T0190106

Thrust Tectonics

Thrust Tectonics

Edited by

K.R. McClay
Department of Geology
Royal Holloway and Bedford New College
University of London

SPRINGER-SCIENCE+BUSINESS MEDIA, B.V.

Additional material to this book can be downloaded from http://extras.springer.com

First edition 1992

Originally published by Chapman & Hall in 1992

A catalogue record for this book is available from the British Library

Library of Congress Cataloging-in-Publication data
Thrust tectonics / edited by K.R. McClay.
p. cm.
Papers presented as part of the Thrust Tectonics Conference held at Royal Holloway and Bedford New College, University of London, from April 4th–7th, 1990.
Includes bibliographical references and index.
ISBN 978-0-412-43900-1 ISBN 978-94-011-3066-0 (eBook)
DOI 10.1007/978-94-011-3066-0
1. Thrust faults (Geology)—Congresses. I. McClay, K.R. (Kenneth R.) II. Thrust Tectonics Conference
(1990 : Royal Holloway and Bedford New College, University of London)
QE606.T57 1991
551.8′7—dc20 91–34275
CIP

CONTENTS

CONTRIBUTORS

J.L. Alonso, Departamento de Geologia, Universidad de Oviedo, Oviedo, Spain
T.G. Apotria, Department of Geology and Center for Tectonophysics, Texas A&M University College Station, Texas, USA
C. Beaumont, Oceanography Department, Dalhousie University, Nova Scotia, Canada
E.G. Bombolakis, Department of Geology and Geophysics, Boston College, Chestnut Hill, Massachusetts, USA
S. Boyer, Department of Geological Sciences, University of Washington, Washington, USA
R.L. Brown, Department of Earth Sciences, Carleton University and Ottawa-Carleton Geoscience Centre, Ontario, Canada
P.G. Buchanan, 30 Marler Road, Forest Hill, London, UK
R.W.H. Butler, Department of Earth Sciences, The Open University, Milton Keynes, UK
S.D. Carr, Department of Earth Sciences, Carleton University and Ottawa-Carleton Geoscience Centre, Ontario, Canada
A.F. Chambers, School of Geological Sciences, Kingston Polytechnic, Surrey, UK
G.T. Chou, Texaco USA, 4601 DTC Boulevard, Denver, Colorado, USA
V.J. Coleman, Department of Earth Science, Carleton University and Ottawa-Carleton Geoscience Centre, Ontario, Canada
F.A. Cook, Department of Geology and Geophysics, The University of Calgary, Alberta, Canada
M. Cooper, BP Resources Canada Limited, Calgary, Alberta, Canada
M.P. Coward, Department of Geology, Imperial College, London, UK
J. Dinares, Department de Geologia Dinamica, Universitat de Barcelona, Zona Universitaria de Pedralbes, Barcelona, Spain
J.M. Dixon, Experimental Tectonics Laboratory, Department of Geological Sciences, Queen's University, Kingston, Ontario, Canada
C. Doglioni, Dipartimento di Scienze Geologiche e Paleontologiche, Universita di Ferrara, 44100, Italy
C.A. Evenchick, Geological Survey of Canada, Vancouver, British Columbia, Canada
M.P. Fischer, Department of Geosciences, The Pennsylvania State University, University Park, Pennsylvania, USA
P. Fullsack, Oceanography Department, Dalhousie University, Nova Scotia, Canada
J. Hamilton, Oceanography Department, Dalhousie University, Nova Scotia, Canada
R.D. Hatcher, Jr, Department of Geological Sciences, University of Tennessee, Tennessee, USA
S.C. Hook, Texaco Inc, PO Box 70070, Houston, Texas, USA
R.J. Hooper, Department of Geology, University of South Florida, Florida, USA
A.T. Hsui, Department of Geology, University of Illinois, Urbana, Illinois, USA
C.N. Izatt, International Exploration, British Gas, 59 Bryanston Street, London, UK
K.C. Jackson, Department of Geology, Imperial College, London, UK
I.A.K. Jadoon, Department of Earth Sciences, Quaid-i-Azam University, Islamabad, Pakistan
W.R. Jamison, Amoco Production Research, Tulsa, Oklahoma, USA
B.J. Johnson, Department of Earth Sciences, Carleton University and Ottawa-Carleton Geoscience Centre, Ontario, Canada
P. Jordan, Geological Institute of Basel University, Bernoullistrasse 32, Basel, Switzerland
R. Kligfield, Department of Geological Sciences, University of Colorado, Boulder, Colorado, USA
H. Laubscher, Geological Institute of Basel University, Bernoullistrasse 32, Basel, Switzerland
R.D. Lawrence, Department of Geosciences, Oregon State University, Oregon, USA
R.J. Lillie, Department of Geosciences, Oregon State University, Oregon, USA
H. Liu, Department of Geology, Royal Holloway and Bedford New College, University of London, Surrey
S. Liu, Experimental Tectonics Laboratory, Department of Geological Sciences, Queen's University, Kingston, Ontario, Canada
S. Marshak, Department of Geology, University of Illinois, Urbana, Illinois, USA
A. Martinez, Servei Geologic de Catalunya, Parallel, Barcelona, Spain
A.M. McCaig, Department of Earth Sciences, University of Leeds, UK
E. McClelland, Department of Earth Sciences, University of Oxford, UK
J. Mosar, Department of Geological and Geophysical Sciences, Guyot Hall, Princeton University, New Jersey, USA

J.A. Munoz,Department de Geologia Dinamica, Universitat de Barcelona, Zona Universitaria de Pedralbes, Barcelona, Spain
T. Noack, Geological Institute of Basel University, Bernoullistrasse 32, Basel, Switzerland
D. Powell, Brunary House, Glen Moidart by Lochailort, Invernesshire, Scotland
C. Puigdefabregas, Servei Geologic de Catalunya, Parallel, Barcelona, Spain
J.G. Ramsay, Geologisches Institut, ETH Zentrum, Zurich, Switzerland
M. Rowan, Alastair Beach Associates, 11 Royal Exchange Square, Glasgow, Scotland
C.M. Rubin, Division of Geological and Planetary Sciences, California Institute of Technology, Pasadena, California, USA
F.J.B. Saleeby, Division of Geological and Planetary Sciences, California Institute of Technology, Pasadena, California, USA
P. Santanach, Department de Geologia Dinamica, Universitat de Barcelona, Zona Universitaria de Pedralbes, Barcelona, Spain
W. Sassi, Institut Francais de Petrole, 92506 Rueil Malmaison Cedex, France
G. Schonborn, Geological Institute of Basel University, Bernoullistrasse 32, Basel, Switzerland
W.T. Snedden, Texaco E&P Technology Division, Houston, Texas, USA
J.H. Spang, Department of Geology and Center for Tectonophysics, Texas A&M University College Station, Texas, USA
J. Suppe, Department of Geological and Geophysical Science, Guyot Hall, Princeton University, New Jersey, USA
P.W.G. Tanner, Department of Geology and Applied Geology, University of Glasgow, Glasgow, Scotland
A. Teixell, Division de Geologia, ITGE, Jaca, Spain
P.J. Treloar, Department of Geology, Imperial College, London, UK
J.L. Varsek, Department of Geology and Geophysics, The University of Calgary, Alberta, Canada
J. Verges, Servei Geologic de Catalunya, Parallel, Barcelona, Spain
M.S. Wilkerson, Department of Geology, University of Illinois, Urbana, Illinois, USA
S.D. Willett, Department of Oceanography, Dalhousie University, Nova Scotia, Canada
D.V. Wiltschko, Department of Geology and Center for Tectonophysics, Texas A&M University College Station, Texas, USA
S. Wojtal, Department of Geology, Oberlin College, Oberlin, Ohio, USA
N.B. Woodward, 26117 Viewland Drive, Damascus, Maryland, USA
R. Zoetemeijer, Institut Francais du Petrole, 92506 Rueil Malmaison Cedex, France

ACKNOWLEDGEMENTS

The papers presented in this volume were part of the Thrust Tectonics Conference held at Royal Holloway and Bedford New College, University of London from 4–7 April 1990. The Conference and production of this volume were greatly assisted by the generous sponsorship of the Geology Department, Royal Holloway and Bedford New College, Brasoil UK Ltd, BP Petroleum Development PLC, Chevron Petroleum UK Ltd, Clyde Petroleum, Enterprise Oil, Esso Exploration and Production Ltd and Shell Petroleum UK.

The staff and students of the Geology Department, Royal Holloway and Bedford New College, in particular Tim Liverton, Luis D'el Rey Silva, Martin Insley, James Dobson, Kevin D'Souza and Sandra Muir, are thanked for their assistance and support in running the conference.

The authors who submitted manuscrips for consideration for publication in this volume are thanked for their efforts and particularly for following the editor's guidelines and submitting in Macintosh format. Minor grammatical changes in the texts remain the responsibility of the editor.

The manuscrips were ably and promptly reviewed by the persons listed below, many of whom kindly reviewed two or more papers:

J. Alvarez-Marron
A. Baird
A. Bally
C. Banks
A. Barber
S. Boyer
M. Brandon
K. Brodie
R. Brown
R. Butler
T. Chapman
P. Cobbold
M. Cooper
J. Cosgrove
D. Cowan
M. Coward
F. Dahlen
M. Daly
I. Davison
D. De Paor
J. Dewey
D. Dietrich
P. Ellis
P. Fermor
M. Ford
H. Gabrielse
R. Gayer
D. Gee
P. Geiser
R. Graham
R. Groshong
S. Hall
W. Higgs
R. Holdsworth
W. Horsfield
D. Hutton
M. Insley
C. Izatt
C. Jackson
W. Jamison
G. Karner
R. Knipe
P. Kooms
G. Lloyd
G. Manby
N. Mancktelow
A. McCaig
A. Meneilly
G. Mitra
S. Mitra
J. Muñoz
A. Pfiffner
R. Price
A. Roberts
G. Roberts
M. Rowan
E. Rutter
D. Sanderson
W. Sassi
J. Suppe
C. Talbot
G. Tanner
R. Thompson
P. Treloar
J. Warburton
P. Weston
J. Wheeler
N. White
G. Yeilding
R. Zoetemeijer.

Finally a special vote of heartfelt thanks goes to Andrew D. Scott and Ruth Durrell for their invaluable and cheerful assistance in organizing the conference and in the production of this volume. This book was compiled and typeset by R. Durrell, A. Scott and K. McClay, Geology Department, Royal Holloway and Bedford New College, University of London using Microsoft Word 4.0® and Pagemaker 4.0®. Kevin D'Souza kindly produced the PMTs. Phototypesetting was done at the Typesetting Bureau of the University of London Computer Centre.

Ken McClay
Geology Department
Royal Holloway and Bedford New College
University of London

Thrust Tectonics: An introduction

K.R. McClay

Department of Geology, Royal Holloway and Bedford New College, University of London, Egham, Surrey, England TW20 0EX.

Since the first Thrust and Nappe Tectonics Conference in London in 1979 (McClay & Price 1981), and the Toulouse Meeting on Thrusting and Deformation in 1984 (Platt *et al.* 1986) there have been considerable advances in the study of thrust systems incorporating new field observations, conceptual models, mechanical models, analogue and numerical simulations, together with geophysical studies of thrust belts. *Thrust Tectonics 1990* was an International Conference convened by the editor and held at Royal Holloway and Bedford New College, University of London, Egham Surrey, from April 4th until April 7th 1990. There were one hundred and seventy participants from all continents except South America. The conference was generously sponsored by Brasoil U.K. Limited, BP Exploration, Chevron U.K. Limited, Clyde Petroleum, Enterprise Oil, Esso Exploration and Production UK Limited, and Shell U.K. Exploration and Production.

One hundred and five contributions were presented at the meeting, - seventy six oral presentations (together with poster displays) and an additional twenty nine posters without oral presentation (McClay 1990, conference abstract volume).

The thirty six papers presented in this volume represent a distillation of the material presented at the conference and are grouped into thematic sections that follow the main topics of discussion at the meeting. To an extent the grouping of papers into specific thematic sets is somewhat arbitrary as there are many common linkages between the various sections. In particular many papers focus upon both the geometries of thrust systems and their kinematic evolution - a recurring theme throughout the volume. Space limitations prevented publication of all of the papers presented at the meeting but it is hoped that this volume will provide the reader with an appreciation of the current research themes and concepts of thrust systems. The contents of the volume are summarized below.

The first section of this volume is headed '*Theoretical aspects and thrust mechanics*'. **Beaumont *et al.*** describe an elegant finite element model of the evolution of the Southern Alps of New Zealand. The model incorporates tectonic data, erosional and orthographic data and demonstrates surprisingly that the climate, erosion and fluvial systems are coupled to the orogenic evolution such that the structural and metamorphic style of this thrust belt may be controlled by the fluvial erosion and climate on the western flank of the Southern Alps. **Willett** presents a finite-element model of Coulomb wedges in order to simulate the growth of accretionary prisms and foreland fold and thrust belts. The model demonstrates self-similar growth of the Coulomb wedge and investigates the kinematics and particle velocities of critical wedges growing by accretion. Changes in the boundary conditions of the wedge (e.g. a change in the basal friction) changes the wedge geometry and thus may permit out of sequence thrusting (increase in basal friction) or extensional collapse of the wedge whilst in compression (decrease in basal friction). **Bombolakis** examines the partitioning of elastic strains and strain-rate dependent ductile deformation in the emplacement of the frontal parts of thrust systems. He demonstrates that significant thrust movement in these environments can occur by aseismic processes. **Wojtal** presents a one-dimensional model for plane and non-plane power-law flow in shortening and elongating thrust zones. He concludes that viscous creep is significant along external thrust faults and that fluid assisted diffusive mass transfer processes (pressure solution) are important deformation mechanisms in the fine-grained banded cataclasites of such thrust systems.

The second section of this volume titled '*Physical modelling*' contains four papers on analogue modelling of thrust structures. **Dixon & Liu** investigate the formation of foreland fold and thrust belts using scaled centrifuge models. They demonstrate that folding precedes thrusting, and that the thrust periodicity is inherited from the buckling wavelength of the antecedent fold systems. The models also demonstrate the 3D coalescence of small thrusts to form larger thrust systems. **Liu *et al.*** present a detailed study of sandbox models of thrust systems found in accretionary wedges and foreland fold and thrust belts. The boundary conditions of the models were systematically varied and the results are in broad agreement with Coulomb wedge theories (cf. Davis et al. 1983; Dahlen 1990). Low basal friction models are characterised by both foreland-vergent and hinterland-vergent thrust faults (i.e. backthrusts) as well as low critical tapers of the resultant Coulomb wedge. Intermediate basal friction models show a dominance of foreland-vergent thrust systems whereas models in which the coefficient of friction at the base is equal to that within the wedge, depart from ideal Coulomb wedges and show greater tapers than predicted by theory. **Marshak *et al.*** investigate the generation of curved thrust belts using viscous analogue models, finite difference models and sandbox models. They conclude that curved fold and thrust belts may be generated by differential displacement such as that resulting from the interaction with obstacles or along strike variations in sediment thicknesses. **McClay & Buchanan** investigate the geometries of thrust systems formed by inversion of extensional basins. Sandbox models demonstrate the complexities of thrusts produced by inversion - steep angle thrusts, footwall shortcut thrusts, backthrusts and both 'forward-breaking' and 'break-back' thrust sequences.

The third section of this volume is titled - *'Thrust geometries and thrust systems'* and focuses upon geometric and kinematic problems of thrust systems. **Suppe *et al.*** set out the theory for the geometry of growth folding associated with kink band folding, fault-bend folding and fault-propagation folding. They demonstrate that the stratal geometries of growth sequences may be used to determine fault slip rates and illustrate their results with both seismic and map examples of growth fold systems. **Mosar & Suppe** introduce the theory for fault-propagation folding with variable shear. They show how the geometries of fault-propagation or tip-line folds may be significantly altered by layer parallel shear. **Zoetemeijer & Sassi** describe a computer method for 2D reconstruction of thrust systems using the fault-bend fold model. The method can incorporate syntectonic sedimentation and erosion. The problems of three dimensional thrust geometries developed over oblique ramps are tackled by **Apotria *et al.*** They describe a detailed kinematic analysis of fault-bend fold interactions at frontal and oblique ramps and illustrate their results with both cross-sections and map views. The stress controls on thrust related folding are addressed by **Jamison.** He proposes that the style of folding in a thrust belt is a result of the competition between buckling and faulting and as such is dependent upon the depth of burial and tectonic regime. The hypothesis is illustrated with examples from the Rocky Mountain thrust belt. **Rowan & Kligfield** present a kinematic analysis of asymmetric detachment folds from the Wildhorn nappe, Helvetic Alps. They propose that overthrust shear between the nappe boundaries generated buckle folds and rotational strains within the nappe system. **Alonso & Teixell** describe an analysis of fault-propagation folds in which the forelimbs are thinned by heterogeneous simple shear localized at ramps and not penetrative through the whole of the thrust sheet. The progressive geometric evolution of thrust faults and related folds is examined by **Fischer & Woodward.** They propose, using examples from both Wyoming and the Appalachians, that fold-thrust systems do not evolve in a self-similar fashion but argue that variations of fold-thrust structures along strike in a thrust belt do not represent temporal variations in evolution but rather reflect intrinsic variations in the mechanical response of the stratigraphic package to contraction.

Ramsay argues against unquestioning acceptance and wholesale application of preconceived geometric models for thrust-fold systems - particularly those where the deformation is concentrated in the hangingwalls of thrust faults. He describes examples where footwall deflection is important and advocates many more detailed studies of natural thrust systems. The geometries and kinematics of duplex structures are addressed by **Tanner** who proposes that the Boyer & Elliott (1982) model does not universally apply and that many natural duplex systems show indications of synchronous fault movement. The use of palaeomagnetic studies to determine rotations about vertical axes in thrust belts is reviewed by **McCaig & McClelland** whereas **Hatcher & Hooper** review the main characteristics of thrust sheets that involve crystalline basement in the interior parts of mountain belts.

The fourth section on the *'Pyrenees'* concerns recent research which integrates structural, sedimentological and geophysical results. An overview and balanced crustal cross-section of the Pyrenean continental collision belt is presented by **Munoz.** He points out the discrepancy between the cross-sectional length of the upper crust and that of the lower crust and argues for subduction of Iberian lower crust northwards under the European crust. The structural, stratigraphic and sedimentological evolution of the southern Pyrenean foreland basins is synthesized by **Puigdefabregas *et al.*** The evolution of the foreland basin sequences is described with reference to the patterns of thrusting. **Verges *et al.*** investigate the role of the foreland evaporitic sequences in controlling the evolution and emplacement of the frontal thrust systems and demonstrate their influence on the strike of the thrust system and on the location of oblique ramps. The palaeomagnetic study of **Dinares *et al.*** demonstrates that up to 45° rotation about vertical axes occurs in thrust sheets associated with oblique ramps in the southern Pyrenean thrust belt.

The fifth section of this volume contains five papers on the *'Alps'*. **Laubscher** addresses the problems of linked compression and strike-slip tectonics in the central Alps and quantifies the E-W strike-slip components. **Butler** describes the western Chartreuse fold and thrust system and emphasizes the role of folding and preexisting extensional structures in the evolution of the belt. The detailed geometries and kinematics of a transverse zone in the southern Alps are analysed by **Schonborn. Jordan & Noak** develop models for the evolution of the hangingwall geometries of thrust systems developed above ductile decollements and apply these to examples in the Jura. The final paper in this section by **Doglioni** describes the structure of the Venetian Alps.

The sixth section of this book contains two papers on the *'Himalayas'*. **Treloar *et al.*** summarize the kinematic evolution of the northwest Himalayas linking displacement patterns, thrust sequences with the analysis of rotations of the thrust sheets using palaeomagnetic data. **Jadoon *et al.*** describe the geometry and kinematics of the Sulaiman fold belt, Pakistan.

The final section contains five papers on the *'North American Cordillera'*. **Brown *et al.*** describe the Monashee decollement in the southern Canadian Cordillera. Using new LITHOPROBE data they are able to demonstrate the existence of a linked detachment from the Rocky Mountain foreland fold and thrust belt through to the accreted terranes on the western margin of the Canadian Cordillera. **Evenchick** describes the newly recognized Skeena fold belt and demonstrates a kinematic linkage between the Coast Plutonic complex and the Omineca and Rocky Mountain fold and thrust belts. The geometric and kinematic evolution of duplex systems is addressed by **Boyer** with reference to duplexes in the southern Canadian Rocky Mountains and in the Rocky Mountains of Montana. He argues that these duplexes evolved through synchronous thrust movement - not by a sequential, forward-breaking 'piggy-back' sequence. **Cooper** describes a detailed subsurface fracture analysis of thrust-related folds in the Foothills of the Canadian Rocky Mountains. The last paper in this section by **Rubin & Saleeby** describes the thrust belt features in the interior of the orogen between the Intermontane and Insular terranes of the North American Cordillera.

It will be evident to a reader of this volume that the theoretical and conceptual models for the geometries and kinematic evolution of high-level foreland fold and thrust belt systems are well developed and highly sophisticated (Suppe *et al.* 1991, this volume; Mosar & Suppe 1991, this volume). However, detailed field observations are still needed to test their applicability particularly with respect to the question of whether thrust structures evolve in a self-similar fashion (Fischer & Woodward 1991, this volume) (cf.fault related folds) or whether they evolve through different structural forms (e.g. from detachment folds - to fault propagation folds - to fault bend folds). It is also apparent that whilst the external zones of thrust belts are relatively well studied, the interior parts of orogens are less well understood and require much more detailed research on the more complex thrust systems found therein. Field work will need to be integrated with geophysical data and sections of the whole crust and lithosphere will need to be constructed. Petrofabric, strain and P-T-t data are needed in order to interpret the processes operating during thrust emplacement. The mechanics of thrust belt evolution are now being addressed using Coulomb wedge models (cf. Willett 1991, this volume), variations of which may provide answers for the problems of fault sequencing in thrust belts. It is apparent that, from both analogue modelling (Dixon & Liu 1991, this volume; Liu *et al.* 1991, this volume) and from field studies (Boyer 1991, this volume), synchronous thrust movement is a significant element in the kinematic evolution of foreland fold and thrust belts. This therefore will have important implications for thrust sequences and the balancing and restoration of cross-sections in thrust terranes.

Field based research, however, is the ultimate test of the theoretical, conceptual, numerical and analogue models. It is clear that the better understanding of thrust belts - for example that of the Canadian Rocky Mountains (e.g. Price 1981; Evenchick 1991, this volume; Brown *et al.* 1991, this volume) and of the Pyrenees (e.g. Munoz 1991, this volume; Puigdefabregas *et al.* 1991, this volume), stems from detailed fieldwork that integrates structural geology, stratigraphy, sedimentology and geophysical data. Future research on thrust tectonics will still need to be firmly based upon multidisciplinary field studies if progress is to be made in the understanding of the geometries, kinematics and mechanics of thrust systems.

References

Boyer, S.E. 1991. Geometric evidence for synchronous thrusting in the southern Alberta and northwest Montana thrust belts, this volume.

Boyer, S.E. & Elliott, D. 1982. Thrust systems. *American Association of Petroleum Geologists Bulletin*, **66**, 1196-1230.

Brown, R. L., Carr, S. D., Johnson, B. J., Coleman, V. J., Cook, F. A. & Varsek. J. L. 1991. The Monashee decollement of the southern Canadian Cordillera: a crustal-scale shear zone linking the Rocky Mountain Foreland Belt to lower crust beneath accreted terranes, this volume.

Dahlen, F.A. 1990. Critical taper model of fold-and-thrust belts and accretionary wedges. *Annual Reviews of Earth and Planetary Sciences*, **18**, 55-99.

Davis, D., Suppe, J. & Dahlen, F.A. 1983. Mechanics of fold-and-thrust belts and accretionary wedges. *Journal of Geophysical Research*, **94**, 10,347-54.

Dixon, J. M. & Shumin Liu. 1991. Centrifuge modelling of the propagation of thrust faults, this volume.

Evenchick, C. A. 1991. The Skeena fold belt: a link between the Coast Plutonic Complex, the Omineca Belt and the Rocky Mountain fold and thrust belt, this volume.

Fischer, M. P. & Woodward, N. B. 1991. The geometric evolution of foreland thrust systems, this volume.

Liu Huiqi, McClay, K. R. & Powell, D. 1991. Physical models of thrust wedges, this volume.

McClay, K.R. 1990. (ed.) *Thrust tectonics 1990*. Programme with abstracts volume, Royal Holloway and Bedford New College, University of London, 106p.

McClay, K.R. & Price, N.J. 1981. (eds). *Thrust and Nappe Tectonics*. Geological Society of London Special Publication, **9**, 544p.

Muñoz, J. A. 1991. Evolution of a continental collision belt: ECORS-Pyrenees crustal balanced cross-section, this volume.

Mosar, J. & Suppe, J. 1991. Role of shear in fault-propagation folding, this volume.

Platt, J.P., Coward, M.P., Deramond, J. & Hossack, J. 1986. (eds) *Thrusting and Deformation*, Journal of Structural Geology, **8**, 215 - 483.

Price, R.A. 1981. The Cordilleran foreland thrust and fold belt in the southern Canadian Rocky Mountains. In: McClay, K.R. & Price, N.J. (eds). *Thrust and Nappe Tectonics*, Geological Society of London Special Publication, **9**, 427-448.

Puigdefabregas, C., Muñoz, J. A. & Vergés, J. 1991. Thrusting and foreland basin evolution in the Southern Pyrenees, this volume.

Suppe, J., Chou, G. T. & Hook, S. C. 1991. Rates of folding and faulting determined from growth strata, this volume.

Willett, S. D. 1991. Dynamic and kinematic growth and change of a Coulomb wedge, this volume.

PART ONE
Theoretical aspects and thrust mechanics

Erosional control of active compressional orogens

Christopher Beaumont, Philippe Fullsack & Juliet Hamilton

Oceanography Department, Dalhousie University, Halifax, Nova Scotia, Canada B3H 4J1

Abstract: Denudation has long been acknowledged as a process that contributes to the unroofing of compressional orogens. It has, however, mainly been considered as a passive process, not one that can dictate or control the tectonic evolution. This view prevails despite the knowledge that the style of deformation is controlled by the interplay of gravitational and tectonic stresses: an interplay that is sensitive to the mass removed by denudation.

This paper presents a model designed to investigate the syn-tectonic style of a compressional orogen that is subject to large-scale denudation. The model couples tectonics and erosion in the following manner. The rate of lithospheric deformation, including surface uplift, is calculated by a plane-strain rigid plastic and/or viscous model of the lithosphere's response to tectonic compression. A digital erosion model acts on the current planform surface of the orogen to redistribute mass by a combination of short-range (hillslope) diffusion and long-range (fluvial) transport.

The fluvial network is recharged by a model of orographic rainfall in which the distribution is related to the current model topography. Tectonic uplift is added to the planform surface at the beginning of each erosional timestep. The current planform topography after erosion is averaged along the direction normal to compression and is then used to define the surface for the next step of the tectonic model.

Preliminary results compare favourably with the orogen-scale precipitation, morphology and tectonics of the Southern Alps of South Island, New Zealand. The potential importance of atmospheric coupling to orogenic evolution is illustrated by results from contrasting models of the Southern Alps in which: (1) there is no erosion; (2) rainfall is derived from a westerly source, and; (3) rainfall is derived from an easterly source. If the model is correct, the structural and metamorphic style of this orogen is controlled by fluvial denudation of its western flank charged by the orographic precipitation.

Can erosion control the evolution of compressional orogens at the scale of the orogen? This paper investigates the way in which orographic rainfall and consequent erosion of a mountain belt may be linked to the tectonics of the underlying deforming orogen. If this analysis is correct, palaeoclimatic information may be required before the evolution of ancient orogens can be fully understood.

The basis of the coupling between erosion and deformation was discussed by Jamieson & Beaumont (1988, 1989). They linked the concept of critical topography (Chapple 1978; Stockmal 1983; Davis *et al.* 1983; Dahlen 1984, among others) applied at the orogen scale, which they termed critical shape, to syn-tectonic deformation within the orogen when the critical shape is destroyed by erosion. The purpose of this paper is to quantify these concepts. A critical orogen will respond to erosion by a perturbation to the internal velocity distribution that will act to restore the critical shape. Persistent erosion leads to exhumation of material from within the orogen in proportion to the mass removed by erosion. The process may control how rocks from deep within an orogen are exposed at the surface.

By considering erosion of a critical Coulomb wedge in steady state, Dahlen & Barr (1989) and Barr & Dahlen (1989) illustrated the consequences of the dynamic steady state for the velocity distribution, and pressure-temperature-time (PTt) paths of rocks as they pass through the wedge. Erosion destroys the critical state of the deforming wedge by reducing the component of the stress tensor within the wedge that is related to the weight of the overburden. Additional deformation follows because the yield strength of the Coulomb (frictional) material decreases as overburden is removed.

It may, alternatively, be argued that a viscous, not Coulomb or plastic, rheology is more appropriate for orogenic mechanics at the largest scale. If this is true, the viscous sheet model (England & McKenzie 1982; Vilotte *et al.* 1986, among others) demonstrates the relative importance of gravitational and tectonic forces on the velocity and strain-rate distributions within the sheet through their ratio, the Argand number. For finite Argand numbers gravitational forces limit orogenic thickening, a plateau develops and high strain rate regions are limited to the periphery of the plateau. When erosion removes mass from the plateau or its flanks, gravitational stabilization is reduced and strain rates correspondingly increase where erosion has occurred. The vertical strain, coupled with isostatic adjustment, induces enhanced uplift as a response to the erosion. If erosion persists, the coupled tectonic response induces exhumation of the crust in the viscous sheet model that is equivalent to that of the eroding Coulomb wedge.

The strain within a *deforming* compressional orogen is related to the internal state of stress, irrespective of the rheology. At crustal and lithospheric scales gravitationally induced stress is as important as tectonic stress. It follows that the syn-tectonic deformation of an orogen will be modified by erosion.

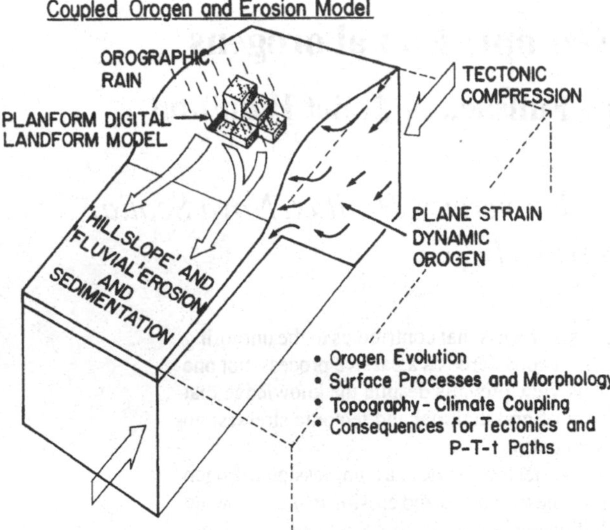

Figure 1. Conceptual illustration of the coupled plane-strain tectonic and surface processes model. Orogenic evolution is controlled by the interplay of deformation, driven by tectonic compression, with surface transport by hillslope and fluvial processes. Potential applications of the model are listed at bottom right.

Dahlen & Barr (1989) solve the problem of exhumation in a steady-state Coulomb critical wedge. The corresponding exhumation for a steady-state viscous sheet can be derived from the zero Argand number plane-stress solution (England & McKenzie 1982) if it is also assumed that the sheet deforms by pure shear. This approach is different because a steady state in which the geometry remains static, and uplift and erosion are equal is not assumed a priori. Instead, this paper considers the dynamic interaction between a deforming orogen and the erosion of its topography by mass diffusion and fluvial transport. The model (Fig. 1) links a planform surface mass transport model with a cross-sectional plane-strain model of crustal tectonics in which the plane is normal to strike of the compressional orogen. It is assumed that the orogen responds to an areal average of the surface mass redistribution. If this is correct, the complete three-dimensional solution of the tectonic problem is not required. Instead, the plane-strain tectonic model is coupled with the strike average of the surface mass redistribution. The evolution of the surface therefore represents the erosional response to orogenic uplift which does not vary along strike.

The formulation of the tectonic and surface mass transport models is first discussed and preliminary model results are then compared with the Southern Alps of central South Island, New Zealand, which independent evidence suggests is a near steady-state orogen.

Tectonic model

When modelling the dynamics of lithospheric deformation assumptions have to be made about the constitutive relationship between the stress field, and the deformation and its rate. Elastic-plastic, rigid-plastic and viscous rheologies are commonly used to represent brittle-frictional and ductile regions. Here the assumption is made that when the deformation is large, elastic strains are negligible, and a rigid-plastic rheology is used to model brittle behaviour of an orogen. The theory is briefly described (see also Willett 1991, this volume) and it is shown to be formally analogous to a viscous rheology. On the basis of this analogy the Stokes equation for a viscous fluid is used to solve incrementally for the velocity and strain rate for each step of the deformation.

Rigid-plastic rheology

A rigid-plastic material does not deform until stresses reach a yield level, at which point it deforms according to a plastic flow rule, the stresses never exceeding the yield level. In the Levy-Mises theory (Malvern 1969), the rigid and plastically deforming regions are governed by:

Rigid branch

$$J_2(\sigma') < \sigma_y^2 \qquad (1)$$

$$\dot{\varepsilon} = 0$$

Plastic branch

$$J_2(\sigma') = \sigma_y^2 \qquad (2)$$

$$\dot{\varepsilon} = k\sigma'$$

where, $\dot{\varepsilon} = \frac{1}{2}(\nabla \underline{v} + \nabla^T \underline{v})$ is the instantaneous strain rate for velocity field $\underline{v}$, σ' is the deviatoric stress tensor, $J_2(\sigma') = \sigma'_{ij}\sigma'_{ij}/2$ is the second invariant of σ', and σ_y is the yield stress. k is a positive scalar which varies spatially so that the constraint equations of the rigid and plastic branches and the boundary conditions will be satisfied.

The simple model also assumes that plastic deformation is incompressible, that is $\nabla \cdot \dot{\varepsilon} = 0$, which, when coupled with the equilibrium equations, is used to determine the pressure.

Pseudo-viscous analogy

For a linear viscous rheology, viscosity η,

$$\dot{\varepsilon} = \frac{1}{\eta}\sigma' \qquad (3)$$

It can be seen that $J_2(\sigma') < \sigma_y^2$ when rigid behaviour (eqn. 1) can be approximated, $\dot{\varepsilon} \sim 0$, by letting η take a very large value, η_∞, in equation (3). Correspondingly, taking an effective viscosity η_e in equation (3) such that $\eta_e^2 = \sigma_y^2 / J_2(\dot{\varepsilon})$ simultaneously satisfies both requirements of a plastically deforming region (eqns 2). An equivalence therefore exists between viscous and Levy-Mises plastic flows such that a proper choice of effective viscosity can always be made to approximate the rigid plastic equations.

Velocity and deformation calculations

Having shown that the rigid plastic constitutive equation can be represented by an effective viscosity, the deformation can be found by solving the conservation of momentum equation for viscous fluid flow, the Navier-Stokes equation. The inertial term is not required because large, long-term deformation of the lithosphere is a quasistatic process. The governing equations are therefore:

$$\eta_e \nabla^2 \underline{v} - \nabla p + \rho g = 0 \qquad \text{(equilibrium)} \qquad (4)$$

$$\nabla \cdot \underline{v} = 0 \qquad \text{(incompressibility)} \qquad (5)$$

subject to the boundary conditions. Equations (4) and (5) are solved for υ in the current configuration which is then updated before the next step of the solution.

The equilibrium distribution of the pseudo viscosity, η_e, is found by iteration. A high level of stress, generated by assuming $\eta_e = \eta_\infty$ everywhere and applying the boundary conditions, is relaxed progressively by 'softening' locations where this level of stress exceeds the yield. Both plastic and viscous regions of the lithosphere can be included in the solution by assigning a fixed 'true' viscosity where appropriate. This technique is used to examine the effect of a basal viscous boundary layer.

Computations were made using a modified form of the finite element code ADINA (ADINA 1987).

Typical solutions on a CRAY-XMP require 40s to deform a crustal scale 30 x 60 element grid of 4-noded elements by 400 m.

Surface transport model

The surface volume transport and the evolution of topography and morphology at the orogenic scale is considered in terms of a quantitative, yet simple, model of the interaction between short range and long range transport on a cellular topographic grid. The model is similar to Chase's (1988, 1989, pers. comm.) precipitation model of fluvial landsculpting, but differs because this paper considers distributed orographic rainfall and the cumulative effect of fluvial discharge down a drainage network. While Chase's approach is appropriate for an arid environment subject to small-scale isolated storm events, our model includes large-scale distributed discharge in continuously flowing rivers. Sediment transport is the sum of the two independent processes, which are assumed to be in a dynamic steady state during each of the model steps of length Δt.

Short range transport

The short range transport model represents the cumulative effect of processes (soil creep, landslides, rainsplash, surface and subsurface wash) that remove material from hill and mountain sides and transport it to the valleys (Carson & Kirkby 1972). Here Culling (1960, 1965), Coleman & Watson (1983) and Flemings & Jordan (1989), for example, are followed by representing the sum of these processes as linear downslope diffusion of material volume. The horizontal flux, $\underline{q}_s$ is related to the local slope, ∇h, by

$$\underline{q}_s = -u_s h_s \nabla h \qquad (6)$$

where u_s is the transport speed, h_s is the vertical height scale of the erodable surface boundary layer and subscript s denotes short range slope processes. u_s and h_s may be combined as a single transport coefficient $K_s = u_s h_s$, but it is useful to distinguish between them, partly because they represent different processes, u_s the ease of material transport once it has become fragmented and, h_s the steady-state thickness of the boundary layer in which cohesion has been destroyed by weathering. As Carson & Kirkby (1972) explain, h_s is governed by the rock type, the surface processes at work and the relative timescales of downslope transport versus weathering. In our model the poorly understood problem of weathering is avoided and the thickness of the weathered layer is represented by parameters h_c for crustal rocks that have not been previously transported and h_a for alluvium created by the model. An improved model will include the processes which determine the spatial dependence of h_s.

If it is also assumed that there is no tectonic transport of the surface material, that volume is conserved and that the effects of solution are negligible, the transport equation can be combined with the continuity equation,

$$\frac{dh}{dt} = -\nabla \cdot \underline{q}_s \qquad (7)$$

to give the linear diffusion equation for the rate of change of local height, the denudation, in response to erosion by the short range processes,

$$\frac{dh}{dt} = u_s h_s \nabla^2 h \qquad (8)$$

The short-range transport operates on a regular square cellular topographic grid, cell side length c_L (Fig. 2). Transport to each cell j from its surrounding eight neighbours is considered in turn on the assumption that during the interval Δt the flux is in a dynamically steady state; that is, it does not

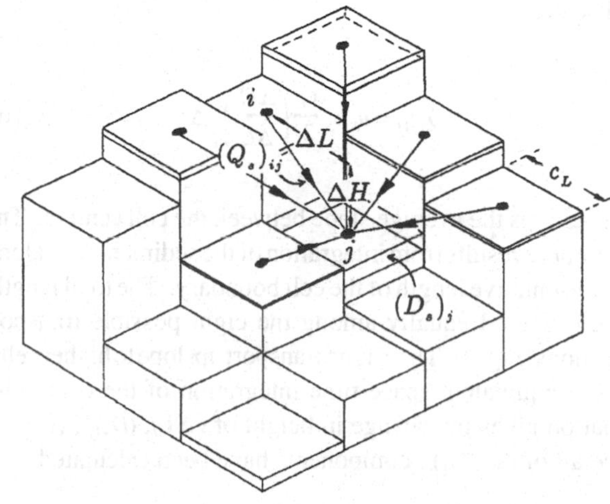

Figure 2. Illustration of the diffusive transport model on part of the cellular topography.

Table 1. Model Parameters, Dimensions, and Values

Model parameters		Dimensions	Values
Tectonic model			
Internal angle of friction	ϕ	-	20°
Pore Pressure ratio	λ	-	0.36
Density of crust	ρ	ML^{-3}	2800 kg m^{-3}
Viscosity of basal shear zone	η	$ML^{-1}T^{-1}$	3 x 10^{18} Pa·s
Depth of basal shear zone	d_s	L	≤ 250 m
Velocity of boundary	υ_0	LT^{-1}	2 x 10^{-2} myr^{-1}
Increment in boundary displacement per timestep		L	400 m
Duration of tectonic timestep	ΔT	T	2 x 10^4 yr
Surface Processes Model: Models 2 & 3			
Size of grid		L x L	104 x 72 km
Cell size	c_L x c_L	L x L	(2 x 10^3) x (2 x 10^3) m^2
Duration of erosion substeps	$\Delta T/m$	T	2 x 10^2 yr
Short Range Processes			
Transport speed	u_s	LT^{-1}	1 x 10^{-2} myr^{-1}
Erodability height scale	h_s	L	
Crust	h_c	L	1m
Alluvium	h_a	L	5 m
Diffusion coefficient	$K_s = u_s h_s$	L^2T^{-1}	
Crust	$h_c u_s$		1 x 10^{-2} m^2yr^{-1}
Alluvium	$h_a u_s$		5 x 10^{-2} m^2yr^{-1}
Precipitation			
Water flux (in form of water vapour) incident on model	$s_W(0)$	L^2T^{-1}	3 x 10^4 m^2yr^{-1}
Topographic scale height	h_R	L	4 x 10^3 m
Length scale	l_R	L	4 x 10^4 m
Extraction efficiency	$a_R = h_R l_R$	L^2	1.6 x 10^7 m^2
Source Model 2 from west (strictly, northwest) Model 3 from east (strictly, southeast)			
Long Range Processes			
Long range transport coefficient	K_f	-	10^{-2}
Erosion/deposition length scale	l_f	L	
Crust	l_c	L	10^5 m
Alluvium	l_a	L	2 x 10^4 m
Deposition	$l_f = c_L/2$	L	10^3 m

vary during Δt. If neighbour i is higher than j, the volume of material transported between the two cells $(Q_s)_{ij}$ in time Δt is the space-time integrated form of equation 6 in which all quantities in capital letters are assumed to be averaged at the cell scale,

$$(Q_s)_{ij} = u_s h_s \frac{c_L}{2}\left(\frac{\Delta H}{\Delta L}\right)_{ij} \Delta t \tag{9}$$

$(\Delta H/\Delta L)_{ij}$ is the average slope between the cell centres. The factor $c_L/2$ results from integration of the sediment flux along a representative length of the cell boundary. The total length, $4c_L$, is shared equally among the eight possible transport directions (Fig. 2). There is no transport upslope to higher cells.

The equivalent space-time integration of the continuity equation gives the change in height of cell j, $(D_s)_j$, (Fig. 2) once all of the $(Q_s)_{ij}$ components have been calculated,

$$(D_s)_j = \frac{1}{c_L^2} \sum_{i=1}^{8} (Q_s)_{ij} \tag{10}$$

This explicit method of calculation is valid if the change in height is small by comparison with the overall topography, a constraint that places an upper bound on $\nabla t << \tau = c_L^2 / u_s h_s$.

The effects of 'dry' and 'wet' (that is, water enhanced) transport can be considered separately in the short-range transport model. The overall model remains diffusive with equation 8 modified to

$$\frac{dh}{dt} = u_s \left(1 + \frac{\upsilon_R}{\upsilon_R^c}\right) h_s \nabla^2 h \tag{11}$$

where υ_R is the average rainfall accumulation rate and υ_R^c is the rainfall rate required to keep the diffusing layer saturated. The $\upsilon_R / \upsilon_R^c$ enhancement term can be interpreted as contributing to increase the downslope transport velocity and/or to increase the steady-state value of h_s.

The relative rates of 'wet' and 'dry' transport are not well known, therefore this paper does not attempt to distinguish them but instead uses only one average transport coefficient for each of crustal rocks and alluvium (Table 1). It should, however, be remembered that the transport velocity or the thickness of the eroding boundary layer may vary considerably with aridity and elevation on the scale of a mountain belt.

Long range transport

The long-range transport model represents the cumulative effect of fluvial transport either as bedload or in suspension. No attempt is made to distinguish the river type, or any of its dynamic characteristics. Instead, the local equilibrium sediment carrying capacity, q_f^{eqb}, of the long range transport system is considered proportional to its local power (Begin *et al.* 1981; Armstrong 1980),

$$q_f^{eqb}\ \hat{l} = -K_f q_r \hat{l}\frac{\partial h}{\partial l} \quad (12)$$

where $\hat{l}$ is the local direction of the river and the product of the discharge, q_r, and the slope, $\partial h/\partial l$, is its power. While equation 12 obviously simplifies the true nature of sediment transport in a particular river, it does represent the first-order property of rivers to transport sediment. The form of equation 12 is also amenable to improvement by making q_f^{eqb} a more general function of q_r and $\partial h/\partial l$. The simple form is also consistent with the design of the cellular surface transport model; that it be as simple as possible, involve the least number of parameters and yet represent transport on timescales $\geq 10^3$ years and spatial scales $\geq 10^3$ m.

Fluvial discharge is related to the net precipitation accumulation rate, υ_R by conservation of water over the upstream watershed area,

$$q_r(x,y)\hat{l}\ (x,y) = \int_{\text{upstream}\,(x,y)} \upsilon_R(x_1,y_1)\ \hat{l}\ (x_1,y_1)dA, \quad (13)$$

where $\hat{l}$ is the drainage direction assumed to be down the steepest slope. υ_R represents the precipitation that contributes to surface runoff and that recharges the river system. Evaporation and infiltration are assumed to be negligible, although their effect could be added. The dynamically steady-state nature of the model means that there is no storage of water and that precipitation is balanced by discharge.

In order to calculate the river discharge on the cellular grid, a model is required that predicts the precipitation (Fig. 3) and the calculation of the collection of runoff by the fluvial network (Fig. 4). For the applications discussed here the precipitation is taken to be orographic in response to uplift and adiabatic decompression of moist air as it passes over the model topography (Fig. 3). This paper considers cases where the prevailing direction of incidence of the air masses is normal to one of the boundaries of the grid, taken here to be the x direction, and that the speed is uniform over the grid and does not vary with time. These restrictions can be removed by integrating the equations along flux tubes if a more general model is required. Let $s_w(x=0)$ be the flux of water, in the form of water vapour, per unit length normal to the direction of flow. Then, conversion of this flux to precipitation rate $\upsilon_R(x)$ is assumed to follow the relationship,

$$\upsilon_R(x) = -\frac{ds_w(x)}{dx} = \frac{h(x)}{h_R l_R}\ s_w(x) \quad (14)$$

where $a_R = h_R l_R$ is an extraction efficiency parameter governed by the topographic scale height, h_R, and the length scale, l_R, over which $1/e$ of the available precipitation would be extracted were $h(x)=h_R$. The physics of orographic precipitation is more complex than this simple model and the extraction efficiency will vary with geographic location and climate, therefore a_R is considered to be a free parameter in the model. Its value is chosen so that the precipitation agrees with observations.

The equivalent space-time integrated precipitation, R_i, that contributes to the river system on cell i in time Δt (Fig. 3) is given by,

$$R_i = \frac{H_i}{h_R l_R}(S_W)_i c_L^2\ \Delta_t = (V_R)_i c_L^2\ \Delta_t \quad (15)$$

where H_i is the height of the cell, V_R the average precipitation accumulation rate, and

$$(S_W)_i = S_W(0) - \frac{1}{c_L\Delta t}\sum_{n=1}^{i=1} R_n \quad (16)$$

is the average water flux in the form of water vapour over cell i. The flux $S_w(0)$ can also be considered as the effect of a water column, height h_w (0), that is advected over the

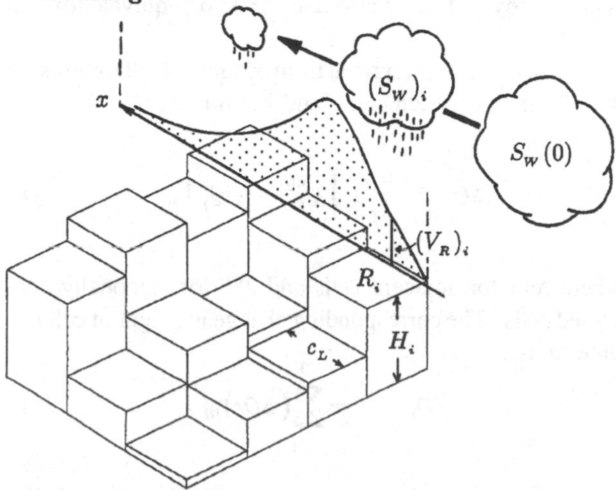

Figure 3. Illustration of the orographic rainfall model in which the clouds diagrammatically represent a long-term integrated precipitation source in the form of water vapour that is progressively depleted as it passes over one row of the cellular topography. The corresponding precipitation distribution is shown in the diagrammatic graph.

Long Range (Fluvial) Transport Model

Distributed Rainfall Collected by Drainage Network

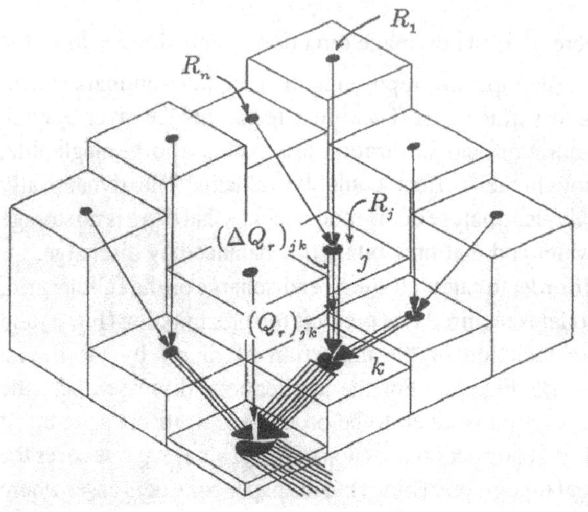

Figure 4. Illustration of the steepest descent drainage network on part of the cellular topography when precipitation is distributed.

topography with velocity $\upsilon_W \hat{\underline{x}}$. The summation $n = 1, \ldots i - 1$ is over the upwind direction from cell i.

Having determined the precipitation, the space-time integrated discharge, from equation 13, for cell j (Fig. 4) is

$$(\Delta Q_r)_{jk} = R_j \tag{17}$$

where k is the cell among the surrounding eight to which the descent is steepest and $(\Delta Q_r)_{jk}$ is the increment in total discharge from precipitation on cell j. The total discharge along jk in time Δt (Fig. 4) is

$$(Q_r)_{jk} = \sum_{n=1}^{j} R_n \tag{18}$$

where the summation is over the cells that are upstream in the drainage network. The river network (Fig. 4) is determined by evaluating the steepest descent path between cells as a function of decreasing elevation.

The space-time integrated equilibrium sediment carrying capacity $(Q_f^{eqb})_{jk}$ of river segment jk in time Δt is, from equation 12,

$$(Q_f^{eqb})_{jk} = -K_f\,(Q_r)_{jk} \left(\frac{\Delta H}{\Delta L}\right)_{jk} \tag{19}$$

Capital letters again represent quantities averaged at the cell scale.

The fluvial erosion-deposition relationships are based on the simple physical assumption that the ability of a river to entrain sediment, both as suspended and bed load, is equivalent to a thermodynamic potential (P. Koons pers. comm. 1990) and is governed by a first-order reaction in which the reaction rate is proportional to a rate constant, t_f, and the degree of disequilibrium in the fluvial sediment flux as it moves in a Lagrangian sense along a river,

$$\frac{dq_f}{dt}\hat{l} = \frac{1}{t_f}\left(q_f^{eqb} - q_f\right)\hat{l} \tag{20}$$

where q_f^{eqb} and q_f are local equilibrium and local, integrated fluxes and $\hat{\underline{l}}$ is the local river direction. This reaction equation is equivalent to the more usual form written on the concentrations because the integrated flux, $q_f = c\upsilon_f A_f$, is proportional to the local mean sediment concentration, c, in a dynamically steady-state river which has constant sediment transport velocity, υ_f, and constant cross section A_f. In a stationary (Eulerian) coordinate system,

$$\frac{dq_f}{dt} = \frac{\partial q_f}{\partial t} + \underline{\upsilon}_f \cdot \nabla q_f \tag{21}$$

which reduces to $\upsilon_f \hat{l}\, dq_f / dl$ in our steady-state model which assumes that fluxes do not vary with time during the interval Δt. Under these circumstances the time dependence of erosion or deposition takes the form of a spatial dependence,

$$\frac{dq_f}{dl}\hat{l} = \frac{1}{l_f}\left(q_f^{eqb} - q_f\right)\hat{l} \tag{22}$$

where $l_f = \upsilon_f\, t_f$. For constant υ_f, l_f is a material property, the erosion-deposition length scale required for the disequilibrium to be reduced by $1/e$ when q_f^{eqb} is constant.

When $q_f < q_f^{eqb}$, the river either entrains alluvium created by the model, in which case $l_f = l_a$, or erodes the channel bedrock, in which case $l_f = l_c$, where $l_c >> l_a$. When $q_f > q_f^{eqb}$, sediment is deposited from the overcapacity river with a small $l_f < l_a$, as discussed below. These relationships illustrate the asymmetry between the control of erosion by a long l_c, equivalent to a large t_f, and control of deposition by a short l_f, equivalent to a small t_f. An implicit assumption is that steady-state is achieved in $t << \Delta t$ or, equivalently that $t_f << \Delta t$.

The space-time integrated form of equation 22 for erosion or deposition between cells j and k in time Δt is

$$(\Delta Q_f)_{jk} = \frac{\alpha\, c_L}{l_f}\left[Q_f^{eqb} - Q_f\right]_{jk} \tag{23}$$

where α=1 for adjacent cells and $2^{1/2}$ for diagonally connected cells. The corresponding change in height of cell j in time Δt is,

$$(D_f)_j = \frac{1}{c_L^2}\sum_{n=1}^{m}\left(\Delta Q_f\right)_{jn} \tag{24}$$

where the summation is over river segments entering and leaving cell j, and $(\Delta Q_f)_{jn} = -(\Delta Q_f)_{nj}$. By taking $l_f = c_L/2$ for deposition, it is ensured that excess sediment is deposited within the cell where the overcapacity developed. These explicit integrations are valid when $c_L << l_f$, but deposition at a scale less than c_L is not resolved. In addition, by

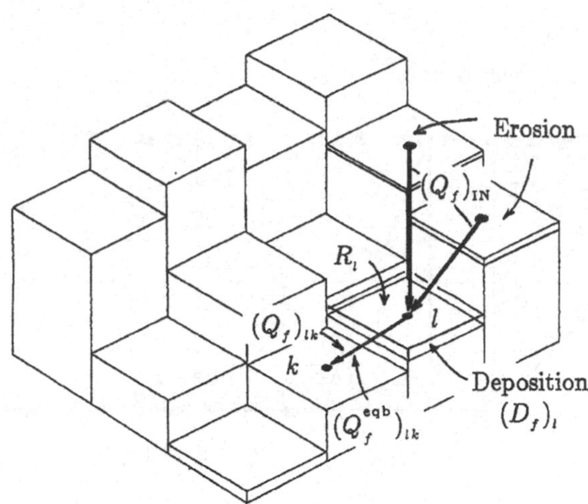

Figure 5. Illustration of the difference between the integrated fluvial transport capacity onto and away from a topographic cell.

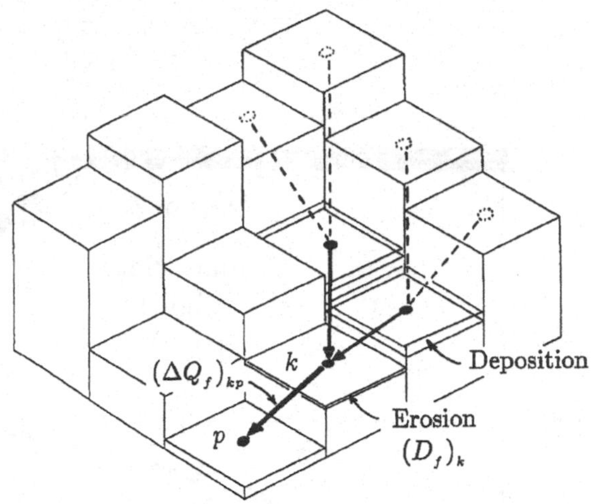

Figure 6. Illustration of the consequence for erosion or sedimentation of disequilibrium in sediment transport onto and away from a topographic cell.

evaluating $(Q_r)_{jk}$ in the order of cells ranked by descending elevation and simultaneously calculating $(Q_f^{eqb})_{jk}$, $(Q_f)_{jk}$, and $(\Delta Q_f)_{jk}$ avoids the need to calculate and store any information about the overall pattern of the drainage network.

Calculation of fluvial erosion on cell l (Fig. 5) requires summation of the sediment transported onto the cell by the n rivers draining higher cells in the network,

$$(Q_f)_{IN} = \sum_n (Q_f)_{nl} \tag{25}$$

The maximum transport capacity of the river draining l to k is $(Q_f^{eqb})_{lk}$. Figure 5 illustrates the example where the inflowing rivers onto l were at or near capacity and precipitation on l, R_l, did not significantly increase the discharge. There is also a large decrease in the downstream slope from l to k, therefore, sediment, thickness $(D_f)_l$, is deposited on cell l. It follows that all sediment is deposited at a local minimum in the topography. This paper does not, however, consider the formation of a lake or its outflow. Figure 6 illustrates the next step in the drainage between cells k and p. Even if there is no precipitation on k, erosion equal to $(D_f)_k$ will occur because the downstream carrying capacity increases by $(\Delta Q_f)_{kp}$ with the increase in downstream slope.

Model of the Southern Alps, New Zealand

The Southern Alps of South Island, New Zealand are the topographic expression of a young (~ 5 Ma old) collisional orogen that is a good example of the processes that we wish to investigate (Fig. 7). Space limitations restrict our discussion of its development but Norris *et al.* (1990) describe the current state of knowledge of this orogen. Furthermore, Koons (1989, 1990) and Norris *et al.* (1990) discuss models of the tectonics and erosional setting of the Southern Alps. While their models differ from our quantitative coupled model, many of the underlying concepts, which were in part developed during discussions with Peter Koons, have the same basis.

The plane-strain finite element model (Fig. 8) represents a 700 km wide region that is normal to and spans the Pacific-

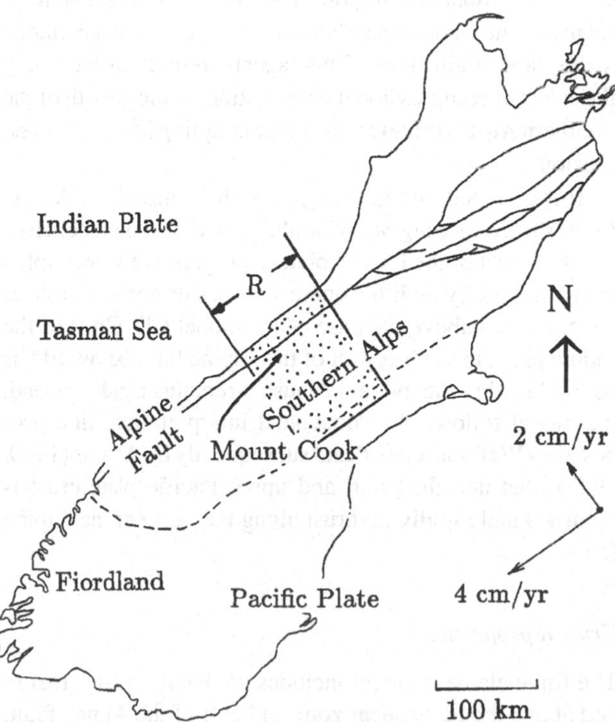

Figure 7. Geographic and plate tectonic setting of South Island, New Zealand. The velocities are the estimates for the current average motion of the Pacific plate with respect to the Indian plate. The relative motion is absorbed within the deforming plate boundary. Region labelled R and shown stippled is the one considered in the modelling.

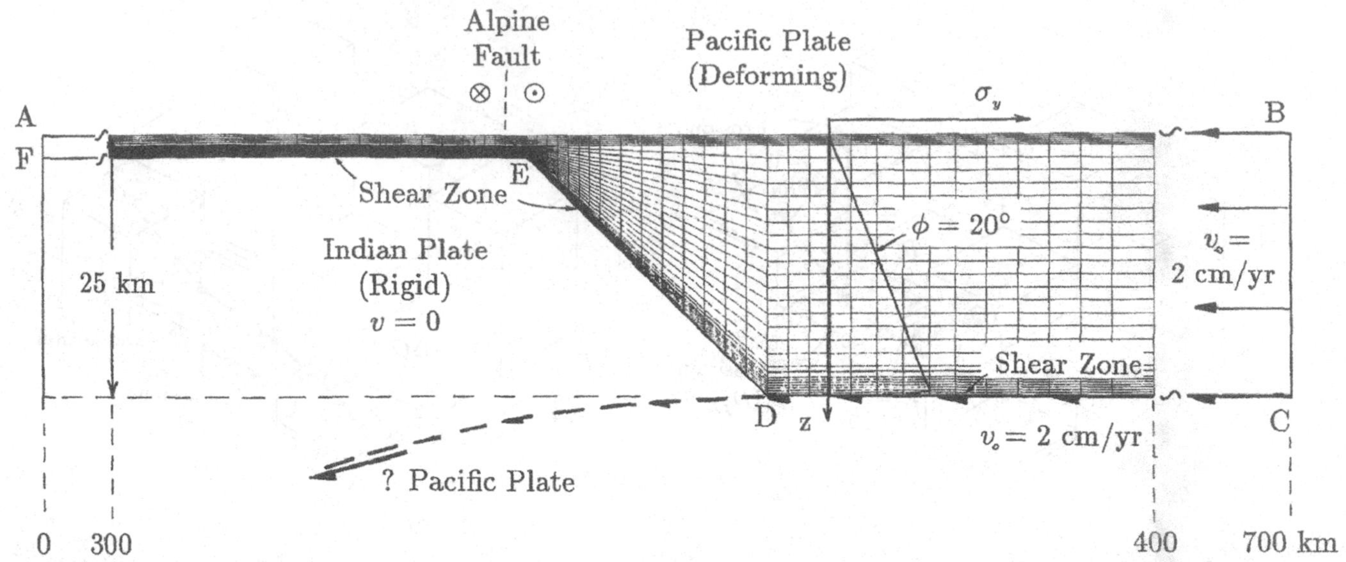

Figure 8. Plane-strain finite element model used to represent the tectonics of compressional deformation of central South Island, New Zealand. The grid has 59 x 31, 4-noded quadrilateral elements.

Indian plate boundary through central South Island. The width of the model was chosen to be greater than the plate boundary zone (Walcott 1984), therefore, the relative plate velocity component normal to the Alpine fault (Fig. 7), presently ~ 2 cm/yr, is absorbed by crustal deformation within the modelled region. The corresponding ~ 4cm/yr relative plate velocity parallel to the boundary is not included in the plane-strain model. This paper therefore considers only the effect of compressional deformation on the growth of the Southern Alps in this example of the coupling of tectonics and erosion.

It is assumed that to the right of the ramp of the Alpine fault, shown dipping at 45° at the centre of the model (ED, Fig. 8), crust of the Pacific plate is progressively decoupled from the underlying lithosphere along a sub-horizontal shear zone, located above the base of the model CD. Crust of the Indian plate, in the footwall of the Alpine fault below ED, is not included in the model because it remains rigid. Instead, the model follows the conceptual interpretation, first proposed by Wellman (1979) and subsequently by Adams (1980, 1985), that detached mid and upper Pacific plate crust is upturned and rapidly upthrust along the ramp of the Alpine fault.

Crustal properties

The finite element model includes the Pacific plate crust in and above the basal shear zone and east of the Alpine fault. The depth of the model (BC, Fig. 8) was chosen on the assumption that continued deformation and erosion has exposed rocks from as deep as the basal shear zone at the surface in the hanging wall adjacent to the Alpine fault; above E. The highest grade metamorphic rocks from this location are garnet oligoclase gneiss, which on the basis of the phase assemblage kyanite-quartz-talc indicates an equilibration depth ~25km (Cooper 1980). This depth was therefore taken as the depth of the model shear zone. Although this depth remains poorly determined, the choice is consistent with interpretations that typical continental lithosphere tends to delaminate at mid-crustal levels. The choice of model depth can also be tested *a posteriori* for consistency with observations because the predicted width of the zone of deformation and uplift is related to the assumed geometry and can be compared with the width of the Southern Alps.

The 2km thick region of the model above the rigid Indian plate (i.e. above region FE) represents crust and cover rocks that may thicken and detach to form a thin-skinned fold and thrust belt when subject to stress transmitted across the Alpine fault. This region is included in order to investigate how deformation might propagate westward with the growth of the Southern Alps were there no erosion. That this region has not deformed requires explanation.

A plastic-viscous rheology is used to model a frictional crustal rheology with a basal shear zone. When the internal angle of friction of the Drucker-Prager plastic material is taken to be 20°, the yield strength in compression, in crust density $\rho = 2800$ kg/m^3, increases linearly with depth to approximately 300 MPa at 20km. This yield strength with depth relationship approximates a uniform gradient form of Byerlee's Law (Brace & Kohlstedt 1980) in a thrust environment for a pore pressure ratio, $\lambda \approx 0.36$.

The transition from brittle-frictional behaviour to the ductile regime, represented by a linear viscous (Newtonian) rheology, is assumed to occur in a thin, basal shear zone, thickness d_S = 250m along DC, of effective viscosity, η_e = 3 x 10^{18} Pa.s. It could be argued that the model should include an ~ 5 km thick transition zone of decreasing effective viscosity below the brittle-ductile transition, which may be as deep as 20 km in a quartzo-feldspathic crust with a 20°C/km geothermal gradient (Ord & Hobbs 1989). This transition would require more, poorly determined, model parameters and is itself only inferred on the basis of laboratory data. Instead, the crustal rheology is purposely simplified. The present model focuses on the interplay between integrated frictional strength of the crust, specified by ϕ, and the opposing strength of the basal shear zone, characterized by the shear stress, $\tau_s \leq \eta_e \upsilon / d_s$, that it transfers from the boundary to the base of the frictionally dominated crust; where $\upsilon < \upsilon_o = 2$cm/yr is the relative velocity across the shear zone and is the specified boundary velocity. The model results are therefore sensitive to the ratio $\eta_e / d_s = 3 \times 10^{18} /250$ which determines τ_s and not to the independent values of η_e and d_s, to which no particular importance should be assigned.

The shear zone thickness decreases linearly with decreasing depth along the ramp DE and is 40 m thick beneath the region of the compressed finite elements along FE. This reduction in d_s could be compensated by a corresponding reduction in η_e but this was not done in order to increase the shear coupling in the colder, near surface, part of the crust. An improved model will consider a frictional base in the shallow parts of the model.

Boundary conditions

The behaviour of the lower crust and mantle below the modelled region dictates the basal boundary conditions. Simple buoyancy calculations suggest that typical continental lithosphere below ~ 25 km is negatively buoyant and will subduct if detached from the mid and upper crust. In this model it is assumed that the lower part of the Pacific plate has detached and, following Adams (1985), that it has passively underthrust the Indian lithosphere to the west in a similar way to the oceanic lithosphere further north except that the process is aseismic. This interpretation requires that the basal horizontal velocity along DC be equal to the Pacific plate velocity and equal to the uniform horizontal velocity along the boundary BC. These velocities are both taken to be $\upsilon_o = -2$cm/yr and the boundary on the Indian plate along AFED is fixed with $\upsilon = 0$. The model therefore considers compression of the Pacific plate crust as it is 'pushed' by the net normal force along BC and is 'dragged' by the shear, τ_s, that develops along DC. This interpretation is entirely equivalent to that which takes υ along BC to be zero and considers the indentation of the Pacific plate by the rigid Indian plate, $\upsilon_o = 2$ cm/yr, which acts as a tectonic wedge in the manner described by Price (1986). The former boundary conditions are preferred only because the location of the Alpine fault remains fixed in the model. It may actually be rotating in an absolute reference frame (Allis 1986).

Points along boundaries AF and BC are free to move vertically but this is not important because they are far enough removed from the plate boundary that deformation does not propagate to them. The upper boundary of the model is a free surface on which the shear stress is zero.

This tectonic interpretation implies that the sense of shear will be dextral along FED but sinistral along DC. In the absence of any additional net plate tectonic forces transmitted through the crust as a stress guide, i.e. along BC and AF, it is the *opposing* basal shear stresses that give an overall initial force balance to the model. An alternative kinematic interpretation (Allis 1980, 1986) holds that the underthrusting is opposite in sense; that the Indian plate underthrusts the Pacific plate as it does south of New Zealand beneath the Puysegur trench and probably beneath Fiordland. This choice of polarity reverses the sense of shear along DC and will, everything else being equal, tend to transfer deformation eastward away from the Alpine fault. It also requires additional forces, probably transmitted as shear from out of the plane of the model, to maintain an overall force balance. Although Allis's interpretation cannot be precluded and may be correct for the southern part of South Island we choose the simpler one to illustrate our coupled tectonic-erosion model. This model may therefore apply only to the central Southern Alps.

Surface processes model and coupling with the tectonic model

It is assumed for the purposes of this illustrative model that the current morphology and drainage pattern of the Southern Alps is not significantly influenced by the ancestral drainage pattern that predated the Late Cenozoic uplift as transpression developed. Model experiments show that this assumption is valid for random height topography of ~ 20 m but the maximum ancestral topography that is consistent with this assumption has not yet been determined. The initial surface was therefore assumed to be flat and at sea level. Small random height perturbations were then added to remove the degeneracy in the cellular grid heights and to create a random drainage pattern.

The model is progressively stepped in a coupled way, alternately calculating the tectonic velocity and displacements, followed by the effect of the two components of erosion acting on the current topography. Specifically, the uplift distribution normal to strike, $u_n(x)$, predicted by the tectonic model for timestep n, length ΔT, is divided into m equal substeps, $u_n^m(x)$, length Δt, for the surface processes calculation. The surface transport was then calculated over the m substeps, each step involving uplift of the topography by an increment $u_n(x)/$m, such that uplift varies in the cross-strike direction but is uniform along strike. The substeps are required to ensure that fluxes for each step are small by comparison with the total topography. The precipitation distribution was calculated separately for each substep because the topography changes with substep.

Feedback coupling between the surface processes model and the tectonics is calculated prior to the next step, $n + 1$, of the tectonic model. The current average along-strike topography from the surface model defines the new surface for the tectonic model. The tectonic model, therefore, incorporates the changing average mass of the growing mountains. Their weight influences the gravitational component of the subsur-

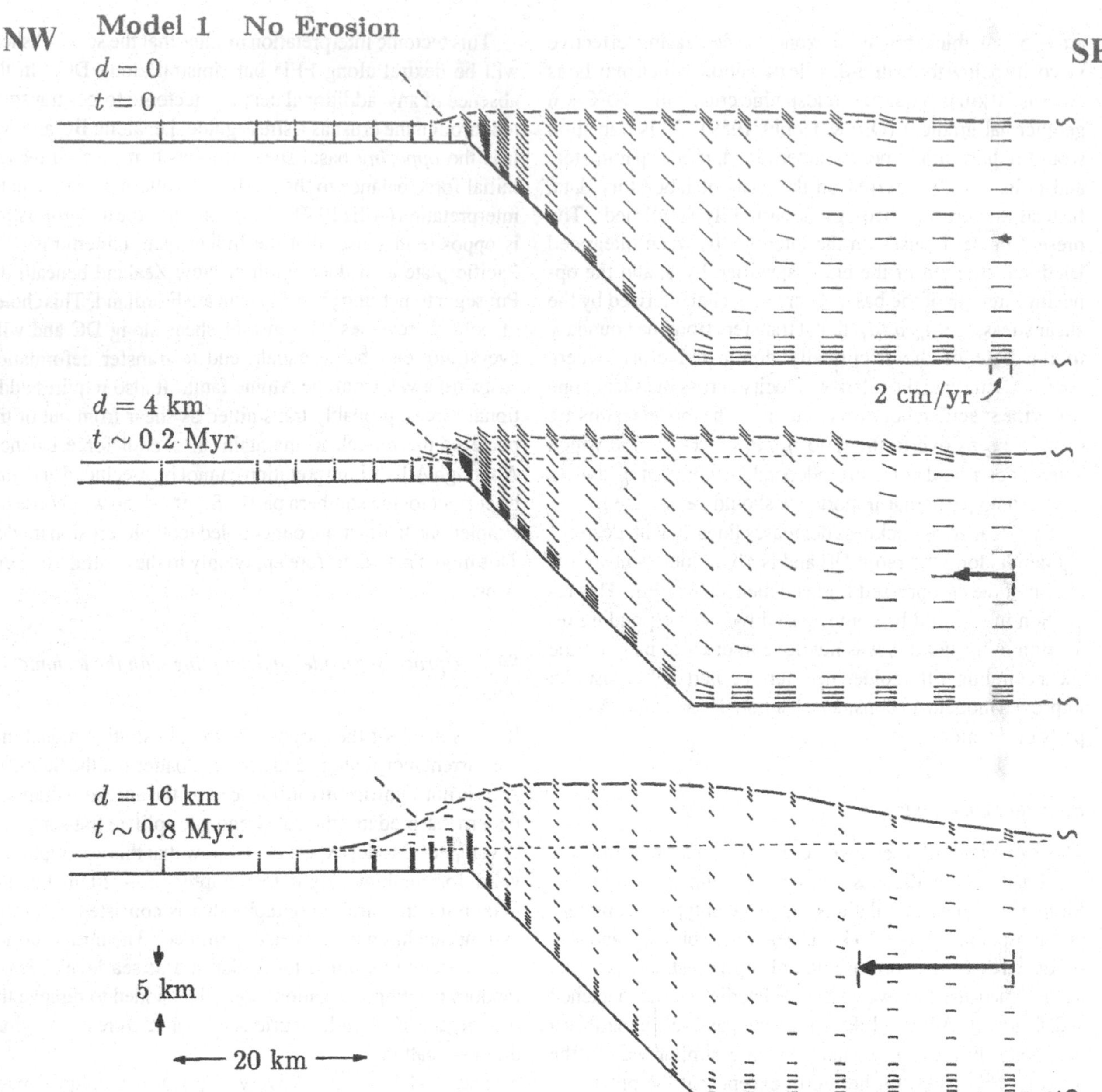

Figure 9. Model 1 results. Evolution of the plane-strain model in the absence of erosion. Velocity vectors are relative to a fixed Indian plate and show the progressive transformation of horizontal to vertical motion in the deforming plate boundary. Thrusting over the ramp of the Alpine fault, which acts as a finite height 'backstop', produces a westerly vergent tectonic wedge. Similarly, easterly vergent 'backthrusting' creates a second outward facing tectonic wedge on the eastern flank of the orogen. Bold horizontal arrows show the total amount of shortening, *d*.

face stress and ultimately modifies the pattern of deformation and uplift during the $n+1$ 'th tectonic step by comparison with the n 'th step.

Model results

The three models discussed here are only preliminary results and were chosen to illustrate how climate coupling and denudation may control the evolution of the Southern Alps for one particular interpretation of the tectonic style and properties of the orogen. The tectonic properties of all three models (Table 1) are identical. The evolution when there is no climate coupling or erosion is first examined and then two cases where denudation is dominated by fluvial transport with orographic rainfall derived from either a western or eastern source are considered.

Model 1: No erosion

The geometry and velocity distribution of model 1 are shown (Fig. 9) at the onset of deformation and again after ~ 0.2 and ~ 0.8 Ma, at which time the overall shortening is 4 and 16 km. The results illustrate that the model does respond by upturning the crust and thrusting it along the ramp of the Alpine fault. The initial velocity field comprises a region of passive hori-

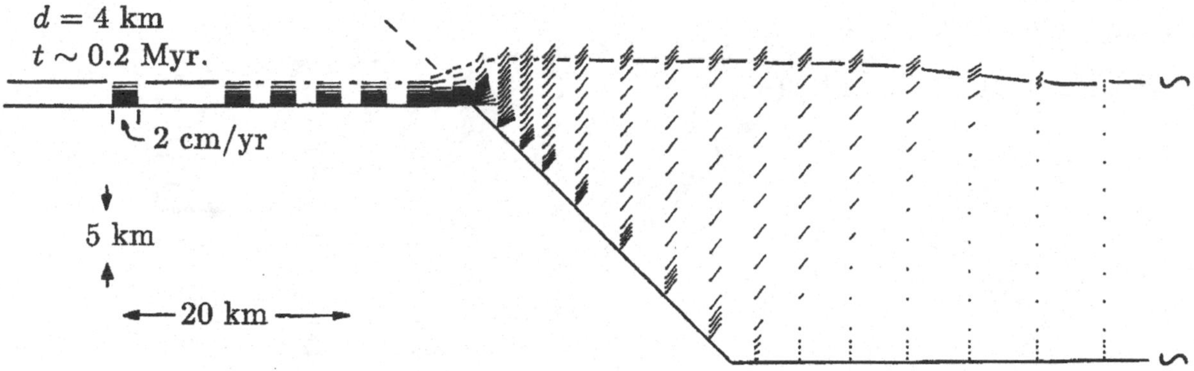

Figure 10. The same model result as shown in Figure 9 (middle panel) but with velocity vectors shown relative to a fixed Pacific plate. The easterly vergent tectonic wedge caused by backthrusting is seen clearly in this reference frame.

zontal transport of the Pacific plate crust without deformation and a region of uplift and deformation in a triangular wedge or 'plug' bounded by the ramp and its reflection about a vertical axis through D (Fig. 8). The initial width of the uplifting region is therefore approximately twice its depth when the dip of the ramp is 45°.

The initial style of deformation resembles the first of Wellman's (1979, Fig. 4) conceptual models of possible faulting patterns for the Southern Alps. His concern that this style would lead to an unrealistically steep eastern margin for the Southern Alps was unfounded as is seen from the topography after finite deformation (Fig. 9, middle panel).

The central region of the orogen continues to uplift in a relatively uniform manner, while the leading (western) and trailing (eastern) zones take the form of outwardly propagating tectonic wedges which increase in size with the uplift. The taper of the wedges depends on the internal frictional strength of the model crust, the basal dip, and the shear stress applied to the base of the frictional crust by the viscous shear zone. The results differ from the critical Coulomb wedge models (e.g. Dahlen 1984) because the model considers the development of a two-sided orogen and because the base of the model is viscous not frictional. The equilibrium topography therefore depends on the convergence velocity, in addition to the current geometry. The results have properties in common with Koons' (1990) conceptual models but differ because no significant basal decollement develops for this amount of shortening. The importance of the eastward propagating wedge becomes clear when velocities are plotted with respect to the Indian plate (Fig. 10).

It can be seen that only a small amount of shortening, ≤ 6 km, is required to create a mountain system with the same average height and width as the present Southern Alps. Shortening by a more realistic amount, 50 km for example, would have produced a much larger orogen by the addition of approximately 1250 km^2 per unit length of crust to the Southern Alps orogen, enough to thicken the crust by 20 km over a 60 km wide zone normal to strike. Although some of this excess mass almost certainly resides in thickened crust located to the east of the Alpine fault (Woodward 1979; Allis 1986), a significant proportion has apparently been eroded. Note that the model must be improved to include isostasy before firm conclusions can be made. Further evidence of erosional control of structure is the absence, except for a few nappes, of a thin-skinned fold and thrust belt west of the Alpine Fault. The model indicates that this westward propagating wedge would develop rapidly and achieve a significant size after only 16 km of shortening. It would be a fully fledged fold and thrust belt after 50 km of shortening.

Model 2: wind from the west

The second model (Figs 11 & 12) includes the effects of surface erosion by slope processes and a fluvial system recharged with orographic precipitation from a western source. Model parameter values are given in Table 1. The evolution of the tectonic model (Fig. 11) shows the strike average response as the surface uplifts and the orographic precipitation develops. The volume of sediment transported by denudation equals the rate of tectonic influx at ~ 0.5 Ma, d = 10 km of shortening, after which the orogen remains on average in a dynamically near steady state as discussed below.

Erosion on the western flank of the mountains dominates because the precipitation is greatest on that side and because the boundary condition on the surface transport model represents the Tasman Sea ($x = 332$km, Fig. 9) as an infinite sink for eroded sediment. This choice is appropriate because most of the sediment that enters the Tasman is transported to the south. The boundary condition on the east is the same, while the north and south boundaries are reflective.

The strike-averaged precipitation distribution has the same character as that shown by Griffiths & McSaveney (1983, their Fig. 2; see also Whitehouse 1988) a maximum to the west of the main divide but some precipitation on the eastern

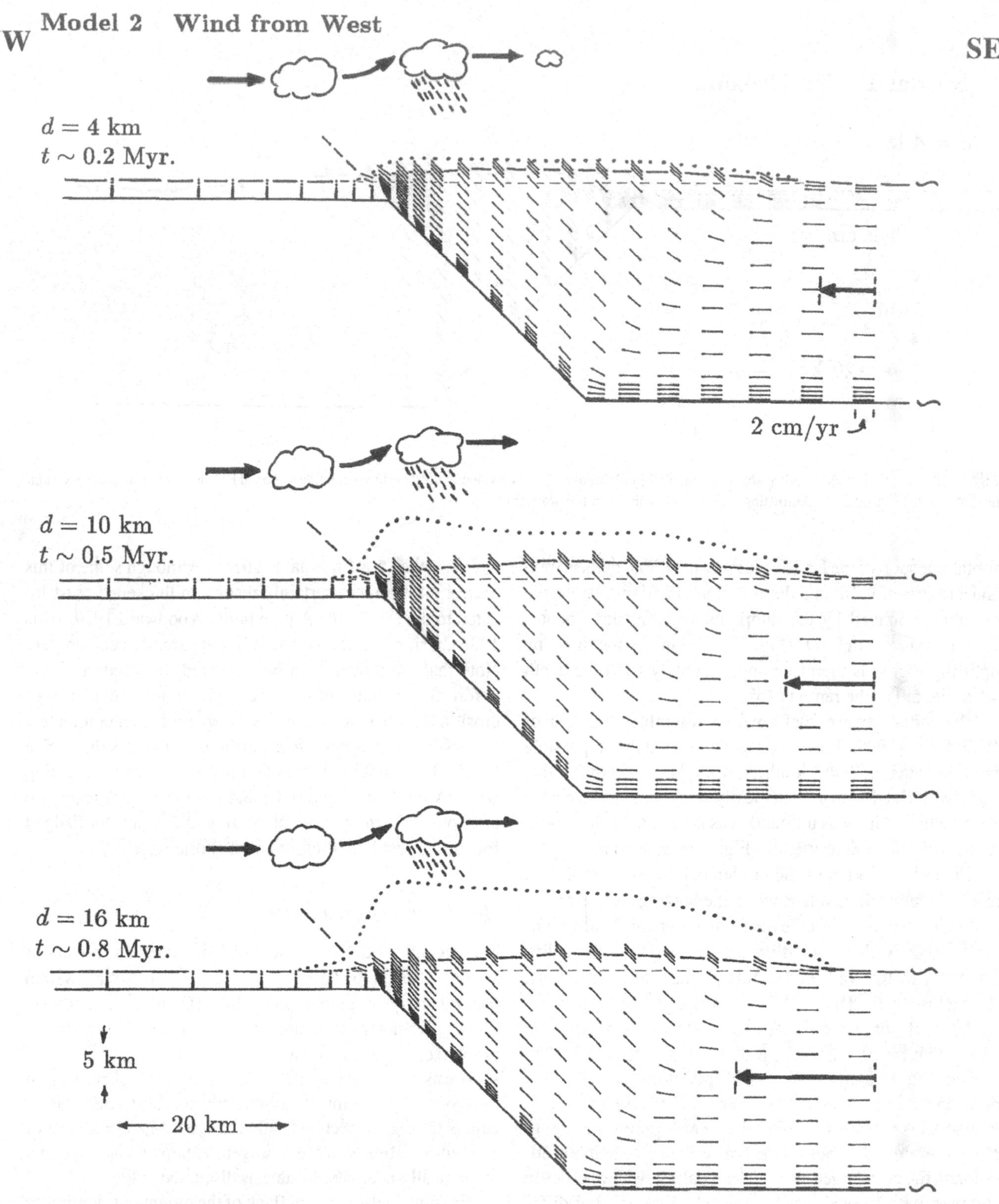

Figure 11. Model 2 results. Evolution of the plane-strain model when orographic precipitation is derived from a westerly source. Velocity vectors are relative to a fixed Indian plate. The dotted lines show the projected position of the initial surface. The westerly vergent tectonic wedge of Figure 9 has been suppressed by erosion. The western sink for mass also stunts the growth of the orogen as a whole and a near steady state is reached with only a small easterly vergent tectonic wedge. It is clear that erosion has modified structural and metamorphic style throughout the model orogen, not just locally. Bold horizontal arrows show the total amount of shortening, *d*.

flank. Agreement of the model precipitation with short-term observations is not independent confirmation that the model is correct because the timescales are totally different. The model rainfall must represent processes at > 10^2 yrs.

The dotted lines (Fig. 11) show the position of the original surface projected by its cumulative vertical displacement. The distance between the model surface and the dashed line is, therefore, a measure of the total erosion in the spatially fixed model coordinate system. It would equal the exhumation experienced by rocks currently at the surface were the uplift

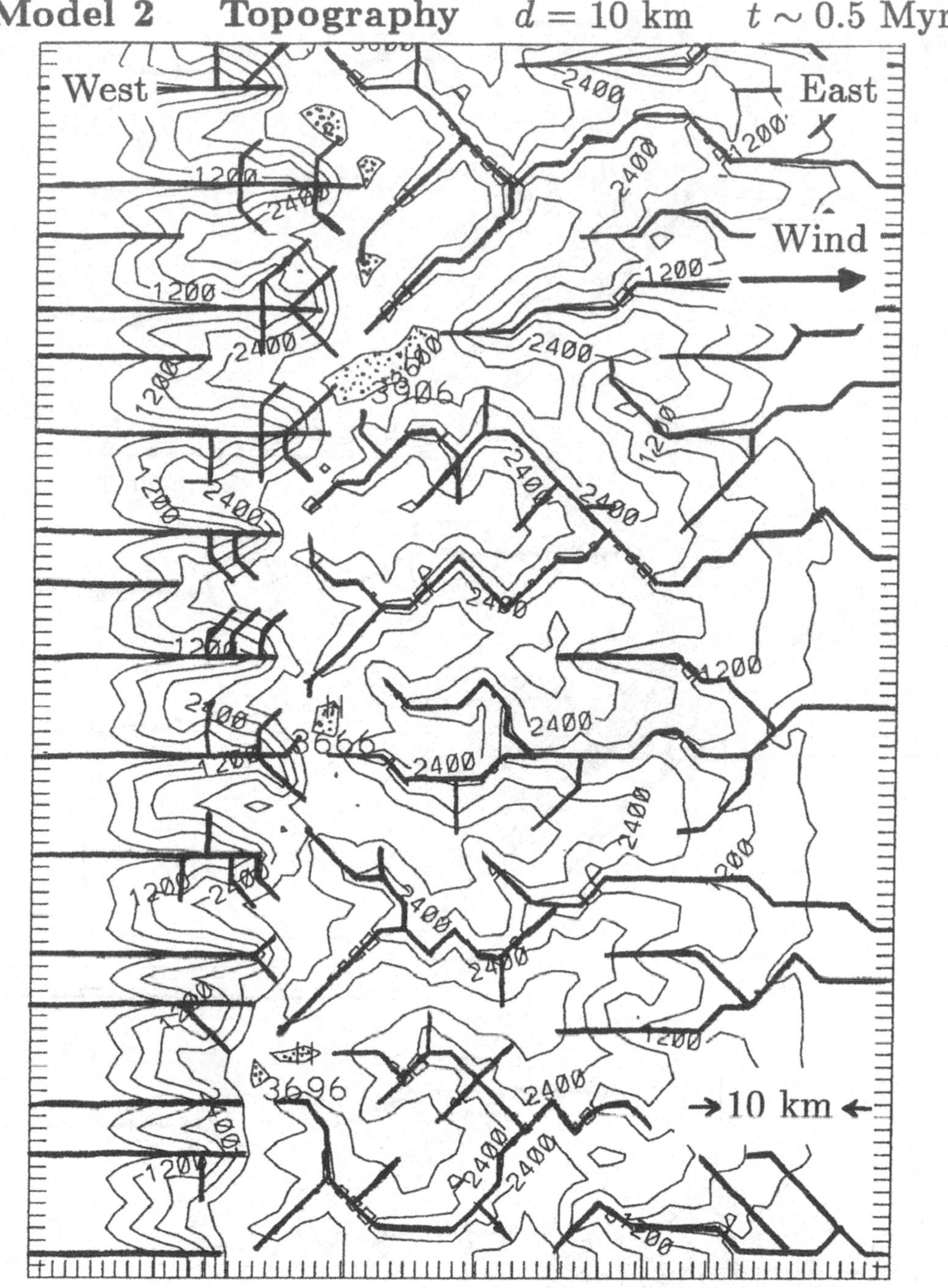

Figure 12. Map of Model 2 topography at 0.5 Ma after 10 km of shortening. The contrast in the drainage pattern and morphology east and west of the divide is also characteristic of the central Southern Alps. The contour interval is 600 m, heights above 3600 m are stippled, and the area is 104 x 72 km. The cell size for the calculations was 2 x 2 km. Only the major rivers are shown.

vertical. However, the horizontal velocity advects the uplifting crust progressively to the west and rocks reach the surface horizontally offset from their original position. Consequently, once steady state has been achieved, the highest grade metamorphic rocks, those exhumed from the greatest depth, will reach the surface of the model adjacent to the Alpine Fault. This result agrees with the observed distribution of metamorphic grade, which is a maximum at the Alpine fault and decreases to the east. It also follows that rocks which have undergone the greatest exhumation do not reach the surface where the erosion rate is greatest even in a steady-state orogen. This is an important point to remember when interpreting the tectonics of ancient orogens on the basis of distribution of metamorphic grade. The metamorphic grade is, however, a measure of the integrated erosion in the tectonic 'upstream' direction and the horizontal gradient of exhumation is the quantity most closely related to the spatial distribution of surface erosion.

The contour map of the topography (Fig. 12) for model 2 after 0.5 Ma has a number of features that can be related to the topography of central South Island in the area shown in Figure 7. It is important to remember that the morphology and river system distribution has been self-selected by the model as it evolves from the initial low amplitude random

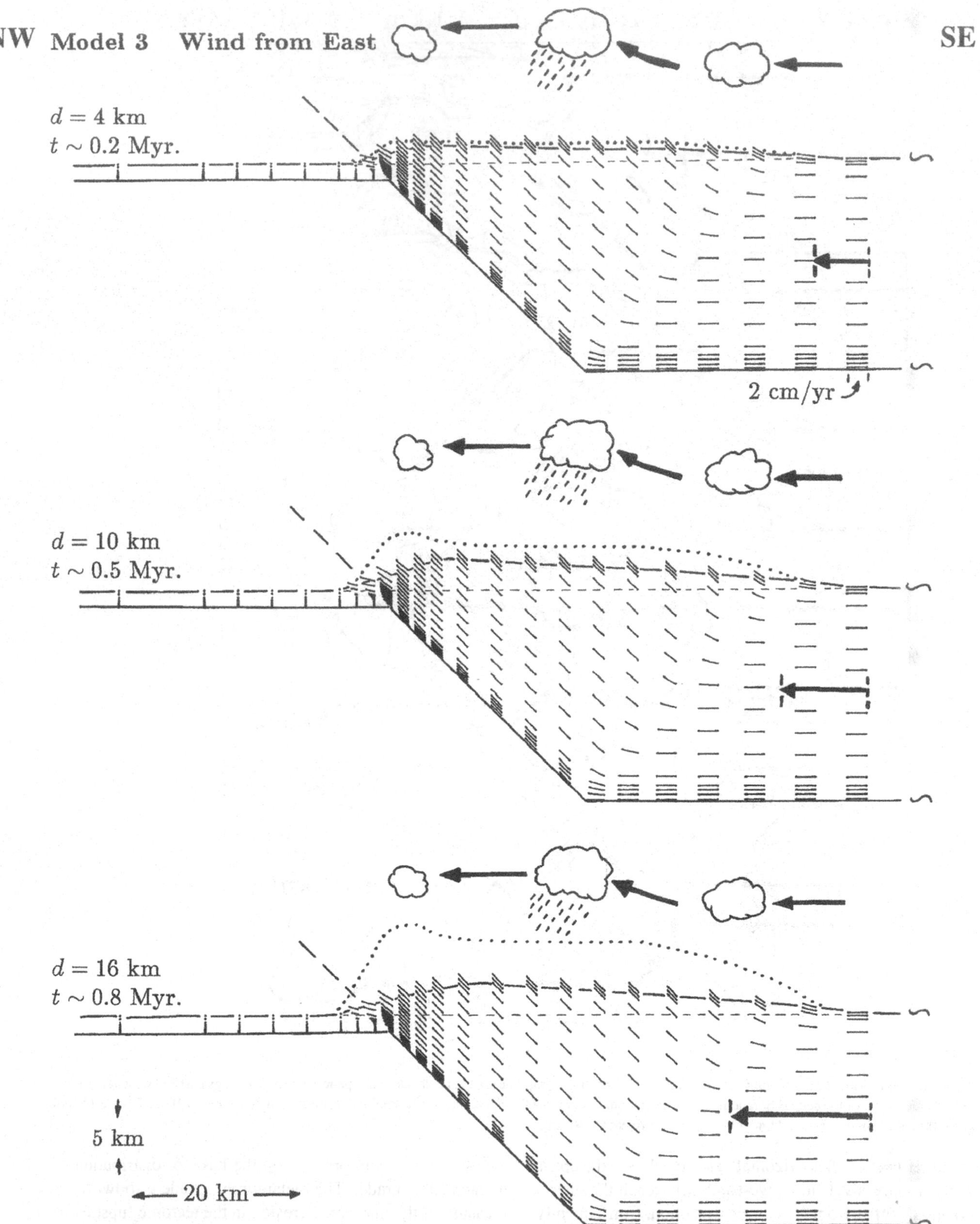

Figure 13. Model 3 results. Evolution of the plane-strain model when orographic precipitation is derived from an easterly source. The source has the same water vapour flux at its eastern boundary as Model 2 had at its western boundary. Velocity vectors are relative to a fixed Indian plate. The dotted lines show the projected position of the initial surface. Bold horizontal arrows show the total amount of shortening, *d*.

topography. The topography therefore contains information on the dynamic interaction of uplift with the surface processes which will be analyzed in a future paper.

West of the drainage divide, the front of the model orogen is cut by linear streams, spaced from 6-12 km apart, that have created an along-strike ridge-valley system. The ridges have steep noses, lower gradient centres and rise again to the east, where they climb to the divide. The valleys have strongly concave profiles. These are the same features that Koons (1989) recognized as characteristic of the western face of the

Southern Alps and reproduced with a planform diffusive mass transport model. In this model he specified the river spacing and valley topography. Here the river spacing is self-selected, and need not be imposed, when fluvial transport is included in the model. The spacing of these small rivers, which is similar to that of the central Southern Alps, results from the competition between hillslope and fluvial processes during uplift. Initially, all of the one hundred rivers equally drain and erode the western face of the model orogen, but an instability occurs with increasing slope and fluvial discharge which leads to finite incision of rivers with the spacing shown in Figure 12. This instability may be related to slope concavity (Smith & Bretherton 1972) but our model is complicated by the additional effects of diffusion and erosion/deposition kinetics. At least one aspect of the river selection, growth and incision requires that when a cell height is perturbed, the change in ΔQ_f from the cell be larger than the corresponding change in Q_s onto the cell. The same competition may explain why the main rivers on the eastern slope of the model orogen are more widely spaced than their westward-drainage counterparts. The eastward-draining rivers have smaller equilibrium sediment carrying capacities for equivalent sized watersheds. Therefore, only rivers that acquire large discharge by developing large watersheds can achieve the same degree of incision as their western equivalents.

The main peaks that resemble Mount Cook are located on or close to the drainage divide (Fig. 12). The true counterparts of the model ridges east of the divide and normal to it may be the southerly trending ridges of the Southern Alps. It is suggested that the distributed plate-boundary shear strain has rotated what were initially ridges trending perpendicular to the orogen and propose to test this hypothesis with a complete planform model of thin-sheet tectonics coupled with erosion (Ellis *et al.* in press).

The near steady-state nature of the model orogen (Fig. 11) for $t > 0.5$ Ma was judged qualitatively from the change in strike-averaged topography by comparison with the same measure of the erosion between $t = 0.5$ Ma and $t = 0.8$ Ma (Fig. 11). The results indicate that the volume or mass of the orogen is approximately in dynamic equilibrium. Individual topographic features are not, however, in equilibrium. For example, the westward draining rivers become longer with increasing time.

The present model cannot provide a quantitative assessment of the morphological evolution because the topography is neither advected nor strained by the tectonic deformation in the horizontal plane. The results are therefore no longer valid when the effects of horizontal advection become significant. For example, the rate of advection of the topography toward the Alpine fault is almost equal to the predicted rate of increase in the length of the westward-draining rivers. Once advection of the topography has been added to the model, it will be possible to assess whether a statistically steady-state topography can exist and track the birth, growth and death of the Mount Cook-like mountains as they are advected across the model.

Model 3: wind from the east

The third model (Figs 13 & 14) is identical to model 2 except that the precipitation is now derived from an eastern source, which is exactly opposite to the prevailing conditions in the Southern Alps. Model 3 is therefore a hypothetical example designed to indicate how the Southern Alps may have developed with reversed symmetry in the climate.

As the surface uplifts and orographic rainfall develops, the eastern flank of the orogen is subject to the greatest precipitation and erosion. Erosion on the tectonic divide and on the west flank of the orogen is therefore reduced by comparison with model 2. Consequently, the western flank of the orogen continues to uplift and a major range of peaks is created which sit astride the divide. All of these peaks exceed the height of Mount Cook within 0.5 Ma of the onset of compression (Fig. 14). They continue to grow and begin to assume Himalayan stature after approximately 1 Ma of convergence.

River incision of the western flank is less well developed (Fig. 14) than in model 2 (Fig. 12) because the fluvial discharge is smaller. The instability that leads to incision is therefore delayed until the slopes are steep enough for the rivers to achieve sufficient power. In Figure 14, the instability has not fully developed along the entire western flank. The steepness of the western face is, however, preserved by the flux boundary along the western side which mimics the effects of longshore transport in the Tasman Sea.

Creation of Himalayan-size mountains occurs with increasing speed as the extraction efficiency of the precipitation is increased. For example, when a_R is increased by a factor of three, precipitation is nearly exhausted east of the drainage divide which now develops only 10 km from the model coastline (Fig. 15). The divide is crowned by 8000 m peaks within 0.8 Ma of the onset of compression.

Correspondingly, as a_R increases in models with an eastern precipitation source, the implied belt of highest grade metamorphic rocks migrates eastward away from its position against the Alpine fault and may even cross the drainage divide. With declining precipitation on the western flank, neither erosion nor horizontal advection are sufficient to expose deeply buried rock against the model Alpine fault. The influence of erosion patterns on exhumation raises the possibility that the present location of the Alpine fault is partly controlled by the surface processes.

Conclusions

Both the model and the results as preliminary and only general conclusions are valid. The results do, however, indicate that in instances where the surface mass transport occurs over distances and at rates comparable to the tectonic mass flux, orogenic evolution may be significantly modified by climatic conditions at the scale of the orogen. One implication is that because precipitation is often asymmetric across an orogen, given sufficient time, the orogen may adopt the same asymmetry in its tectonic style.

Proof of climate influence on the tectonics of ancient

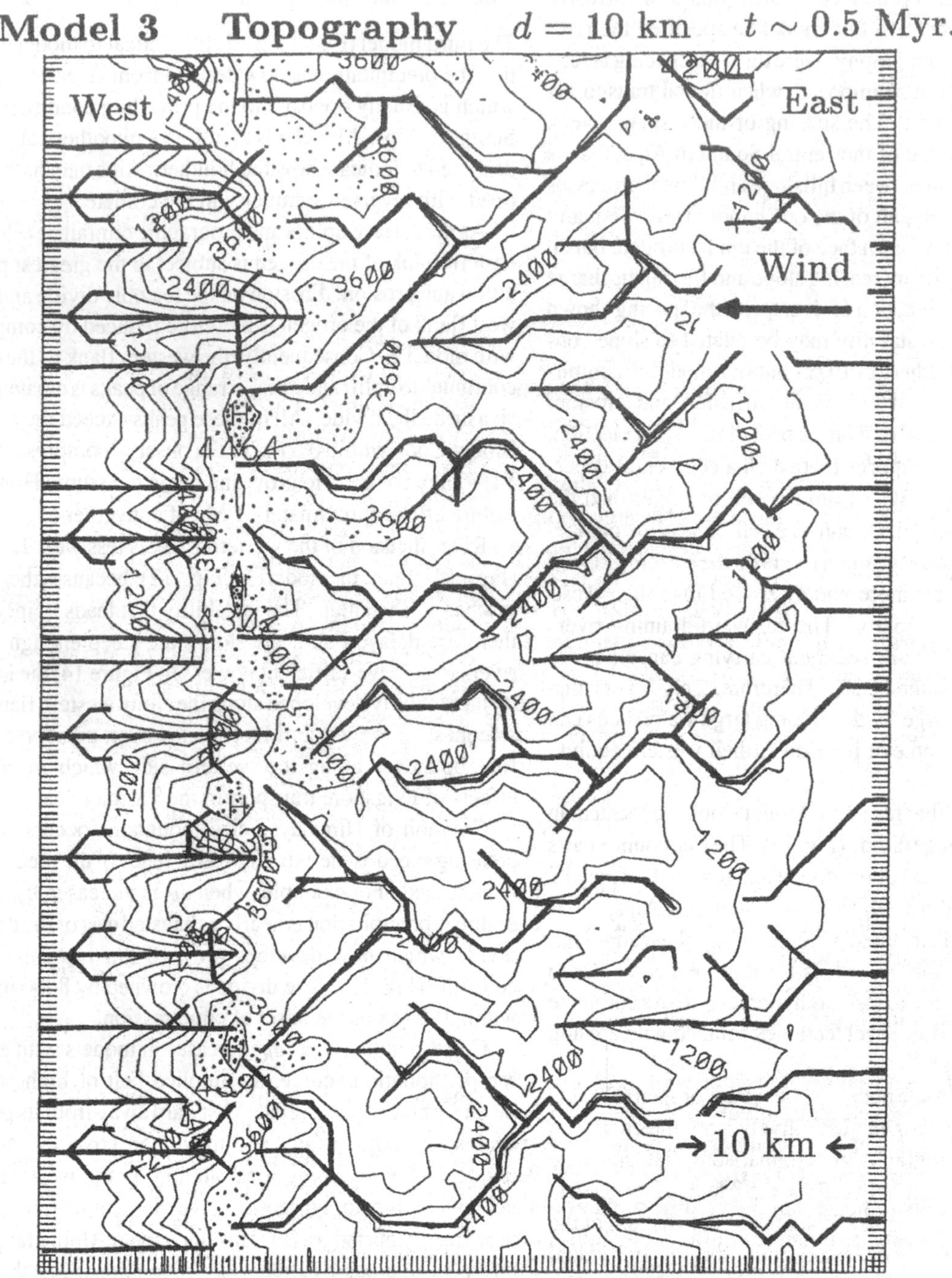

Figure 14. Map of Model 3 topography at 0.5Myr after 10 km of shortening. The contour interval is 600 m, heights above 3600 m are stippled, and the area is 104 x 72 km. The cell size for the calculations was 2 x 2 km. Only the major rivers are shown.

orogens will necessarily be based on circumstantial evidence. South Island, New Zealand, however, offers a dramatic contemporary example of these process at work. Control of this orogen is shared between plate tectonic and surface processes. The dynamic linkage between tectonics and erosion is provided by the atmosphere. Prevailing climatic conditions ensure a sufficient source of orographic rainfall which, when coupled with the tectonically induced steep western slopes, locks the orogen in a dynamically near steady-state mass balance.

These models are only first-order approximations of the evolution of orogens in general, and the Southern Alps, in particular. If it is accepted that model 2 does reproduce some of the major characteristics of the southern Alps, then comparison with models 1 and 3 illustrates the significance of the coupling between the atmosphere and the solid earth. Timescales of 0.5 - 5.0 Ma may be sufficient for profound climatic effects in collisional orogens.

Not all orogens are, or were, as strongly modified by climate coupling as the Southern Alps - an extreme example chosen to demonstrate this thesis. It is possible to suggest that, if the Himalayan orogen was subject to intervals of steady-state evolution, these may have been responsible for the exhumation of the belt of highest grade metamorphic rocks.

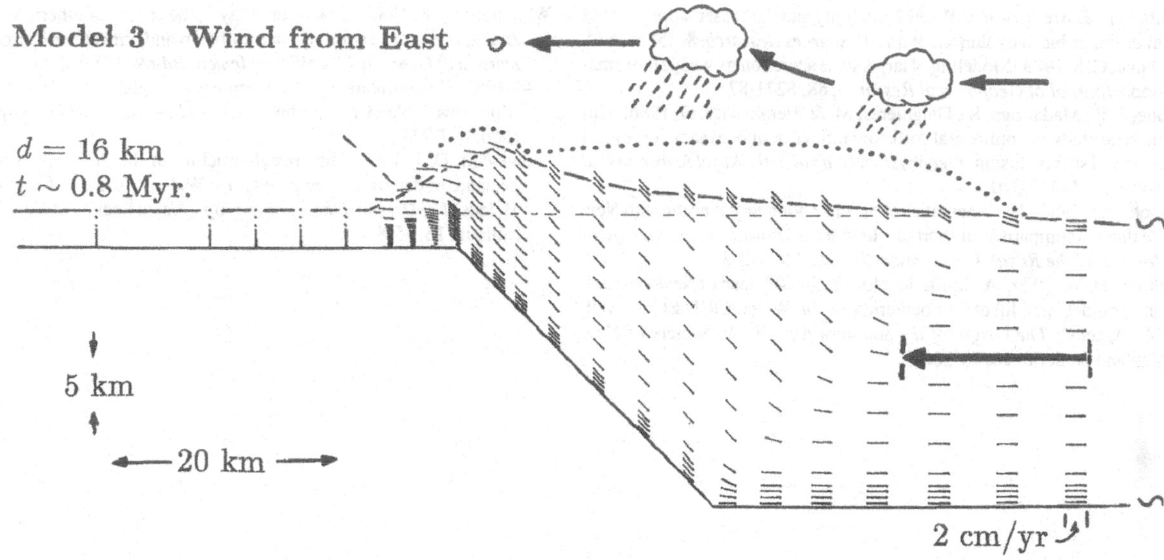

Figure 15. Model 3 results. Evolution of the plane-strain model when orographic precipitation is derived from an easterly source. The model is identical with the one shown in Figure 13 (bottom panel) except that the rainfall extraction efficiency has been increased by a factor of 3.

This work benefited greatly from discussions with Sean Willett, Peter Koons and Becky Jamieson. We would also like to thank Clem Chase for providing us with information on his precipitation landsculpting model prior to publication and for his and Peter Koons' comments on a preliminary version of the manuscript. The research has been funded by the Natural Sciences and Engineering Research Council of Canada through Operating and Lithoprobe Supporting Geoscience grants to C. Beaumont.

References

Adams, J. 1985. Large scale tectonic geomorphology of the Southern Alps, New Zealand. *In*: Morisawa, M. and Hack, J. T. (eds). *Tectonic Geomorphology*, Allen and Unwin, 105-28.

—— 1980. Contemporary uplift and erosion of the Southern Alps, New Zealand. *Geological Society of America Bulletin, Part II*, **91**, 1-114.

Adina. 1987. ADINA - a finite element program for automatic dynamic incremental nonlinear analysis. *Report ARD 87-1, ADINA R and D Inc.*

Allis, R. G. 1986. Mode of crustal shortening adjacent to the Alpine Fault, New Zealand. *Tectonics*, **5**, 15-32.

—— 1981. Continental underthrusting beneath the Southern Alps of New Zealand. *Geology*, **9**, 303-7.

Armstrong, A. C. 1980. Soils and slopes in a humid environment. *Catena*, **7**, 327-38.

Barr, T. D. & Dahlen, F. A. 1989. Brittle frictional mountain building 2, thermal structure and heat budget. *Journal of Geophysical Research*, **94**, 3923-48.

Begin, S. B., Meyer, D. F. & Schumm, S. A. 1981. Development of longitudinal profiles of alluvial channels in response to base-level lowering. *Earth Surface Processes and Landforms*, **6**, 49-68.

Brace, W. F. & Kohlstedt, D. L. 1980. Limits on lithospheric stress imposed by laboratory experiments. *Journal of Geophysical Research*, **85**, 6248-52.

Carson, M. A. & Kirkby, M. J. 1972. *Hillslope Form and Processes*, Cambridge University Press, 475p.

Chapple, W. A. 1978. Mechanics of thin-skinned fold and thrust belts. *Geological Society of America Bulletin*, **84**, 1189-98.

Chase, C. G. 1989. Fluvial landsculpting: why topography is like a fractal. *Geological Society of America, Abstracts with Programs*, A 38.

—— 1988. Fluvial landsculpting and the fractal dimension of topography. *EOS*, **69**, 1207.

Coleman, S. M. & Watson, K. 1983. Ages estimated from a diffusive equation model for scarp degradation. *Science*, **221**, 263-5.

Cooper, A. F. 1980. Retrograde alteration of chromian kyanite in metachert and amphibolite whiteschist from the Southern Alps, New Zealand, with implications for uplift on the Alpine Fault. *Contributions to Mineralogy and Petrology*, **75**, 153-64.

Culling, W.E.H. 1965. Theory of erosion on soil-covered slopes. *Journal of Geology*, 73, 230-54.

—— 1960. Analytical theory of erosion. *Journal of Geology*, **68**, 336-44.

Dahlen, F. A. & Barr, T. D. 1989. Brittle frictional mountain building 1, deformation and mechanical energy budget. *Journal of Geophysical Research*, **94**, 3906-22.

—— 1984. Noncohesive critical Coulomb wedges: An exact solution. *Journal of Geophysical Research*, **89**, 10125-33.

Davis, D., Suppe, J. & Dahlen, F. A. 1983. Mechanics of fold-and-thrust belts and accretionary wedges. *Journal of Geophysical Research*, **88**, 1153-72.

Ellis, S., Fullsack, P. & Beaumont, C. 1990. Incorporation of erosion into thin sheet numerical models of continental collision, *EOS* (in press).

England, P. C. & McKenzie, D. P. 1982. A thin viscous sheet model for continental deformation. *Geophysical Journal of the Royal Astronomical Society*, **70**, 295-321.

Flemings, P. & Jordan, T. E. 1989. A synthetic stratigraphic model of foreland basin development. *Journal of Geophysical Research*, **94**, 3851-66.

Griffiths, G. A. & McSaveney, M. J. 1983. Distribution of mean annual precipitation across some steepland regions of New Zealand. *New Zealand Journal of Science*, **26**, 197-209

Jamieson, R. A. & Beaumont, C. 1989. Deformation and metamorphism in convergent orogens: a model for uplift and exhumation of metamorphic terrains. *In*: Daly, J.S., Cliff, R.A. & Yardley, B.W.D. (eds) *Evolution of Metamorphic Belts*, Geological Society of London Special Publication, **43**, 117-29.

—— & —— 1988. Orogeny and metamorphism: a model for deformation and pressure-temperature-time paths with application to the central and southern Appalachians. *Tectonics*, **7**, 417-45.

Koons, P. O. 1989. The topographic evolution of collisional mountain belts: a numerical look at the Southern Alps, New Zealand. *American Journal of Science*, **289**, 1041-1069.

——1990. The two-sided orogen: collision and erosion from the sand box to the Southern Alps, New Zealand. *Geology*, **18**, 679-682.

Malvern, L.E. 1969. *Introduction to the Mechanics of a Continuous Medium*. Prentice-Hall, New Jersey.

Norris, R. J., Koons, P. O. & Cooper, A. F. 1990. The obliquely-convergent plate boundary in the South Island of New Zealand: implications for ancient collision zones. *Journal of Structural Geology*, **12**, 715-725.

Ord, A. & Hobbs, B. E. 1989. The strength of the continental crust, detachment zones and the development of plastic instabilities. *Tectonophysics*, **158**, 269-89.

Smith, T. R. & Bretherton, F. P. 1972. Stability and the conservation of mass in drainage basin evolution. *Water Resources Research*, **8**, 1506-1529.

Stockmal, G. S. 1983. Modeling of large-scale accretionary wedge deformation. *Journal of Geophysical Research*, **88**, 8271-87.

Vilotte, J. P., Madariaga, R., Daignières, M. & Zienkiewicz, O. 1986. Numerical study of continental collision: influence of buoyancy forces and an initial stiff inclusion. *Geophysical Journal of the Royal Astronomical Society*, **84**, 279-310.

Walcott, R. I. 1984. The kinematics of the plate boundary zone through New Zealand: a comparison of short-and long-term deformations. *Geophysical Journal of the Royal Astronomical Society*, **79**, 613-33.

Wellman, H. W. 1979. An uplift map for the South Island of New Zealand, and a model for uplift of the Southern Alps. *In*: Walcott, R. I. & Cresswell, M. M. (eds) *The Origin of the Southern Alps*, Royal Society of New Zealand Bulletin, **18**, 13-20.

Whitehouse, I. E. 1988. Geomorphology of the central Southern Alps, New Zealand: the interaction of plate collision and atmospheric circulation. *Zeitschrift Geomorphologie Supplement Bd-69*, 105-16.

—— 1987. Geomorphology of a compressional plate boundary, Southern Alps, New Zealand. *In*: Gardiner, V. (ed.) *International Geomorphology 1986*, 897-923.

Woodward, D. J. 1979. The crustal structure of the Southern Alps, New Zealand, as determined by gravity. *In*: Walcott, R. I. and Cresswell, M. M. (eds) *The Origin of the Southern Alps*. Royal Society of New Zealand Bulletin, **18**, 95-8.

Dynamic and kinematic growth and change of a Coulomb wedge

Sean D. Willett

Department of Oceanography, Dalhousie University, Halifax, Nova Scotia, B3H4J1, Canada

Abstract: Deformation and structural relationships in accretionary prisms and fold and thrust belts are the result of dynamic changes in the size, geometry, or strength of the deforming wedge and its boundary conditions. The concepts of critical slope or taper that have been successful in explaining the static geometry and state of stress in a Coulomb wedge can be expanded through the use of finite element models to consider the kinematics and dynamics of a deforming Coulomb wedge. The numerical technique adopts a Coulomb failure criterion and isotropic plastic flow in a velocity-based Eulerian formulation. This formulation allows for very large deformation to be accommodated by a numerical mesh that remains fixed in space, deforming only to follow the movement of the upper surface.

Critical wedge theory defines deformational domains bounded by the critical wedge solutions. Imposed changes in boundary conditions or geometry can move the mechanical state of a wedge off a critical line into either the sub-critical or stable domain, in which a wedge is unstable during accretion, leading to transient deformation as the wedge adjusts to a new critical geometry.

With steady boundary conditions the accretion process leads to self-similar growth. A zone of high strain rate representing the frontal step-up thrust and the decollement develops and separates underthrust sediment from the deforming wedge. An increase in basal strength produces large internal deformation as the wedge increases its taper. A decrease in basal strength concentrates deformation at the toe of the wedge. A large decrease in basal strength may lead to extensional collapse. A complex geological history involving repeated cycles of growth and collapse could produce tectonic exhumation of the deeply buried interior of the wedge, even in the absence of erosion.

In compressional tectonic regimes much of the convergence, at least at upper crustal depths, is manifested by shortening and thickening of a fold and thrust belt. However, at less than crustal scale, styles of deformation and the kinematics of the development of a fold and thrust belt can vary greatly. The classic model of a forward propagating sequence of thrust faults remains an important component of most or all thrust belts and plate margin accretionary wedges, but other large scale deformational processes including out-of-sequence faulting (Morley 1988), sediment underplating (Moore *et al.* 1982; Platt *et al.* 1985; Platt 1986, 1987; Westbrook *et al.* 1988) and even extension (Platt 1986) have been recognized as contributing to the observed structure of thrust belts or compressional orogens as a whole. Along with a recognition of the complexity of the large scale structural and temporal relationships in thrust belts has been an acceptance of the fact that thrust belts, and even entire orogens, behave as a mechanical entity. This has led to the development of a number of continuum models for the mechanics of thrust belts.

Continuum models have generally addressed one of two aspects of thrust belt mechanics; (1) the geometry and state of stress within the thrust belt; or (2) the kinematics of material transport through the thrust belt. Chapple (1978) presented a model for the state of stress in a thrust belt assuming it acted as a plastic material with the stress everywhere at the yield stress. Stockmal (1983) used slip line theory to expand this model and to calculate both stress and internal velocities based on the thrust belt geometry and assumptions regarding the slip velocity on the underthrusting plate. Davis *et al.* (1983) argued that a Coulomb yield stress was more appropriate for thrust belts than the depth-independent von Mises yield criterion used by Chapple (1978) and Stockmal (1983). They presented solutions for the state of stress in both non-cohesive (Davis *et al.* 1983) and cohesive (Dahlen *et al.* 1984) Coulomb thrust belts based on approximately wedge shaped cross-sectional geometries, modified to include the effects of pore fluid pressures. Subsequent work addressed changes in cohesion associated with compaction and lithification (Zhao *et al.* 1986). Dahlen (1984) presented an exact analytic solution for the state of stress in a non-cohesive Coulomb wedge. This Coulomb wedge theory has been widely applied as it provides a simple explanation for many features of accretionary wedges and fold and thrust belts.

The other class of continuum models was designed to address the kinematics of the deformation internal to a thrust belt or convergent orogen. A dynamic model of flow of a linear viscous fluid inside a rigid corner was proposed by Cowan & Silling (1978) to explain scale models and exhumation in convergent settings. This corner flow model was developed further by Cloos (1982, 1984) who proposed viscous flow as a mechanism for emplacing exotic blocks in melange terranes, and by Shi & Wang (1988) who calculated pressure-temperature paths in the Barbados accretionary prism. Cloos & Shreve (1988a, 1988b) developed a similar model for viscous flow in a confined channel under an accretionary wedge. These linear viscous models are attractive because, in contrast to the plasticity solutions, particle velocities and flow paths are easy to calculate. However, rocks are not likely to deform in a linear viscous fashion, at least not at near surface pressure and temperature conditions, so the stresses and even the kinematics of these viscous models are suspect.

Other less quantitative models for the kinematic development of tectonic wedges have been proposed. Silver *et al.* (1985) proposed that growth of an accretionary wedge is best accomplished by systematic underplating and duplexing such that the taper predicted by critical wedge theory is maintained. This model is consistent with the observation that sediments draped over the slopes of accretionary wedges are often undeformed, implying the internal wedge is not undergoing significant horizontal shortening. Platt (1986) also proposed underplating as an important mechanism in compressional orogens, but claimed that this leads to oversteepening of the wedge and hence extensional deformation in the near surface. In Platt's (1986) model this extension serves as a mechanism for unroofing and can explain the rapid exhumation of high pressure metamorphic terranes without invoking high rates of erosion.

Erosion has often been proposed as a driving mechanism for deformation or as a process controlling the kinematics and pressure-temperature-time (PTt) conditions experienced by rocks in a compressional orogen. Erosion was implicit as a driving force in the viscous flow model of Shi & Wang (1988). Beaumont *et al.* (1991, this volume) demonstrate that erosion is capable of controlling the PTt paths internal to an orogen, and that erosion can determine the spatial characteristics and distribution of deformation. Dahlen & Barr (1990) presented a model of the kinematics of the Taiwan fold and thrust belt that is consistent with critical Coulomb wedge theory by explicitly coupling the internal strain rate to the theoretical stress field. The system is then driven by the distribution and magnitude of erosion. This model was used to explain PTt paths and conditions in the wedge as well as the system energetics (Barr & Dahlen 1990), but the model remains kinematic in that the geometry and stress are assumed rather than calculated as part of the solution.

Numerical techniques can potentially include the best aspects of each of these models. Finite element methods, developed for plasticity problems, are based on deformational formulations; displacements or velocities are obtained together with the stresses as part of the solution. This approach eliminates the need to choose between a plastic formulation in terms of stress and a kinematic formulation in terms of displacement or velocity. For this reason, and because of the generality of the method in allowing specification of rheology, geometry or boundary conditions, finite elements have become increasingly popular for modelling lithospheric deformation (e.g. Bird & Piper 1980), including accretionary wedge mechanics (Borja & Dreiss 1989).

In this paper the Coulomb plastic rheology is adopted in a finite element technique that yields the particle velocities and hence kinematics of deformation along with the stress solution. Results of this model predict the dynamic velocity field and deformation in a growing Coulomb wedge that remains at yield as in the critical wedge theory. The generality of the numerical technique allows for modelling of the stresses and deformation in a wedge that is changing its taper geometry in response to changes in driving boundary stresses and this case is also considered. The kinematic implications for movement of material in these dynamic velocity fields are discussed.

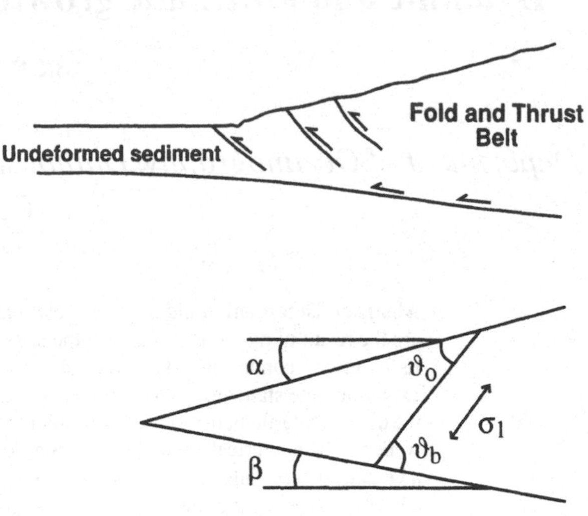

Figure 1. Critical Coulomb wedge model for a fold and thrust belt. Orientation of principal compressive stress with respect to surface slope and basal decollement.

Critical wedge theory

Before presenting the numerical model, a review of critical wedge theory is useful to establish the conceptual framework of the problem and provide a context for the numerical models. In the non-cohesive theory, a dry, subaerial thrust belt or accretionary wedge is assumed to have a triangular cross-section (Fig. 1) with characteristic surface slope angle α and basal dip β (Dahlen 1984). Stresses in the wedge are assumed to be at the Coulomb yield stress, so that

$$\tau = \sigma_n \tan\phi \qquad (1)$$

where τ and σ_n are the shear stress and normal stress on a plane; the internal angle of friction ϕ defines the strength of the material. The boundary condition imposed at the base of the wedge is of the same form as equation (1), with an angle of friction of ϕ_b. If the wedge can be treated as infinite in extent, a simple geometric solution is obtained, following Dahlen (1984), in terms of the orientation of the maximum compressive stress which is constant throughout the wedge and forms angles of ϑ_o and ϑ_b with the upper surface and basal surface respectively. The angles ϑ_o and ϑ_b are functions of the strength of the wedge and of the base,

$$\vartheta_o = \frac{1}{2}\sin^{-1}\left[\frac{\sin\alpha}{\sin\phi}\right] - \frac{\alpha}{2} \qquad (2)$$

$$\vartheta_b = \frac{1}{2}\sin^{-1}\left[\frac{\sin\phi_b}{\sin\phi}\right] - \frac{\phi_b}{2} \qquad (3)$$

The geometric relationship requires that,

$$\alpha + \beta = \vartheta_b - \vartheta_o \qquad (4)$$

A simple relationship therefore holds between the geometry of the wedge and the strength parameters through equations (2), (3) and (4).

The critical wedge solutions define a four dimensional surface in terms of the parameters α, β, φ, and ϕ_b. Two sections of this surface are shown in Figure 2 for φ = 30° (Figs 2a & 2b), ϕ_b = 10° (Fig. 2a) and β = 0° (Fig. 2b). Curve I gives the minimum slope geometry for a compressional wedge and is presumably the geometry of wedges that have grown from a flat surface. Continuous deformation keeps the wedge at this critical geometry as new material is accreted. This solution is consistent with the sandbox experiments of Davis *et al.* (1983). A wedge with geometry and strength state that lies on line II is also critical, but differs in its state of stress in that it is under horizontal tension and will deform by horizontal extension and downslope flow. The two lines I and II are sections of surfaces that are the boundaries of domains in α-β-φ-ϕ_b space that are not addressed by critical wedge theory. A wedge with geometry and strength that place it in domain III in Figure 2 is not steep enough for its internal strength and basal strength. It is therefore unstable and will increase its surface slope to reach a stable critical configuration on line I. A wedge with geometry and strength that place it in domain IV (Fig. 2) is stable, implying that the shear stresses are everywhere less than the yield strength. Such a wedge is capable of sliding stably along its base with no internal deformation. However, if it is forced to accrete new material the deformation will drive it back down to line I, where it is again critical. The other fields shown in Figure 2 but left unlabelled are generally not physically or geologically accessible. For example, if the state of a wedge enters the field in the upper left corner of Figure 2, above line II, it is too steep for its strength and collapses instantaneously to a position on line II.

The transient deformation and interim state of stress experienced by a wedge whose state enters fields III or IV can not be predicted by critical wedge theories that require stresses to remain on yield, yet it is quite conceivable that dynamic changes of the boundary conditions, which determine the wedge geometry, could send the state of a wedge into either field III or IV. Changes in fluid pressure, for example could result in an apparent strengthening or weakening of the basal strength, moving the state of the wedge on a horizontal line off line I in Figure 2b into domain III or IV.

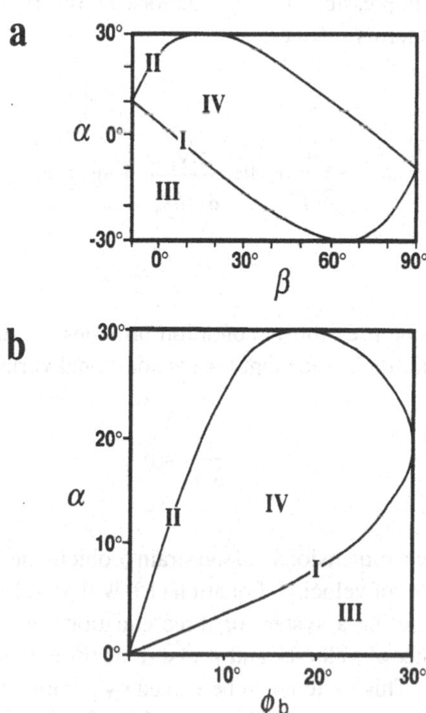

Figure 2. Critical Coulomb wedge solution (**a**) as function of surface slope α and decollement dip β (after Dahlen, 1984) with φ = 30° and ϕ_b = 10°; (**b**) as function of surface slope α and basal strength ϕ_b with φ = 30° and β = 0°. I and II represent the critical wedge solutions. III is the region of subcritical wedges whose taper is less than critical. IV is the region of stable wedges whose state of stress is less than the Coulomb yield stress.

Finite element model

Finite element techniques are increasingly used in nonlinear problems such as plasticity, where the inherent advantages of the numerical techniques have been valuable, particularly in engineering applications where complicated geometries and material properties are common. The advantages in modelling thrust belt wedges include the coupling between kinematics and stress and the ability to include non-linear plastic rheologies. The extremely large strains that result from the accretion process present a problem even for finite element techniques. This problem is addressed here through the use of a rate formulation of the plasticity problem that combines the Eulerian viscous flow equations with a plastic constitutive law. Although the rate theory was originally developed for pressure-independent von Mises yield criteria, the theory is easily modified to include the pressure-dependent Coulomb criterion.

The most common formulation of plasticity problems incorporates an elastic-plastic material in which the total strain is partitioned between elastic and plastic strain components (Zienkiewicz 1977). Under loading, stresses are maintained elastically until the material yield stress is reached after which the material deforms according to a plastic flow rule. For problems in which the plastic strains are much larger than the elastic strains, it may be advantageous to neglect the elastic component of strain. A rate theory of deformation can then be used to solve the problem in velocity and deformation rate (strain rate) rather than displacement and strain, providing important increases in efficiency. The cost of the numerical efficiency is in the loss of elastic strains and displacements.

One such rate formulation is the Levy-Mises theory (Malvern 1969) that describes a rigid-plastic material. Finite element techniques based on this theory have been developed for engineering applications in metal forming (Zienkiewicz & Godbole 1974) and for geodynamic models of lithosphere deformation (Vilotte *et al.* 1982, 1984, 1986). In the Levy-Mises theory an increment of plastic strain or, equivalently, the rate of deformation, is assumed to be proportional to the deviatoric stress tensor(σ'_{ij}),

$$D_{ij} = \frac{d\lambda}{dt} \sigma'_{ij} \tag{5}$$

where D_{ij} is the rate of deformation tensor, defined in terms of the velocity components v_i as,

$$D_{ij} = \frac{1}{2}\left[\frac{\partial v_i}{\partial x_j} + \frac{\partial v_j}{\partial x_i}\right] \tag{6}$$

The proportionality constant $(\frac{d\lambda}{dt})$ is a scalar that is defined by the plastic yield criterion. The standard theory incorporates the von Mises yield criterion,

$$\sqrt{J'_2} = \sigma^Y \tag{7}$$

where J'_2 is the second invariant, or norm, of deviatoric stress, defined as,

$$J'_2 = \frac{1}{2} \sigma'_{ij} \sigma'_{ij} \tag{8}$$

with summation implied over repeated indices, and σ^Y is the von Mises yield stress, a scalar. In order that the constitutive law (eqn. 5) gives a state of stress consistent with the von Mises yield criterion (eqn. 7) the proportionality constant must be,

$$\frac{d\lambda}{dt} = \frac{\sqrt{I_2}}{\sigma^Y} \tag{9}$$

I_2 is the second invariant or norm of the rate of deformation, and takes the same form as equation (8),

$$I_2 = \frac{1}{2} D_{ij} D_{ij} \tag{10}$$

Substituting equation (9) into equation (5) and inverting gives a constitutive law,

$$\sigma'_{ij} = \frac{\sigma^Y}{\sqrt{I_2}} D_{ij} \tag{11}$$

The second invariants of stress, J'_2, and rate of deformation, I_2, defined by equations (8) and (10), respectively, are L_2 norms. By taking the norm of both sides of equation (11) and assuming the material is incompressible, we obtain these invariants such that stress is independent of the rate of deformation and equation (11) becomes simply the von Mises yield criterion (eqn. 7). The constitutive law defined by equation (11) relates deviatoric stress to rate of deformation in the form of a viscous flow constitutive law,

$$\sigma'_{ij} = \mu_e D_{ij} \tag{12}$$

with an effective, non-linear viscosity, μ_e, defined by,

$$\mu_e = \frac{\sigma^Y}{\sqrt{I_2}} \tag{13}$$

The constitutive equation must include the isotropic part of the stress tensor as well as the deviatoric stress which is in equation (11). By including the pressure (or mean stress: $P = \sigma_{ii}/3$) it is possible to obtain the complete constitutive equation for a viscous fluid,

$$\sigma_{ij} = -P\delta_{ij} + \mu_e D_{ij}, \qquad \delta_{ij} = \begin{cases} 1 & \text{if } i = j \\ 0 & \text{otherwise} \end{cases} \tag{14}$$

By applying continuity to equation (14) the Navier-Stokes equation of motion is obtained. For the purposes of this paper, a simplified version of the Navier-Stokes equation that includes incompressible flow, gravitational body forces (ρg), but no inertial terms suffices,

$$-\frac{\delta P}{\delta x_i} + \mu_e \frac{\delta^2 v_i}{\delta x_i \delta x_j} + \rho g = 0 \tag{15}$$

The incompressibility condition provides the additional equation needed to account for the additional variable, P,

$$\frac{\delta v_i}{\delta x_i} = 0 \tag{16}$$

For a two dimensional, plane strain problem there are two components of velocity. Equation (15) with i=1,2 and equation (16) define a system of three equations with the two components of velocity and pressure as three independent variables. This system can be solved by a number of finite element techniques (Zienkiewicz 1977; Taylor & Hughes 1981). The technique used in this study is based on 6-noded triangular elements. Velocity is defined at all 6 nodes and interpolated with quadratic basis functions, but pressure is defined only at the three vertex nodes and is interpolated with linear basis functions. The lower order of interpolation of pressure is consistent with the lower order of the pressure term in the Navier-Stokes equation and is necessary for numerical reasons (Zienkiewicz 1977). Including the non-linear viscosity (eqn.13) to solve the plasticity problem requires a non-linear solver, but standard techniques, including direct iteration, can be used (Zienkiewicz 1977). More efficient iterative techniques (e.g. Newton's Method) are also applicable and may accelerate convergence, but are not necessary. All techniques require an initial solution, obtained by assuming a large constant initial viscosity (as large as is numerically practical), which assures large stresses and plastic 'failure' in part of the problem domain. A large initial viscosity also keeps strain rates small in the part of the domain that is not at plastic failure. This algorithm therefore defines a rheology which, under loading, is approximately rigid (highly viscous) until the yield stress is reached, after which it deforms viscously with stresses remaining at the yield stress. In summary, equations (15), (16) and (13) define a non-linear viscous flow problem that has a solution consistent with the von Mises yield criterion (eqn.7).

One modification to this theory is required to solve the Coulomb wedge problem; the Coulomb yield criterion (eqn. 1) must be used in place of the von Mises criterion. Any yield criterion that can be written in the same form as equation (7) can be included in the above algorithm. The Coulomb criterion must first be cast in terms of the stress invariants, and in the absence of cohesion is (Zienkiewicz 1977),

$$\frac{1}{3}J_1 \sin\phi + \sqrt{J'_2}\left(\cos\theta - \frac{1}{\sqrt{3}}\sin\theta\sin\phi\right) = 0 \qquad (17)$$

J_1 is the first invariant of total stress, ϕ is the internal coefficient of friction, and θ is a parameter giving the orientation of the principal stresses and is defined by,

$$\sin(3\theta) = -\frac{3\sqrt{3}}{2}\frac{J'_3}{\sqrt{J'_2}^{\,3}} \qquad (18)$$

where J'_3 is the third invariant of deviatoric stress. Factoring out $\sqrt{J'_2}$ defines a scalar quantity analogous to the von Mises yield stress,

$$\sigma^Y = \frac{-J_1 \sin\phi}{3\cos\theta - \sqrt{3}\sin\theta\sin\phi} \qquad (19)$$

This equation can now be substituted into equation (13).

To model the development of a Coulomb wedge, a simple parameterization in terms of boundary velocities is used. The initial domain that represents the undeformed sediments is rectangular with the lower boundary resting on the underthrust plate and the right side assumed to be against a vertical 'backstop' (Fig. 3a). The boundary condition imposed on the base is a constant horizontal velocity at the convergence velocity and no vertical velocity. The right side of the rectangular domain has a zero horizontal velocity and a free vertical component. Material is therefore forced by the basal

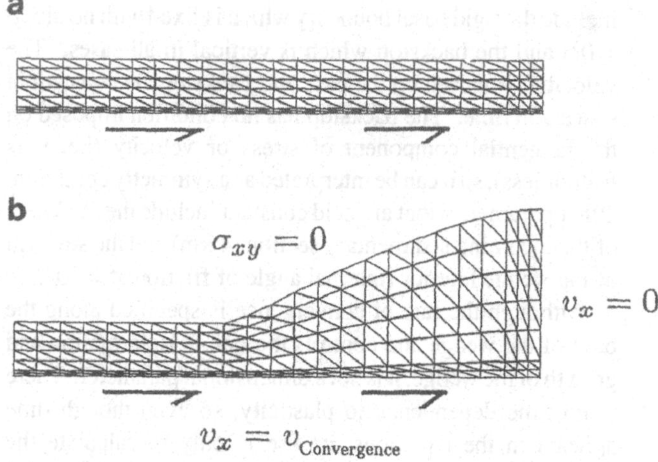

Figure 3. Finite element mesh and boundary conditions used for Coulomb wedge simulations. **(a)** Initial mesh prior to deformation. **(b)** Mesh reconstructed to fit deformed upper surface. Note that mesh does not track internal deformation. Boundary conditions are: fixed horizontal velocity along base and left side (v_x = convergence velocity) and backstop ($v_x = 0$); fixed vertical velocity along base ($v_y = 0$); no shear stress on free surface.

traction against the 'backstop' where it deforms plastically as it develops into a deformed wedge. The upper surface is free to move with the deformation (Fig. 3b), and the mesh is reconstructed at each timestep to track the boundary movement. The finite element solution gives the instantaneous velocity field, the strain rate field, the stress field and the free surface position as deformation progresses through time. Velocity is a relative quantity and an equivalent parameterization would be obtained by fixing the velocity on the base at zero and moving the backstop from right to left (Fig. 3) at the convergence velocity.

Much of this paper discusses the role of the strength of the basal decollement, which is an important parameter in the critical wedge theory. To include a decollement weaker than the wedge as a whole, several rows of elements along the base of the model have been assigned an independent coefficient of friction; they serve as a potential weak basal layer.

Model results

The models presented in this section are designed to address two questions. First, what are the kinematic displacements and particle velocities within a wedge that is growing by accretion while remaining critical, i.e. on the critical line I in Figure 2? Second, what are the kinematic and deformational effects of transient deformation experienced by a wedge with geometry and strength state which place it in one of the other domains in Figure 2? This second question is addressed by imposing changes in the strength of the basal decollement, but it could also be investigated by imposing changes in other parameters, e.g. β or ϕ.

The models are limited to consideration of the deformation and kinematics in Coulomb wedges that are actively growing by accretion. Some aspects of the geometry and boundary conditions remain constant for all models. These include the rigid basal boundary which is fixed with no dip ($\beta = 0°$) and the backstop which is vertical in all cases. The velocities imposed on the base and backstop are constant in space and time. The backstop has no condition imposed on the tangential component of stress or velocity (i.e. it is frictionless), so it can be interpreted as a symmetry condition. Other parameters that are held constant include the thickness of the incoming sedimentary section (4 km) and the strength of the wedge interior (internal angle of friction, $\phi = 30°$).

Although the rate of convergence is specified along the base of the wedge and controls the rate of deformation and growth of the wedge, it is not a dimensional parameter. There is no time dependence to plasticity, so even though time appears in the equations, its use is only to calculate the amount of displacement. Time and the convergence velocity scale each other such that only the product, the total convergence, is important to the deformation. In other words, a decrease in convergence velocity is equivalent to a proportional increase in time. Rate of deformation (strain rate) also scales with time and convergence velocity so that only the relative deformation rates are important.

Constant basal strength model

The strength of the basal decollement is the principal parameter in these models. By holding the basal strength constant with time, a wedge is produced that is always critical and lies somewhere on line I in Figure 2. The critical wedge with the maximum slope for a given basal dip (β) and wedge strength (ϕ) is obtained when the basal strength is equal to the wedge interior strength ($\phi_b = \phi$; Fig. 2). A model of this case is shown in Figure 4. The geometry and instantaneous deformation rates are shown with increasing time, or, equivalently, increasing convergence. The wedge that develops against the backstop shows the self-similar growth characteristic of Coulomb wedges. The upper surface develops a characteristic slope that is maintained as the wedge grows. In contrast to the analytic theory (Dahlen 1984) the upper surface is not linear, but has some curvature to it. This curvature is most likely a response to the vertical, frictionless backstop, which is not present in the analytic theory. The shear stresses on a frictionless surface must be zero which requires the upper surface to be approximately normal to the backstop. The

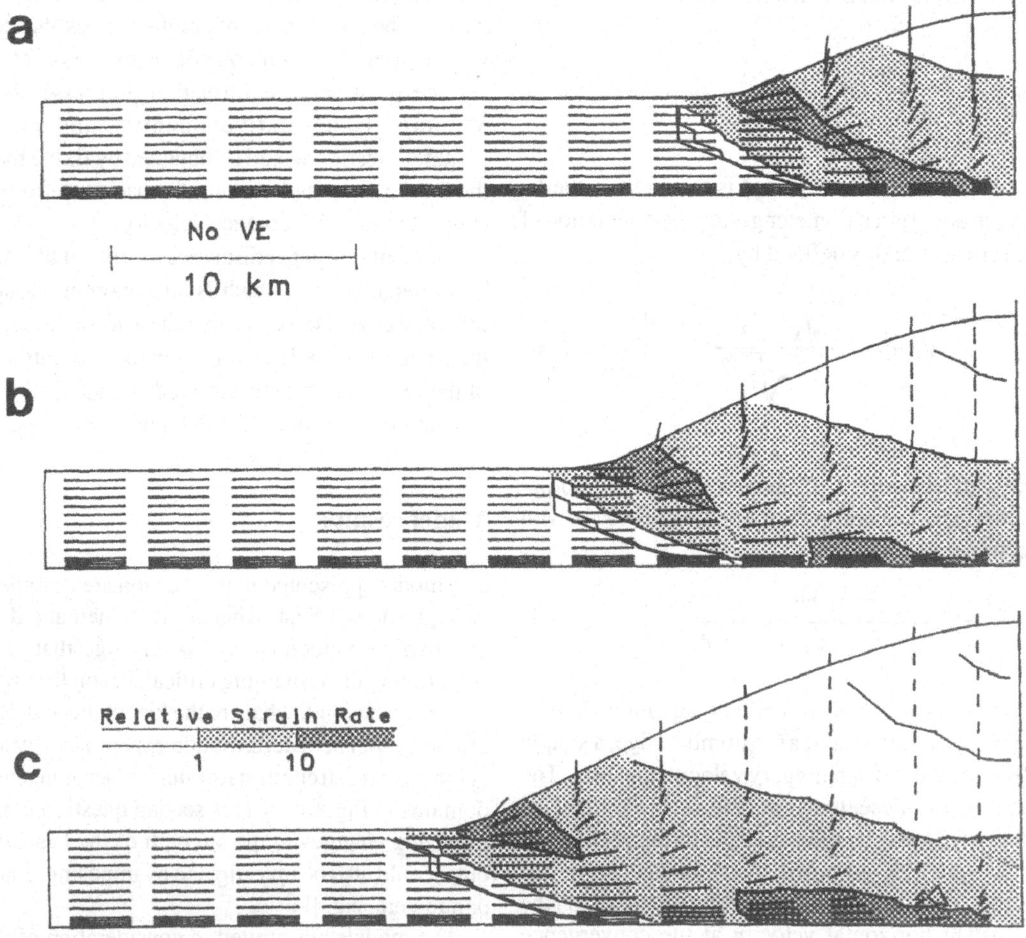

Figure 4. Model of growing Coulomb wedge with basal strength equal to wedge interior strength ($\phi_b = \phi = 30°$). Nodal velocities relative to the overriding plate are shown as vectors. Relative strain rate (second invariant of rate of deformation) contoured with larger relative values shaded. Results are shown for total convergence of: **(a)** 3 km; **(b)** 12 km; **(c)** 24 km.

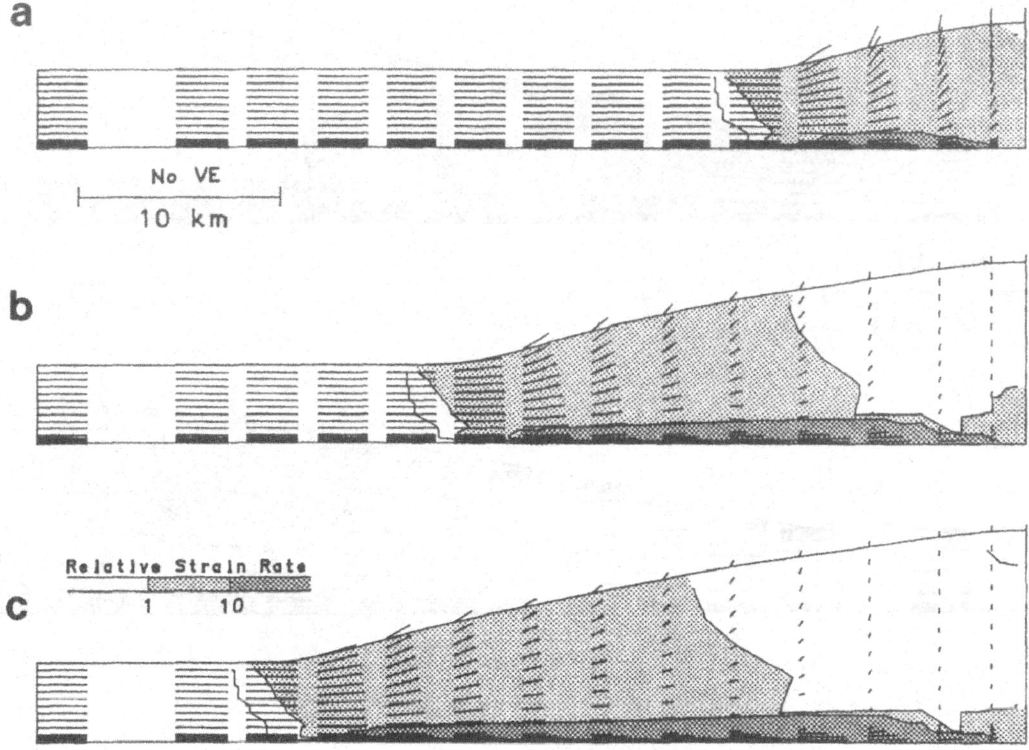

Figure 5. Model of growing Coulomb wedge with basal strength less than wedge interior strength (ϕ_b = 15°; ϕ = 30°). Nodal velocities relative to the overriding plate are shown as vectors. Relative strain rate (second invariant of rate of deformation) contoured with larger relative values shaded. Results are shown for total convergence of: (**a**) 3 km; (**b**) 22 km; (**c**) 35 km.

upper surface is also somewhat steeper than predicted by the analytic theory, but this discrepancy is probably also due to the effect of the backstop. Aside from these differences, the geometry and stress field is in good agreement with the analytic solution.

The kinematics of deformation corresponding to this stress field are shown as the instantaneous velocity field. The velocities show the undeformed sediments moving with constant velocity equal to the convergence velocity until they encounter the toe of the wedge (Fig. 4). Sediments near the surface are immediately deformed as the material turns up into the wedge. Sediments that are progressively deeper in the section pass further under the wedge before being pulled into the deformed wedge. Once into the wedge, material moves nearly vertically upward and experiences little additional deformation. This velocity field produces a distinctive pattern of strain rate with the highest strain rates found in a band from the toe of the wedge back to the corner at the base of the backstop. These high strain rates correspond to the zone where sediment passes from the underthrust plate into the wedge. It is difficult to interpret continuum velocity fields in terms of geological structures and the kinematics associated with discrete faults. However, the kinematics of the model in Figure 4 do resemble the displacements found in the scale models of Cowan & Silling (1978).

The kinematic style is also suggestive of a mechanism that could be interpreted as sediment underplating. The self-similar growth necessary to maintain the critical taper is achieved with very little deformation of the wedge interior. A frontal accretion mechanism requires significant pure shear deformation of the wedge interior in order to maintain the critical taper. This internal deformation is not observed in the model since sediment deformation is localized to the transition between undeformed sediment and the wedge interior. Growth of the back of the wedge is accomplished by 'subducting' undeformed sediment under the wedge and accreting it directly to the base, a pattern that is similar to the deformation style described by Brandon (1984) and Silver *et al.* (1985).

If the base of the wedge is significantly weaker than the wedge interior a distinctly different kinematic pattern emerges (Fig. 5). As predicted by the critical wedge theory a weaker base implies a lower surface slope. The wedge again grows in a self-similar fashion, but with a much smaller taper angle. Changes in the kinematic pattern are more subtle. The maximum strain rates again extend from the toe of the wedge along the lower boundary to the corner at the base of the backstop. However, the zone of high strain rates is much wider, extending far back into the wedge. The velocities do not show the sharp transition from horizontal to vertical apparent in Figure 4. A particle path in the model of Figure 5 changes slowly from horizontal towards vertical, but may never reach vertical. There is also no zone of undeformed, underthrust sediments. This overall kinematic style is suggestive of a frontal accretion mechanism with sediment accreted at the toe of the wedge and pure shear shortening maintaining the critical taper.

The models shown in Figures 4 & 5 are examples of

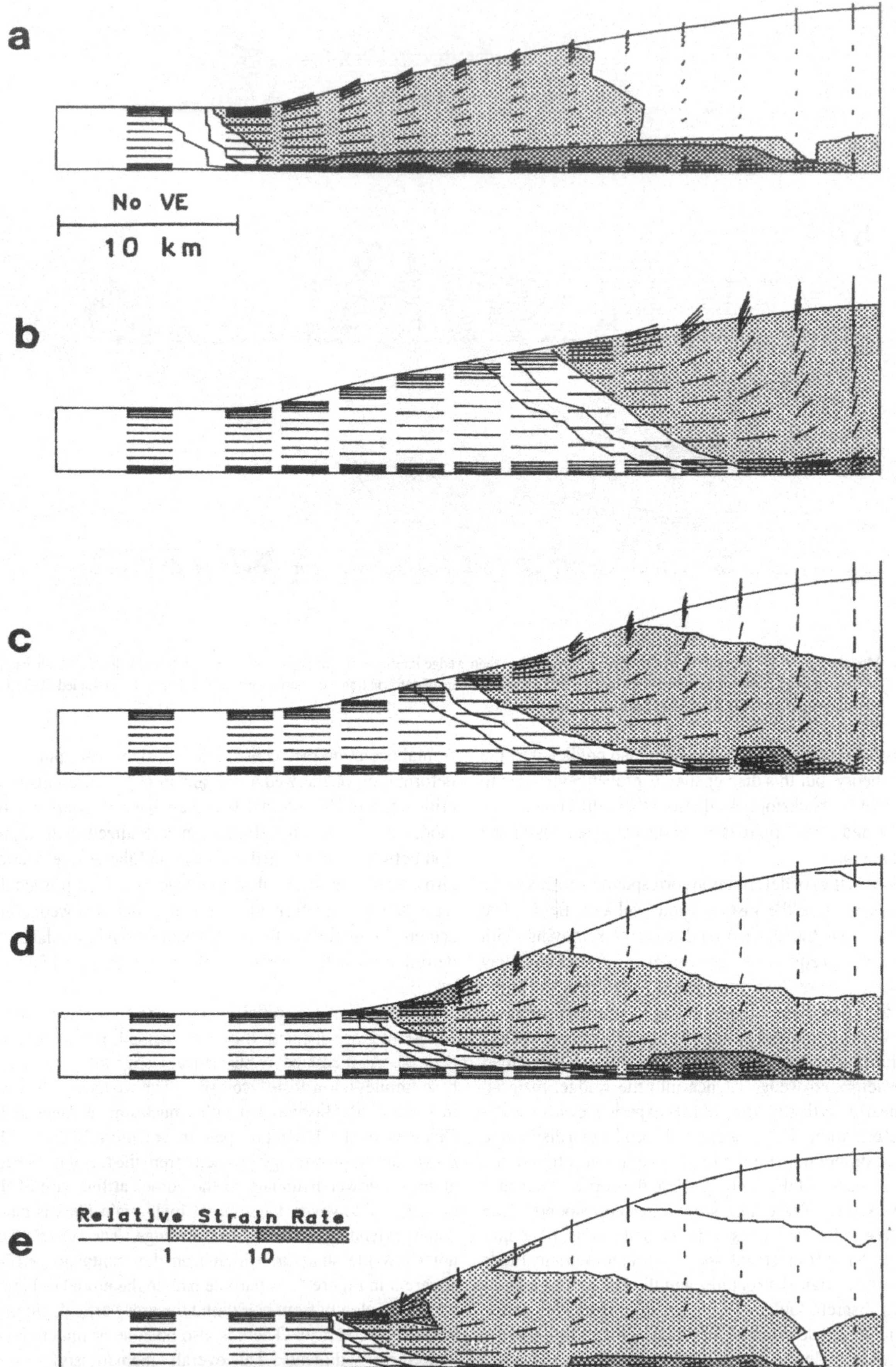

Figure 6. Model of Coulomb wedge experiencing an increase in basal strength. **(a)** Initial critical wedge develops with basal strength $\phi_b = 15°$. Total convergence of 38 km. **(b)** Basal strength increases from $\phi_b = 15°$ to $\phi_b = 30°$ and deformation jumps back to backstop. Total convergence of 38 km. **(c)** Deformation propagates forward. Total convergence of 41 km. **(d)** Total convergence of 44 km. **(e)** New critical taper is achieved. Total convergence of 50 km.

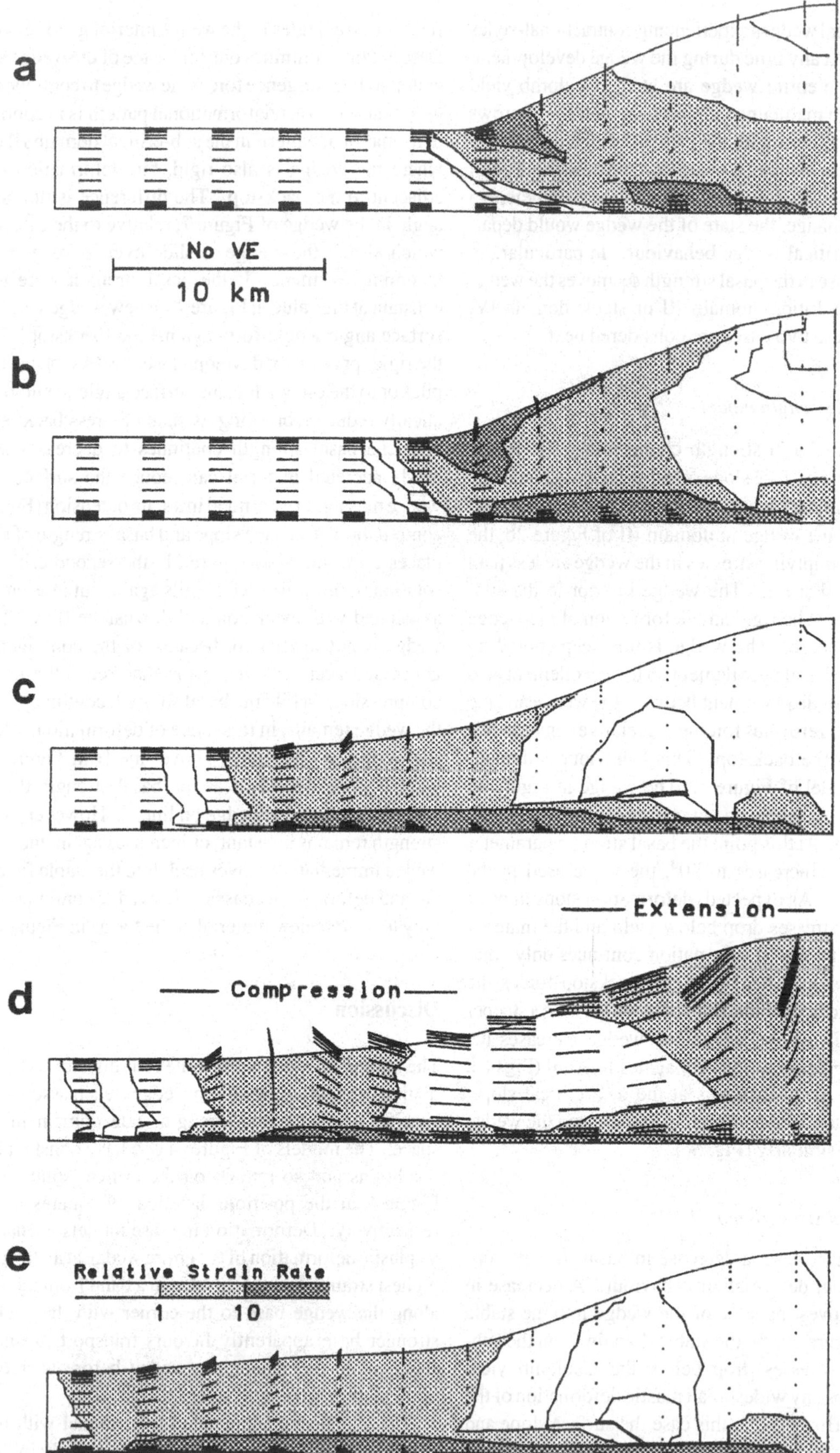

Figure 7. Model of Coulomb wedge experiencing a decrease in basal strength. (a) Initial critical wedge develops with basal strength $\phi_b = 30°$. Total convergence of 12 km. (b) Basal strength decreases linearly to $\phi_b = 20°$. Total convergence of 12.5 km. Most of the wedge is in the stable field and experiences no deformation although continued convergence causes deformation of the toe region. (c) Basal strength decreases linearly to $\phi_b = 10°$. Total convergence of 13 km. (d) With continued weakening of the base ($\phi_b = 6°$), wedge again becomes critical in a tensional mode with horizontal extension. Total convergence of 13.5 km. (e) The basal strength is held constant at $\phi_b = 6°$ and the wedge again becomes stable. Total convergence of 14.5 km. Continued convergence will result in the wedge growing from the toe backward until a new critical taper is achieved.

compressive critical wedges experiencing contractional-styles of deformation. At any time during the wedge development, stresses within the entire wedge are at the Coulomb yield stress. The wedge maintains the same geometry as it grows self-similarly consistent with the critical wedge theory. As such, the state of each model is characterized by a single point on line I in Figure 2b. However, if one of the parameters in Figure 2 should change, the state of the wedge would depart from line I and critical wedge behaviour. In particular, an increase or decrease in the basal strength ϕ_b moves the wedge state into the sub-critical domain III or stable domain IV, respectively. These two cases are considered next.

Increase in basal strength model

An increase in the basal strength changes the state of the wedge such that the surface slope is less than is required to be critical. This combination of surface slope and basal strength place the state of the wedge in domain III of Figure 2b, the sub-critical field, implying stresses in the wedge are less than the Coulomb yield stress. The wedge interior in the sub-critical domain must be rigid and deformation of the wedge as a whole must cease. The wedge is not steep enough to propagate over its basal decollement so the decollement also 'locks' and relative displacement between the wedge and the lower plate goes to zero, thus forcing the relative convergence to be taken up at the backstop. This behaviour is demonstrated by the model of Figure 6. The wedge in Figure 6a develops with a basal strength parameter $\phi_b = 15°$, as in the model of Figure 5. At this point the basal strength parameter ϕ_b instantaneously increases to 30°, the value used in the model of Figure 4. As expected, deformation stops in most of the wedge, as stresses drop below yield and the material becomes rigid (Fig. 6b). Deformation continues only adjacent to the backstop, where the rigid backstop forces the material to fail and begin building a new wedge with a steeper surface. As uplift of the rear of the wedge increases the surface slope, the deformation propagates forward (Figs 6c, 6d). When the entire surface is at the new critical slope, stresses in the entire wedge are again at yield and the wedge again grows self-similarly (Fig. 6e).

Decrease in basal strength model

The complementary case, a decrease in basal strength, implies very different deformation behaviour. A decrease in basal strength moves the state of the wedge into the stable domain IV of Figure 2b. In the stable domain, as in the sub-critical domain, stresses drop below the Coulomb yield stress, which stops any widespread plastic deformation of the wedge interior. However, in this case the surface slope and wedge taper are sufficiently large to permit the wedge to slide stably over its basal decollement forcing relative convergence to be taken up at the toe of the wedge. This behaviour is demonstrated by the model of Figure 7. The wedge in Figure 7a develops with a basal strength parameter $\phi_b = 30°$. Subsequently, the basal strength systematically decreases (Figs 7b, 7c). Most of the wedge quickly enters the stable field and strain rates in the wedge interior go to zero (Fig. 7c). Deformation continues only at the toe of the wedge where the continued convergence forces the wedge to continue accreting new material. This deformational pattern is in contrast to the deformational pattern in the sub-critical domain III (Fig. 6b), where the wedge is also rigid, but deformation continues adjacent to the backstop. The difference is the large taper angle in the wedge of Figure 7, relative to the basal strength, which allows the wedge to slide over its base and accrete incoming sediments. If the basal strength were to remain constant at the value in Figure 7c, a new wedge with a shallow surface angle would form against the 'backstop' formed by the rigid, previously developed wedge. As the new material piles onto the old wedge, the surface angle would be systematically reduced, bringing its state of stress back onto yield.

If the basal strength continues to decrease faster than newly accreted material can reduce the surface slope the wedge may again go critical in a state of tension (Fig. 7d). The combination of surface slope and basal strength of the wedge places it on line II in Figure 2b, the second critical wedge solution. The entire wedge fails again, but in a tensile mode associated with extension and downslope flow. The entire wedge is not in tension, because of the continued convergence and accretion at the toe that keeps the toe in local compression. While the basal strength continues to decrease the wedge remains in this mode of deformation. The state of the wedge can not move above line II in Figure 2, so the surface slope decreases as the basal strength decreases to keep the state of the wedge on line II. However, if the basal strength remains constant, or increases again, the state of the wedge immediately moves back into the stable field IV (Fig. 7e) and deformation ceases. The wedge continues to deform only to accrete new material at the toe as in Figures 7b & 7c.

Discussion

The numerical models illustrate four distinct styles of deformation in growing Coulomb wedges, each associated with a specific geometry and strength state domain in α-β-ϕ-ϕ_b space. The models of Figures 4 & 5 have constant boundary conditions and so remain on the critical solution line I in Figure 8 at the positions labelled by squares (e) and (a), respectively. Deformation in these models is characterized by plastic deformation of the entire wedge at all times with the highest strain rate concentrated in a band from the wedge toe along the wedge base to the corner with the backstop. A stronger base apparently favours transport of undeformed sediment further under the wedge before accretion in an underplating kinematic style.

The evolution of the state of the model with increasing basal strength (Fig. 6) is shown schematically by the square points in Figure 8, each of which corresponds to a stage of development in Figure 6. The initial wedge develops at (a) on the critical line I; the increase in basal strength immediately transfers the wedge state along the horizontal path to square (b) in domain III. Subsequent deformation steepens the surface slope moving the state vertically up to its final state

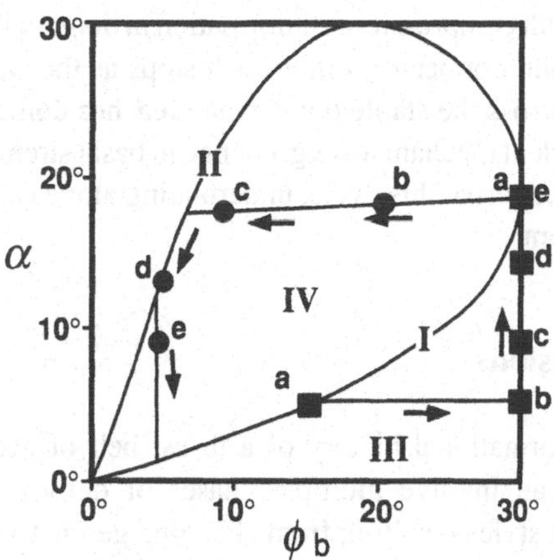

Figure 8. Mechanical state in ϕ_b, α space of finite element models relative to critical wedge solutions. Square points represent path taken by model of Figure (6); points (a) through (e) correspond to Figures (6a) through (6e), respectively. Circles represent path taken by model of Figure (7); points (a) through (e) correspond to Figures (7a) through (7e), respectively.

at square (e). The geometric and strength state of the wedge in this model places it in domain III for the entire transition. With reference to Figures 6b-6d, deformation in domain III is characterized by high strain rates in the rear of the wedge that propagate forward as the surface of the wedge steepens. The style of deformation observed in domain III, the sub-critical field involves the highest compressional strain rates in the interior of a wedge. It is, therefore, the most likely cause of large internal strains as evidenced by out-of-sequence (backstepping) faulting. Internal strain is experienced by the wedge interior in other modes of deformation, in particular, steady growth on the critical line, but the strain is smaller and is experienced at lower rates. As demonstrated in Figure 4, it is possible to maintain the critical taper with no strain in the upper, interior part of the wedge. Large internal, out-of-sequence strains are most likely the result of sub-critical deformation as the wedge deforms to regain its critical taper.

The model of Figure 7 demonstrates two additional styles of deformation, those associated with a wedge state in domains II and IV. The path representing the evolving state of the wedge in Figures 7a through 7e is given by the circles (a-e) in Figure 8. The wedge model in Figure 7a develops initially at point (a) on the critical line. As the basal strength decreases (Fig. 7b), the wedge state moves into the stable domain IV. Continued weakening of the base moves the wedge state across domain IV until it reaches the other critical line, II, above point (d). The base continues to weaken, moving the wedge state along line II until the strength of the base is held constant, at which point the wedge state again enters the stable field as the surface slope decreases by accretion. The wedge state moves down to the final state of the model at point (e), although continued accretion would force the state eventually to reach a new critical state on line I. Figures 7b, 7c & 7e represent the style of deformation in the stable domain IV; Figure 7d shows the deformation occurring while the wedge state is critical on line II. The domain IV deformation is characterized by localized deformation of the wedge interior; the only significant strain rates are in the immediate toe region or are confined to the decollement region where the lower plate shears beneath the rigid wedge. The toe of the wedge continues to deform in order to accrete all the incoming sediment. The vertical segment of the path directly above circular point (e) in Figure 8 represents the decrease of slope resulting from the continued accretion and would be achieved by backward propagation of the new critical slope that is forming at the toe of the wedge in Figure 7e. Dahlen (1984) suggested that a wedge in this stable field would simply slide over the underthrust sediments with no internal deformation and no accretion of new sediments. Sliding without accretion is possible, but not necessary; whether or not new material is accreted likely depends on the detailed mechanics of deformation at the wedge toe.

An example of deformation in domain II on the second critical wedge line is provided by Figure 7d which shows the deformational state of the wedge on the critical line II at circle (d) in Figure 8. The domain II line represents the boundary of the permissible states of a Coulomb wedge. As continued weakening of the base of the wedge attempts to move the wedge state to the left of the critical line, the surface slope decreases at a rate that keeps the state on line II. This slope decrease is achieved by extensional collapse of the wedge. If material continues to be accreted to the toe of the wedge, the toe remains in compression, but the rest of the wedge is in tension and accompanying extension. The extensional strain rates are determined by the rate of weakening of the base; strain rates must be high enough to keep the wedge state on the critical line. If the basal strength stops weakening, the extension immediately ceases as the wedge becomes stable, its state moving into domain IV.

The deformation and kinematic styles observed in these models is associated with the mechanical state of a wedge and the domain implied by that state, not the mechanism or mechanisms responsible for the change in state. The styles of deformation associated with each of the four domains identified in Figure 2 in α-β-ϕ-ϕ_b space are demonstrated by models that change the basal strength, ϕ_b, with time. However, a change in any of the other parameters that causes the state of the wedge to enter another domain produces the same style of deformation. Basal strength was varied in these models because it seems the most likely parameter to experience significant changes. Basal strength can change in response to changes in strength of underthrust material, changes in fluid pressure due to differing porosity and permeability of underthrust material, or changes in convergence rate that could lead to variations in the rate of sediment dewatering and hence fluid pressure generation. The basal dip (β) can also change independently, for example, as an isostatic response to a growing wedge. Surface slope (α) can change as a response to erosion; depending on the distribution of erosion, the surface slope could be either increased or decreased. The surface slope could also be decreased by sedimentation over the wedge, pushing the state down into domain III. Any of these changes would result in the transient deformational states demonstrated by the models of Figures

6 or 7.

A particular thrust belt or accretionary wedge may have a long and complex geological history that involves multiple phases of deformation of different styles as the state of the wedge changes deformational domains. For example, the deformation may shift from the toe of a wedge back to the interior and forward to the toe again as the mechanical state of the wedge shifts between domains III and IV.

The predicted kinematic patterns have implications for the tectonic exhumation of high pressure terranes. All models showed vertical components to the velocity field, but exhumation occurs only where the velocity of a particle at depth is larger than the velocity of the free surface overlying that particle. This condition does not occur in any of the models with the exception of Figure 7d, in which gravitational collapse leads to the exhumation of the rear of the wedge. Otherwise, all models show no exhumation of material at depth in the wedge. This kinematic pattern of a growing Coulomb wedge is in contrast to a wedge with a viscous rheology as proposed in the models of Emerman & Turcotte (1983) and Platt (1986). If the accreted material deforms viscously, the surface slope drives flow down towards the toe in all cases, inevitably leading to exhumation in the rear of the wedge.

The exceptional case in these models, the gravitational collapse of a Coulomb wedge, has limited potential for significant exhumation. A single collapse phase as in Figure 7 is not likely to produce significant exhumation. However, if a wedge passes through several cycles of collapse and growth, cycling its deformational state through domains II, III and IV, as in the full path of square (a) to circle (e) in Figure 8, reproducing the full range of deformational styles observed in Figures 6 and 7, exhumation could be quite significant. The styles of growth (domain III) and collapse (domain II) are very different. Deformation during growth starts from the back and propagates forward; deformation during collapse is ubiquitous with contemporaneous extension and contraction. The corresponding kinematic patterns imply different particle paths during contractural construction and extensional destruction of the wedge leading to a net upward movement of deeply buried rocks as the wedge state cycles between growth and collapse.

This exhumation mechanism is similar to the model discussed by Platt (1986, 1987) in which he proposed that tectonic exhumation was due to extension driven by underplating. However, the Coulomb plastic rheology implies some important differences. A cycling of the wedge through domains II and III would involve underplating during growth in domain III and extension in domain II, and could result in extensional structures associated with underplating. This model would also result in exhumation of high pressure terranes without excessive erosion. However, this mechanism differs from Platt's model in that the extension and underplating would not be contemporaneous nor causally related. Underplating is a growth mechanism associated with compressional deformation in domain III. Extension would not occur until growth stops as the wedge state moves across the stable domain and reaches domain II. An independent mechanism, e.g. change in basal strength, is still required to drive this cycle, underplating alone cannot drive the system.

Conclusions

The deformational history of a thrust belt or accretionary wedge can involve multiple phases of deformation with different styles resulting from changing geometries, boundary conditions and internal strength of the wedge. Critical wedge theory defines distinct domains dependent on the geometric and strength state of a Coulomb wedge; the dynamics and kinematics of these domains can be determined using numerical techniques. Each domain has an associated deformational style. Growth of a wedge on the compressional critical surface in the geometry-strength space implies self-similar growth by either frontal accretion or underplating with frontal accretion favoured by a weak decollement. If changing conditions drive the state of a wedge into the subcritical domain, deformation enters a transient state with no deformation in most of the wedge interior and no displacement along the decollement except at the back of the wedge where the rigid backstop forces the wedge to continue deforming internally, increasing the wedge taper and driving deformation progressively forward to the toe. If the taper of a wedge becomes too large relative to its basal strength, the state of the wedge enters the stable domain and slides over its base with no internal deformation, other than at the toe where continued accretion forces a wedge with a shallower taper to form against the older, now rigid, wedge. The wedge may return to critical in a state of horizontal tension and associated extension if the taper becomes larger than the basal strength can support. In this mode, the entire wedge deforms in extension to reduce the wedge taper, with the exception of the toe region which may remain in compression if accretion continues.

The superposition of phases of these transient and steady deformational styles may help explain complex stuctural relationships. Features such as out-of-sequence faulting and tectonic exhumation of high-pressure terranes can be interpreted in terms of transient phases of non-critical wedge deformation.

The author was supported in this research by a Dalhousie University Killam Post-doctoral Fellowship. This work was also supported by a Natural Sciences and Engineering Research Council Operating Grant to Christopher Beaumont. This work benefited greatly from many discussions with Chris Beaumont and Philippe Fullsack. The manuscript was improved by helpful reviews by Mark Brandon and F. A. Dahlen.

References

Barr, T. D. & Dahlen, F. A. 1989. Steady-State Mountain Building 2. Thermal Structure and Heat Budget. *Journal of Geophysical Research*, **94**, 3923-3947.

Beaumont, C., Hamilton, J. & Fullsack, P. 1991. Erosional Control of Active Compressional Orogens, this volume.

Bird, P. & Piper, K. 1980. Plane Stress Finite Element Models of Tectonic Flow in Southern California. *Physics of the Earth and Planetary Interiors*, **21**, 158-195.

Borja, R. I. & Dreiss, S. J. 1989. Numerical Modelling of Accretionary Wedge Mechanics: Application to the Barbados Subduction Problem. *Journal of Geophysical Research*, **94**, 9323-9339.

Brandon, M. T. 1984. A study of deformational processes affecting unlithified sediments at active margins: A field study and a structural model, Ph.D. thesis, University of Washington, Seattle, 160p.

Chapple, W. M. 1978. Mechanics of thin-skinned fold-and-thrust belts. *Geological Society of America Bulletin*, **89**, 1189-1198.

Cloos, M. 1982. Flow melanges: Numerical modelling and geologic constraints on their origin in the Franciscan subduction complex, California. *Geological Society of America Bulletin*, **93**, 330-345.

—— 1984. Flow melanges and the structural evolution of accretionary wedges. *Geological Society of America Special Paper*, **198**, 71-79.

—— & Shreve, R. L. 1988a. Subduction-Channel Model of Accretion, Melange Formation, Sediment Subduction, and Subduction Erosion at Convergent Plate Margins: 1. Background and Description. *Pure and Applied Geophysics*, **128**, 455-500.

—— & —— 1988b. Subduction-Channel Model of Accretion, Melange Formation, Sediment Subduction, and Subduction Erosion at Convergent Plate Margins: 2. Implications and Discussion. *Pure and Applied Geophysics*, **128**, 501-545.

Cowan, D. S. & Silling, R. M. 1978. A Dynamic, Scaled Model of Accretion at Trenches and Its Implications for the Tectonic Evolution of Subduction Complexes. *Journal of Geophysical Research*, **83**, 5389-5396.

Dahlen, F. A. 1984. Noncohesive Critical Coulomb Wedges: An Exact Solution. *Journal of Geophysical Research*, **89**, 10125-10133.

——, Suppe, J. & Davis, D. 1984. Mechanics of Fold-and Thrust Belts and Accretionary Wedges: Cohesive Coulomb Theory. *Journal of Geophysical Research*, **89**, 10087-10101.

—— & Barr, T. D. 1989. Steady-State Mountain Building 1. Deformation and Mechanical Energy Balance. *Journal of Geophysical Research*, 3906-3922.

Davis, D., Suppe, J. & Dahlen, F. A. 1983. Mechanics of Fold-and-Thrust Belts and Accretionary Wedges. *Journal of Geophysical Research*, **88**, 1153-1172.

Emerman, S. H. & Turcotte, D. L. 1983. A fluid model for the shape of accretionary wedges. *Earth and Planetary Science Letters*, **63**, 379-384.

Malvern, L.E. 1969. *Introduction to the Mechanics of a Continuous Medium*. Prentice-Hall, New Jersey.

Moore, J. C., Watkins, J. S., Shipley, T. H., McMillen, K. J., Bachman, S. B., & Lundberg, N. 1982. Geology and tectonic evolution of a juvenile accretionary terrane along a truncated convergent margin: Synthesis of results from Leg 66 of the Deep Sea Drilling Project, southern Mexico. *Geological Society of America Bulletin*, **93**, 847-861.

Morley, C. K. 1988. Out-Of-Sequence Thrusts. *Tectonics*, **7**, 539-561.

Platt, J. P., Leggett, J. K., Young, J., Raza, H. & Alam, S. 1985. Large-scale sediment underplating in the Makran accretionary prism, southwest Pakistan. *Geology*, **13**, 507-511.

—— 1986. Dynamics of orogenic wedges and the uplift of high-pressure metamorphic rocks. *Geological Society of America Bulletin*, **97**, 1037-1053.

—— 1987. The uplift of high-pressure-low-temperature metamorphic rocks. *Philosophical Transactions of the Royal Society of London A*, **321**, 87-103.

Shi, Y. & Wang, C. 1988, Thermal Structure of the Barbados Accretionary Complex. *Pure and Applied Geophysics*, **128**, 749-766.

Silver, E. A. , Ellis, M. J., Breen, N. A. & Shipley, T. H., 1985. Comments on the growth of accretionary wedges. *Geology*, **13**, 6-9.

Stockmal, G. S. 1983. Modelling of Large Scale Accretionary Wedge Deformation. *Journal of Geophysical Research*, **88**, 8271-8287.

Taylor, C. & Hughes, T. G. 1981. *Finite Element Programming of the Navier-Stokes Equations*. Pineridge Press Ltd., Swansea, UK, 244p.

Vilotte, J. P., Daignieres, M. & Madariaga, R. 1982. Numerical Modelling of Interplate Deformation: Simple Mechanical Models of Continental Collision. *Journal of Geophysical Research*, **87**, 10709-10728.

——, ——, —— & Zienkiewicz, O. 1984. The role of a heterogeneous inclusion during continental collision. *Physics of the Earth and Planetary Interiors*, **36**, 236-259.

——, ——, —— & ——. 1986. Numerical study of continental collision: influence of buoyancy forces and an initial stiff inclusion. *Geophysical Journal of the Royal Astronomical Society*, **84**, 279-310.

Westbrook, G. K., Ladd, J. W., Buhl, P., Bangs, N. & Tiley, G. J. 1988. Cross section of an accretionary wedge: Barbados Ridge complex, *Geology*, **16**, 631-635.

Zhao, W.-L., Davis, D. M., Dahlen, F. A. & Suppe, J. 1986. Origin of Convex Accretionary Wedges: Evidence From Barbados. *Journal of Geophysical Research*, **91**, 10246-10258.

Zienkiewicz, O.C. & Godbole, P. N. 1974. Flow of Plastic and Visco-Plastic Solids with Special Reference to Extrusion and Forming Processes. *International Journal for Numerical Methods in Engineering*, **8**, 3-16.

—— 1977. *The Finite Element Method*. McGraw-Hill, New York, 851p.

A developmental stage of a foreland belt

E. G. Bombolakis

Department of Geology and Geophysics, Boston College, Chestnut Hill, MA 02167, USA

Abstract: Active foreland-type belts indicate that recurring seismically related modes of deformation played a fundamental role in ancient foreland belts. Earthquakes along ramps (?) or listric segments of thrusts apparently can be larger in magnitude than along associated decollements. When this situation exists, additional styles of fault slip must occur along decollements to generate the larger earthquakes at the 'locked' inclined fault segments. Fault-slip styles in active tectonic terranes include typical earthquakes, slow earthquakes, and recurring forms of fault creep. They indicate that velocity-dependent damping is a key element in several faulting processes. Fault models presented here take this parameter into account, and they are applied to the Høgsback sheet in the Kemmerer region of the Wyoming Salient. They indicate that appreciable long-term fault creep displacement cannot occur along decollements at average geological slip rates, with few possible exceptions such as extensive layer-parallel shortening due to pressure solution. Data analysis for Kemmerer indicates that the Hogsback sheet probably was emplaced by the more rapid forms of recurring fault slip. The partitioning of elastic tectonic strain imposed by typical and slow earthquakes accordingly may depend to a considerable extent on the relation between relaxation times of deformational processes and the recurrence intervals of these fault-slip events.

Active continental tectonic terranes illustrate that upper crustal deformation is strongly influenced by coseismic faulting and short-term fault creep events (e.g. Allen 1981; Savage 1983; Schwartz & Coppersmith 1986; Wesson 1988; Sibson 1989; Stein & Yeats 1989). Consequently, the objectives here are (1) to present some key evidence of seismically related deformation in foreland-type belts, (2) to quantify potential modes of dynamic fault slip along a decollement aft of a frontal ramp or listric segment of an imbricate, (3) to illustrate applications with a field example from the Wyoming Salient, and (4) to suggest how some of the elastic tectonic strains imposed by fault slip may be partitioned between brittle deformation and strain-rate dependent ductile deformation.

Seismically related deformation in foreland belts

Seismic moment tensor analyses indicate that moderate to large earthquakes account for 25% to 70% of the upper crustal fault slip occurring in several major continental zones where active deformation is distributed over large distances from major plate boundaries (Ekström & England 1989). Evidence that seismically related modes of deformation are important in foreland-type belts within these zones is illustrated by the Himalayan foreland belt, the Shotori belt in eastern Iran, the Kopah Dagh belt northeast of Iran, the El Asnam thrust terrane in the Tell Atlas of Algeria, the western Transverse Ranges of California, and by the Coalinga region of the fold-and-thrust belt along the southern Coast Ranges bordering the San Joaquin Valley.

The May 2, 1983 M 6.7 Coalinga mainshock and its aftershock sequence were associated principally with blind thrusts and a major fold. Controversy has existed as to which nodal plane of the mainshock was the active fault, but a recent analysis of the three-dimensional velocity structure and seismicity provides very strong evidence that the mainshock occurred along a 30° SW dipping portion of a low-angle blind thrust system at some 10 km depth (Eberhart-Phillips 1989). Relevant interpretations are that the mainshock occurred along an inclined fault segment of a fault-propagation fold (Stein & Yeats 1989), along a listric segment of an imbricate thrust system beneath a major fold (Wentworth & Zoback 1989), or along a frontal ramp associated with a fault-bend fold (Namson & Davis 1988a). The Coalinga anticline was either uplifted or had its amplitude increased by half a metre or more (Stein 1987).

The Coalinga region is not unique. Pronounced seismicity reportedly is related spatially to major ramps (or listric segments of blind thrusts) associated with major folds in the western Transverse Ranges (Namson & Davis 1988b) and in the fold belt of the Los Angeles region where the Oct. 1, 1987 M_L 5.9 Whittier Narrows mainshock occurred at about 14 km depth (Lin & Stein 1989). In both the Whittier Narrows and Coalinga mainshocks, the preferred nodal planes dip about 30°, consistent with balanced cross sections of these regions (Namson & Davis 1988a; Davis *et al.* 1989). In the Tell Atlas of Algeria, the axis of a well developed anticline in the hanging wall of a complex thrust sheet is parallel to the 30 km surface rupture produced by the Ms 7.3 1980 El Asnam earthquake. Palaeoseismic studies indicate that two previous large earthquakes had occurred along this zone with a recurrence interval of about 450 years (Swan 1988).

Coseismic faulting also appears to be associated with flats and decollements. Low-angle nodal planes in the eastern half of the Himalayan foreland belt lie either on a basal decollement or on flats associated with the Main Boundary thrust system

(Baranowski *et al.* 1984). Examples are the Oct. 21, 1964 M 5.9 mainshock and the March 24, 1974 M 5.4 mainshock, both of which had slip vectors plunging 2° or 3°. Similarly, south of Coalinga and east of the North Kettleman Hills Dome, the M 5.5 Avenal earthquake is consistent with low-angle thrusting along a flat (Namson & Davis 1988a).

Most of the catalogued continental thrust-fault mainshocks ($M > 5$) have preferred nodal planes dipping 20° to 60° (e.g. Ekström & England 1989). Instructive examples are the 1980 Ms 7.3 El Asnam mainshock on a complex fault dipping ≈ 45° NW at some 6 km depth in the Tell Atlas (Nabalek 1985; Yielding 1985), and the 1978 Ms 7.4 Tabas-e-Golshan mainshock nodal plane dipping 30° NE at about 10 km depth in the Shotori belt (Berberian 1982). An important feature in both cases is that detailed aftershock studies indicate that the major structures include listric imbricate faults that converge downward into low-angle decollements. And in both cases, like Coalinga, some of the smaller aftershock focal mechanisms are indicative of low-angle slip on flats or decollements.

Observations for formulation of decollement fault-slip models

The analysis of seismic moment rates shows that a few large earthquakes produce considerably more cumulative fault slip along a fault than fairly numerous small earthquakes along the same fault (Brune 1968; Schwartz & Coppersmith 1986). Seismic data, discussed above, indicate that earthquakes along ramps or inclined thrust segments can be larger than along flats or decollements. Therefore, for this type of situation, additional styles of fault slip along a decollement, such as fault creep, need to be considered; e.g. in order that sufficient elastic tectonic strains are stored at 'locked' listric fault segments to generate the larger earthquakes at those segments.

In a review of the paradox of overthrust faulting, Price (1988) emphasized that traditional models of fault slip, such as the Hubbert-Rubey model, cannot be applied to the entire subhorizontal base of a thrust sheet, and that the fault displacements are described more realistically by a Somigliana dislocation fault model. This model is one of the elastic dislocation models frequently employed in geodetic analyses of faulting (e.g. Savage 1983; Lin & Stein 1989). However, these models do not take into account inertial forces of the fault blocks, nor inelastic behaviour along the faults. Consequently, the models presented here are formulated to take these parameters into account with respect to the following observations in earthquake seismology and seismotectonics.

Active faults are associated with barriers defined as features that interrupt or terminate rupture propagation (Aki 1984). Examples of geometric barriers are a tear fault and an imbricate zone that sequentially interrupted rupture propagation of the main thrust during the 1980 Ms 7.3 El Asnam earthquake (Yielding 1985). An example of a rheologic barrier is the thick evaporites in the Zagros (Jackson 1983). And in the case of the 1983 Coalinga mainshock, the upward extent of the mainshock rupture ended at the approximate boundary between the Franciscan and Great Valley sequence strata (Eberhart-Phillips 1989), a potentially important stratigraphic type of barrier.

A fundamental phenomenon related to barriers is the recurrence of earthquakes of similar size along specific segments of a fault zone (see Schwartz & Coppersmith 1986), as illustrated by the Oued Fodda thrust fault in the Tell Atlas (Swan 1988). The 'characteristic' earthquakes apparently recur along the same fault segments until the geometry, rheology, or boundary conditions of the fault zone are altered sufficiently to accommodate changes in the style of deformation.

Instrumental monitoring of active faults has revealed that the modes of fault slip include typical earthquakes of seconds duration, slow earthquakes of minutes to tens of minutes duration (Sacks *et al.* 1981; Beroza & Jordan 1989), accelerating-decelerating fault creep episodes each of an hour or so duration, and slower forms of fault creep (Wesson 1988). Potentially important mechanisms of long-term fault creep include the LPS (layer-parallel shortening) thrusting described by Geiser (1988), and thrusting along evaporite beds But despite more than 30 years of intensive study, 'long term' fault creep has been documented in relatively few active continental locales; notably, along the central section of the San Andreas fault zone, the Imperial fault to the south, one locale along the North Anatolian fault in Turkey (Wesson 1988), the Nahan thrust in the Himalayas (Sinvhal *et al.* 1973), and the Buena Vista thrust in California (Wilt 1958). Attempts to discover long term fault creep in New Zealand have been unsuccessful thus far (Scholz 1989).

Velocity-dependent damping along faults accordingly seems to be a key element in several faulting processes (Bombolakis 1989b). For example, analyses in mechanical engineering show that velocity-dependent damping along a slip surface controls the rate, amount, and time duration of slip (e.g. Den Hartog 1958; Timoshenko *et al.* 1974). The analyses incorporate various forms of velocity-dependent damping in terms of a damping parameter δ with respect to inertial forces. The damping forces can be very complex, and so δ is evaluated such that the dissipative energy associated with δ is equivalent to the dissipative energy produced by the actual damping forces (Timoshenko *et al.* 1974, p. 64, 81-88). Consequently, relevant concepts from mechanical engineering should have an important bearing on faulting in the upper continental crust.

Fault slip models

Figure 1 provides a reference for discussion. Recurring fault slip occurs along length L of an active thrust-belt segment of unit width W, bounded by the topographic slope and vertical sections h_1 and h_2. The barrier shown in Figure 1 is a frontal ramp, but it alternatively could be represented by a listric segment of an imbricate fault. The other barrier in the vicinity of a-b is not shown because various types of barriers are possible. In the case of a coseismic net slip event along L, the slip equations of Bombolakis (1989a) indicate that fault slip X is given approximately by

$$X \approx \frac{(P - S)LW}{k_1 + k_2}\left[1 - \cos \omega_n t\right] \quad (1)$$

where P is the peak shear strength at onset of dynamic slip, S is the average shear resistance during dynamic slip, (P - S) is the stress drop, k_1 and k_2 are the respective elastic composite stiffnesses calculated from seismic reflection data for the sedimentary packages along h_1 and h_2 (Bombolakis 1989a), and

$$\omega_n = \sqrt{\frac{k_1 + k_2}{\rho \cdot (\frac{L}{2} \tan \alpha + h_2) \cdot L \cdot W}} \quad (2)$$

is the fundamental frequency of the thrust-belt segment. Tan α is the topographic slope and ρ is average bulk density. Each composite stiffness constant is analagous to the spring constant of a spring-mass system. The stiffer the spring, the larger the spring constant. Competent strata usually are elastically stiffer than incompetent strata. The resulting fundamental frequency characterizes the simplest mode of motion that the active mass would have if it were not affected by dissipative forces. This frequency therefore is one of the fundamental parameters inherent in the elastic rebound theory of earthquake faulting. An example of how the elastic stiffnesses are calculated from seismic profiling data is illustrated in Table 2 of Bombolakis (1989a).

The time duration of a coseismic net slip in equation (1) is $t = \pi/\omega_n$. For this t, equation (1) becomes identical in form to the net slip equations of traditional crack-growth models employed in earthquake seismology (see equation 26 of Bombolakis 1989a). But like the traditional crack-growth models, equation (1) neglects velocity-dependent damping.

If velocity-dependent damping is incorporated in the derivation of equation (1), we obtain (Bombolakis 1989b)

$$X \approx \frac{(P - S)LW}{k_1 + k_2}\left[1 - e^{-\frac{\delta}{2}t}(\cos \omega_d t + \frac{\delta/2}{\omega_d} \sin \omega_d t)\right] \quad (3)$$

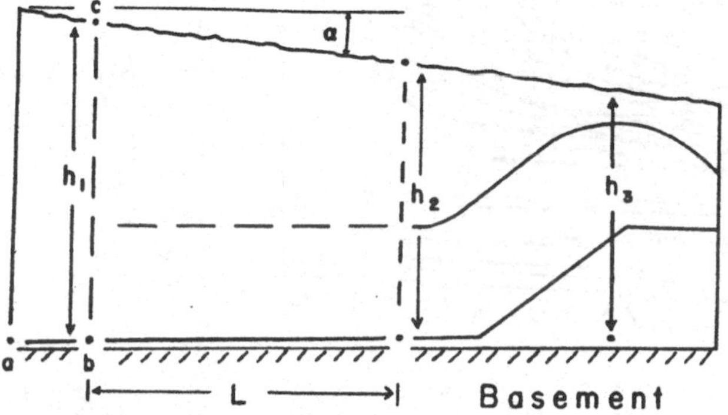

Figure 1. A developmental stage of a foreland belt. The temporally active thrust-belt segment is bounded by L, h1, h_2, the topographic slope, and unit width W perpendicular to diagram.

where δ is the damping parameter expressed in frequency units (Timoshenko *et al.* 1974), $\omega_d = \sqrt{\omega_n^2 - \delta^2/4}$ is the damped fundamental frequency, and $t = \pi/\omega_d$ is the duration of a net slip event. Therefore equation (3) can describe the approximate fault slip resulting from typical earthquakes, slow earthquakes, and accelerating-decelerating fault creep events along an active segment of a decollement when $\delta/2 < \omega_n$.

In contrast, when $\delta/2 > \omega_n$ we have the case of heavy damping (Timoshenko *et al.* 1974). The fault-slip equation for this situation is transformed to

$$X \approx \frac{(P - S)LW}{k_1 + k_2}\left[1 - e^{-\frac{\delta}{2}t}(\cosh \beta t + \frac{\delta/2}{\beta} \sinh \beta t)\right] \quad (4)$$

where $\beta = \sqrt{\delta^2/4 - \omega_n^2}$. The importance of the fundamental frequencies of active fault blocks with respect to velocity-dependent damping along faults is illustrated nicely by the case of $\delta >> \omega_n$. For this case, $\delta/2 \approx \beta$, causing the bracketed expression to approach zero because of a mathematical identity between hyperbolic functions and exponential functions. Therefore, $X \approx 0$ irrespective of time t, with the consequence that no significant fault creep can occur along L. This physical condition accordingly might be one of the basic reasons why so few examples of 'long term' fault creep have been documented in the upper continental crust since instrumental monitoring of active fault zones was initiated circa 1960. And it is a physical condition that can be evaluated in analyses of thrust-sheet emplacement, as illustrated in the following field example.

Hogsback thrust sheet in the Kemmerer region, Wyoming

In general, damping includes both a constant component of damping and a velocity-dependent component. S in equations (3) and (4) is the shear resistance associated with the constant component (Bombolakis 1989a). If we denote τ_d as the average shear resistance associated with both components, then (τ_d - S) represents the average increment of shear resistance associated with velocity-dependent damping along length L of the decollement in Figure 1. For these conditions, the approximate value of δ is

$$\delta \approx \frac{(\tau_d - S)}{\rho \cdot \dot{X}avg. \cdot (\frac{L}{2} \tan \alpha + h_2)} \quad (5)$$

where $\dot{X}$avg. is the average slip rate of a fault slip event along L, and h_2 is the thickness of the sedimentary package indicated in Figure 1.

The Hogsback thrust sheet illustrates the problem of long-term fault creep of thrust sheets. Extensive deformation in its frontal ramp region near Kemmerer is shown in Figure 2. (See Figs 1 & 2 in Delphia & Bombolakis 1988 for detailed stratigraphic section and location in the Wyoming Salient). The predominantly competent sequence encompassing the

Ordovician Bighorn Dolomite (O_{bh}) and the Triassic Nugget sandstone (TR_n) had undergone considerable imbrication in the break back mode. In contrast, the same stratigraphic sequence just above the basal decollement extends westward from c-d for 30 km with no significant deformation apparent within the resolution of seismic reflection profiles and other subsurface data of Dixon (1982) and Lamerson (1982), until the trailing edge contact with a major ramp of the Absaroka thrust is approached.

Suppose that this section of the Hogsback plate is assumed to have undergone long-term fault creep at an average geological slip rate along L in Figure 1. The estimated rate for thrusts in the Wyoming Salient (Wiltschko & Dorr 1983) is similar to the 10^{-7} to 10^{-8} cm/sec. rates calculated by Elliott (1976) for external thrusts of the Canadian Rockies foreland belt. L ≈ 30 km for the Hogsback sheet, and $h_2 \approx 10$ km with $\alpha \approx 4°$ prior to post-thrusting erosion. Utilizing elastic stiffness data for the Kemmerer region (Bombolakis 1989a), equation (2) indicates that the fundamental frequency of this part of the Hogsback plate is less than 1 Hertz. A conservative estimate of (τ_d - S) is ≈ 0.1 bar in equation (5) because estimated stress drops of fault creep along the central section of the San Andreas are in the range of 0.4 bar to several bars (Wesson 1988). Consequently, for average geological slip rates, equation (5) yields values of δ in the range of 10^5 to 10^6 Hertz. These values would increase by a factor of 10, for example, if (τ_d - S) ≈ 1 bar, instead of 0.1 bar. Therefore equation (4) stipulates that no significant fault creep displacement possibly could have occurred under these conditions, even if the shear resistance associated with velocity-dependent damping were small.

The net slip of this section of the Hogsback sheet exceeds 12 km (Delphia & Bombolakis 1988). How, then, could the fault slip have occurred? Consider again fault slip data from active tectonic terranes, keeping in mind that equations (2) and (5) show that the fundamental frequency is more sensitive to length L of the active thrust-belt segment than the velocity-dependent damping parameter.

The 'long term' fault creep along the central section of the San Andreas actually appears to be dominated by sequences of short-term fault creep events, each affecting a few or more km of fault length along the traces (Wesson 1988). Many of the creep events are interpreted to be associated with yield point phenomena (Nason & Weertman 1973; Bombolakis *et al.* 1978; Wesson 1988), which is consistent with (P - S) of equations (3) and (4). In the case of thrust belts, field evidence of repeated cycles of work softening is reported for several low-angle thrust faults in the foreland zone of the Appalachians (Wojtal & Mitra 1986, 1988; Woodward *et al.* 1988).

Slip rates of fault creep events along the San Andreas frequently range from 10^{-3} to 10^{-6} cm/sec. (Wesson 1988; Bilham 1989). Applying these data to the Hogsback sheet for length L ≈ 5 - 10 km, equation (5) yields δ in the range of 10 - 10^5 Hertz, whereas equation (2) yields a fundamental frequency of only a few Hertz for the same length segments. Consequently, for the more rapid fault creep events, more cumulative fault slip is possible than for the slower fault creep events. For example, slip rates of slow earthquakes in the range of 0.1 - 1 cm/sec. yield values of δ of about 10^{-1} to 10^{-2} Hertz. Therefore, equations (2), (3), (4), and (5) indicate that it is only slow earthquakes and relatively rapid fault creep events, as well as typical earthquakes, that could have produced significant propagating fault slip between the trailing edge and the Hogsback frontal ramp. Since no significant fault displacement can result from recurring slow fault creep of work-hardening/work-softening material, a potentially

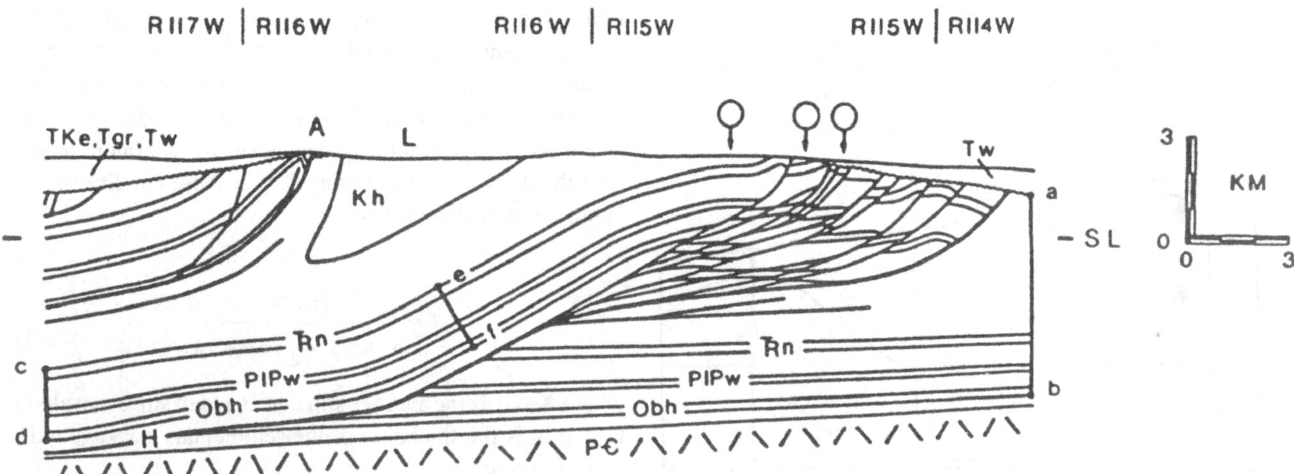

Figure 2. Balanced cross-section of the Absaroka (A) and Hogsback (H) thrust sheets 15 km north of Kemmerer in the Wyoming Salient. The Lazeart syncline (L) is demarcated by a solid-line curve drawn along contact between the Frontier formation and the Cretaceous Hilliard clastic sequence (K_h). Open circles indicate locations of three deep wells close to the section. Three key marker beds within the predominantly competent stratigraphic sequence are the Ordovician Bighorn Dolomite (O_{bh}), the Permo-Pennsylvanian Weber sandstone (PIP_w), and the Triassic Nugget sandstone (TR_n). From Delphia & Bombolakis (1988).

important effect of such events may be temporal thickening of the fault zone in accordance with conservation of energy principles.

Partitioning of elastic tectonic strains - a suggestion

Whether fault slip occurs by typical earthquakes, slow earthquakes, or relatively rapid fault creep events, the elastic tectonic strains imposed along h_2 of the active thrust-belt segment in Figure 1 are 'instantaneous' with respect to typical strain rates of flow processes involved in layer-parallel shortening and layer-parallel shear. A conceptual way of partitioning is illustrated by the Maxwell model in Figure 3. During coseismic fault slip, for example, the imposed tectonic strain is taken up by the elastic spring. Therefore, the amount of subsequent strain-rate dependent deformation depends on how the relaxation time of the dominant flow process compares with the recurrence interval of the fault-slip events. By definition, the relaxation time is the time it takes to decrease the elastic strain by 1/e of its value. The recurrence intervals of moderate to large magnitude thrust earthquakes thus far seem to be in the 100-500 year range (e.g. Swan 1988; Stein & Yeats 1989). Consequently, if the relaxation time were of the same magnitude as the recurrence interval, then relatively little strain-rate dependent ductile deformation would occur, and considerable elastic tectonic strains potentially would be available for brittle and semi-brittle deformation.

Much of the current fold theory is based on linear viscosity η (Hudleston 1986), and so it is worthwhile at this time to consider the relaxation times of layer-parallel shortening and layer-parallel shear based on linear visco-elasticity theory. This theory shows that the relaxation time T for layer-parallel shortening, without the constraint of plane strain, is the equation shown in Figure 4, where E is Young's modulus. This equation differs from the plane strain case by a factor of only $1/(1-\nu^2)$, which is negligible for the customary range of Poisson's ratio ν. For comparison, the relaxation time T' for layer-parallel shear is given also in Figure 4, where G is the shear modulus. If we calculate the ratio of these two relaxation times for each bed of a thrust sheet, then the ratios reduce to the elastic constants of each bed, which can be calculated from seismic reflection profiling data (Bombolakis 1989a). For some beds, T' can be 2/3 smaller than T, in which case layer-parallel shear should be more dominant than layer-parallel shortening for appropriate values of τ/σ. Consequently, the idealized scheme in Figure 4 illustrates how these concepts can be tested with deformation profiles of the simpler thrust sheets described by Sanderson (1982) and Mitra & Protzman (1988).

The implications of these concepts are illustrated by the Hogsback structure in Figure 2 with respect to Figure 1. Elastic stiffness contrasts between the sedimentary packages along h_2 and h_3 in Figure 1 indicate that some or much of the sequential imbrication in Figure 2 probably had been associated with seismic deformation in the ramp region (Bombolakis 1989a). If this is correct, then the recurrence intervals of the fault-slip events must have been appreciably shorter than the relaxation times of layer-parallel shortening and layer-parallel shear of the competent strata. For example, the Twin Creek limestone lies stratigraphically above the Nugget sandstone in Figure 2, but there is no evidence of significant layer-parallel shortening of either of these formations within the resolution of subsurface data. Layer-parallel shortening of the Twin Creek limestone due to pressure solution is minor in this part of the Wyoming Salient (Mitra *et al.* 1984).

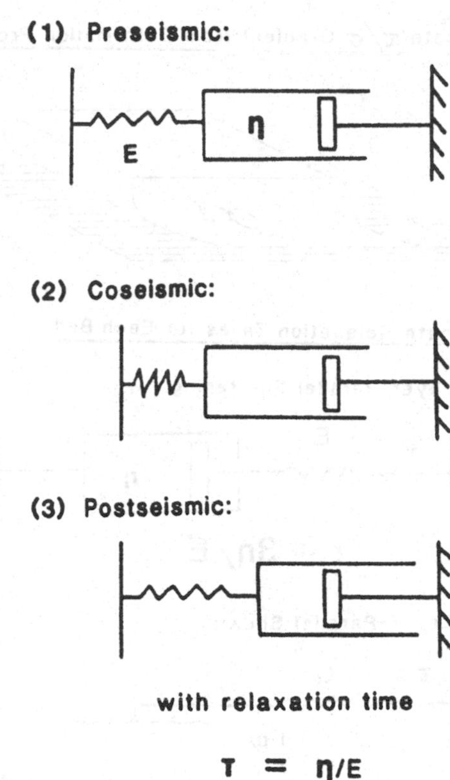

Figure 3. Conceptual Maxwell model, illustrating relaxation of coseismic elastic tectonic strain imposed on sedimentary package along h_2 in Figure 1.

The deformation of the competent sequence in Figure 2 is dominated by flexural-slip folding and imbrication (Delphia & Bombolakis 1988). And the fact of the matter is that modern examples of seismically induced flexural-slip have been recognized in recent years; notably, bedding slip in a syncline due to the April 7, 1981 ML 2.5 Lompac earthquake in the northwest Transverse Ranges (Yerkes *et al.* 1983), bedding slip in a syncline due to the 1980 El Asnam earthquake in Algeria (Philip & Meghraoui 1983), and a stage of flexural-slip folding of an anticline along the Elmore Ranch fault during the 1987 Superstition Hills earthquake sequence in the southern San Andreas (Klinger & Rockwell 1989).

Concluding remarks

Recurring earthquakes along a specific fault segment can be viewed as elastic rebound with respect to 'asperities' and

Partitioning Seismic Strains :

1. Estimate τ/σ Gradient from Deformation Profile:

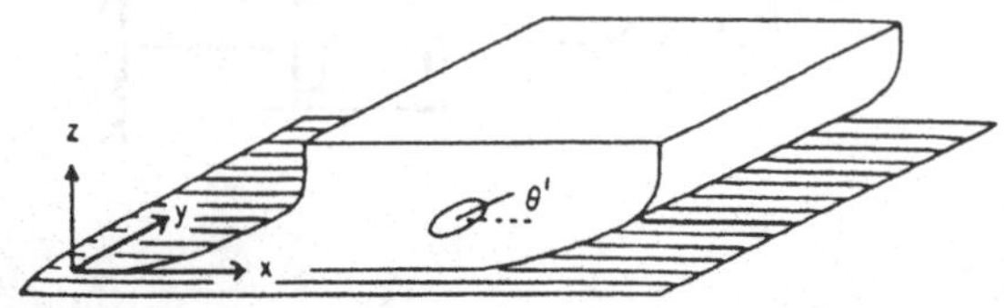

2. Estimate Relaxation Times for Each Bed:

(a) Layer-Parallel Shortening:

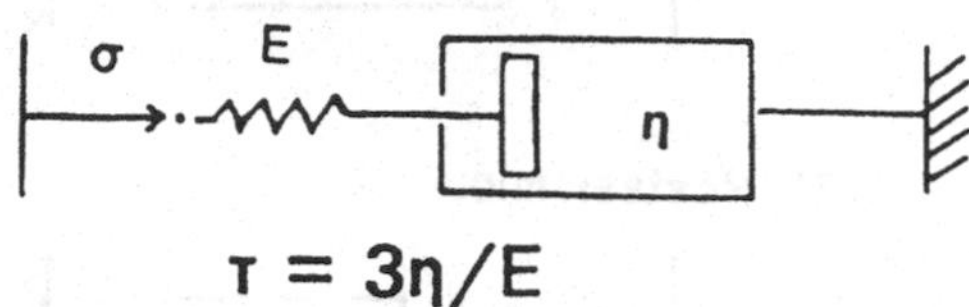

(b) Layer-Parallel Shear:

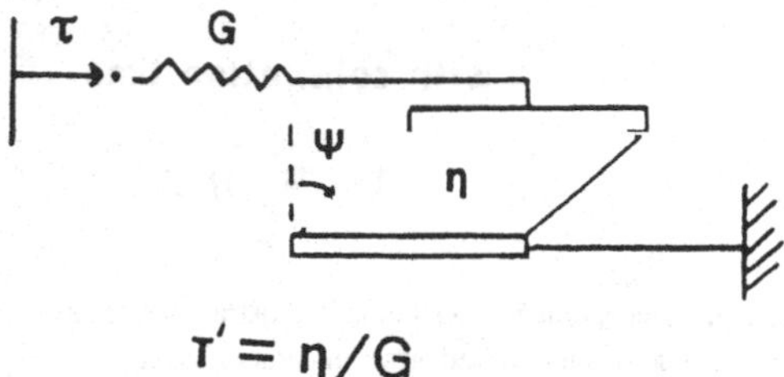

Figure 4. Idealized example of partitioning layer-parallel shortening and layer-parallel shear within a thrust sheet in terms of relaxation times T and T '. See text.

barriers (Aki 1984). In the case of fault creep events, Wesson (1988) finds that they can be viewed as arising from (1) stress applied from external sources, (2) stress caused by geometry and distribution of displacement on the fault, arising from the elastic response of the surrounding medium to the displacements within the inelastic fault zone itself, and (3) constitutive relations and yield point phenomena that characterize slip resistance along the fault. (Wesson's models currently do not incorporate inertial forces).

These parameters are incorporated here in prototype models that include inertial effects. In the case of a typical earthquake, (P - S) in equation (3) is directly proportional to stress drops calculated in earthquake seismology (Bombolakis 1989a). The alteration of fault-zone material, however, can alter fault-slip style along specific segments of a decollement. The constitutive relations for the segments probably are very complex, and so it is more convenient to characterize several effects of fault-rock behaviour by damping parameters. Damping factor δ, for example, can be calculated directly from basic field parameters in equation (5). In the case of episodic fault creep associated with yield point phenomena, P is the peak shear strength equal to or greater than the yield point, depending on work hardening. The result is that (P - S) is the stress drop associated with work softening, where S can be calculated from a steady-state creep flow law when appropriate. (P - S) therefore can be larger than the stress drop associated with the initial phase of a fault-creep event. This is one reason why the approximate sign is employed in equations (3) and (4); other reasons are discussed elsewhere (see discussion and relevant references in Bombolakis 1989a). For very heavy damping, the fault displacement is very small, regardless of whether (P - S) is larger than the stress drop associated with the early phase of the motion. Consequently, because of opposing damping and inertial effects, the relation between δ and the fundamental frequencies of active thrust-block segments is one of the critical relations that determine whether fault slip may occur as a typical earthquake, slow earthquake, or a relatively rapid or slow fault creep event.

The analysis of the Hogsback thrust sheet in the Kemmerer region illustrates potential modes of emplacement of external thrust sheets in a foreland belt. For each style of fault slip event analysed here, the increment of shear resistance associated with velocity-dependent fault damping is $(\tau_d - S) < (P - S)$. And for $(\tau_d - S)$ as small as 0.1 bar, data analysis for the Kemmerer region demonstrates that the 12 or more km of cumulative net slip could not have developed at average geological slip rates. Instead, the Hogsback sheet would have had to be emplaced via more rapid types of recurring fault slip. Potential exceptions permitting emplacement at average geological slip rates include thrusting along salt and the LPS thrusting described by Geiser (1988), which do not apply to the Hogsback sheet (Delphia & Bombolakis 1988). Furthermore, fault-rock studies of several foreland thrusts in the Appalachians indicate that fault slip involved cycles of work hardening and work softening (Wojtal & Mitra 1986, 1988; Woodward *et al.* 1988). Hence, in view of the relatively rapid forms of fault slip operating in active zones of the upper continental crust, it is unlikely that the development of foreland belts can be analysed exclusively in quasi-static terms. The mechanical partitioning of elastic tectonic strains accordingly would need to include analyses based on relations between relaxation times and recurrence intervals.

An important observation in recent years is that syntectonic extensional faulting is associated with compressional faulting in several thrust belts (e.g. Platt & Leggett 1986; Wojtal & Mitra 1986; Coward 1988). In Figure 1, the sedimentary package of the active thrust-belt segment along vertical section h_1 can undergo interludes of extension during dynamic fault slip. However, it is not clear yet whether extensional faulting would occur in the vicinity of h_1, as was observed during the 1941 Hartford dyke detachment faulting event (Bombolakis 1981). Syntectonic extensional faulting in thrust belts therefore constitutes another potentially important problem for future study.

Constructive suggestions for revision of the preliminary manuscript were made by P. D'Onfro and W. Rizer of Conoco Inc., R. H. Martin III of New England Research Inc., an anonymous reviewer, and the editor of this volume. Special thanks to Jeffrey Townsend for the arrangement and editing of a technically difficult manuscript.

References

Aki, K. 1984. Asperities, barriers, characteristic earthquakes, and strong motion prediction. *Journal of Geophysical Research*, **89**, 5867-5872.

Allen, C. R. 1981. The modern San Andreas Fault. *In*: Ernst, W. G. (ed.) *The Geotectonic Development of California, Rubey Volume 1*. Prentice-Hall, Englewood Cliffs, New Jersey, 223-249.

Baranowski, J., Armbruster, J., Seeber, L. & Molnar, P. 1984. Focal depths and fault plane solutions of earthquakes and active tectonics of the Himalayas. *Journal of Geophysical Research*, **89**, 6918-6928.

Berberian, M. 1982. Aftershock tectonics of the 1978 Tabas-e-Golshan (Iran) earthquake sequence: a documented active 'thin-and-thick skinned tectonic' case. *Geophysical Journal of the Royal Astronomical Society*, **68**, 499-530.

Beroza, G. C. & Jordan, T. H. 1989. Comparison of free-oscillation excitation events with global seismicity: evidence for silent earthquakes. *Transactions of the American Geophysical Union (EOS)*, **70**, 397.

Bilham, R. 1989. Surface slip subsequent to the 24 November 1987 Superstition Hills, California, earthquake monitored by digital creepmeters. *Bulletin of the Seismological Society of America* , **79**, 424-450.

Bombolakis, E. G. 1981. Analysis of a horizontal catastrophic landslide. *In*: Carter, N. L., Friedman, M., Logan, J. M. & Stearns, D. W. (eds) *Mechanical Behaviour of Crustal Rocks, the Handin Volume*. American Geophysical Union. Geophysical Monograph, **24**, 251-257.

—— 1989a. Thrust-fault mechanics and dynamics during a developmental stage of a foreland belt. *Journal of Structural Geology*, **11**, 439-455.

—— 1989b. Fault slip and partitioning of elastic tectonic strains during dynamic low-angle thrusting. *Geological Society of America Abstracts with Programs*, **20**, A134-A135.

——, Hepburn, J. C. & Roy, D. C. 1978. Fault creep and stress drops in saturated silt-clay gouge. *Journal of Geophysical Research*, **83**, 818-829.

Brune, J. N. 1968. Seismic moment, seismicity, and rate of slip along major fault zones. *Journal of Geophysical Research*, **73**, 777-784.

Coward, M. 1988. The Moine thrust and the Scottish Caledonides. *In*: Mitra, G. & Wojtal, S. (eds) *Geometries and Mechanisms of Thrusting, with Special Reference to the Appalachians*. Geological Society of America Special Paper, **222**, 1-16.

Davis, T.L. , Namson, J. & Yerkes, R. F. 1989. A cross section of the Los Angeles area: seismically active fold and thrust belt, the 1987 Whittier Narrows earthquake, and earthquake hazard. *Journal of Geophysical Research*, **94**, 9644-9664.

Delphia, J. & Bombolakis, E. G. 1988. Sequential development of a frontal ramp, imbricates, and a major fold in the Kemmerer region of the Wyoming thrust belt. *In*: Mitra, G. & Wojtal, S. (eds) *Geometries and Mechanisms of Thrusting, with Special Reference to the Appalachians*. Geological Society of America Special Paper, **222**, 207-222.

Den Hartog, J. P. 1956. *Mechanical Vibrations*. McGraw-Hill, New York.

Dixon, J. S. 1982. Regional structural synthesis, Wyoming Salient of western overthrust belt. *American Association of Petroleum Geologists Bulletin*, **66**, 1560-1580.

Eberhart-Phillips, D. 1989. Active faulting and deformation of the Coalinga anticline as interpreted from three-dimensional velocity structure and seismicity. *Journal of Geophysical Research*, **94**, 15565-15586.

Ekström, G. & England, P. 1989. Seismic strain rates in regions of distributed continental deformation. *Journal of Geophysical Research*, **94**, 10231-10257.

Geiser, P. 1988. Mechanisms of thrust propagation: some examples and implications for the analysis of overthrust terranes. *Journal of Structural Geology*, **10**, 829-845.

Hudleston, P. J. 1986. Extracting information from folds in rocks. *Journal of Geological Education*, **34**, 237-245.

Jackson, J. A. 1983. The use of earthquake source studies in continental tectonic geology. *In*: Kanamori, H. & Boschi, E. (eds) *Earthquakes: Observation, Theory, and Interpretation, Course LXXXV*. Proceedings of the International School of Physics 'Enrico Fermi', Italian Physical Society. North-Holland, Amsterdam, 456-478.

Klinger, R. E. & Rockwell, T. K. 1989. Flexural-slip folding along the Elmore Ranch fault in the Superstition Hills earthquake sequence of November 1987. *Bulletin of the Seismological Society of America*, **79**, 297-303.

Lamerson, P. R. 1982. The Fossil Basin and its relationship to the Absaroka thrust system, Wyoming and Utah. *In*: Powers, R. B. (ed.) *Geologic Studies of the Cordilleran Thrust Belt, Volume 1*. Rocky Mountain Association of Geologists, Denver, Colorado, 279-340.

Lin, J. & Stein, R. S. 1989. Coseismic folding, earthquake recurrence, and the 1987 source mechanism at Whittier Narrows, Los Angeles Basin, California. *Journal of Geophysical Research*, **94**, 9614-9632.

Mitra, G., Yonkee, W. A. & Gentry, D. J. 1984. Solution cleavage and its relationship to major structures in the Idaho-Utah-Wyoming thrust belt. *Geology*, **12**, 354-358.

—— & Protzman, M. 1988. Strain trajectories and deformation profiles in thrust belts and extensional terranes. *Geological Society of America Abstracts with Programs*, **100**, A107.

Nabalek, J. 1985. Geometry and mechanism of faulting of the 1980 El Asnam, Algeria earthquake from inversion of teleseismic body waves and comparison with field observations. *Journal of Geophysical Research*, **90**, 12713-12728.

Namson, J. & Davis, T. L. 1988a. Seismically active fold and thrust belt in the San Joaquin Valley, central California. *Bulletin of the Geological Society of America*, **100**, 257-273.

—— & Davis, T. L. 1988b. A structural transect of the western Transverse Ranges, California: implications for lithospheric kinematics and seismic risk evaluation. *Geology*, **16**, 675-679.

—— & Weertman, J. 1973. A dislocation theory analysis of fault creep events. *Journal of Geophysical Research*, **78**, 7745-7751.

Philip, H. & Meghraoui, M. 1983. Structural analysis and interpretation of the surface deformations of the El Asnam earthquake of October 10, 1980. *Tectonics*, **2**, 17-49.

Platt, J. P. & Leggett, J. K. 1986. Stratal extension in thrust footwalls, Makran accretionary prism: implications for thrust tectonics. *American Association of Petroleum Geologists Bulletin*, **70**, 191-203.

Price. R. A. 1988. The mechanical paradox of large overthrusts. *Bulletin of the Geological Society of America*, **100**, 1898-1908.

Sacks, I. S., Linda, A. T., Snoke, J. A. & Suyehiro, S. 1981. A slow earthquake sequence following the Izu-Oshima earthquake of 1978. *In*: Simpson, D. W. & Richards, P. G. (eds) *Earthquake Prediction, An International Review*. American Geophysical Union, Maurice Ewing Series, **4**, 617-628.

Sanderson, D. J. 1982. Models of strain variation in nappes and thrust sheets: a review. *Tectonophysics*, **88**, 201-233.

Savage, J. C. 1983. Strain accumulation in the western United States. *Annual Review of Earth and Planetary Sciences*, **11**, 11-43.

Scholz, C. H. 1989. Mechanics of faulting. *Annual Review of Earth and Planetary Sciences*, **17**, 309-334.

Schwartz, D. P. & Coppersmith, K. J. 1986. Seismic hazards: new trends in analysis using geologic data. *In*: *Active Tectonics*. National Academy of Sciences Press, Washington D.C., 215-230.

Stein, R. S. & Yeats, R. S. 1989. Hidden earthquakes. *Scientific American*, June, 48-57.

—— 1987. Contemporary plate motion and crustal deformation. *American Geophysical Union, Reviews in Geophysics*, **25**, 855-863.

Sibson, R. H. 1989. Earthquake faulting as a structural process. *Journal of Structural Geology*, **11**, 1-14.

Swan, F. H. 1988. Temporal clustering of Palaeoseismic events on the Oued Fodda fault, Algeria. *Geology*, **16**, 1092-1095.

Synvhal, H., Agrawal, P. N., King, G.C.P. & Guar, V. K. 1973. Interpretation of measured movement at a Himalayan (Nahan) thrust. *Geophysical Journal of the Royal Astronomical Society*, **34**, 203-210.

Timoshenko, S., Young, D. H. & Weaver, W. 1974. *Vibration Problems in Engineering*. John Wiley and Sons, New York.

Wentworth, C. M. & Zoback, M. D. 1989. The style of late Cenozoic deformation at the eastern front of the California Coast Ranges. *Tectonics*, **8**, 237-246.

Wesson, R. L. 1988. Dynamics of fault creep. *Journal of Geophysical Research*, **93**, 8929-8951.

Wilt, J. W. 1958. Measured movement along the surface trace of an active thrust fault in the Buena Vista Hills, Kern County, California. *Bulletin of the Seismological Society of America*, **48**, 169-176.

Wiltschko, D. V. & Dorr, J. A. Jr. 1983. Timing of deformation in overthrust belt and foreland of Idaho, Wyoming and Utah. *American Association of Petroleum Geologists Bulletin*, **67**, 1304-1322.

Wojtal, S. & Mitra, G. 1986. Strain hardening and strain softening in fault zones from foreland thrusts. *Bulletin of the Geological Society of America*, **97**, 674-687.

—— & —— 1988. Nature of deformation in some fault rocks from Appalachian thrusts. *In*: Mitra, G. & Wojtal, S. (eds) Geometries and Mechanisms of Thrusting, with Special Reference to the Appalachians. *Geological Society of America, Special Paper*, **222**, 17-33.

Woodward, N. B., Wojtal, S., Paul, J. B. & Zadins, Z. Z. 1988. Partitioning of deformation within several external thrust zones of the Appalachian orogen. *Journal of Geology*, **96**, 351-361.

Yerkes, R. F., Ellsworth, W. L. & Tinsley, J. C. 1983. Triggered reverse fault and earthquake due to crustal unloading, northwest Transverse Ranges, California. *Geology*, **11**, 287-291.

Yielding, G. 1985. Control of rupture by fault geometry during the 1980 El Asnam (Algeria) earthquake. *Geophysical Journal of the Royal Astronomical Society*, **81**, 641-670.

PART TWO

Physical modelling

One-dimensional models for plane and non-plane power-law flow in shortening and elongating thrust zones

Steven Wojtal

Department of Geology, Oberlin College, Oberlin, OH 44074, USA

Abstract: The movement of a thrust sheet, even in external parts of fold-thrust belts, occurs by deforming rocks within a zone of finite thickness. A significant fraction of thrust-zone strains may accrue slowly as viscous deformation. This contribution uses a simple one-dimensional model of power-law flow to analyse the viscous deformation in thrust zones. The model generates profiles of stress and velocity for plane and non-plane flows that shorten or elongate during shearing. These stress and velocity profiles are, in essence, 'snapshots' of stresses and velocities in real thrust zones that sequentially shorten and elongate during thrust emplacement. A comparison of stress and velocity solutions for flows with different stress exponents gives a way to examine structural settings where a single weak layer develops and affects stresses throughout the sheet. Using data on the orientations of stress principal directions during episodes of shortening and elongation derived from the mesoscopic fault array in an external thrust zone from the southern Appalachian fold-thrust belt, the plane-flow solution suggests that differential stresses in the thrust zone did not exceed 20 MPa. Moreover, body forces due to the dipping upper surface of the sheet were apparently not the primary source of the tractions during thrust zone shortening or elongation episodes. The additional tractions that other portions of the thrust sheet exerted on this thrust-zone segment had comparable magnitudes during shortening and elongation episodes. Comparing plane and non-plane solutions indicates that (1) shearing may localize at shallower depths or occur with different orientations for principal stresses relative to the thrust-zone boundaries, and (2) shortening or elongation may occur at smaller magnitudes of in-transport compression or tension, in non-plane structural settings like the growing tips of thrusts.

The traditional view of fold-thrust belts has each thrust sheet sliding over its footwall on a discrete fault surface (Hubbert & Rubey 1959; Hsu 1969; Price 1988). This view does not conform with the movement mechanisms known for thrust sheets in the internal portions of fold-thrust belts (cf. Schmid 1975; Gilotti & Kumpulainen 1986). It is at odds increasingly with observations in the external portions of fold-thrust belts, where studies of well-exposed thrust zones indicate that thrust sheets move by deforming rocks within a zone several metres to several tens of metres thick (Harris & Milici 1977; Wojtal 1986; Platt & Leggett 1986). Prevailing models for thrust belt evolution, on the other hand, specify that wedges exhibiting uniform mechanical behaviour yield throughout while sliding above a basal detachment (Chapple 1978; Davis *et al.* 1983; Emerman & Turcotte 1983). The intensities and geometries of strains within foreland thrust sheets do not conform with the predictions of yielding-wedge models (Woodward 1988). Moreover, rocks at different positions within orogenic wedges and along distinct segments of thrust zones within the wedges may exhibit distinct material properties (Platt 1986; Wojtal & Mitra 1988). These factors may limit the applicability of elegant analyses of orogenic deformation, and they make exact solutions for realistic orogenic wedges difficult to attain (Fletcher 1989).

A more tractable task is to examine limited segments of a thrust sheet where a single deformation mechanism prevails. Even then, one must tailor the analysis to fit the desired focus. One intent of this contribution is to present a simple mathematical model, based on field studies of several external fold-thrust belt thrust zones, that gives stresses and velocities in plane strain as functions of position below the surface of a thrust sheet. The model provides insight into the forces that drove the emplacement of an external thrust sheet in the southern Appalachian fold-thrust belt. It also guides inferences on the effects of changes in material behaviour due to initial differences in rock rheology or to the mesoscopic and microscopic structural changes that accompany strain accumulation. A second intent of this contribution is to extend this analysis to three-dimensions, considering non-plane flow in thrust sheets. Preliminary examination of the results of an added degree of freedom, even in the unsophisticated manner considered here, suggests that non-plane strain treatments are warranted.

The model, derived from glacier flow analyses (Nye, 1957; Paterson, 1981), considers the flow of a power-law sheet between parallel-sided boundaries. The formulation (described in greater detail below) does not specify a priori the stress exponent n for the sheet, so it is sufficiently general to treat linearly viscous sheets ($n = 1$) and perfectly plastic sheets ($n \rightarrow \infty$). Stress and velocity distributions for flows with different stress exponents n provide some insight to the effects of different rheologies. More importantly, the model provides explicit solutions for stresses and velocities in laminar flows (i.e. heterogeneous simple shear flows) as well as solutions for compressing and extending flows (flows that shorten or elongate as they shear). This contribution uses those solutions to examine the elongation and shortening that accompanies shearing in many fold-thrust belt thrust zones.

Structure of foreland fold-thrust belt thrust zones

Mesoscopic structure

Immediately above and below many foreland fold-thrust belt thrust surfaces, pervasive arrays of mesoscopic minor faults occur within thin (relative to the thrust sheet's thickness) layers (Harris & Milici 1978; Wojtal 1986; Platt & Leggett 1986; Mitra 1986). Faults in the arrays typically make either low (<45°) or high (>45°) angles to the thrust zone boundaries (Fig. 1). Faults at low angles to the thrust zone boundaries usually (1) have lineations whose pitches are close to 90°, (2) also cut the enveloping bedding at low (<45°) angles, and (3) exhibit reverse offsets of bedding. Faults at high angles to the thrust zone boundaries have less consistent kinematics. Whereas these faults generally also cut enveloping bedding at high (>45°) angles, the pitches of lineations on them may be close to 0°, close to 90°, or in between. Normal offsets of bedding on high-angle faults are more common, but reverse offsets occur. Where both low-angle and high-angle faults occur together, high-angle faults usually postdate low-angle faults.

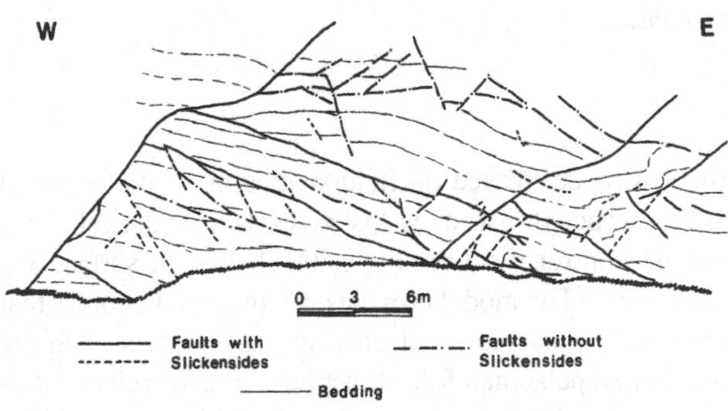

Figure 1. Downplunge projection (onto a vertical plane with a strike of 110°) of a portion of the mesoscopically faulted rock in the Cumberland Plateau thrust zone, southern Appalachians (after Harris & Milici 1977; Wojtal 1986). The primary decollement in a coal layer lies parallel to the base of the drawing, just below the edge of the drawing. Irregular line at base represents rubble that locally covers the detachment. Solid bold lines and dashed bold lines show faults with slickensides preserved. Dash-dot bold lines show faults with no slickensides. Fine lines show bedding traces. The dashed fault traces are those faults with slickensides that do not conform with either of two well-defined palaeostress tensors (Wojtal & Pershing 1991).

This pervasive mesoscopic faulting accommodates deformation that is continuous at the scale of a thrust sheet (Wojtal 1989). The distribution of displacement among faults in these arrays indicates that small numbers of faults account for most of the slip in the zones of pervasive faulting (Wojtal & Mitra 1986). Moreover, the mesoscopic faults that account for most of the total slip constitute subsets of faults that conform with well-defined palaeostress tensors (Fig. 2) (Wojtal & Pershing 1991). Bulk deformation in these pervasively-faulted zones was, then, 'block-boundary sliding' in the sense that bulk shortening, bulk elongation, and bulk shear occurred by moving relatively undeformed but fault-bounded blocks of rock past each other. From the standpoint of how the faults in the arrays together accommodated the strains associated with movement on major faults and from the standpoint of how stresses were transmitted through the pervasively-faulted rocks in the vicinity of major faults, it is therefore sensible to take a larger view and examine this deformation at the scale of the whole thrust sheet.

Some foreland fold-thrust belt thrust zones possess thin layers of pervasively-cleaved rock along movement surfaces (Platt *et al.* 1987; Figs. 6 & 7 in Woodward *et al.* 1988). The pervasive cleavage in these zones usually forms by localized pressure solution in rocks with mixed clay-quartz or clay-calcite compositions (Woodward *et al.* 1988), and it accommodates thrust sheet movement by inhomogeneous ductile shearing in narrow layers like that envisioned by Kehle (1970). This type of ductile shearing might not persist within an individual layer throughout the life of a thrust zone, however. Continued shearing accommodated by diffusive mass transfer, particularly in thrust zones that exchange pore fluids with the surrounding rocks, could remove entirely the relatively soluble phases. The resulting narrow layer of clay-rich selvage might continue to accrue slip, or the localization of strain elsewhere in the thrust zone could generate another zone of pervasive cleavage. The occurrence of individual and multiple selvage bands within several large displacement thrust zones (e.g. Hunter Valley and Saltville thrusts in the southern Appalachians and McConnell thrust in the Canadian Rockies) suggests that inhomogeneous shearing of this type may commonly contribute to thrust sheet movement.

Microscopic structure

In many fold-thrust belt thrust zones, cataclastic rocks (1) decorate individual fault surfaces and bed-parallel slip surfaces, and (2) separate hangingwall rocks from footwall rocks (Engelder 1974; House & Gray 1982; Wojtal & Mitra 1986; Platt *et al.* 1987). Thrust-zone deformation may generate either penetratively microfractured rocks with little bulk strain or distinctive banded cataclasites and ultracataclasites that accommodated very high strains (Mitra 1984; Wojtal & Mitra 1986). The magnitude of strain accommodated in cataclastically deformed rocks varies, of course, with rock type and position within individual thrust zones, and from one thrust zone to another. In thrust zones where temperatures during deformation remained below about 200°C, siliciclastic sandstones are penetratively fractured and/or pervasively faulted, but cataclasis does not normally account for large strains and banded cataclasites are absent. Banded carbonate-matrix cataclasites or zones of pervasive cleavage in shales are common in these low temperature thrust zones, suggesting that shearing occurred in units other than siliciclastic sandstones (Wojtal & Mitra 1986; Woodward *et al.* 1988; Arboleya 1989). Where temperatures during deformation exceeded 200°C, banded quartz-rich cataclasites are common in thrust zones (Mitra 1984; Wojtal & Mitra 1988). Regardless of thrust-zone temperature, banded cataclasites occur most often along primary thrust surfaces,

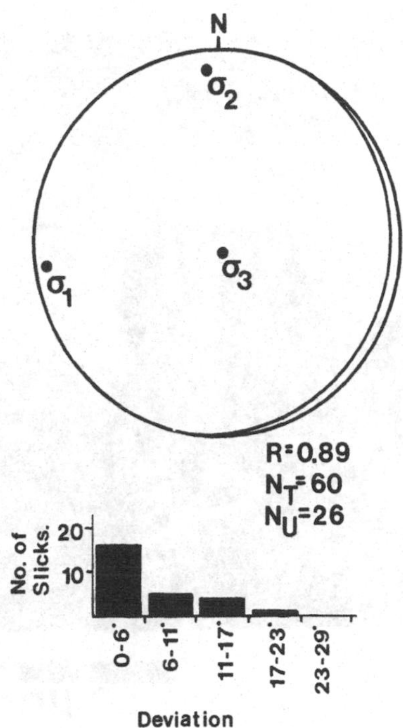

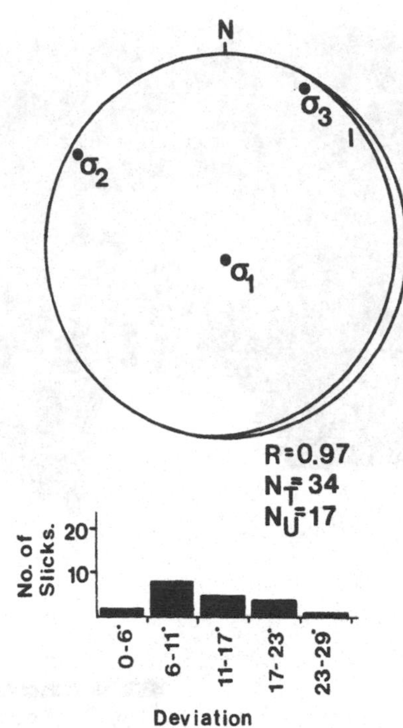

Figure 2. Stereographic projections giving the orientations of palaeostress principal directions calculated, using the attitudes of mesoscopic faults and their slickensides, for the portion of the thrust zone illustrated in Figure 1. This figure shows the two palaeostress tensors calculated for the Cumberland Plateau thrust zone (Wojtal & Pershing 1991). The great circle in each stereographic projection gives attitude of the basal detachment in this thrust zone. The designations 'first' and 'second' for the stress tensors come from cross-cutting relationships among faults. Histograms beneath each stereographic projection give the angular deviations of the fault/slickenside pairs used to calculate each tensor. $R = (\sigma_2 - \sigma_3)/(\sigma_1 - \sigma_3)$. NU is the number of fault slickenside pairs actually used to calculate each palaeostress tensor. NT is the total number of fault-slickenside pairs that could have constrained tensors during each program iteration.

where they constitute layers centimetres to metres thick that separate hangingwall and footwall rocks. The significance of banded cataclasites in such settings is the primary focus here.

In cataclasite to ultracataclasite layers between hangingwall and footwall rocks, rounded and angular fragments of wall rock, mineral-filled veins, and pre-existing cataclasite sit in a matrix of equant grains with diameters less than 10 microns (Fig. 3). Matrix grains usually exhibit no discernible preferred orientation. Alternating light and dark bands aligned parallel to thrust-zone boundaries have respectively greater and lesser matrix fractions. Nearly planar bands with high matrix fractions often extend 10s of centimetres across a hand sample or outcrop.

Planar to smoothly-curved shear surfaces and mineral-filled fractures, most with orientations that correspond to *R*, *R'*, *P*, *Y*, *X*, or *e* (Logan *et al.* 1981; Rutter *et al.* 1986; Platt *et al.* 1987; Woodward *et al.* 1988), cut across matrix and fragments in banded cataclasites. The cataclasites also typically possess folded veins, stylolite-connected vein segments that resemble boudins formed by localized pressure solution (Mullenax & Gray 1984), or otherwise distorted remnants of mineral-filled veins or shear surfaces. Strains inferred from distorted veins are consistent with shear distributed throughout the fine-grained matrix of banded cataclasites (Figs 11 & 12 in Wojtal & Mitra 1986). In addition, many fragments contain fractures that do not extend into the surrounding matrix. These intragranular fractures exhibit weak preferred orientations parallel to *R*, *Y*, and *e* (Crossland & Wojtal 1989). Offset markers within fragments, and orientations roughly

Figure 3. Photomicrograph (plane-polarised light) of the banded cataclasite from the Copper Creek thrust zone, southern Appalachians. The thrust zone boundaries parallel the top and bottom of the photograph. Fragments of the brecciated limestone in the footwall are displaced slightly from each other; elsewhere in the section fragments of limestone are entrained into banded cataclasite. The banding itself is irregular near the irregular footwall contact, but is nearly planar away from the footwall rocks. Some bands contain fragments of footwall limestone and calcite single crystals that resemble vein filling. The hangingwall (not shown in this photograph) moved to the left.

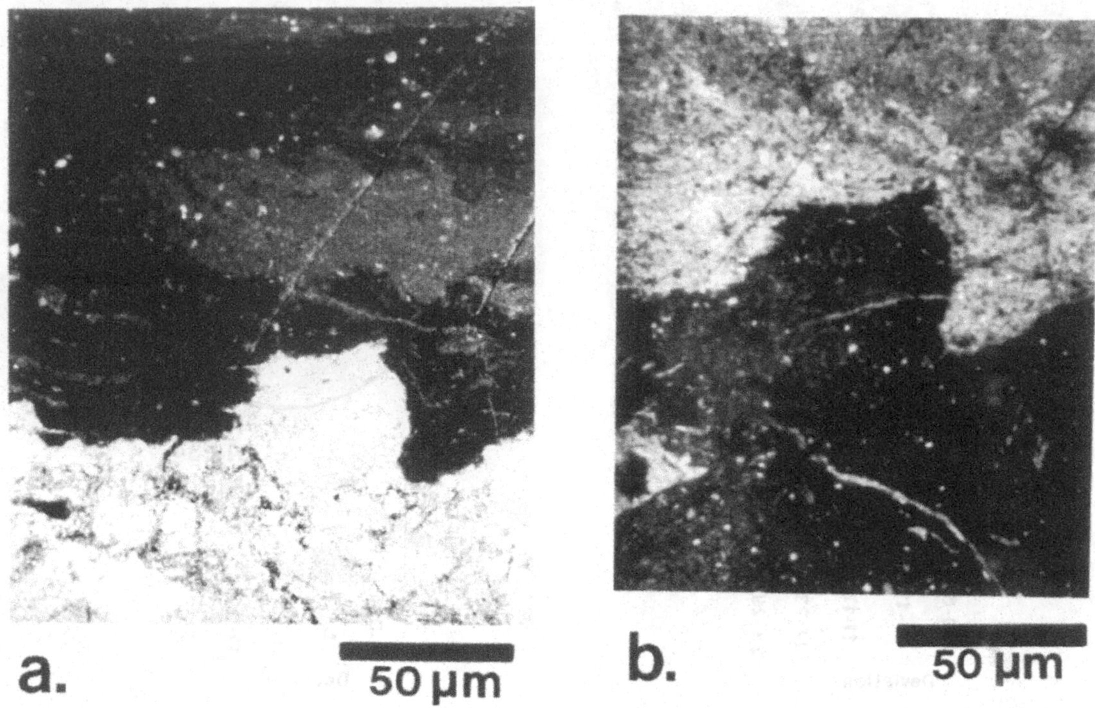

Figure 4. **(a)** Plane polarized light photomicrograph of a portion of Figure 3. **(b)** Approximately the same field of view as (a) under cathodo-luminescence. The cataclasite matrix and vein material on one hand and wall rock and wall-rock fragments on the other hand exhibit significant differences in the intensity and colour of their luminescence.

parallel to *R* and *Y*, support the inference that some intragranular fractures originated as intergranular slip surfaces. Mineral-filled intragranular fractures are more common, however. Most mineral-filled intragranular fractures resemble, in thickness and in the microstructure of their mineral fillings, the late veins found elsewhere in the cataclasites. Others have distinctive, thin mineral fillings and orientations roughly appropriate for *e* microfractures (Fig. 11c in Wojtal & Mitra 1986). These intragranular fractures may be extensional microcracks that never propagated past the fragment boundaries because the matrix blunted the growing fracture tip. Such a situation requires the fine-grained matrix to have been sufficiently ductile to dissipate fracture tip stresses. The flow of fine-grained matrix into opening fractures (Figs 11e, f & 12c in Wojtal & Mitra 1986) supports the inference that the fine-grained matrix was, at times, highly ductile.

Matrix ductility in cataclasites might result from either cataclasis or flow by diffusive mass transfer (Rutter 1984). Determining, if possible, which deformation mechanism prevailed is important in order to know what are the mechanical implications of banded cataclasites along thrusts. Cathodo-luminescence provides one way to assess the contribution of diffusive mass transfer to deformation in some rocks. Under cathode-ray illumination, the luminescence of the fine-grained matrix in banded carbonate cataclasites is often qualitatively quite different from the luminescence of its presumed parent rocks (Fig. 4). The matrix may luminesce intensely while wall rocks and wall-rock fragments in the cataclasites barely luminesce, or vice versa. Where the cataclasite matrix and wall rocks luminesce differently, the cataclasite matrix and mineral-filled veins cutting the wall rocks usually exhibit similar luminescence. Differences in calcite luminescence result from slight differences in concentrations of activators like Mn and suppressors like Fe (Marshall 1988). Different trace element chemistry in host rocks and veins, like different isotopic composition in hosts and veins (Rye & Bradbury 1988; Wiltschko & Budai 1988), suggest that host rocks did not buffer solutions during mineral precipitation, perhaps because fluid to rock ratios were high. Comparable trace element chemistry for veins and cataclasite matrix suggests that both equilibrated with a single fluid phase. Matrix grains may, because of small diameters, have been highly reactive, and their chemical equilibration with a fluid phase may have occurred statically. This would not explain, however, why fine-grained carbonates in the host rocks did not also equilibrate with that fluid or why no reacted carbonate rims occur on host fragments in the matrix. Dissolution and reprecipitation of the fine-grained matrix during diffusion-accommodated flow, as Wojtal & Mitra (1986) proposed, could cause differences in trace element chemistry (Beach 1982) that are restricted to the fine-grained matrix. Thus, cathodo-luminescence observations suggest that diffusive mass transfer was a significant deformation mechanism in banded cataclasites (Mitra 1984; Wojtal & Mitra 1986, 1988), even at temperatures below 200°C. If diffusion creep accommodated matrix flow, cataclasites would have exhibited a linear stress-strain rate relation and strong grain size sensitivity.

The planar to smoothly-curved extensional veins and shear surfaces occur in subparallel sets that cut fragments, banding, and the distorted remnants of earlier subparallel veins where all occur together. In some cases, these intergranular features abut or curve asymptotically into matrix-rich bands aligned parallel to the thrust zone boundaries, suggesting that they formed during episodes of shear by localized matrix flow in matrix-rich bands. Sets of veins and/or shear surfaces cutting through entire samples indicate that episodes of intergranular fracturing postdate matrix flow in some cataclasites. Most cataclasite samples possess both undistorted and uniformly-distorted intergranular features, suggesting that episodes of intergranular fracturing repeatedly interrupted ductile flow in thrust zone cataclasites (Wojtal & Mitra 1986). Inasmuch as matrix flow accounts for nearly all of the shear strain within these banded cataclasites, shear displacement apparently accumulated mainly by diffusive mass transfer. In thrust zones segments where banded cataclasites constitute the only rocks along planar boundaries between deformed hangingwall and footwall rocks (e.g. segments of the southern Appalachian Copper Creek thrust described in Wojtal & Mitra 1986, or segments of the southern Appalachian Saltville thrust described by Woodward *et al.* 1988), diffusive mass transfer accommodated the last sizeable increments of sheet movement. Since bulk flow by diffusive mass transfer probably exhibits a linear relationship between differential stress and strain rate (Elliott 1973; Rutter 1976), sheet movement on such a thrust zone would, in essence, occur by shear of a viscous fluid (Kehle 1970).

In other thrust zone exposures, banded cataclasites occur adjacent to pronounced selvage bands or pervasively cleaved shales (Fig. 5). Cataclasites in these settings possess the entire complement of microstructures indicating that ductile flow accommodated large shear strains. The concern most pertinent here is to assess the magnitude of displacement by ductile flow within layers of cataclasite, cleaved rock, or selvage relative to the displacement by slip on cataclasite-selvage or cataclasite-cleaved rock contacts. Wall-rock deformation does not provide unequivocal answers. O'Hara *et al.* (1990), citing increased coal rank along mesoscopic faults, inferred that fault slip occurred in discrete, temporally abrupt, steps. Crack-seal fibres in mineral lineations on mesoscopic faults also suggest episodic slip on individual mesoscopic faults. These observations suggest that discrete, probably seismic, slip contributes to thrust sheet movement, an inference supported by analytical treatments (Bombolakis 1989; pers. comm.). Lineations on mesoscopic faults in many thrust zones are, however, continuous fibres, suggesting that mesoscopic faults may move in part by continuous slip. Moreover, the microstructures in thrust zone cataclasites indicate that a sizable fraction of the total shear strain in these rocks occurred by continuous, i.e. viscous, flow. At present, there are no general ways to ascertain *from rock structures alone* whether viscous creep or discrete, seismic slip events predominated. One contention of this contribution is that viscous deformation of rocks along external thrust faults is significant, and that it provides important additional insight to the overall mechanics of thrust sheet movement.

Interrelationship of mesoscopic and microscopic structures

Microstructures in cataclasite layers and mesoscopic structures in the layers of pervasively faulted or cleaved rocks evolve concurrently. Using cross-cutting relationships in pervasively faulted strata (Wojtal 1986; Platt & Leggett 1986), cross-cutting vein patterns, or microfracture orientations in fragments from cataclasites (Woodward *et al.* 1988), one may infer deformation histories for different segments of foreland thrust zones. Thrust zone deformation histories commonly include shortening and elongation episodes, with elongation following shortening (Platt & Leggett

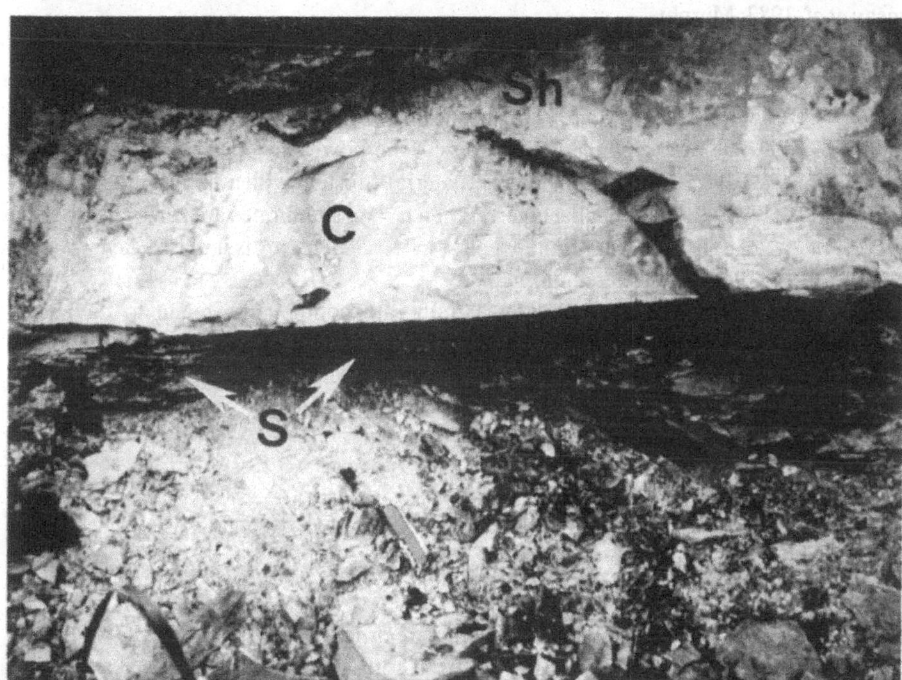

Figure 5. Field photograph of a portion of the Hunter Valley thrust zone of the southern Appalachians showing the shaly carbonate of the hangingwall strata (Sh), a banded cataclasite derived primarily from hangingwall limestones (C), and a selvage band (S) in the footwall shales. Note pocket knife in foreground for scale. The hangingwall moved out of the plane of the photograph toward the viewer.

1986; Woodward *et al.* 1988). The individual mesoscopic faults by which wall rocks shorten or elongate often extend directly into distinct intergranular slip surfaces in cataclasites, suggesting that both thrust zone components shortened or elongated coevally. This extension-then-contraction deformation sequence could result from moving a sheet past an asperity or through an area of rough slip along its thrust zone (Platt & Leggett 1986; Erickson & Wiltschko 1986).

Alternatively, this structural sequence may have evolved during thrust surges (Coward 1982) akin to surging glaciers (Sharp *et al.* 1988). Perhaps an analogy more precise than that with surging valley glaciers is the switching on and off of ice streams within the Antarctic ice-sheet (MacAyeal *et al.* 1988; MacAyeal 1989a & b). Ice stream initiation occurs when a portion of an ice-sheet becomes decoupled from its base, either due to increased subglacial water pressure or due to weakening of subglacial till (Bentley 1987; MacAyeal 1989b). The flow strengths of and stress distributions within external thrust zones are not known precisely, but two observations suggest that the basal layers of thrust sheets are weaker than the sheets themselves. First, the localization of strain in cataclasite layers along thrusts suggests that they became weaker than the surrounding rocks even if they were not originally weaker than them (Mitra 1984; Wojtal & Mitra 1986). Second, shortening and elongation episodes in pervasively faulted rock layers apparently occurred in response to principal stresses aligned nearly parallel to thrust zone boundaries (Wojtal & Pershing 1991). Principal stress directions aligned parallel to thrust zone boundaries rather than oblique to them suggests that the sheets moved over zones considerably weaker than the sheets themselves (cf. Chapple 1978, p. 1196; Stockmal 1983). In this respect, external thrust zones apparently differ from internal thrust zones. In the latter, thrust-parallel shear is distributed throughout significant volumes of sheets, suggesting that principal stresses were oblique to the thrust-zone boundaries near the thrust sheet base (Mitra & Elliott 1980; Ramsay *et al.* 1983; Murphy 1987). To be completely analogous to ice-stream flow, thrust surges would occur when a sheet had pre-existing, a sufficiently deformable stratum composing its footwall, or if such a deformable layer were generated during emplacement by strain accumulation or the influx of fluids. Inferring that processes like these occur during thrusting is reasonable, but this inference cannot be tested directly.

Mesoscopic and microscopic structures in external thrust zones indicate that sizeable fractions of the total deformation accrued by quasi-viscous creep. It is, therefore, appropriate to examine analytically viscous deformation under conditions appropriate to thrust emplacement. Such an analysis will provide important insight to the overall conditions by which thrust sheets move.

Model for continuous flow in external thrust zones

To examine the viscous flow within external thrust sheets, consider a model derived from the treatment of power-law glacier flow by Nye (1957). In this treatment, a uniform sheet with planar and parallel top and bottom surfaces flows in response to surface forces that act at its trailing edge and body forces due to its dipping upper surface (Fig. 6). The driving stresses in the treatment are those that operate in thrust systems (cf. Siddans 1984; Hudleston 1991), and the formulation is general enough to consider compressing and extending flows.

Consider that the sheet's upper surface dips in the direction the sheet moves. Cartesian coordinates in the sheet have x_1 parallel to the upper boundary and pointed in the direction of movement, x_2 normal to the upper boundary and directed upwards, and x_3 perpendicular to the x_1-x_2 plane and directed in such a manner to make the coordinate frame right-handed (Fig. 6). The x_1-x_2 plane is a principal plane for stress and strain rate tensors in the sheet.

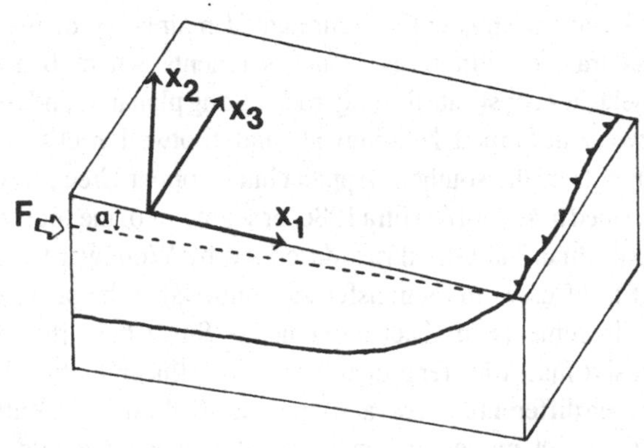

Figure 6. Sketch showing the orientation of the coordinate frame used in the mathematical model described in the text, and showing the general geometry of the stresses acting on rocks in a thrust zone considered in the model.

The flow law at all points in the layer, written in terms of the second invariants of the stress and strain rate tensors τ and $\dot{\varepsilon}$, is

$$\dot{\varepsilon} = r\left[\tau / \ \tau^*\right]^n \quad (1)$$

where r is a unit strain rate, and τ^* is a material constant with the units of stress. The power law stress exponent, n, is a dimensionless material constant with a single value throughout the layer. When $n = \infty$, strain rates are zero when $\tau \leq \tau^*$ and finite when $\tau = \tau^*$. Thus, τ^* is a plastic layer's yield strength.

Individual strain rate components relate to deviatoric stress components (where i is the coordinate direction in which a traction vector acts and j the direction of the outward normal to the surface on which the traction vector acts) according to

$$\dot{\varepsilon}_{ij} = (\partial v_i/\partial x_j + \partial v_j/\partial x_i)/2 = \lambda\sigma'_{ij} \quad (2)$$

v_i are velocity components, σ'_{ij} are deviatoric stress components, and $\lambda = r \ \ \tau^{(n-1)}/(\tau^*)^n$.

In a plane-strain treatment of sheet movement (Nye 1957; Wojtal 1991), stress components in the layer, taken to vary only with x_2, are

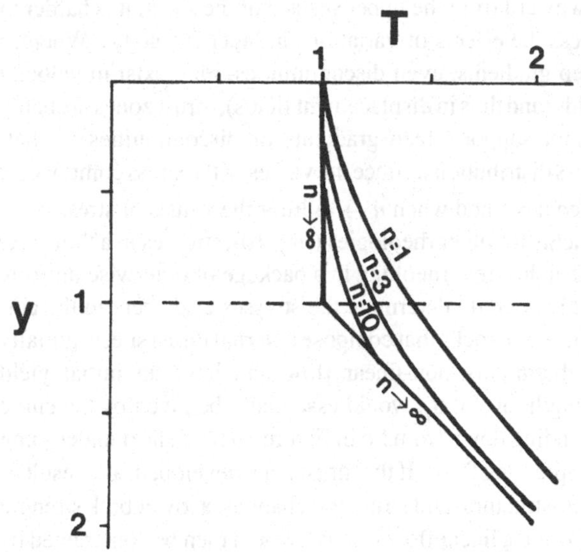

Figure 7. Plot of a dimensionless second invariant of the stress tensor, T, versus scaled depth, y, for sheets with different stress exponents, *n*, undergoing plane flow. Taken from Nye (1957).

$$\sigma_{12} = \rho g x_2 \sin\alpha \quad (3a)$$

$$\sigma_{22} = \rho g x_2 \cos\alpha \quad (3b)$$

$$\sigma_{11} = \rho g x_2 \cos\alpha \pm 2F \quad (3c)$$

ρ is the average density of the sheet, α the dip of its upper surface, and F, the traction applied at the sheet's trailing edge, varies only with x_2. The upper sign in equation (4c) puts σ_{11} greater than σ_{22}, and yields extending flows. The lower sign puts σ_{11} less than σ_{22}, and yields compressing flows. This contribution uses the lower sign, i.e. considers solutions for compressing flow, in all examples; the results for extending flow are comparable.

These stress components satisfy the equilibrium equations in this coordinate frame and, if F = 0, the stress system degenerates to the down-surface slope system of stresses defined by Elliott (1976a & b) for thrust emplacement. Nye (1957) showed that viable solutions result when

$$[\tau/\tau(0)]^{(n-1)} = Fo/F = (Fo)/[\tau^2 - (\rho g x_2 \sin\alpha)2]^{(1/2)} \quad (4)$$

where $\tau(0)$ and Fo are the values of τ and F, respectively, at x_2=0. Using the value of the second invariant for stress at the upper surface to define dimensionless parameters (see Nye 1957 and Appendix), equation (4) becomes

$$T^{(n-1)} = [T^2 - y^2]^{-(1/2)} \quad (5)$$

where T is the dimensionless second invariant of the stress tensor, and y is the dimensionless distance below the upper surface. When $n = 1$, F = Fo at all depths, and τ versus x_2 is a simple quadratic equation. When $n \longrightarrow \infty$, $\tau = \tau(0)$ for all depths, and F versus x_2 is a simple quadratic equation. If all other factors are unchanged, the stress distributions for linearly viscous and perfectly plastic sheets are end member cases that define the range of stress intensity with depth (Fig. 7). Figure 8 gives velocity profiles in laminar flows with different stress exponents for depths of $x_2 \leq l$, the depth at which the shear stress due to the dipping upper surface equals the plastic yield stress (after Nye 1957). Figure 9 shows displacement profiles for laminar flows with different stress exponents at successive times during flow. Individual profiles predict the attitudes, at successive times during flow, of lines originally perpendicular to the thrust surface (these lines are, then, *loose lines* as defined by Geiser 1988 after work by D. Elliott).

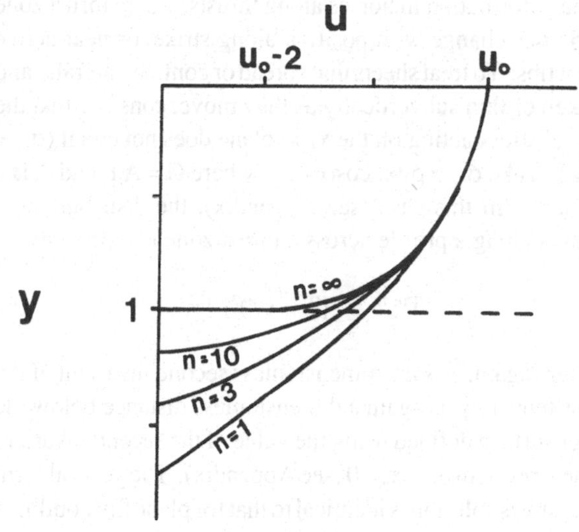

Figure 8. Velocity component u, parallel to x1, versus dimensionless depth, y, for compressing plane flows with different stress exponents, *n* (taken from Nye 1957). u_o = velocity parallel to x_1 at the upper surface. Note that velocities for perfectly plastic sheets are undefined for depths below *l*.

Some thrust sheets spread laterally as they move forward and their thrust zones shorten (Price 1967; Bielenstein 1969). Non-plane strain must occur in surge zones in thrust zones (Coward 1982). Moreover, a non-plane analysis may pertain

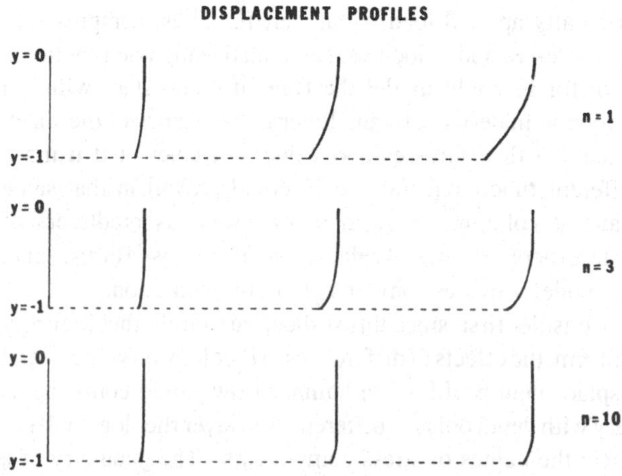

Figure 9. Displacement profiles for laminar flows with different stress exponents, *n*. Each sequence of curves from left to right represents the successive shapes taken by initially straight lines, i.e. loose lines.

to the deformation in horses along thrusts, where thrust zone properties change with position along strike, or near active thrust tips. To treat sheets that spread or contract laterally and thicken or thin subvertically as they move, consider that the normal stress acting on the x_1-x_2 plane does not equal $(\sigma_{11} + \sigma_{22})/2$. Take $\sigma_{33} = \rho g x_2 \cos\alpha - 2G$, where $G = A F$ and A is a constant. In this case (see Appendix), the distribution of stresses along a profile across a thrust zone is limited by

$$T^{(n-1)} = [T^2 - (y')^2]^{-(1/2)} \quad (6)$$

Here again, T is the dimensionless second invariant of the stress tensor. y' is again a dimensionless distance below the upper surface defined using the value of the second invariant of the stress tensor at $x_2 = 0$ (see Appendix). The general form of the stress solution is identical to that for plane flow outlined by Nye (1957), but the dimensionless distances are not identical ($y' = [3/B^2]^{(1/2)} y$, where $B^2 = 4A^2 - 4A + 4$). Well-defined stress solutions exist for all $x_2 \geq l'$ $(= [3/B^2]^{(1/2)} l)$. $y' = y$ when $A = 1/2$ and $\sigma_{33} = (\sigma_{11} + \sigma_{22})/2$. The value of the factor $[3/B^2]^{(1/2)}$ when $A = 1/2$, unity, is its minimum value. When A is either greater than or less than 1/2, the factor $[3/B^2]^{(1/2)}$ is less than one. Thus, for perfectly plastic flows with equal yield strengths, stress solutions for non-plane flows are well-defined for shallower depths than those for plane flows.

Analyzing flow in external thrust zones

Due to their simple formulation, the plane and non-plane flow solutions for stress and velocity presented here cannot model precisely how stresses and velocities vary throughout moving thrust sheets. Flow in thrust sheets is rarely laminar, and episodes of compressing or extending flow will alter the boundary conditions, which, by thickening or thinning the sheet, alter in turn the driving forces, which further alters the flow, etc. The solutions are, moreover, one-dimensional, i.e. stresses and velocities vary only with distance below the upper surface. One could, however, join solutions sequentially, incorporating variations in boundary conditions into the externally applied load parameter, F. Thus, compressing-flow stresses and velocities, calculated using one functional form for F, could model the flow at one instant within a particular material column. Later in the history of the sheet, extending-flow stresses and velocities, calculated using a different functional form of F, could prevail in that same material column. Viewed in this way, as predictors of instantaneous stress and velocity profiles across a thrust zone, the model provides some insight to thrust motion.

Consider first, since thrust sheets are rarely rheologically uniform, the effects of differences in rheology on velocity and displacement profiles. In laminar flow, stress components vary with depth only, so differences in layer rheology will not affect the values of stress components. The geometry of a loose line in a sheet composed of layers with different stress exponents *n* will, however, depend on the layer's stress exponent (Fig. 9). In compressing or extending flows where stresses and velocities must be calculated by integrating downward from the upper surface of the sheet, it is harder to assess the effects of variations in layer rheology. Whereas steep gradients, even discontinuities, may exist in velocity fields (and thus in displacement fields), thrust zones probably cannot support steep gradients or discontinuities in their stress distributions. Since the values of the stress components when n = 1 and when $n \rightarrow \infty$ limit the values of stress components for other rheologies (Fig. 10), stresses in a thin layer with a different rheology in a package of otherwise uniform rheology could determine the stresses elsewhere in the column. If the rocks that compose external thrust sheets initially exhibit highly non-linear flow and have an initial yield strength of Fo, σ_{11} would essentially be given by the curve extending down from Fo in Figure 10 for a sheet undergoing compressing flow. If the thrust zone developed, as a result of microstructural and structural changes, a lower bulk strength (F_1o) and a linear flow law, σ_{11} would then be constrained by the curves extending downward from F_1o. In the vicinity of the thrust zone (near depth = y), σ_{11} could become more compressive as a result of the change in rheology. Since σ_{22} and σ_{12} are unchanged, the orientation of the principal compression would become more nearly parallel to the thrust zone boundaries. The shear stress at depth y would, in a layer with low viscosity, localize shear strains due to a larger value of the velocity gradient $\partial v_i/\partial x_2$. At shallower depths, σ_{11} would become less compressive than the value required for compressing flow, and nearly plug flow in the nearly plastic sheet would prevail. The overall evolution of the thrust zone would resemble, in this case, a surging glacier. A change in the rheology of thrust-zone rocks may then have profound implications for the stress and velocity distributions throughout the sheet.

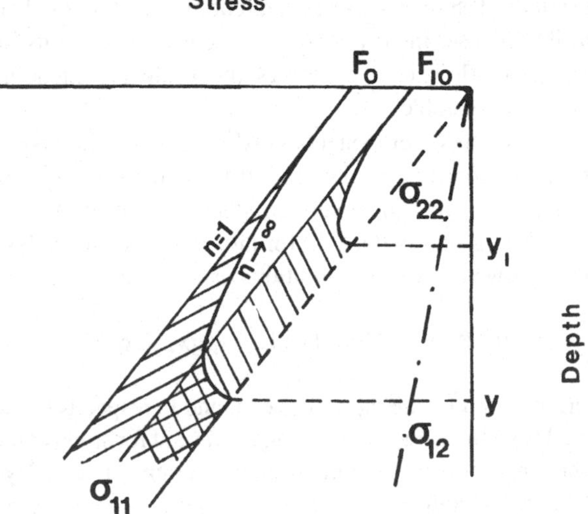

Figure 10. Plot showing a variation of stress components in compressing flow. σ_{12} and σ_{22} vary only with depth. σ_{11} depends on the value of the stress exponent for the sheet, but it is must lie between the straight line (when n = 1) and quadrant of an ellipse (when $n \rightarrow \infty$) emanating downward from Fo (region shown by right-to-left diagonal ruling). Velocities in perfectly plastic sheets are well-defined for all depths shallower than y. If the strength of the sheet drops to F_1o, σ_{11} will fall in the region outlined by the left-to-right diagonal ruling, and velocities in plastic sheets will be defined for all depths shallower than y_1. See text for more explanation.

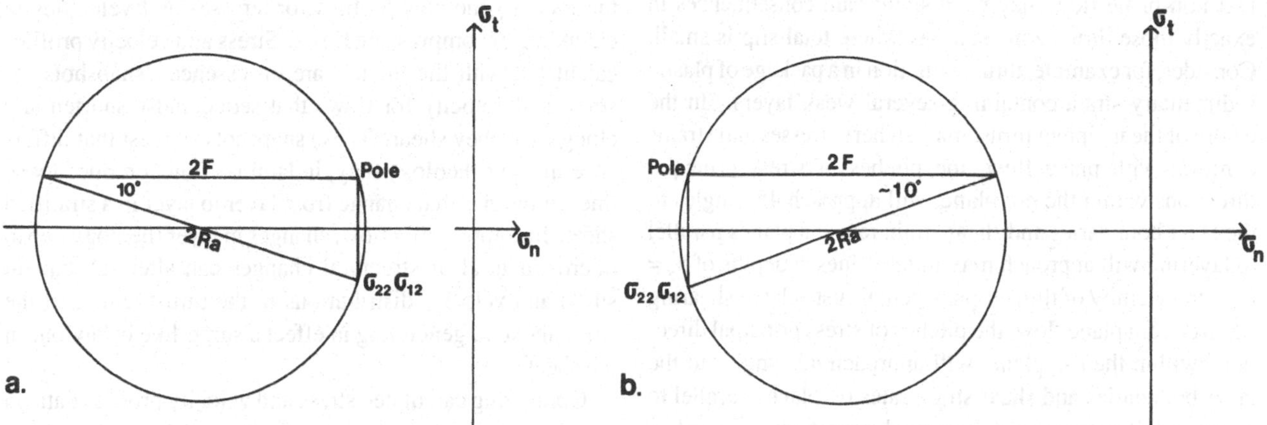

Figure 11. Mohr circles for stress for the Cumberland Plateau thrust zone, showing only the geometric relationships between σ_{12}, the parameter F from the plane flow model, and the radius of the stress Mohr circle, Ra, during thrust-zone shortening episodes (**a**) and thrust zone elongation episodes (**b**). Compressive stresses are negative. Palaeostress tensors calculated from mesoscopic fault surface and slickenside measurements fix the angle between the principal stress directions and the thrust surface. As outlined in the text and in Wojtal (1991), one can calculate that Ra = 20 MPa by estimating overburden at 1.2 km and estimating surface slope at 1°.

Combining field data on thrust zone stresses with this analysis provides an estimate of bulk sheet properties during thrust emplacement. At exposures of the base of the southern Appalachian Cumberland Plateau thrust sheet along the west edge of the southern Appalachian fold-thrust belt, a zone of faulted strata nearly 100 m thick composes the thrust zone (Wojtal 1986). The thrust zone shortened and elongated successively as the sheet moved. Principal stress directions during shortening and elongation episodes were essentially parallel to the thrust zone boundaries (Fig. 2). The pitch of the stress principal directions in the a-c plane and the relationships between σ_{11}, σ_{22}, and σ_{12} derived above, fix the relative magnitudes of F and the radius of a two dimensional stress Mohr circle for the thrust zone (Fig. 11) (Wojtal 1991). Using conodont colour alteration index trends to infer an approximate sheet thickness of 1 to 2 km and an approximate surface slope of 1° to 2°, these geometric relationships suggest that at the base of this sheet (1) the parameter F had a magnitude of 3 to 10 MPa, and (2) the maximum differential stresses were of the order of 6 to 20 MPa. The geometry of stresses indicates that the parameter F and the maximum differential stresses had comparable magnitudes during shortening and elongation episodes (Fig. 11).

This argument indicates that the magnitude of shear stress resolved parallel to the base of this sheet was small relative to the parameter F. The pitches of stress principal directions change little with distance above the base of the sheet. Both observations suggest, not surprisingly, that the lateral movement and attendant deformation of this sheet are not consistent with uniform sheet rheology. Shearing was localized in a laterally-continuous coal bed, which is now the base of the sheet, and thrust-related deformation accrued mainly in response to stresses transmitted to this region by the surrounding sheet. These stresses include the 'push from behind' responsible for the thrust-zone shortening episodes, and a 'pull forward' responsible for the thrust-zone elongation episodes. The inferred strength of this sheet was approximately equal in shortening and elongation. Such strength isotropy for layered sedimentary rocks is, in the author's view, curious. The inferred sheet strength, 6 to 20 MPa, is roughly equal to the stress drops observed in shallow earthquakes (Sibson 1989).

If this thrust-zone deformation reflected 'far-field' stresses, and if those stresses varied smoothly across this thrust zone, stresses in strata beneath this local detachment should conform generally with this simple plane flow solution for all points above a depth l, calculated using the 6 to 20 MPa yield stress here to be 6 to 12 km. Provided the plane flow solution held, detachments along subjacent weak horizons could occur, with stress principal directions in the vicinity of successively lower detachments approaching 45° angles to thrust zone boundaries as the depths of detachments approach the limiting depth, l = 6 to 12 km.

The southern Appalachian Hunter Valley thrust zone originated at greater depths and accommodated larger net displacement than the Cumberland Plateau thrust zone, yet principal stress directions during later shearing increments were apparently also aligned nearly parallel to thrust zone boundaries (Wojtal & Pershing 1991). The deformation during later movement increments in this, or any, large displacement thrust zone need not resemble that during the early history of the zone. Large thrust displacement may go hand in hand with the generation of a weak basal layer, like a banded cataclasite, that allows stress principal directions to rotate parallel to the thrust zone boundaries and allows resolved shear stresses across thrust zones to decrease. Moreover, erosional unroofing may reduce the overburden and alter the position of a thrust zone relative to the upper surface of the sheet. Only by examining the thrust-zone structures in the vicinity of more deeply-buried thrust tips, where total slip is small, can one determine whether external thrusts ever are, like their higher temperature and pressure counterparts, surfaces of high resolved shear stress.

The analysis of non-plane flow presented here suggests

that non-plane flow may have significant consequences in exactly those thrust zone settings where total slip is small. Consider, for example, thrust initiation in a package of plastic sedimentary strata containing several weak layers. In the centre of the incipient thrust mass, where stresses and strains conform with plane flow, the pitches of stress principal directions within the *a-c* plane, will approach 45° angles to the layer boundaries and shear strain rates on planes parallel to layering will approach maximum values at depths of $x_2 = l$. In the vicinity of thrust tips, where thrust-related shearing departs from plane flow, the pitches of stress principal directions within the *a-c* plane, will approach 45° angles to the layer boundaries and shear strain rates on planes parallel to layering will approach maximum values at depths of $x_2 = l' = [3/B^2]^{(1/2)} l$, where $[3/B^2]^{(1/2)} < 1$. Decollements might, then, initiate at shallower depths where non-plane flow conditions prevail. As a detachment grows in length and the thrust tip migrates away from a particular rock column, local flow conditions could evolve toward plane flow. Stress principal directions in the vicinity of the active detachments would rotate toward orientations parallel and perpendicular to the active detachment during such an evolution. Lateral changes in rock strength like those due to facies changes, even local increases in rock strength, could cause then the maximum rate of shearing parallel to layering to localize at a shallower level within a particular stratigraphic column. If the strains associated with the early shear increments led to a weakening of rock along the active detachment, an initially stronger rock mass would have led indirectly to a localization of shear strain and an eventual weakening now associated with the detachment. A comparable sequence of changes in the pitches (in the *a-c* plane) of stress principal directions from 45° to the thrust zone boundaries toward parallel to and perpendicular to the boundaries could occur as the active portion (i.e. slipped area) of a detachment grows in size during the surge. Restricting analyses to plane flow may then, may severely limit our understanding of thrust zone evolution.

Conclusions

Thrust sheet movement, even in the external portions of fold-thrust belts, occurs by deforming rocks within a zone of finite thickness, suggesting that analyses of viscous deformation in zones with finite thickness will help elucidate the processes by which thrust sheets move. The simple one-dimensional model of power-law flow between parallel top and bottom surfaces presented here provides insight to some deformation within thrust zones. With driving forces that include a dipping upper surface and supplemental horizontal stresses, the model generates profiles for stresses and velocities in extending or compressing flows. Stress and velocity profiles calculated with the model are, in essence, 'snapshots' of stress and velocity for flows that sequentially shorten and elongate as they shear. These snapshots suggest that difference in layer rheology may, in laminar flow, produce loose line geometries that change from layer to layer in a stratified sheet. In compressing flow, changes in layer rheology due to microstructural or structural changes can alter subsequent stress and velocity distributions in the thrust zone and the sheet above it, generating in effect a surge-like behaviour in the sheet.

Comparing calculated stress and velocity profiles data on the stress orientations during discrete episodes of shortening or elongation episodes in the southern Appalachian Cumberland Plateau thrust zone suggests that differential stresses in the thrust zone did not exceed 20 MPa. Moreover, body forces due to the dipping upper surface of the sheet were not the primary source of these tractions. Additional tractions that neighbouring portions of the thrust sheet exerted on this segment of the thrust zone were on the order of 10 MPa, with comparable magnitudes during shortening and elongation episodes.

The analysis further suggests that non-plane flow differs from plane flow in significant ways, and that these differences may affect the distribution of stresses and velocities in structural settings like the growing tips of thrusts. In highly non-linear sheets, maximum rates of shear strain parallel to layering may occur at shallower depths in non-plane flows than in plane flows. If, as a thrust tip migrates past a rock column, a non-plane flow evolves into a plane flow and if early shear strain increments facilitate a weakening of the layer experiencing maximum shear strain rates, the later evolution of the detachment may be characterized by stress principal directions in the a-c plane that are nearly parallel and perpendicular to the detachment. Both more data on the nature of stresses and strains in the vicinity of thrust zones and more sophisticated three-dimensional analyses of thrust system deformation are needed to more fully understand the nature of thrust zone evolution.

I thank Wendy for her patience and support in this project, Mike Bombolakis for kindly sharing his views on thrust zone mechanics, and Peter Hudleston for giving me a preprint of his comparison of glaciers and thrust sheets. Thought-provoking criticism by Dave Wiltschko, Andy Schedl, Terry Engelder, and Shankar Mitra helped me to organize this analysis. Comments by Rob Knipe and an anonymous reviewer helped me to clarify further the manuscript. Finally, I thank Ken McClay both for his comments on this work and his efforts in organizing the Thrust Tectonics 1990 Conference and publishing the conference proceedings.

References

Arboleya, M. L. 1989. Fault rocks of the Esla Thrust (Cantabrian Mountains, N Spain - an example of foliated cataclasites. *Annales Tectonicae*, **3**, 99-109.

Beach, A. 1982. Deformation mechanisms in some cover thrust sheets from the external French Alps. *Journal of Structural Geology*, **4**, 137-149.

Bentley, C. R. 1987. Antarctic Ice streams: A review. *Journal of Geophysical Research*, **92**, 8843-8858.

Bielenstein, H. U. 1969. The Rundle thrust sheet, Banff, Alberta. Ph.D. Dissertation, Queen's University, Kingston, Ontario.

Bombolakis, E. G. 1989. Thrust fault mechanics and dynamics during a developmental stage of a foreland belt. *Journal of Structural Geology*, **11**, 439-456.

Chapple, W. M. 1978. Mechanics of thin-skinned fold and thrust belts. *Geological Society of America Bulletin*, **89**, 1189-1198.

Coward, M. P. 1982. Surge zones in the Moine thrust zone of NW Scotland. *Journal of Structural Geology*, **4**, 247-256.

Crossland, A. & Wojtal, S. 1989. Rheological implications of intragranular fractures in cataclasites from foreland fold-thrust belt thrusts (abst.). *Program of the Geological Society of America*, **21**, A135.

Davis, D., Suppe, J. & Dahlen, F. A. 1983. Mechanics of fold and thrust belts and accretionary wedges. *Journal of Geophysical Research*, **88**, 1153-1172.

Elliott, D. 1973. Diffusion flow laws in metamorphic rocks. *Geological Society of America Bulletin*, **84**, 2645-2664.

—— 1976a. The motion of thrust sheets. *Journal of Geophysical Research*, **81**, 949-963.

—— 1976b. The energy balance and deformation mechanisms of thrust sheets. *Philosophical Transactions of the Royal Society of London*, **A283**, 289-312.

Emerman, S. H. & Turcotte, D. L. 1983. A fluid model for the shape of accretionary wedges. *Earth and Planetary Science Letters*, **63**, 379-384.

Engelder, J. T. 1974. Cataclasis and the generation of fault gouge. *Geological Society of America Bulletin*, **85**, 1515-1522.

Erickson, S. G. & Wiltschko, D. V. 1986. Analytical models of an asperity on a thrust surface (abst.). *Program of the Geological Society of America*, **18**, 595.

Fletcher, R. C. 1989. Approximate analytical solutions for a cohesive fold-and-thrust wedge: Some results for lateral variation in wedge properties and for fine wedge angle. *Journal of Geophysical Research*, **94**, 10,347-10,354.

Geiser, P. A. 1988. The role of kinematics in the construction and analysis of geological cross sections in deformed terranes. *In:* Mitra, G. & Wojtal, S. (eds) Geometries and mechanisms of thrusting, with special reference to the Appalachians. *Special Paper of the Geological Society of America*, **222**, 47-77.

Gilotti, J. & Kumpulainen, R. 1986. Strain-softening induced ductile flow in the Sarv thrust sheet, Scandinavian Caledonides. *Journal of Structural Geology*, **8**, 44-56.

Harris, L. D. & Milici, R. C. 1977. Characteristics of thin-skinned style of deformation in the southern Appalachians, and potential hydrocarbon traps. *Professional Paper US Geological Survey* 1018.

House, W. M. & Gray, D. R. 1982. Cataclasites along the Saltville thrust, USA and their implications for thrust-sheet emplacement. *Journal of Structural Geology*, **4**, 257-269.

Hsu, K. 1969. A preliminary analysis of the statics and kinetics of the Glarus overthrust. *Eclogae Geologicae Helvetiae*, **62**, 143-154.

Hubbert, M. K. & Rubey, W. W. 1959. Role of fluid pressure in mechanics of overthrust faulting. *Geological Society of America Bulletin*, **70**, 115-166.

Hudleston, P. J. 1991. A comparison between glacial movement and thrust sheet or nappe emplacement and associated structures. *In*: Mitra, S. & Fisher, G. W. (eds) *Structural Geology of Fold and Thrust Belts (Elliott Volume)*. Johns Hopkins University Press, Baltimore (in press).

Kehle, R. O. 1970. Analysis of gravity sliding and orogenic translation. *Geological Society of America Bulletin*, **81**, 1641-1664.

Logan, J. M., Higgs, N. G. & Friedman, M. 1981 Laboratory studies on natural gouge from the US Geological Survey Dry Lake Valley No. 1 well, San Andreas fault zone. *American Geophysical Union of Geophysics Monograph*, **24**, 121-134.

MacAyeal, D. R. 1989a. Large-scale ice flow over a viscous basal sediment: Theory and application to Ice Stream B, Antarctica. *Journal of Geophysical Research*, **94**, 4071-4087.

—— 1989b. Ice-shelf response to Ice-stream discharge fluctuations: III. The effects of Ice-stream imbalance on the Ross Ice Shelf, Antarctica. *Journal of Glaciology*, **35**, 38-42.

——, Bindschadler, R. A., Jezek, K. C. & Shabtaie, S. 1988. Can relict crevasse plumes on Antarctic ice shelves reveal a history of ice-stream fluctuation? *Annals glaciology*, **11**, 77-82.

Marshall, D. J. 1988. *Cathodoluminescence of geological materials*. Unwin Hyman, Boston.

Mitra, G. 1984. Brittle to ductile transition due to large strains along the White Rock thrust, Wind River Mountains, Wyoming. *Journal of Structural Geology*, **6**, 51-61.

—— & Elliott, D. 1980. Deformation of basement in the Blue Ridge and the development of the South Mountain cleavage. *In:* Wones, D. R. (ed.) *The Caledonides in the USA*. Department of Geological Sciences Virginia Polytechnic Institute and State University Memoirs, **2**, 307-311.

Mitra, S. 1986. Duplex structures and imbricate thrust systems: Geometry, structural position, and hydrocarbon potential. *Bulletin of the American Association of Petroleum Geologists*, **70**, 1087-1112.

Mullenax, A. C. & Gray, D. R. 1984. Interaction of bed-parallel stylolites and extension veins in boudinage. *Journal of Structural Geology*, **6**, 63-72.

Murphy, D. C. 1987. Suprastructure/infrastructure transition, east-central Cariboo Mountains, British Columbia: geometry, kinematics, and tectonic implications. *Journal of Structural Geology*, **9**, 13-29.

Nye, J. F. 1957. The distribution of stress and velocity in glaciers and ice sheets. *Proceedings of the Royal Society of London*, A239, 113-133.

O'Hara, K., Hower, J. C. & Rimmer, S. M. 1990. Constraints on the emplacement and uplift history of the Pine Mountain thrust sheet, eastern Kentucky: Evidence from coal rank trends. *Journal of Geology*, **98**, 43-51.

Paterson, W. S. B. 1981. *The physics of glaciers*. 2nd ed. Pergamon Press, Oxford.

Platt, J. P. 1986. Dynamics of orogenic wedges and the uplift of high-pressure metamorphic rocks. *Geological Society of America Bulletin*, **97**, 1037-1053.

—— & Leggett, J. K. 1986. Stratal extension in thrust footwalls, Makran accretionary prism: implications for thrust tectonics. *Bulletin of the American Association of Petroleum Geologists*, **70**, 191-203.

——, Leggett, J. K. & Alam, S. 1987. Slip vectors and fault mechanics in the Makran accretionary wedge, southwest Pakistan. *Journal of Geophysical Research*, **93**, 7955-7973.

Price, R. A. 1967. The tectonic significance of mesoscopic subfabrics in the southern Canadian Rockies of Alberta and British Columbia. *Canadian Journal of Earth Science*, **4**, 39-70.

—— 1988. The mechanical paradox of large overthrusts. *Geological Society of America Bulletin*, **100**, 1898-1908.

Ramsay, J. G., Casey, M. & Kligfield, R. 1983. Role of shear in the development of the Helvetic fold-and-thrust belt of Switzerland. *Geology*, **11**, 439-442.

Rutter, E. H. 1976. The kinetics of rock deformation by pressure solution. *Philosophical Transactions of the Royal Society of London*, A283, 203-219.

——, Maddock, R. H., Hall, S. H. & White, S. H. 1986. Comparative microstructures of natural and experimentally produced clay-bearing fault gouges. *Pure and Applied Geophysics*, **124**, 3-30.

Rye, D. M. & Bradbury, H. J. 1988. Fluid flow in the crust: An example from a Pyreanean thrust ramp. *American Journal of Science*, **288**, 197-235.

Schmid, S. M. 1975. The Glarus overthrust: Field evidence and mechanical model. *Eclogae Geologicae Helvetiae*, **68**, 247-280.

Sharp, M., Lawson, W. & Anderson, R. S. 1988 Tectonic processes in a surge-type glacier. *Journal of Structural Geology*, **10**, 499-516.

Sibson, R. H. 1989. Earthquake faulting as a structural process. *Journal of Structural Geology*, **11**, 1-14.

Siddans, A. W. B. 1984. Thrust tectonics, a mechanistic view from the West and Central Alps. *Tectonophysics*, **104**, 257-281.

Stockmal, G. S. 1983. Modeling of large scale accretionary wedge deformation. *Journal of Geophysical Research*, **88**, 8271-8287.

Wiltschko, D. V. & Budai, J. M. 1988. A model for fluid motion through layered rock: Evidence from the Idaho-Wyoming thrust belt (abst.). EOS, *Transactions of the American Geophysical Union*, **69**, 484.

Wojtal, S. 1986. Deformation within foreland thrust sheets by populations of minor faults. *Journal of Structural Geology*, **8**, 341-360.

—— 1989. Measuring displacement gradients and strains in faulted rocks. *Journal of Structural Geology*, **11**, 669-678.

—— 1991. Shortening and elongation of thrust zones within the Appalachian foreland fold-thrust belt. *In*: Mitra, S. & Fisher, G. W. (eds) *Structural Geology of Fold and Thrust Belts (Elliott Volume)*. Johns Hopkins University Press, Baltimore (in press).

—— & Mitra, G. 1986. Strain hardening and strain softening in fault zones from foreland thrusts. *Geological Society of America Bulletin*, **97**, 674-687.

—— & Mitra, G. 1988. Nature of deformation in fault rocks from Appalachian thrusts. *In*: Mitra, G. & Wojtal, S. (eds) Geometries and mechanisms of thrusting, with special reference to the Appalachians. *Special Paper of the Geological Society of America*, **222**, 17-33.

—— & Pershing, J. 1991. Paleostresses associated with faults of large offset. *Journal of Structural Geology*, **13**, 49-62.

Woodward, N. B., 1988. Geological applicability of critical-wedge thrust-belt models. *Geological Society of America Bulletin*, **99**, 827-832.

——, Wojtal, S., Paul, J. B. & Zadins, Z. Z. 1988. Partitioning of deformation within several external thrust zones of the Appalachians. *Journal of Geology*, **96**, 351-361.

Appendix

To narrow the considerable latitude in choices for the functional form of F, choose $\partial v_2/\partial x_1 = 0$. Invoking this condition in equations (2) and combining them to eliminate λ, Nye (1957) showed that

$$\lambda = a/F = a/[\tau^2 - (\rho g x_2 \sin\alpha)^2]^{(1/2)} \quad \text{(I1)}$$

and, from equation (3),

$$r\,\tau^{(n-1)}/(\tau^*)^n = a/F = a/[\tau^2 - (\rho g x_2 \sin\alpha)^2]^{(1/2)} \quad \text{(I2)}$$

Following Nye (1957), choose dimensionless parameters based on a unit strain rate $a = r(Fo/\tau^*)^n$, that is tied to the value of the second invariant of the stress tensor at $x_2 = 0$,

$$\tau = T\,\tau(0) \qquad x_2 = y\,l \qquad Fo = \rho g\, l \sin\alpha$$

Substituting those dimensionless parameters into equation (I2) yields

$$T^{(n-1)} = [T^2 - y^2]^{-(1/2)} \quad (5)$$

In the case of a three-dimensional sheet where $\sigma_{33} = \rho g x_2 \cos\alpha - 2G$, equations (2) are

$$\partial v_1/\partial x_1 = \lambda\sigma'_{11} = \lambda[-4F + 2G]/3 \quad \text{(I3a)}$$

$$\partial v_2/\partial x_2 = \lambda\sigma'_{22} = \lambda[2F + 2G]/3 \quad \text{(I3b)}$$

$$\partial v_3/\partial x_3 = \lambda\sigma'_{33} = \lambda[-4G + 2F]/3 \quad \text{(I3c)}$$

$$\partial v_1/\partial x_2 = 2\lambda\sigma'_{12} = 2\lambda\rho g x_2 \sin\alpha \quad \text{(I3d)}$$

Combining (I3b) and (I3d) to eliminate λ yields

$$\partial v_1/\partial x_2 = [3\rho g x_2 \sin\alpha/(F + G)]\,\partial v_2/\partial x_2 \quad \text{(I4)}$$

Differentiate with respect to x_1

$$\partial^2 v_1/\partial x_1 \partial x_2 = [3\rho g x_2 \sin\alpha/(F + G)]\,\partial^2 v_2/\partial x_1 \partial x_2 = 0 \quad \text{(I5)}$$

Volume remains constant during deformation, so $\partial v_1/\partial x_1 + \partial v_2/\partial x_2 + \partial v_3/\partial x_3 = 0$.

Differentiating with respect to x_2 indicates that

$$\partial^2 v_2/\partial x_2^2 = -\,\partial^2 v_3/\partial x_2 \partial x_3 \quad \text{(I6)}$$

and

$$\partial/\partial x_2\,[\lambda(F + G)]/3 = \partial/\partial x_2\,[\lambda(2G - F)]/3 \quad \text{(I7)}$$

or

$$(1/\lambda)(\partial\lambda/\partial x_2) = -(2F' - G')/(2F - G) \quad \text{(I8)}$$

where primes denote derivatives. Integrating with respect to x_2,

$$\lambda/\lambda_o = (2Fo - Go)/(2F - G) \quad \text{(I9)}$$

where the subscript 'o' denotes a parameter's value at $x_2 = 0$. If G = AF, where A is a constant, we have

$$\lambda/\lambda o = Fo/F \quad \text{(I10)}$$

and, from equation (3)

$$[\tau/\tau(0)]^{(n-1)} = Fo/F \quad \text{(I11)}$$

The second invariant of the stress tensor at any point in a sheet where G=AF is

$$\tau = \{[(4A^2 - 4A + 4)/3]F^2 + (\rho g x_2 \sin\alpha)^2\}^{(1/2)} \quad \text{(I12)}$$

or

$$[\tau/\tau(0)]^{(n-1)} = (B\,Fo)/[\{3t^2 - [3(\rho g x_2 \sin\alpha)^2]\}^{(1/2)}] \quad \text{(I13)}$$

where $B^2 = (4A^2 - 4A + 4)$. Equation (I13) is similar in form to equation (26) in Nye (1957). Using dimensionless parameters

$$\tau = T\,\tau(0) \qquad x_2 = y\,l' \qquad Fo = [B^2/3]^{(1/2)} \rho g\, l \sin\alpha = \tau(0)$$

equation (I13) becomes

$$T^{(n-1)} = [T^2 - (y')^2]^{-(1/2)} \quad (6)$$

Centrifuge modelling of the propagation of thrust faults

John M. Dixon & Shumin Liu

Experimental Tectonics Laboratory, Department of Geological Sciences, Queen's University, Kingston, Ontario, Canada K7L 3N6

Abstract: Three-dimensional propagation of thrust faults in duplex structures and the mechanism of formation of large thrusts have been investigated by analogue modelling using the centrifuge technique. Plasticine and silicone putty are used to simulate rocks such as limestone and shale, respectively. The models contain a stratigraphic succession composed of six units with alternating bulk competency (beginning with low competency at the base).

The models are subjected to horizontal, layer-parallel compression from one end. They exhibit three mechanisms of shortening: layer-parallel shortening (LPS), buckling and thrust faulting. LPS is homogeneously distributed and accumulates throughout the deformation. Folds nucleate as isolated periclines which propagate and link together along strike.

Sections cut through the models after successive stages of compression show that in the lowest competent unit folds propagate from the hinterland to foreland, and the front (foreland-dipping) limbs of these low-amplitude folds are soon cut by foreland-verging thrust faults. As shortening continues, this unit is thrusted into a duplex structure with floor thrust in the underlying (lowest) incompetent unit and roof thrust in the overlying (middle) incompetent unit. Thrusts develop, one by one, in time and space from hinterland to foreland, but not a single thrust dies during the deformation.

Observation of the model deformation sequence shows that the thrust ramps are localized solely by earlier-stage low-amplitude folds. Therefore, the pronounced systematic spacing of thrust ramps is inherited from the equal-wavelength buckling instability in competent units. Fault-related folds nucleate as detachment folds, develop into fault-propagation folds as thrust ramps propagate, and finally are modified by fault-bend folding as thrust displacement increases.

Thrusts nucleate at different points along a single fold which was formed by along-strike propagation of several doubly-plunging small folds. In the same way that the folds join together, the thrusts localized by them also join to form a major thrust. The thrust displacement in such cases varies from smaller values at linkage points to larger values at nucleation points.

In fold-thrust belts and accretionary prisms the horizontal shortening and vertical thickening of the stratigraphic pile is accommodated by layer-parallel shortening, folding and low-angle overthrust faulting. The relative importance of these mechanisms depends largely on the rheological properties of the stratigraphic sequence, and the interference among them plays a major role in the structural evolution of fold-thrust belts. The relationship between folds and thrusts has in particular received a great deal of attention. For example, Rogers & Rogers (1894), Heim (1878), Willis (1894), and other workers of that era viewed thrust faults as having formed from pre-existing buckle folds caused by layer-parallel compression, the faults often propagating from breaks through fold limbs in competent beds. In contrast, Buxtorf (1915) concluded that low-angle reverse faulting can produce associated drag folds.

Folds can also be formed passively by translation of a thrust sheet over a ramp. Rich (1934) first developed the 'ramp-anticline' or 'fault-bend fold' theory to interpret the Powell Valley anticline (Virginia, USA) as a product of displacement of a thrust sheet on a fault with a trajectory that is in part parallel to bedding and in part ramping upwards across bedding. This mechanism has been analysed geometrically (Boyer & Elliott 1982; Suppe 1983) and mechanically (Berger & Johnson 1980). It has been applied very widely to fold-thrust belts and accretionary prisms and became for a time the leading fold-thrust paradigm. In this model, the thrust is clearly implied to develop first, and the fold is a product of passive bending. The model has the advantage of being easy to analyse in a rigorous geometric way, but it involves a large number of implicit assumptions (Ramsay 1990) that are valid in natural fold-thrust systems.

Another genetic link between folds and thrusts is represented by the fault-propagation model (Williams & Chapman 1983; Suppe & Medwedeff 1984; Suppe 1985). A fault-propagation fold develops simultaneously with propagation of the ramp portion of a stepped thrust fault; the fold directly overlies the ramp and grows in amplitude as the thrust propagates across the bedding.

A detachment fold (Jamison 1987) also develops at the termination of a propagating thrust, but differs from a fault-propagation fold in that it is not related to a ramp. This kind of fault-related fold represents localized shortening of the thrust plate above the tip of a bedding-parallel fault or zone of decollement.

In all of these relationships between folding and thrusting, with the exception of the first, folding is either a consequence of or concurrent with thrusting; the faulting (be it localized on a discrete planar fracture or distributed in a zone of shear or decollement) is inferred to be the primary phenomenon, and

Table 1. Model ratios applicable to models TH16, TH18, TH20, TH22, TH23 and TH24

quantity	ratio	equivalence (model = prototype)
length	$l_r = 1.0 \times 10^{-6}$	10 mm = 10 km
specific gravity	$\rho_r = 0.6$	1.60 = 2.67 (bulk value for whole stratigrapraphic column)
acceleration	$a_r = 4.0 \times 10^{3}$	4000 g = 1 g
time	$t_r = 1.0 \times 10^{-10}$	1 hour = 1.15 Ma (for example)
stress	$\sigma_r = \rho_r\, l_r\, a_r = 2.4 \times 10^{-3}$ (calculated from other ratios)	
viscosity	$\mu_r = \sigma_r\, t_r = 2.4 \times 10^{-13}$ (calculated from other ratios)	

folding a secondary consequence. One problem with this hierarchy is that it provides no obvious mechanical explanation for the localization of thrust ramps.

In the authors' opinion, the role of faulting has been overemphasized. Furthermore, recent theoretical (e.g. Davis *et al.* 1983 and others) and analogue (e.g. Mulugeta & Koyi 1988) model work has tended to focus on faulting and ignore the role of folding because thrust belts have been assumed to consist of a homogeneous material and bedding anisotropy has been neglected. On the basis of analogue model studies of the interaction between buckling and faulting (and layer-parallel shortening), Dixon & Tirrul (1991) and Liu & Dixon (1990, 1991) have proposed that early buckling plays a fundamental role in localizing thrust ramps and in guiding along-strike propagation of thrusts, and thus may control the formation of duplex structures. In this regard, a return to the hierarchy proposed by Rogers & Rogers (1894), Heim (1878) and others is proposed.

Folds and thrusts nucleate serially and propagate spatially from hinterland to foreland, and the conventional view has been that early-formed structures lock up as younger ones develop at the toe of a growing fold-thrust belt (e.g. Boyer & Elliott 1982). On the other hand, the critical Coulomb wedge analysis of Davis *et al.* (1983) predicts that in order to maintain a critical taper while new material is accreting to the toe of the fold-thrust belt, the wedge must continue to deform internally (for example by slip on thrust faults throughout its volume). The model studies of Liu & Dixon (1990) confirm

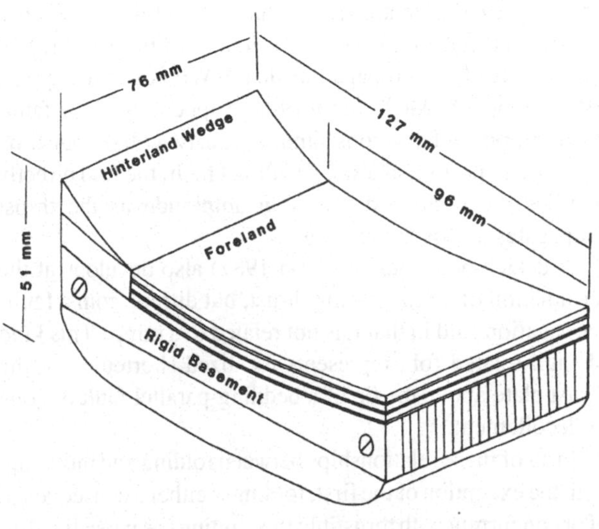

Figure 1. Dimensions and initial configuration of the models.

that early-formed faults continue to accumulate displacement even though later-formed faults have developed closer to the foreland; thus the fold-thrust wedge deforms throughout while the toe is propagating.

Two previous papers (Liu & Dixon 1990, 1991) have discussed the sequence of thrusting and along-strike structural variation (displacement transfer) in fold-thrust belts. This paper summarizes the observations of the three-dimensional interaction between folds and thrusts and the propagation of thrust faults in duplex structures. The results from models TH16, TH18, TH20, TH22, TH23, and TH24 are discussed. The last four models have been discussed in the previous papers but here more details of the structural relationships in new three-dimensional diagrams are shown. The stratigraphic sequences in models TH16 and TH18 differ in detail from that of the other models, but the relationships are consistent.

Experimental procedure

The centrifuge in the Experimental Tectonics Lab at Queen's University is capable of subjecting two specimens to accelerations as high as 20,000 g (see Dixon & Summers (1985) for a detailed description of the machine). Readers are referred to Ramberg (1981) for a detailed discussion of the theory of scale modelling and the centrifuge technique.

The models are constructed of analogue materials. Plasticine (Harbutt's Gold Medal Brand) and silicone putty (Dow Corning dilatant compound 3179) have the appropriate rheological properties to represent, respectively, competent rocks such as limestone and sandstone, and incompetent rocks such as shale, under natural conditions typical of the upper part of the Earth's crust. The rheological similitude achieved with these materials and the method of construction of finely-laminated multilayers of the two materials have been discussed in detail elsewhere (Dixon & Summers 1985; Dixon & Tirrul 1991). It will suffice to repeat here that an internally-laminated stratigraphic unit of a desired bulk competency can be prepared by repeated rolling and stacking of an initial couplet of layers of the two materials whose thickness ratio corresponds to the thickness ratio desired for the unit. Competent stratigraphic units have a higher proportion of plasticine, whereas less-competent units contain more silicone putty. Scale-model ratios chosen for the experiments described here are listed in Table 1. Note that 1 mm in the model represents 1 km in the prototype.

The dimensions and initial geometric configuration of the

models are illustrated in Figure 1, although the foreland stratigraphic sequence varies from one model to another. The models all consist of a six-part stratigraphic sequence (Fig. 2), three competent and three incompetent units. For convenience these will be referred to as Units I—VI from top to bottom. Competent units I, III and V each contain internal laminae of plasticine without silicone putty interlayers. The competent units in model TH16 (Fig. 2a) are each 1.67 mm thick and contain 10 layers of blue and red plasticine, and in model TH18 (Fig. 2b) they are each 1.33 mm thick and contain 10 layers of yellow and blue plasticine. The competent units in models TH22, TH23 and TH24 (Fig. 2c) contain four laminae of plasticine of four different colours (yellow, red, black and blue from top to bottom), while in model TH20 (Fig. 2d) they contain 8 laminae, with a total unit thickness of 1.0 mm. In all six models the incompetent units II, IV and VI each contain four laminae, two of silicone putty and two of black plasticine in a 2:1 ratio of thickness. The total thickness of the model stratigraphic pile is 6 mm in TH16; 5 mm in TH18; and 4 mm in models TH20, TH22, TH23 and TH24 (see Fig. 2). There is minor variance in strength between plasticines of different colour, but no structural manifestations of this variance have been detected.

The model stratigraphic units simulate rock units such as limestone and shale with prototype thicknesses of 1670, 1330 or 1000 m, and 330 m, respectively. The units are internally laminated so that they simulate to some extent the bedding anisotropy of natural sedimentary units. The model stratigraphic pile represents a prototype which has 6-fold stratigraphy with alternating competent and incompetent units. An example of such a sequence is provided by the succession in the Appalachian Valley and Ridge Province in West Virginia, Virginia and Maryland (Parry 1978; Kulander & Dean 1986).

The models are subjected to horizontal compression from one end by gravitational collapse and spreading of a 'hinterland wedge' of plasticine (see Fig. 1). The wedge begins to fail when the acceleration in the centrifuge reaches about 2500 g. As the wedge collapses, it loses gravitational potential; the acceleration must be gradually increased (to 4000 g in these experiments) to maintain the spreading. This increase in acceleration causes a proportional (minor) change in the model scaling. The experiments are run in stages so that the progressive development of structures can be monitored. Some models are sectioned after each stage of deformation, parallel to the shortening direction, are photographed and

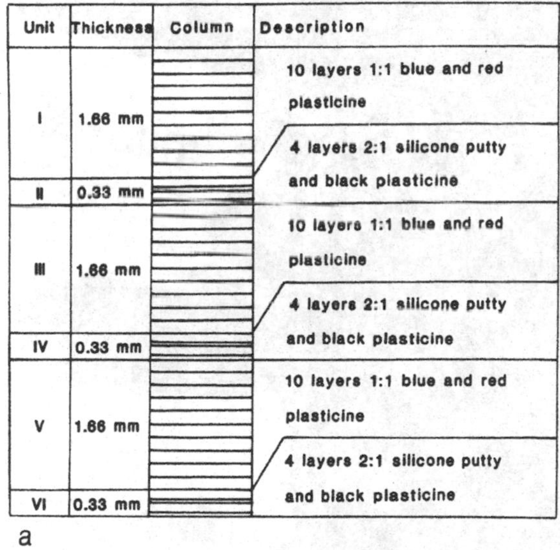

Unit	Thickness	Column	Description
I	1.33 mm		10 layers 1:1 yellow and blue plasticine
II	0.33 mm		4 layers 2:1 silicone putty and black plasticine
III	1.33 mm		10 layers 1:1 yellow and blue plasticine
IV	0.33 mm		4 layers 2:1 silicone putty and black plasticine
V	1.33 mm		10 layers 1:1 yellow and blue plasticine
VI	0.33 mm		4 layers 2:1 silicone putty and black plasticine

b

Unit	Thickness	Column	Description
I	1 mm		4 layers, 1:1:1:1 yellow, red, black and blue plasticine
II	0.33 mm		4 layers , 2:1 silicone putty and black plasticine
III	1 mm		4 layers, 1:1:1:1 yellow, red, black and blue plasticine
IV	0.33 mm		4 layers, 2:1 silicone putty and black plasticine
V	1 mm		4 layers, 1:1:1:1 yellow, red, black and blue plasticine
VI	0.33 mm		4 layers, 2:1 silicone putty and black plasticine

c

Unit	Thickness	Column	Description
I	1.0 mm		8 layers 1:1:1:1 yelloy, black, blue and red plasticine
II	0.33 mm		4 layers 2:1 sl silicone putty and black plasticine
III	1.0 mm		8 layers 1:1:1:1 yellow, black, blue and red plasticine
IV	0.33 mm		4 layers 2:1 silicone putty and black plasticine
V	1.0 mm		8 layers 1:1:1:1 yellow, black, blue and red plasticine
VI	0.33 mm		4 layers 2:1 r silicone putty and black plasticine

d

Figure 2. Stratigraphic sequences in the foreland portions of models TH16 (**a**), TH18 (**b**), H22, TH23, TH24 (**c**) and TH20 (**d**).

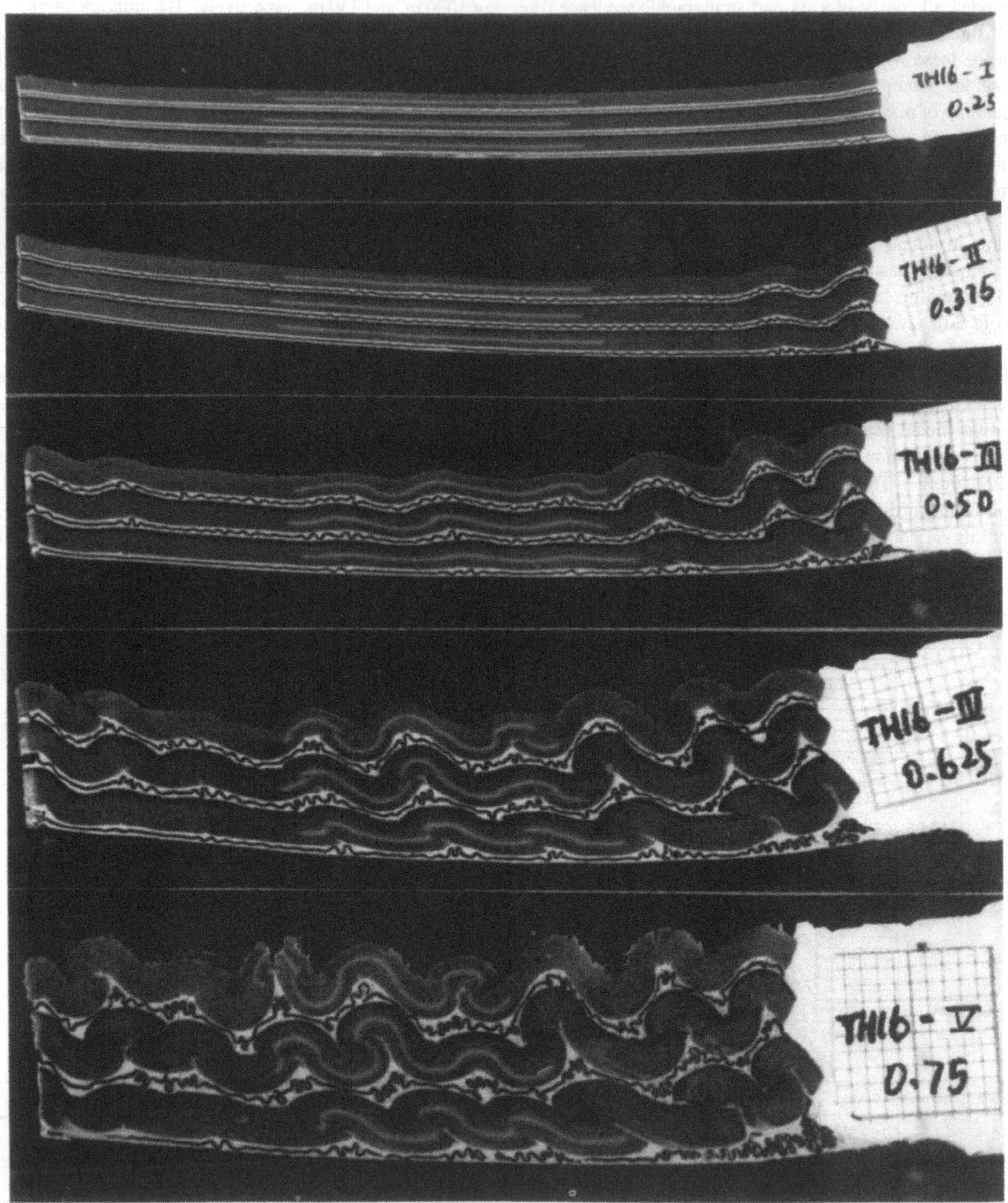

Figure 3. Profile sections through model TH16 at stage I through stage V. The positions of sections relative to one edge of the model are indicated in inches on the labels. The labels measure 10 mm x 10 mm.

then reassembled for a subsequent stage of deformation. Other models are run through several stages of deformation without sectioning, and then sectioned horizontally or transversely to the shortening direction at the final stage. For each new stage the driving wedge is rejuvenated by addition of a tapered slice of plasticine to its surface. This process is repeated after each stage until the whole deformation sequence is finished.

An analysis of the magnitude of the stress applied to the edge of the model fold-thrust belt in similar models has been presented elsewhere (Dixon & Tirrul 1991). With a more gently-tapered 'hinterland wedge' and with a 10 mm foreland stratigraphic pile the models commence to yield under horizontal compression estimated at about 2.3 x 10^4 Pa, which corresponds to a prototype stress of about 3.0 MPa. This value is below, but probably within one order of magnitude of, realistic levels of differential stress required to drive deformation of thrust belts (see Dixon & Tirrul (1991) for detailed discussion). The stress levels in the models described here are similar.

Nucleation and across-strike propagation of thrusts

Sequence of thrust nucleation

The progressive evolution of model TH16 was monitored by cutting profile sections parallel to the shortening direction after each stage of the shortening. These are shown in photographs in Figure 3 and in line drawings in Figure 4a. For each section the Roman numeral refers to the model stage, and the Arabic number indicates the position of the line of section relative to one edge of the model (distance in inches). As demonstrated in other models (Dixon & Tirrul 1991; Liu & Dixon 1990), the early stages of deformation (stages I and II) are characterized by harmonic buckle-fold trains in the competent units and localized groups of small folds in the incompetent units. The small folds in the incompetent units are localized beneath the anticlines in the competent units. The folds nucleate serially, from hinterland towards foreland, and grow progressively in amplitude. By stage III, one thrust (labelled 'a' in Fig. 4) has ramped upward across the competent unit V in the position of the front limb of an earlier-stage low-amplitude fold (fold 'a' in Fig. 4, stage II), and an incipient thrust ('b') has begun to propagate through the lower part of unit V in the position of the front limb of another fold ('b') that had formed in stage II. This suggests the serial nucleation and progressive growth of thrust faults. By this stage (III), in addition to the two thrusts, more low-amplitude folds (folds 'c', 'd' and 'e') have formed in unit V in front of the thrusts.

During stage IV (Fig. 4), the folds continue to increase amplitude in unit I and unit III. New thrusts propagate across unit V in the front limbs of low-amplitude folds (folds 'c', 'd' and 'e' of stage III) on the foreland side of those ramps formed at earlier stages, but the old thrusts continue to increase their displacements. Note that thrust 'e', like thrust 'b' at stage III, is only an incipient thrust; this, as well, suggests the sequence of thrusting.

At stage IV, Figure 3 shows that the groups of small folds in the incompetent units are evenly spaced and are clearly localized beneath the anticlines of the overlying competent units. Coincident with this (Figs 3 & 4a), the spacing of thrusts in competent unit V is also constant. By following the deformation sequence, it is easy to see that the thrusts have a constant spacing because they are localized by the earlier folding (Dixon & Tirrul 1991; Liu & Dixon 1990) which has a buckling wavelength dependent on the material properties and thicknesses of the mechanical units involved in the folding process (e.g. Ramberg 1963).

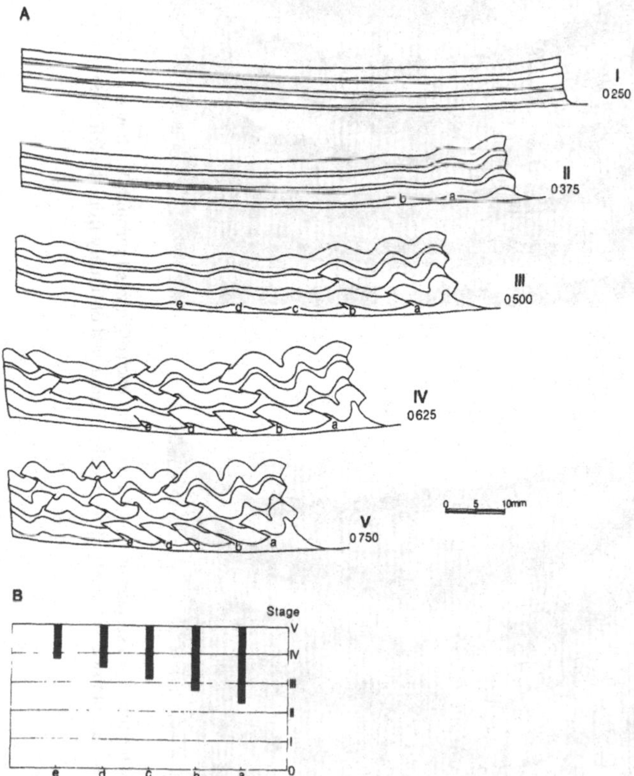

Figure 4. (a) Line drawings of the sections of model TH16 shown in Figure 3. The folds and thrusts in the lowest competent unit are indicated by letters which are referred to in the text. (b) Bar graph showing the stage of initiation and duration of displacement on faults in unit V of model TH16. Thrusts are lettered as in (a). Note that the fault ramps nucleate serially from hinterland to foreland, and not a single thrust dies during the deformation.

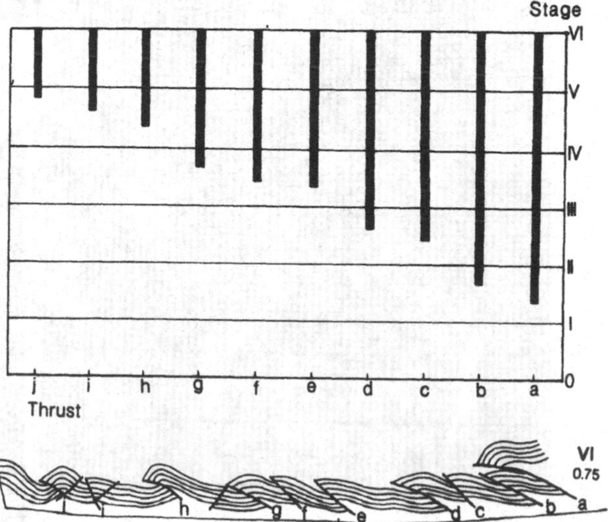

Figure 5. Bar graph showing the stage of initiation and duration of displacement on faults in unit V of model TH24. The positions of thrusts 'a' to 'j' are shown in the line drawing of section 0.75, stage VI, at the bottom of the figure. As in TH16 (Fig. 4), the fault ramps nucleate serially from hinterland to foreland, and not a single thrust dies during the deformation.

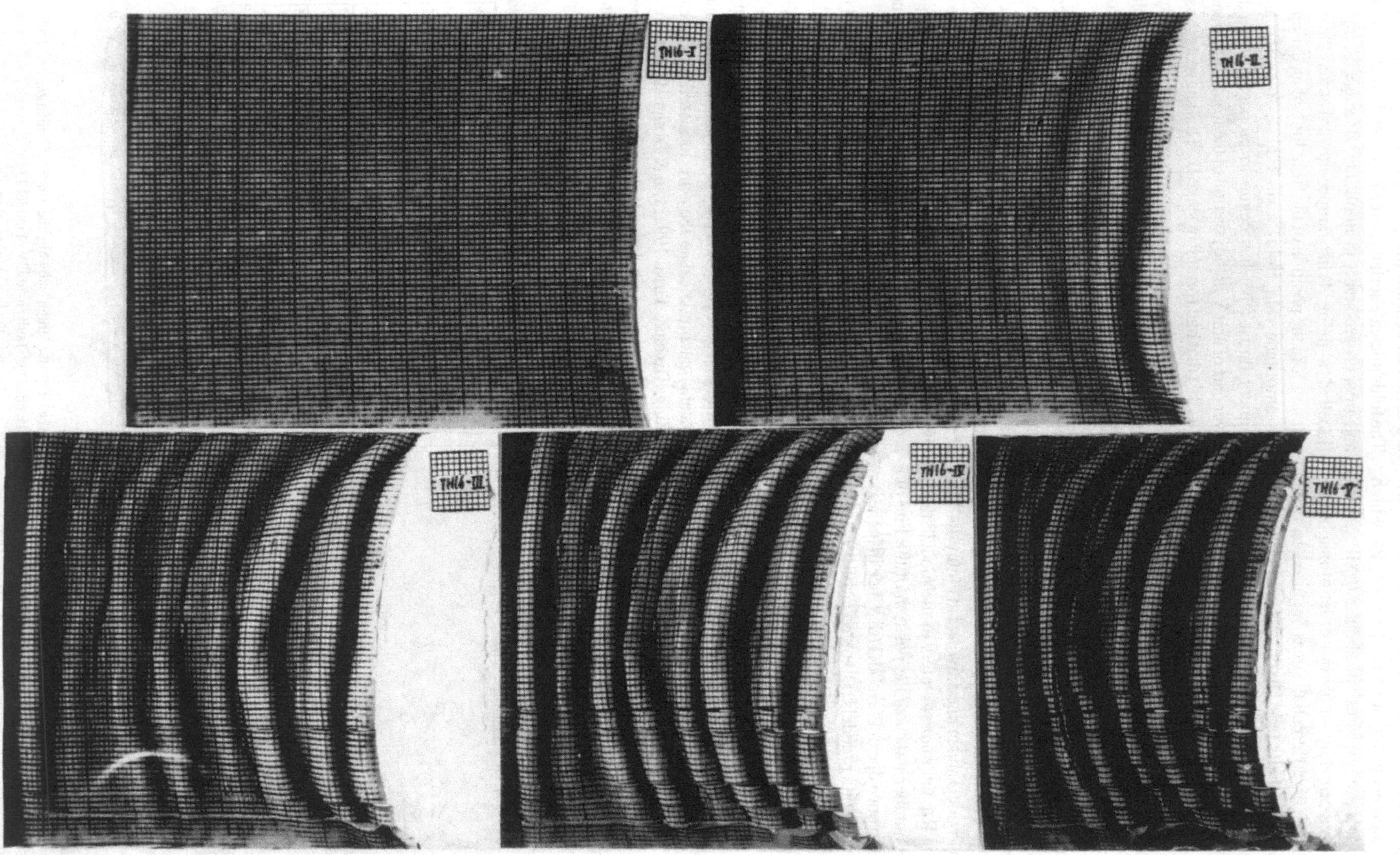

Figure 6. Photographs of the top (free) surface of model TH16 after each of stages I through V, showing the pattern of serial nucleation and foreland propagation of folds. The 'hinterland wedge' of plasticine (white, not shown in its entirety) advanced towards the left. The model is illuminated obliquely from the left to enhance surface relief. The surface of the model is marked with a print of metric graph paper (cm, mm) for scale and as a strain marker,

The competent units in model TH16 (Fig. 3) were constructed with short strips of lighter-coloured (yellow) plasticine embedded within them in the central portion of the model. These strips extend across the width of the model (that is, along strike) and serve as passive markers which can be used to measure relative displacement between the competent units. At stage I the ends of the light-coloured markers in units I, III and V are situated in a vertical row. As shortening proceeds, there is progressive shearing of the upper layers towards the foreland due to drag against the rigid base of the model. The differential displacement is accommodated by decollement in the incompetent units II, IV and VI.

The time of nucleation and the duration of displacement of all the thrusts developed in unit V of model TH16 is documented in Figure 4b. The data are obtained by careful measurement of the displacement of competent unit V by thrusts after each stage. The plot shows that thrusts nucleate successively, one by one, toward the foreland, but that an older thrust does not die out or cease its displacement as a new thrust forms in front of it. Model TH24 (Fig. 5) evolved in a similar way (although it contains more thrusts), and shows the same pattern of serial nucleation but long-lived displacement of the thrusts. This is not in agreement with the simple conception that imbricate thrusts develop serially with early-formed thrusts becoming inactive as later-formed ones develop towards the foreland. On the other hand, the continued accumulation of displacement on early-formed faults within the older, hinterland end of the fold-thrust belt is in agreement with the critical Coulomb wedge model of Davis *et al.* (1983). The nucleation sequence of thrusts from hinterland to foreland is inherited from the sequence of development of folds from hinterland to foreland. The propagation of folds (of competent unit I) across the top surface of model TH16 is shown in Figure 6.

The evolution of a fold into a thrust can be observed in the sections at stages II, III, IV and V of model TH24 (Fig. 7). Note the structure in the location marked by '*' in the sections. At stage II, the structure is a low-amplitude fold; it could be classified as a 'detachment' fold (Jamison 1987) for it overlies a zone of layer-parallel shear or decollement in the underlying incompetent unit VI. At stage III, the lower part of unit V is displaced by a propagating thrust ramp; the fold is by now a 'fault-propagation' fold (Suppe & Medwedeff 1984; Suppe 1985) which grows in amplitude as the fault propagates. The fault is shown as a solid line in unit V, but it extends back towards the hinterland as a decollement horizon within unit VI to the right of the ramp. By stage IV, the thrust has propagated upward so as to ramp all the way through unit V and into the overlying incompetent unit IV. Now the fault follows a flat-ramp-flat trajectory and the ramp-flat geometry of the thrust begins to influence the shape of the fold. By stage V the fold has evolved into a well-developed 'fault-bend' fold (Suppe 1983) or 'ramp anticline' (Rich 1934; Dahlstrom 1970). Therefore, there is an evolution of the relationship between fold and thrust: the fold nucleates first and localizes the thrust ramp in its foreland-dipping limb; the fold and thrust dominate alternately during the deformation process.

In the models, fault ramps clearly propagate upward through the competent units (see Fig. 4, fault 'b' at stage III and fault 'e' at stage IV). These ramps form at the tip of a bedding-parallel decollement zone in the underlying incompetent unit. This pattern of upward propagation, from the incompetent unit into and through the overlying competent unit, is contrary to Eisenstadt & DePaor's (1987) suggestion that ramps nucleate within competent units and propagate both upward and downward into neighbouring weak units.

Strain partitioning

Natural fold-thrust belts undergo shortening by three different deformation mechanisms: layer-parallel shortening (LPS), buckling and faulting. We have previously analysed the partitioning of strain among these three mechanisms in the competent units of model TH24 (Liu & Dixon 1991). Using the same methods, it has been possible to reconstruct restored

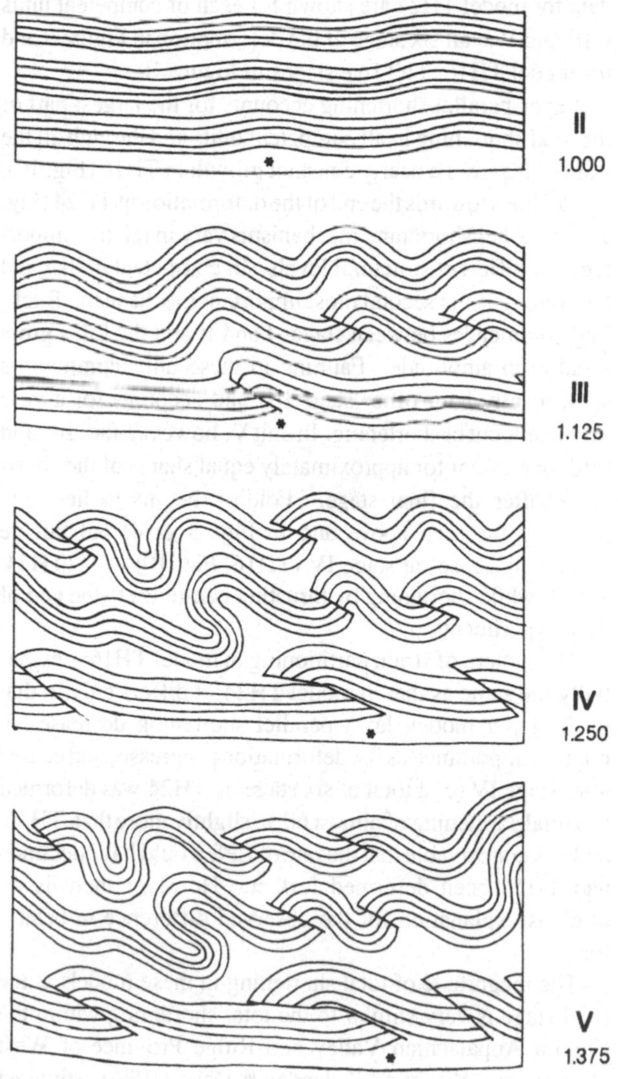

Figure 7. Line drawing from magnified photographs of sections through model TH24 after the stages II, III, IV and V. The structure marked by '*' is discussed in detail in the text.

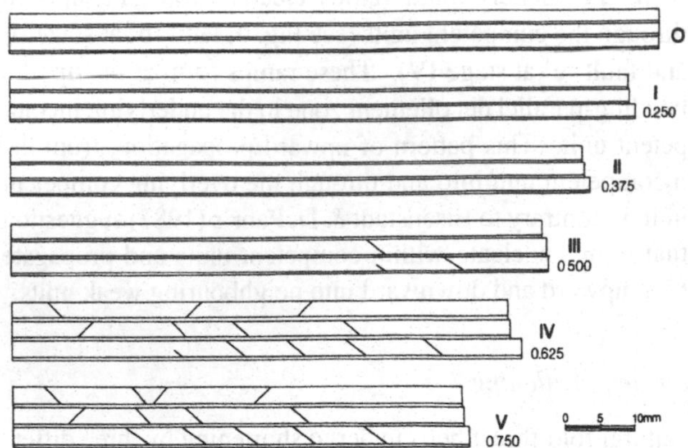

Figure 8. Palinspastically reconstructed sections of model TH16, stages O (initial) and stages I to V restored by unslipping thrust-fault displacement and unfolding buckle folds. The difference between the restored and initial length shows the effect of layer-parallel shortening.

sections of model TH16 (Fig. 8) and to calculate the total shortening and to partition it into the three mechanisms. The data for model TH24 are shown for each of competent units I, III and V at all six stage of the deformation in Figure 9 and for model TH16 at all five stages in Figure 10.

Layer-parallel shortening accounts for the largest part of the total shortening in all competent units and through all the stages. It shows a nearly constant growth in TH16 (Fig. 10), but declines towards the end of the deformation of TH24 (Fig. 9). The other shortening mechanisms vary in relative importance at different structural levels. In competent units I and III, folding is the second most important mechanism. Buckling commences between stages I and II and the folds grow steadily in amplitude. Faulting in these units commences significantly later (after stage III) and accounts for a very minor amount of shortening. In unit V, however, faulting and folding account for approximately equal shares of the shortening after the final stage. Folding begins earlier than thrusting (as seen above and in Figs 3 & 4) but buckle shortening ceases at stage IV (TH16, Fig. 10) or V (TH24, Fig. 9) while shortening by thrusting continues to the end of both experiments.

The pattern of strain partitioning in model TH16 is essentially the same as that in model TH24; it differs only in that in the latter model, layer-parallel shortening decreases in relative importance as the deformation progresses and ceases after stage IV (of a total of six stages). TH24 was deformed to a total shortening of almost 60%, slightly more than TH16, and it is conceivable that the rate of LPS would have declined had TH16 been deformed further. However there is no obvious explanation for the observed difference in behaviour.

The magnitude of total shortening in these models at the final stage is very similar to the total shortening within the Central Appalachian Valley and Ridge Province of West Virginia and Virginia. Kulander & Dean (1986) estimated shortening of Cambro—Ordovician carbonates (modelled by unit V in models TH16 and TH24) by thrusting and folding in the range of 24% to 32%, and not including layer-parallel shortening by plastic or pressure-solution deformation

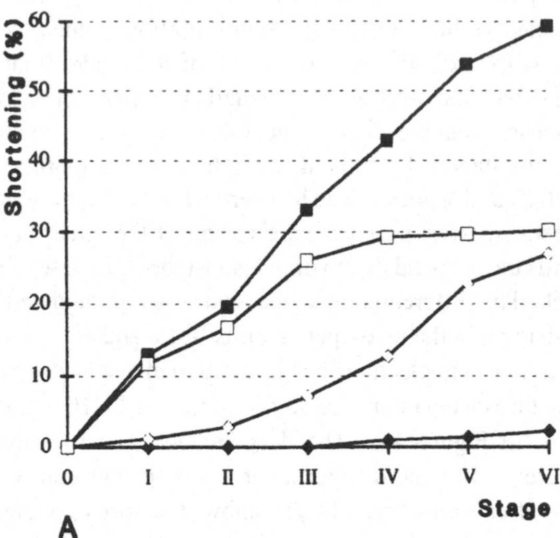

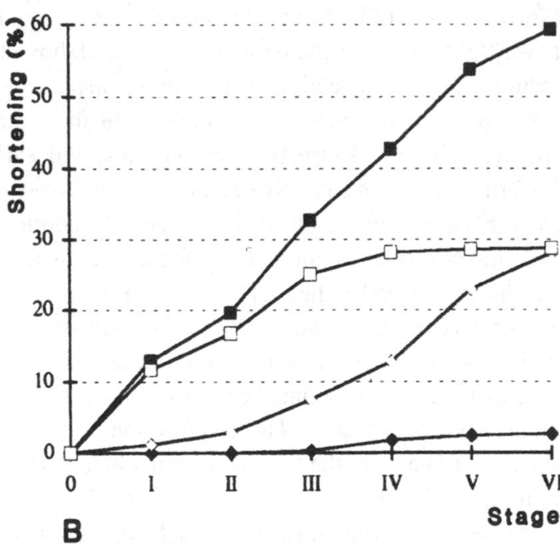

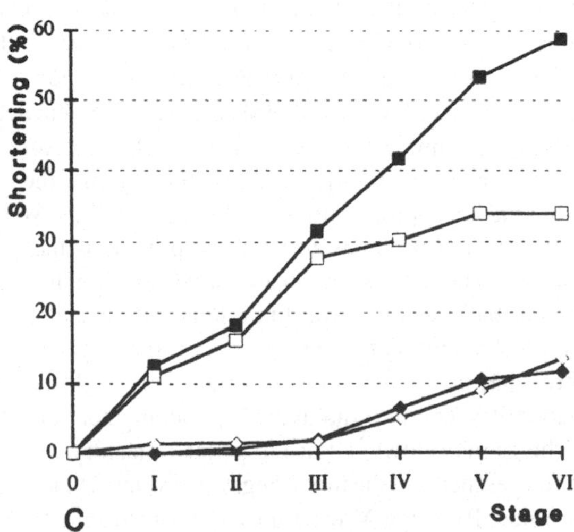

Figure 9. Graphical representation of horizontal shortening of the three competent units ((A) unit I, (B) unit III, and (C) unit V) of model TH24 through the deformation sequence, stage I to VI. The total strain (solid square) is partitioned into layer-parallel shortening (open square), shortening by faulting (solid diamond) and shortening by folding (open diamond).

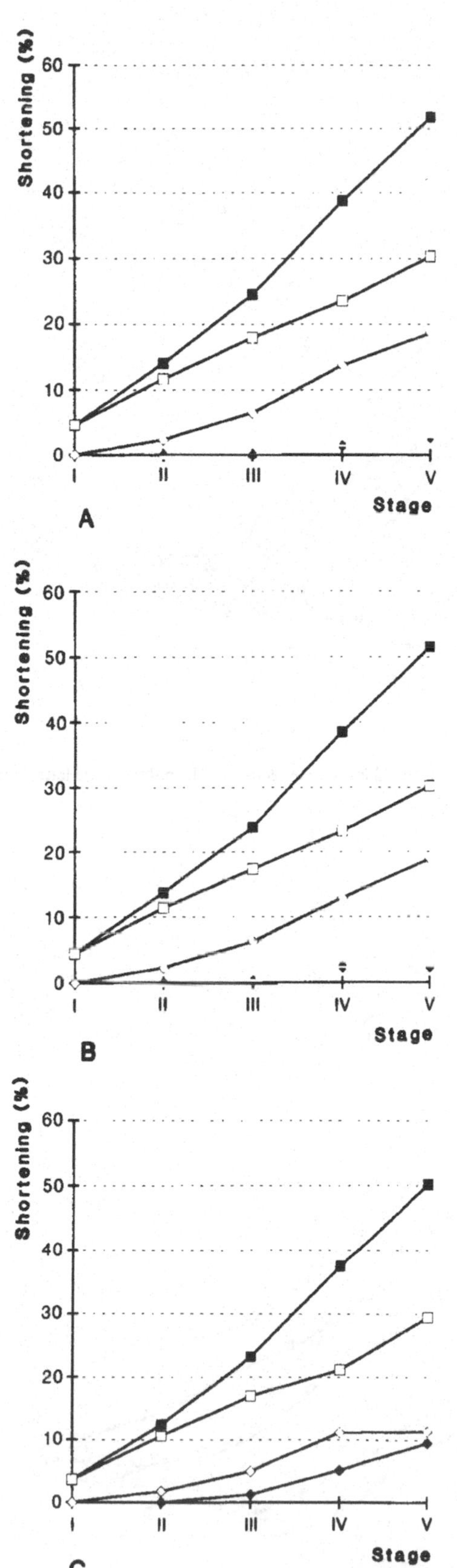

Figure 10. Graphical representation of horizontal shortening of the three competent units ((A) unit, (B) unit III, and (C) unit V) of model TH16 through the deformation sequence, stages I to V. The total strain (solid square) is partitioned into layer-parallel shortening (open square), shortening by faulting (solid diamond) and shortening by folding (open diamond).

mechanisms. They stated that layer-parallel shortening reaches 30% at some locations.

Along-strike propagation

Thus far, the discussion has focused on how thrusts propagate across strike. It is the earlier-stage low-amplitude folds that control the localization of thrust ramps, and because the folds propagate from hinterland to foreland, the thrusts develop in the same sequence. But how does a thrust develop along strike? The fold-thrust models provide insight into this problem because the progressive evolution of individual structures can be monitored as the models are deformed in stages, and the three-dimensional character of structures can be documented in detail from closely-spaced serial sections cut in different orientations. Liu & Dixon (1991) previously discussed along-strike propagation and displacement transfer in models TH22 and TH23. These phenomena are examined based on these and several other models, and new three-dimensional diagrams reveal the relationships more clearly.

Three-dimensional character of the structures

Figure 11 is an orthographic representation of the structures in unit V of model TH22 after the final stage of deformation based on the data collected from 22 serial sections cut through the model parallel to the shortening direction (that is, across strike) and at 0.125-inch spacing. Photographs and line drawings of twelve of these sections were shown in Plate I and Figure 3 of Liu & Dixon (1991); the perspective reconstruction shown here gives a better impression of the along-strike continuity and variability of structures. The main structural features of unit V are a series of thrusts which branch up from a lower decollement (within incompetent unit IV), forming a duplex structure; and thrust-related hangingwall folds. The diagram (Fig. 11) shows the top surface of this duplex.

A most remarkable feature is a local overlapping of thrusts: for example, thrust 1 overlaps thrust 2 in the central part of the model; thrust 7 overlaps thrust 8; and thrust 8 overlaps thrust 9. Examination of these localities shows that they coincide with positions at which the overlapping thrusts change their displacement. This phenomenon of along-strike displacement variation will be discussed below.

The figure also shows that a thrust can pass along strike into a fold. For example, two segments of thrust 6 are linked by a fold in the central portion of the model. This relationship is examined in more detail below.

Another feature, which is also found in natural thrust belts, is that the leading edges of thrusts are curvilinear in detail. However, in model TH22 the total transverse shortening of the model does not vary along strike, for the sides of the model were lubricated to reduce drag against the side walls of the centrifuge model chamber. Thus the faults and folds have a generally straight form. This is in contrast to model TH18.

The three-dimensional character of the structures in unit V of model TH18 is shown in Figure 12. As in TH22 (Fig. 11),

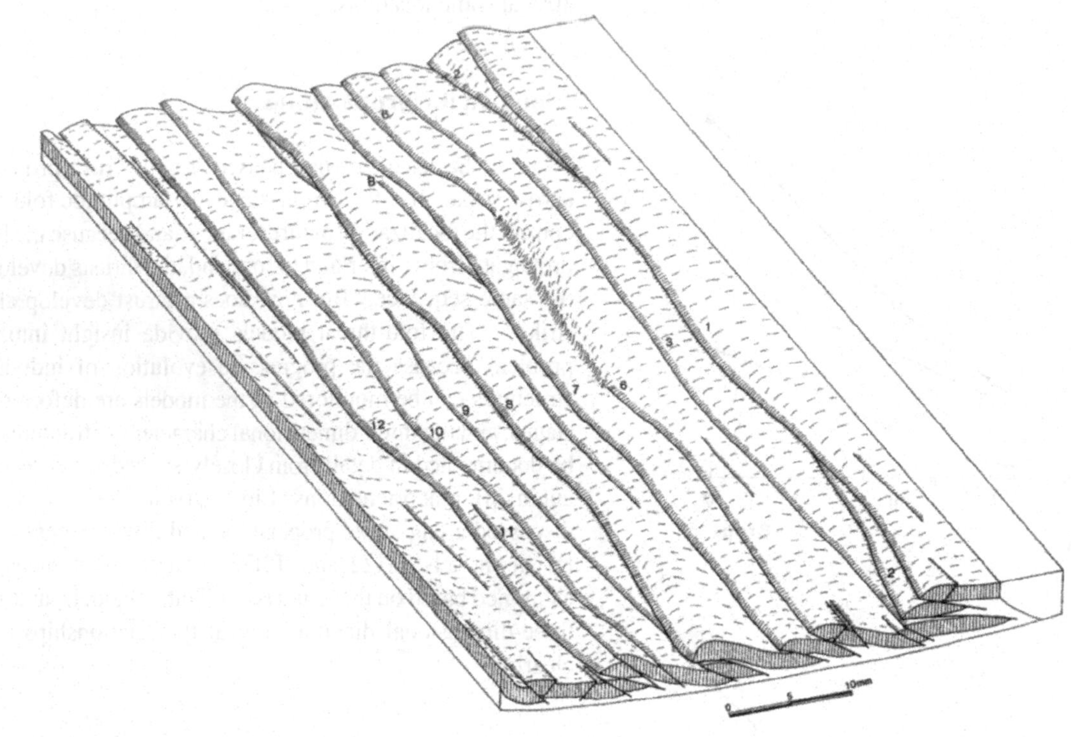

Figure 11. Three-dimensional view of structures in unit-V duplex of model TH22, drawn from 22 transverse sections. The numbers and letters marking individual structures (faults and folds) are referred in the text.

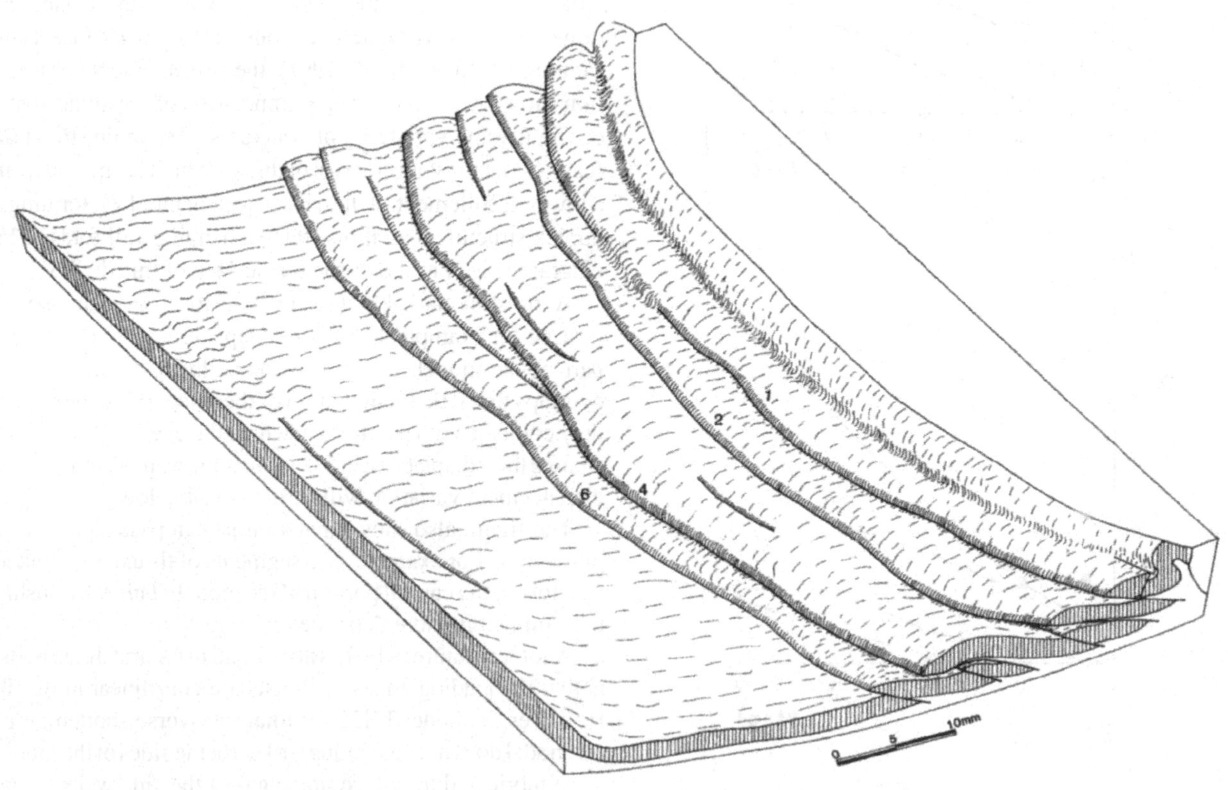

Figure 12. Three-dimensional view of structures in unit-V duplex of model TH18, drawn from 22 transverse sections. The numbers marking individual structures (faults and folds) are referred in the text.

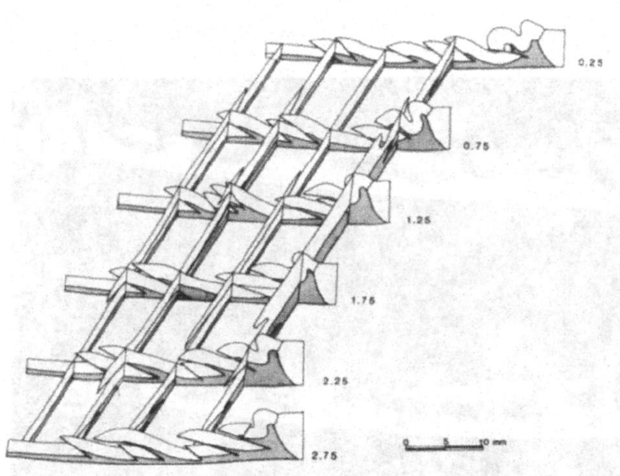

Figure 13. Across- and along-strike sections of unit V in model TH18. The numbers at the right end of the transverse sections indicate the positions of these sections (in inches) relative to the far edge of the model. Note that some ramps are arcuate (oblique relative to the right-to-left shortening direction). See text for discussion.

this lowest competent unit has been thrusted into a duplex structure with its floor thrust in incompetent unit VI and its roof thrust in incompetent unit IV. The leading edges of thrusts in this model are also curvilinear in detail, but are fan shaped, convex toward the foreland, at the scale of the whole model. This curvature reflects a gradient in total shortening and thrust shortening from minima at the edges of the model to a maximum in the centre, resulting from side drag because neither side of this model was lubricated. As a result, thrusts in this model developed oblique ramps (ramps which strike at an angle to overall transport direction of the fold-thrust belt). The oblique ramps can be seen in Figure 13, a fence-diagram representation of the structure of the unit-V duplex in model TH18. As in model TH22, overlapping thrusts are present in model TH18 (Figs 12 & 13) and reflect along-strike variation of thrust displacement. On the far side of the model the thrusts have approximately equal displacement and there is no overlapping. Towards the near side of the model, the overlapping begins to appear (e.g. thrust 4 overlaps thrust 5; see Fig. 12).

Deformation at the tip of a thrust

Deformation in the region of thrust terminations has been studied by various authors (e.g. Dahlstrom 1970; Gardner & Spang 1973; Elliott 1976; House & Gray 1982; O'Keefe & Stearns 1982; Geiser 1988 and others). House & Gray (1982) amplified on two mechanisms described by Dahlstrom (1970): (1) folding in the hangingwall of the thrust, and (2) development of en-echelon overlapping thrusts. Our models reproduce both mechanisms, as can be seen in closely-spaced serial sections cut parallel to the shortening direction near thrust terminations.

The phenomenon of termination of a thrust in the core of a fold is observed in model TH22. Figure 14 shows a series of closely-spaced sections in the vicinity of the point 'B' in Figure 11. The section interval is 0.06 inches. The thrust that can be seen to terminate near 'B' in Fig. 11 is marked '*' in Fig. 14. The thrust exhibits decreasing displacement from section 0.81 to 0.75. In sections 0.69 and 0.63 the thrust cuts only the lower part of unit V and then dies into the core of a small fold in section 0.56. This mechanism is also evident in models constructed by Gardner & Spang (1973).

The phenomenon of transfer of displacement between overlapping, en-echelon thrusts is also present in our models. Figure 15 shows 8 sections cut through model TH39 (which had a stratigraphic column similar to that of models TH22, TH23 and TH24 (see Fig. 1)) after the final stage of deformation. While thrust 'a' increases its displacement along strike from section A to section E, thrust 'b' decreases its displacement in the same direction and dies into a fold in section E. From section F to section H, the sense of displacement transfer reverses. Thrusts 'c' and 'd' have the same kind of relationship: from section A to section F, thrust 'c' decreases its displacement gradually, finally dying into a fold in section F where thrust 'd' picks up its displacement.

Along-strike displacement gradients

The displacements of five large thrusts which have propagated across the full width of model TH20 have been measured. Figure 16a shows a plan view of the traces of all faults on the top surface of unit V after the final stage of deformation. The variations of displacement of five selected faults are plotted in Figure 16b. There is more variability of displace-

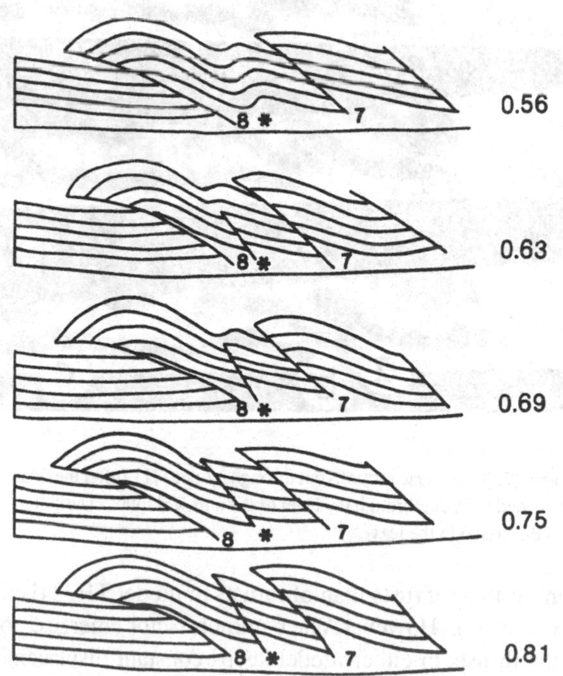

Figure 14. Closely-spaced transverse sections through part of unit V, model TH22, showing detailed structure at the tip line of a thrust (in the area marked by letter b in Fig. 11). The thrust marked by '*' decreases its displacement from section 0.81 to section 0.56 while it dies into the core of a fold. The numbers on the right give the location (in inches) of the sections relative to one edge of the model. (After Liu & Dixon in press).

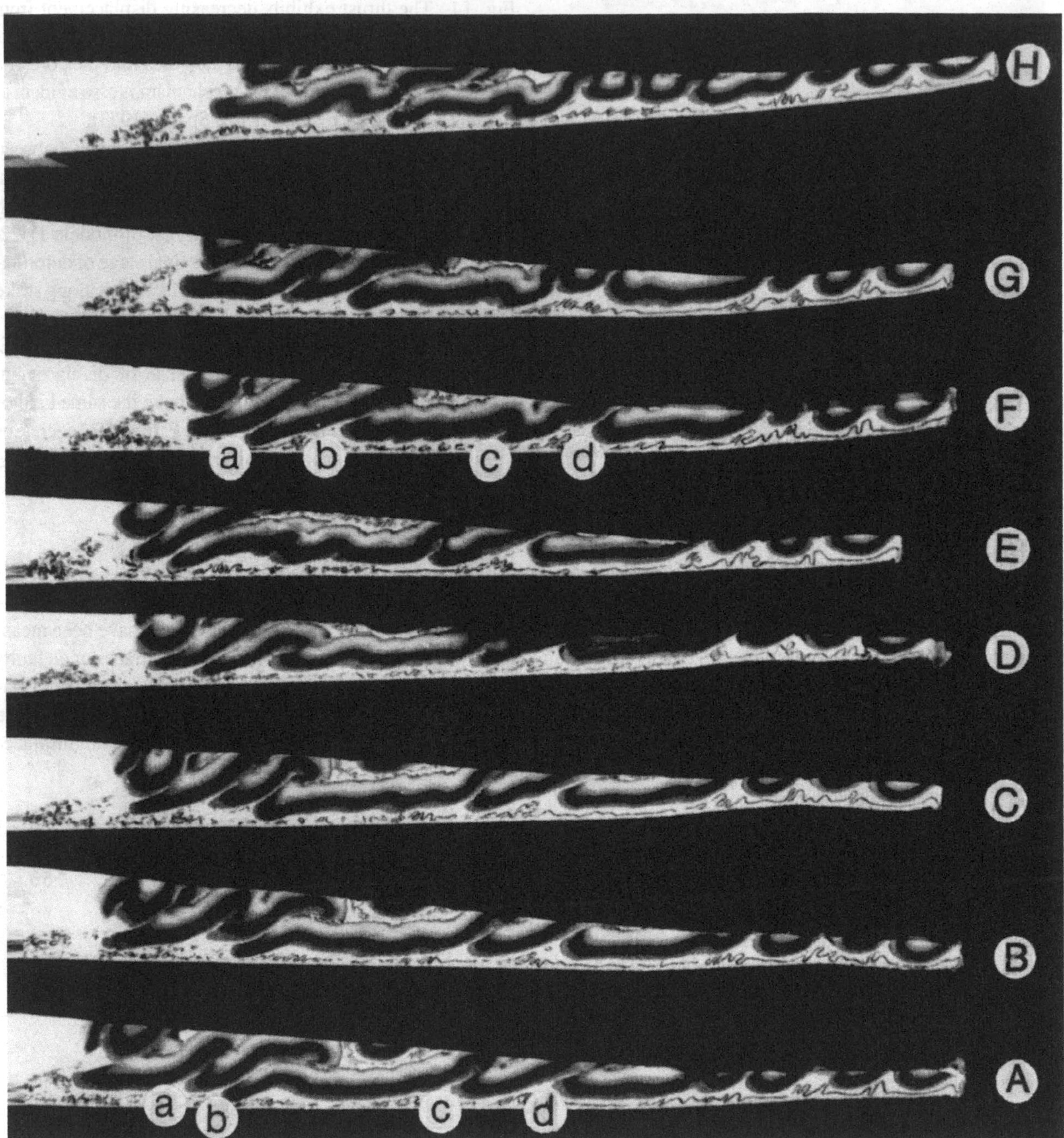

Figure 15. Transverse sections cut through model TH39 after the final stage of deformation. Only the bottom portion of the model is shown, as a horizontal section had been cut through the level of the unit-V duplex. The along-strike transfer of displacement between faults 'a' and 'b', and between faults 'c' and 'd' is discussed in the text.

ment of these thrusts than of thrusts in model TH22 (Liu & Dixon 1991). However, one common factor emerges: none of the thrusts in either model keeps constant displacement along its full strike length. As was pointed out above, the faults have curvilinear traces; but it is apparent that variable displacement is not directly related to sinuousity of trace. Like those of thrusts 1 and 7 in model TH22 (Fig. 10 of Liu & Dixon 1991), the displacement profile of thrust d in model TH20 is also roughly bow-shaped with its maximum located near the middle of the model. This fault confirms the point made by Elliott (1976) and in Liu & Dixon (1991): a large thrust may nucleate at one point and propagate in both directions along strike. For the other thrusts, the displacement profiles each show several maxima and minima. In this model thrust 'c' has a particularly large displacement of 8.7 mm along section 0.50, and its displacement decreases towards the other side of the model. In that portion of the model where thrust 'c' has large displacement, thrusts 'b', 'd' and 'e' all have small displacements; these faults grow in the same direction that 'c' decreases. This suggests displacement

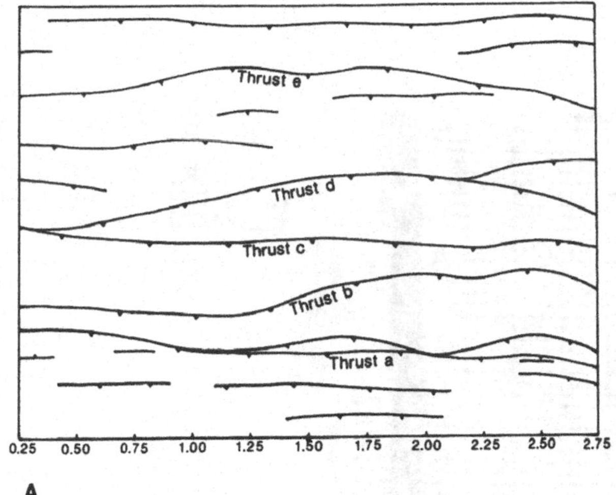

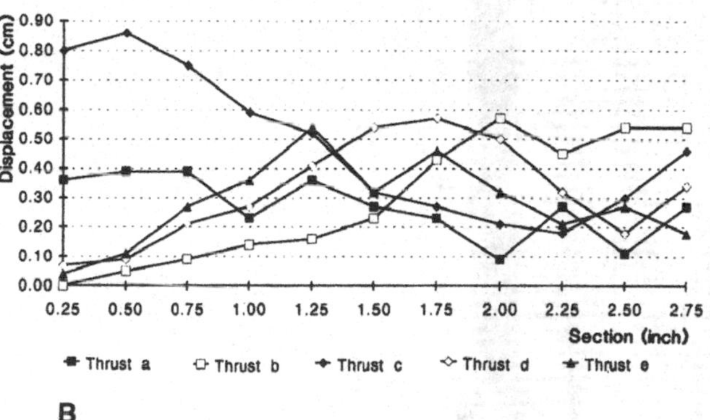

Figure 16. (a) Plan view of the top surface of unit V, model TH20, showing the positions of faults which ramp through this unit. The model was shortened from the bottom of the figure. (b) Graphical representation of along-strike variation of thrust displacement for thrusts 'a' to 'e' in unit V, model TH20. See (a) for locations of the faults.

transfer along strike among this bundle of thrusts.

Variation of fault displacement along strike can be demonstrated by plotting fault-plane maps. Figure 17 is a vertical projection of four fault ramps that cut unit V in model TH18. The thrusts are numbered in the same sequence as in Figure 12. The displacement of a single thrust can change along strike very rapidly (e.g. thrust 1 has a displacement of 3 mm at section 2.75, dies into a fold at section 2.00, and regains displacement towards section 0.25). The ramp angle also varies along strike, especially for fault 4, as shown by the variation in width of the projected stratigraphic cut-off of unit V against the fault plane. Another feature shown in this diagram is that the ramp angle of thrusts tends to decrease from hinterland to foreland. This is principally a result of steepening of early-formed ramps by passive rotation due to continued layer-parallel shortening of the model.

Mechanism of formation of a major thrust

Ellis & Dunlap (1988) studied the along-strike variation of dip-slip displacement of thrust faults. They related displacement minima to points of merger of separate, coplanar faults, or to barriers to fault movement. Here the models contain no obvious barriers to fault movement and the possibility that major thrusts which exhibit displacement variations may result from along-strike propagation and merger of smaller thrusts can be examined.

Observation of the top (free) surface of models during deformation reveals that folds nucleate as isolated periclinal culminations which grow in amplitude and propagate along strike as the shortening accumulates (see Dixon & Tirrul 1991). As they propagate along strike, these isolated anticlines may intersect or pass each other. If they intersect, they merge to form a longer anticline which has crestal culminations at the sites of nucleation and crestal depressions at sites of merger.

Figure 18 shows the free surface (top of unit I) of model TH24 after stages II and III. After stage II, only one fold has developed across the full width of the model, directly in front of the spreading wedge. All other folds are doubly-plunging, localized culminations. These folds join together to form longer anticlines after stage III. Note that the two folds

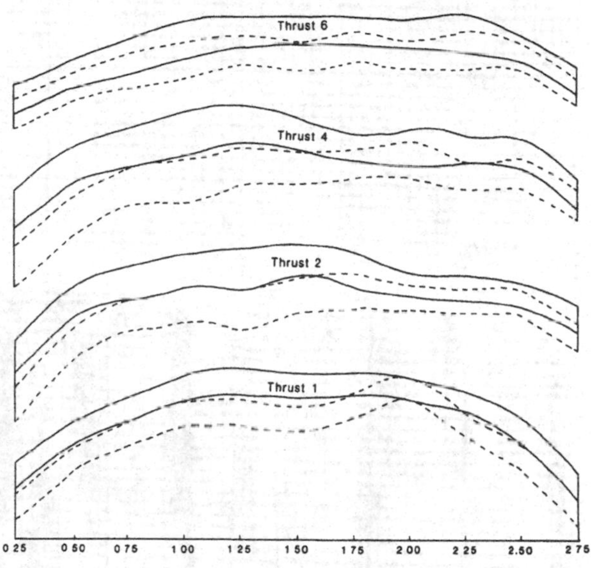

Figure 17. Fault-plane projection maps for four thrusts which ramp through unit V, model TH18. Solid lines outline the hangingwall cutoffs of unit V, and the dashed lines outline the corresponding footwall cutoffs. See Figure 12 for the positions of the faults.

marked with letters A and B at stage II have propagated towards the middle of the model and joined together by stage III (fold B'). Similarly, three individual folds (marked with C, D and E) at stage II have merged by stage III to form a single fold (marked C').

It has been shown that thrust ramps in the models are nucleated in foreland-dipping limbs of earlier-formed folds, and that the faults propagate once the folds have grown to sufficient amplitude (see above, Figs 3 & 4). It is possible to infer that along-strike propagation of thrusts follows along-strike propagation of folds; therefore, as periclines propagate and merge to form an elongate anticline with crestal culminations, the related thrusts may also merge to form a major thrust with displacement that exhibits maxima and minima along strike (as for example thrusts 'a' and 'e' of model TH20

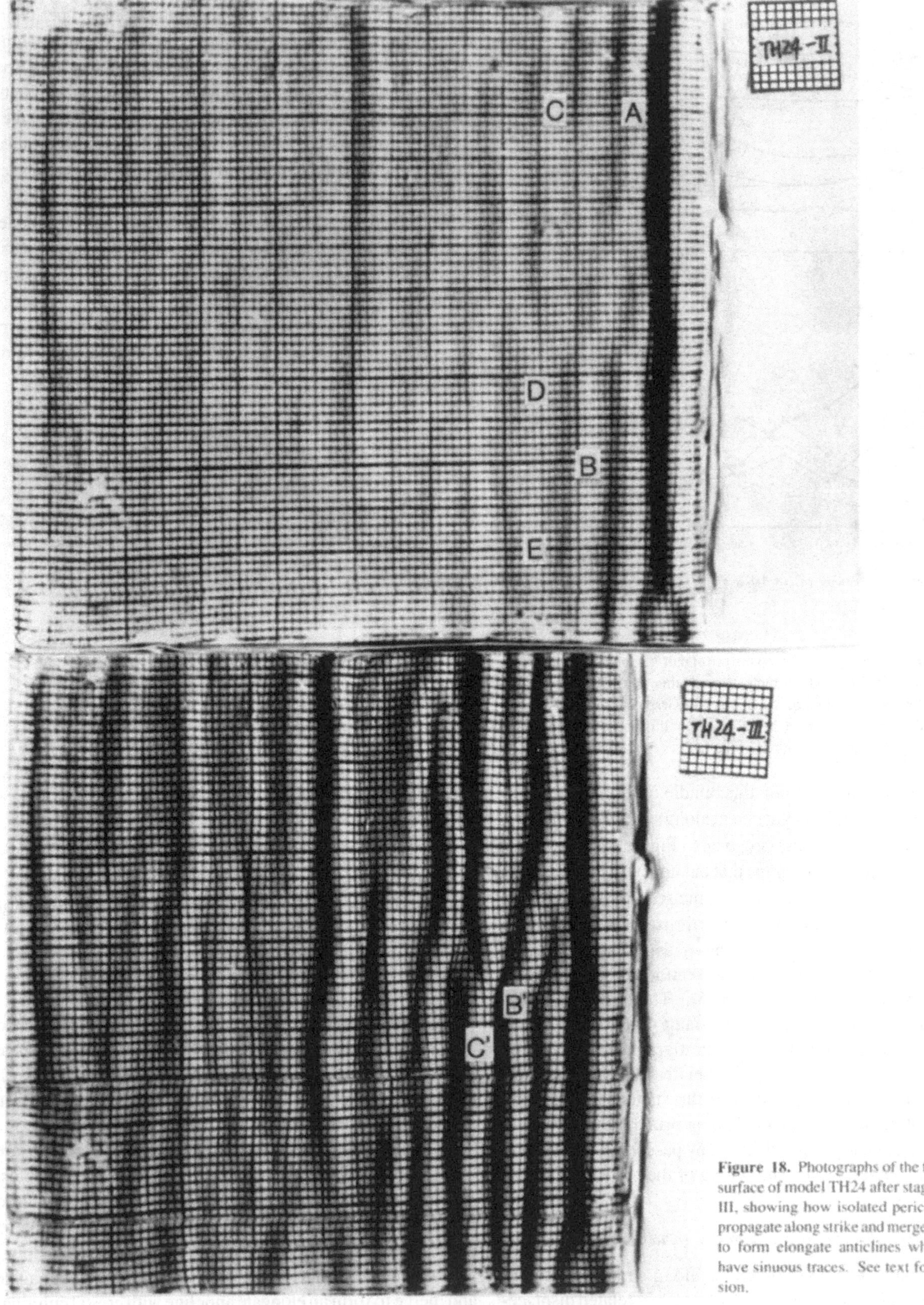

Figure 18. Photographs of the top (free) surface of model TH24 after stages II and III, showing how isolated periclines can propagate along strike and merge together to form elongate anticlines which may have sinuous traces. See text for discussion.

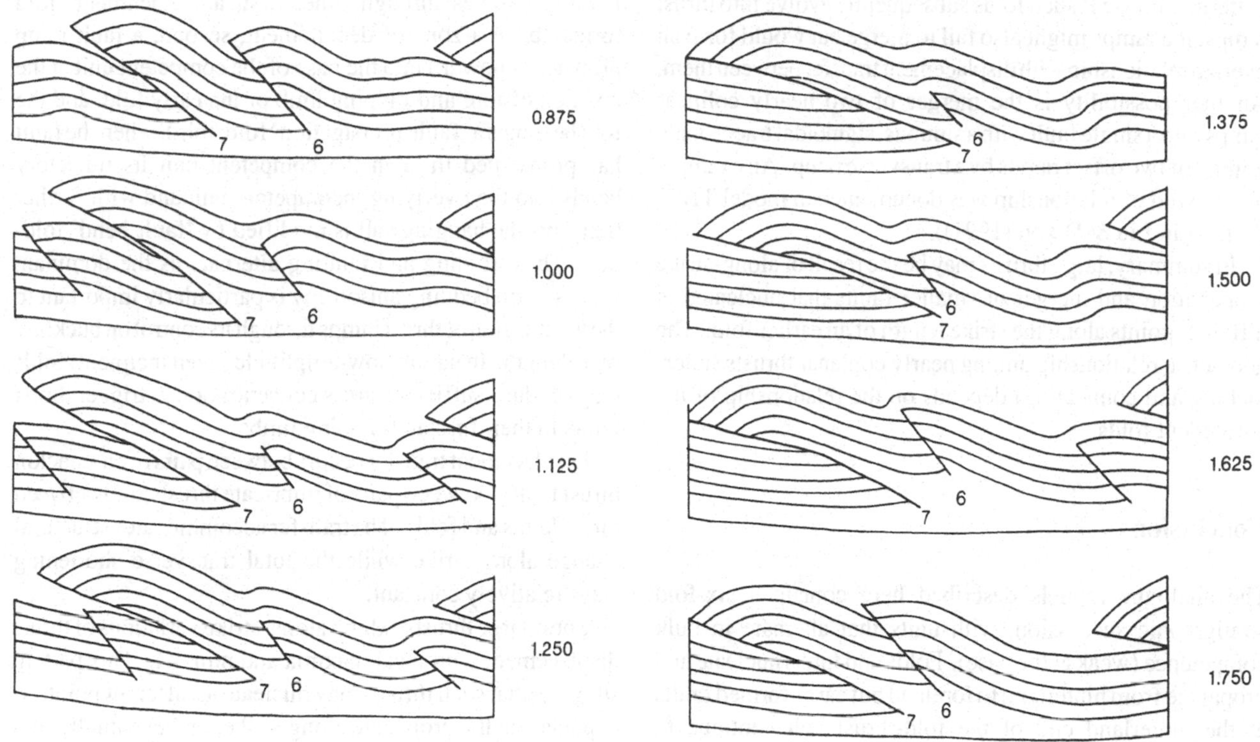

Figure 19. Line drawings of enlarged photographs of serial transverse sections cut through centre part of model TH22 (Fig. 11). The numbers on the right show the positions of the sections (in inches) relative to the far side of the model. Fault 6 cuts the full thickness of unit V in section 0.875 but its displacement decreases along strike to section 1.375 and then increases to section 1.750 where it again cuts the full thickness of unit V. The fault is overlain by a fold from section 1.00 to section 1.625.

(Fig. 16b)). The points of maximum displacement should correspond to points where small faults nucleated in the limbs of isolated folds; the points of minimum displacement should correspond to points of merger of the propagating, coplanar faults to form a through-going larger thrust.

As was mentioned earlier, in model TH22 (Fig. 11) it has been observed that, as seen on the top surface of unit V, thrust 6 occurs in two segments separated by a fold in the centre part of the model. Detailed study of sections cut across strike through the portion of the model between the fault segments shows that there is an incipient thrust ramp in the lower part of unit V beneath the fold. These sections, drawn from enlarged photographs, are shown in Figure 19. In section 0.875, thrust 6 cuts the whole thickness of unit V; between section 1.000 and section 1.625 it cuts only the lower part of unit V; finally, in section 1.750 it again cuts the full thickness of unit V. Another example of this relationship is found in model TH23. Figure 20 shows a horizontal section, at the level of the unit V duplex, of part of model TH23, together with nine closely-spaced transverse sections. Thrust 'b' appears as two segments joined by a fold in the centre part of the horizontal (map view) section, but in profile the fault is seen to extend continuously from section A to section I, although it is 'blind', and overlain by a fold, between sections B and G. If deformation were to continue further, the fault would propagate through the remainder of unit V.

An alternative evolutionary scenario is that isolated folds may propagate along strike towards each other, but may not be colinear. The result will be an en-passant or en-echelon

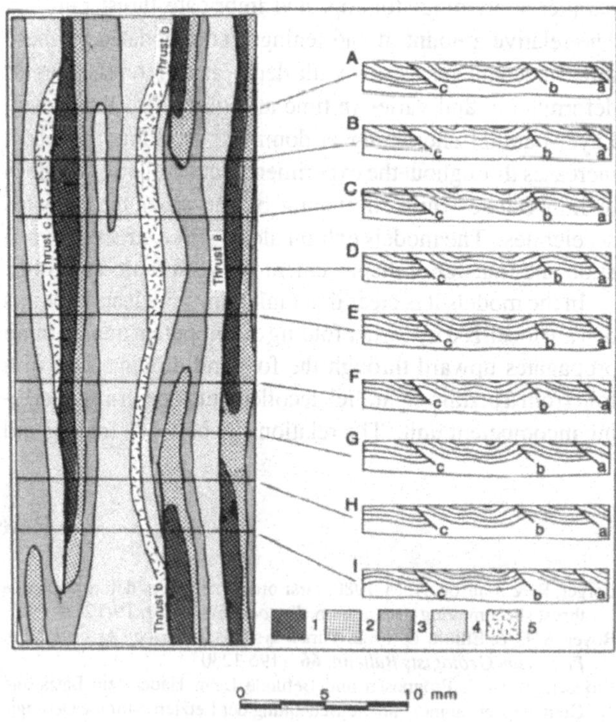

Figure 20. The map on the left shows part of a horizontal section cut through the unit-V duplex in model TH23. Profile sections A through I at the right show how thrust b is connected in the subsurface and overlain by a fold between section B and H. Map units: 1-blue plasticine, 2-black plasticine, 3-red plasticine, 4-yellow plasticine (all are part of unit V); 5-unit VI. See text for discussion.

pattern of folds. If such folds subsequently evolve into thrust ramps, the ramps might also fail to merge but would form an overlap relationship with displacement transfer between them. Another possibility is the merger of two nearly colinear ramps into a single fault with a sinuous, sigmoidal trace, or the joining of two offset ramps by a transverse ramp. An example of this kind of relationship was documented in model TH23 (Fig. 14 in Liu & Dixon (1991)).

In summary, large thrusts may be the result of along-strike propagation and merger of smaller faults that nucleated at different points along the strike length of an earlier fold. The geometric relationship among nearly coplanar thrusts (interlocking and connecting) depends on the relationship of the antecedent folds.

Conclusions

The analogue models described here contain a six-fold stratigraphic succession, with units that alternate in bulk competence (weak at the base). Folds and faults nucleate and propagate from hinterland to foreland but early-formed faults at the hinterland end of the fold-thrust belt continue to accumulate displacement as the deformation progresses. Thus the model fold-thrust belt undergoes continuous thickening in its hinterland portion while the toe or deformation front propagates towards the foreland.

Horizontal shortening strain in the models is partitioned among three deformation mechanisms: homogeneous, layer-parallel shortening; folding; and imbricate thrust faulting. The relative amount of shortening accommodated by these three mechanisms varies with depth at different stages of deformation, and varies in time at a given structural level. Layer-parallel shortening is dominant; in some models it increases throughout the experiment, but in others it declines in importance while shortening by the other mechanisms accelerates. The models exhibit along-strike structural variation and transfer of displacement between faults and folds.

In the models it is clear that fault ramps nucleate at points that are localized by earlier folding of competent units: a ramp propagates upward through the foreland-dipping limb of a fold from a bedding-parallel decollement zone in an underlying incompetent unit. The relationship between folding and faulting evolves through time: first, a 'detachment' fold forms above a zone of decollement; second, a fault ramp propagates upward into the base of the competent unit at the position of foreland-dipping limb of the early fold, and the fold becomes a 'fault-propagation' fold; third, when the fault has propagated through the competent unit its trajectory bends into the overlying incompetent unit and with further transport the hangingwall is modified by 'fault-bend' folding. Thus, folding and faulting alternate as the dominant process, and bedding anisotropy is particularly important to the localization of thrust ramps through its control on buckling wavelength. In nature, low-amplitude (even incipient) folds may produce sufficient stress concentration to trigger thrust faults in their foreland-dipping limbs.

Displacement transfer occurs between pairs of en-echelon thrust faults, between pairs of imbricate thrusts, and between thrust faults and folds. The transfer accommodates structural change along strike while the total transverse shortening stays relatively constant.

Some large thrusts exhibit along-strike variations of thrust displacement with local maxima and minima. This pattern suggests that such thrusts may nucleate at different points as coplanar faults, propagate along strike, and eventually join together because the nucleation points are all under an anticline which formed by along-strike propagation and merger of colinear, earlier-formed periclinal folds. The thrust displacement in such cases varies from smaller values at linkage points to larger values at nucleation points. Displacement transfer through transverse ramps and overlapping thrusts may occur as a means of linking non-colinear propagating thrusts which nucleated beneath en-echelon folds.

Our analogue-modelling investigation of fold-thrust tectonics is supported in part by a grant from Exploration Research, Arco Oil and Gas Company. The construction and operation of the Experimental Tectonics Laboratory at Queen's University has been supported by grants from the Natural Sciences and Engineering Research Council of Canada to J. M. Dixon. S. Liu gratefully acknowledges scholarship support in the course of his Ph.D. studies and a travelling grant to the Thrust Tectonics 1990 Conference granted by Queen's University.

References

Berger, P. & Johnson, A. M. 1980. First-order analysis of deformation of a thrust sheet moving over a ramp. *Tectonophysics*, **70**, T9-T24.

Boyer, S. E. & Elliott, D. 1982. Thrust systems. *American Association of Petroleum Geologists Bulletin*, **66**, 1196-1230.

Buxtorf, A. 1915. Prognosen und Befunde Beim Hauenstein Basis und Grenchenbergtunnel und die Bedeutung der Letzteren fur die Geologie des Juragebirges. Basel: *Verhandlungen des Naturforschunden die Gesellschaft*, **27.**

Dahlstrom, C.D.A. 1969. Balanced cross sections. *Canadian Journal of Earth Sciences*, **6**, 743-757.

—— 1970. Structural geology in the eastern margin of the Canadian Rocky Mountains. *Canadian Society of Petroleum Geologists Bulletin*, **18**, 332-406.

Davis, D., Suppe, J. & Dahlen, F. A. 1983. Mechanics of fold-and-thrust belts and accretionary wedges. *Journal of Geophysical Research*, **88,w** 1153-1172.

Dixon, J. M. & Summers, J. M. 1985. Recent developments in centrifuge modelling of tectonic processes: equipment, model construction techniques and rheology of model materials. *Journal of Structural Geology*, **7**, 83-102.

—— & Tirrul, R. 1991. Centrifuge modelling of fold-thrust structures in a tripartite stratigraphic succession. *Journal of Structural Geology*, **13**, *3-20.*

Eisenstadt, G. & DePaor, D. C. 1987. Alternative model of thrust-fault propagation. *Geology*, **15**, 630-633.

Ellis, M. A. & Dunlap, W. J. 1988. Displacement variation along thrust faults: implications for the development of large faults. *Journal of Structural Geology*, **10**, 183-192.

Gardner, D.A.C. & Spang, J. H. 1973. Model studies of the displacement transfer associated with overthrust faulting. *Canadian Society of Petroleum Geologists Bulletin*, **21**, 534-552.

Geiser, P. A. 1988. Mechanisms of thrust propagation: some examples and implications for the analysis of overthrust terranes. *Journal of Structural Geology*, **10**, 829-845.

Heim, A. 1921. *Untersuchungen uber den Mechanismus der Gebirgsbildung*. Basel: Schwabe.

House, W. M. & Gray, D. R. 1982. Displacement transfer at thrust terminations in Southern Appalachians - Saltville thrust as example. *American Association of Petroleum of Geologists Bulletin*, **66**, 830-842.

Jamison, W. R. 1987. Geometric analysis of fold development in overthrust terranes. *Journal of Structural Geology*, **9**, 207-219.

Kulander, B. R. & Dean, S. L. 1986. Structure and tectonics of Central and Southern Appalachian Valley and Ridge Provinces, West Virginia and Virginia. *American Association of Petroleum Geologists Bulletin*, **70**, 1674-1684.

Liu, S. & Dixon, J. M. 1990. Centrifuge modelling of thrust faulting: strain partitioning and sequence of thrusting in duplex structures, *In:* Knipe, R. J. (ed.) *Deformation Mechanisms, Rheology and Tectonics*, Geological Society of London Special Publication, **54**, 431-434.

—— & —— 1991. Centrifuge modelling of thrust faulting: Structural variation along strike in fold-thrust belts. *Tectonophysics*, **188**, 39- 62.

Mulugeta, G. & Koyi, H. 1987. Three-dimensional geometry and kinematics of experimental piggyback thrusting. *Geology*, **15**, 1052-1056.

O'Keefe, F. X. & Stearns, D. W. 1982. Characteristics of displacement transfer zones associated with thrust faults, in Powers, R.B. (ed.) *Geologic Studies of the Cordilleran Thrust Belt*. Rocky Mountains Association of Geologists, **1**, 219-233.

Perry, W. J. Jr. 1978. Sequential deformation in the central Appalachians. *American Journal of Science*, **278**, 518-542.

Ramberg, H. 1963. Fluid dynamics of viscous buckling applicable to folding of layered rocks. *American Association of Petroleum Geologists Bulletin*, **47**, 484-505.

—— 1982. *Gravity, Deformation and the Earth's Crust*, Second Edition. London: Academic Press.

Ramsay, J. G. 1990. Some geometrical problems of ramp-flat thrust models. *In:* McClay, K. R. (ed.) *Thrust Tectonics 1990, Programme with Abstracts*, Royal Holloway and Bedford New College, University of London (4—7 April, 1990), 71.

Rich, J. L. 1934. Mechanics of low-angle overthrust faulting as illustrated by Cumberland thrust block, Virginia, Kentucky, and Tenessee. *American Association of Petroleum Geologists Bulletin*, **1**, 1584-1596.

Rogers, W. B. & Rogers, H. D. 1894. On the physical structure of the Appalachian chain, as exemplifying the laws which have regulated the elevation of great mountain chains generally. *Association of American Geologists, Nature Reports*, 474-531.

Suppe, J. 1983. Geometry and kinematics of fault-bend folding. *American Journal of Science*, **283**, 684-721.

—— 1985. *Principles of Structural Geology*. Englewood Cliffs: Prentice-Hall.

Suppe, J. & Medwedeff, D. A. 1984. Fault-propagation folding. *Geological Society of America Program with Abstracts*, **16**, 670.

Williams, G. & Chapman, T., 1983. Strains developed in the hangingwalls of thrusts due to their slip/propagation rate: a dislocation model. *Journal of Structural Geology*, **5**, 563-571.

Willis, B. 1894. The mechanics of Appalachian structure. *United States Geological Survey Annual Report*, Part II: 211-282.

Physical models of thrust wedges

Liu Huiqi, K. R. McClay & D. Powell

Department of Geology, Royal Holloway and Bedford New College, University of London, Egham, Surrey TW20 0EX, UK

Abstract: Scaled sandbox models have been used to simulate the growth and sequential development of critical thrust wedges in isotropic cohesionless and anisotropic cohesionless materials. Variations in the initial thickness of the layered sequence, the friction of the basal detachment, and the anisotropy of the layered system have been systematically investigated. Imbricate fans of dominantly foreland-vergent thrust systems are developed similar to those found in accretionary prisms and in foreland fold and thrust belts. Critical taper wedges close to theoretically predicted geometries are developed for intermediate values of basal friction $\mu_b = 0.47$ whereas for the lower value of basal friction low-taper wedges are formed with tapers less than predicted by theory. Supra-critical wedges are formed when the basal friction equals or is greater than the coefficient of friction in the wedge and the wedge has a high taper closer to the angle of rest for the modelling material. The spacing/thickness ratio of foreland-vergent thrusts increases as the layer thickness increases. The spacing of thrust faults increases with increased basal friction. Higher basal friction or anisotropy within the layered systems favours displacement along foreland-vergent thrusts and suppresses backthrusts.

Physical modelling of thrust structures dates from the earliest days of structural geology. Cadell (1889) constructed the first documented experiment in a squeeze box using clay material to simulate the formation of structures found in the Northwest Scottish Highlands. This was closely followed by Willis (1892) who modelled the structures of the Appalachians. Despite their limitations, physical models have provided many insights into the study of thrust tectonics (e.g. Hubbert 1937, 1951; Mulugeta & Koyi 1987; Mulugeta 1988a, b). Recent physical modelling of thrust systems has been largely concerned with the development of critical wedges, with experiments performed under normal gravity (e.g. Davis *et al.* 1983; Dahlen *et al.* 1984; Malavieille 1984) and under higher gravitational loading in a centrifuge (e.g. Mulugeta & Koyi 1987; Mulugeta 1988 a, b).

The critical-tapered, Coulomb wedge model proposed by Davis *et al.* (1983), Dahlen (1984) and Dahlen *et al.* (1984), together with the subsequent modifications and developments (e.g. Davis & Engelder 1985; Davis & von Huene 1987; Zhao *et al.* 1986; Dahlen & Barr 1989; Dahlen 1990) provides a coherent explanation for the geometry and sequential propagation of thrust sheets in accretionary prisms and in foreland fold and thrust belts. According to this model, thrusts propagate sequentially until an overall critical taper is attained such that subsequent deformation involves transport of the whole wedge along the basal decollement. The state of internal stress within such a critical wedge is such that everywhere the wedge is on the verge of brittle failure according to the Coulomb failure criterion, and the basal shear stress is at a level for frictional sliding along the basal detachment.

The critical taper of a steady-state geological wedge such as an accretionary prism or foreland fold and thrust belt, or that of an appropriate analogue model, is defined by the surface slope of the wedge topography (α) and the dip of the basal decollement (β). The steady state wedge shape is controlled by the relative magnitudes of the strength of the wedge material and the frictional strength of the basal detachment. Increasing the basal friction increases the critical taper, whereas increasing the internal strength of the wedge material decreases the critical taper (Davis *et al.* 1983; Dahlen *et al.* 1984). Other factors that affect the critical taper are pore fluid pressure, erosion and synchronous sedimentation at the top of the wedge, and underplating of material at the base of the wedge (Dahlen & Suppe 1988; Roure *et al.* 1990).

In the critical-tapered wedge theory the wedge material has been treated as homogeneous with an average rheology (Coulomb behaviour). Variations in cohesion, coefficients of internal friction and basal friction have not been considered, nor have the roles of mechanical anisotropy, competency contrasts, and variations in local stress fields (Price 1988). Such parameters have not been systematically examined in previous model studies (Mulugeta 1988a, b; Mulugeta & Koyi 1987). This paper presents the results of a detailed series of analogue experiments which investigate the influences of model thicknesses, basal friction variations and anisotropy within the critical-taper Coulomb wedge. The experiments illustrate how increased basal friction dramatically affects the wedge geometry, how enhanced mechanical strength due to increased thickness affects the final geometries of thrust wedges, as well as the spacing of the thrust sheets, and how anisotropy modifies the structural style of a critical-taper Coulomb wedge.

Experimental method

Four series of experiments are described in this paper (Table

Table 1. Experimental Details.

Expt. No.	Thickness (cm)	μ_b	$\alpha + \beta$	l / h	Shortening
Homogeneous sand models					
Series I. Plastic sheet detachment					
C.56	3.0	0.37	2°	5.1	21.0%
C.63	2.5	0.37	2°	4.9	26.5%
C.62	2.0	0.37	2°	4.2	32.0%
C.54	1.5	0.37	2.5 - 3.5°	4.1	46.0%
C.59	1.0	0.37	3 - 4°	3.6	55.0%
Series II. Drafting film detachment					
C.102	3.0	0.47	3.5 - 4°	5.1	37.5%
C.103	2.5	0.47	4 - 5°	4.7	44.0%
C.104	2.0	0.47	5 - 6°	4.6	51.5%
C.105	1.5	0.47	6 - 8°	4.3	58.0%
C.106	1.0	0.47	8 - 10°	3.8	70.0%
Series III. Sand paper detachment					
C.64	3.0	0.55	17 - 22°	5.6	62.0%
C.73	2.5	0.55	20 - 22°	5.0	66.0%
C.68	2.0	0.55	20 - 23°	4.7	66.5%
C.69	1.5	0.55	20 - 24°	4.2	71.5%
C.71	1.0	0.55	20 - 24°	3.2	81.5%
Anisotropic sand/mica models					
Series IV. Constant thickness					
C.113	2.5	0.37	2 - 2.5°	4.4	31.0%
C.112	2.5	0.47	8 - 10°	4.5	53.0%
C.114	2.5	0.55	21 - 24°	4.5	67.0%

μ_b = basal friction; l = spacing of foreland-vergent thrust; h = layer thickness.

1). Series I-III used homogeneous sand models to investigate the effects of variations in both basal friction and initial model thicknesses on the geometries and kinematics of the resultant Coulomb wedge. Series IV experiments used sand-mica models of a constant initial thickness (2.5 cm) to investigate the effects of variations in the basal friction on an anisotropic wedge system.

The apparatus used in this experimental programme was a glass-sided, rectangular deformation box (Fig. 1) similar to that used by Davis *et al.* (1983). The motor, through the roller system, pulls a sheet of detachment material at a constant rate underneath the sand model, and as a result of the buttressing force of the rigid 90° backstop, generates a Coulomb wedge that simulates an accretionary prism or fold and thrust belt. The internal dimensions of the box were 100-150 cm x 20 cm x 20 cm. The sheet of detachment material moved at a constant displacement rate of 0.6 cm min^{-1}. In all of the experiments described in this paper, the base of the deformation box was horizontal (β = 0°) and the initial lengths of the models were all 100 cm.

Models were constructed by carefully sieving alternating layers of coloured and white sand (isotropic models) into the deformation box taking care to minimize local inhomogeneities in layer thickness and/or packing. Anisotropic models were constructed by sieving thin layers of mica flakes (1 mm thick) between the alternating sand layers. The glass side-walls were cleaned using an anti-static cleaner to minimize the side-wall friction. Sequential deformation of the sedimentary succession was monitored and photographed at constant time intervals. Each experiment was run until a steady state was reached, i.e. an overall critical-taper wedge was attained and visible internal deformation had ceased. At this stage the whole wedge either slid along the basal detachment (low basal friction experiments), or material was continuously underplated at the base and the wedge continued to grow with the surface slope approaching the angle of rest of the modelling material (high basal friction experiments). The completed models were impregnated and serially sectioned in order to examine the resultant structures in three dimensions. Side-wall friction generated thrusts which were curved in plan view but severe curvature was limited to about 2 cm from the edges of the glass, and the structures formed in the centre of the deformation box were laterally consistent within the model. The deformation observed through the glass side-walls was shown to be similar to the internal parts of the model. The serial sections shown in this paper were taken from the centre of the models and are representative of the deformation of the models as a whole. Each series of experiments have been duplicated and the results were found to be reproducible.

Physical properties of the modelling materials

Dry, cohesionless quartz sand has been commonly used as a physical modelling material as it satisfactorily simulates the brittle Coulomb behaviour of shallow crustal rocks in scaled laboratory experiments (Hubbert 1937; Horsfield 1977; McClay & Ellis 1987a, b; Ellis & McClay 1988; Mandl 1988; Mulugeta 1988a, b). Dry quartz sand has a Navier-Coulomb rheology and an angle of friction of 30° similar to that for

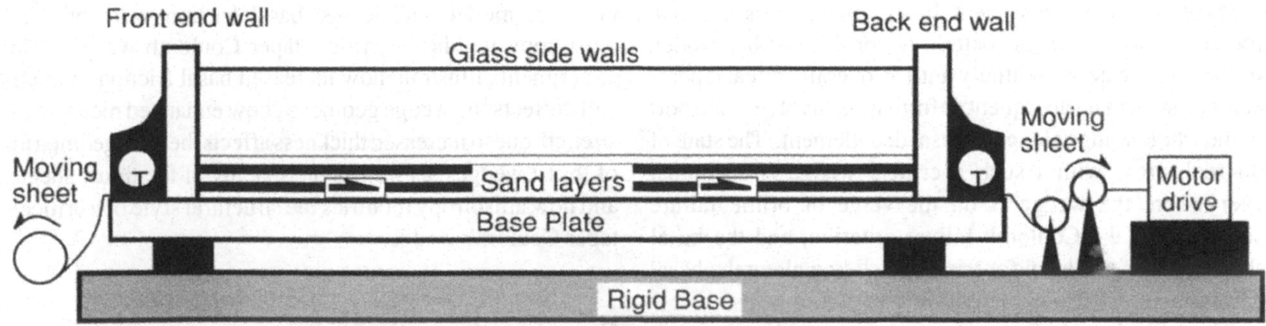

Figure 1. Schematic diagram of the deformation apparatus.

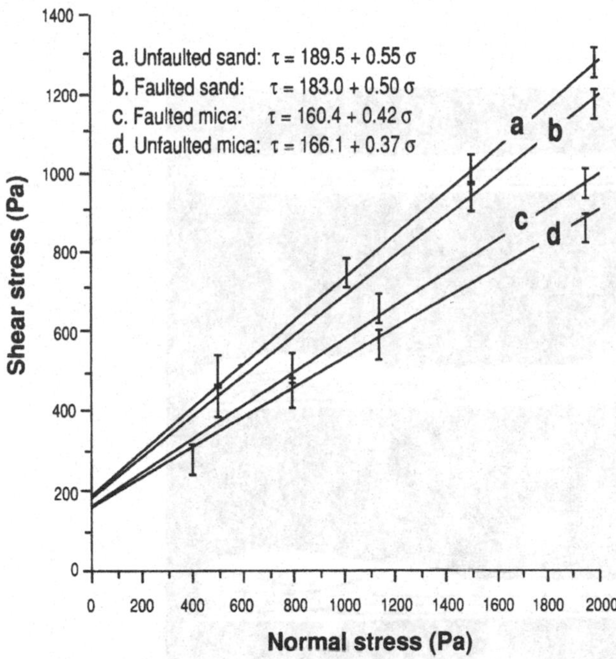

Figure 2. Mechanical properties of the modelling materials (sand and mica) at failure.

many sedimentary rocks (cf. McClay 1990). In terms of geometric similarity (Ramberg 1981), the length ratio of the model experiments to natural examples is 10^{-4}-10^{-5}, i.e. a 1 cm sand layer corresponds to 100 m to 1 km of sedimentary strata in nature. However, the packing, distribution, and size of the grains of the modelling material can not be accurately scaled and consequently the fault structures in the analogue experiments are not discrete failure planes but are dilatant shear zones of varying width (Horsfield 1977; McClay 1990).

Dry quartz sand (grain size 200-300 μm, average density 1.58 ± 0.1 g cm^{-3}) was used to simulate the brittle deformation of isotropic sedimentary sequences (experiment series I-III, Table 1) and thin layers of dry vermiculite mica (grain size 500 μm (long axis), average density 1.0 g cm^{-3}) were introduced between the sand layers to provide a mechanical anisotropy and hence simulate the deformation of anisotropic sediments which deform by interlayer slip (experiment series IV, Table 1).

The mechanical properties of the sand and mica at very low normal stresses (Fig. 2) have been investigated using a similar apparatus to that described by Krantz (1991). Shear strength was measured for both the unfaulted and faulted materials (after fault displacement of 5 mm) (Fig.2). Linear Mohr envelopes were found for all materials with coefficients of friction ranging from $\mu_i = 0.55$ for unfaulted sand to $\mu_i = 0.37$ for unfaulted mica (Fig. 2). Low apparent cohesive strengths were found for all materials ($\tau_0 = 166 - 189.5$ Pa, Fig. 2). Faulted sand shows a strength reduction of approximately 10% (Fig. 2) whereas faulted mica gained strength (14%) after 5 mm of fault displacement (Fig. 2). The latter feature is probably a result of reorientation of mica flakes so that they obstruct the sliding plane at low displacements. However, at larger displacements, it is anticipated that entrainment of a large number of mica grains into the shear plane would lower the effective coefficient of friction in the fault zone

Frictional properties of detachment materials

Three detachment materials were used: (a) plastic sheet, (b) drafting film and (c) industrial sand paper. The frictional characteristics between the basal detachments and the sand packs were measured at low values of normal stress (Fig. 3). The plastic sheet has the lowest coefficient of sliding friction $\mu_b = 0.37$; the drafting film has an intermediate value of sliding friction $\mu_b = 0.47$, whereas the sandpaper has a high value of sliding friction $\mu_b = 0.55$ which is equal to the coefficient of friction for quartz sand. The vertical backstop had a coefficient of friction for sand sliding against it of $\mu = 0.55$.

Results

The results of four series of experiments (15 models, Table 1) are presented in this paper. (Duplicate models have produced similar results (Liu unpubl. data)). In all models a foreland propagating sequence of thrust slices formed a Coulomb wedge-like geometry with a taper dependent upon both the basal friction and, to a lesser degree, on the initial thickness of the model (Table 1). In all of these models a rigid vertical backstop was used in order to produce wedges that closely approximate the theoretical geometries (cf. Davis *et al.* 1983; Dahlen 1990). Further research is currently being carried out using variable backstop geometries that may be more geologically realistic.

Thrust sequences and thrust geometries

The progressive evolution of the thrust system in a typical

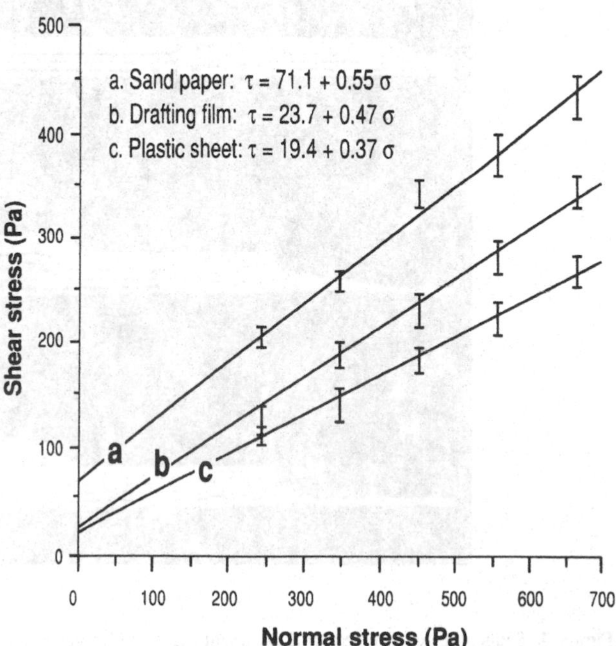

Figure 3. Frictional properties of the three basal detachment materials.

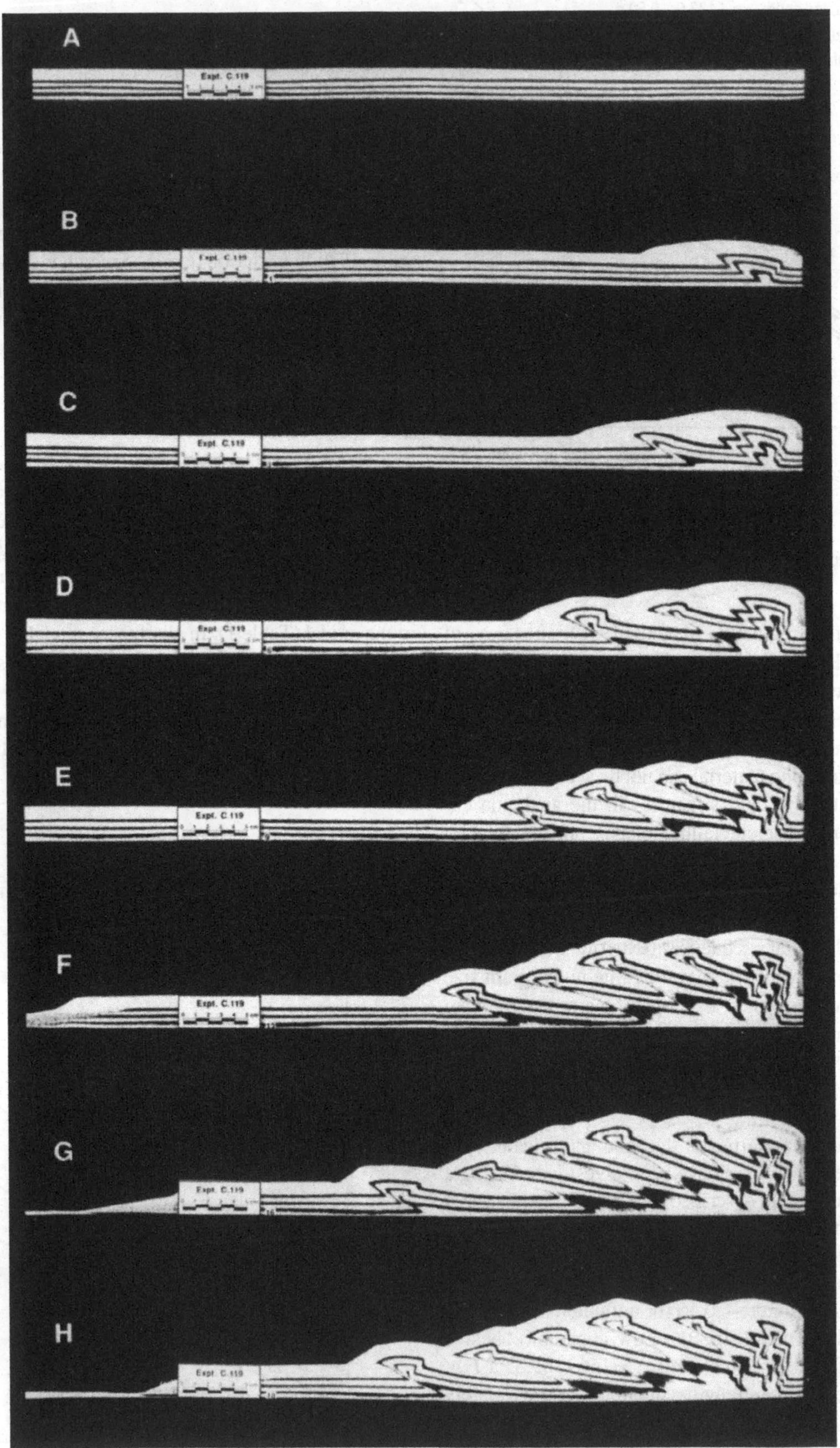

Figure 4. Sequential development of Experiment C.119, with homogeneous sand, $\mu_b = 0.47$ and a 2.5 cm thick initial parallel layered sequence. **(a)** Initial model 0% shortening; **(b)** 3% shortening; **(c)** 9% shortening; **(d)** 18% shortening; **(e)** 27% shortening; **(f)** 39% shortening; **(g)** 48% shortening; **(h)** 49% shortening - end of the deformation run with a stable wedge formed. Slight underplating of sand causes the wedge to grow vertically from (g) above.

critical-taper Coulomb wedge model is illustrated in Figure 4 for an homogeneous 2.5 cm thick sand model with an intermediate basal detachment friction of $\mu_b = 0.47$. The first thrust appeared at about 2.5% shortening and is followed almost immediately by a steeper 'backthrust' (Fig. 4b). Another foreland-vergent thrust developed in the immediate footwall of the first thrust and was followed by a second major foreland-vergent thrust approximately five cms in front of thrust 1 (Fig. 4c). At 18% shortening thrust 2 has developed a small backthrust at the tip and thrust 3 had nucleated in front of thrust 2 (Fig. 4d). The older, higher thrusts became inactive and are passively back rotated once a new foreland-vergent thrust nucleated in the footwall. A similar pattern of nucleation and displacement on the most foreland-ward thrust combined with back-rotation of the hinterland thrusts, is continued throughout the progressive development of the Coulomb wedge (Figs 4 e-h). The topographic slope and hence the wedge taper decreases throughout the deformation from 15° at the start of thrusting to 7-8° when a stable critical taper has been achieved (Fig. 5a).

Foreland-vergent thrusts nucleate at angles of 20-25° and are always accompanied by a backthrust (initial angles 35-40°) at their tips (Fig. 4). The majority of the displacement is concentrated on the foreland-vergent thrust. As deformation proceeds older, higher thrusts are progressively back-rotated so that their dip increases until approximately 45° where they become involved in the vertical uplift zone at the back of the wedge (Fig. 5b). The overall geometry of the Coulomb wedge evolves from a convex geometry (Fig. 4b) to progressively more triangular geometries (Figs 4 c-h) but there is still an element of convexity in the shape of the wedge. The rear of the wedge has a near horizontal surface and is progressively elevated by displacement on both the most foreland-ward thrust and on the backthrusts adjacent to the vertical backstop (Figs 4 d-g). The position of the deformation front (i.e. the frontal thrust) moves progressively into the foreland with new thrusts nucleating at an approximately uniform spacing (Fig. 5c). At the end of the experiment (Fig. 4h) a stable wedge has formed and this slides on the basal detachment with no further internal deformation.

Initial thickness variations

In the experiments described in this paper a uniform parallel sided sedimentary prism was used in order to invesigate the effect of stratigraphic thickness on thrust spacing and wedge geometries. Typical results are shown in Figure 6 for a constant basal detachment friction of $\mu_b = 0.47$ and values of initial thicknesses ranging from 1.0-3.0 cm. For $\mu_b = 0.37$, initial thicknesses of 3.5 cms and above did not undergo thrusting and form a Coulomb wedge but rather slid undeformed on the basal detachment. For initial thicknesses of 3.0 cm and below three principal effects were found -

(1) With decreasing initial thickness there is an increase in the amount of shortening that occurs before a stable wedge is developed;

(2) With decreasing initial thickness there is a decrease in the spacing of the thrusts and a corresponding increase in the

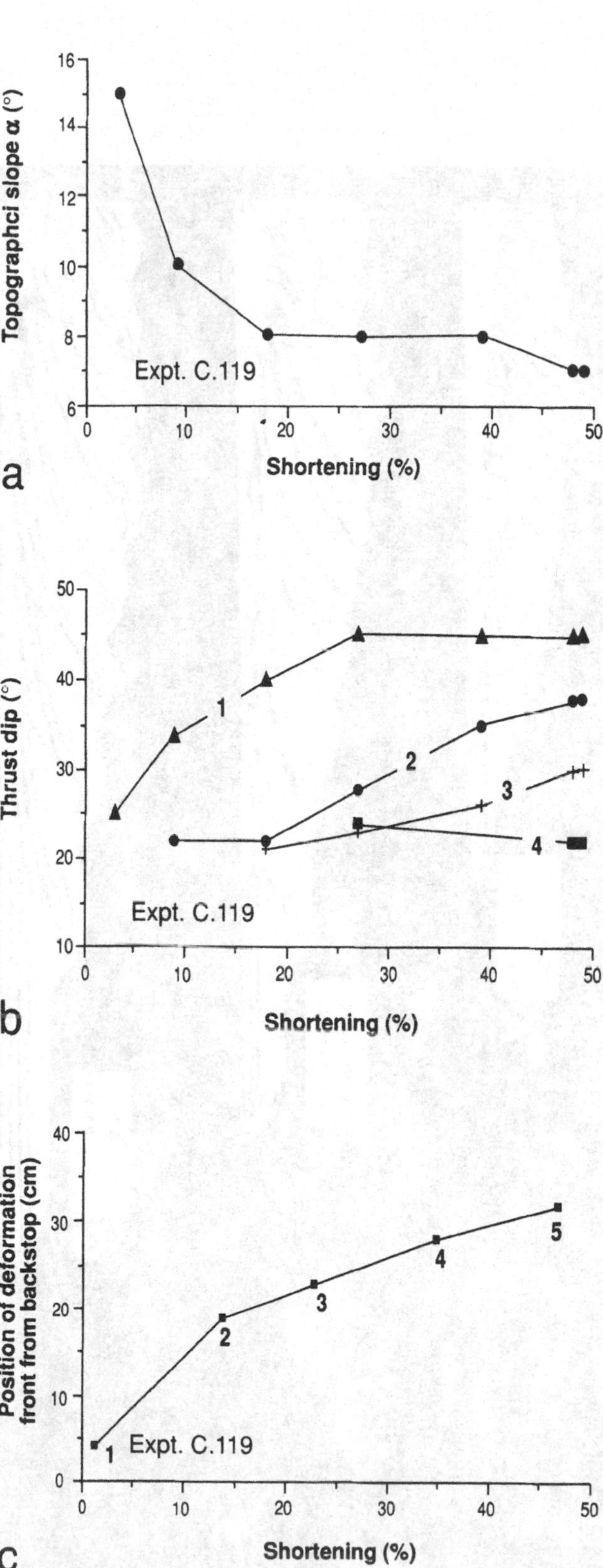

Figure 5. Detailed measurements taken from Experiment C. 119 (Fig. 4). (**a**) Evolution of the topographic slope (wedge taper) with percentage shortening. A stable wedge slope of 7 - 8° is developed after 20% shortening. (**b**) Dip of foreland-vergent thrusts with progressive shortening. Thrusts 1 - 4 developed in sequence from the back of the wedge (Fig. 4). (**c**) Evolution of the thrust front with progressive shortening. Each point marks the formation of a new foreland-vergent thrust (1 - 5).

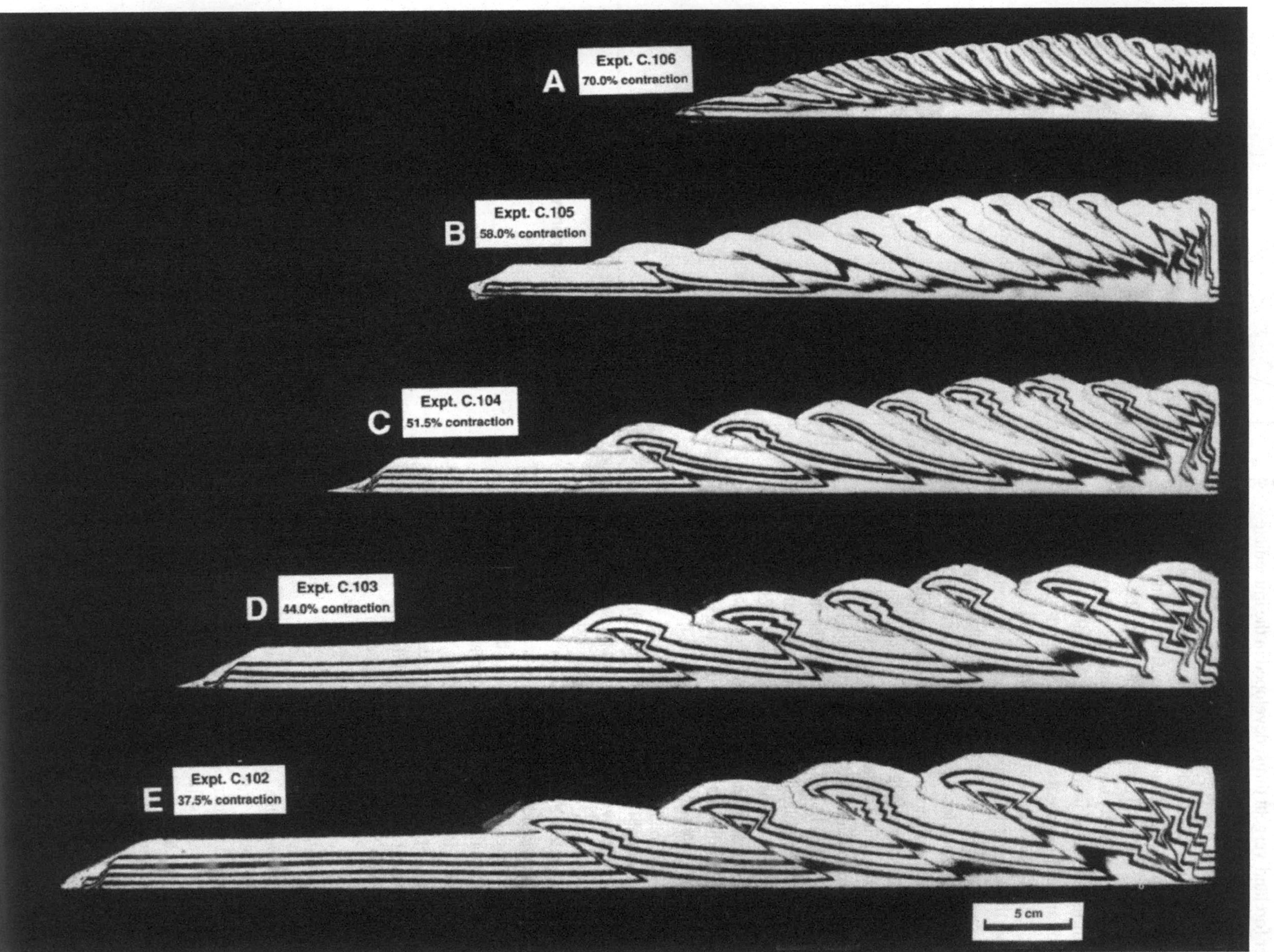

igure 6. Five sections of sand models with different thickness, all using intermediate basal friction. C.102, 3 cm model, C.103, 2.5 cm model, C.104, 2 cm model, C.105, 1.5 cm model and C.106, 1 cm model.

number of thrusts that form the final stable wedge; and (3) The critical taper of the wedge only decreases slightly with an increase in initial thickness from 9.5° for an initial thickness of 1.0 cm to 7° for an initial thickness of 3.0 cms (Fig. 6).

In all experiments except C.106 (initial thickness 1.0 cm) an early backthrust commonly formed at the tip of each foreland-vergent thrust (Fig. 6). In the experiments with thinner initial thicknesses more foreland-vergent thrusts formed with the result that the more hinterland thrusts have undergone greater back rotation than for the wedges formed from greater initial thicknesses (Fig. 6).

Variation in basal friction

The effects of varying the basal friction on final wedge geometry are shown in Figure 7. An increase in friction produces a dramatic increase in the critical taper from 1.5° for a basal friction of $\mu_b = 0.37$ (Fig. 7c), to a wedge taper of 7.0° for a basal friction of $\mu_b = 0.47$ (Fig. 7b), and to a wedge taper of 18.5° for a basal friction of $\mu_b = 0.55$ (Fig. 7a). Concomitant with the increase in wedge taper there are corresponding increases in the lengths of thrusts in the wedge (Fig. 7). The number of backthrusts associated with each foreland-vergent thrust decreases with an increase in basal friction. For the lowest value of basal friction ($\mu_b = 0.37$) the deformation is characterized by asymmetric 'pop-up' structures that result in a low critical taper for the wedge (Fig. 7c).

The effects of anisotropy

The introduction of anisotropy into the system by placing thin layers of mica flakes between the sand layers produces critical tapers (Fig. 8) which are slightly increased when compared to the isotropic models (cf. Fig. 7). The styles of deformation in the anisotropic models are very similar to the isotropic examples (Figs 7 & 8) with possibly slightly more folding in the anisotropic models (the mica flakes facilitate layer parallel slip).

Discussion

Sandbox analogue models have successfully simulated the formation of critical taper Coulomb wedges comparable to those found in accretionary prisms and foreland fold-thrust belts (Davis *et al.* 1983; Dahlen 1990). In all models a foreland nucleating and foreland-vergent thrust system developed with varying degrees of backthrusting. Active foreland-vergent deformation was concentrated on the last-formed foreland-vergent thrust such that the older, higher thrusts were passively back-rotated to steeper angles in a manner analogous to that found in many thrust belts (Bally *et al.* 1966; Dahlstrom 1970; Boyer & Elliott 1982). In many models the tips of individual thrusts have an associated small backthrust (Figs 4, 6-8) similar to those found both in fold-thrust belts (Alonso & Teixell 1991 - this volume) and in accretionary prisms (Lewis *et al.* 1988).

Critical taper Coulomb wedges

According to Coulomb wedge theory the critical taper ($\alpha + \beta$) of a dry cohesionless wedge can be approximated by the equation (e.g. Dahlen 1990, eq. 16) -

$$\alpha + \beta \approx \mu_b \left(\frac{1 - \sin\ \phi}{1 + \sin\ \phi} \right) \qquad (1)$$

where α = topographic slope of the wedge, β = dip of the basal detachment, μ_b = coefficient of friction of the basal detachment, ϕ = angle of internal friction of the material in the wedge.

For the analogue model experiments, substituting for μ_b and ϕ gives theoretical critical wedge tapers of 7.4° for $\mu_b = 0.37$ (plastic sheet), 9.4° critical taper for $\mu_b = 0.47$ (drafting film), and a critical taper of 11° for $\mu_b = 0.55$ (sandpaper). Comparison of the theoretical values with actual values (Table 1) shows that the best results were obtained for thin initial thicknesses (1.0-1.5cms) and for an intermediate basal detachment friction of $\mu_b = 0.47$ - the drafting film detachment. For low values of basal friction low-taper wedges (taper less than that predicted by eq. 1) were formed because the deformation partitioned approximately equally between foreland-vergent thrusts and backthrusts (Figs 6 & 7) such that symmetric uplift occurred thus reducing the wedge taper. Wedges produced when the basal detachment had a frictional coefficient equal to that of the modelling material produced critical tapers of 18-24°, greatly in excess to that expected from theory (11°) and unaffected by initial thickness of the model (Table 1). In these cases a critical point is not reached - the wedge continues to build up as long as new material is fed into the system and the wedge angle is closer to a dynamic angle of rest for the granular modelling material. Such wedges are supra-critical and the Coulomb wedge theory is no longer applicable (it was an assumption in the theoretical analysis that the basal detachment is weaker than the internal material). For a basal friction of $\mu_b = 0.37$ there is an initial thickness above which a Coulomb wedge is not formed but rather the whole model slides on the basal detachment with no internal deformation. Such models are termed supercritical and it has been found that the supercritical thickness increases with an increase in the basal friction. In this case the model is significantly stronger than the basal detachment and no internal deformation occurs.

Thrust geometries

In the analogue models (Figs 4, 6-8) imbricate thrust fans are produced which have similar forms to those found in accretionary prisms (e.g. Seely 1974; Westbrook 1982; Lewis *et al.* 1988; Westbrook *et al.* 1988) and in the frontal parts of foreland fold and thrust belts (e.g. Bally *et al.* 1966; Dahlstrom 1970; Lillie *et al.* 1987; Stanley 1990). In the models described in this paper the spacing/thickness ratio of the thrust sheets varies from 3.2-5.7 and increases with an increase in basal friction in contrast to the findings of Mulugeta (1988b) who found that this ratio increased with a decrease in basal friction. The length of individual thrust slices increased with

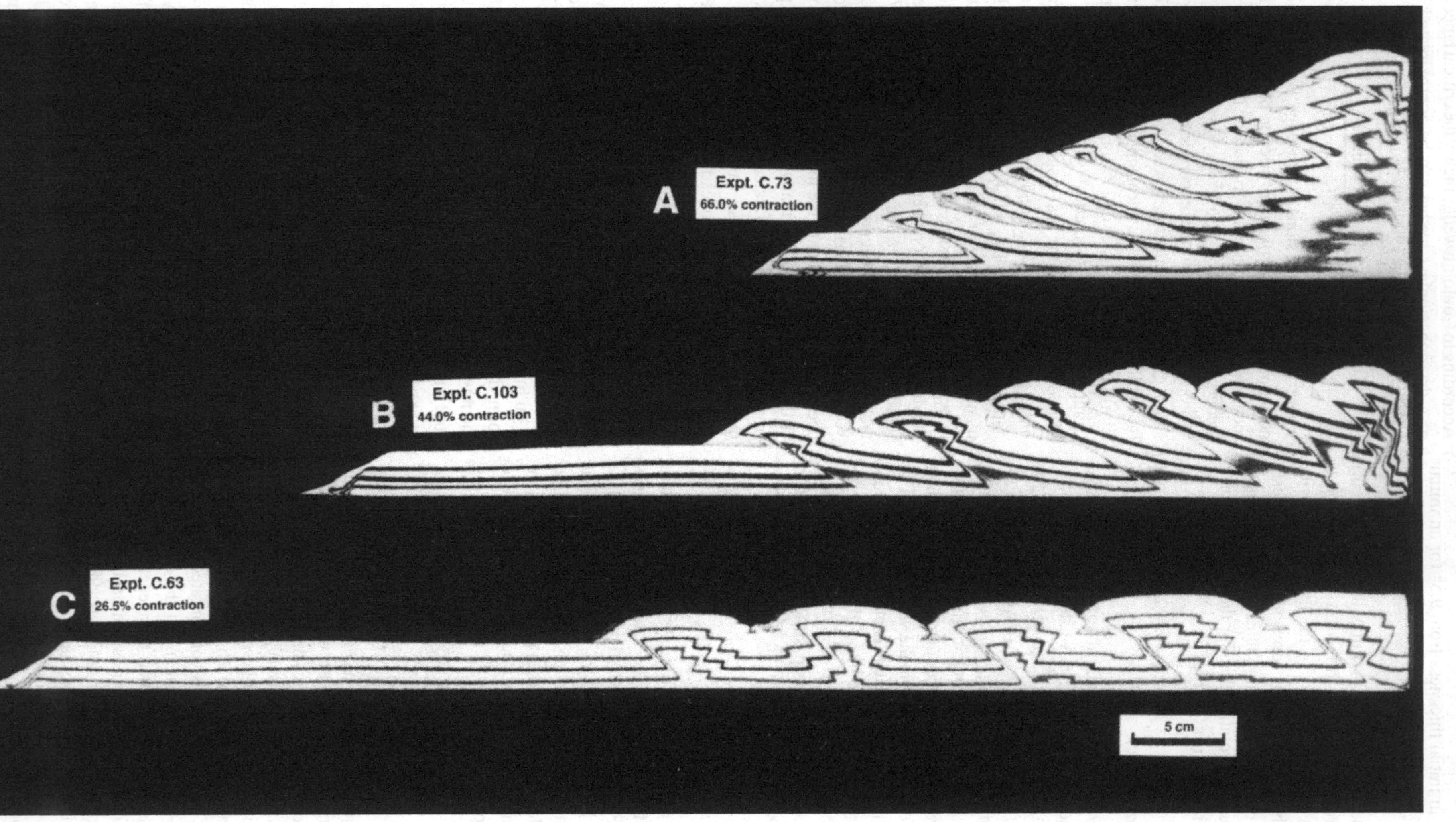

Figure 7. Three representative sections of selected homogeneous sand models with the same layer thickness (2.5 cm); **(a)** High basal friction model, $\mu_b = 0.55$; **(b)** Intermediate basal friction model, $\mu_b = 0.47$; and **(c)** Low basal friction model, $\mu_b = 0.37$.

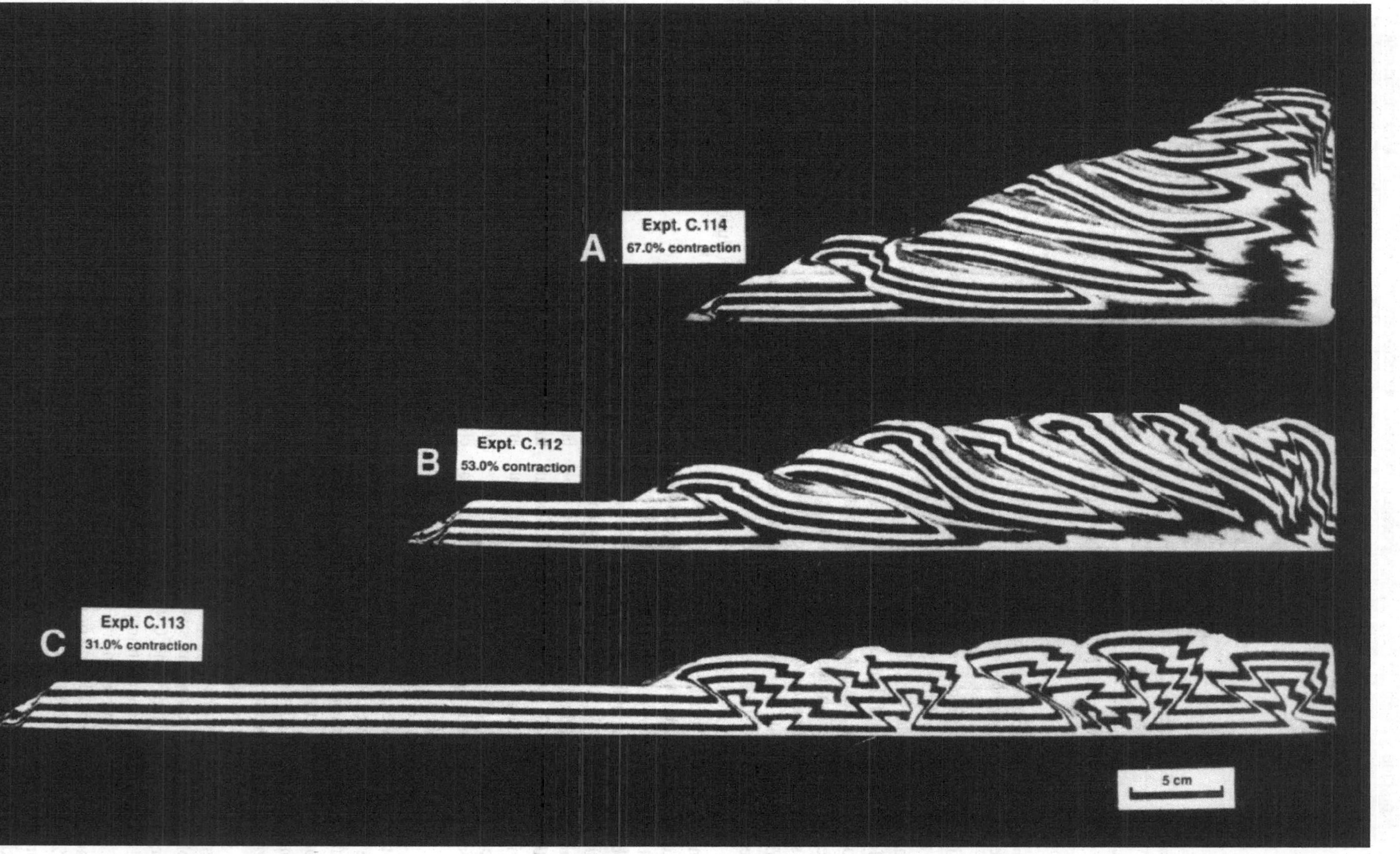

Figure 8. Three representative sections of sand/mica models with the same layer thickness (2.5 cm); **(a)** High basal friction model, $\mu_b = 0.55$; **(b)** Intermediate basal friction model, $\mu_b = 0.47$; and **(c)** Low basal friction model, $\mu_b = 0.37$.

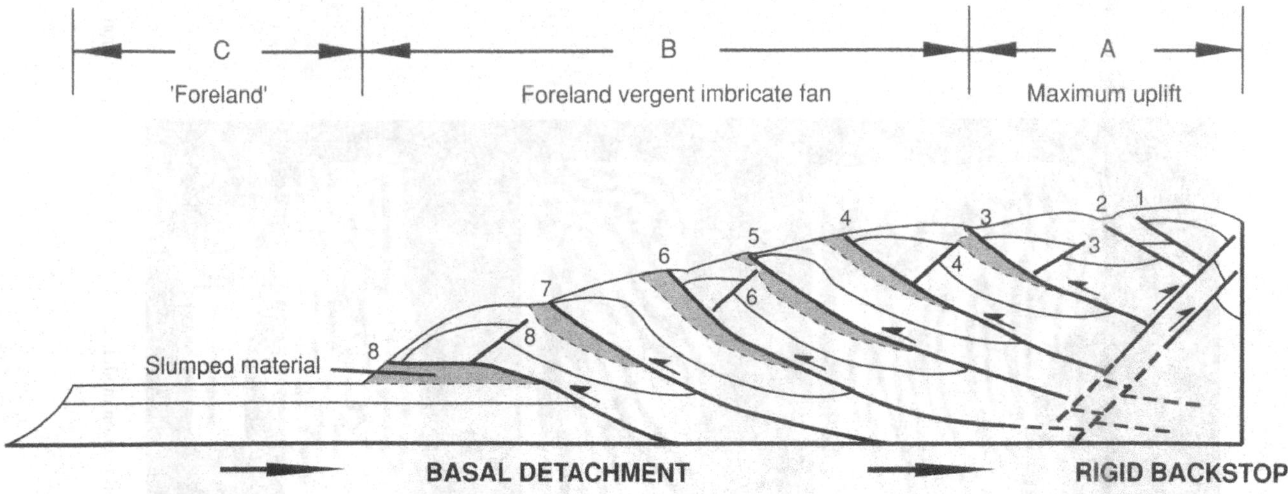

Figure 9. Schematic diagram of a deformed wedge showing the three deformation zones and the general sequence of thrust faults. Note that the back thrusts at the rear of the wedge are active at various stages of wedge growth.

an increase in initial model thickness (Fig. 6) and similar features are found in foreland fold and thrust belts e.g. in cross-sections of the Canadian Rocky Mountains (Price 1981) and of the Moine thrust zone (Elliott & Johnson 1981). The analogue models show that a decrease in the basal friction increases the number of backthrusts in the system (Fig. 7) with a more symmetric uplift and a low critical taper similar to that found in foreland fold and thrust belts with a weak basal decollement (e.g. Pakistan, Davis & Engelder (1985), Jadoon *et al.* (1991, this volume)). In the anisotropic models more shortening is required to achieve a stable wedge system (Fig. 8) and deformation of the hangingwall above the foreland-vergent thrusts is accommodated by folding with suppression of backthrusts thus indicating that the anisotropic models are mechanically weaker than the isotropic sequences.

The initial ramp angles of the foreland-vergent thrusts (25-30°) and of the backthrusts (35-40°) do not show significant variations with changes in basal friction thus indicating that care must be exercised when attempting to deduce the coefficient of basal friction from thrust fault angles (cf. Mulugeta 1988b).

Figure 9 is a summary diagram illustrating the main features of the Coulomb wedge models described in this paper. Zone A delimits an area adjacent to the back end wall which has suffered maximum bulk shortening and thickening and where foreland-vergent thrusts have suffered a maximum amount of passive rotation. This section undergoes vertical uplift by synchronous movement on the foreland-vergent detachment and on the backthrusts adjacent to the vertical backstop. In zone B the foreland-vergent thrusts have the largest displacements, most regular spacing and form a foreland-vergent imbricate fan; zone C (the 'foreland') is undeformed and has been passively transported on the basal detachment. The wedges have a concave upwards topography which suggests a strain hardening process at the rear of the system (Zhao *et al.* 1986; Mulugeta 1988b). The slumped material from each of the thrust fronts is later over-ridden by the foreland-vergent thrust faults which is similar to the 'erosional thrusts' mentioned by Burchfiel *et al.* (1982). Similar slumping was also been reported by Mulugeta & Koyi (1987) in their experiments, and can be compared to that reported by Moore & Shipley (1988) in the Middle America Trench accretionary prism. Whereas the experiments described in this paper produce Coulomb wedges that show many similarities to foreland fold and thrust belts and to accretionary prisms, it must be remembered that the experiments use dry material and pore-fluid pressures have not been simulated - a factor important in many accretionary prisms and in fold-thrust belts (Hubbert & Rubey 1959).

Conclusions

Physical models have been successfully constructed to simulate the progressive deformation of brittle Coulomb wedges. Imbricate fans of dominantly foreland-vergent thrust systems are developed similar to those found in accretionary prisms and in foreland fold and thrust belts. Critical taper wedges close to theoretically predicted geometries are developed for intermediate values of basal friction $\mu_b = 0.47$ whereas for the lower value of basal friction low-taper wedges are formed with tapers below that predicted by theory. Supra-critical wedges are formed when the basal friction equals or is greater than the coefficient of friction in the wedge and the wedge has a high taper closer to the angle of rest for the modelling material. In sandbox critical wedges, the spacing/thickness ratio of foreland-vergent thrusts increases as the layer thickness increases, suggesting a process which is not geometrically self-similar. The experimental data also indicate that the spacing of thrust faults increases with increased basal friction. Higher basal friction or anisotropy within the layered systems favours displacement along foreland-vergent thrusts and suppresses backthrusts. In the anisotropic models enhanced interlayer slip arising from the introduction of mica layers allows accommodation in the hangingwalls to be taken up by folds.

Liu Huiqi gratefully acknowledges receipt of a research studentship from the Sino-British Friendship Scholarship Scheme administered by the British Council. This work was in part supported by the Industry Association of the Geology Department, Royal Holloway and Bedford New College, University of London, and by the Structural Geology Modelling Laboratory, RHBNC. Thanks are due to Brian Adams for constructing the deformation apparatus, to Martin Insley for helping with the shear tests and to Ian Davison and Tony Dahlen for helpful comments on an early version of the manuscript.

References

Alonso, J. L. & Teixell A. 1991. Forelimb deformation in some natural examples of fault-propagation folds (this volume).

Bally, A. W., Gordy, P. L. & Stewart, G. A. 1966. Structure, seismic data, and orogenic evolution of the southern Canadian Rocky Mountains. *Bulletin of Canadian Petroleum Geology*, **14**, 337-381.

Boyer, S. E. & Elliott, D. 1982. Thrust systems. *Bulletin of the American Assocation of Petroleum Geologists*, **66**, 1196-1230.

Burchfiel, B. C., Wernicke, B., Willemin, J. H., Axen, G.J . & Cameron, C. S. 1982. A new type of decollement thrusting. *Nature*, **300**, 513-515.

Cadell, H. M. 1889. Experimental researches in mountain building. *Transactions of Royal Society of Edinburgh*, **35**, 337-357.

Dahlen, F. A. 1984. Noncohesive critical Coulomb wedges: an exact solution. *Journal of Geophysical Research*, **89**, 10125-10133.

—— 1990. Critical taper model of fold-and-thrust belts and accretionary wedges. *Annual Review of Earth and Planetary Sciences*, **18**, 55-99.

—— & Barr, T. D. 1989. Brittle frictional mountain building, 1, Deformation and mechanical energy budget. *Journal of Geophysical Research*, **94**, 3906-3922.

—— & Suppe, J. 1988. Mechanics, growth, and erosion of mountain belts. *In:* Clark, S. P., Jr., Burchfiel, B. C. & Suppe, J. (eds) *Processes in Continental Lithospheric Deformation*. Geological Society of America Special Paper, **218**, 161-178.

——, ——, J. & Davis, D. M. 1984. Mechanics of fold-and-thrust belts and accretionary wedges: Cohesive Coulomb theory. *Journal of Geophysical Research*, **89**, 10087-10101.

Dahlstrom, C.D.A. 1970. Structural geology of the eastern margin of the Canadian Rocky Mountains. *Bulletin of Canadian Petroleum Geology*, **18**, 332-406.

Davis, D. M. & Engelder, T. 1985. The role of salt in fold-and-thrust belts. *Tectonophysics*, **119**, 67-89.

—— & von Huene, R. 1987. Inferences on sediment strength and fault friction from structures at the Aleutian Trench. *Geology*, **15**, 517-522.

——, Suppe, J. & Dahlen, F. A. 1983. Mechanics of fold-and-thrust belts and accretionary wedges. *Journal of Geophysical Research*, **88**, 1153-1172.

Elliott, D. & Johnson, M.R.W. 1980. Structural evolution in the northern part of Moine thrust belt, NW Scotland. *Transactions of Royal Society of Edinburgh (Earth Sciences)*, **71**, 69-96.

Ellis, P. G. & McClay, K. R. 1988. Listric extensional fault systems - results of analogue model experiments. *Basin Research*, **1**, 55-70.

Horsfield, W. T. 1977. An experimental approach to basement controlled faulting. *Geologie en Mijinbouw*, **56**, 363-370.

Hubbert, M. K. 1937. Theory of scale models as applied to the study of geologic structures. *Geological Society of America Bulletin*, **48**, 1459-1520.

—— 1951. Mechanical basis for certain familiar geologic structures. *Geological Society of America Bulletin*, **62**, 355-372.

—— & Rubey, W. M. 1959. Role of fluid pressure in mechanics of thrust faulting: I. Mechanics of fluid-filled porous solids and its application to overthrust faulting. *Geological Society of America Bulletin*, **70**, 115-166.

Jadoon, I.A.K., Lawrence, R. D. & Lillie, R. J. 1991. Balanced and retrodeformed geological cross-section from the frontal Sulaiman Lobe, Pakistan: Duplex development in thick strata along the western margin of the Indian Plate (this volume).

Krantz, R. W. 1991. Measurements of friction coefficients and cohesion for faulting and fault reactivation in laboratory models using sand and sand mixtures. *Tectonophysics*, **188**, 203-207.

Lewis, S. D., Ladd, J. W. & Bruns, T. R. 1988. Structural development of an accretionary prism by thrust and strike-slip faulting: Shumagin region, Aleutian Trench. *Geological Society of America Bulletin*, **100**, 767-782.

Lillie, R. J., Johnson, G. D., Yousuf, M., Zamin, A.S.H. & Yeats, R. S. 1987. Structural development within the Himalayan foreland fold-and-thrust belts of Pakistan. *In*: Beaumont, C. & Tankard, A. J. (eds) *Sedimentary Basins and Basin-Forming Mechanisms*. Canadian Society of Petroleum Gelogists Memoir, **12**, 379-392.

Malavieille, J. 1984. Modélisation expérimentale des chevauchements imbriqués: Application aux chaînes de montagnes. *Geological Society of France Bulletin*, **7**, 129-138.

Mandl, G. 1988. *Mechanics of Tectonic Faulting: Models and Basic Concepts. Developments in Structural Geology*, **1**, Elsevier, Amsterdam, 407p.

McClay, K.R. 1990. Deformation mechanics in analogue models of extensional fault systems. *In*: Knipe, R. J. & Rutter, E. H. (eds), *Deformation Mechanisms, Rheology and Tectonics*. Geological Society of London Special Publication, **54**, 445-453.

—— & Ellis, P. G. 1987a. Analogue models of extensional fault geometries. *In*: Coward, M. P., Dewey, J. F. & Hancock, P. L. (eds) *Continental Extensional Tectonics*. Geological Society of London Special Publications, **28**, 109-125.

—— & —— 1987b. Geometries of extensional fault systems developed in model experiments. *Geology*, **15**, 341-344.

Moore, G. F. & Shipley, T. H. 1988. Mechanics of sediment accretion in the Middle America Trench off Mexico. *Journal of Geophysical Research*, **93**, 8911-8927.

Mulugeta, G. 1988a. Squeeze box in a centrifuge. *Tectonophysics*, **148**, 323-335.

—— 1988b. Modelling the geometry of Coulomb thrust wedges. *Journal of Structural Geology*, **10**, 847-859.

—— & Koyi, H. 1987. Three-dimensional geometry and kinematics of experimental piggyback thrusting. *Geology*, **15**, 1052-1056.

Price, R. A. 1981. The Cordilleran foreland thrust and fold belt in the southern Canadian Rocky Mountains. *In:* McClay, K. R. & Price, N. J. (eds) *Thrust and Nappe Tectonics*, Geological Society of London Special Publications, **9**, 427-448.

—— 1988. The mechanical paradox of large overthrusts. *Geological Society of America Bulletin*, **100**, 1898-1908.

Ramberg, H. 1981. *Gravity, Deformation, and the Earth's Crust: In Theory, Experiments and Geological Application* (2nd ed.). Academic Press, London. 452p.

Roure, F., Howell, D. G., Guellec, S. & Casero, P. 1990. Shallow structures induced by deep-seated thrusting. *In:* Letouzey, J. (ed.) *Petroleum and Tectonics in Mobile Belts*, Éditions Technip, Paris, **43**, 15-30.

Seely, D. R., Vail, P. R. & Walton, G. G. 1974. Trench slope model. *In*: Burk, C. A. & Drake, D. L. (eds) *The Geology of Continental Margins*. Springer-Verlag, New York. 249-260.

Stanley, R. S. 1990. The evolution of mesoscopic imbricate thrust faults—an example from the Vermont Foreland, USA. *Journal of Structural Geology*, **12**, 227-241.

Westbrook, G. K. 1982. The Barbados Ridge complex: Tectonics of a mature forearc system. *In*: Leggett, J. K. (ed.) *Trench-Forearc Geology*. Geological Society of London Special Publication, **10**, 275-290.

——, Ladd, J. W., Buhl, P., Bangs, N. & Tiley, G. J. 1988. Cross section of an accretionary wedge: Barbados Ridge complex. *Geology*, **16**, 631-635.

Willis, B. 1892. The mechanics of Appalachian structure. US Geological Survey 13th Annual Report, 271-281.

Zhao, W.-L., Davis, D. M., Dahlen, F. A. & Suppe, J. 1986. Origin of convex accretionary wedges: Evidence from Barbados. *Journal of Geophysical Research*, **91**, 10246-10258.

Generation of curved fold-thrust belts: Insight from simple physical and analytical models

S. Marshak, M. S. Wilkerson & A. T. Hsui

Department of Geology, University of Illinois, Urbana, IL, 61801, USA

Abstract: Viscous-flow and sand-wedge models provide insight into factors that determine the location, shape, and temporal evolution of fold-thrust belts whose traces are curved in plan view. These models suggest that curves are localized by shear between segments of fold-thrust belts and adjacent regions, and by lateral variations in sediment thickness. Furthermore, they indicate that the shape of antitaxial (convex toward the foreland) belts may reflect the rheology and/or strength of rock comprising a belt, as well as regional variations in sediment thickness, topography, and backstop geometry. Finally, the models help distinguish between settings in which curvature is secondary (i.e. curves are oroclines) and settings in which curvature is primary.

Many major orogens, and their component fold-thrust belts, contain pronounced plan-view curves defined by lateral variation in structural trends. For example, the Appalachians contain two major *antitaxial* curves (convex in the direction of vergence) and two major *syntaxial* curves (concave in the direction of vergence). Work during the past few decades (e.g. Carey 1955; Ries & Shackleton 1976; Beutner 1977; Burchfiel 1980; Eldredge *et al.* 1985; Lowrie & Hirt 1986; Marshak & Tabor 1989) has clarified our understanding of structural and palaeomagnetic patterns within curved belts, and has emphasized the persistence of three outstanding problems: (1) Why do curves form? (2) What factors control the shape of individual curves? and, (3) Do curves initiate with their present shape, or do they represent plan-view bending and tangential extension of pre-existing straight structures?

The purpose of this paper is to describe results of viscous-flow and sand-wedge modelling experiments which help constrain solutions to these problems. Curved fold-thrust belts form in a variety of different tectonic settings, but this paper examines only curves formed as a consequence of fold-thrust belt interaction with obstacles in the foreland, curves formed in advance of an indenter, and curves formed adjacent to strike-slip faults. In this context, the term *obstacle* refers to any feature that restrains thrust-sheet movement at its lateral margins - physically, obstacles may be basement massifs, pinch-outs of stratigraphic glide horizons, or strike-slip faults. An *indenter* is a rigid mass or backstop of finite length (measured perpendicular to the transport direction) that pushes into the deforming medium.

Modelling procedures and observations

Several different types of models were used to investigate the generation of curved fold-thrust belts. The first was an analytical model of viscous flow describing the interaction of a fluid layer with obstacles to test the effects of rheology on curve geometry. To gain a clearer image of how obstacle geometry controls flow trajectories, a physical model of viscous flow that used cooled glycerine as the deforming medium was constructed. The need to have more quantitative constraints on the shape of deforming lobes led the authors to construct a finite-difference model describing crustal deformation. Finally, to gain insight into the kinematic evolution of curved fold-thrust belts, three types of sand-wedge models that allowed simulation of collision with obstacles, impact of an indenter, and superposition of strike-slip shear were constructed.

Analytical model of fluid flow

Model setup. This model uses the evolution of a passive marker line on the surface of a fluid sheet to describe curve development accompanying interaction of a growing fold-thrust belt with obstacles in the foreland. The fluid sheet represents a thrust wedge, and the marker line represents the structural trend as a function of location. All points on the marker line are constrained to flow along parallel lines. To simulate interaction with obstacles, the endpoints of the marker line are assigned a velocity of zero. Initially, the marker line is straight and is perpendicular to the flow direction, but as flow progresses, it arches into a parabola, as described by:

$$x = \{(h/2)^{n+1} - [y-(h/2)]^{n+1}\} / F(n+1) \qquad (1)$$

In this equation, which is similar to the equation defining one-dimensional channel flow, x defines the distance between the initial and final positions of a point on the marker line as measured in the direction of flow, y defines the distance between a point and the left endpoint of the marker line prior to flow, and h is the distance between endpoints of the marker line measured along the y-axis (Fig. 1; Wilkerson *et al.* 1988; modified from Turcotte & Schubert 1982, p. 316-318, and Hsui & Youngquist 1985; see also Johnson 1970).

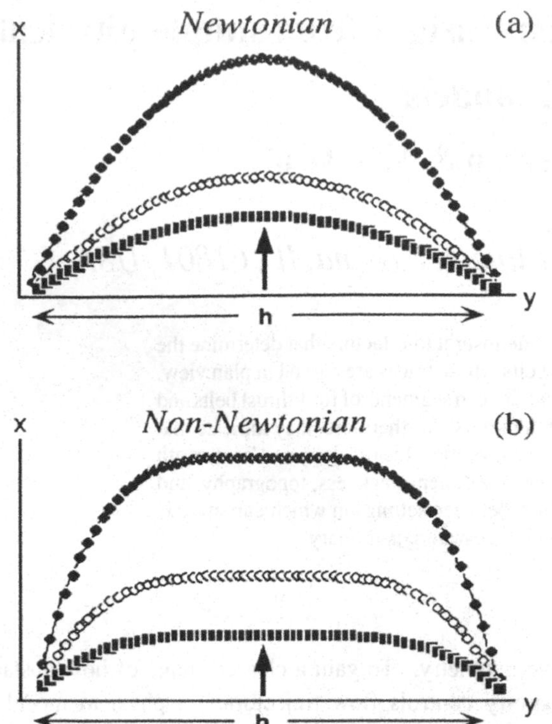

Figure 1. Results from the analytical model of parallel fluid flow described by equation 1. Flow is parallel to the x axis (as indicated by arrow). The distance between the endpoints measured along the y-axis is h. At the endpoints, flow velocity is 0. The marker line is originally straight and lies on the y axis. Movement stages are distinguished by patterns: squares = early, open diamonds = middle, and filled diamonds = late. Displacement fields for successive stages of Newtonian flow (n = 1) are parabolic curves (**a**), whereas displacement fields for successive stages of non-Newtonian flow (n = 3) are flat-crested curves (**b**).

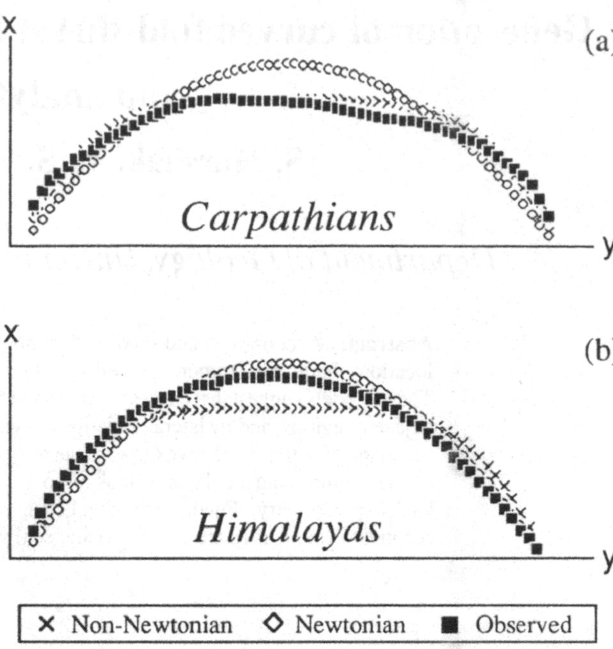

Figure 2. Comparison of antitaxial fold-thrust belt curves (defined by the map-view shape of the deformation front) to model curves produced by the analytical flow equation. On these plots, x is the flow direction (= structural vergence), and y is the distance from the endpoints. The model curves are generated by choosing a best-fit value for F in equation 1. Each plot shows the Newtonian (diamonds) and non-Newtonian (overlapping X's) curves for the specified F, and the simplified map trace of a real deformation front (filled squares). The Carpathians (**a**) match a non-Newtonian curve, and the Himalayas (**b**) match a Newtonian curve.

F, the only free parameter in the equation, is a constant described by $F^{-1} = C(dp/dx)^n t$, where n is the power-law exponent, t is the time since initiation of flow, dp/dx is the pressure gradient in the flow direction generated by a horizontal stress imbalance, and C is a proportionality constant related to fluid viscosity. Equation 1 generates curves (Fig. 1) for Newtonian flow if n = 1, or for non-Newtonian flow if n = 3 (only positive odd integer values produce real numbers). Movement of all points is parallel to the x-axis.

Observations. The purpose of this model is to illustrate how curve shape depends on the flow law (specified by the value of n in equation 1) describing the behaviour of the deforming material. For Newtonian flow, generated curves are parabolic, whereas for non-Newtonian flow, generated curves are flat-crested (Fig. 1; cf. Johnson 1970). With these two end-member curve geometries in mind, the shape of real antitaxial fold-thrust belts were examined to determine if actual orogenic curves are 'Newtonian' or 'non-Newtonian' in plan. To do this, the shape of an orogen from a published map was traced and this shape was used to specify a best-fit value for F. This value is inserted into equation 1 to generate Newtonian and non-Newtonian model curves which can then be compared to real curves. Both Newtonian and non-Newtonian types of curves appear to exist - the Carpathians, for example, fit a non-Newtonian curve, whereas the Himalayas fit a Newtonian curve (Fig. 2).

Glycerine models

Model setup. Two physical models of fluid flow were constructed to help understand how obstacle geometry might influence curve geometries (Fig. 3). Both models utilize a 2 cm-thick sheet of chilled glycerine (-78°C) in a 16 x 23 x 5 cm plastic tank; obstacles were cemented to the floor of the tank. In one model, obstacles were parallel channel walls composed of 1.2 cm-thick plexiglass sheets spaced 12.5 cm apart (Fig. 3a). In the second model, obstacles were made from 1.2 cm-diameter wooden cylinders (Fig. 3b). Flow was initiated by tilting the tank toward the foreland, and flow trajectories (particle paths) were defined by tracking the movement of rows of 0.7 cm-diameter paper circles placed on the surface of the glycerine.

Observations. The pattern of displacement in the channel model (Fig. 3c) closely resembles the pattern created by the analytical model; flow trajectories are parallel to the channel walls, and the marker lines arch into parabolas. The pattern of displacement displayed by the cylindrical-obstacle model, however, is markedly different (Fig. 3d). In this model, a trough develops on the foreland side of the obstacles as the glycerine is pushed aside by the obstacles during flow. A point on the fluid surface, therefore, is subjected to two topographic potentials - a regional topographic potential caused by tipping the tank, and a local topographic potential caused by the troughs that exist in the wake of the obstacles. Each paper circle moves both toward the foreland and laterally down into one of the two troughs, thereby describing a

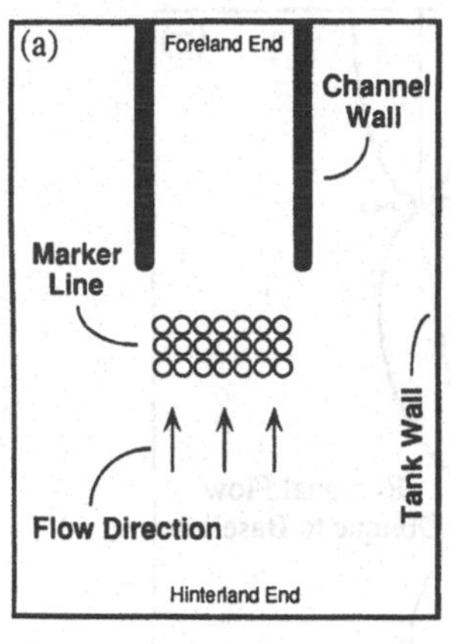

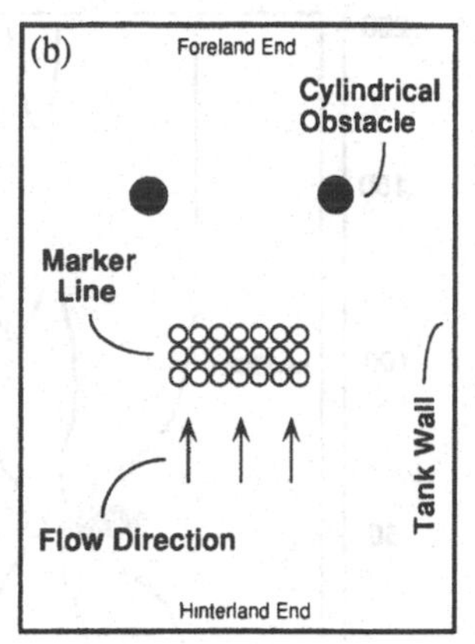

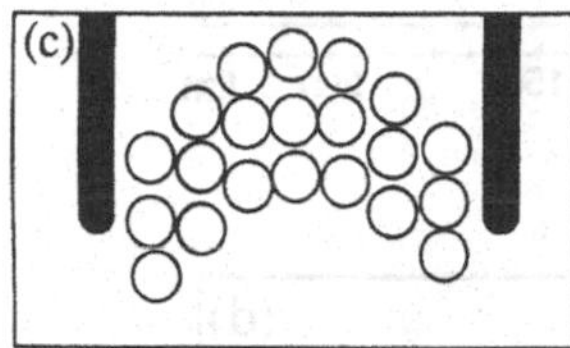

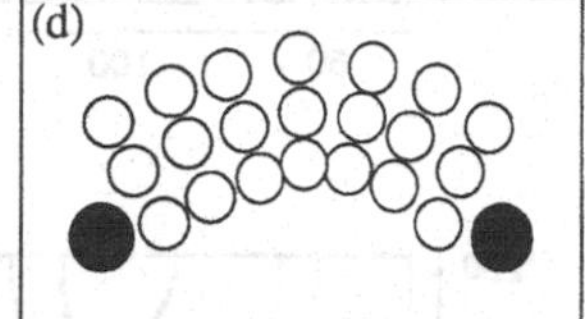

Figure 3. Map sketches of glycerine-tank models (approximate scale). Open circles represent paper markers on the surface of the 2 cm-thick glycerine layer. **(a)** Map of the tank configured for channel-flow simulation; **(b)** Map of the tank configured for cylindrical-obstacle interaction; **(c)** Close-up map showing the pattern of paper circles after interaction with the channel mouth. Note the parallel flow; **(d)** Close-up map showing the pattern of paper markers after interaction with the cylinders. Note the divergent flow.

curved trajectory. The marker line not only develops into a curve, but also undergoes tangential extension (Fig. 3d), and movement of a point along the line is roughly perpendicular to the curve at any time.

Finite-difference model of obstacle interaction

Model setup. To determine whether differences between flat-crested and parabolic curves are necessarily a consequence of rheology, as suggested by the analytical model, a finite-difference method was used to model crustal flow driven by a topographic gradient toward foreland obstacles (for model details, see Hsui *et al.* 1990a,b). At time = 0, the gradient is represented by an 800 m-high vertical step (Fig. 4a). Immediately after initiation of flow, the step transforms into a foreland-dipping slope and the fluid wedge migrates toward the foreland ultimately intersecting and flowing around two cylindrical obstacles that project above the surface of the fluid. While flow is occurring, the material behaves like a Newtonian fluid. This model allows us to determine if there is a relationship between curve shape resulting from obstacle interaction and strength of the deforming wedge. To simulate strength, the material composing the sheet is allowed to flow only if the slope of the wedge exceeds a *critical angle*; the larger the critical angle, the stronger the material.

Observations. After flow begins, the trace of the flow front is straight until it intersects the obstacles, at which time the front bends and the region between the two obstacles becomes a curved lobe (Fig. 4b). The curved lobe in the finite-difference model, as with the glycerine model (Fig. 3d), represents the interaction of a regional topographic potential with a local potential caused by obstacle interaction. The relative contribution of these two potentials at any given point along a curve controls the local flow trajectory and the overall curve shape. Results of calculations for different critical angles indicate that the shape of the wedge, as depicted by contour maps (Fig. 5) depends on the critical angle (i.e. strength). If the angle is small (< 2°), then the flow-front curve is flat-crested (Fig. 5a), whereas if the angle is large (5°), then the curve is parabolic (Fig. 5b). This contrast illustrates that both parabolic and flat-crested curves can be generated by flow of a material with a Newtonian rheology, and thus it is not necessary to call upon an *ad hoc* change of rheology to explain differently shaped curves (Hsui *et al.* 1990a,b). For a given critical angle, contours at the flow front are more parabolic than contours up the wedge slope, a pattern observed in real fold-thrust belts such as the Verkhoyansk Range (Fig. 6). The finite-difference model

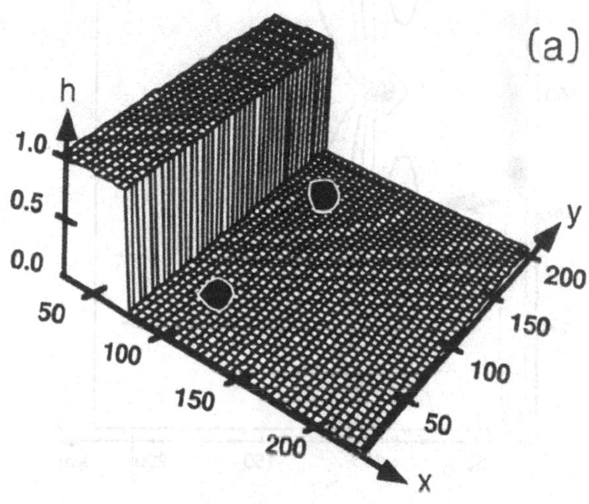

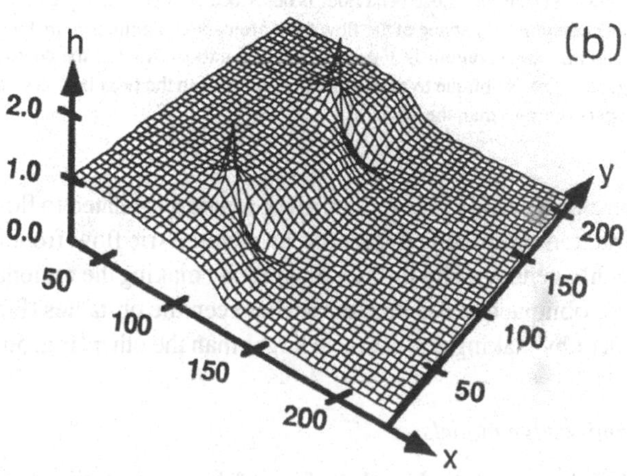

Figure 4. Perspective grid diagrams representing a finite-difference model of the interaction between a Newtonian material that flows toward the obstacles in the foreland. On the diagrams, all numbers refer to distances in kilometres (vertical exaggeration about 50X). h is elevation, x and y are horizontal directions. **(a)** Diagram of the initial state, when the material has a vertical step. Black dots represent the obstacles; **(b)** Diagram showing the shape of the flow front after about 3 million years. The front is curved (lobe-shaped) between the two obstacles.

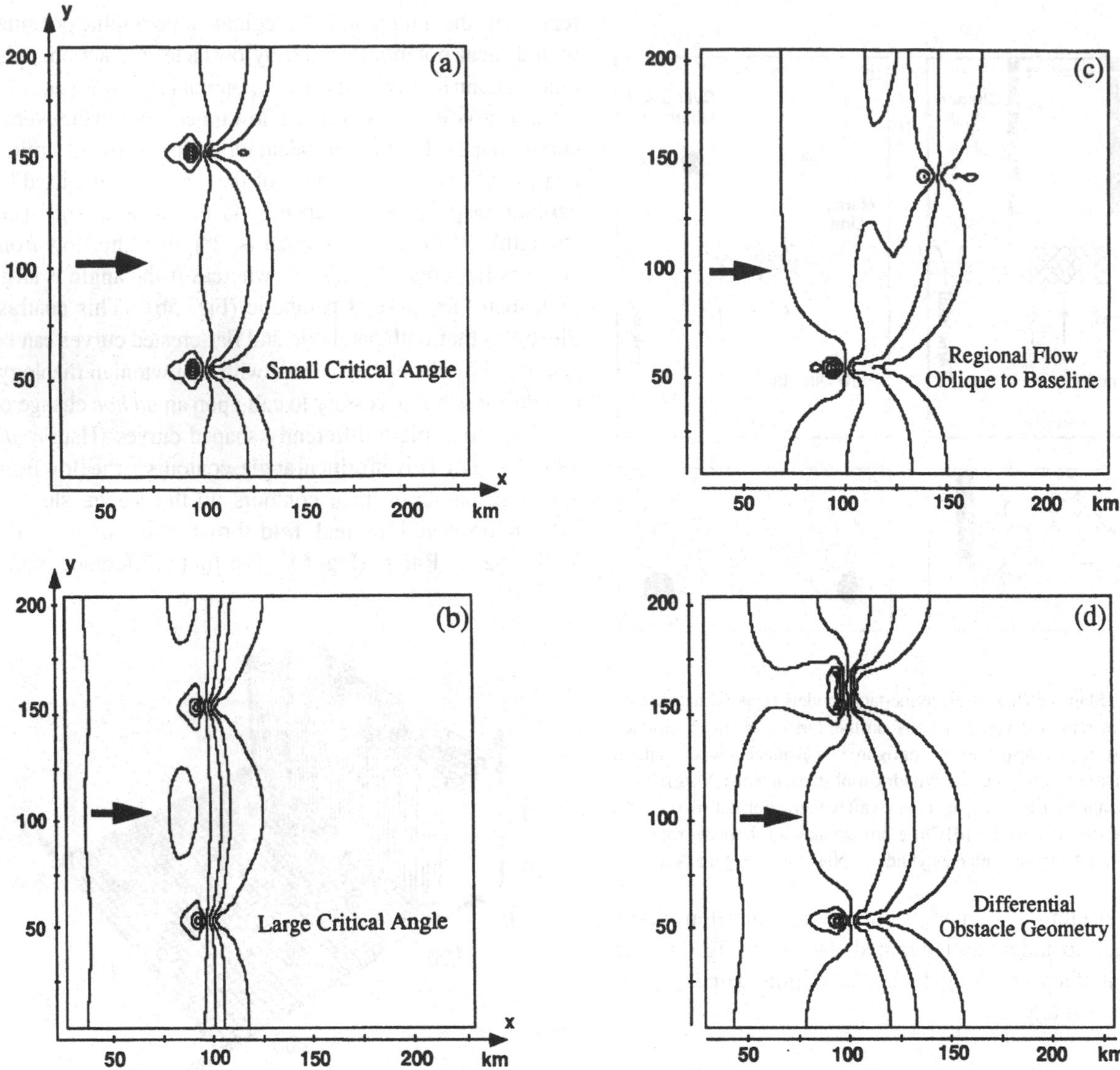

Figure 5. Form-line contour maps showing the shape of flow fronts generated by the finite-difference model. (**a**) Shape of the flow front after about 3 million years for a material whose behaviour is described by a critical angle of 0°. Contour interval is 333 m, lowest contour is 533 m. Note the flat-crested trace of the contours; (**b**) Shape of the flow front after about 3 million years for a material whose behaviour is described by a critical angle of 5°. Contour interval is 160 m, lowest contour is 360 m. Note the parabolic trace of the contours. (**c**) Shape of the flow front after about 3 million years for a model where the regional flow is oblique to a base line drawn between the two obstacles. (**d**) Shape of the flow front after about 3 million years for flow where one obstacle (upper) is larger than the other.

can also be used to show that flow geometry is related to flow direction or obstacle geometry; an asymmetric flow front is readily generated by the model either by making the regional flow oblique to a baseline drawn between the obstacles (Fig. 5c), or by making one obstacle larger than the other (Fig. 5d).

Sand-wedge models

Model setup. Sand wedges formed by compression have proven to be successful tools in developing an intuitive understanding of fold-thrust belt generation (e.g. Davis *et al.* 1983; Mulugeta & Koyi 1987; Mulugeta 1988). This paper focuses on models that illustrate development of fold-thrust belt geometries in map view (see also Burg *et al.* 1987; Davy & Cobbold 1988). The models employ commercial silica sand (0.3 - 0.6 mm diameter grains, bulk density of 1.57 g/ cm^3, internal friction angle of about 30°) that behaves as a Coulomb material. Sand is spread in a smooth 2 cm-thick layer across the base of a 75 x 120 x 30 cm wooden deformation box, so the dimensional scale factor is between 10^{-5} and 10^{-6}, similar to values used by Horsfield (1977) and McClay & Ellis (1987). Slightly dampening the sand made fault traces developed during shortening visually more distinct, though their shape was identical to those formed in dry sand. The capillary effect of the water films on the sand grains effectively creates negative pore pressure (see Lambe & Whitman 1979), a factor that does not affect the slope of the failure envelope. In order to maintain constant sand dampness during the runs, sand wedges were generated at a strain rate of about 0.5 cm/s. Though this rate is significantly faster than rigorous scaling dictates, studies suggest that the yield envelope of Coulomb materials is practically independent of

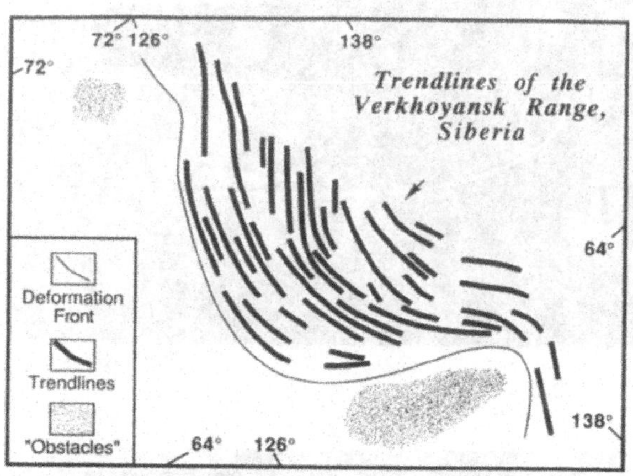

Figure 6. Sketch map of trendlines in the antitaxial Verkhoyansk Range, Siberia. The trendlines represent the orientation of structural grain, as indicated by traces of folds represented on the Geological Map of the Soviet Union (Nalivkin 1965). The deformation front represents the boundary between folded and unfolded strata. The obstacles are geological domes. Tectonic vergence, indicated by an arrow, is to the west-southwest.

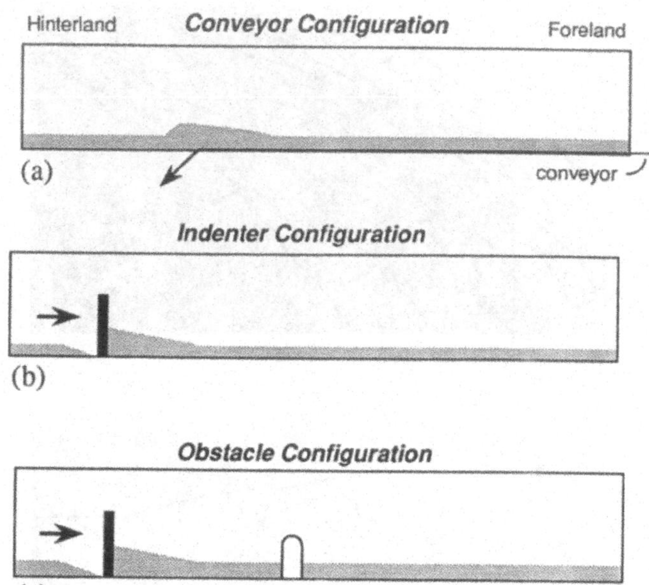

Figure 7. Schematic sketches (not to scale) depicting the three experimental configurations used for sand-wedge experiments. Each figure shows a cross section of the deformation box drawn parallel to the length of the box. Sand is represented by the stipple pattern, and the indenters are represented by thick vertical black lines. **(a)** Conveyor configuration. The sand wedge is generated by moving a mylar conveyor underneath the sand and out of a slit in the floor of the box. Arrows indicate movement direction. **(b)** Indenter configuration. The sand wedge is generated in front of the indenter moving toward the foreland (movement direction is indicated by the arrow). **(c)** Obstacle configuration. A pair of cylindrical obstacles are fixed to the floor of the box on a line oriented perpendicular to movement. This cross section passes through one of these obstacles (open column), and depicts the sand wedge prior to collision.

strain rate (e.g. McClay & Ellis 1987), and thus the geometries generated in these experiments can provide insight into the geometries of real fold-thrust belts. The box was tilted so that its floor sloped at about 4° in the direction opposite to thrust vergence. For purposes of reference, the higher end of the box is called the *foreland end* and the lower end of the box is called the *hinterland end.* Three different model configurations (Fig. 7) were used.

Basal conveyor configuration. In this configuration (Fig. 7a), a 2 cm-thick sand layer is placed over a 75 cm-wide by 0.003 mm-thick mylar conveyor belt which lines the floor of the box. The belt enters through a slit at the base of the foreland wall and exits through a slit in the box floor, 45 cm from the hinterland wall. A fold-thrust wedge is generated by moving the mylar conveyor toward the hinterland, so that the unmoving sand of the hinterland serves as the backstop (cf. Davis *et al.* 1983). The deformation box also has a 20 cm-long slit cut into its floor. By pulling a mylar strip through this slit, a narrow belt of sand moves differentially with respect to the sand on either side.

Indenter configuration. To simulate the evolution of a fold-thrust wedge in advance of an indenter, the sand layer was pushed with a vertical plexiglass sheet in contact with the box floor (Fig. 7b). Experiments were repeated with three indenters (18 cm, 25.5 cm, and 31 cm wide by 20 cm high). To examine the effects of sediment-thickness variation on the localization of curves, 20 x 1.5 cm notches were cut from each side of a 55 cm-wide indenter, thereby leaving only a 15 cm segment in the centre that remained in contact with the base of the box (Fig. 8).

Obstacle configuration. This configuration simulated the interaction between a growing sand wedge and stationary obstacles in the foreland (Fig. 7c). In this model, two 5 cm-diameter glass cylinders spaced 22 cm apart represented obstacles. A straight sand wedge was pushed toward the obstacles by a 40 cm-wide plexiglass indenter.

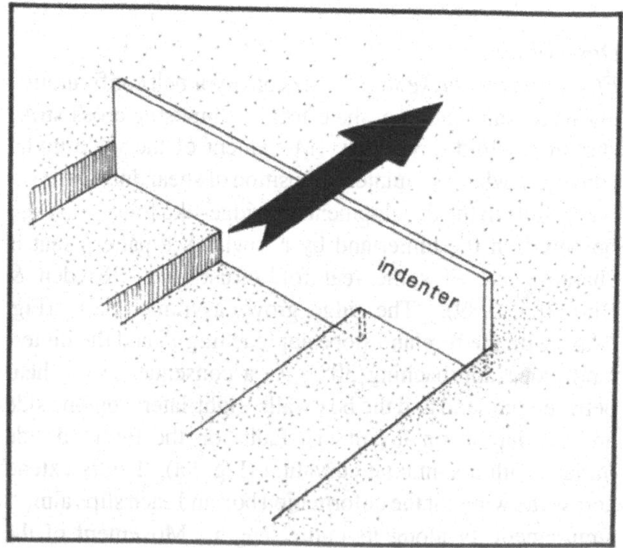

Figure 8. Sketch depicting the notched indenter as it is displaced toward the foreland (motion is indicated by arrow). This view is from the rear. In this diagram, steps in the sand are depicted with vertical faces to emphasize the shape of the indenter; in reality, they slump into the central trough.

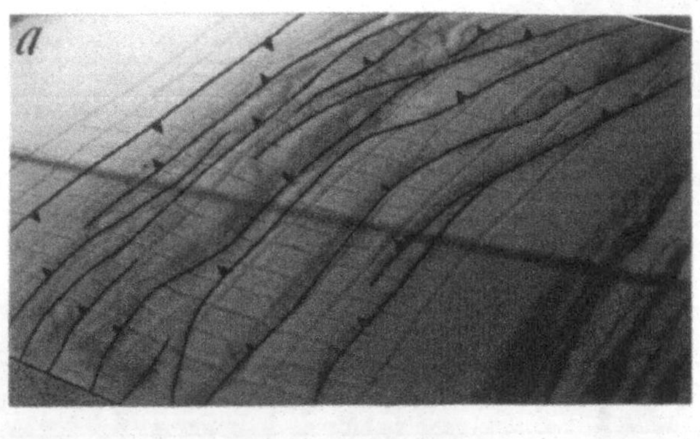

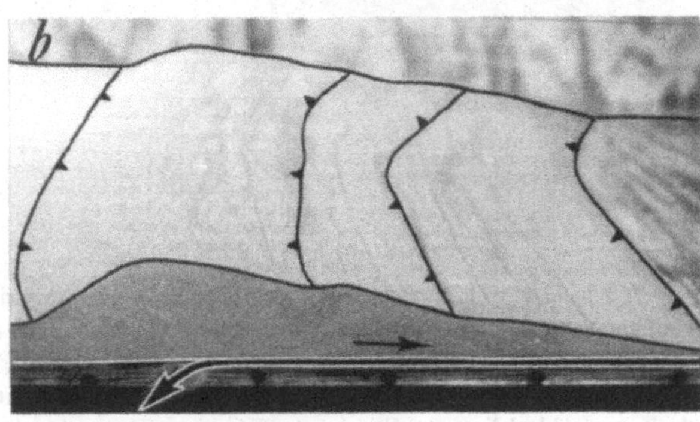

Figure 9. Photos of sand-wedges generated by movement of the conveyor. Toothed lines emphasize thrust traces. The wedge is about 50 cm wide. (a) Oblique view of the surface of the sand after displacement of the conveyor. The foreland direction is to the right and faults verge toward the foreland. The dark grey line across the middle of the picture is the top of the plexiglass wall of the box. Fault traces define a syntaxial curve; (b) Cross section of the sand wedge. The exit slit is indicated by the line and arrow, and the foreland direction is to the right. Note the asymmetry of the wedge.

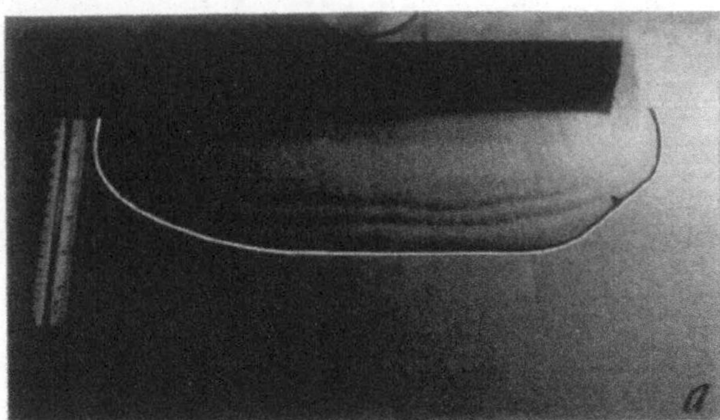

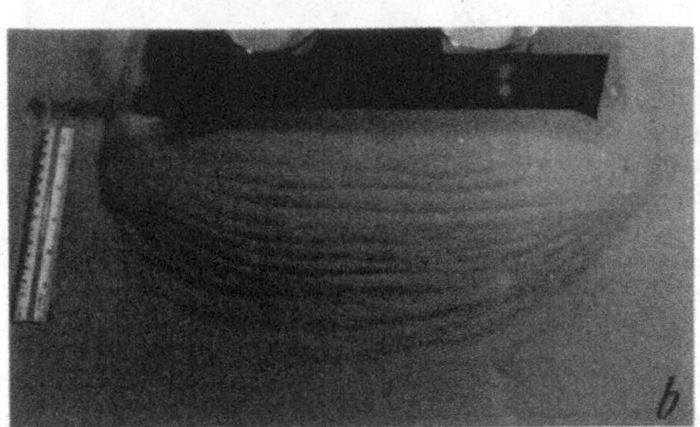

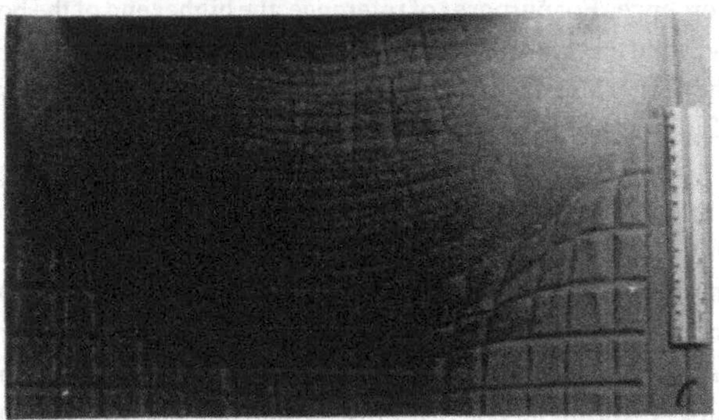

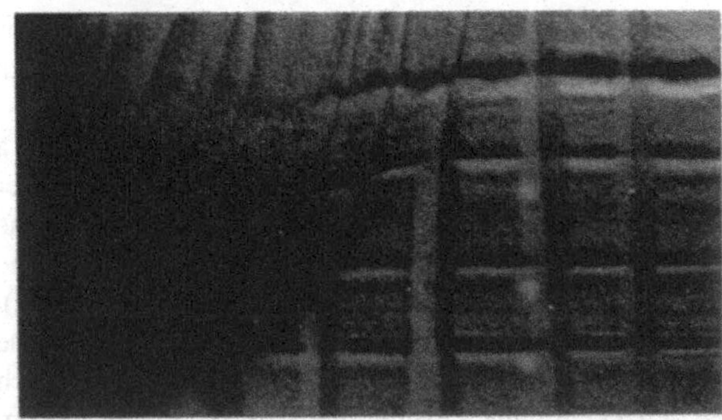

Figure 10. Photos of sand wedges developed in advance of an indenter. The indenter is the dark face near the top of each photo. Scale is 15 cm long. All photographs are map view. (a) Early stage in wedge development. Note that the early formed thrusts (indicated by steps in the sand surface) are elliptical. (b) Intermediate stage in wedge development. Note that the younger thrusts are less eccentric than the older thrusts. (c) Late stage in wedge development. Note the overall parabolic shape of the deformation front. Oldest faults (adjacent to the indentor face) are flat-crested. The grid roughly shows strain distribution in the map plane. (d) Close up of the strain grid along the side of the wedge.

Observations

Effect of cross-strike shear. The conveyor belt configuration allowed examination of the effects of imposing cross-strike shear on a fold-thrust belt. Movement of the wide mylar conveyor, which simulates imposition of shear during thrusting, results in the development of a ridge-like fold-thrust belt bordered on the hinterland by a lowland, a pattern that is characteristic of some real fold-thrust belts (Royden & Burchfiel 1989). The ridge grows asymmetrically (Fig. 9a,b), with the foreland slope approaching 7° and the hinterland slope approaching 30°. As a consequence of shear between the sand and the box walls (right-lateral on one side and left-lateral on the other), faults on the foreland side initiate with a syntaxial curvature (Fig. 9a). Faults extend across the width of the deformation box and each slips almost simultaneously along its entire length. Movement of the narrow mylar conveyor beneath a linear sand wedge simulates imposition of shear on a pre-existing fold-thrust belt. After movement, the straight fault traces were bent into antitaxial curves, one above each edge of the mylar conveyor. The curve adjacent to the left-lateral shear zone is similar in appearance to the curve of the Makran fold-thrust belt where it intersects the Chaman fault in Pakistan (see Fig. 15 in

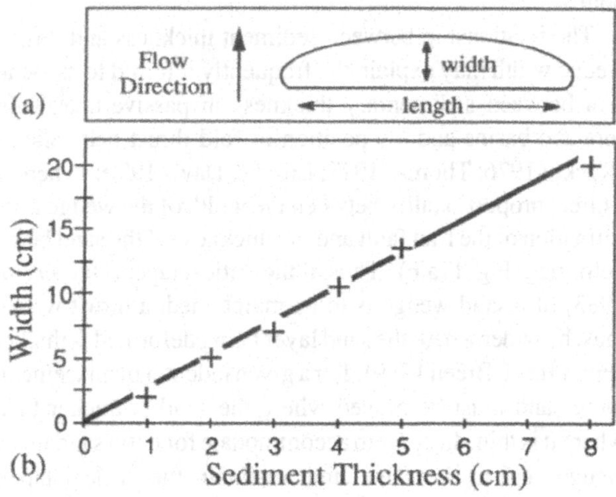

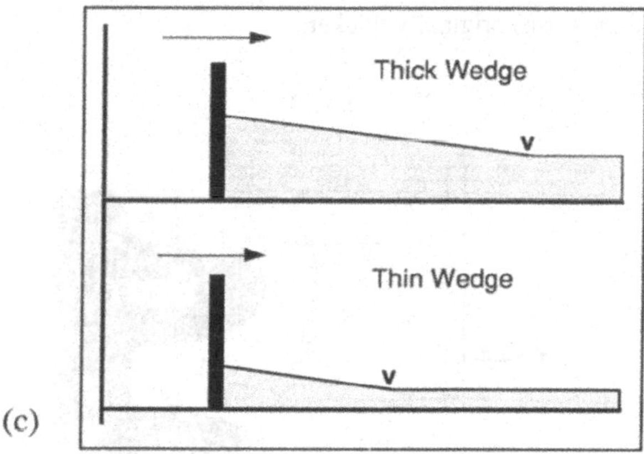

Figure 11. Relationship between sand thickness and 'width' (the distance between the indenter face and the trace of the first formed thrust). **(a)** Schematic map showing the definition of width. On the right side of the diagram is an elliptical fault trace in front of a straight indenter face. Length is the distance between endpoints. The movement direction is indicated by the arrow; **(b)** Graph of width as a function of sediment thickness for six experiments; **(c)** Schematic cross section emphasizing why the distance between the toe of the wedge (marked by 'v') and the indenter face depends on sand thickness. Shaded area is sand; the thick black line is the indenter, which moves in the direction indicated by the arrow. The slope of both wedges is the same.

Marshak 1988).

Indenter-generated curves. Pushing into the sand with an indenter leads to the development of an antitaxial fold-thrust wedge, and allowed observation of how curved structures initiate and evolve. Following the first increment of movement, a fault develops in advance of the indenter (Fig. 10a). The shape of the fault trace depends on indenter length (measured perpendicular to movement direction) relative to sand thickness. In experiments with a 2 cm-thick sand layer, the first fault formed in front of the 18 cm-long indenter initiates as an elliptical arc (spoon-shaped in three dimensions; cf. Davy & Cobbold 1988), whereas the first fault formed in front of the 31 cm-long indenter was curved near the ends of the indenter, but was parallel to the indenter face away from the ends. With progressive displacement into the foreland, the width (measured parallel to the movement direction) of the thrust wedge increases and its interior thickens as sand is added to the wedge toe. For the 18 cm-long indenter, each successively younger fault trace initiates with a progressively more curved trace, defining the arc of an ellipse with less eccentricity (Fig. 10b), until eventually, the curve shape defined by the thrust front resembles a parabola (Fig. 10c). For the 31 cm-long indenter, the initial flat-crested curve evolved into an ellipse as the sand in the wedge thickened (thereby decreasing the length to thickness ratio) and then ultimately into a parabola.

The initial displacement associated with fault slip results in a slight component of radial movement (Figs 10c,d; see also Fig. 2 in Davy & Cobbold 1988). Grid lines that were originally oriented perpendicular to the indenter face bow outwards and diverge at the wedge toe (cf. Coward & Potts 1983). As new thrust sheets are incorporated at the wedge toe, older sheets shorten in the direction of vergence and undergo extension parallel to the indenter face. This strain is indicated in the central part of the wedge by an increase in the spacing of cross-strike grid lines and a decrease in the spacing of strike-parallel grid lines (Fig. 10c). Curvature of thrust faults, therefore, effectively decreases as the faults migrate up the slope of the wedge. Close to the face of the indenter and away from its ends, cross-strike grid lines do not diverge, because these faults were originally flat-crested in plan.

Effect of sand-thickness variation. Study of the relationship between initial sand thickness (H) and the width (W) of the thrust wedge immediately after initiation of the first fault indicates that the two parameters are directly proportional to one another (Fig. 11a,b) as defined empirically by the relation: $W = 2.75H$. This proportionality suggests that the location of a curve is affected by lateral variations of sand thickness in the foreland and that, once formed, the evolution of a curve is controlled in part by lateral variations in the thickness of the wedge itself.

To physically illustrate how lateral variation in sand thickness localizes curves, sand was pushed with an indenter that had matching notches removed from both sides along its base (Fig. 8). The full thickness of sand was displaced, therefore, only adjacent to the central part of the indenter, where the indenter contacted the box floor; adjacent to the notches, only three-quarters of the sand thickness was displaced. When the indenter moves, the sand wedge formed adjacent to the notches is relatively narrow (as measured

Figure 12. Photo of a sand lobe formed in front of the notched indenter sketched in Figure 8. This photo is taken looking vertically down on the surface, and movement is from the top to the bottom of the photo. The lobe, which is 15 cm-long measured parallel to the indenter face, forms where the sand being pushed by the indenter is thicker. Note that the spacing of faults depends on wedge width.

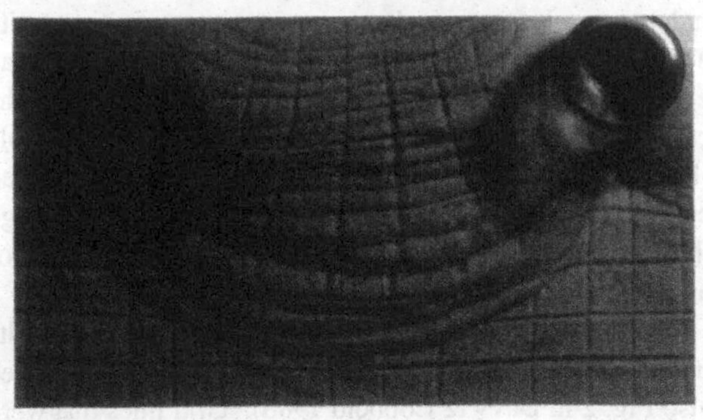

Figure 13. Photo showing the sand lobe developed on the foreland side of two glass obstacles. The curvature of originally straight faults to the hinterland of the obstacles was imposed after collision, and the curvature of the faults to the foreland of the obstacles is primary. Movement is from the top to the bottom of the photo.

perpendicular to the wedge) and contains closely spaced faults, whereas the sand wedge formed adjacent to the central part of the indenter is wider and contains more widely spaced faults (Fig. 12). An antitaxial lobe developed, therefore, where the fold-thrust belt propagated into a band of thicker sediment. Fault traces in the lobe are curved and connect to traces of the narrow wedges.

Collision with obstacles. Upon collision of an initially straight thrust wedge with paired obstacles, pre-existing faults bend into antitaxial curves bounded by strike-slip shear zones along the inner sides of the obstacles (Fig. 13). Continued foreland displacement of the indenter results in fault generation to the foreland side of the obstacles. These new faults form with an initial curvature, just like faults in advance of an indenter.

Discussion and conclusions

What controls the location of curves?

Modelling emphasizes that curves form in response to: (1) differential displacement between a portion of a fold-thrust belt and neighbouring regions, caused, for example, by collision with obstacles, impact of an indenter, or interaction with strike-slip faults; (2) an inhomogeneous stress field characterized by smoothly curving stress trajectories (Fig. 14; see Laubscher 1972; Beutner 1977; Jaeger & Cook 1979); (3) along-strike variations in sediment thickness - where the sediments are thicker, thrusts initiate further to the foreland and the thrust wedge is wider; and (4) topographic and/or thickness variations in the fold-thrust belt hinterland, which affect the gravitational-potential component of the fold-thrust belt driving force. All of the above circumstances cause variation in slip trajectory and magnitude as a function of location in a fold-thrust belt. Real orogenic curves, as well as curves illustrated by the models in this paper, are smooth and continuous from end to end (i.e. they lack irregular steps and are not composed of multiple sub-lobes), suggesting that variation in slip trajectory and magnitude are continuous at both scales.

The relationship between sediment thickness and thrust-wedge width may explain the frequently referred to association between sedimentary thickness in passive-margin or foreland basins and the position of fold-thrust belt salients (Rankin 1976; Thomas 1977; Lillie & Davis 1990). There is a direct proportionality between the width of the wedge after formation of the first fault and the thickness of the sand being deformed (Fig. 11a,b). Thus, if the critical taper (Davis *et al.* 1983) of a sand wedge is to be maintained, a thrust wedge must be wider where the sand layer being deformed is thicker (Fig. 11c; cf. Breen 1989). For a given indenter displacement, more sand must be moved where the sand is thicker than where it is thin. In order to accommodate for extra sand in the wedge, and at the same time maintain the critical taper, faulting propagates further into the foreland (Fig. 11c). Therefore, an antitaxial lobe develops over the region where sediment was originally thicker.

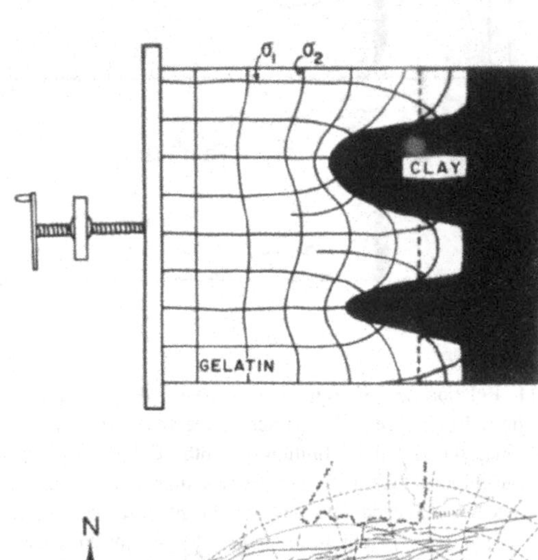

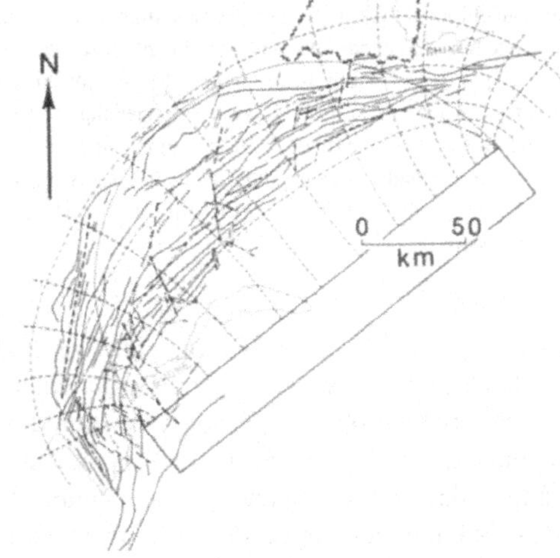

Figure 14. Elastic models of fold-thrust belt curvature. (**a**) Obstacle model of Beutner (1977) which shows curved stress trajectories where an elastic sheet (gelatin) pushes against obstacles (clay); (**b**) Indenter model of Laubscher (1972) which shows curved stress trajectories resulting from indentation of an elastic half space by a block of finite length. The stress trajectories (dashed lines) are superimposed on a map of structural features in the Jura Mountains. The maximum principal stress trajectories are perpendicular to structural trends.

What controls the shape of curves?

Deformation fronts of antitaxial curves range in shape from parabolic (e.g. the Himalayas) to flat-crested (e.g. the Carpathians) in plan. Furthermore, within a given belt, the foreland fault traces tend to be more parabolic than hinterland fault traces (Figs 6, 10c, 14b). Elastic models suggest that such a pattern reflects the geometry of the instantaneous stress field (Fig. 14). The analytical parallel-flow model suggests that the shape of a curve (parabolic vs. flat-crested) may also depend on the rheology of the material being deformed (cf. Johnson 1970) which could reflect the deformation mechanism contributing to strain development in a belt (e.g. pressure-solution mechanisms in shallow belts, dislocation glide/climb in deeper belts; Spratt 1983).

A major difficulty with the analytical model is the requirement that displacement trajectories of all points on the marker line be straight and mutually parallel. Since the work of Seuss (1904) and Argand (1924; Fig. 15), it has been recognized that movement trajectories in real orogens may be divergent. Modern kinematic data from real orogens suggest that movement at any given point along the curve is typically perpendicular to the curve; for example, the P-axes of fault-plane solutions for earthquakes in the Himalaya are perpendicular to the range front (Seeber 1984) and slip lineations on thrust faults in the Wyoming salient are typically perpendicular to the regional map trace of the faults (Crosby 1969). Also, considering the structural complexity that is locally characteristic of fold-thrust belts (e.g. Coward & Potts 1983; Marshak 1986), motion of discrete blocks within a belt may be somewhat chaotic, thereby leading to non-parallel movement especially near the endpoints of curves.

Finite-difference models suggest that the contrast between parabolic and flat-crested curves need not reflect rheology, but alternatively may reflect the overall strength of the fold-thrust belt (perhaps controlled by lithology, degree of fracturing, or deformation conditions). It was found that curve shape changes as a function of the critical angle, a term representing strength in the model. As the critical angle increases, the flow front becomes more parabolic. This model suggests that antitaxial curve shape reflects interaction between a regional topographic potential driving the orogen into the foreland (e.g. caused by plate convergence or overthickening of crust) and local topographic potentials caused by restraints on movement (e.g. obstacles, strike-slip faults).

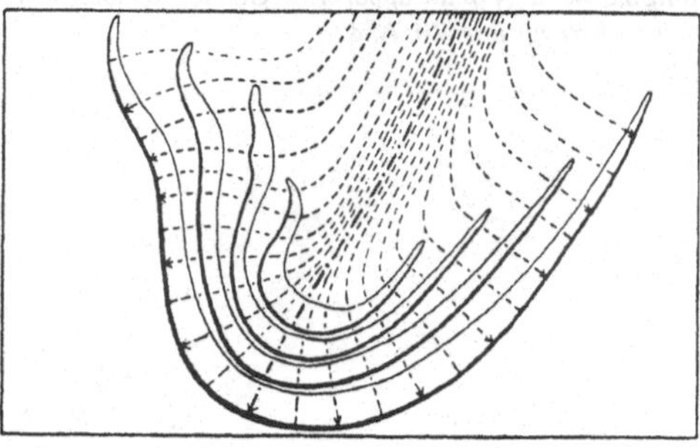

Figure 15. Reproduction of a sketch from Argand (1924) showing his concept of a 'double virgation of the first type' (i.e. divergent flow in a curved orogen). The arrows indicate assumed flow lines.

The shape of the deformation front of an antitaxial curve changes as the fold-thrust belt grows into the foreland. All of the models (e.g. Figs 3 & 10) show that the interlimb angle of the curves decreases as the deformation front propagates into the foreland. The sand model, in addition, suggests that during early stages in the formation of antitaxial curves, curve geometry is controlled by the ratio between the length of the indenter and the thickness of the sediment being deformed. Qualitatively, if the ratio is very small, then the curve initiates as a parabola, if it is intermediate, then the curve initiates as an arc of an ellipse, and if the ratio is large, the fault is curved near the ends of the indenter but is parallel to the indenter face away from the ends. Initially flat-crested deformation fronts evolve into arcs of ellipses as the thrust wedge grows. Each new fault at the toe initiates with progressively lower eccentricity, until ultimately the deformation front resembles a parabola. Once formed, a thrust sheet is subjected to pure shear as new thrust sheets add to the toe of the wedge. Therefore, the curvature of a given fault decreases with time. This effect coupled with change in fault shape as a function of distance from the indenter face means that curves towards the hinterland are flatter than curves toward the foreland, a phenomenon that is observed in many antitaxial fold-thrust belts (e.g. the Verkhoyansk and Jura Ranges; Figs 6 & 14b). The finite-difference model also predicts such variation in curve geometry (Fig. 5) as a function of position in the wedge.

Do faults initiate with curved traces, or are curves secondary?

One of the principal debates concerning curved fold-thrust belts concerns the question of whether curves are primary (i.e. initiate with their present structural trends) or are secondary (i.e. are oroclines). The models in this paper emphasize that true oroclines can form due to superposition of cross-strike shear on an originally straight fold-thrust belt, and due to collision with obstacles in the foreland (see Carey 1955 and Marshak 1988 for discussion of other settings in which oroclines form). In the case of collision with obstacles, only structures that had formed prior to collision are bent. Primary curves can form in advance of an indenter, to the foreland of obstacles in collisional zones, or as a consequence of along-strike variations of sediment thickness in the foreland. Curves like the Pennsylvania salient, for example, which are spatially associated with thicker strata deposited in a continental re-entrant, may have initiated with a curved trajectory and need not reflect significant oroclinal bending. Palaeomagnetic data so far have not demonstrated large amounts of bending in this salient, and structural data have not shown major along-strike extension (Schwartz & Van der Voo 1984; Miller & Kent 1986). The initial movement on a curved fault at the toe of the thrust wedge results in some tangential extension and some rotation, but the amount is less than would accompany the development of a curved fault trace from an originally straight fault and may explain the lack of reported tangential extension in many natural orogenic curves.

Partial support for this work was provided by NASA grant NAGS-1312 to Hsui, NSF grant EAR-84-07785 to Marshak, and an NSF graduate fellowship to Wilkerson. We wish to thank P. R. Cobbold and K. R. McClay for helpful reviews of the manuscript, R. D. Hatcher, Jr. for emphasizing the importance of sediment thickness to us, C. Hedlund for help with the sand-wedge modelling, and G. P. Salisbury for suggestions concerning curve nomenclature.

References

Argand, E. 1924 (1977 translation by A. V. Carozzi). *Tectonics of Asia*. Hafner Press, New York, 218p.

Beutner, E. C. 1977. Causes and consequences of curvature in the Sevier orogenic belt, Utah to Montana. *In: Wyoming Geological Association Twenty-Ninth Annual Field Conference Guidebook*. Wyoming Geological Association, 353-365.

Breen, N. A. 1989. Structural effect of Magdalena fan deposition on the northern Columbia convergent margin. *Geology*, **17**, 34-37.

Burchfiel, B. C. 1980. Eastern European Alpine system and the Carpathian orocline as an example of collision tectonics. *Tectonophysics*, **63**, 31-61.

Burg, J. P., Bale, P., Brun, J. P. & Girardeau, J. 1987. Stretching lineation and transport direction in the Ibero-Armorican arc during the Siluro-Devonian collision. *Geodinamica Acta*, **1**, 71-87.

Carey, S. W. 1955. The orocline concept in geotectonics. *Papers and Proceedings of the Royal Society of Tasmania*, **89**, 255-288.

Coward, M. P. & Potts, G. L. 1983. Complex strain patterns developed at the frontal and lateral tips to shear zones and thrust zones. *Journal of Structural Geology*, **5**, 383-400.

Crosby, G. W. 1969. Radial movements in the western Wyoming salient of the Cordilleran overthrust belt. *Geological Society of America Bulletin*, **80**, 1061-1078.

Davis, D., Suppe, J. & Dahlen, F. A. 1983. Mechanics of fold-and-thrust belts and accretionary wedges. *Journal of Geophysical Research*, **88**, 1153-1172.

Davy, P. & Cobbold, P. R. 1988. Indention tectonics in nature and experiment. 1. Experiments scaled for gravity. *Bulletin of the Geological Institute of the University of Uppsala, N.S.*, **14**, 129-141.

Eldredge, S., Bachtadse, V. & Van der Voo, R. 1985. Paleomagnetism and the orocline hypothesis. *Tectonophysics*, **119**, 153-179.

Horsfield, W. T. 1977. An experimental approach to basement controlled faulting. *Geologie en Mijnbouw*, **56**, 363-370.

Hsui, A. T., Wilkerson, M. S. & Marshak, S. 1990a. Generation of non-Newtonian-like mountain belt curvatures within a Newtonian crust (abst.). *Transactions of the American Geophysical Union (EOS)*, **71**, 637.

——, —— & —— 1990b. Topographically driven crustal flow and the generation of curved fold-thrust belts. *Geophysical Research Letters*, (in press).

—— & Youngquist, S. 1985. A dynamic model of the curvature of the Mariana Trench. *Nature*, **318**, 455-457.

Jaeger, J. C. & Cook, N.G.W. 1979. *Fundamentals of rock mechanics*. Chapman & Hall, London, 593p.

Johnson, A. M. 1970. *Physical processes in geology*. Freeman, Cooper & Company, San Francisco, 577p.

Lambe, T. W. & Whitman, R. V. 1979. *Soil mechanics, SI version*. John Wiley & Sons, New York, 553p.

Laubscher, H. P. 1972. Some overall aspects of Jura dynamics. *American Journal of Science*, **272**, 293-304.

Lillie, R. J. & Davis, D. M. 1990. Structure and mechanics of the foldbelts of Pakistan. *In:* McClay, K. R. (ed.) *Thrust Tectonics 1990, Programme with Abstracts*, Royal Holloway and Bedford New College, University of London, (4-7 April, 1990), 69.

Lowrie, W., & Hirt, A. M. 1986. Paleomagnetism in arcuate mountain belts. *In:* Wezel, F.-C. (ed.) *The Origin of Arcs, Developments in Geotectonics 21*. Elsevier, 141-158.

Marshak, S. 1986. Structure and tectonics of the Hudson Valley fold-thrust belt, eastern New York State. *Geological Society of America Bulletin*, **97**, 354-368.

—— 1988. Kinematics of orocline and arc formation in thin-skinned orogens. *Tectonics*, **7**, 73-86.

—— & Tabor, J. R. 1989. Structure of the Kingston orocline in the Appalachian fold-thrust belt, New York. *Geological Society of America Bulletin*, **101**, 683-701.

McClay, K. R. & Ellis, P. G. 1987. Analogue models of extensional fault geometries. *In:* Coward, M. P., Dewey, J. F. & Hancock, P. L. (eds) *Continental Extensional Tectonics*. Geological Society Special Publication, 109-125.

Miller, J. D. & Kent, D. V. 1986. Paleomagnetism of the Upper Devonian Catskill Formation from the southern limb of the Pennsylvania salient: Possible evidence of oroclinal rotation. *Geophysical Research Letters*, **13**, 1173-1176.

Mulugeta, G. 1988. Modelling the geometry of Coulomb thrust wedges. *Journal of Structural Geology*, **10**, 847-859.

—— & Koyi, H. 1987. Three-dimensional geometry and kinematics of experimental piggyback thrusting. *Geology*, **15**, 1052-1056.

Nalivkin, D. V. (editor-in-chief) 1965. *Geological map of the Union of Soviet Socialist Republics*, sheet 7. Ministry of Geology of the USSR, 1:2,500,000.

Rankin, D. W. 1976. Appalachian salients and recesses: Late Precambrian continental breakup and the opening of the Iapetus Ocean. *Journal of Geophysical Research*, **81**, 281-288.

Ries, A. C. & Shackleton, R. M. 1976. Patterns of strain variation in arcuate fold belts. *Philosophical Transactions of the Royal Society of London A*, **282**, 281-288.

Royden, L. & Burchfiel, B. C. 1989. Are systematic variations in thrust belt style related to plate boundary processes? *Tectonics*, **8**, 51-62.

Schwartz, S. Y. & Van der Voo, R. 1984. Palaeomagnetic evaluation of the orocline hypothesis in the central and southern Appalachians. *Geophysical Research Letters*, **10**, 505-508.

Seeber, L. 1984. Large components of rotation and/or longitudinal stretching are predicted for the Himalayan accretionary wedge (abst.). *International Association of Seismology and Physics of the Earth's Interior, Regional Assembly*.

Seuss, E. 1904. *The Face of the Earth*. Clarendon Press, 604p.

Spratt, D. A. 1983. Deformation mechanisms associated with the initiation and propagation of thrust faults (abst.). *Transactions of the American Geophysical Union (EOS)*, **64**, 318.

Thomas, W. A. 1977. Evolution of Appalachian-Ouachita salients and recesses from reentrants and promontories in the continental margin. *American Journal of Science*, **277**, 1233-1278.

Turcotte, D. L. & Schubert, G. 1982. *Geodynamics applications of continuum physics to geological problems*. John Wiley & Sons, New York, 450p.

Wilkerson, M. S., Marshak, S. & Hsui, A. T. 1988. Dynamic modelling of an orocline in the Idaho-Wyoming fold-thrust belt: Implications for modelling the rheology of the upper crust. *Geological Society of America Abstracts with Programs*, **20**, A58.

PART THREE

Thrust geometries and thrust systems

Thrust faults in inverted extensional basins

K. R. McClay & P. G. Buchanan

Department of Geology, Royal Holloway and Bedford New College, University of London, Egham, Surrey, England TW20 0EX, UK

Abstract: Thrust faults in inverted extensional basins may not exhibit the simple low-angle, ramp-flat thrust trajectories that are characteristic of many foreland fold and thrust belts, but are expected to show more complex geometries controlled by the architecture of the earlier extensional fault system. The results of sandbox analogue models are reviewed as guides to possible thrust fault architectures in contracted extensional basins. High-angle thrust faults and thrust faults with convex-upwards shapes are characteristic of inverted extensional fault systems. Footwall-vergent shortcut thrusts and hangingwall-vergent backthrusts are characteristic of the inversion models. Fault sequences are complex and do not follow simple 'footwall-nucleating - footwall-vergent' rules. The architecture of the pre-existing extensional fault system exerts a strong control on the character of thrust faults in inverted basins and must be taken into account when analysing such regions.

Inversion tectonics involves a switch in tectonic mode from extension to compression such that extensional basins are contracted and become regions of positive structural relief. This process is generally accepted to involve the reactivation of pre-existing extensional faults such that they undergo reverse slip and eventually may become thrust faults (Cooper & Williams 1989; Williams *et al.* 1989). Pre-existing extensional basins have been incorporated into many fold and thrust belts - in the Alps (Butler 1989; de Graciansky *et al.* 1989; and others), in the Pyrenees (Puigdefàbregas *et al.* 1986; Berastegui *et al.* 1990; Martinez *et al.* 1989) and the Canadian Cordillera (McClay *et al.* 1989). The architecture of thrust systems in inverted extensional basins will be influenced and in some cases strongly controlled by the pre-existing extensional fault geometries (cf. Cooper & Williams 1989; Coward *et al.* 1991; McClay *et al.* 1989; Hayward & Graham 1989). In such situations thrust faults have geometries, trajectories and sequences that are significantly different to those found in many foreland fold thrust belts which deform platformal strata (e.g. Dahlstrom 1970; Boyer & Elliott 1982).

Inversion geometries have been commonly modelled as involving complete reactivation of the basin forming extensional faults such that the extensional sedimentary wedge is pushed up - 'inverted' above the pre-extensional 'regional elevation' (Fig. 1, Bally 1984). Such conceptual models of inversion ignore the mechanics of the inversion process and do not necessarily resemble natural inversion geometries (Fig. 2). It is clear that the range in structural styles and geometries resulting from the inversion and contraction of previously distended crust is likely to be far more complex than the relatively simple imbricate thrust geometries in areas of layer cake stratigraphy (Boyer & Elliott 1982). This paper aims to illustrate thrust geometries in inverted extensional basin systems at various scales and to demonstrate, using sandbox analogue models, the geometries and mechanics of thrusting during inversion.

Thrust systems in inverted basins

Thrust systems in inverted basins vary from small, high-angle thrust faults in basins which have undergone only small amounts of contraction to large, low-angle thrust systems which override and obliterate the pre-existing extensional architecture (Fig. 2). Extensional fault systems which have undergone low amounts of contraction (inversion) commonly show reverse movement on only a few of the faults that could possibly have undergone reactivation (Fig. 2a). For example the inversion structures across the South Hewett fault zone in the southern North Sea show a growth anticlinal structure with the development of a number of small footwall shortcut thrusts and only localized reactivation of extensional faults in the main domino array within the basin (Fig. 2a). Growth folding (i.e. the fold develops in strata that are being deposited at the same time as the fold grows, e.g. Suppe *et al.* (1991, this volume)) is typically developed with thinned growth strata (Suppe *et al.* 1991, this volume) deposited above the reactivated basin margin fault and thicker stratigraphic intervals deposited on the flanks of the structure (Fig. 2a). Within the inverted domino fault array the thrust faults are typically small, high-angle faults which show net extensional displacement at depth. In examples where listric extensional faults have been reactivated during inversion, hangingwall anticlines are developed and 'pop-up' structures in the region of the crestal collapse graben (McClay 1990) are characteristic (Fig. 2b). At greater values of contraction reactivated extensional faults propagate into the post-rift and syn-inversion sequences, flattening upwards as they do so (Fig. 2c). Complex 'arrowhead' or 'harpoon' geometries result from the wedge of syn-rift sediments being elevated above regional. Where extensional half-grabens have been subjected to greater contraction, thrust faults may either transect across the half-graben thus dissecting it (Fig. 2d) or new thrust faults will develop out of the half-graben (Fig. 2d). In cases of strong contraction pre-existing extensional faults may also be rotated

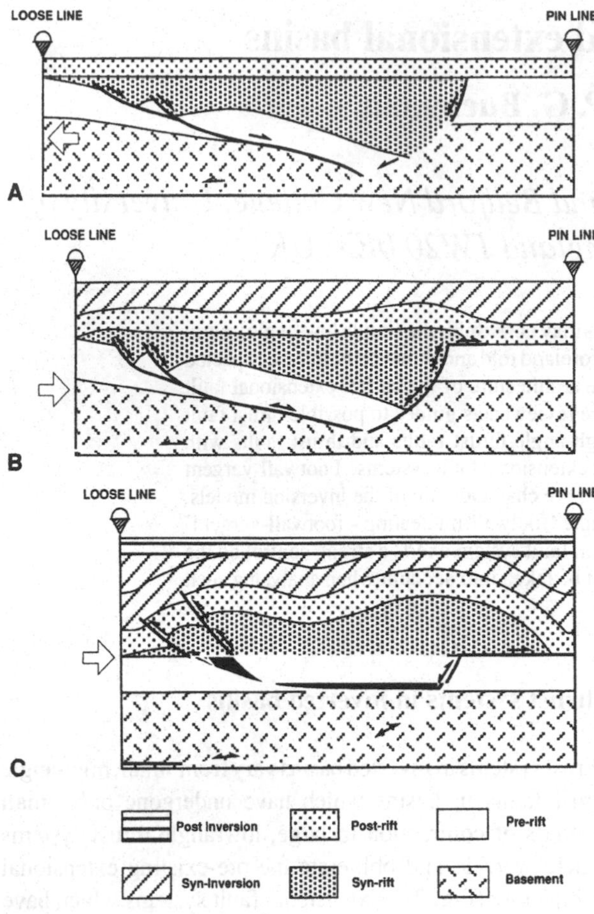

Figure 1. Conceptual model of an inverted half-graben system (modified after Bally 1984). (**a**) Extension; (**b**) Partial inversion; (**c**) Total inversion.

and re-utilized as thrust faults (Fig. 2d). Basement blocks may be incorporated into the thrust system via footwall 'shortcut' thrusts thus creating exotic horses within the thrust complex (Fig. 2e). Extensional faults which are unfavourably oriented for reactivation remain bypassed by the contractional tectonics (Fig. 2e). In areas of very strong thrusting such as the Alps (Butler 1989; de Graciansky *et al.* 1989) the thrust faults appear to bypass pre-existing extensional faults which display little or no reactivation. The thrust trajectories are, however, strongly controlled by the extensional fault architecture with thrust ramps located at pre-existing extensional fault steps.

In the examples shown in Figure 2 it is obvious that the architecture of the extensional basin fault system exerts a strong influence on the thrust systems developed during inversion and thrust faulting. If the thrust faults nucleate or grow from reactivated extensional faults they may be shallow-dipping in their upper sections but steepen downwards, thus not following a more conventional ramp-flat trajectory. The volume problem caused by contracting an extensional half-graben may generate new thrusts which nucleate and propagate out of the 'half-graben' not necessarily in a footwall propagating sequence. The stratigraphic sequences undergoing thrusting are not 'layer cake' but exhibit significant thickness changes with the resultant implications for thrust trajectories. Hangingwall anticlines in extensional basin systems are generated by roll-over and exist prior to inversion. The later contractional deformation may serve to tighten and amplify these pre-existing structures. Thus, in fold thrust belts which are developed upon earlier extensional fault regimes, complex thrust architectures and sequences may be expected. In order to understand some of the possible thrust geometries that may develop when extensional fault systems are contracted, an extensive analogue modelling programme has been carried out (Buchanan 1991; Buchanan & McClay 1991) and the results are summarized below.

Analogue models of inversion

Experimental method

Two types of sandbox experiment are briefly reviewed in this paper. The first type, the simple listric and the ramp-flat listric fault shapes, consisted of a fixed footwall geometry (wooden block) over which a deformable hangingwall was first extended and then contracted by horizontal recompression (see also Buchanan & McClay 1991). The second type of model experiment was designed to simulate a series of rigid domino blocks that firstly undergo extension and then are inverted or pushed back to their pre-extensional configuration.

The simple listric and ramp-flat listric experiments were carried out in a glass walled deformation rig (Fig.3a) 150 cm long, 20 cm wide and up to 20 cm deep. Initial model dimensions were typically 30 cm long, 20 cm wide and 10 cm deep. The sand models were constructed between the end walls of the deformation rig and consisted of alternating layers of white and coloured sand (average grain size 300 μm). Dry cohesionless quartz sand has an angle of friction of 31° (Liu *et al.* this volume) and successfully simulates the brittle deformation of sedimentary rocks in the upper crust (McClay 1990; Horsfield 1977). Faulted sand has a lower strength than unfaulted sand (Liu *et al.* 1991, this volume) and as such one would expect fault zones formed in extension to at least partly reactivate during inversion. The models are broadly scaled such that 1 cm in the model represents between 100 m and 1 km (McClay 1990).

Deformation was achieved by moving one of the end walls at a constant displacement rate of 4.16 x 10^{-3} cm sec^{-1}. A plastic sheet between the sand model and the footwall block / base of the deformation rig simulates a low friction ($\mu_b = 0$) basal detachment. For extension the hangingwall is pulled down and along the basal detachment, whereas for inversion the hangingwall is pushed back up the detachment. During extension syn-rift layers were incrementally added to the model in order to simulate syn-rift sedimentation. At the end of the extensional phase, a thin post-rift sequence was added to the models and during the contraction, syn-inversion layers were introduced in order to prevent the formation of unstable surface slopes as thrust faults propagated to the surface of the model.

The second type of experiment was designed to simulate

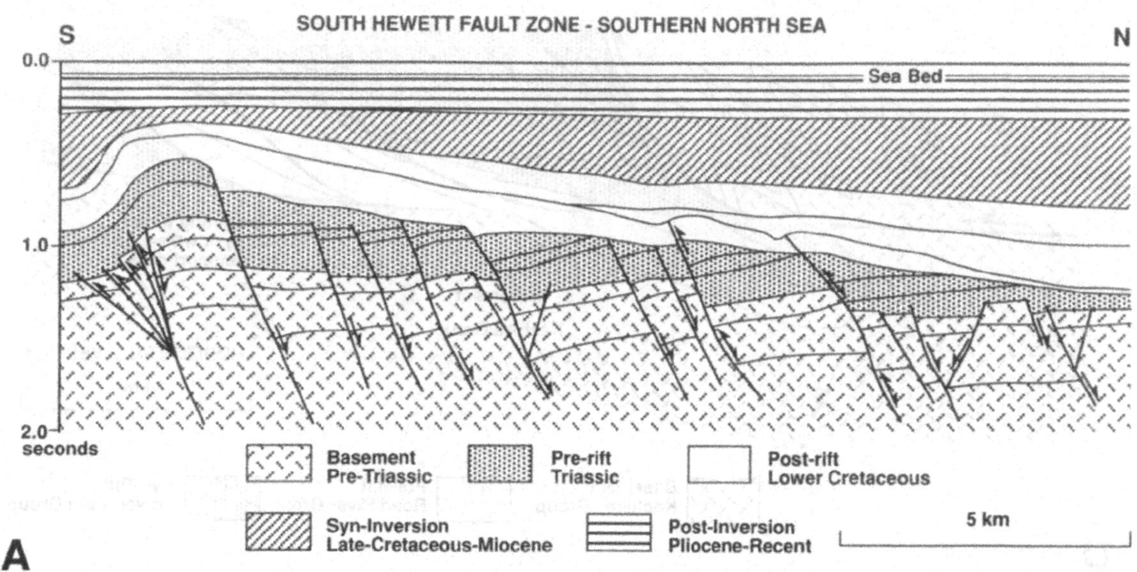

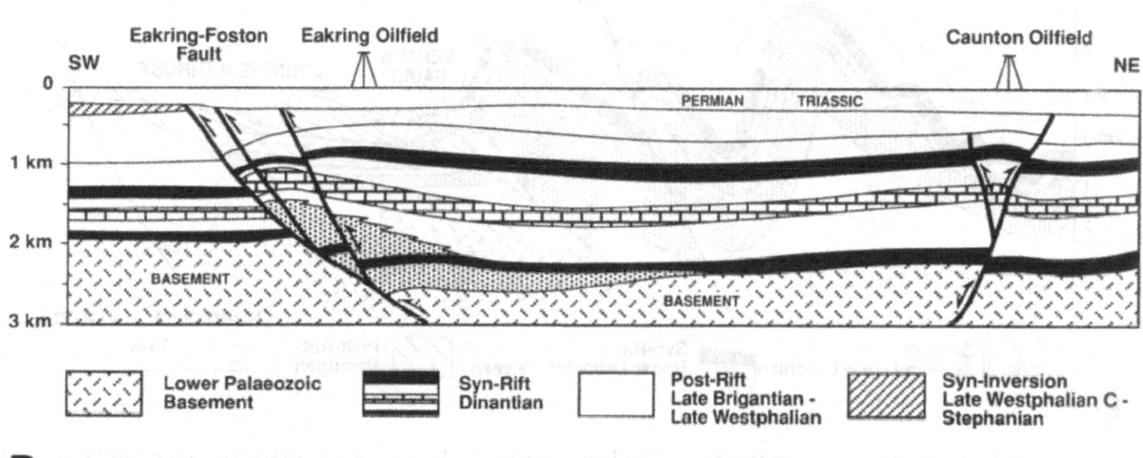

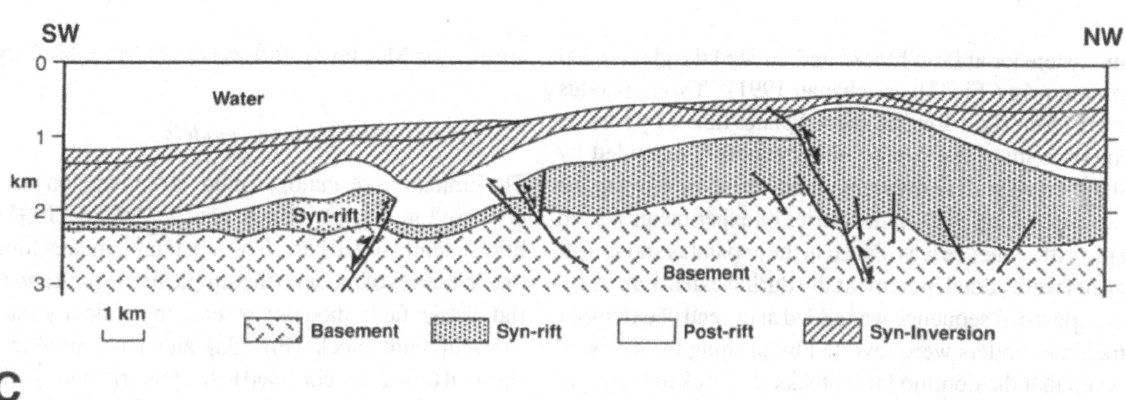

Figure 2. Examples of thrust faults in inverted basin systems. **(a)** Geoseismic section across the South Hewett fault zone, southern North Sea. The inverted domino fault array shows a major growth anticline at the left hand end of the section and high angle thrust faults have produced arrowhead or 'harpoon' structures at the right hand end of the section; **(b)** Cross-section through the Eakring oilfield, England (modified after Fraser *et al.* 1990). The extensional half-graben system has undergone partial inversion with contractional displacement on the main extension fault and a 'pop-up' at the location of the Caunton oil field; **(c)** Cross-section through part of the Sunda Arc (modified after Letouzey 1990) showing inversion of half-graben fault systems to produce characteristic arrowhead and harpoon geometries.

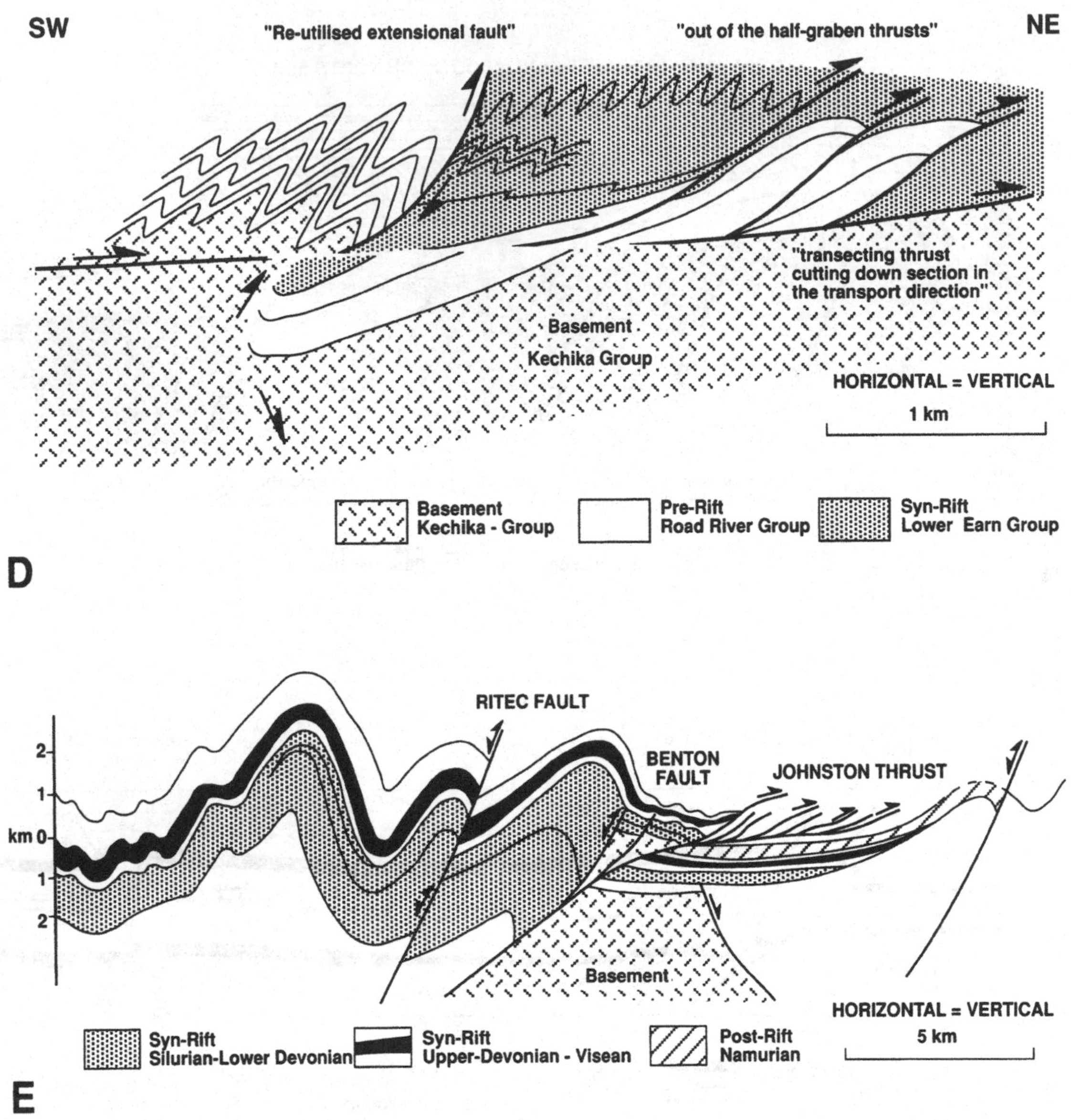

Figure 2. (d) Strongly inverted half-graben system from the Kechika Trough, northern Rocky Mountains, Canada (adapted after McClay *et al.* 1989). Pre-existing extensional faults have been rotated and re-utilised. 'Out of the half-graben' thrusts occur together with a through going thrust that cuts down stratigraphy in the direction of transport; **(e)** Inversion structures in the South Wales coalfield (modified after Powell 1989). Note the development of short-cut faults in the footwall to the reactivated Benton fault system.

domino extensional fault blocks and utilized the glass-sided deformation rig (Fig. 3b, Buchanan 1991). This apparatus consists of four rigid domino shaped plates linked by a trellis system such that they can be simultaneously extended by moving the end wall to the right and inverted by moving the end wall back (Fig. 3b). In this series of experiments a 3 cm thick pre-rift sequence was placed in the apparatus and synrift sediments were added incrementally during extension. A 1.5 cm thick post-rift sequence was added at the end of extension and then the models were inverted by pushing the end wall back such that the domino fault blocks slid back to their pre-extensional configuration.

In both series of experiments, progressive deformation during both the extension and inversion phases was monitored using time lapse photography. The completed models were impregnated and serially sectioned to study the final deformation geometry. Details of the experimental methods are given in McClay (1990) and Buchanan & McClay (1991).

Limitations of the analogue models

The limitations of analogue models of inversion tectonics are discussed in detail by Buchanan & McClay (1991) and are briefly summarized here. The most fundamental limitation of the first series of models (the simple listric fault and the ramp-flat listric fault models) is that they incorporate a rigid footwall fault block (Fig. 3a) and any possible footwall shortcut thrusts are confined to the post-rift and syn-inversion strata. In addition, the rigid nature of the footwall block in these experiments generates a marked mechanical contrast between the sandpack and the footwall such that strong buttressing effects are to be expected (see below). The sandbox modelling also cannot simulate compaction, thermal and pore fluid pressure effects, all of which may affect the

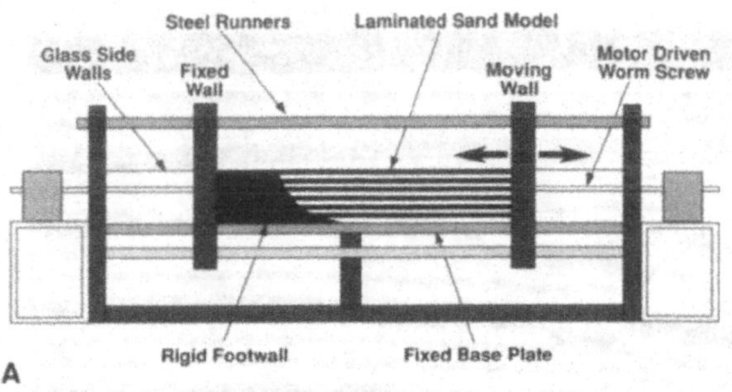

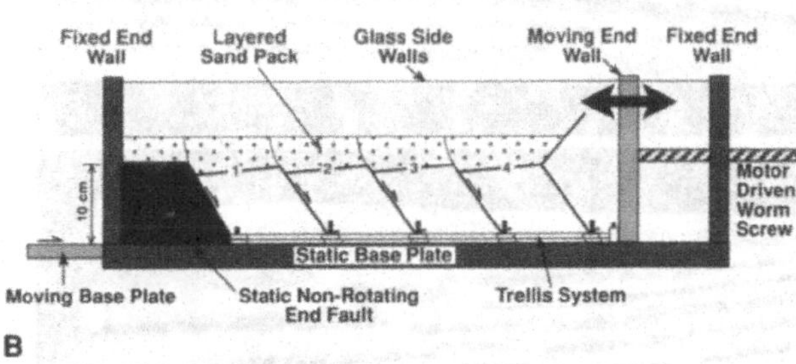

Figure 3. Experimental apparatus. **(a)** 2D deformation rig for extension and inversion of listric fault systems; **(b)** 2D deformation rig for the extension and inversion of domino fault systems.

geometries and mechanics of both the extension and inversion phases of deformation. In particular pore fluid pressure may be expected to strongly control whether pre-existing high-angle extensional faults reactivate (Sibson 1985; Etheridge 1986). However, if the above limitations are borne in mind, the analogue models do provide useful guides to the geometries and mechanics of dip-slip inversion of extensional fault systems (Buchanan & McClay 1991).

Results

The geometries and sequences of inversion structures that are developed in the analogue models are described below with reference to the architecture of the antecedent extensional fault system. Thus thrust faults which verge towards the footwall of the pre-existing extensional fault system are termed 'footwall-vergent' (rather than 'foreland-vergent'), and thrusts which verge towards the hangingwall of the pre-existing extensional fault system are termed 'hangingwall-vergent' (rather than 'hinterland-vergent'). The latter thrusts may also be termed 'backthrusts' as they verge backwards against the direction of compression. The sequence of thrust fault development (i.e. the direction in which successive new thrust faults nucleate) is described in terms of 'forward-breaking' sequences in which successive thrust faults nucleate in front of older thrusts and verge in the same direction as the nucleation sequence, and in terms of 'break-back' sequences in which successive new thrust faults nucleate behind older thrust faults and thus 'back step' in the direction opposite to the vergence of the faults.

Listric faults

Extension above a listric detachment fault produces a characteristic roll-over structure with an associated crestal collapse graben (Fig. 4a). The amount of bed rotation in the roll-over is proportional to the amount of extension and the shape of the detachment (e.g. Ellis & McClay 1988). The crestal collapse graben is generated by arc stretching with displacement on the bounding synthetic and antithetic faults dying out downwards towards the main detachment surface (Fig. 4a).

Inversion was achieved by horizontal recompression and the model was shortened 7 cm - 1 cm more than the extensional deformation. Upon compression the main detachment surface was reactivated (and remained active throughout the inversion) such that the synrift sequence was pushed back up the extensional fault. The resultant inversion structure shows a characteristic harpoon geometry with footwall shortcut faults, out of the crestal collapse graben thrusts, hangingwall-vergent backthrusts, and a growth anticline structure (Fig. 4b). These features are described below in the order of their development.

Footwall shortcut thrusts. At the early stages of inversion convex-upwards footwall shortcut thrusts developed in the post-rift sequence in front of the main detachment (Fig. 4b). These formed a break-back sequence until most of the displacement was transferred onto the steeper main detachment which propagated upwards into the post-rift and syn-inversion strata.

Out of graben thrusts. The next stage of inversion is characterised by rotation of the hangingwall back up the main detachment surface together with the nucleation of small thrust faults from the tips of the crestal collapse graben faults (Fig. 4b). These thrusts show growth features where they propagate upwards into the post-rift and syn-inversion sequences. They pass downwards to lower tips lines where they do not link with the reactivated basal detachment.

Backthrusts. Towards the latter stages of inversion large, hangingwall-vergent backthrusts develop in a forward-breaking sequence. These thrusts dissect and ignore the pre-existing extensional structures. The backthrusts are located in the undeformed hangingwall of the antecedent extensional fault system and follow more closely the rules of classic thin-skinned tectonics - i.e. forward-breaking. The backthrusts result from the buttressing of the contractional deformation against the concave upwards segment of the listric extensional detachment. As a result, complex fault patterns are produced by the superposition of the contractional structures upon the earlier extensional deformation. The backthrusts become progressively more numerous at the latter stages of the inversion and accommodate most of the contractional displacement.

Growth anticlines. During inversion the main detachment propagates upwards through the post-rift and syn-inversion strata towards the upper free surface of the model. As a consequence growth geometries are formed in the syn-inversion sequence (Fig. 4b) where syn-inversion strata are thicker on the footwall to the active fault and thin onto the hangingwall ramp-anticline above the fault (Fig. 4b). Growth anticlines

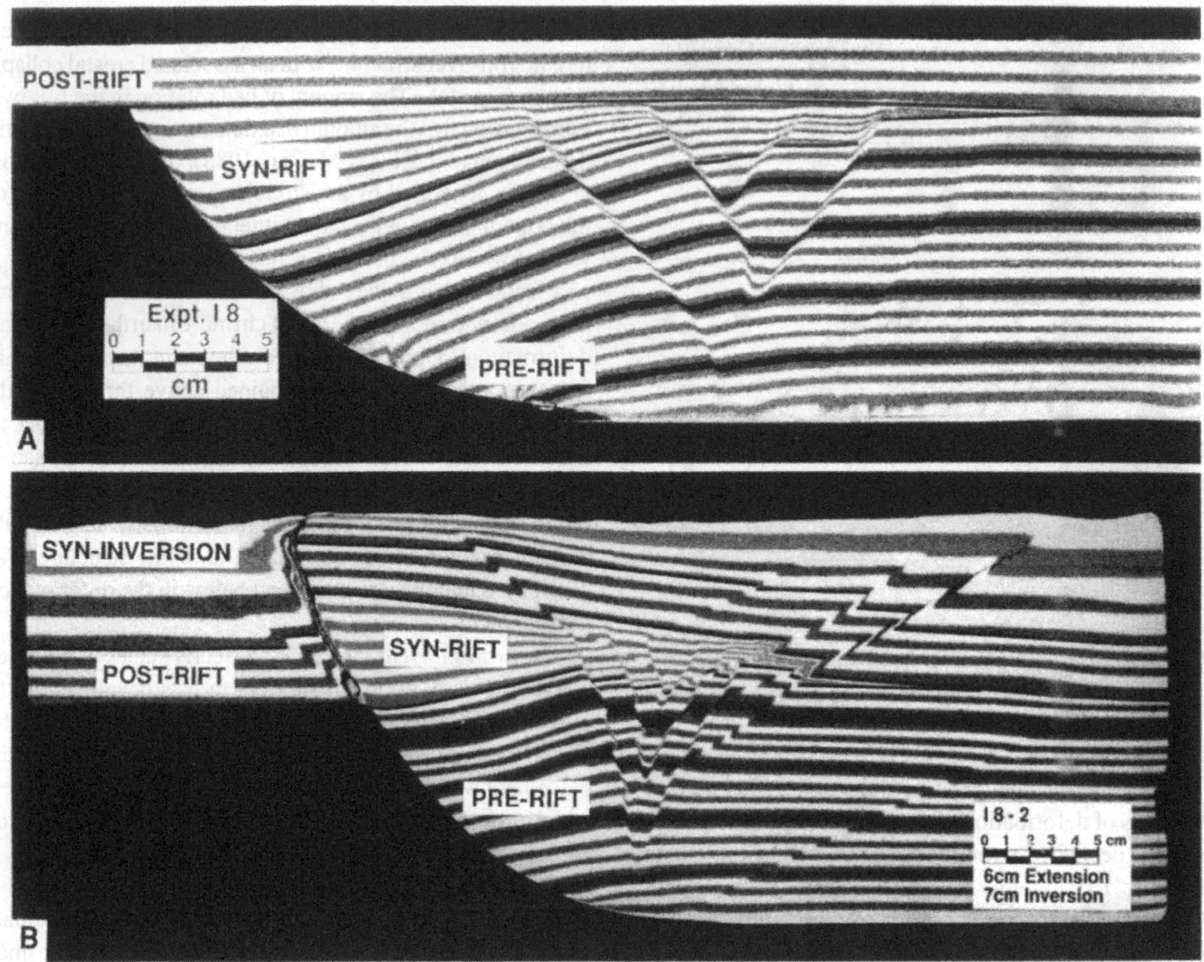

Figure 4. Simple listric fault - Experiment I8. **(a)** Model at the end of 6 cm extension. Top of the pre-rift sequence is marked by the upper grey-black-grey sequence. Syn-rift grey and white layers were added incrementally up to the top of the footwall block. A 2 cm thick, horizontal post-rift sequence has been added prior to inversion; **(b)** Serial section of the model at the end of 7 cm inversion. See text for details.

are generated by the reactivated main detachment, by the out of graben thrusts and by the hangingwall-vergent backthrust system (Fig. 4b).

For the simple listric fault system the maximum uplift as indicated by the upper syn-rift, lower post-rift sequences, is focused above the upper part of the original listric extensional detachment, and forms the crest of a large growth anticline. The sequence of thrusting is complex with reactivation of the basal detachment throughout the inversion. Footwall shortcut thrusts form a break-back sequence early in the inversion history whereas forward-breaking backthrusts dominate the latter stages of inversion (see also Buchanan & McClay 1991).

Ramp-flat listric faults

The structural architecture which evolves above an extensional ramp-flat detachment geometry has been discussed in detail by Ellis & McClay (1988) and McClay & Scott (1991). The basic structural elements are similar to those seen above the listric fault described above. An upper roll-over and associated crestal collapse graben develops above the upper listric portion of the detachment (Fig. 5a). This is separated by a hangingwall syncline related to the ramp in the detachment, from a larger roll-over with superimposed graben structures associated with the lower listric segment of the detachment.

Inversion was achieved in a similar manner to that for the listric fault model and the magnitude of contraction exceeded extension (by 12 cm). Reactivation of the lower listric segment of the basal detachment resulted in the propagation of a major, footwall-vergent thrust fault upwards and bypassed the higher and smaller of the two crestal collapse graben (Fig. 5b). This footwall-vergent thrust becomes a growth fault with significant thickness variations in the syn-inversion sequence. Minor thrusts parallel to the major thrust in its hangingwall were also generated. The geometry of this footwall-vergent thrust is relatively planar with almost no deformation in its footwall (Fig. 5b). Thus the upper extensional roll-over structure is preserved as an exotic, fault-bounded, lozenge-shaped block unaffected by the contraction. As in the case of the listric fault model (Fig. 4b), the curvature of the basal detachment at the lower listric section results in the development of a series of 'forward-breaking', hangingwall-vergent 'backthrusts'. These have a convex-upwards profile geometry caused by the presence of the non-

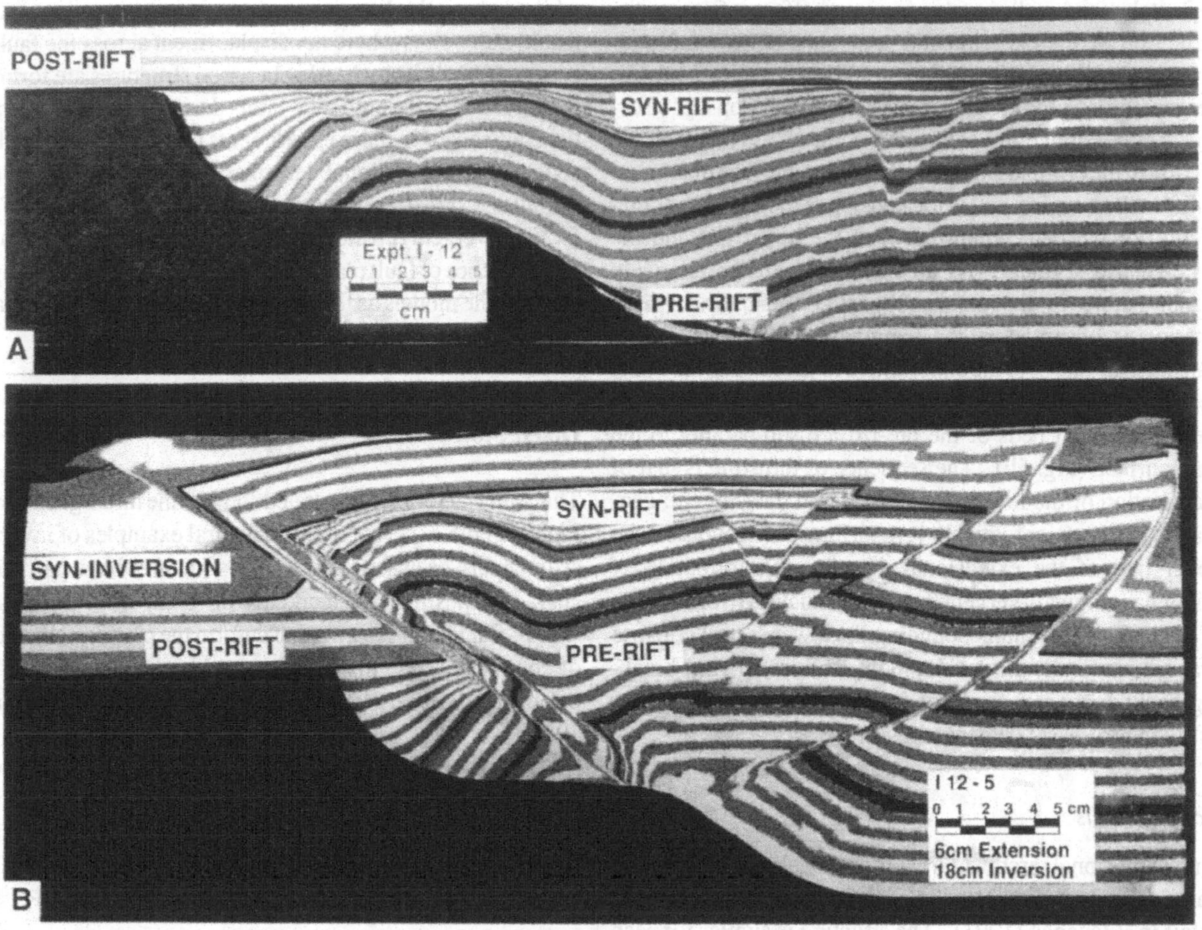

POST-RIFT
SYN-RIFT
Expt. I - 12
0 1 2 3 4 5
cm
PRE-RIFT
A

SYN-RIFT
SYN-INVERSION
POST-RIFT
PRE-RIFT
I 12 - 5
0 1 2 3 4 5 cm
6cm Extension
18cm Inversion
B

Figure 5. Ramp-flat listric fault - Experiment I 12. **(a)** Model at the end of 6 cm extension. The model consists of alternating layers of coloured sand with thin mica layers between them in order to facilitate layer parallel slip (McClay & Scott 1991). The top of the pre-rift sequence is marked by the upper grey-black-grey sequence. Syn-rift grey and white layers were added incrementally up to the top of the footwall block. A 2 cm thick, horizontal post-rift sequence has been added prior to inversion; **(b)** Serial section of the model at the end of 18 cm inversion. See text for details.

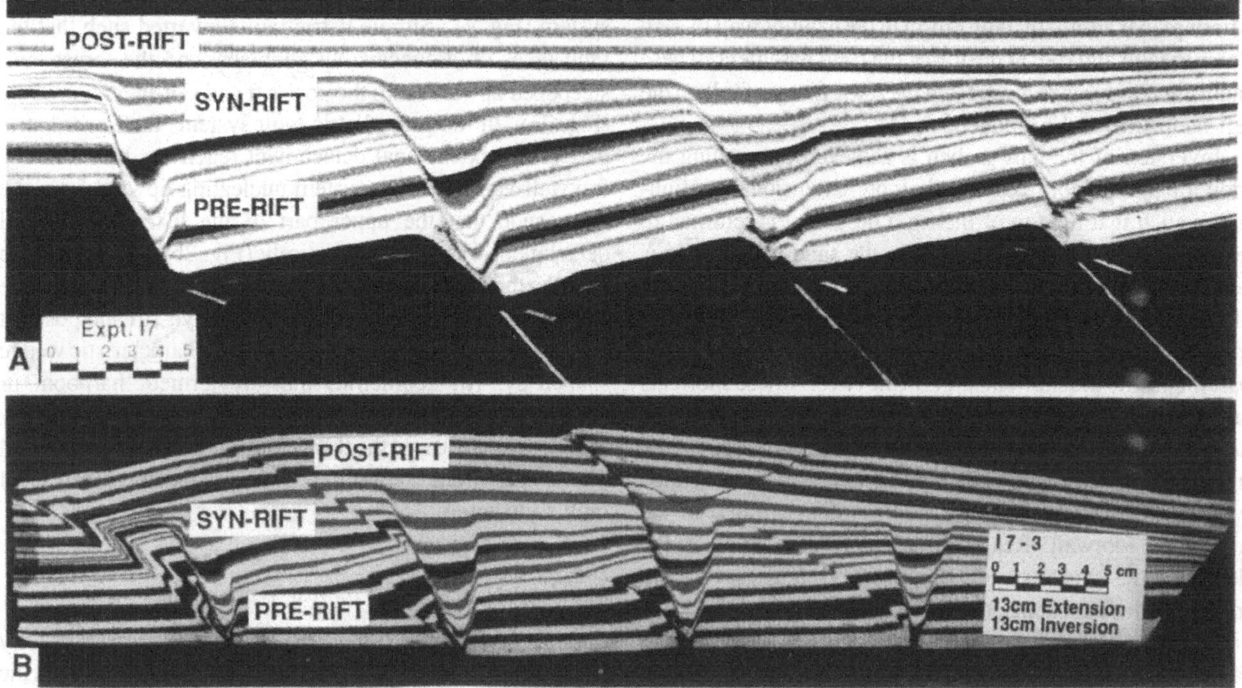

Figure 6. Domino fault system. **(a)** Model at the end of 13 cm extension. Top of the pre-rift sequence is marked by the upper grey-black sequence. Syn-rift grey and white layers were added incrementally up to the top of the footwall block. A 1.5 cm thick, horizontal post-rift sequence has been added prior to inversion; **(b)** Serial section of the model at the end of 13 cm of inversion. See text for details. No syn-inversion strata have been added to this model

deflectable end wall, i.e. this is an end effect reflecting the limitation of the experimental apparatus. These forward-breaking backthrusts, in conjunction with the major footwall-vergent thrust, produce a characteristic broad 'pop-up' structure centred above the soling-out point of the lower listric section of the original extensional detachment. The maximum uplift in the hangingwall is located above the top of the ramp section of the detachment between the upper flat and the lower listric segment. As a consequence of the relatively planar nature of the major footwall-vergent thrust, little rotation is observed in the hangingwall which is simply translated up and over the undeformed footwall. The footwall-vergent thrust becomes 'thin-skinned' in this region where the hangingwall beds of the pre-rift are sub-parallel to the fault surface (Fig. 5b). The sequence of thrusting is that of reactivation of the lower part of the basal detachment; generation of the major footwall-vergent thrust (that bypasses the upper part of the extensional detachment); formation of minor hangingwall-vergent thrusts that nucleate from the tips of the extensional faults bounding the lower crestal collapse graben; and finally the development of major forward-breaking backthrusts as a result of buttressing against the lower listric part of basal detachment surface.

Domino faults

The extensional architecture produced above a series of domino faults is illustrated in Figure 6a and is described in detail in Buchanan (1991). The structure is similar for each domino fault and consists of a half graben filled with a syn-rift wedge in which the layering is planar. The final profile geometry of the domino faults has a slight listric shape acquired during extension as a result of the interaction between sedimentation and rotation (Vendeville & Cobbold 1988). Minor planar antithetic faults form in the hangingwall to each domino fault and root at the top of the rigid basement. These faults increase in size away from the static margin fault on the left-hand side of the model.

Inversion of the half graben is achieved by pushing the individual fault blocks back to their pre-extensional configuration such that the top of the rigid basement is returned to horizontal. Contractional strain in the sand pack above is accommodated by reactivation of the domino faults and along several footwall shortcut faults (Fig. 6b). Displacement is generally transferred from the domino faults in the basement onto the shortcut faults in the cover. These shortcuts are planar to slightly convex upwards in profile and define downward tapering wedge-shaped blocks at the margin of the individual half grabens. The contractional strain associated with each reactivated domino fault is concentrated in the immediate footwall. Reactivation of the entire length of the domino faults causes steeply dipping thrust faults to develop in the post-rift sequence (Fig. 6b). All the extensional and contractional faulting in the cover relating to each domino fault, links into a single master fault in the rigid basement below. Reactivation of individual domino faults causes post-rift and upper syn-rift layers to be put into net contraction whilst below, the layering remains in net extension. As a result, null points where net displacement across the fault is zero and classic arrowhead or harpoon structures are formed. Reactivation of the antithetic faults during inversion does not generally occur in these experiments (Buchanan 1991). The style of the contractional deformation is strongly asymmetric, all the thrust faults verging in the same direction (i.e. footwall-vergent) up the dip of the reactivated domino faults. The sequence of fault reactivation generally occurs from right to left in the models as the right hand end wall is pushed back to its pre-extensional position.

Discussion

This paper has briefly illustrated some of the thrust geometries found in both experimental and natural examples of inverted sedimentary basins. Clearly, such structures have particular characteristics that are markedly different from the classic 'ramp - flat' thrust systems that deform the platformal regions of continental margins (e.g. Rocky Mountains and the Appalachians, Bally *et al.* 1966; Dahlstrom 1970; Boyer & Elliott 1982).

The experimental models of inverted extensional fault systems illustrate the sequential development of reactivation of pre-existing extensional faults as thrust faults, the formation of growth structures and the nucleation of new thrust faults.

Reactivation of pre-existing extensional faults as thrust faults

In the experimental models pre-existing extensional faults were reactivated as high-angle thrust faults whose architecture was controlled by the shape of the extensional detachment system (Fig. 7). Simple listric faults inverted such that the fault takes on a concave-up geometry and the thrust fault steepened upwards into the post-rift and syn-inversion strata (Fig. 7a). In ramp-flat listric fault systems the lower listric part of the extensional detachment reactivated with a new footwall-vergent thrust system nucleating upward from the ramp such that the upper part of the listric extensional detachment was bypassed entirely (Fig. 7b). In the domino models the extensional faults can reactivate and became convex upwards in the upper part of the syn-rift and post-rift strata (Fig. 7c). The inversion produces characteristic wedge shaped syn-rift geometries that are termed 'harpoon' or arrowhead structures (Badley *et al.* 1989; Buchanan & McClay 1991). In many cases the thrust faults are typically steep and their shapes strongly controlled by the pre-existing extensional fault architecture.

Reactivation of crestal collapse graben faults

Crestal collapse graben faults (Figs 4, 5, 7a & b) show only partial reactivation with new thrust faults nucleating from the upper tips of both the antithetic and synthetic crestal collapse graben faults. Extensional architecture is preserved at depth

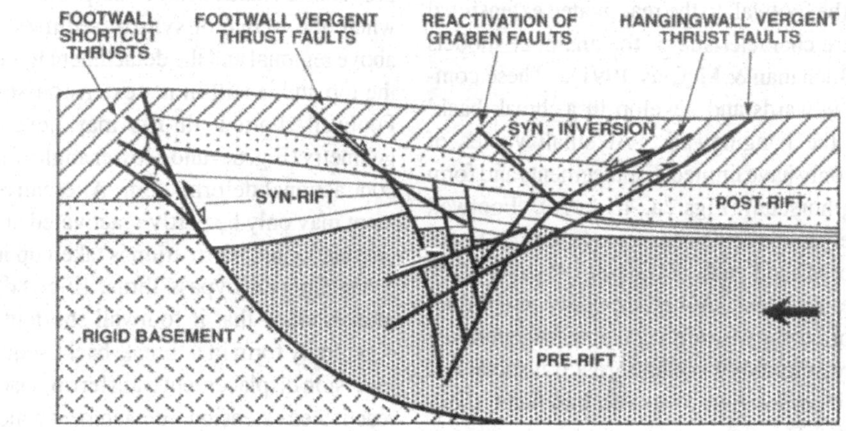

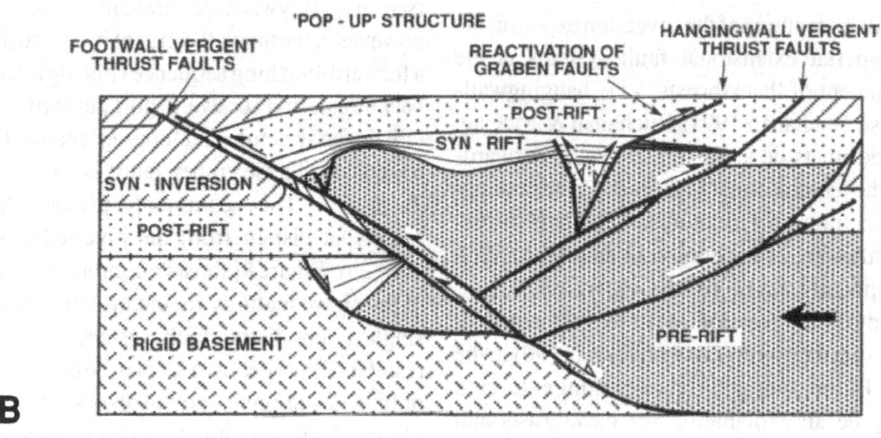

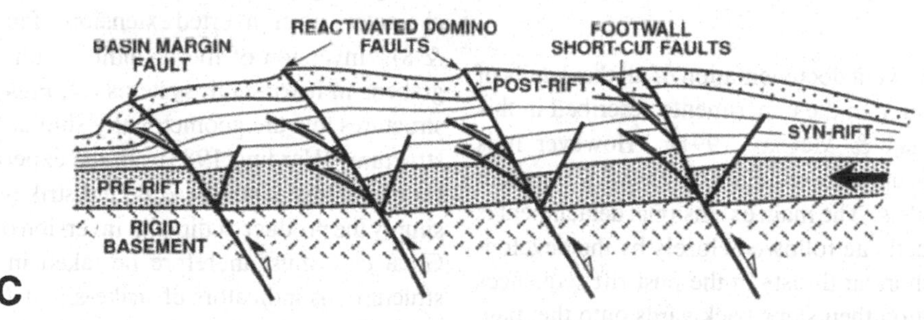

Figure 7. Summary diagrams of inverted extensional fault systems. **(a)** Inverted listric extensional fault showing the rigid basement footwall block, pre-rift layers, syn-rift layers added incrementally during extension, post-rift layers and syn-inversion strata. **(b)** Inverted ramp-flat listric extensional fault system showing the rigid basement footwall block, pre-rift layers, syn-rift layers added incrementally during extension, post-rift layers and syn-inversion strata. Note the development of a prominent hangingwall bypass thrust that leaves the upper roll-over unaffected by the inversion. Major backthrusts lead to the development of a triangular 'pop-up' structure. **(c)** Diagram for the inversion of a domino fault array showing reactivation of the domino faults and the development of footwall shortcut thrusts in the pre-rift and syn-rift strata.

(Figs 7a & b). The thrust faults typically have a convex upward trajectory flattening into the post-rift sequence.

Growth structures

Growth folds are characteristic of the inversion of basin bounding extensional faults (Figs 2a, 4b & 7a). They are fault-propagation growth folds, typically asymmetric with onlap sequences in the syn-inversion strata. The growth asymmetry is footwall-vergent (Figs 2a & 4b).

Footwall shortcut faults

Shortcut thrusts in the footwall to the reactivated extensional faults (cf. Fig. 7) are characteristic of the analogue models (Buchanan 1991; Buchanan & McClay 1991). These commonly are convex upwards and develop in a 'break-back' sequence towards the hangingwall. At small values of inversion they may only have limited displacements and form an upward fanning horsetail (Figs 2a, 4b & 7a). Footwall shortcut thrusts are geometrically necessary in order to accommodate the excess bed length of the syn-rift strata in the extensional rollover anticline. At higher values of contraction the footwall shortcut thrusts are responsible for generating a lower angle, more smoothly varying thrust trajectory and may result in the incorporation of exotic basement slices within thrust belts (cf. Fig. 2e).

'Backthrusts'

One of the most striking features of the inversion experiments on listric and ramp flat extensional fault systems is the ubiquitous development of 'backthrusts' - i.e. hangingwall-vergent thrusts (also Buchanan 1991; Buchanan & McClay 1991). These backthrusts develop 'in sequence' forward-breaking towards the hangingwall (Figs 7a & b) and result in significant uplift in 'pop-up' structures. The backthrusts result from the buttressing of the contractional deformation by the concave-upwards shape of the extensional fault. Backthrusts related to the inversion of an extensional fault system have been described by Berastegui *et al.* (1990) in the southern Pyrenees. Buttressing against pre-existing extension fault systems may be an explanation for backthrusts and hangingwall-vergent thrust and nappe systems found in the interior portions of many fold and thrust belts.

Fault sequences

Space limitations prevent documentation of all the details of sequential deformation in the experiments described in this paper (see Buchanan & McClay 1991). However it is possible to comment upon the sequence of thrust development observed in the models. The main extensional detachment is the first fault to reactivate followed closely by the development of footwall shortcut thrusts in the post-rift sequences (Fig. 7). Deformation then steps backwards onto the main detachment via other footwall shortcut thrusts. Reactivation of crestal collapse graben faults occurs at this time. After appreciable contractional deformation (approximately 25%) forward-breaking, hangingwall-vergent back thrusts develop. The net result of the above is a complex sequence of thrusts that is not simply footwall-nucleating but is governed by the architecture of the pre-existing extensional detachment and by the amount of shortening.

Conceptual models for the inversion of extensional fault systems may be erected based upon the experimental models and natural examples (e.g. Fig. 2). A series of schematic models for possible geometries that could arise upon the inversion of a simple listric fault system are shown in Figure 8. Dip-slip inversion of the main extensional detachment produces a characteristic 'harpoon' or arrowhead geometry where the wedge of syn-rift sediments is partially elevated above regional and the detachment has thrust displacement at the top and is still in net extension at the bottom (Fig. 8b). Footwall shortcut thrusts may develop with a sigmoidal geometry (Fig. 8c) and a lower angle trajectory that facilitates contractional deformation. Alternatively, the main detachment may only be partly reactivated and then stick such that contractional deformation is taken up in the hangingwall via a hangingwall bypass thrust (Fig. 8d). The experimental models show that if footwall shortcut thrusts develop they commonly form in a 'break-back' sequence with most of the inversion displacement transferring backwards onto the main reactivated extensional detachment which propagates upwards into the post-rift and syn-inversion stratigraphy (Fig. 8e). Similarly, if hangingwall bypass thrusts develop one might expect them to also form in a break-back sequence (Fig. 8f) (see also Hayward & Graham 1989). More commonly, however, hangingwall-vergent backthrusts would develop in a forward-breaking sequence (Fig. 8g). Combination of both backthrusts and footwall short cut thrusts produce a characteristic fan structure and 'pop-up' feature (Fig. 8h). Inversion of crestal collapse graben faults is commonly observed (Figs 2b & 8i). They commonly produce 'pop-up' structures somewhat distant from the inverted basin bounding fault (Fig. 2b). Figure 8j shows a combined conceptual model if all of the above features are incorporated into the inversion of a single major listric fault system. It is apparent that the resultant thrust geometries may be complex and the sequences of fault development do not obey simple footwall nucleation - footwall propagation rules that seem to apply in many thin-skinned foreland fold and thrust belts (cf. Boyer & Elliott 1982).

High-angle thrust faults and footwall shortcut thrusts are characteristic of inverted extensional fault systems (Figs 2, 7 & 8). Inversion of the bounding faults to crestal collapse grabens in listric fault systems (cf. Figs 4b & 7a) produces structures that are geometrically similar to positive 'flower' structures (Harding 1985). In the experimental models described in this paper these are not strike-slip features but are simply the product of dip-slip inversion of extensional faults. Great care must therefore be taken in using such flower structures as indicators of strike-slip deformation.

Analysis of thrust systems in terranes where inverted earlier extensional faults might be expected (e.g. McClay *et al.* 1989) should take into account the possibilities of thrusts deforming already folded and faulted strata (cf. rollover anticlines and crestal collapse grabens), complex fault architectures with both forward-breaking and break-back thrust sequences (Fig. 8), together with the difficulties of distinguishing between similar structural geometries such as thrust-related ramp anticlines and extension-related rollover anticlines (Fig. 7). These complexities have significant implications for the theories and methods of balancing sections of thrust systems developed in previously extended terranes.

A EXTENSION - HALF-GRABEN AND ROLLOVER ANTICLINE

Regional
Regional
SYN - RIFT
PRE - RIFT BASEMENT

B INVERSION - 'HARPOON' STRUCTURE

Regional
Regional

C INVERSION - FOOTWALL SHORTCUT THRUST

Regional
Regional

D INVERSION - HANGINGWALL BYPASS THRUST

Regional
Regional

E INVERSION - IMBRICATE FAN OF 'BREAK-BACK' FOOTWALL SHORTCUT THRUSTS

Regional
Regional
1
2
3
4

F INVERSION - 'BREAK-BACK BYPASS' THRUSTS

1
2
3
4
5
Regional

G INVERSION - IMBRICATE FAN OF 'FORWARD-BREAKING' BACK THRUSTS

Regional
Regional
1
2
3
4

H INVERSION - FAN STRUCTURE OF BACK - THRUSTS AND FOOTWALL SHORTCUT THRUSTS

1 - 6
1
2
3
4
5
6
Regional

I INVERSION - CRESTAL COLLAPSE 'POP - UP' STRUCTURE

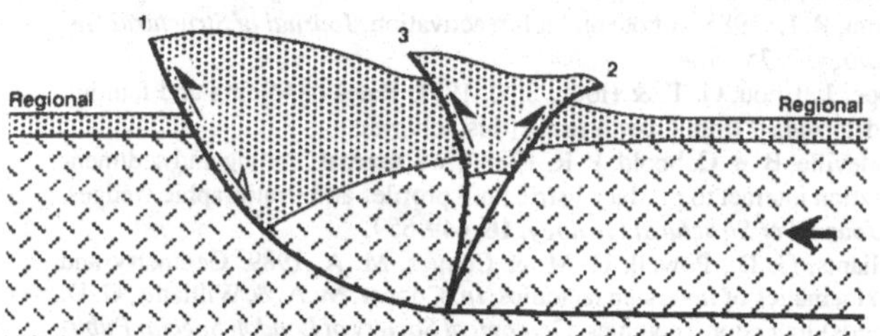

J INVERSION - COMBINED MODEL

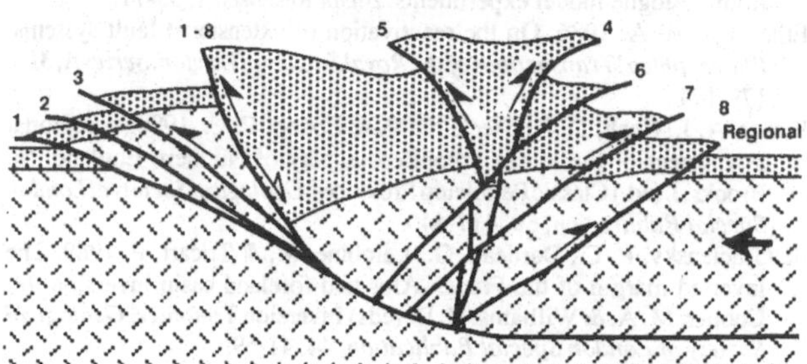

Figure 8. (a- j) Conceptual models for thrust faults developed by the dip-slip inversion of a listric fault system. See text for detailed discussion. Note the regional at the top of the syn-rift strata in (a).

Conclusions

The experimental models and examples cited in this paper illustrate some of the possible complexities of thrust fault systems that are found when extensional basin systems are subjected to dip-slip inversion. Inversion of extensional fault systems is characterised by high-angle thrust systems that may be convex upwards, steepen downwards and join pre-existing extensional detachments. Footwall shortcut faults and backthrusts are common features of both the experimental models and are also found in nature. Thrust fault nucleation is not necessarily in towards the footwall in a 'forward-breaking' sequence but is controlled by the pre-existing extensional architecture and buttressing effects. The analogue sandbox models serve simply as a guide to some of the possible geometries and mechanics of inversion features. These, however, need to be tested by detailed field analysis of thrust systems in inverted extensional basins.

At low values of shortening many of the inversion features described above may be recognized but one would expect that, at higher amounts of shortening, many of these features may be obliterated and more complex structures will result such that the 'normal' rules of thrust systems are no longer obeyed. The extensional architecture of basins incorporated in fold and thrust belts must be carefully considered if the thrust geometries of such contractionally deformed basins are to be fully understood (McClay *et al.* 1989) and meaningful balanced sections constructed.

This paper is in part based upon Ph.D research by Peter Buchanan funded by BP Petroleum who are thanked for permission to publish. Ken McClay acknowledges support by the Fault Dynamics consortium - Arco British Limited, BP Exploration, Brasoil U.K. Ltd., Conoco (U.K.) Limited, Mobil North Sea Limited and Sun Oil. Critical reviews by Bob Thompson, Peter Ellis and Ian Davison are gratefully acknowledged. Sandra Muir is thanked for help with drafting the diagrams and Kevin D'Souza for photographic assistance.

References

Badley, M. E., Price, J. D. & Blackshall, L. C. 1989. Inversion, reactivated faults and related structures - seismic examples from the southern North Sea. *In*: Cooper, M. A. & Williams, G. D. (eds), Inversion Tectonics. *Geological Society of London Special Publication*, **44**, 201-219.

Bally, A. W. 1984. Tectogenése et sismique réflexion. *Bulletin Societe Geologique France*, **7**, 279-285.

Bally, A.W., Gordey, P.L. & Stewart, G.A. 1966. Structure, seismic data, and orogenic evolution of the southern Canadian Rocky Mountains. *Bulletin of Canadian Petroleum Geology*, **14**, 337 - 381.

Berastegui, X., Garcia-Senez, J. M. & Losantos, M. 1990. Tectono-sedimentary evolution of the Organya extensional basin (central south Pyrenean unit, Spain) during the Lower Cretaceous. *Bulletin Societe Geologique France*, **8**, 251 - 264.

Boyer, S. M. & Elliot, D. W. 1982. Thrust systems. *Bulletin American Association of Petroleum Geologists*, **66**, 1196-1230.

Buchanan, P. G. 1991. Geometries and kinematic analysis of inversion tectonics from analogue model studies. Ph.D thesis (unpubl.), University of London.

Buchanan, P. G. & McClay, K. R. 1991. Sandbox experiments of inverted listric and planar faults systems. *In*: P. R. Cobbold (ed.) Experimental and Numerical Modelling of Continental Deformation. *Tectonophysics*, **188**, 97-115.

Butler, R.W.H. 1989. The influence of pre-existing basin structure on thrust system evolution in the Western Alps. *In*: Cooper, M. A. & Williams, G. D. (eds) Inversion Tectonics, *Geological Society of London Special Publication*, **44**, 105-122.

Cooper, M. A. & Williams, G. D.(eds) 1989. Inversion Tectonics. *Geological Society of London Special Publication*, **44**, 376p.

Dahlstrom, C.D.A. 1970. Structural geology in the eastern margin of the Canadian Rocky Mountains. *Bulletin of Canadian Petroleum Geology*, **18**, 332-406.

Ellis, P. G. & McClay, K. R. 1988. Listric extensional fault systems: results from analogue model experiments. *Basin Research* **1**, 55-71.

Etheridge, M. A. 1986. On the reactivation of extensional fault systems. *Philosophical Transactions of the Royal Society of London, Series A*, **317**, 179-194.

Fraser, A. J., Nash, D. F., Steele, R. P. & Ebdon, C. C., 1990. A regional assessment of the intra-Carboniferous play of Northern England. *In*: Brooks, J. (ed.) Classic Petroleum Provinces. *Geological Society of London Special Publication*, **50**, 417-440.

de Graciansky, P. C., Dardeau, G., Lemoine, M. & Tricart, P. 1989. The inverted margin of the French Alps and foreland basin inversion. *In*: Cooper, M. A. & Williams, G. D. (eds) Inversion Tectonics. *Geological Society of London Special Publication*, **44**, 41-59.

Harding, T.P. 1985. Seismic characteristics and identification of negative flower structures, positive flower structures, and positive structural inversion. *Bulletin American Association of Petroleum Geologists*, **69**, 582-600.

Hayward, A. B. & Graham, R. H. 1989. Some possible characteristics of inversion. *In*: Cooper, M. A. & Williams, G. D. (eds) Inversion Tectonics. *Geological Society of London Special Publication*, **44**, 17-39.

Horsfield, W.T. 1977. An experimental approach to basement controlled faulting. *Geologie en Mijnbouw*, **56**, 363 - 370.

Letouzey, J. 1990. Fault reactivation, inversion and fold-thrust belt. *In*: Letouzey, J. (ed.) *Petroleum and Tectonics in Mobile Belts*. IFP Editions Technip, Paris, 101-128.

Liu, Huiqui, McClay, K. R. & Powell, D. P. 1991. Physical models of thrust wedges (this volume).

McClay, K. R. 1990. Extensional fault systems in sedimentary basins. A review of analogue model studies. *Marine and Petroleum Geology*, **7**, 206 - 233

McClay, K. R., Insley, M. W. & Anderton, R. 1989. Inversion of the Kechika Trough, Northeastern British Columbia, Canada. *In*: Cooper, M. A. & Williams, G. D. (eds) Inversion Tectonics. *Geological Society of London Special Publication*, **44**, 235-257.

McClay, K. R. & Scott, A. D. 1991. Experimental models of hangingwall deformation in ramp-flat listric extensional fault systems. *In*: P. R. Cobbold (ed.) Experimental and Numerical Modelling of Continental Deformation. *Tectonophysics*, **188**, 85-96.

Martinez, A., Verges, J., Clavell, E. & Kennedy, J. 1989. Stratigraphic framework of the thrust geometry and structural inversion in the southeastern Pyrenees: La Garrotxa area. *Geodinamica Acta*, **3**, 185-194.

Powell, C. M. 1989. Structural controls on Palaeozoic basin evolution and inversion in southwest Wales. *Journal of the Geological Society of London*, **146**, 439-446.

Puigdefàbregas, C., Munoz, J. A. & Marzo, M. 1986. Thrust belt development in the eastern Pyrenees and related depositional sequences in the southern foreland basin. *In:* Allen, P. A. & Homewood, P. (eds) Foreland Basins. *International Association of Sedimentologists Special Publication*, **8**, 229-246.

Sibson, R. H. 1985. A note on fault reactivation. *Journal of Structural Geology*, **7**, 751-754.

Suppe, J., Chou, G. T. & Hook, S. C. 1991. Rates of folding and faulting determined from growth strata (this volume).

Vendeville, B. & Cobbold, P. R. 1988. How normal faulting and sedimentation interact to produce listric fault profiles and stratigraphic wedges. *Journal of Structural Geology*, **10**, 649-659.

Williams, G. D., Powell, C. M. & Cooper, M. A. 1989. Geometry and kinematics of inversion tectonics. *In*: Cooper, M. A. & Williams, G. D. (eds) Inversion Tectonics. *Geological Society of London Special Publication*, **44**, 3-15.

Rates of folding and faulting determined from growth strata

John Suppe[1], George T. Chou[2] & Stephen C. Hook[3]

[1]Department of Geological and Geophysical Sciences, Guyot Hall, Princeton University, Princeton, New Jersey 08544-1003, USA
[2]Texaco USA, 4601 DTC Boulevard, Denver, Colorado 80237, USA
[3]Texaco Inc, PO Box 70070, Houston, Texas 77215-0070, USA

Abstract: Many upper crustal folds on the scale of 1-10 km in compressive mountain belts grow by kink-band migration as a result of fault-bend, fault-propagation, or box folding. One or both kink-band boundaries sweep through the rock as the kink bands widen during fold growth. The kink bands typically have a constant width in *pregrowth strata*, which are strata that existed before deformation, whereas we predict - and observe on seismic lines - an upward decrease in kink-band width within the stratigraphic sequence deposited during fault slip and associated fold growth - here called *growth strata*. In fact this growth stratigraphic sequence provides a complete, decipherable record of the kinematics of deformation, much in the same way that sea-floor magnetic anomalies provide a decipherable record of plate kinematics. It is the continual addition of material that provides the detailed record of motion in both cases. In the simplest folds the kink band has a constant width within the pregrowth strata, narrows upward through the growth strata, and finally has a zero width at the top of the growth stratigraphy.

The rate of folding can be described geometrically as the rate of kink-band migration divided by the sedimentation rate, which is dimensionless. Furthermore, if key beds can be dated, absolute fault-slip rates can be computed. Examples are presented from the Gulf of Mexico, the Philippines, Venezuela, California, and Oklahoma. Folding rates in many of these examples are observed to be 1-2 mm/y lasting for 1-8 Ma. Translation of folding rates into fault-slip rates requires knowledge of the fault shape and its relationship to fold shape. Nevertheless, even if fault geometry is poorly constrained, fault slip rates generally can be estimated to within narrow limits based on rates of kink-band migration, a fact that makes growth structures valuable for assessing earthquake hazards.

The common upper-crustal folding mechanisms of fault-bend folding, fault-propagation folding and box folding all predict that folds grow by kink-band migration (Suppe 1983, 1985; Suppe & Medwedeff 1984, 1990). Specifically, these primary folding mechanisms do not predict gradual amplification by limb rotation; rather they predict that beds only change dip as a result of rolling through axial surfaces*. For this reason fold limbs are predicted by these mechanisms to increase in length as the fold grows, whereas limb dips remain constant. Therefore, if sedimentation takes place during the growth of the fold, the limb length will decrease upward within the stratigraphic interval deposited during deformation (here called *growth strata*). It is the purpose of this paper to show that upward narrowing kink bands are indeed observed as predicted, thus validating these theories of folding. Furthermore, if key beds within the growth strata can be dated then rates of folding - that is limb lengthening - can be determined. Finally, quantitative relationships between limb length and fault slip are derived for fault-bend folding and fault-propagation folding that allow fault slip rates to be determined from rates of limb lengthening.

Constant-dip kink-band migration

Let us put aside for the moment the specific mechanisms of fault-related folding and simply consider the possible kinematics of constant-dip kink-band migration. Five kinematic classes exist because one or both of the axial surfaces bounding a kink band can move with respect to the material (Fig. 1).

---

*It should be noted that we are not arguing that constant-dip kink-band migration is the only important folding mechanism in the upper crust, rather that the predictions of these theories of fault-related folding are in fact borne out in a number of specific cases. Some upper crustal folding does involve progressive limb rotation. For example, some folds appear to lock up and then amplify at a later stage in deformation (e.g. Suppe & Medwedeff 1990, Fig. 1). The case of progressive limb rotation is beyond the scope of this paper.

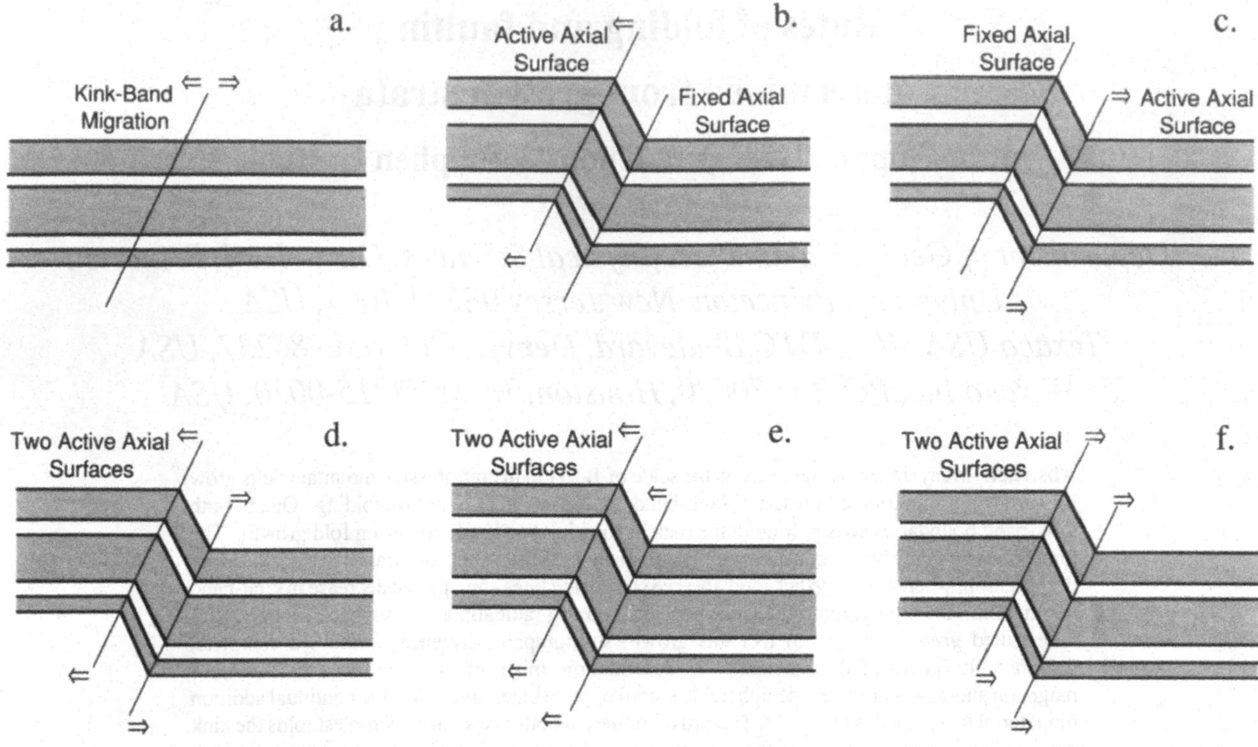

Figure 1. The five possible kinematic classes of constant-dip kink-band migration. The motion of the axial surfaces relative to the material is indicated. All five folds appear the same because they are composed of layers that predate the deformation, that is *pregrowth strata*. Only layers deposited during deformation, that is *growth strata*, record the deformational history (Fig. 2).

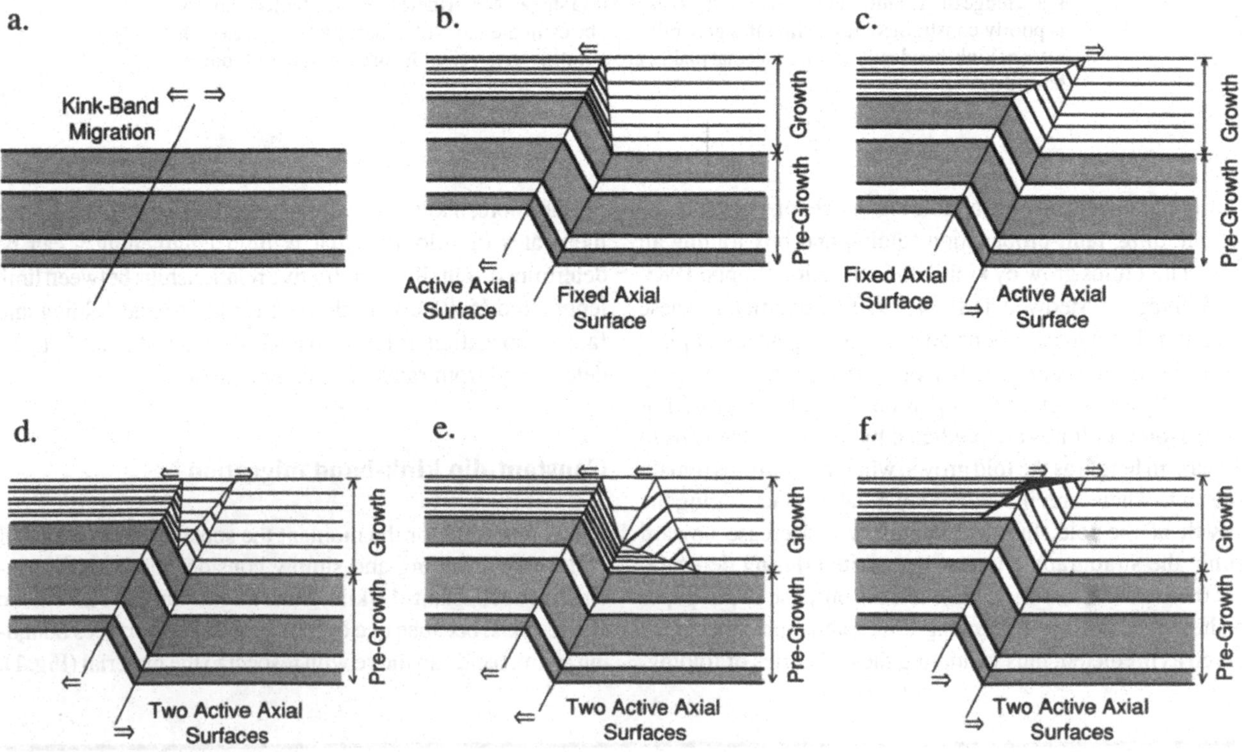

Figure 2. The five possible classes of constant-dip kink-band migration with strata deposited during deformation, called *growth strata*. The five folds only differ within the growth strata, which record the kinematics of deformation. It is assumed in these models that sedimentation rate is constant relative to fold growth rate and that sedimentation keeps ahead of deformation such that there is never bathymetric or topographic expression of the fold. The motion of the axial surfaces is indicated relative to the material.

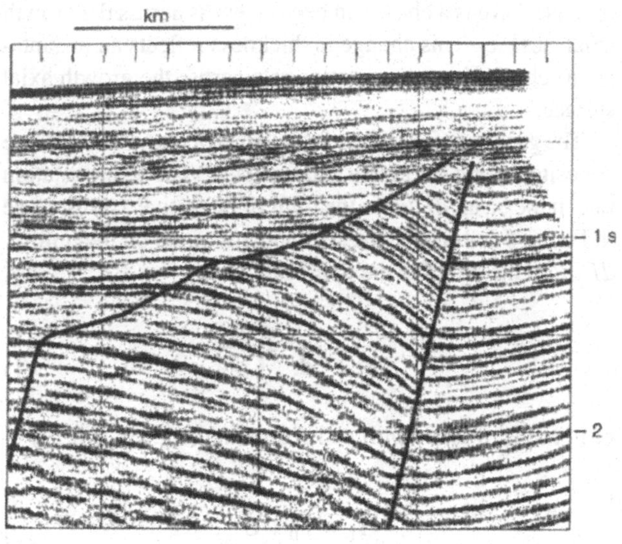

Figure 3. Seismic profile of an upward narrowing kink band within growth strata, Santa Barbara Channel, California. The growth structure is analogous to the model in Figure 2c with the fixed or inactive axial surface to the left and the active axial surface to the right. The boundary between pregrowth and growth strata can be recognized as the top of the constant-width kink band at a depth of about 1.5 s on the fixed axial surface (compare Fig. 2c). The apex of the growth triangle is obscured by horizontal multiples at depths less than 1 s. The average vertical exaggeration in the upper 1.5 s is about 1.7. Seismic profile provided by Texaco USA, BP Exploration Inc., and Sun Operating LP.

An axial surface that moves with respect to the rock is here called an *active axial surface* whereas one fixed in the material is called a *fixed* or *inactive axial surface*. The possible kinematic classes are: one active axial surface moving to the left or right (Figs 1b & 1c), two active axial surfaces moving outward relative to the material (Fig. 1d), and two active axial surfaces moving in the same direction (Figs 1e & 1f).

It is important to note that all the folds in Figure 1 look the same. The final shape does not record the kinematics because all the layers predate the folding; that is, all the strata are *pregrowth strata.* If however sedimentation takes place during deformation the growth strata record the kinematics of kink-band migration. Figure 2 shows the same five folds with the addition of growth strata; now each case appears distinct and in fact contains enough information for bed-by-bed, cinematic retrodeformation.

The specific geometry of the fold within the growth strata strongly depends on the history of sedimentation relative to the history of deformation. The simple models of Figure 2 assume that the rate of sedimentation is constant relative to the rate of folding and that sedimentation keeps ahead of deformation such that bathymetric or topographic relief never develops. Note that in all cases the kink bands narrow upward from a constant width in the pregrowth strata to zero width at the current surface of sedimentation.

It goes without saying that some of the fold shapes in Figure 2 are quite bizarre at first blush. Nevertheless we need to take them seriously because all five possibilities are

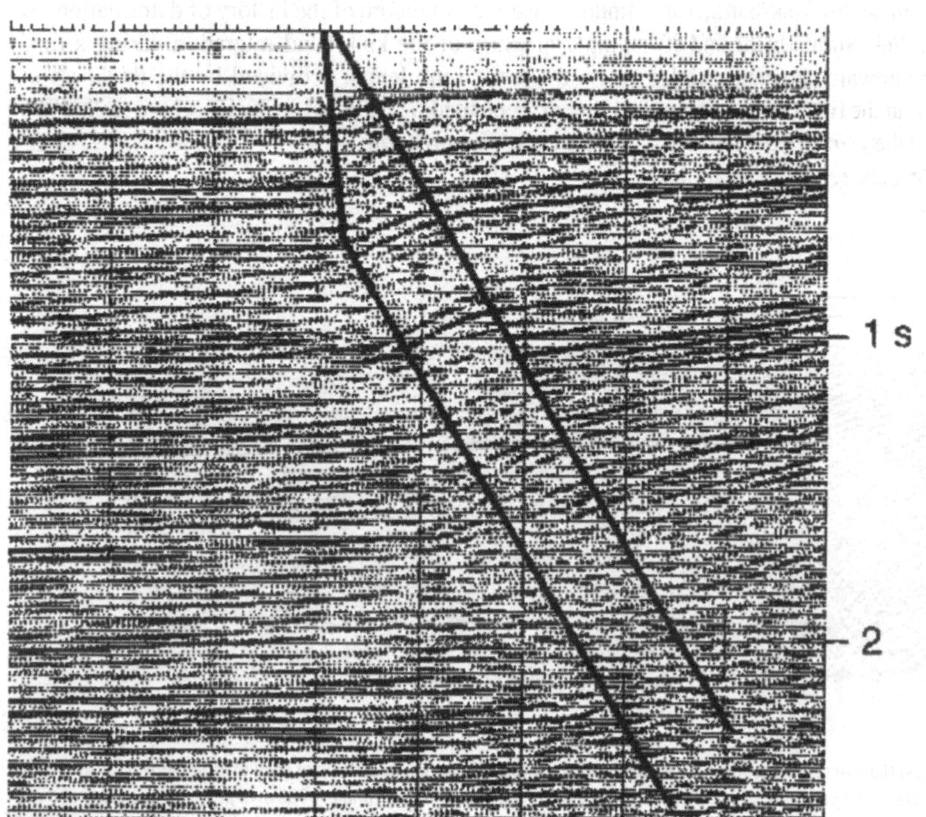

Figure 4. Seismic profile of an upward narrowing kink band within growth strata, Los Angeles basin, California. The growth structure is analogous to the model in Figure 2b with the fixed axial surface to the left and the active axial surface to the right. The beginning of growth can be recognized as the top of the constant-width kink band at a depth of about 0.7 s on the fixed axial surface (compare Fig. 2b). Growth appears to be continuing today because the apex of the growth triangle is the present ground surface, which is depositional. The average vertical exaggeration in the upper 0.9 s is about 1. Seismic profile provided by Texaco USA and Nippon Western US Company, Ltd.

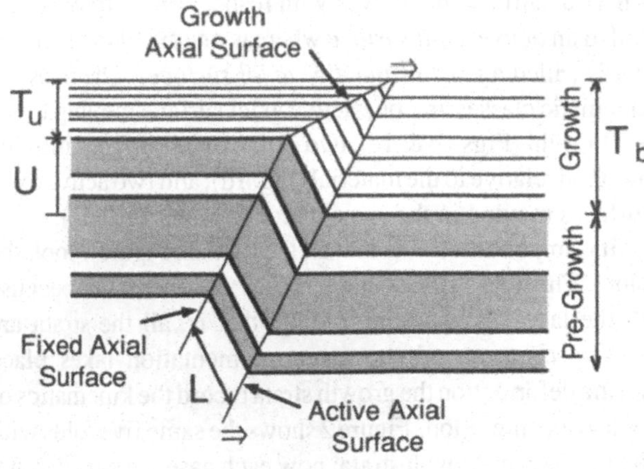

Figure 5. The fixed axial surface within the growth strata is called a *growth axial surface,* which is the locus of particles that were deposited along the active axial surface. The growth axial surface does not bisect the angle between the fold limbs because there is a change in thickness that reflects a primary change in sedimentation rate across the the active axial surface (eq. 1).

predicted by the standard fault-related folding mechanisms (although case 2e is unusual, it is known to exist in normal faulting). All five kinematic possibilities have been observed at map-scale in nature. Figure 3 is a seismic example similar to Figure 2c in which the right-hand axial surface is active and the left-hand axial surface is inactive. Figure 4 is a seismic example similar to Figure 2b.

It should be noted in both the models and the seismic examples (Figs 2b, 2c, 3 & 4) that the beginning of deformation - that is, the pregrowth-growth boundary - can easily be identified as the top of the constant-width kink-band, for which the two axial surfaces are parallel. Similarly the end of growth is marked by the apex of the upward narrowing kink band or *growth triangle.* Also note that the fixed axial surface within the growth strata - here called the *growth axial surface* (Fig. 5) - does not bisect the angle between the fold limbs because there is a change in bed thickness across the growth axial surface. This change in thickness reflects an instantaneous change in sedimentation rate across the growth axial surface.

The growth axial surface is the locus of particles that were deposited along the active axial surface. The sedimentation rate in the basin dT_b/dt to the right of the active axial surface in Figure 5 is equal to the sedimentation rate on the uplift dT_u/dt to the left plus the uplift rate dU/dt

$$\frac{dT_b}{dt} = \frac{dT_u}{dt} + \frac{dU}{dt} \quad (1)$$

or in terms of total uplift and sediment thickness (Fig. 5)

$$T_b = T_u + U \quad (2)$$

In general the sedimentation and uplift rates could be continuously varying; therefore, in the general case the orientation of a growth axial surface will be variable and its shape will be curved or bent. However, in the seismic examples (Figs 3 & 4) the growth axial surfaces are relatively straight overall indicating a relatively constant ratio of deformation rate to sedimentation rate. Examples are given later in this paper that record major variation in this ratio.

Measurement of kink-band migration rates

Growth axial surfaces and growth triangles provide a continuous quantitative record of the history of deformation. A simple measure of the kink-band migration during a time interval t is the limb length L divided by the thickness of sediment deposited T_b (Fig. 5), which is here called the *dimensionless growth rate G.*

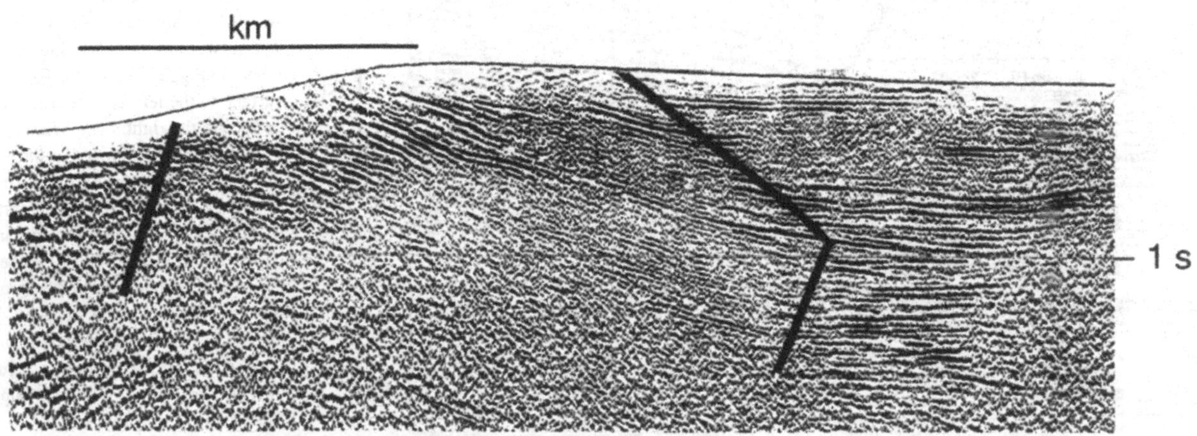

Figure 6. East-west seismic profile of the east flank of the Allegria anticline, western Cebu, Philippines. The beginning of growth is marked by the inflection and reversal of dip on the right-hand axial surface, indicating a kink band of the class shown in Figure 2b. Approximately 900 m of growth strata are preserved which record 600 m of limb lengthening; therefore the dimensionless growth rate G is about 0.7. The average vertical exaggeration in the upper 1 s is about 0.8.

$$G = \frac{L}{T_b} \tag{3}$$

The dimensionless growth rate has the advantage that it is directly observable in the fold geometry and that it can be measured for each identifiable stratigraphic interval δT_b

$$G = \frac{\delta L}{\delta T_b} \tag{4}$$

Furthermore if key beds can be dated to determine a time interval δt, then the dimensionless growth rate provides the relationship between the sedimentation rate $\delta T/\delta t$ and the *absolute fold-growth rate* $G_a = \delta L/\delta t$.

$$G_a = \frac{\delta L}{\delta t} = G\frac{\delta T}{\delta t} \tag{5}$$

Thus the dimensionless growth rate is a geometric measure of the rate of folding.

A seismic example of a measurement of a dimensionless growth rate is given in Figure 6, which is the east limb of the 6 km wide Allegria anticline in Cebu. The axial surfaces are simply located as the boundaries between regions of homogeneous dip (Suppe & Chang 1983). The kinematic class of kink band is that of Figure 2b. The inflection on the right-hand axial surface marks the beginning of growth; approximately 900 m of growth strata are preserved above the inflection, which were deposited during 600 m of limb lengthening. Therefore, the dimensionless growth rate is about 0.66. The time involved is no more than about 4Ma, yielding an absolute fold-growth rate G_a of not less than 0.15 mm/y.

The seismic example from the Santa Barbara Channel, California in Figure 3 has a limb length of about 2.5 km with an associated 2.1 km of growth strata, yielding a dimensionless growth rate G of 1.2. The time involved is 2-4Ma therefore the absolute fold-growth rate G_a is about 0.6-1.3 mm/y. A similar growth structure in the City of Los Angeles, California (Fig. 7) has a dimensionless growth rate G of 0.6 and an absolute fold-growth rate G_a of about 1 mm/y.

Effects of variation in sedimentation rate

The examples and models of growth structures shown so far are probably not typical of the full scope of growth structures because they all involve high sedimentation rates relative to the rates of fold growth (dimensionless growth rates less than 1). In particular sedimentation keeps ahead of deformation in these models and geological examples such that there is no bathymetric or topographic expression of folding, which is expected not to be true in many cases. Expected absolute fold-growth rates in tectonic environments are not well known but probably do not range much below one or two orders of magnitude less than plate motions, that is perhaps fold-growth rates are in the range 0.1 mm/y-1 cm/y. In contrast, sedimentation rates are known to range over many

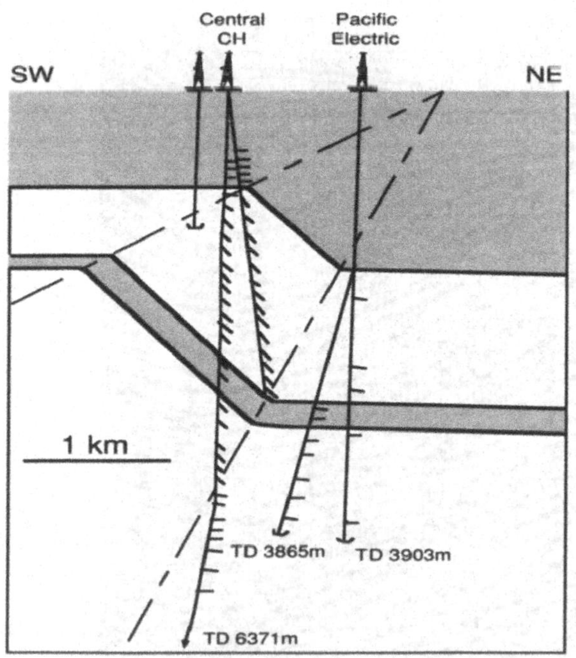

Figure 7. Cross section through a growth structure in the Los Angeles basin, California, based on well and seismic data. The right-hand axial surface is the active axial surface. A dimensionless growth rate G of about 0.6 is indicated for approximately the last 2 Ma, implying an absolute fold-growth rate of about 1 mm/y. This apparently active structure extends at least 40 km along strike in the Los Angeles metropolitan area.

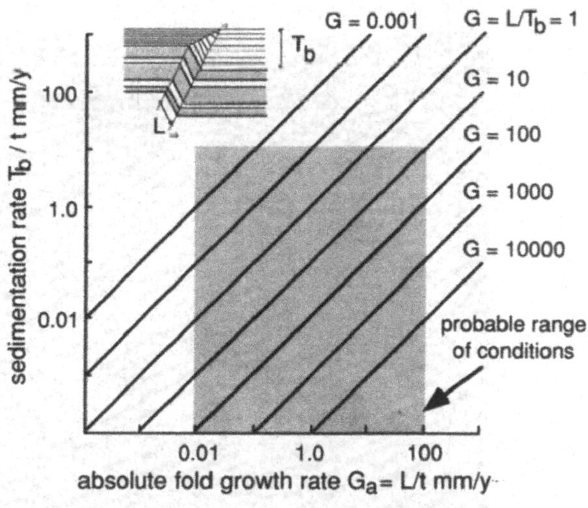

Figure 8. Graph showing the expected geological range of absolute fold-growth rates $G_a = L/t$ and sedimentation rates T_b/t in relation to the dimensionless growth rate $G = L/T_b$. Tectonic folding rates in some cases might be as fast as plate-tectonic rates (1 cm/y-10 cm/y) and probably are never much less than 0.1 mm/y, whereas sedimentation rates rarely exceed 1 mm/y and can be several orders of magnitude less. Therefore sedimentation rates are expected to be commonly one or two orders of magnitude less than absolute fold-growth rates, leading to dimensionless growth rates $G = L/T_b$ of 10-100 or greater. Dimensionless growth rates less than 1, which are shown in the models of Figure 2, require high sedimentation rates relative to fold-growth rates. Models with lower sedimentation rates are shown in Figure 9.

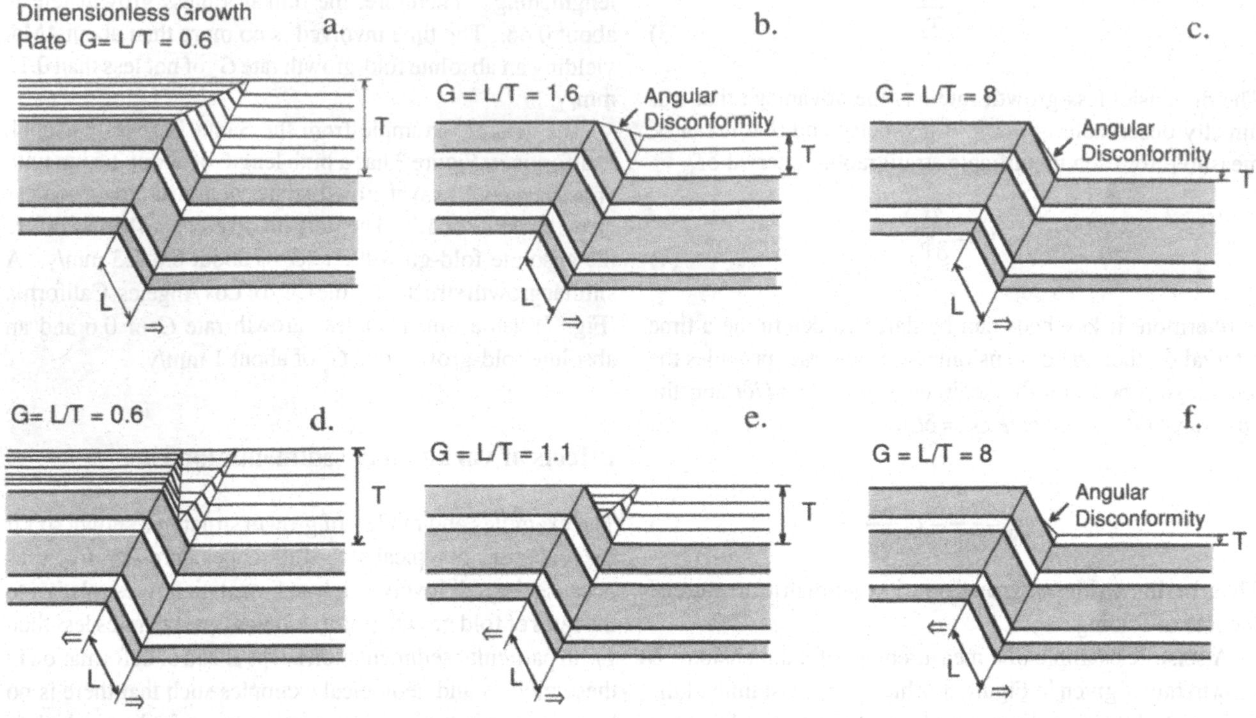

Figure 9. Models showing the effect of reducing the sedimentation rate in a growth structure. The upper models are for one active axial surface moving to the right. The lower models are for two active axial surfaces moving outward relative to the material. Angular disconformities develop when the boundary between deposition and non-deposition, which is the edge of topographic or bathymetric relief, lies along an active axial surface.

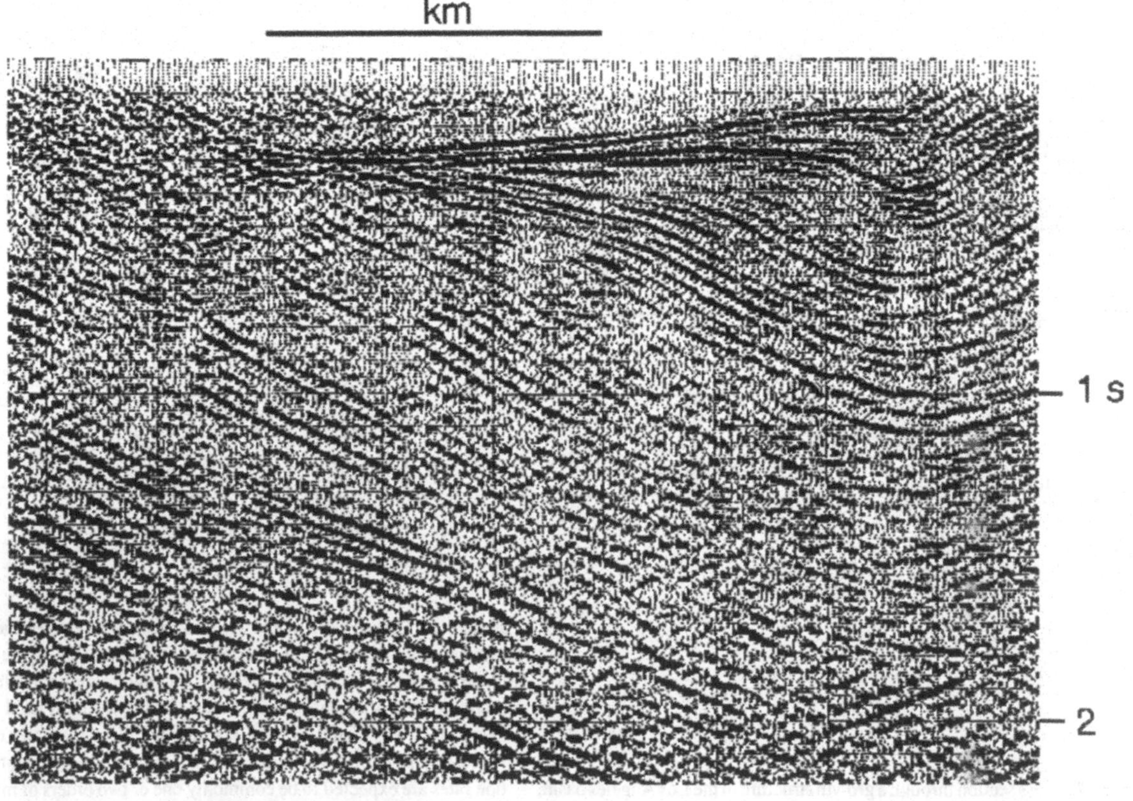

Figure 10. An east-west seismic profile of a growth structure similar to Figure 9e from Antelope Hills, San Joaquin basin, California. Vertical and horizontal scales are approximately equal.

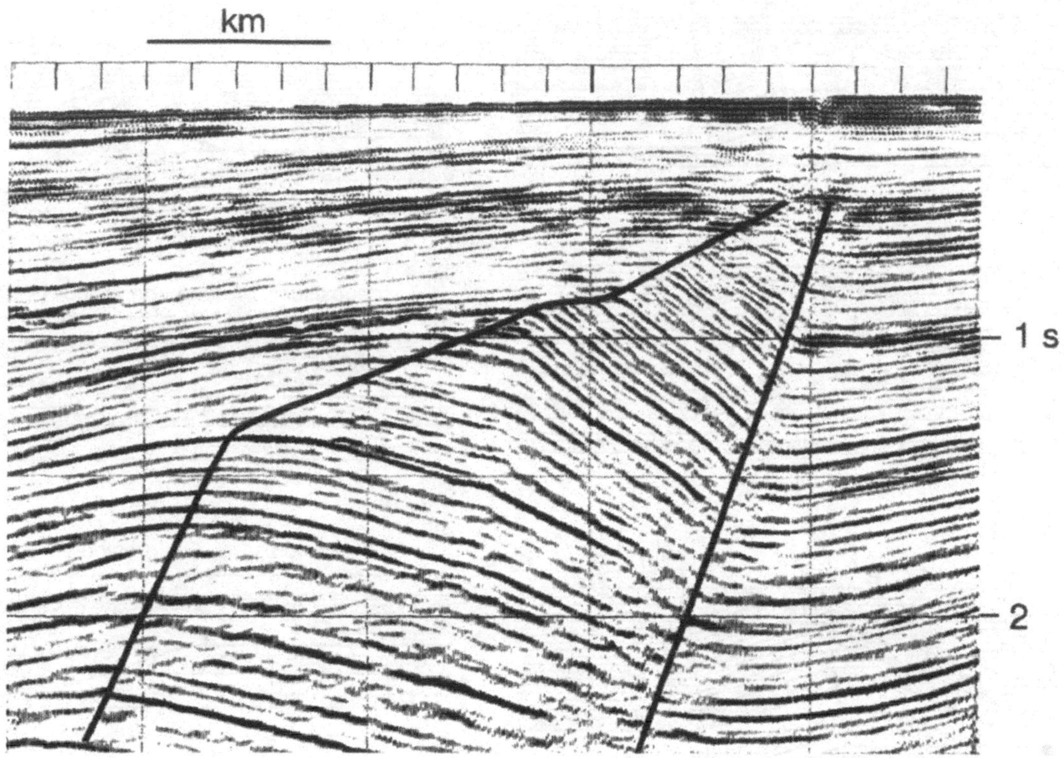

Figure 11. Seismic profile showing a short-term decrease in sedimentation rate relative to the rate of limb lengthening at a depth of about 0.9 s; Santa Barbara Channel, California. The average vertical exaggeration in the upper 1.5 s is about 1.7. The apex of the growth triangle is obscured by horizontal multiples at depths less than 0.7 s. Seismic profile provided by Texaco USA, BP Exploration Inc., and Sun Operating LP.

orders of magnitude but are expected to be commonly one or two orders of magnitude less than deformation rates, leading to dimensionless growth rates $G = L/T_b$ of 10-100 or greater (Fig. 8). A fast deformation rate is expected to be on the order of centimetres per year whereas fast sedimentation rates are on the order of millimetres per year.

Growth structures appear drastically different with lower sedimentation rates; for example Figure 9 shows that even a modest one-half reduction of sedimentation rate makes the uplifting block non-depositional in the models of Figure 2. This gives rise to time-transgressive, cross-cutting surfaces of nondeposition, here called *angular disconformities*. An example of an onlapping time-transgressive disconformity similar to Figure 9e is shown in Figure 10. If the uplifting block undergoes erosion, then time-trangressive angular unconformities are predicted, examples of which are shown in later sections.

The simple models at high sedimentation rate in Figures 2 and 9 assume that sedimentation is independent of deformation. In contrast we may expect that the bathymetric or topographic relief that is produced at low sedimentation rates will have a direct impact on the process of sedimentation. Both sedimentary facies and sedimentation rates will be affected by structure - for example pelagic drape, downslope transport, and shoreline facies - which give rise to more complex fold and reflector geometry within the growth strata. The differences between the model in Figure 9e and the data in Figure 10 are thought to be caused by heterogeneous sedimentary facies and associated heterogeneous compaction. These complexities are largely ignored in this paper.

Relatively short term fluctuations in sedimentation rate - for example caused by climactic variation - are expected to be reflected in the fine structure of the growth axial surfaces. For example the seismic line in Figure 11 shows a local fluctuation in dimensionless growth rate at a depth of about 0.9 s whereas the long-term dimensionless growth rate is quite constant. It is of course possible that these changes in dip of the growth axial surface could reflect short-term fluctuations in deformation rate, but sedimentation rate is a more likely cause given known climatic fluctuations.

Growth fault-bend folding

So far we have considered growth structure only from the perspective of kink-band migration, ignoring the fundamental causes of the motion of active axial surfaces. This approach is useful because growth structure is observed commonly at shallow depths on seismic profiles, whereas the deep structure causing the axial surfaces to move commonly is not imaged or is poorly imaged. Insight into these deep structural causes is obtained most rapidly by constructing models of the growth structures that are predicted for the known common mechanisms of fault-related folding. Fault-bend folding is discussed in this section and fault-propagation folding in the next. The growth structure associated with normal faulting is presented in detail elsewhere (Xiao & Suppe 1990), but is essentially analogous to compressive

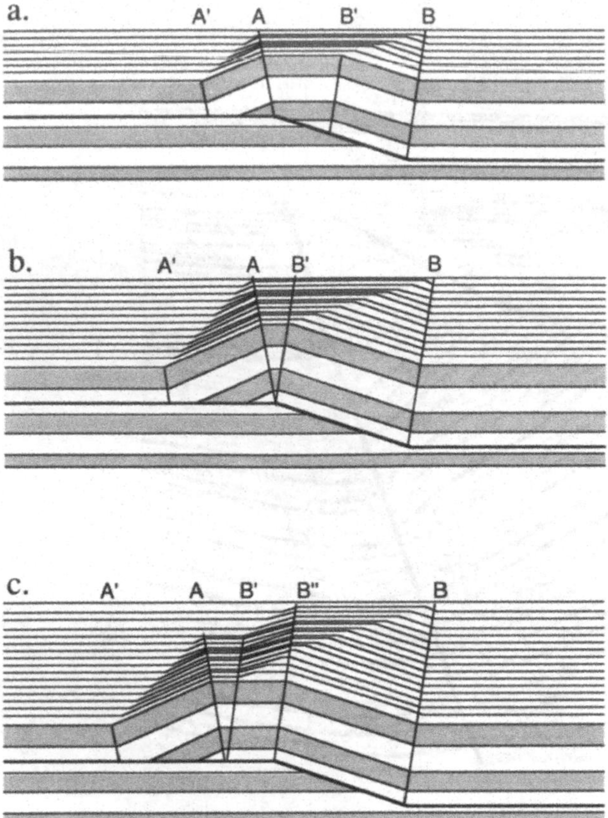

Figure 12. Model of the sequential development of a fault-bend fold at high sedimentation rate such that no bathymetric or topographic relief develops. The active axial surfaces are *A* and *B* from the beginning of deformation until instant of Figure *b*, at which point *A* is released from the footwall bend and becomes inactive. At the same instant axial surface *B'* spawns a new axial surface *A''* that is pinned to the footwall bend and begins to form a new kink band bounded by inactive axial surface *B'* and active axial surface *B''* (Fig. *c*). Kink band *B'*-*B''* is only visible in the growth strata because the folding of pregrowth strata along *B'* at the instant of Figure *b* unfolds the fold that was present in pregrowth strata along *B'* at the stage of Figure *a*. At the stage of Figure *c* axial surfaces *B* and *B''* are active (analogous to Fig. 2f) and *A'*, *A*, and *B'* are inactive.

fault-bend folding except that the active axial surfaces normally do not bisect interlimb angles, which is the condition for conservation of layer thickness.

The simplest fault-bend fold involves a thrust fault stepping up from one decollement horizon to another, for example the classic Cumberland or Pine Mountain thrust sheet of the southern Appalachians (Rich 1934). Figure 12 shows a sequential set of models of a simple-step fault-bend fold with a high but constant sedimentation rate relative to the slip rate, such that no bathymetry or topography develops. Figure 12a shows the fold at an early stage of development. The front side of the anticline shows a kink band *A*-*A'* with the left-hand axial surface inactive and the right-hand active, analogous to the kink-band model of Figure 2b and the seismic lines of Figures 4 & 6. The back side of the anticline shows a kink-band *B*-*B'* analogous to the kink-band model of Figure 2c, the seismic lines of Figures 3 & 11, and the cross section of Figure 7. Note that the two active axial surfaces are pinned to bends in the footwall - that is, footwall cutoffs - whereas the inactive axial surfaces terminate at the corresponding hangingwall

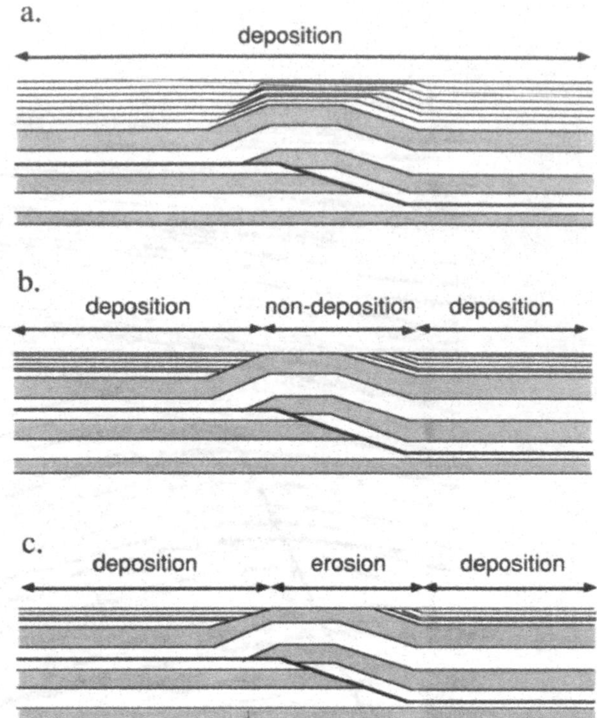

Figure 13. Models showing the effect of changing sedimentation rate on geometry of growth fault-bend folding. Figure *a* is for conditions similar to Figure 12a, Figure *b* is for uplift on the ramp equal to sedimentation rate in the basin, and Figure *c* is for uplift greater than sedimentation. The upper surface in these models is constrained to be horizontal such that no relief develops.

cutoffs. Thus the total slip on the fault is the width of the kink-band in the pregrowth strata and the history of slip is recorded in the upward-narrowing growth triangles. Note that even if the fault were not imaged, the overall asymmetry of the growth structure would betray the leftward vergence of the thrusting - a fact that would not be known from the shape of the fold in pregrowth strata alone.

A sudden change in kinematics takes place when the slip equals the length of the ramp (Suppe 1983), that is when the hangingwall cutoff associated with *B'* reaches the footwall cutoff associated with *A* (Fig. 12b). At this instant axial surface *A* is released by the footwall cutoff while axial surface *B'* spawns a new axial surface *B''* that is pinned to the footwall cutoff (Fig. 12c). At this final stage both axial surfaces of the

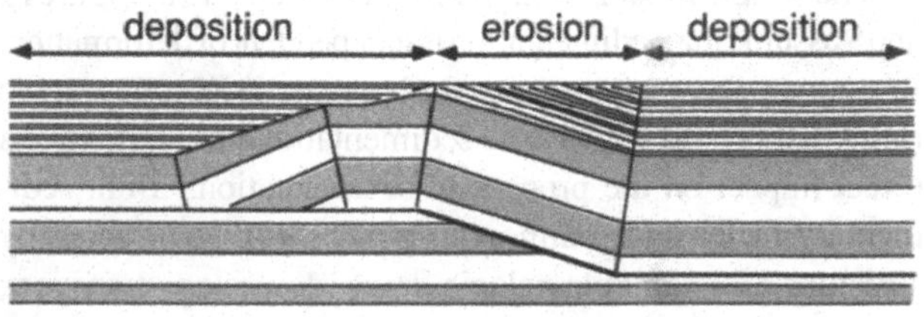

Figure 14. Model of fault-bend fold under erosive conditions that has moved beyond the top of the ramp. An earlier stage in fault slip is shown in Figure 13c. The upper surface in this model is constrained to be horizontal such that no relief develops.

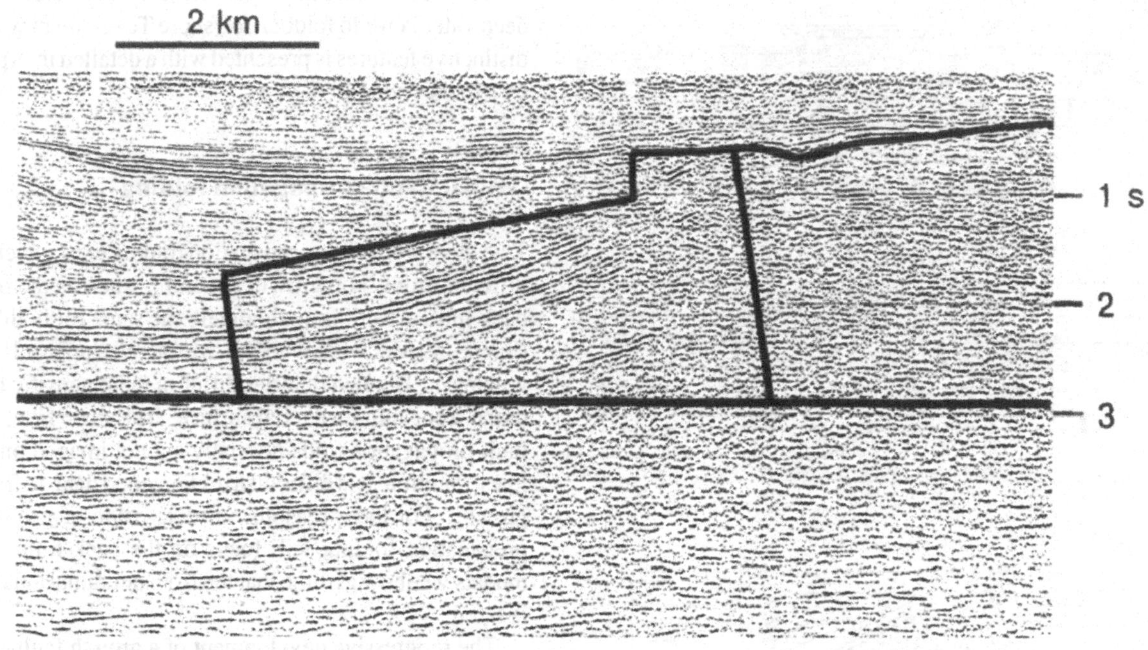

Figure 15. Seismic example of an erosive growth fault-bend fold analogous to the front flank and crest of the model fault-bend fold in Figure 14; eastern Venezuelan basin. Notice that the synclinal axial surface terminates at the unconformity above and the fault below. The steep axial surface connecting two segments of unconformity in the front flank of the anticline records an interval of higher sedimentation rate relative fold growth rate. The poorly imaged nearly flat fault is the Pirital thrust. The average vertical exaggeration in the upper 1.0 s of syncline is approximately 0.8.

back kink band are active and are analogous to the axial surfaces in the kink-band model of Figure 2f.

It is important to realize that in fault-bend folding the particles are moving everywhere parallel to the local orientation of the fault. Uplift takes place instantaneously only over the ramp, therefore thin sedimentation takes place over the ramp in the zone bounded by the two active axial surfaces (*A* and *B* in Fig. 12a and *B* and *B''* in Figs 12b & 12c). The sedimentation rate over the ramp is equal to the sedimentation rate in the basin minus the uplift rate over the ramp. Therefore the swath of thin beds running diagonally upwards from left to right in Figure 12c are the beds that were deposited over the ramp and the growth axial surfaces bounding the thin sediments record the locus of particles deposited on the active axial surfaces that bound the region of thin sedimentation over the ramp.

If we lower the sedimentation rate in our models the

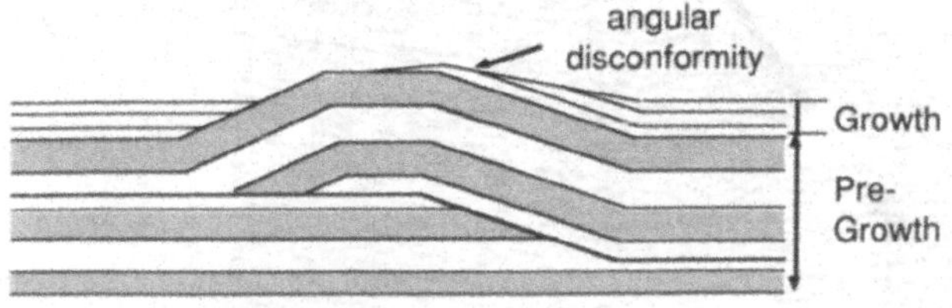

Figure 16. Model of growth fault-bend fold with sedimentation rate less than uplift rate and non-deposition over bathymetric or topographic relief. The structurally highest part of the fold is along the back anticlinal axial surface where the angular disconformity is being folded. Note that growth strata are missing at the front anticlinal axial surface. A similar fold from the Perdido foldbelt of the Gulf of Mexico is modelled in detail by Mount *et al.* (1990).

thickness of strata deposited over the ramp decreases and the dip of the growth axial surfaces decreases. In the limit of nondeposition over the ramp (Fig. 13b) the growth axial surface on the back limb is horizontal and is an angular disconformity whereas the growth axial surface on the front limb is parallel to the front-limb dip and is a surface of disconformable onlap. For any horizon within the growth sequence, the region in which a bed is missing records the region of non-deposition over the ramp. If we move into the erosive regime we observe down cutting into the pregrowth strata (Fig. 13c). The magnitude of erosion is proportional to the time spent over the ramp; for example note the progressively deeper erosion updip on the front flank of Figure 13c.

Figure 14 is a model of a fault-bend fold that has slipped beyond the top of the ramp in the erosive regime. Figure 15 is a seismic example from eastern Venezuela that is analogous to the front flank and crest of the model. Notice the progressive down cutting of the pregrowth strata and the progressive onlap of the growth strata. Also notice in the model that the unconformity has been folded on the crest such that the unconformity is dipping but the beds above and below are horizontal. This relationship is observed at the right edge of the seismic example and in other lines along strike. This same model has been applied to Lost Hills anticline, California (Medwedeff & Suppe 1986a, 1986b; Medwedeff 1989).

If the uplifting block does not erode then substantial topography or bathymetry can develop, as shown in Figure 16. Notice that once the fault slips beyond the top of the ramp the angular disconformity is folded such that the bathymetrically highest point on the fold is along the back anticlinal axial surface whereas growth strata are missing

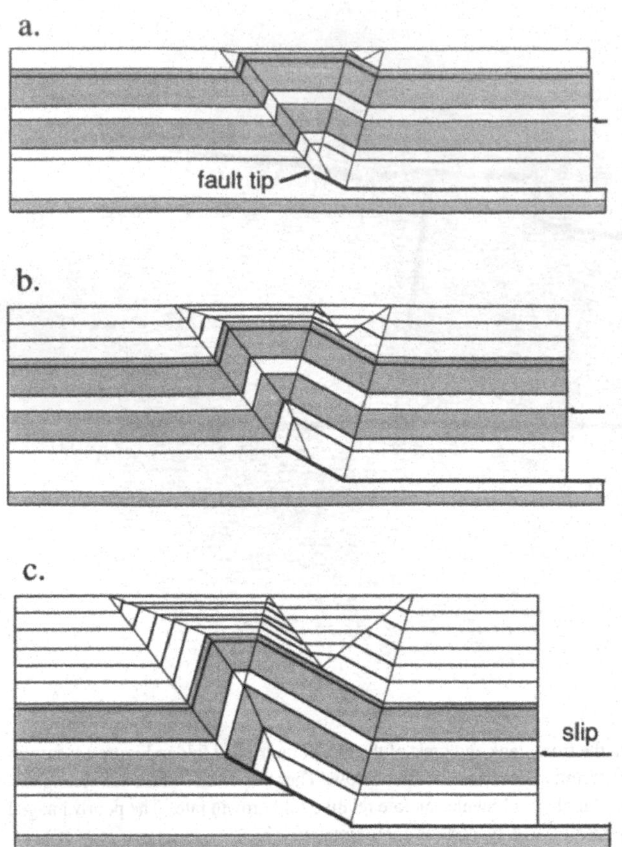

Figure 17. Progressive development of a fault-propagation fold at high sedimentation rate such that no bathymetric or topographic relief develops. Both axial surfaces are active on the back limb and move outward relative to the material (see Fig. 2d); the triangle of flat-lying beds were deposited over the back limb and record its progressive widening. The front synclinal axial surface is active and the front anticlinal axial surface is fixed (see Fig. 2c).

over the front anticlinal axial surface. A structure from the deepwater Perdido foldbelt, offshore Texas, displaying these distinctive features is presented with a detailed interpretation by Mount *et al.* (1990).

Growth fault-propagation folding

Two theories of fault-propagation folding exist which predict different front-limb growth structure, whereas the predicted back-limb growth is qualitatively the same for both theories (Suppe & Medwedeff 1990; Mosar & Suppe (1991, this volume)). The two theories differ in the kinematic behavior of the front anticlinal axial surface. The steep front limbs of fault-propagation folds normally are not imaged on seismic lines, therefore it is difficult to obtain good data on front-limb growth. For these reasons our discussion of growth fault-propagation folding will focus on the back limb. All the models we show assume that the front anticlinal axial surface is fixed.

The progressive development of a growth fault-propagation fold at high sedimentation rate relative to slip rate is shown in Figure 17. Both axial surfaces bounding the back limb are active and moving outward, therefore the kink-band migration is analogous to Figure 2d. The inverted triangle of horizontal beds over the back limb records the progressive widening of the kink-band during growth. A seismic example from the Oak Ridge anticline, Ventura basin, California is shown in Figure 18. The steep front limb is known from drilling but is not imaged on the seismic profile (for a well section along strike see Suppe & Medwedeff 1990). A growth fault-propagation fold from the Perdido foldbelt showing the characteristic flat dips over the long back limb and the unimaged front limb is shown in Figure 19. Another example from the same region is interpreted in detail by

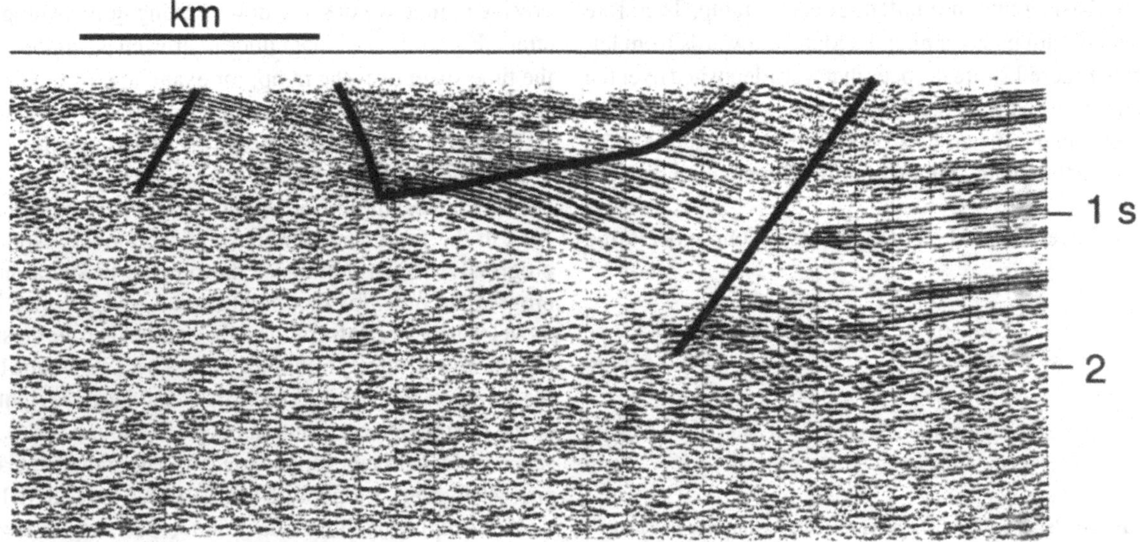

Figure 18. Seismic profile of the back limb of a growth fault-propagation fold showing an upward widening triangle of gently dipping beds similar to Figure 17, Oak Ridge anticline, Ventura basin, California. A well section of this structure along strike is given by Suppe & Medwedeff (1990), which shows it to be a fault-propagation fold less than a million years old.

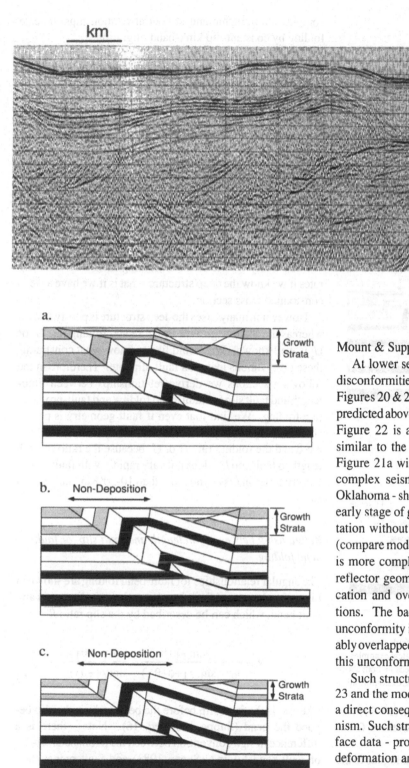

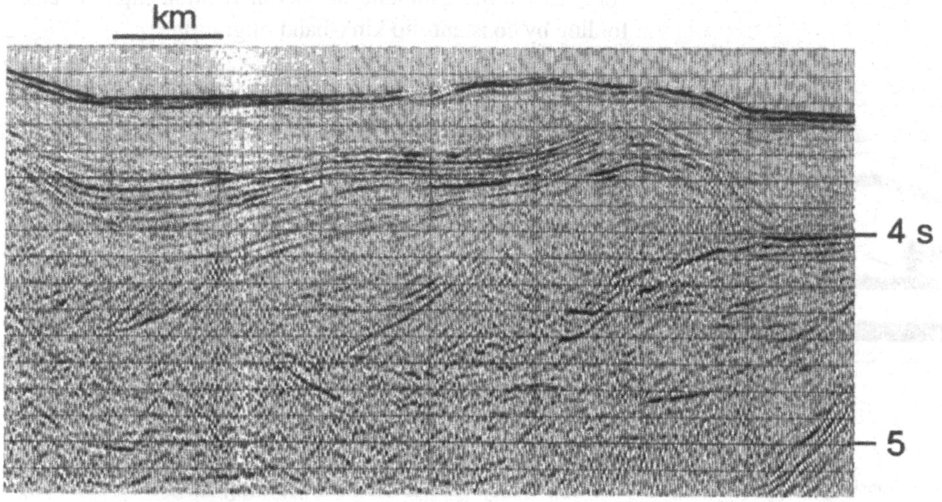

km

4 s

5

Figure 19. Seismic profile of growth fault-propagation fold, Perdido foldbelt, offshore Texas.

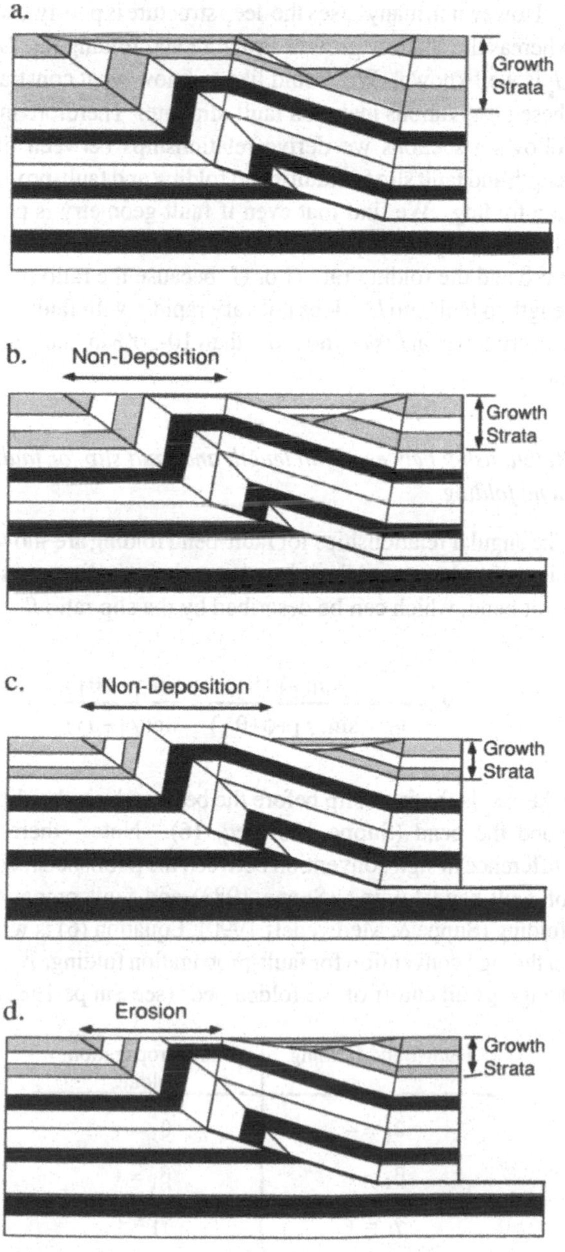

a.

Growth
Strata

b.

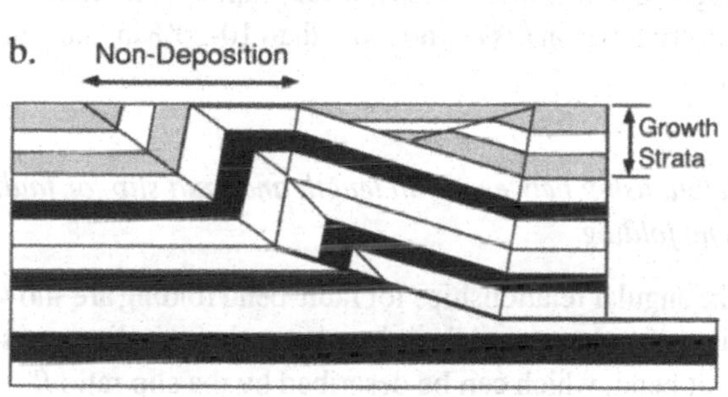

c.

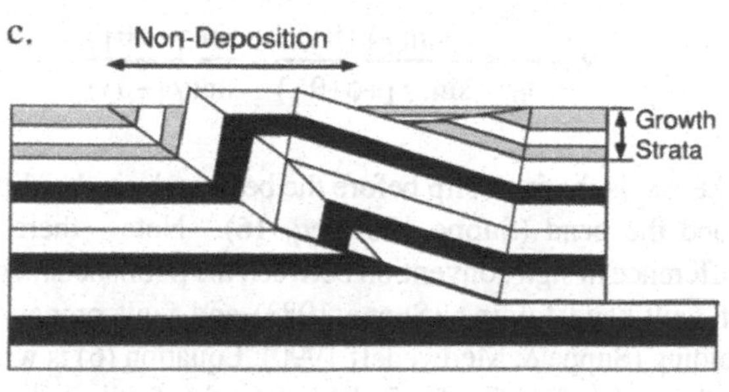

d.

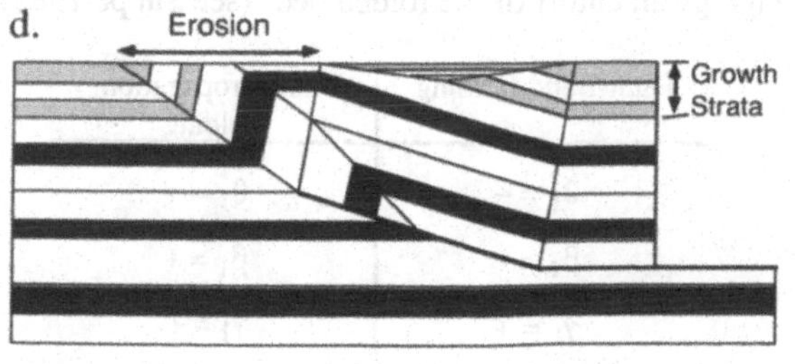

Mount & Suppe (1990).

At lower sedimentation rates time-transgressive angular disconformities and unconformities are predicted as shown in Figures 20 & 21. Note in Figure 21 that the same age beds are predicted above and below the unconformity or disconformity. Figure 22 is a seismic example from southern Oklahoma similar to the back limb of the balanced growth model in Figure 21a with non-deposition on the fold crest. A more complex seismic example (Fig. 23) - also from southern Oklahoma - shows erosive growth similar to Figure 20d at an early stage of growth followed by a long period of sedimentation without much deformation followed by deformation (compare model Fig. 21b). In detail the first period of growth is more complex than the model as shown by the detailed reflector geometry, which shows alternation between truncation and overlap - possibly recording sea-level fluctuations. The back-limb growth predicts a nearly orthogonal unconformity in the front limb (Fig. 21b) which is conformably overlapped by the base of the upper stratigraphic sequence; this unconformity is observed as predicted in drilling.

Such structures as shown in the seismic profile of Figure 23 and the model of Figure 21b are of course bizarre but are a direct consequence of the kinematics of the folding mechanism. Such strange structures require very extensive subsurface data - properly interpreted - to be documented or later deformation and erosion to allow them to be mapped at the surface. The fact that these structures have the same age beds above and below the unconformities and that beds abruptly change from conformable to unconformable relationships suggests that such structures would be widely misinterpreted in the subsurface. Similar time-transgressive angular

Figure 20. Models of growth fault-propagation folds showing the effect of varying sedimentation rate. In all but Figure *c* the upper surface is constrained to be horizontal. Figures *c* and *d* have the same sedimentation in the basin, whereas *c* is non-erosive and *d* is completely erosive. Note that the growth axial surfaces flatten as the sedimentation rate decreases.

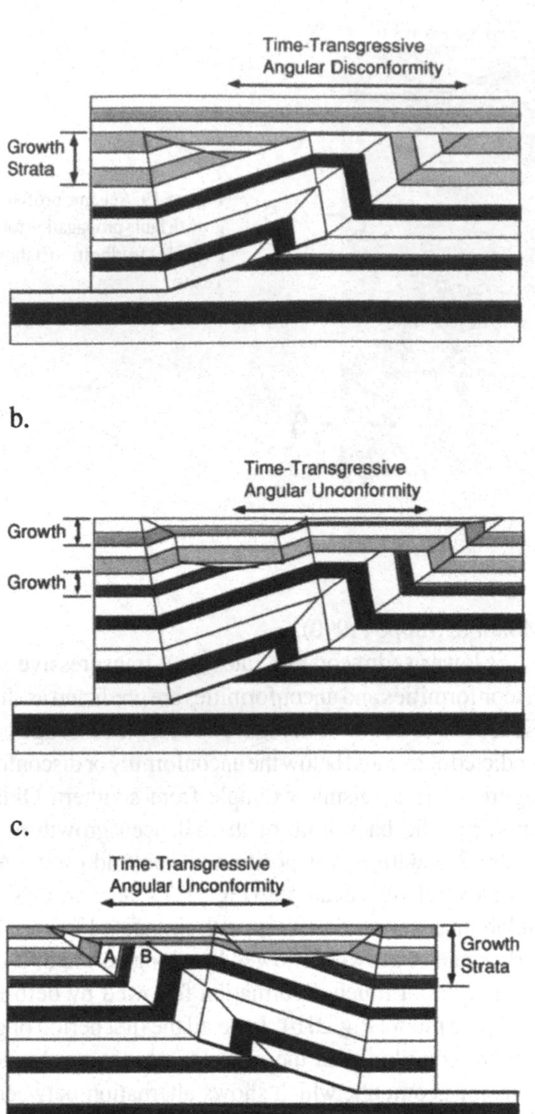

Figure 21. Three models of growth fault-propagation folds showing the effect of varying sedimentation history. Analogous geological examples are shown in Figures 22, 23 & 24. Notice that the same age beds lie both above and below the time-transgressive angular unconformities and disconformities.

unconformities were documented by Tomlinson (1952) in his classic paper 'Odd Geologic Structures in Southern Oklahoma'.

Dibblee (1966, 1986) mapped a structure analogous to the predicted front-limb structure of Figures 21b & 23 in the Transverse Ranges of California as shown in Figure 24. Notice that on the left-hand side of Figure 24 - below point *A* - the Juncal Formation lies conformably on Cretaceous shale. At point *A*, where the anticlinal axial surface disappears, the contact suddenly makes a right-angle bend to become a nearly orthogonal angular unconformity (compare point *A* in Fig. 21). The angular unconformity extends to the right to point *B* where the anticlinal axial surface reappears, similar to point *B* in Figure 21. To the right of point *B* the Juncal Formation lies on the Espada Formation - roughly parallel but with several kilometres of section cut out between *A* and *B*. These complex stratigraphic and structural relationships indicate folding by constant-dip kink-band migration.

Relationships between limb length and fault slip

We would like to transform our observations of rates of growth of folds into fault-slip rates for both scientific and practical reasons, for example to constrain the kinematics and dynamics of deformation in the Earth and for assessing earthquake hazards. These observations of growth structure at shallow depths based on seismic profiles and wells allow us to measure directly the rates of folding or limb lengthening by way of the dimensionless growth rate $G = \delta L/\delta T$ or, if we can date key beds, the absolute fold-growth rate $G_a = \delta L/\delta t$. We can easily transform these folding rates into fault-slip rates if we know the deep structure - that is if we have a well-constrained cross section.

However in many cases the deep structure is poorly known whereas the shallow growth structure and folding rate G or G_a is well known. We would like to know what constraints these observations make on fault-slip rate. Therefore in the following sections we derive relationships between limb-length and fault slip for fault-bend folding and fault-propagation folding. We find that even if fault geometry is poorly constrained, fault-slip rates are well constrained by the limb dip δ and the folding rate G or G_a because the ratio of limb length to fault slip L/s does not vary rapidly with fault shape. Uncertainties in L/s are no more than 10-20% in many actual cases.

Relationship between limb length and fault slip for fault-bend folding

The angular relationships for fault-bend folding are shown in Figure 25. In general there is a change in fault slip across any fault bend, which can be described by the slip ratio R

$$R = \frac{s}{s_o} = \frac{\sin(\gamma_1+\theta_1)}{\sin(\gamma_1+\phi+\theta_1)} = \frac{\sin(\gamma_1+\theta_1)}{\sin(\gamma_1+\theta_2)} \quad (6)$$

where s_o is the fault slip before the bend and s is the slip beyond the bend (Suppe 1983, eq. 16). Note - there is a difference in sign convention between the published theories of fault-bend folding (Suppe 1983) and fault-propagation folding (Suppe & Medwedeff 1990): Equation (6) is written in the sign convention for fault-propagation folding. β_1 is the hangingwall cutoff of the folded beds (see Suppe 1983).

Fault-Bend Folding	Fault-Propagation Folding
$\theta_1 = -$	$\theta_1 = +$
$\beta_1 = -$	$\beta_1 = +$
$\gamma_1 = +$	$\gamma_1 = +$
$\phi = +$	$\phi = +$

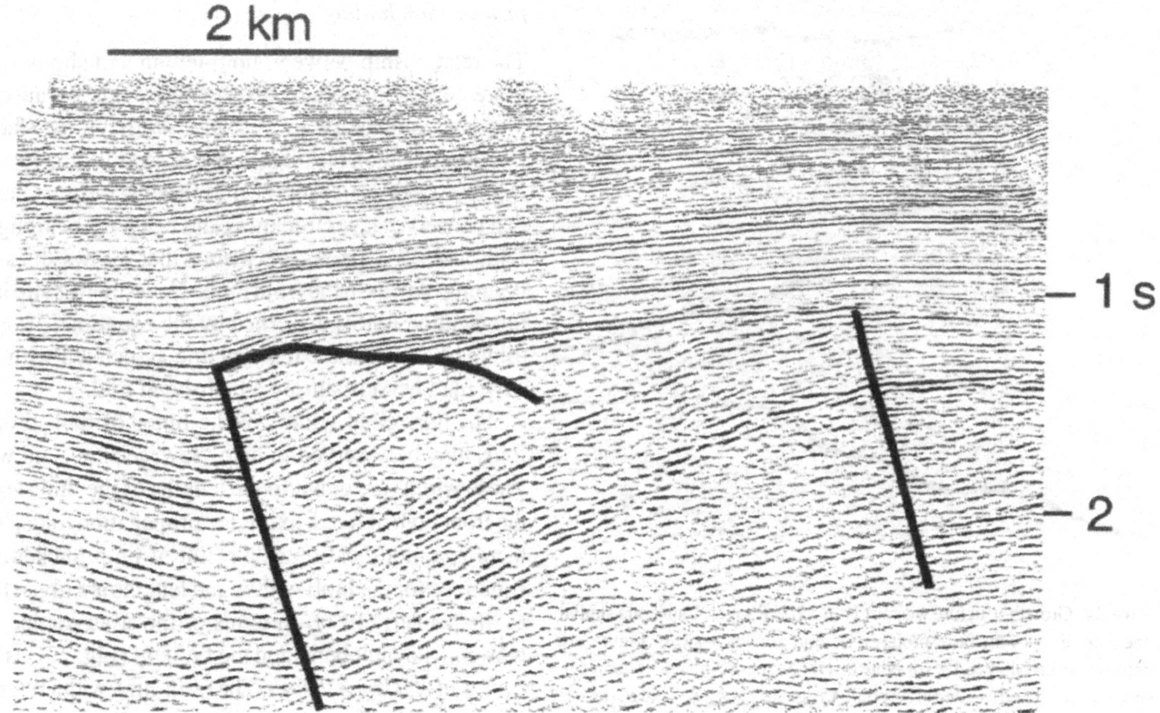

Figure 22. Seismic profile of the back limb and crest of a growth fault-propagation fold from southern Oklahoma. The growth axial surface terminates in the middle of the back limb at a time-transgressive disconformity similar to the model of Figure 21a. Average vertical exaggeration in the upper 1.5 s is about 0.9. Seismic profile provided by Professional Geophysics Inc. (PGI).

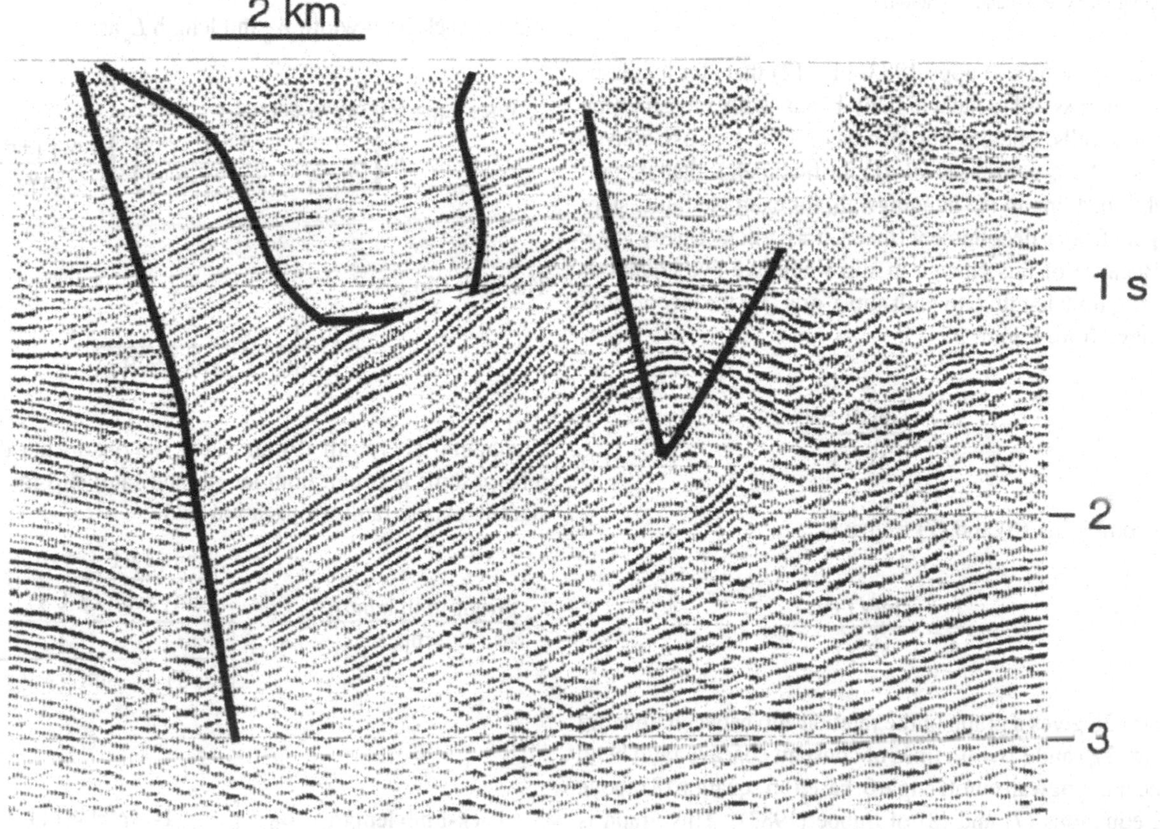

Figure 23. Seismic profile of a growth fault-propagation fold from southern Oklahoma, which is similar to the model of Figure 21b. The structure grew under erosive conditions at an early stage, after which it stopped growing or grew slowly relative to the sedimentation. The structure was reactivated late in the depositional sequence. The strong angular unconformity that is predicted for the front limb (Fig. 21b) has been encountered in drilling but is not imaged on the seismic profile. A similar unconformity is shown in Figure 24. Average vertical exaggeration in the upper 1.5 s is about 1.5. Seismic profile provided by Halliburton Geophysical Services, Inc.

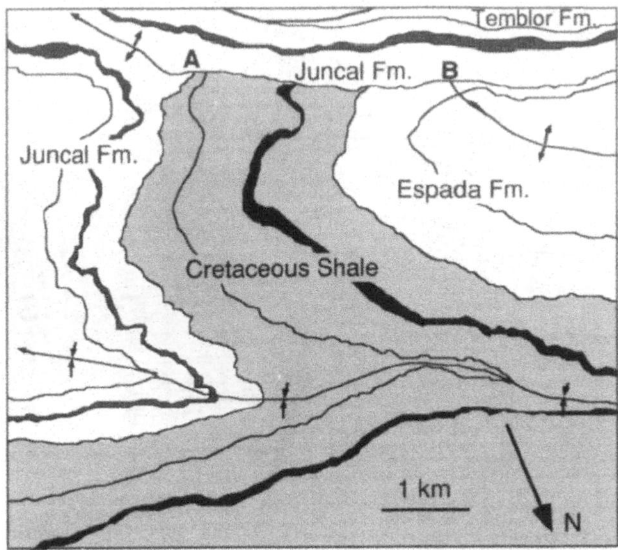

Figure 24. Growth structure western Transverse Ranges, California similar to the front limb of the fault-propagation fold growth model of Figure 21c (simplified from Dibblee 1966). Note that the Juncal Formation is conformable to the Cretaceous shale below point *A* whereas between *A* and *B* a nearly orthogonal discordance exists. The Juncal Formation is approximately conformable to the Espada Formation to the right of *B*. The anticlinal axial surface that disappears at point *B* reappears at point *A*, as predicted in the model of Figure 21c. These complex stratigraphic and structural relationships are indicative of growth during sedimentation and erosion with deformation by kink-band migration.

It can be shown (Suppe 1983, Fig. 12) that slip increases (R>1) across all concave fault bends with cross-cutting hangingwalls whereas slip decreases (R<1) across convex bends. Slip is conserved (R =1) for fault bends that are parallel to bedding in the hangingwall, for example faults stepping up from a decollement (that is $\phi = \theta_2$, note π-[$\phi+\gamma_1$]=γ_1).

Because of this change in slip we derive two equations relating limb length L to fault slip s_0 or s. By the law of sines we have from Figure 25

$$\frac{L}{s_0} = \frac{\sin(\gamma_1+\theta_1)}{\sin\gamma_1} \tag{7}$$

and combining (6) and (7) we have

$$\frac{L}{s} = \frac{\sin(\gamma_1+\theta_2)}{\sin\gamma_1} \tag{8}$$

Figure 26 gives a graph of the relationship between limb dip $\delta = (\pi-2\gamma_1)$ and L/s with lines of constant concave fault bend ϕ and hangingwall cutoff angle θ_1 based on equation (8) above and equations (7) and (8) of Suppe (1983). This graph is discussed after derivation of the corresponding equations for fault-propagation folding.

Relationship between limb length and fault slip for fault-propagation folding

The relationship between limb-length and slip is somewhat more complex for fault-propagation folds because of their intrinsic complexity and because we have two theories of fault-propagation folding (Suppe & Medwedeff 1990; Mosar & Suppe (1991, this volume)): (1) a constant-thickness theory that conserves layer thickness and bed length throughout the fold, and (2) a fixed-axis theory that allows thickening or thinning of beds in the front limb. Nevertheless slip in a fault-propagation fold is governed by the equations of fault-bend folding because the back syncline associated with the concave fault bend is simply a fault-bend fold.

The horizontal widths of the limbs of fault-propagation folds may be observed in map-view, seismic, and well data. These widths are equal to the lengths of the limbs except for the front limb of fixed-axis fault-propagation folds, for which there is a modification as a result of limb thickening. The equations for the limb widths and lengths are given by Suppe & Medwedeff (1990) in terms of the stratigraphic height h between the fault bend and the fault tip (Fig. 27), as follows: *Constant-thickness theory*. The front-limb width W_f and length L_f are given by the constant-thickness theory as

$$W_f = L_f = h\left[\frac{\sin(2\gamma-\theta_2-\gamma^*)}{\sin\gamma^*\sin(2\gamma-\theta_2)}\right] \tag{9}$$

and the back-limb width W_b and length L_b as

$$W_b = L_b = h\,[\cot\gamma^*-\cot\gamma_1] + h\left[\frac{1}{\sin\theta_2} - \frac{1}{\sin(2\gamma-\theta_2)}\right]\left[\frac{\sin(\gamma_1+\theta_2)}{\sin\gamma_1}\right] \tag{10}$$

The fault slip s is

$$s = h\left[\frac{1}{\sin\theta_2} - \frac{1}{\sin(2\gamma-\theta_2)}\right] \tag{11}$$

Combining (11) with (9) and (10) we obtain the desired relationships between limb length and fault slip for a constant-thickness fault-propagation fold

$$\frac{\text{front-limb length}}{\text{slip}} = \frac{L_f}{s} = \frac{\left[\dfrac{\sin(2\gamma-\theta_2-\gamma^*)}{\sin\gamma^*\sin(2\gamma-\theta_2)}\right]}{\left[\dfrac{1}{\sin\theta_2} - \dfrac{1}{\sin(2\gamma-\theta_2)}\right]} \tag{12}$$

and

$$\frac{\text{back-limb length}}{\text{slip}} = \frac{L_b}{s} = \left[\frac{[\cot\gamma^*-\cot\gamma_1]}{\left[\dfrac{1}{\sin\theta_2} - \dfrac{1}{\sin(2\gamma-\theta_2)}\right]}\right] + \left[\frac{\sin(\gamma_1+\theta_2)}{\sin\gamma_1}\right] \tag{13}$$

Fixed-axis theory. The front-limb length in the fixed-axis theory is equal to the front-limb width times the stretch:

$$W_f = L_f \left[\frac{\sin\gamma_e}{\sin\gamma_i}\right] \tag{14}$$

The front-limb width W_f is

$$W_f = h\left[\frac{\sin(\gamma_e-\theta_2)}{\sin\gamma_e \sin\theta_2}\right] \tag{15}$$

Combining (14) and (15) we obtain L_f

$$L_f = h\left[\frac{\sin\gamma_e}{\sin\gamma_i}\right]\left[\frac{\sin(\gamma_e-\theta_2)}{\sin\gamma_e \sin\theta_2}\right] \tag{16}$$

The back-limb width W_b and length L_b equal

$$W_b = L_b = h\,[\cot\gamma_e^* - \cot\gamma_1] + h\left[\frac{1}{\sin\theta_2} - \frac{\sin\gamma_i/\sin\gamma_e}{\sin(\gamma_e+\gamma_i-\theta_2)}\right]\left[\frac{\sin(\gamma_1+\theta_2)}{\sin\gamma_1}\right] \tag{17}$$

Fault slip s in a fixed-axis fault-propagation fold is

$$s = h\left[\frac{1}{\sin\theta_2} - \frac{\sin\gamma_i/\sin\gamma_e}{\sin(\gamma_e+\gamma_i-\theta_2)}\right] \tag{18}$$

Combining (18) with (15), (16), and (17) we obtain the desired relationships between slip and limb lengths and widths

$$\frac{\text{front-limb width}}{\text{slip}} = \frac{W_f}{s} = \frac{\left[\frac{\sin(\gamma_e-\theta_2)}{\sin\gamma_e \sin\theta_2}\right]}{\left[\frac{1}{\sin\theta_2} - \frac{\sin\gamma_i/\sin\gamma_e}{\sin(\gamma_e+\gamma_i-\theta_2)}\right]} \tag{19}$$

$$\frac{\text{front-limb length}}{\text{slip}} = \frac{L_f}{s} = \frac{\left[\frac{\sin\gamma_e}{\sin\gamma_i}\right]\left[\frac{\sin(\gamma_e-\theta_2)}{\sin\gamma_e \sin\theta_2}\right]}{\left[\frac{1}{\sin\theta_2} - \frac{\sin\gamma_i/\sin\gamma_e}{\sin(\gamma_e+\gamma_i-\theta_2)}\right]} \tag{20}$$

and

$$\frac{\text{back-limb length}}{\text{slip}} = \frac{L_b}{s} = \frac{[\cot\gamma_e^* - \cot\gamma_1]}{\left[\frac{1}{\sin\theta_2} - \frac{\sin\gamma_i/\sin\gamma_e}{\sin(\gamma_e+\gamma_i-\theta_2)}\right]} + \frac{\left[\frac{1}{\sin\theta_2} - \frac{\sin\gamma_i/\sin\gamma_e}{\sin(\gamma_e+\gamma_i-\theta_2)}\right]\left[\frac{\sin(\gamma_1+\theta_2)}{\sin\gamma_1}\right]}{\left[\frac{1}{\sin\theta_2} - \frac{\sin\gamma_i/\sin\gamma_e}{\sin(\gamma_e+\gamma_i-\theta_2)}\right]} \tag{21}$$

The above equations give relationships between fault slip and limb length and width for fault-propagation folds. Inasmuch as back-limb growth is most easily observed, Figure 28 gives a graph of the relationship between back limb dip $\delta_b = 2(\pi-\gamma_1)$ and L_b/s with lines of constant fault bend ϕ or hangingwall cutoff angle θ_2 based on the above equations and numerical solution of the fundamental equations of fault-propagation folding of Suppe & Medwedeff (1990). This graph is discussed below.

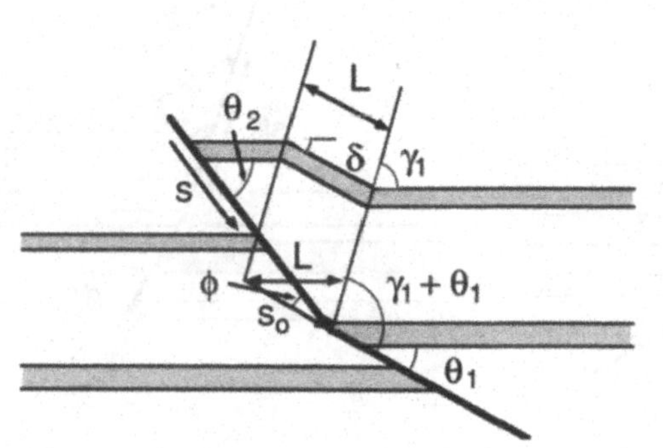

Figure 25. Fault-bend fold showing angular relationships for derivation of relationship between limb length and fault slip.

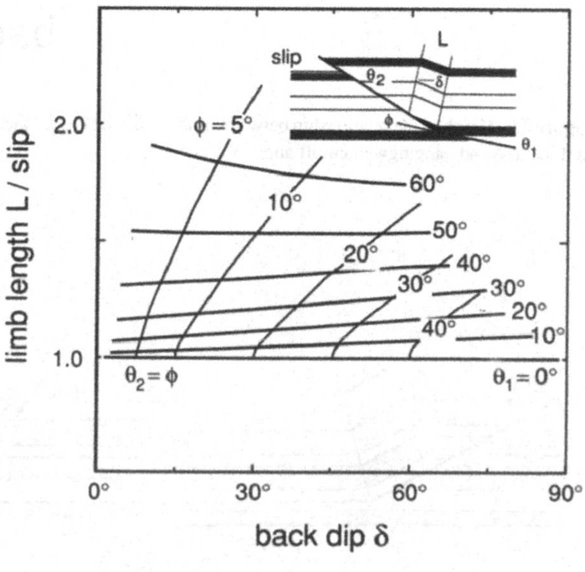

Figure 26. Graphs of the relationship between limb dip δ and limb length *L/slip* for synclinal fault-bend folds with lines of constant fault bend ϕ or hangingwall cutoff angle θ_1.

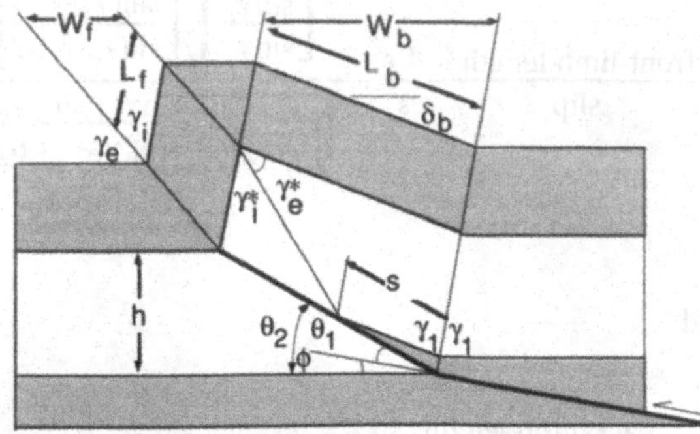

Figure 27. Fault-propagation fold showing angular relationships for derivation of the relationships between fault slip s and limb length L and width W. The axial angles are shown for the fixed-axis theory. The constant thickness theory simplifies to $\gamma = \gamma_e = \gamma_i$ and $\gamma^* = \gamma_e{}^* = \gamma_i{}^*$

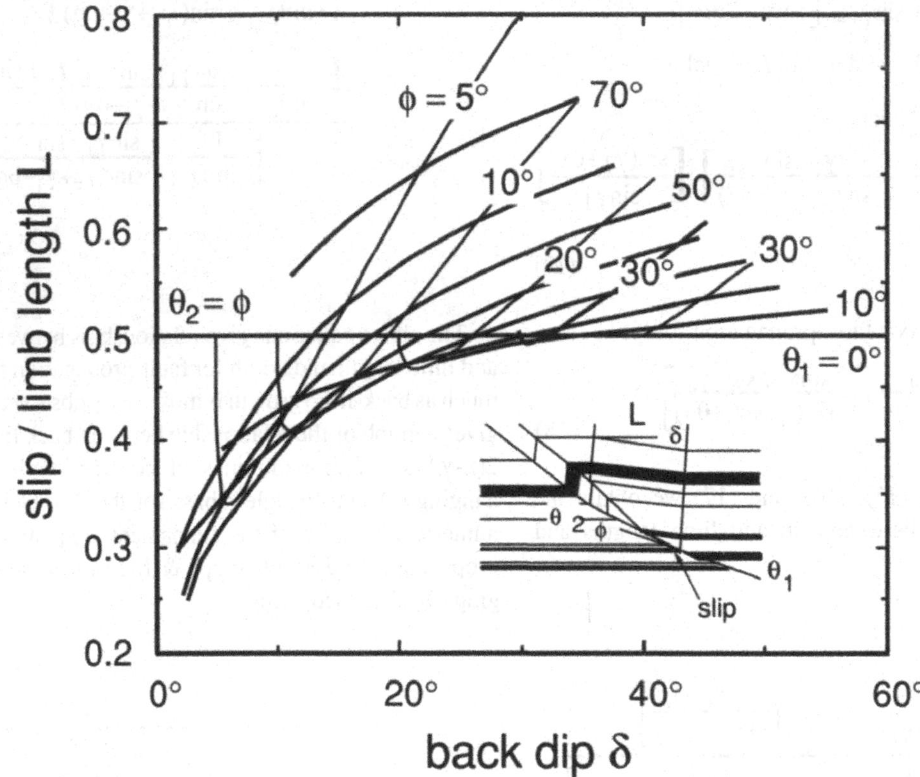

Figure 28. Graph of the relationship between limb dip δ and *slip* / *L* (back limb length) for constant-thickness fault-propagation folds with lines of constant fault bend ϕ and hangingwall cutoff angle θ_1.

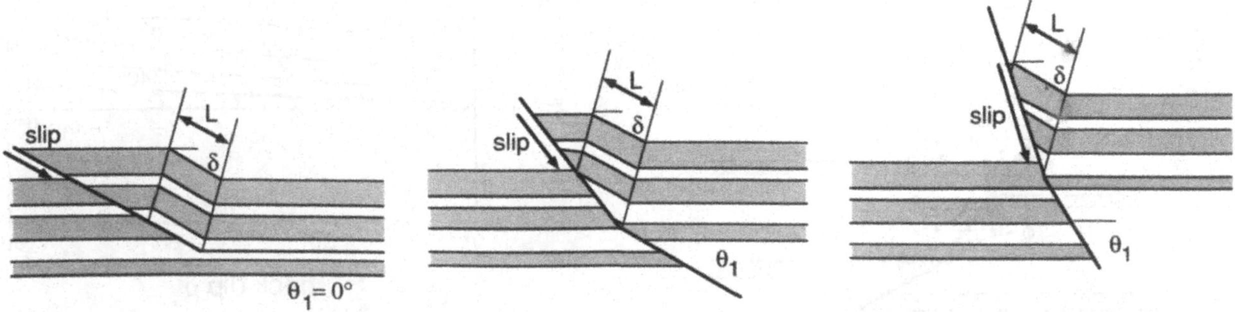

Figure 29. Models of a kink band with a 30° limb dip showing the range of fault shapes that can produce the same fold. Note the variation in fault slip with constant fold geometry.

Discussion of relationships between limb length and slip

We would like to estimate fault-slip rates from observation of shallow growth structure and to assign uncertainties based upon our uncertainties in the underlying fault geometry. For example Figure 27 shows schematically how the same kink band might form by slip on a variety of faults. The equations and associated graphs obtained in the previous sections allow us to assign these uncertainties in fault-slip rate based on our uncertainty in fault geometry. This discussion assumes a two-dimensional geometry - essentially dip-slip motion. If the folding is produced by oblique slip through a fault bend, such that there is a strike-slip component, then the kink band records the dip-slip component.

In the schematic example of Figure 29 a limb dip of 30 degrees is displayed. If based on growth structure we believe the kink band is produced by a synclinal fault bend then Figure 26 shows that the ratio of limb length to fault slip L/s ranges between 1.0 and about 2.0. If we can constrain the problem further, for example regional geological relationships might indicate that the upper fault dip θ_2 must be less than about 50°, then we can further limit L/s to the range 1.0 to 1.25. Thus with only rather modest information on fault geometry we can constrain fault slip rather closely from growth structure.

Similar conclusions are reached for fault-propagation folds (Fig. 28); namely that the relationship between limb length and fault slip does not vary rapidly with fault geometry at a constant limb dip. For example with a back-limb dip of 15-25°, which is quite common, the fault slip ranges between 0.4 and 0.6 times the back-limb length for almost all fault geometries. Fault geometry of fault-propagation folds commonly can be constrained rather closely from shallow geological data such as ratios of limb widths (Suppe & Medwedeff 1990). Therefore fault slip commonly can be closely constrained from fold geometry.

Conclusions

The well known processes of fault-related folding such as fault-bend and fault-propagation folding specifically predict fold growth by constant-dip kink-band migration. This implies specific fold geometries within the stratigraphic sequences deposited during fold growth, which are observed as shown by seismic examples. Specifically, kink bands decrease in width upward through the growth strata. Rates of kink-band migration are observed in a few structures to be on the order of 1-2 mm/y. Straightforward relationships are derived between limb length and fault slip which commonly allow fault slip to be determined within rather narrow limits from shallow growth structure, even if fault geometry is poorly constrained - a fact that makes growth structures valuable for assessing earthquake hazards.

Suppe is grateful to the Guggenheim Foundation for support during a sabbatical leave in 1978-79 when the initial theory of growth fault-bend folding was developed (Fig. 12). However most of the theory of growth structure was developed during Fall of 1985 when Chou and Hook were visiting professors at Princeton, supported by Texaco USA. We are all grateful to Texaco for the continued support and encouragement of this work and to the many sources of the seismic examples, including Texaco USA; Professional Geophysics Inc. (PGI); BP Exploration Inc.; Sun Operating Limited Partnership; Nippon Western US Company, Ltd; Halliburton Geophysical Services, Inc.; R. Prieto, R. del Pilar, and A. Saldivar-Sali. We thank Steve Boyer and Shankar Mitra for reviews of the manuscript. The work on the relationship between fold growth and fault slip and on the examples from California were supported by USGS NEHRA grant 14-08-0001-G1699.

References

Dibblee, T. W. Jr. 1966. Geology of the central Santa Ynez Mountains, Santa Barbara County, California. *California Division of Mines and Geology Bulletin*, **186**, 1-99.

—— 1986. Geologic map of the Hildreth Peak Quadrangle, Santa Barbara County, California. *Thomas W. Dibblee Jr. Foundation Map*, **DF-07**.

Medwedeff, D. 1989. Growth fault-bend folding at southeast Lost Hills, San Joaquin Valley, California. *American Association of Petroleum Geologists Bulletin*, **73**, 54-67.

—— & Suppe, J. 1986a. Growth fault-bend folding - precise determination of kinematics, timing, and rates of folding and faulting from syntectonic sediments. *Geological Society of America Abstracts with Programs*, **18**, 692.

—— & —— 1986b. Kinematics, timing, and rates of folding and faulting from syntectonic sediment geometry. *EOS Transactions of the American Geophysical Union*, **68**, 1223.

Mosar, J. & Suppe, J. 1991. Role of shear in fault-propagation folding (this volume).

Mount, V.S. & Suppe, J. 1990. Seismic structural analysis of the deep-water Perdido fold belt, Aliminos Canyon, Northwest Gulf of Mexico. *In*: Future Petroleum Provinces of North America. *American Association of Petroleum Geologists Memoir* (in press).

——, Suppe, J. & Hook, S.C. 1990. A forward modelling strategy for balanced cross sections. *American Association of Petroleum Geologists Bulletin*, **74**, 521-531.

Rich, J. L. 1934. Mechanics of low-angle overthrust faulting as illustrated by Cumberland thrust block, Virginia, Kentucky, and Tennessee. *American Association of Petroleum Geologists Bulletin*, **18**, 1584-1596.

Suppe, J. 1983. Geometry and kinematics of fault-bend folding. *American Journal of Science*, **283**, 684-721.

——1985. *Principles of Structural Geology*. Prentice-Hall Inc., Englewood Cliffs, New Jersey, 1-537.

—— & Chang, Y. L. 1983. Kink method applied to structural interpretation of seismic sections, western Taiwan. *Petroleum Geology of Taiwan*, **19**, 29-49.

——& Medwedeff, D. A. 1984. Fault-propagation folding. *Geological Society of America Abstracts with Programs*, **16**, 670.

—— & —— 1990. Geometry and kinematics of fault-propagation folding. *Eclogae Geologicae Helvetiae*, **83** (Laubscher volume), (in press).

Tomlinson, C. W. 1952. Odd geologic structures in Southern Oklahoma. *American Association of Petroleum Geologists Bulletin*, **36**, 1820-1840.

Xiao, H. B. & Suppe, J. 1990. Origin of rollover. *American Association of Petroleum Geologists Bulletin* (submitted).

Role of shear in fault-propagation folding

Jon Mosar[1] & John Suppe

Department of Geological and Geophysical Sciences, Guyot Hall, Princeton University, Princeton, New Jersey 08544-1003, USA

Abstract: The effect of layer-parallel shear on the shapes of fault-propagation folds is explored for the two theories of fault-propagation folding of Suppe & Medwedeff (1990): (1) constant layer thickness and (2) variable front-limb layer thickness; the range of possible fold shapes is significantly expanded relative to the case of no shear. In this analysis, a homogeneous differential layer-parallel shear is applied to the beds that are cut by the thrust fault in the cores fault-propagation folds. This shear may be applied in three ways: (1) self-similar fold growth during fault propagation with constant shear applied instantaneously along the bed of the fault tip, (2) progressively increasing shear during fault propagation, and (3) shear after the fault is locked modifying the existing fault-propagation fold. The final shape of the fold is independent of the history of shearing in relation to fault slip. The fold shapes are largely governed by the fault steepness and the amount of imposed shear. Strong differential shear profiles and shallow faults produce overturned or thickened front-limbs. Little differential shear and steep faults result in upright or thinned front-limbs.

The concept of fault-propagation folding, as well as other concepts of fault-related folding, has proven successful in unravelling the development of numerous folds (Laubscher 1977; Suppe 1983, 1985; Suppe & Medwedeff 1984, 1990; Suppe *et al.* 1990; Jamison 1987; Mitra 1988, 1990; Mount *et al.* 1990). Fault-propagation folds form contemporaneously with a propagating fault tip. An anticline develops over the blind thrust and a syncline at the fault tip (Fig.1), which lies along the same layer as the branching point of the anticlinal axial surfaces. Slip along the thrust surface goes to zero at the fault tip.

Material transport during folding is achieved by translation parallel to the thrust surface. Changes in transport direction occur when material passes through axial surfaces. The geometric and kinematic constraints of the model arise largely from applying the laws of conservation of volume (Goguel 1952; Laubscher 1962; Dahlstrom 1969, 1990) and the kink method for construction of cross sections (Coates 1945; Faill 1973; Suppe 1983, 1985). Two models of fault-propagation folding are discussed here. Both models are balanced forward models, one based on conservation of layer thickness and bed length, the other allowing thickness changes in beds in the front limb (Suppe & Medwedeff 1990).

The quantitative fault-propagation fold model is appropriately applied only in the domain of upper crustal brittle folding, where rocks have suffered little internal deformation under low grade or non-metamorphic conditions - for example the Jura Mountains (Laubscher 1962, 1976) or the Prealps (Mosar 1989). The interpreter should evaluate carefully whether the model should be applied - either quantitatively or qualitatively - to folds from areas that have undergone significant flow or if other models would be invoked more fruitfully.

Model description

The theory describing fault-propagation folding is based on the kink method for constructing cross sections (Coates 1945; Suppe 1985) and incorporates line-length, balanced models. The kink method is based on the observation that fold limbs are frequently straight and hinge zones relatively small compared to the limb lengths (Faill 1973; Suppe 1983; Boyer 1986). If layer thickness is conserved, the kink method implies that axial surfaces bisect adjacent fold limbs. If layer thickness changes across the fold hinge, the axial surfaces no longer bisect adjacent limbs (Suppe 1983, 1985; Marshak & Mitra 1988). Applying the kink method results in angular fold hinges, some angles being rather tight (Fig.1). More rounded hinges can be obtained by introducing numerous closely spaced axial surfaces. Although this does change the details of the fold geometry, it does not, however, change the overall kinematic and geometric features involved in fault-propagation folding. In order to understand implications arising from the models and not to complicate unnecessarily the models, this analysis uses simple geometries, which involve only as many axial surfaces as required to describe the fold shape.

The mathematical theory of fault-propagation folding is simply a compatibility, conservation relationship, such that material particles are preserved between the undeformed and deformed states with no voids or overlaps and such that layers are continuous except where cut by the fault. There is no scale to the theory and therefore no explicit history. However, in the case of no shear it is conveniently considered that the folds grow self-similarly as the fault propagates (Fig. 1) because in this case fault shape and slip are the only variables. In

[1]Present address: Musée Géologique, Université de Lausanne - BFSH2, CH-1015 Lausanne, Switzerland

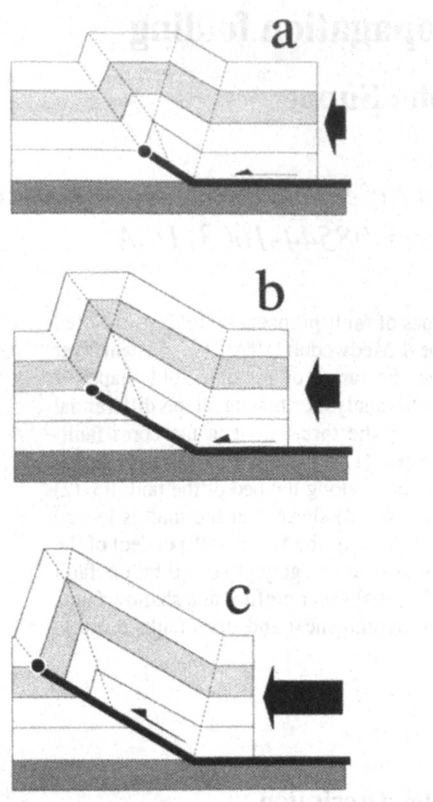

Figure 1. Fault-propagation fold development (simple step-up, constant layer thickness model, fault ramp angle = 30°; anticlinal interlimb angle in fold core = 78°; interlimb angle of syncline at fault tip = 108°).

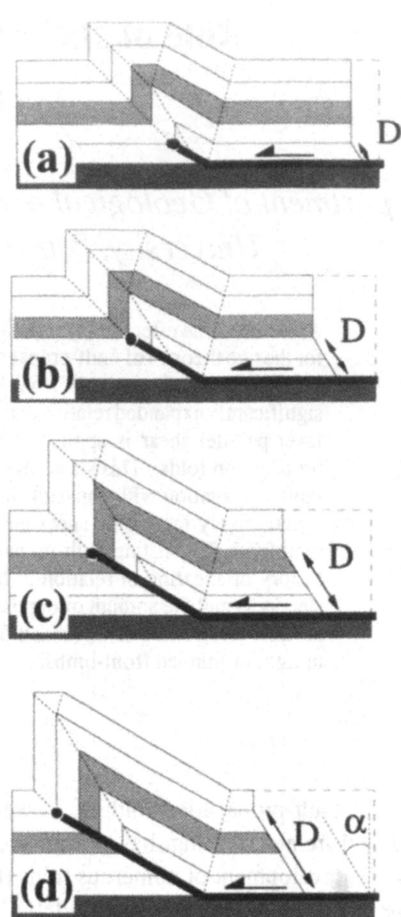

Figure 2. Self-similar growth of a simple-step fault-propagation fold. The differential layer parallel shear (angular shear α) is determined by the original fold geometry. Once this geometry is acquired, α stays the same as the fault propagates from (a) through (d), only the width D of the shear profile increases. (Fault ramp angle = 30°; anticlinal interlimb angle in fold core = 60°; interlimb angle of syncline at fault tip = 90°; α = 31.8°).

contrast, the deformational history of a final fold shape becomes clearly non-unique in the case of non-zero shear. In general the shear may be applied in three ways: (1) self-similar fold growth during fault propagation with constant shear applied instantaneously along the bed of the fault tip, which is shown in Figure 2, (2) progressively increasing shear applied in various possible ways during fault propagation and fold growth, and (3) shear applied after the fault is locked modifying the existing fault-propagation fold. The final gross shape of the fold in pre-existing strata is independent of the history of shearing in relation to fault slip, but the fold shape in growth strata will record the deformational history (Suppe *et al.* 1990).

The fold develops kinematically by consuming layer parallel slip and the overall deformation is one of simple shear. The amount of layer parallel shear across the fold is determined by the fold geometry (interlimb angles as well as angle and height of the ramping thrust plane). Thus some fold geometries may require a differential bedding parallel shear to be transmitted through the structure (see also Mitra 1990, p. 925). A displacement profile of single layers through the flat lying beds in the trailing part of the fault-propagation fold, may then be a straight line perpendicular to bedding or a more complex line if excess layer parallel shear is implied by a different fold geometry. Excess layer parallel shear can, for example, be induced by bending of beds along a supplementary axial surface (Suppe 1985, p. 318). Although excess layer parallel shear appears to be minor in many examples of fault-propagation folding (Suppe & Medwedeff 1990), shear occuring within specific stratigraphically controlled weak horizons may be important.

The conditions for conservation of line-length and area can be expressed quantitatively in a mathematical formulation using only angles describing the fold shape (Fig. 3). Exact solutions involving shear have been derived for the general case, not involving a decollement (Suppe & Medwedeff 1990). This phase will only be concerned with simple step-up fault-propagation folds, that is, when the fault steps up at a given angle θ_2 from a bedding plane decollement (Fig. 3). The results indicate that in this simple step case, the fold shape (viz. the various internal axial angles) is a function of two independent variables only: S_p, the *shear* associated with layers in the fold core; and θ_2, the *fault step-up angle*.

Two different fault-propagation folding theories have been developed (for details on the quantitative derivations see Suppe & Medwedeff 1990).

Constant-thickness theory

In the constant-thickness model, bed length and layer thickness are conserved throughout the fold. The quantitative geometric relationships are shown in equation (1). Note that

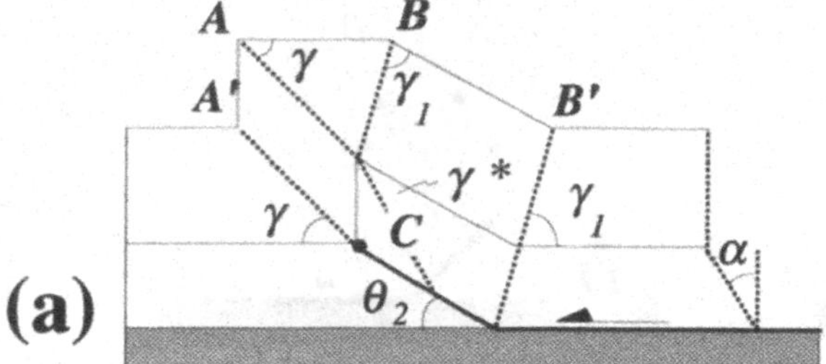

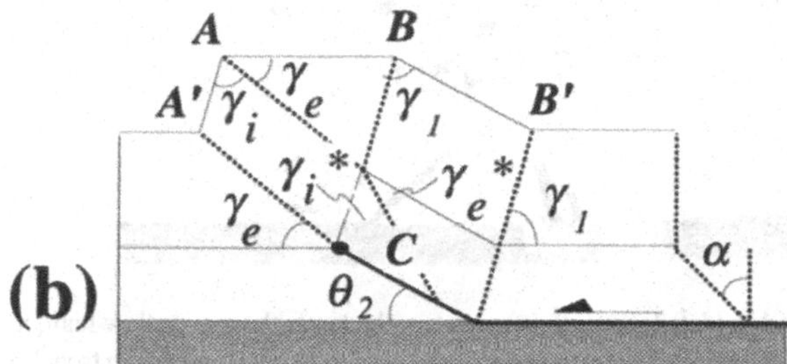

Figure 3. Simple step-up fault-propagation fold model. θ_2 step-up angle, γ_1 back-limb axial surface angle, γ, γ_e, γ_i front limb axial surface angles, γ^*, γ_e^*, γ_i^* fold core inter-limb angles, α 'shear angle'. A, A' front limb axial surfaces (axial surface A is fixed in the fixed axial surface theory); B, B' back-limb axial surfaces; C fixed fold core axial surface. (**a**) constant-thickness theory, (**b**) fixed axial surface theory. In both theories $S_p = \tan \alpha$.

$\theta_2 = 2(\gamma - \gamma^*)$. Axial surfaces always bisect adjacent fold limbs. Overturned frontal limbs occur at angles 2γ less than 90°.

$$S_p = \frac{1 + 2\cos^2\gamma^*}{\sin 2\gamma^*} + \frac{\cos\theta_2 - 2}{\sin\theta_2} \qquad (1)$$

Fixed axial surface theory

In the fixed axial surface model we allow for thickening or thinning in the front limb of the fold, with the constraint of conservation of area. This thickening or thinning is caused by the front limb anticlinal axial surface (Fig. 3b, axial surface A) being fixed relative to the rock. No material moves through axial surface A and bed thickness is conserved in the crestal and back-limb portions of the fold. In this case axial surfaces in the front limb and the fold core do not bisect adjacent limbs. The quantitative geometric relationships are presented in equations (2) and (3).

$$S_p = 2\cot\gamma_e^* + \frac{2\cos\theta_2 - 3}{\sin\theta_2} \qquad (2)$$

The change in bed thickness in the front limb is given by the ratio T_e/T_i, where T_e is the unchanged bed thickness external to the front limb and T_i the altered layer thickness in the front limb (thickening $1 > T_e/T_i > 1$ thinning). Note that $\theta_2 = (\gamma_i + \gamma_e) - (\gamma_i^* + \gamma_e^*)$, which requires that the front limb cannot be overturned. The front-limb dip is always parallel to the back-limb axial surfaces!

$$\frac{T_e}{T_i} = \frac{\sin\gamma_e^*}{\sin\gamma_i^*} = \frac{\sin\gamma_e}{\sin\gamma_i} \qquad (3)$$

Kinematic aspects

The kinematic aspects and characteristics of fault-propagation folding can be addressed in several different ways, only a few of which are briefly described here. Once the fold is initiated, it grows self-similarly (except for later possible imposition of additional shear, as discussed above). Structural relief increases during propagation and the flat anticlinal top progressively disappears (Fig. 1).

Material transport direction

Displacement directions of material with respect to the fault-propagation fold footwall (Fig. 4) are the same for both the constant-thickness and the fixed axial surface theories. Instantaneous velocities are constant within regions bounded by the axial surfaces through which the thrust sheet passes. In the trailing part of the structure and above the basal thrust, material moves parallel to the decollement. A change in translation direction occurs along the synclinal axial surface B' associated with bend in the fault plane (Fig. 4). Between the back-limb axial surfaces and in the fold core the transport direction is parallel to the ramping fault plane. In the anticline crest and the front-limb the translation is parallel to the two front-limb axial surfaces A and A' (Fig. 4).

The bending stresses arising from these rotations are released by layer parallel slip on discrete surfaces (Suppe 1983, Kilsdonk & Wiltschko 1988), by fracture formation (faults and veins) and dissolution along stylolites (Laubscher 1976, Srivastava & Engelder 1990, Pfiffner 1990).

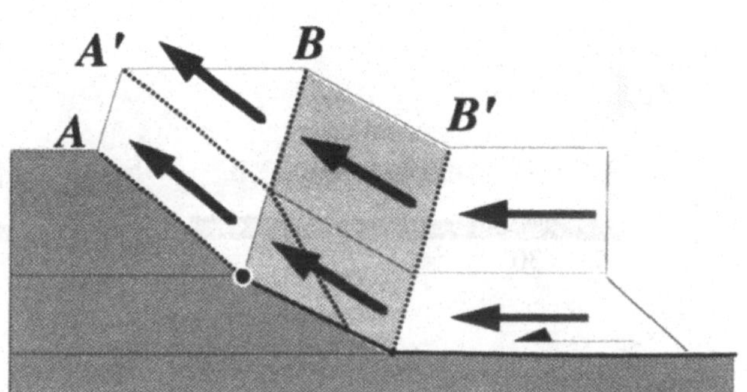

Figure 4. Kinematic properties of simple step-up fault-propagation folds. Material transport directions (no velocities given) during self-similar fold growth with respect to footwall (fixed axial surface theory with front-limb thickening).

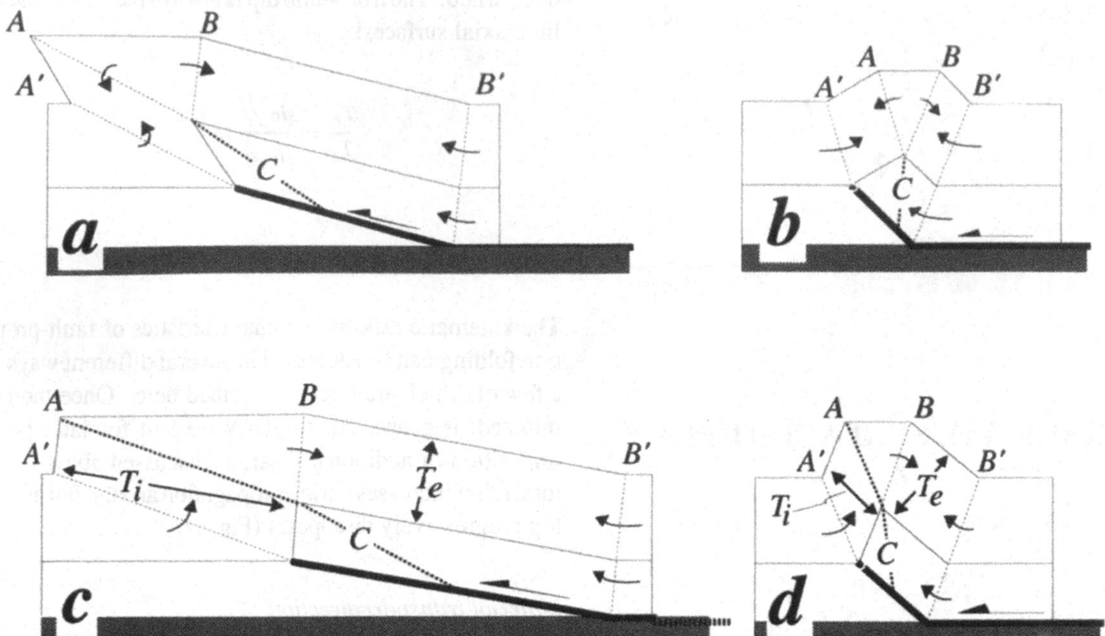

Figure 5. Kinematic properties of different simple step fault-propagation fold models (detail see text). **(a)** constant-thickness theory, shallow fault, low $\theta_2 = 15°$; $\gamma^* = 21.6°$; $\gamma = 29.1°$, **(b)** constant-thickness theory, steep fault, high $\theta_2 = 45°$; $\gamma^* = 51.2°$; $\gamma = 73.7°$, **(c)** fixed axial thickness model with front limb thickening: $T_e / T_i = 0.35$, shallow fault, low $\theta_2 = 10°$; $\gamma_e^* = 18.6°$; $\gamma_e = 19.7°$. **(d)** fixed axial thickness model with front limb thinning: $T_e / T_i = 1.37$ steep fault, high $\theta_2 = 40°$; $\gamma_e^* = 41.2°$; $\gamma_e = 67.5°$. Arrows indicated sense of material moving through active axial surfaces.

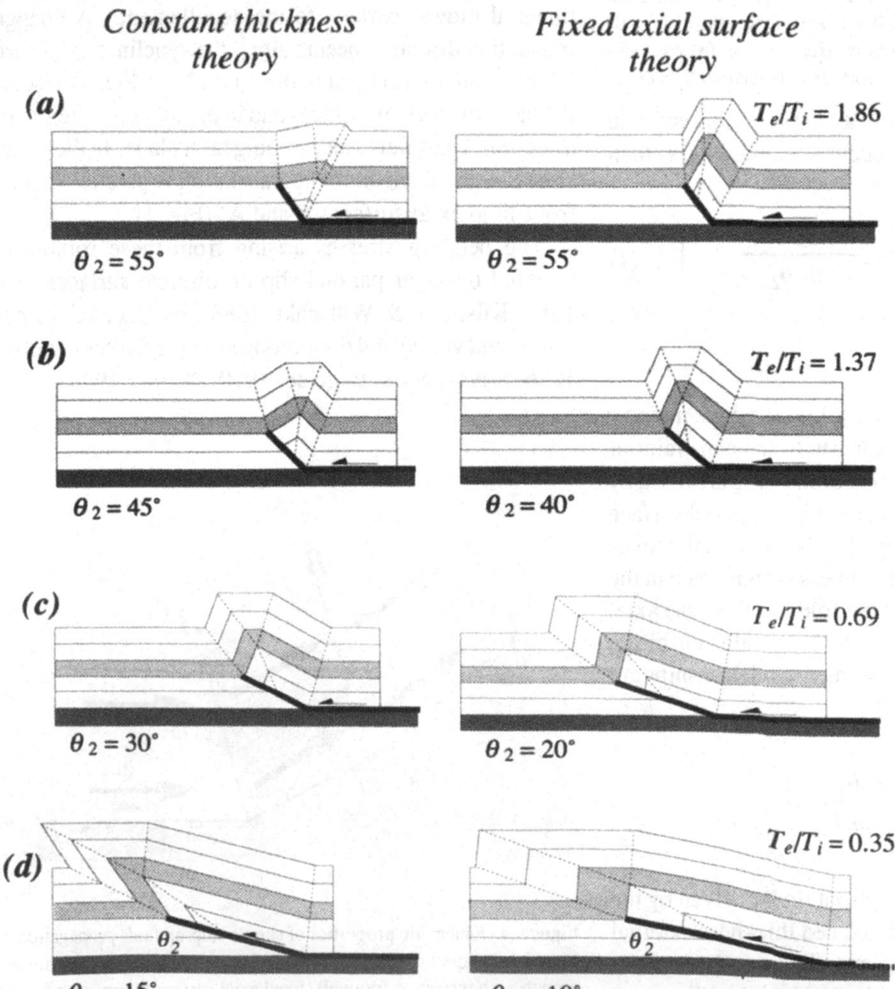

Constant thickness theory

Fixed axial surface theory

(a)

$\theta_2 = 55°$

$\theta_2 = 55°$

$T_e/T_i = 1.86$

(b)

$\theta_2 = 45°$

$\theta_2 = 40°$

$T_e/T_i = 1.37$

(c)

$\theta_2 = 30°$

$\theta_2 = 20°$

$T_e/T_i = 0.69$

(d)

$\theta_2 = 15°$

$\theta_2 = 10°$

$T_e/T_i = 0.35$

Figure 6. Influence of fault-ramp angle θ_2 on fold shape in simple step fault-propagation folds *with constant ramp height*. Shear is zero; step-up angle changes from **(a)** through **(d)**. The total displacement at the fold back end of case (d) in the fixed axial surface theory could not be represented accurately.

Fold axial surfaces

Different types of axial surfaces can be recognized in fault-propagation folds (Fig. 5):

(1) *Active axial surfaces* in which material moves through the axial plane as the fold grows. They are attached to the footwall as in the case of bends in the fault surface (Fig. 5, axial surface B') or to the fault tip (Fig. 5 axial surface A'). Whereas the former active axial surface is fixed relative to the footwall, the latter is translated as the fault tip propagates. In the constant-thickness theory both the anticlinal axial surfaces A and B (Fig. 5a & b) are active. In the fixed axial surface theory, only the back-limb anticlinal axial surface B is active (Fig. 5c & d). It is also worth noting that from a kinematic point of view, an additional axial surface exists parallel to bedding along the line connecting the fault tip and the anticlinal branching point. In the fixed-axis theory this 'hidden' axial surface is identical to B (Figs 5c & d), whereas in the constant-thickness theory it is distinct (Figs 5a & b).

(2) *Fixed, or inactive axial surfaces* are planes fixed relative to material in adjacent fold limbs, but not with respect to the footwall. No material rolls through fixed axial surfaces. In all of the models the axial surface C in the fold core is fixed (Fig. 5) although this surface extends upwards as the fold propagates. In the fixed axial surface theory the front-limb anticlinal axial surface A (Fig. 5c & d) is also fixed. As no material moves through surface A, the kinematics require thickness changes in the front limb.

Fault steepness and fold shape

If no differential displacement profile exists in the layers cut by the fault - that is the fold geometry does not require excess layer parallel shear - then the fold shape is a function of the fault geometry only. In a fault-propagation fold stepping up from a decollement, the fault shape is given by the step up angle θ_2 (Fig. 5). For a fixed stratigraphic height of fault-propagation both slip and structural relief decrease with increasing fault dip θ_2. With constant displacement, but increasing fault dip, the structural relief increases and the fault tip climbs into higher stratigraphic levels. Front limbs are overturned for shallow fault angles and upright for higher step-up angles. Also the anticlinal axial surface of the fold core dips in the same direction as the ramping fault for shallow values of θ_2 (Fig. 6c, d). With higher step-up angles, however, this axial surface is steeper and eventually has a reversal in dip direction (Fig. 6a, b). At the same time the fold asymmetry is reversed and faces in the opposite direction of the transport sense on the thrust.

In fixed axial surface models the structural relief at constant ramp height is not significantly altered by changes in the fault angle θ_2. As in the constant-thickness models the axial surface of the fold core is rather steep at high step-up angles and eventually has a dip direction opposite of that of the ramping fault plane (Fig. 6). The front-limb is always upright, but the overall width of the beds is thickened for low fault angles and thinned for high θ_2 values.

Shear in fault-propagation folds

Differential bedding parallel shear in fault-propagation folds is associated with each bend in the fold and with the termination of the fault. In general, bending of beds along a synclinal or anticlinal axial surface produces a specific amount of layer parallel simple shear that is a function of the change in dip across the axial surface (Suppe 1983, 1985). In simple-step fault-propagation folds that ramp from a decollement, we have the possibility of shear occuring in layers that form the fold core (that is, layers that are truncated by the fault), as defined by equations (1) and (2) (Figs 2 & 3). Layer parallel shear in beds above the fold core (above beds lying along the line connecting the fault tip and the branching point forming the anticlinal axial surfaces) can also be expected. This shear may, for example, be accounted for by branching fold axial surfaces. The bending shear associated with these axial planes can account for shear imposed at the trailing edge of the fold (Namson 1981, Mitra & Namson 1989).

Here the focus is only on the implications of shear applied to the core of simple-step fault-propagation folds. It is suggested that structures analogous to this type of situation are to be found in the Jura Mountains and the Prealps (Swiss and French Alps). Structures in these fold and thrust belts develop above a weak basal decollement associated with evaporite-rich layers. These layers frequently form the fold cores and may be able to accommodate large amounts of differential shear.

The differential bedding parallel angular shear α is taken as positive in a counter-clockwise sense. Negative angular shear cannot be ruled out and can be considered as a wedge shaped structure affecting the fold core layers. The axial surface C (Fig. 3) in the fold core dips at a steep angle when the angular shear is negative or very low (Fig. 7). With increasing shear this axial plane has a shallower dip angle, close to that of the ramping fault plane. In the constant-thickness models, increasing shear causes the frontal limb to overturn. Larger amounts of shear in fixed axial surface fault-propagation folds are associated with front-limb thickening whereas low shears are associated with front-limb thinning. The no-thickness-change situation $S_p = 0$ occurs for a step-up angle $\theta_2 = 29°$. This also represents the unique case where both the constant-thickness and the fixed axial surface theories coincide. Note that for step-up angles less than 10° and for fold core inter-limb angles less than 20-25°, small changes of these angles induce dramatic changes in shear. Conversely large changes in shear only produce minor changes in fold shape if step-up angles and fold core inter-limb angles are small (Figs 8 & 9).

Thus applying shear to the fold core layers enlarges the range of possible geometries for fault-propagation folds. These families of configurations can be represented graphically as a function of shear S_p, the fault step-up angle θ_2 and the angle between the axial surface of the fold core and the fold limb γ^* and γ_e^* (Fig. 8). In the case of the fixed axial surface theory it is necessary also to consider the impact of thickness changes in the front limb as given by T_f/T_i (Fig. 9). Not all geometries predicted by the mathematical solutions

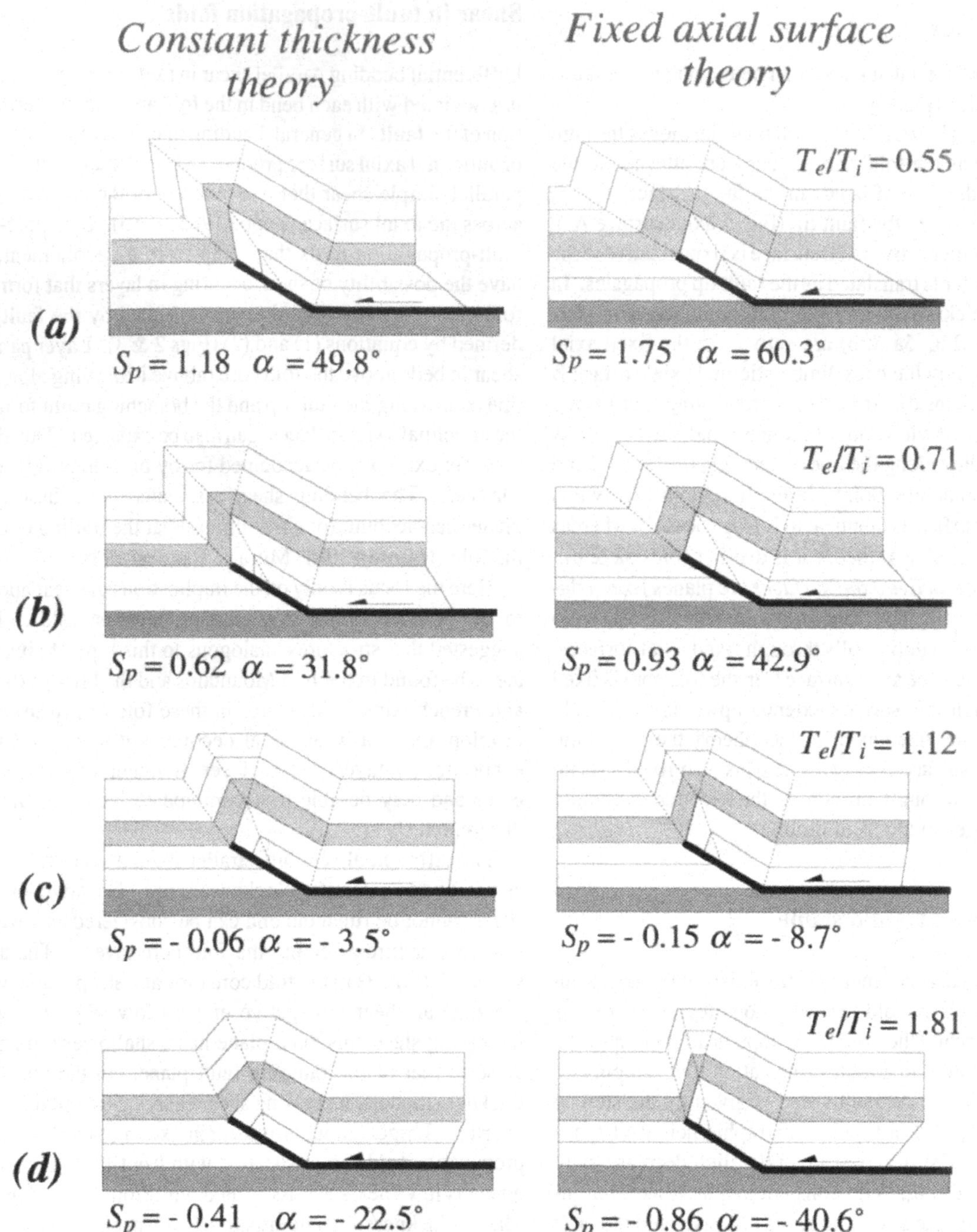

Figure 7. Influence of shear in fold core layers in simple-step fault-propagation folds *with constant ramp height*. Shear decreases from **(a)** through **(d)**; step-up angle for all four cases is 30°. Fold core angles γ^* respectively γ_e^* are: **(a)** 25°, **(b)** 30°, **(c)** 40°, **(c)** 50°.

are geometrically consistent with fault-propagation folding. Three types of solutions can be found (Fig.10):

(a) first, a set of situations where geometries (sets of fold internal angles) are consistent with the fault-propagation fold models. That is, displacement occurs along a fault surface - a decollement portion and a ramp portion in simple step-up models - which implies that the axial surface C of the fold core intersects the ramping fault segment between the fault tip and the point where the fault steps up from the decollement (Fig. 10a);

(b) second, a set of solutions in which simple parallel-kink folding without faulting is produced (simple shear folding, not to be confused with detachment folds such as box folds and lift-off folds because displacement goes to zero at the basal layer, rather than a finite value as in detachment folds). In this case no thrust surface exists and no slip has occurred other than layer parallel slip. The axial plane C of the fold core intersects the back-limb axial surface B' at the basal horizon of shear (Fig. 10b);

(c) third, a set of geometries which are inconsistent with fault-propagation folding by thrust faulting. The displacement sense along the ramp segment of the fault appears to be a normal fault and the axial plane C in the fold core intersects the back-limb axial surface B' (Fig. 10c).

The geometries expected by the third type of solutions are not considered here. They fall within the upper right of the

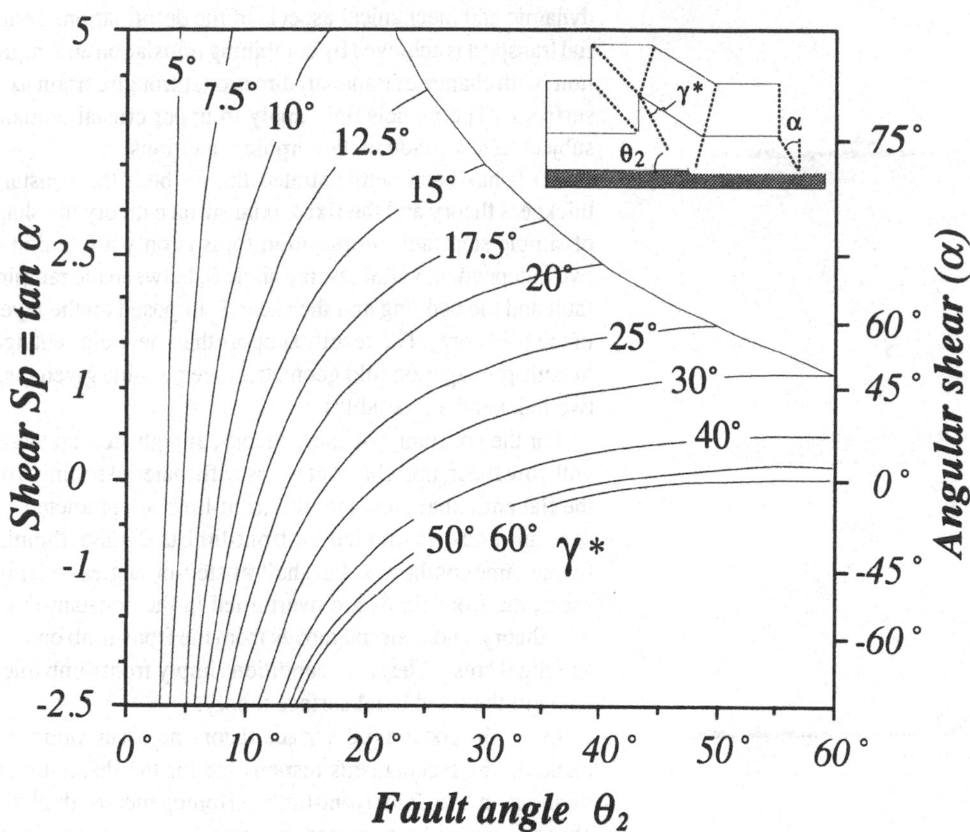

Figure 8. Inter-limb half angle γ^* as a function of fault-ramp angle θ_2 and shear S_p for constant-thickness fault-propagation fold models (solutions from eq. (1)).

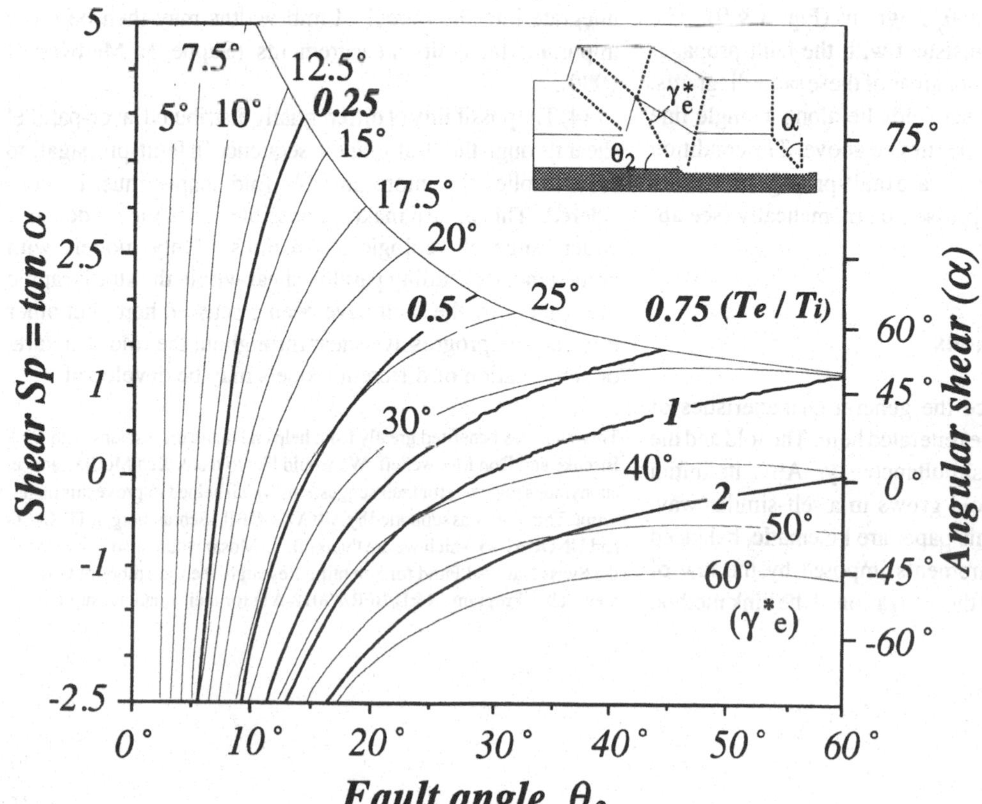

Figure 9. Inter-limb half angle γ_e^* as a function of fault-ramp angle θ_2 and shear S_p for fixed axial surface fault-propagation fold models (solutions from eq. (3)).

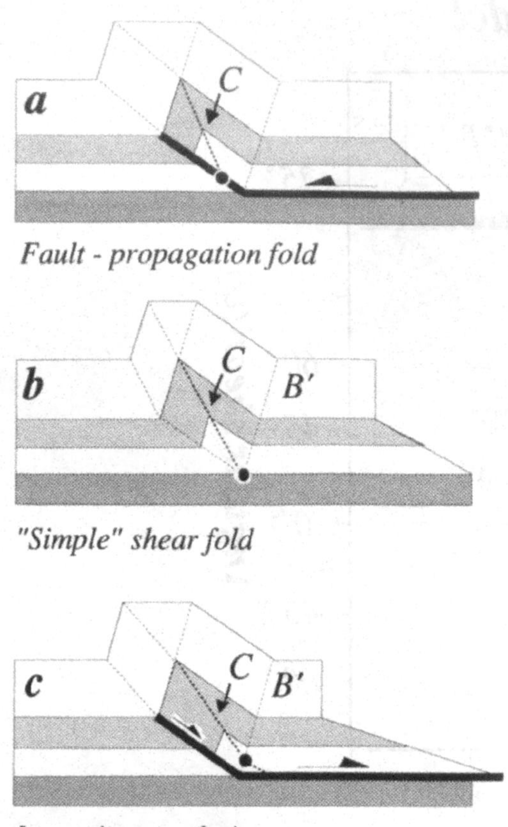

Figure 10. Solutions of consistent and inconsistent geometries for fault-propagation model as obtained from the general mathematical formulations. Models are fixed axial surface, simple step fault-propagation folds. **(a)** $S_p = 1.09$, $\alpha = 47.5°$, $\theta_2 = 35°$; **(b)** $S_p = 1.93$, $\alpha = 59.9°$, $\theta_2 = 35°$; **(c)** $S_p = 3.12$, $\alpha = 72.2°$, $\theta_2 = 35°$.

shear S_p versus step-up angle θ_2 diagrams (Figs 8 & 9). The solutions for geometries consistent with the fault-propagation models represent the major areas of these same diagrams. The solutions for simple shear folds lie along a single line separating the two domains mentioned above. The condition for consistency with the general case fault-propagation model (no decollement) can be expressed mathematically (see appendix).

Discussion and conclusions

(1) Given their importance the general characteristics of fault-propagation folding are reiterated here. The fold and the propagating fault develop simultaneously. After its initial geometry is acquired the fold grows in a self-similar way. The models developed in this paper are kinematic, balanced models based on the requirements imposed by the law of conservation of volume and the utilization of the kink method for cross section construction. They do not attempt to explain dynamic and mechanical aspects of the deformation. Material transport is achieved by combining translation and migration (with change of transport direction) through certain axial surfaces. The models only apply to upper crustal domains subject to low grade metamorphic conditions.

(2) It has been demonstrated that in both the constant-thickness theory and the fixed axial surface theory the shape of simple step fault-propagation folds is only a function of two independent variables: the angle θ_2 between the ramping fault and the bedding and the shear S_p imposed on the layers in the fold core. The results support that the major changes in fault-propagation fold geometries are possible given these two independent variables.

For the constant-thickness theory, at high step-up angles with low shear, upright front limbs with material rolling from the flat anticlinal crest into the front-limb are predicted. In constant-thickness models the front-limb undergoes thinning for the same conditions. For shallow step-up angles with high shear, the front-limbs are overturned in the constant-thickness theory and material moves from the front-limb onto the anticlinal crest. The same conditions imply front-limb thickening in the fixed axial surface theory.

(3) In the fixed axial surface theory no assumptions are made on the mechanisms responsible for the thickening or thinning in the fold front-limb. Homogeneous thickness changes cannot be expected in domains subjected to brittle deformation. Small scale folds, imbrications, faults and other similar structures are likely to play a dominant role in achieving these results. As a consequence, homogeneous dips can no longer be expected in the front limb, and accurate dip measurements obtained from field data may be difficult to integrate into the model. Limb widths may then be more important than dip measurements (Suppe & Medwedeff 1990).

(4) The possibility of differentially distributed layer-parallel shear through the stratigraphic sequence in fault-propagation folds implies that more possible fold shapes must be considered. This in turn makes it possible to use the models in a wider range of geological situations. Only models with homogeneous bedding parallel shear within the stratigraphic interval cut by the fault have been discussed here, but other models with progressive shear throughout the fold structure, or combination of different models may be developed.

This study has benefited greatly from help and many discussions with Dick Bischke and Don Medwedeff. We would like to thank Ken McClay and an anonymous reviewer for their suggestions which helped improve this manuscript. This work was supported by ARCO, MOBIL, Schlumberger, TEXACO and UNOCAL to which we are thankful. J. Mosar would also like to thank the Swiss National Fund for Scientific Research, the Luxembourg Government R&D Program (R&D/BFR/89012-A1) for their financial support.

References

Boyer, S. E. 1986. Styles of folding within thrust sheets: examples from the Appalachian and Rocky Mountains of the USA and Canada. *Journal of Structural Geology* **8**, 325-339.

Coates, J. 1945. The construction of geological sections. *Quarterly Journal of the Geological, Mineralogical and Metallurgical Society of India*, **17**, 1-11.

Dahlström, C.D.A. 1969. Balanced cross sections. *Canadian Journal of Earth Science*, **6**, 743-757.

—— 1990. Geometric constraints derived from the law of conservation of volume and applied to evolutionary models for detachment folding. *American Association of Petroleum Geologists Bulletin*, **74**, 336-344.

Faill, R. T. 1973. Kink-band folding, Valley and Ridge Province, Pennsylvania. *Geological Society of America Bulletin*, **84**, 1289-1341.

Goguel, J. 1952. Traité de tectonique. Masson, Paris, 383p.

Jamison, W.R. 1987. Geometric analysis of fold development in overthrust terranes. *Journal of Structural Geology* **9**, 207-219.

Kilsdonk, B. & Wiltschko, D. V. 1988. Deformation mechanisms in the southeastern ramp region of the Pine Mountain block, Tennessee. *Geological Society of America Bulletin*, **100**, 653-664.

Laubscher, H. P. 1962. Die Zwei-phasenhypothese der Jurafaltung. *Eclogae Geologicae Helvetiae*, **55**, 1-22.

—— 1976. Geometrical adjustments during rotation of a Jura fold limb. *Tectonophysics*, **36**, 347-365.

—— 1977. Fold development in the Jura. *Tectonophysics*, **37**, 337-362.

Marshak, S. & Mitra, G. 1988. *Basic methods of structural geology*. ed. Prentice Hall, 446p.

Mitra, G. 1988. Area balanced models of fault propagation folds. *Geological Society of America Abstracts with Program*, A56.

—— 1990. Fault-propagation folds: geometry, kinematic evolution, and hydrocarbon traps. *American Association of Petroleum Geologists*, **74**, 921-945.

—— & Namson, J. 1989. Equal-area-balancing. *American Journal of Science*, **289**, 563-599.

Mosar, J. 1989. Déformation interne dans les Préalpes Médianes (Suisse). *Eclogae Geologicae Helvetiae*, **82**, 765-793.

Mount, V.S., Suppe, J. & Hook, S. C. 1990. A forward modelling strategy for balancing cross sections. *American Association of Petroleum Geologists*, **74**, 521-531.

Namson, J. S. 1981. Structure of the western foothills belt, Miaoli-Hsinchu area, Taiwan: I southern part. *Petroleum Geology of Taiwan*, **18**, 31-51.

Pfiffner, O. A. 1990. Kinematics and intra-bed strain in mesoscopically folded limestone layers: examples from the Jura and the Helvetic zone of the Alps. *In:* 7. ann. meeting Swiss Tectonic Studies Group, Basel, Abstract vol.

Srivastava, D. C. & Engelder, T. 1990. Crack-propagation sequence and pore-fluid conditions during fault-bend folding in the Appalachian Valley and Ridge, central Pennsylvania. *Geological Society of America Bulletin*, **102**, 116-128

Suppe, J. 1983. Geometry and kinematics of fault bend folding. *American Journal of Science*, **283**, 648-721.

—— 1985. *Principles of Structural Geology*. Prentice-Hall, Inc., New-Jersey, 537p.

——, Chou, G. T., & Hook, S. C. 1991. *Rates of folding and faulting determined from growth strata* (this volume)

—— & Medwedeff, D. A. 1984. Fault propagation folding. *Geological Society of America Abstracts with Program*, **16**, 670.

—— & Medwedeff, D. A. 1990. Geometry and kinematics of fault-propagation folding. *Eclogae Geologicae Helvetiae*, **83** (Laubscher volume, in press).

Appendix

Although various angles, and hence fault-propagation fold shapes are consistent with the mathematical formulation derived from volume conservation, not all solutions are geometrically feasible.

Several boundary conditions impose limiting constraints on the variety of fault-propagation fold shapes (Suppe & Medwedeff 1990). (1) Internal fold angles cannot exceed certain upper limits. The step-up angle θ_2 has to be less than

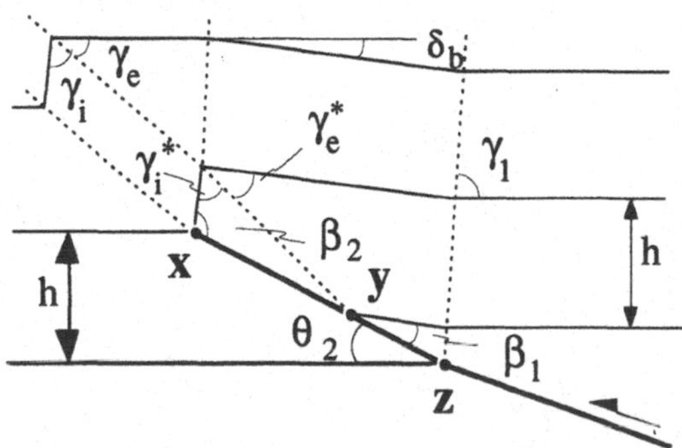

Figure 11. Outline of parameters involved in deriving the general consistency conditions. General case fault-propagation fold, fixed axial surface model.

60°, hence γ_1 and ($\gamma_e^* + \gamma_i^*$) have to be smaller than 90° and larger than 60°. (2) To be consistent with the fault-propagation fold model, the distance xy between the fault tip and the intersection of the axial surface of the fold core has to be smaller than the length xz of the ramping fault segment (Fig. 11). Expressed differently, the angles γ^* (constant-thickness theory) and γ_i^* (fixed axial surface theory) have to be in a range such that the axial surface of the fold core always intersects the ramping fault segment. Conversely, the axial surface of the fold core cannot intersect the back limb axial surface associated with the fault bend at the intersection of the decollement and the ramping fault segment (Fig. 10c).

The consistency conditions for the general (not only simple step-up from a decollement) fault-propagation fold model is given by : xy ≤ xz. Using trigonometric relationships, xy and xz can than be expressed as:

$$xz = \frac{h}{\sin\theta_2} \tag{4}$$

and

$$xy = \frac{\sin\gamma_i^*}{\sin\beta_2}\,\frac{h}{\sin\gamma_e^*} \tag{5}$$

From internal angular relationships (Suppe & Medwedeff 1990) we know that:

$$\beta_2 = 180 - \left(\gamma_i^* + \gamma_e^*\right) + \beta_1 \tag{6}$$

$$\beta_1 = \theta_2 - \delta_b = \theta_2 + 2\gamma_1 - 180 \quad (7)$$

Replacing and substituting we then obtain the following relationship:

$$\frac{h}{\sin\theta_2} \geq \frac{\sin\gamma_i^*}{\sin\gamma_e^*} \frac{h}{\sin\left[\left(\gamma_i^* + \gamma_e^*\right) - \beta_1\right]} \quad (8)$$

$$\sin\theta_2 \leq \frac{\sin\gamma_e^* \sin\left[\left(\gamma_i^* + \gamma_e^*\right) - \beta_1\right]}{\sin\gamma_i^*} \quad (9)$$

And finally we obtain the following general condition for consistency with the fault-propagation fold model (general case, both theories)

$$\sin\theta_2 \leq \frac{\sin\gamma_e^*}{\sin\gamma_i^*} \sin\left[\left(\gamma_i^* + \gamma_e^*\right) - \left(\theta_2 - \delta_b\right)\right] \quad (10)$$

For the constant-thickness fault-propagation fold model, where $(\gamma_i^* + \gamma_e^*) = 2\gamma^*$ we thus will have the following limiting condition:

$$\sin\theta_2 \leq \sin\left(2\gamma^* - \theta_2 + \delta_b\right) \quad (11)$$

In a simple-step fault-propagation fold with a fixed front anticlinal axial surface, we have $\theta_2 = \delta_b$. Thus the former equation simplifies to:

$$\sin\theta_2 \leq \sin\left(\gamma_i^* + \gamma_e^*\right) \quad (12)$$

and for a simple-step fault-propagation fold conserving layer thickness this simplifies to:

$$\sin\theta_2 \leq \sin\left(2\gamma^*\right) \quad (13)$$

Equations (12) and (13) represent the domains shown in the upper right corners of Figures 7 and 8.

2-D reconstruction of thrust evolution using the fault-bend fold method

Reini Zoetemeijer* & William Sassi

Institut Français du Pétrole, BP 311, 92506 Rueil Malmaison Cedex, France

Abstract: This paper presents a computer-aided method for analysing geological cross sections in compressive terranes. The deformation of thrust sheets is simulated using the geometrical model of fault-bend folding. The modelling focuses on complex thrust configurations like multiple imbrications and duplexes. This paper discusses the geometrical constraints of the fault-bend fold model to highlight the applicability and methodology of the kinematic reconstruction method. An application to a section from the Jura Mountains shows that complex thrust geometries can be analysed. Improvements on earlier work concern the treatment of transported piggy-back thrust sheets and the introduction of sedimentation and erosion processes. With the current computer method it is possible to investigate the stratigraphic patterns that develop in the syn-orogenic basins. This can provide insight in understanding the kinematic evolution of thrust systems.

The conceptual model of fault-bend fold deformation may be used for structural interpretation of geological cross-sections as a method of restoration and may serve as such, as an aid to seismic interpretation. With fault-bend fold deformation one assumes conservation of bed length and orthogonal thickness throughout deformation and formation of kink-band type of folding. This model belongs to a class of geometric models of finite deformation, like the vertical simple shear which is also used in section balancing (Moretti & Larrére 1989). The model of fault-bend folding was described in detail by Suppe (1983), who demonstrated its applicability for the Pine Mountain thrust sheet in the southern Appalachians, and the fold- and thrust belt of western Taiwan. Suppe & Medwedeff (1984) and more recently Medwedeff (1989) have further extended the analysis describing the kinematic relationships which may prevail between fault-related folds and contemporaneous sedimentation processes.

The fault-bend fold model may also be used to reconstruct the forward evolution of thrust emplacement (Endignoux 1989). Although a number of forward computer methods making use of this model have been proposed (Usdansky & Groshong 1989; Chou & Suppe 1987; Charlesworth &McLellan 1986; Endignoux 1989), very few case history studies have been reported in the literature (Endignoux *et al.* 1989). This paper attempts to demonstrate that such a modelling approach may be applied successfully to rather complex thrust geometries. An example of modelling the progressive evolution of thrusting is presented using a geological section taken from the Jura Mountains, western Alps, studied by Guellec *et al.* (1990).

The work has been done using a computer program that was developed by Endignoux (1989) and described in detail by Endignoux & Mugnier (1990). This program has been modified and extended in order to enable analyses of complex fault geometries (e.g. to allow faults to be activated in any order) and to incorporate sedimentation and erosion processes in the simulation.

The main limitations of the geometric model of deformation are discussed in the paper. These concern the treatment of 'high-angle fault-ramp geometry' and the fusion of two kink-axes during deformation. Earlier assumptions introduced to simplify passive deformation are abandoned and it is demonstrated that passively transported thrust sheets can be deformed without production of back shear deformation.

Finally, sedimentation and erosion processes are introduced so that the stratigraphic patterns that develop in the sedimentary overburden during thrust displacement can be studied.

Geometric method

The computer method (Endignoux & Mugnier 1990) is based on the equations of Suppe (1983). The solutions of these equations strictly satisfy the conservation of bed-thickness and bed-length, and therefore volume remains constant during the deformation. The mathematical problem is described by equations (1) and (2) and involves four parameters:

$$\tan\varphi = \frac{-\sin(\gamma-\theta)[\sin(2\gamma-\theta)-\sin\theta]}{\cos(\gamma-\theta)[\sin(2\gamma-\theta)-\sin\theta]-\sin\gamma} \quad (1)$$

$$\beta = \theta - \varphi + (\pi - 2\gamma) \quad (2)$$

θ is the cut-off angle of the fault with stratigraphic layering, φ is the change in fault dip, β is the cut-off angle after bending and γ is the axial angle, between kink-axis and layering (Fig. 1). The preservation of layer thickness is guaranteed when taking as the kink-axis, the bisectrix between the two bedding surfaces, before and after bending. In the case where two solutions of γ exist (mode I and mode II of Suppe (1983)), one has to chose one of the two possibilities. Figure 1 shows two schematic kink folds, one for an increase of fault dip (anti-

*Present address: Department of Earth Sciences, Vrije Universiteit, PO Box 7161, 1007 MC Amsterdam, The Netherlands

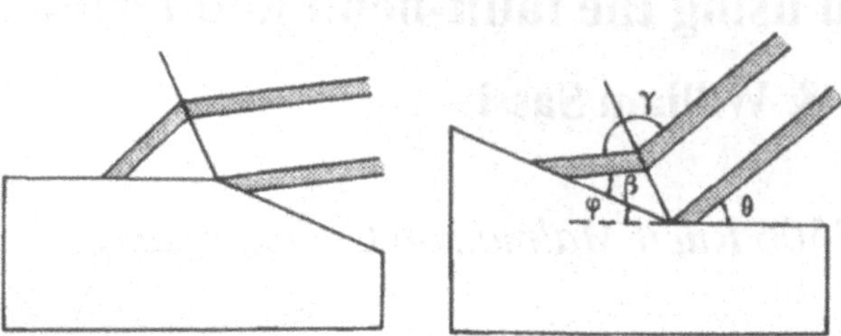

Figure 1. Fault-bend fold method showing (a) anticlinal fold in case of decrease in fault dip and (b) synclinal fold in case of increase in fault dip. Parameters: θ is the cut-off angle of the fault relative to bedding attitude, φ is the change in fault dip, β is the cut-off angle after bending and γ is the axial angle, between kink and layering.

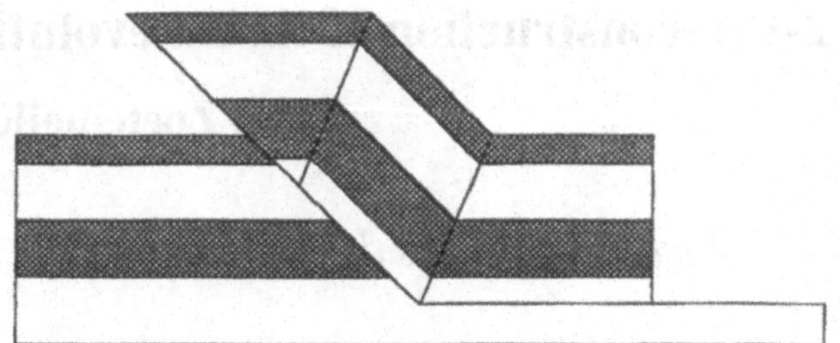

Figure 2. High-angle fault-ramp geometry. For this situation, no kink-angle γ exists for which bed-length and thickness remain conserved, when trying to bend the toe towards the surface.

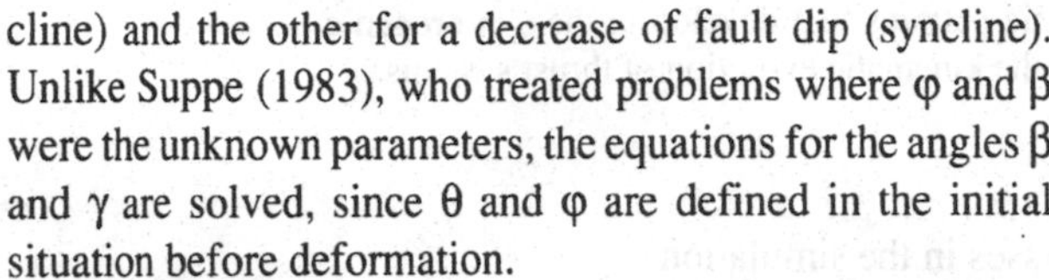

cline) and the other for a decrease of fault dip (syncline). Unlike Suppe (1983), who treated problems where φ and β were the unknown parameters, the equations for the angles β and γ are solved, since θ and φ are defined in the initial situation before deformation.

Starting from an initial geological configuration, thrust bodies will deform by addition of an incremental displacement along the fault surface. The resulting folds are defined by the parameters β and γ, which are calculated from equations (1) and (2) using the Newton method. The angle β is the newly obtained cut-off angle that is used for the creation of folds during further deformation. Also calculated are the amount of bed-parallel shear due to folding, and displacement variations along the fault, given by the slip ratio, see Suppe (1983, eq. 16).

The process of incremental displacement is stopped when the total shortening is achieved and all kink-axes are defined in the model. At this stage a thrust geometry is built, taking into account the local effects of bed-parallel shear. The bed-parallel shear modifies the thrust geometry (cut-off angles) towards the foreland. However, when geometric problems occur (i.e. high-angle fault-ramp geometry, discussed below) the bed-parallel shear modifies the thrust geometry towards the hinterland (rear boundary).

Geometric constraints of the model

The fault-bend folding hypotheses: conservation of orthogonal layer thickness, preservation of bed length and pure layer-parallel shear, described mathematically by equations (1) and (2) are very strong constraints. The assumed behaviour of fault-bend folding is not applicable to all situations (Moretti *et al.* 1990), and as with other geological models of deformation, it has limited range of applicability. However, numerous examples in the foreland of thrust belts could be investigated. If one assumes the fault-bend fold modelling as a valid approximation, it must be emphasized that during the reconstruction, geometric problems may arise for only two reasons. The first reason is the high-angle fault ramp geometry which cannot be solved in terms of bed-parallel shear and conservation of both bed length and bed thickness, and the second reason is when two kink-axes undergo fusion.

High-angle fault-ramp geometry

A solution for γ and β only exists for certain pairs of values of φ and θ (eq. 1 & 2). For example, geometric problems will occur for a single ramp-flat situation in which both ramp angle φ and cut-off angle θ exceed 30°. In such a case (Fig. 2), it is not possible to construct a symmetric kink-fold, while maintaining constant the local volume of the material that is folded. This implies that situations such as reactivation of normal faults (fault ramp of 60°) cannot be treated with this geometric model of deformation.

Multiple solutions may be envisaged to allow folding in case of high-angle fault-ramp geometry (Fig. 3 a-d). One can slacken the constraints of the fault-bend folding hypotheses, for example by assuming that bed-length (Fig. 3b), or bed-thickness (Fig. 3c) are no longer preserved. However, these solutions should not be adopted, because the assumptions imply respectively an increase in volume (Fig. 3b) or a loss of volume (Fig. 3c). An alternative would be to combine fault-bend folding with a simple shear deformation that is oblique to the stratigraphy of bedding. The simple shear preserves local volume throughout the deformation. The shear direction could be freely chosen, but must be parallel to the axis that separates the area where simple shear is imposed, from the remaining part of the model (Fig. 3d). It seems difficult, however, to work out a model in which the involved mechanisms of deformation vary laterally for the same material.

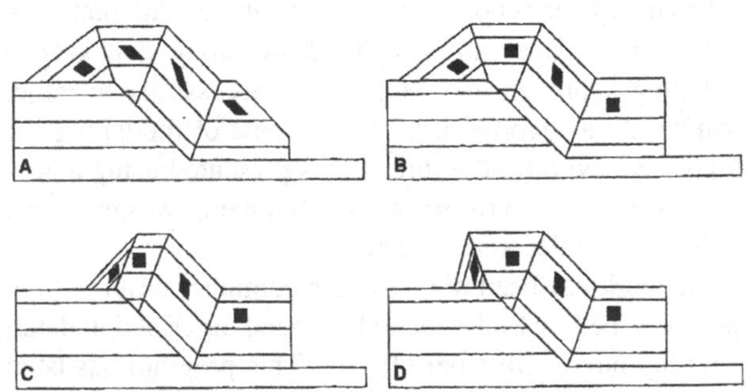

Figure 3. Different forced solutions for the treatment of the high-angle fault-ramp geometry. (a) solution with layer-parallel shear as extra parameter (conservation of bed-length and thickness). (b) conservation of bed-thickness, but not bed-length, (c) conservation of bed-length, but not bed-thickness and (d) oblique simple shear solution (conservation of volume).

Finally, in order to satisfy the local lack of material that is necessary for the further deformation of high-angle fault-ramp geometry, a certain amount of extra bed-parallel shear is introduced to ensure preservation of overall volume (Suppe 1983; Jamison 1987), (Fig. 3a). Due to length preservation, the change in shape at the rear boundary or trailing edge of the model represents the amount of required layer-parallel shear (also called 'negative' back shear). In the proposed approach the value of the kink-angle γ is determined such that the amount of extra bed-parallel shear is minimized (see also Endignoux & Mugnier 1990; Moretti *et al.* 1990).

Fusion of kink-axes

Back shear may also occur when the constructed kink-axes come to intersection during the kinematics of the deformation. An important increase in fault dip can bend the overlying sediments in a synclinal shape, in such a way that kink-axes intersect. The kink-axes will undergo fusion and, if one treats the problem in the mathematically correct way, another kink-axis has to split up elsewhere in the model on the same shear-level. The problem of kink-axis fusion and splitting up is difficult to solve. The solution is no longer unique; several kink-axes can be considered for splitting up. However, the volume of material lost in the reconstruction is usually very small compared to the volume of material needed to allow folding in the case of high-angle fault-ramp geometry. In this method the loss of material during the fusion of kink-axes is not compensated, and as a result, this will produce a (positive) backshear towards the hinterland of the models.

The deviation from a vertical rear boundary (positive or negative back shear) measures the amount of material that is needed (high-angle fault-ramp geometry) or is in excess (kink-axis fusion) because the assumption of volume preser-

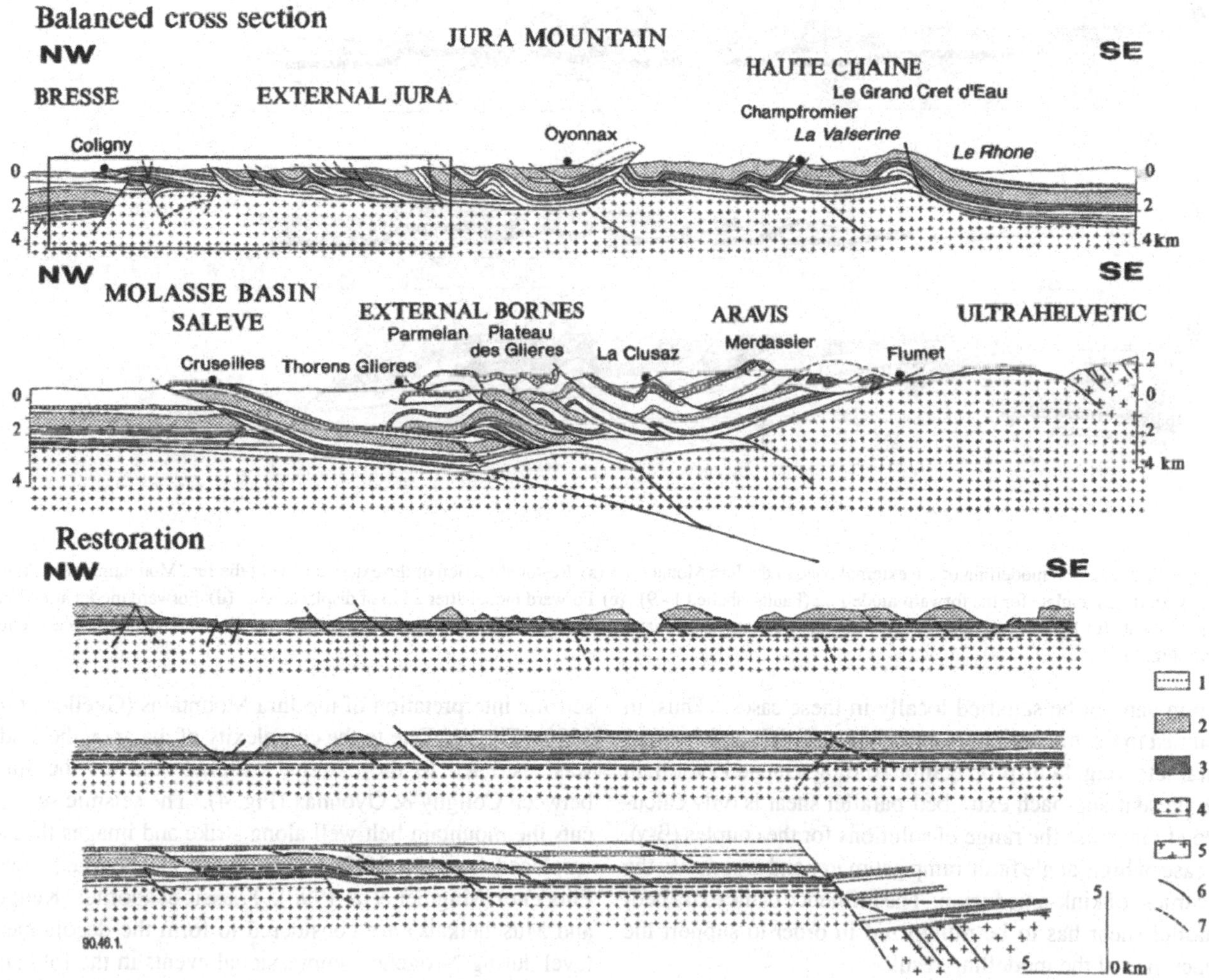

Figure 4. Balanced cross section along the ECORS profile with restoration before Cenozoic deformation. 1: Urgonian limestone; 2: Subalpine Tithonian limestone and Upper Jurassic platform carbonates; 3: Middle Jurassic limestone; 4: Basement of Jura Mountains and External Belledonne; 5: Basement of Internal Belledonne; 6: Thrusts and faults; 7: Potential thrusts. I: Cenozoic of Bresse graben and Molasse basin; II Oligocene flysch; III, IV and V: Cretaceous (III Bresse graben, Jura Mountains and Molasse basin, IV: Presubalpine domain, V: Subalpine domain); VI and VII: Jurassic (VI: Bresse graben, Jura Mountains and Molasse basin, VII: Presubalpine and Subalpine domain); VIII: Basement. After Guellec *et al.* (1990). The frame indicates the position of the area tested by numerical modelling in Figure 5.

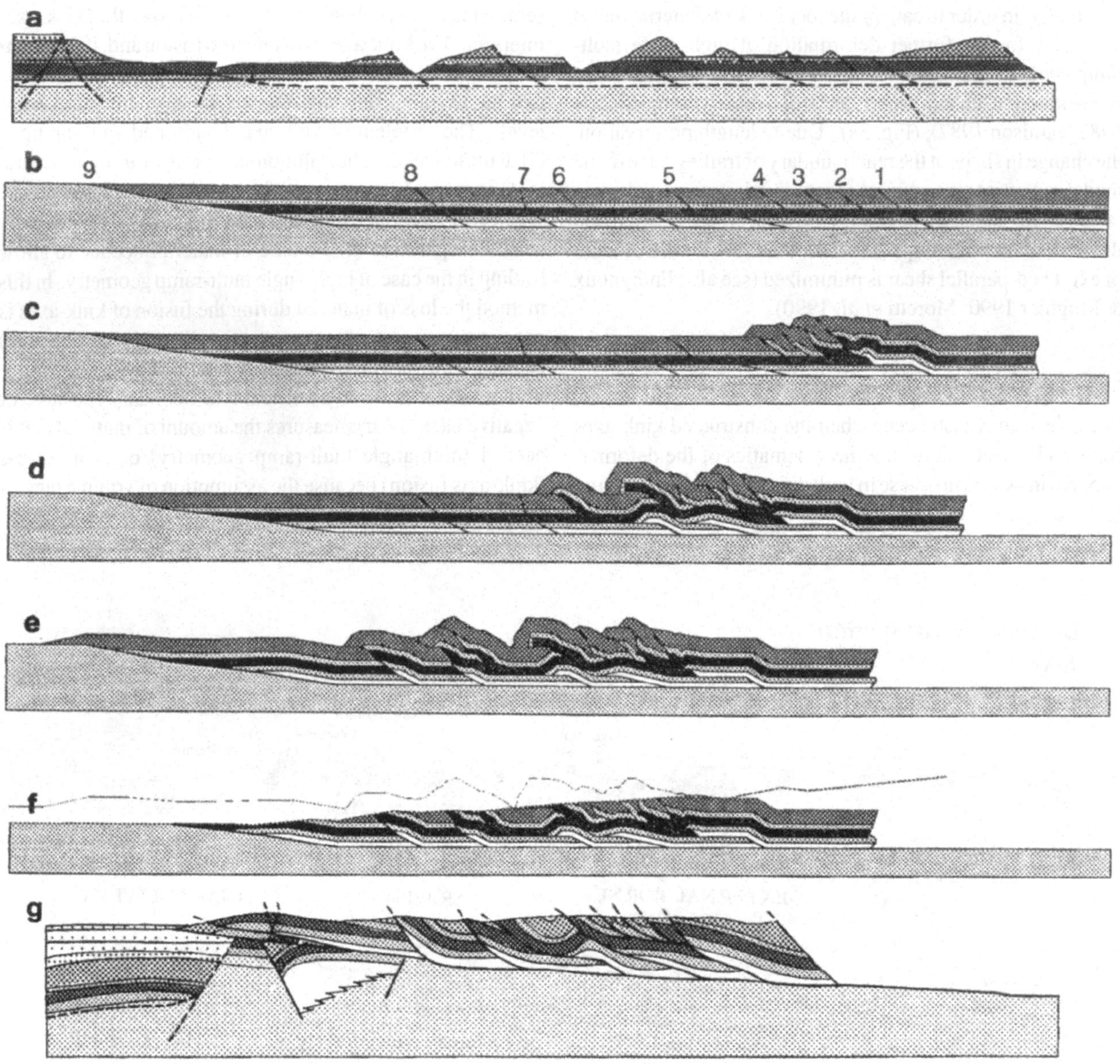

Figure 5. Numerical modelling of the external zone of the Jura Mountains. **(a)** Restored section of the external zone of the Jura Mountains from Figure 4. **(b)** Starting template for the forward modelling (Faults labelled 1 - 9). **(c)** Forward model after 2 km of displacement. **(d)** Forward model after 4 km displacement. **(e)** Forward model after 6 km of displacement. **(f)** Forward model, 6 km of displacement with erosion. **(g)** Comparative geological section (from Fig. 4).

vation can not be satisfied locally in these cases. Thus, in contrast to treatments where extra bed-parallel shear is a free parameter (e.g. fault-propagation folding (Suppe 1990a)), in the present approach extra bed-parallel shear is only calculated to increase the range of solutions for the couples (β-γ), in case of high-angle fault-ramp geometry and to simplify the treatment of kink-axis fusion. The influence of the extra bed-parallel shear has to be minimized in order to support the coherence of the modelling results.

The following example of the Jura Mountains illustrates a solution with minimum bed-parallel shear.

Jura Mountains

The code was tested on a real case example taken from a seismic interpretation of the Jura Mountains (Guellec *et al.* 1990) (Fig. 4). Due to the complexity of the area, the study was restricted to the extreme external zone of the Jura, between Coligny & Oyonnax (Fig. 4). The seismic section cuts the mountain belt well along strike and images thrusts emplaced between 10 Ma (Tortonian) and 4 Ma (Lower Pliocene) (Mugnier *et al.* 1990). Triassic evaporites (Keuper and Muschelkalk) are considered to form the decollement level during Neogene compressional events in the fold and thrust belt. The situation before the compressional events is described in the restoration of this section (Fig. 5a). It is obtained using a section balancing technique assuming conservation of length. Due to the extensive erosion three reference horizons are chosen: Urgonian, Upper Jurassic and Dogger limestones.

Layer-thicknesses remain constant all along the section,

and are locally parallel to the basement interface. The regional basement dip towards the hinterland can be interpreted as the lithospheric deflection due to nappe loading (foreland flexure). However, the wavelength of the flexure is too large to affect the thrust evolution in the studied area. The basement dip has been neglected in the simulation and a horizontal sole thrust assumed. A piggy-back thrust propagation sequence has also been assumed (cf. Fig. 5b). The balanced reconstruction of the interpreted section shows several steep fault geometries with dip values in the range of 55°. So, the assumption of high-angle fault-ramp geometry is put into practice.

The consequence of extra bed-parallel shear for the treatment of high-angle fault-ramp geometries is displayed in the modelling by the irregular rear boundary (Fig. 5c-f). However, the resulting backshear, at the final deformational stage, is found to be very small relative to the 6 km of shortening. The final state obtained with post-tectonic erosion, is in good agreement with the interpreted geological section of Guellec *et al.* (1990) (Fig. 5g). Although it will not influence the results of the geometric modelling, it should be noted that it is more likely that the erosion of 1,000 m locally, has taken place gradually during thrust emplacement.

The good fit of the final state suggests that calculated intermediate stages may be adopted as a reliable geological model of the thrust evolution in this part of the thrust belt. Figures 5c-d show the intermediate stages of the thrusting after 2 and 4 km of displacement.

Multiple imbrications

To simplify the deformation of imbricate fault-bend folding, Suppe (1983) made the assumption of constant cut-off angles during the passive deformation. That is the thrust deformation due to a gradual change in fault-shape by later deformations. This assumption was initially followed in the program. However, computer results have shown that geometric problems arise in case of multiple imbrications, when earlier deformed thrusts are transported piggy-back on newly formed thrusts (Endignoux 1989). When the thrusts were passively deformed, an additional layer-parallel shear occurred along the entire bed-length. This is shown in Figure 6c by a simple model simulating a duplex structure. The change in fault geometry occurs in the upper layer and induces the back shear at the rear boundary of the model.

The problem of back shear caused by fault deformation can be solved when the fault-bend fold equations are applied to the passively deformed thrusts. This is demonstrated in Figure 6d, for the same stage as in Figure 6c. However, instead of the back shear an apparent forward displacement along the passive fault is observed. Figure 6d shows the resulting thrust simulation that can be compared to the simplified treatment in Figure 6c. It is interesting to note that the fault 1 has been partially 'reactivated' during its passive transport on top of the fault 2, with an amount of displacement that balances the backshear shown in Figure 6c.

Thus, when adequately applied, the fault-bend fold method

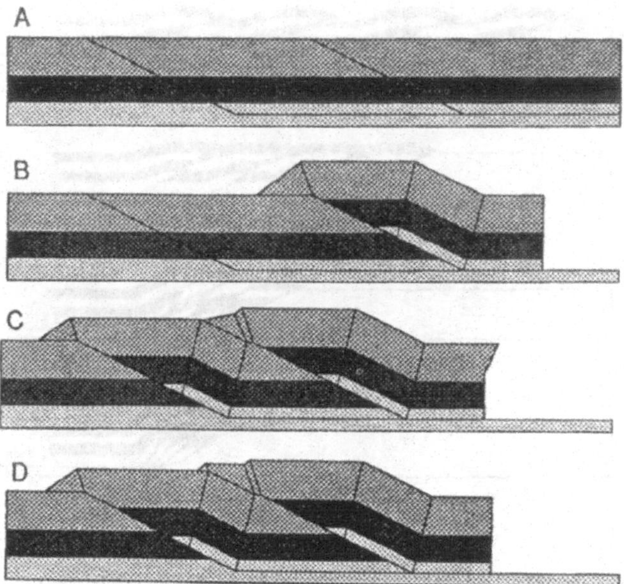

Figure 6. Simple model simulation of duplex structure, showing the back-dip of the fault that is passively deformed on the back of the new thrust. (**a**) Initial template. (**b**) Movement on the first fault - fault 1. (**c**) Backshear resulting from the assumption of constant cut-off angles, and (**d**) solution of the modified modelling, where the fault-bend fold equations are also applied to passive deformation (note the straight back edge).

can provide an additional compatibility condition for the deformation. The fault reactivation described above could be considered as a consequence of an homogeneous horizontal push at the rear end of the model.

In Figures 7a & b, the modelling results of duplex imbrications for various initial spacing of faults are given. In Figure 7a the original version of the program is used, which assumes constant cut-off angle with passive deformation. The back shear induced by the change in fault geometry is very different from one case to another. This results from the difference in change of the fault geometry (back-dip) for each initial fault spacing (Fig. 7a). In Figure 7b, the simulation shows the attitude of the duplexes when displacement on earlier faults is re-adjusted to maintain a system with vertical back boundary.

A remarkable feature is that the order in which the thrusts are activated was important for the initial version of the program, but not for the current program. If the most external fault is activated first, the second thrust will deform along an already deformed fault, and the same final geometry will be obtained. This implies that the geometric modelling of imbricate fault-bend folding is not able to distinguish the order in which faults have been activated.

Sedimentation and erosion processes

So far it has been demonstrated that the model can calculate correctly the geometry of a thrust belt at any stage of deformation. The next step in the development of the method is the introduction of sedimentation and erosion processes which are important elements in the analyses of thrust evolution (Roure *et al.* 1990; Medwedeff 1989).

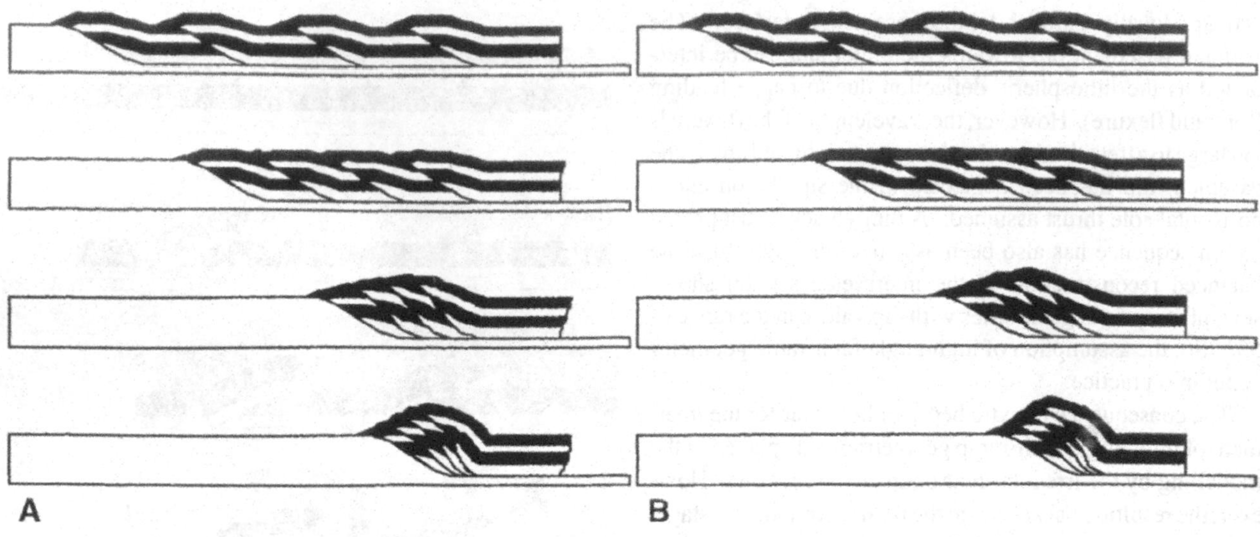

Figure 7. Reproduction of duplex imbrications. 4 duplexes with various initial fault spacing, respectively 15, 10, 5 and 3 units. (**a**) results of initial model with passive deformation assuming constant cut-off angles, and (**b**) same model as in a, but with adjustment of cut-off angles according to the fault-bend fold method.

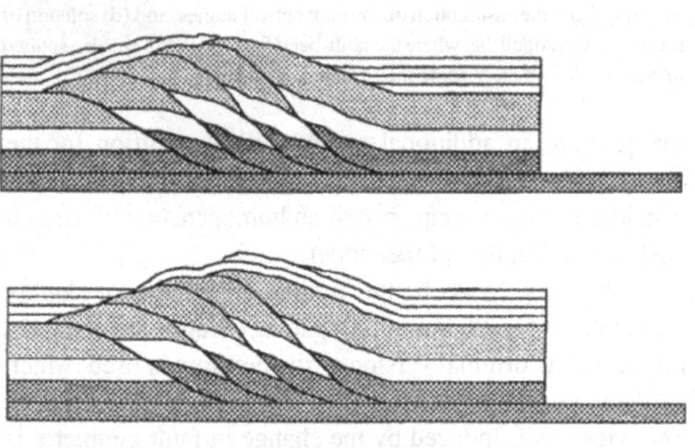

Figure 8. Synthetic model of imbricate fault-bend folding. The order in which thrusts are activated is not visible from the duplex geometry, but the overlying sediments have recorded the evolution and the different situations are therefore clearly distinguishable.

Erosion and sedimentation processes contemporaneous with thrusting are simulated by numerically following the evolution of time-lines which may be defined at any given stage of deformation. For kinematic purposes one may control independently thrust velocities and rate of sedimentation and erosion. The continual addition of 'material' on top of the model will fossilize a stratigraphic pattern in the newly formed basin.

As mentioned above, the geometric description of imbricate fault-bend folding cannot distinguish the order of faulting in the deformation history. However, basin stratigraphic analyses of the sedimentary overburden may reveal information on the kinematics of deformation, which should be used to constrain the reconstruction (Fig. 8).

An example of how sediments can register thrust evolution is described by Roure *et al.* (1990), in their study of the southern Apennines accretionary wedge. The observed piggy-back basins, formed by accumulation of syn-tectonic depos-

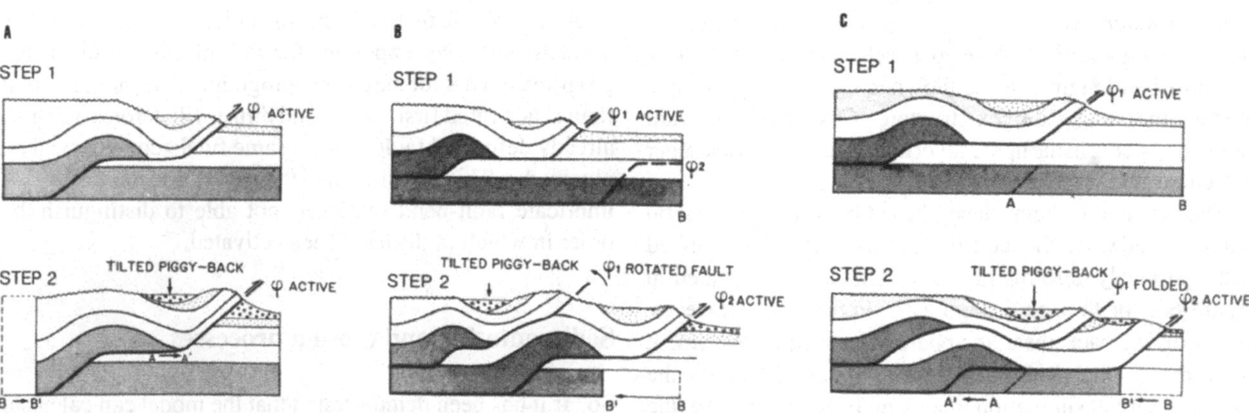

Figure 9. Development of geometry in piggy-back basins in relation to deep-seated structure. (**a**) depocenters migrate away from the front of the underlying fault when the displacement along the thrust is sufficiently slow. (**b**) The same general geometry of deposits is observed when the thrust is being tilted by younger thrusts. (**c**) Depocenters migrate toward the front of the underlying fault when a previously flat portion of thrust becomes folded. (after Roure *et al.* 1990).

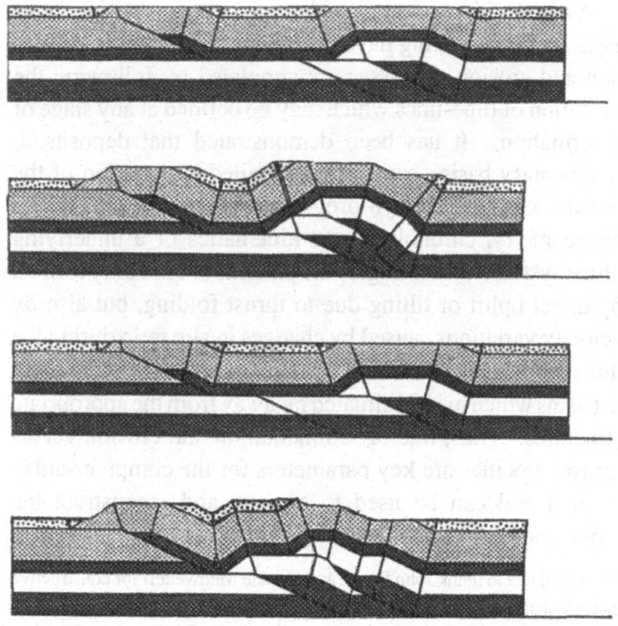

Figure 10. Model simulation of piggy-back basin migration. Note how the the variation in local position of the deep structures affects the shape of the basins and erosional surfaces.

its of Pliocene and Early Quaternary age, indicate locally different directions of depocenter migration (Roure *et al.* 1990). Typical relationships that one may expect between syn-orogenic sedimentation and erosional processes are illustrated in Figure 9. Similar thrust configurations may be reconstructed using this numerical method. In Figure 10 two schematic examples of such reconstruction for piggy-back basins being deformed by thrusting on a duplex structure are shown. Note how variation in the local position of the deep structures affects the shape of the basins and also that of the erosional surfaces.

Relationships between rate of sedimentation and thrust velocities are key factors to the kinematics (Suppe 1990b). When sedimentation rate decreases, the syn-orogenic deposits are deformed and important internal angular unconformities and erosion-surfaces will occur in the stratigraphy. When deformation slows down, important onlaps are found in the basin stratigraphy. These features can be reconstructed using the fault-bend fold method.

Medwedeff & Suppe (1986) have shown that the development of the stratigraphic patterns are controlled by two processes: (1) the instantaneous uplift above the ramp that creates a crest with less sedimentation or even erosion, and (2) when deformation continues the sedimentary beds roll from the back flank onto the the ramp and crest, where they are covered by younger sediments (Fig. 11)

Furthermore, when the allochthonous material has reached its maximum vertical uplift at the crest, not only the asymmetric shape of the rolling sediment package emerges, but also the slip along the fault is changed, since the slip is no longer taken up in the growth of the fold. This initiates changes in thrust rates versus sedimentation rates that will be visible in the overlying material deposited farther along the fault (Fig. 11). Thus the kinematics of deep thrust may be recorded laterally by the deposits in the syn-tectonic foreland basins, even if the distance between these basins and the deep thrust is relatively large.

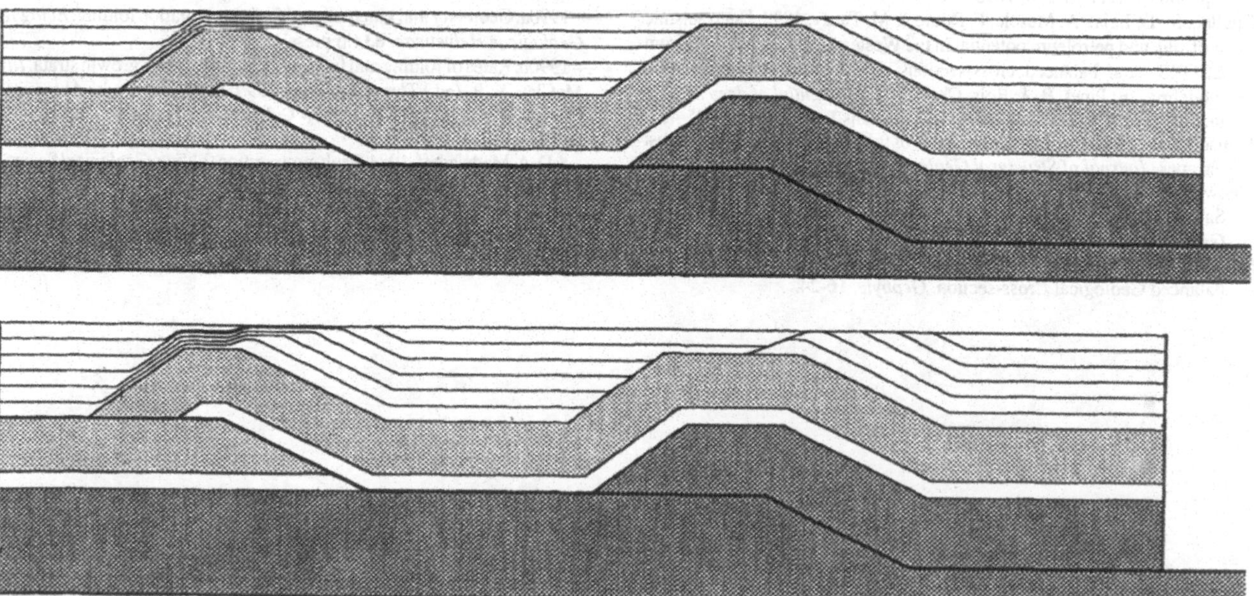

Figure 11. Two stages in the simulation of double ramp thrust structure in which time-lines record the evolution of thrusting. The velocity of the rear boundary is taken constant. The simulated sedimentation has a constant velocity equal to the initial vertical growth of the deepest fold ($v_s = v_t \cdot \sin\varphi$). However, displacement along the fault changes as a function of the slip ratio and the kinematics of the deep fold are recorded in the basins, not only close to the fold but also more towards the foreland.

Conclusions

This paper has examined the fault-bend fold model as a method to simulate the kinematics of thrust evolution. The computer model is shown to be an efficient tool for building models of thrust evolution, especially when a large number of faults with complex geometries are involved in the analysis of a geological cross section.

Although the model has important limitations (2-D approach and unsolvable high-angle fault ramp geometries in terms of local volume preservation), it is shown that rather complicated situations can be treated successfully. This is illustrated by the reconstruction of thin-skinned thrusting applied to the Jura Mountains, western Alps, using data of an interpreted geological section from Guellec *et al.* (1990). The example serves to demonstrate that cases of high-angle fault-ramp geometry may be treated satisfactorily if one minimizes the required amount of extra bed-parallel shear to accommodate deformation.

As an additional feature sedimentation and erosion processes in the modelling have been included. The sedimentation and erosion processes are simulated by following the evolution of time-lines which may be defined at any stage of deformation. It has been demonstrated that deposits in sedimentary basins can show a detailed registration of the thrust evolution. The syn-orogenic deposits give indications on geometry, chronology and kinematics of a underlying thrust system. The stratigraphic patterns are not only formed by direct uplift or tilting due to thrust folding, but also by velocity variations caused by changes in slip ratio during the thrust emplacement. These velocity variations are recorded in basins which may be situated far away from the appropriate structures. Thus, rate of sedimentation and erosion versus thrust velocities are key parameters for the computer-aided method and can be used to analyse and reconstruct the kinematic evolution of thrust systems.

We would like to thank John Suppe and Donald Medwedeff for constructive discussions. Ken McClay and François Roure reviewed earlier versions of the manuscript and we thank them for their thoughtful comments.

References

Charlesworth, H.A.K. & G. C. McLellan. 1986. Refold; Fortran 77 program to construct model block diagrams of multiply folded rocks. *Computers and Geosciences*, **12**, 349-360.

Chou, T. G. & J. Suppe. 1987. Balanced computer-modeling of complex thrust structures. *Eos abs.*, **68**, **44**, 1451.

Endignoux, L. 1989. Une modélisation numérique bidimensionelle de l'évolution cinématique et thermique des structures chevauchantes. Ph.D., Univ. Joseph Fourier, Grenoble, 243p.

Endignoux, L. & J. L. Mugnier. 1990. The use of a forward kinematical model in the construction of a balanced cross-section. *Tectonics*, **9**, 1249-1262.

——, I. Moretti & F. Roure. 1989. Forward modelling of the Southern Apennines. *Tectonics*, **8**, 1095-1104.

Guellec, S., D. Lajat, A. Mascle, F. Roure & M. Tardy. 1990. Deep seismic profiling and petroleum potential in the Western Alps: Constraints from ECORS data, balanced cross-sections and Hydrocarbon maturation modeling. *In:* Pinet, B. & Bois, Ch. (eds) *The potential of deep seismic profiling for hydrocarbon exploration*, Editions Technip, Paris, 425-437.

Jamison, J. W. 1987. Geometrical analysis of fold development in overthrust terranes. *Journal of Structural Geology*, **9**, **2**, 207-219.

Medwedeff, D. A. 1989. Growth Fault-bend folding at Southeast Lost Hills, San Joaquin Valley, California. *American Association of Petroleum Geologists Bulletin*, **73**, **1**, 54-67.

Moretti, I. & M. Larrère. 1989. LOCACE: Computer-aided construction of balanced Geological Cross-section. *Geobyte*, 16-24.

——, S. Triboulet, & L. Endignoux. 1990. Some remarks on the geometrical modeling of geological deformations, *In:* Letouzey, J. (ed.) *Petroleum and tectonics in mobile belts*. Editions Technip, Paris, 155-162.

Mugnier, J. L., S. Guellec, G. Ménard, F. Roure, M. Tardy, & P. Vialon. 1990. Crustal balanced cross-sections through the external Alps deduced from the ECORS profile. *In:* Roure, F., Heizman, P. & Polino, R. (eds) *Deep structures of the Alps*. Memoirs de la Société. Géologique de France **156**, *Suisse* **1**, *Italiana* **1** (in press).

Roure, F., D. Howell, S. Guellec, & P. Casero. 1990. Shallow structures induced by deep-seated thrusting, *In:* Letouzey, J. (ed.) *Petroleum and tectonics in mobile belts*. Editions Technip, Paris, 15-30.

Suppe, J. 1983. Geometry and kinematics of fault-bend folding, *American Journal of Science*, **283**, 684-721.

—— 1990a. Geometry and kinematics of fault-propagation folding. *Eclogae Geologicae Helveticae*, **83** (in press).

—— 1990b. Rates of folding and faulting determined from growth strata. *In:* McClay, K. R. (ed.) *Thrust Tectonics 1990, Programme with Abstracts*, Royal Holloway and Bedford New College, University of London, (4-7 April, 1990), 16.

—— & D. A. Medwedeff. 1984. Fault-propagation folding, *Geological Society of America Program with Abstracts*, **16**, 670.

Usdansky, S. I., & R. H. Groshong. 1989. Thrustramp 1 and 2, *Rockware Inc., Geological Software Catalog, Spring*, 65-66.

Kinematic models of deformation at an oblique ramp

Theodore G. Apotria[1*], William T. Snedden[2], John H. Spang[1], David V. Wiltschko[1]

[1]Department of Geology and Center for Tectonophysics, Texas A&M University College Station, Texas 77843, USA
[2]Texaco E&P Technology Division, Houston, Texas 77042, USA

Abstract: Kinematic models for the deformation of hangingwall material moving over a footwall oblique ramp are developed by considering two end members of assumed mechanical behaviour, vertical shear and layer-parallel shear. In the former case, material is sheared vertically and displacements remain within the tectonic transport plane; the deformation is accommodated by thinning of the hangingwall over the ramp. In the later case, material is deflected out of the transport plane such that the pitch angle of the hangingwall particle path in the plane of the oblique ramp is equal to the angle between the transport direction and the strike of the oblique ramp. As a result, shear strains above the oblique ramp are non-zero in both the transport and transport-normal planes. The deflection and transport-normal shear strains are a minimum for the special cases of pure frontal and lateral ramps, and maximum at an intermediate orientation, depending on oblique ramp dip. Fault-bend folds are similar in most respects for both vertical shear and layer-parallel shear mechanisms. At frontal ramp - oblique ramp intersections, synformal or antiformal multiple bends in the footwall generate, respectively, second order hangingwall synclines or anticlines, which terminate along strike into simple fault-bend folds. For the layer-parallel shear mechanism along the pure oblique ramp, deflected hangingwall material passes through the transport plane, conserving area and volume. At the rearward intersection zone (concave toward the transport direction), hangingwall material diverges resulting in local strike-parallel extension. This extension may be a mechanism for the generation of transverse faults (or 'tear faults') in the hangingwall. At the forward intersection zone (convex toward the transport direction), displacement paths converge resulting in local strike-parallel shortening. The attitude of the oblique ramp and the amount of displacement significantly affect the map geometry and magnitude of lateral strains.

A characteristic of thin-skinned thrust faults is that they are parallel to the stratigraphy along 'flats' and cut across the stratigraphic section along 'ramps'. Ramps are termed frontal, oblique (or transverse), or lateral, depending on their orientation with respect to the regional transport direction (Fig. 1). Similar ramp - flat geometries have been documented in extensional terranes as well (e.g. Gibbs 1984; Bosworth 1985; Wernicke & Burchfiel 1982), and these are termed normal fault ramps. Ramps are a special case of fault-bends. In general, a fault-bend is a change in fault surface attitude.

The purpose of this paper is to establish a criterion for the kinematic development of a hangingwall moving over a footwall oblique ramp. As the hangingwall negotiates an oblique ramp, do material particles remain entirely within the tectonic transport plane (xz, Fig. 1), or are they deflected resulting in out-of-plane deformation?

Two-dimensional kinematic models of hangingwall deformation over fault-bends include those with a preferred direction of simple shear (e. g. Sanderson 1982; Suppe 1983; Jamison 1987; Groshong & Udansky 1988, in thrust terranes; Verral 1981; Gibbs 1983; White *et al.*, 1986; Groshong, 1988, in extensional terranes). Mechanical models have addressed ramp deformation of isotropic half-spaces (Berger & Johnson 1982; Kilsdonk & Fletcher 1989) as well as isotropic and anisotropic finite layers (Wiltschko 1979, 1981). Here, a kinematic model is developed bounded by two end members of assumed mechanical behaviour, vertical shear and layer-parallel shear. The two end members bound a range of behaviour where a simple shear is imposed at an arbitrary angle to bedding, a range also investigated in two dimensions by White *et al.* (1986). However, this discussion is focused on vertical shear and layer-parallel shear only. The three-dimensional displacement and strain behaviour of these two mechanisms due to motion over an oblique ramp are outlined, and discussed with respect to the implications on the development of fault-bend folds. Here, the analysis is restricted to the motion of hangingwall material by developing simple kinematic models, with no regard for the state of stress due to irregular footwall topography. The mechanics of this process

* Present address: Shell Development Company, PO Box 481, Houston, Texas 77001, USA

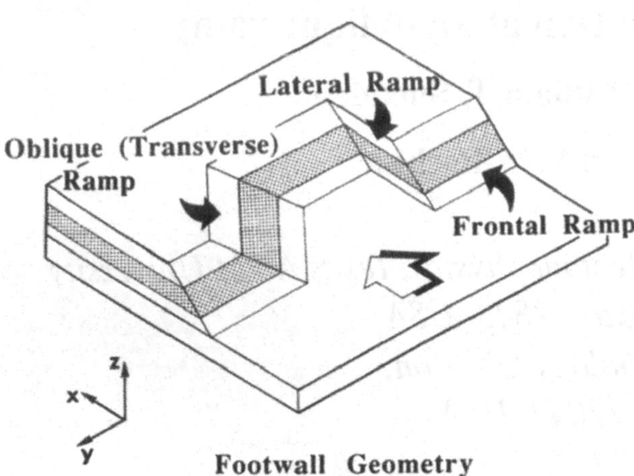

Figure 1. Footwall geometry. Frontal ramps strike perpendicular to the transport direction (shadowed arrow). Lateral ramps strike parallel to the transport direction. Oblique (transverse) ramps strike oblique to the transport direction. Similar footwall geometries exist in extensional terranes.

is the subject of parallel research (Apotria 1988, 1990).

Review of selected previous work

Oblique ramps and fault-bends appear to be scale independent and have several structural associations. Oblique ramps inhibit the motion of thrust sheets by macroscopic folding, as do frontal ramps, because they represent topography along otherwise subplanar faults. Oblique ramps serve as a mechanism of displacement transfer by linking en-echelon frontal thrusts (Gardener & Spang 1978; Pfiffner 1981; Goldberg 1984; Mitra 1988) and of displacement termination as in the bounding faults to duplexes (Boyer & Elliot 1982; Butler 1982; Lageson 1984). Oblique ramps have been associated with presumed regional stress concentration mechanisms. Woodward (1987a) suggests that lateral ramps in the northern Absaroka thrust system are localized by facies changes that occur across the strike of the thrust belt. Oblique ramps may be associated with basement buttresses where thrust belts interact with foreland style deformation (Schmidt & O'Neil 1982; Schmidt *et al.* 1988; Couples & Lewis 1988).

Documentation of the occurrence of oblique ramps is common in the literature, yet work concerning the details of deformation processes involved is limited. The following studies suggest that oblique ramps are places where the common assumption of plane deformation is violated. Butler (1982) presented a model involving the formation of culminations and depressions along the strike of a thrust sheet as a result of footwall lateral ramps. He proposed that the hangingwall above the bounding faults of duplexes may extend during ramping, and shorten during footwall imbrication. This idea was quantified by Petrini & Wiltschko (1986) in which they determined the thinning or shear required in the hangingwall moving over a promonitory structure as a function of lateral ramp dip and ramp height. Boyer (1985) described hydrocarbon trap styles related to hangingwall deformation associated with oblique ramps. He suggested that strike- parallel extension and dip-oriented porosity occur as a result of hangingwall collapse on the upper footwall flat, as well as the occurrence of strike closure of ramp anticlines and tear faults. Mitra & Yonkee (1985) attribute cleavage development to multiple phases of deformation in the Idaho-Wyoming-Utah thrust belt. An early cleavage, formed axial planar to early buckle folds in the Crawford thrust sheet, is fanned as folds move over major frontal ramps. A second cleavage, axial planar to accompanying cross folds, developed by motion over a lateral ramp in the Crawford thrust. Wheeler (1980) noted zones of anomalous folding and fracturing associated with cross-strike structural discontinuities in the Appalachian overthrust belt. These discontinuities are the loci of bends or style changes of detached folds associated with lateral ramps, transverse faults with differential displacement, enhanced fracture development, and facies or thickness changes of the stratigraphic units. Sanderson (1982), Coward & Potts (1983), and Coward & Kim (1981) developed models of differential transport within thrust sheets and nappes where strains are factorized into wrench, thrust, and pure shear components. A wrench shear (differential slip in the plane of the detachment) occurs in the vicinity of a lateral thrust tip or oblique ramp. Principal shortening strains calculated from the model of Sanderson (1982) agreed with those inferred from oblique cleavages and stretching lineations around lateral ramps near Mayo, Ireland. Based on the above work, it appears that out-of-plane deformation is to be expected near oblique ramps. The goal is to develop a context

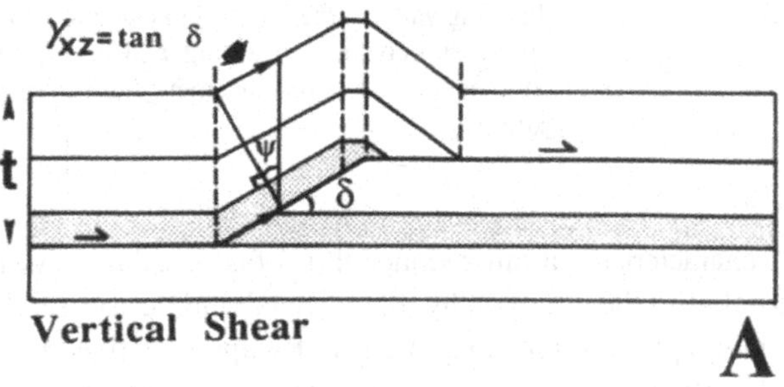

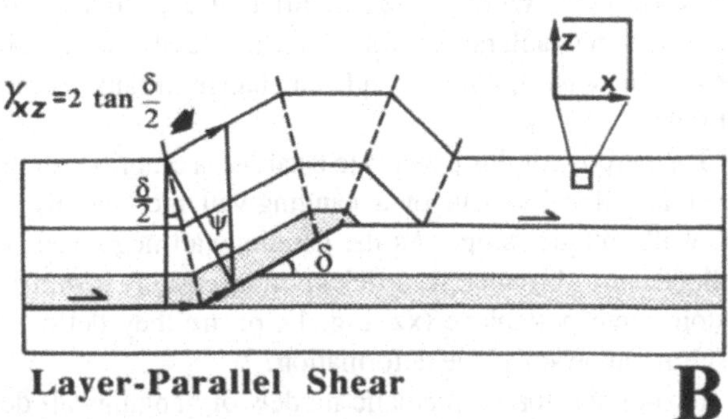

Figure 2. Fault-bend folds demonstrating the mechanisms of and the differences in shear strain between vertical shear (a) and layer-parallel shear (b) for a unit thickness (t) of the folded layer. (a) shear strain $\gamma_{xz} = \tan\ \psi = \tan\ \delta$. (b) $\gamma_{xz} = \tan\ \psi = 2\ \tan\ (\delta/2)$. Note also the variation in width of horizontal beds and forelimb dip for identical ramp dips and displacements.

based on kinematic end members within which observed behaviour can be evaluated, and unobserved behaviour can be predicted.

End members of mechanical behaviour

The kinematic development of a hangingwall negotiating a ramp or fault bend can be placed in the context of two end members of assumed mechanical behaviour. These two end member models were discussed by Sanderson (1982) for two-dimensional deformation and were termed 'bending fold' and 'flexural flow'. To emphasize the primary mechanism involved in each of these, they will be referred to here as the 'vertical shear' and 'layer-parallel shear' models, respectively. These two end members are herein considered macroscopic mechanisms and do not necessarily refer to actual deformation mechanisms at the grain or outcrop scale. Although the examples here illustrate thrust fault-bends, the analysis is quite general. Normal fault-bends or ramps are evaluated by reversing the transport direction. For simplicity we assume that the footwall is rigid, ramps and flats have a planar geometry, and that the regional transport direction is perpendicular to the strike of the frontal ramp (parallel to the x-direction in a Cartesian coordinate system, see Fig. 1).

Vertical shear

As hangingwall material passes through the lower hinge of a frontal ramp, it may be sheared vertically such that the thickness measured in a vertical direction remains constant, analogous to similar folding (Fig. 2a). The bedding-normal thickness of the hangingwall is reduced over the ramp, and the boundary between sheared and unsheared material (axial surface) is vertical. The magnitude of the shear strain in the forelimb and backlimb in the xz plane is a function of the ramp angle, δ (Sanderson, 1982) :

$$\gamma_{xz} = \tan \delta \tag{1}$$

where the change in angle between two lines originally perpendicular and parallel to bedding on the lower flat (y, Fig. 2) is equivalent to the ramp dip. If, at some time *t*, the next increment of strain at a ramp - flat hinge is infinitesimal, expressions for tensorial shear strain in the xz and yz plane can be used to illustrate how displacements are related to shear strains (Fig. 3).

$$\varepsilon_{xz} = \frac{1}{2}\left(\frac{\partial u}{\partial z} + \frac{\partial w}{\partial x}\right) \tag{2}$$

$$\varepsilon_{yz} = \frac{1}{2}\left(\frac{\partial v}{\partial z} + \frac{\partial w}{\partial y}\right) \tag{3}$$

For two-dimensional plane deformation, $\varepsilon_{yz} = 0$. In the case of vertical shear, $\partial u/\partial z = 0$ in equation (2), displacements are vertical and in the xz plane for a reference frame fixed to bedding in the folded layer. Upon moving from the ramp to the upper flat, material undergoes a shear of opposite sign, and, if the flats are parallel, equal magnitude, resulting in zero

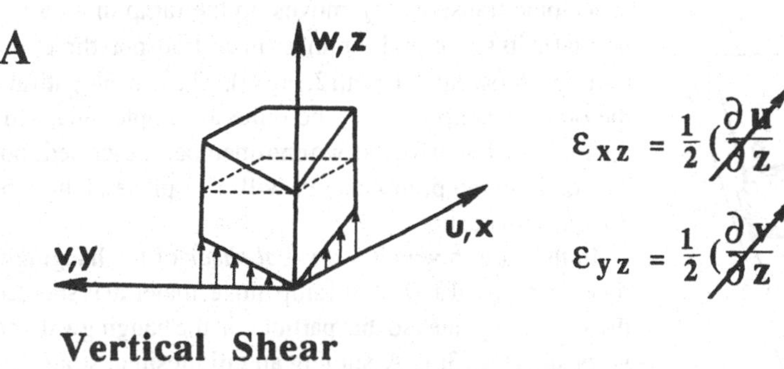

$$\varepsilon_{xz} = \frac{1}{2}\left(\frac{\cancelto{0}{\partial u}}{\partial z} + \frac{\partial w}{\partial x}\right)$$

$$\varepsilon_{yz} = \frac{1}{2}\left(\frac{\cancelto{0}{\partial v}}{\partial z} + \frac{\partial w}{\partial y}\right)$$

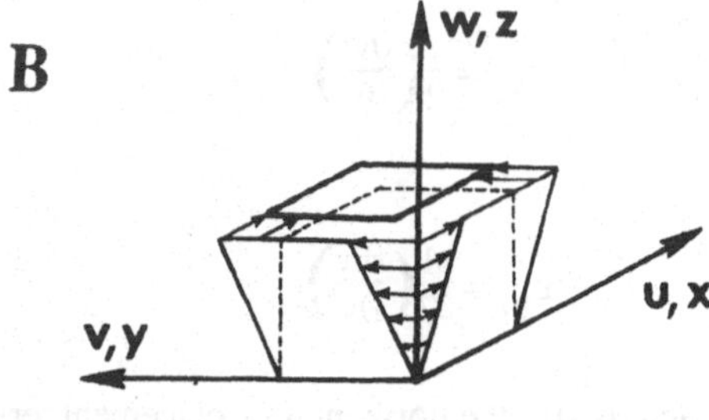

$$\varepsilon_{xz} = \frac{1}{2}\left(\frac{\partial u}{\partial z} + \frac{\cancelto{0}{\partial w}}{\partial x}\right)$$

$$\varepsilon_{yz} = \frac{1}{2}\left(\frac{\partial v}{\partial z} + \frac{\cancelto{0}{\partial w}}{\partial y}\right)$$

Figure 3. If the increment of strain at a ramp-flat hinge is infinitesimal, expressions for tensorial shear strain in the xz and yz plane can be used to illustrate how displacements are related to shear strains for vertical and layer-parallel shear. u, v, and w are respectively small displacements in the x, y, and z direction. The x-y plane of the reference frame indicated corresponds to bedding in the folded layer. The dashed lines outline the undeformed cube, the solid lines represents the deformed cube. **(a)** For vertical shear, displacements are parallel to z with zero out-of-plane displacement. **(b)** Layer-parallel shear admits displacements in the x and y directions. Deformed cube consists of components resolved parallel to x and y.

net strain. This mechanical behaviour is likely to occur if the hangingwall is isotropic with respect to mechanical layering, if the rocks are vertically jointed, if the deformation occurs under conditions where ductile mechanisms predominate, or perhaps where rocks are unlithified or unconsolidated.

Layer-parallel shear

Alternatively, shear may occur parallel to bedding such that bedding-normal thickness remains constant throughout the deformation, analogous to parallel folding (Fig. 2b). This mechanism is also called flexural slip if the shear occurs along bedding planes, or flexural flow if the shear is otherwise distributed through the rock (i.e. at the grain scale). To maintain constant bed thickness, the axial surface bisects the angle between the flat and the ramp. The magnitude of the shear strain at each ramp - flat intersection is (Sanderson 1982):

$$\gamma_{xz} = 2 \tan\left(\frac{\delta}{2}\right) \qquad (4)$$

Using the same analogy with tensorial shear strain, for plane deformation $\varepsilon_{yz} = 0$ and in the case of layer-parallel shear, $\partial w/\partial x = 0$ in equation (2). Infinitesimal displacements are parallel to bedding in the xz plane. If the upper and lower footwall flats are parallel, the shear strains at the lower and upper hinges are opposite in sign, resulting in zero net shear strain. The layer-parallel shear mechanism is likely to occur in the external, 'colder' levels of layered hangingwall sheets where interbed slip due to mechanical layering is prominent. This mechanism provided the basis of folding models developed by Suppe (1983), Suppe & Medwedeff (1984), and Jamison (1987).

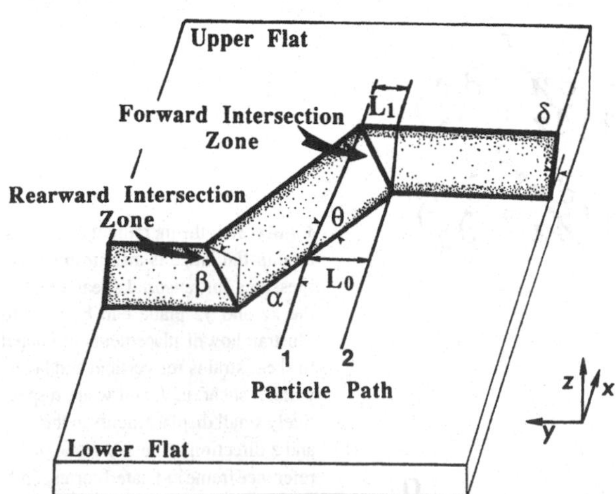

Figure 4. Footwall geometry including the salient footwall parameters. α is the angle between the transport direction and the strike of the oblique ramp. β is the dip of the oblique ramp. δ is the dip of the frontal ramp. θ is the pitch of the particle path in the plane of the oblique ramp. If the strike of the frontal ramp is perpendicular to the transport direction, then the angle between the oblique ramp and the frontal ramp is 90-α. L_0 is the initial distance between two adjacent particles on the lower flat, L_1 is the final distance on the upper flat.

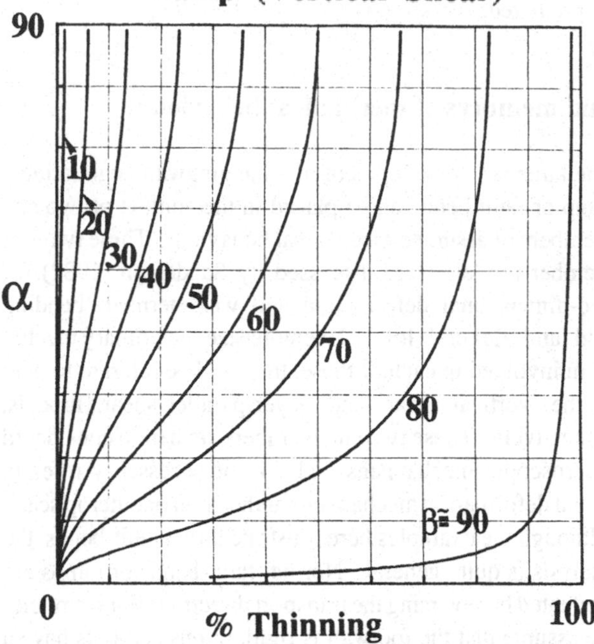

Figure 5. Folding over a ramp is accommodated by thinning of the hangingwall for the case of vertical shear. Thickness is measured perpendicular to bedding before (t) and after (t') passing through a ramp-flat hinge (Fig. 2). % thinning = (1- t'/t) • 100. The percentage of thinning in the xz plane increases with both a and b, since each corresponds to an increase of the component of ramp dip in that plane.

Application to oblique ramps

To extend the analysis to three dimensions, initially consider the path of a particle passing over a footwall oblique ramp (Fig. 4). For the case of a frontal ramp, where two-dimensional plane deformation is assumed, a particle moves in the direction of tectonic transport (x), moves up the ramp in a direction normal to its strike, and continues in the transport direction on the upper footwall flat (path 2, Fig. 4). The particle path along the oblique ramp may not be quite as simple, and will be distinctly different for the two end-members described above. The oblique ramp, in general, will be a plane along which oblique slip occurs.

In the case in which a *vertical shear* of the hangingwall rocks is imposed at the flat-ramp hinge, material is sheared in the xz and yz planes so that particles in the hangingwall move vertically (Fig. 3a). Assuming an infinitesimal strain increment to illustrate the displacements, the tensor shear strain components reduce to:

$$\varepsilon_{xz} = \frac{1}{2}\left(\frac{\partial w}{\partial x}\right) \qquad (5)$$

$$\varepsilon_{yz} = \frac{1}{2}\left(\frac{\partial w}{\partial y}\right) \qquad (6)$$

where $\partial u/\partial z$ and $\partial v/\partial z$, the horizontal displacement terms, are zero. Therefore, there are only vertical displacements in

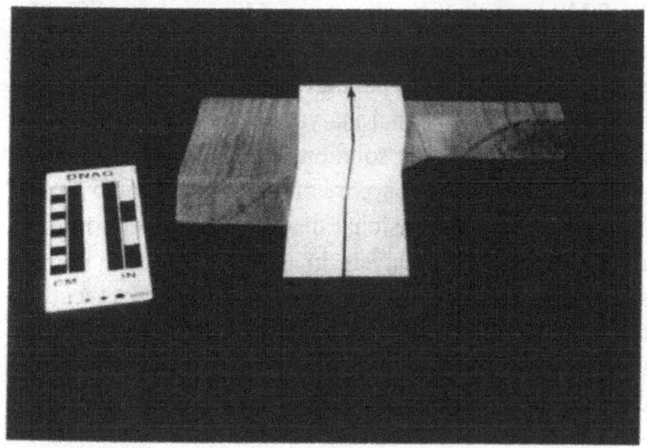

Figure 6. The relation $\theta = \alpha$ for the case of layer-parallel shear can be demonstrated using a simple physical model. The rigid footwall is represented by a wooden block and the hangingwall by a thickness of paper sheets, which is a useful analogue for folding by layer-parallel shear. Here, all but one sheet of paper has been removed and marked with a line representing the transport direction. As the sheet is moved up the oblique ramp (Fig. 6b), the particle path represented by the line is deflected out of the transport plane such that $\theta = \alpha$. In general, the particle paths on the lower and upper flats are parallel, but no longer contained in the same vertical plane. (Camera is directed parallel to the transport direction).

proportional to apparent dips of the oblique ramp:

$$\gamma_{xz} = \tan \beta_x \tag{7}$$

$$\gamma_{yz} = \tan \beta_y \tag{8}$$

where β_x and β_y are apparent dips of the oblique ramp in the x and y directions, respectively (see also Fig. 7). There is no displacement of the particle out of the transport (xz) plane for the case of vertical shear. The two components of shear strain are accommodated by thinning of the hangingwall over the oblique ramp (Fig. 5). If there is no displacement of a hangingwall particle path out of the transport plane, its path may be specified in the plane of the oblique ramp. In this case, an expression is derived for the relationship between α and the angle q, where q is the pitch of the particle path in the plane of the oblique ramp (see Fig. A1 - Appendix 3):

$$\sin \theta = \frac{\tan \alpha}{\cos \beta} \tag{9}$$

For small values of oblique ramp dip (β), the pitch of the particle path (θ), is nearly equal to the transport direction-oblique ramp angle (α). This is also true at low and high values for α, where the oblique ramp approaches a simple frontal or lateral ramp, respectively.

The case corresponding to an end member model in which *layer-parallel shear* is imposed at the flat-ramp intersection results in a slight modification of equations (2) and (3) above (Fig. 3b) such that :

$$\varepsilon_{xz} = \frac{1}{2}\left(\frac{\partial u}{\partial z}\right) \tag{10}$$

$$\varepsilon_{yz} = \frac{1}{2}\left(\frac{\partial v}{\partial z}\right) \tag{11}$$

where ∂w/∂x and ∂w/∂y, the vertical displacements terms, are now zero. In general, there are displacements in the x and y directions. The displacements are such that the pitch angle of the particle path in the plane of the oblique ramp (θ) is equal to the angle between the transport direction and the oblique ramp (α), regardless of the dip of the oblique ramp (compare eq. (9)):

$$\theta = \alpha \tag{12}$$

The condition that θ=α for layer-parallel shear is a result of the condition that there is no deformation in the plane of

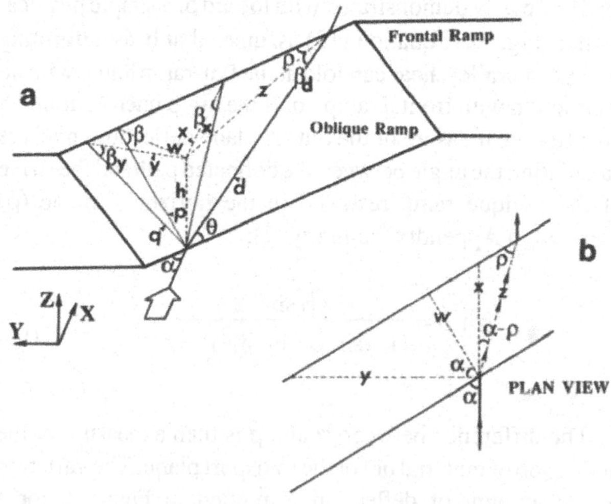

Figure 7. Parameters used to calculate the out-of-plane deflection for the layer-parallel shear mechanism. (a) β_d is the apparent dip of the oblique ramp in the displacement direction. ρ is the angle between the deflected path and the strike of the ramp measured in the horizontal plane. z, d, x, y, p, w, q are leg lengths of right triangles, h is the height of the ramps. (b) In plan view, the deflection is α-ρ. Other symbols are listed in Appendix 1.

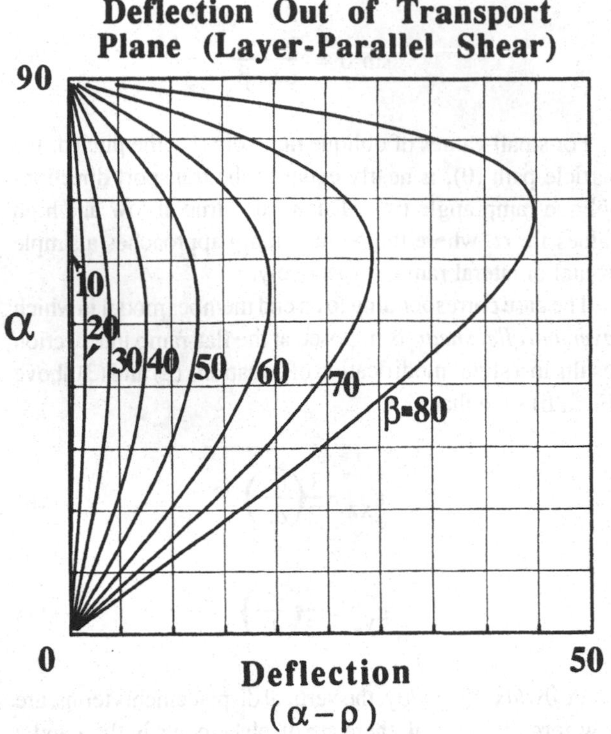

Figure 8. Variation in the amount of deflection for a spectrum of ramp geometries. The magnitude of the deflection increases with the dip of the oblique ramp (β). For any value of β, the deflection is zero at α=0° and α=90°, and reaches a maximum at an intermediate value. In other words, there is no out-of-plane deflection for pure lateral and frontal ramps. The location of the deflection maximum increases from α= 45° to 90° as β increases from 0° to 90°.

bedding. Therefore, each 'bedding plane' is rotated about an axis parallel to the strike of the oblique ramp, an amount equal to the ramp dip. The transport direction is a line in the bedding plane, and is deflected. This rotation is perhaps best visualized and easily demonstrated with the aid of a simple physical model (Fig. 6). Equation (12) assumes that beds deforming by layer-parallel shear can fold at the flat-ramp hinge with no interaction with frontal ramp - oblique ramp intersections. A quantitative measure of the out-of-plane deflection involves calculating the angle between the deflected path and the strike of the oblique ramp resolved in the horizontal plane (ρ) (derived in Appendix 2 and Fig. 7):

$$\sin \rho = \frac{\cos\beta \sin \alpha}{(1-(\sin \alpha \sin \beta)^2)^{1/2}} \tag{13}$$

The difference between α and ρ is then a measure of the deflection of material out of the transport plane. The variation in the amount of deflection is plotted in Figure 8 for a spectrum of ramp geometries. The magnitude of the deflection increases with the dip of the oblique ramp (β). For any value of β, the deflection is zero at α=0° and α=90°, and reaches a maximum at an intermediate value. In other words, there is no out of plane deflection for lateral and frontal ramps. We will illustrate the significance of this deflection in a later section.

Given the displacement path of particles in the hangingwall for the layer-parallel shear mechanism (eq. 12), the resultant shear strain can be calculated, analogous to that in equation (4), using the apparent dip of the oblique ramp in the direction of displacement (see β_d in Fig. 7). The shear strain in the displacement direction can then be resolved into x and y components.

This analysis has utilized two kinematic end member models to describe the displacement paths and shear strains of hangingwall material at an oblique ramp. These displacement paths can be applied to determine fault-bend fold forms over oblique ramps by making a transition from particle paths to three dimensional volumes of rock.

Fault-bend folding over oblique ramps

The fold forms that develop over oblique ramps differ for each of the two end members of assumed mechanical behaviour, vertical and layer-parallel shear. For layer-parallel shear, the two-dimensional analytic solution of Suppe (1983) is used, while a graphic solution is utilized for the vertical shear mechanism. The purpose of this paper is not to present a complete three-dimensional theory of fault-bend folding, but to perform the analysis in two dimensions, initially assuming plane strain, and constructing serial cross sections across two frontal ramp - oblique ramp intersections (Fig. 9). It is then suggested as to how the theory must be modified to account for three dimensional geometry and kinematics.

Footwall geometry

The footwall geometry in Figure 9 depicts two frontal ramps linked by an oblique ramp in map view. Note that the

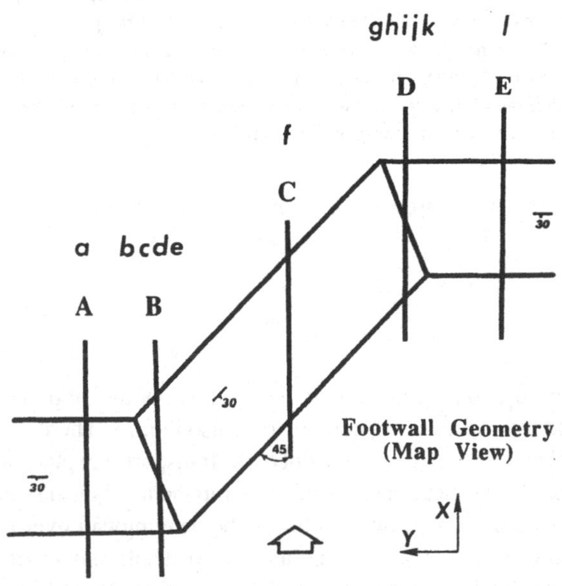

Figure 9. Footwall geometry in map view showing the location of serial cross sections *A-E* in Figures 10 and 11, and sections *a-l* in Figure 12. All ramps dip 30° (α=β=30°) and α, the angle between the transport direction and the strike of the oblique ramp, is 45°.

intersection between frontal and oblique ramps include a zone which has a multiple fault-bend. To illustrate the hangingwall geometry, serial cross sections have been constructed, labelled *A* through *E* (Figs 10 & 11). Sections *A* and *E* are simple fault-bend folds over ramps which dip 30°. Section *C* is a simple fault-bend fold over a ramp which has an apparent dip of 22° in the transport direction. Section *B* is a antiformal multiple fault-bend with ramps dipping consecutively 30° and 22° in the transport direction. Section *D* is a synformal multiple fault-bend with ramps dipping consecutively 22° and 30° in the transport direction. The footwall geometry is the same for both vertical and layer-parallel shear mechanisms. In the cross sections that follow, the displacement is small to avoid complicated geometries. Complications that arise due to interference between axial surfaces are treated in the following section.

Vertical shear

Fault-bend folds produced by this mechanism were constructed graphically by displacing the hangingwall up the ramp, maintaining constant vertical thickness (Fig. 10). Like parallel fault-bend folding (Suppe 1983), two axial surfaces are generated at each fault-bend (dashed lines), which unlike parallel fault-bend folding, are vertical. One remains fixed to the footwall through which material in the hangingwall passes, whereas the other is a passive marker in the hangingwall. This mechanism requires thinning of the hangingwall over the ramp (see Fig. 5), and bedding line length is not, in general, conserved before and after slip. Because there is no interbed slip in the hangingwall, the forelimb has the same magnitude of dip in the foreland direction as the ramp dip, in this case 30°. Section *C* is similar to sections *A* and *E*. However, the shallower ramp dip generates a ramp anticline with less structural relief and a shallower forelimb dip (22°). The difference in relief results in frontal ramp anticlines that plunge with three-way closure toward the oblique ramp. The amount of structural relief and closure changes with displacement. Section *B* has an antiformal bend in the footwall. This additional bend results in a third pair of vertical axial surfaces and an anticlinal bend in the hangingwall. Similarly, section *D* has a synformal bend in the footwall and resultant synclinal bend in the hangingwall bound by a third set of axial surfaces.

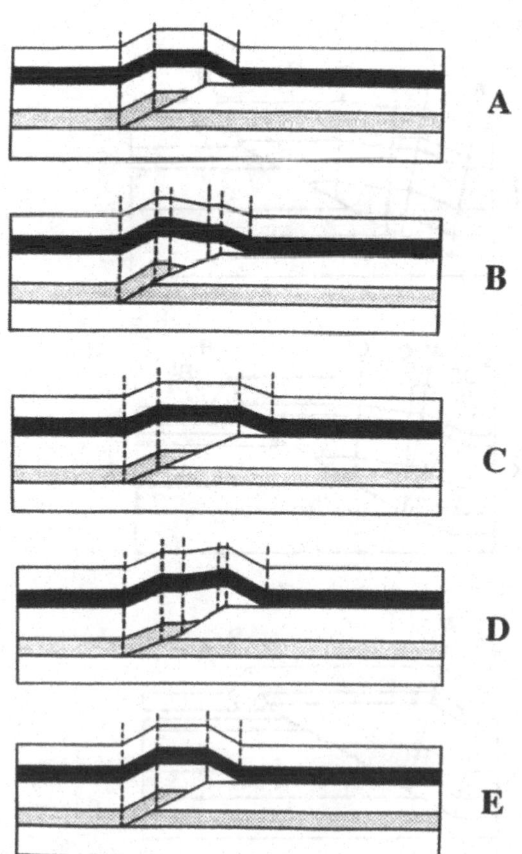

Figure 10. Fault-bend folds produced by vertical shear. Section position corresponds to Figure 9. See text for explanation.

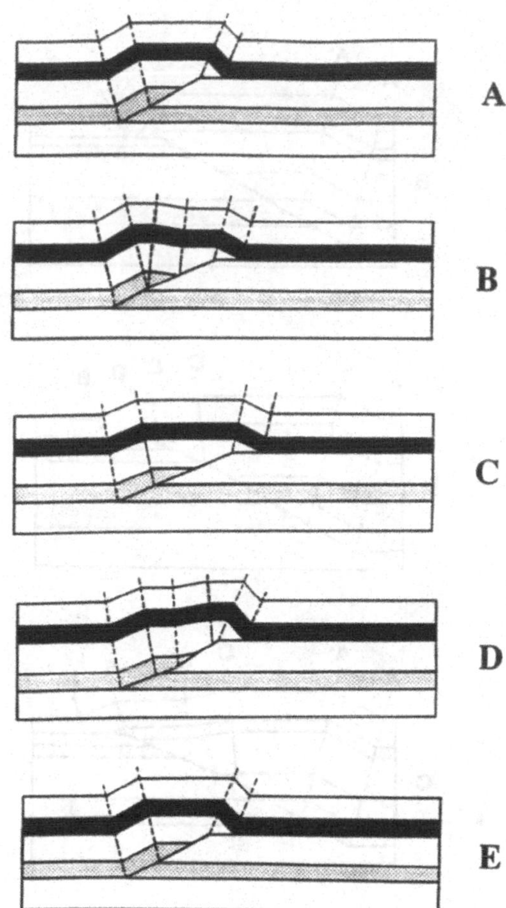

Figure 11. Fault-bend folds produced by layer-parallel shear. Section position corresponds to Figure 9. See text for explanation

Layer-parallel shear

Fold forms for this mechanism are described in Suppe (1983). Ramp dip is a trigonometric function of forelimb dip and axial angle (see Suppe's Fig. 7, 1983). In this case, there is no

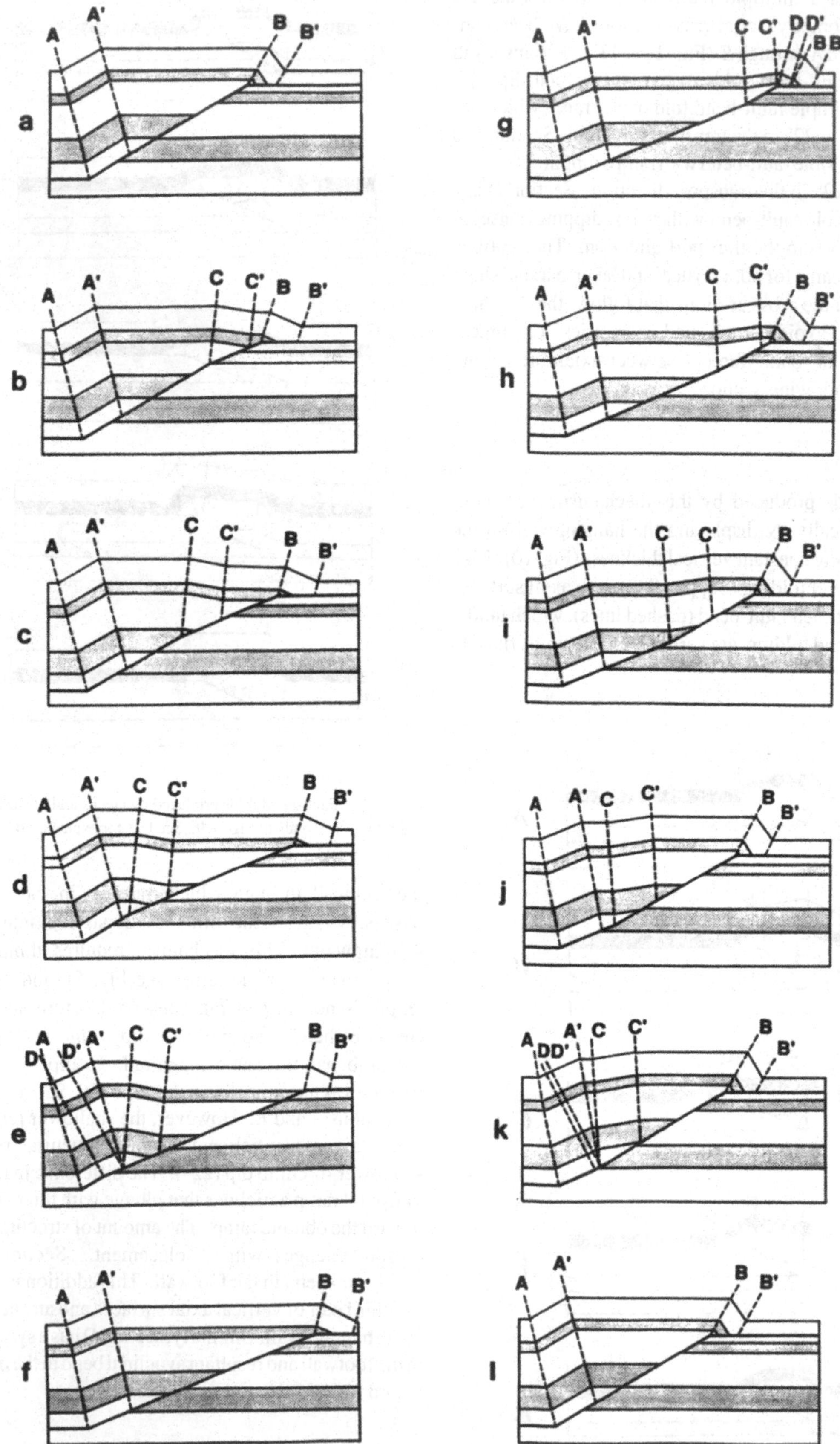

Figure 12. Detailed serial cross sections across ramp intersection zones for the layer-parallel shear mechanism. These sections are labelled *a, b, c, d, e, f, g, h, i, j, k, l*, and have a displacements indicated by the distance along the ramp between axial surfaces A and A' in section *a*. The horizontal erosion level in section *a* is the datum for the map in Figure 13. The location of these cross sections is given in Figure 9. See text for explanation.

deformation in the plane of bedding so that bedding line length is conserved. Due to slip parallel to bedding, slip along the fault is lost to folding through each bend (Suppe 1983, eq. 16). Interbed slip and resultant angular shear also results in a forelimb dip that is steeper, in general, than in the case of vertical shear (Fig. 11). Axial surfaces bisect bedding in the hangingwall, thereby maintaining constant bed thickness. Sections *A* and *B* are folds over frontal ramps dipping 30°, resulting in forelimb dips of 60°. In section *C*, an apparent dip of 22° yields a forelimb which dips about 23°. Clearly, as ramp dip decreases, the amount of interbed slip decreases. Therefore, at small values of ramp dip, the forelimb dip approaches the same dip as the ramp. As in the case of vertical shear, the shallower, oblique ramp apparent dip results in a difference in structural relief between frontal ramp anticlines and the oblique ramp anticline which changes with displacement. The antiformal bend in section B produces a second order anticline in the hangingwall. Similarly, there is a second order syncline in the hangingwall of section *D*. Note that ramp anticlines produced by layer-parallel shear are broader than those produced by vertical shear.

Interference of axial surfaces at multiple bends

In cross sections *A* through *E* of the previous discussion, the amount of displacement and the section locations were chosen for simplicity such that no interference of axial surfaces occurred in the hangingwall. For layer-parallel shear, with increased displacement and/or close spacing of fault-bends, interference generates additional pairs of axial surfaces which bound new dipping panels of rock.

In Figure 12, the detail at each ramp intersection zone is analysed by constructing several serial, two-dimensional fault-bend folds under the assumption of layer-parallel shear. The footwall geometry is identical to that in Figure 9; here section *A* is replaced with *a*, section *B* with *b*, *c*, *d*, and *e*, section *C* with *f*, section *D* with *g*, *h*, *i*, *j*, and *k*, and section *E* with *l*. The cross sections reflect one instant of time at a specified displacement which is constant along strike. Sections *a* and *l* are simple fault-bend folds over frontal ramps dipping 30° with two pairs of axial surfaces labelled A-A' and B-B'. In sections *b*, *c*, *d*, and *e*, the additional fault-bend due to the ramp intersection generated axial surfaces C-C'. In sections *b* to *e*, the position of the fault-bend and C-C' is lower in the stratigraphy toward the basal detachment. Section *e* is an example of interference close to the lower flat. In this case, axial surface A' is at the fault-bend, at which point axial surfaces D-D' bisect the angle between the two ramp segments. The orientation of surfaces D-D' is determined by the cutoff angle between beds within A-A' (in this case zero), instead of the horizontal beds between A' and C (see Suppe 1983, Fig. 7). The units bounded by D-D' then dip parallel to the ramp above the bend. Were displacement to increase, D would remain fixed to the footwall, D' and A' passively move with the hangingwall. Material passes only through axial surfaces that are fixed to the footwall at a bend, therefore C and C' also passively move with the hangingwall.

At the synformal footwall bend near the upper flat (section *g*, Fig. 12), axial surface C' is at the upper bend, localizing surfaces D-D'. The orientation of the D-D' and the dip of beds within is determined from the cutoff angle of beds between C-C', rather than the cutoff angle which determined the dip of axial surface B-B' (horizontal beds between C' and B'). Because this new cutoff angle is less, the dip of axial surfaces D-D' is correspondingly less than B-B' (see Suppe 1983, Fig. 7). Section *k* shows similar interference structures.

From these serial cross sections, a map of the various dip domains bounded by axial surfaces was generated at a specified horizontal erosion level (Fig. 13). In general, the hangingwall geometry is simple over pure frontal and oblique ramps with two pairs of axial surfaces corresponding to each fault-bend. Complexities arise at the intersection zones due to multiple fault-bends and interfering axial surfaces. Within the intersection zones, multiply oriented dipping panels of rock contribute an additional component of three-way closure between frontal and oblique ramp anticlines. Domain dips in the intersection zones are the sum of two components: (1) dip in the plane of the cross section predicted by fault-bend

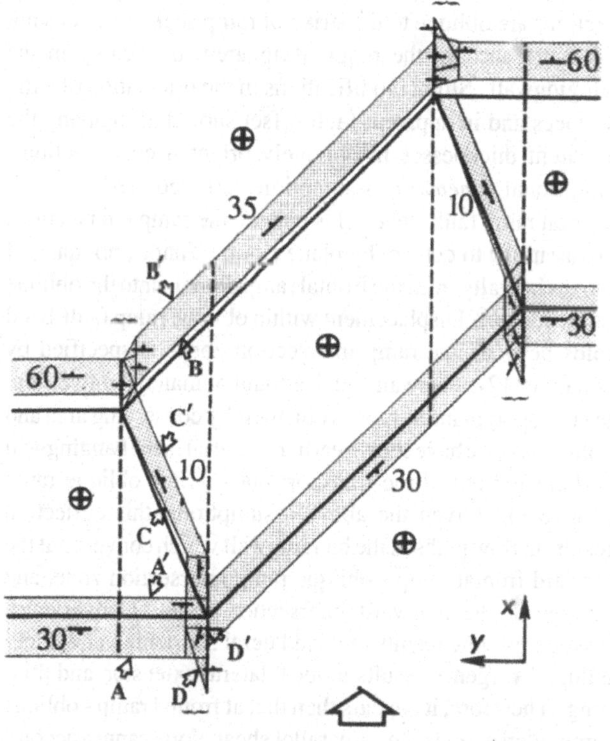

Figure 13. Map of dip domains generated from serial cross sections at a specified horizontal erosion level (Fig. 12a). The bold solid line represents the trace of the footwall geometry in plan view. The solid lines represent axial surfaces which bound dip domains. Dashed lines mark the boundaries of the ramp intersection zones. Between these lines, the footwall consists of a multiple fault-bend. Regions referred to by the brackets depict areas of interference between axial surfaces. In map view, these appear as discrete panels of dipping beds bounded by axial surfaces A-A', B-B', C-C', and D-D'. Some dip magnitudes are omitted for simplicity but can be obtained from the cross sections. The dip domain geometry will change significantly as the amount of displacement and footwall geometry change.

fold theory, (2) dip in the strike direction due to variations in structural relief. The map pattern will significantly change if the footwall geometry (α,β,δ) and/or the amount of displacement changes. For example, cross section *e* (Fig. 12) will show the following changes by increasing the displacement. Panel D-D' will widen, panels between D' and C' will translate, and the horizontal panel between C' and B will narrow as B-B' widens. At a greater displacement, C' will intersect the upper fault-bend and interfere with axial surface B, generating a new pair of axial surfaces and panel of dipping beds These changes in cross section geometry will have corresponding changes in map geometry.

Modifications of fault-bend folds due to 3-D geometry and kinematics

The cross sections and map generated in the previous section assumed two-dimensional deformation, resulting in a first order approximation of the geometry and kinematics. Extending the results to three dimensions requires additional considerations and assumptions. With regard to *geometry*, the use of two-dimensional fault-bend fold theory to generate folds over oblique footwall ramps, in cases where cross sections are oblique to the strike of ramp segments, does not take into account the required apparent thickening in the hangingwall. Slight modifications in the orientation of axial surfaces and in apparent fault offset should accompany the apparent thicknesses in obliquely oriented cross sections. Additional *kinematic* assumptions are required as well. Frontal ramp fault-bend folds outside the ramp intersections are assumed to deform by plane strain. That is, no material moves laterally from the frontal ramp domain into the oblique ramp domain. Displacement within oblique ramp fault-bend folds between the ramp intersection zones is specified by equation (12), where an equal amount of material moves into the transport plane as moves out, thereby conserving area and volume. As we have shown earlier, material in the hangingwall is deflected out of the transport plane at the oblique ramp (Figs 6-8). Given the above assumptions, this deflection results in flow paths in the hangingwall which converge at the forward frontal ramp - oblique ramp intersection zone, and diverge at the rearward intersection zone. Convergence within this zone results in local lateral shortening and thickening, divergence results in local lateral extension and thinning. Therefore, it appears then that at frontal ramp - oblique ramp intersections, layer-parallel shear alone cannot accommodate the deformation.

In Figure 14, two different ramp geometries are depicted in map view. The deflected path of hangingwall material is shown by the flow lines, as derived from equation (13) and Figure 8. A calculation of the longitudinal elongation (e_y) at each ramp intersection has been made graphically by measuring the line length change between two particles, originally a distance L_0 apart (see Fig. 4). One particle undergoes pure frontal ramp deformation, while the adjacent particle encounters the oblique ramp and is deflected out of the transport plane. On the upper footwall flat, the two particles are now separated by a distance L_1. In case (a), the oblique ramp dip (β) is 30° and the deflection (α–ρ) is 4°. At the front intersection zone, the resultant elongation across the entire intersection zone (ΔL/L) is -0.25. At the rear zone, the elongation is 0.33. In case (b), β=70° and the deflection angle is 26°. The elongation at the front hinge is -0.58, and at the rear hinge is 1.4. With α and δ constant, an increase in the oblique ramp dip increases the longitudinal strain. Although the magnitude of the displacement path deflection (Fig. 8) is relatively small for oblique, dips under 40°, they should result in longitudinal strains at ramp intersections that may be measurable in the field.

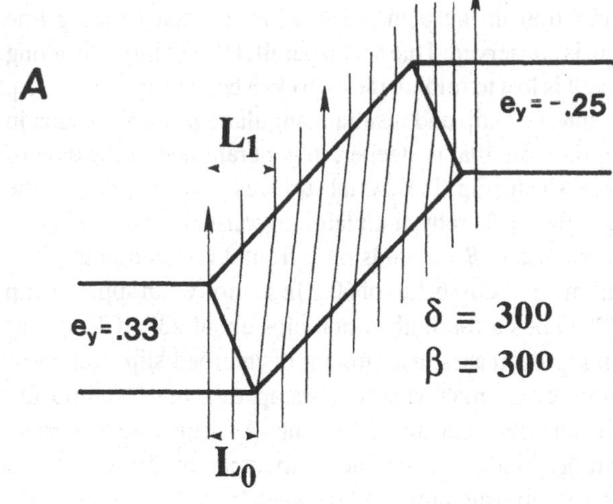

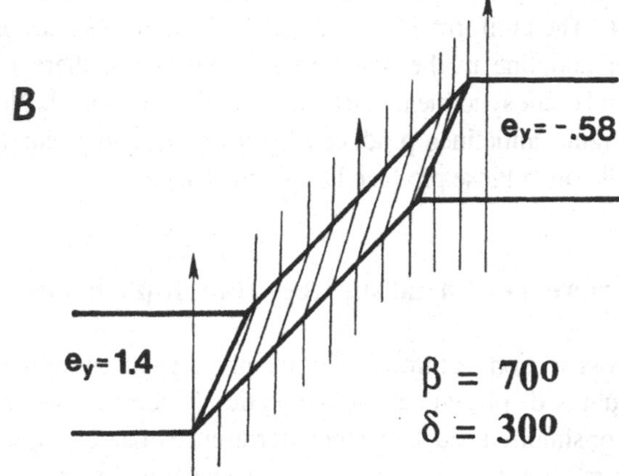

Figure 14. Two different ramp geometries are depicted in map view. In both cases, the frontal ramp dip (δ) is 30°, and the angle between the transport direction and the strike of the transverse ramp (α) is 45°. The deflected path of hangingwall material is shown by the flow lines, as derived from equation (12) and Figure 8. **(a)** Oblique ramp dip (b) = 30°. **(b)** β = 70°. See text for further explanation.

Figure 15 is a schematic block diagram depicting how the lateral strains might be manifest in the field. At the rearward intersection zone, rocks originally on the lower flat will be strained (γ_{xz}) at the lower ramp hinge. Above the line of intersection of the two ramps, two shear strains are imposed (γ_{xz},-γ_{yz}) as well as a lateral extension due to divergent deflection (e_y). The extension may be manifest as fracturing or normal faulting in the hangingwall. If the extension is accommodated by a normal fault, hinterland dipping rocks of the backlimb will show an apparent left-lateral strike-slip offset in map view. Deflection and lateral extension may be

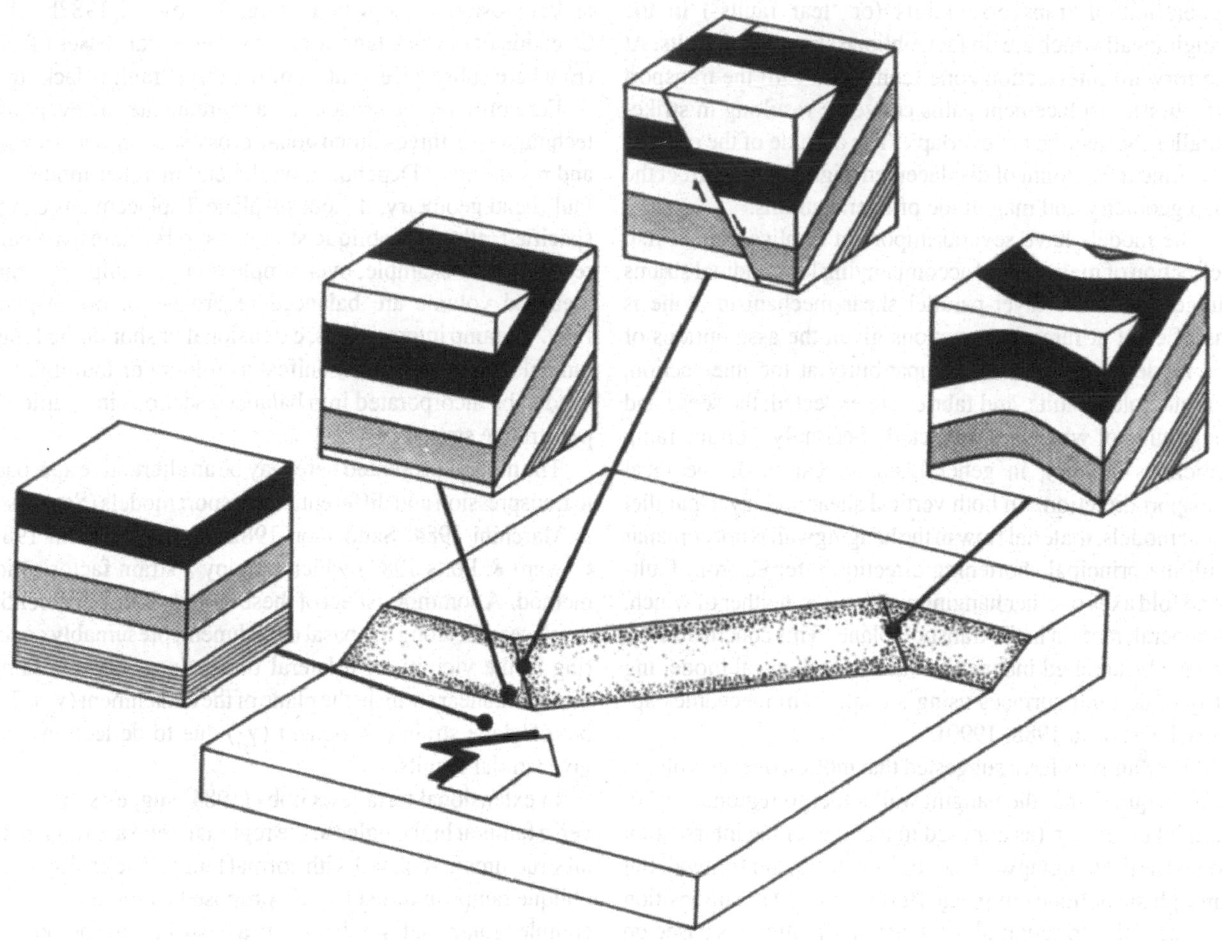

Figure 15. Schematic block diagram depicting how the lateral strains may be manifest in the field. Each cube of rock represents a small element of the hangingwall moving up the frontal ramp-oblique ramp intersection zone. See text for further explanation.

a mechanism for 'tear faulting' in the hangingwall, which, in this case, is normal faulting with an oblique slip component. At the forward intersection zone, a component of lateral shortening (e_y) should result in folds, faults, or cleavage at a high angle the regional transport direction.

Discussion

On the basis of previous work, oblique ramps are suspected to be structures where the common assumption of plane strain is violated. They may be the locus of oblique folds, faults, cleavages, and fractures, and tend to be avoided when constructing balanced cross sections. One component in understanding the geometry and kinematics of oblique ramp deformation requires a three-dimensional modelling approach.

Kinematic models for the deformation of hangingwall material moving over a footwall oblique ramp are developed by considering two end members of assumed mechanical behaviour, vertical shear and layer-parallel shear. The observed behaviour may be expected to lie between the two idealized end members, depending on the intrinsic and extrinsic conditions during thrusting. In the former case, material is sheared vertically and displacements remain within the tectonic transport plane; the deformation is accommodated by thinning of the hangingwall over the ramp. In the later case, material is deflected out of the transport plane such that the pitch angle of the particle path in the plane of the oblique ramp is equal to the initial angle between the transport direction and the strike of the oblique ramp. This deflection results in shear strains imposed at ramp - flat hinges. The deflection and out-of-plane shear strain are zero for the special cases of pure frontal and lateral ramps, and maximum at an intermediate oblique orientation, depending on ramp dip. Fault-bend folds are grossly similar for both vertical shear and layer-parallel shear mechanisms. At frontal ramp - oblique ramp intersections, synformal or antiformal multiple bends in the footwall generate second order hangingwall synclines or anticlines. These terminate along strike into simple fault-bend folds. For the layer-parallel shear mechanism, the deformation is more complex than vertical shear. Along the simple oblique ramp, deflected hangingwall material passes through the transport plane, conserving area and volume. Within the intersection zones, axial surface interference generates additional panels of dipping beds At the rearward intersection zone (concave toward the transport direction), hangingwall material diverges resulting in strike-parallel extension, or the development of a 'gap'. This extension may be a mechanism for the generation of transverse faults (or 'tear faults') in the hangingwall which are, in fact, oblique slip normal faults. At the forward intersection zone (convex toward the transport direction), displacement paths converge resulting in strike-parallel shortening, or 'overlap'. The attitude of the oblique ramp and the amount of displacement significantly affect the map geometry and magnitude of lateral strains.

The models have several important implications. First, deflection of material and accompanying longitudinal strains suggests that the layer-parallel shear mechanism alone is insufficient at ramp intersections given the assumptions of the model. To maintain compatibility at the intersection, oblique folds, faults, and fabrics are expected, the sense and magnitude of which are predicted. Secondly, oblique ramp anticlines are not, in general, an indication of the local transport direction. In both vertical shear and layer-parallel shear models, material flow in the hangingwall is not coplanar with the principal shortening direction inferred from fault-bend fold axes or other hangingwall fabrics, neither of which, in general, remain in the transport plane. This conclusion has been substantiated independently by mechanical modelling of oblique fault surfaces using a continuum mechanics approach (Apotria 1988, 1990).

Some authors have suggested that motion over an oblique ramp requires that the hangingwall undergo regional strike-parallel extension (as opposed to local, over the intersection zones) as the hangingwall moves from the lower footwall flat through the oblique ramp (e.g. Butler 1982). The supposition that regional strike-parallel extension is required is based on conservation of line length in a direction parallel to the strike of the frontal ramp, but is dependent on the bulk deformation mechanism of the hangingwall. For example, in the case of layer-parallel shear, the apparent extension is accommodated by layer-parallel slip and out-of-plane shear strains. If one relaxes the assumption that deformation is plane over pure frontal ramp segments, out-of-plane displacement of material from the frontal ramp domain into the oblique ramp domain may accommodate the apparent regional extension suggested by Butler, and perhaps distribute the local lateral strains due to deflection over a greater area. A component of strike-parallel extension could occur if the hangingwall deformed as an isotropic bending beam, with extensional strains in the outer arc and shortening strains in the inner arc.

The model may be useful in deciphering oblique footwall ramping from other along-strike variations. In map view, a change in stratigraphic separation along the trace of a thrust is often interpreted as an oblique ramp. However, nearly identical map geometries can result from: (a) a change in hangingwall displacement along strike, and (b) uplift, plunge, and/or erosion of a thrust fault (e.g. Woodward, 1987b). The detection of out-of-plane strains may eliminate cases (a) and (b) where subsurface control of the thrust fault is lacking.

Furthermore, the models are a prerequisite to developing techniques of three-dimensional cross section construction and restoration. Depending on the end member model and fault-bend geometry, the out-of-plane displacements can be specified, allowing oblique structures to be palinspastically restored. For example, over simple oblique ramp segments, area and volume are balanced regardless of out-of-plane flow. At ramp intersections, extensional or shortening longitudinal strains, perhaps manifest as folding or faulting, will need to be incorporated into balanced sections in addition to penetrative strains.

The models presented here may be an alternative approach to transpression and differential transport models (Sanderson & Marchini 1984; Sanderson 1982; Coward & Kim 1981; Coward & Potts 1983) which employ a strain factorization method. A common aspect of these models is that differential displacement along the basal detachment, presumably occurring in the vicinity of a lateral thrust tip or oblique ramp, imposes a shear strain in the plane of the detachment (γ_{xy}). The out-of-plane strain component (γ_{yz}) due to deflection may give similar results.

In extensional terranes, Gibbs (1984) suggests that transverse faults at high angles to the regional trend are fundamental structures associated with normal faults. The analogy with oblique ramps in thrust belts is proposed by Gibbs, in which complex rotational, synthetic dip, and strike slip components allow extension to transfer style and displacement along a graben. Similarly, both shortening and extensional structures can develop depending on the relative orientations of the transverse fault and normal faults. The models proposed herein may accommodate regional extension by simply reversing the transport direction in Figure 1. For the layer-parallel shear model, equation (12) holds for extensional oblique ramps. More appropriate models in extensional terranes may employ inclined simple shear, rather than vertical or layer-parallel shear. Models of this type are in progress.

Preparation of this manuscript benefited from discussions with D. Goff, S. G. Erickson, D. DePaor, and D. Medwedeff. Financial support was contributed by Texaco E&P Technology Division, ARCO Oil and Gas Co., Shell Western E&P Inc., Chevron USA Inc., Unocal Science and Technology Division, Geological Society of America and Sigma Xi grants-in-aid of research. We appreciate the constructive reviews by D. J. Sanderson, K. R. McClay, and an anonymous reviewer.

References

Apotria, T. G. 1988. A 3-D solution for hangingwall motion over a wavy fault surface. *Geological Society of America Abstracts with Programs*, **20**: 7, A57.

—— 1990. The Kinematics and Mechanics of Oblique Ramp Deformation within Fold and Thrust Belts. Ph.D. dissertation, Texas A & M University.

Berger, P. & Johnson, A. M. 1980. First order analysis of deformation of a thrust sheet moving over a ramp. *Tectonophysics*, **70**, T9-T24.

Bosworth, W. 1985. Geometry of propagating continental rifts. *Nature*, **316**, 625-627.

Boyer, S. E. 1985. Hydrocarbon trap styles in fold and thrust belts and related terranes. *Offshore Technology Conference*, OTC 4873, 297-305.

Boyer, S. & Elliott, D. 1982. Thrust systems. *American Association of Petroleum Geologists Bulletin*, **66**, 1196-1230.
Butler, R.W.H. 1982. Hangingwall strain: a function of duplex shape and footwall topography. *Tectonophysics*, **88**, 235-246.
Couples, G. D. & Lewis, H. 1988. 'Thrust belt' structures with 'foreland' influence: the interaction of tectonic styles. *In:* Schmidt, C. J. & Perry, W.J. (eds), Interaction of the Rocky Mountain Foreland and the Cordilleran Thrust Belt. *Geological Society of America Memoir, 171*, 99-110.
Coward, M. P. & Kim, J. H. 1981. Strain within thrust sheets. *In:* McClay, K. R. & Price, N. J. (eds), *Thrust and Nappe Tectonics*. Geological Society of London Special Publication, **9**, 275-292.
—— & Potts, G. J. 1983. Complex strain patterns developed at the frontal and lateral tips to shear zones and thrust zones. *Journal of Structural Geology*, **5**, 383-395.
Gardener, D.A.C. & Spang, J. H. 1973. Model studies of the displacement transfer associated with overthrust faulting. *Bulletin of Canadian Petroleum Geologists*, **21**, 534-552.
Gibbs, A. D. 1983. Balanced cross section construction from seismic sections in areas of extensional tectonics. *Journal of Structural Geology*, **5**, 153-160.
—— 1984. Structural evolution of extensional basin margins. *Journal of the Geological Society of London*, **141**, 609-620.
Goldburg, B. L. 1984. Displacement transfer between thrust faults near the Sun River in the Sawtooth Range, northwestern Montana. *Montana Geological Society 1984 Field Conference*, 211-220.
Groshong, R. H. 1988. Half-graben structures: balanced models of extension fault-bend folds. *Geological Society of America Bulletin*, **101**, 96-105.
—— & Udansky, S. I. 1988. Kinematic models of plane-roofed duplex styles. *In:* Mitra, G. & Wojtal, S. (eds), *Geometries and Mechanisms of Thrusting, with special reference to the Appalachians*. Geological Society of America Special Paper **222**, 197-206.
Jamison, W. R. 1987. Geometric analysis of fold development in overthrust terranes. *Journal of Structural Geology*, **9**, 207-219.
Kilsdonk, B. & Fletcher, R. C. 1989. An analytical model of hangingwall and footwall deformation at ramps on normal and thrust faults. *Tectonophysics*, **163**, 153-168.
Lageson, D. R. 1984. Structural geology of the Stewart Peak Culmination, Idaho-Wyoming thrust belt. *American Association of Petroleum Geologists Bulletin*, **68**, 401-416.
Mitra, G. & Yonkee, W. A. 1985, Relationship of spaced cleavage to folds and thrusts in the Idaho-Utah-Wyoming thrust belt. *Journal of Structural Geology*, **7**, 361-373.
Mitra, S. 1988. Three-dimensional geometry and kinematic evolution of the Pine Mountain thrust system, southern Appalachians. *Geological Society of America Bulletin*, **100**, 72-95.
Petrini, H. & Wiltschko, D. V. 1986. Some simple geometrical models for strains due to motion over lateral ramps. *Geological Society of America, Abstracts with Programs*, **18**, 717.
Pfiffner, O. A. 1981. Fold and thrust tectonics in the Helvetic Nappes (E. Switzerland). *In:* McClay, K. R. & Price, N. J. (eds), *Thrust and Nappe Tectonics*. Geological Society of London Special Publication, **9**, 319-328.
Sanderson, D. J. 1982. Models of strain variation in nappes and thrust sheets: a review. *In:* G. D. Williams (ed.), Strain within Thrust Belts, *Tectonophysics*, **88**, 201- 233.
—— & Marchini, W.R.D. 1984. Transpression. *Journal of Structural Geology*, **6**, 449-458.
Schmidt, C. J., O'Neil, J. M. & Brandon, W. C. 1988. Influence of Rocky Mountain foreland uplifts on the development of the frontal fold and thrust belt, southwestern Montana. *In:* Schmidt, C. J. & Perry, W. J. (eds), Interaction of the Rocky Mountain Foreland and the Cordilleran Thrust Belt, *Geological Society of America Memoir 171*, 171-202.
—— & O'Neil, J. M. 1982. Structural evolution of the southwest Montana transverse zone. *In:* R. B. Powers (ed.), *Geologic Studies of the Cordilleran Thrust Belt, vol. 1*. Rocky Mountain Association of Geologists, 193-218.
Suppe, J. 1983. Geometry and kinematics of fault-bend folding. *American Journal of Science*, **283**, 684-721.
—— & Medwedeff, D. A. 1984. Fault-propagation folding. *Geological Society of America, Abstracts with Programs*, **16**, 670.
Verral, P. 1981. Structural interpretation with applications to North Sea problems: *Joint Association of Petroleum Exploration Courses*, London, Course Notes No. 3.
Wernicke, B. & Burchfiel, B. C. 1982. Modes of extensional tectonics. *Journal of Structural Geology*, **4**, 104-115.
Wheeler, R. L. 1980. Cross-strike structural discontinuities: possible exploration tool for natural gas in the Appalachian overthrust belt. *American Association of Petroleum Geologists Bulletin*, **64**, 2166-2178.
White, N. J., Jackson, J. A. & McKenzie, D. P. 1986. The relationship between the geometry of normal faults and that of the sedimentary layers in their hangingwalls. *Journal of Structural Geology*, **8**, 897-909.
Wiltschko, D. V. 1979. A mechanical model for thrust sheet deformation at a ramp. *Journal of Geophysical Research*, **84**, 1091-1104.
—— 1981. Thrust sheet deformation at a ramp: summary and extensions of an earlier model. *In:* McClay, K. R. & Price, N. J. (eds), *Thrust and Nappe Tectonics*. Geological Society of London Special Publication, **9**, 55-63.
Woodward, N. B. 1987a. Primary and secondary basement controls on thrust sheet geometries. *In:* Schmidt, C. J. & Perry, W. J. (eds), *Interaction of the Rocky Mountain Foreland and the Cordilleran Thrust Belt*. Geological Society of America, Memoir **171**, 353-366.
—— 1987b. Stratigraphic separation diagrams and thrust belt structural analysis. *Wyoming Geological Association Guidebook*, Thirty-Eighth Field Conference, 69-77.

Appendices

Appendix 1: List of symbols

α: angle between transport direction and strike of oblique ramp measured in horizontal plane.
δ: dip of frontal ramp
β: dip of oblique ramp
β_d: apparent dip of oblique ramp in the direction of displacement for the layer-parallel shear mechanism
β_x: apparent dip of oblique ramp in x-direction
β_y: apparent dip of oblique ramp in y-direction
θ: pitch of hangingwall particle path in the plane of the oblique ramp
ρ: angle between particle path in the plane of the oblique ramp and the strike of the oblique ramp measured in the horizontal plane
γ_{xz}: shear strain in the xz (transport) plane
γ_{yz}: shear strain in the yz plane
γ_d: shear strain in the plane containing the displacement path for the layer-parallel shear mechanism

Appendix 2: Derivation of r for case of layer-parallel shear (Fig. 7)

Ramp height h is assumed to be 1. p, d, z, and w are arbitrary triangle leg lengths to be eliminated leaving a final expression relating angles.

$$p = \frac{h}{\sin \beta} \qquad \text{(A2-1)}$$

$$d = \frac{p}{\sin \alpha} \qquad \text{(A2-2)}$$

$$\sin \beta_d = \frac{h}{d} \qquad \text{(A2-3)}$$

$$z = \frac{h}{\tan \beta_d} \qquad \text{(A2-4)}$$

$$w = \frac{h}{\tan \beta} \qquad \text{(A2-5)}$$

$$\sin \rho = \frac{w}{z} \qquad \text{(A2-6)}$$

By back substitution, and a trigonometric identity:

$$\sin \rho = \frac{\cos \beta \sin \alpha}{(1 - (\sin \alpha \sin \beta)^2)^{1/2}} \qquad \text{(A2-7)}$$

Appendix 3: Derivation of θ for case of vertical shear (Fig. A1)

$$\text{Ramp height, } h = 1 \qquad \text{(A3-1)}$$

x, y, z, and L are arbitrary lengths to be eliminated leaving final expression in terms of angles

$$x = \frac{h}{\sin \beta} \qquad \text{(A3-2)}$$

$$y = \frac{h}{\tan \beta} \qquad \text{(A3-3)}$$

$$z = y \cos \alpha \qquad \text{(A3-4)}$$

$$L = \frac{z}{\sin \alpha} \qquad \text{(A3-5)}$$

$$\tan \theta = \frac{x}{L} \qquad \text{(A3-6)}$$

By back substitution:

$$\tan \theta = \frac{\tan \alpha}{\cos \beta} \qquad \text{(A3-7)}$$

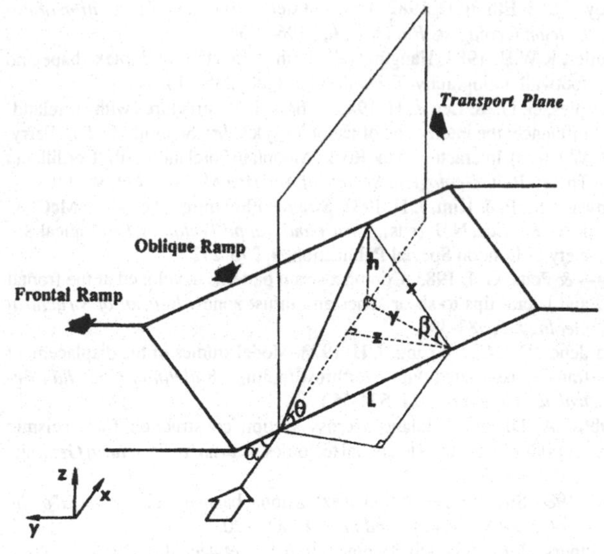

Figure A1. Footwall geometry used to derive the pitch angle of the particle path in the plane of the oblique ramp (θ) for vertical shear. See derivation in Appendix 3.

Stress controls on fold thrust style

William R. Jamison*

Amoco Production Research, PO Box 3385, Tulsa, Oklahoma 74102, USA

Abstract: Fold-fault relationships in overthrust terranes can usually be placed in the spectrum of fold/thrust styles ranging from fault-bend to fault-propagation to detachment folds. The specific fold/thrust styles of three major folds in the Wyoming-Idaho-Utah thrust belt have been assessed geometrically. Whitney Canyon-Carter Creek anticline is a fault-bend fold, Haystack Peak anticline is a detachment fold, and Hunter Creek anticline is probably a fault-propagation fold. All three structures involve essentially the same mechanical stratigraphy. It is postulated that the preferential development of a specific fold/thrust style reflects a fundamental competition between buckling and faulting in the layered rock package. Both deformational processes may be represented by instability envelopes in three-dimensional stress space. The form and location of these two envelopes is a function of the mechanical stratigraphy. The fold/thrust style that develops depends upon which instability surface is intersected first by the stress path. Depth of burial and regional tectonics are the major factors determining the stress path. In a thrust belt setting, the buckling instability surface is likely to be the initial intersect of the stress path only at relatively shallow depths of burial. At greater depths, the faulting instability surface is the initial intersect. This suggests that detachment folds develop most readily in the shallow subsurface, whereas fault-bend folds dominate in the deeper subsurface. Consequently, outcropping structures may not always be appropriate analogues for the structures occurring at depth.

Geologists working in overthrust terranes have recognized for over a century that the folds they observed were inherently linked to thrusts, either bedding-parallel thrusts (decollements) or thrust ramp segments (e.g. Willis 1890; Rich 1934). The specific nature of the fold/thrust interaction can usually be categorized as one of three basic styles (Fig. 1), viz. fault-bend folding (e.g. Rich 1934; Suppe 1983), fault-propagation folding (e.g. Faill 1973; Williams & Chapman 1983; Suppe & Medwedeff 1984), and detachment folding (e.g. Willis 1890; Laubscher 1977). All three styles exist, and are common in nature. The focus of this paper is on some of the mechanical factors that influence the selective development of these various fold/thrust styles.

The mechanical characteristics of the rock package, i.e. the mechanical stratigraphy, are a major factor affecting the macroscale deformational characteristics of layered rock. It is an important factor in determining fold/thrust style, but it is not the dictating influence. To support this contention, examples of different fold/thrust styles (all from the Wyoming-Idaho-Utah thrust belt) that have developed in the same rock package are presented.

The three end-member styles of fold/thrust interaction (Fig.1) are categories of fold/ramp timing. In the fault-bend fold, ramping precedes folding; in the detachment fold, folding precedes ramping; and in the fault-propagation fold, ramping and folding are synchronous events. The different fold/thrust styles thus reflect the relative dominance of buckling vs. faulting (folding vs. ramping). A stress space approach is used to investigate the competition between these two processes. The choice between buckling and faulting, and the choice of fold/thrust style, appears to be a function of the state of stress at the time of fold initiation as well as the mechanical stratigraphy.

Wyoming thrust belt examples

The Wyoming-Idaho-Utah thrust belt (Fig. 2a) has a gentle but persistent southerly plunge. Numerous major hangingwall anticlines are well exposed in the northern portion of the thrust belt. To the south, the correlative structures, many of which are highly constrained by development drilling, reside in the subsurface. The detailed cross-sectional geometries of three of these folds have been determined, and are used to infer fold/thrust style. Hunter Creek and Haystack Peak anticlines are both surface exposed structures, located in the Prospect and Absaroka thrust sheets respectively, whereas Whitney Canyon-Carter Creek anticline is a subsurface structure in the Absaroka thrust sheet (Fig. 2a).

Mechanical stratigraphy

As these are Laramide-age structures, they may have involved the entire Mesozoic and Palaeozoic section, a sequence of sedimentary rocks ~5 km thick (Fig. 2b). The mechanical response of this sequence was likely dominated by units within the Palaeozoic and lower Mesozoic section. The major competent struts in this stratigraphic sequence are the Bighorn, Madison, and Thaynes carbonates, each 150 m to 300 m thick. The Weber-Wells and Nugget Formations, both

*Present address: Centre for Earth Resources Research, Department of Earth Sciences, Memorial University of Newfoundland, St. John's, Newfoundland, Canada A1B 3X5

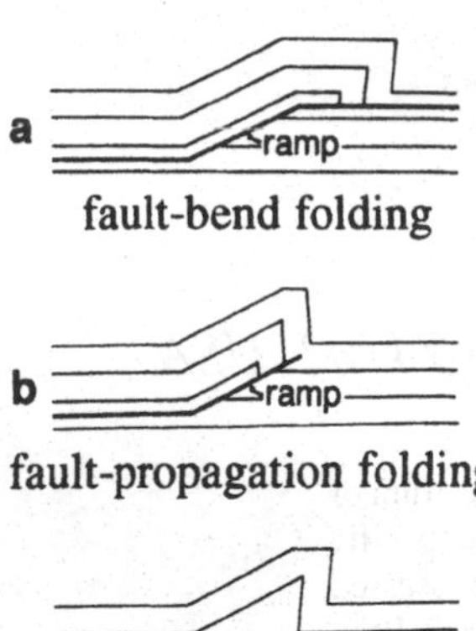

Figure 1. End-member styles of fold/thrust interaction: (**a**) fault-bend folding, (**b**) fault-propagation folding, (**c**) detachment folding.

fairly massive aeolian sandstones (each 150 m to 200 m thick), probably serve as secondary struts. The Cretaceous section, largely composed of shale-dominated marine clastics, is a very thick (>2 km), incompetent succession at the top of the sequence. The Cambrian shale sequence, ~325 m thick, is a relatively incompetent unit at the base of the stratigraphic sequence.

The bulk of the Mesozoic and Palaeozoic section is directly seen to be involved in the folding in the subsurface example, but erosion has removed all of the Cretaceous and much of the Jurassic section from the surface examples. If the erosion completely post-dates the folding, then these surface folds also involved the same 4 to 5 km sedimentary sequence. However, if the bulk of the erosion preceded fold development, the thickness of the section actually involved in the folding may have been considerably reduced, to perhaps 2 to 2.5 km. The mechanical stratigraphic response of the rock sequence should be relatively insensitive to this potential erosion because it involves only the mechanically incompetent upper Mesozoic strata. However, the state of stress at the time of fold initiation would have been significantly affected by the thickness of the overburden.

Haystack Peak anticline

Haystack Peak anticline lies within the Absaroka thrust sheet, roughly 9 km west of the surface trace of the Absaroka thrust (Figs 2a & 3). There is also a second major anticline, McDougal Pass anticline, between Haystack Peak anticline and the Absaroka thrust trace (Fig. 3). Both anticlines have a chevron fold configuration, with planar limbs and a sharp, angular hinge. The intervening syncline is a comparatively broad, rounded feature. There are a few minor faults and folds in the Absaroka sheet east of the McDougal Pass structure, but no major faults are mapped between Haystack Peak anticline and the Absaroka thrust trace (e.g. Rubey 1973).

In both fault-bend and fault-propagation folding, the fold height (measured from anticlinal crest to synclinal trough) is roughly equal to the stratigraphic height of the associated ramp. Haystack Peak and McDougal Pass anticlines each have fold heights in excess of 1 km. Along the Absaroka thrust trace, Ordovician Bighorn Dolomite in the hangingwall overrides Cretaceous strata in the footwall. Cambrian shales are mapped in the core of Haystack Peak anticline (Rubey 1973). Consequently, the Absaroka thrust has cut upsection through the hangingwall rocks about 400 m, at most, between

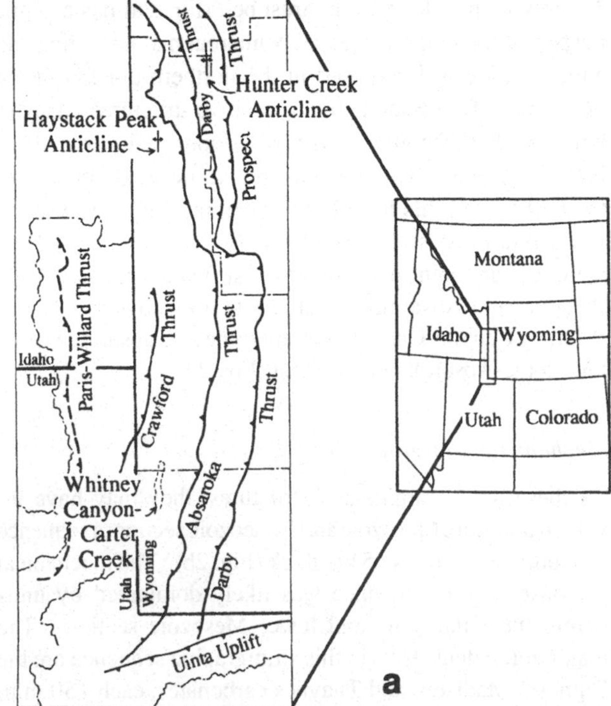

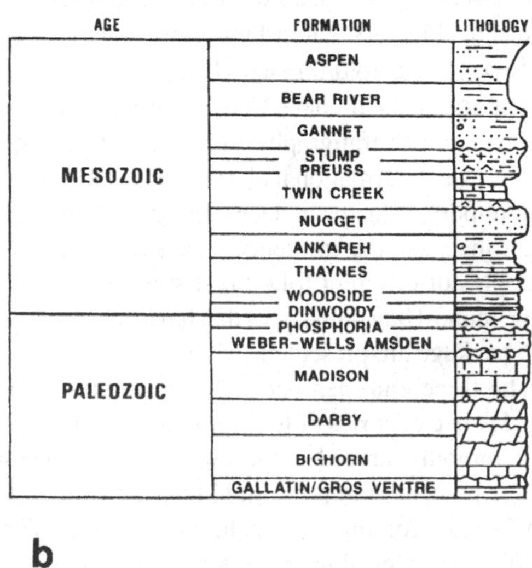

AGE	FORMATION	LITHOLOGY
MESOZOIC	ASPEN	
	BEAR RIVER	
	GANNET	
	STUMP	
	PREUSS	
	TWIN CREEK	
	NUGGET	
	ANKAREH	
	THAYNES	
	WOODSIDE	
	DINWOODY	
PALEOZOIC	PHOSPHORIA	
	WEBER-WELLS AMSDEN	
	MADISON	
	DARBY	
	BIGHORN	
	GALLATIN/GROS VENTRE	

b

Figure 2 (**a**) Reference map for the Wyoming-Idaho-Utah thrust belt showing the location of the thrust belt in the western USA, the surface traces of the major thrusts, and the locations of the three anticlines discussed in this paper. (**b**) Generalized stratigraphy of the Wyoming-Idaho-Utah thrust belt. Standard lithological symbols used. Units not shown to relative scale.

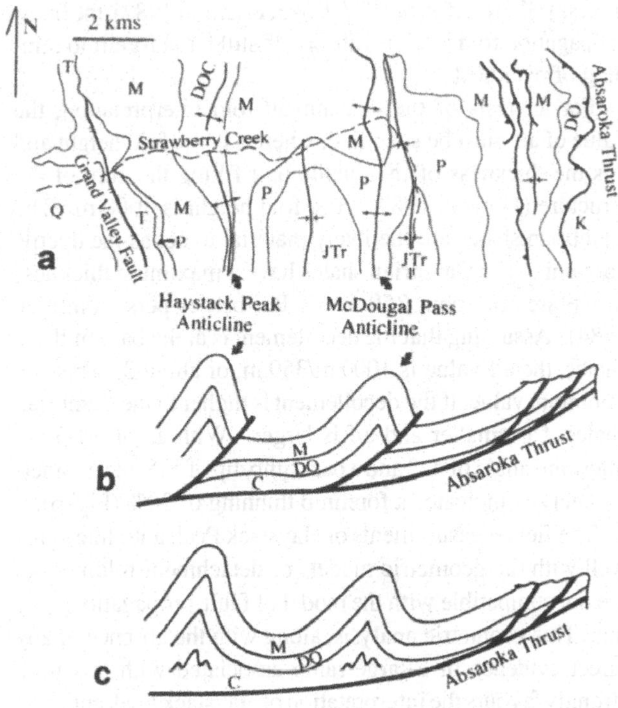

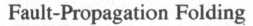

Figure 3. (**a**) Surface map of major structural features along the Strawberry Creek drainage, Salt River Range, Wyoming. Modified from Rubey (1973). Rock units depicted by age groups: Q-Quaternary; T-Tertiary; K-Cretaceous; J-Jurassic; JTr-Jurassic plus Triassic; P-Permian plus Pennsylvanian; M-Mississippian; DO-Devonian plus Ordovician; DOC-Devonian plus Ordovician plus Cambrian; C-Cambrian. Schematic cross-section interpretation of Haystack Peak and McDougal Pass anticlines along the Strawberry Creek drainage as either (**b**) fault-propagation folds or (**c**) detachment folds.

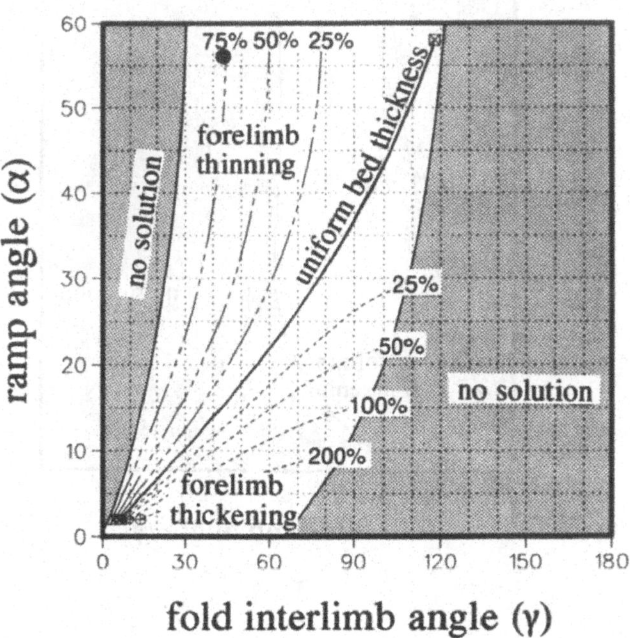

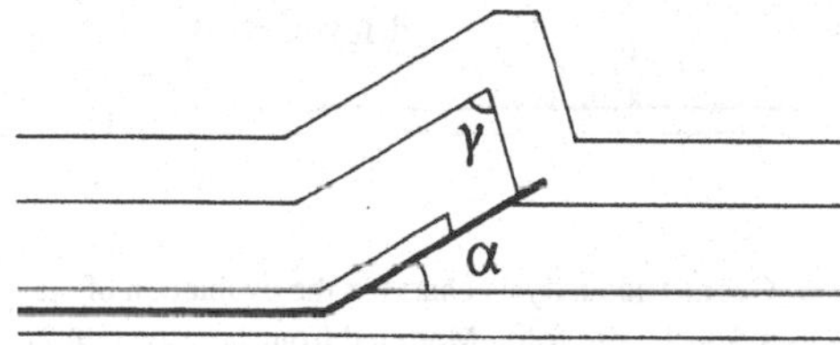

Figure 5. Geometric analysis chart for the evaluation of Haystack Peak anticline as a fault-propagation fold. Modified from Jamison (1987).

Haystack Peak anticline and the Absaroka thrust trace. The amount of hangingwall truncation is equal to the cumulative height of the associated ramps. The very large amplitudes of both Haystack Peak and McDougal Pass anticlines relative to potential ramp heights precludes their being either fault-bend or fault-propagation folds related to the Absaroka thrust. These two folds must be either (1) fault-propagation folds associated with some imbricate thrusts of the Absaroka thrust (Fig. 3b) or (2) detachment folds above the Absaroka thrust (Fig. 3c). These alternative interpretations are assessed for Haystack Peak anticline via analysis of the fold geometry.

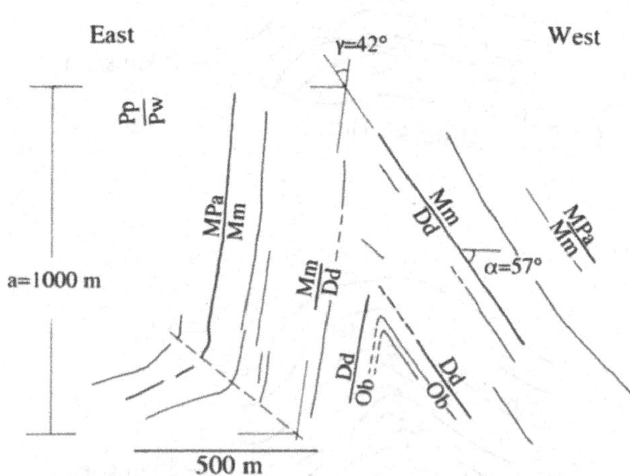

Figure 4. Survey-controlled cross section of Haystack Peak for exposure on the south side of the Strawberry Creek drainage. Pp-Permian Phosphria Fm.; Pw-Pennsylvanian Wells Fm.; MPa-Pennsylvanian-Mississippian Amsden Fm.; Mm-Mississippian Madison Fm.; Dd-Devonian Darby Fm.; Ob-Ordovician Bighorn Fm. Angular measurements and fold height (a) derived from base of Madison Fm.

The exposure of Haystack Peak anticline along the south side of Strawberry Creek drainage (Fig. 4) has been surveyed and projected into a cross-sectional plane perpendicular to the fold axis (18°/176°). The interlimb angle (γ) is 42° and the backlimb dip is 57°. If this is a fault-propagation fold, it is, in this case, still located directly above its associated ramp. Thus, the ramp angle (α) must be very close to the backlimb dip, or 57°. The Devonian through Mississippian section is about 20% thinner in the forelimb than in the backlimb. (This thinning is not uniformly distributed though the involved units. The relative forelimb thinning in the Devonian Darby Formation is about 40%, whereas thinning in the Mississippian Madison carbonates is only about 4%.) A fault-propagation fold with γ=42° and α=57° should exhibit relative forelimb thinning of 75% (Fig. 5). Even if a smaller, more conventional ramp angle could be rationalized for this struc-

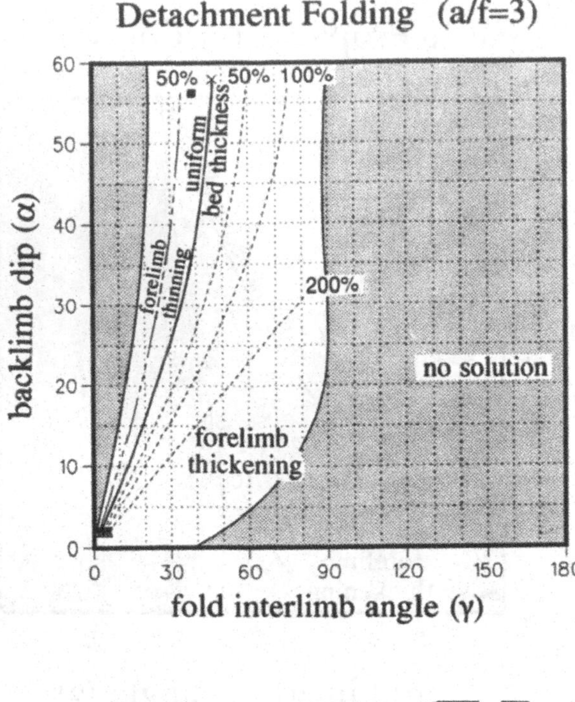

Figure 6. Geometric analysis chart for the evaluation of Haystack Peak anticline as a detachment fold. Modified from Jamison (1987).

ture, say 30° (e.g. Serra 1977; Boyer & Elliott 1982), the fault-propagation fold model with γ= 42° still has a forelimb thinning of over 60%.

For analysis of the detachment fold interpretation, the value of a/f must be specified, where a is the fold height and f is the thickness of the ductile unit filling the core of the structure (Jamison 1987). The fold height is 1000 m. The Cambrian shales are the likely material to act as the ductile core unit. The Cambrian shales have a maximum thickness in the area of about 350 m (C. Bartberger pers. commun. 1984). Assuming that the decollement is at the base of these shales, the a/f value is 1000 m/350 m, or about 3. This is a minimum value; if the decollement is higher in the Cambrian shales, f is smaller and a/f is larger. With an a/f of 3, an interlimb angle of 42° and a backlimb dip of 57°, the geometric analysis indicates a forelimb thinning of 20% (Fig. 6).

The field measurements of Haystack Peak anticline agree well with the geometric models of detachment folding, but are not compatible with the model of fault-propagation folding. This geometric analysis, along with the absence of any direct evidence of a large ramp associated with this fold, strongly favours the interpretation of Haystack Peak anticline as a detachment fold. Though McDougal Pass anticline has not been studied in comparable detail, its strong similarity in form and setting to the Haystack Peak structure suggests it is also a detachment fold.

Whitney Canyon-Carter Creek anticline

Whitney Canyon-Carter Creek anticline is an oil-field structure in the southern portion of the Wyoming-Idaho-Utah

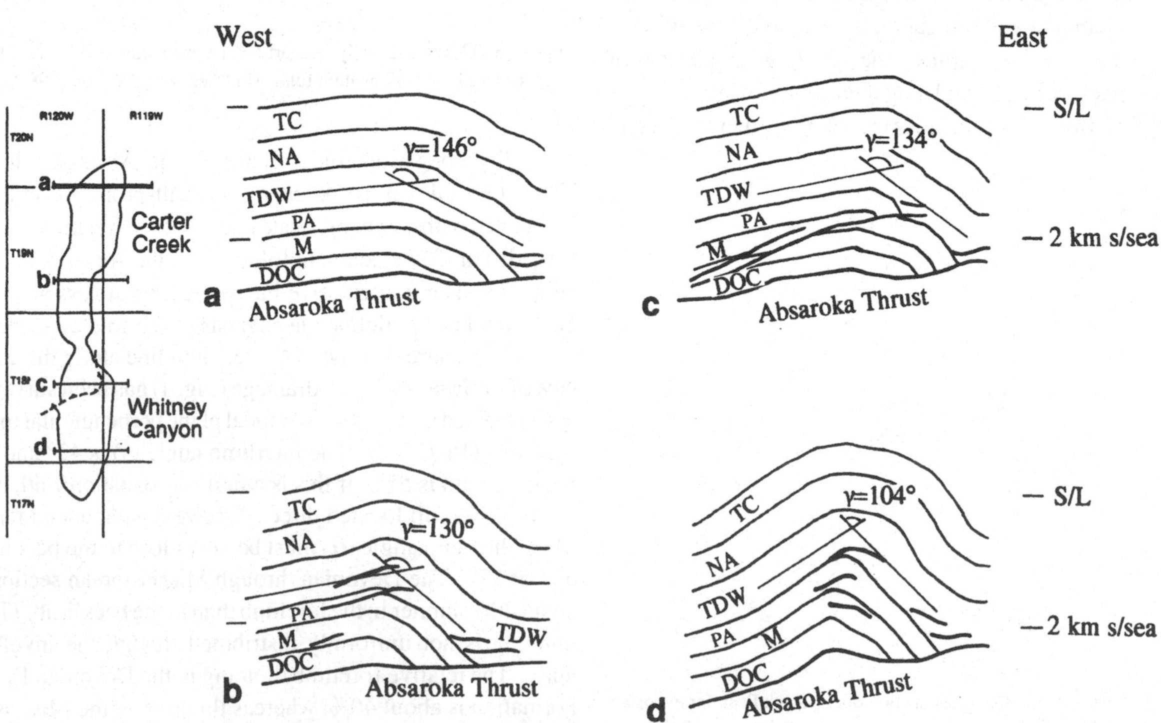

Figure 7. Cross-section imterpretations of the Whitney Canyon-Carter Creek oil field structure. Modified from Weir (1983). TC-Twin Creek Fm.; NA-Nugget plus Ankareh Fms.; TDW-Thaynes plus Dinwoody plus Woodside Fms.; PA-Phosphoria plus Weber & Amsden Fms.; M-Madison Fm.; DOC-Devonian plus Ordovician plus Cambrian units. Angular measurements made using top of the Dinwoody Fm. Wells penetrating to the top of the footwall of the Absaroka thrust beneath this structure record only Cretaceous rocks.

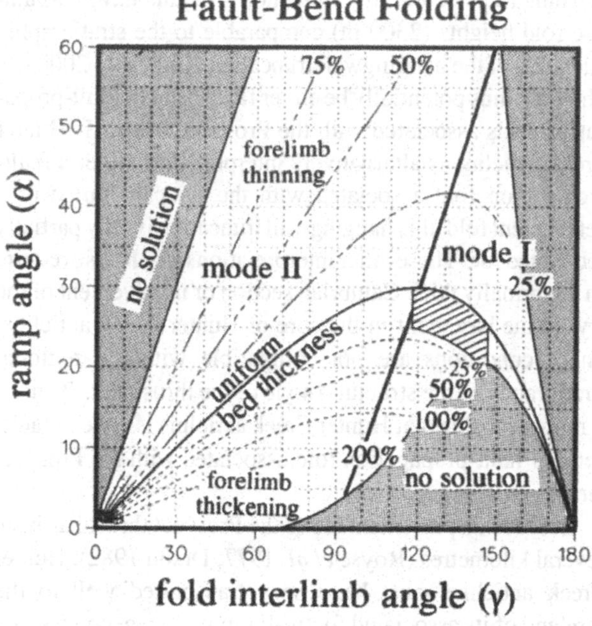

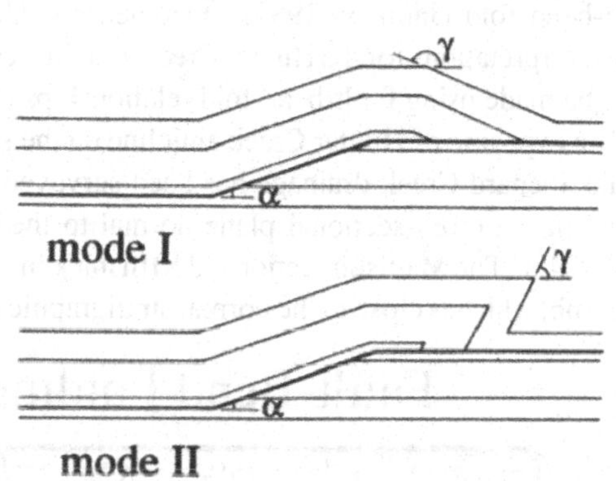

Figure 8. Geometric analysis chart for the evaluation of Whitney Canyon-Carter Creek anticline as a fault-bend fold. Cross-lined region encompasses range of interlimb angles and bedding thickness changes obtained from the four cross sections of Figure 7. Modified from Jamison (1987).

thrust belt (Fig. 2a). A series of cross sections, integrating seismic and well data, have been constructed through the Whitney Canyon-Carter Creek structure (Fig. 7) by Gary Weir (Weir 1983; Couples *et al.* 1987). Numerous wells penetrating the forelimb, hinge, backlimb, and the transporting (Absaroka) thrust provide good constraints on the overall fold geometry. The hangingwall truncation of this structure extends from the Cambrian section into the lower Triassic section, a stratigraphic thickness of 1200 m to 1250 m. The presence of the Cretaceous section in the footwall of this structure implies that this fold has been transported foreland from its originating ramp and upsection, through a second ramp, to its present location.

In the northern two-thirds of this field the Absaroka thrust is a relatively planar surface beneath the fold, but in the southern third the thrust is folded up into the core of the fold (Fig.7). In the northern cross sections, where the thrust is nearly planar, the interlimb angle (γ) ranges from 118° to 146°. In the southernmost cross section, where the thrust is quite non-planar, γ=104°. However, if this warp in the fault surface is removed, geometrically, the fold opens to γ=119°.

Intra- and interformational thrusts produce structural thickening in the backlimb, hinge and forelimb. These contractional faults thicken the affected units up to 60%. However, when distributed through the entire section, the maximum thickening is on the order of 15% to 20%. In some of the cross sections there is negligible structural thickening or thinning across the fold (e.g. Fig. 7a). The observed structural thickening is not restricted to the fold forelimb, as in the models of Jamison (1987). Thickness variations in the hinge and backlimb, though, carry geometric implications very similar to forelimb thickness changes.

The large interlimb angles observed in this structure cannot be obtained by fault-propagation folding (see Fig. 5). The Whitney Canyon-Carter Creek anticline must be either a detachment fold with a/f<1 (see Jamison 1987) or a mode I fault-bend fold. The amplitude of the fold is about 1175 m. The absence of a thick ductile unit (f) that would potentially fill the core of the fold (and provide a low a/f) discounts the detachment fold interpretation. A fault-bend fold with γ=118° to 146° and forelimb thickness increases of 0% to 20% should have an associated ramp angle of 25°±5° (Fig. 8). The ramp angle here cannot be directly assessed because the associated ramp lies somewhere, unconstrained, in the hinterland. Boyer & Elliott (1982) record ramp angles ranging from 25° to 40° and Serra (1977) measures ramp angles ranging from 10° to 30°. The implied ramp angle for fault-bend folding here is, thus, a quite reasonable value. Geometrically, Whitney Canyon-Carter Creek anticline is compatible only with the mode I fault-bend fold model.

Hunter Creek anticline

Hunter Creek anticline lies within the hangingwall of the Prospect thrust, about 6.5 km west of the surface trace of this thrust (Figs 2a & 9). The intervening Game Creek and Shepard thrusts, which have much smaller displacements that the Prospect thrust, are probably imbricates of the Prospect fault. Hunter Creek anticline is very well exposed in a series of valleys cross-cutting this structure. On the south side of the Shepard Creek drainage (Fig.10) the Darby and the lower two-thirds of the Madison formations are well exposed through the hinge and backlimb. The forelimb has reasonably good expression in the Madison, Amsden and Wells formations. The main anticlinal flexure is narrow and sharp.

At this location, and in the other transecting valleys, the Shepard thrust extends from the core of the fold into the forelimb. This thrust is parallel to bedding in the backlimb and cuts through bedding at a very high angle in the forelimb. The maximum measurable displacement on the Shepard thrust is about 600 m, with displacement diminishing as the fault penetrates upsection through the forelimb. In fact, displacement on the Shepard thrust apparently goes to zero before it intersects any of the ridgelines separating the several valleys (Fig. 9a). Although this thrust is quite visible in the valleys, it is a blind thrust along the ridgelines.

Along the trace of the Prospect thrust, hangingwall Nug-

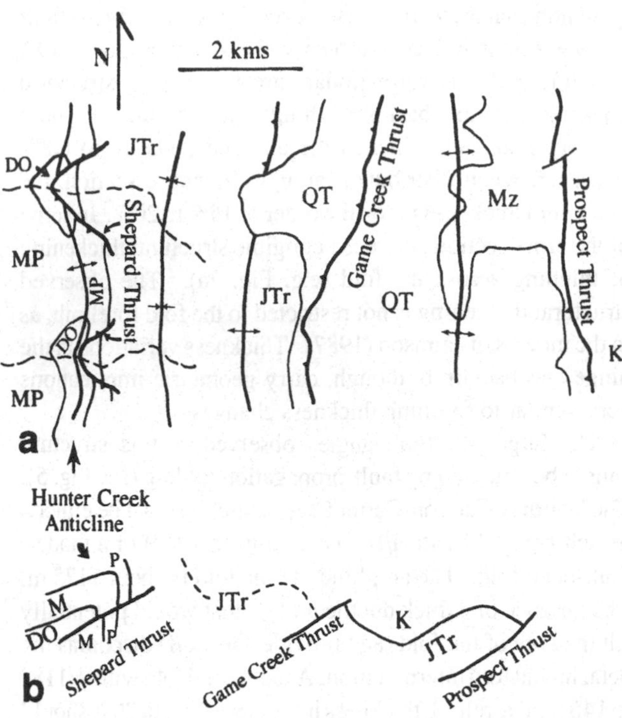

Figure 9. (a) Surface map of the major structural features between Shepard Creek and the Prospect thrust trace, Clause Peak quadrangle, Wyoming. Modified from Schroeder (1973). Rock units depicted by age groups: QT-Quaternary plus Tertiary; K-Cretaceous; JTr-Jurassic plus Triassic; Mz-Mesozoic; P-Permian plus Pennsylvanian; M-Mississippian; PM-Permian plus Pennsylvanian plus Mississippian; DO-Devonian plus Ordovician. (b) Schematic cross-section of major structures along east-west line from Shepard Creek to the Prospect thrust trace.

get Sandstone overrides footwall Cretaceous. The Ordovician Bighorn Dolomite is exposed in the core of Hunter Creek anticline. Cambrian rocks may also be involved in this structure below the limits of erosion. Thus, between Hunter Creek anticline and the Prospect fault trace, the Prospect thrust cuts upsection 1600 m to 2000 m in the hangingwall. The fold height of Hunter Creek anticline is, by extrapolation, about 1300 m. A second anticline, located between Hunter Creek anticline and the Prospect fault trace, has a fold height

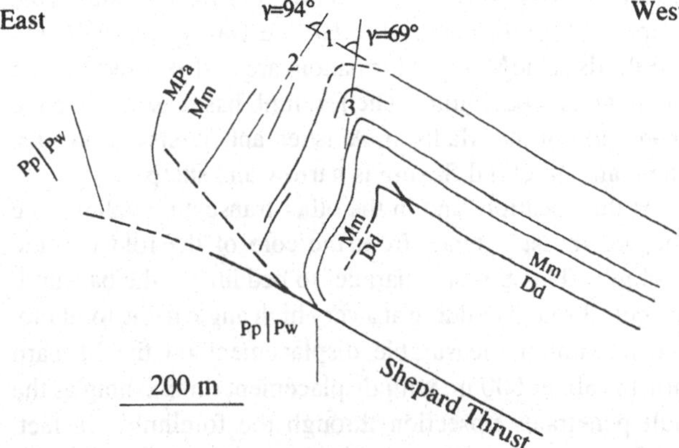

Figure 10. Survey-controlled cross section of Hunter Creek anticline for exposure south of Shepard Creek drainage. Angular measurements made using marker horizon within the Madison Fm. Line 1 projects from backlimb; line 2 from central forelimb; line 3 from upper forelimb.

of about 1000 m.

Hunter Creek and its companion anticline have cumulative fold heights (2300 m) comparable to the stratigraphic thickness of the hangingwall truncation (1600 m to 2000 m). They can thus potentially be either fault-bend or fault-propagation folds associated with the Prospect thrust. If Hunter Creek anticline is alternatively interpreted as either a fault-propagation fold associated with the Shepard thrust or a detachment fold, this hangingwall truncation is only partially accounted for. These latter interpretations would also require an abnormally thick Cambrian section or involvement of the crystalline basement in the core of Hunter Creek anticline. These conditions are not compatible with the regional stratigraphy or the structural style of the thrust belt. Thus, it is more probable that Hunter Creek anticline is either a fault-bend or fault-propagation fold associated with the Prospect thrust.

Because displacement along the Prospect thrust has been several kilometres (Royse *et al.* 1977; Dixon 1982), Hunter Creek anticline must have been transported well to the foreland of its associated footwall ramp. The geometry of a transported fault-propagation fold is very similar to a mode II fault-bend fold (Jamison 1987). Geometric evaluation of both interpretations for the Hunter Creek anticline can, therefore, be made using fault-bend fold relationships (Fig. 11).

The exposure of Hunter Creek anticline on the south side of the Shepard Creek drainage has been surveyed and projected into a cross-sectional plane normal to the fold axis (354°/04°). The Madison section is 310 m thick in the central forelimb, which is close to the normal stratigraphic thickness

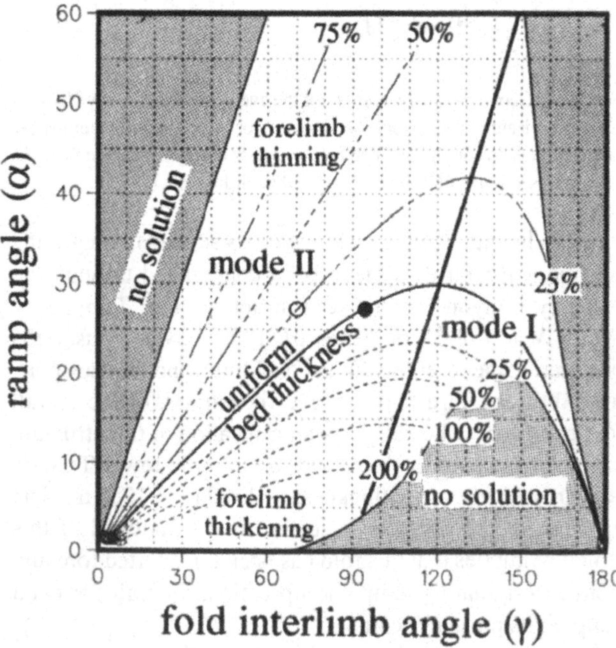

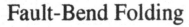

Figure 11. Geometric analysis charts for the evaluation of Hunter Creek anticline as a fault-bend fold (or a transported fault-propagation fold). Solid circle is for interlimb angle measured using central forelimb. Open circle is for interlimb angle measured using thinned upper forelimb. Modified from Jamison (1987).

of the Madison Formation in this region. The fold interlimb angle (γ) determined using the mid-forelimb dip is 94° (Fig. 10). No forelimb thickness change and γ=94° suggests that Hunter Creek anticline is either a mode II fault-bend fold or a transported fault-propagation fold with a ramp angle (α) of 27° (Fig.11). There is no direct control on α, but this inferred α=27° is very compatible with the generalized observations of Serra (1977) and Boyer & Elliott (1982). In the lower Madison Formation, right in the hinge region, γ=69° and the forelimb has locally thinned about 20% (Fig. 10). For α=27° and γ=69°, a thinning of 25% is indicated (Fig. 11), providing further support for the inferred α= 27° and the fault-bend/ fault-propagation fold interpretation.

The geometric analysis does not provide a direct method to distinguish between the mode II fault-bend fold and the transported fault-propagation fold. From kinematic considerations, though, it may be argued that a mode II fault-bend fold should exhibit substantial deformation in the forelimb (Jamison 1987). The absence of such deformation suggests that this structure is probably a fault-propagation fold.

Stress controls

Haystack Peak anticline is definitely a detachment fold. Whitney Canyon-Carter Creek anticline is definitely a fault-bend fold. Hunter Creek anticline is probably a fault-propagation fold. All three of these structures involve the same fundamental mechanical stratigraphy. This inconsistency in the nature of the fold/thrust interaction in the same rock package argues that some factor beyond mechanical stratigraphy influences variations in fold/thrust style. In the following, the potential control by the state of stress at the inception of fold development is explored. The underlying theme is that the observed variations in fold/thrust style reflect the competition between folding and faulting, and that this competition can be assessed by examining the form of the instability envelopes for buckling and faulting in terms of stress.

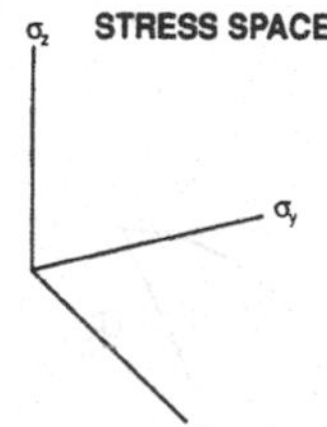

Figure 12. Stress space is a three-dimensional reference system with axes representing the principal stress magnitudes.

The working stress space

Stress space (Fig. 12) is a three-dimensional reference space whose axes are the principal stress values (e.g. Jaeger & Cook 1967). It is a non-physical space; only the principal stress magnitudes are recorded. There is no reference to any orientation data in stress space, including the principal stress orientations. This is not of particular concern when dealing with a mechanically isotropic material. However, the folds discussed above involve a distinctly stratified rock sequence, which is mechanically anisotropic. The orientation of the principal stress axes relative to the material layering is important in determining the response of this stratified material to a given stress state. In order to address the layering effects, a very specialized stress condition is assumed for the following discussions. Specifically, it is assumed that the principal stress axes, indicated as σ_{h1}, σ_{h2}, and σ_v, are parallel and perpendicular to layering (Fig. 13). There is no restriction on the relative or absolute magnitudes of the principal stresses, only their orientation relative to bedding. This special set of conditions is referred to as the 'working stress state'.

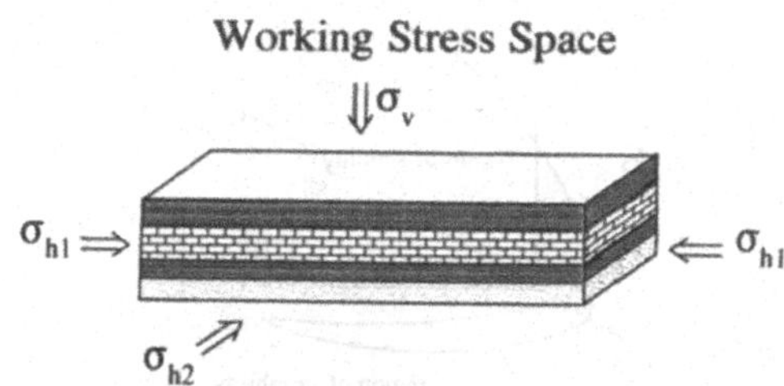

Figure 13. The 'Working Stress Space' is the special case of principal stress axes oriented parallel and perpendicular to bedding, referenced as shown.

Faulting instability envelope

The stress conditions for the initiation of brittle faulting depend on both the mean stress and the differential stress as described, for example, by the Mohr-Coulomb or Drucker-Prager failure criteria. In stress space, the condition of zero differential stress is described by the hydrostat (H in Fig. 14), the line defined by the condition $\sigma_x=\sigma_y=\sigma_z$. (E is the projection of H onto the σ_{h1}-σ_{h2} plane.) A given state of stress plots as a point in stress space, and any point not on H indicates a non-zero differential or deviatoric stress. The farther the point is from H, the larger the differential stress. The faulting instability envelope for an isotropic material is symmetric about H. The conical faulting instability surface shown in Figure 14a is defined by the Drucker-Prager failure criteria (Drucker & Prager 1952):

$$\sigma_f = \beta J_1 + \sqrt{J_2'} \tag{1}$$

where σ_f is the failure stress, J_1 and J_2' are the first stress invariant and the second invariant of deviatoric stress, respectively, and β is a coefficient (≥0 for compression positive) that may vary as a function of J_1. As long as the state of stress plots as a point internal to this cone the material is stable, but when the stress state moves to contact with this conical failure surface the material will fault of fracture (under conditions of brittle deformation).

Laboratory tests of rock samples with definitive layering have found that the failure stresses are (1) relatively reduced in value for σ_1 oriented less than 45° to layering at low confining pressures and (2) independent of principal stress orientations at high confining pressures (e.g. McGill & Raney 1970; Donath 1961). In the working stress state, these anisotropic effects can be incorporated, at least qualitatively. The layering-related anisotropy will move the faulting instability

Faulting Instability

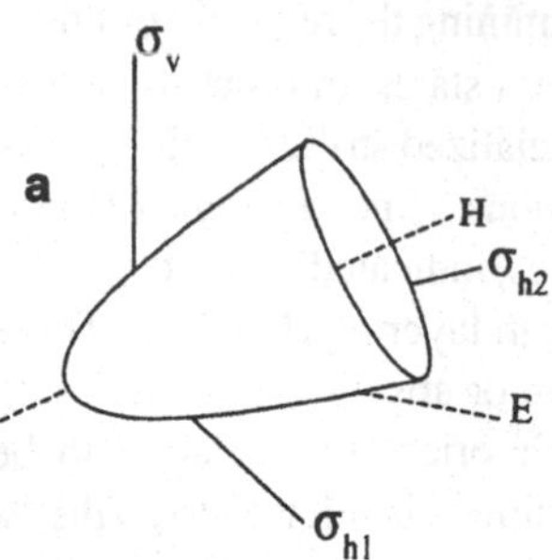

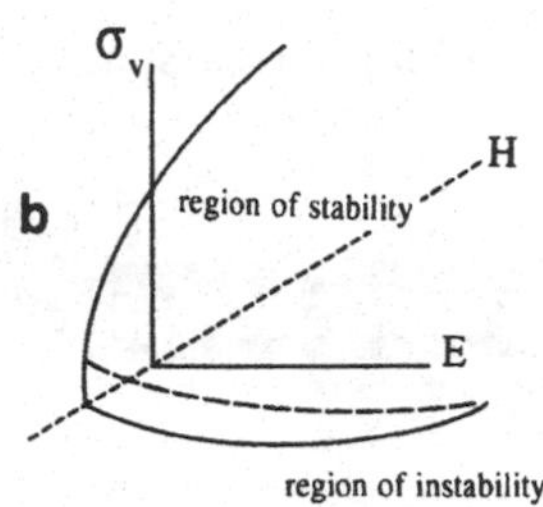

Figure 14. The faulting instability envelope for the Drucker-Prager failure criterion in the working stress space is **(a)** a conical surface symmetric about the hydrostat (H). E is the projection of H onto the σ_{h1}-σ_{h2} plane. A sectional view in the σ_v-E plane **(b)** cuts through the centre of the conical failure surface. The region internal to the cone is the area of stability for faulting, and the region external to the cone is the area of instability for faulting. The solid curved lines depict the faulting instability envelope for an isotropic material. The dashed curved line is the suggested modification of the faulting instability envelope for an anisotropic material in the working stress space.

envelope towards H along the σ_{h1} and σ_{h2}, axes (as suggested by the dashed line in Fig. 14b). This shift becomes less pronounced as the stress values increase. The net effect is to make the faulting instability surface for layered rock a somewhat lopsided cone.

Buckling instability envelope

If an elastic layer is compressed in a layer-parallel direction to a stress level exceeding the Euler load (σ_{cr}, the critical load for buckling), it is in a state of unstable equilibrium. In response to a perturbation, the layer will deflect to a sinusoidal waveform (e.g. Timoshenko 1936; Johnson & Page 1976). This is elastic buckling. For a layer of thickness h and width b, loaded parallel to its length, the critical stress for buckling is:

$$\sigma_{cr} = \frac{E}{\left(1-\nu^2\right)} \frac{k\pi^2 h^2}{12b^2} \qquad (2)$$

where k is a shape factor and E and ν are Young's modulus and Poisson's ratio, respectively, for the layer. For a multilayered system, the expression for σ_{cr} is a bit more complicated, but still has basically the same ingredients. For example, for layers of uniform geometry and elastic properties, and frictionless interfaces (after Johnson & Page 1976):

$$\sigma_{cr} = \frac{E}{(1-\nu^2)}\left[\frac{t^2k^2}{12} + \left(\frac{\pi}{n^2tk}\right)^2\right] \qquad (3)$$

where n is the number of layers and t is the individual layer thickness.

In equations (2) & (3), layer-parallel load is differential stress (i.e. $\sigma_{cr}=\sigma_1-\sigma_3$), and the equations are for the case of plane stress (i.e. $\sigma_2=0$). In the working stress state, these conditions translate to $\sigma_{cr}=\sigma_{h1}-\sigma_v$ and $\sigma_{h2}=0$. The critical stress (using either eq. 2 or 3) plots simply as a line of slope m=1 in the σ_v-σ_{h1} plane, intersecting the σ_{h1} axis at σ_{cr} (Fig. 15a). This is the buckling instability surface. The layer shape and elastic properties will alter the intersect but not the slope of the line.

Rock layer interfaces are not frictionless, as assumed for equation (3). As the stress normal to the layer surfaces (σ_v) increases, the frictional resistance will suppress potential slippage between the layers. In essence, the layers no longer behave as independent units, but rather as fewer beams of greater individual thickness. Effectively, t becomes larger and n becomes smaller in eq. (3), with the net result that σ_{cr} becomes larger with increasing σ_v. The buckling instability envelope becomes convex in a positive σ_v direction, as suggested, qualitatively, by the dashed line of Figure 15a.

Finally, the critical load for layer buckling is not independent of σ_{h2}. In fact, as σ_{h2} increases, the value of σ_{h1} decreases by an equivalent amount (Timoshenko 1936), i.e. $\sigma_{cr}=\sigma_{h1}+\sigma_{h2}-\sigma_v$. In the σ_{h1}-σ_{h2} plane, the buckling instability envelope is simply the straight line intersecting both of these axes at σ_{cr} (Fig. 15b). Putting the instability surface together from the σ_{h1}-σ_v and σ_{h1}-σ_{h2} planes yields a buckling instability surface for the working stress space of cylindrical form (Fig. 15c).

Buckling Instability

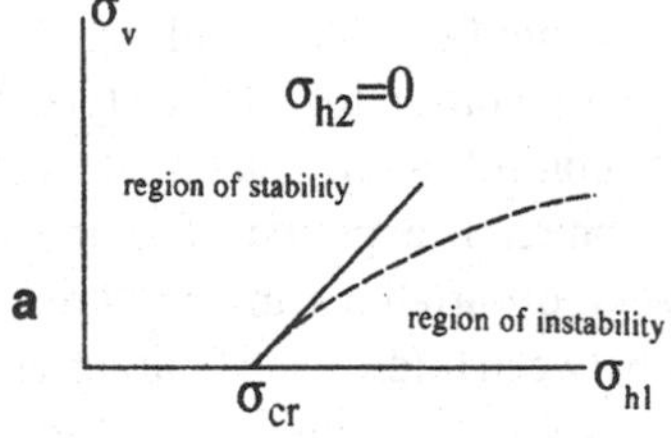

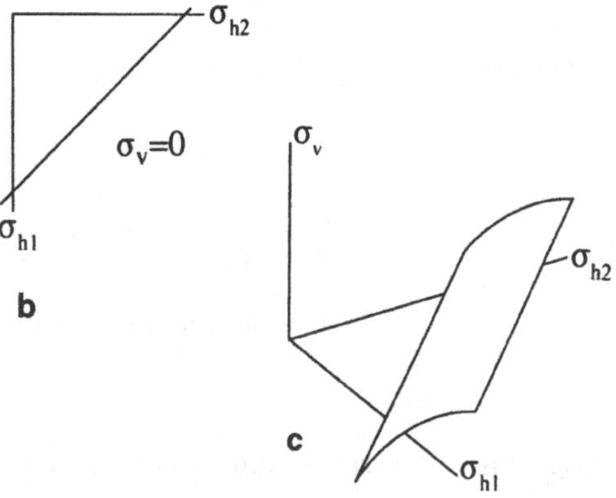

Figure 15. The buckling instability envelope in the working stress space viewed **(a)** in the σ_v-σ_{h1} plane and **(b)** in the σ_{h1}-σ_{h2} plane. The solid line of slope m=1 in (a) is the instability envelope for a free layer or multilayers with frictionless interfaces. The dashed curve is the suggested modification for frictional resistence at layer interfaces. **(c)** The buckling instability envelope in the working stress space for a multilayered material with frictional layer interfaces.

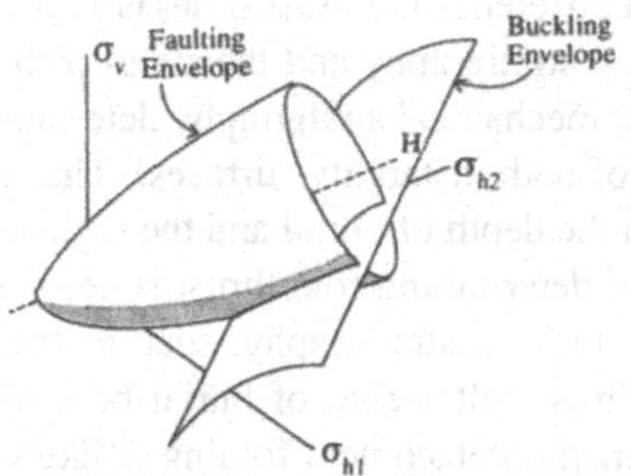

Figure 16. The faulting and buckling instability envelopes viewed together in the working stress space.

Competing failure surfaces

The faulting and buckling instability envelopes delimit areas of stability and instability for these separate deformational responses, and they define competing failure surfaces in the working stress space (Fig. 16). If the state of stress for a while in the region of stability for buckling, it will fail by faulting or fracturing. In regards to the issue of fold/thrust style, fault-bend faulting is expected. Alternatively, if the state of stress reaches the buckling instability envelope while within the stability field for faulting, it will buckle. Detachment folding is the probable fold/thrust style. For some materials, the buckling instability envelope may reside entirely outside the stability field for faulting, in which case buckling (and detachment folding) is not a viable deformation mechanism. The form and positioning of the two instability envelopes in the macroscale geological situation is determined by the mechanical characteristics of the rock package, i.e. the mechanical stratigraphy.

For the case where the faulting and buckling instability envelopes do, in fact, overlap, the choice of fold/thrust style is dictated by the stress path of the rock to the point of fold or fault initiation. Depth of burial plays a major role in determining the stress path. For horizontal strata buried in a subsiding basin with normal fluid pressures and no lateral stresses of tectonic origin, the overburden defines the vertical stress, i.e.

$$\sigma_v = (\rho_r - \rho_w)gh, \tag{4}$$

where h is the thickness of the overburden, g is gravitational acceleration, and ρ_r and ρ_w are the densities of rock and fluid in the overburden, respectively. Lateral stresses are

$$\sigma_{h1} = \sigma_{h2} = s\sigma_v, \tag{5}$$

where s varies between $\nu/(1-\nu)$ and 1, depending on whether a condition of uniaxial strain or lithostatic pressure, respectively, is assumed. In the working stress space, the burial stress path lies in the σ_v-E plane and is positive definite (Fig. 17a). It is in the stability field for both faulting and buckling. Subsequent to burial, increases in the lateral stresses associated with the thrusting (tectonic loading) will move the stress path toward the buckling and faulting instability envelopes (Fig. 17b). The stress paths are projected into a plane of constant σ_{h2} for the convenience of illustrating the concept. However, σ_{h2} will, in general, increase during the buildup of this tectonic load.) For any particular rock package, the two instability envelopes may vary in location and form from the example of Figure 17, but they will maintain their opposing curvatures. Where the instability surfaces overlap, the buckling instability envelope will be the first instability surface intersected by the stress path only at low values of σ_v (relatively shallow burial). The faulting surface will be the initial intersect at large values of σ_v (greater burial depths). Different stress paths may, thus, explain the development of Haystack Peak anticline as a detachment fold versus Whitney Canyon-Carter Creek anticline as a fault-bend fold. At Whitney Canyon and Carter Creek the Madison is buried beneath 3 km of structurally conformable overburden, whereas at Haystack Peak the Madison is at the surface. If Haystack Peak (and McDougal Pass) anticline formed very late in the displacement history of the Absaroka thrust, then a great deal of overburden may have been eroded before this folding initiated. There was certainly more overburden than currently exists at these outcrops, but perhaps little enough to result in a stress preference for detachment folding (i.e. low σ_v). In contrast, the more deeply buried Whitney Canyon-Carter Creek structure (with higher σ_v) initiated preferentially by faulting, and formed as a fault-bend fold.

The geometric analysis of Hunter Creek anticline suggests that it developed as a fault-propagation fold, i.e. that folding and faulting occurred concurrently. Fault-propagation folding is intermediate between detachment and fault-bend folding. Thus, it might be expected that fault-propagation folding occurs when the stress path contacts the instability envelopes somewhere near their points of intersection. Alternatively, the development of a fault-propagation fold may be explicitly linked to the composition of the rock package. Even if the stress conditions favour detachment folding, the growth of the detachment fold can proceed only if there is a supply of ductile material to fill the core of the fold as it increases amplitude. If that material is not available, faulting is

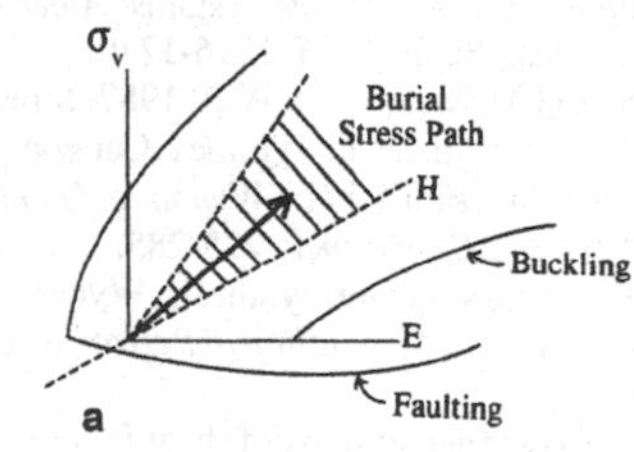

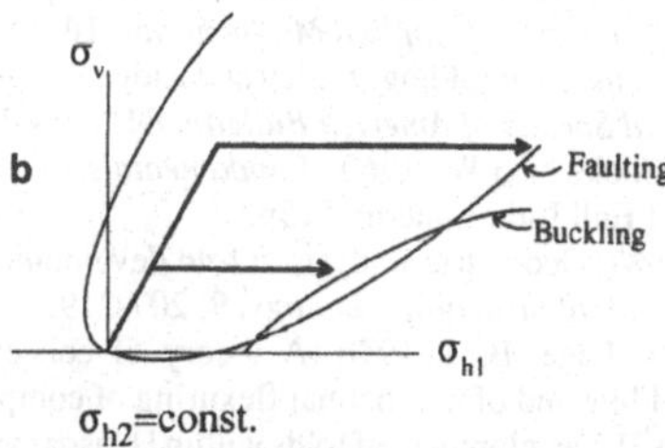

Figure 17. **(a)** The stress path for burial within a basin with no lateral stresses of tectonic origin (thus, with $\sigma_{h1}=\sigma_{h2}$) will fall in the σ_v-E plane and within the cross-lined region. **(b)** The application of regional horizontal stresses will change the horizontal, but not the vertical stresses. The stress path will move toward intersection with the instability envelopes.

necessitated for continued fold growth. This fault development during the incipient stages of fold growth may yield a fault-propagation fold.

The development of Hunter Creek anticline as a fault-propagation fold may be rationalized using either explanation. There is no Cambrian strata exposed in the core of this fold. It is, thus, possible that the Prospect thrust beneath Hunter Creek anticline carries little or none of the Cambrian shales, the only lithology that would serve as the critical core-filling material for detachment folding. Fault-propagation folding would then evolve as per the second scenario above. Alternatively, Hunter Creek anticline may have developed at a burial depth intermediate between Haystack Peak and Whitney Canyon-Carter Creek anticlines. Although roughly the same units outcrop at Hunter Creek anticline as at Haystack Peak anticline, the former structure has probably been transported several kilometres foreland and over a kilometre upsection from its developing location (Dixon 1982). The stress conditions at the inception of the Hunter Creek fold may have been located near the intersection of the two instability envelopes.

Conclusions

Geometric analysis provides a methodology for distinguishing different fold/thrust styles. The application of this procedure to three folds in the Wyoming-Idaho-Utah thrust belt indicates that fault-bend, detachment, and, possibly, fault-propagation folds can all occur in the same stratigraphic sequence. To explain the implication that fold/thrust style is at least partially independent of mechanical stratigraphy, the concept of competing buckling and faulting instability surfaces in stress space has been developed. In this approach, the occurrence of different fold/thrust styles is a function of both the mechanical stratigraphy and the stress path of the rock package. The mechanical stratigraphy determines the form and location of both instability surfaces. The stress path is dependent on the depth of burial and the regional tectonics.

In terms of determining fold/thrust style in a thrust belt with a fairly uniform stratigraphy, such as the Wyoming-Idaho-Utah thrust belt, depth of burial becomes a critical factor. In general, detachment folding is likely to develop only at relatively shallow depths, and fault-bend folding will become the dominant fold/thrust style at greater depths. Because fault-bend folds can have larger interlimb angles than fault-propagation folds and most detachment folds (Jamison 1987), folds in thrust belts might be expected to be more open structures at depth than in outcrop or in the shallow subsurface. An important consequence of this concept is that the structures observed in outcrop may not always be appropriate analogues for subsurface structures.

The concepts of stress controls on fold/thrust style presented here have assumed (1) a very special set of principal stress orientations and (2) the deformation is appropriately treated via brittle and elastic descriptions. However, these assumptions must be dropped for the generalized application of stress control and deformational style. If the principal stress axes deviate from the working stress space conditions, both the configuration and the location of the instability envelopes, especially for buckling, will change, but the basic concepts and conclusions presented here should remain intact. As conditions of deformation shift into the semi-brittle and ductile regimes, deformation becomes progressively more strain-rate dependent. Consequently, the use of discrete instability surfaces in stress space may not be not appropriate for conditions of ductile deformation.

References

Boyer, S.E. & Elliott, D. 1982. Thrust systems. *American Association of Petroleum Geologists Bulletin*, **66**, 1196-1230.

Couples, G.D., Weir, G.M. & Jamison, W.R. 1987. Structural development and hydrocarbon entrapment at Whitney Canyon and Yellow Creek Fields, Wyoming overthrust belt. *Wyoming Geological Association Guidebook, 38th Field Conference*, 275-285.

Dixon, J.S. 1982. Regional structural synthesis, Wyoming salient of western overthrust belt. *American Association of Petroleum Geologists Bulletin*, **66**, 1560-1580.

Donath, F.A. 1961. Experimental study of shear failure in anisotropic rocks. *Geological Society of America Bulletin*, **76**, 985-999.

Drucker, D.C. & Prager, W. 1952. Soil mechanics and plastic anaylsis for limit design. *Quarterly of Applied Mathematics*, **10**, 157-165.

Faill, R.T. 1973. Kink-band folding, Valley and Ridge Province, Pennsylvania. *Geological Society of America Bulletin*, **84**, 1289-1314.

Jaeger, J.C. & Cook, N.G.W. 1969. *Fundamentals of rock mechanics*. Chapman and Hall Ltd., London, 515p.

Jamison, W.R. 1987. Geometric analysis of fold development in overthrust terranes. *Journal of Structural Geology*, **9**, 207-219.

Johnson, A.M. & Page, B.M. 1976. A theory of concentric, kink and sinusoidal folding and of monoclinal flexuring of compressible, elastic multilayers. VII. Development of folds within Huasna syncline, San Luis Obispo County, California. *Tectonophysics*, **33**, 97-143.

Laubscher, H.P. 1977. Fold development in the Jura. *Tectonophysics*, **37**, 337-362.

McGill, G.E. & Raney, J.A. 1970. Experimental study of faulting in an anisotropic, inhomogeneous dolomitic limestone. *Geological Society of America Bulletin*, **81**, 2949-2958.

Rich, J.L. 1934. Mechanics of low-angle overthrust faulting as illustrated by Cumberland thrust block, Virginia, Kentucky, Tennessee. *Bulletin of the American Association of Petroleum Geologists*, **18**, 1584-1596.

Rubey, W.W. 1973. Geologic map of the Afton quadrangle and part of the Big Piney quadrangle, Lincoln and Sublette Counties, Wyoming. 1:62,500. *United States Geological Survey, Map I-686*.

Schroeder, M.L. 1973. Geologic map of the Clause Peak quadrangle, Lincoln, Sublette and Teton Counties, Wyoming. 1:24,000. *United States Geological Survey, Map GQ-1092*.

Serra, S. 1977. Styles of deformation in the ramp regions of overthrust faults. *Wyoming Geological Association Guidebook, 29th Field Conference*, 487-498.

Suppe, J. 1983. Geometry and kinematics of fault-bend folding. *American Journal of Science*, **283**, 684-721.

—— & Medwedeff, D.A. 1984. Fault-propagation folding. *Geological Society of America Annual Meeting Program with Abstracts*, **16**, 670.

Timoshenko, S. 1936. *Theory of elastic stability*. McGraw-Hill, New York, 518.

Weir, G. 1983. Structural analysis of Whitney Canyon-Carter Creek field, Uinta and Linclon Counties, Wyoming. *Amoco Production Company Report WN-22-83R*.

Williams, G. & Chapman, T. 1983. Strain developed in the hangingwalls of thrust sheets due to their slip/propagation rate: a dislocation model. *Journal of Structural Geology*, **5**, 563-571.

Willis, B. 1890. Mechanics of Appalachian structure. *United States Geological Survey 13th Annual Report, pt. II*, 211-281.

Kinematics of large-scale asymmetric buckle folds in overthrust shear: an example from the Helvetic nappes

Mark G. Rowan* & Roy Kligfield

Department of Geological Sciences, Campus Box 250, University of Colorado, Boulder, CO 80309, USA

Abstract: Kinematic analysis of asymmetric detachment folds from the Wildhorn nappe, central Helvetics, Switzerland, supports models of buckle fold formation in overthrust shear between nappe boundaries. The overall geometry shows a 2500 m thick multilayer of competent Jurassic limestones deflected into super- and subjacent less competent marls and shales by buckling above the basal decollement. Detailed investigation and measurement of the distribution, orientation, and relative timing of mesoscale structures, primarily dilatant veins and solution cleavage, allow documentation and quantification of the following kinematic development: (1) the nappe boundaries originated at a low angle (ca. 10°) to undeformed bedding; (2) local stress reorientation and buckling instabilities initiated symmetric buckle folds possibly characterized by tangential longitudinal strain; (3) the far-field stress directions imposed by the shear zone became dominant, and mesoscale structures developed asymmetrically on backlimbs and forelimbs; and (4) buckling ceased when fold interlimb angles reached 90-100°, and further shearing caused the fold limbs and axial planes to rotate into their current orientations. The total shear strain required is 2.9, equivalent to ca. 8 km of displacement of the overlying Cretaceous with respect to the base of the nappe. It is suggested that only an overthrust shear model can explain all the observations, and that other models, such as gravitational sliding and buttressing by normal faults, are incompatible with the data.

Large-scale asymmetric folds are commonly associated with thrust faults. Several popular kinematic models have been advanced to explain their origins: the fault-bend fold model relates fold development to movement over thrust ramps (Rich 1934; Suppe 1983), and the fault-propagation fold model links fold amplification to growth and propagation of blind thrust faults (Suppe & Medwedeff 1984; Suppe 1985). Many folds, however, belong in neither of these categories, and are best described as detachment folds, cored by relatively incompetent lithologies, that develop due to shortening and/or shearing of a multilayer above a basal decollement.

The Helvetic nappes of Switzerland offer a classic example of detachment folds. Some kinematic models suggest these folds formed due to either gravitational sliding (Lugeon 1943; Trümpy 1969, 1973) or gravitational spreading (Milnes & Pfiffner 1977; Merle 1986, 1989). Another proposes a major component of pure-shear compression due to buttressing against pre-existing normal faults during nappe movement (Lemoine *et al.* 1986; Gillcrist *et al.* 1987), and a fourth model is based on Rocky Mountain-style thin-skinned thrust tectonics (Boyer & Elliott 1982; Butler 1985). Finally, in the model supported by the evidence presented here, the folds developed in simple shear between thrust faults bounding the nappes (Laubscher 1983; Ramsay *et al.* 1983; Casey & Huggenberger 1985; Dietrich & Casey 1989). The latter authors stressed that only this model is capable of explaining the wedge-shaped nappe geometries, the orientation and variation in finite and incremental strains, and the observed metamorphic gradients.

Simple shear has been invoked as the dominant factor in fold development in other areas as well. Sanderson (1979) proposed that variations in fold geometries in southwest England, from open, upright folds to tight, asymmetrical, recumbent folds, were caused by increasing amounts of simple shear. Similar explanations have been applied to account for spatially varying fold styles from other thrust belts (Bruhn 1979; Kligfield *et al.* 1981; Bosworth & Vollmer 1981; Tanner & Macdonald 1982; Gibson & Gray 1985). While the proposed role of simple shear offers an attractive explanation for the observed deformations, it remains a model based solely on geometries and, occasionally, strain patterns; evidence for the kinematic development of a single fold in overthrust shear has not been provided. Furthermore, there is some question as to the mechanism of fold initiation: while some models (Ghosh 1966; Manz & Wickham 1978; Sanderson 1979; Gibson & Gray 1985) require only simple shear, Ramsay *et al.* (1983) proposed early ramp anticlines, Casey & Huggenberger (1985) invoked a pre-existing gentle warping of bedding, and Gillcrist *et al.* (1987) suggested compression due to buttressing.

In this paper, data are presented which document the kinematic history of large-scale asymmetric detachment folds from an area of the central Helvetic nappes in the Bernese Oberland, Switzerland (Fig. 1). It is shown that overthrust

*Present Address: Alastair Beach Associates, 11 Royal Exchange Square, Glasgow G1 3AJ, Scotland

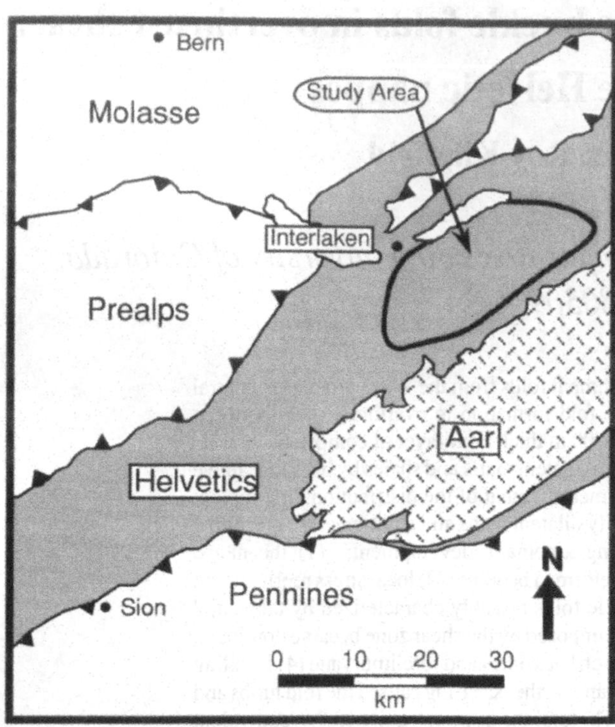

Figure 1. Regional tectonic map.

shear between nappe boundaries oriented at an originally low angle to bedding (as proposed by Ramsay *et al.* 1983, and Casey & Huggenberger 1985) created initially open, upright, symmetric folds, and that further shearing subsequently rotated these folds into their current asymmetric geometries. The amount of required shear strain and associated displacement is calculated by application of standard equations for simple shear deformation.

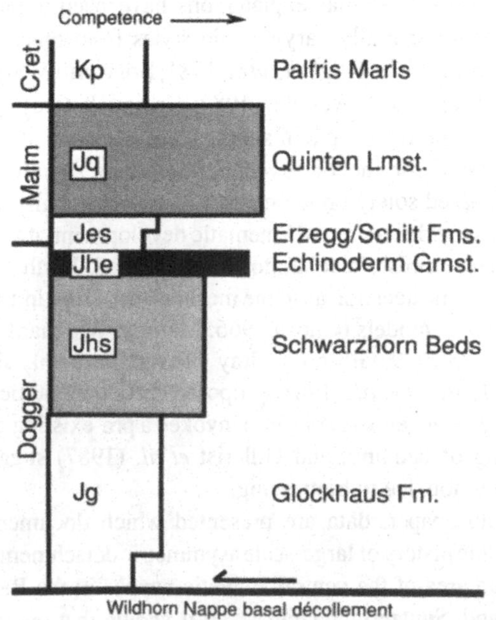

Figure 2. Generalized stratigraphy, roughly to scale, (approx. 2500 m), with competency increasing with unit width. See text for lithological descriptions.

Geological setting

The area of investigation forms part of the Wildhorn nappe, the uppermost of the Helvetic nappes, located directly northwest of the Aar massif culmination (Fig. 1). Although the nappe contains rocks of latest Triassic through early Tertiary age, the folds in the study area are defined by Jurassic formations. Figure 2 is a simplified stratigraphic column depicting a roughly 2500 m thick multilayer riding on a major thrust fault with the sedimentary cover of the Aar massif in its footwall. Immediately above the basal decollement are incompetent shales and interbedded sandstones of the Dogger Glockhaus Formation (Jg), and at the top of the section are thick marls of the Cretaceous Palfris Formation (Kp). This latter unit serves as the upper detachment, as the overlying Cretaceous folds are largely disharmonic with respect to those involving Jurassic rocks (Ramsay 1981, 1989; Ramsay *et al.* 1983; Burkhard 1988). The more competent portion of the multilayer is formed by: the Schwarzhorn Beds of the Dogger Hochstollen Formation (Jhs), fine-grained clastic limestones interbedded with thin marls; the Echinoderm Member (Jhe), a medium-bedded, fine- to coarse-grained echinoderm grainstone; thin shales and marls of the Erzegg and Schilt Formations (Jes); and the Malm Quinten Limestone (Jq), a thick-bedded to massive micrite. Approximate relative competencies are indicated in Figure 2; the Malm limestone is the thickest and most competent unit and thus asserts the most control over fold wavelength and geometry.

The field area was initially mapped in the early part of the century (Heim 1919; Stauffer 1920; Günzler-Seiffert 1924). Results of more detailed mapping by the authors are illustrated in a simplified map (Fig. 3) and a northwest-southeast cross section (Fig. 4). The area is dominated by a series of northwest-vergent, overturned folds which may have strike lengths of over forty kilometres (cf. Ramsay 1989) and which show cuspate-lobate shapes characteristic of strong competency contrasts (Ramsay 1982; Ramsay & Huber 1987). The two largest-amplitude folds are the focus of this paper, and are termed here the Birg (BG) and Schynige Platte (SP) folds. Each is characterized by a moderately dipping backlimb, a domain of low dips, and a steeply dipping, overturned forelimb. The folds are not cored by thrust ramps and are thus interpreted as undisrupted detachment folds.

Kinematic analysis

Although both the observed fold geometry and total finite strain must be correctly predicted by any kinematic model, neither imparts much information on fold evolution. The most useful criteria in deciphering the kinematic development are mesoscale structures and measured incremental strains. As pressure fringes and curved fibrous vein filling were not observed in samples from the Helvetic folds, the analysis depends on small-scale faults, bedding-plane fibres indicative of flexural slip, solution cleavage, and dilatant veins. Examination and measurement of the distribution, orientation, and relative timing of these features, especially

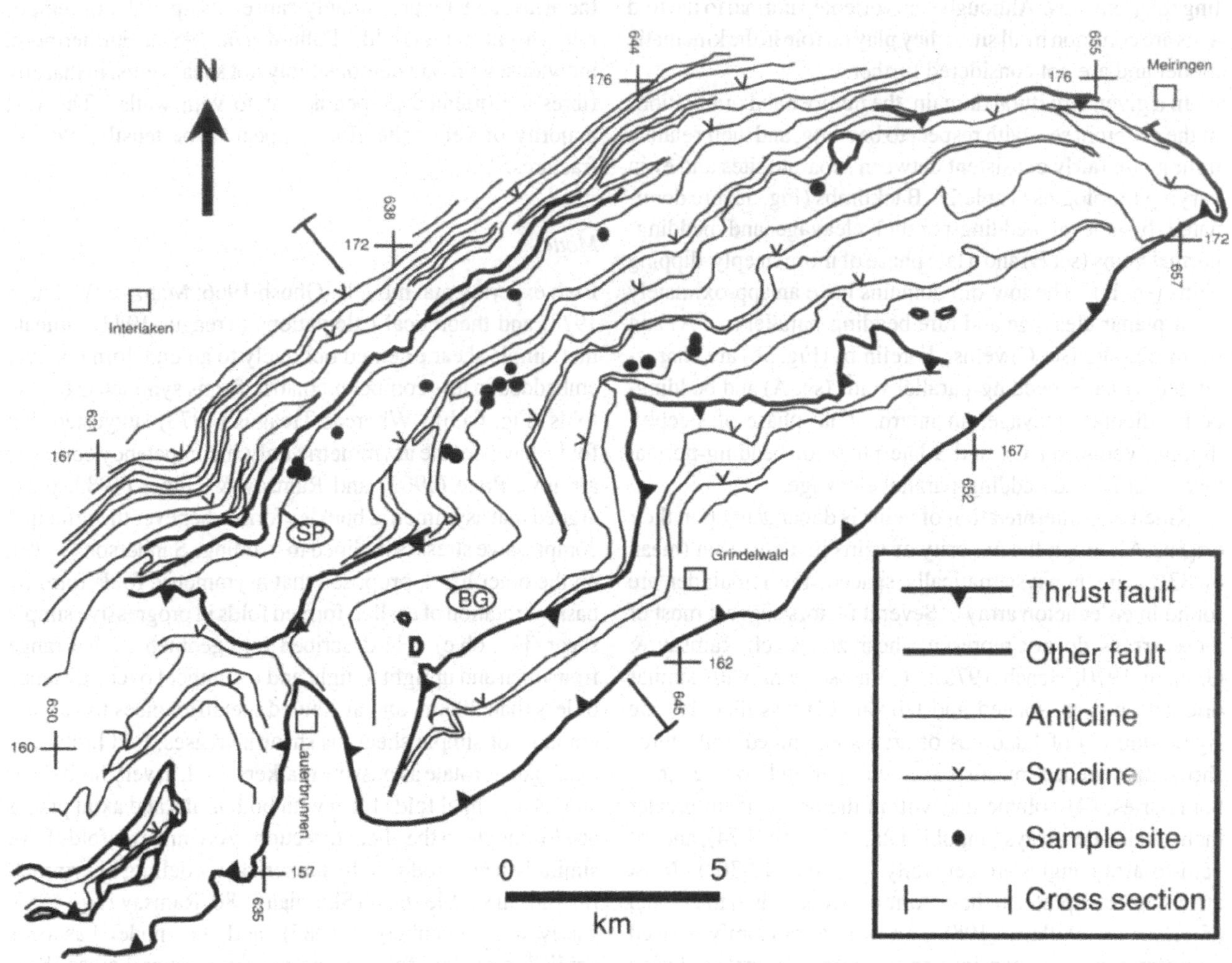

Figure 3. Simplified structural map of study area. The northwestern boundary is the Brienzer See, and the southeastern boundary is the basal detachment of the Wildhorn Nappe. Crosses refer to Swiss national map grid coordinates.

the veins and cleavage, form the foundation for documenting the proposed role of simple shear in fold initiation, amplification, and modification.

Vein and cleavage orientations were measured at nineteen representative sites (Fig. 3, Table 1). The vein data for each site were divided into sets (Fig. 5) using cluster-analysis (Shanley & Mahtab 1975) and field observations. Mean orientations were determined by eigenvector analysis. Each set is defined relative to site bedding, and does not necessarily have the same orientation as the equivalent set at another site. The relative timings of the vein and solution cleavage populations (Table 1) were determined by examining cross-cut-

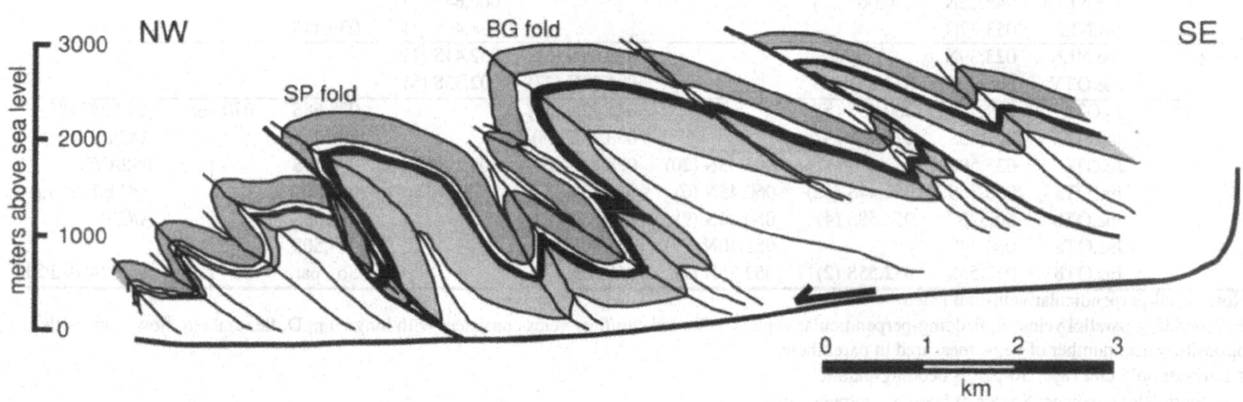

Figure 4. Simplified structural cross section across center of study area (line of section shown in Fig. 3). Thick lines are faults, thin lines are fold hinge lines, stratigraphic patterns as in Figure 2. Topographic elevations in vicinity of cross section range from 700 m to 2500 m above sea level.

ting relationships. Although veins oriented normal to the fold axes are common in all sites, they play no role in the kinematic model and are not considered further.

In a given structural domain, the number and orientations of the different sets with respect to bedding, and their relative timing, are fairly consistent between separate sites and even varying lithologies (Table 1). Backlimbs (Fig. 5a) are dominated by coeval bedding-parallel cleavage and bedding-normal veins (set B) and a late phase of more steeply-dipping veins (set D). The low-dip domains have an approximately axial-planar cleavage and rare bedding-parallel (set A) and north-dipping (set C) veins. Forelimbs (Fig. 5b) are characterized by early bedding-parallel veins (set A) and bedding-perpendicular cleavage, an intermediate phase of steeply-dipping veins (set C), and a late phase of bedding-normal veins (set B) and bedding-parallel cleavage.

Kinematic interpretation of veins is dependent upon their origin. Although the majority of veins in a given set (mean of 83%) are nonsystematically spaced, the remainder are found in en-echelon arrays. Several factors suggest most of these arrays do not represent shear zones (cf. Ramsay & Graham 1970; Beach 1975): (1) most veins with similar orientations are isolated and tabular; (2) less than 1% are sigmoidal; (3) of hundreds of arrays examined, only three showed evidence of displacement parallel to the array boundaries; (4) volume loss within the arrays is no greater than outside the arrays (unpubl. data; cf. Beach 1974); and (5) vein-to-array angles are generally very low (10-30°). Most arrays are interpreted to have formed according to the model of Olson & Pollard (1988), in which randomly spaced microfractures develop into en-echelon vein arrays during growth due to mechanical interaction around vein tips. Others can be seen to merge into a primary vein oriented parallel to the array, and are presumably caused by spatially or temporally varying stress fields (Pollard *et al.* 1982). Furthermore, individual veins are demonstrably not shear veins, in that any fibres are oriented perpendicular to vein walls. The vast majority of veins, therefore, appear to be tensile, mode-1 fractures.

Model

Both experimental models (Ghosh 1966; Manz & Wickham 1978) and theoretical calculations (Treagus 1973) indicate that simple shear oriented obliquely to an undeformed layer embedded in less competent matrix forms symmetric buckle folds (Fig. 6a,b). Whereas Treagus (1973) suggested that folds may become asymmetric only if competency contrasts are low, Price (1967) and Ramsay & Huber (1983, p.27) argued that asymmetric buckles form whenever the principal compressive stress is inclined to bedding. Sanderson (1979), on the other hand, proposed that asymmetric folds form by passive rotation of earlier-formed folds in progressive simple shear (Fig. 6b,c). He described fold geometries that range from open and upright to tight and recumbent over a distance of less than 20 km, and attributed the differences to varying amounts of simple shear: as shear increases, fold limbs and axial planes rotate as passive markers until, at very high shear strains, isoclinal folds form with both limbs and axial planes at a low angle to the shear direction. Asymmetric folds have similarly been produced by numerical modelling of symmetric folds in simple shear (Skjernaa 1980; Ramsay *et al.* 1983; Casey & Huggenberger 1985), and the model has been applied to explain fold geometries from several areas (Bosworth & Vollmer 1981; Tanner & Macdonald 1982; Gibson & Gray 1985).

Table 1. Mean orientations and relative timing of veins sets and cleavage

Domain	Site	Bedding	A*	B*	C*	D*	PS1†	PS2‡	Timing§
Backlimb	Jq. NLa	054,19S		069,72N (24)			043,27S		b,ps1/B,PS1
	Jq.NLb	092,10S		080,74N (2)	050,52N (9)	050,88S (15)	043,17S		B,c/C,PS1,b/D
	Jhe.NLa	080,24S		068,71N (69)		090,82S (25)	059,23S		B/D
	Jhe.NLb	052,11S		098,74N (10)		088,82S (22)	091,03N		B/D
	Jhe.NLc	090,24S		076,66N (20)		110,87S (21)	So - par.		B,ps1/PS1,b/D
	Jhs.NLa	076,26S		080,74N (61)	057,30N (6)	044,90 (102)	So - par.		B,c/C,b,d/D
	Jhs.NLb	069,26S		085,67N (36)	120,48N (9)	068,88N (15)	079,37S		B,c/C,b,d/D,c
Crest	Jhe.NLFa	046,12N	048,13N (5)		024,84N (11)		030,10S		
	Jhe.NLFb	046,12N	050,06N (2)			060,64S (1)			
	Jhs.NLF	053,27N				038,46S (1)	039,44S		
Hinge	Jhe.NLN	023,39N			058,64N (6)	032,41S (19)			
	Jhe.OTV	035,90	073,81N (1)		094,39N (27)	002,35S (5)			
Forelimb	Jq.OTa	059,46S	060,45S (5)		044,72N (64)		028,68S	070,36N	A,PS2/C/PS1
	Jq.OTb	071,45S	073,43S (3)		050,78S (28)		So - par.		A/C/PS1
	Jq.OTc	033,59S		095,43N (20)	098,82N (20)	105,10N (15)	046,53S	So - per.	PS2/C/PS1
	Jhe.OTa	050,36S	059,44S (32)	060,45N (6)	039,88N (23)		066,50S		A/C,b/B,PS1,c
	Jhe.OTb	50,52S	055,58S (4)	050,36N (9)	070,73N (12)		042,55S		A/C/B
	Jhs.OTa	062,44S		055,40N (79)	065,80N (3)		064,50S		
	Jhs.OTb	052,55S	052,55S (2)	062,35N (7)	066,73N (7)		So - par.	So - per.	A,PS2/C/B,PS1

Note: axial-perpendicular veins not listed.

* A, bedding-parallel veins; B, bedding-perpendicular veins; C, flexural slip/flow veins consistent with hinge pin; D, flexural slip/flow veins with opposite sense; number of veins measured in parentheses.

† Early or only cleavage; So-par. is bedding-parallel.

‡ Second, later cleavage; So-per. is bedding-perpendicular.

§ Timing relationships, with different phases separated by slashes (oldest on left, youngest on right). Commas separate mutually cross-cutting sets; capital letters designate dominant or only occurrence of a set; lower-case letters designate minor occurrence of a set.

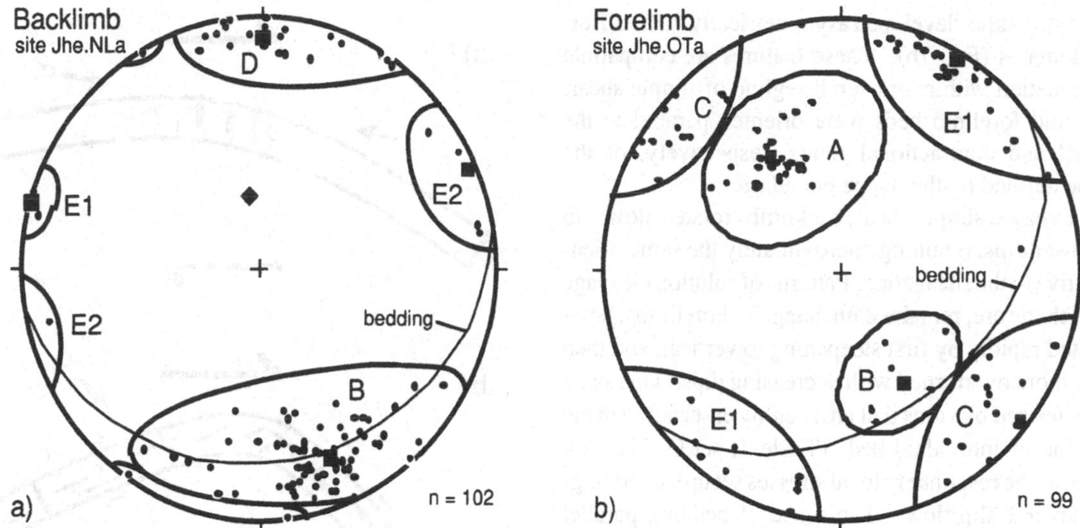

Figure 5. Equal-area, lower-hemisphere plots of dilatant veins from representative sites (both in Echinoderm Member): (a) backlimb, (b) forelimb. Dots - poles to veins, diamond - pole to site bedding, squares - poles to mean vein orientation for each set. Capital letters refer to different vein sets: A - bedding-parallel, B - bedding-perpendicular, C - veins related to flexural slip/flow during folding, D - late flexural slip/flow veins, E - axial extension veins. n - number of veins.

It is suggested that the Wildhorn nappe folds formed in a simple shear zone defined by nappe boundaries which were originally oriented at a small angle to bedding. Initiation of overthrust shear formed symmetric buckle folds (Fig. 7a) which subsequently developed asymmetrically distributed mesoscale structures (Fig. 7b). Further shearing modified the folds primarily by rotation of the forelimbs through the vertical (Fig. 7c) to their present orientations (Fig. 7d).

Initial shortening appears to have been accommodated immediately by buckling of the competent units of the multilayer into the surrounding incompetent marls and shales. Evidence for early layer-parallel shortening is almost nonexistent: small-scale contractional wedge faults are very rare, and bedding-perpendicular pressure solution is confined to forelimbs. The earliest-formed veins and cleavage have relationships to bedding that are opposite on the two limbs, resulting in extension and thinning of backlimbs and shortening and thickening of forelimbs (Fig. 7b).

Evidence that the initial folds were symmetric rather than asymmetric comes primarily from the low-dip domains. These regions have none of the features of the backlimbs, but do have veins and cleavage orientations analogous to those of the earliest forelimb structures. In addition, backlimb bed lengths (measured from cross sections and including layer-parallel strains) are roughly equal to the sum of the forelimb and low-dip domain bed lengths. Thus, the initial folds were symmetric, with hinges located at the boundaries between the low-dip domains and backlimbs. As there is no evidence for progressive hinge migration through the low-dip domains, it is suggested that the fold hinges jumped to their present locations prior to significant forelimb rotation.

The early symmetric shape of the folds implies that σ_1 was locally reoriented parallel to bedding at interfaces between competent beds and incompetent matrix (Fig. 7a; Treagus 1973). However, there are no observed mesoscale structures symmetrically developed on both fold limbs. Instead, layer distortion may have been manifested by tangential longitudinal strain (Ramsay 1967), with deformation confined to fold hinges. It is emphasized that this is only a proposed model, as no unequivocal evidence for tangential longitudinal strain has been documented. Although fold hinges are sometimes characterized by more intense fracture development, including veins both parallel and perpendicular to bedding, the relative timing of these features is ambiguous, and any finite neutral surface cannot be defined. At some point during fold amplification, bedding was distorted enough so that it could no longer act effectively as a stress guide, and the far-field stress direction imposed by the shear zone became dominant. At this stage, the fold geometry was still symmetric, but

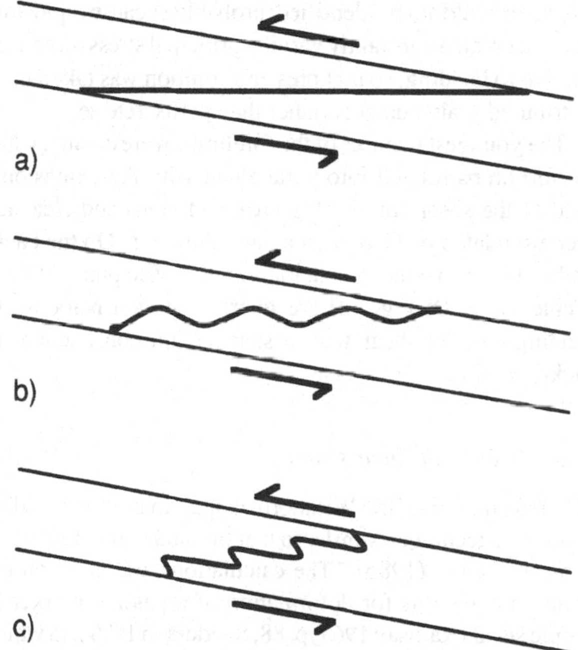

Figure 6. Kinematic model of asymmetric fold development in simple shear: (a) undeformed geometry; (b) initial generation of symmetric buckle folds (after Ghosh 1966, Manz & Wickham 1977, Treagus 1973); (c) development of asymmetry by passive rotation (after Sanderson 1979).

mesoscale structures developed asymmetrically in the different fold domains (Fig. 7b). These features are compatible with deformation within an overall regime of simple shear: backlimb and forelimb beds were oriented parallel to the extensional and contractional fields, respectively, of the shear zone defined by the nappe boundaries.

With increased simple shear, backlimbs rotated slowly to slightly lower dips, retaining approximately the same orientation relative to the shear zone. Patterns of solution cleavage and veins, therefore, remained unchanged. Forelimbs, however, rotated rapidly by first steepening to vertical, and then becoming more overturned with decreasing dips. This stage was characterized by veins that are roughly en-echelon in the profile plane of individual beds (Table 1, set C; Fig. 7c), interpreted as the response to local stresses set up by bedding-parallel flexural slip/flow. The sense of bedding-parallel shear is compatible with that indicated by common bedding plane fibres. Relative lack of similar features on the backlimbs indicates little flexural slip or flow in this domain, and thus suggests approximate backlimb pins during fold rotation and tightening. Forelimb solution cleavage associated with this stage could not be identified, probably because rapid limb rotation created constantly varying principal stress directions relative to bedding, so that pressure solution was taken up at distributed grain contacts rather than at discrete seams.

The youngest features of the forelimbs were created when the forelimbs rotated into parallelism with the extensional field of the shear couple (Fig. 7d), and veins and cleavage accommodated bed lengthening and thinning. On the backlimbs, a set of veins en-echelon in the profile plane of beds (Table 1, set D; Fig. 7d) are interpreted as a response to bedding-parallel shear with a sense again consistent with backlimb pins.

Quantification of shear strain

The deformation of the Wildhorn nappe can be quantified by applying a technique similar to that of Sanderson (1979) and Gibson & Gray (1985). The calculations are based on the standard equations for deformation of a passive marker in simple shear (Ramsay 1967, p. 88; Sanderson 1979), in which the final length ($\int$) and orientation (α) of a line are dependent upon its original length ($\int_0$), its original orientation with respect to the shear zone (α_0), and the amount of shear strain (γ):

$$\cot \alpha = \cot \alpha_0 + \gamma \qquad (1)$$

$$(\int/\int_0)^2 = 1 + \gamma^2 \sin^2\alpha_0 + \gamma \sin 2\alpha_0. \qquad (2)$$

In the present analysis, several assumptions are made. First, the deformation is divided into two separate phases: an initial period of symmetric folding with rotation only of the fold envelope, followed by passive rotation of fold limbs, axial plane, and fold envelope. Second, it is assumed that the shear zone boundaries are parallel, and that there is no change in shear zone thickness (i.e. no volume change) during deformation. Finally, the deformation is modelled with a constant shear strain profile, although shear strains are generally higher near the nappe boundaries (Ramsay 1981, Casey & Huggenberger 1985, Dietrich & Casey 1989).

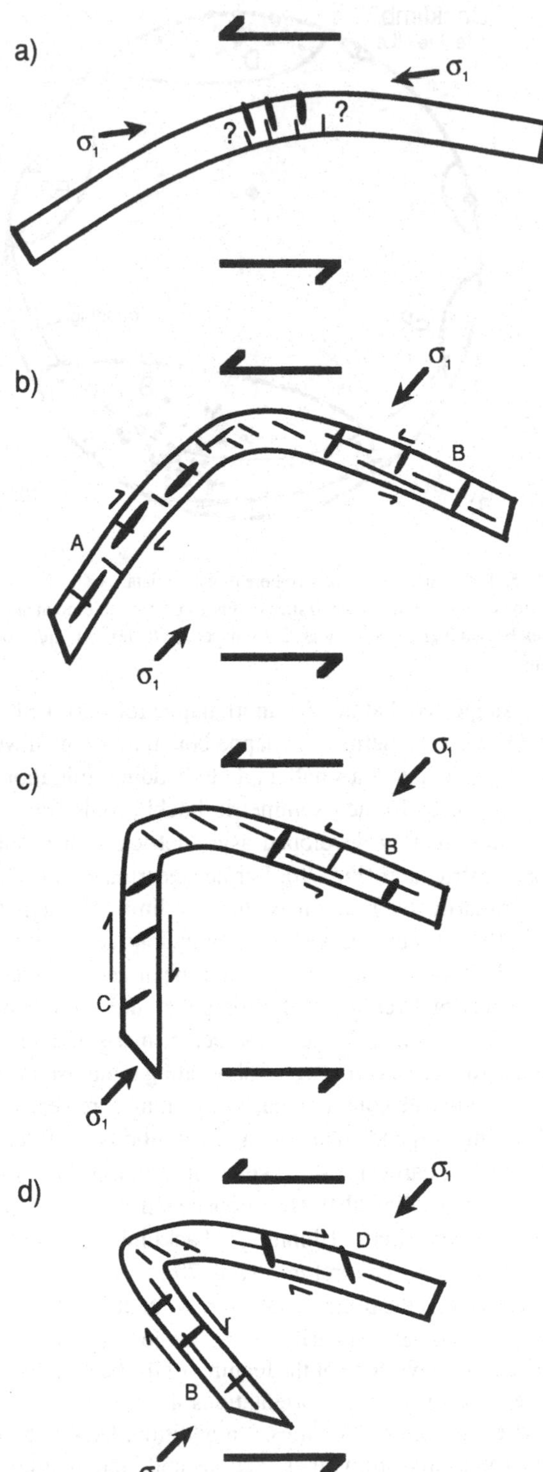

Figure 7. Schematic diagrams showing kinematic development of Wildhorn nappe folds: (**a**) initiation stage, with σ_1 locally reoriented parallel to original bedding, (**b**) symmetric stage, (**c**) rotation stage, (**d**) final stage (present geometry). Nappe boundaries are horizontal, with sense of overthrust shear shown by large arrows determining the regional orientation of σ_1 in (**b**), (**c**), (**d**). Dilatant veins are indicated by filled gashes, solution cleavage by thin lines. Small arrows parallel to bedding show sense and relative magnitude of flexural slip, and edges of each bed represent lines originally perpendicular to bedding. Letters A,B,C,D refer to vein sets of Table 1.

A number of independent parameters can be used as

Table 2. Quantification of simple shear deformation (see text).

Run	γ_s	α_0	ILA	γ_r	Env	AP	BL	FL	Err	γ_t
1	1.2	167	95	1.30	151	32	16	54	11	2.50
2	1.2	168	99	1.48	154	30	15	51	6	2.68
3	1.2	169	102	1.69	156	27	15	47	11	2.89
4	1.3	167	92	1.20	151	33	17	56	13	2.50
5	1.3	168	95	1.38	154	31	16	52	5	2.68
6	1.3	169	99	1.59	156	28	15	48	9	2.89
7	1.4	168	92	1.28	154	32	17	54	7	2.68
8	1.4	169	95	1.49	156	30	16	50	4	2.89
9	1.4	170	99	1.76	158	27	15	44	16	3.16
10	1.5	168	89	1.18	154	34	18	56	10	2.68
11	1.5	169	92	1.39	156	31	17	51	3	2.89
12	1.5	170	96	1.66	158	28	16	46	12	3.16
13	1.6	168	85	1.08	154	36	19	58	15	2.68
14	1.6	169	89	1.29	156	32	18	53	5	2.89
15	1.6	170	93	1.56	158	29	17	48	8	3.16
16	1.7	169	86	1.19	156	34	19	56	11	2.89
17	1.7	170	91	1.46	158	30	18	50	4	3.16
18	1.7	171	95	1.78	161	27	16	43	19	3.48
19	1.8	169	84	1.09	156	36	20	58	16	2.89
20	1.8	170	88	1.36	158	32	19	52	7	3.16
21	1.8	171	93	1.68	161	28	17	45	15	3.48
measured*					155	31	18	50		

γ_s, symmetric stage shear strain; α_0, original angle between bedding and shear zone; ILA, inter-limb angle of symmetric fold; γ_r, rotation stage shear strain; Env, final orientation of fold envelope; AP, final orientation of axial plane; BL, final orientation of backlimb; FL, final orientation of forelimb; Err, degrees of cumulative mismatch; γ_t, total shear strain.
* mean values measured from cross sections relative to horizontal shear zone.

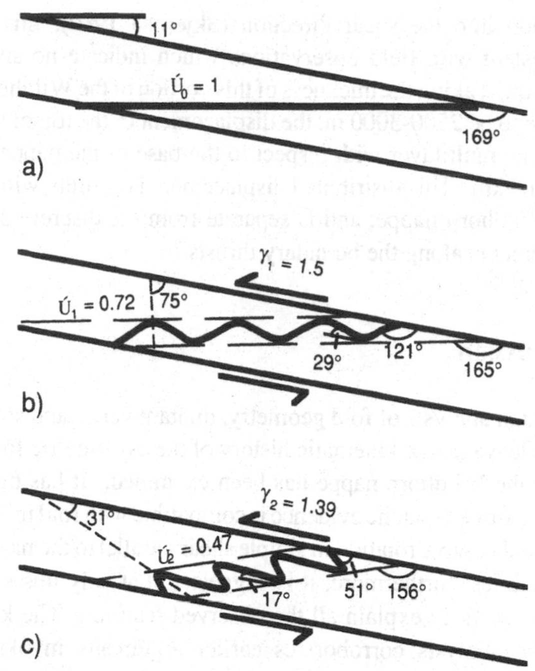

Figure 8. Quantification of overthrust shear deformation (see text): (a) undeformed state, (b) symmetric stage, (c) final geometry. Values from Run 11 of Table 2.

variables in equations (1) and (2), namely, the total measured shortening (53%) and the average orientations of the fore-limbs, backlimbs, axial planes, and fold envelope. The equations were applied to over 200 different scenarios with varying combinations of shear strains, original shear zone orientations relative to bedding, and current shear zone orientations. Selected results are listed in Table 2, and the best solution is illustrated in Figure 8. The procedure for each run is as follows: (1) an orientation of the shear zone with respect to bedding (α_0) is chosen (Fig. 8a), and an initial shear strain (γ_s) is imposed; (2) the bedding rotation and shortening are calculated from equations (1) & (2); (3) the new bedding orientation is equivalent to the fold envelope of the symmetric folds, and the shortening determines the fold interlimb angle (ILA, equal to $180°-2[\cos^{-1}\int_1]$); (4) from these values, symmetric fold axial plane, backlimb, and forelimb orientations relative to the shear zone can be computed (Fig. 8b); (5) the shortening from the symmetric stage is subtracted from the total measured shortening to give that of the second, rotation, stage, and the equivalent shear strain (γ_r) is determined from equation (2); (6) this shear strain is applied to the symmetric fold elements, yielding the calculated final orientations (Env, AP, BL, FL, Table 2; Fig. 8c); (7) the present shear zone dip is assumed, and the observed orientations of the various parameters measured relative to it; and (9) the measured and calculated final orientations are compared, and the cumulative degrees of mismatch are calculated. The best solution is that which minimizes the cumulative mismatch.

The calculations are remarkably well constrained: while other combinations of shear strain and orientation may accurately reproduce one or several aspects of the deformation, only a narrow range can account for the measured rotation of all fold elements and the total shortening. Table 2 lists the results of scenarios in which the shear zone was assumed to be currently horizontal. Other scenarios with north- and south-dipping shear zones yielded the same best total shear strain and original shear zone orientation, but the cumulative mismatches were greater. The best solution is run 11, with a cumulative mismatch of only 3° compared with measured values. With bedding originally horizontal, the shear zone had an original dip of 11° SE (Fig. 8a). Its current horizontal orientation implies a regional rotation of this part of the Wildhorn nappe of approximately 11° to the northwest. An initial shear strain of $\gamma = 1.5$ caused bedding (or the fold envelope) to rotate 4° and shorten by 28%. The shortening was accommodated by creation of symmetric buckle folds with interlimb angles of about 92° and axial planes at 75° to the shear zone boundaries (Fig. 8b). A further shear strain of $\gamma = 1.39$ then modified the folds by passive rotation into the presently observed geometries (Fig. 8c).

The division of the deformation into two separate phases, although a simplification, may be mechanically justifiable. Ramsay *et al.* (1983) have pointed out that folds of this type are mechanically active until interlimb angles approach 100°, at which point the folds lock and further tightening occurs by passive rotation in simple shear. The results presented here are in excellent agreement, with rotation commencing when interlimb angles reached 92°.

The total shear strain of $\gamma = 2.9$ compares well with that determined for the Diablerets Nappe ($\gamma = 3$) by Dietrich & Casey (1989). In addition, it falls below the value of $\gamma = 3$, at which structural elements such as fold axes start to show rotation into the shear direction (Skjernaa 1980); this is consistent with field observations which indicate no such rotation. Taking the thickness of this portion of the Wildhorn nappe to be 2500-3000 m, the displacement of the top of the Jurassic multilayer with respect to the base of the nappe is about 8 km. This distributed displacement is entirely within the Wildhorn nappe, and is separate from the discrete displacements along the boundary thrusts.

Discussion

Through analysis of fold geometry, dilatant veins, and solution cleavage, the kinematic history of the asymmetric folds from the Wildhorn nappe has been examined. It has been demonstrated that the evidence is compatible with fold initiation and passive rotation in simple shear parallel to the nappe boundaries; furthermore, it is suggested that only this scenario can fully explain all the observed features. The kinematic analysis corroborates earlier arguments invoking overthrust shear in the Helvetic nappes (Laubscher 1983; Ramsay *et al.* 1983; Casey & Huggenberger 1985; Ramsay 1989; Dietrich & Casey 1989), and supports proposed explanations for asymmetric fold development in other areas (Sanderson 1979; Bruhn 1979; Kligfield *et al.* 1981; Bosworth & Vollmer 1981; Tanner & Macdonald 1982; Gibson & Gray 1985).

It could be argued that the early symmetric folds were caused not by simple shear at an angle to bedding, but instead by buttressing against pre-existing normal faults, as in the models of Lemoine *et al.* (1986) and Gillcrist *et al.* (1987). These authors clearly demonstrated the effects of buttressing in the western Alps, and suggested that similar deformation styles, modified by intense simple shear, can explain the geometry and strain of the Morcles Nappe. Dietrich & Casey (1989), however, pointed out the inadequacies of this model: first, that two of the three Helvetic nappe boundaries are marked by clearly visible isoclinal synclines rather than by normal faults; and second, that buttressing implies a significant pure shear component and corresponding steep axial planes at the front of the nappe, features not characteristic of the Helvetic nappes. Furthermore, the present study has shown that early mesoscale structures are not symmetric on both limbs, as would be expected if buttressing were a significant factor.

Although the driving mechanism for fold development was overthrust shear, the actual deformation processes that accommodated buckling and rotation were varied. Certainly flexural slip played a major role during forelimb rotation, as indicated by the abundant bedding-plane fibres. A component of flexural flow is also suggested, both by the en-echelon veins on the forelimbs and by finite strain patterns (unpubl. data). But folding was not by flexural mechanisms alone, as some evidence exists for deformation by tangential longitudinal strain (Ramsay 1967). Small parasitic folds sometimes show radial dilatant veins and solution cleavage on the outside and inside, respectively, of the folds. Cores of the major folds can also show more extensive fracturing, although this is not developed everywhere. Hinge areas of the early symmetric folds are broad and poorly defined, so that the role of tangential longitudinal strain in the initial buckling is difficult to determine. It is probable, but not proven, that initial buckling was manifested primarily by tangential longitudinal strain, and that later rotation was accompanied principally by flexural slip and flow.

Brittle and chemical processes were important components of the Wildhorn nappe deformation (see also Groshong *et al.* 1984), and were primarily responses to the regional and local stresses imposed by shearing along the nappe boundaries. Vein formation accommodated both limb lengthening and thickening, depending on orientation, as well as axial extension, with total volume gains of up to 10% (unpubl. data). Dissolution, however, was the major factor in creating the thickness variations apparent in Figure 4. Although maximum volume losses in both the Malm limestone and Echinoderm Member are less than 10%, limestones of the Schwarzhorn Beds, where thinning is most pronounced on overturned limbs, have volume losses of up to 35% (unpubl. data). Dissolution in the interbedded marls may be even greater, as marls make up 45% of the Schwarzhorn Beds thickness on backlimbs, but only 15% on forelimbs. This appears contradictory at first, as backlimbs have always been properly oriented for limb thinning, whereas forelimbs did not reach appropriate dips until late in the folding history. It may be that intense shear associated with flexural slip/flow in the weak marls of the overturned limbs facilitated the dissolution process.

In closing, it is important to emphasize the differences in the fold kinematics documented here versus the fault-bend and fault-propagation models which dominate the recent literature. Widely varying structural styles exist in the high-level, low-temperature portions of separate, and even individual, fold-and-thrust belts, and different explanations are required. For example, models appropriate to the Rocky Mountains should not necessarily be applied to Alpine nappes. The variable styles are caused by many factors. The most critical include: tectonic setting (collision vs. subduction), presence and orientation of older faults, depth of burial and geothermal gradient, strain rate, and relative thicknesses and competencies of the sedimentary layers. It is the unique combination of such elements in each fold-and-thrust belt that determines the response to regional compression through folding, faulting, or more penetrative strain.

We wish to thank R. Ratliff, J. Cosgrove, and an anonymous reviewer for constructive comments, and K. McClay and A. Scott for their tireless efforts in organizing the conference and putting together this publication. The research was supported by NSF grant EAR 86-16640 to R. Kligfield.

References

Beach, A. 1974. A geochemical investigation of pressure solution and the formation of veins in a deformed greywacke. *Contributions to Mineralogy and Petrology*, **46**, 61-68.

—— 1975. The geometry of en-echelon vein arrays. *Tectonophysics*, **28**, 245-263.

Bosworth, W. & Vollmer, F. W. 1981. Structures of the Medial Ordovician Flysch of eastern New York: deformation of synorogenic deposits in an overthrust environment. *Journal of Geology*, **89**, 551-568.

Boyer, S. & Elliott, D. 1982. Thrust systems. *American Association of Petroleum Geologists Bulletin*, **66**, 1196-1230.

Bruhn, R. L. 1979. Rock structures formed during back-arc basin deformation in the Andes of Tierra del Fuego. *Geological Society of America Bulletin*, **90**, 998-1012.

Burkhard, M. 1988. L'Helvétique de la bordure occidentale du massif de l'Aar (évolution tectonique et métamorphique). *Eclogae geologicae Helvetiae*, **81**, 63-114.

Butler, R.W.H. 1985. The restoration of thrust systems and displacement continuity around the Mont Blanc massif, NW external Alpine thrust belt. *Journal of Structural Geology*, **7**, 569-582.

Casey, M. & Huggenberger, P. 1985. Numerical modelling of finite-amplitude similar folds developing under general deformation histories. *Journal of Structural Geology*, **7**, 103-114.

Dietrich, D. & Casey, M. 1989. A new tectonic model for the Helvetic nappes. *In*: Coward, M. P., Dietrich, D. & Park, R. G. (eds) *Alpine Tectonics*, Geological Society of London Special Publication, **45**, 47-63.

Ghosh, S. K. 1966. Experimental tests of buckling folds in relation to strain ellipsoid in simple shear deformations. *Tectonophysics*, **3**, 169-185.

Gibson, R. G. & Gray, D.R. 1985. Ductile-to-brittle transition in shear during thrust sheet emplacement, Southern Appalachian thrust belt. *Journal of Structural Geology*, **7**, 513-525.

Gillcrist, R., Coward, M. P. & Mugnier, J.-L. 1987. Structural inversion and its controls: examples from the Alpine foreland and the French Alps. *Geodinamica Acta*, **1**, 5-34.

Groshong, R. H. Jr., Pfiffner, O. A. & Pringle, L. R. 1984. Strain partitioning in the Helvetic thrust belt of eastern Switzerland from the leading edge to the internal zone. *Journal of Structural Geology*, **6**, 5-18.

Günzler-Seiffert, H. 1924. Der geologische Bau der östlichen Faulhorngruppe im Berner Oberland. *Sonderabdruck aus Eclogae geologicae Helvetiae*, **19**.

Heim, A. 1919. Das helvetische Deckengebirge. *Sonderabdruck aus Geologie der Schweiz, Bd. II*. Tauchnitz, Leipzig.

Kligfield, R., Carmignani, L. & Owens, W. 1981. Strain analysis of a northern Apennine shear zone using deformed marble breccias. *Journal of Structural Geology*, **3**, 421-436.

Laubscher, H. P. 1983. Detachment, shear, and compression in the central Alps. *Geological Society of America Memoir*, **158**, 191-211.

Lemoine, M., Bas, T., Arnaud-Vanneau, A., Arnaud, H., Dumont, T., Gidon, M., Bourbon, M., De Graciansky, P.-C., Rudkiewicz, J.-L., Megard-Galli, J. & Tricart, P. 1986. The continental margin of the Mesozoic Tethys in the western Alps. *Marine and Petroleum Geology*, **3**, 179-199.

Lugeon, M. 1943. Une nouvelle hypothèse tectonique: la diverticulation. *Bulletin de la Societé Vaudoise des Sciences Naturelles*, **42**, 301-303.

Manz, R. & Wickham, J. 1978. Experimental analysis of folding in simple shear. *Tectonophysics*, **44**, 79-90.

Merle, O. 1986. Patterns of stretch trajectories and strain rates within spreading-gliding nappes. *Tectonophysics*, **124**, 211-222.

—— 1989. Strain models within spreading nappes. *Tectonophysics*, **165**, 57-71.

Milnes, A. G. & Pfiffner, O. A. 1977. Structural development of the Infrahelvetic complex, eastern Switzerland. *Eclogae gelogicae Helvetiae*, **70**, 83-95.

Olson, J. & Pollard, D. D. 1988. Interpretation of joint sets based on fracture mechanics (abst). *Geological Society of America Abstracts with Programs*, **20**, A318-A319.

Pollard, D. D., Segall, P. & Delaney, P. T. 1982. Formation and interpretation of dilatant echelon cracks. *Geological Society of America Bulletin*, **93**, 1291-1303.

Price, N. J. 1967. The initiation and development of asymmetrical buckle folds in non-metamorphosed competent sediments. *Tectonophysics*, **4**, 173-201.

Ramsay, J. G. 1967. *Folding and Fracturing of Rocks*. McGraw-Hill, New York.

—— 1981. Tectonics of the Helvetic nappes. *In:* McClay, K. R. & Price, N. J. (eds) *Thrust and Nappe Tectonics*, Geological Society of London Special Publication, **9**, 293-309.

—— 1982. Rock ductility and its influence on the development of tectonic structures in mountain belts. *In:* Hsü, K. (ed) *Mountain Building Processes*, 111-127. Academic Press, London.

—— 1989. Fold and fault geometry in the western Helvetic nappes of Switzerland and France and its implication for the evolution of the arc of the western Alps. *In:* Coward, M. P., Dietrich, D. & Park, R. G. (eds) *Alpine Tectonics*, Geological Society of London Special Publication, **45**, 33-45.

——, Casey, M. & Kligfield, R. 1983. Role of shear in development of the Helvetic fold-thrust belt of Switzerland. *Geology*, **11**, 439-442.

—— & Graham, R. H. 1970. Strain variation in shear belts. *Canadian Journal of Earth Sciences*, **7**, 786-813.

—— & Huber, M. 1983. *The Techniques of Modern Structural Geology, Vol. 1, Strain Analysis*. Academic Press, London.

—— & —— 1987. *The Techniques of Modern Structural Geology, Vol. 2, Folds and Fractures*. Academic Press, London.

Rich, J. L. 1934. Mechanics of low-angle overthrust faulting as illustrated by the Cumberland thrust block, Virginia, Kentucky and Tennessee. *American Association of Petroleum Geologists Bulletin*, **18**, 1584-1596.

Sanderson, D. J. 1979. The transition from upright to recumbent folding in the Variscan fold belt of wouthwest England: a model based on the kinematics of simple shear. *Journal of Structural Geology*, **1**, 171-180.

Skjernaa, L. 1980. Rotation and deformation of randomly oriented planar and linear structures in progressive simple shear. *Journal of Structural Geology*, **2**, 101-109.

Shanley, R. J. & Mahtab, M. A. 1975. FRACTAN: a computer code for analysis of clusters defined on the unit hemisphere. *US Bureau of Mines Information Circular*, **8671**.

Stauffer, H. 1920. *Geologische Untersuchung der Schilthorngruppe im Berner Oberland*. Unpublished Ph.D thesis, Universität Bern.

Suppe, J. 1983. Geometry and kinematics of fault bend folding. *American Journal of Science*, **283**, 648-721.

—— 1985. *Principles of Structural Geology*. Prentice-Hall, New Jersey.

—— & Medwedeff, D. A. 1984. Fault-propagation folding. *Geological Society of America Abstracts with Programs*, **16**, 670.

Tanner, P. W. G. & Macdonald, D.I.M. 1982. Models for the deposition and simple shear deformation of a turbidite sequence in the South Georgia portion of the southern Andes back-arc basin. *Journal of the Geological Society, London*, **139**, 739-754.

Treagus, S. H. 1973. Buckling stability of a viscous single-layer system, oblique to the principal compression. *Tectonophysics*, **19**, 271-289.

Trümpy, R. 1969. Die helvetischen Decken der Ostschweiz: Versuch einer palinspastischen Korrelation und Ansätze zu einer kinematischen Analyse. *Eclogae geologicae Helvetiae*, **62**, 105-142.

—— 1973. The timing of orogenic events in the central Alps. *In:* De Jong, K. A. & Scholten, R. (eds) *Gravity and Tectonics*, Wiley, New York, 229-251.

Forelimb deformation in some natural examples of fault-propagation folds

Juan Luis Alonso[1] & Antonio Teixell[2]

[1] *Departamento de Geología, Universidad de Oviedo, 33005 Oviedo, Spain*
[2] *División de Geología, ITGE, Mayor 20, 22700 Jaca, Spain*

Abstract: An analysis of two natural examples of fault-propagation folds from the Pyrenees and the Cantabrian zone (N Spain) is presented. These folds possess common features despite their different geological settings, i.e. related to thrust tectonics in the Pyrenean example and to strike-slip faulting in the Cantabrian zone. Transition from fault-bend to fault-propagation folding along individual thrusts is related to changes in lithology. The fault-propagation folds studied are tight and markedly asymmetric with respect to limb thickness and shape. The forelimbs are curvilinear and display progressive thinning of beds towards the fault surface, which can be interpreted in terms of a characteristic heterogeneous strain superimposed onto flexural shear, and restricted to them. This strain has been factorized by means of grid construction, and is found to be a combination of thrust-parallel heterogeneous simple shear and pure shear, attributed here to a shear zone with diverging walls and extrusion. In contrast with previous models, it is shown that thrust-parallel simple shear is not necessarily distributed above the entire fault surface but may instead be restricted to regions where motion was inhibited such as tip zones or ramps.

Anticline-syncline fold pairs which accommodate fault displacement ahead of tip lines are commonly referred to as fault-propagation folds (Suppe & Medwedeff 1984). These folds are often broken as a result of further tip propagation. A kinematic model for fault-propagation folding was presented by Suppe (1985), in which displacement was accommodated by kink folds developed as a result of layer-parallel shear. Other models to account for displacement variations along faults, and the differences in thrust-parallel shortening between hangingwall and footwall have been both theoretically invoked (Elliott 1976; Berger & Johnson 1980; Sanderson 1982; Williams & Chapman 1983; Coward & Potts 1983), and documented from nature (Geiser 1988; Hyett 1990).

In this paper an analysis of two natural examples of fault-propagation folds from non-metamorphic terranes of both the Pyrenees and the Cantabrian zone of N Spain is presented. These folds display common geometric features even though they are from different geological settings (related to thrust tectonics in the Pyrenees and to strike-slip faulting in the Cantabrian zone). This paper shows that the deformation occurring in the fold forelimbs can be factorized into a flexural shear component, and a superimposed heterogeneous strain, both accommodating displacement variations along the faults.

Geological setting

The Pyrenean examples of fault-propagation folds presented belong to an imbricate fan developed in Upper Cretaceous sandstones and Palaeogene carbonates located just south of the Axial zone of the Central Pyrenees (Figs 1 & 2), in the Aragüés valley. The individual thrusts sole into a decollement located at the base of the sandstones (Larra floor thrust, Teixell 1990). This thrust system is folded into overturned folds with local cleavage development (S_2 in Fig. 1c), related to the emplacement of lower basement thrusts.

The Tejerina fault-propagation fold, located within the Cantabrian zone, is related to a strike-slip fault developed in the northern, subvertical limb of the earlier Tejerina syncline (Figs 3 & 4a). The stratigraphic succession involved consists of clastic rocks of Carboniferous age which unconformably overlie the Esla nappe (Alonso 1985, 1987). This subvertical limb was folded into a broad, open fold in which various structures were formed depending on the lithology and thickness of the beds. In the northern, inner arc of the fold, where thick-bedded alternations of quartzitic and carbonate conglomerates predominate, fault-propagation folds were formed. By contrast, in the south, where a thin-bedded carbonate conglomerate and shale multilayer exists, only smaller, conjugate kink folds and faults occur, involving less shortening as would be expected in the outer arc of the fold (Fig. 3b).

Geometry

The Aragüés and Tejerina structures are tight, fault-related folds of hectometric scale. Their detailed geometry is shown in Figures 2 and 4. In Aragüés, the folds are inclined, with subhorizontal axes. In Tejerina, the fold axes are vertical, and therefore the map pattern constitutes a true fold profile. In the Aragüés examples, the hangingwall anticlines, well-exposed, have been chosen for analysis. In Tejerina, the vertical syncline occurring in the north wall of the strike-slip fault is tighter than the south wall anticline, also better exposed, and

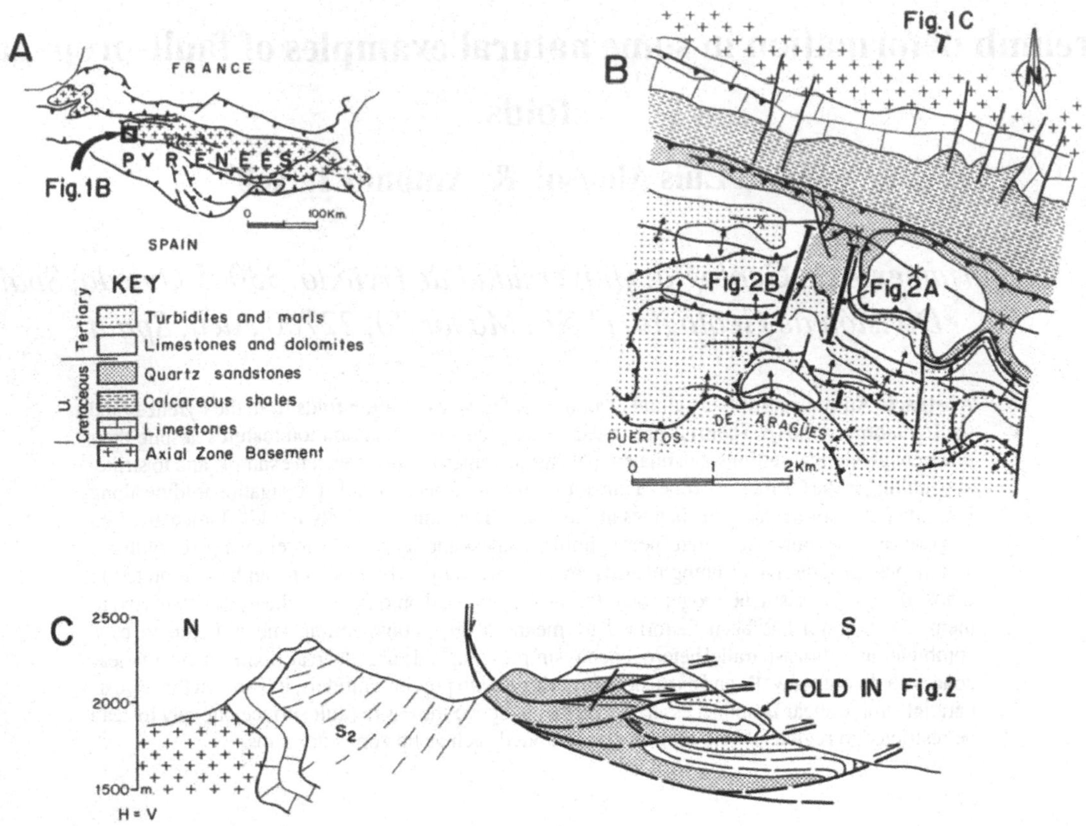

Figure 1. (**a**) and (**b**) Location maps for the fault-propagation folds in the Aragüés valley (south central Pyrenees). (**c**) Simplified geological cross-section. See (**b**) for location.

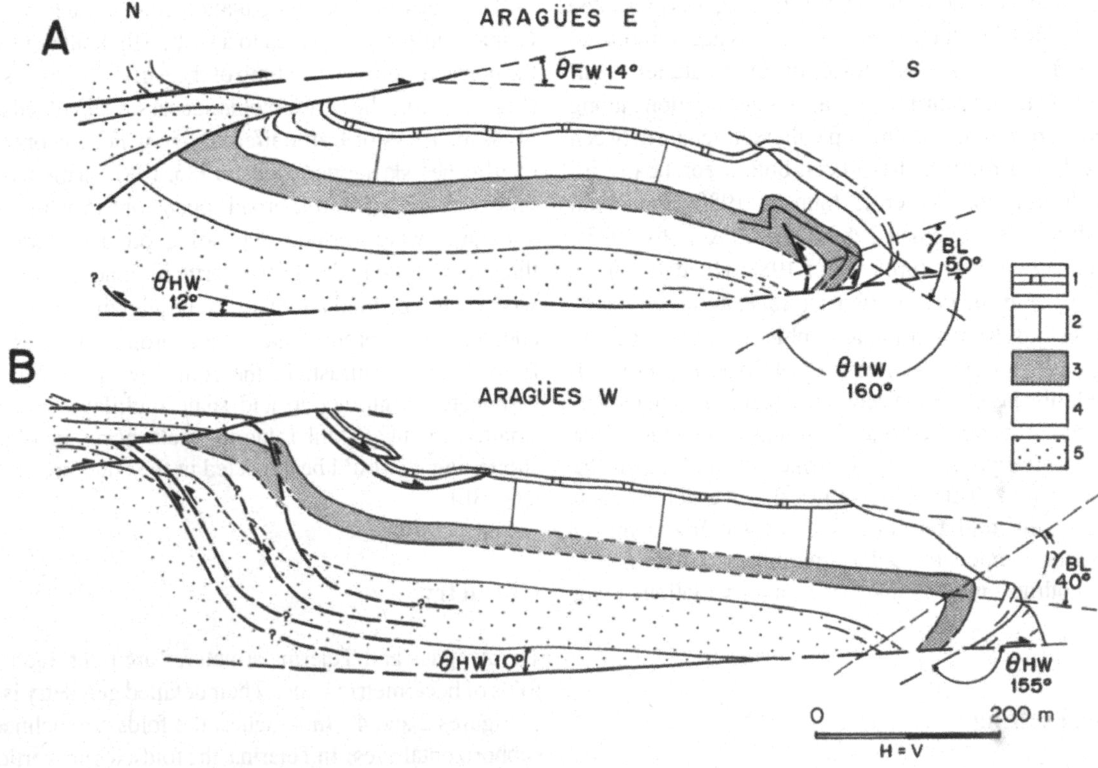

Figure 2. Detailed cross-sections of the same structure on both sides of the Aragüés valley, showing the geometry of thrust-related folds and backthrusts (see Fig. 1 for location). γ_{BL}, backlimb axial angle; θ_{HW}, hangingwall cutoff angle; θ_{FW}, footwall cutoff angle; Key: 1, well-bedded limestones; 2, massive limestones; 3, fine-grained limestones and dolomites; 4, dolomites; 5, quartz sandstones (Marboré formation).

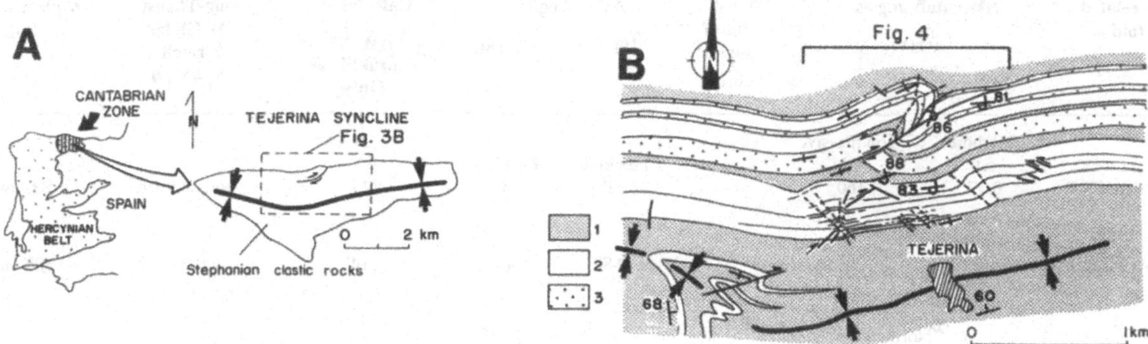

Figure 3. (**a**) and (**b**). Location maps for the Tejerina fault-propagation fold, in the Cantabrian zone. Key: 1, shales; 2, calcareous conglomerates; 3, quartz conglomerates.

was thus selected. The tighter nature of the north wall fold is probably related to the existence of a thick shaly packet, easily deformed by contact strain, immediately north of the fault-propagation fold.

There are several similarities between the Aragüés and Tejerina fault-propagation folds. Both examples consist of asymmetric folds with a straight backlimb in which the bed thicknesses remain unchanged, except for the hinge zones. In contrast, the forelimbs display a characteristic curvilinear form, which results in markedly asymmetric axial angles (angles between the fold limbs and the axial plane). In addition, the forelimb regions show a progressive thinning towards the fault plane, where the beds show quasi-asymptotic attitudes (Figs 2 & 4). This deformation took place mainly by pressure solution mechanisms, evidenced by the occurrence of stylolites and rough cleavage in Aragüés, and pitted pebbles in Tejerina. Small fractures and veins are also present.

In the Tejerina example, the fold shape changes, in the direction of displacement reduction, from chevron-like through rounded to box forms. This transition is similar to that occurring in the kink folds of the theoretical model for fault-propagation folds of Suppe (1985).

In Tejerina, the fold axial plane of the northern fault block originates from the trailing flat to ramp transition (Fig. 4). In contrast, in the Aragüés example the axial plane originates from the upper part of the ramp, near the boundary between the quartz sandstones and the overlying carbonates, leaving an unfolded section of sandstones in the lower part of the ramp (Fig. 2). This feature can be related to a lithologically-controlled upward change in slip/propagation rate, being lower within the sandstones than in the carbonates, where fault-propagation folds develop. Angular values and other parameters of the structures are presented in Figures 2 and 4, and in Table 1.

Strain factorization

In order to explain the curvilinearity and thickness variations in the forelimbs of the fault-propagation folds studied, consider a model in which a symmetric fold shape that would result from the common folding mechanisms (layer-parallel shear and tangential longitudinal strain) is modified by a

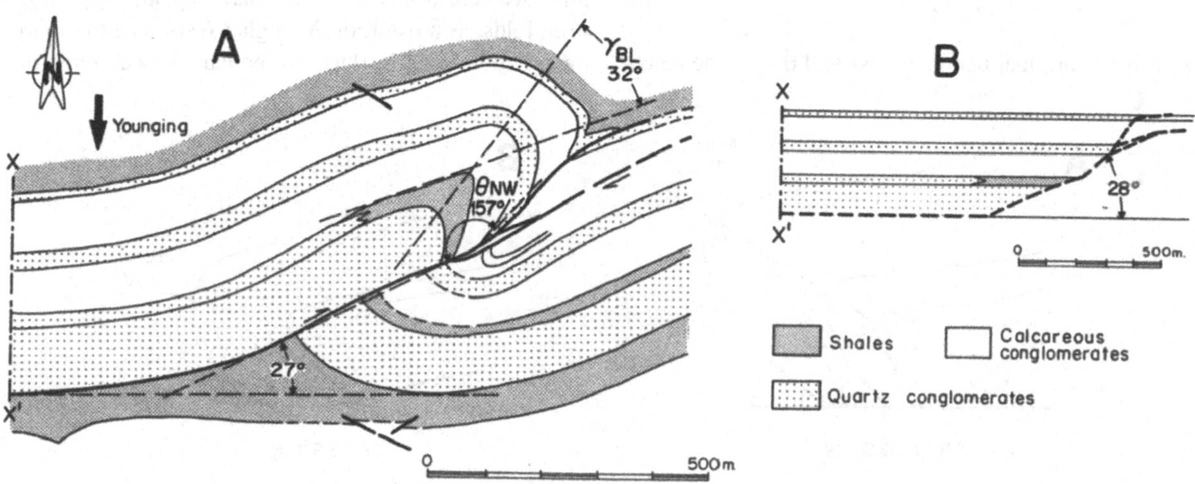

Figure 4. (**a**). Detailed geological map constituting the profile of the Tejerina fault-propagation fold. γ_{BL}, backlimb axial angle; θ_{NW}, cutoff angle in north wall of the fault (see Fig. 3 for location). (**b**) Restoration of the Tejerina structure.

Fault-related fold	NW cutoff angles θ_{HW}		FW cutoff angles θ_{FW}	Axial Angles γ_{FL}	γ_{BL}	Calculated θ_{HW} for Flexural Shear Only	Along-Thrust NW Global Stretch $S_t = l_1/l$	Displacement (m)
	Marboré sst.	Top massive lst.		Forelimb	Backlimb			
ARAGÜES E	0-12°	160°	---	21°	50°	117°	not recorded	500 min
ARAGÜES W	10°	155°	---	0-8°	40°	114°	0.79	500 min
	θ_{North} Wall		θ_{South} Wall					
TEJERINA	157° (Top lst.Conglom.)		20-90°	0°	32°	133°	0.86	170

Table 1. Angular values and other parameters of the fault-propagation folds analysed in this study.

deformation restricted to the forelimb. This restriction is supported by the constant thickness and the absence of significant strain which characterizes the straight backlimbs. The procedure is used solely for convenience in the factorization analysis, and a particular deformation sequence is not implied.

If an ideal symmetric fold is constructed using the backlimb axial angles measured in the folds studied, keeping the length of the axial plane constant, the forelimb obtained has the same cross-sectional area as that of the actual fold (Fig. 5). This suggests plane strain deformation with the intermediate strain axis perpendicular to the fold profile.

Furthermore, it can be observed that the cutoff lengths of the individual layers in the actual fold forelimb (l_1 in Figs 6b & d) are shorter than the orthogonal thicknesses measured in the backlimb, implying that thrust-parallel shortening occurred in the forelimb. The along-thrust stretch which modified the ideal symmetric fold into the present shape is defined as

$$S = l_1/l_0 \quad (1)$$

where

$$l_0 = t_0/\sin\theta'_{HW} \quad (2)$$

where t_0 is the original bed thickness and θ'_{HW} is the reconstructed hangingwall cutoff angle (Fig. 6). If an orthogonal grid is drawn onto the forelimb of the symmetric fold (Figs 6a & c), and if it is subsequently overlain on the actual fold shape with a displacement $d = l_0 - l_1$ (axial plane displacement), a deformed grid can be constructed using the distortion of bedding (Figs 6b & d). The superimposition of both the undeformed and deformed grids is shown in Figure 7. This figure illustrates a shear zone with diverging walls and heterogeneous strain, a model which is comparable to the extrusion effect of Ramsay & Huber (1987, Fig. 26.35) and Dietrich & Casey (1989). In the deformation zone shown here, one of the diverging walls (axial plane) is translated parallel to itself along the fault slip direction, and the other wall (fault plane) constitutes a discontinuity in the displacement field which experiences shortening. The resulting extrusion accounts for the characteristic curvilinearity of the forelimbs studied (Figs 6b, d & 7). The strain distribution within the deformation zone (Fig. 7) indicates the occurrence of thrust-parallel stretch (pure shear) combined with heterogeneous simple shear to maintain strain compatibility.

It is possible that strain extrusion models similar to the one proposed here also apply to true hangingwall ramps of fault-bend folds, as a result of the higher resistance to motion in ramp regions than in flats. In the hangingwall ramp of the

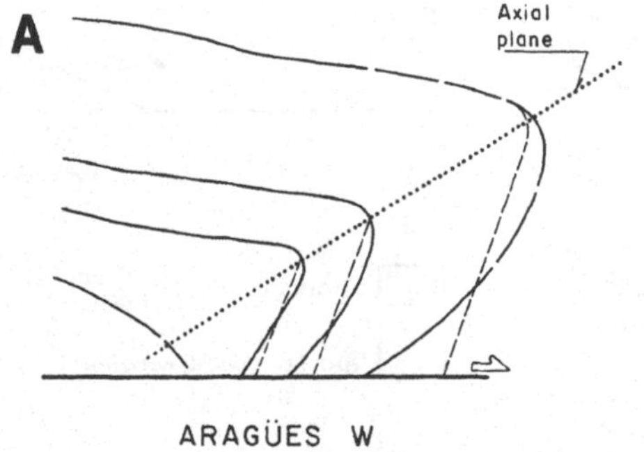

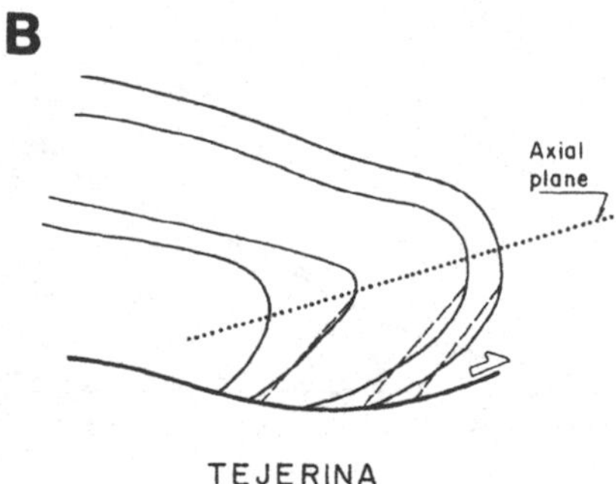

Figure 5. Reconstruction of symmetric folds from the backlimb axial angles measured in the Aragüés W and Tejerina structures. The actual fold profile is represented by continuous lines, whilst the dashed lines represent the bedding traces of the reconstructed symmetric folds. In the Aragüés example the reconstruction has been made taking into account the mean thickness of the entire backlimb.

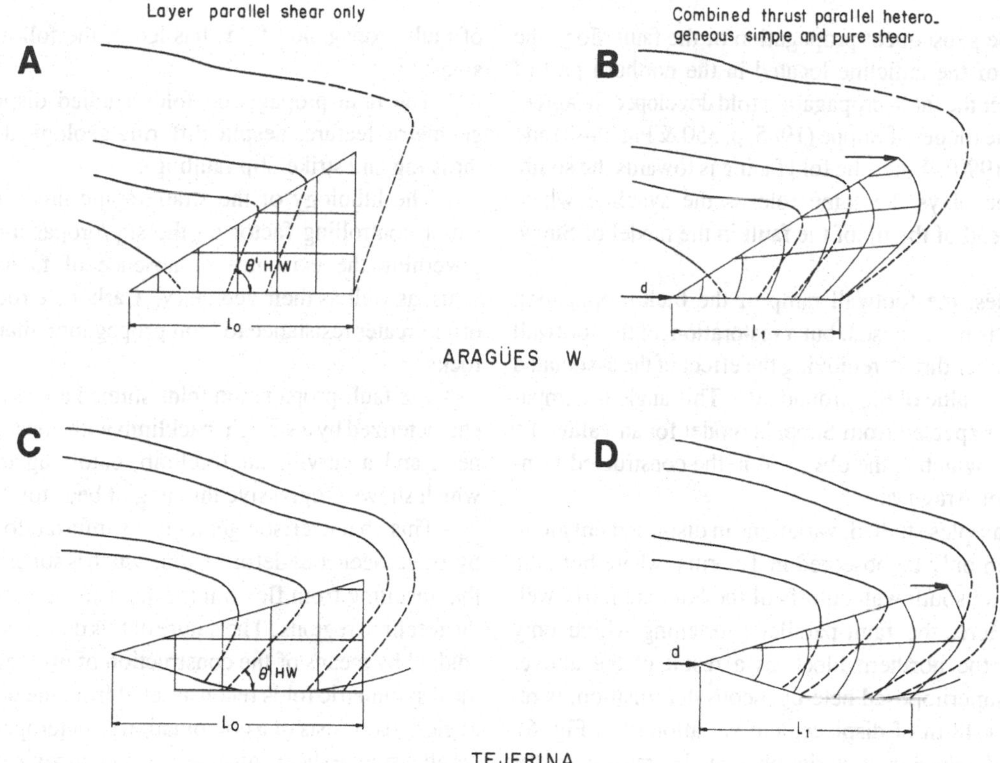

Figure 6. Modification of the symmetric folds into the actual fold forms. (**a**) and (**c**) Reconstructed symmetric folds, with undeformed orthogonal grids. (**b**) and (**d**) Presently observed fold forms showing the superimposed heterogeneous deformation by means of the deformed grid. l_0, initial cutoff length; l_1, present cutoff length; d, displacement.

Esla nappe (Cantabrian zone), some incompetent stratigraphic units show progressive thinning towards the thrust surface, with cutoff lengths much smaller than the original orthogonal thicknesses; by contrast, in the hangingwall flat, thrust-parallel shortening is negligible (Alonso 1985). Hence heterogeneous simple shear, supported by a well-developed fold train in the hangingwall ramp, would not necessarily affect the whole thrust sheet but instead would occur only in the frontal culmination wall. This differs from previous models in which simple shear superimposed on fault-bend folds is envisaged to affect the whole thrust sheet, i. e. both hangingwall ramp and flat regions (Elliot 1976; Ramsay *et al.* 1983; Suppe 1983). In our model, simple shear is a result of heterogeneous thrust-parallel stretch, necessary for the compatibility of deformation.

Comparison with previous models

The symmetric folds obtained by removing the effect of the superimposed heterogeneous strain as proposed in our model (Fig. 5) show quantitative relationships θ–γ which are in good agreement with those predicted by Suppe's (1985) model for fault propagation folding.

In Tejerina, the restoration of the structure profile (Fig. 4b), assuming a pin line (X–X´) perpendicular to the regional trend of bedding, yields a ramp angle of 28°. This value is similar to the actual angle between the present fault surface and the regional trend of bedding (27° in Fig. 4a). According to Suppe's model, an axial angle of 32° as that of the symmetric fold would imply θ_{FW} =24. This departure in θ_{FW} angle can be explained by the fact that the folds in Tejerina are rounded, implying thus less shortening than the kink folds of Suppe's model. The different geometries of the fault trajectories shown in Figure 4b and in the south wall ramp in Figure 4a may be attributed to the deformation of the fault by heterogeneous layer-parallel shear related to the development of the south wall fold. In this southern wall, the cutoff angles are now around 80°. To the north, the fault branches into two splays, the northwesternmost splay shows a higher ramp angle than the southern part of the ramp. This can be

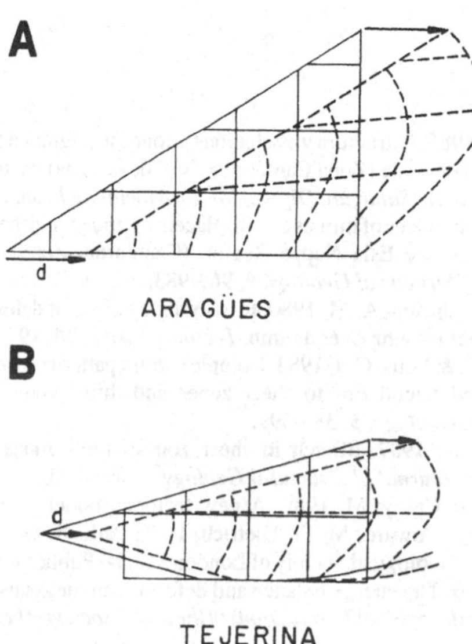

Figure 7. Transformation of the orthogonal grids, illustrating the translation of the axial plane, the shortening along the thrust plane, and the strain distribution.

related to the subsequent propagation of the fault along the axial plane of the anticline located in the northern part of Figure 4, after the fault-propagation fold developed, in agreement with the model of Suppe (1985, p. 350 & Fig. 9-47) and also Mitra (1990). Since the fold facing is towards the south, this anticline plays the same role as the syncline which develops ahead of the tip of the fault in the model of Suppe (1985).

In Aragüés, the footwall ramp of the fault-propagation fold studied is not exposed, but a restoration of the footwall ramp of a higher thrust, removing the effect of the associated folds, gives a value of θ_{FW} around 30°. This angle is comparable to that expected from Suppe's model for an value of γ equal to 40°, which is the observed in the constructed symmetric fold of Aragüés.

In the examples studied, variations in displacement along the faults can only be observed in Tejerina, where both the north wall and south wall cutoffs of the fault are fairly well exposed. Here, the fault-parallel shortening which only occurred in the northern block as a result of the above-mentioned superimposed heterogeneous deformation, is responsible for 48 m of displacement variation (d in Fig. 6). Additional displacement variation may be related to the differences in cutoff lengths caused by flexural shear processes as in the models of Suppe (1985) and De Paor (1987).

Comparable geometries to those of the forelimbs of the folds studied, with progressive bed thinning towards the fault, have been obtained experimentally by Mulugeta & Koyi (1987) and Lui *et al.* (this volume). These experimental models also contain backfolds, whose geometry is strikingly similar to the hangingwall box fold of Aragüés E (Fig. 2a).

Conclusions

The geometric and kinematic analysis of two field examples of fault-propagation folds has led to the following conclusions:

- The fault-propagation folds studied display common geometric features despite differing geological settings (i.e. thrusting and strike-slip faulting).
- The lithology of the stratigraphic units involved is a major controlling factor on the slip/propagation rate, thus governing the existence or absence of fault-propagation folds, as well as their geometry. Carbonate rocks appear to offer greater resistance to fault propagation than siliciclastic rocks.
- The fault-propagation folds studied are asymmetric and characterized by a straight backlimb with constant bed thickness, and a curvilinear forelimb, cutoff against the fault, which shows progressive thinning of beds towards the fault.
- This characteristic geometry is inferred to be produced by a heterogeneous deformation, which is superimposed onto that resulting from flexural mechanisms and is restricted to the forelimb regions. The nature of this deformation has been studied by means of the construction of orthogonal grids on ideal symmetric folds reconstructed from the backlimb axial angles. It consists of a combination of heterogeneous thrust-parallel simple shear and pure shear, occurring in a shear zone with diverging walls and extrusion.
- The folds obtained by removing the effect of the superimposed heterogeneous deformation display angular relationships (ramp angles and axial angles) which are in good agreement with previous theoretical models of fault-propagation folds.

We appreciate the reviews by Ken McClay and two anonymous referees which substantially helped to improve the manuscript. The comments by the colleagues of the Structural Geology Group at the University of Oviedo are also acknowledged.

References

Alonso, J. L. 1985. Estructura y evolución tectonoestratigráfica de la Región del Manto del Esla (Zona Cantábrica, NW de España). Institución Fray Bernardi *no de Sahagún. Diputación Provincial de León*, 276p.

—— 1987. Sequences of thrusts and displacement transfer in the superposed duplexes of the Esla Nappe Region (Cantabrian Zone, NW Spain). *Journal of Structural Geology*, **9**, 969-983.

Berger, P. & Johnson, A. M. 1980. First-order analysis of deformation of a thrust sheet moving over a ramp. *Tectonophysics*, **70**, T9-T24.

Coward, M. P. & Potts, G. J. 1983. Complex strain patterns developed at the frontal and lateral tips to shear zones and thrust zones. *Journal of Structural Geology*, **5**, 383-399.

De Paor, D. G. 1987. Stretch in shear zones: implications for section balancing. *Journal of Structural Geology*, **9**, 893-895.

Dietrich, D. & Casey, M. 1989. A new tectonic model for the Helvetic nappes. *In:* Coward, M. P., Dietrich, D. & Park. R. G. (eds), *Alpine Tectonics*. Geological Society of London Special Publication, **45**, 47-63.

Elliot, D. 1976. The energy balance and deformation mechanisms of thrust sheets. *Philosophical Transactions of the Royal Society of London*, **A283**, 289-312.

Geiser, P. A. 1988. Mechanisms of thrust propagation: some examples and implications for the analysis of overthrust terranes. *Journal of Structural Geology*, **10**, 829-845.

Hyett, A. J. 1990. Deformation around a thrust tip in Carboniferous limestone at Tutt Head, near Swansea, South Wales. *Journal of Structural Geology*, **12**, 47-58.

Mitra, S. 1990. Fault-Propagation Folds: Geometry, Kinematic evolution, and Hydrocarbon traps. *American Association of Petroleum Geologists Bulletin*, **74**, 921-945.

Mulugeta, G. & Koyi, H. 1987. Three-dimensional geometry and kinematics of experimental piggy-back thrusting. *Geology*, **15**, 1052-1056.

Ramsay, J. G., Casey, M. & Kligfield, R. 1983. Role of shear in development of the Helvetic fold-thrust belt of Switzerland. *Geology*, **11**, 439-442.

—— & Huber, M. I. 1987. *The techniques of modern Structural Geology. Volume 2: Folds and Fractures*. Academic Press, London, 309-700.

Sanderson, D. J. 1982. Models of strain variation in nappes and thrust sheets: a review. *Tectonophysics*, **88**, 201-233.

Suppe, J. 1985. Principles of Structural Geology. Prentice-Hall, New Jersey.

—— & Medwedeff, D. A. 1984. Fault-propagation folding. *Geological Society of America Abstracts with Programs*, **16**, 670.

Teixell, A. 1990. Alpine thrusts at the western termination of the Pyrenean Axial Zone. *Bulletin de la Societé Géologique de France*, **8 (VI)**, 241-249.

Williams, G. & Chapman, T. 1983. Strains developed in the hangingwalls of thrusts due to their slip/propagation rate: a dislocation model. *Journal of Structural Geology*, **5**, 563-571.

The geometric evolution of foreland thrust systems

M. P. Fischer[1] & N. B. Woodward[2]

[1]*Department of Geosciences, The Pennsylvania State University, University Park, PA 16802, USA*
[2]*26117 Viewland Drive, Damascus, MD 20872, USA*

Abstract: Thrust system models which consider the observed variation in structural geometry from minimum displacement regions to maximum displacement regions to be equivalent to the temporal variation in structural geometry for any cross section through the thrust sheet are incompatible with observations of lateral variations in ramp angles and fold geometry in well-exposed thrust systems in Wyoming and Tennessee. The self-similar model of thrust system evolution assumes ductile bead strain is always accommodated by tip-line folding, and suggests that variations in structural geometry observed anywhere in the thrust system are the product of the continued deformation of currently observed lateral tip structures. These variations, however, are interpreted to reflect the influence of the intrinsic physical properties of the thrust system on thrust propagation. Rather than representing the progressive deformation of tip folds, the structural geometry of many thrust systems is most likely indicative of spatial and/or temporal variations in these intrinsic properties.

Since direct observation of changes in the geometry of an evolving thrust system is not possible, assumptions must be made when reconstructing thrust system kinematics. The assumption that presently observed spatial variations in deformation (i.e. thrust system geometry) are equivalent to inferred temporal variations in deformation (i.e. geometry) is widely employed in geological modelling (see Means 1976, p. 27-30 for a discussion of this technique; Dickenson & Snyder 1979; Delaney *et al.* 1986; Wojtal 1986) and has been called an 'Eulerian' approach (Wojtal & Mitra 1988). This approach can yield perfectly viable kinematic models provided there is sound evidence of sequential overprinting of deformation (Wojtal & Mitra 1988). Because such overprinting relationships are often difficult to discern in macroscopic studies of the evolution of thrust systems, however, this paper discusses an approach wherein the deformation path (i.e. temporal geometric evolution) of individual portions of the thrust system is inferred based on an understanding of how the local physical properties of system constituents affect fault kinematics.

Self-similar modelling techniques

A commonly used, but rarely explicitly outlined technique (hereafter called a self-similar model) used to describe the geometric and kinematic evolution of thrust systems is based on the observation that over geological time spans the displacement at any particular location on a single thrust surface is proportional to the duration of thrusting in that region (Elliott 1976). Hence, since total thrust displacement (i.e. displacement of the lowest hangingwall cutoff) goes to zero at lateral thrust tip lines, the structures observed in these regions should represent the earliest stages of deformation (i.e. developing thrust system geometry). Likewise, since maximum total thrust displacement is suggested to occur beneath the internal portions of thrust sheets (Elliott 1976), structures in these regions should represent the final stages of deformation (i.e. thrust system geometry). It follows from the self-similar model that many of the fundamental geometric characteristics of a thrust system are determined at the earliest stages of thrusting, and that some record of this early deformation (i.e. geometry) should be observed in older parts of the system (e.g. 'background strain' of Williams & Chapman 1983; overprinting of deformation described by Wojtal 1986; Mitra 1987; Blenkinsop & Drury 1988; Wojtal & Mitra 1988). To examine the factors which might influence initial thrust geometry, it is therefore logical to study the deformation (i.e. structural geometry) at lateral thrust tip lines.

Faults are commonly modelled as individual large-scale fractures (Segall & Pollard 1980; Aydin & Nur 1982; Aydin 1988; Rybicki 1989), which in turn are equated with displacement discontinuities or dislocations (Bilby & Eshelby 1968). Thrust fault surfaces are often conceptualized as Somigliana dislocations (Eshelby 1973) where displacement is zero at the edges of the dislocation (i.e. thrust tip lines; Elliott 1976; Williams & Chapman 1983; Chapman & Williams 1985; Pfiffner 1985; Price 1988). Elliott (1976) recognized the strain inherent at the edges of such dislocations and suggested this strain was accommodated by a 'ductile bead' of deformation propagating just ahead of the moving thrust tip line in much the same way deformation in a crack tip process zone accommodates the strain developed at fracture terminations (Broek 1978).

The consistent observation of tip folds (asymmetric, blind fold complexes, e.g. Elliott 1976; Boyer & Elliott 1982) at thrust terminations (Buxtorf 1916; Spang & Brown 1981; Thompson 1981; Berger & Johnson 1982) has led to the

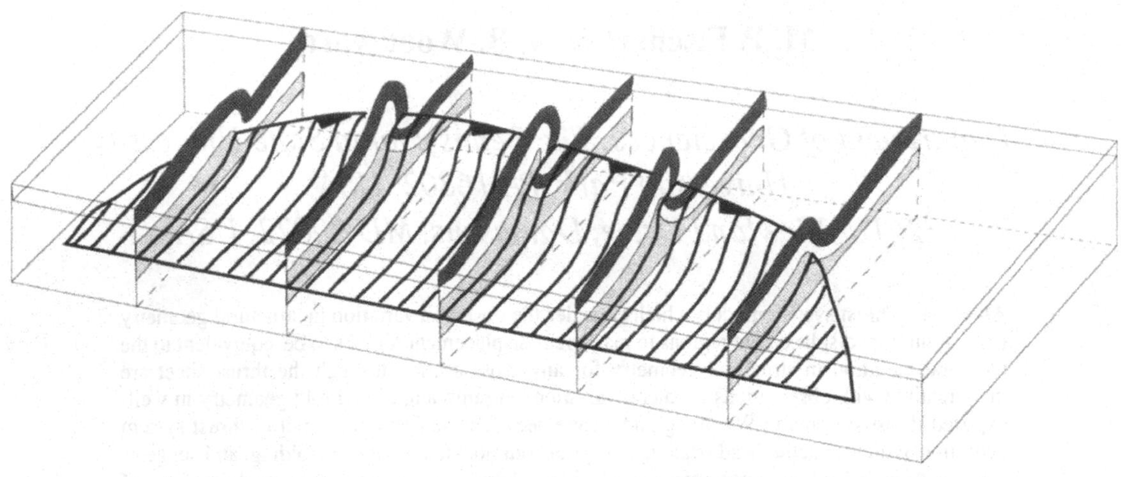

Figure 1. Conceptual model of an ideal self-similar thrust sheet. Portion of thrust fault (lined surface) between teeth is exposed at surface. Deformed strata (shaded layers) have been reconstructed above an ideal horizontal erosion surface.

hypothesis that thrust faults normally die out into these structures (Elliott 1976; Laubscher 1976; Williams 1980; Williams & Chapman 1983; Boyer 1986), and implies that deformation within the ductile bead is consistently manifested as folding in front of a thrust tip. Figure 1 is a conceptual model of a simple thrust system consisting of a single thrust sheet that evolved according to the self-similar model. In this system a thrust fault deforms strata during forward and lateral propagation, and dies out laterally into tip folds. Along strike towards the centre of the thrust sheet these anticlines are truncated and increasingly offset, preserving a syncline in the footwall and an asymmetric, fold-first (e.g. Dahlstrom 1970) anticline in the hanging wall.

Case studies

Since along-strike changes in the geometry of a self-similarly propagating thrust sheet represent the geometry of the system at different times during the kinematic evolution, the practical applicability of the self-similar model can be tested using relatively simple thrust systems in which the three-dimensional geometry and/or kinematics is well understood.

Baldy thrust system

The Baldy thrust system is located in the northern portion of the Idaho-Wyoming thrust belt in the Snake River Range and

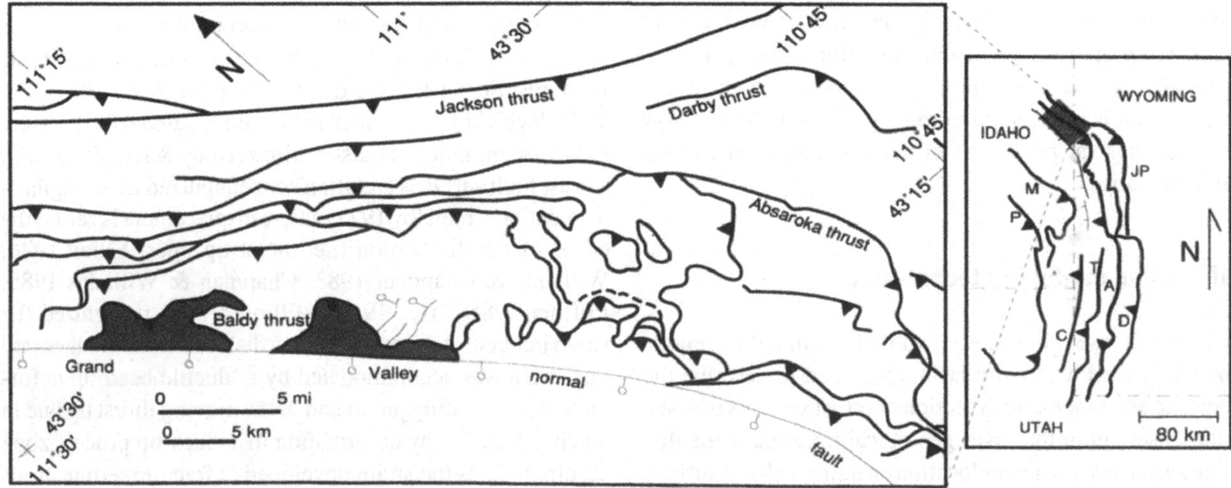

Figure 2. (a) Maps depicting the principal faults of the Idaho-Wyoming thrust belt and the Snake River Range (shaded box). Baldy thrust system is shaded on large-scale map. A= Absaroka thrust, C= Crawford thrust, D= Darby thrust, JP= Jackson-Prospect thrust, M= Meade thrust, P= Paris thrust, T= Tunp thrust (Modified from Woodward, 1986).

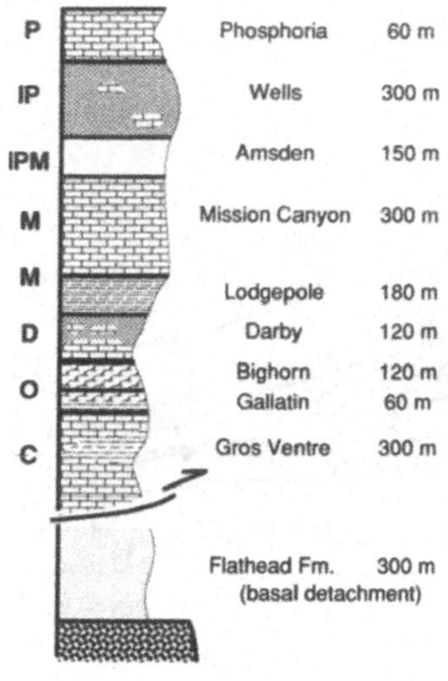

Figure 2. (**b**) Cambrian through Pennsylvanian stratigraphic units exposed in the Baldy thrust system. (Modified from Woodward, 1986). Standard stratigraphic ornamentation used.

is exposed for roughly 30 km as several klippen along the western margin of the frontal imbricate fan of the Absaroka thrust system (Fig 2a; Woodward 1986). The Baldy thrust sheet involves Cambrian through Mississippian strata, and is the uppermost fault slice in the imbricate stack of thrust sheets comprising the Snake River Range in this area (Fig 2a, b; Woodward 1986). NE-SW trending normal faults locally cut the Baldy thrust sheet, and the NW-SE trending Grand Valley normal fault truncates the trailing edge of the sheet, preventing accurate estimates of thrust displacement from restoration of balanced cross sections.

Figure 3 combines a map of the surface trace of the Baldy thrust, a generalized structure contour drawn on the thrust surface, a hangingwall cutoff map (e.g. Diegel 1986), and three schematic cross sections. The structure contour was constructed by recording and contouring points at various elevations where the thrust was observed (~1000 m of local relief). Where not preserved or observed, hangingwall cutoffs were reconstructed based on known stratigraphic thicknesses and fault displacements inferred from nearby locations in the thrust sheet where displacement estimates were better constrained.

In the northwest part of the thrust sheet (section A-A') the Baldy thrust dips roughly 15° west and places a hangingwall flat in Cambrian strata over a footwall flat in Mississippian strata (Fig. 3). Along strike to the southeast (section B-B') the thrust dips from 30°-40° west and places a hangingwall ramp in Cambrian strata over a footwall ramp in Cambrian through Pennsylvanian strata with displacement less than in section A-A' (Fig. 3). Further southeast along strike (section C-C') the Baldy thrust dips at nearly 40° to the west and places a hangingwall ramp in Cambrian strata over a footwall ramp in the same strata, indicating a continued southeastward decrease in displacement (Fig. 3). In addition, there is significant along-strike variation in Baldy thrust-related folding. At the southeast end, Mississippian strata in the footwall are folded, whereas no folding is observed in these footwall strata farther north, beneath the more internal portions of the thrust sheet (Woodward 1990). This simple geometric analysis of the Baldy thrust system suggests the observed lateral variations in structural geometry do not reflect the temporal evolution of the system.

Because there is a decrease in total thrust displacement from northwest to southeast, it is reasonable to infer that serial cross sections A-A' through C-C' represent a traverse from the interior toward a lateral tip of the Baldy thrust. In a self-similar model, early, intermediate and late stages in the geometric evolution of the Baldy thrust system are respectively represented by cross sections C-C', B-B' and A-A'. Furthermore, although section C-C' does not specifically represent a lateral tip structure, the model suggests this

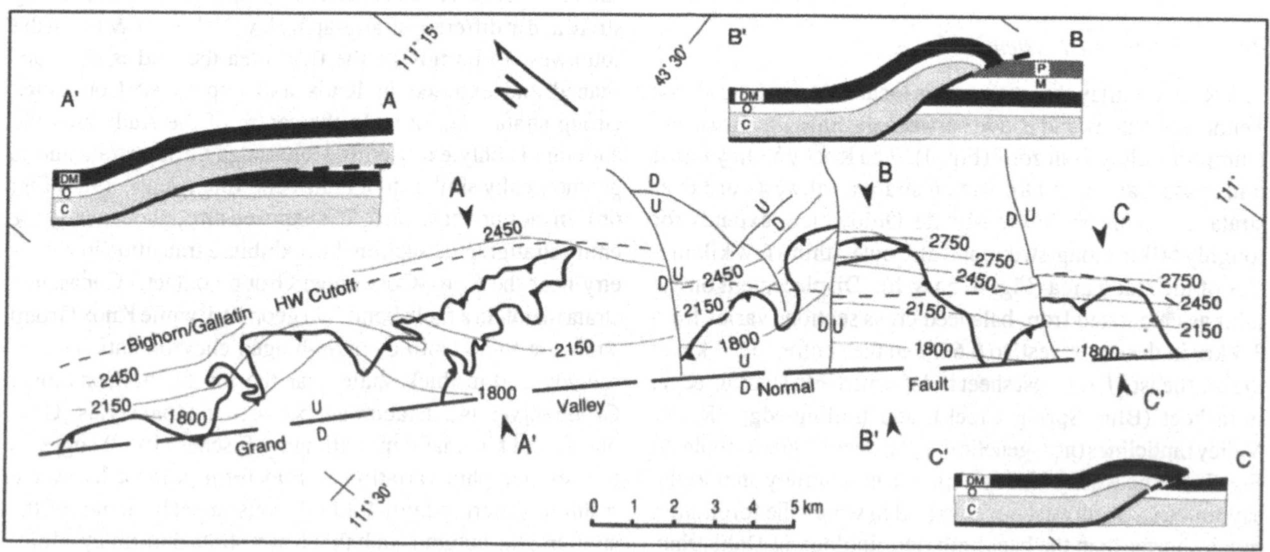

Figure 3. Map of the surface trace of the Baldy thrust combined with a hangingwall cutoff map and structure contour drawn on the thrust surface. Contours are in metres. Schematic cross sections A—A' through C—C' are intended to show generalized hanging wall and footwall geometries only. Sections are not to scale, but relative thicknesses of units are correct. Note the decrease in thrust displacement from section A—A' to section C—C', the increase in fault dip towards the southeast and the along-strike variation of thrust geometry in Devonian-Mississippian strata (black layer). (Modified from Woodward, 1986).

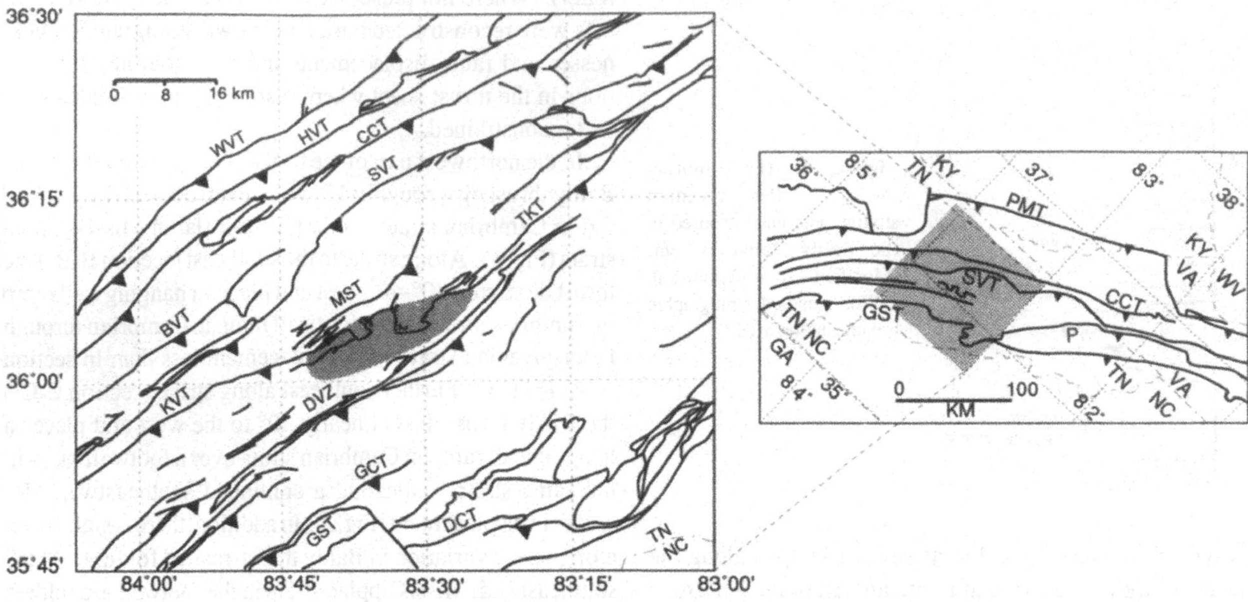

Figure 4. Maps depicting principal faults of the central east Tennessee Valley and Ridge province. Symbols: BVT= Beaver Valley thrust, CCT= Copper Creek thrust, DCT= Dunn Creek thrust, DVZ= Dumplin Valley fault zone, GCT= Guess Creek thrust, GST= Great Smoky thrust, HVT= Hunter Valley thrust, KVT= Knoxville thrust, MST= Mill Spring thrust, P= Pulaski thrust, SVT= Saltville thrust, TKT= Town Knobs thrust, WVT= Wallen Valley thrust. Rocky Valley thrust system is shaded on larger-scale map. (Modified from Woodward *et. al* 1988).

section represents an earlier stage of deformation (relative to the other sections), and that all 'later' (i.e. more internal) geometries are the product of progressive deformation of a structure like that depicted in section C-C'. In this case, an early-formed footwall ramp through Devonian-Mississippian and Mississippian strata (sections C-C' and B-B') must change cutoff angle through time and eventually be transformed into a footwall flat (section A-A'). Similar aged footwall strata must be folded and unfolded during evolution of the thrust system.

Rocky Valley thrust system

The Rocky Valley thrust system is located in the central east Tennessee Valley and Ridge between the Saltville thrust and Dumplin Valley fault zone (Fig. 4). The Rocky Valley thrust transports Cambrian-Ordovician strata northwestward over strata as young as lower Middle Ordovician, extends for roughly 50 km along-strike, and dies out within a few kilometres of the study area (Figs 4, 5a & b). Displacement on the fault as calculated from balanced cross sections varies from 3.4 km in the southwest, to 6.6 km in the centre, to 5.3 km in the northeast. The thrust sheet is deformed into leading-edge, intrasheet (Blue Spring Creek), and trailing-edge (Rocky Valley) anticlines (non-genetic use of Boyer 1986 terminology; Fig. 5c). The Rocky Valley anticline is generally an upright, asymmetric, northwest-verging fold in which the forelimb is much shorter than the backlimb (terminology of Dahlstrom, 1970). Interlimb angles of this fold vary from 160° in the northeast to 80° in the southwest, where the anticline is locally overturned. The leading-edge fold is generally an asymmetric, northwest-verging fold in which the forelimb is overturned to the south. Interlimb angles of this fold average near 100°, but may be as small as 50°. The Rocky Valley anticline, well developed throughout the thrust sheet, and the leading-edge anticline, well developed only in the centre of the sheet, merge northwestward to form one fold. The Blue Spring Creek anticline is interpreted as the product of later, local out-of-sequence thrusting which deforms the southwestern portion of the Rocky Valley thrust sheet (Fischer 1989).

The Rocky Valley anticline overlies the crest of a footwall ramp through Cambrian-Ordovician strata of the Conasauga and Knox Groups, and exhibits varying geometry both along strike and at different stratigraphic levels (Figs 5c & 6). In the southwestern portion of the field area the fold is chevron-shaped and exposed in Knox and uppermost Conasauga Group strata (Fig. 6a). In the centre of the study area the anticline is only exposed in Conasauga Group strata and is geometrically similar to a fault-bend fold (Suppe 1983; Fig. 6b). In the northeast the fold is exposed throughout nearly the entire stratigraphic section, but exhibits a transition in geometry near the Knox-Conasauga Group contact. Conasauga strata maintain a fault-bend fold geometry while Knox Group strata are folded into a single-hinged chevron anticline. A roughly 150 m thick shale near the top of the Conasauga Group serves as a detachment between the fold styles. Geometric and kinematic modelling by Fischer (1989) suggests this stratigraphic variation in fold form is most likely the result of differing fault-fold relations in each section of the stratigraphic column such that Knox strata deform by a fold-first (Dahlstrom 1970), fault-propagation fold (Suppe 1985) mechanism, while Conasauga strata concurrently deform by a fault-first (Dahlstrom 1970), fault-bend fold mechanism. This model explains the transition in exposed fold geometry between the central and southwest portions of the study area

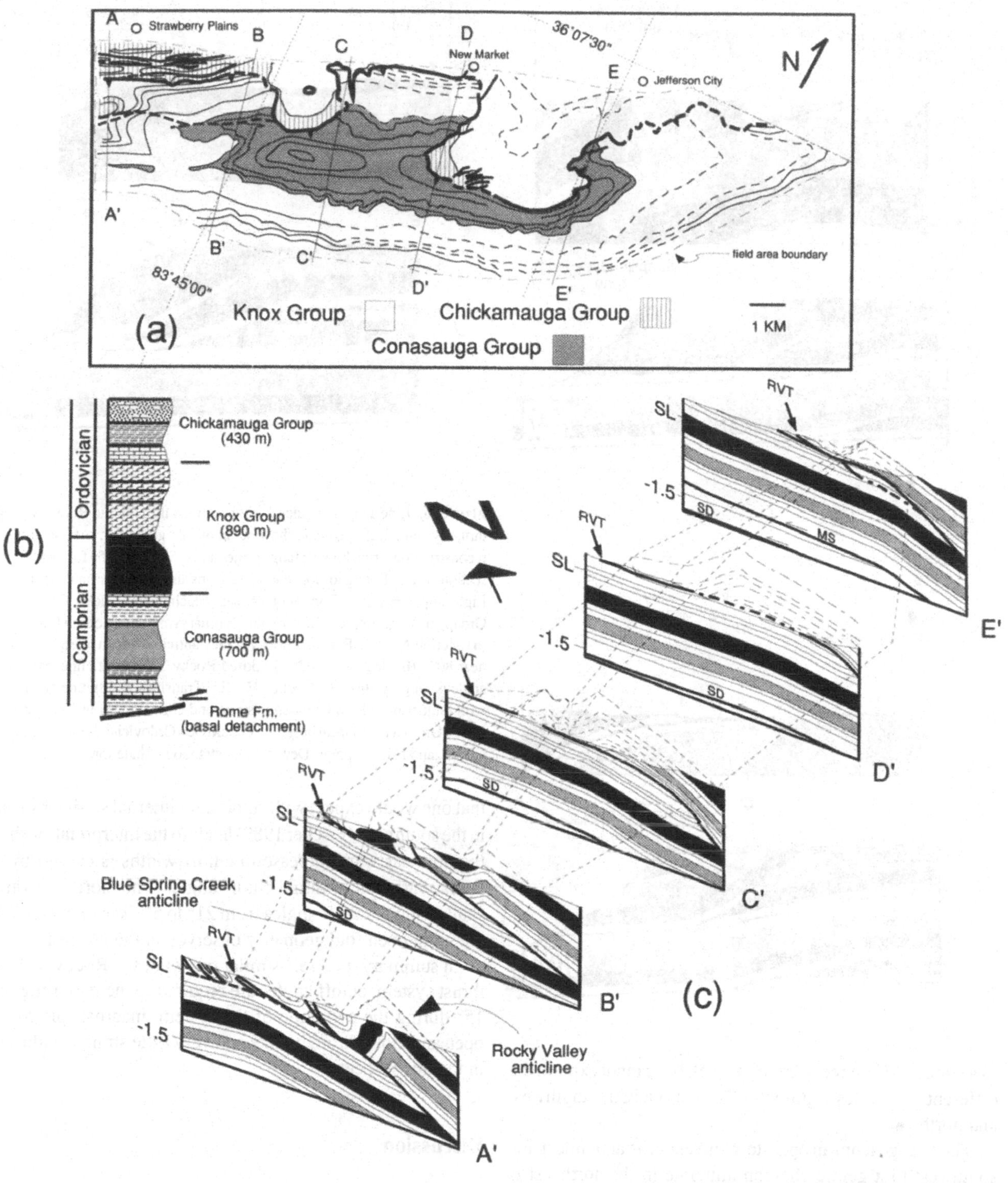

Figure 5. (a) Generalized geological map of the Rocky Valley thrust system in the vicinity of New Market, Tennessee. Rocky Valley thrust shown with teeth on hangingwall. Other faults in the area are moderate to low angle reverse faults of lesser displacement. (b) Stratigraphic column depicting units exposed in the Rocky Valley thrust system. Thicknesses of units are averages. Standard lithological symbols are used. Note that the entire Chickamauga Group is not exposed in the field area, and that the Rome Formation is never exposed in the field area. Groups correspond roughly to incompetent (Conasauga and Chickamauga Groups) and competent (Knox Group) structural lithic units (Woodward & Rutherford, 1989). (c) Block diagram of the Rocky Valley anticline showing along-strike variations in fold geometry. Vertical scale is in kilometres below sea level (SL). Horizontal scale along section lines equal to vertical scale. Distances between sections are not to scale. Black layer is lowermost massive dolostone in Knox Group, stippled layer is thick limestone unit in Conasauga Group (Fig. 5b). See Figure 5a for locations of section lines. MS= Mill Spring thrust, RVT= Rocky Valley thrust, SD= Saltville detachment.

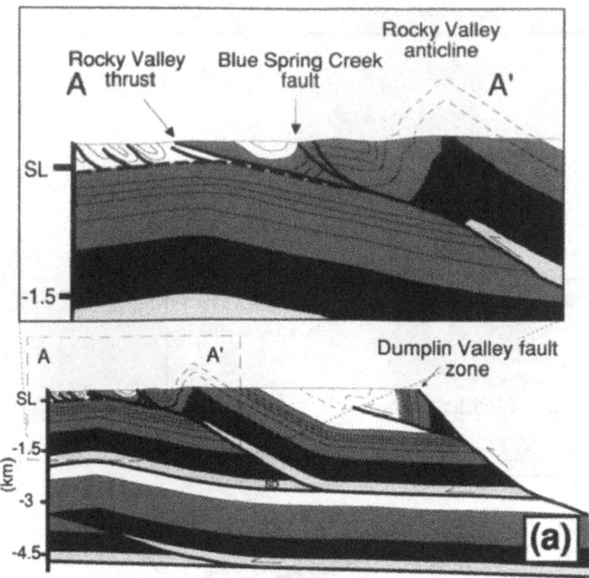

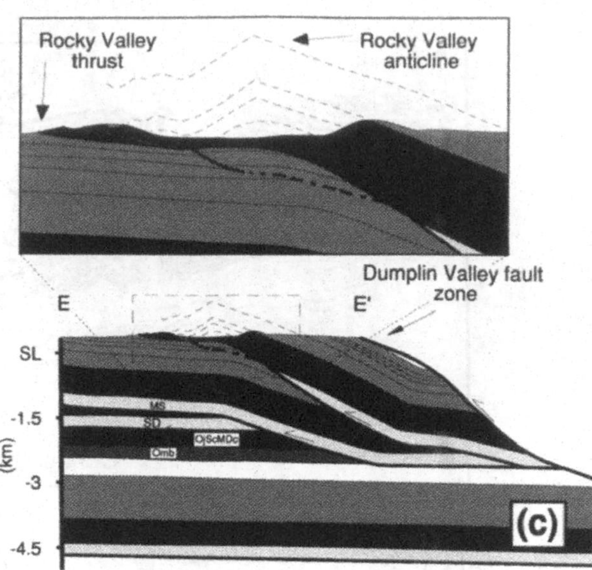

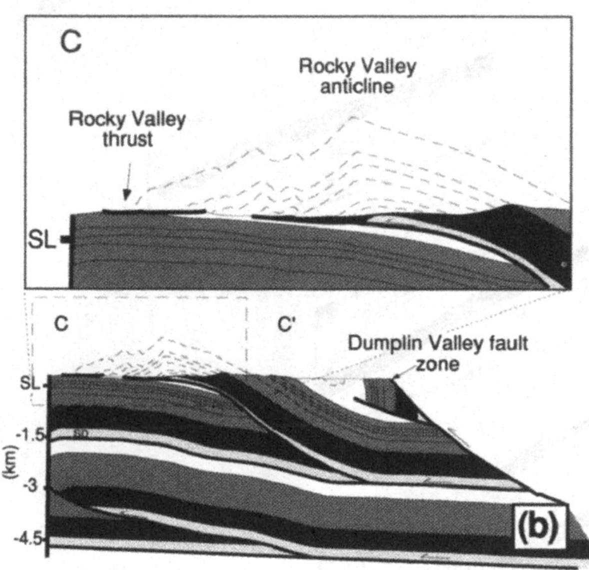

Figure 6. Line-length balanced cross sections through the Rocky Valley thrust system. Enlargements show detail of fold geometry. Above ground reconstructions from down-plunge projections. See Figure 5a for location of section lines. Enlarged portions of sections used to construct Figure 5c. Light stipple= Rome Formation (basal detachment), black= Conasauga Group, medium stipple= Knox Group. Thrust symbols same as in Figure 5c. (**a**) Section A—A'. Exposed Rocky Valley anticline has an interlimb angle near 80°. (**b**) Section C—C'. Exposed Rocky Valley anticline interlimb angle is roughly 160°. (**c**) Section E—E'. Transitional anticline geometry with fault-bend fold in Conasauga strata and chevron fold in Knox strata. Omb=Ordovician Martinsburg Fm. OjScMDc=Ordovician Juniata, Silurian Clinch, and Mississippian-Devonian Chattanooga Shale formations.

as a simple difference in erosion level, but cannot explain the different fold styles exposed in Knox strata in the southwest and northeast.

The change from an open-to-close chevron anticline in the southwest to a gentle chevron anticline in the northeast is attributed to a variation in thrust geometry. It can be seen from Figure 7 that the width of hangingwall cutoffs increases toward the northeast. Given that cutoff width is a function of the initial angle at which the fault cuts through strata and the amount of internal strain and shear induced in the thrust sheet during transport, there are two possible explanations for the observed variation in cutoff widths: either the fault cut through strata at a much shallower angle in the northeast, or strain and shear in the southwestern portions of the thrust sheet have reduced cutoff widths there. The lack of well-developed interbed shear, extensive mesoscopic faulting, penetrative cleavage, mesoscopic folding or stylolitization that one would expect with significant internal strain or shear in the southwest (Fischer 1989) leads to the interpretation that the northeastward increase in cutoff widths is caused by a lateral variation in initial cutoff angle. This northeastward shallowing of cutoff angles from 21° to 6°, is responsible for the more open fold geometry observed in the northeast.

In summary, in a self-similar model of the Rocky Valley thrust system, cutoff angles are required to increase roughly 15° during the evolution of the system, macroscopic folds open and close through time, and very little strain is induced in the thrust sheet.

Discussion

Along-strike variations in macroscopic structural geometry are nearly ubiquitous in the thrust sheets which comprise foreland fold and thrust belts (Rodgers 1953; Dahlstrom 1970; Harris 1970; Jacobeen & Kanes 1974, 1975; King & Yielding 1984; Kulander & Dean 1986; Woodward 1986; Bartholomew 1987; Mitra 1988; Evans 1989; Woodward 1990). The self-similar model for the geometric evolution of thrust systems explains along-strike changes in thrust system geometry as the product of the progressive deformation of structures which form in a thrust-tip ductile bead and is based on two assumptions: (1) displacement decreases from the interior of a thrust sheet toward the lateral tips, and (2) tip folding precedes thrusting. This paper points out, however, that although tip folds are often presumed to be the most common physical manifestation of ductile bead strain (Elliott 1976; Williams 1980; Williams & Chapman 1983; Boyer 1986), other types of deformation may also accommodate this strain (e.g. distributed internal strain, localized shear zones; Elliott 1976; Chapman & Williams 1985; vertical strike-slip faulting and fracturing; King & Yielding 1984). The observation that tip folds are not the only structures which form to accommodate ductile bead strain suggests a ductile bead may not propagate self-similarly, and that many thrust system geometries may not be explained by the self-similar model. If a ductile bead does not propagate in self-similar fashion, structures currently preserved at lateral thrust terminations only represent the ephemeral last stage of thrusting (Knipe 1989; Woodward 1990), and should have no kinematic relation to structures in the more internal parts of the system.

A structural geometry observed anywhere in a thrust system is most likely the product of the kinematics of thrusting in that portion of the system. Thrust kinematics are in turn controlled by the extrinsic (e.g. temperature, pressure) and intrinsic (e.g. pore fluid pressure, stratigraphy, structural position) characteristics of the system. In most foreland thrust systems, where extrinsic properties do not vary significantly within the system, stratigraphy is generally accepted as the controlling influence on macroscopic thrust geometry (Rich 1934). Furthermore, thrust sheet geometry is largely controlled by cutoff angle, displacement and overall thrust geometry (Suppe 1983; 1985; Jamison 1987), and is therefore also largely a function of stratigraphy. However, while numerous researchers have noted the influence of stratigraphy on both thrust and thrust sheet geometry (Berger & Johnson 1982; Tippet *et al.* 1985; Woodward 1990), exactly how this influence is exerted in determining which of the two end member thrust sheet geometries (i.e. fault- and fold-first) develops in a particular area is unknown.

In a simplistic sense, the relative rate of brittle versus plastic deformation around a thrust tip is the limiting factor in determining which of the two geometric forms develops (Coward & Potts 1983). Fold-first geometries should develop when significant plastic deformation occurs before failure, whereas fault-first geometries develop in regions where failure occurs before significant plastic deformation accumulates. Williams & Chapman (1983) quantitatively described this relation in terms of a fault slip versus propagation ratio (S/P). If a thrust sheet is transported a long distance over a fault which has propagated a much shorter relative distance, S/P is large and the theoretical strain gradient at the fault tip is similarly quite large. Areas with large S/P values are more likely to form fold-first geometries because more strain can be accommodated by plastic deformation.

In the Rocky Valley thrust system, although different structural geometries are observed throughout the system, different structural lithic units (Woodward & Rutherford 1989) exhibit characteristic fold geometries in various locations in the thrust sheet. Macroscopic folds in competent structural lithic units consistently exhibit fold-first geometries while folds in incompetent units exhibit fault-first geometries (e.g. Coward & Potts 1983; Boyer 1986). These observations suggest competent structural lithic units in the area may cause a locally increased S/P ratio, and thereby inhibit fault propagation while intensifying deformation around the thrust tip. In contrast, the S/P ratio in incompetent units may be small, such that the thrust tip advances relatively rapidly, and there is little time for strain to accumulate at the fault tip. Other field studies (Morley 1987) have similarly shown an increased tendency for thrusts to die out in competent sections of a stratigraphic column, and further suggest these strata may cause a locally increased S/P ratio. This relationship inferred from field studies is also in agreement with the results of numerous theoretical, quantitative studies of the influence of material competency on fracture propagation.

In a situation analogous to the local interaction between a thrust fault tip and strata of different mechanical properties, Dundurs & Mura (1964) and Dundurs (1967) reported that both screw and edge dislocations were 'repelled' by inclusions of a higher rigidity modulus than the medium which contains the dislocation. Rybicki (1989) examined the interaction of two colinear cracks with an intervening high

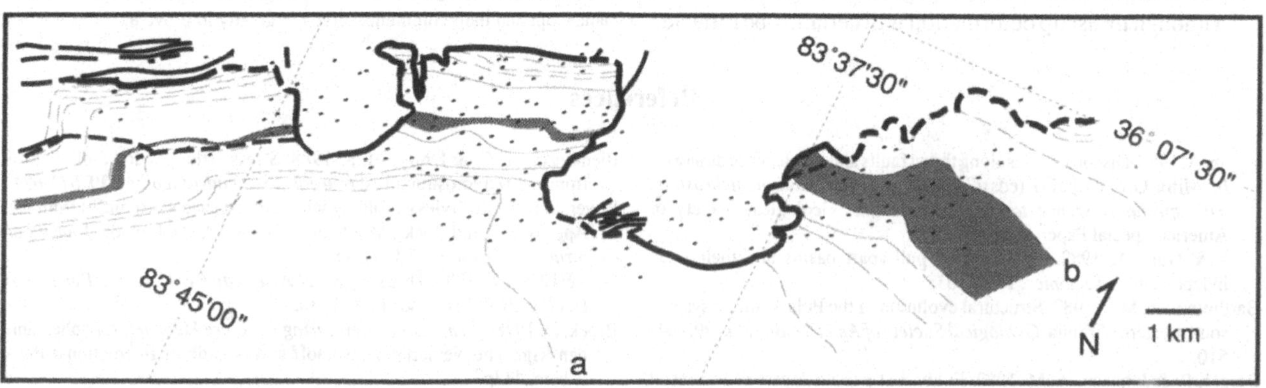

Figure 7. Hangingwall cutoff map for the Rocky Valley thrust. Dots show locations of 218 shallow drill holes used to constrain the map. Cutoff widths increase from southwest to northeast by an average of 200%. Shaded pattern shows the cutoff for the Maynardville Limestone portion (uppermost formation) of the Conasauga Group. Cutoffs: a= Rome Fm.-Conasauga Group, b= Conasauga Group-Knox Group.

rigidity inclusion and showed that the mere presence of the inclusion inhibited the fractures from propagating toward one another. In addition, the stress induced in the inclusion by the cracks is higher than the corresponding stress in a homogeneous medium, and this induced stress increases with increasing rigidity of the inclusion. The fracture mechanics study of Lemiszki & Landes (1989) reported that rocks of higher competency will shorten (by folding) significantly before a mode III fracture will propagate through the hinge of the fold, while in less competent strata, the fracture will propagate before significant fold shortening occurs. These studies further suggest competent strata may cause a locally increased S/P ratio and an increased tendency to form fold-first geometries.

Although the spatial variation in stratigraphy appears to have been the dominant influence on the geometric evolution of the Rocky Valley thrust system, other intrinsic properties certainly played a role. Because the self-similar model pre-determines many fundamental characteristics of thrust system geometry and ignores any relation between the intrinsic properties of the system and thrust kinematics, interpretations based on the model are severely limited. Instead of the self-similar model, we believe it is more beneficial to examine the variation in the intrinsic properties of a thrust system and how these variations affect thrust kinematics.

Conclusions

The self-similar model for the geometric evolution of thrust systems uses an 'Eulerian' approach and is based on the assumption that the ductile bead of deformation present at a thrust tip line propagates as a tip fold throughout the active history of the thrust. Since structures observed at lateral tips should therefore represent the deformation associated with the earliest stages of thrusting, this assumption suggests the geometry of structures observed in the interior part of the thrust sheet can be explained by the progressive deformation of lateral-tip folds. Because neither the Baldy nor Rocky Valley thrust system geometry can be explained in this manner, it is suggested (for these two well-exposed cases) that a ductile bead does not propagate self-similarly.

Although thrust tip deformation will certainly be reflected in the overall structural style of a thrust sheet, the spatio-temporal variation in the style and geometry of ductile bead structures suggests there is little value in explaining the geometry of a thrust sheet by the progressive deformation of currently observed lateral tip structures. Consequently, this paper questions the usefulness of the self-similar model, and to a degree, the 'Eulerian' approach upon which it is based, as an untested general assumption which has not been adequately documented. At one scale of frontier investigation the model was a useful advance over all previous (generally non-viable, inadmissible and unbalanceable) approaches. However, as more data become available, modern investigations of thrust kinematics cannot accept *a priori* the self-similar approach without much greater scrutiny.

Thrust system geometry is most likely determined by the kinematics of thrusting which occurred throughout the region. The geometry of the Rocky Valley thrust system is explained by a stratigraphic variation in the mode of thrust propagation and a lateral variation in cutoff angle. Thrust kinematics in this system appear to have been most strongly influenced by local stratigraphic (facies) and structural (competency) variations. Spatial variations in thrust system geometry do not necessarily reflect the evolution of the system, but may instead reflect a spatial or spatio-temporal variation in the intrinsic properties of the system. It is this variation of intrinsic properties which ultimately determines thrust kinematics of thrusting and hence the geometric evolution of thrust systems.

This paper represents a portion of Fischer's Master's thesis at the University of Tennessee. R. E. Williams, K. R. Walker and W. M. Dunne provided helpful insight and comments throughout the completion of this thesis. R. D. Hatcher Jr. provided advice, editorial comments, and unparalleled field assistance throughout Fischer's stay at Tennessee. Drill data from the Rocky Valley area were provided by the Tennessee Division of Geology, American Smelting and Refining Company, and New Jersey Zinc Mining Company. The personal assistance of Bob Fulweiler and Vance Green is gratefully acknowledged in obtaining these data. Seismic data through the Rocky Valley area were provided by Amoco with the help of Larry Knox. This project was funded by grants from the American Association of Petroleum Geologists, Southeastern Section of the Geological Society of America, Professor's Honor Fund of the University of Tennessee, and a University of Tennessee Center of Excellence fellowship awarded to Fischer. D. Fisher, P. Lemiszki, S. Marshak and S. Wojtal provided careful review of this manuscript and their comments served to greatly improve it.

References

Aydin, A. 1988. Discontinuities along thrust faults and the cleavage duplexes. *In:* Mitra, G. & Wojtal, S. (eds) *Geometries and mechanisms of thrusting, with special reference to the Appalachians.* Geological Society of America Special Paper, **222**, 223-232.

—— & Nur., A. 1982. Evolution of pull-apart basins and their scale independence. *Tectonics*, **1**, 91-105.

Bartholomew, M. J. 1987. Structural evolution of the Pulaski thrust system, southwestern Virginia. *Geological Society of America Bulletin*, **99**, 491-510.

Berger, P. & Johnson, A. M. 1982. Folding of passive layers and forms of minor structures near terminations of blind thrust faults - application to the central Appalachian blind thrust. *Journal of Structural Geology*, **4**, 343-354.

Bilby, B. A. & Eshelby, J. D. 1968. Dislocations and the theory of fracture. *In:* Leibowitz, H. (ed) *Fracture, an Advanced Treatise*, **1**, 99-182.

Blenkinsop, T. G. & Drury, M. R. 1988. Stress estimates and fault history from quartz microstructures. *Journal of Structural Geology*, **10**, 673-684.

Boyer, S. E. 1986. Styles of folding within thrust sheets: examples from the Appalachian and Rocky Mountains of the USA and Canada. *Journal of Structural Geology*, **3/4**, 325-339.

—— & Elliott, D. 1982. Thrust Systems. *American Association of Petroleum Geologists Bulletin*, **86**, 1196-1230.

Broek, D. 1978. *Elementary Engineering Fracture Mechanics.* Alphen ann den Rijn, The Netherlands, Sijthoff and Noordhoff International Publishers, 437p.

Buxtorf, A. 1916. Prognosen und Befunde beim Hauensteinbasis und Grenchenberg tunnel und die Bedeutung der letzteren für die Geologie des Juragebirges. *Verhandlungen des Naturforschunden die Gesellschaft* Basel, **27**, 185-254.

Chapman, T. J. & Williams, G. D. 1985. Strains developed in the hanging walls of thrusts due to their slip-propagation rate: a dislocation model: Reply. *Journal of Structural Geology*, **7**, 759-762.

Coward, M. P. & Potts, G. J. 1983. Complex strain patterns developed at the frontal and lateral tips to shear zones and thrust zones. *Journal of Structural Geology*, **5**, 383-399.

Dahlstrom, C.D.A. 1970. Structural geology in the eastern margin of the Canadian Rocky Mountains. *Bulletin of Canadian Petroleum Geology*, **18**, 332-406.

Delaney, P.T., Pollard, D. D., Ziony, J. T., & McKee, E. H. 1986. Field relations between dikes and joints: Emplacement processes and paleostress analysis. *Journal of Geophysical Research*, **91**, 4,920-4,938.

Dickenson, W. R. & Snyder, W. S. 1979. Geometry of triple junctions related to San Andreas transform. *Journal of Geophysical Research*, **84**, 561-572.

Diegel, F. A. 1986. Topological constraints on imbricate thrust networks, examples from the Mountain City window, Tennessee, USA. *Journal of Structural Geology*, **8**, 269-280.

Dundurs, J. 1967. On the interaction of a screw dislocation with inhomogeneities. *In*: Eringen, A.C. (ed) *Recent Advances in Engineering Sciences*, **2**, 223-233.

—— & Mura, T. 1964. Interaction between an edge dislocation and a circular inclusion. *Journal of the Mechanics and Physics of Solids*, **12**, 177-189.

Elliott, D. 1976. The energy balance and deformation mechanisms of thrust sheets. *Philosophical Transactions of the Royal Society of London*, **283**, 289-312.

Eshelby, J. D. 1973. Dislocation theory for geophysical applications. *Philosophical Transactions of the Royal Society of London*, **274**, 331-338.

Evans, M. 1989. The structural geometry and evolution of foreland thrust systems, northern Virginia. *Geological Society of America Bulletin*, **101**, 339-354.

Fischer, M. P. 1989. Structural geometry of the Rocky Valley thrust system in the vicinity of New Market, Tennessee (MS thesis): Knoxville, University of Tennessee, 345p.

Harris, L. D. 1970. Details of thin-skinned tectonics in parts of Valley and Ridge and Cumberland Plateau provinces of the southern Appalachians. *In*: Fisher, G. W., Pettijohn, F. J., Reed, J. C., Jr. & Weaver, K. N. (eds) *Studies of Appalachian Geology: Central and Southern*. New York, 161-173.

Jacobeen, F. & Kanes, W. H. 1975. Structure of Broadtop synclinorium, Wills Mountain anticlinorium, and Alleghany frontal zone. *American Association of Petroleum Geologists Bulletin*, **59**, 1136-1150.

—— & —— 1974. Structure of Broadtop synclinorium and its implications for Appalachian structural style. *American Association of Petroleum Geologists Bulletin*, **58**, 362-375.

Jamison, W. R. 1987. Geometric analysis of fold development in overthrust terranes. *Journal of Structural Geology*, **9**, 207-219.

King, G.C.P. & Yielding, G. 1984. The evolution of a thrust fault system: Processes of rupture initiation, propagation and termination in the 1980 El Asnam (Algeria) earthquake. *Geophysical Journal of the Royal Astronomical Society*, **77**, 915-933.

Knipe, R. J. 1989. Deformation mechanisms - recognition from natural tectonites. *Journal of Structural Geology*, **11**, 127-146.

Kulander, B. R. & Dean, S. L. 1986. Structure and tectonics of the central and southern Appalachian Valley and Ridge and Plateau provinces, West Virginia and Virginia. *American Association of Petroleum Geologists Bulletin*, **70**, 1674-1684.

Laubscher, H. P. 1976. Fold development in the Jura. *Tectonophysics*, **37**, 337-362.

Lemiszki, P. J. & Landes, J. D. 1989. Mixed-mode fracture propagation induced by folding ahead of a fault: Analogy to fault-propagation folds. *Geological Society of America Abstracts with programs*, **21**, 136.

Means, W.D. 1976. *Stress and Strain*. New York, Springer-Verlag, 339p.

Mitra, S. 1988. Three dimensional geometry and kinematic evolution of the Pine Mountain thrust system, southern Appalachians. *Geological Society of America Bulletin*, **100**, 72-95.

—— 1987. Regional variations in deformation mechanisms and structural styles in the central Appalachian orogenic belt. *Geological Society of America Bulletin*, **98**, 569-590.

Morley, C. K. 1987. Lateral and vertical changes of deformation style in the Osen-Røa thrust sheet, Oslo region. *Journal of Structural Geology*, **9**, 331-343.

Pfiffner, O. A. 1985. Displacements along thrust faults. *Eclogae Geologae Helvetiae*, **78**, 313-333.

Price, R. A. 1988. The mechanical paradox of large overthrusts. *Geological Society of America Bulletin*, **100**, 1098-1908.

Rich, J. L. 1934. Mechanics of low-angle overthrust faulting illustrated by Cumberland thrust block, Virginia, Kentucky and Tennessee. *American Association of Petroleum Geologists Bulletin*, **18**, 1584-1596.

Rodgers, J. 1953. Geologic map of east Tennessee with explanatory text. *Tennessee Department of Conservation, Division of Geology Bulletin*, **58**, parts I and II, 168p.

Rybicki, K. R. 1989. On fault slip zones in the external shear-field separated by a high rigidity asperity (barrier) - inplane and antiplane strain models. *Geophysical Journal*, **96**, 101-115.

Segall, P. & Pollard, D. D. 1980. Mechanics of discontinuous faults. *Journal of Geophysical Research*, **85**, 4,337-4,350.

Spang, J. H. & Brown, S. P. 1981. Dynamic analysis of a small imbricate thrust and related structures, Front Ranges, southern Canadian Rocky Mountains. *In*: McClay, K. R. & Price, N. J. (eds) *Thrust and Nappe Tectonics*. Geological Society of London Special Publication, **9**, 143-149.

Suppe, J. 1985. *Principles of Structural Geology*: Englewood Cliffs, New Jersey, Prentice-Hall Inc., 537p.

—— 1983. Geometry and kinematics of fault-bend folding. *American Journal of Science*, **283**, 684-721.

Tippett, C. R., Jones, P. B. & Frey, F. R. 1985. Strains developed in the hangingwalls of thrusts due to their slip/propagation rate: A dislocation model: Discussion. *Journal of Structural Geology*, **7**, 755-758.

Thompson, R. I. 1981. The nature and significance of large 'blind' thrusts within the northern Rocky Mountains of Canada. *In:* McClay, K. R. & Price, N. J. (eds) *Thrust and Nappe Tectonics*. Geological Society of London Special Publication, **9**, 449-462.

Williams, G. & Chapman, T. 1983. Strains developed in the hanging walls of thrusts due to their slip-propagation rate: A dislocation model. *Journal of Structural Geology*, **5**, 563-571.

Williams, R. E. 1980. Creation and geometry of the thrust fracture. *Geological Society of America Abstracts with programs*, **12**, 549.

Wojtal, S. 1986. Deformation within foreland thrust sheets by populations of minor faults. *Journal of Structural Geology*, **8**, 341-360.

—— & Mitra, G. 1988. Nature of deformation in some fault rocks from Appalachian thrusts. *In:* Mitra, G. & Wojtal S. (eds) *Geometries and mechanisms of thrusting, with special reference to the Appalachians*. Geological Society of America Special Paper, **222**, 17-34.

Woodward, N. B. 1990. Deformation styles and geometric evolution of some Idaho-Wyoming thrust belt structures. *David Elliott Memorial Volume*, The Johns Hopkins University, (in press).

—— 1986. Thrust fault geometry of the Snake River Range, Idaho and Wyoming. *Geological Society of America Bulletin*, **97**, 178-193.

—— & Rutherford, E. Jr. 1989. Structural lithic units in external orogenic zones. *Tectonophysics*, **158**, 247-267.

——, Walker, K. R. & Lutz, C. T. 1988. Relationships between early Paleozoic facies patterns and structural trends in the Saltville thrust family, Tennessee Valley and Ridge, southern Appalachians. *Geological Society of America Bulletin*, **100**, 1758,1769.

Some geometric problems of ramp-flat thrust models

John G. Ramsay

Geologisches Institut, ETH Zentrum, CH-8092 Zürich, Switzerland

Abstract: Various classic models used to relate fold geometry to thrust fault geometry are critically discussed and it is shown how modifications might be made which could provide better fits to geological data. A region on the South coast of Dorset, UK, where well developed small scale thrust faults are developed, is described and the geometric features of the fold and fault geometry related to the previously proposed modified models.

The Appalachian 'Rocky Mountain' thrust model, originally described by Rich (1934) and refined to a high level of sophistication by many subsequent workers (Boyer & Elliott 1982; Suppe 1983), seems to have become a ruling guide to much recent section construction in orogenic zones. In the author's opinion these fault bend- and fault propagation-models are being applied rather indiscriminately, and to situations where they are geologically inappropriate. Thus, published sections of highly water impregnated sediments at the toe of an accretionary wedge (Davis *et al.* 1983; Knipe & Needham 1986) are often geometrically indistinguishable from those of foreland thrust belts developed in lithified rocks of varying competence (Price 1981), and almost identical reconstructions have been applied to highly deformed, folded and regionally metamorphosed rocks from the central parts of orogens and to crystalline basement deformed at deep crustal levels (Coward 1983; Butler 1983).

Although today there is available an enormous amount of excellent seismic data from fault zones and this provides very sound guide lines to certain aspects of thrust and fold geometry, perhaps there is a tendency to fit these data too readily to previously selected geometric models. The fold geometry of the hangingwalls of thrust faults is usually clear and relatively easy to interpret, the data from footwall sectors is often much less clear and should be analysed with circumspection. In the sections below a number of modified kinematic models for the interrelationships of fold and fault geometry will be discussed.

The relationships of folds and faults in thrust zones

Two models are usually proposed to account for the interactive development of folds and faults in thrust systems; the fault bend- and fault propagation-models. In both, folding is essentially a consequence of the progressive development of the fault. The individual layers are deflected and folded as the overlying thrust sheet passes over irregularities in the thrust surface or develop as a fault tip propagates through unfaulted rock. The rheological properties of individual layers play no guiding role in controlling the fold geometry. This assumption, even though convenient for the purposes of computing the interlimb geometry of the folds, is either unrealistic, or at least needs some justification in the particular geological region under investigation. It does seem improbable that rock rheology, which is generally considered to control the initial locations of ramps (where thrusts cut across more competent members of the succession) or flats (generally sub-parallel to the incompetent layers), should play no further role in controlling the subsequent development of the overall structure. In the classic region of the Helvetic nappes of Switzerland rock lithologies do have a considerable influence of the development of the structures in the nappes and fold develop which are not just the result of passive layer bending above a variably inclined thrust surface. Because thrust surfaces cut upwards across the stratigraphy in the overthrust sense, shear distortions in the transported nappe block inevitably lead to layer parallel shortening (Ramsay *et al.* 1983; Ramsay & Huber 1987, p. 377-378). This shortening sets up buckle folds in the more competent layers with initial wavelengths controlled by layer thickness and competence contrast. The amplitudes of these folds depends upon the extent of the internal shear and also upon the inclination of the bedding surfaces to the thrust sheet boundaries (Fig. 1).

In the classic model of fault bend folding the folds are only developed in the hangingwall of the structure (Fig. 2). In natural examples of such structures which can be observed in surface exposures folds may also be found in the footwall. One interpretation of these footwall folds is that they result from the development of other fault bend folds beneath the observed structure (Suppe 1983). In such situations Suppe shows how such features can be used to extrapolate downwards so as to predict the locations of the ramps and flats of the underlying thrusts. However, this explanation might not offer a universal model, especially where a well developed footwall synform is present. Several alternative models should be considered. Heim (1921) developed a model based on many years of field work in the Helvetic nappes that suggested that thrust surfaces developed as a secondary feature of folds, and were the result of the shearing out of the middle limbs of antiform-synform pairs (Fig. 3). Although this theory is not in fashion today, it should be considered as a possibility in regions of strongly developed regular folds, especially where the form of footwall synforms mirrors that of hangingwall antiforms.

Another well known interpretation to account for folds in the footwall of a fault is that localized shear displacements

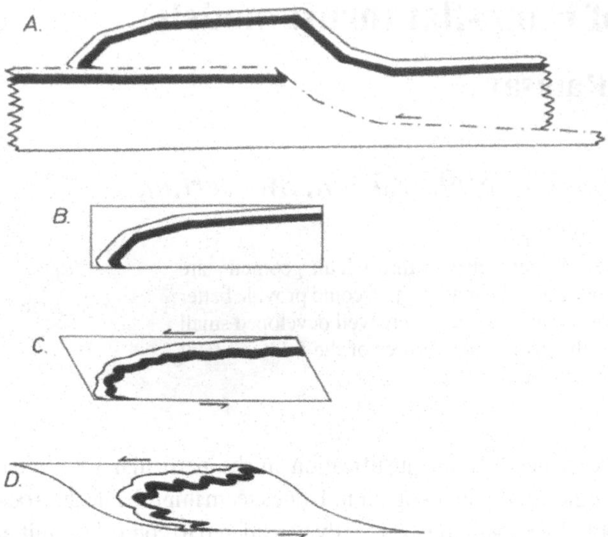

Figure 1. A. Model of a fault bend developed over a ramp of varying inclination. Local shearing near the thrust plane can lead to overturning of the hangingwall near the fault surface, and develop a footwall syncline. B indicates a part of this model and C illustrates how this part can be modified by homogeneous simple shear with development of buckle folds in the competent black layer. D illustrates how heterogeneous simple shear could further modify the fold pattern (after Ramsay & Huber 1987).

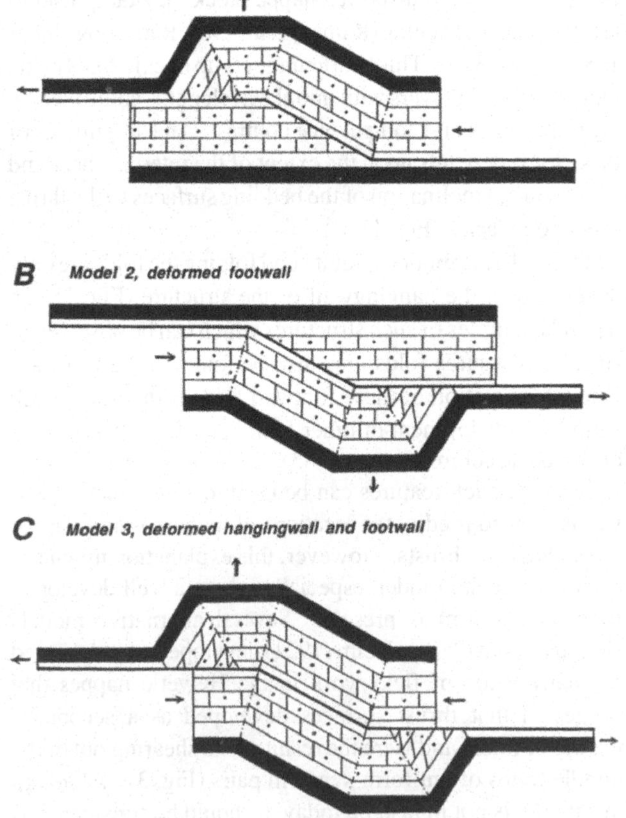

Figure 2. (a) Model 1. The classic fold-bend fault model with actively deforming hanging wall (after Suppe 1983). Models 2 (b) and 3 (c) are geometric equivalents with active footwall or both hanging and footwalls respectively.

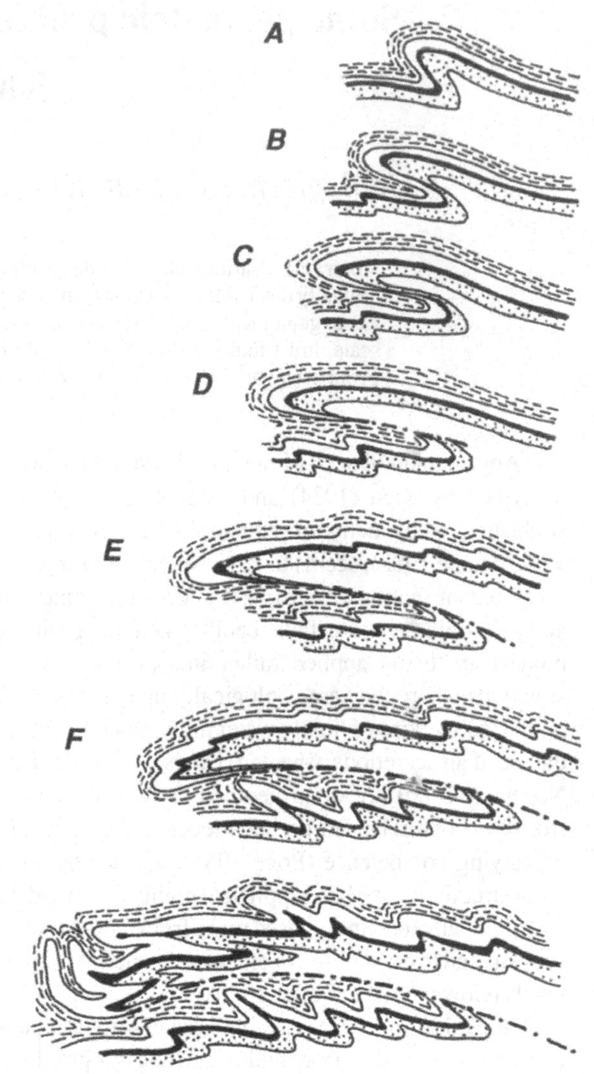

Figure 3. The development of thrusts as a result of shearing of the common limbs between an antiform-synform pair (After Heim 1921). The stages recognized by Heim were: A. Inclined fold with middle limb thickness not reduced, B Recumbent fold with reduced thickness of beds in the middle limb, C Fold with strongly reduced middle limb, D Fold with middle limb broken by a fault, E Thrust fold with the start of secondary folding in the normal limb, F Overthrust fold nappe, G Plunging nappe with frontal fold digitations and frontal blocking.

might occur in the rocks directly beneath the thrust surface as a result of strong shear strains in the proximity of the thrust surface (see Fig. 23.33A in Ramsay & Huber 1987).

Other possibilities of modifying the usual fault-bend fold model are shown in Figures 2B and 2C. Published versions of the fault bend model show that the structures have a polarity, that is to say that the two sides of the fault surface behave differently; the passive footwall contrasts with a kinematically active and structurally elevated hanging wall (Fig. 2A). The author has never seen any discussion why this should always be so. It seems to be assumed that the possibilities of a downwards deflection of the footwall into the substratum are limited, presumably because of the volumetric or space constraints arising during such a displace-

Figure 4. Thrust structures seen in a quarry face of limestones and marls at Holderbank (30 km WNW Zürich, Switzerland) with passive hangingwall and deformed footwall (cf Fig. 2B, Model 2).

ment. However, if the rocks above and below the thrust had more or less similar rheologies, and if the ground surface was fairly high above the thrust plane then geometrically there might be other solutions to the displacement field than that shown in Figure 2A. For example it would be possible to take up the same overall shortening between the two sides of the thrust with a passive hangingwall and a folded footwall (Fig. 2B), or it would be possible to take up shortening with mirror like deflections on both sides of the thrust (Fig. 2C). Although these possibilities might shock the advocates of a unique solution to the fault-bend model, nature provides examples of both. Figure 4 shows a 50 m high quarry face in Jurassic limestones (competent) and marls (incompetent) in the folded Jura of Switzerland which seems in perfect accord with the geometry of Figure 2B, and in the later discussion of thrusting at Kimmeridge Bay the model of Figure 2C seems to be predominant (Fig. 13). Why should downward displacements of a fault footwall occur? Such a geometry seems to imply that the fault walls do not 'feel' the effects of gravitational force (because they are embedded deeply in the rock mass, or the overthrust is small relative to the total system size), or that the footwall displacements are unconstrained by rigid undeformable material (e.g. a competent crystalline basement).

Probably the geometric possibilities of these models should be extended when considering the structure of duplexes (Boyer & Elliott 1982; Mitra1986). Again there seems to be no inevitability of always lifting up the duplex roof and, in certain terranes, the possibility of duplexes being formed with opposing sides having mirror symmetry (Fig. 5) should at least be considered.

Another feature of the standard fault bend- and fault propagation-models is the kink-like nature of the hangingwall folds. Natural folds with kink band geometry are well known,

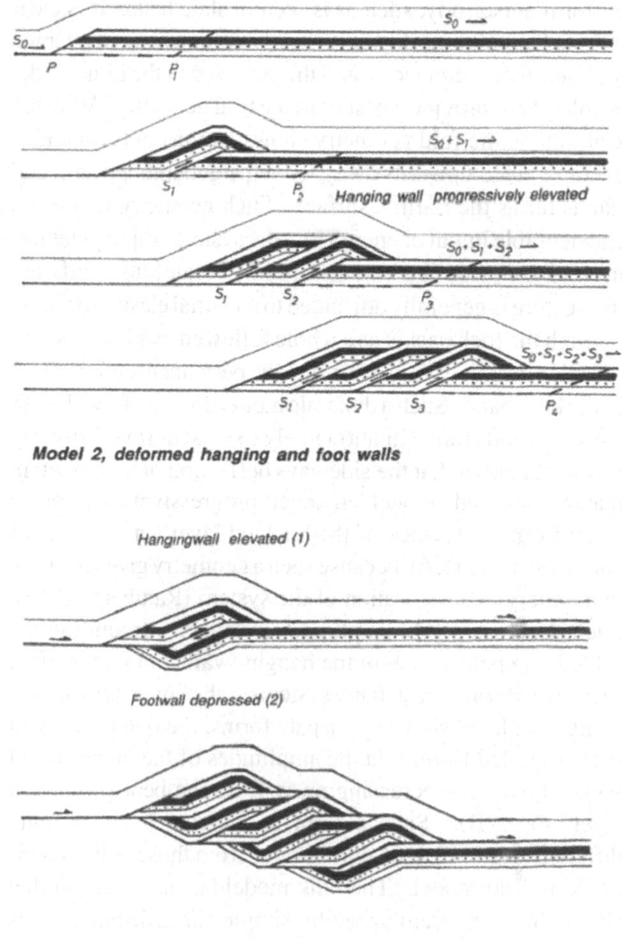

Figure 5. Model 1. Classic development of a duplex structure with active hangingwall. Model 2 shows how the third stage of Model 1 might develop with the development of an active footwall.

Figure 6. Typical active buckle fold developed by shortening of a competent layer (black) in an incompetent rock showing progressively deamplifying folds in the zone of contact strain of the buckle.

especially on a centimetric or decimetric scale. Although large kink folds are frequently mentioned in the literature, the fold geometry is rarely analysed in detail to see if it is really in accord with this model. In fact perfect kink band geometry on a large scale is extremely rare (one excellent example has been described by Collomb & Donzeau 1974). Kink folds develop only in terranes where the rocks are very uniformly bedded with regular thickness periodic alternations of competent and incompetent layers, or where there is a rather uniform anisotropy, such as is seen in slate belts. It is clear that such conditions do not generally pertain in most average rock sediment sequences, and this means that the kink model is unlikely to provide a best fit to the fold geometry. With the kink model the fold geometry shows a uniform form and is unconstrained upwards away from the generating ramp and flat as far as the Earth's surface. Such geometry requires a considerable imput of energy into the system. Long penetrating kink zones are characteristic of small scale kink bands, but this feature is generally attributed to an initial elastic straining through the rock packet as a whole followed by rapid kinking redeploying this elastic strain into the permanent deformation of the kink band. Such a dynamic model does not fit well with the stress and strain situation likely to exist during thrusting. It is well known that the sideways deflection of incompetent material around a buckled layer progressively decreases away from the surface of the buckled layer (in the zone of contact strain, Fig. 6), because such a geometry gives the least work energy configuration of the system (Ramberg 1963). Such decreasing fold amplitudes in layers surrounding the folded competent rock in the hangingwall anticline is often seen in natural thrust fault systems: the inner arcs of the competent layer showing cuspate forms, the outer arcs with more rounded forms and the amplitudes of the incompetent rocks above the decreasing away from the bent competent layer (Fig. 7B). Such a form is associated with a quite different distribution of finite strains from those of the standard kink fold model. The kink model has no layer parallel shortening and strain arises by simple shear displacements predominantly in the kinked sectors of the structure. This model predicts that initially unstrained or slightly strained strata should pass into a kink zone, and as they pass through the sharply defined kink axial plane they immediately develop a finite strain, the value and orientation of which is a function of the change of angle between the flat and ramp sections of the underlying fault. Subsequently this layer on the back limb of the hangingwall anticline passes through another kink axial plane to take on a flat orientation and, in so doing, it becomes unstrained (Elliott 1976; Ramsay & Huber 1987, p. 538). At present little data exist to back up this model, although small scale structural features of cleavage and pressure shadows have been described in the Appalachian fold and thrust belt which fit this plan quite well (Beutner *et al.* 1988). The strain plan of the model of Figure 7B is quite different from that of the kink model and should show closer correspondence with contact strain buckle models (Ramberg 1963; Dieterich 1970; Ramsay 1966, 1976). Future research might be effectively oriented to obtain more detail of the fold geometry, finite strain and incremental strain history of well exposed hangingwall anticlines.

Special problems arise when investigating the geometric features of thrust zones located in the deeper, more internal parts of orogenic belts. In such regions the thin-skin concept of an easily deformable cover overlying a rigid, undeformable basement are untenable. The basement rocks are ductile or

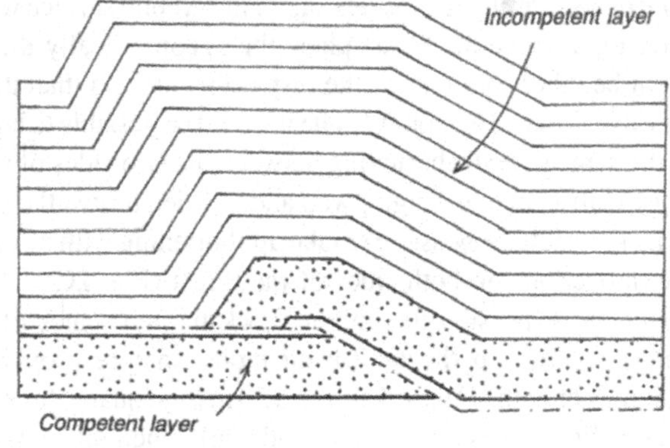

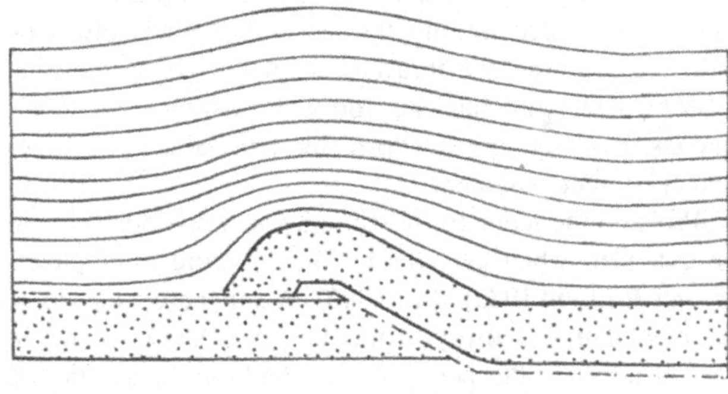

Figure 7. Comparison of the classic fault bend fold model with kink folds in the hangingwall and a possible modification where the competent layer shows buckle fold geometry with deamplifying folds in the zone of the contact strain.

semi-ductile and become involved in the structural edifice of the cover strata, and competence contrasts between rocks of differing composition may be modified as a result of mineralogical changes during metamorphism (Ramsay 1983). The models used to describe near surface structural features in the crust are mostly constrained by plane strain with no material passing in or out of a specific section plane. Such an outlook is perfectly acceptable in this environment, but it would be unwise to use such a model without very careful thought in internal orogenic zones. Metamorphic rock fabrics, backed up with strain measurements, often indicate that flattening or constrictional types of strain ellipsoids predominate, and it is often surprising how such different types of strain can vary from place to place. The total displacement pattern in these regions is commonly built up by the superposition of several phases of deformation, superposed fold geometry and multiple cleavages clearly evidence this. In such situations, even where the individual deformations are of the plane strain type, their combination can only be a plane strain under exceptional circumstances. Successive and differently oriented shear displacements on individual fault planes can bring together rocks which were originally far separated and the footwall and hangingwall of a fault in any one cross section cannot be simply matched. Seismic sections are a great help in interpreting structural geometry in foreland regions but they have severe limitations in basement rocks because the methods only allow us to identify reflectors with low inclinations, steeply inclined reflectors give rise to transparent data windows. Even in foreland regions seismic records still require interpretation and it should not be forgotten that, without borehole data, the reflection data can sometimes be made to fit several kinematic models equally well. There is a tendency today for some geologists to browbeat their colleagues into acceptance of a particular cross-section construction, being of the opinion that, just because they possess seismic data, their views represent perfect truth.

Thrust tectonics at Kimmeridge Bay, Dorset

In this section of the paper the geometric features of extremely well exposed sets of conjugate thrusts seen in dolostones and shales exposed in the Upper Jurassic Kimmeridgian of the type locality at Kimmeridge Bay in South Dorset are described.

The sediments consist of regularly bedded dolostones of uniform thickness separated by dark brown or black organic shales (House 1969). On the west side of Kimmeridge Bay the foreshore is dominated by the wide exposure of a very gently inclined dolostone layer of about 50 cm thickness, known as the Flats Stone Band, situated near the core of a regionally important anticline, the Purbeck Anticline (Donovan & Stride 1961; House 1969). This structure is overturned to the north, having a steeply or vertically inclined northern limb and a gently southward dipping southern limb, and is generally interpreted as resulting from the reactivation of underlying pre-Albian extensional fault planes as compressional thrusts during the late Tertiary. The Flats Stone Band dolostone shows the development of complicated interferring thrust systems (Bellamy 1977; Leddra *et al.* 1987).

The dolostone shows the development of four distinct types of deformation structures which, from their intersection and interference relationships, appear to be formed at different times during the structural evolution of these rocks:
1. Low angle synsedimentary normal faults with displacements of a few centimetres. The fault surfaces are injected by the

Figure 8. Thrust surface in the dolostone of the Flats Stone Band, Kimmeridge Bay showing overlapping white calcite fibres on the thrust surface.

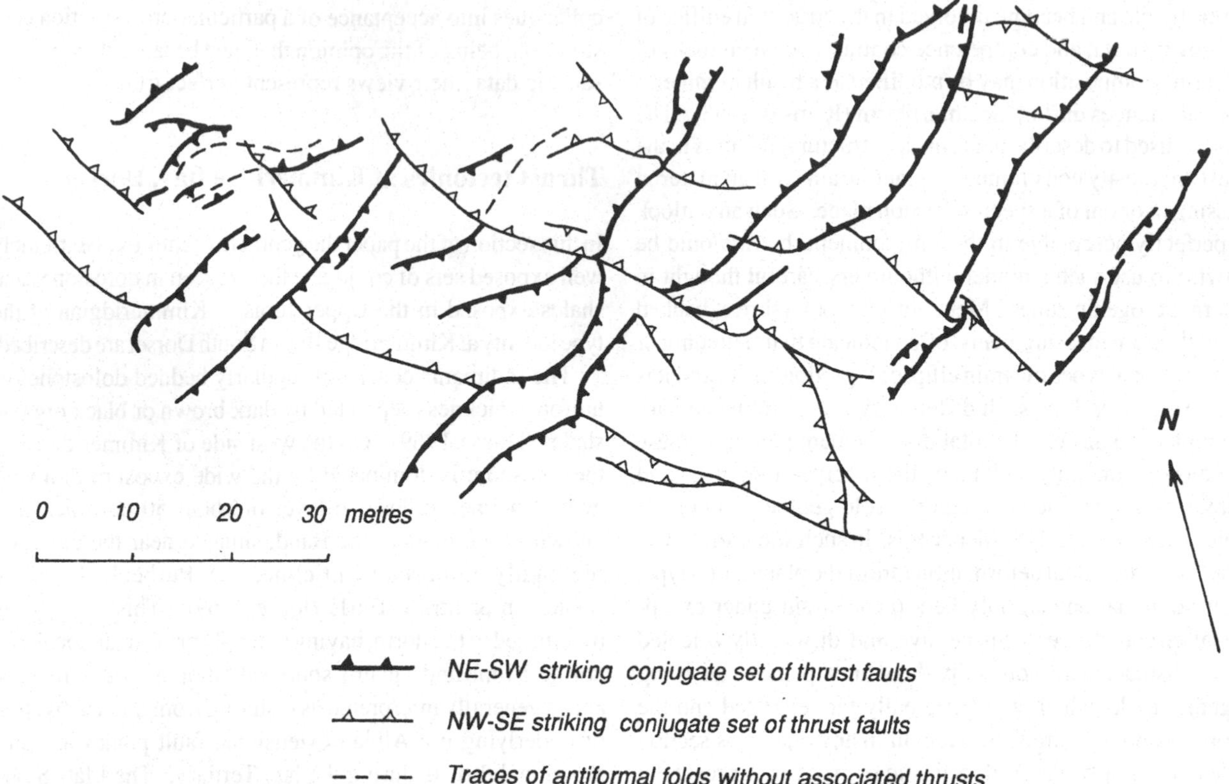

Figure 9. Geological map of part of the surface of the Flats Stone Band, Kimmeridge Bay showing the thrust faults.

surrounding shaly material.

2. Low angle thrusts forming two groups of conjugate sets with displacements varying from zero to around 50 cm. The fault surfaces are quite distinct from those of the synsedimentary faults, being coated with overlapping fibres of white calcite (Fig. 8). The overlap sense and lengths of these fibres is clearly in accord with the thrust displacements indicated by the displacements on the dolostone-shale contacts (Durney & Ramsay 1973).

3. Conjugate normal faults with displacements of up to a few metres with general N-S or NNE-SSW strikes dipping east or west at angles of about 50°. These clearly cut and displace the thrust structures (2).

4. Regular sub-vertical planar or curved sets of joint surfaces cut through all other structures. At least three major trends are seen (striking 80, 0 and 150°), and all show plumose and parabolic fracture surface markings indicative of generation by rapid brittle fracture mechanisms.

The geometric features of the conjugate thrusts (3) are now discussed further. On the exposed beach platform each thrust shows a local uplift of the hangingwall above the regional level of the upper dolostone surface very similar to that of the classic fault bed fold model, although the outer arcs of the hangingwall anticlines are more rounded than are the inner arcs, conforming better to the model shown in Figure 7. Individual antiforms can be traced along their hinges and change amplitude and form in accordance with the extent of the displacements on the underlying thrust surface. They can deamplify to zero, or they can meet and merge with a differently oriented fault bend antiform (Fig. 9). The thrusts, movement fibres and their hangingwall antiforms can be grouped into two differently oriented conjugate sets with average strikes of 110 and 45° respectively (Figs 9, 10 & 11). At some localities the triangular shaped block between two parallel conjugate thrusts is lifted up as a type of horst mass (Fig. 12 and Leddra *et al.* 1987, Figs 3D & 4D). The positions where differently oriented hangingwall antiforms and their underlying thrusts meet show geometric relations suggesting

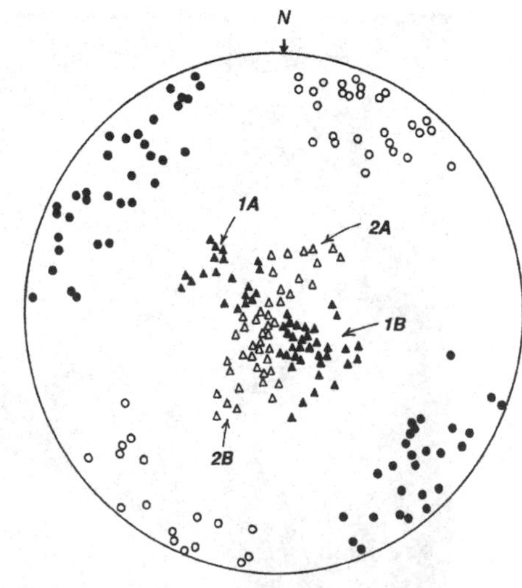

Figure 10. Equal area projection illustrating the orientations of thrust planes and calcite fibre orientations from Figure 9. The faults occur in conjugate sets with poles situated in four main groups: 1A, 1B and 2A, 2B.

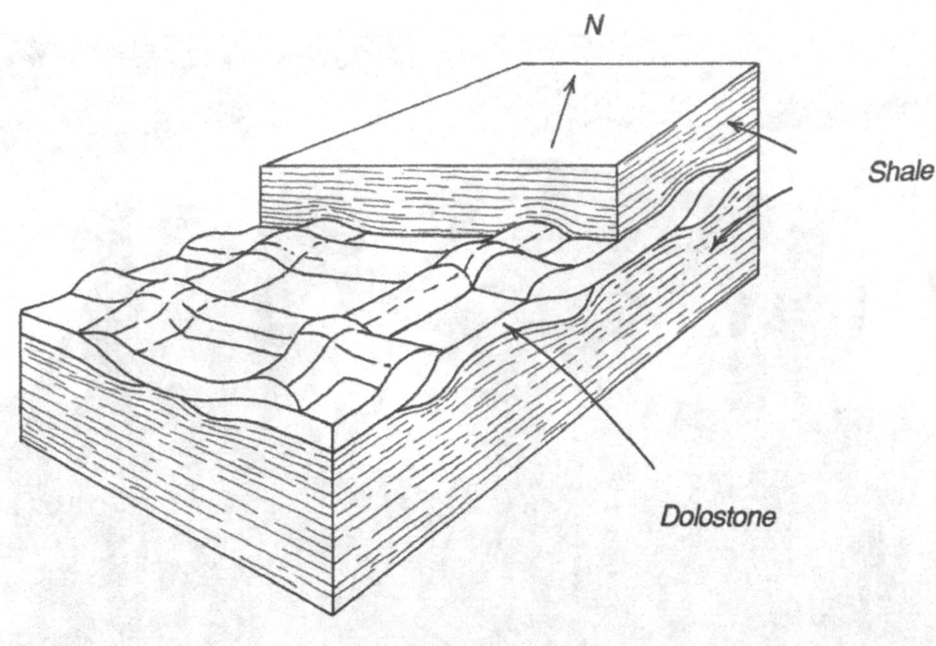

Figure 11. Schematic diagram of the geometric features of fold and faults developed in the Flats Stone Band dolostone of Kimmeridge Bay.

that the two sets of conjugate structures were formed at the same time. Although the resulting overall interference geometry has been decribed as polygonal (Leddra *et al.* 1987) it would be better described as lozenge shaped because the polygons are not as regular and equidimensional as, for example, seen in mud crack- or basalt cooling-polygons. Leddra *et al.* also suggested that there was a slight orientation effect (1987, Figs 2 & 5) and both data sets show a significant anisotropy with a shorter axis in a N-S direction than that in the E-W direction.

In the cliffs behind the foreshore good profile sections of these fold and thrust structures can be observed, and the two sides always show a mirror image downbending of the footwall and elevation of the hangingwall (Fig. 13) in accord with the model shown in Figure 2C. The folding seen in the shale layers around the dolostone is not kink like and the folds progressively deamplify around the folded dolostone layer in the manner shown in Figure 7.

Figure 12. Thrust elevated triangular wedge of dolostone between conjugate thrust planes. Note the arcuate fractures developed in the frontal zones of both thrusts, and the concave form of these fractures toward the relative forward movement direction of the upper thrust plate. The fold and fault structures are cut across by late joints.

Figure 13. Ramp thrust fault in the Flats Stone Band dolostone showing mirror symmetric folds in the hangingwall and footwall (cf. Fig. 2, Model 3).

The overlapping calcite fibres, when studied in thin section, show inclusion bands and small screens of wall rock material characteristic of the crack-seal mechanism of fibre growth (Ramsay 1980). This structure is generally interpreted as arising during tectonic fibre development where vein opening and fibre growth occur periodically over long time periods. At some positions along the thrust faults isolated angular fragments of dolostone occur in the form of dolomite-calcite cemented fault breccias. The structure of this carbonate cement is very different from that of the fault surface fibres, the crystals forming radiating aggregates growing sub-perpendicularly to the walls of the dolostone fragments, and with a comb-structure typical of vug fillings (Bateman 1942).

Figure 14. Thrust of the Flats Stone Band dolostone showing calcite filled fault surface and calcite fibre lineations of the leading edge of the hangingwall antiform (parallel to the pencil). Sub horizontal calcite veins branch off from the calcite filling of the thrust surface and appear to represent echelon extension veins formed contemporaneously with the thrust development.

The frontal parts of each hangingwall anticline and footwall syncline are striated and also may sometimes show fibrous calcite fibres indicative of thrust sense movements (Fig. 14). Such localized sliding could be the result of the bedding plane slip on the steepened bedding surface according to the classic fault bend fold model, however it might also arise from the effect of wedging of the cut off tips of the hanging and footwalls into the surrounding shales. Other special features seen in hanging wall anticlines are curving sets of fractures clearly related to the bending processes set up in these regions (Fig.12). These fractures are always concave towards the shear sense direction. They appear to be extension structures, and their curved form seems to suggest that various sectors of the overthrust advance preferentially while adjacent parts of the thrust sheet are held back from thrust advance.

The displacement relationships between hanging and footwalls of the thrusts are clear and unambiguous in the steep ramp sections within the dolostone layer. However, when the faults are followed into the bedding plane parallel flats they become less well marked, and eventually disappear (Fig. 15B). The thrusts appear to be generated only in the dolostone layer, and this implies either that other layer parallel shortening mechanisms must be present in the shale material or that the shortening is confined to the dolostone layer. The second of these possibilities was proposed by Bellamy (1977), and accepted by Leddra *et al.* (1987). They suggested that local shortening was the result of dolomitisation processes in the dolostone which had resulted in a localized volumetric increase only in the carbonate rock. It is well known that dolomitisation can take place with either a volumetric increase or decrease, depending upon the nature of the dolomitisation reaction (see summary in Morrow 1982). This explanation is rejected in favour of a tectonic shortening mechanism for the following reasons:-

1. Dolostone bands of comparable composition and thickness which lie to the south east of Kimmeridge Bay (The Yellow Ledge Stone Band, Grey Ledge Stone Band, Basalt Stone Band and White Band, exposed in the cliffs of the Kimmeridge Ledges, see House 1969) do not show any of the contraction features seen in the Flats Stone Band. It would be surprising if the dolomitisation processes were of different type in these lithologically identical dolostones. The only differences in the locations of these layers is their proximity to the hinge zone of the Purbeck Anticline. The layers without contraction features all lie further south from the hinge zone of the fold than that of the Flats Stone Band.
2. The lozenge structures of unfolded dolostone statistically show longer axes in an E-W direction than in a N-S direction, and this is the result of their formation by interference of two fairly distinct sets of conjugate thrusts with specific orientations. Dolomitisation processes, being volumetric changes, should give rise to isotropic structural features with uniform contractions in all directions within the dolostone bed.
3. The carbonate fibre growths along individual fault planes all consist of crack-seal calcite. If the faults had been developed during a dolomitisation process it is difficult to explain the absence of vein dolomite when the carbonate bed consists of 98 per cent dolomite.

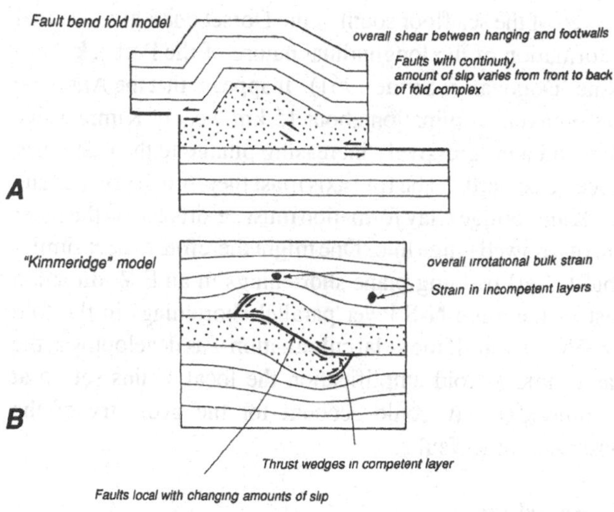

Figure 15. Comparison of the classic fault bend fold model (**a**) with the geometry of the thrust ramp faults and folds seen at Kimmeridge Bay (**b**).

4. The overall regional shortening of the dolostone by the thrust development is not large, probably of the order of 3 to 5 per cent. At several localities in the shales immediately above or below the ramp thrusts the ammonites have distorted forms indicative of local strains of the order of 25 per cent bedding plane parallel shortening, with the short axis if the strain ellipse located in a direction perpendicular to the local fault bend fold. Where such strains were detected there was a tendency for the shales to develop a crude pencil fabric parallel to the fold axes. Linear fabrics of this type are known to be characteristic of the imprinting of small tectonic strains on pre-existing diagenetic flattening fabrics (Graham 1978; Reks & Gray 1982; Ramsay & Huber 1983). At distances exceeding about 50 cm from the folded dolostone surface no traces of folding can be seen and no shape changes were detected in the ammonite forms. The lack of folding must imply either an absence of strain or that the strain is homogeneous. If there was a homogeneous bedding parallel shortening in the shales it would be so small as to be practically undetectable.

It is suggested that these thrusts are the result of tectonic shortening associated with the formation of the Purbeck Anticline. The Kimmeridge Clay formation is an incompetent group of rocks and the outcrops at Kimmeridge Bay are situated in the core of this anticline and located on the inner arc of formations comprised of competent rocks - the Upper Jurassic Portland and Purbeck Limestones. Structural positions such as this are well known to be associated with the strongest rock strains in a fold system, the maximum shortenings oriented parallel to the bedding surfaces and at right angles to the fold hinge. The overall geometry of the Kimmeridge Bay thrust lozenges, with maximum shortening in a N-S direction fits well with this model, but a problem arises: why are their two distinct sets of contemporaneously formed conjugate thrusts, a geometry which also implies that there has been an E-W shortening in the system, albeit much smaller than the N-S shortening? A geophysical accoustic survey of the sea floor south of the Dorset coast gives further information of the longitudinal nature of the Purbeck Anticline (Donovan & Stride 1961). It appears that the Anticline has an axial culmination about 15 km west of Kimmeridge Bay and a progressively increasing plunge to the east when traced eastwards. The fold axis must therefore be curved and the Kimmeridge Clay formation must be situated on the inner arc of a bowed hinge line. One might therefore expect similar (but lesser) bedding plane shortenings in an E-W direction just as there are N-S layer parallel shortenings in the fold profile section. If the axial culmination was developing at the same time as fold amplification the local strains set up at Kimmeridge bay could account for the geometry of the observed thrust faults.

Conclusions

In this paper, it has been suggested that there has been a tendency to overplay the geometry of the classic fold and fault models and to force too readily the geometry of naturally developed thrust systems to fit these models. It has been pointed out that some geometric features of these models are difficult to explain in terms of mechanical behaviour of rock materials and, probably more important for the practical geologist, many well exposed examples of such structures do not fit too well the predicted geometry. Seismic data is not in every case a useful tool for resolving the dilemma because so often a model forms the skeleton for interpreting the geological flesh. It would be appropriate for geologists to show a little more flexibility in the model choice and, in the future, to make much more accurate investigations of well exposed examples of these types of structures to discover the exact geometric forms of hangingwall antiforms and how these forms change at distance from the thrust surface. Although realising that strain data is not always obtainable in natural systems it could provide a critically valuable key for the kinematic and dynamic interpretation of thrusts and their associated fold features.

References

Bateman, A. M. 1942. *Economic mineral deposits.* Wiley & Sons, New York.

Bellamy, J. 1977. Surface expansion megapolygons in Upper Jurassic dolostone (Kimmeridge, UK). *Journal of Sedimentary Petrology,* **47**, 973-978.

Beutner, E. C., Fisher, D. M. & Kirkpatrick, J. L. 1988. Kinematics of deformation at a thrust fault ramp (?) from syntectonic fibres in pressure shadows. *Geological Society of America Special Paper* **222**, 77-88.

Boyer, S. E. & Elliott, D. 1982. Thrust systems. *American Association of Petroleum Geologists,* **66**, 1196-1230.

Butler, R.W.H. 1983. Balanced cross-sections and their implications for the deep structure of the northwest Alps. *Journal of Structural Geology,* **5**, 125-137.

Collomb, P. & Donzeau, M. 1974. Relations entre kink-bands décamétriques et fractures de socle dans l'Hercynian des Monts d'Ougarta (Sahara occidental, Algérie). *Tectonophysics* **24**, 213-242.

Coward, M. P. 1983. Thrust tectonics, thin skinned or thick skinned, and the continuation of thrusts to deep in the crust. *Journal of Structural Geology,* **5**, 113-123.

Davis, D., Suppe, J. & Dahlen, F. A. 1983. Mechanics of fold-and-thrust belts and accretionary wedges. *Journal of Geophysical Research,* **88**, 1153-1172.

Dieterich, J. H. 1970. Computer experiments on mechanics of finite amplitude folds. *Canadian Journal of Earth Science,* **7**, 467-476.

Donovan, D. T. & Stride. 1961. An acoustic survey of the sea floor south of Dorset and its geological interpretation. *Philosophical Transactions of the Royal Society of London,* **B 244**, 299-330.

Durney, D. W. & Ramsay, J. G. 1973. Incremental strains measured by syntectonic crystal growths. *In:* De Jong, K. A. & Scholten, R. (eds) *Gravity and Tectonics,* John Wiley & Sons, 67-96.

Elliott, D. 1976. The energy balance and deformation mechanisms of thrust sheets. *Philosophical Transactions of the Royal Society of London,* **A 283**, 289-312.

Graham, R. H. 1978. Quantitative deformation studies in the Permian rocks of the Alpes-Maritimes. *Goguel Symposium, B.R.G.M.* 220-238.

Heim, A. 1921. *Geologie der Schweiz.* Band II/I. Tauchnitz, Leipzig.

House, M. R., 1969. The Dorset coast from Poole to the Chesil Beach. *Geologists Association Guide,* **22**.

——, 1989. Geology of the Dorset Coast. *Geologists Association Guide.*

Knipe, R. J. & Needham, T. 1986. Deformation processes in accretionary wedges - examples from the SW margin of the Southern Uplands, Scotland. *In:* Coward, M. P. & Ries, A. C. (eds) *Collision Tectonics.* Geological Society of London Special Publication, **19**, 51-65.

Leddra, M. J., Yassir, N. A., Jones, C. & Jones M. E. 1987. Anomalous compressional structures formed during diagenesis of a dolostone at Kimmeridge Bay, Dorset. *Proceedings of the Geological Association,* **8**, 145-155.

Mitra, S. 1986. Duplex structures and imbricate thrust systems: geometry, structural position, and hydrocarbon potential. *American Association of Petroleum Geologists Bulletin,* **70**, 1087-1112.

Morrow, D. W. 1982. Diagenesis 1. Dolomite, The chemistry of dolomitization and dolomite precipitation. *Geoscience Canada,* **9**, 5-13

Price, R. A. 1981. The Cordilleran foreland thrust and fold belt in the southern Canadian Rocky Mountains. *In:* McClay, K. R. & Price, N. J. (eds) *Thrust and Nappe Tectonics.* Geological Society of London Special Publication, **9**, 427-448.

Ramberg, H. 1963. Strain distribution and geometry of folds, *Geological Institute of the University of Uppsala Bulletin,* **42**, 1-20.

Ramsay, J.G. 1966. *Folding and fracturing of rocks.* McGraw Hill Book Co.

—— 1976. Displacement and strain. *Philosophical Transactions of the Royal Society of London.* **A 283**, 3-25.

—— 1980 The crack-seal mechanism of rock deformation. *Nature,* **284**, 135-139.

—— 1983. Rock ductility and its influence on the development of tectonic structure in mountain belts. *In:* Hsü, K. (ed.) *Mountain building processes.* Academic Press, 111-127

——, Casey M. & Kligfield, R. 1983. Role of shear in the development of the Helvetic thrust belt of Switzerland. *Geology* **11**, 439-442.

—— & Huber, M. I. 1983. *The techniques of modern structural geology, Vol 1, Strain analysis.* Academic Press, 1-307.

—— & Huber, M. I., 1987. *The techniques of modern structural geology, Vol.2, Folds and faults.* Academic Press, 309-700.

Reks, I. J. & Gray, D. R. 1982. Pencil structure and strain in weakly deformed mudstone and siltstone. *Journal of Structural Geology,* **4**, 161-176.

Rich, J. L. 1934. Mechanics of low-angle overthrust faulting as illustrated by Cumberland thrust block, Virginia, Kentucky and Tennessee. *American Association of Petroleum Geologists Bulletin,* **18**, 1584-1596.

Suppe, J. 1983. Geometry and Kinematics of fault-bend folding. *American Journal of Science,* **283**, 684-721.

The duplex model: Implications from a study of flexural-slip duplexes

P.W.G. Tanner

Department of Geology & Applied Geology, University of Glasgow, Glasgow, G12 8QQ, UK

Abstract: Criteria for the recognition of duplexes developed on the limbs of flexural-slip folds are summarized and the morphology of these flexural-slip duplexes is compared with that of well-documented, fully-exposed examples of duplexes from thrust belts. The two groups of structures have many features in common regardless of scale or of the rock type in which they occur. They exhibit flat or gently curved floor and roof thrusts; low-angle thrusts in both the tip and rear of the structure; and limited rotation of horses in the middle portion. In marked contrast to current geometrical models for duplex formation, the development of ramp anticlines above newly-formed thrusts at the tip of the duplex is relatively unimportant in natural examples. A revised model for duplex formation is proposed in which new horses are accreted to the tip of the wedge-like structure on low-angle thrusts, are then rotated between active floor and roof thrusts as the structure grows, and become sigmoidal in the interior part of the duplex. The duplex maintains its streamlined profile by internal adjustments, some oblique to the main movement direction, on both link thrusts and backthrusts; all movements are accompanied by the growth of quartz or calcite fibre veins on thrust surfaces.

The first imbricate structures were reported by Peach *et al.* (1907) from the Moine Thrust zone, NW Scotland but the term 'duplex' was not applied to such structures until much later (Dahlstrom 1970). Since then duplexes have been described from a wide range of geological environments including thrust belts (cf. Boyer & Elliott 1982), strike-slip fault zones (Woodcock & Fischer 1986), extensional fault systems (Gibbs 1984), and rocks affected by soft-sediment deformation (Shanmugam *et al.* 1988).

This paper is concerned with contractional duplexes, which are rather brittle, easily eroded structures generally found in gently dipping strata in thrust belts. As a result, the complete duplex structure is seldom exposed and well-documented examples of duplexes are rare. They include the Basse Normandie duplex (Cooper *et al.* 1983); the cleavage duplexes described by Nickelsen (1986); internal duplexes from the Lewis Thrust zone (McClay & Insley 1986); and a small duplex from the Moine Thrust zone (Bowler 1987). Partially exposed, large-scale duplexes whose geometry is well known include the Haig Brook duplex (Fermor & Price 1976); the Foinaven (Elliott & Johnson 1980) and Lighthouse (Coward 1984) duplexes from the Moine Thrust zone; the Mountain City window duplex (Boyer & Elliott 1982), and a duplex from the Makran accretionary prism (Platt & Leggett 1986).

Due to this scarcity of natural examples, our conception of what a duplex looks like, i.e. the mental template used to guide section construction in poorly exposed ground or on seismic reflection profiles, is much influenced by theoretical models, particularly that of Boyer & Elliott (1982). This model was based on the ramp anticline concept of Rich (1934) whereas later modelling (Mitra 1986; Groshong & Usdansky 1988) used the Suppe kinematic model (Suppe 1983) to generate the ramp-related folds and hence the duplex geometry.

The aim of this short contribution is to summarize the main features of newly-discovered flexural-slip duplexes and, by using this information in combination with that from previously-described duplexes, to suggest a revised model for the formation of these structures in general.

Flexural-slip duplexes

These are duplexes which form when, after the rocks have become lithified, packets of welded beds slip over one another during the development of flexural-slip folds (Tanner 1989). They have been found recently in turbidite facies sandstones and shales of Upper Carboniferous age from coastal cliff sections in North Devon and North Cornwall between Hartland Quay and Bude (SW England) (Fig. 1). At Hartland Quay the rocks have been folded into a single generation of upright, slightly curvilinear chevron folds as a result of high-level deformation during the later stages of the Variscan (end-Carboniferous) orogeny (see Tanner 1989 and references therein for details). Farther south these folds have been affected by S-directed simple shear (Sanderson 1979) and, a few kilometres south of Widemouth Bay, become tight, recumbent structures (Fig. 1). Nearly all of the duplexes reported here occur on the limbs of folds which are either unaffected (Hartland Quay), or little modified (Maer Cliff), by the later deformation. There is some evidence that an early phase of N- directed thrust movement took place in these rocks prior to the development of the chevron folds (Enfield *et al.* 1985; Whalley & Lloyd 1986) and the possibility that the duplexes described here are of pre-folding origin has been fully considered by the author, but rejected for the reasons outlined below. It should be stressed however that it is not the structural setting in which these duplexes formed but their mode of formation and morphology which are of most

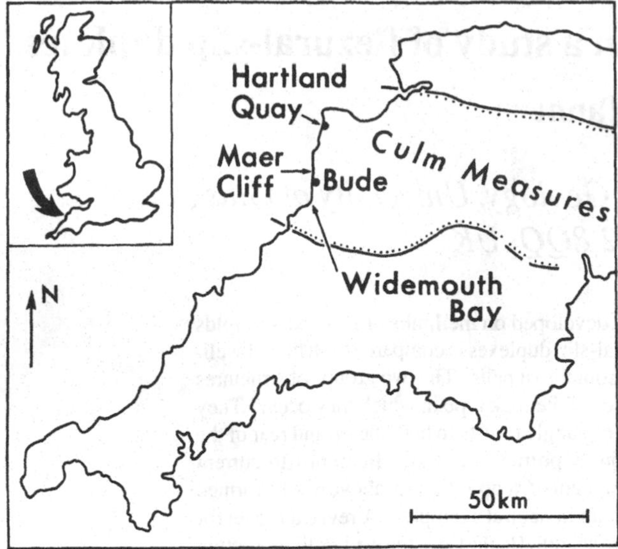

Figure 1. Localities in SW England from which flexural-slip duplexes are reported.

importance here.

The duplexes range in scale from a centimetre to a metre thick; involve thin bands of slate, single sandstone beds, or packages of beds; and are associated with other structures such as simple ramps, blind thrusts, and imbricate structures (Fig. 2). The detailed geometry and morphology of the duplexes, including slate duplexes (Nickelsen 1986), will be described elsewhere but the main criteria for their recognition can be summarized as follows:

1. They always show a shear sense which corresponds to the inferred flexural-slip displacement for the fold limb on which they occur. In several cases duplexes from the same group of beds on two adjacent fold limbs show a reversal of shear sense across the fold axis.

2. The direction of slip given by lineations on the floor and roof thrusts is statistically parallel to the mean orientation of identical lineations on flexural-slip surfaces (movement horizons) between packets of beds on the same fold limb.

3. The floor, roof and link thrusts (McClay & Insley 1986) of the duplexes are marked by laminated quartz-carbonate fibre veins up to a centimetre thick which are identical in morphology to those on movement horizons found throughout the fold structure. Fibre steps on the link thrusts in some of the duplexes confirm the sense of shear inferred from the morphology of the duplex, and the reversal of shear sense shown by fibre steps on the movement horizons on alternate fold limbs shows that these surfaces were active during flexural slip.

4. Slickenfibre lineations (Tanner 1989) on the fibre veins both within and bounding the duplexes, and on the adjacent movement horizons, consist mainly of quartz and/or calcite fibres with parallel slickolite development. Individual mineral fibres are oriented at a low angle to the margins of the fibre sheets and some contain crack-seal inclusion bands and trails (Ramsay 1980) which indicate that they have grown by incremental extension parallel to their length, and hence to the macroscopic lineation on the surface. The morphology of these structures has been illustrated by Tanner (1990).

The duplexes have therefore grown as a result of aseismic, incremental movements on the various internal and bounding thrust surfaces. These movements occurred in harmony with progressive flexural-slip movements which were taking place on bedding-parallel surfaces throughout the fold structure as it developed. Duplexes develop on movement horizons which, because of minor perturbations, are not precisely planar for the complete length of a fold limb. Flexural-slip chevron-style folds affecting distal turbidites or other planar-bedded sequences do not appear to develop duplexes on their limbs. However in more proximal and irregularly-bedded sequences, duplex formation is triggered by local variations in bed thickness, by lateral facies changes, and by the presence of large-scale soft-sediment structures. These 'sticking points' (Knipe 1985) cause the flexural-slip surface to ramp up to a higher easy-glide horizon and collapse of the resulting footwall ramp gives rise to a duplex.

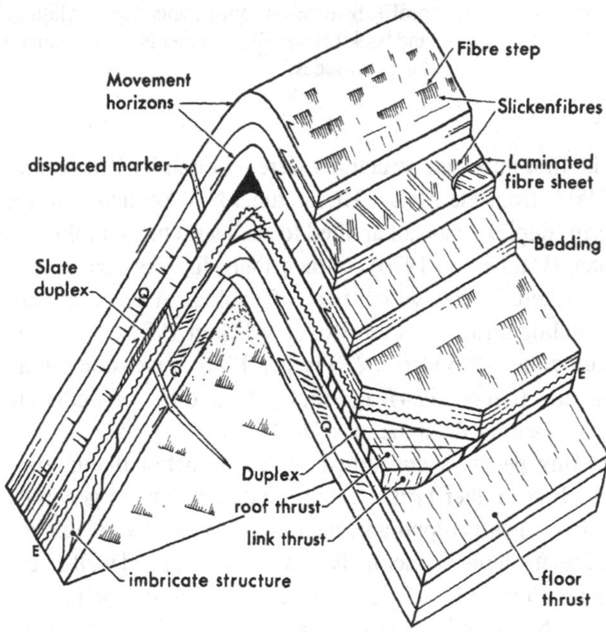

Figure 2. Geometry of duplexes in relationship to other features of the flexural-slip model. E, erosion surface; Q, bedding-normal and en-echelon sets of quartz veins.

The main morphological features of flexural-slip duplexes, based on 27 examples located in North Devon and North Cornwall, are summarized in Figure 3. All of the duplexes have planar or gently curved floor and roof thrusts marked by fibrous quartz-carbonate veins, and slickenfibre and other lineations on the two sets of veins are generally parallel to one another. Differently oriented slickenfibres are sometimes found on successive laminations within these veins and reflect changes in the slip vector with time. In these cases, similar sets of multiple fibre orientations are commonly found on adjacent flexural-slip surfaces between packets of beds on the same fold limb. The duplexes are asymmetric (Fig. 3) when viewed in cross-section normal to the movement direction shown by the lineations on the floor and roof thrusts: they have a pointed 'toe' region with gently dipping

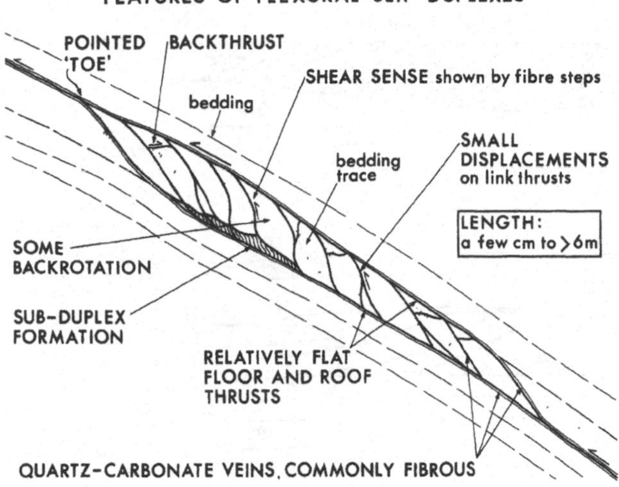

Figure 3. Summary in cartoon form of the main features of flexural-slip duplexes.

thrusts, and a rear part with gently dipping planar thrusts. Curved bedding traces are seen within the horses in the central parts of some duplexes, as shown diagrammatically in Figure 3.

Displacements on individual link thrusts are small and consistently show a thrust sense. The dip directions of the link thrusts and the trends of the slickenfibre lineations on their surfaces are commonly oblique to the movement directions seen on the floor and roof thrusts. As this obliquity varies from place to place in the duplex the structure has developed under conditions of variable 3-D strain and no cross-section can be area balanced. Small backthrusts are common in some duplexes; they have slickenfibre lineations which lie approximately in the same plane as those on the link thrusts and appear to be kinematically related to them. Stacked duplexes are often found in which duplexes on different scales formed from, for example, a single sandstone bed several centimetres thick, and from a laminated siltstone-shale unit a few millimetres thick, are intimately associated and form a single structural unit (Fig. 3).

Comparison with other duplexes

The profiles of typical flexural-slip duplexes are compared with those of well-documented duplexes from various thrust belts, starting with the simplest (? least evolved) of these structures (Fig. 4). The Foinaven duplex (Fig. 4a), first described from the Moine Thrust zone by Peach *et al.* (1907), has the Moine Thrust as a roof thrust. It is a foreland-propagating structure and the normal stratigraphic thickness of Pipe Rock and An t-Sron Formation - 122 m, is represented by a structural thickness of 510 m in the duplex, a total shortening by area balance of -49% (Boyer & Elliott 1982).

In overall morphology the Foinaven duplex closely resembles the foreland portion of the Basse Normandie duplex from Boulonnais, France (Cooper *et al.* 1983), which was probably the first fully-exposed duplex to be described and analysed. In this duplex (Fig. 4b) a 2-m thick sequence of thin-bedded limestones with two marker horizons has been affected by layer-parallel shortening (-22%), followed by a -

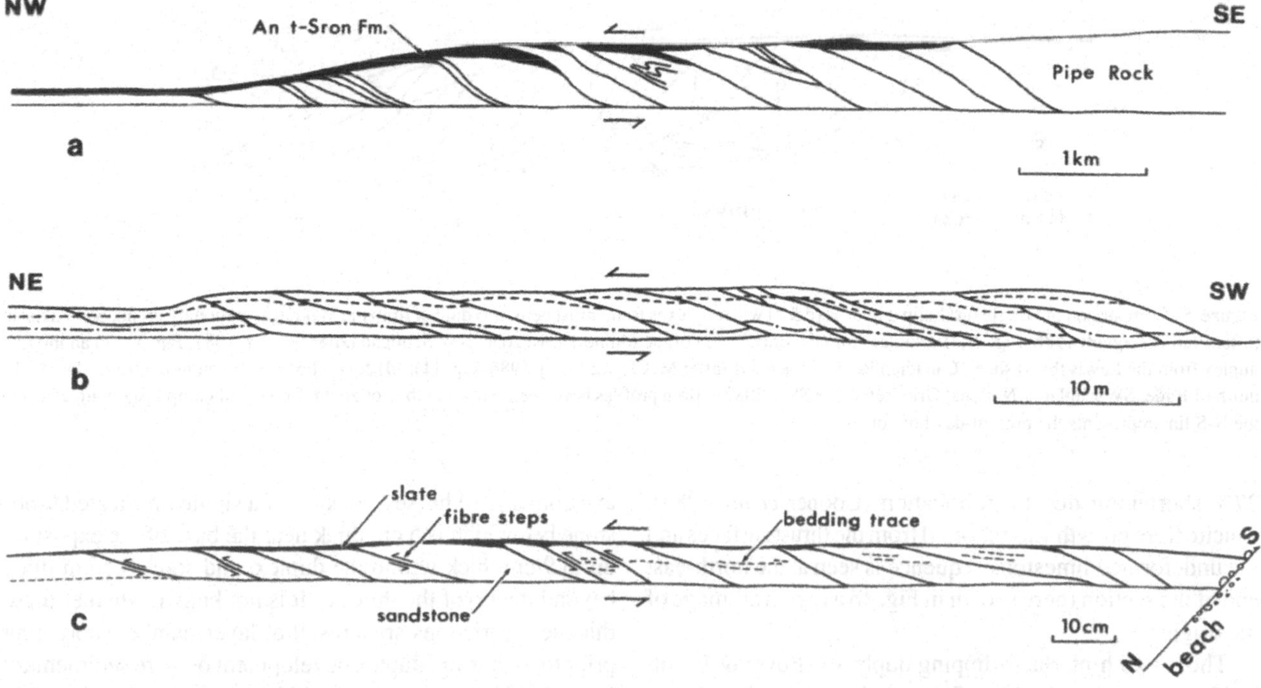

Figure 4. Comparison of a simple flexural-slip duplex (c) with two duplexes of similar morphology from thrust belts. **(a)** the Foinaven duplex, Moine Thrust zone, NW Scotland (after Elliott & Johnson 1980, Fig. 8); **(b)** the lower duplex, Basse Normandie, Boulonnais (after Cooper *et al.* 1983, Fig. 6); **(c)**, a flexural-slip duplex from near Berry Cliff, north of Hartland Quay, N Devon (SS 225258). The structure dips at 49°S but has been rotated to the horizontal in this figure for comparison with a and b.

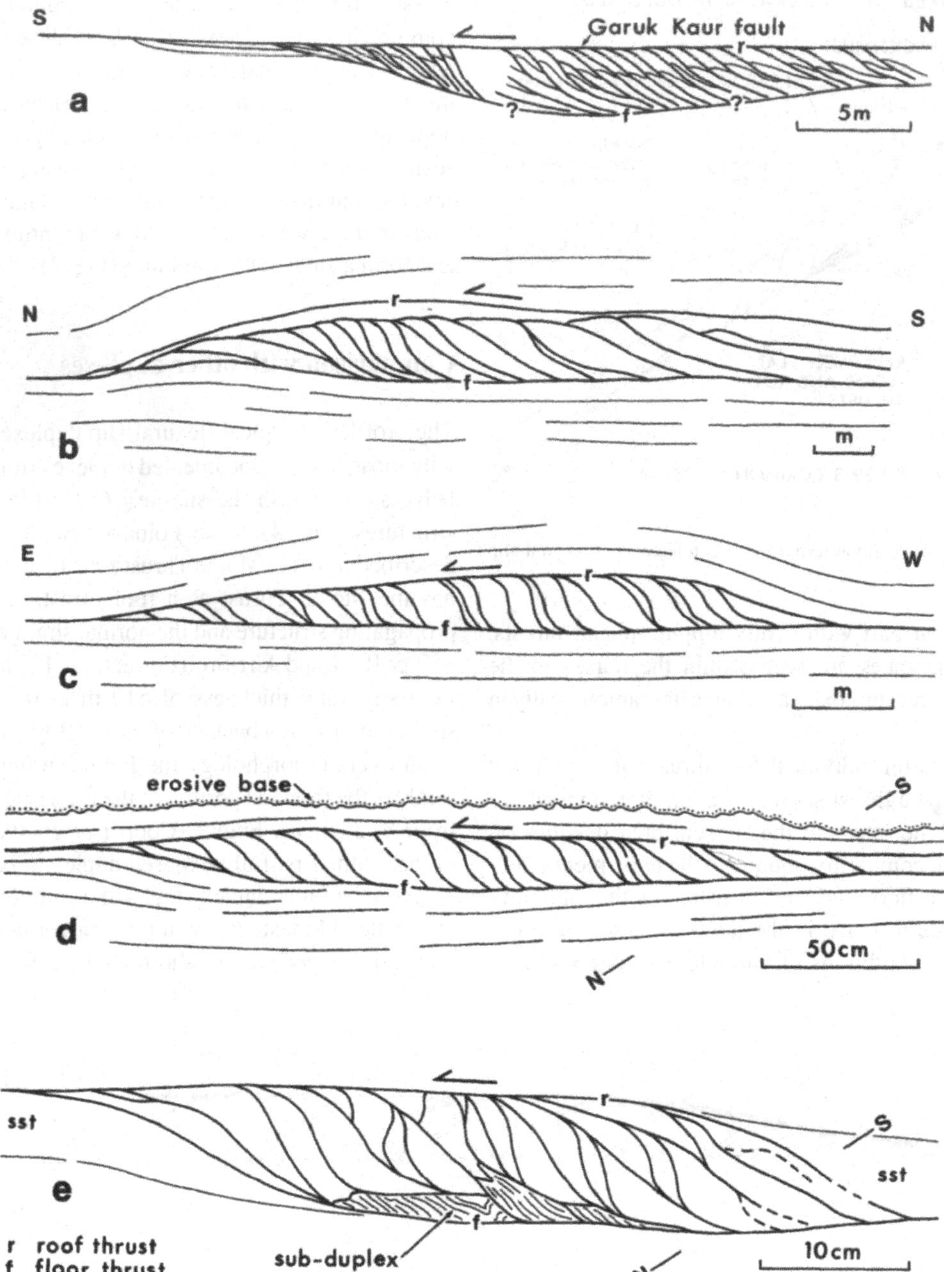

Figure 5. Comparison of two flexural-slip duplexes (d & e) with duplexes from thrust belts. **(a)** duplex from the Makran accretionary prism, SW Pakistan (after Platt & Leggett 1986, Fig. 7); **(b)** duplex from the Eriboll area of the Moine Thrust zone, NW Scotland (after Bowler, 1987, Fig. 3); **(c)** an internal duplex from the Lewis thrust sheet, Canadian Rocky Mountains (after McClay & Insley 1986, Fig. 11); **(d)** & **(e)** flexural-slip duplexes from Maer Cliff, north of Bude, SW England (National Grid Reference SS 202082). Both profiles have been rotated to the horizontal for ease of comparison with a - c and the N-S line represents the present-day horizontal.

27% shortening due to imbrication (Cooper *et al.* 1983). Calcite fibre growth was reported from the thrust surfaces and the undeformed limestone sequence is seen at the north-east end of the section (here shown in Fig. 4b as a mirror image of the original).

These two hinterland-dipping duplexes (Boyer & Elliott 1982) are compared with a flexural-slip duplex found near Berry Cliff (SS 225258), north of Hartland Quay, North Devon, which dips at 49°S on the south limb of an anticline (Fig. 4c). The duplex, which is about 2m long (only the lower part is accessible and it has been drawn from photographs) and contains 14 horses, consists of a single imbricated sandstone bed which is 5 cm thick near the base of the exposure, up to 9 cm thick within the duplex, and some 4.5 cm thick beyond the tip of the duplex. It is not known whether these thickness variations are a result of layer-parallel shortening prior to, or during, duplex development or were sedimentary in origin. Traces of bedding within the horses show that little rotation of these has occurred, and displacements are on the order of a few millimetres to 1 cm on each link thrust. Fibre steps on 5 link thrusts give the sense of displacement shown in Figure 4c and confirm the direction of propagation of the duplex. It is of significance to the later discussion on the origin of duplexes that the roof thrust of this particular example continues to the limit of exposure at the foot of the cliff and that it is marked throughout the exposure by a thick laminated quartz fibre vein up to 9 mm thick. From a study of adjacent rocks in this section (Tanner 1989) it can be inferred that a bedding-parallel vein of this thickness is indicative of a considerable amount of movement parallel to the roof thrust.

Features of these simple duplexes which are of particular significance are the (a) flat floor and roof thrusts, (b) constant ramp angles and lack of any ramp anticline development, and (c) lack of evidence of significant rotation of the horses within the duplex.

Well-exposed examples of duplexes showing a greater degree of internal rotation of horses are now described and are compared with two examples of flexural-slip duplexes from the Bude area of North Cornwall (Fig. 1). They are all internal duplexes (McClay & Insley 1986) in that they are contained within one stratigraphic formation and repeat a single bedding unit.

The first example (Fig. 5a) is of an exceptionally well exposed duplex reported from the Makran Coast Ranges of southwest Pakistan by Platt & Leggett (1986), which occurs in well-bedded turbidite facies rocks of Tertiary age. It lies beneath a major bedding-parallel thrust, the Garuk Kaur fault, is 30m long, and was formed from a 30-cm thick 'turbidite bed' shortened by -66% (Platt & Leggett 1986). The frontal ramp dips at about 15°, the bed shows progressive back-rotation away from the tip of the duplex and the roof thrust has remained planar (Fig. 5a). The front portion of this duplex was figured on the cover of the *Journal of Structural Geology* for 1986 and it is the best example described so far of such a structure. Calcite fibre growth was reported from thrust surfaces in this area but was not described specifically from the duplex.

The next example (Fig. 5b) is of a small, completely exposed duplex from the north end of the Moine Thrust zone near Loch Eriboll described by Bowler (1987). It has formed from the imbrication of a 30-cm thick quartzite unit, and is 13m long and up to 1.2m thick (Fig. 5b), but it is not clear from the description why a slice of quartzite apparently intervenes between the roof thrust (Bowler 1987, Figs 3 & 4b) and the 'roof' defined by the aligned upper ends of the stacked horses. This duplex shows the internal features which here are considered to be characteristic of duplexes in general: a gently dipping frontal thrust; back rotation of horses in the front-to-mid part of the duplex; and gently inclined thrust surfaces in the rear portion. The link thrusts are mostly curved and somewhat oblique to the movement direction and the horses are lensoid in 3-D. It is noteworthy that two or three sets of slickenfibre lineations occur on some link thrusts and that fibre orientations in general are rather variable in orientation (Bowler 1987).

A series of small-scale imbricates and duplex structures were described from the Lewis thrust sheet in the Canadian Rocky Mountains by McClay & Insley (1986). That reproduced in Figure 5c is a completely exposed duplex in argillaceous limestone which is 14 m long and contains some 27 horses. It has an overall geometry very similar to that described above and shows a progressive increase, then decrease, in the maximum degree of clockwise rotation of each link thrust from the floor thrust, as measured from the tip to the rear of the duplex (McClay & Insley 1986, Fig. 11). The slip vectors marked by grooving, slickenside lineations and slickolites on the floor and roof thrusts are consistent in direction, whereas the slip directions on the link thrusts vary considerably.

For comparison with these duplexes described from thrust belts, two duplexes of flexural-slip origin from Maer Cliff, near Bude in North Devon (Fig. 1) are shown in Figure 5, d & e and Figure 6. Both were drawn from detailed photographs supplemented by field sketches and occur in interbedded sandstones and slates dipping at 30-35° to the south, on the southern limb of an anticline. Example d has 24 horses, is approximately 2 m long, and formed by the imbrication of a 5-cm thick package of beds. Duplex e consists of over 18 horses, is 54 cm long and formed from the imbrication of an homogeneous 4-cm thick sandstone bed. All of the thrust surfaces in d and e are marked by quartz fibre veins. Both duplexes are exposed on flat joint surfaces but another duplex of similar size nearby is seen in 3-D and slickenfibres show that movement on the link thrusts was oblique to that on the floor and roof thrusts. In the latter duplex some of the internal thrusts which dip in the opposite direction to the main link thrusts preserve fibre steps indicating a backthrust sense of movement.

In the examples described so far, crucial information on the form of the bedding surfaces within the horses in each duplex is generally lacking; in the Bude-Hartland Quay area the best evidence of this seen so far is given by a partially eroded duplex found north of Widemouth Bay (Fig. 1) (National Grid Reference SS 198037). The duplex consists of a bedded sandstone unit repeated by imbrication and is over 6 m long and up to a metre thick, but only the hindward portion is preserved (Fig. 7). The shear sense and other features shown by this duplex are consistent with a flexural-slip origin. A series of photographs, taken at right angles to the parallel fibre lineations seen on the floor and roof thrusts, were used in the field to record details of the internal structure. The resultant profile is only approximate in places because of perspective problems but it shows that bedding in individual horses is generally only slightly sigmoidal, and there is only one clear example (at a, Fig. 7) of a possible ramp-related anticline. As with the other duplexes described above, slickenfibre lineations on the link thrusts indicate that some out-of-section movement has taken place but a line-balanced restored section using two stratigraphic marker horizons (contacts between beds A, B and B, C in Fig. 7) within the duplex shows that the total shortening is -45% and that the link thrusts were initially approximately parallel to one another and had a hindward dip of <20°. Because of uncertainty in the positions of the bottom of bed A and the top of bed C in the unrestored section it is not known whether the link thrusts were planar or slightly concave upwards.

All of these carefully documented duplexes have a similar

Figure 6. The flexural-slip duplexes from Maer Cliff (National Grid Reference SS202082) shown in Figure 5d & e. The scale in (**a**) is 25 cm long and that in (**b**) is 5 cm long.

external shape and internal geometry regardless of the fact that they have developed in rock types as diverse as limestone, sandstone, and quartzite, and that there is a 60-fold variation in scale for those shown in Figure 5. It should be stressed that these examples have not been selected: to the best of the author's knowledge, they represent the complete set of well-documented examples of fully exposed duplexes (apart from some flexural-slip duplexes similar to those described here) and any omissions are unintentional.

Based on these examples, the features that have to be explained by any model for the formation of a duplex are:

1). Planar or gently curved floor and roof thrusts bound the duplex and it has an asymmetric profile.

2). Back rotation of horses to steeply dipping attitudes is restricted to the middle portion of the structure.

3). Link thrusts at the tip of a duplex have a low angle of dip, as do those at the rear end of the structure.

4). Displacements on individual link thrusts are oblique to the main transport direction and backthrusts are commonly developed within the duplex.

5). All thrust surfaces are marked by calcite- or quartz-fibre growth, depending upon host rock composition.

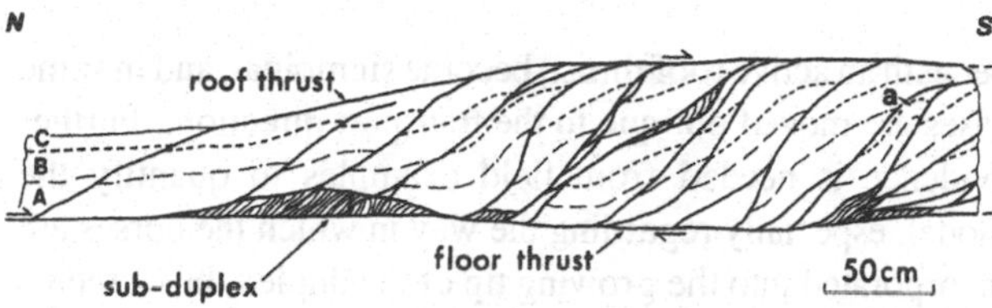

Figure 7. Flexural-slip duplex from north of Widemouth Bay, N Cornwall, SW England (National Grid Reference SS198037).

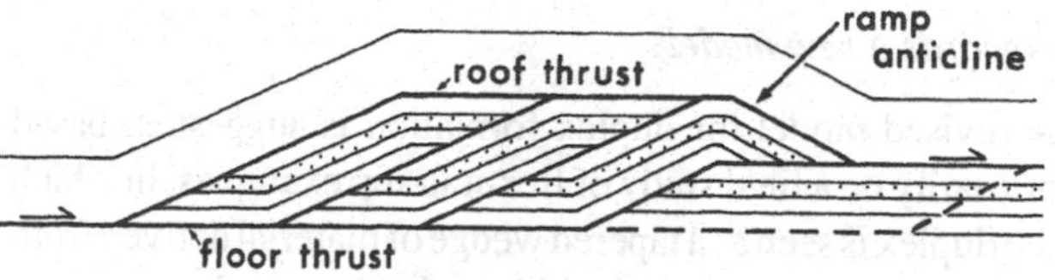

Figure 8. The Boyer & Elliott (1982) duplex model.

The duplex model

The basic model for duplex formation, which has dominated thinking on the subject since it was published in 1982, is that of Boyer & Elliott (Fig. 8). In this forward-propagating model each horse in sequence is transported up a footwall ramp, forms a ramp anticline as it passes over the top of the ramp, and is partially unfolded as the next horse is emplaced beneath it. Later theoretical models for duplex development have confirmed the main features of the Boyer & Elliott (1982) model but this has been a self-fulfilling prophecy as these models are all based upon the assumption that the addition of each horse to the duplex is accompanied by the formation of a new ramp anticline. The models vary in using either the Rich (1934) geometry for the ramp anticline (Boyer & Elliott 1982; Mitra & Boyer 1986) or the Suppe (1983) fault-bend construction (Mitra 1986; Groshong & Usdansky 1988; Cruikshank *et al.* 1989).

Several problems exist with the Boyer & Elliott (1982) model. Firstly it requires that each horse is folded, and then partially unfolded to give an open S-shape. In discussing the mode of development of the Basse Normandie duplex, Cooper *et al.* (1983, p.150) commented that such a process is unlikely and 'can be avoided if several horses move at the same time'. They also demonstrated that the shortening in the duplex (Fig. 4b) increases progressively from front to back of the structure and not in the discrete jumps predicted by the Boyer & Elliott model. The second problem is that no natural example of a duplex has been *reported* (as opposed to modelled) which preserves the last-formed ramp anticline at its tip. Finally, a duplex is considered to form as a result of footwall collapse and thus facilitate the passage of a thrust over an obstacle or 'sticking point', yet the model duplex which has been proposed has a profile which would mechanically impede, rather than assist, this process. Thus either all of the small and medium scale duplexes illustrated in Figures 4-7 are atypical because of some feature such as the relatively small displacement between adjacent horses, or they are *representative* of the typical duplex and the theoretical models proposed so far over-emphasize the importance of the formation of ramp anticlines in duplex development. Despite differences in the mechanisms which operated during the relatively low temperature deformation of the sandstone, quartzite, and limestone beds in which these structures occur, the morphologies of the resultant duplexes are remarkably similar.

The most striking feature of naturally-occurring duplexes is their flat or gently curved roof. This is seen in the examples described earlier and also in well-exposed portions of other duplexes such as the Cate Creek window duplex (Dahlstrom 1970), the Haig Brook duplex (Fermor & Price 1976), the Mountain City window duplex (Boyer & Elliott 1982), and in the duplexes beneath the Copper Creek detachment (Mitra 1986). This feature has been commented upon by several authors including McClay & Insley (1986) who suggested that the roof thrust of duplexes like that in Figure 5c may have been active throughout the period of duplex formation and, as one of several possibilities, that slip could have taken place on all of the link thrusts simultaneously with the greater slip occurring on those in the central, most rotated, portion of the duplex. Bowler (1987) also invoked repeated imbrication of the footwall, and rejuvenation of movement on previously formed link thrusts within the duplex shown in Figure 5b, together with a late movement on the roof thrust.

A flat roof is created in the Boyer & Elliott (1982) model but requires special geometric conditions for its generation when the ramp anticlines are produced using the Suppe (1983) construction (Groshong & Usdansky 1988). It is highly significant that the closest approach to a plane-roofed duplex is given by the computer models of Groshong & Usdansky (1988) when the thrusts within the duplex are initiated as low-angle, nearly planar surfaces. Other attempts to produce plane-roofed duplexes propose geometrical solutions (Geiser 1988) which may be valid in individual cases but do not constitute general models for the formation of such duplexes.

Other features common to many naturally occurring examples are the sigmoidal shapes of the rotated horses and the way in which the hindward dip angle of the thrusts becomes less at the tip of the duplex. This geometry precludes the formation of the duplex by out-of-sequence or break-back collapse of the footwall: it is a forward-propagating structure and the low-angle thrusts at the tip are the last to form. The small amount of field evidence regarding the orientation of bedding in some of the rotated horses found in the central parts of duplexes suggests that very shallow ramp anticlines may have formed on some of the link thrusts as they developed as low-angle features at the tip of the duplex. Subsequent back-rotation of the horses formed in this way gave them a sigmoidal profile and accentuated the curvature of the internal bedding surfaces to an open S-shaped profile (Figs 3 & 5). However, these features are a minor aspect of the internal geometry of the duplex and have not made an important contribution to the external morphology as in the case of the Boyer & Elliott (1982) model. The planar and relatively low angle thrusts at the rear of some duplexes may be late thrusts (cf. McClay & Insley 1986) which have cut up-section in the transport direction after the main development of the duplex structure, but this is not certain.

Towards a new model

A revised model for duplex formation is suggested, based primarily on a field study of flexural-slip examples, in which the duplex is seen as a tapered wedge of material derived from the footwall and enclosed within active floor and roof thrusts (Fig. 3). It has low-angle thrusts at the tip (although those developed at an earlier stage of development of the structure may have had slightly steeper dips) and incremental adjustments throughout the duplex on floor, roof, link and backward-propagating thrusts enable compatibility to be maintained between the constituent horses as the duplex grows and allow the roof thrust to remain planar. This is analogous to the situation in which the Garuk Kaur fault (Fig. 5a), the roof to a duplex, has remained planar whilst the footwall structure takes up the deformation (Platt & Leggett 1986). These movements are accompanied by the incremental development of fibre veins, some at least by the crack-seal mechanism, on all movement surfaces. The duplex is driven forward by the thrust movement, picks up slices of footwall on gently inclined thrust planes and as further material is incorporated into the tip region these slices are back-rotated beneath an active roof thrust, become sigmoidal, and in some cases are moved oblique to the transport direction. Further evidence is needed from field examples to quantify the model, especially regarding the way in which the horses are incorporated into the growing tip of the duplex, but it seems clear that, because of the low angle of the leading thrusts and the way they curve asymptotically into the active roof thrust, the formation of ramp anticlines does not play a significant role in the development of the duplexes described in this paper. As further examples of duplexes from contractional settings are documented it may be found that the type of duplex described here has a geometry more typical of natural situations than that of the Boyer & Elliott (1982) model, and this would require a change in the way these structures have been represented on many interpretive cross-sections in the past decade.

I wish to thank Judith Tanner for finding several of the duplexes and acting as devil's advocate during many discussions as to their origin; Martin Insley and John Lloyd for reviewing the manuscript; and Sheila Brown for the word-processing.

References

Bowler, S. 1987. Duplex geometry: an example from the Moine Thrust belt. *Tectonophysics*, **135**, 25-35.

Boyer, S. E. & Elliott, D. 1982. Thrust systems. *American Association of Petroleum Geologists Bulletin*, **66**, 1196-1230.

Cooper, M. A., Garton, M. R. & Hossack, J. R. 1983. The origin of the Basse Normandie duplex, Boulonnais, France. *Journal of Structural Geology*, **5**, 139-152.

Coward, M. P. 1984. A geometrical study of the Arnaboll and Heilam thrust sheets, NW of Ben Arnaboll, Sutherland. *Scottish Journal of Geology*, **20**, 87-106.

Cruikshank, K. E., Neavel, K. E. & Zhao, G. Z. 1989. Computer simulation of growth of duplex structures. *Tectonophysics*, **164**, 1-12.

Dahlstrom, C.D.A. 1970. Structural geology in the eastern margin of the Canadian Rocky Mountains. *Bulletin of Canadian Petroleum Geology*, **18**, 332-406.

Elliot, D. & Johnson, M.R.W. 1980. Structural evolution in the northern part of the Moine thrust belt, NW Scotland. *Transactions of the Royal Society of Edinburgh: Earth Sciences*, **71**, 69-96.

Enfield, M. A., Gillcrist, J. R., Palmer, S. N. & Whalley, J. S. 1985. Structural and sedimentary evidence for the early tectonic history of the Bude and Crackington Formations, north Cornwall and Devon. *Proceedings of the Ussher Society*, **6**, 165-172.

Fermor, P. R. & Price, R. A. 1976. Imbricate structures in the Lewis thrust sheet around the Cate Creek and Haig Brook windows, southeastern British Columbia. *Geological Survey of Canada Paper*, **76-1B**, 7-10.

Geiser, P. A. 1988. The role of kinematics in the construction and analysis of geological cross sections in deformed terranes. *In*: Mitra, G. & Wojtal, S. (eds) *Geometries and Mechanisms of Thrusting*. Geological Society of America Special Paper, **222**, 47-76.

Gibbs, A. D. 1984. Structural evolution of extensional basin margins. *Journal of the Geological Society, London*, **141**, 609-620.

Groshong, R. H. & Usdansky, S. I. 1988. Kinematic models of plane-roofed duplex styles. *In:* Mitra, G. & Wojtal, S. (eds) *Geometries and Mechanics of Thrusting*. Geological Society of America Special Paper, **222**, 187-206.

Knipe, R. J. 1985. Footwall geometry and the rheology of thrust sheets. *Journal of Structural Geology*, **7**, 1-10.

McClay, K. R. & Insley, M. W. 1986. Duplex structures in the Lewis thrust sheet, Crowsnest Pass, Rocky Mountains, Alberta, Canada. *Journal of Structural Geology*, **8**, 911-922.

Mitra, S. 1986. Duplex structures and imbricate thrust systems: geometry, structural position, and hydrocarbon potential. *American Association of Petroleum Geologists Bulletin*, **70**, 1087-1112.

Mitra, G. & Boyer, S. E. 1986. Energy balance and deformation mechanisms of duplexes. *Journal of Structural Geology*, **8**, 291-304.

Nickelsen, R. P. 1986. Cleavage duplexes in the Marcellus Shale of the Appalachian foreland. *Journal of Structural Geology*, **8**, 361-371.

Peach, B. N., Horne, J., Gunn, W., Clough, C. T. & Hinxman, L. W. 1907. The geological structure of the north-west Highlands of Scotland. *Memoir Geological Survey of Great Britain.*

Platt, J. P. & Leggett, J. K. 1986. Stratal extension in thrust footwalls, Makran accretionary prism: implications for thrust tectonics. *American Association of Petroleum Geologists Bulletin*, **70**, 191-203.

Ramsay, J. G. 1980. The crack-seal mechanism of rock deformation. *Nature*, **284**, 135-139.

Rich, J. L. 1934. Mechanics of low-angle overthrust faulting as illustrated by Cumberland thrust block, Virginia, Kentucky and Tennessee. *American Association of Geologists Bulletin*, **18**, 1584-1596.

Sanderson, D. J. 1979. The transition from upright to recumbent folding in the Variscan fold belt of southwest England: a model based on the kinematics of simple shear. *Journal of Structural Geology*, **1**, 171-180.

Shanmugam, G., Moiola, R. J. & Sales, J. K. 1988. Duplex-like structures in submarine fan channels, Ouachita Mountains, Arkansas. *Geology*, **16**, 229-232.

Suppe, J. 1983. Geometry and kinematics of fault-bend folding. *American Journal of Science*, **283**, 684-721.

Tanner, P.W.G. 1989. The flexural-slip mechanism. *Journal of Structural Geology*, **11**, 635-655.

—— 1990. The flexural-slip mechanism: Reply. *Journal of Structural Geology*, **12**, 1085.

Whalley, J. S. & Lloyd, G. E. 1986. Tectonics of the Bude Formation, north Cornwall - the recognition of northerly directed décollment. *Journal of the Geological Society, London*, **143**, 83-88.

Woodcock, N. H. & Fischer, M. 1986. Strike-slip duplexes. *Journal of Structural Geology*, **8**, 725-735.

Palaeomagnetic techniques applied to thrust belts

A. M. McCaig[1] & E. McClelland[2]

[1] *Department of Earth Sciences, University of Leeds, LS2 9JT, UK*
[2] *Department of Earth Sciences, University of Oxford, OX1 3PR, UK*

Abstract: Palaeomagnetic data can be used to determine rotations of thrust sheets which cannot be detected by conventional structural techniques. In basement rocks this may be the only way to constrain rotations. A detailed knowledge of rotations is necessary for accurate section balancing and can help in understanding the evolution of structures and the mechanisms of thrust sheet deformation. Previously unsuspected rotations about steeply plunging axes have recently been identified in many thrust belts around the world, and we present examples of the use of both pre- and syntectonic remanences in both sedimentary and basement rocks. However, not all areas are suitable for palaeomagnetic studies, and it is vital that appropriate structural and palaeomagnetic tests are used to establish the ages of remanence components, and that the uncertainties inherent in palaeomagnetic data are fully appreciated.

The study of the structural evolution of thrust belts is in large measure the study and interpretation of finite rotations. This is normally accomplished by referring folded and tilted sedimentary rocks to an assumed initial horizontal orientation. This assumption is implicit in the construction of balanced and restored cross-sections (Dahlstrom 1969; Hossack 1979), where all rotations are normally assumed to take place about axes normal to the plane of section. The occurrence of other rotations will automatically invalidate the assumption of plane strain within the section.

Conventional structural studies in sedimentary rocks cannot detect rotations which do not fold or tilt the bedding, and in polydeformed or intrusive basement rocks thrust-related rotations are usually impossible to constrain. Palaeomagnetic studies have the power to identify rotations not previously suspected in sedimentary rocks, and may be the only way of detecting rotations in basement rocks. Palaeomagnetic studies are an important tool in the understanding of the structural evolution of both internal and external parts of mountain belts. However, because of the time consuming nature of palaeomagnetic work and difficulties in establishing the age of remanence, areas for such study need to be selected with some care.

In this paper, the principles and techniques of palaeomagnetic study in deformed rocks will first be reviewed. Examples will then be given of the application of palaeomagnetism to thrust belts in both sedimentary and basement rocks.

Palaeomagnetic techniques and methods

Data collection

The stages in the collection of a palaeomagnetic dataset can be summarized as follows: (1) selection of field area; (2) selection of sample sites; (3) collection and orientation of samples; (4) demagnetisation of samples; (5) establishment of stable remanence components, and rejection of sites with no meaningful remanence components; (6) statistical comparisons within and between sites; (7) interpretation of data.

There is potential for error in all stages of this process, and palaeomagnetic programmes must be both planned and executed with great care. It is inevitable that some sites will not yield meaningful data, and it is particularly important that sites be included in the dataset on purely rational criteria. Unfortunately, this has not always been the case, with the result that palaeomagnetism has acquired an unjustifiably bad reputation amongst many geologists.

The most suitable lithologies for palaeomagnetic work are volcanic rocks, shallow intrusives and iron-rich sediments, but with the advent of cryogenic magnetometers practically any rock may yield significant results. For structural studies, young (Tertiary) rocks are generally best since the range of possible field directions is small. This means that any significant deviations from the present day field direction are almost certainly due to local tectonic rotations.

Selection of sample sites should always be accompanied by field mapping and structural measurements with a view to using palaeomagnetic tests to establish the age of remanence as discussed below. Best results are obtained by using a rock drill to collect core samples in the field. Use of oriented blocks introduces an additional source of error and it is difficult to collect sufficiently fresh material. Cores can be very accurately oriented using a specially adapted sun compass (Collinson 1983), eliminating problems associated with magnetic rocks. At least six and preferably eight samples must be collected at each site if statistically significant results are to be obtained.

It is important that as many samples as possible be fully demagnetized to determine the vector structure of the remanence. Whether Alternating Field or Thermal Treatment or a combination of both is used depends on the magnetic mineralogy and the correct treatment is determined by pilot experiments (Collinson 1983). Magnetic anisotropy may deflect remanence components and should be tested for in deformed samples (Lowrie *et al.* 1986).

Methods for establishing statistically significant remanence components both within a sample and within a site were established by Fisher (1953). Sites can be compared using standard fold and contact tests as described below.

Establishing the age of remanence

Most rocks contain several components of remanence acquired at different times and by different mechanisms. Depositional remanence (DRM) arises from the settling of magnetic minerals in the Earth's field. Thermal remanence (TRM) results from cooling of igneous or metamorphic minerals through the Curie point, while chemical remanence (CRM) components can arise whenever new minerals grow within a rock during diagenesis, metamorphism or deformation. During thermal demagnetisation, different components held in different minerals will be stripped off at different temperatures. However, it cannot be assumed that the highest temperature components are the earliest, or even that they formed at higher temperatures (O'Reilly 1984). It is often very hard to relate a particular component to a particular mineral growth episode in the rock, and indirect tests must be used to establish the age of magnetisation.

The most useful test in structural studies is the *fold test* (McElhinny 1964). In a positive *fold test*, a remanence component on the two limbs of a fold is grouped significantly more closely after unfolding about the fold axis than in situ, showing that remanence acquisition predated folding. A negative *fold test* indicates that remanence acquisition postdated folding. In some cases, one component may predate folding while another may be acquired during folding, and become most closely grouped after partial unfolding (McClelland-Brown 1983).

The *contact*, or *dyke test* is useful wherever sedimentary or basement rocks are cut by later intrusions (e.g. Allerton & Vine 1987; McClelland & McCaig 1989). If a particular remanence direction is present in both a dyke and its immediate wall rocks, but disappears further away, then that component almost certainly dates from the time of dyke intrusion. Contact tests are usually more successful with small intrusions than with large ones.

Other useful tests include the *reversal test* and the *conglomerate test* (cf. McElhinny 1973). In the former, the presence of reversals of magnetization within a sedimentary sequence indicates that the sequence has not been remagnetized since the original remanence was acquired. In the latter, the presence of randomly oriented remanence directions within blocks of the same type indicates that this remanence, which may be identifiable within the source formations of the conglomerate, must pre-date deposition. If the remanence in blocks and underlying formation is uniform in direction, then both conglomerate and source have been remagnetized.

In more complex fold structures, it is useful to test whether the angle between a given remanence and the bedding remains constant across the structure. If it does, the remanence must predate folding, and can be used to constrain the evolution of the structure even where two phases of folding have occurred (cf. Bonhommet *et al.* 1981).

Palaeomagnetic studies of thrust belts in sedimentary rocks

Previous work.

On the scale of an orogenic belt, much palaeomagnetic effort has been directed towards the question of 'oroclinal bending' (e.g. Irving & Opdyke 1965; Van der Voo & Channell 1980; Lowrie & Hirt 1986). Smaller scale studies of rotations in thrust belts have been less common, but nevertheless rotations about previously unsuspected axes have been recognised from several areas.

One of the earliest such studies was by Kotasek & Krs (1965), who were able to refute arguments for the autochthonous nature of parts of the western Carpathians on the basis of palaeomagnetic evidence. Local rotations of thrust sheets around salients in the Wyoming-Idaho thrust belt have been attributed to the buttressing effects of basement highs (Grubbs & Van der Voo 1976; Schwartz & Van der Voo 1984; Eldredge & Van der Voo 1988). Around the Wyoming salient, the rotations vary more or less smoothly, while large variations in rotation around the Helena salient imply that here the thrust sheets broke and rotated as individual pieces. Kent (1988) has suggested that similar results around the Pennsylvania salient in the Appalachians are due to a tightening of originally arcuate structures. In Nevada, studies by Geissman *et al.* (1982), Hudson & Geissman (1984) and Geissman *et al.* (1984) indicate that rotation has occurred due to the oblique incidence of thrusting onto a shelf margin, compounded by buttressing effects of irregularities in the shelf margin. Opdyke *et al.* (1982) and Klootwijk *et al.* (1986) have presented evidence that the northwestern and central Himalayan regions have been thrusted over the Indian shield with a coherent large-scale clockwise rotation. In Mallorca, palaeomagnetic data has been used to rule out differential rotation as an explanation for different orientations of structures in different thrust sheets (Freeman *et al.* 1989). Channell *et al.* (1990) describe palaeomagnetic data from a thrust pile in western Sicily which show up to 140° of clockwise rotation in the uppermost unit, and decreasing amounts of rotation in structurally lower units, providing important new information for palinspastic reconstructions of the western Mediterranean region. Finally, Dinarès *et al.* (1991, this volume) have used palaeomagnetic data to define lateral variability in rotations of carbonate thrust sheets from the southern Pyrenees.

Scope and limitations of the method

Pre-deformational remanences. If pre-deformational remanences of known age are present, the evolution of complex structures can be constrained, since reversal of the deformation path must restore both bedding to a horizontal attitude and the remanence to its reference direction (cf. Bonhommet *et al.* 1981). A useful concept here is that of the *finite rotation pole* (Bates 1989). This is the pole about which the bedding rotates to horizontal and simultaneously a remanence direction of known age rotates into coincidence with the reference palaeopole for this age. Figure 1 illustrates

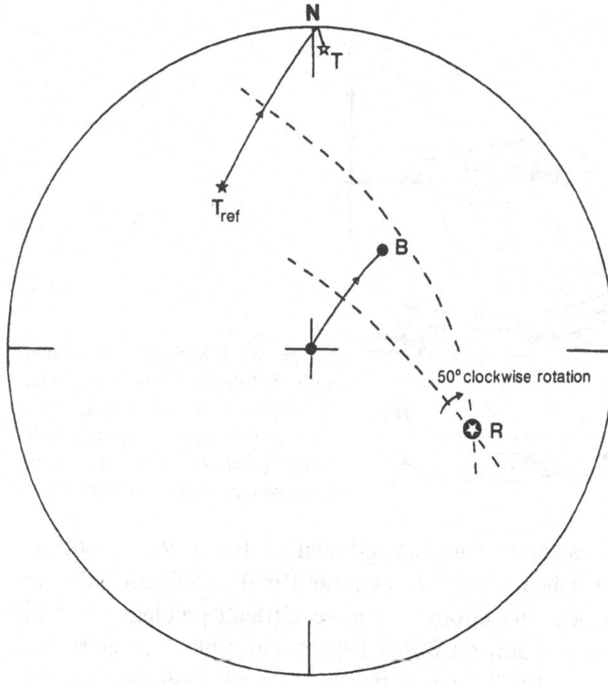

Figure 1. Determination of finite rotation pole (R) which simultaneously rotates in situ bedding pole (B) to vertical, and insitu Triassic (T) remanence direction to coincidence with the Triassic (Tref) reference direction for 'stable' Iberia. After Bates (1989). This example shows a 50° clockwise rotation about a rotation pole plunging 30° towards 115°. The dotted lines are the great circles which are the perpendicular bisectors of the small circles joining T to Tref, and B to the vertical. These are the loci of all possible rotation poles for Triassic remanence and bedding respectively. Their intersection defines the finite rotation pole.

a simple stereographic construction for finding finite rotation poles. The finite rotation matrix can be the sum of any number of finite rotations about different axes. If the amount of rotation about a measured fold axis is known from structural measurements, then the orientation of an additional rotation axis can be found by subtraction from the finite rotation (although the additional rotation may also be made up of several components). In some circumstances two pre- or syntectonic remanence components may be rotated during the same deformation event (Bates 1989), giving very tight constraints on possible rotation poles.

It is commonly supposed that because of the non-commutativity of finite rotations (e.g. Ramsay 1967), palaeomagnetic data might be capable of resolving the sequence of rotation events about different axes. However, this is not usually possible in thrust belts because identifiable rotation axes are not mutually independent. In Figure 2 a thrust sheet has been rotated about a vertical axis while beds within it have also been rotated about a subhorizontal fold axis. In this case the order of rotation makes no difference to the final orientation of a remanence direction since the fold axis will be rotated along with the thrust sheet if folding occurs first.

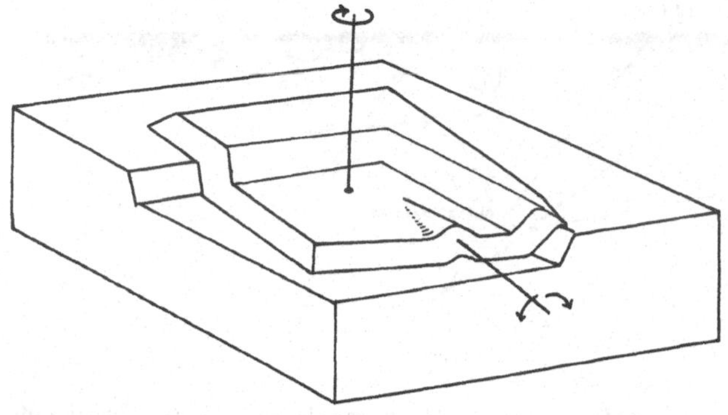

Figure 2. An example of non-independent rotations in a thrust sheet rotating about a vertical axis and a fold developing with a subhorizontal fold axis.

The orientation of the palaeomagnetic vector is an important limitation. If the vector was normal to the bedding at the time of acquisition, no additional information can be obtained from palaeomagnetic work. This means that sedimentary rocks deposited at low latitudes are likely to be best for palaeomagnetic studies of deformation. Additional restrictions are imposed by the 10-15° error limit typical of most palaeomagnetic site means. Any rotation must separate the remanence vectors before and after deformation by at least 10° if it is to be detected.

Syn-deformational remanences. Where a component of remanence is acquired during deformation or between deformation events, additional information can be obtained. For example Courtillot *et al.* (1986) determined that magnetization in Devonian limestones from a polyphase Hercynian fold was acquired between two major phases of folding, allowing them to reconstruct the shape of the fold at the end of the first phase, and to determine that the shortening direction rotated some 30° between the two phases.

Bates (1989) used multicomponent remanences to constrain the evolution of thrust sheets containing Triassic redbeds from the Nogueras Zone of the southern Pyrenees. He suggests that an intermediate blocking temperature (Tb) component was acquired in some cases while a thrust sheet was climbing an footwall ramp and the beds were tilted steeply up to the north, and in other cases, later in the evolution of the structural pile when the beds had been tilted down towards the south. A high Tb component is believed to be primary. Calculation of finite rotation poles using both intermediate and high Tb directions allowed Bates to investigate local deformation and folding; for example, several sites were interpreted as having undergone rotation about a vertical axis as a result of lateral cutoffs and imperfect lateral transfer of displacement.

A simple example of the use of syn-deformational remanences is shown in Figure 3, and comes from Triassic redbeds overlying basement rocks just north of the area studied by Bates (1989) in the Pyrenees. Intermediate and high Tb components are each well grouped at site level but vary in direction around a syncline. Application of stepwise unfolding techniques to the data demonstrates that the high Tb component was acquired before the folding, and its magnetic characteristics indicate that it is a primary Triassic remanence. After unfolding, this component gives a mean direction which is rotated about 15° clockwise from the expected Trias reference direction. The intermediate Tb component was acquired halfway through the development

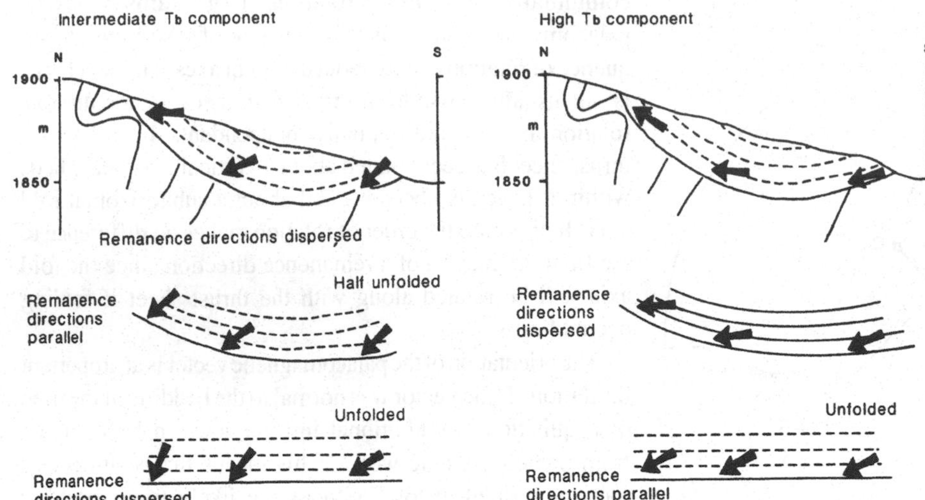

Figure 3. Schematic cross-section through fold in Triassic redbeds, Isabena valley, Pyrenean Axial Zone, showing how intermediate and high blocking temperature components group on incremental unfolding.

of the fold and after the rotation of the thrust sheet, as the partially unfolded mean is indistinguishable from the Alpine age reference direction. This data can be used to show that the rotation of this area finished before folding was complete.

Internal strain of rocks can cause problems in interpreting syn-deformational remanences (Kligfield *et al.* 1983; Cogne *et al.* 1986). Kodama (1988) has suggested that some remanences interpreted to be 'syn-folding' may have resulted either through deflection of earlier remanence directions due to internal strain or deflection of newly formed remanences away from the field direction due to magnetic anisotropy. Where two or more components are present, deflection due to strain should affect all components which are affected by the folding. If the strain can be measured independently, its effects can be removed and total rotation poles can still be established (Cogne & Perroud 1985). Deflection due to magnetic anisotropy is a more difficult problem. Recent work in southwest Wales (Ogden *et al.* 1990; Stearns & Van der Voo 1987) suggests that this may have affected some of the Devonian redbeds studied by McClelland-Brown (1983) and McClelland (1987), where syn-deformational remanences were identified. Further work is in progress to try to resolve this problem, but clearly anisotropy of magnetic susceptibility must be tested for, and results from anisotropic rocks must be viewed with caution. Syn-folding remanences certainly do occur however, as shown by the very detailed study of Hudson *et al.* (1989).

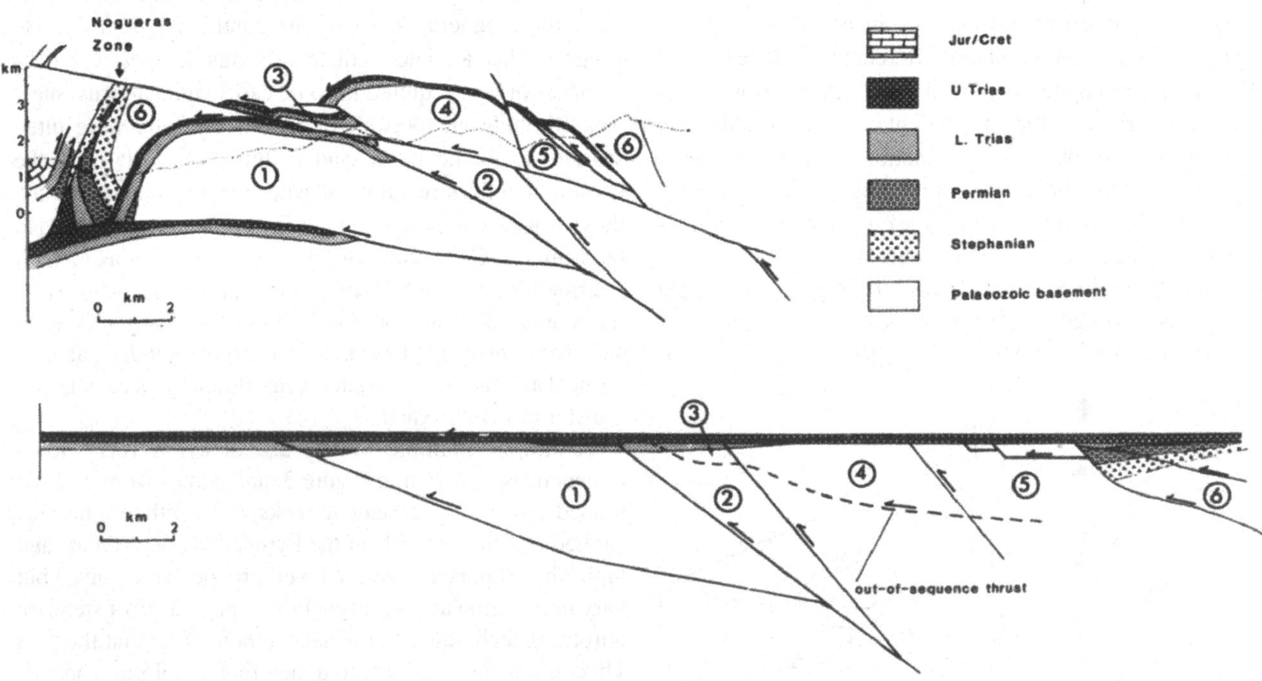

Figure 4. Cross-section 3 km west of line A-B in Fig. 5(a), constructed using the map of Mey (1969) and field data from this study. Note reactivation of normal faults to explain absence of lower Trias, Permian and Stephanian in sheet 5, and out-of-sequence thrust which truncates fold structures in Triassic footwall rocks.

Basement rocks

Scope and limitations

In basement rocks, both the scope and limitations of palaeomagnetic studies are greater than in sedimentary rocks. The scope is greater because no bedding is available to provide a palaeohorizontal datum, and palaeomagnetism may provide the only constraint on finite rotations. On the other hand, fold tests cannot be easily used, the range of useful rock types is much more limited, and rotations about axes close to the remanence direction cannot be detected. Best results are obtained when a suitable suite of contemporaneous igneous rocks such as a dyke swarm is available. Contact tests can be used to establish that remanence dates from the age of intrusion of the dykes and the orientations of remanence vectors can be compared between fault blocks. In the case cited below, the presence of overlying sedimentary rocks allows a fold test to be performed on dykes cutting basement rocks.

Results from Triassic redbeds from the southern Axial Zone of the Pyrenees have already been presented in Figure 3. In the underlying basement, late Hercynian dykes show excellent remanence characteristics and can be used to demonstrate significant clockwise rotations of thrust sheets (McClelland & McCaig 1989). Figure 4 is a balanced cross section through the area constructed using field data, which shows that six separate thrust sheets are present in an antiformal stack.

In Figure 5, finite rotations have been partitioned into a rotation about a vertical axis and tilt about a horizontal axis compared with the late Carboniferous reference direction. The horizontal axis corresponds to the dominant fold axis seen in overlying Triassic redbeds (McClelland & McCaig 1989). Figure 5(a) shows the true declination of the remanences in situ; the horizontal component of rotation is the difference between this direction and the reference direction. Figure 5(b) shows the estimates of tilt, i.e. the difference between the reference inclination and the inclination obtained at the site or group of sites shown.

In the lowermost thrust sheet (sheet 1), the tilts conform closely to the fold shape defined by the post-Hercynian unconformity. In addition, this sheet has experienced clockwise rotation (about a vertical axis) which increases from 15° in the west to 30° in the east. Thrust sheets 2 to 5 show increasing amounts of northward tilting, and clockwise rotations of between 30° and 100°. Remanence directions from sheet 6 are significantly different from those in underlying sheets, and lie close to the reference direction for the late Carboniferous, indicating either that this sheet has not been tilted or rotated, or that any rotations have cancelled each other out.

The clockwise rotations which have been documented are in agreement with palaeomagnetic results from Triassic redbeds of the Nogueras Zone immediately to the south (Bates 1989). The significant backward tilting of rocks in sheets 3 and 4 was not predicted by the structural model shown in Figure 4, and neither was the lack of tilting in sheet 6. Work is in progress on a unified structural and palaeomagnetic model incorporating all the data (McClelland & McCaig 1989; and unpubl. data).

This is the first comprehensive palaeomagnetic study of structural rotations in a thrusted basement terrane. The data illustrate the considerable scope for palaeomagnetic work in such terranes, provided that good remanence carriers exist which are not affected by folding during previous orogenic events. The studies of folded Hercynian metasediments in the same area have only rarely yielded interpretable results.

Discussion

Rotation mechanisms

Rotations within thrust belts can arise from a variety of causes: (1) movement of thrust sheets over combinations of frontal, lateral and oblique ramps; (2) folding and tilting due to internal deformation of thrust sheets; (3) differential movement on thrusts, or transfer of deformation from one thrust to another; (4) penetrative strain within thrust sheets; (5) strain taken up on networks of domino faults.

Combined palaeomagnetic and structural work has the potential to distinguish between these mechanisms. For example, movement over ramps should lead to folding about axes parallel to the lines of intersection between ramps and flats. Careful field mapping should reveal the orientations of such structures and the predicted rotations can be tested against the palaeomagnetic data. It may be possible to constrain structural geometrics either at depth or further back down the thrust sheet's trajectory (Bates 1989), particularly if syntectonic remanences can be found.

Differential movement on thrusts is one of the most commonly cited reasons for rotations about steep axes detected by palaeomagnetic studies (Schwartz & Van der Voo 1984; Bates 1989). Such models depend on pinning of thrust sheets and should result in arcuate movement directions about a local rotation pole. The radius of the arc is a simple function of the amount of rotation and the displacement on the thrusts. For small displacements, very tight arcs are required which should lead to arcuate patterns of lineation on the faults concerned and be readily detectable by structural studies.

Domino type models have been proposed to explain palaeomagnetic rotations seen in strike slip terranes (e.g. Ron *et al.* 1984). Jackson & McKenzie (1989) have shown how coupled rotations of fault blocks about vertical and horizontal axes can take up both extensional and strike-slip motions within large volumes of crust. There seems no reason why similar rotation mechanisms should not occur in thrust belts. A feature of this model is that the movement vectors on the fault array maintain a constant angle with the boundaries of the zone, and no arcuate pattern of movement directions should be expected.

In the example shown in Figure 5, many of the thrusts die out westwards into fold structures affecting the Triassic unconformity. At least part of the clockwise rotation can probably be explained by pinning of the thrust sheets in the west. However lineations on minor faults within the entire thrust stack show a remarkably consistent trend towards the

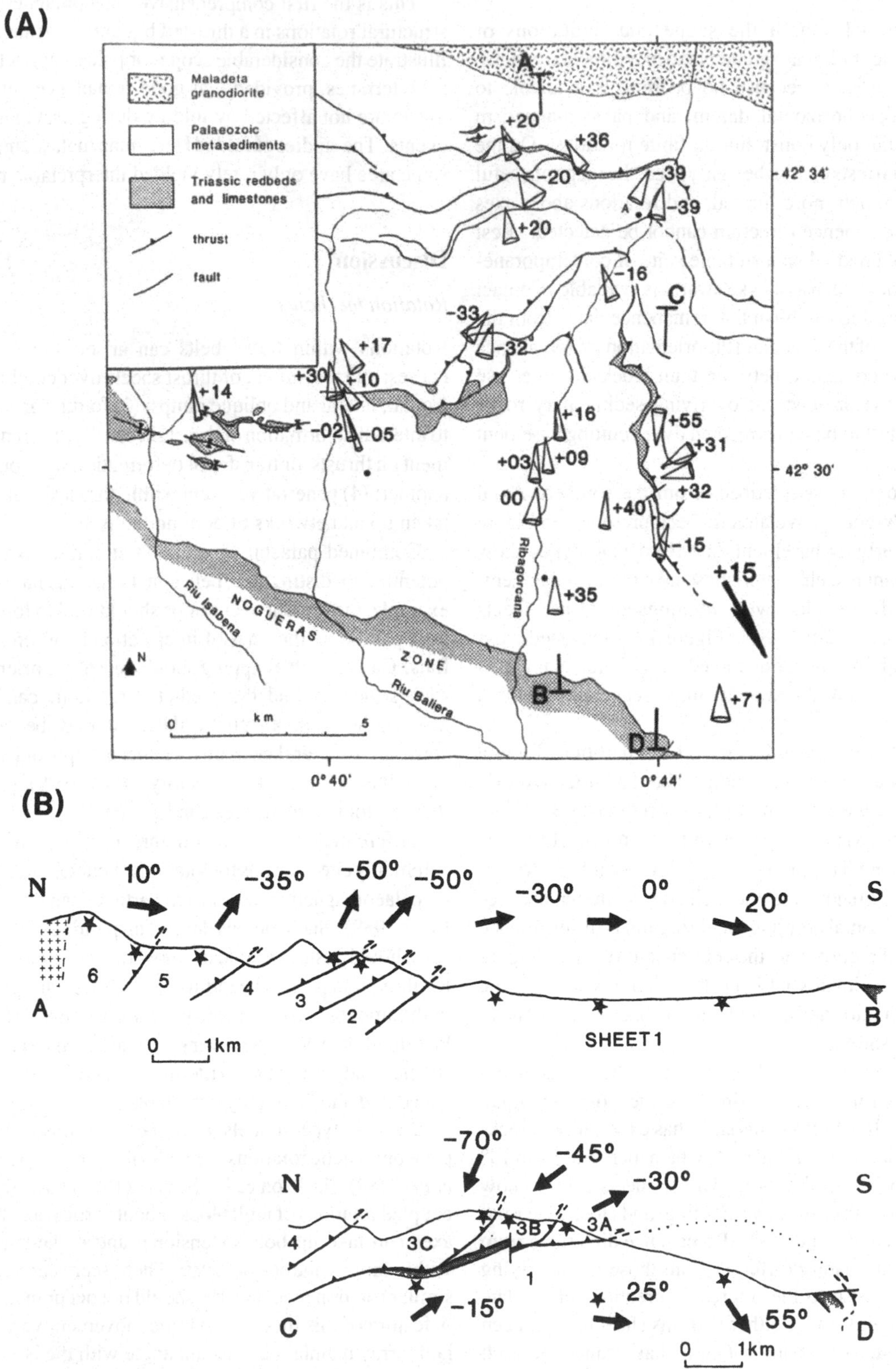

Figure 5. (a) Mean dyke remanence directions from sites or groups of sites in an antiformal stack cutting basement rocks in the southern Pyrenean Axial Zone, updated after McClelland & McCaig (1989). The cone opens towards the actual declination of the remanence (point of cone is at site), width of cone is twice alpha 95 (the cone of confidence), number by cone indicates inclination (positive numbers indicate polarity down plunge). Large arrow indicates Late Carboniferous reference field direction. A-B and C-D are lines of sections in Fig 5(b). (b) Schematic cross-sections showing palaeomagnetic estimates of tilt about a horizontal axis. Arrows are oriented at the estimated angle of tilt obtained from the group of dykes immediately below the arrow, the number next to the arrow is the tilt estimate. Note that both normal and reversed polarities exist in the dykes.

SSW, with no sign of any arcuate pattern (McCaig & McClelland unpubl. data), and it seems that part of the rotation may have occurred through domino faulting.

Section balancing

Balanced cross-sections are only valid if all strain is confined to the plane of section. Rotations about axes not perpendicular to the plane of section automatically invalidate the technique, although the extent to which errors are introduced depends on the amount of rotation and the orientation of bedding when rotation occurs. For example, rotation of a flat lying thrust sheet about a vertical axis need not introduce any excess area into the plane of section, whereas rotation of a steeply dipping or irregular sheet almost certainly will.

The area studied in the Pyrenees lies astride the section line studied and balanced by Williams (1985) and Williams & Fischer (1984). These authors did not suspect the rotations documented here, and there is no doubt that small errors in their shortening estimate will have resulted. However, greater errors were probably associated with the unrealistic geometry they proposed for the Nogueras Zone antiformal stack, in particular the failure to document out of sequence thrusts which have been identified by field mapping (Fig. 4).

Ultimately, the only way to incorporate palaeomagnetic data into such studies is by volume balancing (Butler 1990). This requires an exceedingly precise 3-D model of the present-day geometry of a portion of a thrust belt which is not generally possible where data is only available from surface mapping. It is recommended that future studies of volume balancing should concentrate on areas where extensive sub-surface data is available and where palaeomagnetic studies can also be undertaken. It should always be borne in mind, however, that the main purpose of such studies is not to endlessly refine shortening estimates but to understand the mechanical processes involved in the evolution of thrust belts and thrust geometries.

Evolution of structures

As discussed earlier, syntectonic remanences may provide valuable information about intermediate points on the strainpath leading to a particular structure. Lateral propagation of folds may lead to tilting and then untilting of beds along the fold axis, and similar effects may occur as thrust sheets move over ramps and flats. A number of studies have attempted to address this type of problem (McClelland-Brown 1983; Bates 1989), but although syntectonic remanence components have been identified, little useful data for constraining thrust sheet evolution has so far emerged, which was not already available from study of pretectonic remanences. The problem is that extremely detailed studies are required to constrain the relationship between remanence components which may have been introduced at slightly different times at different locations in the structure.

Fluid movement during thrusting

New remanence components are introduced into rocks as a result of new mineral growth or the recrystallisation of existing minerals. Numerous studies in the fields of diagenesis and metamorphism have shown that such mineral growth often occurs as a result of fluid infiltration. Palaeomagnetism therefore offers a means of constraining fluid movement pathways through deforming rocks, as well as the age of fluid movement relative to particular tilting events (cf. Bates 1989). Such studies may prove particularly valuable in the petroleum industry, where the age of a diagenetic event which affects reservoir permeability relative to the age of formation of structural traps can be of great importance.

This work was supported by NERC grant GR3/5933 to A. McCaig and E. McClelland. Reviews by P. Geiser and C. Jackson helped to improve the manuscript.

References

Allerton, S. & Vine, F. J. 1987. Spreading structure of the Troodos ophiolite, Cyprus: some palaeomagnetic constraints. *Geology*, **15**, 583-597.

Bates, M. P. 1989. Palaeomagnetic evidence for rotations and deformation in the Nogueras Zone, Central Southern Pyrenees, Spain. *Journal of the Geological Society of London*, **146**, 459-476.

Bonhommet, N., Cobbold, P. R., Perroud, H. & Richardson, A. 1981. Palaeomagnetism and cross-folding in a key area of the Asturian Arc (Spain). *Journal of Geophysical Research*, **86**, 1873-1887.

Butler, R.W.H. 1990. Evolution of fold-thrust complexes: a linked kinematic approach. *In:* Mitra S. (ed.) *Structural geology of fold and thrust belts*. Johns Hopkins University Press, Baltimore (in press).

Channell, J.E.T., Oldow, J. S., Catalano, R. & D'Argenio, B. 1990. Palaeomagnetically determined rotations in the western Sicilian fold and thrust belt. *Tectonics*, **9**, 641-660.

Cogne, J. P. & Perroud, H. 1985. Strain removal applied to palaeomagnetic directions in an orogenic belt: the Permian redbeds of the Alpes Maritimes, France. *Earth and Planetary Science Letters*, **72**, 125-140.

—— Perroud, H., Texier, M. P. & Bonhommet N. 1986. Strain reorientation of hematite and its bearing upon remanent magnetization. *In:* McClelland, E. A., Courtillot, V. & Tapponier, P. E. (eds). *Magnetotectonics*. Tectonics, **5**, 753-767.

Collinson, D. W. 1983. *Methods in Rock Magnetism and Palaeomagnetism*. Chapman and Hall, London, 503p.

Courtillot, V., Chambon, P., Brun, J. P., Rochette, P. & Matte, P. 1986. A Magnetotectonic study of the Hercynian Montagne Noire (France). *In:* McClelland, E.A., Courtillot, V. & Tapponier, P. E. (eds) *Magnetotectonics*. Tectonics, **5**, 733-751.

Dahlstrom, C. D. 1969. Balanced cross-sections. *Canadian Journal of Earth Sciences*, **6**, 743-757.

Dinarès, J., McClelland, E. & Santanach, P. 1991. Contrasting rotations within thrust sheets and kinematics of thrust tectonics as derived from palaeomagnetic data: an example from the Southern Pyrenees (this volume).

Elderedge, S. & Van der Voo, R. 1988. Palaeomagnetic study of thrust sheet rotations in the Helena and Wyoming salients of the Northern Rocky Mountains. *Geological Society of America Memoir*, **171**, 319-332.

Elliot, D. & Johnson M.R.W. 1980. Structural evolution in the northern part of the Moine thrust belt of NW Scotland. *Transactions of the Royal Society of Edinburgh, Earth Sciences*, **71B**, 69-96.

Fisher, R. A. 1953. Dispersion on a sphere. *Proceedings of the Royal Society of London, A*, **217**, 295-305.

Freeman, R., Sabat, F., Lowrie, W. & Fontbote, J.-M. 1989. Palaeomagnetic results from Mallorca (Balearic Islands, Spain). *Tectonics*, **8**, 591-608.

Geissman, J. W., Van der Voo, R. & Howard, K. L. 1982. A Palaeomagnetic study of the structural deformation in the Yerington district, Nevada, II. Mesozoic basement units and their total and pre-Oligocene tectonism. *American Journal of Science*, **282**, 1080-1109.

—— Callian, J.T., Oldow, J. S. & Humphries, S. E. 1984. Palaeomagnetic assessment of oroflexural deformation in west-central Nevada and relations to emplacement of allochthonous assemblages. *Tectonics*, **3**, 179-200.

Grubbs, K. L. & Van der Voo R., 1976. Structural deformation of the Idaho-Wyoming overthrust belt (USA), as determined by Triassic palaeomagnetism. *Tectonophysics*, **33**, 321-336.

Hossack, J. R. 1979. The use of balanced cross-sections in the calculation of orogenic contraction: a review. *Journal of the Geological Society of London*, **136**, 705-711.

Hudson, M. R. & Geissman, J. W. 1984. Preliminary palaeomagnetic data from the Jurassic Humboldt lopolith, west-central Nevada: evidence for thrust belt rotation in the Fencemaker allochthon. *Geophysical Research Letters*, **1**, 828-831.

—— Reynolds, R. L. & Fishman, N. S. 1989. Synfolding magnetization in the Jurassic Preuss sandstone, Wyoming-Idaho-Utah thrust belt. *Journal of Geophysical Research*, **94**, 13,681-13,705.

Irving, E. & Opdyke, N. D. 1965. The palaeomagnetism of the Bloomsberg redbeds and its possible application to the tectonic history of the Appalachians. *Geophysical Journal of the Royal Astronomical Society*, **9**, 153-167.

Jackson, J. & McKenzie, D. P. 1989. Relations between seismicity and palaeomagnetic rotations in zones of distributed continental deformation. *In:* Kissel, C. & Laj, C. (eds) *Palaeomagnetic Rotations and Continental Deformation*. NATO ASI Series C, **254**, 33-42.

Kent, D. V. 1988. Further palaeomagnetic evidence for oroclinal rotation in the central folded Appalachians from the Bloomsburg and Mauch Chunk formations. *Tectonics*, **7**, 749-759.

Kligfield, R., Lowrie, W., Hirt, A. & Siddans, A.B.W. 1983. Effect of progressive deformation on remanent magnetization of Permian redbeds from the Alpes Maritimes. *In:* McClelland Brown, E. & Vandenberg, J. (eds) *Palaeomagnetism of Orogenic Belts*. Tectonophysics, **98**, 43-57.

Klootwijk, C. T., Sharma, M. L., Gergan, J., Shah, S. K. & Gupta, B. K. 1986. Rotational overthrusting of the north western Himalaya: further palaeomagnetic evidence from the Riasi thrust sheet, Jammu foothills, India. *Earth and Planetary Science Letters*, **80**, 375-393.

Kodama, K. P. 1988. Remanence rotation due to rock strain during folding and the stepwise application of the fold test. *Journal of Geophysical Research*, **93**, 3357-3371.

Kotasek, J. & Krs, M. 1965. Palaeomagnetic study of tectonic rotation in the Carpathian mountains of Czechoslovakia. *Palaeogeography, Palaeoclimatology and Palaeoecology*, **1**, 39-49.

Lowrie, W. & Hirt, A. M. 1986. Palaeomagnetism in arcuate mountain belts. *In:* F.-C. Wezel (ed.) *The Origin of Arcs*. Elsevier, Amsterdam.

——,—— & Kligfield, R. 1986. Effects of tectonic deformation on the remanent magnetization of rocks. *In:* McClelland, E.A., Courtillot, V. & Tapponier, P.E. (eds) *Magnetotectonics*. Tectonics, **5**, 713-722.

McClelland-Brown, E. 1983. Palaeomagnetic studies of fold development and propagation in the Pembrokeshire Old Red Sandstone. *In:* McClelland-Brown, E. & Vandenberg, J. (eds) *Palaeomagnetism of Orogenic Belts*. Tectonophysics, **98**, 131-149.

McClelland, E. 1987. Palaeomagnetic results from the Lower Devonian Llandstadwell formation, Dyfed, Wales - Discussion. *Tectonophysics*, **143**, 335-336.

—— & McCaig, A. M. 1989. Palaeomagnetic estimates of rotations in compressional regimes and potential discrimination between thin-skinned and deep crustal deformation. *In:* Kissel, C. & Laj, C. (eds) *Palaeomagnetic Rotations and Continental Deformation*. NATO ASI Series C, **254**, 365-379.

McElhinney, M. W. 1964. Statistical significance of the *fold test* in palaeomagnetism. *Geophysical Journal of the Royal Astronomical Society*, **8**, 338-340.

—— 1973. *Palaeomagnetism and plate tectonics*. Cambridge University Press.

Mey, P.H.W. 1969. Geology of the upper Ribagorzana and Tor valleys, central Pyrenees, Spain. *Leidse Geologische Mededelungen*, **41**, 229-292.

Molnar, P. & Lyon-Caen, H. 1989. Fault plane solutions of earthquakes and active tectonics of the Tibetan Plateau and its margins. *Geophysical Journal International*, **99**, 123-153.

Ogden, K. L., Torsvik, T. H. & McClelland, E. 1990. The relationship between syn-deformational remanences and magnetic fabric in the Old Red Sandstone of SW Dyfed, Wales. *Geophysical Journal International*, **101**, 283 (abst.).

Opdyke, N. D., Johnson, N. H., Lindley, E. H. & Tahirrkhehi, L. 1982. Palaeomagnetism of the Middle Siwalik formation of Northern Pakistan and rotation of the Salt Range decollement. *Palaeogeography, Palaeoclimatology and Palaeoecology*, **37**, 1-16.

O'Reilly, W. 1984. *Rock and Mineral Magnetism*. Blackie, London, 220p.

Ramsay, J. G. 1967. *Folding and fracture of rocks*. McGraw Hill, 568p.

Ron, H., Freund, R., Garfunkel, Z. & Nur, A. 1984. Block rotation by strike-slip faulting: structural and palaeomagnetic evidence. *Journal of Geophysical Research*, **89**, 6256-6270.

Schwartz, S. Y. & Van der Voo, R. 1984. Palaeomagnetic study of thrust sheet rotation during foreland impingement in the Wyoming-Idaho overthrust belt. *Journal of Geophysical Research*, **89**, 10,077-10,086.

Van der Voo, R. & Channell, J.E.T. 1980. Palaeomagnetism in Orogenic Belts. *Revues of Geophysics and Space Physics*, **18**, 455-481.

Williams, G. D. 1985. Thrust tectonics in the south central Pyrenees. *Journal of Structural Geology*, **7**, 11-18.

—— & Fischer, M.W. 1984. A balanced cross-section across the Pyrenean orogenic belt. *Tectonics*, **3**, 773-780.

Evolution of crystalline thrust sheets in the internal parts of mountain chains

Robert D. Hatcher, Jr.[1] & Robert J. Hooper[2]

[1]Department of Geological Sciences, University of Tennessee, Knoxville, Tennessee 37996–1410 USA, and
Environmental Sciences Division, Oak Ridge National Laboratory, PO Box 2008, Oak Ridge, Tennessee 37831 USA;*
[2]Department of Geology, University of South Florida, Tampa, Florida 33620–5200 USA

Abstract: Two kinds of crystalline thrust sheets form in the internides of mountain chains. Type C megathrust sheets are internally brittle slabs of intact crust (composite basement) that detach within the thermally softened ductile–brittle transition (DBT) and, once formed, behave as thin-skinned thrust sheets. Type F thrust sheets are fold–related lobe-shaped thrust sheets that form below or within the DBT by attenuation of the common limb between antiforms and synforms in passive- or flexural–flow folding; transport is controlled by ductile flow. Type C megathrust sheets form by continent–continent or arc–continent collision accompanying A–subduction; Type F sheets form via A- or B-subduction below the DBT. Both result in crustal thickening. Individual Type C megathrusts are very strong (compared to large foreland sheets), with maximum size and displacement attained where crystalline thrusts ramp into weak zones in platform sedimentary rocks. Here crustal thickness may be duplicated, but α (basal detachment) and β (surface slope) angles remain constant (and near zero) because of slab geometry. Coefficient of internal friction along the base of nascent Type C megathrust sheets would be low (<0.3), while the sheet itself would be very strong and coherent, with a high coefficient of internal friction ($\geq$0.85). Foreland thrusts are driven ahead of Type C sheets as crystalline and foreland thrusts merge into the Coulomb wedge of the foreland. Behaviour modes also merge here, because, once formed, Type C sheets commonly ramp onto the platform and propagate along the basal detachment of the deforming platform wedge. Thin platform assemblages on continental promontories restrict the size and displacement of Type C and foreland sheets. Crustal duplexes form in ramp zones as thrusts can no longer propagate along the DBT. Duplexes of platform sedimentary rocks ($\pm$ basement) may also form beneath the crystalline sheet and arch the sheet above. Late macro- and meso–scale structures (isolated domes, out of sequence thrusts, open folds, some crenellations, and ductile shears) related to thrust emplacement may form within the sheet. Thrust–related meso- and microfabrics are mostly concentrated in or near the fault zone. Character of fault rocks varies with location, rate of motion on the fault, availability of fluid, and ambient T–P conditions in the fault zone.

Fault rocks vary from cataclasite formed near the surface or at depth at high strain rate, to retrograde mylonite formed in low–T or high strain rate zones, to annealed mylonite formed in high–T zones or zones where recrystallization/recovery rates exceed strain rate, to absent along the base of some Type F thrust sheets. Type F thrust sheets form in the middle to lower crust in upper greenschist to granulite facies conditions. They may be thicker (or thinner) and have less regular shape than Type C sheets because of association with folding, but are not as extensive because of lower overall strength. Mesofabrics within the sheets are commonly more penetrative and representative of transport and deformation of the entire sheet, with sheath folds a common meso- and macro– style, but cylindrical folds with orogen–parallel lineations normal to transport or mixed–mode linear fabrics may be present.

Crystalline and foreland thrust systems were once thought to be separate and independent structures, thus mechanical models were developed that characterized them separately (e.g. Armstrong & Dick 1974; Chapple 1978; Boyer & Elliott 1982; Davis *et al.* 1983). Certain characteristics - maximum size, location in orogens, and initial detachment properties - are certainly different; other properties, however, including geometry, appear quite similar. Thus the importance of both the similarities and differences should therefore be emphasized in any mechanical model (Hatcher & Williams 1986).

Early divisions centered around the involvement of crystalline rocks (or basement) in thrust systems. If crystalline rocks or basement were involved, the thrust system was thick-skinned; if crystalline rocks were not involved, the system

*Operated by Martin Marietta Energy systems, Inc., for the US Department of Energy under contract No. DE-AC05-840R21400.

was thin-skinned. As shown herein, however, particular kinds of crystalline thrust sheets exhibit thin-skinned behaviour. The division of thrust systems into thick- or thin-skinned based solely on the involvement of basement thus has only historical significance and should probably be abandoned for attempting to understand thrust mechanics. Interestingly, several classic studies where many of the fundamental geometric properties of thrusts were first observed were in fact conducted on crystalline thrusts - for example, the Moine thrust in the Northwest Highlands of Scotland and the thrust at Åreskutan in Sweden (see references in Hatcher & Williams 1986). Whereas they may be geometrically similar, and crystalline and foreland thrust sheets evolve behaviourally toward each other during transport, there are important mechanical differences in the way several kinds of crystalline thrusts are generated and are transported. It has been suggested that differences occur within foreland thrust systems; the mechanical characteristics of foreland fold-thrust belt thrust systems also differ significantly from those formed in accretionary wedge complexes because of fundamental differences in the mechanical properties of the materials being deformed (Woodward 1987).

The purpose here is to explore the processes related to the formation and evolution of crystalline thrust sheets that form in the internal parts of mountain chains (Fig. 1). Most kinds of crystalline thrust sheets may be grouped into two fundamental types. Type C crystalline megathrust sheets form at the ductile–brittle transition (DBT), and possibly the base of the lithosphere, and transport slabs of previously consolidated upper crust and cover sedimentary rocks. Type F crystalline sheets form below the DBT in the realm of ductile flow. These two kinds comprise most of the crystalline thrust

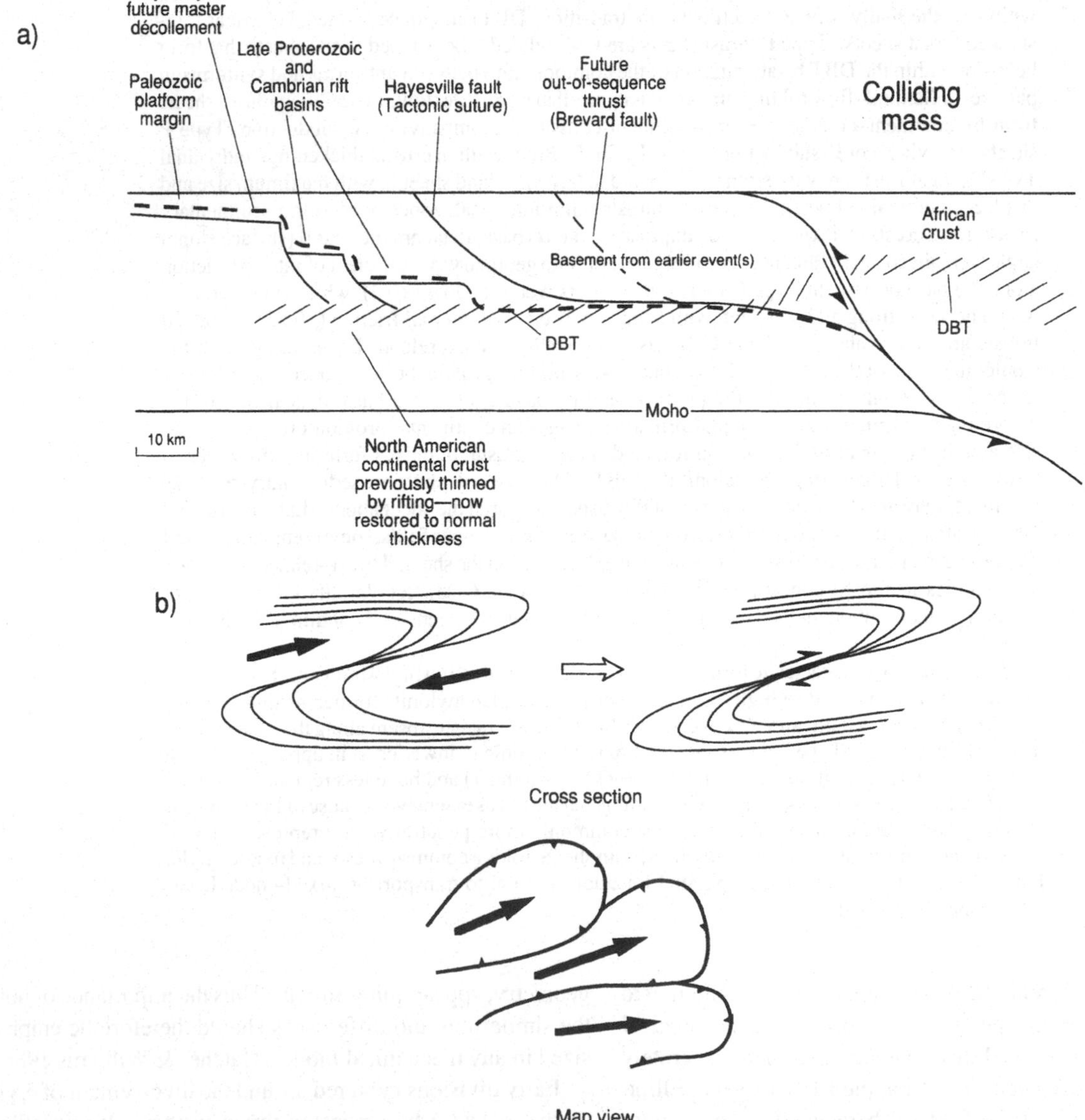

Figure 1. **(a)** Ideal Type C sheet prior to formation. Dashed line indicates the projected location of detachment as it propagates along the ductile–brittle transition (DBT) and onto the continental platform. Note the similarity of this diagram for a contractional regime to Figure 1 of Gibbs (1990), which shows a linked fault model for an extensional regime. **(b)** Formation of a Type F sheet by differential flow in a ductile mass forming a fold, then shearing out of the common limb between the antiform and synform forming a thrust. Cross section view is above; map view of lobate thrust sheets is below.

sheets observed in orogenic belts worldwide. Ophiolite sheets mechanically involve delamination of composite upper mantle, the oceanic crust, and cover sediments at the oceanic DBT, and are thus a subset of Type C thrust sheets.

Nature of the ductile–brittle transition

The ductile-brittle transition (DBT) - the brittle-plastic transition of Rutter (1986) - is the zone in the crust where thermal softening begins and brittle deformation is inhibited (Fig. 2a). Below the DBT, deformation is dominated by ductile (plastic *and* viscous) flow; above the DBT, deformation occurs mostly by brittle (elastic) failure. The depth to the DBT varies and depends amongst other things on heat flow and the amount of fluid present: the DBT may be close to the surface in zones of high heat flow where fluids may be abundant (e.g. zones of crustal extension and active magma systems coeval with faulting), and deep in the crust in zones of low heat flow where rocks contain only small amounts of fluid or are dry (e.g. continental interiors). Armstrong & Dick (1974) suggested that crystalline thrusts form by detachment along the DBT, producing thrust sheets with thicknesses ranging from 3 to 10 km. The DBT may be locally considered a planar surface. The location of this transition is largely controlled by the geothermal gradient and composition of crustal rocks, but it may also be influenced by fluid pressure and strain rate. In addition, at least two transitions probably exist, one for the crust, and a second for the upper mantle where differences exist in composition and fluid content between the quartz-rich, wet upper continental crust to the dry lower crust and the olivine-rich dry upper mantle.

Plastic flow below the DBT may be described by power laws such as

$$\dot{\varepsilon} = A\sigma^n \qquad (1)$$

where $\dot{\varepsilon}$ is strain rate, A is a material constant, σ is normal stress, and n depends on the flow mechanism. A creep law that describes shear strain rate, $\dot{\gamma}$ can be expressed as

$$\dot{\gamma} = A\tau^n e^{\frac{\psi P}{RT}} \qquad (2)$$

where τ is shear stress, ψ is another material constant (or the activation energy), P is pressure in MPa, R is the gas constant 1.987 cal/deg, and T is temperature in °K (Nicolas 1987).

Brittle and ductile behaviour are not always restricted to one or the other side of the DBT. Ductile flow may be induced above the DBT in the upper crust by local concentrations of fluids or perturbations of the geothermal gradient caused by igneous intrusion. Brittle deformation may likewise occur beneath the DBT if strain rates are high enough to exceed the rate of ductile flow - a possible explanation for deep-focus earthquakes in the lower crust and mantle. Type C thrusts that are initiated as ductile faults within the DBT may ramp into the cooler upper crust and foreland and exhibit brittle behaviour up dip.

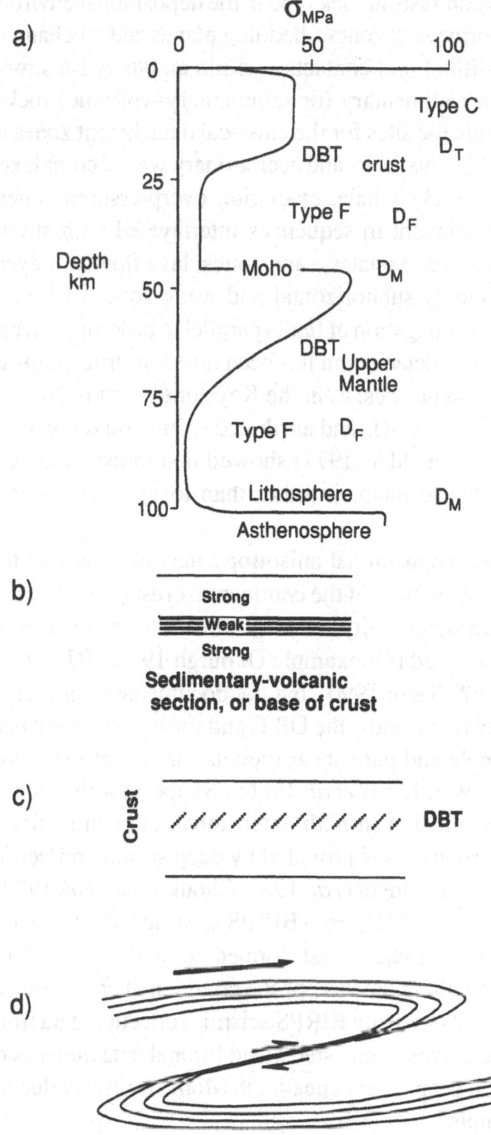

Figure 2. (a) Depth-stress diagram showing the variation in properties of crustal and mantle materials, assuming the crust is wet and quartz-bearing in the upper part, dry and quartz-bearing in the lower part and the mantle is composed of olivine and pyroxene. DBT - ductile-brittle transition. D_T - thermal detachment. D_F - differential flow-related detachment. D_M - mechanical/compositional detachment. (b) Mechanical/compositional detachment. (c) Thermal detachment. (d) Detachment related to differences in flow rate.

Detachments (decollements)

Detachments are zones of mechanical weakness zones along which faults may propagate. Those in thrust- and normal-fault systems are formed by anisotropy that may be produced in several ways (Fig. 2): (1) original compositional planar anisotropy; (2) thermally induced anisotropy; and (3) locally induced anisotropy produced by differences in ductile flow (or strain) rate.

Original compositional planar anisotropy (Fig. 2a) may

result from contrasts in rock type in the depositional environment that form weak zones - bedding planes and mechanical (structural-lithic) unit contacts - within an otherwise strong sequence of sedimentary (or sedimentary–volcanic) rocks. These provide the sites for the classical detachment zones in foreland fold-thrust belts and accretionary wedge complexes where weak rocks (shale, evaporite, overpressured zones) localize detachment in sequences interlayered with strong rocks (massive carbonates, sandstones, lava flows). Layering is commonly subhorizontal and weak zones easily accommodate propagation of faults parallel to bedding. Exceptions do occur: detachment has been noted in strong units in sedimentary sequences, as in the Keystone thrust in Nevada (Burchfiel *et al.* 1974), and attributed to high fluid pressure, but Brock & Engelder (1977) showed that thrust sheets are too permeable to maintain other than local overpressured zones.

Original compositional anisotropy may also arise in the lithosphere at the base of the continental crust (Fig. 2b), thus providing an opportunity for detachment. This mechanism has been proposed (for example Oxburgh 1972, 1974; Price 1986; Sacks & Secor 1990), but not documented, for crustal thrust detachment below the DBT, and forms an integral part of both simple and pure shear models for crustal extension (Wernicke 1985; Lister *et al.* 1986). Support for the idea of crustal delamination in both extensional and contractional crustal environments is provided by deep seismic reflection data (e.g. Allmendinger *et al.* 1987; Choukroune *et al.* 1990). Soper & Barber (1982), from BIRPS seismic reflection data, suggested the Moine thrust formed as a detachment by delamination along the base of the continental crust. Butler & Coward (1984), using BIRPS seismic reflection data from the MOIST survey, suggested an additional detachment occurred in the deep crust beneath the Moine thrust producing a crustal duplex.

Thermally induced anisotropy may be the principal mechanism worldwide for detachment in crystalline rocks in orogenic belts and zones of crustal extension (Fig. 2c). The zone of maximum anisotropic contrast occurs where thermal softening of the crust reaches a threshold where ductile flow is initiated - the DBT.

The third detachment mechanism - anisotropy induced by differences in flow or strain rate - occurs below the DBT in rocks being deformed by uniform (passive) ductile flow (Fig. 2d). Evidence for this mechanism is abundantly preserved in rocks from the internal parts of mountain chains formed during at least the past 3 Ga of Earth history. Faults are formed either compressionally by excision and simple shear producing a thrust fault along the common limb between an antiform and synform, or tensionally by ductile necking and simple shear producing ductile normal faults. Differences in flow or strain rate must be present to produce folds, or structures other than foliation, in the realm of uniform flow. Detachment initiated from instability created by folding propagates as displacement increases and thrusts with tens of kilometres of displacement may be produced. This mechanism is discussed in greater detail below.

Fault rocks

The character of fault rocks developed along thrusts vary with the thermal regime of the fault zone, the amount of fluid available in the fault zone, the shear strain rate, and the nature of the rocks involved, including their composition, grain size, and strength. While no unifying theory has been developed to fully explain the processes related to the formation of fault rocks, several recent papers address specific aspects of the problem. The relationships between formation of mylonite and cataclasite, and static and dynamic recrystallization of

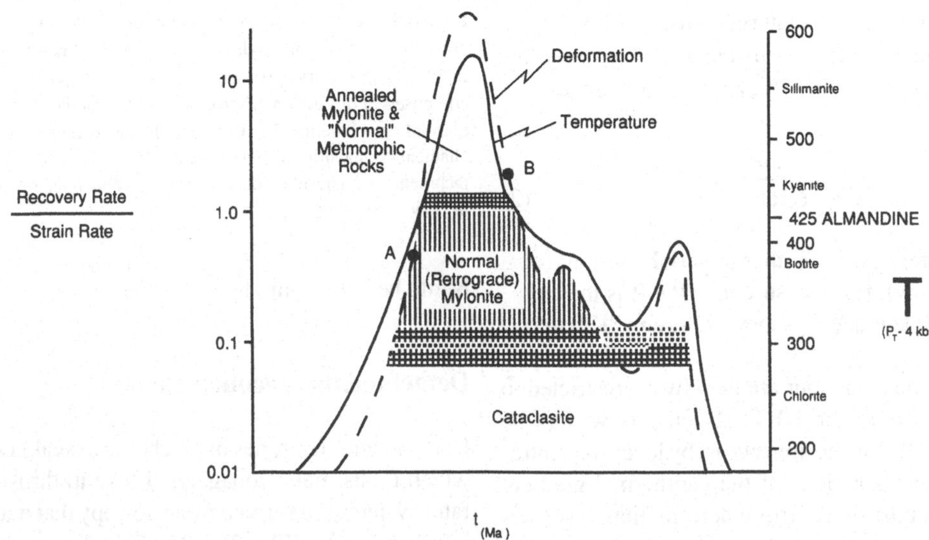

Figure 3. Relationships between recrystallization-recovery rate, strain rate, temperature and total strain and the products of deformation along a fault. Point A represents a rock that was deformed under conditions of increasing temperature. Mylonitic fabric would probably be annealed following cessation of deformation, unless the rock mass cooled without increasing the temperature further so the existing textures would be preserved. Point B represents a rock deformed following the thermal peak. Mylonitic (or metamorphic) texture would be preserved, and much of the strain in the rock may not be annealed.

mylonite and cataclasite have been discussed by Sibson (1977, 1983) and Wise *et al.* (1984); Tullis & Yund (1985) have discussed the relationships between feldspar deformation, dynamic recrystallization, and the development of ductile shear zones. Significant progress has been made in understanding the development of microfabrics in ductile fault rocks. Lister & Snoke (1984) addressed the development of S-C fabrics (shear bands) and devised a classification of mylonites that is rock–type dependent: quartzofeldspathic rocks produce type I S-C mylonite, whereas mica–rich rocks produce type II S-C mylonite. Hatcher (1978) and Hatcher & Hooper (1981) first suggested that the ratio of recovery (or recrystallization) rate to deformation (strain rate) (R/D) is important in determining the resultant fault rock type (Fig. 3). Wise *et al.* (1984) considered this ratio a determining factor. Passchier & Simpson (1986) and Hooper & Hatcher (1987), in an attempt to explain the development of porphyroclast systems, considered the R/D ratio a controlling factor in determining the type of system that develops. The R/D ratio is possibly the single most important parameter affecting the character of fault rocks produced because the ratio incorporates most of the variables affecting rock deformation. The recrystallization rate, for example, is controlled by mineralogy, temperature, prior deformation history, fluid pressure, confining pressure, and deviatoric stress. If the R/D ratio is very small, as might occur in the upper crust, brittle faults would develop. Ductile faults - and fault rocks - will form under lower crustal conditions where the R/D ratio is larger (Fig. 3).

The timing of thrusting relative to the thermal peak also affects the character of fault rocks that form. Thrusting may occur prior to, during, or after the thermal peak in an orogen affecting the degree (and rate) of recrystallization of fault rocks that form along a crystalline thrust. The character of fault rocks (texture, degree of recrystallization, incorporation of ductile fragments in cataclasite) also provides clues about the movement history of the fault.

Most discussions of fault rocks deal with retrograde mylonite without considering their relationships to the processes that form prograde regional metamorphic rocks. A clear interrelationship exists between ambient temperature and pressure (including fluid pressure) and strain rate along a moving fault and the fault rocks produced (Fig. 3).

Mylonites formed prior to a thermal peak would initially have the same retrograde character as those formed after the peak, but available thermal energy would increase the rate of dynamic or static recrystallization so that the accumulated lattice strain and microtexture are annealed. Mylonites formed during the thermal peak may also anneal and produce a microfabric that contains little evidence of shear zone deformation. The megascopic fabric would, however, survive providing most of the evidence of ductile shear. Retrograde mylonites preserve both a mega- and microfabric that is indicative of ductile deformation.

Propagation of a thrust from the more ductile to the more brittle parts of the crust at the base of a Type C megathrust sheet may involve a long history of recurrent movement as well as the movement of rocks that were at one time within or below the DBT to more cool parts of the crust. For that reason, mylonite may be superposed by cataclasite along the same fault; alternatively, recurrent motion separated by long time periods may erosionally unload a thrust sheet so that it may deform brittlely when it is reactivated. Type F sheets form below the DBT so that fault rocks are commonly prograde (annealed) or retrograde mylonite. Occasionally, both types of crystalline sheets develop with few or no fault rocks along the contacts for significant distances. Either they never developed or fault rocks formed, ceased to form, and were removed by tectonic erosion along the fault as it moved.

Types of crystalline thrust sheets

Two kinds of crystalline thrust sheets form in the internides of mountain chains (Fig. 1): Type C - internally brittle slabs of intact crust (composite basement) that detach within the thermally softened ductile–brittle transition (DBT) and, once formed, behave as thin–skinned thrust sheets; and Type F - fold–related lobe–shaped thrust sheets that form below or within the DBT by attenuation of the common limb between antiforms and synforms in passive- or flexural–flow folding where transport is controlled by ductile flow. Type C thrust sheets are commonly thought to form by continent–continent collision - or possibly arc–continent collision - accompanying A-subduction, whereas Type F sheets form via A- or B-subduction below the DBT. Ophiolite sheets, a subset of Type C thrust sheets, are probably the products of B-subduction. Crustal thickening accompanies formation of both types of thrust sheets.

Type C sheets - crystalline megathrust sheets

Individual Type C (composite) thrust sheets are very strong compared to thrust sheets in foreland fold-thrust belts, and, as a consequence, are generally much larger than the latter, with area to (restored) thickness ratios of from ~3,000:1 (Austroalpine, Moine) to ~30,000:1 (Blue Ridge–Piedmont). Maximum size and displacement are attained where detachment occurs over a wide area and the crystalline thrust ramps into one or more weak zones in platform sedimentary rocks. Here crustal thickness increases significantly during thrusting, but basal detachment (α) and surface slope (β) angles remain constant (near zero) - except in ramp zones - because of the slab geometry (Figs 1a & 4) and non-Coulomb behaviour of the thrust mass. Foreland thrusts are driven ahead of a Type C sheet where crystalline and foreland thrust geometry merge and Coulomb wedge behaviour prevails above the DBT - as predicted by Davis *et al.* (1983).

Thrust–related mesofabrics are mostly concentrated in or near the fault zone at the base of the thrust sheet. One of the best examples is the Jotun thrust sheet in southern Norway where Middle Proterozoic structures and textures are preserved except near the base of the sheet where all earlier fabrics become transposed into the strong L-S fault–related fabrics formed during emplacement of the sheet (Milnes &

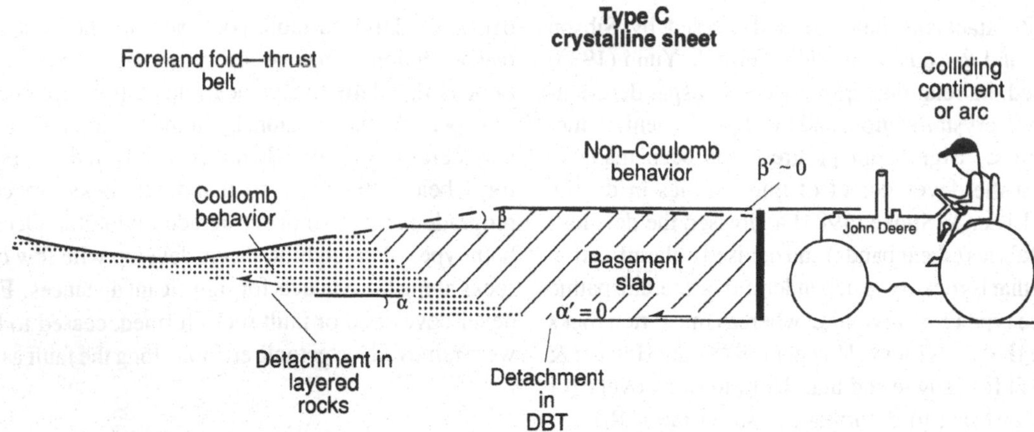

Figure 4. Diagrammatic representation of the relationships between a Type C thrust sheet that exhibits non-Coulomb behaviour and the transition to Coulomb behaviour in the foreland fold–thrust belt being pushed ahead of the crystalline sheet. α – dip of basal decollement. β – surface slope angle.

Koestler 1985). More subtle late macro- and meso-scale structures (isolated domes, out-of-sequence thrusts, open folds, some crenellations, and ductile shears) related to thrust emplacement may occur within Type C sheets.

Thin platform assemblages restrict the size and displacement of Type C sheets and foreland fold-thrust belts in general. The total width and displacement of the US Cordilleran thrust belt narrows as the aggregate thickness of the stratigraphic section decreases southward. The width of the thrust belt is at the minimum in southeastern California where the Winters Pass thrust formed as one of the few basement thrusts in this part of the orogen. The width of the thrust belt increases again southward paralleling the increase in thickness of the platform assemblage (Burchfiel & Davis 1971, 1975). A similar phenomenon occurs in the central Appalachians. Displacement in the foreland fold–thrust belt diminishes from the south to the Reading Prong and Hudson Highlands in the north where the basement and thin cover become involved in thrusting (Drake *et al.* 1988; Hatcher 1989)

The overall geometry of the fault systems carrying Type C crystalline thrust sheets is similar to that of crustal-scale fault systems recognized in extensional terranes. Gibbs (1990) provided a general model for large-scale linked fault systems in extensional terranes where discrete brittle faults above the DBT, forming asymmetrical half-graben basins, are linked to detachments in the middle to lower crust at or below the DBT. Strain is taken up in the lower crust either along a shear zone (or detachment) for throughgoing simple shear of the crust, or in some cases by more continuous deformation in a broad zone of ductile pure shear. The overall geometry of the extensional systems (see Gibbs 1990, Fig. 1) is remarkably similar to the fault systems in contractional orogens associated with Type C thrust sheets where ductile faults at or below the DBT may merge with a foreland fold-thrust belt in the upper crust (Fig. 1).

Examples

Blue Ridge–Piedmont sheet, Southern and Central Appalachians. The Blue Ridge–Piedmont (BRP) thrust sheet (Fig. 5) is one of the largest intact Type C thrust sheets in the world. It was emplaced as a product of Alleghanian (Late Carboniferous–Permian) collision between Africa and North America. The BRP thrust sheet extends through the southern and central Appalachians from Alabama to Pennsylvania and contains a number of earlier thrusts, as well as a series of Alleghanian strike–slip faults. Once the crystalline sheet ramps onto the platform, foreland deformation is driven in front of and beneath the crystalline sheet (Fig. 5). The central Piedmont suture was once considered the eastern boundary of the thrust sheet, but attempts to balance and palinspastically restore the BRP sheet (Hatcher 1989; Hatcher *et al.* 1989) require that it root east of the central Piedmont suture, so the eastern edge of the BRP is thus probably the Alleghanian suture now buried beneath the Coastal Plain. Palinspastic reconstruction yielded a minimum displacement of 350 to 400 km for the part of the sheet with the greatest displacement.

The western edge of the BRP thrust sheet consists of the eastern part of the foreland fold-thrust belt. This segment exhibits classic thin-skinned behaviour, as the master detachment ramped from the basement and older (Upper Proterozoic–Lower Cambrian) rift-drift facies sedimentary rocks into the Lower Cambrian master decollement of the foreland. Most of the BRP consists of Palaeozoic basement generated during earlier orogenies (Grenville, Taconian, Acadian) that mostly behaved as an intact homogeneous crust during the Alleghanian event. Structures and plutons contained within it were therefore cut through by the BRP master decollement. The BRP detachment propagated westward along the ductile-brittle transition in a thickened softened crust of Grenville and pre–Alleghanian Palaeozoic metamorphic rocks before it ramped from the more western (cooler? & stronger?) basement into weak units in the sedimentary cover (Fig. 5b).

Most of the internal parts of the BRP thrust sheet contain little evidence that it is a huge intact thin (2–12 km thick) sheet of crystalline rock, largely because few obviously Alleghanian structures are present in the internal parts of the sheet. The Brevard fault, a complex fault zone having a possible earlier history of dip-slip motion and an Alleghanian history of both dip and strike–slip motion, is a major Alleghanian out-of-sequence structure that resides in the

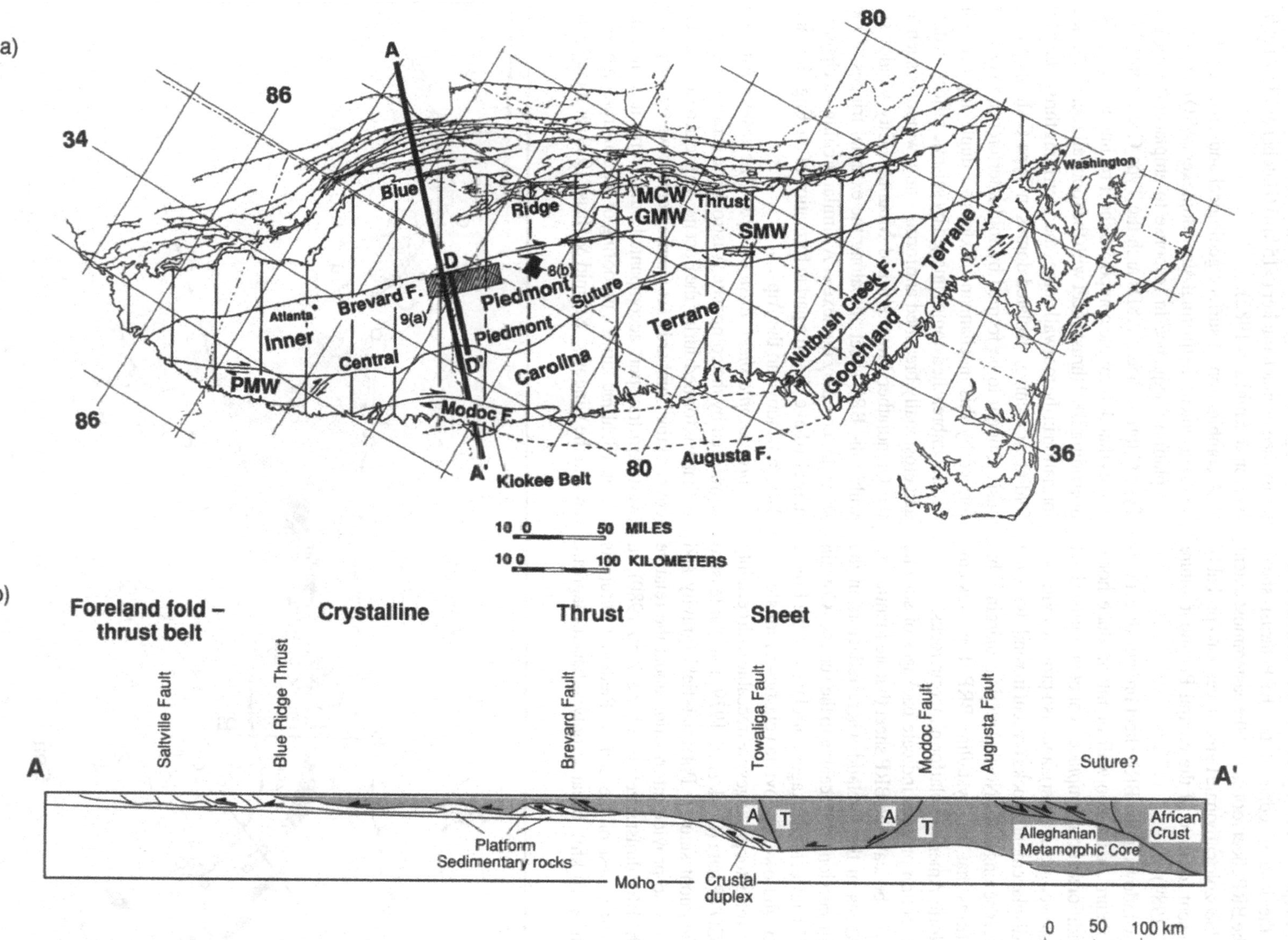

Figure 5. **(a)** Outline map of the southern and central Appalachians showing the extent of the Blue Ridge–Piedmont thrust sheet. MCW - Mountain City window. GMW - Grandfather Mountain window. PM - Pine Mountain window. Blue Ridge-Piedmont thrust sheet is shown by a diagonal line pattern. A-A' is the location of the section in (b). D-D' is the line of section in Figure 8(a). The labelled boxes indicate the locations of Figures 8(b) & 9(a). **(b)** Section through the southern Appalachians at the widest point of the Blue Ridge-Piedmont thrust sheet (patterned) showing only Alleghanian (Late Carboniferous–Permian) structures and the dominance of the Type C crystalline sheet in this part of the orogen.

BRP sheet. The carbonate slices from the Brevard fault zone provide strong evidence that the Palaeozoic carbonate platform extended at least as far east as the western Piedmont. This evidence, coupled with the presence of Lower Cambrian rocks within the leading edge of the BRP thrust sheet, requires that the BRP sheet root east of the easternmost extent of the early Palaeozoic carbonate bank succession - probably east of the present location of the central Piedmont suture (Hatcher *et al.* 1989).

Numerous windows in the BRP sheet provide direct evidence of the nature of the footwall or intermediate horse rocks, P-T conditions during emplacement of the thrust sheet (anchizone to lower greenschist), and also provide evidence for minimum displacement. Rocks present in windows in the BRP thrust sheet belong to the North American margin. This evidence confirms that the crystalline BRP sheet was emplaced onto North American platform margin rocks.

Seismic reflection profiles indicate packages of seismic reflectors present beneath the BRP sheet that are identical to reflector packages in the foreland where reflectors in the sedimentary succession are traceable to the surface (Costain *et al.* 1989). The reflector packages correlate with the Lower Cambrian through Lower Ordovician clastic–carbonate succession. These reflector packages are traceable with certainty using the COCORP data (Cook *et al.* 1979, 1983) at least to the central Piedmont suture. Potential–field (gravity and magnetic) data also provide information about the relative thickness of the BRP thrust sheet (Hatcher & Zietz 1980). In general, the BRP west of the central Piedmont suture is relatively thin (< 10 km), because only subdued magnetic anomalies may be recognized that are related to surface geological features. East of the central Piedmont suture numerous surface geological features are directly correlative with magnetic and gravity anomalies indicating the BRP thrust sheet is much thicker (>15 km) (Hatcher & Zietz 1980; Iverson & Smithson 1982).

A combination of surface geological and seismic reflection data indicate antiformal-stack duplexes (ASD) formed in the platform sequence in response to emplacement of the Type C thrust sheet. ASDs arch the Type C sheet into late domes that deform the composite sheet by interaction of the overriding BRP thrust sheet with the footwall rocks. Duplex formation in the footwall domes the overriding thrust sheet by folding or arching. Isolated domes in Type C sheets previously thought to be formed by fold interference may have formed by this mechanism. Extensive duplexing beneath several unbreached domes imaged by seismic reflection data and erosionally breached antiforms - now windows such as the Grandfather Mountain and Mountain City windows - within the BRP sheet and along the external fringes of the sheet are cored by duplexes. A similar arching mechanism may also have played a part in formation of the Engadine and Tauern windows in the Alps.

Austroalpine sheet, Eastern Alps. The eastern Alps are dominated by the composite Austroalpine thrust sheet (Fig. 6) that was formed during the Eocene. Variscan basement and cover of the Lombardy (or Carnic) terrane were transported westward transport accompanied emplacement of internal sheets (Schliniq thrust), unroofing of the Austroalpine sheet around the windows. Schmid & Haas (1989) associated the

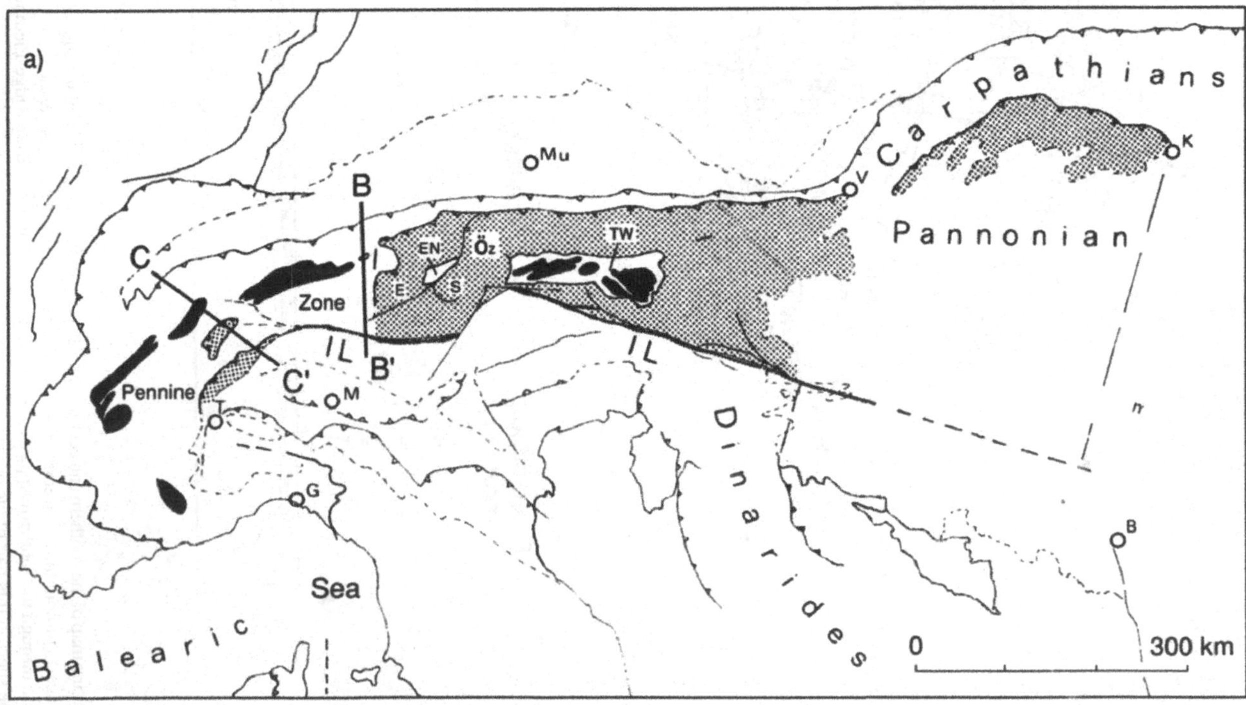

Figure 6. (a) Outline map of the Alps with the patterned area showing the distribution of the Austroalpine sheet and related structures (patterned) and external basement massifs (black). B-B' is the section in Figure 6(b). C-C' is the section in Figure 7. (After Laubscher 1988.) IL - Insubric fault zone. E - Engadine fault. EN - Engadine window. TW - Tauern window. S - Schlinig thrust. Öz - Ötztal nappe. T - Turin. M - Milan. G - Genoa. Mu - Munich. V - Venice. B - Belgrade. K - Kishinev.

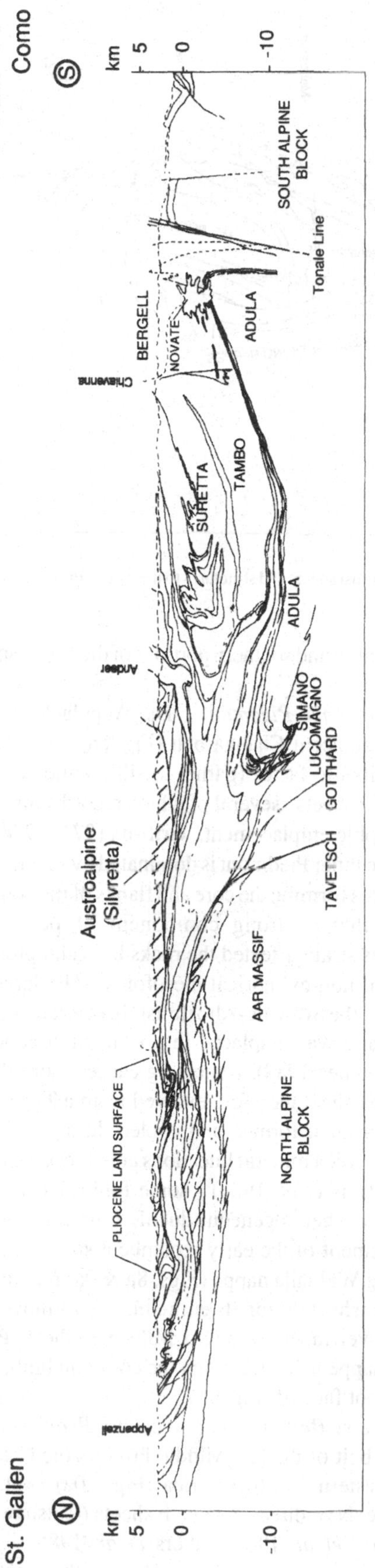

Figure 6. (b) Structural style of the Alps along a section from St. Gallen to Como. (From Milnes & Pfiffner 1980.)

sinistral Engadine fault with later deformation of the sheet. They also demonstrated that simultaneous ductile extension occurred as cover and basement were extruded beneath the over the Penninic rocks and the European foreland (Trümpy 1975). The Engadine and Tauern windows expose the European basement and cover of the Pennine zone and attest to the thin character of the sheet. Selverstone (1988) and Schmid & Haas (1989) have shown, using kinematic indicators and P-T relationships, that a major component of the Schliniq thrust and higher Ötztal nappe were compressionally emplaced during assembly of the Austroalpine sheet. All of these events would have occurred before the Austroalpine sheet was detached and emplaced above the European platform and deforming Pennine nappes. Laubscher (1988) suggested the sheet was further deformed and partially dismembered during late Eocene to early Miocene and later (Pliocene–Pleistocene) extension. Butler (1983) attempted a palinspastic reconstruction of the western Alps that would occur below the projected western extent of the Austroalpine sheet and estimated that a minimum of 100 km of shortening had occurred on these structures. The Austroalpine sheet probably had a minimum displacement of 200 km, although not all of this would have been north directed.

Yukon–Tanana sheet, Canadian–Alaskan Cordillera. The Yukon–Tanana terrane is a Type C sheet composed of a stack of thrust sheets composed of Palaeozoic sedimentary, arc (?) volcanic, and plutonic rocks formed on a continental (non-North American?) basement (Templeman-Kluit 1974; Dusel-Bacon & Aleinikoff 1985). It was imbricated with volcanic and sedimentary rocks of the Campbell Range terrane (Slide Mountain terrane of Monger 1984) and had been thrust onto the North American craton by mid–Cretaceous time (Templeman–Kluit 1979; Mortensen & Jilson 1985). It has subsequently been partially dismembered by dextral motion on the Tintina fault. This terrane is interesting because it is a sizable Type C sheet that is clearly not a product of continent–continent collision.

Type F thrust sheets - ductile, fold–related sheets

Type F thrust sheets are the same as the tectonic slides described by Fleuty (1964), and may be considered geometric analogues of the fault–propagation folds of Suppe (1985) or the break thrusts of Willis (1893). They form as recumbent structures by anisotropy induced by differences in flow (strain) rate in rocks being deformed by uniform penetrative ductile flow (Figs. 1b & 2d). They form below the DBT in the middle to lower crust under upper greenschist to granulite facies conditions. They may be thicker (or thinner) and have less regular shape than Type C sheets because of the association with folding, but are commonly not as extensive as the largest Type C sheets in the same orogen because of lower overall strength. In plan view Type F thrust sheets tend to have a lobate shape owing to a strong component of penetrative inhomogeneous simple shear producing strongly noncylindrical to sheath folds in the direction of transport.

Mesofabrics within the sheets are commonly penetrative

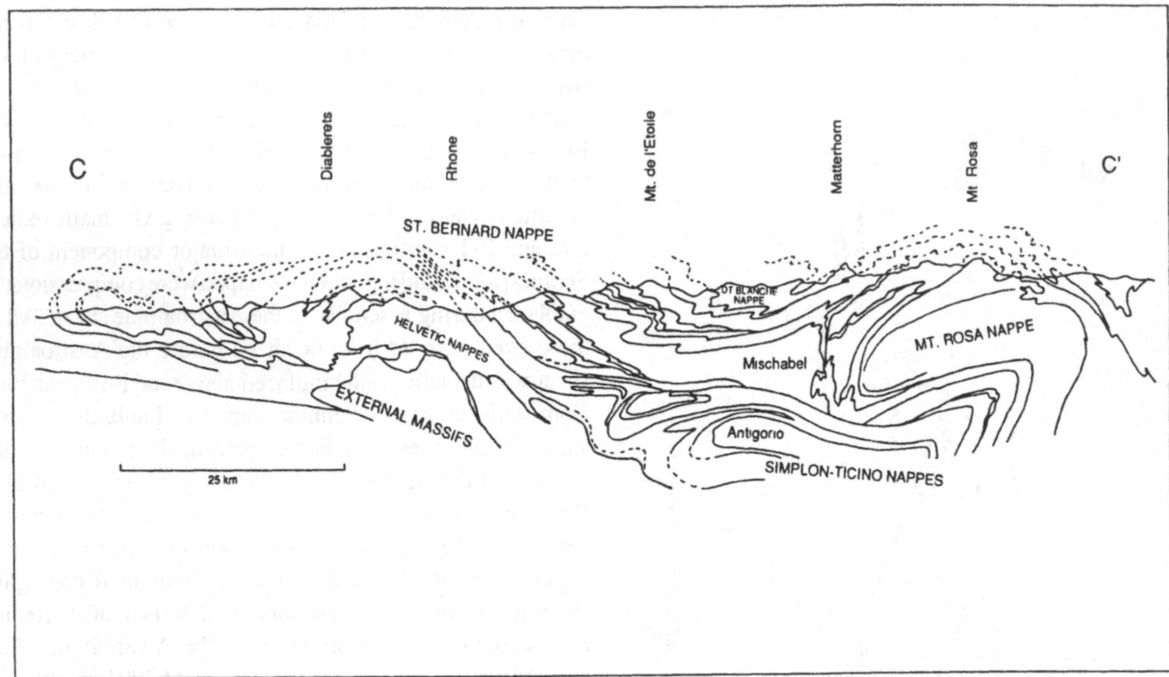

Figure 7. Section through part of the western Alps showing the structural style of Type F Pennine thrust sheets and structures to the west. (From Homewood *et al.* 1980.)

and more representative of transport and deformation of the entire sheet than mesofabrics in Type C sheets, with sheath folds a common meso- and macro-style, although cylindrical folds with orogen–parallel lineations normal to transport or mixed–mode linear fabrics may be present (Hatcher & Tabor 1989). The close relationships of internal fabrics to displacement on large faults and formation of strongly noncylindrical macroscopic folds is restricted to faults that are emplaced at or near the thermal peak. Fault fabrics may be totally annealed or nonexistent because the entire moving rock mass is in the same P–T state.

Post-metamorphic faults in these internal zones will truncate the more penetrative fabrics and form retrograde, albeit frequently ductile, assemblages and new fault fabrics along the later faults. These effectively behave as Type C sheets because they have cooled appreciably, although frequently only to lower amphibolite or greenschist facies conditions. Fault rocks may contain stable garnet-biotite that retrogrades middle to upper amphibolite facies - or higher grade - assemblages.

Examples

Pennine Alps. The structure of the Pennine zone in the Alps (Fig. 7) is dominated by several large recumbent Type F fold-thrust structures that have the typical map (e.g. Spicher 1972) and cross–section structure described above. They consist mostly of bulbous fold nappes of isoclinally folded sediments, remobilized Variscan basement and Mesozoic cover sediments, and deep ocean sedimentary and ophiolitic assemblages (Milnes 1974; Trümpy 1980). All units thin markedly into the root zone (Fig. 7). Transport directions are strongly imprinted in the mesofabrics, particularly in the higher grade central and southern portions of the zone (Spicher 1972).

Appalachian Inner Piedmont. The Appalachian Inner Piedmont and adjacent Chauga belt (Fig. 8a) consists of a stack of amphibolite facies (garnet to sillimanite zone) recumbent Type F sheets, several of which record a history of post–metamorphic emplacement. Griffin (1971, 1978) recognized that the Inner Piedmont is dominated by several large Type F structures forming the core and flanks of the zone. He also showed that a strong component of penetrative inhomogeneous strain affected the rocks here and produced a dominance of non-cylindrical mesofolds. The largest of these structures, the Six Mile-Alto thrust sheet is late- or post-metamorphic and was emplaced in a semi-rigid condition (Hopson & Hatcher 1988), truncating earlier more ductile structures. This sheet therefore evolved from a Type F to a Type C sheet after it formed and cooled during transport. Because quartz-mica mineral lineations are of mixed character (Fig. 8b, Davis *et al.* 1989) - some fold-related, others transport-related - heterogeneous ductile flow was involved in the emplacement of the early recumbent structures here. The underlying Walhalla nappe (Figs. 8a & 9a) remained as a Type F thrust sheet during its entire history of movement, and was later overridden by the Six Mile-Alto sheet. Part of the Walhalla nappe is faulted along the common limb, while part of it was not faulted (Fig. 9b).

Central Gneiss Belt Domain, Grenville Province. The central gneiss belt of the late Middle Proterozoic Grenville province of eastern North America (Fig. 10a) contains a series of lobate NW-directed Type F sheets (Culshaw *et al.* 1983; Davidson *et al.* 1982; Rivers *et al.* 1989). These structures, like many in similar zones in other mountain chains, are probably large noncylindrical to sheath structures

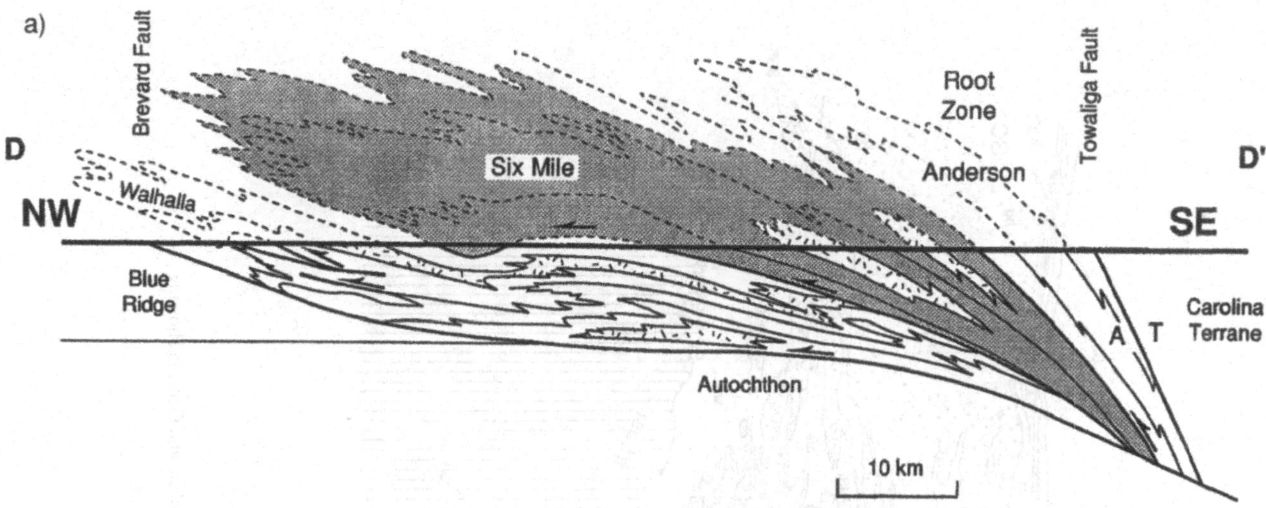

Figure 8. (**a**) Section through the Inner Piedmont in South Carolina. Type F sheets have been modified by later open folding and movement on the Alleghanian Brevard, Towaliga, and Blue Ridge-Piedmont faults. The Walhalla, Six Mile, and Anderson nappes were named by Griffin (1978).

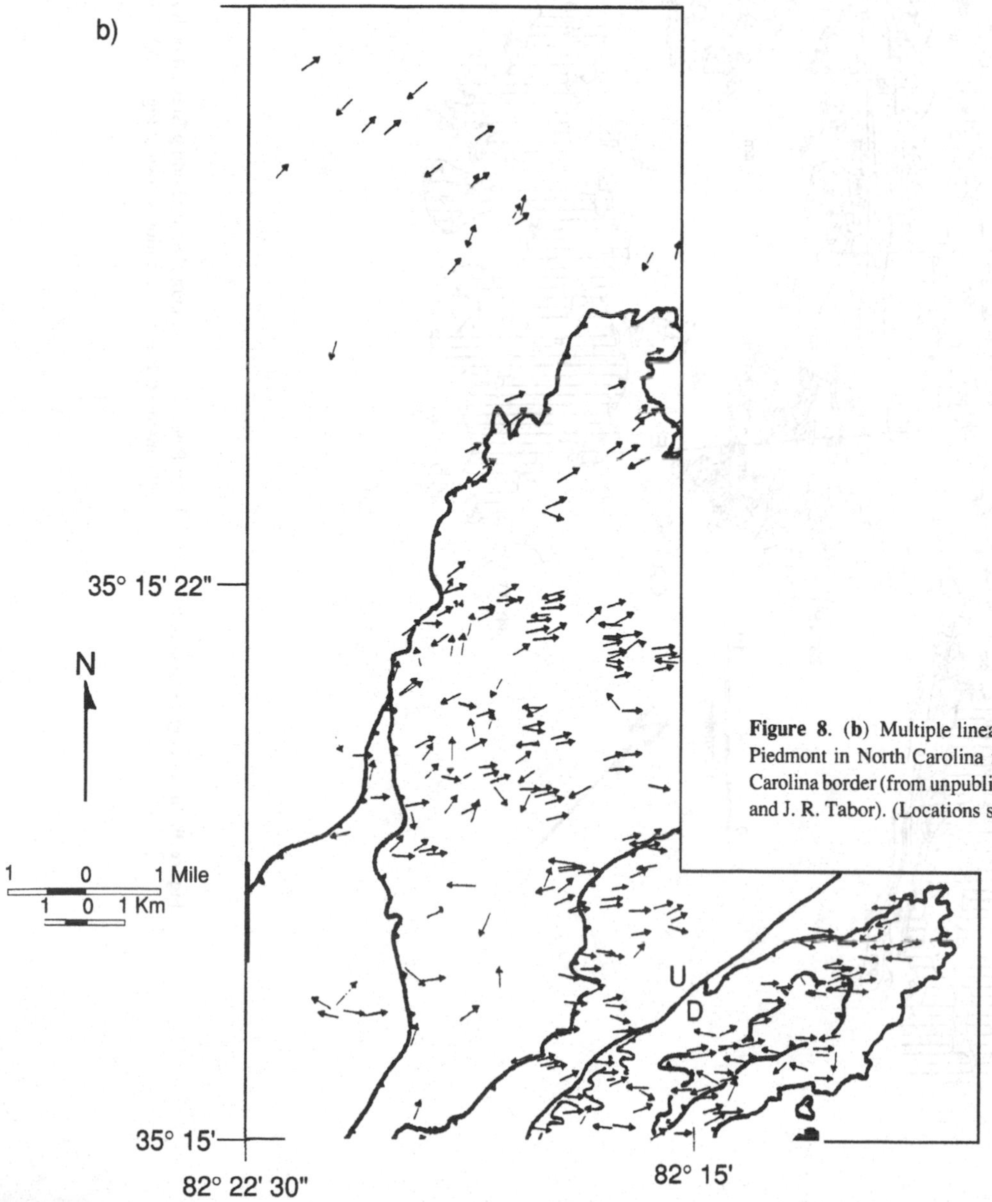

b)

35° 15' 22"

N

1 0 1 Mile

1 0 1 Km

U

D

35° 15'

82° 22' 30"

82° 15'

Figure 8. (**b**) Multiple lineation trends in the western Inner Piedmont in North Carolina immediately north of the South Carolina border (from unpublished data courtesy of T. L. Davis and J. R. Tabor). (Locations shown in Fig. 5(a).)

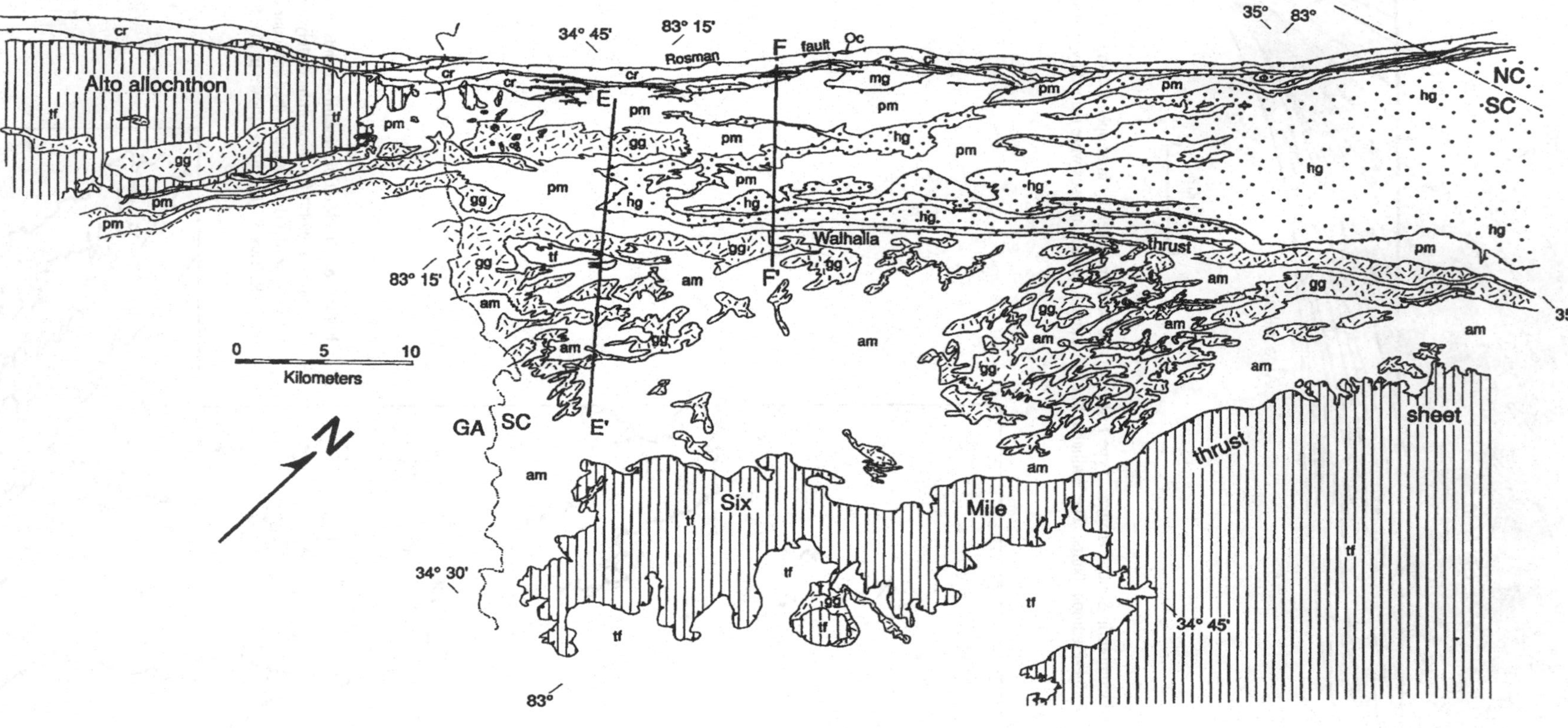

Figure 9. (a) Geological map of the western Inner Piedmont and Chauga belt (including Brevard fault zone) in South Carolina and Georgia showing the major structures. (Location shown in Fig. 5(a).)

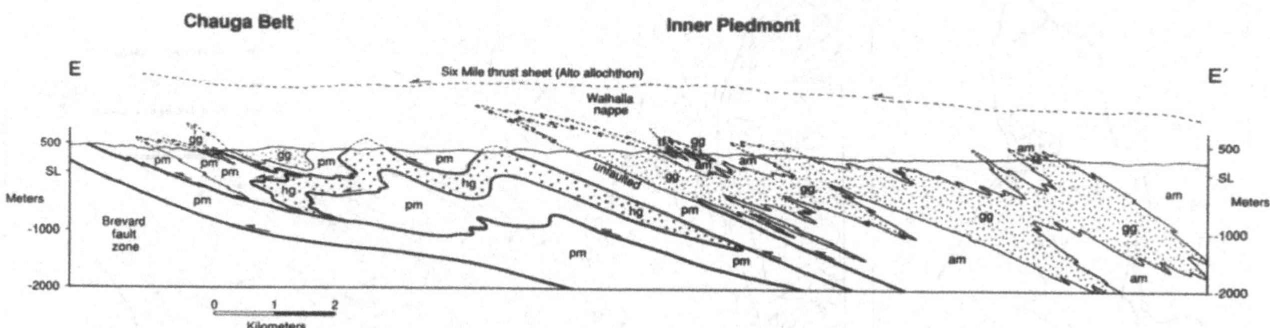

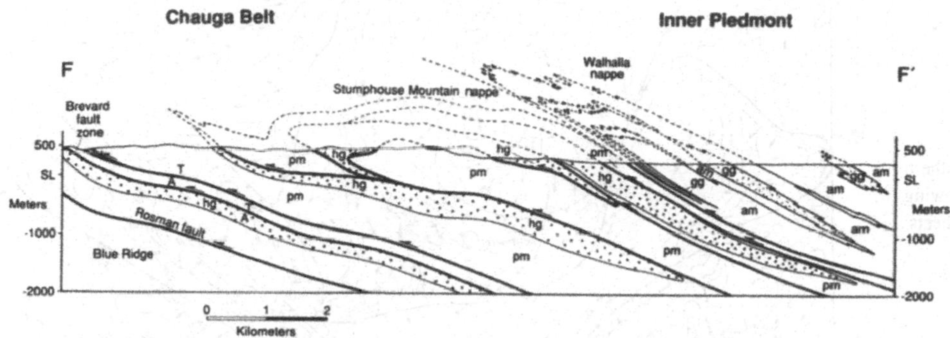

Figure 9. (**b**) Cross sections E-E' and F-F' through the Walhalla nappe (locations shown in Fig. 9(a)). am - amphibolite, hornblende gneiss, and biotite gneiss. tf - Tallulah Falls Formation. gg - Palaeozoic granitoids (cross-hatch pattern). hg - Henderson Gneiss (stippled). pm - Poor Mountain Formation. cr - Chauga River Formaiton. mg - mylonite gneiss. Oc - Ordovician platform carbonate horse (in Brevard fault zone). Six Mile thrust sheet and Alto allochthon are vertical lined. Toothed lines are thrusts (teeth on upthrown side).

dominated by mesofabrics that indicate non-uniform ductile flow of the entire rock mass as the sheets were being emplaced (Fig. 10b).

Crustal thickening processes

The processes of generation and emplacement of Types C and F thrust sheets are responsible for much of the crustal thickening and duplication accompanying orogenesis (Fig. 11). Much of this occurs during plate collision. Crustal thickening by formation of crystalline sheets is thus more a product of subduction and collision than a product that produces thickening.

An additional phenomenon that may produce crustal thickening is the formation of crustal duplexes. Structures of this kind have been proposed to explain seismic reflector packages in the Appalachians (Ando *et al.* 1983; Hatcher 1989) and British Caledonides (Butler & Coward 1984), and crustal thickening in the Himalaya (Brunel & Kienast 1986). Duplexes form in crustal rocks where faults at the base of Type C thrust sheets can no longer propagate along the DBT, so they ramp into the upper crust or, if available, into the foreland platform sequence. Crustal duplexes may be a mechanism for uplift of some high plateaus. Duplexes of platform sedimentary rocks (plus basement) may also form beneath Type C crystalline sheets and arch the sheet above, as discussed in the Blue Ridge–Piedmont sheet example above.

Discussion

Elaborate theory has evolved during the past several decades to describe the mechanics of thrust faults, principally those that form in accretionary wedges and foreland fold–thrust belts. Probably the most significant contribution was the attempt by Chapple (1978) to model foreland fold-thrust belts by assuming a perfectly plastic rheology of a deforming mass with a very weak basal layer. More recently, much of this theory has been derived using sand-box Coulomb wedge models (e.g. Davis *et al.* 1983; Dahlen *et al.* 1984; Muguleta 1988). These models approximate thrust geometry by deforming internally along predicted slip lines. Attempts have been made to apply the theory by analogy to real accretionary wedges, notably Taiwan (although it too is a non-ideal accretionary wedge), and to foreland fold-thrust belts. The theory suffers from being developed largely assuming the material is noncohesive (dry sand) and that the equilibrium critical wedge shape attained in the models may be applied at all to accretionary wedges and foreland fold-thrust belts. Dahlen *et al.* (1984) attempted to remedy the problem by extending the theory to cohesive materials, but difficulties remain with assumed coefficients of friction and the behaviour of rocks at the base of and inside the sheet. Woodward (1987) was one of the first to point out some of the problems with the necessity of internal deformation in a wedge and the lack of evidence from field studies that it actually occurs. A possible solution to the problem is by formation of out of

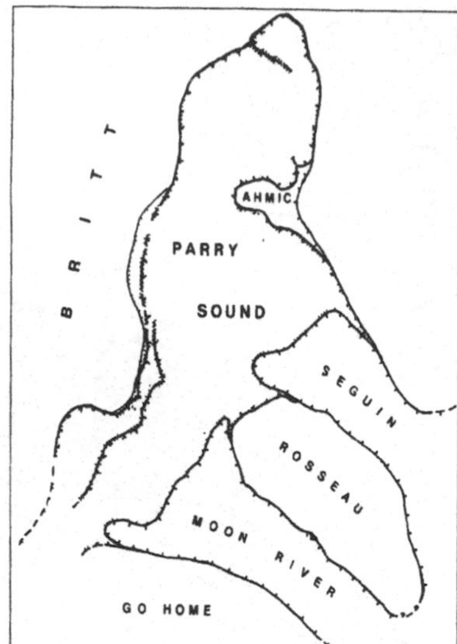

Figure 10. (a) Part of the southern Grenville Province in Ontario showing the shapes of several large Type F sheets. (From Davidson *et al.* 1982.)

Figure 10. (b) Structural trends reflecting strong body deformation in parts of the Algonquin Seguin, Rousseau, and Moon River Type F sheets in the southern Grenville Province in Ontario. (From Culshaw *et al.* 1983.)

sequence thrusts, but the number that exist falls far below the number required by the sand–box models and theory. Part of the inability to apply Coulomb theory to real thrust systems resides in the difficulty in estimating coefficients of friction for the materials involved and the strengths of the different elements that compose the thrust system. Dahlen *et al.* (1984) discussed this problem and noted that the coefficient of friction at the base of a moving thrust sheet, μ_b may be as low as 0.3 to 0.6, based on experiments with clays and clay–rich fault gouge, whereas the coefficient of friction of the sedimentary rocks of the sheet averages 0.85, as predicted by Byerlee's law.

The applicability of Coulomb wedge theory to crystalline thrust sheets has many of the same difficulties as for foreland fold–thrust belts. Type C megathrusts are generated along the DBT as non–Coulomb detachments with low basal coefficients of friction (probably <0.3), but with high strengths and coefficients of internal friction (probably >0.85) within the sheet. The latter make it impractical to ignore the cohesion term, S_0, in the Coulomb equation,

$$|\tau| = S_0 + \mu(\sigma_n - \sigma_f) \quad (3)$$

where τ is shear stress, μ is the coefficient of friction, σ_n is normal stress, and σ_f is pore fluid stress. As the thrust moves and crosses the first ramp into the rifted margin and platform sedimentary rocks, the sheet enters the wedge assemblage and will assume the behaviour of foreland sheets, except for their dimensions. The footwall at this point will be localized in either an ordinary weak unit in the platform succession, or will climb over footwall ramps like foreland sheets.

The geometry of Type C sheets could therefore be modelled like foreland sheets, as Elliott & Johnson (1980) have assumed for the Moine thrust. Butler (1986) suggested that for the Moine, there is thrust-related internal deformation (ductile shears and thrusts) in the basement rocks thereby

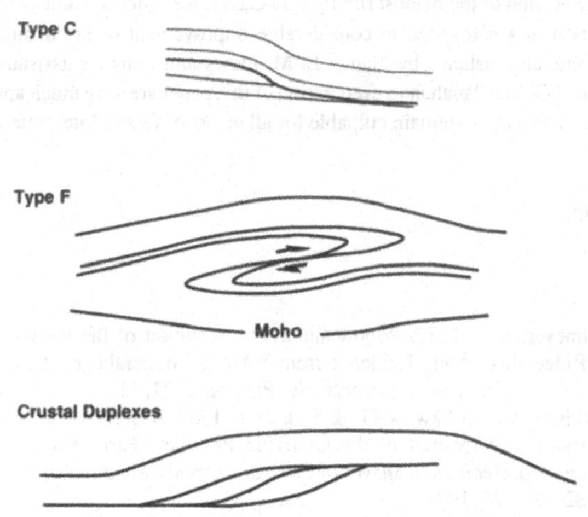

Figure 11. Crustal thickening processes involving (**a**) Type C, (**b**) Type F, and (**c**) crustal duplex structures.

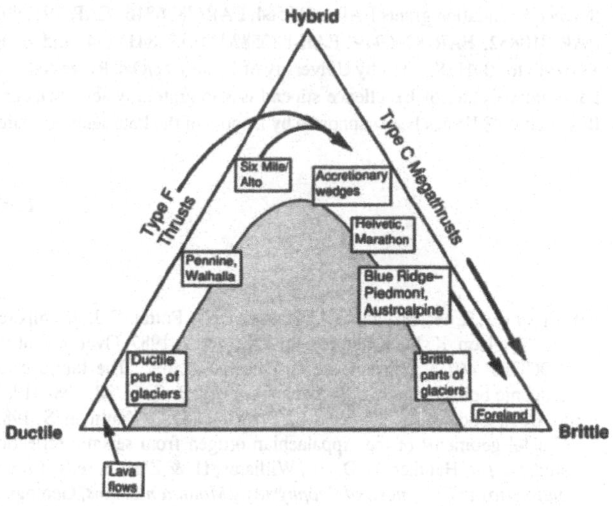

Figure 12. Rheological interrelationships between different kinds of thrust sheets. Note that some thrust sheets may evolve toward more brittle behaviour during transport and emplacement. Hybrid behaviour includes elasticoviscous and elastic–plastic behaviour; ductile includes both viscous and plastic.

indicating greater displacement than was previously thought. Wojtal & Mitra (1986) related internal deformation to strain softening and strain hardening of thrust sheets, and used the occurrence of ductile shears and fracture sets coated with low-temperature mineral assemblages as evidence of internal deformation in the western part of the Appalachian Blue Ridge–Piedmont Type C megathrust sheet. This kind of deformation is also present in the basement of the Appalachian Reading Prong in Pennsylvania and New Jersey, but here it is more difficult to find and to show that it is related to the master Type C megathrust in the internal parts of larger sheets, like the Blue Ridge–Piedmont sheet. Away from the thinner leading edges of such sheets they appear to have remained intact with less internal deformation, although ductile deformation zones with young $^{40}Ar/^{39}Ar$ ages (Dallmeyer 1989) have been recognized in the internal parts of the Blue Ridge-Piedmont megathrust sheet.

Crystalline thrust sheets can be related to each other and to other kinds of thrusts in a rheological continuum that ranges from brittle behaviour at one end to ductile (plastic or viscous) behaviour at the other (Fig. 12). Foreland thrust sheets plot closer to the elastic end, while ideal Type F sheeets would plot near the ductile end. Lava flows could be considered a kind of superductile Type F thrust sheet, because they move over a discontinuity of simple shear (the ground surface) and are internally deformed by inhomogeneous simple shear - hence their lobate shape in map view that resembles the shapes of flattened sheath folds and Type F thrust sheets. Flow patterns are similar, but the rheology is obviously different.

Many thrusts that form as Type C or Type F thrust sheets undergo a change in behaviour mode during progressive deformation and transport. This change is probably driven by decreasing temperature and pressure as the sheet moves upward in the crust. Type C megathrust sheets may become more elastic as they ramp onto a continental platform and become geometrically indistinguishable from thrust sheets in the foreland formed by fault–bend folding. Except for their size and composition, Type F thrust sheets may form in a more ductile realm as ductile internally deforming sheaths, then evolve during progressive deformation and transport into more coherent sheets that exhibit Type C hybrid behaviour.

Conclusions

1). Crystalline thrust sheets form as products of A- or B-subduction processes.
2). Crystalline sheets form mostly by non–Coulomb behaviour, but may evolve during transport toward Coulomb behaviour.
3). Detachment occurs along zones of original mechanical weakness, thermally induced anisotropy, and by variation in rates of ductile flow.
4). Type C sheets are generally slabs of consolidated basement that detach from the DBT or the base of the lithosphere as rigid masses that drive foreland deformation.
5). Type C sheets are always larger than the largest foreland sheets in the same orogen, and thus are some of the largest structures on Earth.
6). Type F sheets are lobate, sheath structures that form below the DBT by variation in rate of inhomogeneous ductile flow.
7). Type C and F sheets function in crustal thickening and underplating processes. Crustal duplexes form as byproducts of the formation of Type C sheets and also function in the thickening process.
8). Crystalline thrust sheets can be related to each other and to other kinds of thrusts in a rheological continuum that ranges from elastic behaviour at one end to ductile (plastic or viscous) behaviour at the other. They may form in one mode and evolve into another during progressive deformation and transport.

Research on crystalline thrust sheets has been supported by the US National Science Foundation grants EAR-7615564, EAR-7826316, EAR-7911802, EAR-810852, EAR-8206949, EAR-8305832, EAR-8417894, and EAR-8816343 to RDH. Support by University of Tennessee/Oak Ridge National Laboratory Center of Excellence stipend is also gratefully acknowledged. Research by RJH has been supported by Donors of the Petroleum Research Fund, administered by the American Chemical Society. Comments on an early version of the manuscript by Rod Gayer, Ken McClay and an anonymous reviewer resulted in considerable improvement of the manuscript. Secretarial assistance by Nancy L. Meadows and graphics assistance by Donald G. McClanahan in preparation of this paper are very much appreciated. The authors remain culpable for all errors of fact or interpretation.

References

Allmendinger, R. W., Hauge, T. A., Hauser, E. C., Potter, C. J., Klemperer, S. L., Nelson, K. D., Kneupfer, P. & Oliver, J. 1987. Overview of the COCORP 40° N transect, western United States: The fabric of an orogenic belt. *Geological Society of America Bulletin*, **98**, 308–319.

Ando, C. J., Cook, F. A., Oliver, J.E., Brown, L. D. & Kaufman, S. 1983. Crustal geometry of the Appalachian orogen from seismic reflection studies. *In:* Hatcher, R. D., Jr., Williams, H. & Zietz, I. (eds) *Contributions to the Tectonics and Geophysics of Mountain Chains*, Geological Society of America Memoir **158**, 83–102.

Armstrong, R. L. & Dick, H. J. B. 1974. A model for the development of thin overthrust sheets of crystalline rock. *Geology*, **1**, 35–40.

Boyer, S. E. & Elliott, D. 1982. Thrust systems. *American Association of Petroleum Geologists Bulletin*, **66**, 1196–1230.

Brock, W. G. & Engelder, T. 1977. Deformation associated with the movement of the Muddy Mountains overthrust in the Buffington window, southeastern Nevada. *Geological Society of America Bulletin*, **88**, 1667–1677.

Brunel, M. & Kienast, J–R. 1986. Étude pétro-structurale des chevauchements ductiles himalayens sur la transversale de l'Everest - Makalu (Népal) oriental. *Canadian Journal of Earth Sciences*, **23**, 1117–1137.

Burchfiel, B. C. & Davis, G. A. 1971. Clark Mountain thrust complex in the Cordillera of southeastern California: Geological summary and field trip guide. *In*: Elders, W. A. (ed.) *Geological excursions in southern California*, University of California Riverside Campus Museum Publications No. 1, 1–38.

—— 1975. Nature and controls of Cordilleran orogenesis, western United States: Extensions of an earlier synthesis. *American Journal of Science*, **275-A**, 363–396.

——, Fleck, R. J., Secor, D. T., Vincelette, R. R. & Davis, G. A. 1974. Geology of the Spring Mountains, Nevada. *Geological Society of America Bulletin*, **85**, 1013–1022.

Butler, R.W.H. 1983. Balanced cross–sections and their implications for the deep structure of the northwest Alps. *Journal of Structural Geology*, **5**, 125–137.

—— 1986. Structural evolution of the Moine thrust of northwest Scotland: a Caledonian linked thrust system? *Geological Magazine*, **123**, 1–11.

—— & Coward, M. P. 1984. Geological constraints, structural evolution, and deep geology of the northwest Scottish Caledonides. *Tectonics*, **3**, 347–365.

Chapple, W. M. 1978. Mechanics of thin–skinned fold–and–thrust belts. *Geological Society of America Bulletin*, **89**, 1189–1198.

Choukroune, P., Roure, F., Pinet, B. & ECORS Pyrenees Team 1990. Main results of the ECORS Pyrenees profile. *Tectonophysics*, **173**, 411–423.

Cook, F. A., Albaugh, D. S., Brown, L. D., Kaufman, S., Oliver, J. E. & Hatcher, R. D., Jr. 1979. Thin–skinned tectonics in the crystalline southern Appalachians: COCORP seismic–reflection profiling of the Blue Ridge and Piedmont. *Geology*, **7**, 563–567.

——, Brown, L. D., Kaufman, S. & Oliver, J. E. 1983. The COCORP seismic–reflection traverse across the southern Appalachians. *American Association of Petroleum Geologists Studies in Geology*, **14**, 1–61.

Costain, J. K., Hatcher, R. D., Jr. & Çoruh, C. 1989. Appalachian ultradeep core hole (ADCOH) Project site investigation regional seismic lines and geological interpretation. *In:* Hatcher, R. D., Jr., Viele, G. W. & Thomas, W. A. (eds) *The Appalachian–Ouachita orogen in the United States*. Geological Society of America, Boulder, The Geology of North America, **F–2**, Plate 8.

Culshaw, N. G., Davidson, A. & Nadeau, L. 1983. Structural subdivisions of the Grenville province in the Parry Sound - Algonquin region, Ontario. *Geological Survey of Canada Current Research Part A*, Paper 83–1B, 243–252.

Dahlen, F. A., Suppe, J. & Davis, D. 1984. Mechanics of fold-and–thrust belts and accretionary wedges: Cohesive Coulomb theory. *Journal of Geophysical Research*, **89**, 10,087–10,101.

Dallmeyer, R.D. 1989. Polymetamorphic evolution of the western Blue Ridge allochthon: Evidence from $^{40}Ar/^{39}Ar$ mineral ages. *Geological Society of America Abstracts with Programs*, **21**, 11.

Davidson, A., Culshaw, N. G. & Nadeau, L. 1982. A tectono–metamorphic framework for part of the Grenville Province, Parry Sound region, Ontario. *Geological Survey of Canada Current Research Part B*, Paper 82–1A, 175–190.

Davis, D., Suppe, J. & Dahlen, F. A. 1983. Mechanics of fold–and–thrust belts and accretionary wedges. *Journal of Geophysical Research*, **88**, 1153-1172.

Davis, T. L., Tabor, J. R. & Hatcher, R. D., Jr. 1989. Orogen–parallel to orogen–oblique ductile deformation and possible late Palaeozoic (?) ductile deformation of the western Piedmont, southern Appalachians. *Geological Society of America Abstracts with Programs*, **21**, A–65.

Drake, A. A., Jr., Hall, L. M. & Nelson, A. E. 1988. Basement and basement–cover relation map of the Appalachian orogen in the United States. US Geological Survey Map, **I-1655**, scale 1:1,000,000.

Dusel–Bacon, C. & Aleinikoff, J. N. 1985. Petrology and tectonic significance of augen gneiss from a belt of Mississippian granitoids in the Yukon-Tanana terrane of east–central Alaska. *Geological Society of America Bulletin*, **96**, 411–425.

Elliott, D. & Johnson, M.R.W. 1980. Structural evolution in the northern part of the Moine thrust belt, NW Scotland. *Transactions of the Royal Society of Edinburgh: Earth Sciences*, **71**, 69–96.

Fleuty, M. J. 1964. Tectonic slides. *Geological Magazine*, **101**, 452–456.

Gibbs, A. D. 1990. Linked fault families in basin formation. *Journal of Structural Geology*, **12**, 795–801.

Griffin, V. S., Jr. 1971. The Inner Piedmont of the southern crystalline Appalachians. *Geological Society of America Bulletin*, **82**, 1885–1898.

—— 1978. Detailed analysis of tectonic levels in the Appalachian Piedmont. *Geologische Rundshau*, **67**, 180–201.

Hatcher, R. D., Jr. 1978. Reply. Eastern Piedmont fault system: Speculations on its extent. *Geology*, **6**, 580–582.

—— 1989. Tectonic synthesis of the US Appalachians, Chapter 14, *In*: Hatcher, R. D., Jr., Thomas, W. A. & Viele, G. W. (eds) *The Appalachian–Ouachita orogen in the United States*. Geological Society of America, Boulder, The Geology of North America. **F-2**, 511–535.

—— & Hooper, R. J. 1981. Controls of mylonitization processes: Relationships to orogenic thermal/metamorphic peaks. *Geological Society of America Abstracts with Programs*, **13**, 469.

—— & Tabor, J. R. 1989. Orogen–parallel coaxial mineral lineations and folds - Products of buckling and passive folding in coaxial and noncoaxial flow. *Geological Society of America Abstracts with Programs*, **21**, A–177.

—— & Williams, R. T. 1986. Mechanical model for single thrust sheets Part I: Crystalline thrust sheets and their relationships to the mechanical/thermal behaviour of orogenic belts. *Geological Society of America Bulletin*, **97**, 975–985.

—— & Zietz, I. 1980. Tectonic implications of regional aeromagnetic and gravity data from the southern Appalachians. *In:* Wones, D. R. (ed.) *The Caledonides in the USA*. Virginia Polytechnic Institute Department of Geological Sciences Memoir, **2**, 235–244.

—— Thomas, W. A., Geiser, P. A., Snoke, A. W., Mosher, S. & Wiltschko, D. V. 1989. Alleghanian orogen. *In:* Hatcher, R. D., Jr., Thomas, W. A. & Viele, G. W. (eds) *The Appalachian-Ouachita orogen in the United States*. Geological Society of America, Boulder, The Geology of North America, **F-2**, 233–318.

Homewood, P., Gosso, G., Escher, A. & Milnes, A. 1980. Cretaceous and Tertiary evolution along the Besançon–Biella traverse (Western Alps). *Eclogae geologicae Helvetiae*, **73**, 635–649.

Hooper, R. J. & Hatcher, R. D., Jr. 1988. Mylonites from the Towaliga fault zone, central Georgia: Products of heterogeneous non–coaxial deformation. *Tectonophysics*, **152**, 1–17.

Hopson, J. L. & Hatcher, R. D., Jr. 1988. Structural and stratigraphic setting of the Alto allochthon, northeast Georgia. *Geological Society of America Bulletin*, **100**, 339–350.

Iverson, W. P. & Smithson, S. B. 1982. Master decollement root zone beneath southern Appalachians and crustal balance. *Geology*, **10**, 241–245.

Laubscher, H. 1988. Material balance in Alpine orogeny. *Geological Society of America Bulletin*, **100**, 1313– 1328.

Lister, G. S. & Snoke, A. W. 1984. S–C mylonites. *Journal of Structural Geology*, **6**, 617–638

——, Etheridge, M. A. & Symonds, P. A. 1986. Detachment faulting and the evolution of passive continental margins. *Geology*, **14**, 246–250.

Milnes, A. G. 1974. Structure of the Pennine zone (central Alps): A new working hypothesis. *Geological Society of America Bulletin*, **85**, 1727–1732.

—— & Koestler, A. G. 1985. Geological structure of Jotunheimen, southern Norway (Sognefjell–Valdres cross section). *In:* Gee, D. G. & Sturt, B. A. (eds) *The Caledonide orogen - Scandinavia and related areas*. John Wiley & Sons, London, 457–474.

Milnes, A. G. & Pfiffner, O. A. 1980. Tectonic evolution of the Central Alps in the cross section St. Gallen–Como. *Eclogae geologicae Helvetiae*, **73**, 619-633.

Monger, J.W.H. 1984. Cordilleran tectonics: A Canadian perspective. *Société Géologique de France Bulletin*, **26**, 255–278.

Mortensen, J. K. & Jilson, J. A. 1985. Evolution of the Yukon-Tanana terrane: Evidence from southeastern Yukon Territory. *Geology*, **13**, 806–810.

Mulugeta, G. 1988. Modeling the geometry of Coulomb thrust wedges. *Journal of Structural Geology*, **10**, 847–859.

Nicolas, A. 1987. *Principles of Rock Deformation*. D. Reidel Publishing Company, 208p.

Oxburgh, E. R. 1972. Flake tectonics and continental collision. *Nature*, **239**, 202–204.

—— 1974. The plain man's guide to plate tectonics. *Proceedings of the Geological Association*, **85**, 299–357.

Passchier, C. W. & Simpson, C. 1986. Porphyroclast systems as kinematic indicators. *Journal of Structural Geology*, **8**, 831–843.

Price, R. A. 1986. The southeastern Canadian Cordillera: Thrust faulting, tectonic wedging, and delamination of the lithosphere. *Journal of Structural Geology*, **8**, 239–254.

Rivers, T., Martignole, J., Gower, C. F. & Davidson, A. 1989. New tectonic divisions of the Grenville province, southeast Canadian Shield. *Tectonics*, **8**, 63–84.

Rutter, E. M. 1986. On the nomenclature of mode of failure in rocks. *Tectonophysics*, **122**, 381–387.

Sacks, P. E. & Secor, D. T., Jr. 1990. Delamination in collisional orogens. *Geology*, **18**, 999–1002

Schmid, S. M. & Haas, R. 1989. Transition from near–surface thrusting to intrabasement decollement, Schliniq thrust, eastern Alps. *Tectonics*, **8**, 697–718.

Selverstone, J. 1988. Evidence for east–west crustal extension in the eastern Alps: Implications for the unroofing history of the Tauern window. *Tectonics*, **7**, 87–105.

Sibson, R. H. 1977. Fault rocks and fault mechanisms. *Geological Society of London Journal*, **133**, 191–213.

—— 1983. Continental fault structure and the shallow earthquake source. *Geological Society of London Journal*, **140**, 741–767.

Soper, N. J. & Barber, A. J. 1982. A model for the deep structure of the Moine thrust zone. *Geological Society of London Journal*, **139**, 127–138.

Spicher, A. 1972. *Tektonische Karte der Schweiz*. Wepf and Company, Basel. Schweizerische Geologische Kommission. scale 1/500,000.

Suppe, J. 1985. *Principles of Structural Geology*. Prentice–Hall, Englewood Cliffs, New Jersey, 537p.

Templeman–Kluit, D. J. 1974. The Yukon crystalline terrane: Enigma in the Canadian Cordillera. *Geological Society of America Bulletin*, **87**, 1343–1357.

—— 1979. Transported cataclasite, ophiolite and granodiorite in Yukon: Evidence of arc–continent collision. Geological Survey of Canada Paper, **79-14**, 27.

Trümpy, R. 1975. Penninic–Austroalpine boundary in the Swiss Alps: A presumed former continental margin and its problems. *American Journal of Science*, **275-A**, 209–238.

—— (ed.) 1980. *Geology of Switzerland, a guidebook. Part A: An outline of the geology of Switzerland*. Wepf & Company, Basel, Schweizerische Geologische Kommission. 104p.

Tullis, J. & Yund, R. A. 1985. Dynamic recrystallization of feldspar: A mechanism for ductile shear zone formation. *Geology*, **13**, 238–241.

Wernicke, B. 1985. Uniform–sense normal simple shear of the continental lithosphere. *Canadian Journal of Earth Sciences*, **22**, 108–125.

Willis, B. 1893. The mechanics of Appalachian structure. *US Geological Survey 13th Annual Report 1891-1892, Part 2*, 212–281.

Wise, D. U., Dunn, D. E., Engelder, J. T., Geiser, P. A., Hatcher, R. D., Jr., Kish, S. A., Odom, A. L. & Schamel, S. 1984. Fault–related rocks: Suggestions for terminology. *Geology*, **12**, 391–394.

Wojtal S. & Mitra, G. 1986. Strain hardening and strain softening in faults zones from foreland thrusts. *Geological Society of America Bulletin*, **97**, 674–687.

Woodward, N. B. 1987. Geological applicability of critical-wedge thrust–belt models. *Geological Society of America Bulletin*, **99**, 827–832.

PART FOUR

Pyrenees

Evolution of a continental collision belt: ECORS-Pyrenees crustal balanced cross-section

Josep Anton Muñoz

Departament de Geologia Dinàmica, Geofísica i Paleontologia, Facultat de Geologia, Universitat de Barcelona, Zona Universitària de Pedralbes, 08028, Barcelona, Spain

Abstract: Construction of a crustal balanced cross-section across the Pyrenean chain shows a minimum shortening of 147 km, 112 km of which are related to stacking of basement thrust sheets in the southern Pyrenees. Metamorphic conditions of the basement rocks, as well as thrust geometry, indicate the maximum depth for the detachment level to be at 15 km. In the restored cross-section, the upper crust is 110 km longer than the lower layered crust. The lower crust was subducted together with the lithospheric mantle into the asthenospheric mantle and has not been imaged by geophysical data probably because an increase of density through eclogitic metamorphism. The upper crust constitutes an orogenic lid mainly deformed by thrust structures. The balanced cross-section has been constrained by the ECORS deep reflection seismic profile as well as detailed surface data and available commercial seismic and oil well data. The restored cross-section provides a better picture of the middle Cretaceous combined strike-slip and extensional fault system as well as of the Hercynian crust. Hercynian geological features have been used as an additional tool for the restoration of the basement thrust sheets.

The Pyrenees form an Alpine collision belt located in between the Iberian and European plates. Formation of the Pyrenean orogen is related to the kinematics of the Iberian plate, largely dependent on the motion of the larger neighbouring plates (Eurasia and Africa). Knowledge of the past positions of the plates on both sides of the Pyrenees would constrain the amount of convergence and the amount of shortening involved in the collision. Convergence occurred from Campanian to Early Miocene time, and resulted in a partial closure of the Bay of Biscay along the North Spanish subduction zone (Boillot & Capdevila 1977), and into continental collision in the Pyrenean chain. The exact separation distance between the Iberian and European plates at the beginning of the convergence is not very well constrained, although several reconstructions of Iberia deal with the amount of rotation and sinistral displacement of the Iberian plate with respect to Europe during Cretaceous times. Different solutions point to a separation between 100 km and 150 km (Grimaud *et al.* 1982; Olivet *et al.* 1984; Boillot 1986).

Thrust structures and related synorogenic materials are very well preserved and, as a consequence, the Pyrenean chain has recently been the focus of much detailed structural work. During 1985-1986, a deep seismic survey across the Pyrenees was completed adding information about this orogen (ECORS Pyrenees team 1988; Choukroune *et al.* 1989). The 250 km long deep reflection seismic profile (ECORS-PYRENEES) traverses the main Pyrenean structural units across the central Pyrenees (Fig. 1). Knowledge of thrust structure at the surface and at moderate depths from available commercial seismic and oil-well data, together with the ECORS profile permits construction of a reasonably well constrained crustal cross-section.

Crustal balanced cross-sections can be constructed similarly to the sections of the frontal parts of the orogenic belts (Dahlstrom 1969; Hossack 1979), but the assumptions normally adopted in the balancing methodology strongly limit the restoration of complete orogenic belts. In most of the collision belts, the deep crustal structure is poorly constrained and the internal parts are thermally remobilized and highly deformed by ductile structures. In the case of the Pyrenees, the ECORS profile provides data to constrain the crustal geometry at depth. Moreover, deformation and thermal processes in the internal parts were not strong enough to destroy the pre-Alpine features completely. In fact, the Pyrenees are one of the few mountain chains where no metamorphic or plutonic processes have occurred during collision. All these peculiar characteristics of the Pyrenean chain allow us to construct a complete crustal balanced cross-section which has provided further insight into the tectonic evolution of the chain.

Complete crustal balanced cross-sections have been drawn during the last decade for the Alps (Beach 1981; Butler 1986; Menard & Thouvenot 1987), the Himalayas (Butler & Coward 1989) and the Pyrenees (Deramond *et al.* 1985; Seguret & Daignières 1986). Most of them are based on the determination of the depth and geometry of the sole thrust, deduced from conventional balanced cross-sections of the external parts of the orogenic belt together with the depth of the Moho below the chain, determined by refraction seismic data. These sections are not very well constrained by deep seismic data and the geometry of the crustal thrust system is poorly determined. As a result, some of the initial solutions expressed in these sections have been reconsidered, once deep reflection seismic profiles have been obtained. Crustal balanced

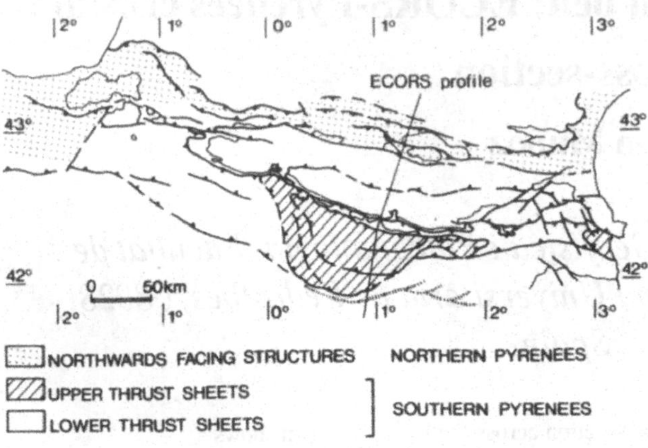

Figure 1. Structural sketch of the Pyrenees and location of the ECORS profile.

cross-sections across the Alps pointed out a paradox in the difference in the amount of apparent shortening for the surface (basement-cover contact) versus the lower crust and Moho. Two possibilities were proposed: either the difference was the result of continental subduction (Butler 1986) or the thickening of the crust was the result of the shortening of a previously thinned crust (Beach 1981). New information on the deep structure furnished by the deep seismic profiles across the Pyrenees and the Alps (Choukroune *et al.* 1989; Bayer *et al.* 1987) can be used for updating earlier attempts at crustal balanced cross-sections and to elucidate new ideas about thrusting kinematics at crustal scale. Crustal balanced cross-sections have already been proposed along the new deep seismic profiles across the Pyrenees (Roure *et al.* 1989) and across the Alps (Mugnier *et al.* 1990).

The main goal of this contribution is to discuss the geometry of the Pyrenean thrust system at a crustal scale, as deduced from the construction of a crustal balanced cross-section along the ECORS deep seismic profile. Questions concerning the evolution of the Pyrenean continental collision belt that are discussed include the influence of the pre-collision crustal geometry, location of detachment levels and the amount of orogenic contraction. The ECORS balanced cross-section has been constructed from the combination of field data, available commercial subsurface seismic and well data and the deep seismic data furnished by the ECORS seismic profile (Choukroune *et al.* 1989).

Main structural features: A cross-section from the Ebro foreland to the North Pyrenean frontal thrust.

The ECORS cross-section traverses the main structural units of the Pyrenean chain which are from south to north: the Cover Upper thrust sheets, the basement-involved Lower thrust sheets, the North Pyrenean fault zone and the North Pyrenean thrust sheets, which also involve Hercynian basement rocks (Figs 1 & 2).

Cover Upper thrust sheets

The Cover Upper thrust sheets, also known in the Central Pyrenees as South Pyrenean Central Unit (Seguret 1972), are made up by Mesozoic, mainly platform series, and Palaeogene rocks. They form three thrust sheets which are, from south to north, Serres Marginals, Montsec and Bóixols (Fig. 2). These thrust sheets have been imbricated southwards over an autochthonous Palaeogene and very reduced Mesozoic series which directly overlies the basement of the Ebro basin (Camara & Klimovitz 1985). The South Pyrenean front is characterized by a triangle zone structure (Fig. 2). In its northern limb, Mesozoic rocks have been thrust onto Upper Eocene-Lower Oligocene gypsum and fine-grained clastic rocks of the Ebro foreland basin. Southwards, a backthrust and related structures deform the northern foreland sequences. Below and south of the thrust front, a detachment is located in the Upper Eocene-Lower Oligocene gypsum as evidenced by open folds at the surface above flat lying basement and Lower Palaeogene rocks (Fig. 2).

The Serres Marginals thrust sheets are located between the southern thrust front (South Pyrenean Main thrust) and the Montsec thrust, to the north (Pocovi 1978; Martinez & Pocoví 1988). They consist of several small imbricate units characterized by a thin Mesozoic sequence, which becomes thicker and more complete to the north (Pocoví 1978). In the Serres Marginals, the Upper Eocene-Lower Oligocene conglomerates, sandstones and gypsum unconformably overlie Lower Eocene Alveolina limestones, Palaeocene and Mesozoic rocks, onlapping previously imbricated structural units. The Upper Eocene-Lower Oligocene rocks also display syntectonic relationships with thrust structures, the younger Oligocene sequences also being affected by thrusts. All these relationships imply an initial emplacement of the Serres Marginals units between the Early Eocene and the Late Eocene and a later deformation in the Late Eocene-Oligocene. The latter thrust structures have been developed in a break-back sequence synchronous to the development of the South Pyrenean Main thrust and related backthrusts, as clearly demonstrated more to the east of the section, in equivalent structural units (Vergés & Muñoz 1990).

Northwards, the Montsec thrust sheet (Fig. 2) contains a Mesozoic and Cenozoic sequence. The Mesozoic sequence is about 2000 m thick, mainly Upper Cretaceous limestones. The Cenozoic sequence is represented by Palaeocene and Lower and Middle Eocene clastic rocks (Tremp-Graus basin). The Montsec thrust sheet presents a simple structure, mainly consisting of a broad syncline which supports the Tremp-Graus basin. The Montsec thrust sheet is bounded to the south by the Montsec thrust which displays a cartographic arc geometry, as does the South Pyrenean Main thrust (Fig. 1). In the ECORS traverse, the age of the Montsec thrust emplacement was determined to be Ypresian as recorded by the syntectonic sediments at both sides of the thrust, especially to the south, in the Ager basin (Williams & Fischer 1984; Mutti *et al.* 1985; Farrell *et al.* 1987). A minimum displacement of about 10 km along the Montsec thrust is well

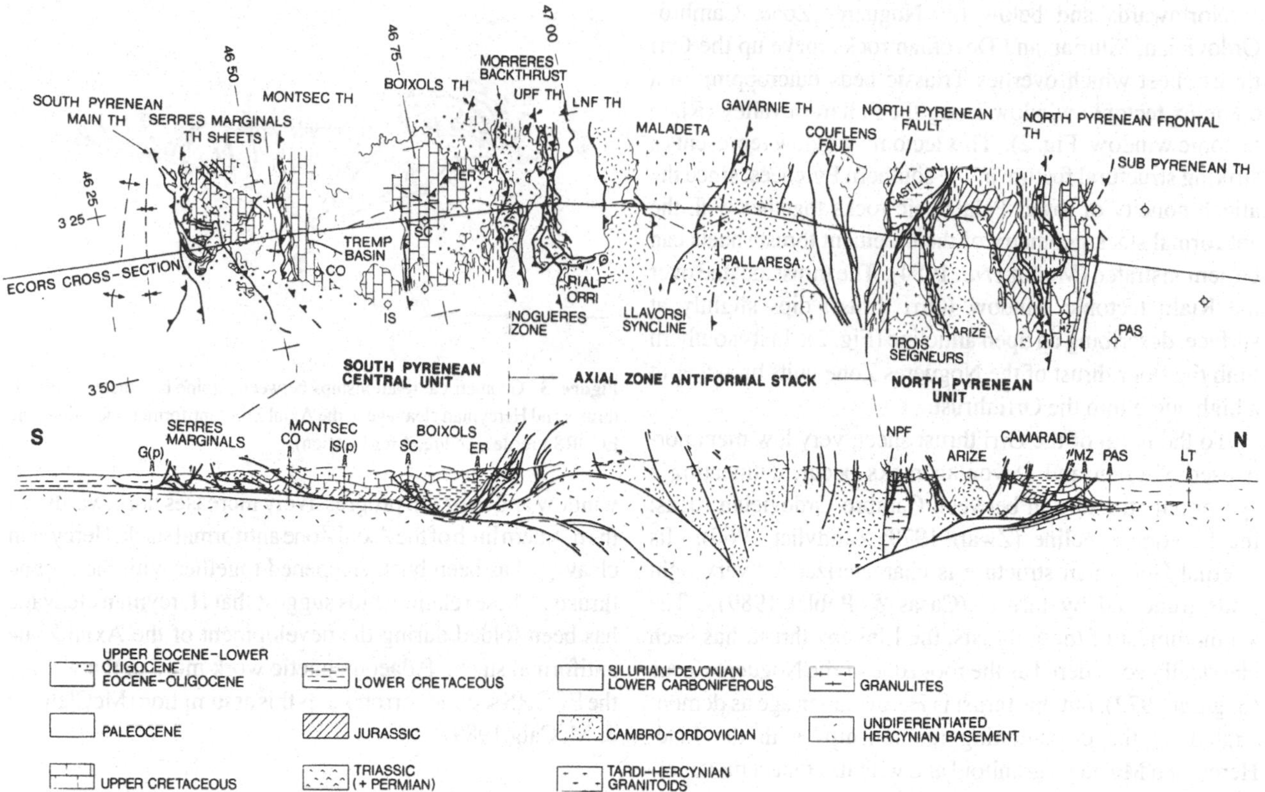

Figure 2. Strip map and cross-section along the ECORS profile. Dashed broken line in the strip map corresponds to the trace of the seismic line.

constrained from the Comiols oil well data and from the cutoff points in its hangingwall and footwall (Fig. 2). The Upper Cretaceous carbonate series is significantly thicker (1500 m) in the Montsec thrust sheet than in the northernmost outcrops of the Serres Marginals thrust sheets, below the Ager basin (Martinez & Pocoví 1988).

The Bóixols thrust sheet is located between the Tremp basin and the northern thrust of the Upper Thrust sheets (Fig. 2). This thrust sheet consists of a thick (over 5000 m) Mesozoic series, mainly Lower Cretaceous in age, the thickness of the Lower Cretaceous marls and limestones being the main stratigraphic difference with respect to the Montsec thrust sheet. The southern boundary of the Bóixols thrust sheet is along most of its cartographic expression, a buried thrust which is overlapped by the Maastrichtian Arén sandstone formation (Souquet 1967; Garrido-Megías & Ríos 1972). The syntectonic character of the Arén sequence is clearly demonstrated by the fan attitude of the sandstone beds and by their onlap disposition over the southern limb of the San Corneli anticline (Simó & Puigdefàbregas 1985). The frontal part of the Bóixols thrust sheet presents a complicated structure, resulting from the inversion of previous Lower Cretaceous extensional faults. The northern fault contact of the Bóixols thrust sheet corresponds to a backthrust (Morreres backthrust). This thrust truncates previously developed folds and very often causes an omission of the Mesozoic stratigraphy. Nevertheless, cleavage and shear zones demonstrate its thrust character. The Morreres backthrust represents a passive-roof thrust (Banks & Warburton 1986) related to the stacking of basement thrust sheets below.

Hercynian basement thrust sheets

North of the South Pyrenean Central Unit, Hercynian basement rocks have been involved in Alpine thrusts. Knowledge of the Alpine structure of the basement is the cornerstone in deciphering the Pyrenean thrust system at a crustal scale. This structure is dominated by thrusting, Alpine cleavage and folds being very subordinate. Basement rocks constitute thrust sheets which are piled on top of each other forming the Axial Zone antiformal stack (Fig. 2).

The uppermost of these thrust sheets is the Nogueres Zone, (Dalloni 1913, 1930) where thrusts affect basement and Triassic cover rocks. In the Nogueres Zone thrusts and Triassic beds are steepened and, as a result, hangingwall anticlines display a downward facing fold geometry. Two units can be differentiated: the Upper Nogueres units and the Lower Nogueres units. The basement of the Upper Nogueres units consists of Silurian, Devonian and Lower Carboniferous series (Fig. 2). This series is affected by Hercynian thrusts and folds evidenced by the unconformity disposition of the Triassic red beds. Thrusts are the main Hercynian structural feature in this unit. Triassic red beds and Devonian limestones are affected by a north-dipping Alpine cleavage associated with E-W trending open folds. The Lower Nogueres units are composed of several small thrust sheets with a hectometric to kilometric displacement to the south. These units form a duplex at the southern limb of the Axial Zone antiformal stack. The Devonian series and its Hercynian structure differ significantly from those of the Upper Nogueres units.

Northwards and below the Nogueres Zone, Cambro-Ordovician, Silurian and Devonian rocks make up the Orri thrust sheet which overlies Triassic beds outcropping in a complex tectonic window along the Pallaresa valley (Rialp tectonic window, Fig. 2). This tectonic window represents a striking structural feature of the southern Pyrenees, since the allochthoneity of lower Palaeozoic rocks together with the antiformal stack geometry of the basement thrust sheets can be demonstrated (Muñoz *et al.* 1984). The thrust located over the Rialp tectonic window (Orri thrust) dips slightly at surface, describing an open anticline (Fig. 2). In its southern limb the floor thrust of the Nogueres Zone units branches at a high angle into the Orri thrust.

To the north of the Orri thrust sheet, very low metamorphosed Silurian and Devonian rocks occur, delineating a synformal structure in between Cambro-Ordovician rocks: the Llavorsí syncline (Zwart 1979; Zandvliet 1960). Its internal Hercynian structure is characterized by very tight folds truncated by thrusts (Casas & Poblet 1989). The southernmost of these thrusts, the Llavorsí thrust, has been classically considered as the root zone of the Nogueres Zone (Seguret 1972), but this thrust is Hercynian in age as demonstrated by the cross-cutting relationships with the Late-Hercynian Maladeta granitoid and with its contact metamorphic aureole (Casas & Poblet 1989). Moreover, the Devonian sequences and Hercynian structure of the Llavorsi syncline are similar to those of the Orri thrust sheet and completely different to those shown by the Nogueres Zone. As a consequence, the Nogueres Zone units have to be rooted more to the north.

The Llavorsí syncline is bounded to the north by a minor thrust which separates the Devonian limestones from the low metamorphic grade Cambro-Ordovician rocks, which occupy a wide area to the north (Pallaresa dome, Zandvliet 1960). This thrust, together with related thrusts located in between the Cambro-Ordovician rocks, has been considered the eastward continuation of the Gavarnie thrust and the root zone of the Nogueres Zone (Muñoz *et al.* 1986). This thrust involves Triassic rocks in its footwall only 20 km to the west of the ECORS section.

The northern exposed rocks of the Axial Zone antiformal stack are deformed by steep faults. The northernmost of these faults (Couflens fault) involves Triassic rocks, demonstrating its Alpine age. Other equivalent and probably Alpine steep faults away from the ECORS section, as for example, the Merens fault and the Bossost fault, have been described (McCaig & Miller 1986; Zwart 1979). These faults show an upthrown northern block with displacements of up to 4-5 km and can be interpreted as Alpine thrusts back-steepened in the northern limb of the Axial Zone antiformal stack (Fig. 3).

Hercynian cleavage and Alpine thrusts of the basement thrust sheets of the Axial Zone antiformal stack are subparallel or form a small angle regardless of the dip of thrusts. The arrangement of Alpine thrusts and faults as well as the attitude of Hercynian cleavage suitably define the geometry of the Axial Zone antiformal stack (Fig. 3). Hercynian cleavage and Alpine thrusts are steep or southerly dipping in the Nogueres Zone, display a flat lying attitude over the Rialp tectonic window and their dip progressively increases northwards. In the northern limb of the Axial Zone antiformal stack, Hercynian cleavage has been back-steepened together with the Alpine thrusts. These relationships suggest that Hercynian cleavage has been folded during the development of the Axial Zone antiformal stack. Palaeomagnetic work more to the west of the ECORS section corroborates this assumption (McClelland & McCaig 1989).

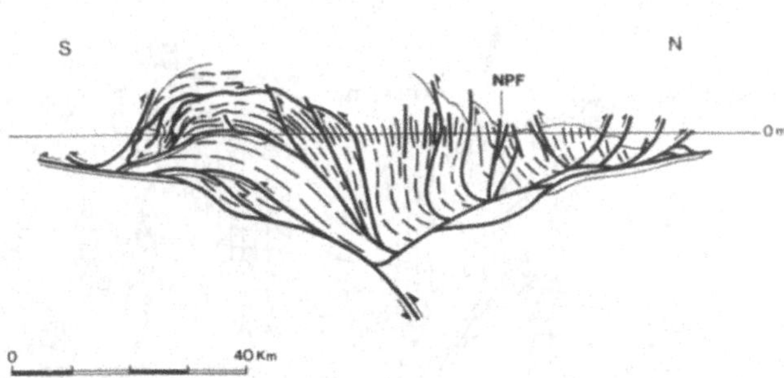

Figure 3. Geometrical relationships between Alpine basement-involved-thrusts and Hercynian cleavage in the Axial Zone antiformal stack along the ECORS profile (see Fig. 4 for location).

The North Pyrenean fault zone

The Axial Zone antiformal stack is bounded to the north by the North Pyrenean fault (Fig. 2). This and related faults are steep and define a narrow belt characterized by the presence of Jurassic and Lower Cretaceous rocks which were affected by a thermal metamorphism and a strong deformation during Middle Cretaceous times.

The North Pyrenean fault developed during the sinistral displacement of Iberia. The age of this displacement is determined by the age of flysch pull-apart basins formed synchronously with the strike-slip movement along the North Pyrenean fault from Middle Albian to Early Cenomanian (Debroas 1987, 1990). High temperature metamorphism developed as a result of crustal thinning synchronously or immediately after the Albo-Cenomanian basin formation. Radiometric data of metamorphic minerals have yielded ages around 95 Ma (105-87 Ma, Montigny *et al.* 1986; Goldberg & Maluski 1988). A syn-metamorphic cleavage developed locally during Turonian and Early Senonian times, still in a dominant strike-slip regime (Debroas 1990). Low-grade metamorphism also affects Senonian rocks, and was thus still active during the first stages of the N-S convergence.

Lower crustal granulitic rocks, as well as ultrabasic upper mantle rocks (lherzolites) occur in the North Pyrenean fault zone (Choukroune 1976; Vielzeuf & Kornprobst 1984). These rocks were carried at upper crustal levels by strike-slip faulting affecting a thinned crust.

The North Pyrenean fault has been considered the main structure of the Pyrenean chain, the axis of the collision belt and the present boundary between the Iberian and European plates (Choukroune 1976; Choukroune *et al.* 1973; Mattauer 1985). These authors considered that the North Pyrenean fault remained subvertical throughout the convergence. The

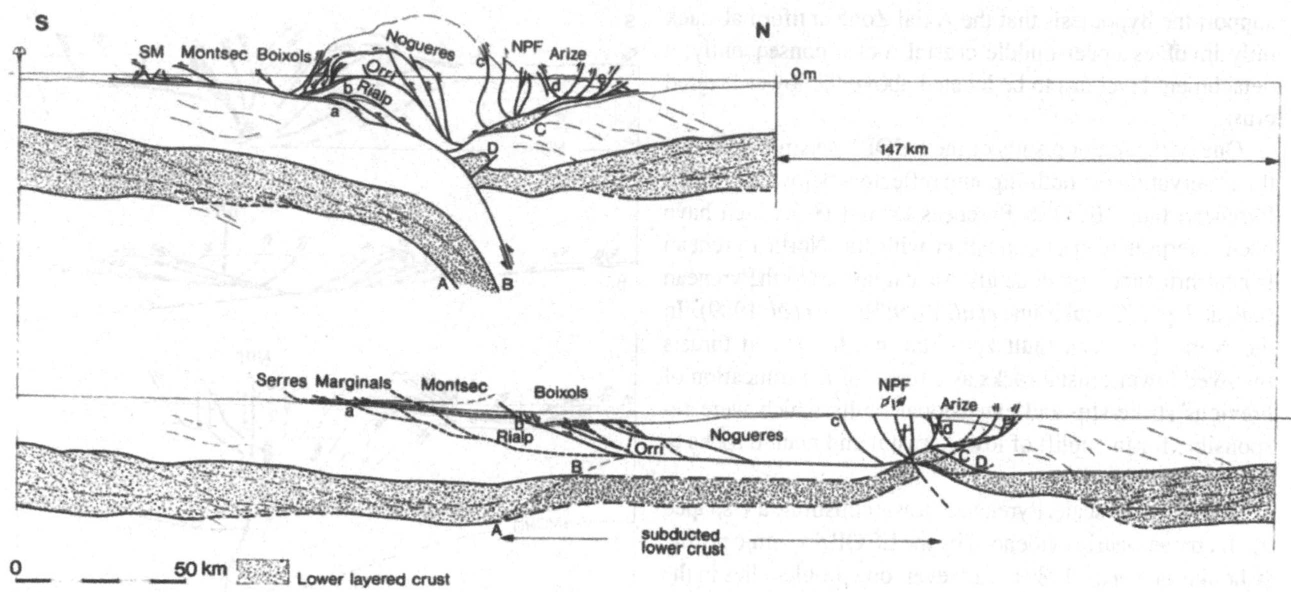

Figure 4. ECORS crustal balanced and restored cross-sections.

offset of the Moho below the North Pyrenean fault revealed by refraction seismic data (Daignières *et al.* 1982) would be in agreement with this hypothesis. The Moho step would be the result of differential thickening of the Iberian crust, with a normal thickness, and the European crust, thinned before the collision at both sides of the North Pyrenean fault.

North Pyrenean thrust sheets

North of the North Pyrenean fault zone, north-directed thrusts involve basement and cover rocks (Fig. 2). The Hercynian basement forms culminations (Fischer 1984), the so called North Pyrenean massifs (Trois Seigneurs and Arize massifs along the ECORS section). The non metamorphic and weakly deformed character of the Upper Cretaceous flysch series which unconformably overlie basement rocks of the Trois Seigneurs massif contrasts with the strongly deformed Jurassic and Lower Cretaceous metamorphic rocks outcropping in the North Pyrenean Fault zone.

The Arize massif (Fig. 2) forms a pop-up structure overthrusting, at both sides, Lower Cretaceous turbiditic series. This massif is allochthonous over Mesozoic rocks at depth as corroborated by commercial seismic data (Souquet & Peybernes 1987) and by the ECORS seismic line (Choukroune *et al.* 1989).

Northwards the cover rocks of the Arize massif, the Albo-Cenomanian Camarade basin occurs (Fig. 2). This basin, filled by breccias and turbiditic series several thousands of metres thick, formed as a result of the sinistral displacement of Iberia. The northern fault boundaries of the Camarade basin controlled the location of the North Pyrenean frontal thrust (Baby *et al.* 1988). This thrust follows on the surface the initial geometry of the basin. Basement short-cuts are observed in the hangingwall of the North Pyrenean frontal thrust as a result of the inversion of the previous steeper strike-slip and extensional Early Cretaceous faults.

The northernmost Pyrenean structure corresponds to an anticline interpreted at depth as a duplex involving Upper Cretaceous flysch series (Deramond *et al.* 1990).

Pyrenean thrust system at crustal scale

Deformation of the Pyrenean belt is best explained by a crustal thrust system which forms an orogenic lid displaying a characteristic geometry: a basement-involved antiformal stack in the middle of the chain (Axial Zone), bounded by imbricated thrust systems, mainly south-directed thrusts to the south (Cover Upper thrust sheets in the Central Pyrenees) and mainly north-directed thrusts to the north (Fig. 4). The basement antiformal stack forms a crustal wedge that moved to the south as it developed. The southern cover succession has been wedged northwards and up the antiformal stack. The result has been a tectonic delamination (Price 1986) between the basement and the cover rocks along a bedding parallel detachment zone in the Middle Triassic evaporites and lutites (Keuper facies).

The Axial Zone antiformal stack only involves upper crustal rocks. In the ECORS cross-section non-metamorphic to low grade metamorphic rocks are exposed at surface. The deepest outcropping basement rocks in the Axial Zone consist of high grade gneisses, but still correspond to middle crustal domains (formation depths about 15 km). Alpine metamorphism is lacking and only some retrogradations at greenschist facies related to Alpine faults have been described (McCaig & Miller 1986). In the ECORS deep seismic profile, the antiformal geometry of the Axial Zone is evidenced by well defined reflectors at upper crustal levels (Choukroune *et al.* 1989). Below these reflectors the Iberian lower layered crust remains undisturbed by thrusts, at least underneath the southern half of the Axial Zone, and dips increasingly below the European crust (Fig. 4). All these data support the hypothesis that the Axial Zone antiformal stack only involves upper-middle crustal rocks, consequently, a detachment level has to be located above the lower layered crust.

One of the major results of the ECORS seismic profile is the observation of south-dipping reflectors below the North-Pyrenean fault (ECORS Pyrenees team 1988) which have been interpreted in continuation with the North Pyrenean frontal thrust and, consequently, truncating the North Pyrenean fault at depth (Choukroune *et al.* 1989; Roure *et al.* 1989). In the North Pyrenean fault zone, the north-directed thrusts involved lower crustal rocks as a result of the truncation of previous strike-slip and extensional faults which were responsible for the uplift of lower crustal and mantle rocks to upper crustal levels.

At the crustal scale, Pyrenean thrusts constitute a v-shaped geometry as clearly evidenced by the ECORS seismic profile (Choukroune *et al.* 1989). However, one problem lies in the geometric and age relationships between the southerly-directed thrusts and the northerly-directed ones. This question, together with the geometric relationships between the north-Pyrenean fault and thrusts, gives rise to major discrepancies between the different crustal structural models proposed for the Pyrenees (Fig. 5). Vertical tectonic models (Seguret & Daignières 1986; Mattauer 1986, 1990) consider the north-Pyrenean fault as the axial plane of a fan thrust system which involves the whole crust (Fig. 5c). Thin-skinned tectonic models (Williams & Fischer 1984) and thick-skinned tectonic models (Déramond *et al.* 1985) show the North-Pyrenean fault truncated by a linked thrust system in which the northerly directed thrusts are considered backthrusts of the southerly directed ones (Fig. 5a, b). In the former, thrusts only involve upper crustal rocks, whereas the latter model supposes that thrusts merge downwards into a detachment level located at the Moho (Fig. 5a). These models were poorly constrained by deep data because they were proposed before the ECORS seismic line was obtained. Recently, and from the ECORS data, the North Pyrenean fault has been interpreted as little deformed by wedging of the Iberian crust by the European plate (Roure *et al.* 1989). This model (Fig. 5d) implies more shortening for the northerly directed thrusts than for the southerly directed ones and thrust kinematics becomes complicated due to the geometric relationships between the upper thrusts of the Axial Zone antiformal stack and the sole thrust of the crustal system. An upper crustal orogenic lid (Fig. 5e) better explains the surface and deep seismic data as well as the kinematics of thrusts at both sides of the chain.

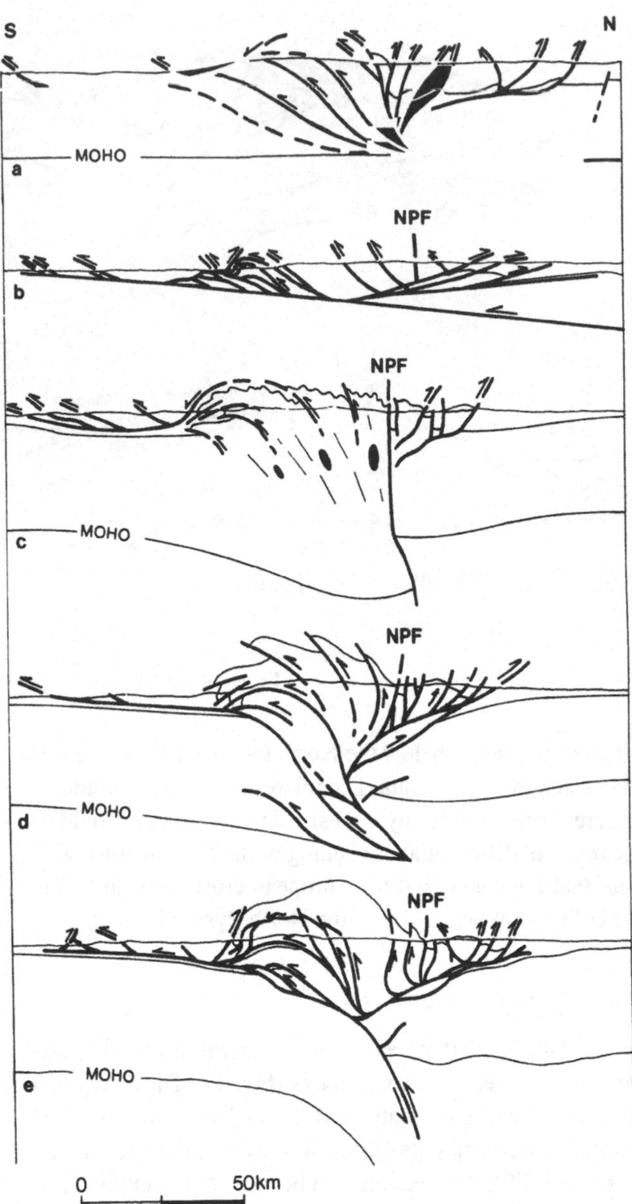

Figure 5. Proposed models for the Pyrenean crustal thrust structure. (**a**) Deramond *et al.* 1985 (Western Pyrenees); (**b**) Williams & Fischer 1984 (Central Pyrenees, west of the ECORS profile); (**c**) Seguret & Daignières 1986; (**d**) Roure *et al.* 1989 and (**e**) model herein proposed. Models c, d and e correspond to the ECORS profile.

Balanced and restored cross-sections

Crustal balanced and restored cross-sections were drawn in order to better constrain the interpretation of the ECORS seismic profile and to calculate the orogenic contraction. The reflectors imaged by the ECORS profile have been converted to depth using a velocity crustal model consistent with gravity data (Torné *et al.* 1989). Migration of the main reflectors which define the boundaries of the seismic fabrics (Banda & Berastegui 1990) along its complete length was not carried out. Only the migration of selected reflectors was established (Roure *et al.* 1989). In the cross-section presented, the results of this migration have only been partially taken into account. Reflectors which, after the migration, shifted from one seismic facies to another or those which, while being part of a set of time markers defining the boundary of a seismic facies, lost this property after migration, have been rejected.

Any balanced cross-section has to be drawn parallel to the regional thrust transport direction. Fortunately, the trace of the ECORS seismic line coincides quite well with the thrust transport direction deduced for the cover thrust sheets. The cartographic pattern of frontal, oblique and lateral structures of the cover South Pyrenean Upper thrust sheets defines a transport direction N-S to NNE-SSW. This transport direction has recently been corroborated by rotations of the footwall and hangingwall along oblique ramps, as deduced by palaeomagnetic results (Burbank *et al.* 1991). Palaeomagnetic results (Dinarès *et al.* 1991, this volume) have also demonstrated the absence of relative rotations both between the Main Upper thrust sheets along the ECORS profile and between these thrust sheets and the foreland. Orientations of thrust structures in the North Pyrenean units demonstrate a transport direction to the NNE, also close to the direction of the northern ECORS profile (Fig. 2).

The cartographic arrangement of the Pyrenean basement thrust sheets also suggests a transport direction to the S or to the SSW, a direction which has been corroborated in other southern Pyrenean sections (in the eastern and western Pyrenees) where no detachment level exists between the basement and cover rocks (Muñoz 1985; Muñoz *et al.* 1986). Rotations of the Nogueres Zone units (Bates 1989) may have been the consequence of a strong longitudinal structural variation which resulted in the relay of small units and the existence of several oblique and lateral structures.

The ECORS cross-section has been restored using line-length balancing techniques for the cover thrust sheets and for the basement units with an attached lower Triassic and Permian series. The basement thrust sheets have been areally balanced, but always taking into account their internal Alpine and Hercynian features. The restored section not only has to conserve the areas of the balanced one, but more importantly, should present a coherent section concerning the precollisional geological features. This methodology provides a powerful tool for discriminating between the different possible restored cross-section solutions. The ECORS restored cross-section will mainly depend on the proposed solution for the restoration of the Axial Zone antiformal stack. The final restored section for the northern Pyrenees will be determined by the interpretation of the reflectors located below the North Pyrenean fault, and by the geometry of the previously thinned European crust adopted in the restored section.

The south Pyrenean Upper thrust sheets have been restored independently of the Axial Zone antiformal stack as imposed by the existence of the detachment level into the Keuper evaporites. The Serres Marginals and Montsec thrust sheets have been restored taking the first synorogenic deposits as a horizontal stratigraphic reference level. As the thrusts mainly developed in a piggy back fashion, the horizontal reference level gets progressively older and becomes asymptotic upwards to the hinterland (Fig. 4). To the north of the Bóixols thrust the horizontal reference corresponds to the base of the Upper Cretaceous succession. The break-back reactivation of the Serres Marginals and Montsec thrusts during Late Eocene-Early Oligocene times has to be taken into account in the restoration (Vergés & Muñoz 1990), mainly in the cross-section thrust sheets, where structural units with intermediate stratigraphies can be thrusted completely out-of-sequence, resulting in strong differences in the stratigraphy across the thrusts (Martinez & Pocoví 1988).

Restoration of the Bóixols thrust sheet reveals the geometry of the Early Cretaceous extensional basin (Fig. 4). The northernmost outcropping Lower Cretaceous rocks reflect a progressive northwards shallowing of the Bóixols basin (Berastegui *et al.* 1990). This shallowing is corroborated by the northern provenance of fine-grained clastics which characterized the last sequence of the basin, as exposed more to the west of the ECORS section. There is no evidence to assume the existence of Early Cretaceous basins located between the Bóixols basin and the North Pyrenean fault (Fig. 4).

The Axial Zone antiformal stack consists of three main thrust sheets, all of them outcropping along the ECORS profile: Nogueres, Orri and Rialp thrust sheets (Fig. 4). The restored cross-section obtained will depend on the initial arrangement of these thrust sheets. Two extreme solutions are possible: a pre-collisional geometry with very little or no overlapping of these units and a pre-collisional geometry with maximum imbrication of these thrust sheets.

A restored cross-section with minimum shortening has been constructed (Fig. 4). As a consequence, basement thrust sheets have been initially superposed as much as possible. This superposition, and the maximum depth for the detachment is constrained by the maximum exposed thickness of the upper thrust sheet (Nogueres) in the northern limb of the Axial Zone antiformal stack and by the Hercynian metamorphic conditions shown by the basement rocks. After the Hercynian orogeny the basement rocks experienced a progressive shallowing, as demonstrated by the rocks observed below the post-Hercynian unconformity along the Axial Zone. A post-Hercynian deepening of basement rocks may only have been restricted to the hangingwall of late to post-Hercynian extensional faults (Permian, Triassic and Early Cretaceous) and implies the existence of thick sedimentary basins. Alternatively, there is no evidence of thick sedimentary sequences over the Axial Zone thrust sheets.

The restored cross-section (Fig. 4) has been constructed assuming a maximum overlapping of the Orri thrust sheet over the Rialp one - the two thrust sheets being separated by a low-dipping Early Cretaceous extensional fault (Bóixols fault). The Cambro-Ordovician rocks of the Orri thrust sheet, located over the Rialp window, show anchimetamorphic conditions, with a calculated maximum temperature of 250-300°C (Bons 1989). The maximum reasonable depth by the end of the Hercynian orogeny, taking into account the mineral parageneses shown by these rocks and the Hercynian crustal configuration (Zwart 1979, 1986), must have been about 9 km, which could have increased by 1-2 km during the Early Cretaceous basin formation. Cambro-Ordovician rocks exposed over the floor thrust of the Nogueres thrust sheet (Pallaresa anticlinorium) show low grade greenschist facies metamorphic conditions (Bons 1989) and, with the proposed restoration, these rocks would have been initially located at a depth of 15 km. This depth is excessive because it is the expected depth for the location of high grade gneisses and because the related Hercynian metamorphic gradient would be too low. To avoid this problem, a downwards movement of the Nogueres thrust sheet to the north over Late-Hercynian and Early Cretaceous extensional faults could be invoked. The maximum depth reached by these rocks determines the

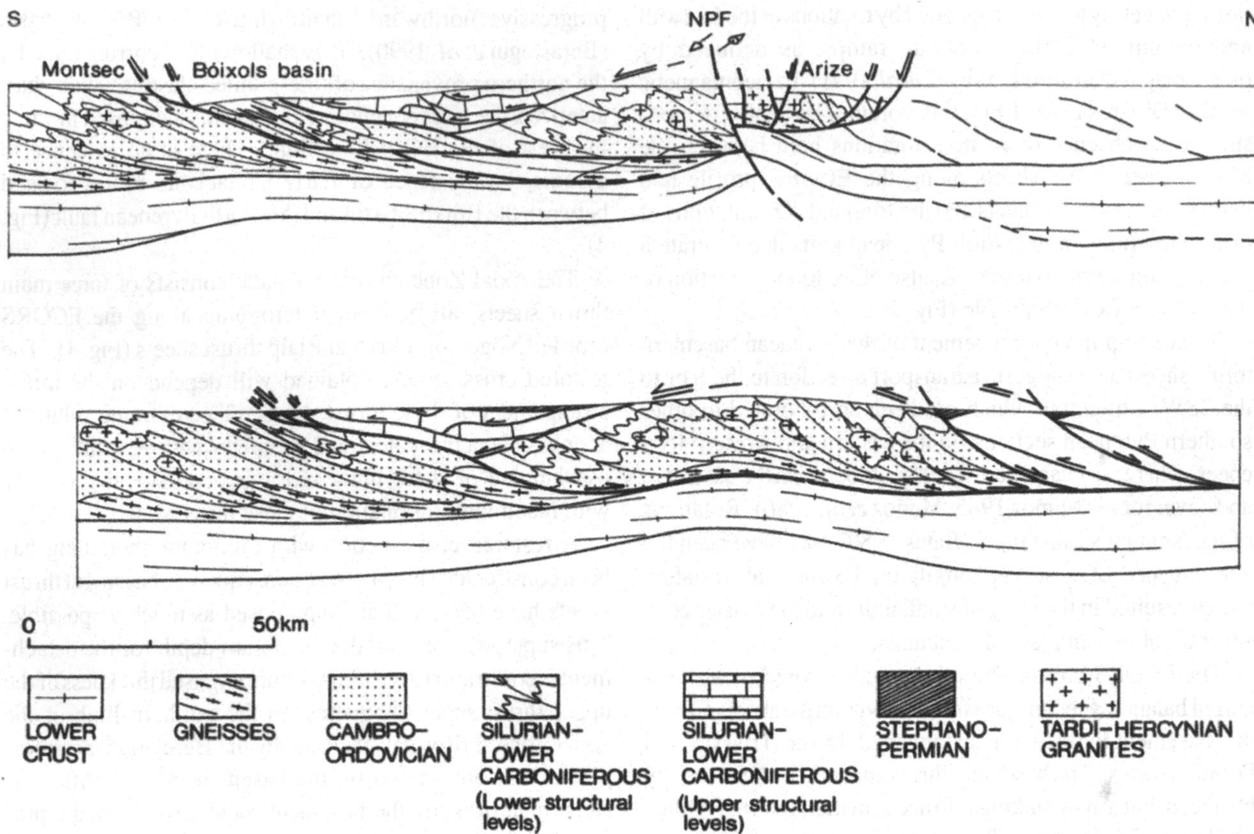

Figure 6. (a) ECORS restored cross-section showing the pre-collisional geological features at Middle Cretaceous times, before the convergence. This section corresponds to the minimum shortening solution. (b) Interpretative cross-section in Permian times once the Early Cretaceous extensional fault system is restored. A Late-Hercynian extensional fault system is inferred from the relationships between thrust sheets adopted in the above restored section in order to minimize the shortening.

maximum depth of the detachment, which in the restored cross-section has been located at 15 km depth (Fig. 4). A deeper location is not possible because the sole thrust climbs northwards, thus defining the wedge geometry of the North Pyrenean thrust sheets. In any case, a deeper detachment level would only be restricted to a narrow area southwards of the North Pyrenean fault.

Devonian rocks of the Nogueres Zone have been located in the restored section above the Devonian and Cambro Ordovician rocks of the Orri thrust sheet in order to minimize the displacement of the Upper Nogueres units (supposed to be only 11 km). As a result, the Nogueres thrust sheet has been restored by overlapping the Orri one (Fig. 4). This proposed restoration imposes a Late-Hercynian extensional geometry. The unmetamorphosed Devonian rocks deformed by Hercynian thrusts (the shallowest exposed structural level) are directly located over slightly metamorphosed Devonian rocks deformed by very tight syn-cleavage folds (Fig. 6). Both Devonian units are unconformably overlapped by Triassic red beds. This Late-Hercynian extensional fault could be synchronous with the Stephano-Permian basins and their related volcanic activity. Permian basins outcrop close to the proposed Permian extensional fault between the Upper and Lower Nogueres units (Fig. 6). The mid-crustal reflectors observed in the northern ECORS profile (Choukroune *et al.* 1989) cannot only be interpreted as Hercynian thrusts (Choukroune *et al.* 1990) but also as Permian extensional faults, as suggested by well data. A Late-Hercynian extensional system could have been responsible for a crustal thinning and a first shallowing of mantle rocks (lherzolites) in the northern Pyrenees, as proposed by Duee *et al.* (1984). The total calculated shortening for the Axial Zone antiformal stack, taking the North Pyrenean fault as a northern reference, is 112 km.

Restoration of the North Pyrenean thrust sheets depends on the geometry predicted for the thinned crust in middle Cretaceous times. Thinning is deduced by the emplacement of lower crustal and mantle rocks to upper crustal levels, and by the thermal metamorphism along the North Pyrenean fault zone. Restoration also depends on the geometry deduced for the lower crustal rocks that have been assumed to exist at middle crustal depths below the North Pyrenean fault (Choukroune *et al.* 1989). The North Pyrenean fault has to be located in the restored section south of the European crust imaged by the ECORS profile. This implies a minimum displacement backwards of 22 km. Restoring the section, taking this minimum displacement into account, the crust below the North Pyrenean Zone becomes 25 km thick, which is probably too thick to explain the geological features cited above. If the reflectors below the North Pyrenean fault are considered as lower crustal rocks and removed to the south, the crust below the North Pyrenean sole thrust becomes thinner and the total displacement along this thrust increases.

A limitation of the extent of this horse of lower crustal rocks is imposed by the cut-off position of the autochthonous contact between the basement and cover (Fig. 4). In the balanced section presented here, only a small unit of lower layered crust which presents the same seismic fabrics as the autochthonous lower layered crust has been considered in contrast with previous interpretations (Roure *et al.* 1989). The crust below the North Pyrenean massifs has been restored to a lower initial thickness than in the deformed state, because some ductile thickening is expected to have occurred. The calculated shortening for the North Pyrenean thrust sheets is 36 km and the thickness of the crust below the North Pyrenean fault zone in the restored section is 15 km (Fig. 4).

The total calculated shortening for the Pyrenean belt along the ECORS profile from the presented crustal balanced cross-section is 147 km. This value is coherent with the proposed northward displacement of Iberia with respect to the European plate during the Pyrenean collision (Grimaud *et al.* 1982; Boillot 1986). The most remarkable result of the restoration of the Pyrenean belt is the apparent discrepancy between the length of the upper crust and the length of the lower layered crust.

In the restored cross-section, the Iberian crust below the foreland and the more external south Pyrenean thrust sheets, presents a constant thickness of about 34 km (Fig. 4), which is in agreement with the crustal thickness of the non-deformed Hercynian areas of the Iberian Massif (Banda *et al.* 1981). Can this portion of the apparently undeformed Iberian crust be thinned in the restored section in order to fill the 110 km hole of the missing middle and lower crust of northern Iberia (below the Orri and Nogueres thrust sheets)? (Fig. 4). This assumption is not reasonable because the resulting Iberian crust is too thin, both in the foreland (less than 30 km) and in the inner parts (over 20 km). The amount of crustal thinning experienced by the northern Iberian crust before the Pyrenean collision is difficult to determine. This thinning could have taken place during Permian, Late Triassic and Early Cretaceous times. Nevertheless, the cover sediments which overlay the Hercynian basement of the Axial Zone thrust sheets do not demonstrate a significant attenuation of the underlying crust. Moreover, deep seismic reflection profiles (ECORS, BIRPS) of genetically related close areas (during Early Mesozoic stretching events), which have not been reworked by the Alpine collision, show thinning of the crystalline crust below the deepest basins of up to 10 km, and a shallowing of the Moho of only a few kilometres (Cheadle *et al.* 1987). The crust in the restored section (Fig. 4) has been reduced to a thickness of 15 km below the Cretaceous basins north of the North-Pyrenean fault and to 28 km below the Bóixols basin in the southern Pyrenees. To the north of this basin, the crust has been drawn with a constant thickness of 28 km although this is too thin if we consider that Lower Cretaceous extensional basins did not occur over the Nogueres thrust sheet.

In the restored cross-section, the main discontinuities which affected the Pyrenean crust before the collision (Hercynian cleavage, Hercynian thrusts, Late-Hercynian extensional faults and Early Cretaceous extensional system) display a listric geometry over the lower layered crust. This geometry has been observed in the undeformed parts of the ECORS profile, deduced after the restoration or by comparision with other areas (Fig. 6). Most of the BIRPS deep reflection profiles across the Mesozoic extensional basins of northwestern Europe show a highly reflective lower crust which is not penetrated by the extensional faults (Cheadle *et al.* 1987). Although the lower layered crust has been interpreted as the result of extensional processes at depth and the faults do not show listric geometry, they must join a detachment level over the lower crust. The same geometry is observed in the Bay of Biscay and Galicia Banks continental margins where the Early Cretaceous extensional faults merge into a detachment level located over the lower crust (Le Pichon & Barbier 1987; Boillot & Malod 1988). The location of these discontinuities favoured the delamination of the crust, the upper part forming an orogenic lid shortened by an upper crustal thrust system. The crust below this middle crustal detachment was subducted beneath the European crust (Fig. 4).

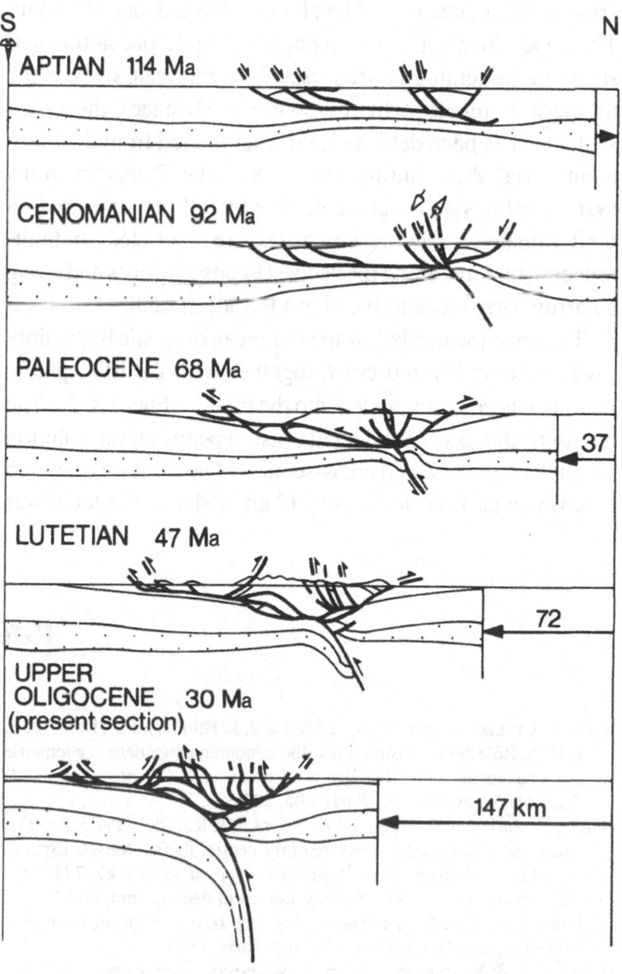

Figure 7. Evolution of the Pyrenean crust along the ECORS profile from the Early Cretaceous to the end of the collision event.

Conclusions

The proposed balanced and restored crustal cross-sections are valid because the lengths of the cover sequences, as well as the area of the basement rocks, are the same (if plane strain is assumed), and admissible because the structural model is in agreement with the observed data. Nevertheless, to obtain a good balanced cross-section, mainly on a crustal scale, at least two more rules have to be followed: the restored section has to furnish a coherent picture of all the previously recorded geological features (Fig. 6) and the passage from the undeformed state to the deformed cross-section has to be kinematically possible (Fig. 7). The restored section shows an Early Cretaceous extensional system which evolved, in Late Albian-Early Cenomanian, to a combined sinistral strike-slip and extensional dip-slip fault system, driven by the North Pyrenean Fault. This fault penetrated the crust and probably the whole lithosphere (Fig. 7).

The geometry of this extensional fault system as well as the geometry of the Late-Hercynian and Hercynian discontinuities, all of them flattening downwards over the lower layered crust, favoured delamination of the crust (Figs 6 & 7). The upper crust has been detached from the lower crust along a detachment level located at a depth of 15 km. The upper crust constitutes a duplex (Axial Zone antiformal stack) in the southern Pyrenees and an imbricate stack in the northern Pyrenees. In the southern Pyrenees the cover sequence has been detached and delaminated from the basement Axial Zone antiformal stack. The Pyrenean thrust system defines an orogenic lid in which thrust deformation predominates. Ductile structures are restricted to fault-related structures along the deepest exposed thrusts and along a narrow strip close to the North Pyrenean fault.

The crust located below the Pyrenean orogenic lid, mainly lower seismic layered crust, together with the lithospheric mantle, has been subducted into the mantle (Figs 4 & 7). The length of the lower subducted crust depends on the solution adopted by the restored cross-section. The solution favoured here involves a shortening of 147 km and a subducted lower crust of 110 km in length (Fig. 4). Restored sections with a shortening less than 120 km have been found to be impossible, implying a subducted slab of 65 km minimum.

A subduction of the lower crust together with the lithospheric mantle can be postulated in order to explain the balance problem that exists at the lithospheric scale. Subduction of the lithosphere in other collision belts, as for example the Himalayas and the Alps, is widely accepted (Mattauer 1986; Butler 1986; Laubscher 1988). In the Himalayas, thrusts have been interpreted affecting the crust and detached from the Moho (Mattauer 1986) whereas, in the Alps, thrusts affect the upper part of the mantle as recently evidenced by the deep seismic profiles (Bayer *et al.* 1987). In the Pyrenees, the lower crust has remained attached with the lithospheric mantle and subducted below the European crust. The lower subducted crust has probably not been imaged in the seismic profile because at this depth its density increases up to mantle values through eclogite facies metamorphism. This interpretation has also been suggested in the Alps for subducted lower crust (Butler 1986, 1990; Laubscher 1988, 1990). Recently, petrological evidence of subducted crustal rocks to depths of more than 100 km has been described in several areas (Chopin 1984, 1987; Schreyer 1988). The balanced and restored cross-sections of the Pyrenees along the ECORS profile represent further evidence supporting subduction of continental crust into the mantle.

The balanced and restored cross-sections, as well as the proposed kinematic evolution, provide a further step towards crustal modelling (flexural and uplift history) and better understanding of the early geological events affecting the Pyrenean domain (Hercynian orogeny and Late Palaeozoic-Mesozoic extensional events).

This paper derives from work carried out at the Servei Geològic de Catalunya in cooperation with C. Puigdefàbregas, X. Berastegui and M. Losantos. Field work, funded by the Servei Geològic, was part of the ECORS-Pyrenees project. I am grateful for discussions with the members of the ECORS-Pyrenees team. I am indebted to Paul Heller for improvement of the original English manuscript. This work has been partially supported by the CICYT project GE089-0254.

References

Baby, P., Crouzet, G., Specht, M., Deramond, J., Bilotte, M. O., Debroas, E. J. 1988. Rôle des paléostructures albo-cénomaniennes dans la géométrie des chavauchements frontaux nord-pyrénéens. *Comptes Rendus de l'Académie des Sciences, Paris,* **306**, 307-313.

Banda, E., Suriñach, E., Aparicio, A., Sierra, J. & Ruiz de la Parte, E. 1981. Crust and upper mantle structure of the central Iberian Meseta (Spain). *Geophysical Journal of the Royal Astronomical Society,* **67**, 779-789.

—— & Berastegui, X. 1990. Seismic fabrics in deep seismic profiles. *In:* Pinet, B. *et al.*, (eds), *The potential of deep seismic profiling for hydrocarbon exploration.* Éditions Technip, Paris, 73-76.

Banks, C. J. & Warburton, J. 1986. 'Passive-roof' duplex geometry in the frontal structures of the Kirthar and Sulaiman mountain belts, Pakistan. *Journal Structural Geology,* **8**, 229-237.

Bates, M. 1989. Palaeomagnetic evidence for rotations and deformation in the Nogueras Zone, Central Southern Pyrenees. *Journal of the Geological Society, London,* **146**, 459-476.

Bayer, R., Cazes, M., Dal Piaz, G. V., Damotte, B., Elter, J., Gosso, G., Hirn, A., Lauza, R., Lombardo, B., Mugnier, J. L., Nicolas, A., Nicolich, R., Polino, R., Roure, F., Sacachi, R., Scarascia, S., Tabacco, I., Tapponnier, P., Tardy, M., Taylor, M., Thouvenot, F., Torreilles, G. & Villien, A. 1987. Premiers résultats de la traversée des Alpes occidentales par sismique reflexion verticale (Programme ECORS-CROP). *Comptes Rendus de l'Académie des Sciences, Paris,* **305**, 1461-1470.

Beach, A. 1981. Thrust tectonics and crustal shortening in the external French Alps based on a seismic cross-section. *Tectonophysics,* **79**, T1-T6.

Berastegui, X., Garcia, J. M. & Losantos, M. 1990. Structure and sedimentary evolution of the Organyà basin (Central South Pyrenean Unit, Spain) during the Lower Cretaceous. *Bulletin de la Société Géologique de France,* **8**, 251-264.

Boillot, G. 1986. Comparison between the Galicia and Aquitaine margins. *Tectonophysics* **129**, 243-255.

—— & Capdevila, R. 1977. The Pyrenees, subduction and collision ? *Earth and Planetary Science Letters,* **35**, 251-260.

—— & Malod, J. 1988. The north and north-west Spanish continental margin: a review. *Revista Sociedad Geológica, España*, **1**, 295-316.

Bons, A.-J. 1989. Very low-grade metamorphism of the Seo Formation in the Orri Dome, South-Central Pyrenees. *Geologie en Mijnbouw*, **68**, 303-312.

Burbank, D. W., Vergés, J., Muñoz, J. A. & Bentham, P. 1991. Coeval hindward- and forward-imbricating thrusting in the Central Southern Pyrenees, Spain, Timing and rates of shortening and deposition. *Geological Society of America Bulletin* (in press)

Butler, R.W.H. 1986. Thrust tectonics, deep structure and crustal subduction in the Alps and Himalayas. *Journal of the Geological Society, London*, **143**, 857-873.

—— 1990. Balancing sections on a crustal scale: a view from the western Alps. Proceedings of the European Geotraverse Sixth workshop, 157-164.

—— & Coward, M. P. 1989. Crustal scale thrusting and continental during Himalayan collision tectonics on the NW Indian plate. *In:* Sengör, A.M.C. (ed.), Tectonic Evolution of the Tethyan Region. *NATO ASI, series C: Mathematical and Physical Sciences*, **259**, 387-413.

Casas, J. M. & Poblet, J. 1989. Essai de restitution de la déformation dans une zone avec plis et chevauchaments, le 'synclinal de Llavorsi' dans le Pyrénées centrales (Espagne). *Comptes Rendus de l'Académie des Sciences, Paris*, **308**, 427-433.

Cámara, P. & Klimowitz, J. 1985. Interpretación geodinámica de la vertiente centro-occidental surpirenaica. *Estudios Geológicos*, **41**, 391-404.

Cheadle, M. J., McGeary, S., Warner, M. R. & Matthews, D. H. 1987. Extensional structures on the western UK continental shelf, a review of evidence from deep seismic profiling. *In:* Coward, M. P., Dewey, J. & Hancock, P. L. (eds) *Continental Extensional Tectonics*. Geological Society of London Special Publication, **28**, 445-465.

Chopin, C. 1984. Coesite and pure pyrope in high-grade blueschists of the Western Alps, a first record and some consequences. *Contributions to Mineralogy and Petrology*, **86**, 197-218.

—— 1987. Very-high-pressure metamorphism in the western Alps, implications for subduction of continental crust. *Philosophical Transactions of the Royal Society of London*, series A, **321**, 183-197.

Choukroune, P. 1976. Structure et évolution tectonique de la zone nord-pyrénéenne (analise de la déformation dans une portion de chaîne à schistosité subverticale). *Mémoires de la Société Géologique de France*, **127**, 116p.

——, Seguret, M. & Galdeano, A. 1973. Caractéristiques et évolution structurale des Pyrénées. *Bulletin de la Société Géologique de France*, **7**, 601-611.

—— & ECORS Team. 1989. The ECORS Pyrenean deep seismic profile reflection data and the overall structure of an orogenic belt. *Tectonics*, **8**, 23-39.

——, Pinet, B., Roure, F. & Cazes, M. 1990. Major Hercynian structures along the ECORS Pyrenees and Biscaye Lines. *Bulletin de la Société Géologique de France*, 8, **VI**, 313-320.

Dahlstrom, C.D.A. 1969. Balanced cross sections. *Canadian Journal Earth Sciences*, **6**, 743-757.

Daignières, M., Gallart, J., Banda, E. & Hirn, A. 1982. Implications of the seismic structure for the orogenic evolution of the Pyrenean Range. *Earth and Planetary Science Letters*, **57**, 88-100.

Dalloni, M. 1913. Stratigraphie et tectonique de la région des Nogueras (Pyrénées centrales). *Bulletin de la Société Géologique de France*, **4**, 243-263.

—— 1930. Étude géologique des Pyrénées Catalanes. *Annales Facultee Sciences Marseille*, **XXVI**, 373pp.

Debroas, E-J. 1987. Modèle de bassin triangulaire à l'intersection de décrochements divergents pour le fossé albo-cénomanien de la Ballongue (zone nord-pyrénéenne, France). *Bulletin de la Société Géologique de France*, **8**, 887-898.

—— 1990. Le Flysch noir albo-cénomanien témoin de la structuration albienne à sénonienne de la zone nord-pyrénéenne en Bigorre (Hautes Pyrénées, France). *Bulletin de la Société Géologique de France*, **8**, 273-285.

Deramond, J., Graham, R. H. Hossack, J. R. Baby, P. & Crouzet, G. 1985. Nouveau modéle de la chaîne des Pyrénées. *Comptes Rendus de l'Académie des Sciences, Paris*, **301**, 1213-1216.

——, Baby, P., Specht, M. & Crouzet, G. 1990. Géométrie des chevauchements dans la zone nord-pyrénéenne ariégoise précisée par le profil ECORS. *Bulletin de la Société Géologique de France*, **8**, 287-294.

Dinares, J., McClelland, E. & Santanach, P. 1991. Contrasting rotations within thrust sheets and kinematics of thrust-tectonics as derived from palaeomagnetic data: an example from the southern Pyrenees (this volume).

Duée, G., Lagabrielle, Y., Coutelle, A. & Fortané, A. 1984. Les lherzolites associées aux chaînons béarnais (Pyrénées occidentales): mise à l'affleurement anté-dogger et resédimentation albo-cénomanienne. *Comptes Rendus de l'Académie des Sciences, Paris*, **299**, 1205-1210.

ECORS Pyrenees team. 1988. The ECORS deep reflection seismic survey across the Pyrenees. *Nature*, **331**, 508-511.

Farrell, S. G., Williams, G. D. & Atkinson, C. D. 1987. Constraints on the age of movement of the Montsech and Cotiella Thrusts, south central Pyrenees, Spain. *Journal of the Geological Society of London*, **144**, 907-914.

Fischer, M. W. 1984. Thrust tectonics in the North Pyrenees. *Journal of Structural Geology*, **6**, 721-726.

Garrido-Megías, A. & Ríos, L.M . 1972. Síntesis geológica del Secundario y Terciario entre los ríos Cinca y Segre (Pirineo central de la vertiente surpirenaica, provincias de Huesca y Lérida). *Boletin Geológico y Minero de España*, **83**, 1-47.

Goldberg, J. M. & Maluski, H. 1988. Données nouvelles et mise au point sur l'âge du métamorphisme pyrénéen. *Comptes Rendus de l'Académie des Sciences, Paris*, **306**, 429-435.

Grimaud, S., Boillot, G., Collette, B. J., Mauffret, A., Miles, P. R. & Roberts, D. B. 1982. Western extension of the Iberian-European plate boundary during the Early Cenozoic (Pyrenean) convergence, a new model. *Marine Geology*, **45**, 63-77.

Hossack, J. R. 1979. The use of balanced cross-sections in the calculation of orogenic contraction: A review. *Journal of the Geological Society of London*, **136**, 705-711.

Laubscher, H. 1988. Material balance in Alpine orogeny. *Geological Society of America Bulletin*, **100**, 1313-1328.

Laubscher, H. P. 1990. Seismic data and the structure of the central Alps. *Proceedings of the European Geotraverse Sixth workshop*, 149-156.

Le Pichon, X. & Barbier, F. 1987. Passive margin formation by low-angle faulting within the upper crust, The Northern Bay of Biscay margin. *Tectonics*, **6**, 133-150.

Martínez, M. B. & Pocoví, J. 1988. El amortiguamiento frontal de la estructura de la cobertera surpirenaica y su relación con el anticlinal de Barbastro - Balaguer. *Acta Geológica Hispànica*, **23**, 81-94.

Mattauer, M. 1985. Présentation d'un modèle lithosphérique de la chaîne des Pyrénées. *Comptes Rendus de l'Académie des Sciences, Paris*, **300**, 71-74.

—— 1986. Intracontinental subduction, crust-mantle décollement and crustal-stacking wedge in the Himalayas and other collision belts. *In:* Coward, M. P. & Ries, A. C. (eds) *Collision Tectonics*. Geological Society of London Special Publication, **19**, 37-50.

—— 1990. Vue autre interprétation du profil ECORS Pyrénées. *Bulletin de la Société Géologique de France*, **VI**, 307-311.

McCaig, A. M. & Miller, J. A. 1986. 40Ar-39Ar age of mylonites along the Merens fault, Central Pyrenees. *Tectonophysics*, **129**, 149-172.

McClelland, E. A. & McCaig, A. M. 1989. Palaeomagnetic estimates of rotation in basement thrust sheets, Axial Zone, Southern Pyrenees. *Cuadernos de Geología Ibérica*, **12**, 181-193.

Menard, G. & Thouvenot, F. 1987. Coupes équilibrées crustales, méthodologie et application aux Alpes occidentales. *Geodinamica Acta*, **1**, 35-45.

Montigny, R., Azambre, B., Rossy, M. & Thuizat, R. 1986. K-Ar study of Cretaceous magmatism and metamorphism from the Pyrenees: age and length of rotation of the Iberian peninsula. *Tectonophysics*, **129**, 257-274.

Mugnier, J. L., Guellec, S., Menard, G. Roure, F., Tardy, M. & Vialon, P. 1990. Crustal balanced cross-section through the external Alps deduced from the ECORS profile. *Bulletin de la Société Géologique de France* (in press)

Muñoz, J. A. 1985. Estructura alpina i herciniana a la vora sud de la zona axial del Pirineu oriental. Tesi doctoral. University of Barcelona, 305p.

——, Puigdefabregas, C. & Fontboté, J. M. 1984. El cicló alpino y la estructura tectónica del Pirineo. *In:* Comba, J. A., *Libro Jubilar J. M. Rios*, Geología de España, 2. IGME, Madrid, 185-205.

——, Martínez, A. & Vergés, J. 1986. Thrust sequences in the eastern Spanish Pyrenees. *Journal of Structural Geology*, **8**, 399-405.

Mutti, E., Rosell, J., Allen, G. P., Fonnesu, F. & Sgavetti, M. 1985. *The Eocene baronia tide dominated delta-shelf system in the Ager Basin*. Exc. Guidebook 6th European Regional Meeting. Lerida, Spain, 579-600.

Olivet, J. L., Bonnin, J., Beuzart, P. & Auzende, J. M. 1984. *Cinematique de l'Atlantique nord et central. Centre National pour l'exploration des oceans*, Rapports scientifiques et techniques, **54**, 108p.

Pocoví, J. 1978. Estudio geológico de las Sierras Marginales Catalanas (Prepirineo de Lérida). *Acta Geológica Hispànica*, **13**, 73-79.

Price, R. A. 1986. The southeastern Canadian Cordillera, thrust faulting, tectonic wedging, and delamination of the lithosphere. *Journal of Structural Geology*, **8**, 239-245.

Roure, F., Choukroune, P., Berastegui, X., Muñoz, J.A ., Villien, A., Matheron, P., Bareyt, M., Seguret, M., Camara, P. & Deramond, J. 1989. ECORS Deep Seismic data and balanced cross-sections, geometric constraints to trace the evolution of the Pyrenees. *Tectonics*, **8**, 41-50.

Schreyer, W. 1988. Subduction of Continental Crust to Mantle Depths, Petrological Evidence. *Episodes*, **11**, 97-104.

Seguret, M. 1972. *Étude tectonique des nappes et séries décollées de la partie centrale du versant sud des Pyrénées*. Publications de l'Université de Sciences et Techniques de Languedoc, série Geologie Structurale. n. 2, Montpellier.

—— & Daignières, M. 1986. Crustal scale balanced cross-sections of the Pyrenees, discussion. *Tectonophysics*, **129**, 303-318.

Simó, A. & Puigdefàbregas, C. 1985. Transition from shelf to basin on an active slope, upper Cretaceous, Tremp area, southern Pyrenees. Exc. Guide-book 6th European Regional Meeting, Lerida, Spain, 63-108.

Souquet, P. 1967. Le Crétacé superieur sud-pyreneen en Catalogne, Aragon et Navarre. Thèse Doctorat Sciences Naturales. Faculté Sciences Toulouse, 529p.

—— & Peybernés, B. 1987. Allochtonie des massifs primaires nord-pyrénéens des Pyrénées Centrales. *Comptes Rendus de l'Académie des Sciences, Paris*, **305**, 733-739.

Torne, M., De Cabissole, B., Bayer, R., Casas, A., Daignières, M. & Rivero, A. 1989. Gravity constraints on the deep structure of the Pyrenean belt along the ECORS profile. *Tectonophysics*, **165**, 105-116.

Vergés, J. & Muñoz, J. A. 1990. Thrust sequences in the Southern Central Pyrenees. *Bulletin de la Société Géologique de France*, **8**, 265-271.

Vielzeuf, D. & Kornprobst, J. 1984. Crustal splitting and the emplacement of Pyrenean lherzolites and granulites. *Earth and Planetary Science Letters*, **67**, 383-386.

Williams, G. D. & Fischer, M. W. 1984. A balanced section across the Pyrenean orogenic belt. *Tectonics*, **3**, 773-780.

Zandvliet, J. 1960. The geology of the Upper Salat and Pallaresa valleys, Central Pyrenees, France/Spain. *Leidse Geologische Mededelingen*, **25**, 1-27.

Zwart, H. J. 1979. The geology of the Central Pyrenees. *Leidse Geologische Mededelingen*, **50**, 1-44.

—— 1986. The Variscan Geology of the Pyrenees. *Tectonophysics*, **129**, 9-27.

Thrusting and foreland basin evolution in the Southern Pyrenees

C. Puigdefàbregas[1], J.A. Muñoz[2] & J. Vergés[1]

[1]*Servei Geologic de Catalunya, Parallel, 71, 08004 - Barcelona, Spain*
[2]*Departament de Geologia Dinàmica, Geofísica i Paleontologia, Facultat de Geologia, Universitat de Barcelona, Zona Universitària de Pedralbes, 08071 Barcelona*

Abstract: The geometry and the infill of the south Pyrenean foreland basin mainly depend on the tectonic subsidence history due to the flexure of the crust, which in turn can be related to the structural evolution of the mountain chain at a crustal scale. Characteristics of the infill of the basin as well as the relationships between structures and synorogenic deposits allow distinction of four stages in the evolution of the south Pyrenean foreland basin. These stages can be related to the structural evolution of the crust as deduced from partial restored cross-section construction through the central Pyrenees. Stage I (Upper Cretaceous) is characterized by strong subsident turbiditic troughs deposited over a thinned crust. Uniform distribution of continental facies and a crust restored to its initial thickness corresponds to the stage II (Palaeocene). During Stage III, turbiditic troughs developed synchronously with the initial subduction of the lower crust (Lower and Middle Eocene). Stage IV (Uppermost Eocene-Oligocene) is characterized by continental deposition coeval with the increase of the crustal cross-sectional area, produced by both piggy-back and break-back thrust sequences.

The development of a thrust system in a mountain chain takes place synchronously with the accumulation of sediments in the related foreland basins. This implies that basin geometry and sedimentation patterns are controlled by the development of the thrust system.

The Pyrenees, mainly in their southern part, are one of the best examples to illustrate the interaction between tectonics and sedimentation because the shallow erosion level is such that the relationships between thrusts and their related deposits are well exposed. The Pyrenean orogen was developed as a consequence of an Alpine age continental collision that ranged from the Upper Cretaceous to Miocene times. The associated thrust system includes (Fig. 1): a basement antiformal-stack bounded by cover imbricated thrust systems (Muñoz, this volume). Thrusts of the central antiformal stack are south-directed and the southern imbricate thrust system involves more shortening than the northern one. Moreover, the floor thrust of the southern thrust system is the sole thrust of the Pyrenean chain, which implies that in the central and eastern Pyrenees, Iberia has been subducted to the north below Europe.

The south Pyrenean foreland basin is a triangular shaped feature between the Pyrenees to the north and the Catalan and Iberian Ranges to the SE and SW respectively. Most of the area represents the last stage of the basin fill known as Ebro foreland basin, whereas the earlier stages involved in the south Pyrenean thrust system occurred as piggy-back basins (Ori & Friend, 1984). An outline of the Ebro basin with the first observations on the relationships between tectonics and sedimentation was given by Riba (1967), who described features such as basin asymmetry and the outward migration of facies depocentres. Riba (1973) also proposed a kinematic model for the progressive unconformities in the northern margin of the Ebro basin and correlated these with the synchronous emplacement of the Pyrenean thrust system. Some authors have attempted to distinguish sedimentary sequences in both the central and eastern southern Pyrenean foreland basins, and relate them to the tectonic evolution (Soler & Puigdefàbregas 1970; Garrido-Megías 1973; Puigdefàbregas 1975; Nijman & Nio 1975; Puigdefàbregas *et al.* 1986; Labaume *et al.* 1987, and Mutti *et al.* 1988).

The aim of this paper is to present a general outline of the geometry and infill of the south Pyrenean foreland basin, its relationship to the thrust system and finally to correlate this with the crustal evolution of the whole crust along the ECORS profile.

Stages in the evolution of the foreland basin

The characteristics of the infill of the Pyrenean foreland basin allow the determination of several stages in its evolution regardless of longitudinal variations of facies, and mainly dependent on the pattern of thrusting. These stages, well established in the central and eastern Pyrenees, have been used in the present paper in order to describe the main features of the foreland basin deposits and their relationships with the evolution of the Pyrenean chain.

Stage I (Upper Santonian-Maastrichtian)

Prior to the Alpine compression of the Pyrenees, a Lower Cretaceous extensional rift basin developed (Puigdefàbregas & Souquet 1986). The Upper Cenomanian transgressive

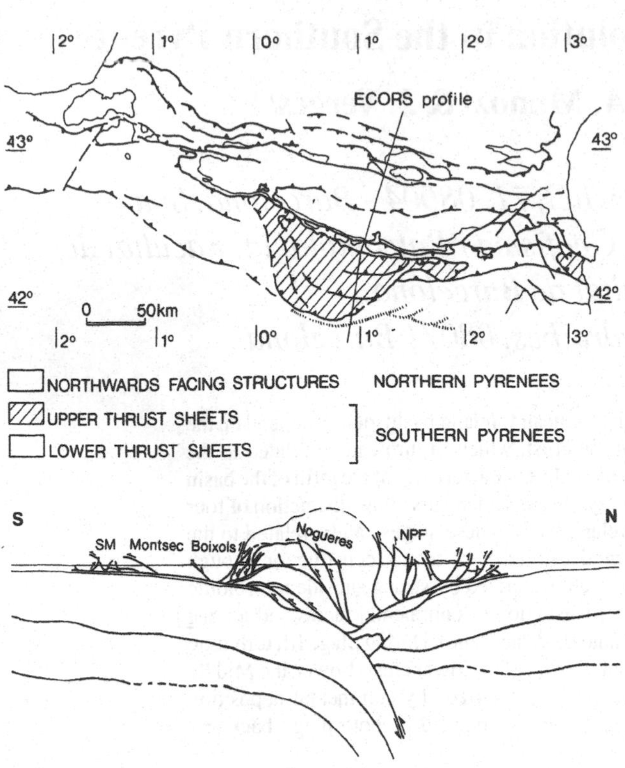

Figure 1. Structural sketch of the Pyrenees and the ECORS crustal cross-section.

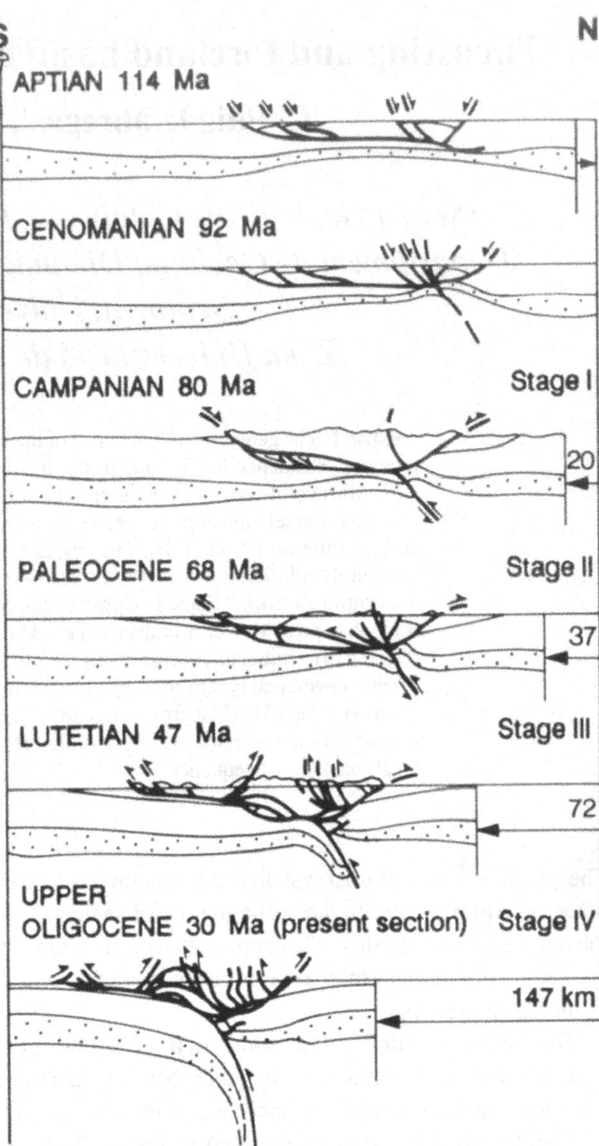

Figure 2. Partially restored cross-sections showing the Lower Cretaceous extensional configuration and thrust geometry at crustal scale during the stages of the foreland basin evolution.

platform deposits unconformably overlie the syn-rift deposits and related normal faults (Berastegui *et al.* 1990). During this period, thermal subsidence was still active as indicated by the superposition of Lower and Upper Cretaceous depocentres, suggesting a pure shear extensional model for the Lower Cretaceous rifting.

The first Pyrenean thrusts developed as a result of the reactivation of the Lower Cretaceous extensional faults during Upper Santonian times. In front of these thrusts, deep E-W elongated basins were formed and infilled with Upper Santonian to Campanian age turbiditic deposits (Vallcarga Formation of Mey *et al.* 1968) which are up to 6000 m in the north (Dubois & Seguin 1978) and 2000 m in the south. These turbidites unconformably overlie the earlier sequences and their deposition seems to be controlled by the development of the first Pyrenean thrust sheets (Bóixols; Fig. 1). Similar relationships were documented westwards for the Turbón thrust sheet which moved from the Uppermost Turonian to Maastrichtian (Souquet & Déramond 1989) and also for thrusts in the northern Pyrenees (Baby *et al.* 1988; Desegaulx *et al.* 1990). As thrusting progressed during the Maastrichtian, progressive unconformities (Garrido-Megías 1973) affected the shallow marine sequences of the Aren Group. The geometry and distinct depositional sequences within this formation have been described by Nagtegaal *et al.* (1983), Simó & Puigdefàbregas (1985), Fondecave *et al.* (1989) and Mutti & Sgavetti (1987) who also discuss eustatic or tectonic models for their formation. A number of unconformity-bounded depositional sequences can be distinguished, in which relative disposition of facies, depocentres and angular unconformities are related to the Bóixols thrust.

The first thrusts developed over an already stretched crust (Fig. 2). As a consequence, both the limited tectonic subsidence induced by the inversion of the lower Cretaceous extensional basins and the remnant thermal subsidence contributed to the formation of turbiditic troughs as suggested by Brunet (1984) in the NW Pyrenees. Moreover, a thin and warm crust would favour the formation of narrow and deep foreland basins in front of the thrust sheets (Karner *et al.* 1983).

Stage II (Uppermost Maastrichtian-Palaeocene)

During this stage the south Pyrenean foreland basin was characterized by the deposition of fluvial and lacustrine red beds, which overlay and in part interfingered with the Aren Group. These deposits are known as Garumnian (Tremp Formation by Mey *et al.* 1968) or Tremp Group (Cuevas 1989). They are up to 1000 m thick in places and extend over the whole central and eastern Pyrenees, in the north as well as in the south (Plaziat 1975). Conglomerate formations unconformably overlie the inverted extensional faults (Bóixols thrust) and are coeval with the southwards piggy-back propagation of the southern thrust system in the central Pyrenees. Thickness and facies distributions demonstrate the contemporaneity of the Garumnian sedimentation with the new-formed thrusts (Montsec thrust; Figs 1 & 2).

The Lower Cretaceous extensional faults were completely inverted during Palaeocene times; thus the stretched upper crust recovered its initial pre-Cretaceous length, and probably the crustal thickness also attained its pre-Cretaceous crustal thickness. This conclusion has also been deduced from the partial restoration of the ECORS balanced cross-section for Palaeocene times (Muñoz 1991, this volume; Fig. 2).

Although thrust structures continuously developed from Stage I, the change in the infill of the foreland basin could be related to the crustal thickness compensation, inferred to occur at the end of the thermal subsidence. As a result, during this stage the foreland basin is characterized by uniform shallow-water deposits occupying a wide area. Involvement of basement rocks in thrust structures was still too restricted to generate strongly subsident troughs by thrust sheet loading (Beaumont 1981; Jordan 1981).

In the western Pyrenees, the equivalent sediments to the Garumnian shallow-water deposits are represented by 350 m of deep-water carbonate and siliciclastic sediments deposited in continuity with the Upper Cretaceous turbidites (Pujalte *et al.* 1989). Thus, the evolution proposed for Stage I, characterized by thrusts superimposed on a thinned crust, continued in the western Pyrenees through the Uppermost Cretaceous-Palaeocene and probably until Early Eocene times. Thickness variations of the series and the occurrence and thickness of volcanic rocks (Montigny *et al.* 1986) demonstrate a greater extension of the western Pyrenean crust during Early Cretaceous times. As a result, during convergence the crust in the western Pyrenees would have recovered its initial thickness later than that in the central and eastern Pyrenees. This concept is in agreement with the proposed crustal and foreland basin evolution in the central Pyrenees.

Stage III (Lower and Middle Eocene)

After the widespread Ilerdian transgression, the south Pyrenean

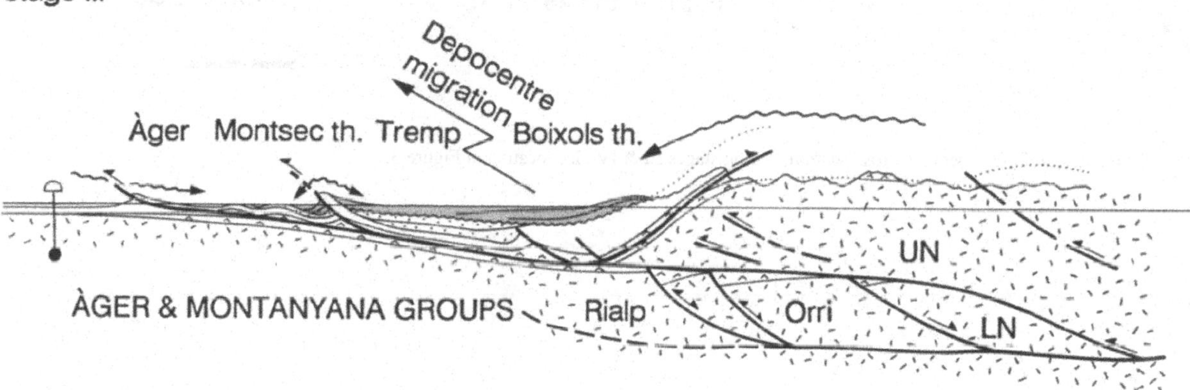

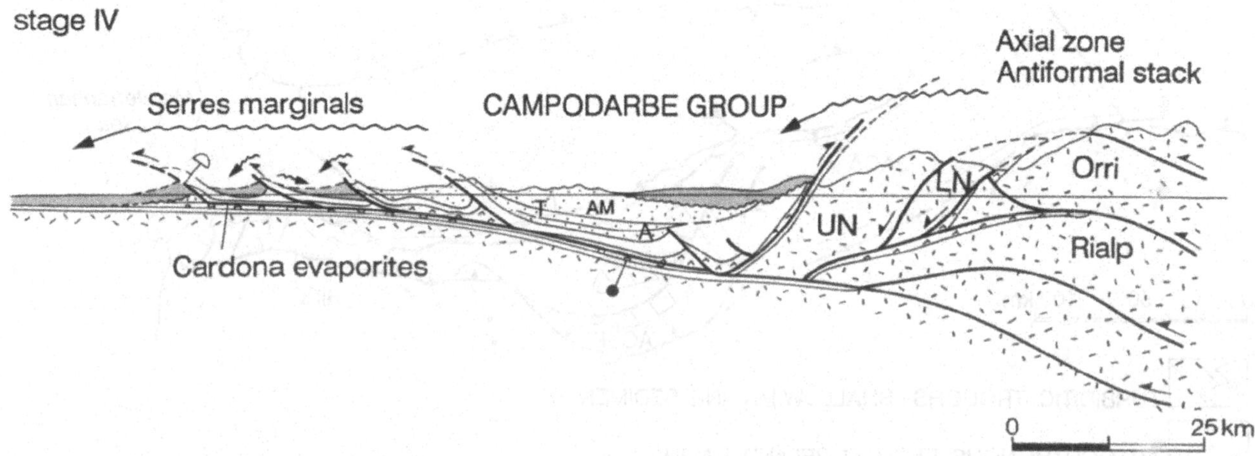

Figure 3. Partially restored cross-sections at the time of stages III & IV. These sections correspond to the southern part of the ECORS profile. The syntectonic sediments related with the stage have been shaded as well as the previous foreland basin deposits stippled. A: Aren Group (Upper Cretaceous); T: Tremp Group; AM: Ager and Montanyana Groups; LN: Lower Nogueres Units; UN: Upper Nogueres Units.

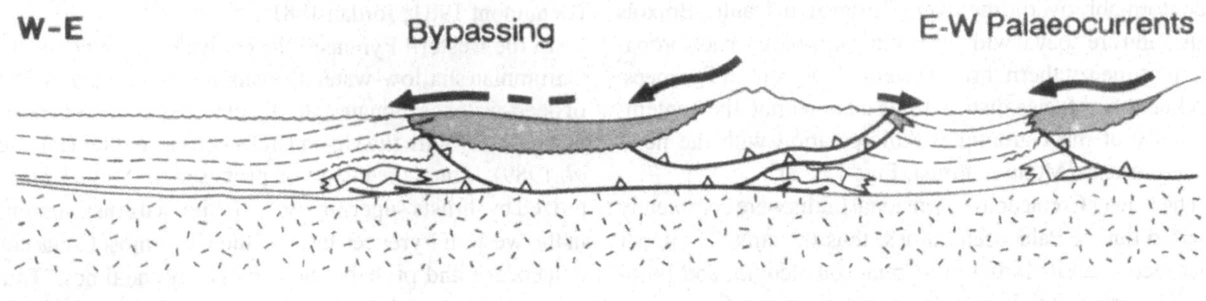

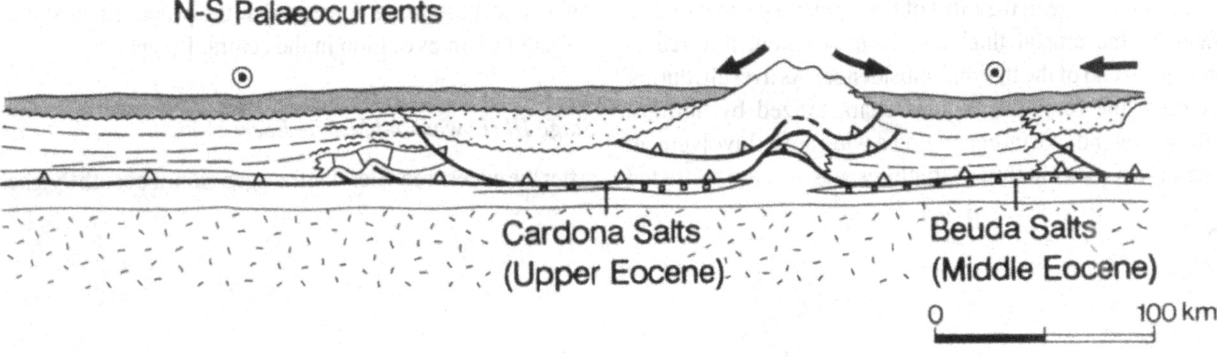

Figure 4 Longitudinal E-W schematic cross-sections during stages III & IV. See location in Figure 3.

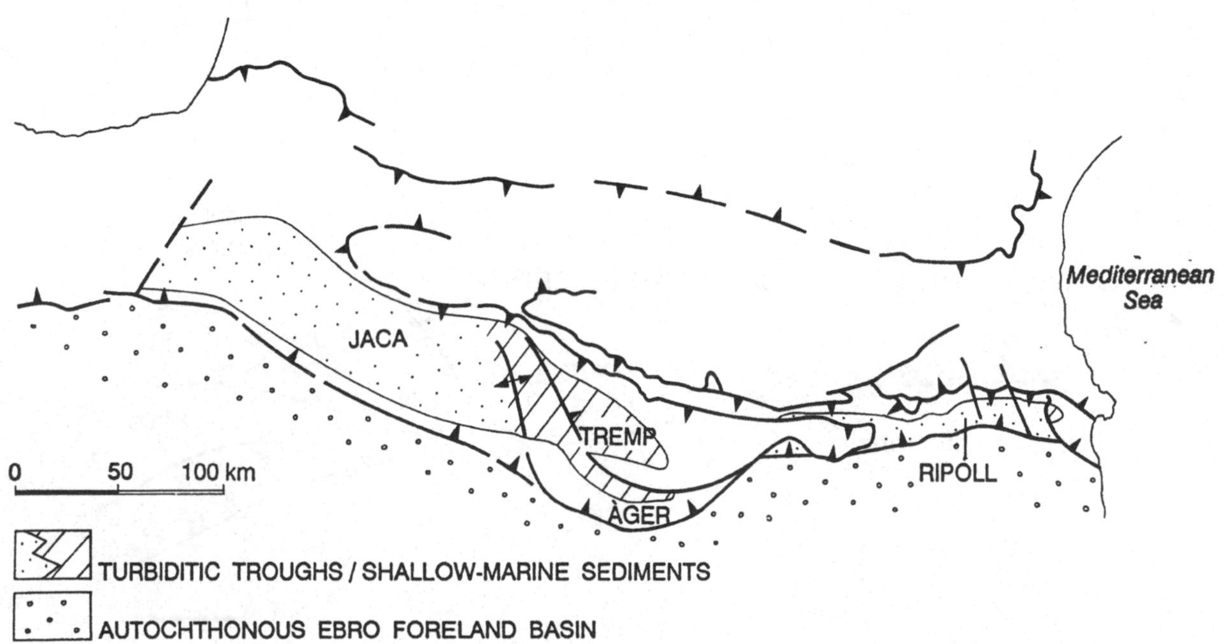

Figure 5. Schematic map showing the facies distribution during stage III.

foreland basin was characterized by the occurrence of strongly subsident troughs infilled by turbiditic sediments. The arrangement of the turbidites and the coeval shallow-marine to continental deposits as well as the facies distribution within them were controlled by the thrust sheet geometries.

During this stage, the southern Pyrenean upper thrust sheets mainly consisted of Mesozoic units, which were previously inverted during the first and second stages, and at this time displaced in a piggy-back fashion to the south over the foreland, characterized by a reduced cover sequence unconformably overlying the basement rocks (Fig. 3). The longitudinal disposition of these thrust sheets, the South Pyrenean Central Unit (SPCU; Séguret 1972), is controlled by the Mesozoic extensional faults. These faults determine the location of oblique and lateral ramps and the arrangement of the foreland facies across them (Figs 4 & 5). On top of the Mesozoic thrust sheets of the SPCU (Serres Marginals, Montsec and Bóixols from S to N) the third stage of the foreland basin infill consisted of the fluvio-deltaic facies of the Montanyana Group (Mutti *et al.* 1988). The basin tended to be elongate and parallel to the structures (cf. the Tremp basin). The sequences within the Montanyana Group are characterized by southward prograding alluvial fans whose fringes were westward axially drained by a river system (Fig. 6). Transverse N-S sections through the basin are asymmetric, with the thicker alluvial fan sequences occupying about the northern two thirds of the basin and the fluvial systems only the remaining southern one third. This basin-fill geometry suggests the synchronous emplacement of the Upper Nogueres thrust sheet, accompanied by passive roof back thrusting, and out of sequence thrusting in the rear, together leading to generation of relief, increase of the erosion rate and clastic supply.

From all of the above one can gather that the tectonic control on sedimentation was also exerted at minor scale within each of the major sequences in response to alternating periods of forward thrust propagation and vertical stacking in the hinterland.

The moderately subsiding Montanyana basin, situated on the hangingwall of the SPCU, was related westwards to the strongly subsiding Jaca basin on the footwall, where the turbidites of the Hecho Group were deposited under a 1000 m water column (Mutti *et al.* 1988). The transition between these two settings occurred through a succession of oblique structures where incised canyons transferred the clastics from the shallow platform to the deeper basin. This strongly subsident trough (Jaca basin) is located over slightly stretched or unstretched Mesozoic crust. The turbiditic trough was a piggy-back basin and its sediments are coeval with a widespread southern carbonate platform which was synsedimentarily deformed by the frontal thrust structures (Puigdefàbregas 1975).

In the eastern Pyrenees, the south Pyrenean foreland basin displays a similar arrangement with fluvio-deltaic deposits in the east, on top of the upper thrust sheet (Pujadas *et al.* 1989), and these grade westwards to slope and turbiditic deposits of the Ripoll basin (Armàncies and Campdevànol sequences; Puigdefàbregas *et al.* 1986). This turbiditic trough was formed

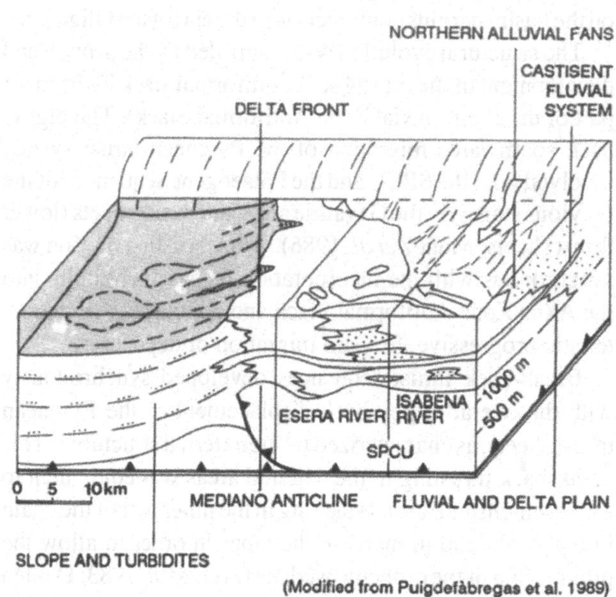

Figure 6. Schematic block diagram of the western oblique ramp zone of the SCU showing the transition from continental and shallow marine deposits (Tremp and Ager basins) to the turbiditic sediments (Jaca basin) of stage III.

in an equivalent structural position to the Jaca basin, over unextended crust, although the Ripoll basin was closed to the west by the oblique boundary structure of the SPCU.

The turbiditic troughs are aligned E-W with their fluvio-deltaic equivalents, independent of the syn-emplacement shape of the cover thrust sheets (SPCU). This suggests that the associated tectonic subsidence is not only due to the loading of the cover thrust sheets, but also due to the subcrustal forces.

Stage IV (Upper Eocene-Oligocene)

This stage was characterized by the final infilling of the earlier turbiditic basins by deltaic systems. The basin became wider and shallower and the generalized deposition of coarse alluvium indicates an increase in the orogenic relief.

The facies distribution, as in the previous stages, was controlled by the geometry of the SPCU. On the hangingwall of this unit mostly alluvial fan systems are found unconformably overlying a substratum which included the depositional sequences of the previous stages of thrust emplacement. Coeval deposits on the footwall consist of thick deltaic to fluvial and alluvial fan sequences and formed the final infill of the Jaca and Ripoll basins. These basins were mostly fed by clastics derived from erosion of the newly created relief in the hinterland, from the alluvial fans overlying the SPCU through the lateral ramp zones, and also from the southern and eastern foreland basin margins where locally thick alluvial fans and fan-deltas accumulated (Montserrat fan-delta; Anadón & Marzo 1986). This configuration defines a basin centre occupied by extensive evaporitic formations (Sáez & Riba 1986) when the basin was finally closed to the marine waters at the end of the Eocene. This arrangement, with lacustrine deposits in the centre and clastic wedges on the basin margins continued until the end of the Oligocene.

The structural evolution was controlled by the growth and development of the crustal scale antiformal stack in the inner part of the chain (Axial Zone Antiformal Stack). The piggy-back southwards migration of the Pyrenean thrust system involved both the SPCU and the Palaeogene sequences of the previous stages, in the Gavarnie and Cadí thrust sheets (lower thrust sheets; Muñoz *et al.* 1986). This thrusting motion was synchronous with the incorporation of basement units into the Axial Zone Antiformal Stack and controlled the characteristic progressive, forward migration of depocentres.

Break-back thrust sequences developed synchronously with the overall piggy-back displacement of the Pyrenean thrust sheets, as characterized by the external structures. This break-back thrusting in the external areas was equivalent to the basement antiformal stacking in the inner part of the chain in that both tend to increase the taper in order to allow the progression of the orogenic wedge (Davis *et al.* 1983; Dahlen & Suppe 1988). This particular structural situation significantly increased the cross-sectional area, so that erosion and supply of coarse clastics was therefore enhanced.

Several examples of break-back thrusting sequence can be unambiguously deduced from the relationships between the structures and the related synorogenic conglomerates (Martínez *et al.* 1988; Vergés & Muñoz 1990). In the eastern oblique boundary of the SPCU (Oliana area) these relationships between structures and clastic deposits are well illustrated (Vergés & Muñoz 1990). The break-back thrust system is constituted by a set of thrusts, each of them covered by a conglomerate formation, which in turn is thrust by the younger thrust formed hindwards of the previous one, in an obvious out-of-sequence disposition (Fig. 7).

The contribution of these out of sequence thrusts in the total displacement of the thrust system was very small. To the contrary, the vertical component of movement was significant and therefore the structural relief increased.

The basement antiformal stack and synchronous with its formation, the cover thrust sheets (SPCU), have been wedged backwards (to the north) over the Morreres passive roof backthrust (Muñoz, this volume). This backthrust together with a set of out of sequence thrusts are related with three successive conglomerate sequences (Mellere, pers. comm.) of the Collegats Fm. (Mey *et al.* 1968). The conglomerates which documented the out-of-sequence thrusts of the Collegats and Oliana area are approximately of the same age.

Discussion and conclusions

The geometry and major infilling of the south Pyrenean foreland basin was mainly dependent upon both the surficial and subcrustal forces which produced tectonic subsidence. The surficial forces were mainly due to thrust sheet loading and the subcrustal forces result from thermal evolution of the lithosphere and from its flexure caused by the traction of the subducting slab (Brunet 1986; Molnar & Lyon-Caen 1988). The geometry of the thrust sheets might have been a factor modifying the tectonic subsidence but this mainly generated longitudinal variations in the infilling of the foreland basin. Finally, as far as the tectonic control on the foreland basin depositional sequences is concerned, the influence of local structures must be taken into account, in addition to other factors such as global sea level changes. Although it is not the

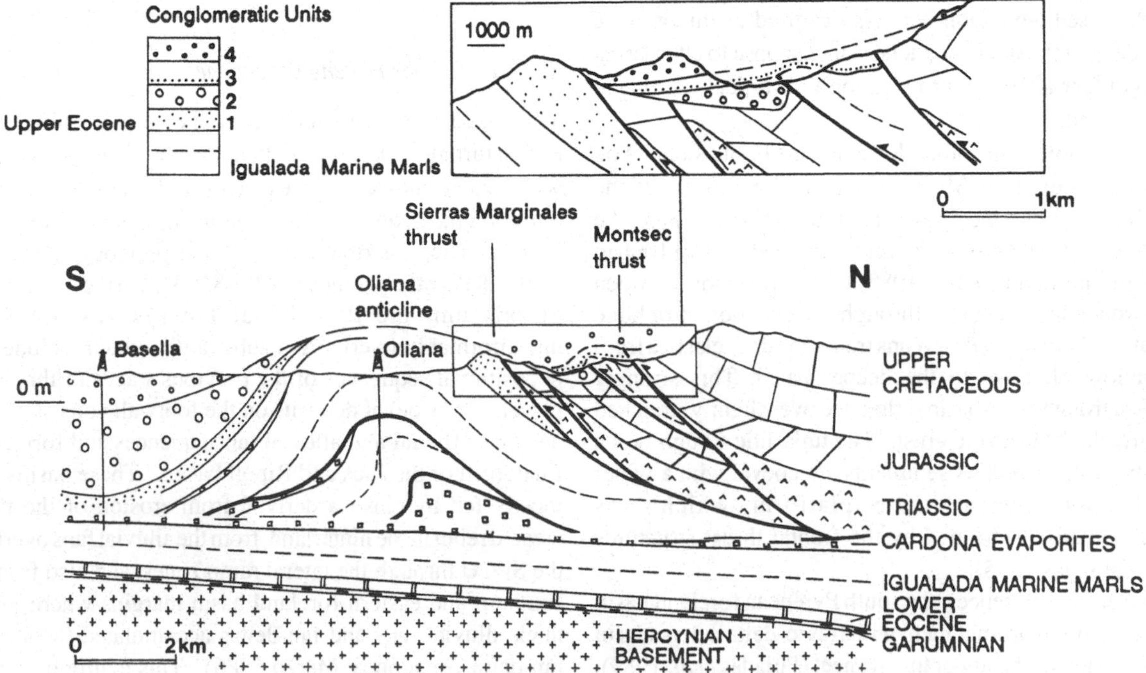

Figure 7. Geological cross-section of the eastern termination of the SCU showing the relationships between out-of-sequence thrusts and synorogenic conglomeratic deposits.

here determine the depositional geometries of the syntectonic conglomerates (Vergés & Muñoz 1990).

Recent papers have modelled the flexure of the lithosphere of the Pyrenees and the geometry of the foreland basin (Brunet 1986; Desegaulx & Moretti 1988; Desegaulx *et al.* 1990) using the physical parameters calculated or deduced from the present state of the lithosphere. The problem of modelling old mountain chains is to know how these physical parameters have evolved during the formation of the orogen and also during the post-tectonic evolution. Partial crustal restored cross-sections can be used as an additional tool. This kind of work through the Central Pyrenees allows investigation of the relationships between the foreland basin geometry and the crustal profile as deduced from partial restored cross-sections (Muñoz 1991, this volume). Crust below the Pyrenean belt has evolved during the formation of the mountain chain from an initial thinned crust to a thickened crust with a subducted slab of about 100 km long (Muñoz 1991, this volume). As a consequence, not only the thickness of the crust below the belt has changed through convergence, but also its elastic parameters, which control the geometry of the foreland basin.

The major subdivisions that have been made in the south central Pyrenean foreland basin are related to stages with deposition of turbidites in strongly subsident narrow troughs and to stages with deposition of shallow-water to continental sediments occupying a wider area. The former indicate a small wavelength of the lithospheric flexure, the latter a large wavelength.

The turbiditic troughs characteristic of Stage I, (Upper Santonian-Maastrichtian) have been interpreted as a result of a restricted surface weight but acting over a warm and thin crust. In addition, subcrustal forces due to the thermal subsidence inherited from the previous extensional history have contributed to the strong tectonic subsidence. During this stage, the crustal shortening has not compensated the initial crustal thickness.

Stage II, (Uppermost Maastrichtian-Palaeocene), characterized by a rather uniform facies distribution, corresponds to the period where the thickening of the crust is compensated by crustal shortening, so that there are insignificant subcrustal forces. The only contribution in the tectonic subsidence is due to the surficial loading by the thrust sheets themselves.

The creation of elongated turbiditic troughs is the main characteristic of Stage III (Lower and Middle Eocene).

The sudden increase of the tectonic subsidence during this period (Burbank, pers. comm.) is due to both a greater influence of the surficial loading and by the existence of subcrustal forces. During this period the basement rocks are involved in thrust sheets and the load increases. In addition the Iberian crust began to subduct below the European crust giving rise to a pull by the subducted slab.

The lithospheric flexure value was probably the same all along the chain but turbiditic troughs only developed in the footwall of the SPCU because the thrust sheet of the SPCU partially fill the trough in the central Pyrenees. During this stage, there is an increase of the thrusting rate with respect to the earlier one (Labaume *et al.* 1985; Vergés & Martínez 1988).

Stage IV (Upper Eocene-Oligocene) is characterized by coarse grained alluvial fan deposition as an erosional response to the increase of the cross-sectional area which resulted from the growth of the basement antiformal stack in the hinterland and to the development of the break-back thrusting sequences in the frontal areas. In spite of the taper increase and consequently the surficial weight, the tectonic subsidence during this stage decreases with respect to previous stages (Burbank *et al.* 1991). As a consequence, a subcrustal explanation such as thermal re-equilibration or collapse of the subducting slab has to be considered.

It must be noted that this evolution in four stages, which lasts 50 Ma, is proposed for this particular transect and is not necessarily synchronous with other transects, first because of differences in structural style and strength of the crust, and secondly because of the diachronous development of the belt (Choukroune 1976).

These four stages may be difficult to distinguish in other mountain chains because of differences in preservation of the foreland deposits or because of a different foreland basin history. As most of the collisional belts evolved from previous stretched crust to a final thickened crust, one might expect that both surficial and subcrustal processes for the evolution of the foreland basin would be similar to those described in this paper.

We are grateful to Ken McClay for comments and revision of the original manuscript. This work has been partially supported by the CICYT project GEO89-0254.

References

Anadón, P. & Marzo, M. 1986. Sistemas deposicionales eocenos del margen oriental de la cuenca del Ebro: sector Igualada - Montserrat. *Libro guia Exc. XI Congreso Español de Sedimentologia*, 4.1-4.59.

Baby, P., Crouzet, G., Specht, M., Déramond, J., Bilotte, M. & Debroas, E. J. 1988. Rôle des paléostructures albo-cénomaniennes dans la géométrie des chevauchements frontaux nord-pyrénéens. *Comptes Rendus de l'Académie des Sciences, Paris*, **306**, 307-313.

Beaumont, C. 1981. Foreland basins. *Geophysical Journal of the Royal Astronomical Society*, **65**, 291-329.

Berastegui, X., Garcia, J. M. & Losantos, M. 1990. Structure and sedimentary evolution of the Organyà basin (Central South Pyrenean Unit, Spain) during the Lower Cretaceous. *Bulletin de la Société Géologique de France*, **8**, VI; 251-264.

Brunet, M. F. 1984. Subsidence history of the Aquitaine basin determined from subsidence curves. *Geological Magazine*, **121**, 421-428.

—— 1986. The influence of the evolution of the Pyrenees on adjacent basins. *Tectonophysics*, **129**, 343-354.

Burbank, D. W., Vergés, J., Muñoz, J. A. & Bentham, P. 1991. Coeval hindward- and forward-imbricating thrusting in the Central Southern Pyrenees, Spain: Timing and rates of shortening and deposition. *Geological Society of America Bulletin* (in press).

Choukroune, P. 1976. Structure et évolution tectonique de la zone nord-pyrénéenne (analise de la déformation dans une portion de chaîne à schistosité subverticale). *Mémoires de la Société Géologique de France*, **127**, 116p.

Cuevas, J. L. 1989. La Formación Talarn: estudio estratigráfico y sedimentológico de las facies de un sistema aluvial en el tránsito Mesozoico-Cenozoico de la Conca de Tremp. *Tesis de Licencitura*, Universitat de Barcelona, 107p.

Dahlen, F. A. & Suppe, J. 1988. Mechanics, growth, and erosion of mountain belts. *Geological Society of America Special Paper*, **218**, 161-178.

Davis, D., Suppe, J. & Dahlen, F. A. 1983. Mechanics of fold-and-thrust belts and accretionary wedges. *Journal of Geophysical Research*, **88**, 1153-1172.

Desegaulx, P. & Moretti, I. 1988. Subsidence history of the Ebro basin. *Journal of Geodynamics*, **10**, 9-24.

——, Roure, F. & Villien, A. 1990. Structural evolution of the Pyrenees: Tectonic heritage and flexural behavior of the continental crust. *In:* Letouzey, J. (ed.) *Petroleum and Tectonics in Mobile Belts*, 31-48.

Dubois, P. & Seguin, J-C. 1978. Les flyschs crétacé et éocène de la zone commingeoise et leur environnement. *Bulletin de la Société Géologique de la France*, **7**, 657-671.

Fondecave-Wallez, M. J., Souquet, P. et Gourinard, Y. 1989. Enregistrement sédimentaire de l'eustatisme et de la tectonique dans la série turbiditique du Crétacé des Pyrénées centro-méridionales (Groupe de Vallcarga, n.gr., Espagne). *Comptes Rendus de l'Académie des Sciences, Paris*, **309**, 137-144.

Garrido-Megías, A. 1973. Estudio geológico y relación entre tectónica y sedimentación del Secundario y Terciario de la vertiente meridional pirenaica en su zona central (prov. Huesca y Lérida). *Tesis Doctoral*, Facultad de Ciencias. Granada, 395p.

Jordan, T. E. 1981. Thrust loads and foreland basin evolution, Cretaceous, Western United States. *American Association of Petroleum Geologists Bulletin*, **65**, 2506-2520.

Karner, G. D., Steckler, M. S. & Thorne, J. A. 1983. Long-term thermomechanical properties of the continental lithosphere. *Nature*, **304**, 250-253.

Labaume, P., Séguret, M. & Seyve, C. 1985. Evolution of a turbiditic foreland basin and analogy with an accretionary prism: Example of the Eocene South-Pyrenean basin. *Tectonics*, **4**, 661-685.

——, Mutti, E. & Séguret, M. 1987. Megaturbidites: A depositional model from the Eocene of the SW-Pyrenean foreland basin, Spain. *Geo-Marine Letters*, **7**, 91-101.

Martínez, A., Vergés, J. & Muñoz, J. A. (1988) Secuencias de propagación del sistema de cabalgamientos de la terminación oriental del manto del Pedraforca y relación con los conglomerados sinorogénicos. *Acta Geològica Hispànica*, **23**, 119-128.

Mey, P.H.W., Nagtegaal, P.J.C., Roberti, K. J. & Hartevelt, J.J.A. 1968. Lithostratigraphic subdivision of post-hercynian deposits in the south-central Pyrenees, Spain. *Leidse Geologische Mededelingen*, **41**, 221-228.

Molnar, P. & Lyon-Caen, H. 1988. Some simple physical aspects of the support, structure, and evolution of mountain belts. *Geological Society of America Special Paper*, **218**, 179-207.

Montigny, R., Azambre, B., Rossy, M. & Thuizat, R. 1986. K-Ar study of Cretaceous magmatism and metamorphism in the Pyrenees: age and length of rotation of the Iberian Peninsula. *Tectonophysics*, **129**, 257-273.

Muñoz, J. A., Martínez, A. & Vergés, J. 1986. Thrust sequences in the eastern Spanish Pyrenees. *Journal of Structural Geology*, **8**, 399-405.

—— 1991. Evolution of a Continental Collision Belt: ECORS-Pyrenees Crustal Balanced Cross-section (this volume).

Mutti, E. & Sgavetti, M. 1987. Sequence stratigraphy of the Upper Cretaceous Aren strata in the Orcau-Aren region, south-central Pyrenees, Spain: Distinction between eustatically and tectonically controlled depo sitional sequences. *Annali dell'Università di Ferrara. Sezione Scienze della Terra*, **1**, 22p.

——, Séguret, M. & Sgavetti, M. 1988. Sedimentation and deformation in the Tertiary sequences of the southern Pyrenees. *American Association of Petroleum Geologists Mediterranean Basins Conference*. Field Trip 7, 153p.

Nagtegaal, P.J.C., Van Vliet, A. & Brouwer, J. 1983. Syntectonic coastal offlap and concurrent turbidite deposition: the Upper Cretaceous Aren Sandstone in the South-Central Pyrenees, Spain. *Sedimentary Geology*, **34**, 185-218.

Nijman, W. & Nio, S. D. 1975. The Eocene Montañana delta (Tremp-Graus Basin, provinces of Lérida and Huesca, Southern Pyrenees, N. Spain). In: *IX Congrés de Sedimentologie. Nice*, 18p.

Ori, G. G. & Friend, P. F. 1984. Sedimentary basins formed and carried piggyback on active thrust sheets. *Geology*, **12**, 475-478.

Plaziat, J. C. 1975. L'Ilerdien a l'intérieur du Paléogène languedocien; ses relations avec le Sparnacien, l'Ilerdien sud-pyrénéen. l'Yprésien et le Paléocène. *Bulletin de la Société géologique de France*, **7**, 168-182.

Puigdefàbregas, C. 1975. La sedimentación molásica en la cuenca de Jaca. *Pirineos*, **104**, 188p.

——, Muñoz, J.A. & Marzo, M. 1986. Thrust belt development in the Eastern Pyrenees and related depositional sequences in the southern foreland basin. *In:* Allen, P. A. & Homewood, P. (eds) *Foreland basins*. Special Publication of the International Association of Sedimentologists, **8**, 229-246.

—— & Souquet, P. 1986. Tectosedimentary cycles and depositional sequences of the Mesozoic and Tertiary from the Pyrenees. *Tectonophysics*, **129**, 173-203.

——, Collinson, J., Cuevas, J. L., Dreyer, T., Marzo, M., Mellere. D., Mercadé, L., Muñoz, J. A., Nijman, W. & Vergés, J. 1989. Alluvial deposits of the successive foreland basin stages and their relation to the Pyrenean thrust sequence. *4th International Conference on Fluvial Sedimentology. Excursion Guidebook*. (Marzo & Puigdefàbregas, eds), 175p.

Pujadas, J., Casas, J. M., Muñoz, J. A. & Sabat, F. 1989. Thrust tectonics and Paleogene syntectonic sedimentation in the Empordà area, southeastern Pyrenees. *Geodinamica Acta*, **3**, 195-206.

Pujalte, V., Robles, S., Zapata, M., Orue-Etxebarria, X. & Garcia-Portero, J. 1989. Sistemas sedimentarios, secuencias deposicionales y fenomenos tectoestratigraficos del Maastrichtiense superior-Eoceno inferior de la cuenca vasca (Guipuzcoa y Vizcaya). *XII Congreso Español de Sedimentología, Bilbao. Guía de Excursiones*, 47-88.

Riba, O. 1967. Resultados de un estudio sobre el Terciario continental de la parte este de la depresión central catalana. *Acta Geològica Hispànica*, **II**, 1-6.

—— 1973. Las discordancias sintectónicas del Alto Cardener (Prepirineo catalán), ensayo de interpretación evolutiva. *Acta Geològica Hispànica*, **8**, 90-99.

Sáez, A. & Riba, O. 1986. Depositos aluviales y lacustres paleogenos del margen pirenaico catalán de la cuenca del Ebro. Libro guia Excursiones. *XI Congreso Español de Sedimentologia*. Barcelona, 6.1-6.29.

Séguret, M. 1972. *Étude tectonique des nappes et séries décollées de la partie centrale du versant sud des Pyrénées*. Publications de l'Université de Sciences et Techniques de Languedoc, série Geologie Structurale n. 2, Montpellier, 155p.

Simó, A. & Puigdefàbregas, C. 1985. Transition from shelf to basin on an active slope, upper Cretaceous, Tremp area, southern Pyrenees. *Exc. Guide-book 6th European Regional Meeting*. Lerida, Spain, 63-108.

Soler, M. & Puigdefàbregas, C. 1970. Lineas generales de la geología del Alto Aragón Occidental. *Pirineos*, **96**, 5-20.

Souquet, P. et Déramond, J. 1989. Séquence de chevauchements et séquences de dépôt dans un bassin d'avant-fosse. Exemple du sillon crétacé du versant sud des Pyrénées (Espagne). *Comptes Rendus de l'Académie des Sciences, Paris*, **309**, 137-144.

Vergés, J. & Martínez, A. 1988. Corte compensado del Pirineo oriental: geometria de las cuencas de antepaís y edades de emplazamiento de los mantos de corrimiento. *Acta Geològica Hispànica*, **23**, 95-106.

—— & Muñoz, J. A. 1990. Thrust sequences in the Southern Central Pyrenees. *Bulletin de la Société géologique de France*, **8**, 265-271.

South Pyrenean fold and thrust belt: The role of foreland evaporitic levels in thrust geometry

J. Vergés[1], J.A. Muñoz[2] & A. Martínez[1]

[1]Servei Geologic de Catalunya, Parallel, 71, 08004 - Barcelona, Spain
[2]Departament Geologia Dinàmica, Geofísica i Paleontología, Universitat de Barcelona, 08071 Barcelona, Spain

Abstract: Combined surface and subsurface data were used to determine the geometry of the foreland fold and thrust belt in the southeastern Pyrenees. In the south Pyrenean foreland basin, three evaporitic horizons were deposited: the Beuda, the Cardona and the Barbastro Formations. These evaporites range from Lutetian to Lower Oligocene in age, and their depocentres shifted successively to the SW in front of the advancing Pyrenean thrust sheets. From Late Eocene to Late Oligocene, a salt-basal foreland fold and thrust belt developed over a detachment level with a staircase geometry controlled by the arrangement of the evaporitic basins. Synchronously, the central Pyrenean thrust sheets were displaced southwards over the same detachment level. The arrangement of the evaporitic basins, mainly the Cardona basin, the N-S transport direction of the south Pyrenean thrust sheets, and the change in trend of the South Pyrenean main thrust (SPMT), were all responsible for the different trend of structures in the foreland fold and thrust belt. The limits of these folded regions coincide with the boundaries of the Cardona evaporitic basin. The southernmost outcropping frontal structure of the Pyrenees is a backthrust related to the edge of the Cardona massive salt level and is located in some places 40 km southwards of the SPMT. Well-constrained timing of deformation permits the interpretation that the shortening occurred from Upper Eocene times and has minimum values of between 21 km and 25 km. These values represent more than 15% of the 147 km of shortening deduced for the overall Pyrenean chain.

The southern Pyrenees are made up of cover and basement thrust sheets which were displaced southwards from the Late Cretaceous to the Oligocene (Early Miocene in the western Pyrenees). The south Pyrenean structural units can be grouped into Upper Thrust Sheets, which consist only of cover rocks, mainly a Mesozoic series, and Lower Thrust Sheets, formed by basement rocks and a Palaeogene series unconformably overlying a reduced and thin cover sequence (Muñoz *et al.* 1986; Muñoz 1991, this volume). In the eastern Pyrenees, Palaeogene sediments involved in the Lower Thrust Sheets (Cadí thrust sheet) occupy a narrow area (Figs 1 & 2). These sediments represent the northern part of the South Pyrenean foreland basin, which has been incorporated into the allochthonous units (Puigdefàbregas *et al.* 1986; Vergés & Martínez 1988). The Cadí thrust sheet is overthrusted by the Pedraforca upper thrust sheet (Fig. 2). Westwards, in the central Pyrenees, upper thrust sheets (Bóixols, Montsec and Serres Marginals) cover a wide area over an autochthonous non-deformed Palaeogene series, which unconformably overlies basement rocks (Muñoz 1991, this volume). The eastern termination of the Upper Thrust Sheets of the central Pyrenees is the oblique Segre thrust zone, which limits with the Ebro foreland basin and, northwards, with the Cadí thrust sheet (Fig. 2). In the central and eastern Pyrenees, the emergent floor thrust of the south Pyrenean thrust sheets, herein called South Pyrenean main thrust (SPMT), describes a step geometry at surface (Figs 1 & 2). The eastern foreland Ebro basin, located southwards and eastwards of the SPMT has been displaced to the south over a detachment level. The southernmost structures affecting the eastern Ebro basin define an approximate E-W front with the SPMT of the central Pyrenees. The displacement and shortening involved in these foreland structures have been probably transferred in the eastern extremity of the Pyrenean chain to dextral NW-SE strike-slip faults, later reactivated by Neogene extensional faults (Fig. 2).

The thrust and fold structures of the eastern Ebro basin are strongly controlled by the geometry of the decollement level, which is largely dependent on the location of weak evaporitic horizons in the foreland series. The structural style of this region resembles the style observed in other detached fold and thrust belts, such as the Appalachian Plateau (Rodgers 1963), and the Salt Range, in Pakistan (Jaumé & Lillie 1988; Lillie *et al.* 1987; Baker *et al.* 1988), amongst other examples.

The main goal of this contribution is the study of the geometry of structures of the eastern foreland Ebro basin, between the Pyrenean thrust sheets and the Catalan Coastal Ranges and their relationships with the arrangement of the different Palaeogene evaporitic basins. Regional work and mapping showing general features of the area has been done (Riba 1967; Wagner *et al.* 1971). The available subsurface data (drill holes and seismic lines) together with detailed field data, enable the construction of accurate cross-sections. In addition, the synchroneity between sedimentation and tectonics permits us to date thrust movements and to calculate the Pyrenean shortening since Late Eocene times.

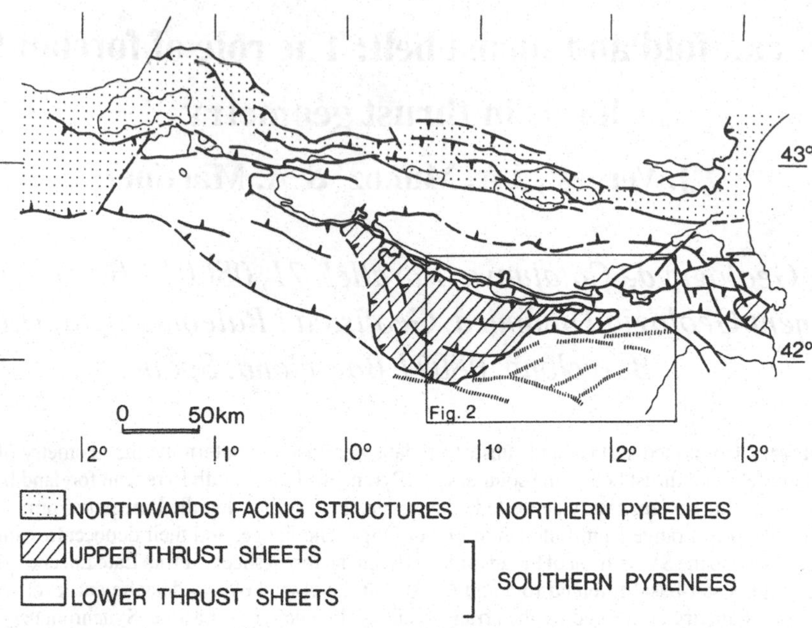

Figure 1. Structural sketch of the Pyrenees and location of the study area (Fig. 2).

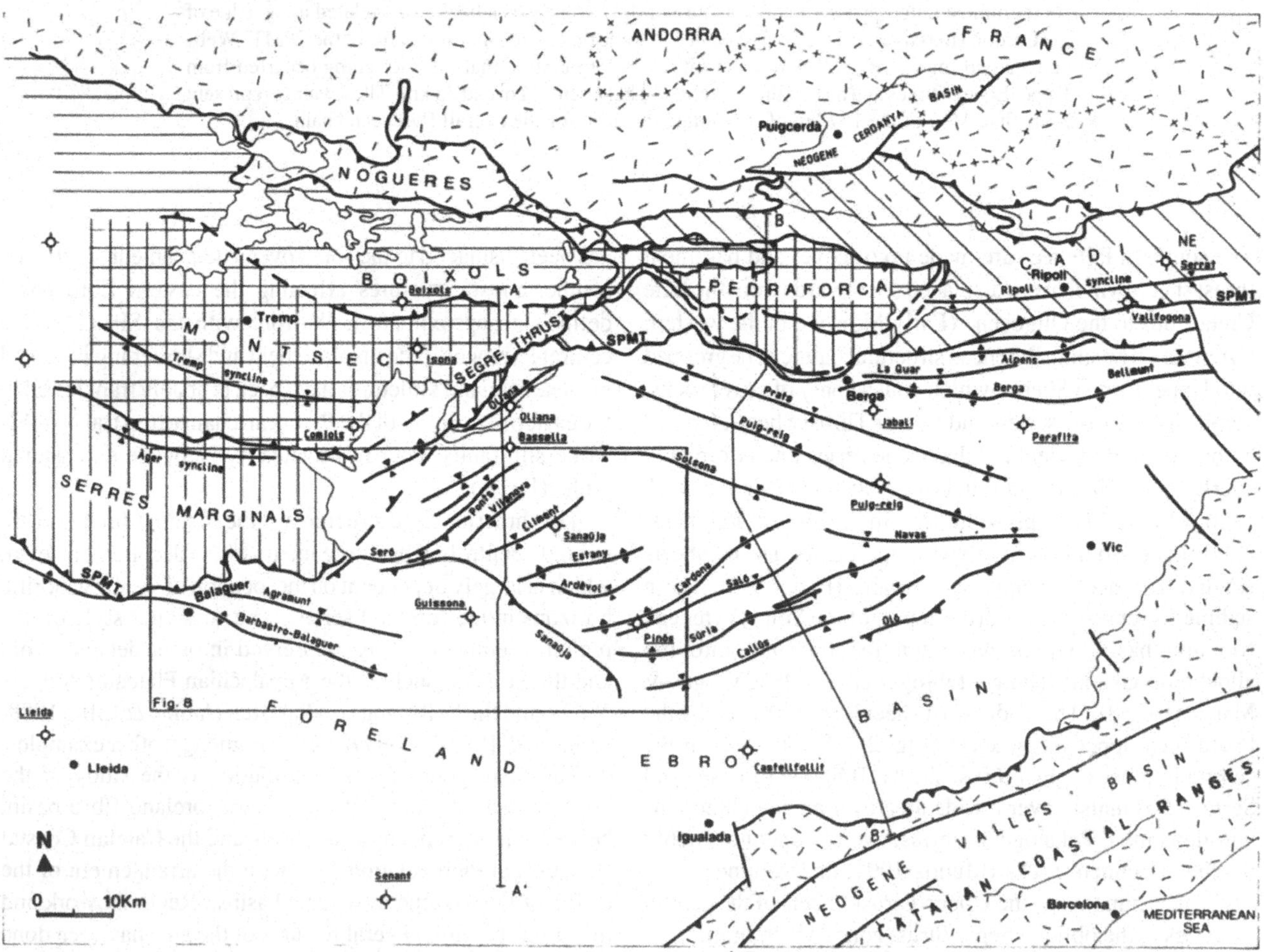

Figure 2. Structural map showing main features of the study area. The South Pyrenean Main thrust (SPMT) represents the southern boundary of the different Mesozoic and Cenozoic cover rocks thrust sheets, which are depicted in different types of continuous lines. The foreland basin is represented in white. Foreland map derived from a compilation of the map by Ramirez & Riba (1975) and from own observations. The crossed circles represent oil wells (Igme 1987).

Stratigraphy and South Pyrenean foreland basin evolution

Palaeogene sedimentation in the southeastern Pyrenean foreland basin evolved, as most of the foreland basins do, from marine to continental. The marine stage is characterized by shallow water and platform sequences, followed by deep-water sequences deposited in narrow troughs in front of the eastern South Pyrenean thrust sheets (Puigdefàbregas *et al.* 1986). Transition from marine to continental deposits occurred at different times in the South Pyrenean foreland basin. Clastic systems, related to emergent thrusts, prograded southwards as the piggy-back thrust system developed. In the northern areas, the first widespread continental deposits, at present outcropping in the Cadí thrust sheet, are Lutetian in age. In the southern areas, now the Ebro basin, continental sedimentation was general in Priabonian times. The boundary between marine and continental deposits is marked by marine evaporites: the Beuda Formation (Lutetian) and the Cardona Formation (Priabonian).

The lower part of the sedimentary pile of the foreland basin has been separated into several depositional sequences (Puigdefàbregas *et al.* 1986). Stratigraphic and sedimentological studies have been carried out in the upper and continental part of the Palaeogene foreland series (Sáez & Riba 1986). The southern margin of the Ebro basin has been studied by Anadón *et al.* (1979). The detailed stratigraphy of evaporites is mainly from Busquets *et al.* (1985) and Martínez *et al.* (1989).

The first Palaeogene sediments (Fig. 3) consist of Garumnian red beds (the Tremp Fm. in the north and the Mediona Fm. in the south), which unconformably overlie Hercynian basement rocks northwards and Triassic sediments southwards (Jurado 1988). The Lower Eocene (Ilerdian) is represented by carbonate platform sediments (Alveolina limestones), which cover most of the basin. To the north, in the Cadí thrust sheet (Fig. 3), limestones grade to deepest marly sediments. During Cuisian and Early Lutetian times slope and basinal sediments were deposited in a narrow trough in front of the Pedraforca thrust sheet (Fig. 4). Southwards of this trough, in the Ebro basin, shallow-water sedimentation persisted (Pontils Group). The trough was progressively closed in the west by the emplacement of the Pedraforca and Montsec thrust sheets. As a result, the Beuda marine evaporites deposited in a restricted basin. The depocentre of the Beuda evaporites is buried and poorly known. This evaporitic sequence, studied from well data, is made up by at least 1000 m of alternating shales and anhydrite, with 100 m of salt in its upper part (Martínez *et al.* 1989).

Continental sedimentation followed the Beuda evaporites in the northern foreland basin, while shallow marine deposition (Santa María Group) occupied the overall extension of the eastern part of the Ebro basin from Late Lutetian to Priabonian (Figs 3 & 4). The upper part of the marine succession consists of the Igualada marls which are overlain by the Tossa reefal limestones (Fig. 5). Over these limestones, a thin marly series shows evidence of progressive restriction (euxinic stage), which precedes the deposition of the Cardona marine

AGE	SEQUENCES	FORMATIONS	GROUPS	LEGEND
RUPELIAN	Solsona	Berga (alluvial) Solsona (fluvial) Barbastro (cont. evaporites)	Santa Maria	
PRIABONIAN	Cardona	Cardona (marine evaporites)		
BARTONIAN	Milany	Tossa (reefal limestones) Milany (delta front) Igualada (prodelta marls)		
LUTETIAN	Bellmunt	Bellmunt (alluvial & fluvial) Banyoles (prodelta marls)		
	Beuda	Beuda (marine evaporites)	Pontils	CADI UNIT
	Campdevànol	Campdevànol (turbidites)		
CUISIAN	Armàncies	Armàncies (slope marls) Penya (carbonate platform)		
	Corones	Corones (outer platform) Cadí (carbonate platform)		
ILERDIAN	Cadí	Sagnari (open marine marls) Orpí (carbonate platform)		
PALEOCENE		Tremp (red beds) Mediona (red beds)		

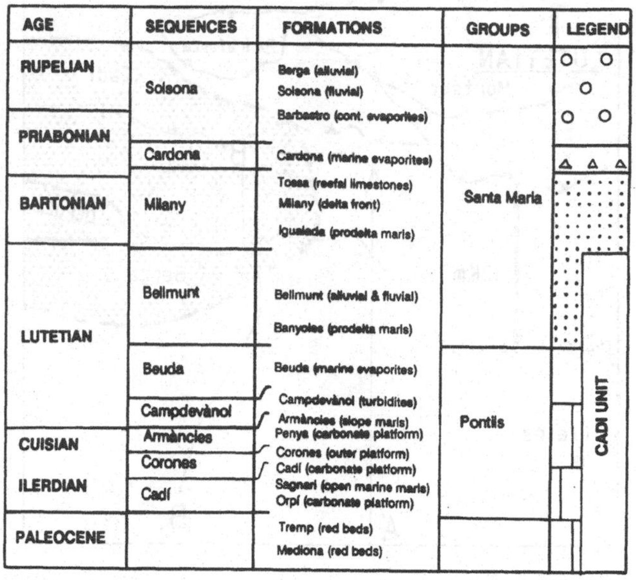

Figure 3. Palaeogene stratigraphic units of the eastern Pyrenees and foreland Ebro basin (from Puigdefàbregas *et al.* (1986) and Anadón *et al.* (1979)).

evaporites (Busquets *et al.* 1985).

The geometry of the Cardona evaporitic basin (Fig. 5) is well documented due to both the existence of extensive potash mining works at the basin centre and the well-exposed marginal facies in the eastern and southern ends of the basin (Reguant 1967; Busquets *et al.* 1985). The northern end of the basin is only visible in the southern flank of the Oliana anticline (Figs 2 & 6, A-A'), where a thin gypsum unit outcrops (Vergés & Muñoz 1990). The Cardona evaporites in the basin centre (Figs 4 & 5), are thin anoxic marls and a 5 m thick basal sulphate member at the bottom, followed by 250 m of massive salt (Lower-Salt Member), overlain by potassium salt with grey lutites at the top (Pueyo 1975). The total thickness in non-deformed areas is 300 m. At the margin of the basin, only anhydrite developed. The top of the Cardona evaporites, according to geochemical data (Rosell & Pueyo 1984), represent the last marine infill of the eastern foreland Ebro basin, although continental evaporitic sedimentation continued.

The Upper Eocene-Lower Oligocene Barbastro gypsums, overlying the grey lutites, are the first continental deposits in the central and western areas of the Ebro basin. The Barbastro evaporites reach more than 1000 m in thickness in the centre of the basin (Fig. 4), and are made up mainly of gypsum (some salt has been drilled at depth). These evaporites developed in the distal parts of fluvial and alluvial systems (Sáez 1987), which deposited in front of the emergent eastern part of the SPMT and along out-of-sequence thrusts in the central Pyrenees in the Oliana zone (Vergés & Muñoz 1990).

Structure of the foreland fold and thrust belt

In the study area the deformed Ebro foreland basin is characterized by different sets of folds and thrusts of varying trends

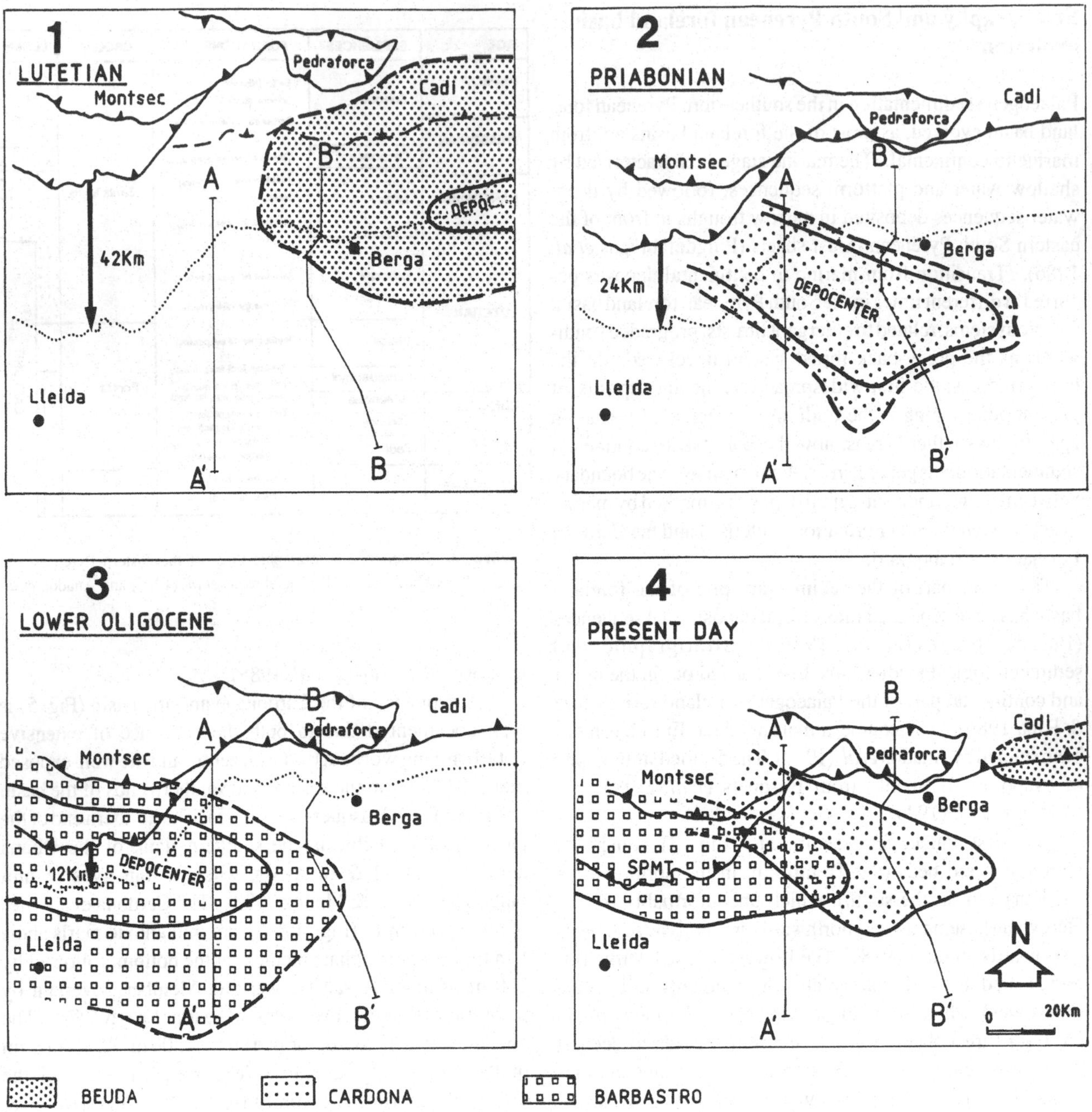

Figure 4. Palinspastic restoration during the deposition of the Beuda (Lutetian), Cardona (Priabonian), and Barbastro evaporites (Later Priabonian-Early Oligocene), with the approximate amount of displacement of the SPMT. Last sketch corresponds to the present position of the evaporitic depocentres. Dashed line represents the present position of the South Pyrenean main thrust (SPMT).

(Riba 1967; Wagner *et al.* 1971; Ramirez & Riba 1975; Malmsheimer & Mensink 1979), which can be divided into three main regions: the NE region with continuous structures varying from E-W to ESE-WNW, the Central region with NE-SW trending structures, and the SW region, which displays roughly E-W trending folds, such as the Barbastro-Balaguer anticline, which continues westwards of the study area (Fig. 2).

The northern structures of the NE region (Alpens, La Quar and Berga) show an E-W trend slightly oblique to the SPMT. Southwards, the Prats and Solsona-Navas synclines and the Puigreig anticline stretch over a distance of 50 km with an ESE-WNW trend (Fig. 2).

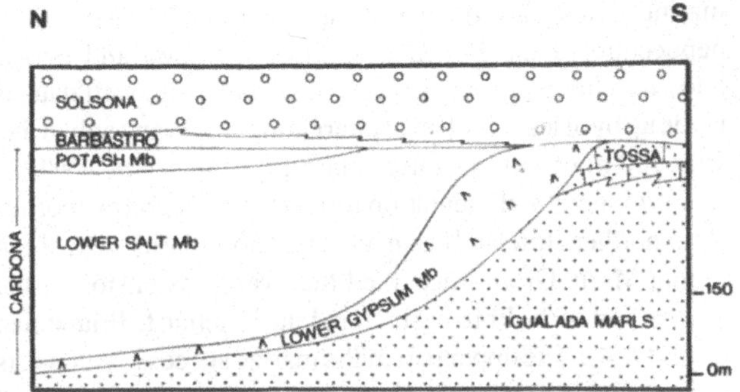

Figure 5. Interpretative distribution of Cardona evaporitic facies and thickness from the basin centre to the southern margin (simplified from Busquets *et al.* (1985). No horizontal scale.

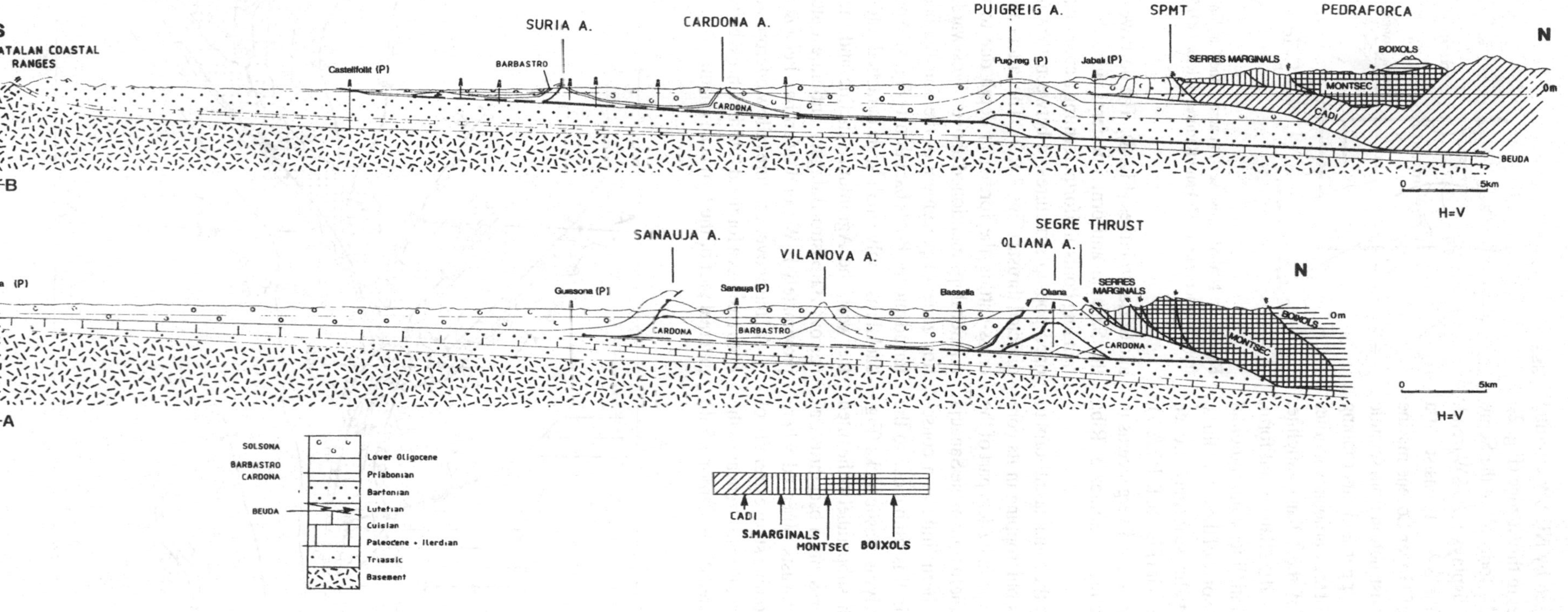

Figure 6. N-S balanced geological cross-sections showing the general features of the entire folded foreland. See Figure 2 for location.

The Central region is characterized by NE-SW trending structures, parallel to the oblique Segre thrust zone (Fig. 2). The Oliana anticline developed in the footwall of the Segre thrust zone. The axis of the anticline displays a NE-SW trend, and plunges both to the NE and to the SW. It consists of a duplex structure at depth made up of Upper Eocene marine marls, as demonstrated by section construction from seismic and well data (Fig. 6, A-A'). The upper and outcropping horse contains the northern pinch-out of the Cardona evaporites (Vergés & Muñoz 1990). South of the Oliana anticline, between the Solsona-Navas syncline and the Sanaüja antiform, a set of broad synclines and narrow anticlines with sigmoidal axial traces can be observed (Fig. 2). Some of these anticlines, such as the Cardona and Súria anticlines, display broken cores due to the diapiric flow of the salt (Figs 6, B-B' & 7). The southernmost structures display a north vergence as in the case of the Súria and Oló structures (Ramirez & Riba 1975), (Figs 2 & 6, B-B').

The Central region is bounded by the Sanaüja antiform to the south with Barbastro evaporites outcropping in its core (Fig. 8). The folds and thrusts of the southeastern part of the Central region interfere with the northern limb of the Sanaüja antiform, but do not affect its southern limb; as a consequence, a detachment must occur in this limb close to the boundary between gypsum and overlying clastic rocks (Fig. 8). The detachment dips to the south, subparallel to the tilted beds of the southern limb, and represents a backthrust, as indicated by a set of related backthrusts observed where clastic rocks are involved. In the southeastern extremity of the Sanaüja antiform, a set of backthrusts truncates the previously developed, forward-directed Pinós-Cardona, Saló and Súria structures (Fig. 8). The northwestern extremity of the Sanaüja antiform interferes with the N-E to NE-SW trending Ponts antiform, forming a culmination which developed pop-up structures in-between forward and hindward directed thrusts (Figs 8 & 9). The latter can be interpreted as cross-cutting the former at depth due to the cross-cutting relationships mentioned above. Westwards of the culmination zone, displacement of the Sanaüja backthrust is transferred into the Ponts backthrust (Fig. 8). South of the Sanaüja backthrust, only small-scale detached folds are present. To the west, the Agramunt syncline and the continuous and narrow Barbastro-Balaguer anticline outcrop with an approximately E-W trend (Martinez-Peña & Pocoví 1988).

All the above structures allow interpretation of the eastern Ebro basin as a foreland fold and thrust belt forming the most external part of the Pyrenean chain.

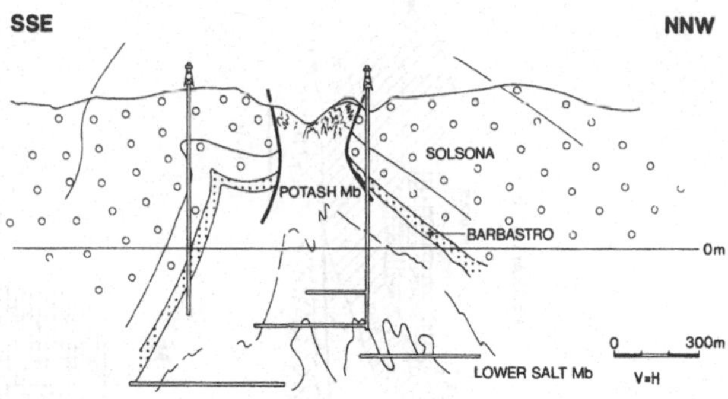

Figure 7. NW-SE cross-section across the Cardona anticline (modified from Wagner *et al.* (1971) and Ramirez & Riba (1975)).

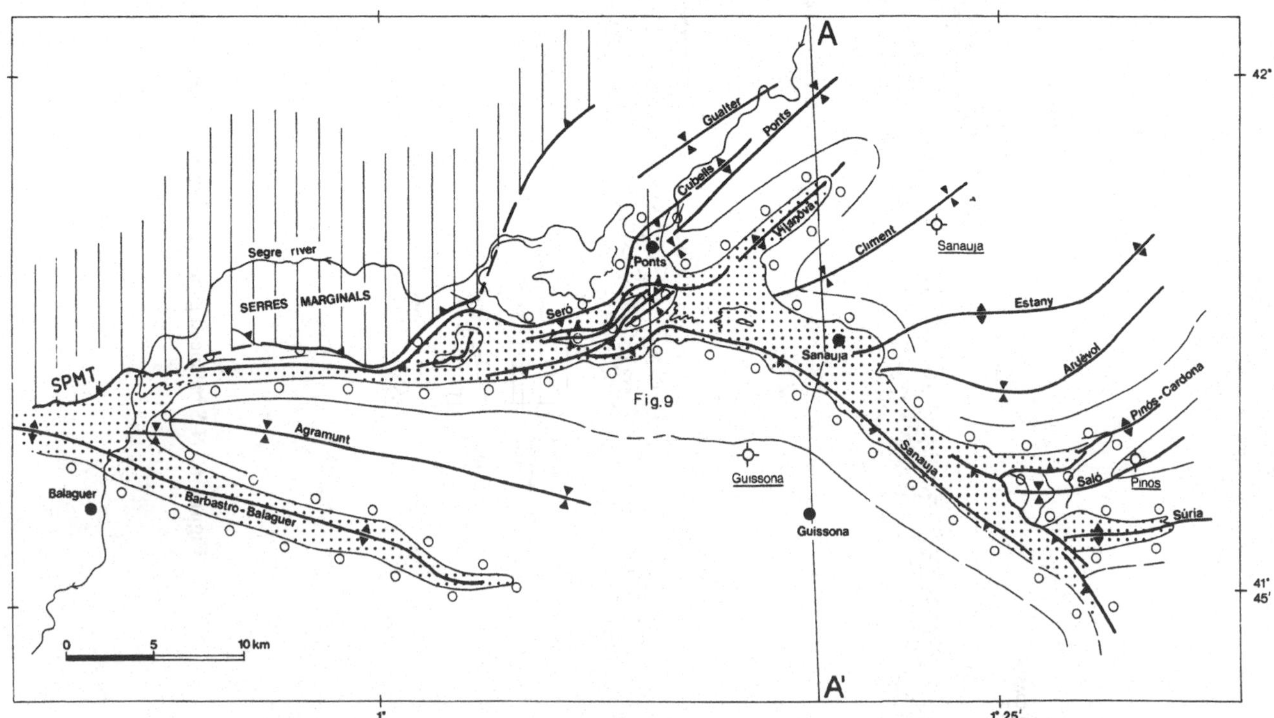

Figure 8. Geological map of the Sanaüja antiform, which separates the folded Central region and the non-deformed SW region. See Figure 2 for location. Crossed circles are oil wells. Black circles are towns. Dotted unit is the Barbastro Formation.

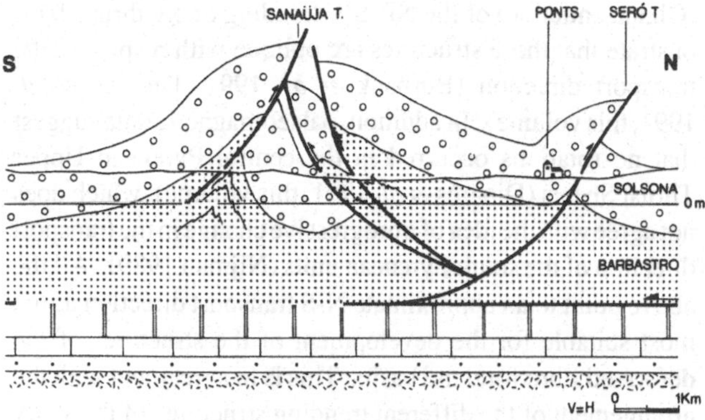

Figure 9. N-S geological cross-section across the culmination zone between the Sanaüja and Vilanova anticlines. See Figure 8 for location.

Geological cross-sections

Two geological cross-sections across the Ebro foreland fold and thrust belt have been included in this paper: the eastern one (Figs 2 & 6, B-B') across the Lower and Upper Thrust Sheets, the deformed Ebro foreland and the Catalan Coastal Ranges: and the western section (Fig. 6, A-A') across the Upper Thrust Sheets and the deformed Ebro foreland. In the latter, basin fold and thrust structures occur above a detachment level, which is located in the different evaporitic horizons of the basin. The geometry of the structures is controlled by the geometry of the detachment level. Palaeogene rocks underlying the decollement remain undeformed (Fig. 6). In the cross-sections, a double vergence can be observed. In the northern part, all structures are south directed (Oliana, Puigreig, and Cardona anticlines), while in the southern part they are north directed (Súria, Oló, and Sanaüja structures). The folds of the middle area, such as the Vilanova anticline, show no clear vergence.

The geometry of the detachment level, the southern continuation of the south Pyrenean sole thrust, displays a flat and ramp geometry controlled by the arrangement of the Beuda, Cardona and Barbastro evaporites. In the eastern section (Fig. 6, B-B'), the detachment climbs up from the Beuda to the Cardona evaporites. In the western section (Fig. 6, A-A'), the decollement climbs up from the Cardona to the Barbastro evaporites in the southernmost extremity. The NE study region studied is detached over the Beuda Formation, the Central region over the Cardona Formation, and the SW region over the Barbastro Formation (Figs 2 & 4). The Cardona evaporites are the most suitable decollement level, due to the existence of a thick salt package (Fig. 5). This staircase geometry contrasts with the flat detachments found in many thin-skinned mountain belts with basal salt decollements.

A duplex structure formed above the ramp that connects the Beuda and Cardona horizons. The Puigreig anticline is the surface expression of this duplex, as can be inferred from seismic lines across the anticline (Fig. 6, B-B'). The Oliana anticline has been described as a duplex at depth formed above a ramp which connected the floor thrust of the Upper Thrust Sheets with the foreland detachment (Fig. 6, A-A'), similar to the Puigreig anticline, later displaced southwards over the Cardona evaporites (Vergés & Muñoz 1990).

In the Central region, over the flat detachment located in the Cardona salt, most of the area is occupied by gently dipping beds corresponding to broad synclines that separate narrow anticlines. These anticlines are cored by salt which flows from the synclines. The load of the synorogenic Upper Eocene-Lower Oligocene clastic rocks (Solsona Formation), deposited in the synclines, contributed to the salt flowage. The synchronism of clastic deposition with salt growth is documented by progressive unconformities in the limbs of the anticlines. Once the salt attained a sufficient amplitude, it continued to flow diapirically, thus breaching the previously-formed anticlines. This can be observed in the Cardona anticline, where salt is still flowing at present (Ramirez & Riba 1975).

The southern part of the deformed Ebro foreland displays north directed structures (backthrusts). These backthrusts (Oló and Sanaüja) coincide with the southern edge of the Cardona evaporitic basin (Figs 2 & 6). They branch with the detachment close to the southern end of the thick Lower Salt Member (Fig. 10). The Oló branch line forms the southern front of the eastern Pyrenees, while westwards of the Sanaüja backthrust, the detachment climbs up to the Barbastro evaporites and deformation continues southwards (Fig. 6). South of the Oló backthrust, a buried tip line can be traced with an approximate E-W trend, merging with the Barbastro-Balaguer anticline tip line to the west (Muñoz *et al.* 1984; Williams 1985).

Timing

Well-constrained geometric relationships between structures and syntectonic sediments, combined with biostratigraphic and magnetostratigraphic data, enable us to calculate the age of the structures.

The Oliana anticline (duplex) developed in Priabonian

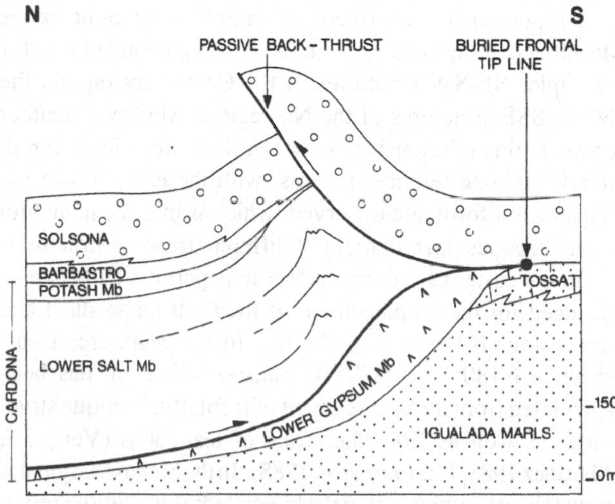

Figure 10. Interpretative tectonic model for the southern boundary of the Cardona evaporitic basin drawn in Figure 5 (with the same legend). The disappearance of the Lower Salt member increases the basal friction of the decollement and, as a consequence, a passive backthrust was developed.

times, between 40Ma and 36Ma, at the same time as the break-back imbricate stack in the hangingwall of the Segre thrust (Burbank *et al.* 1991). The Oliana anticline has rotated 20° anticlockwise (Burbank *et al.* 1991; Dinarès *et al.* 1991, this volume). As a consequence, it formed with a trend of at least N70°E. Eastwards, movement along the E-W trending SPMT south of the Pedraforca thrust sheet and growth of the E-W trending Berga-Bellmunt anticline took place at the same time (Mató & Saula 1991). The Oliana anticline rotated and was displaced southwards over the Cardona evaporites during Earliest Oligocene, synchronous with the growth of the Puigreig anticline. Deformation progressed southwards, as demonstrated by the sedimentological characteristics of the Lower Oligocene Solsona sequence. Both limbs of the Cardona anticline show lower Solsona distal alluvial sediments (coeval with coarser clastics related to the growth of the Oliana anticline) without facies, thickness or palaeocurrent directions (Sáez 1987). Consequently, southern structures of the Central region and Sanaüja antiform are Early Oligocene and younger than the northern ones. In the study area, there are no available criteria to determine the age of the Barbastro anticline, but westwards, it has been dated as Late Oligocene (Pardo & Villena 1979; Riba *et al.* 1983).

Shortening

The calculation of shortening of the Pyrenean chain through its eastern realm should take into account the shortening involved in the structures of the deformed Ebro foreland basin. The construction of balanced and restored cross-sections across this area poses some difficulties due to the different strikes of the structures. As far as the determination of transport direction is concerned, which is crucial for accurate construction of balanced cross-sections, questions arise. Can all these structures be formed by a unique transport direction? If so, which of these structures are oblique and which are frontal?

As previously described, structures of different trends developed at the same time in the deformed foreland, as for example, NE-SW structures of the Central region and the NNW-SSE structures of the NE region. Moreover, neither cross-cutting relationships between structures with different trends nor interference patterns, with the exception of the Sanaüja antiform, are observed, which is in agreement with a synchronous development of different trending structures.

In the eastern Pyrenees, a N-S transport direction can be deduced for the emplacement of the Cadí thrust sheet and lower units (Muñoz *et al.* 1986). In the Pedraforca thrust sheet, a N180°E to N195°E transport direction has been calculated from the arrangement of frontal and oblique structures, as well as from the strike of tear faults (Vergés & Martínez 1988; Martínez *et al.* 1988). In the Sierras Marginales thrust sheets, equivalent to the lower Pedraforca units, similar criteria allow us to deduce a N-S transport direction from structures developed synchronously with the folds and thrusts of the deformed Ebro foreland. Anticlockwise rotations in the structures of both the hangingwall (imbricates) and footwall (Oliana anticline) of the NE-SW trending Segre thrust demonstrate that these structures are oblique with respect to the transport direction (Burbank *et al.* 1991; Dinarès *et al.* 1991, this volume). In addition, palaeomagnetic data suggest that no rotations occurred in the central Pyrenean Upper Thrust Sheets (Dinarès *et al.* 1991, this volume), which does not agree with the idea of a progressive change in the transport direction of the south Pyrenean units (Nijman 1989). All the above point to an approximate N-S transport direction as the most suitable for the development of the structures of the deformed Ebro foreland basin. This is in agreement with the arrangement of the different trending structures of the study area (Fig. 2).

In the study area, where the majority of the trends of the structures are oblique with respect to the thrusting transport direction, an accurate calculation of the shortening would involve a 3-D restoration. Although only two cross-sections have been included, they have been restored, despite the possible errors of such restorations, if only to give an idea of the shortening involved. Bed-length techniques were used for balancing competent layers (Dahlstrom 1969), while evaporitic rocks have been areally restored (taking into account only the N-S migration of salt). The pin lines are located at the end of the deformed zone (black points in cross-sections, Fig. 6). The eastern section (Fig. 6, B-B') is drawn normal to the trend of the structures in each segment of the section, whereas the western one (Fig. 6, A-A') is oblique to the structures, but parallel to the thrust transport direction.

In this paper, only the shortening that occurred after the deposition of Early Priabonian Cardona marine evaporites has been analysed. Both cross-sections show similar amounts of shortening: 25 km in the A-A' and 21 km in the B-B' cross-sections (Fig. 6). This is a minimum value, as neither erosion (the hangingwall ramp of the SPMT is eroded in both the A-A' and B-B' cross-sections) nor internal deformation such as the strong cleavage observed in the overturned Lower and Middle Eocene marls outcropping in front of the SPMT (B-B' cross-section) have been taken into account. The total amount of shortening can be partitioned in two, that is, the shortening that occurred in the foreland structures, and the shortening due to the movement of the SPMT since Priabonian times. In the A-A' cross-section, the shortening south of the SPMT is 21 km while the shortening due to the motion of the SPMT is a minimum of 4 km. In the B-B' cross-section, the deformed foreland southwards of the SPMT shows 11.5 km of shortening, whereas the movement of the SPMT is a minimum of 9.5 km. In the former section, there is a big difference between shortening due to the motion of the SPMT and deformation in foreland deposits, whereas in the latter section, both values of shortening are similar. These differences in the relative amount of shortening between that in the foreland basin and along the SPMT, observed in the two cross-sections, can be explained by the arrangement of the boundaries of the NW-SE Cardona evaporitic basin (Fig. 4). The south Pyrenean sole thrust reached the western part of the northern margin of the Cardona salt (Fig. 6, A-A' cross-section) before reaching the eastern margin (B-B' cross-section). When the sole thrust climbed up to the Cardona decollement level, the movement along the SPMT, in the Segre zone, stopped and a hindward thrusting sequence with no important displacements developed in the hangingwall of the SPMT, while the Oliana anticline developed in its footwall (Vergés & Muñoz 1990). In contrast, in the eastern part (B-B' cross-section), the SPMT overthrusted the overall sedimentary pile, which was involved in the growth of the Oliana anticline (Fig. 6).

Conclusions

The geometry of the structures of the study area is characteristic of foreland fold and thrust belts with a weak evaporitic decollement level at depth (Davis & Engelder 1985). Anticlines are narrow and salt-cored and synclines are broad. Thrusts are mainly blind and backthrusts coexist with forward directed thrusts. Evaporites that accumulated in the anticline cores move diapirically, thus overprinting previously developed structures. This has led all these structures to be interpreted as having developed by diapirism alone and not being related to the Pyrenean contractional structures.

The Central region, located above the Cardona decollement, has undergone shallow erosion as can be deduced from areal distribution of the youngest sediments outcropping in the synclines. The present topographic slope of the Central region is less than 1°, the basal slope is 3°, but more conclusively, syntectonic sediments of the same age are located approximately at the same altitude in the centre of the synclines. This demonstrates a constant narrow taper of the Central region through its evolution, as shown by the salt-basal fold and thrust belts.

A peculiarity of the salt-basal Ebro foreland fold and thrust belt is the existence of structures with different strike, which developed from the Late Eocene to the Early Oligocene along a N-S thrust transport direction. Superposition of the structural map over the evaporitic basin distribution (Figs 2 & 4) clearly shows that there is a close relationship between the defined structural regions and the evaporitic basin distribution. The NE region coincides areally with the Lutetian Beuda evaporites, the Central region with the Priabonian Cardona evaporites and the SW region with the Upper Eocene-Lower Oligocene Barbastro evaporites. The three horizons act as a decollement level, the Cardona being the most suitable due to the existence of a thick salt member. The detachment steps up from the northern and lower (oldest) evaporitic horizon to the southern and upper (youngest) one. Ramps are located at the edge of the Cardona salt basin and correspond at surface to the boundaries of the different structural regions defined by their different strikes. In the northern edge of the Cardona basin, a duplex structure developed above the ramp (Oliana and Puigreig anticlines) while in the southern edge, a set of backthrusts formed, defining a frontal wedge.

The oblique arrangement of the evaporitic basins with respect to the thrust transport direction, which was controlled by the thrusting evolution of the south Pyrenean thrust sheets, determined the existence of numerous oblique structures in the foreland, mainly those of the Central region. The trend of the structures of each region coincides with the direction of the SPMT located hindwards of them, as can be seen in the structural map of Figure 2. In the NE region, structures strike parallel to the E-W segment of the SPMT. In the Central region, NE-SW structures are parallel to the oblique Segre thrust (oblique segment of the SPMT) and, in the SW region fold structures strike parallel to the E-W segment of the SPMT of the central Pyrenees.

The greater extent to the south of the thrust sheets of the central Pyrenees with respect to the eastern Pyrenean units (step geometry of the SPMT) is partially due to a major displacement of the central thrust sheets once they reached in Late Eocene the Cardona salt. The salt was never reached by the eastern part of the SPMT, instead the SPMT climbed up section and developed the structures of the foreland above the evaporitic detachment levels (Fig. 6). The NE-SW structures above the Cardona detachment of the Central region formed by an oblique component of shortening coeval with further movement to the south of the SPMT along the oblique Segre zone.

Well-constrained relationships between tectonics and sedimentation enable one to deduce that the deformation started in the study region later than deposition of the Cardona evaporites (Burbank *et al.* 1991). Shortening produced after the Cardona deposition by both the motion of the SPMT and the deformation in the foreland basin is between 21 km and 25 km, both minimum values. This represents approximately the 15% of the total 147 km of shortening for the whole Pyrenean belt (Muñoz 1991, this volume).

The overall geometry of the southern central and eastern Pyrenees is mainly the result of the disposition of the Cardona salt. The central Pyrenean thrust sheets moved further to the south, synchronously with the development of the foreland fold and thrust belt to the east. The displacement of the central part of the SPMT, over the Cardona salt, is at least of the same magnitude of the shortening involved in the deformed Ebro foreland. Frontal structures, emplaced forward of the Cardona salt, in continuation with the Barbastro-Balaguer tip anticline, are located 40 km southwards of the eastern part of the SPMT trace. Eastwards of the Cardona salt, the shortening is accomplished by folds, thrusts and cleavage normal to the beds, and the southern Pyrenean front is only located 10 km south of the SPMT.

We would like to thank Oriol Riba, Juan José Pueyo and Alberto Sáez for their helpful comments. We have found the observations of our two anonymous referees very useful. The English was revised by Michèle Pereira. This work was supported by the CAICYT project PB85-0098-C04-03 and by the Servei Geològic de Catalunya.

References

Anadón, P., Colombo, F., Esteban. M., Marzo, M., Robles, S., Santanach, P. & Solé Sugrañes, L . 1979. Evolución tectonoestratigráfica de los Catalánides. *Acta Geològica Hispànica*. Homenatge a Lluís Solé i Sabarís. **14**, 242-270.

Baker, D. M., Lillie, R. J., Yeats, R. S., Johnson, G. D., Yousuf, M. & Hamid, A. S. 1988. Development of the Himalayan frontal thrust zone: Salt Range, Pakistan. *Geology*, **16**, 3-7.

Burbank, D. W., Vergés, J., Muñoz, J. A. & Bentham, P. 1991. Coeval hindward- and forward-imbricating thrusting in the Central Southern Pyrenees, Spain: Timing and rates of shortening and deposition. *Geological Society America Bulletin* (in press).

Busquets, P., Ortí, F., Pueyo, J. J., Riba, O., Rosell, J, Sáez, A., Salas, R. & Taberner, C. 1985. Evaporite deposition and diagenesis in the saline (potash) catalan basin, Upper Eocene. *Excursion Guide-book 6th European Meeting*, Lleida, Spain. 13-59.

Dahlstrom, C.D.A. 1969. Balanced cross sections. *Canadian Journal of Earth Sciences*, **6**, 743-757.

Davis, D. M. & Engelder, T. 1985. The role of salt in fold and thrust belts. *Tectonophysics*, **119**, 67-88.

Dinarès, J., McClelland, E. & Santanach, P. 1991. Contrasting rotations within thrust sheets and kinematics of thrust-tectonics as derived from palaeomagnetic data: an example from the southern Pyrenees (this volume).

IGME. 1987. Contribución de la exploración petrolífera al conocimiento de la geología de España. *Instituto Geológico y Minero de España*, 1-465.

Jaumé, S. C. & Lillie, R. J. 1988. Mechanics of the Salt Range-Potwar Plateau, Pakistan: A fold and thrust belt underlain by evaporites. *Tectonics*, **7**, 57-71.

Jurado, M. J. 1988. El Triásico del subsuelo de la cuenca del Ebro. Tesis doctoral. Univ. de Barcelona, 259p.

Lillie, R. J., Johnson, G. D., Yousuf, M., Hamid, A. S. & Yeats, R. 1987. Structural development within the Himalayan foreland fold and thrust belt of Pakistan. *Canadian Society of Petroleum Geologists Memoir*, **12**, 379-392.

Malmsheimer, K. & Mensink, H. 1979. Der geologische Aufbau des Zentralkatalanischen Molassebeckens. *Geologische Rundschau*, **68**, 121-162.

Martínez-Peña, M. B. & Pocoví, J. 1988. El amortiguamiento frontal de la estructura de la cobertera surpirenaica y su relación con el anticlinal de Barbastro - Balaguer. *Acta Geològica Hispànica*, **23**, 81-94.

Martínez, A., Vergés, J. & Muñoz, J. A. 1988. Secuencias de propagación del sistema de cabalgamientos de la terminación oriental del manto del Pedraforca y relación con los conglomerados sinorogénicos. *Acta Geològica Hispànica*, **23**, 119-127.

——,—— Clavell, E. & Kennedy, J. 1989. Stratigraphic framework of the thrust geometry and structural inversion in the southeastern Pyrenees: La Garrotxa area. *Geodinamica Acta*, **3**, 185-194.

Mató, E. & Saula, E. 1991. Los ciclos sedimentarios del Eoceno medio y superior en el sector NE de la depresión del Ebro (zona de Berga-Vic). Caracterización de sus límites. *Geogaceta* (in press).

Muñoz, J. A. 1991. Evolution of a Continental Collision Belt: ECORS-Pyrenees Crustal Balanced Cross-section (this volume).

—— Martínez, A. & Vergés, J. 1986. Thrust sequence in the eastern Pyrenees. *Journal of Structural Geology*, **8**, 399-405.

—— Puigdefàbregas, C. & Fontboté, J. M. 1984. El ciclo alpino y la estructura tectónica del Pirineo. *In:* Comba, J. A., *Libro Jubilar J. M. Rios*, Geológica de España, **2**. IGME, Madrid, 185-205.

Nijman, W. 1989. Thrust sheet rotation? The south Pyrenean Tertiary basin configuration reconsidered. *Geodinamica Acta*, **3**, 17-42.

Pardo, G. & Villena, J. 1979. Aportación a la geología de la región de Barbastro. *Acta Geològica Hispànica*. Homenatge a Lluís Solé i Sabarís. **14**, 289-292.

Pueyo, J. J. 1975. Estudio petrológico y geoquímico de los yacimientos potásicos de Cardona, Súria, Sallent y Balsareny (Barcelona, España). Tesis Doctoral. Universidad de Barcelona, 315p.

Puigdefàbregas, C., Muñoz, J. A. & Marzo, M. 1986. Thrust belt development in the Eastern Pyrenees and related depositional sequences in the southern foreland basin. *In:* Allen, P. A. & Homewood, P. (eds) *Foreland basins*. Special Publication of the International Association of Sedimentologists, **8**, 229-246.

Ramirez, A. & Riba, O. 1975. Bassin potassique catalan et mines de Cardona. IX Congres Inter. de Sédimentologie, Nice 1975. Livret-guide Ex. 20, 49-58.

Reguant, S. 1967. El Eoceno marino de Vic (Barcelona). *Memorias del Instituto Geológico y Minero de España*, **68**, 1-350.

Riba, O. 1967. Resultados de un estudio sobre el Terciario continental de la parte este de la depresión central catalana. *Acta Geològica Hispànica*, **II**, 1-6.

—— Reguant, S. & Villena, J. 1983. Ensayo de síntesis estratigrafica y evolutiva de la cuenca terciaria del Ebro. *In:* Comba, J. A., *Libro Jubilar J. M. Rios*, Geológia de España, **2**, IGME, Madrid, 131-159.

Rodgers, J. 1963. Mechanics of Appalachian foreland folding in Pennsylvania and West Virginia. *American Association of Petroleum Geologists Bulletin*, **47**, 1527-1536.

Rosell, L. & Pueyo, J. J. 1984. Características geoquímicas de la formación de sales potásicas de Navarra (Eoceno superior). Comparación con la cuenca potásica catalana. *Acta Geològica Hispànica*, **19**, 81-95.

Sáez, A. & Riba, O. 1986. Depositos aluviales y lacustres paleogenos del margen pirenaico catalán de la cuenca del Ebro. Libro guia Excursion. XI Congreso Español de Sedimentologia. Barcelona, 6.1-6.29.

—— 1987. Estratigrafía y sedimentología de las formaciones lacustres del tránsito Eoceno - Oligoceno del NE de la cuenca del Ebro. Tesis doctoral. Universitat de Barcelona, 352p.

Vergés, J. & Martínez, A. 1988. Corte compensado del Pirineo oriental: geometria de las cuencas de antepaís y edades de emplazamiento de los mantos de corrimiento. *Acta Geològica Hispànica*, **23**, 95-106.

—— & Muñoz, J. A. 1990. Thrust sequences in the Southern Central Pyrenees. *Bulletin de la Société géologique de France*, **8**, 265-271.

Wagner, G., Mauthe, F. & Mensik, H. 1971. Der Salzstock von Cardona in Nordospanien. *Geologische Rundschau*, **60**, 970-996.

Williams, G. D. 1985. Thrust tectonics in the south central Pyrenees. *Journal of Structural Geology*, **7**, 11-17.

Contrasting rotations within thrust sheets and kinematics of thrust tectonics as derived from palaeomagnetic data: an example from the Southern Pyrenees

J. Dinarès[1,2], E. McClelland[1], P. Santanach[2]

[1]*Department of Earth Sciences, University of Oxford, Parks Rd., Oxford OX1 3PR, UK*

[2]*Departament de Geologia Dinàmica, Geofísica i Paleontologia, Universitat de Barcelona, Zona Universitaria de Pedralbes, 08028 Barcelona, Spain*

Abstract: Determination of palaeomagnetic declinations can help quantify differential rotations of structural units about a vertical axis and can therefore constrain kinematic models. A palaeomagnetic study (105 sites) has been carried out on Jurassic to Eocene lithologies from the structural units in the Pyrenees which constitute the south-central cover thrust sheets (Bóixols, Montsec and Sierras Marginales units) and also from the Cadí and Pedraforca units further east. They developed mainly by piggy-back thrusting sequence between Upper Cretaceous and Oligocene time.

Clockwise rotations ranging from 20° to 45° have been detected at sites associated with dextral oblique thrust ramps whereas those sites related to sinistral oblique thrust ramps exhibit anticlockwise rotation. Furthermore there is palaeomagnetic evidence for non-rotational thrusting in the central area of the Bóixols and Montsec thrust sheets allowing straightforward balancing of structural cross-sections. This means that balancing of 2-D sections in these South-Pyrenean Central Units along the line of the ECORS seismic profile is permissible. Such an arrangement of the rotational components implies a greater total displacement in the central area where no substantial rotation occurred, while rotations along oblique ramps at the edges of the sheets might have occurred as a result of friction during thrust motion since the thrust transport direction (N-S) is not perpendicular with that of the oblique thrust plane.

Rotation of footwall pre- and syntectonic sediments during development of hangingwall imbricate thrust stacks is also reported from the Oliana anticline in the eastern margin of the Montsec thrust sheet. In addition, differential anticlockwise rotation of structural units within the Pedraforca area piggy-back thrust pile related to the piling up of the units are also evaluated.

Palaeomagnetism has been used as a powerful tool for identifying structural rotations about any axis in fold and thrust belts. This technique becomes specially valuable for quantifying structural rotations about the vertical axis and can therefore be used to document thrust sheet rotation in mobile belts. Magnetotectonic studies from several orogenic belts have shown that such rotations are common (e.g. Kotasek & Krs 1965; Schwartz & Van der Voo 1984; McClelland & McCaig 1989) due to complexities such as oblique and lateral ramps, and e.g. the existence of an arcuate foreland margin.

Rotations in the Idaho-Wyoming overthrust belt (USA) have been studied by a number of authors (Grubbs & Van der Voo 1976; Schwartz & Van der Voo 1984; Eldredge & Van der Voo 1988) to attempt to discriminate between various models for the origin of the observed arcuate thrust belt. These authors conclude that buttressing edge effects due to a Precambrian massif in the foreland were the cause of the documented rotations. McClelland & McCaig (1989) have recently postulated that the growth of an antiformal stack in the Axial Zone of the Pyrenees caused differential rotation to occur in this thrust pile. Freeman *et al.* (1989) compared palaeomagnetic directions from two thrust sheets on the island of Mallorca (Spain) which show very different structural patterns (one area associated with frontal ramps and another area associated with oblique-lateral ramps). They found that palaeomagnetic directions after tectonic correction were very similar in both areas, thus indicating no differential rotation between them.

Structural rotation about the vertical axis causes material to move into or out of section, and 2-D sections cannot be expected to balance. Palaeomagnetism is the only way of quantifying such rotations and is thus an essential test before 2-D section balancing is allowable. The South-Pyrenean Central Unit (SCU) as defined by Seguret (1972) is a major trapezoid-shaped thrust sheet consisting of several separate units (Fig. 1). Its western and easternmost boundaries have been interpreted as oblique ramps and therefore a question arises as to whether rotations might have occurred within the unit. A deep seismic profile (French-Spanish ECORS group) has recently been shot in the Pyrenees and this cross-cuts the SCU through its central-eastern part in an approximately N-S section (ECORS Pyrenees Team, 1988; Choukroune *et al.* 1989). The section has also been balanced by Roure *et al.* (1989). Moreover, a putative counterclockwise rotation of

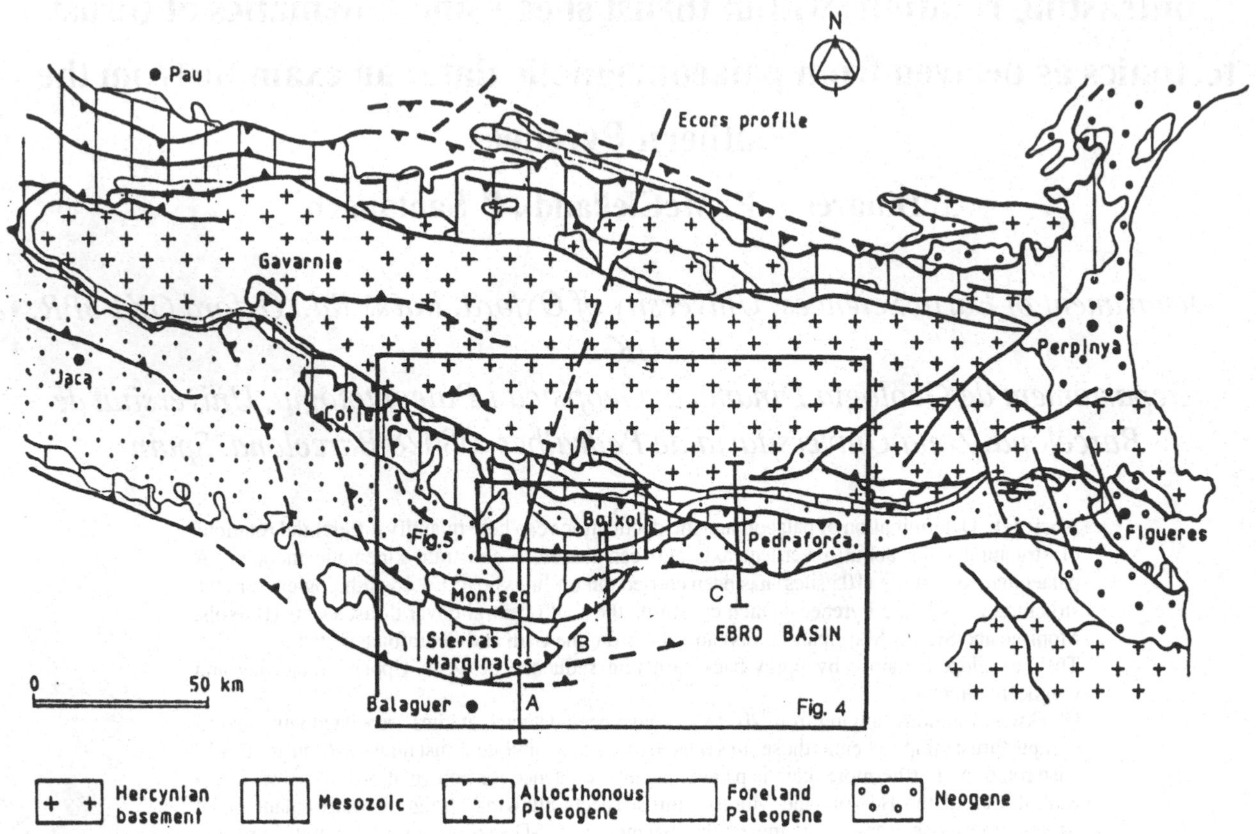

Figure 1. Structural sketch of the central and eastern Pyrenees. Boxes indicate location of Figure 4 & Figure 8. Sections A, B, and C are the location of cross-sections from Figure 2.

the SCU has been postulated by Nijman (1990).

This paper presents an extensive palaeomagnetic study in the south-central cover thrust sheets of the Pyrenees in order to identify and quantify structural rotations within those units. Additional attention has been given to the Pedraforca area piggy-back thrust pile which is one of the constituent parts of the so called Upper Thrust Sheets together with the SCU and the Ampurdà units (Muñoz *et al.* 1986). The foreland Oliana anticlinal stack located in the eastern margin of the SCU has also been studied.

Geological setting and sampling

The South-Pyrenean Central Unit (SCU) includes a succession of south-directed thrust sheets involving Mesozoic-lower Tertiary cover rocks. These are thrust over autochthonous Palaeogene rocks that lie directly over either the Palaeozoic basement of the Ebro foreland basin or over a thin Mesozoic cover (Cámara & Klimovitz 1985; ECORS Pyrenees Team 1988; Roure *et al.* 1989). The major Pyrenean thrust belt developed mainly by foreland propagating thrusting between Upper-Cretaceous and Miocene times (Muñoz *et al.* 1986; Martínez *et al.* 1986; Vergés & Martínez 1988). Coeval break-back and piggy-back sequences have also been reported recently (Vergés & Muñoz 1990).

In the ECORS profile three main thrust sheets can be distinguished (Fig. 2). The uppermost sheet is called the Bóixols unit. It extends southwards from the southern edge of the Nogueres zone. Further to the south the Montsec unit constitutes the second thrust sheet. The Sierras Marginales unit is the lowermost and its floor thrust has followed the Tertiary sediments of the Ebro foreland basin. The southern limit of the Bóixols unit corresponds mostly to a blind thrust overlapped by detritic sediments of the Maestrichtian Aren depositional sequence (Souquet 1967; Garrido-Megias & Rios 1972). The syntectonic character of this sequence is demonstrated by the onlap geometry over the frontal part of the Bóixols unit (Simó & Puigdefàbregas 1985). The age of the Montsec emplacement in the ECORS transverse is Ypresian (Williams & Fischer 1984; Mutti *et al.* 1985; Farrell *et al.* 1987). The Montsec thrust can be connected westwards below the Eocene conglomerates with the Cotiella Thrust along the N-S striking Mediano anticline (Fig. 1). Towards the E the Montsec thrust branches onto the Sierras Marginales thrust and both trend NE-SW (oblique zones) in the so called Segre thrust (Vergés & Muñoz 1990). The Sierras Marginales unit is formed by an imbricate fan thrust system and contains a limited Mesozoic series. The stratigraphic relationships between syntectonic deposits and structures led Vergés & Muñoz (1990) to establish a two-stage evolution of the unit. They postulate an initial emplacement between the Lower Eocene and the Upper Eocene and later development of the structures in the Upper Eocene-Lower Oligocene.

The lithostratigraphy of the SCU sheets is characterized by the almost complete absence of Palaeozoic basement rocks and also by the thinning of the Mesozoic cover from the northern unit towards the south (Puigdefàbregas & Souquet 1986). The Bóixols unit is characterized by a thick Mesozoic series of about 5000 m. The Lower Cretaceous series is thinner in the Montsec unit and is completely absent in the Sierras Marginales unit where Upper Cretaceous rocks directly overlie Jurassic.

In the Pedraforca area three separate thrust sheets can be distinguished (Fig. 2). The lowermost one belongs to the Lower Thrust Sheets of the Pyrenees and is called the Cadí thrust. It is constituted by basement rocks (Devonian and Lower Carboniferous), a thick Stefano-Permian series and a reduced Mesozoic and a thick Palaeogene series (Garumnian facies and Lower and Middle Eocene). The Pedraforca thrust is subdivided into two units, the Lower Pedraforca thrust sheet, and the Upper Pedraforca thrust sheet, structurally equivalent to the Bóixols sheet (Vergés & Martínez 1988).

The purpose of this palaeomagnetic study is to define any structural rotations which have occurred during the Alpine thrusting. The aim was to find rocks which carry remanence components pre-dating the deformation. These magnetization vectors would act as passive markers and record any such rotation. The sampling strategy was, therefore, designed to include the maximum number of fold tests, where the age of remanence can be identified as being pre-, syn- or post-folding (see McCaig & McClelland 1991, this volume, for further discussion). Samples for the palaeomagnetic study were collected with a portable drill and oriented with both magnetic and solar compasses. Typically, five to twelve samples from different horizons were taken per site (105 sites, 748 samples). Sites were chosen to yield spatial and temporal distribution of possible structural rotations in both N-S and E-W transverses. Several sedimentary rock-types ranging from Jurassic to Eocene in age were sampled from different tectonic units, but it was not always possible to find similar and contemporaneous lithologies in each structural unit due to the pre-thrusting basinal facies distribution, specially in a N-S section. 35% of the samples were found to be unsuitable for our purpose due to either a weak intensity of the natural remanent magnetization (NRM) or an erratic behaviour of the magnetic moment upon demagnetization. These unsuitable samples were mostly the continental red-sandstones and lacustrine limestones of the Garumnian facies (uppermost Cretaceous-lowermost Palaeocene) which outcrop widely in the Tremp piggy-back basin in the Montsec thrust sheet and also in the Sierras Marginales unit. Similarly, most of the

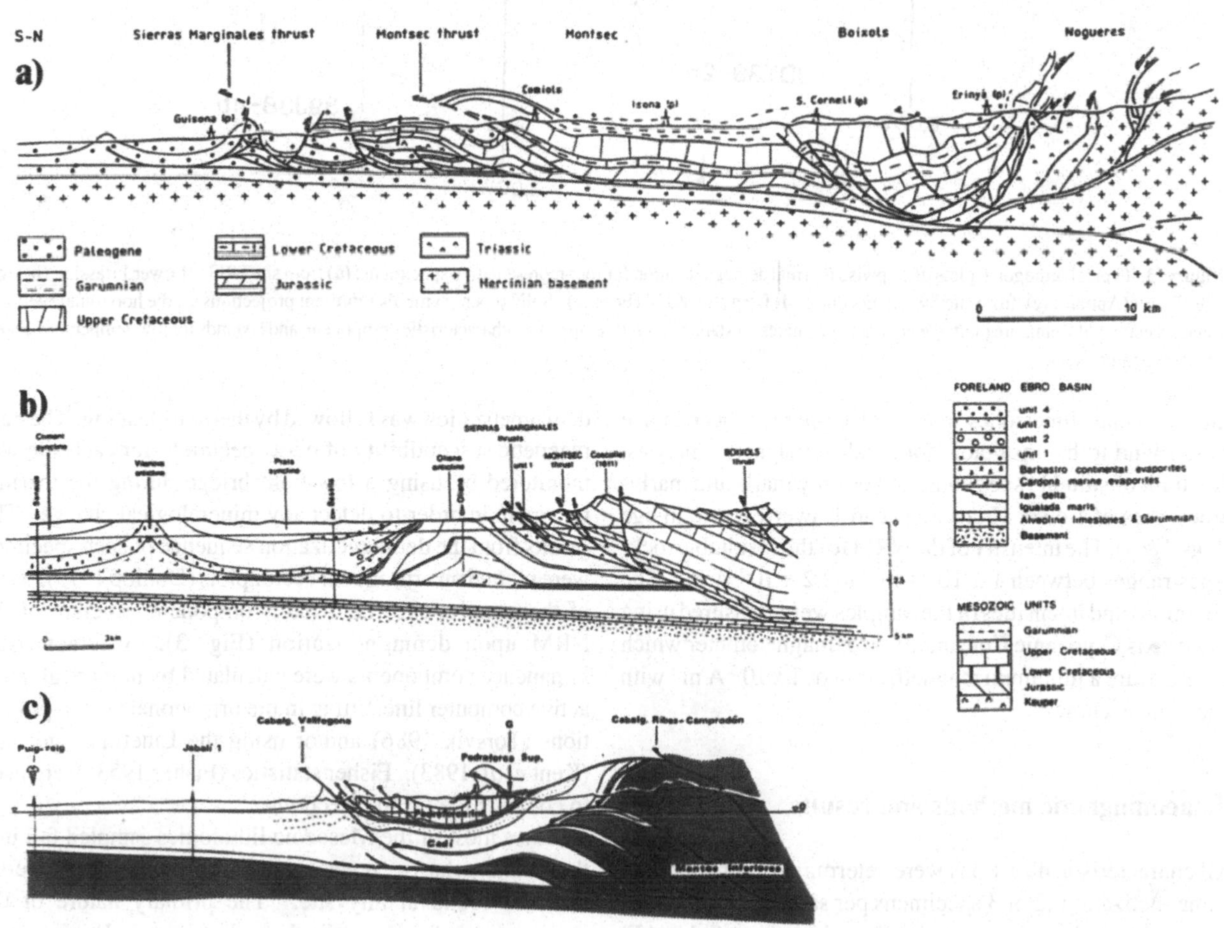

Figure 2. **(a)** Balanced cross-section of the cover thrust sheets across the south central Pyrenees (southern part of the ECORS profile), (Vergés & Muñoz 1990); **(b)** Structural cross-section of the eastern termination of the Montsec and Serres Marginals thrust sheets through the Oliana anticline, (Vergés & Muñoz 1990); **(c)** Balanced cross-section through the eastern Pyrenees (Pedraforca thrust pile), (Vergés & Martínez 1988). See Figure 1 for location.

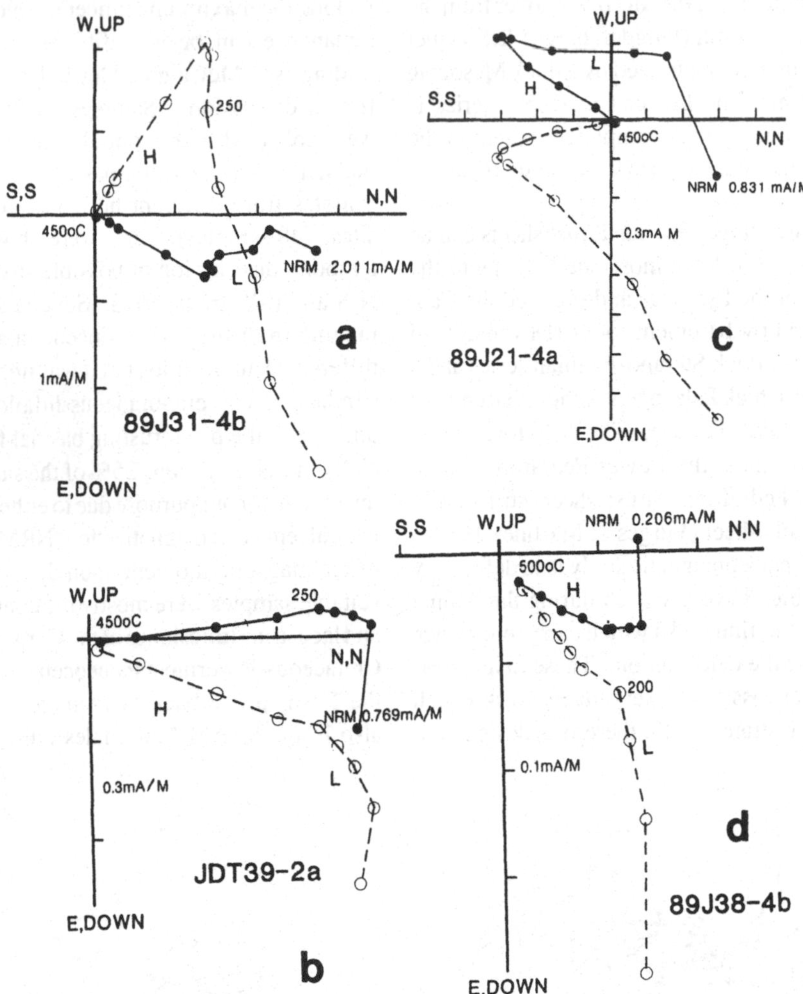

Figure 3. Typical orthogonal plots of stepwise thermal demagnetization for four representative specimens: (**a**) from site 89J31 (Lower Jurassic): (**b**) from site JDT39 (Aptian); (**c**) from site 89J21 (Eocene); (**d**) from site 89J38 (Eocene). Solid (open) symbols represent projections on the horizontal (vertical) plane. Vector end-points are plotted in *in-situ* coordinates. H stands for high-temperature characteristic component, and L stands for low-temperature recent overprint component.

shallow-water limestones mainly of Upper Cretaceous age were found to be unsuitable for a palaeomagnetic analysis. The bulk of suitable sites consisted of grey marls and marly-limestones of Lower Cretaceous and Lower Eocene in age (Figs 3 & 4). The intensity of the NRM for those suitable rock-types ranges between 1×10^{-4} A m^{-1} and 2×10^{-3} A m^{-1}. The directions and intensities of the samples were measured using a two-axis Cryogenic Consultants Ltd. magnetometer which can measure a minimum magnetization of 1×10^{-5} A m^{-1} with adequate accuracy.

Palaeomagnetic methods and results

All characteristic directions were determined after specimen demagnetization (2 or 3 specimens per sample). Specimens were subjected to either progressive alternating field (AF) demagnetization using an apparatus capable of peak fields of 55 mT (550 Oe) or stepwise thermal demagnetization using a non-magnetic furnace. Both instruments were shielded from the Earth's magnetic field. In some instances AF demagnetization was followed by thermal cleaning. The bulk magnetic susceptibility of pilot specimens for each site was monitored by using a low-field bridge during the thermal treatment in order to detect any mineralogical change. The results from the demagnetization sequence of each specimen were plotted on orthogonal vector plots (Dunlop 1979). Most of the samples showed a multicomponent structure of the NRM upon demagnetization (Fig. 3). Characteristic remanence components were calculated by numerical interactive computer line-fitting in the orthogonal vector-projections (Torsvik 1986) and/or using the Linefind algorithm (Kent *et al.* 1983). Fisher statistics (Fisher 1953) were used to compute the mean directions.

Since most of the Mesozoic lithologies sampled fall into the normal-polarity Cretaceous quiet zone, no positive reversal test was found at any site. The primary nature of the characteristic component has been demonstrated by a positive fold test on Lower Cretaceous sites from the Bóixols thrust sheet. The Eocene sites from the Oliana anticline pass both reversal and fold tests. Similar behaviour of the NRM upon demagnetization of the same rock-types in areas where no

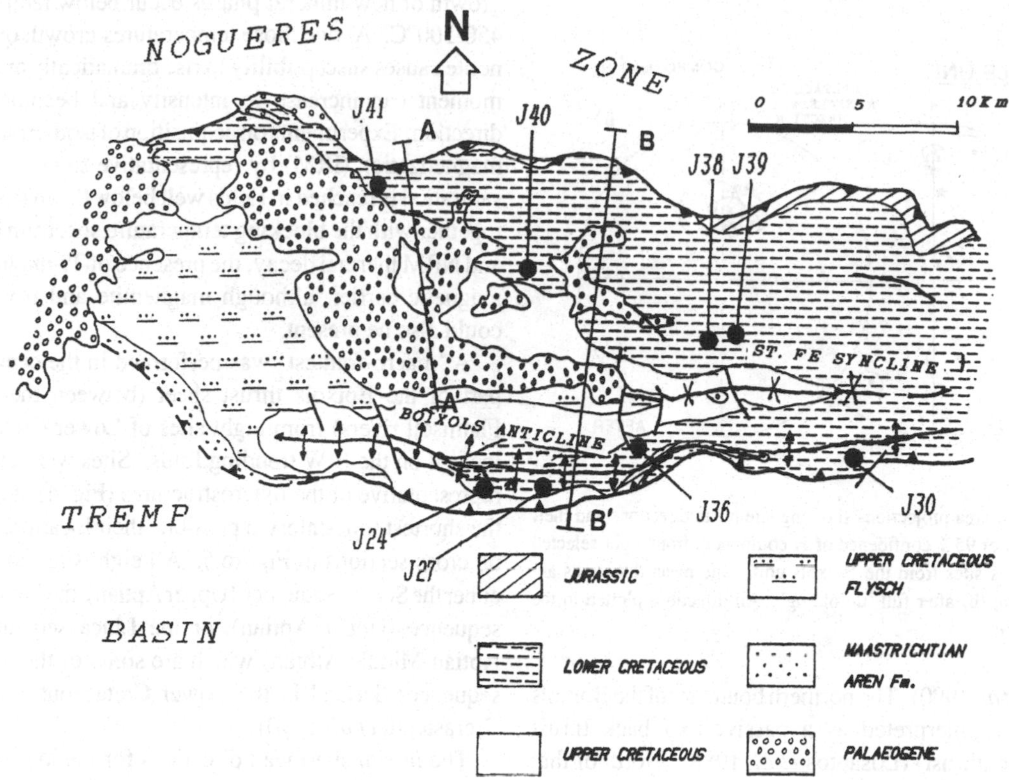

Figure 4. Simplified structural map of the eastern termination of the Boixols thrust sheet and location of the eight sites considered for the fold test (See text) and location of cross sections from Figure 5.

fold test can be performed allows confidence in the assumption of the primary nature of the characteristic component. *In-situ* primary declinations and inclinations have been corrected for dip of the strata by rotation of the bedding plane to the horizontal about the strike. Where plunging structures were observed, the tectonic correction (unfolding) incorporated unplunging followed by unbuckling. The reversed order did not produce substantial differences. This is probably due to the particular geometric relationship of the remanence and structures. In any case the fold axes plunged no more than 15 degrees.

Rock-magnetic experiments were also carried out in order to determine the mineralogy involved in each rock-type. The study of the progressive acquisition of isothermal remanent magnetization (IRM) in increasing magnetic fields and the AF and/or thermal demagnetization of a simple or composite IRM has proved to be a great help in understanding the character of some NRM demagnetization patterns. This will be the subject of a publication elsewhere.

Bóixols thrust sheet

The internal structure of the Bóixols thrust sheet consists of kilometric-scale E-W trending folds (Bóixols and St. Corneli anticlines, Santa Fe syncline), (Fig. 4). These affect a thick Mesozoic series and are the result of the modification by the Alpine compression of extensional structures associated with the formation and the evolution of the Organyà basin

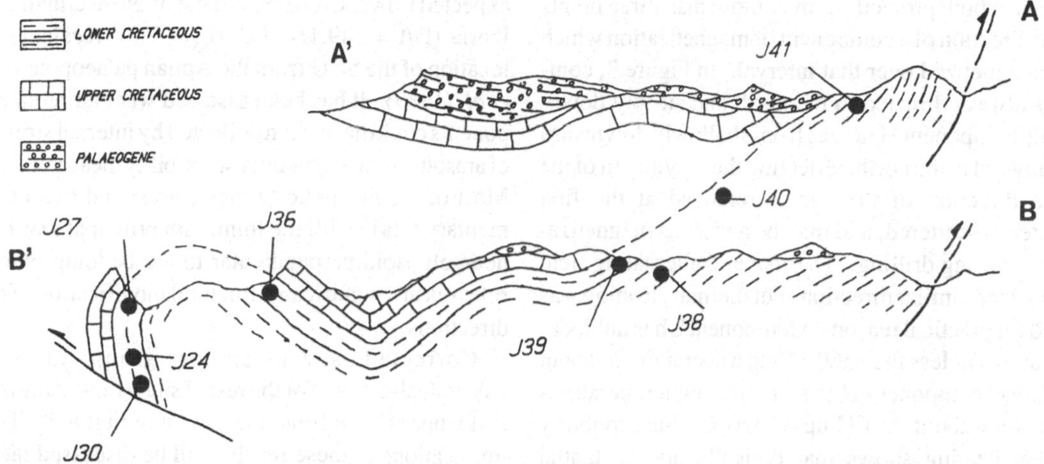

Figure 5. N-S cross-sections with the location of eight Lower-Cretaceous sites from the Bóixols thrust sheet that pass the fold test (see Fig. 4 for location).

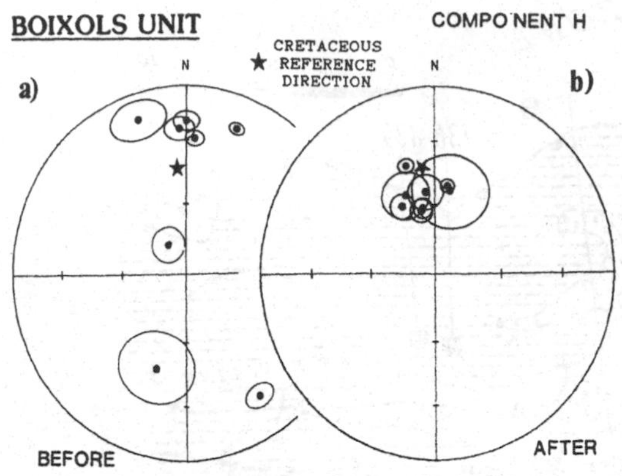

Figure 6. Equal-area projections showing site mean directions and their respective cones of 95% confidence of H component from eight selected Lower-Cretaceous sites from the Boixols unit. Site mean directions are shown (**a**) *in-situ*, (**b**) after full 'unfolding'. All directions plotted in the lower hemisphere.

(Berastegui *et al.* 1990). The northern boundary of the Bóixols thrust sheet is interpreted as a passive-roof back thrust (Morreres backthrust) (Losantos *et al.* 1988). North of that structure, several small thrust slicés showing sequences related to those of the Bóixols thrust sheet, can be recognized. A total of 28 sites with reliable palaeomagnetic results are located in the Bóixols thrust sheet (1 site from the Lower Jurassic, 22 from the Lower Cretaceous and 5 from the Upper Cretaceous).

Palaeomagnetic experiments demonstrate that the magnetization from sites in the Bóixols sheet generally have two components of magnetization, one of which we believe to be primary or at least pre-Alpine (component H), and the other which post-dates Alpine deformation (component L).

Orthogonal plots of some typical demagnetization data are given in Figure 3. In these diagrams, the variation of length and direction of the magnetic vector in a single rock specimen during demagnetization treatment is plotted by projecting the endpoint of the vector at each treatment interval onto both the horizontal (N, E, S, W) plane and a vertical plane (up, N, down, S in Fig. 3) as solid and open symbols, respectively. Straight lines in both projections over more than three points indicate the direction of a component of magnetization which is being demagnetized over that interval. In Figure 3, component L is always directed to the north and steeply downwards, while component H ranges from shallowly downward to steeply upward to the north, reflecting the varying tilt of the beds. The direction of the vector removed at the first treatment step is scattered, and may be a viscous magnetization acquired during drilling. Thermal and alternating field treatment yielded similar directions but thermal cleaning was mostly used for practical reasons. Component L has unblocking temperatures of less than 300°C and a coercivity of about 25mT, whereas component H has unblocking temperatures distributed from about 250°C up to 550°C. Susceptibility control upon heating shows that typically no substantial growth of new mineral phases occur below temperatures of 450-500°C. Above those temperatures growth of new magnetite causes susceptibility to rise dramatically and magnetic moment can increase in intensity and become erratic in direction. Experiments on acquisition of isothermal remanent magnetization (IRM) by representative samples reveal that saturation is reached in fields well below 0.2 mT, suggesting, together with the blocking temperature spectrum of the IRM and NRM thermal decay, the presence of Ti-magnetite as the magnetic carrier, although maghemite and iron sulphides could also be present.

A 'macro-foldtest' was performed in the central-eastern part of the Bóixols thrust sheet (between the Segre and Flamisell rivers) from eight sites of Lower Cretaceous age located on the E-W trending folds. Sites were chosen to be representative of the macrostructures (Fig. 4) and limited to the shortest time-interval possible, their locations are shown on cross-sections in Figure 5. All eight sites correspond to either the Senyús sequence (Upper Aptian), the Font Bordonera sequence (Upper Aptian), or the Lluçà sequence (Upper Aptian-Middle Albian) which are some of the sedimentary sequences defined in the Lower Cretaceous in this sector (Berastegui *et al.* 1990).

The *in-situ* site-mean directions for the low-temperature component cluster around the direction of the present geomagnetic field but become dispersed after tectonic correction. Therefore, component L does not pass the fold test and is interpreted as a viscous remanent magnetization (VRM) acquired in a recent period, possibly during the Brunhes normal polarity interval (< 0.73 Ma) and thus, post-dates folding.

Figure 6 shows the stereographic projection of site-mean directions of the high-temperature component H for the selected eight sites both before and after tectonic correction. It is obvious that the dispersion of site-mean directions is dramatically reduced when directions are rotated into their tilt-corrected configuration. This increase in clustering upon bedding correction results in a positive fold test that is significant at the 99% confidence level (McElhinny 1964) and thus it is likely that component H pre-dates folding. The overall corrected mean direction of component H (D/I = 351/+56, α_{95} = 7.5, k = 55) is not significantly different from the expected Lower Cretaceous palaeofield declination for stable Iberia (D/I = 349.1/+51.9, α_{95} = 7.0), recalculated for the location of the SCU from the Aptian palaeopole of Galdeano *et al.* (1989). It has been assessed whether remanence components could have been deflected by internal strain by means of anisotropy of magnetic susceptibility measurements (ASM). Most of the magnetic fabrics correspond to an oblate sedimentary fabric with the minimum principal axis of the magnetic ellipsoid perpendicular to the bedding plane and this rules out any significant structural modification of remanence directions.

Corrected site-mean or group of sites-mean palaeodeclinations for the rest of sites (Lower Jurassic, Lower and Upper Cretaceous) are shown in Figure 8. The tectonic implications of these results will be discussed later.

Montsec thrust sheet

Ten reliable sites are located in the Montsec thrust sheet (Fig. 4). Five are located in the central-eastern part (one in the Lower Jurassic, one at the Jurassic-Cretaceous boundary, one in the Lower Cretaceous, and two in the Upper Cretaceous). In the western part of the Montsec thrust sheet four sites were sampled (one in Upper Cretaceous rocks, and three in Eocene sediments). Only one site (Upper Cretaceous) is located in the easternmost part of the Montsec thrust sheet.

All sites mentioned above showed a two-component behaviour of the NRM upon demagnetization plus a soft viscous component removed at the first treatment step. The unblocking temperature spectrum for the two characteristic components differs slightly for each site but we can differentiate a low-temperature (150-300°C), low-coercivity (10-25 mT) component (L), and a high-temperature (250-580°C), high-coercivity (20-55mT) component (H). The L component always falls close to the present Earth's magnetic field before any tectonic correction (*in-situ* coordinates) and therefore is interpreted as a recent overprint. Sometimes the soft viscous component can mask the L component. Neither a positive fold test nor a within-site positive reversal test has been found in the Montsec thrust sheet. The lack of a positive fold test is due to the similar tectonic attitude of the strata in most areas of the thrust sheet (Fig. 2). An open syncline (Tremp-Graus piggy-back basin) constitutes the thrust sheet, but no similar suitable lithologies can be sampled from both limbs. Some of the Eocene sites carry a reversely magnetized component H and a between-site positive reversal test thus exists. Rock-magnetic properties, similarity of dèmagnetization patterns and consistency of data from other areas with positive palaeomagnetic field tests allows us to be confident of the primary character of component H.

All 15 sites sampled in the southernmost unit of the SCU (Sierras Marginales Unit) proved unsuitable for palaeomagnetic purposes.

Pedraforca area thrust pile

Three main separate thrust sheets can be distinguished in the Pedraforca area (Cadí, Lower Pedraforca and Upper Pedraforca thrust sheets). As a pilot study one site in each of these thrust sheets was sampled. The age of the lithologies sampled were different in each thrust sheet due to their different lithostratigraphy. Lower Eocene marls were sampled in the Cadí unit (site 89J21), Upper Cretaceous sandy-limestones in the Lower Pedraforca thrust sheet (site 89J20) and Lower Cretaceous marly-limestones in the Upper Pedraforca thrust sheet (site 89J11), proved unsuitable.

Some of the Permian and Triassic rocks from the Cadí unit were the objective of an early study (Van Dongen 1967). At that time the Cadí area was considered autochthonous and the palaeomagnetic results were interpreted in the light of previous results from rocks of the same age in Europe and Iberia. Those results were roughly consistent with other Permian and Triassic palaeodirections from the rest of the Iberian plate. Since the Cadí unit is accepted to be allochthonous it is now possible to conclude that there is no evidence for a post-Triassic rotation of the unit from Van Dongen's data. One site was sampled in the uppermost part of the thrust sheet (Lower Eocene) in order to test the validity of that conclusion and to obtain an Eocene palaeodirection to compare with other Eocene directions from different structural units in the southern Pyrenees. Two component remanences were found, and the site-mean directions of the high-temperature component (H) before and after tectonic correction for the sites and the tectonic-corrected palaeodeclination of the primary components are shown in Figure 8 (see discussion below).

Oliana foreland anticline

A break-back thrust sequence coeval with forward-imbricating thrusting has been postulated in the Oliana area in the easternmost part of the Montsec thrust sheet (Vergés & Muñoz 1990). The break-back sequence is deduced from the relationship between thrusts and four related synorogenic conglomerate units, each unconformably overlying the previous one and truncated by a more hinterland thrust. These unconformities suggest a synchronous growth of the Oliana anticline with the development of the break-back sequence. The anticline is formed by the piling-up of horses over the sole thrust and thus accommodating the foreland deformation (Vergés & Muñoz op. cit.). Timing of the events and rates of deformation has been attempted by means of a magnetostratigraphy study (Burbank *et al.* 1991).

Samples were obtained from the uppermost part of the foreland sequence, the Igualada marls (Bartonian-Priabonian) which outcrop in the core of the anticline (one site from each limb) and from the Uppermost Eocene-Lower Oligocene continental conglomerates (one site in palustrine sandstones from the western limb). The remanence consists of two components, L and H. The component H from the two Igualada marls sites passes the reversal test between sites and thus confirms the primary character of that component otherwise already argued by a positive fold test (Fig. 7).

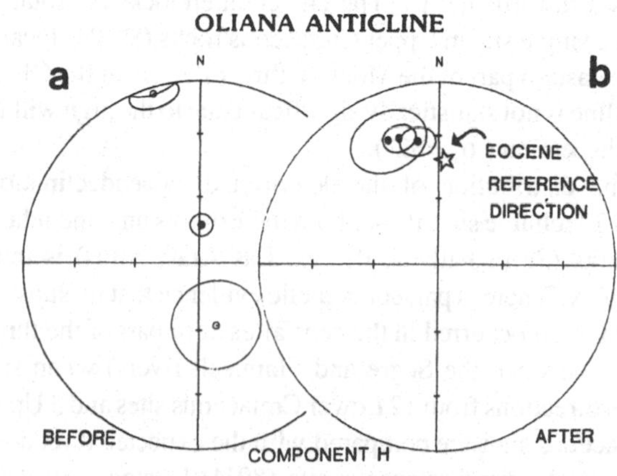

Figure 7. Equal-area projections showing site mean directions and their respective cones of 95% confidence of H component for three sites from the Oliana anticline. Site mean directions are shown (**a**) *in-situ*, (**b**) after tectonic

Discussion

The results of the palaeomagnetic investigation of the SCU and adjacent areas are plotted in Figure 8. These palaeodeclinations have been compared with expected reference directions for stable Iberia, recalculated for the location of the SCU (centred at Tremp) from palaeopoles given by several authors (Jurassic: Shott *et al.* 1981; Hauterivian-Barremian and Aptian: Galdeano *et al.* 1989; Upper Cretaceous: Van der Voo & Zijderveld 1971; Eocene: Westphal *et al.* 1986). A significant structural rotation may be considered to have occurred when there is no overlap of the confidence cones from the reference direction and the local direction from rocks of that age (rotation usually >15°). Alternatively, one can look for relative rotations between sites of the same or similar age from the same or different tectonic units. In this way any uncertainty in the determination of the reference directions can be avoided, as the Jurassic and Lower Cretaceous reference directions are determined from single studies.

It is obvious from Figure 8 that substantial rotation occurs within and between tectonic units. The salient features of the data indicate clockwise and counterclockwise rotation at the western and eastern edges of the units respectively, this rotation decreasing to zero in the central part. This general pattern is seen in both the Montsec and Bóixols units, although the more complex internal structure of the Bóixols unit is reflected in more variable rotations inferred from the palaeomagnetic data. The palaeomagnetic declinations from the central/eastern part of the Montsec thrust sheet (sites JDT48, Jurassic; JDT49, Lower Cretaceous; sites JDT58 and JDT60, Upper Cretaceous) show no significant rotation when compared with the expected directions. This is supported by unpublished data from a magnetostratigraphic study on the Lower Eocene further East in the Montsec thrust sheet (Noguera Pallaresa river) which shows that no rotation occurred there (Pascual, pers.com.). In contrast, the Lower Eocene sites 89J36, 89J38, 89J12 from the western part of the Montsec thrust sheet show a significant clockwise rotation of 31°, 26° and 15° respectively. Note that the rotation decreases towards the E. The 24° counterclockwise rotation from a single site in Upper Cretaceous rocks (89J24) located at the eastern part of the Montsec thrust sheet N of the Oliana anticline is not statistically significant due to the high within-site dispersion (α_{95}=32°).

The distribution of the determined palaeodeclinations from 27 reliable sites throughout the Bóixols unit and related subunits (Serra Fallada, Perves, Las Aras, Alins) is more complex. There is palaeomagnetic evidence that no substantial rotation occurred in the central-eastern part of the thrust sheet (between the Segre and Flamisell rivers) when site-mean directions from 12 Lower Cretaceous sites and 3 Upper Cretaceous sites are compared with the expected directions. A single Lower Cretaceous site (89J4) located east of the Segre river in the eastern edge of the Bóixols thrust sheet shows 23° counterclockwise rotation. It should be noted that site 89J9 (Hauterivian) and site 89J8 (Upper Barremian) both located in the Segre valley show declinations towards the W-NW (D/I = 291/57, α_{95} = 3.2 and D/I = 306/67, α_{95} = 6.8 respectively) which are rotated about 25° and 10° in a counterclockwise sense when compared with the expected local declination calculated from the Hauterivian-Barremian palaeopole from Galdeano *et al.* 1989. The fact that site 89J5, immediately east of sites 89J8 and 89J9, and site 89J7 located south of those two sites and both of Aptian age show no significant rotation when compared with the expected local value, leads one to suspect the apparent rotation of sites 89J8 and 89J9. This would imply that either the single Iberian palaeopole determined for the Lowermost Cretaceous is not adequately constrained or, alternatively, two more possibilities could explain the declination offset between the Hauterivian-Barremian site-mean directions and the Aptian-Albian site-mean directions: (1) effect of the anticlockwise rotation of the Iberian plate as a whole during the lowermost Cretaceous; (2) local rotation prior to both thrusting and sedimentation of the Aptian-Albian sedimentary sequences. Further palaeomagnetic work is needed in both the Pyrenean belt and the rest of the Iberian plate to assess this question properly.

Departures from the general pattern can be related to local structure. The Lower Cretaceous site 89J14 located in a syncline structure (Peracalç syncline) between the Flamisell and Noguera Pallaresa rivers (Fig. 8) shows a 45° clockwise rotation, in contrast to no rotation observed from other sites in the vicinity. The Peracalç syncline is bounded by the Morreres backthrust in the north and by a syn-sedimentary extensional fault to the south (Berastegui *et al.* 1990). The clockwise rotation of site 89J14 is interpreted as a local rotation of the Peracalç syncline during the Alpine compression. Likewise, to the west of the Flamisell river, the Lower Cretaceous site 89J40 shows an anticlockwise rotation of 37°, whereas further to the west, site 89J31 located in Lower Jurassic rocks of the Alins subunit and site 89J32 located in Lower Cretaceous rocks of the Las Aras subunit show 45° and 38° clockwise rotation, respectively. It is interesting to note that the discordant anticlockwise rotation of site 89J40 occurs in an area where macrostructures tend to be oriented N-S (Fig. 8), in contrast to the general E-W trend. This feature and the detailed structure north of this area (Serra Fallada, Perves) are not yet well understood and make detailed interpretation of the palaeomagnetic results difficult. Sites 89J29 and 89J30 located in a thrust slice containing Lower Cretaceous rocks (Fig. 8) show large amounts of rotation (123° anti- and 128° clockwise, respectively). The westernmost site sampled in the SCU (89J35) is located in Upper Cretaceous rocks from a structural subunit related to the Bóixols thrust sheet N of Campo. This site shows a 61° clockwise rotation when compared with the expected palaeodeclination.

Structural interpretation

The construction of balanced sections as discussed by several authors (e.g. Dahlstrom 1969; Hossack 1979) is a widespread practice in orogenic belts especially in their external parts. Balanced and restored sections are particularly valuable in that they can be used to calculate estimates of bulk shortening

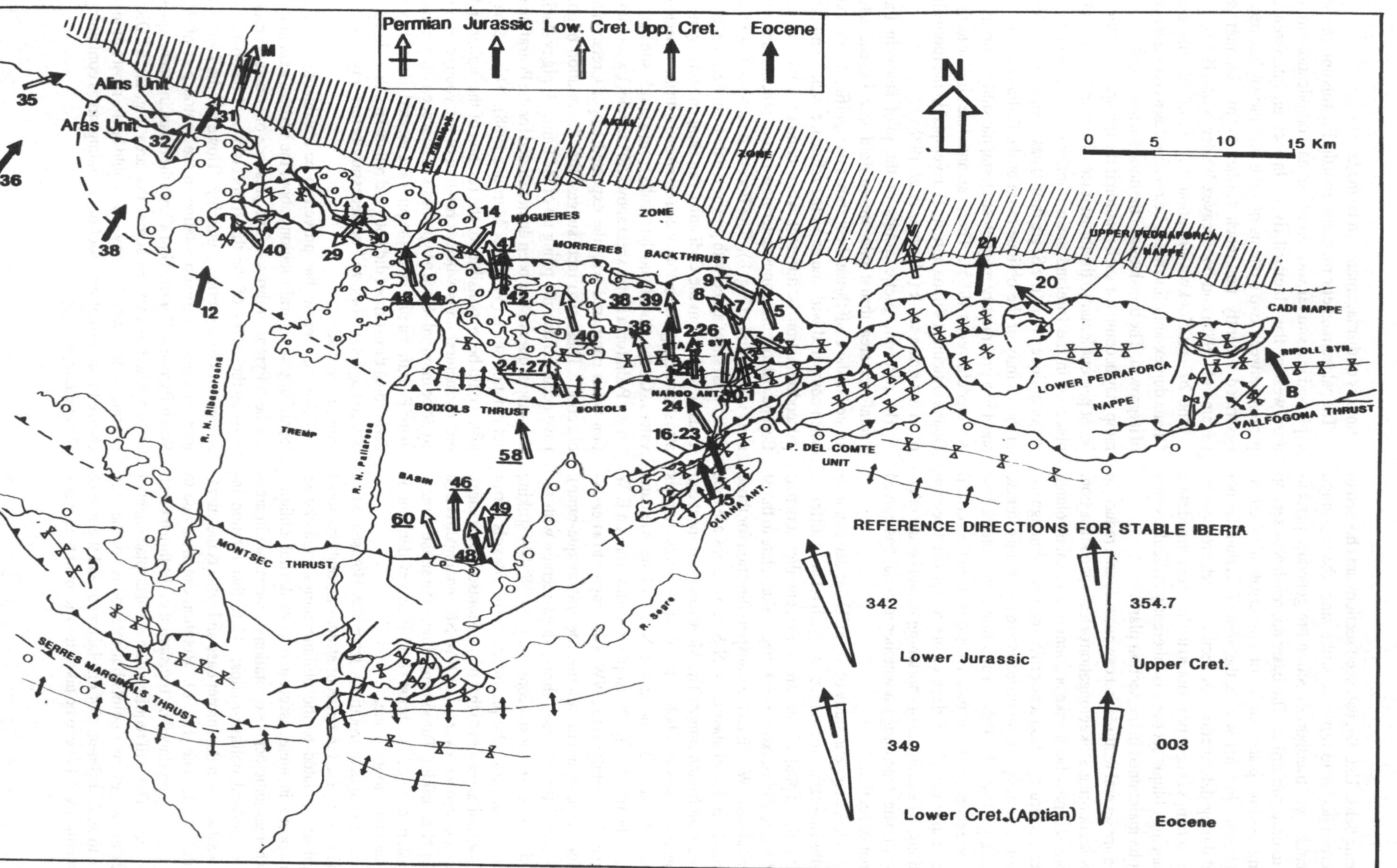

Figure 8. Palaeomagnetic data on structural map of the central eastern Pyrenees. Arrows represent the site-mean or group of sites-mean magnetic declinations (site location at the base of the arrow): numbered directions are sites from this study (underlined sites for prefix JDT otherwise 89J), lettered arrows correspond to data from other studies (V, Van Dongen, 1967; B, Burbank & Puigdefàbregas, 1985; M, McClelland & McCaig, 1989)

in orogenic belts. Usually, balanced sections are choosen to trend perpendicular to orogenic strike since the maximum displacement and transport direction are generally perpendicular to orogenic strike. The basic approaches to section balancing assume plane strain, or conservation of cross-sectional area. In other words, the area of section has not changed during deformation. A section which crosses oblique thrust ramps cannot be restored to the undeformed state and balancing oblique sections is inherently difficult as area may not be maintained in the section plane.

There are several features that can produce area reduction in a cross-section: e.g. strike elongation or tectonic compaction. Therefore, it is possible to take account of estimated volume loss when constructing balanced cross-sections, although the basic assumption of plane strain leads to a minimum estimate of shortening (Hossack 1979). In addition to the volume loss effects, the existence of structural rotations about the vertical or inclined axis can also induce errors in balanced section calculations. The estimate of shortening would be an under- or over-estimate depending on whether material has moved into or out of section.

In the palaeomagnetic study of the SCU one can compare areas within the same structural unit which relate in different ways to the thrust plane. There are areas (central/eastern part of the Montsec and Bóixols thrust sheets) that relate to thrust planes trending E-W. Since the transport direction for those thrusts has been established to be N-S, those areas are thus associated with frontal ramps. The Montsec thrust sheet can be connected westwards below the Eocene conglomerates with the Cotiella thrust along the N-S striking Mediano anticline. The trace of the thrust plane varies from an E-W direction in the frontal part to a NW-SE swinging to a near N-S direction in the Mediano anticline (dextral oblique-ramp). Towards the E the Montsec thrust sheet merges with the Sierras Marginales thrust and constitutes the NE-SW striking Segre thrust (sinistral oblique-ramp). The Bóixols thrust plane has a similar pattern to that of the Montsec thrust sheet. It can be extended westwards in a NW-SE trending direction below the N-S striking Turbón anticline. The sinistral ramp in the eastern edge of the thrust sheet is less evident and is shorter than the Montsec one.

The palaeomagnetic rotations from the Montsec thrust sheet can be explained in terms of structural rotations about a vertical axis related to the oblique ramps. Clockwise rotations occur in areas associated with dextral oblique-ramps whereas anticlockwise rotations concentrate in areas related to sinistral oblique-ramps. Note that rotation increases westwards in the former case and reflects the curvature of the thrust plane (Fig. 8). Rotation can arise due to existence of drag produced by translation of the thrust block over the ramp. This will require differential displacement of the block at the different points but rotation is not the only response to such difference of displacement and the role of local structures will have to be taken into account before any estimates of displacement can be made.

The palaeomagnetic rotations from the Bóixols unit can be explained in a similar way but additional complications may arise west of the Flamisell river. In fact, an interference pattern between two systems of folds and thrusts has been described recently not much further W by Souquet & Déramond (1989) and Fondécavez-Wallez *et al.* (1989).

The 20° anticlockwise rotation (Fig. 8) of the foreland Oliana duplex-anticline has to be regarded as a consequence of its growth. The translation of the horses across the footwall ramp may account for such rotation if the transport direction is not perpendicular to the thrust trace. This kind of rotation observed in the foreland basin immediately adjacent to a major oblique ramp (the Segre thrust) has been also reported in the vicinity of the oblique ramp of the Pedraforca thrust sheet in a preliminary magnetostratigraphic study (Burbank & Puigdefabregas 1985). There, an anticline similar to the Oliana anticline (Vilada anticline) developed in the footwall of an imbricate stack (Martínez *et al.* 1988).

The palaeomagnetic results of the pilot study in the Pedraforca area thrust pile are twofold: the Eocene site located in the Cadí thrust sheet shows no significant rotation when compared with the expected reference value) and the palaeodeclination is different when compared with other Eocene paleodeclinations determined from other structural units in this study (Fig. 8). The Eocene directions from the Cadí thrust sheet agree with previous data from Permian and Triassic rocks from the Cadí unit (Van Dongen 1967) which also show negligible rotation when compared with the expected values. On the other hand the results from a single site in the Lower Pedraforca thrust sheet shows 50° anticlockwise rotation when compared with the expected Upper Cretaceous direction. These two facts suggest that the rotation might have occurred due the piling up of the units. A piggy-back thrusting sequence has been demonstrated for the Pedraforca culmination stack (Vergés & Martínez 1988). This mechanism could be responsible for the rotation of the higher and older structural units whereas rotation of the younger units could be minimal. Such a hypothesis has not been fully tested in the Pedraforca pile since the pilot site located in the Upper Pedraforca thrust sheet did not yield reliable results. Further palaeomagnetic work will be necessary before any conclusion can be made.

Palaeomagnetism has proven to be a valuable tool for identifying structural rotations about the vertical axis in the southern Pyrenean thrusts. The presence of structural rotations at the edges of the thrust sheets would hamper any attempt of cross-section balancing. However, the fact that there is evidence for no rotation in the central part of the South-Pyrenean Central Unit allows the balancing techniques to be applied to sections across that area (e.g. the ECORS seismic profile) which in any case would be the most rational choice for constructing a section, avoiding areas containing oblique ramps.

References

Berastegui X., Losantos, M. & Garcia Senz, J. M. 1990. Structure and sedimentary evolution of the Organya basin (Central South Pyrenean Unit, Spain). *Bulletin de la Societé géologique de France*, **8**, 251-263.

Burbank, D. W. & Puigdefàbregas, C. 1985. *Chronologic investigations of the South Pyrenean basins: preliminary magnetostratigraphic results from the Ripoll Basin*. International Association of Sedimentologists, 6th European Regional Meeting, Abstracts, Lleida, Spain, 66-69.

Burbank, D. W., Vergés, J., Muñoz, J. A. & Bentham, P. 1991. Coeval hindward- and forward-imbricating thrusting in the Central Southern Pyrenees, Spain: Timing and rates of shortening and deposition. *Geological Society America Bulletin* (in press).

Cámara, P. & Klimowitz, J. 1985. Interpretación geodinámica de la vertiente centro-occidental surpirenaica (Cuencas de Jaca-Tremp). *Estudios Geológicos*, **41**, 391-404.

Choukroune, P. & ECORS Team. 1989. The ECORS Pyrenean deep seismic profile reflection data and the overall structure of an orogenic belt. *Tectonics*, **8**, 23-39.

Dahlstrom, C. D. 1969. Balanced cross sections. *Canadian Journal of Earth Sciences*, **6**, 743-747.

Dinarès Turell, J. 1989. Palaeomagnetic evidence for non-rotational thrusting in the South Pyrenean Central Unit. *International Association of Geomagnetism and Aeronomy, (I.A.G.A), 6th Scientific Assembly. Abstracts*, Exeter, UK., 236.

Dunlop, D. J. 1979. On the use of Zijderveld vector diagrams in multicomponent palaeomagnetic studies. *Physics of the Earth and Planetary Interiors*, **20**, 12-24.

ECORS Pyrenees team 1988. The ECORS deep reflection seismic survey across the Pyrenees. *Nature*, **331**, 508-511.

Elderedge, S. & Van der Voo, R. 1988. Paleomagnetic study of thrust sheet rotations in the Helena and Wyoming salients of the northern Rocky Mountains. *Geological Society of America Memoir*, **171**, 319-332.

Farrell, S. G., Williams, G. D. & Atkinson, C. D. 1987. Constraints on the age of movement of the Montsech and Cotiella thrusts, south central Pyrenees, Spain. *Journal of the Geological Society of London*, **114**, 907-914.

Fisher, R. A. 1953. Dispersion on a sphere. *Proceedings of the Royal Society of London*, **A- 217**, 295-305.

Fondecava-Wallez, J. M., Souquet, P. & Gourinard, Y. 1989. Enregistrement sédimentaire de l'eustatisme et de la tectonique dans la série turbiditique du Crétacédes Pyrénées centro-méridionales (Groupe de Vallcarga, n. gr., Espagne). *Comptes Rendus de l'Academie de Sciences, Paris*, **308, Série II**, 1011-1016.

Freeman, R., Sàbat, F., Lowrie, W. & Fontboté, J. M. 1989. Palaeomagnetic results from Mallorca (Balearic Islands, Spain). *Tectonics*, **8**, 591-608.

Galdeano, A., Moreau, M., Pozzi, J. P., Berthou, P. Y., & Malod, J. A. 1989. New palaeomagnetic results from Cretaceous sediments near Lisboa (Portugal) and implications for the rotation of Iberia. *Earth and Planetary Science Letters*, **92**, 95-106.

Garrido-Megias, A. & Rios, L. M. 1972. Síntesis geológica del Secundario y Terciario entre los rios Cinca y Segre (Pirineo central de la vertiente surpirenaica, provincias de Huesca y Lérida). *Boletin Geológico y Minero*, **83**, 1-47.

Grubbs, K. L. & Van der Voo, R. 1976. Structural deformation of the Idaho-Wyoming overthrust belt (USA). *Tectonophysics*, **33**, 321-336.

Hossack, J. R. 1979. The use of balanced cross sections in the calculation of orogenic contraction: A review. *Journal of the Geological Society of London*, **136**, 705-711.

Hudson, M. R., Reynolds, R. L. & Fishman, N. S. 1989. Synfolding Magnetization in the Jurassic Preuss Sandstone, Wyoming-Idaho- Utha Thrust Belt. *Journal of Geophysical Research*, **94**, 13681-13705.

Kent, J. T., Briden, J. C., & Mardia, K.V . 1983. Linear and planar structure in ordered multivariate data as applied to progressive demagnetization of palaeomagnetic remanence. *Geophysical Journal of the Royal Astronomical Society*, **75**, 593-621.

Kotasek, J. & Krs, M. 1965. Palaeomagnetic study of tectonic rotations in the Carpathian Mountains of Czechoslovakia. *Palaeogeography, Palaeoclimatology and Palaeocology*, **1**, 39-49.

Li, Z. X., Powell, McA. & Schmidt, P. W. 1989. Syn-deformational remanent magnetization of the Mount Eclipse Sandstone, central Australia. *Geophysical Journal International*, **99**, 205-222.

Losantos, M., Berastegui, X. Muñoz, J. A. & Puigdefàbregas, C. 1988. Corte geológico cortical del Pirineo Central (perfil ECORS): Evolución geodinámica de la cordillera Pirenaica. *In*: *Simposio sobre cinturones orogénicos*, SGE, 1988, II Congreso Geológico de España, Granada, 1988, 7-16.

Martínez, A., Vergés, J. & Muñoz, J. A. 1986. Secuencias de propagación del sistema de cabalgamientos de la terminación oriental del manto del Pedraforca y relación con los conglomerados sinorogénicos. *Acta Geológica Hispánica*, **23**, 119-128.

McElhinny, M. W. 1964. Statistical significance of the *fold test* in paleomagnetism. *Geophysical Journal of the Royal Astronomical Society of London*, **8**, 338-340.

McClelland Brown, E. 1983. Palaeomagnetic studies of fold developement and propagation in the Pembrokeshire Old Red Sandstone. *Tectonophysics*, **98**, 131-149.

McClelland, E. & McCaig, A. 1989. Palaeomagnetic estimates of rotations in compressional regimes and potential discrimination betwen thin-skinned and deep crustal deformation. *In*: *Palaeomagnetic rotations and continental deformation*. NATO ASI series C, **254**, 365-379.

Muñoz, J. A., Martínez, A. & Vergés, J. 1986. Thrust sequences in the eastern Spanish Pyrenees. *Journal of Structural Geology*, **8**, 399-405.

Mutti, E., Rosell, J., Allen, G. P., Fonnesu, F. & Sgavetti, M. 1985. The Eocene Baronia tide dominated delta-shelf system in the Ager basin. Exc. Guide-book, 6th European Regional Meeting, Lleida, Spain, 579-600.

Nijman, 1990. Thrust sheet rotation ? - The south Pyrenean Tertiary basin configuration reconsidered. *Geodinamica Acta*, **4**, 17-42.

Puigdefàbregas, C. & Souquet, P. 1986. Tecto-sedimentary cycles and depositional sequences of the Mesozoic and Tertiary from the Pyrenees. *Tectonophysics*, **129**, 173-203.

Roure, F., Choukroune, P., Berastegui, X., Muñoz, J. A., Villien, A., Mathernon, P. Baryet, M., Seguret, M., Cámara, P. & Déramond, J. 1989. ECORS deep seismic data and balanced cross-sections: Geometric constraints on the evolution of the Pyrenees. *Tectonics*, **8**, 41-50.

Schott, J. J , Montigny, R., Thuizat, R. 1981. Paleomagnetism and potassium-argon age of the Messejana Dike (Portugal and Spain): angular limitation to the rotation of the Iberian Peninsula since the Middle Jurassic. *Earth and Planetary Science Letters*, **53**, 457-470.

Schwartz, S. Y. & Van der Voo, R. 1984. Paleomagnetic study of thrust sheet rotation during foreland impingement in the Wyoming-Idaho overthrust belt. *Journal of Geophysical Research*, **89**, 10077-10086.

Seguret, M. 1972. *Étude tectonique des nappes et séries décolles de la partie centrale du versant sud des Pyrénées*. Publications de l'Université de Sciences et Techniques de Languedoc, série Geologie Structurale, **2**, Montpellier, 155p.

Simó, A. & Puigdefàbregas, C. 1985. Transition from shelf to basin on an active slope, upper Cretaceous, Tremp area, southern Pyrenees. *Exc. Guide-book, 6th European Regional Meeting*, Lleida, Spain, 63-108.

Souquet, P. 1967. Le Cretace superieur sud-pyreneen en Catalogne, Aragon et Navarre. Thèse Doctorat Sc. Nat., Fac. Sc. Toulouse, 529p.

Souquet, P. & Déramond, J. 1989. Séquence de chevauchements et séquences de depôt dans un bassin d'avant-fosse. Exemple du sillon crétacé du versant sud des Pyrénées (Espagne). *Comptes Rendus de l'Academie de Sciences, Paris*, **309**, 137-144.

Torsvik, T. 1986. IAPD - Interactive Analysis of Palaeomagnetic Data (User-guide and program description). *Internal Publication*, University of Bergen, Institute of Geophysics.

Van der Voo, R. & Zijderveld, J.D.A. 1971. Renewed paleomagnetic study of the Lisbon volcanics and implications for the rotation of the Iberian Peninsula. *Journal of Geophysical Research*, **76**, 3913-3921.

Van Dongen, P. G. 1967. The rotation of Spain: palaeomagnetic evidence from the eastern Pyrenees. *Palaeogeography, Palaeoclimatology and Palaeocology*, **3**, 417-432.

Vergés, J. & Martínez, A. 1988. Corte compensado del Pirineo oriental: geometria de las cuencas de antepais y edades de emplazamiento de los mantos de corrimiento. *Acta Geológica Hispánica*, **23**, 95-106.

—— & Muñoz, J. A. 1990. Thrust sequences in the southern central Pyrenees. *Bulletin de la Societé géologique de France*, **8**, 265-272.

Westphal, M., Bazhenou, M. L., Lauer, J. P., Pechersky, D. M. & Sibuet, J. C. 1986. Paleomagnetic implications on the evolution of the Tethys belt from the Atlantic Ocean to the Pamirs since the Triassic. *Tectonophysics*, **123**, 37-83.

Williams, G. D. & Fischer, M. W. 1984. A balanced section across the Pyrenean orogenic belt. *Tectonics*, **3**, 773-780.

PART FIVE
Alps

The Alps - a transpressive pile of peels

H. Laubscher

Geological Institute of Basel University, Bernoullistrasse 32, CH-4056 Basel, Switzerland

Abstract: One of the striking features of recent reflection seismograph traverses in the Alps is the strong reflection band of the 'Penninic front'. It cuts discordantly through the complex geometry of the Pennine nappes and ties laterally into the Miocene Simplon line. The early Miocene ductile component of this line continues into the base of the roof zone of the Lepontine dome with its retrograde EW shearing and stretching. A kinematic model compatible with the data may be constructed on the concept of an orogenic lid subjected to dextral transpression by the early Miocene Adriatic indenter (Insubric-Helvetic phase). In this model, the Penninic front reflections mark the shear zone at the base of the orogenic lid. The frontal part of this lid is the fold-and-thrust belt of the Helvetic nappes and equivalent contemporaneous thrusts such as those of the Prealps. The lid was deformed and partly eroded during the subsequent Windows phase which pushed up the External Massifs. The Insubric-Helvetic lid was subdivided into a mosaic of sublids which accommodated the divergence of translation at the NW-corner of the early Miocene Adriatic indenter. The movement of this indenter had a component of about 150 km translation to the west (dextral strike-slip along the Insubric line), and 100 km to the north (estimated translation and shortening of the Helvetic nappes). As a consequence, that part of the brittle orogenic lid north of the Insubric line (Silvretta sublid) was pushed in a northerly direction, while that at the western front of the Adriatic indenter (Gran Paradiso sublid) moved more to the west. In between, the Lepontine sublid was stretched axially by normal faulting and uplifted by tectonic underplating due to particularly intensive N-S compression in the narrowest part of the central Alps. It participated as a pull-apart domain in the dextral motion along the Insubric line and was subject to rapid tectonic denudation comparable to that in metamorphic core complexes. Adjacent to the west (external side of the lid), the Prealps sublid accommodated axial stretching largely through complementary strike-slip along its bordering transverse zones (Giffre and Kander lines). Below the orogenic lid, the middle and lower crust were deformed disharmonically. In the western Alps, parts of the lower crust and even the upper mantle were peeled off from the subducted slab and wedged into the middle crust, whereas in the central Alps, during this particular phase, subduction of lower and some middle crust seems to have been dominant, except for the most intensely squeezed Lepontine domain, where middle crust was piled up und produced an uplift of up to 20 km. The pile of Alpine nappes participating in this phase therefore had three components: those inherited from previous phases and forming passively the orogenic lid; those developing at the base and the front of the lid; and those forming disharmonically in the middle and lower crust below the lid.

One of the new and unexpected tectonic features brought to light by the recent reflection seismic traverses of the Alps is that giving rise to the reflection band of the 'Penninic front' (Bayer *et al.* 1987). This band of strong reflections cuts discordantly the complex structures of the Penninic nappes and therefore marks a shear zone that is younger. To the north and east it joins the Simplon-Rhone valley line (Bearth 1956, Steck 1984, Mancktelow 1985) which is equally discordant to the Penninic nappes and is essentially of Miocene age (Mancktelow 1985). The cataclastic Simplon fault was preceded by a ductile shear zone of some thickness which extends to the roof of the Lepontine dome (Merle *et al.* 1989) and is there dated as (latest Oligocene to) early Miocene (Hurford *et al.* 1989). This association of the 'Penninic front' reflection band with the brittle-ductile transition characterizes the base of an 'orogenic lid' (Laubscher 1983), the cold and strong upper 10-20 km of a collision zone that is the most effective stress guide between the colliding segments of continental crust. It acts as 'traîneau écraseur' (Termier 1903): at its base it produces peel nappes through predominantly simple shear, and at its leading edge frontal fold and thrust belts.

This paper focuses upon the quantitative kinematics of this early Miocene orogenic lid in the central Alps and their surroundings, with the aim of arriving at a new perspective of some long-standing problems of Alpine tectonics.

The problem of quantification in Alpine tectonics

Quantification in tectonics, when properly done, proceeds from a set of discrete observations, by interpolation, to continuous geometric bodies, then to a history of these bodies, from an initial position in an undeformed state to their present position and state (kinematics), and finally to an explanation of this historical process in terms of causal relations, mainly applying theoretical mechanics (dynamics). This procedure is necessarily an iterative, trial-and-error

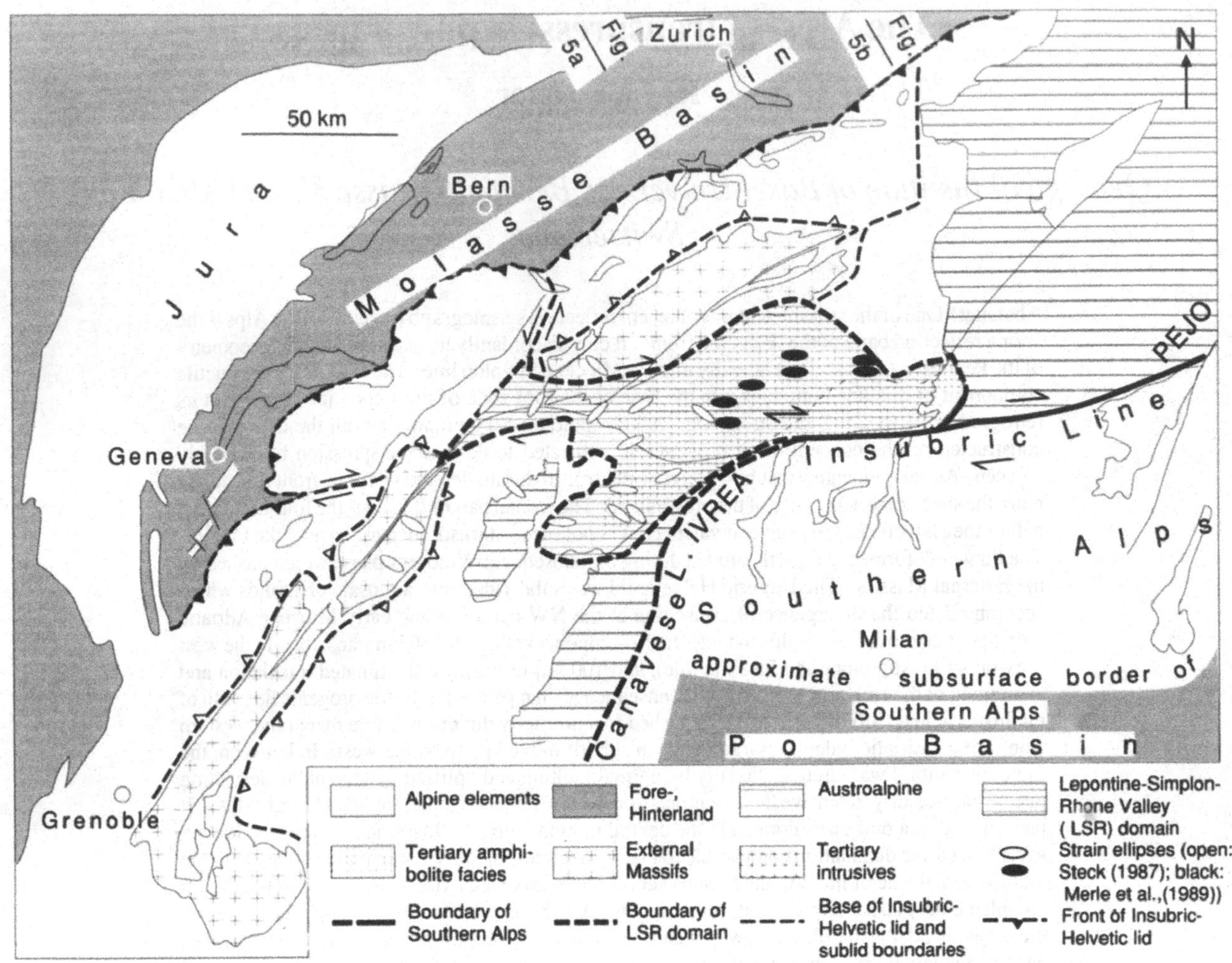

Figure 1. The central Alps and their surroundings.

one. A particular difficulty in this process is the diversity of scale and of the quality and the nature of data: geophysical data must be used, but they involve larger scales and other properties of the rocks than those used by the surface geologist. Microscopic features such as shear indicators must be integrated into domains of regional dimensions (Laubscher 1990a).

The search for plausible initial states is easiest for sedimentary bodies, particularly those involved in foreland thrust belts. Here, the construction of balanced sections as a first step to a quantitative kinematics is feasible with some precision. It relies on conservation principles, general ones such as material conservation or more restrictive ones such as area balancing or even bed-length and bed-thickness balancing (Suppe 1983; Kligfield *et al.* 1986). These by no means trivial but conceptually rather simple procedures fail for complex bodies such as the Alps; for these even a cursory inspection reveals the importance of lateral migration of material and of transformations, e. g. by high pressure, which may disguise material beyond recognition by indirect, geophysical means (Laubscher 1970, 1988, 1990b; Schreyer 1988). The comparative ease in handling thrust belts in sediments led to the realization, about 100 years ago, that the Alps consist of a pile of thin thrust sheets or nappes (Lugeon 1902), involving a series of original basins and intervening highs ('geosynclines' and 'geanticlines'). Even rough quantifications were attempted. As a result, it was inferred that shortening of several hundred kilometres had reduced the original width of the Alpine geosyncline to the present narrow belt. Later, plate tectonics helped by defining the Alpine geosyncline as the Tethys ocean and its margins, and by tying Alpine kinematics to that of the Atlantic (Dewey *et al.* 1973, 1989). Still, there is a marked lack of precision.

In the history of Alpine tectonics, those elements attributable to strike-slip have proved particularly elusive and controversial. A considerable part of this research effort has been aimed at resolving the role of strike-slip in the kinematic puzzle of the Alps.

Identification and quantification of strike-slip components

The concept that strike-slip must have played a considerable role in the kinematics of the narrow arcs of the Alps is readily apparent, and the search for the main elements is not difficult

(Laubscher 1988; Coward & Dietrich 1989). Faults such as the Insubric line or the Giudicarie line that obliquely dissect Alpine structures are obvious candidates. Qualitatively, strike-slip may be verified by any of a number of shear sense indicators. These have been observed all along the Insubric line (Schmid *et al.* 1989) and also along the southern Giudicarie (Laubscher, Schönborn & Schumacher, unpubl. data). Shear-sense indicators, however, have inherent limitations in that they provide sense and direction but not the quantity of translation. In addition, they may be late features that camouflage earlier, possibly more important events.

To obtain quantitative indicators for strike-slip one must either find marker boundaries originally transverse (preferably perpendicular but usually oblique) to the strike-slip (transfer) boundary of a displaced body of rock, or alternatively establish the amount of shortening at the leading edge or extension at the trailing edge of the body. Transverse boundaries are almost invariably fuzzy, it is often uncertain whether they existed before or developed during the strike-slip deformation phase, and correlation across the fault is therefore ambiguous. Under the circumstances trial and error interpretations must be used. Suspected boundaries may be accepted provisionally, and when several of them result in compatible kinematics, they may be integrated into a model *pro tempore*.

Timing in kinematics: The problem of quantifying transfer-linked normal translations for specific time intervals

For transfer-linked transport systems contemporaneity of motions on all the elements must be established, and this again is usually fraught with uncertainties. The timing of events like other geological information is based on a discontinuous set of data of diverse nature (relative timing by structural criteria, stratigraphic timing by means of unconformities and the deposition of erosional products, radiometric dating). Observations are made at discrete points and must be interpolated for the application to entire rock bodies and the transport systems of which they are an element. Again, as in the spatial aspects of kinematics, assignment of times can only be done by trial-and-error procedures, and again, the basic data set is full of pitfalls and may have to be re-examined when it leads to apparent kinematic contradictions.

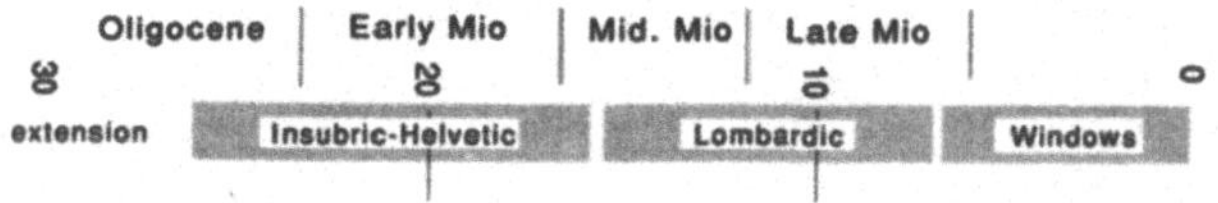

Figure 2. Estimated duration of the Neoalpine phases (grey).

A case in point is the assignation of time intervals to the Neogene motions (Neoalps) discussed in this article, and particularly those of the Helvetic nappes, the Insubric dextral transpression, and the Lombardic thrust belt. Laubscher (1990a, b) argues for the scheme represented in Figure 2: it is compatible with the basic data and provides a viable kinematic scheme. These phases are composed of numerous subphases which, however, are hard to correlate regionally. Different opinions have been voiced. Roure *et al.* (1989) propose an important link between Late Oligocene-Early Miocene Insubric motions and the Lombardic thrust belt, which, from the data alone, not considering their quantitative kinematic relations, seems neither impossible nor necessary. They base their argument on the stratigraphic record of the Sali Vercellese well in the western part of the Po plain (compare Pieri & Groppi 1981), where ditch samples document a gap between Aquitanian and Langhian (D. Bernoulli, written pers. comm., 1990). There is no doubt that the Insubric phase had its repercussions along the arc of the western Alps, but the particular link across the Po basin with the Lombardic thrust belt as proposed by Roure *et al.* (1989) seems arbitrary. As it cannot achieve the apparent allochthoneity of the Iorio-Tonale segment of the Insubric line (Laubscher 1990b), this link would appear to be of doubtful quantitative kinematic significance.

Quantification and timing of strike-slip on the Insubric line

Some of the rather fuzzy boundaries transverse to the Insubric line and its complement at the southern margin of the early Miocene (Insubric-Helvetic phase, Fig. 2) Adriatic indenter, the Villalvernia-Varzi-Levanto line (VVL) between the Ligurian Alps and the Apennines (south of Fig. 1) have been commented on by Laubscher (1988), and Figures 3 & 4 give a summary of estimated displacements.

Of the markers transverse to the Insubric line in the Alps, the western border of the Austroalpine basement units (Rhine line s.s., compare Laubscher 1988,1989) is perhaps the most

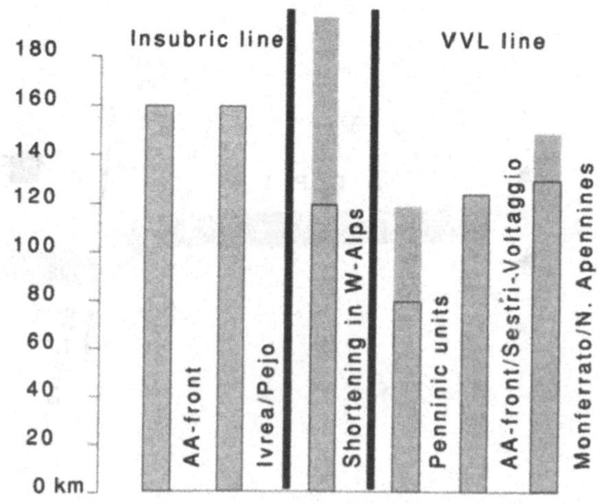

Figure 3. Apparent displacements of possibly correlative structures along the Insubric and Villalvernia-Varzi-Levanto (VVL: between Ligurian Alps and Apennines; Elter & Pertusati 1973, Laubscher 1988) Lines. Outlined columns give estimated minima, shaded columns where exceeding them give estimated maxima.

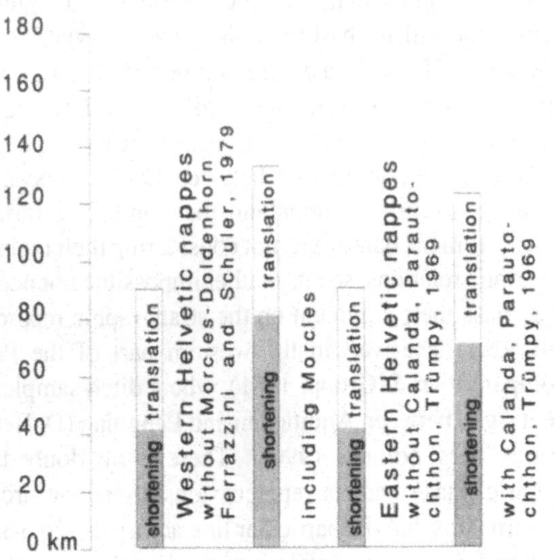

Figure 4. Estimated shortening in the Helvetic domain. Shaded: internal shortening, mainly by thrusting and folding; white: translation from original, autochthonous position.

significant. This line was probably established in the Cretaceous; it certainly predates the Insubric-Helvetic phase and was transported passively on its orogenic lid. In the correlation of nappes east and west of the Lepontine culmination the similarity between the basement rocks of the Dent Blanche and Margna nappes was stressed early on (compare Trümpy, 1980). Although the gap between the two nappes had always been assumed to be erosional, Laubscher (1988) called attention to the fact that they are on different sides of the Lepontine-Simplon-Rhone-branch (LSR, Fig. 1) of the Insubric line, and that consequently some dextral displacement would have to be assumed. The full apparent displacement is about 160 km. Although traditionally the Canavese line (Fig. 1) is considered the main branch in the Western Alps of the Insubric line, nappe correlation suggests that, in the Miocene, the Lepontine-Simplon-Rhone branch was more important.

Another possible marker is suggested by the striking similarities between the highly metamorphosed rocks of the Ivrea and the Pejo zones (Fig. 1). Remarkably, apparent dextral displacement again turns out to be 160 km.

A less satisfactory correlation is that between the Oligocene Bergell intrusion north of the line and boulders derived from it in the Gonfolite Lombarda south of it (Bernoulli *et al.* 1989; Gelati *et al.* 1988; Giger & Hurford 1989). There are some ambiguities in timing, but it would appear probable that before erosion considerable transpression had already occurred, and that any figure obtained by correlating the boulders and their source rock (around 50 km) would only be a part of total Insubric-Helvetic strike-slip (Laubscher 1988).

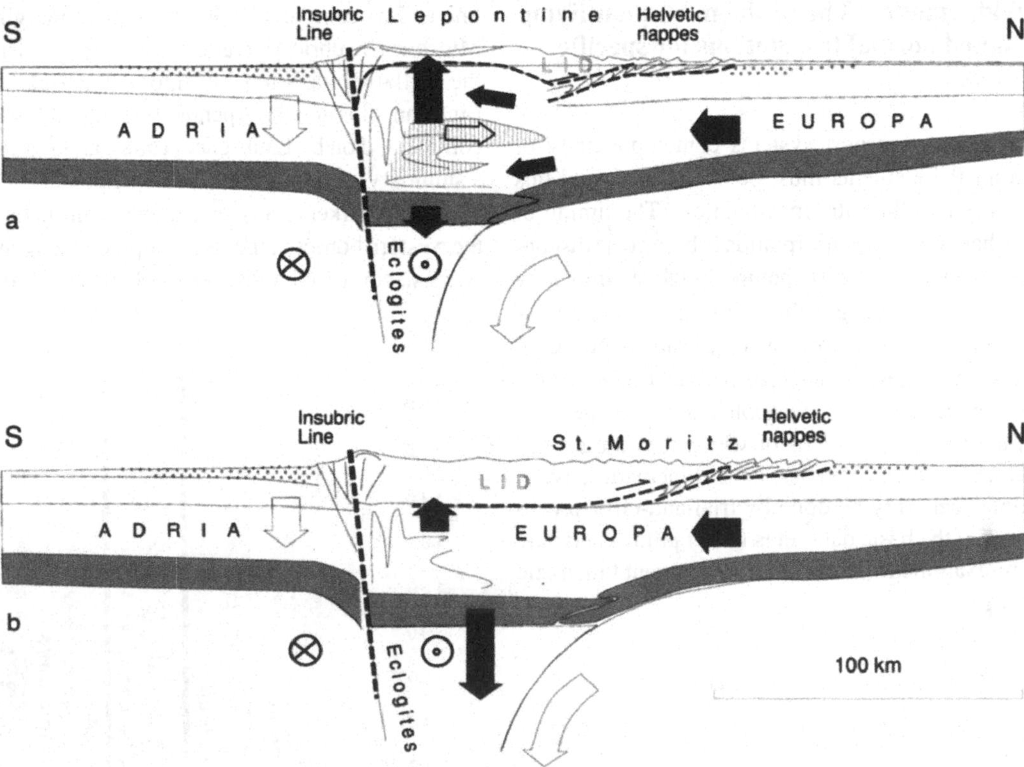

Figure 5. The uplift of the Lepontine domain by tectonic underplating. Light shading - upper (brittle) crust; dark shading - lower (granulitic) crust; dots - Molasse conglomerates of the fore-and hinterland. From the convergence of the Insubric line with the front of the Alps (from (b) to (a), compare Figs 1, 6, 7) it is inferred that the gap between the subducted European plate and the Insubric line becomes narrower. This is actually borne out by the recent reflection seismic surveys (Frey *et al.* 1989). Uplift of Lepontine may be due to piling-up of middle crust north of the Insubric line (tectonic underplating), see text. Cooling assumed somewhat slower than uplift and tectonic denudation by extension of the lid, with steep thermal gradients as in metamorphic core complexes.

Another possibility for the quantification of the westward translation of the Adriatic indenter is that of estimating the shortening of its leading edge in the western Alps.

According to the seismic sections of ECORS-CROP (Bayer *et al.* 1987) a crustal shortening of about 120 km minimum and possibly as much as 200 km seems to be required for the imbrications of lower crust and upper mantle (compare Fig. 7). The average would again be 160 km, and the same amount would have had to be incurred by the orogenic lid. Timing here is difficult, however.

A third group of estimates for the westward translation of the early Miocene Adriatic indenter is based on post-Oligocene sinistral displacement along its southern edge (Villalvernia-Varzi-Levanto line =VVL). This fault zone, originally defined by Elter & Pertusati (1973), though the only fault separating the post-Oligocene Apennines from the pre-Oligocene Ligurian Alps, is often severely underestimated (compare Laubscher 1988). Apparent sinistral displacements vary from about 80 km for the middle Penninic units of the western Alps with respect to those of the Ligurian Alps, and somewhat higher figures for the displacement of the ophiolite and Austroalpine units (Dent Blanche/Sestri-Voltaggio), to between125 km (western tip of the Monferrato Apennines at Turin/Varzi area) and 150 km (Turin/meridian of Levanto). Although the estimates are not unambiguous, they appear to be systematically lower by several tens of kilometres than the estimates along the Insubric line. It is possible that the discrepancy is due to the presence of early (pre-Burdigalian) motions along IL but not VVL. Figures for the Insubric line are probably more appropriate for the kinematics of the Insubric-Helvetic lid.

In spite of these discrepancies, which do not affect the order of the estimates, a remarkably uniform picture emerges, and an assumption of about 150 km post-Rupelian dextral displacement along the Insubric line would appear justified on a *pro tempore* basis.

The Helvetic thrust belt and its orogenic lid

A large part of the motions of the Helvetic nappes thrust belt seems to be coeval with dextral translation along the Insubric line (Insubric-Helvetic phase, Fig. 2; Schmid *et al.* 1989; Laubscher 1990b). Such thrust belts develop at the frontal part of an orogenic lid like that conjectured to overlie the Penninic front reflection band, according to Laubscher (1983). The orogenic lid consists of the upper 10 to 20 km of strong, brittle upper crust in their hinterland (Fig. 5). There is a sort of simple shear at the base of this lid (for a pure shear component see Dietrich & Casey 1989) which presumably becomes somewhat diffuse in the high-T ductile domain (Laubscher 1983; Schmid & Haas 1989). The lid is in need of continuous repair as it gets uplifted and eroded: the cooling, formerly ductile parts are added to its base. The orogenic lid that was active during the Insubric-Helvetic phase was partly destroyed by subsequent uplift, beginning early on in the Lepontine domain, and culminating subsequently during the 'Windows phase' (Fig. 2), when the External massifs were pushed up. Consequently, the base of the deformed and partly destroyed Insubric-Helvetic orogenic lid should at present be exposed immediately behind the 'roots' of the Helvetic nappes, about parallel to the internal border of the External Massifs. This is the site where the 'Penninic front' of the recent reflection seismograph surveys emerges at the surface (Fig. 1). This reflection band may therefore be considered to mark the base of the early Miocene Insubric-Helvetic lid. For this lid, crude estimates suggest about 100 km of N- to NW-translation (Fig. 4).

In map view, the lid appears to be subdivided into a mosaic of sublids (Fig. 6) which seem to have accommodated the divergence of transport direction adjacent to the NW-corner of the Adriatic indenter. This divergence implies lateral stretching. The brittle behaviour of the lid requires a style of accommodation different from that in the ductile zone below where most of the fabric data such as those in Figure 1 were obtained. Lateral stretching in the lid would be expected to be accomplished by systems of strike-slip or normal faults according to the Mohr-Coulomb criteria.

A central role in the mosaic of sublids is played by the Lepontine-Simplon-Rhone domain of Figure 1, which in a first approximation may be represented by one sublid, called 'Lepontine' in Figure 5. In the Simplon area, the base of the lid seems to rise to the top of the Lepontine 'dome', where amphibolite grade rocks were uplifted to the surface (Fig. 1).

Qualitatively, this domain has many of the characteristics of a 'metamorphic core complex' (Davis 1988; Dokka *et al.* 1986). In particular, there are indications of E-W extension of the roof (e.g. Simplon line) as well as the ductile zone underneath (Steck 1984; 1987, Mancktelow 1985) coupled with rapid cooling (Hurford *et al.* 1989) and retrograde west-verging folds and thrusts (Merle & Le Gal 1988; Merle *et al.* 1989). Quantitatively, a maximum vertical uplift of the Lepontine domain of about 20 km may be assumed (upper amphibolite facies adjacent to surface rocks in the south). The amount of E-W stretching is harder to come by but it may have been more than 40 km (40 km original length of the dome stretched by a factor of 2). In the dextral Insubric system this E-W stretching appears to play the role of a pull-apart segment, and the 40 km or so of stretching would be the pull-apart contribution of the 150 km of total (late Oligocene to) early Miocene dextral displacement.

There must be a kinematic link between stretching and rapid uplift in the sublid. Isostatic uplift is unrealistic because the domain is too narrow and the gradient of uplift is too steep. The alternative is some kind of underplating. As the uplift took place at a particular tectonic position within the Insubric-Helvetic system, tectonic underplating seems more plausible than incidental magmatic underplating. The special tectonic position is that in front of the NW corner of the Adriatic plate, because there the E-W trending Insubric fault converges westward with the SW-NE trending northern border of the Alps, thus making the NW corner of the Adriatic plate marks the narrowest section of the Alps (Fig. 6). Here, the problem of crustal escape was the most severe (Figs 5 & 6). Arguably, it was easiest westward and upward. In particular, masses of the middle European crust which were piled up against the

Insubric fault because of the N-S compressive component could plausibly provide the uplift. An element of synergy may be perceived: uplift due to N-S compression would provide the gravity potential to help the lid slide down its slopes and augment E-W extension. Figure 5 illustrates the argument: where the Europa-Adria approach is closest, the geometry is less propitious for downward escape which, at least in the Alps, seems an important way for getting rid of mass excess (Laubscher 1990 a, b).

In the Lepontine sublid, uplift by underplating simultaneously with N-S compression created a situation of the type investigated by Withjack & Scheiner (1982; compare Laubscher 1990a for the Tauern window). In this situation, the largest compressive stress becomes vertical in the uplift, and a field of normal faults develop. Without the uplift, the largest compressive stress would ordinarily be horizontal, and lateral stretching would produce strike-slip faults except for local normal faults within a complex transverse zone. This appears to be the rule outside the Lepontine dome.

Such strike-slip faults and more diffuse transverse zones emanate from the borders of the Lepontine sublid (Fig. 5) and define the other sublids of the Mosaic. The main fault zones are (Fig. 6):

The Neo-Rhine line

The Rhine line (Fig 6) which marks the western border of the Austroalpine nappes (Fig. 1) is joined near its northern end by a Neogene fault zone that cuts off the southern part of the Helvetic nappes and juxtaposes Penninic nappes to their more northerly elements: it is a kind of 'Penninic front'. In the north it penetrates the Helvetic nappes and gradually decays towards their front; to the south it swings into a position parallel to the internal border of the External Massifs. It is hard to trace there but probably splinters into several branches that penetrate the boundaries of the partial massifs shown in Figure 1 before joining the Lepontine dome. The Neo-Rhine line seems to have a sinistral component, and it separates what may be called a 'Gotthard sublid' from a 'Silvretta sublid'.

The Kander line

The Kander line emanates from the NW corner of the Lepontine sublid near Sierre (compare the Geological Map of Switzerland 1:500 000, 1980). It separates the central from the western Helvetic nappes, which from the beginning developed a different style of deformation. Here, the Helvetic front is formed by the Ultrahelvetic to Penninic Préalpes Romandes which overlie the foreland molasse directly. This must be due to a lateral ramp at the Kander line: east of the northern part of the line the basal decollement is in the lower Cretaceous of the Helvetic domain, west of it at the base of the Mesoalpine nappe complex of the Préalpes Romandes which obviously was subjected to important Neoalpine reactivation. Several lines of evidence, moreover, point towards a sinistral displacement of several kilometres (Laubscher 1982; Zwahlen 1986).

The Giffre line

At the western end of the Lepontine sublid its dextral pull-apart demands an outlet in the frontal parts of the lid. There is no well defined fault at the expected location, but there is a very pronounced 'transverse zone' (Thomas 1990). Transverse zones are found frequently in thrust belts where they mark a line-up of structural changes such as ramps from one decollement level to another, lateral ramps, and transfer of shortening. As defined by Thomas (1990): 'Such a cross-strike alignment of points of along-strike change in the structure of frontal ramps has been called a cross-strike stuctural discontinuity (or CSD) (Wheeler 1978) and is here called a transverse zone'. Though sometimes of obvious large-scale kinematic importance, the structural changes are distributed over a wide zone and are not easy to pinpoint. In this sense, a marked transverse zone, here called the Giffre line, characterizes the southern end of the Morcles nappe and its transition into the Subalpine Chains of Savoy (Bornes; compare Gidon 1976). Through the eroded part of the Helvetic lid it connects with the western end of the Lepontine sublid exactly where expected. A dextral component is implied by both the position of the front of the Morcles nappe at Tête de Bonettan (Schweizerische Geologische Kommission 1934, 1937, 1951), the dextral drag of the Subalpine Chains in the Arve valley, and the numerous SW striking dextral faults in the Subalpine Chains (cf. Gidon 1976). The 'cross-strike alignment of points of along-strike change in the structure of frontal ramps' is dramatically demonstrated in the Arve Valley east of Bonneville where in the north the Pennine units of Chablais rest more or less directly on the foreland Molasse, whereas in the south they overlie the Subalpine chains which substitute here for the Morcles nappe (and other Helvetic units?). The Subalpine chains in turn appear to ramp from a basal decollement zone in the Lias through the Valanginian shales onto the same foreland Molasse overridden in the north by the Penninic units. This situation is in many ways the complement of that across the Kander line.

There are, of course, numerous other interpretations based on different perspectives. For instance, Ramsay (1989) advocates continuity of folds throughout the Morcles nappe from north of the Rhone valley through the Dents du Midi into the Chaînes Subalpines of the Bornes. This interpretation is based upon special techniques of projection (structural tie-lines). From the perspective developed in this article, it is possible to interpret Ramsay's Figure 3 somewhat differently. For instance, the shear indicators from the Dent de Morcles to the south are found to be oblique to the tie-lines, leaving a dextral component along the latter. There is a discontinuity of Ramsay's tie-lines south of the Dents du Midi, and new units appear south of Bonneville. Farther south, continuity of units and strike breaks down completely. So far as one can see, even Ramsay's own Figure 3 (1989) leaves the reader at liberty to opt for a discontinuous solution such as the Giffre transverse zone avocated in this paper. Considering, moreover, the facts (1) that the Morcles nappe is the youngest of the Helvetic nappes and was added to

Helvetic kinematics rather late, (2) that south of the Rhone river no trace of a higher and older Helvetic nappe is found ouside the root zone, and (3) that north of Bonneville the Prealps begin with a thrust on the Molasse, it seems to me that a strong case can be made for an important transverse zone.

Obviously the elements of the sublid mosaic moved somewhat independently: the position of ramps and flats changes across their boundaries, and these functioned as transfer zones. The most peculiar sub-lids are those in front of the NW-corner of the Adriatic indenter: the Lepontine and Prealps sublids. Of the border zones of the Prealps sublid, the Kander line has a sinistral, the Giffre line a dextral component. Together, they delimit a domain of lateral stretching at the 'northern hinge of the arc of the Western Alps' (Laubscher 1982). This hinge is more sharply defined on the internal (Lepontine sublid) than on the external (Prealps sublid) side (Fig. 1). The Lepontine sublid was raised differentially and stretched laterally; it was partly incorporated into the dextral Insubric system. The spatial association of the Lepontine and Prealps sublids suggests a causal connection, the details of which, however, remain to be worked out. The position of the Gotthard sublid suggests that it too is associated with the NW corner of the Adriatic indenter. The three together compose the early Miocene Central Alps. The Simplon and Silvretta sublids are the most substantial ones and define the main trunks of the lower Miocene Insubric-Helvetic Alps: the western and eastern Alps, respectively.

Lid kinematics versus middle and lower crust kinematics

Figures 5 & 7 schematically illustrate the disharmony between the tectonics of the upper crust lid and that of the middle and lower crust. The deeper portions of the western Alps seem to be dominated by imbricated peels of lower crust and upper mantle within an interval ordinarily considered as middle crust. In the central Alps, on the other hand, a wedge of high velocity, layered material conjectured to be a peel of Adriatic lower crust, fits better into the kinematics of the middle Miocene Lombardic phase (Laubscher 1990 b) and is consequently not shown in Figures 5 & 7. Therefore, it would seem that, in the Central Alps, deep tectonics in the early Miocene was dominated by eclogitization and subduction of crustal masses, i.e. downward rather than upward escape, except for the Lepontine sublid where upward and westward escape was important. As mentioned before, this may be due to its special position at the NW-corner of the Adriatic indenter.

The transpressive pile of peels of the Insubric-Helvetic Alps thus may be subdivided into three parts: the already existing ones riding passively with the orogenic lid; those formed at the base and front of the moving lid, and those of the middle and lower crust or even upper mantle that escaped subduction and were added to the middle crust.

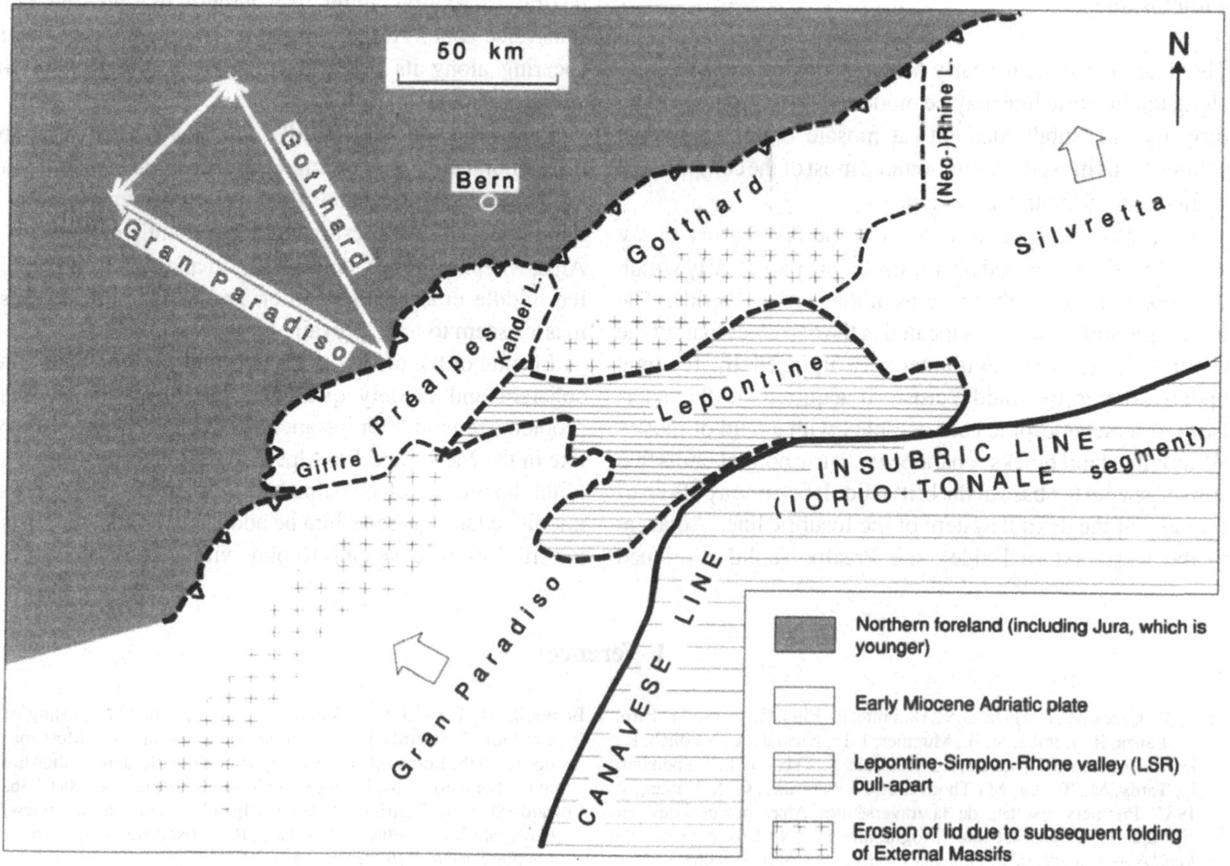

Figure 6. The sublids of the central Alps and their surroundings, Insubric-Helvetic phase.

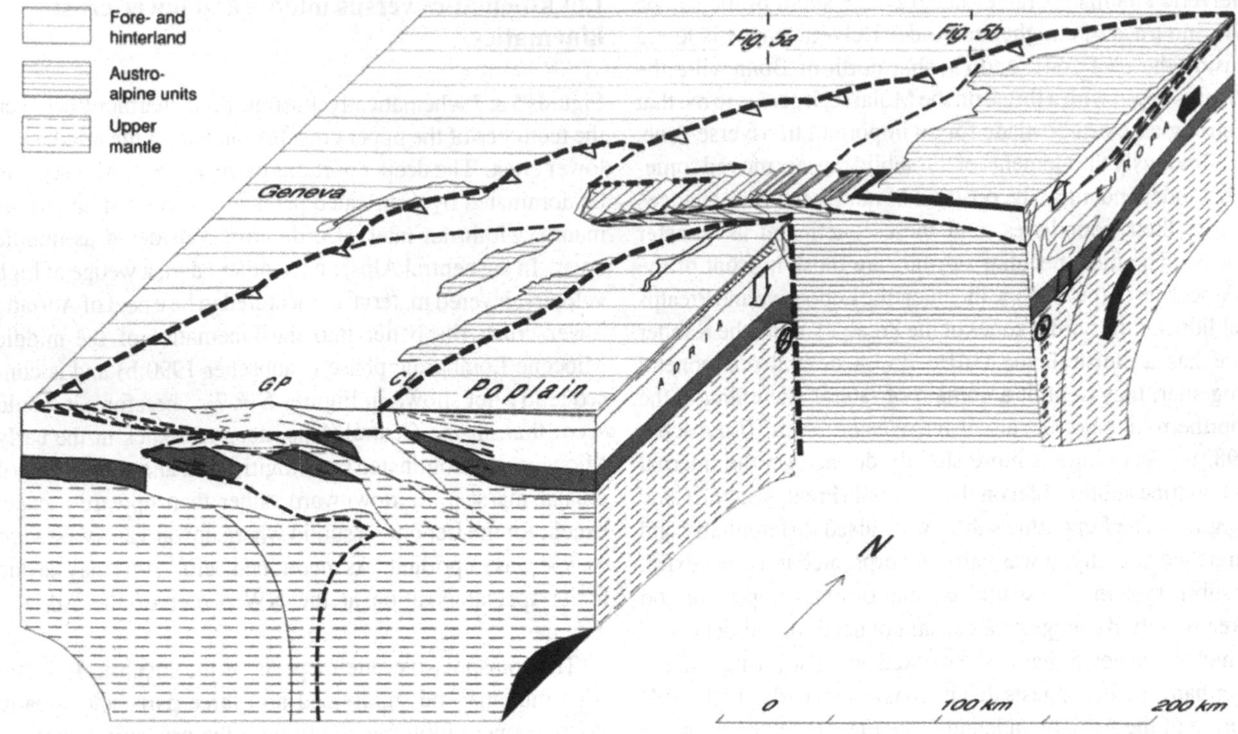

Figure 7. Simplified block diagram illustrating the 3rd dimension of Figures 1 & 6. In particular, the Lepontine-Simplon-Rhone valley (LSR) pull-apart has been greatly simplified. The section through the Gran Paradiso is according to Bayer *et al.* (1988), simplified. The External Massifs are omitted because they did not exist in the lower Miocene. Compare Figures 1 & 5.

Conclusions

The (late Oligocene to) early Miocene dextral transpression along the Insubric line may be modelled on the concept of an orogenic lid, subdivided into a mosaic of four principal sublids. This mosaic accommodated most of the compressive component of the transpression.

Divergent transport directions at the NW corner of the Adriatic indenter caused lateral stretching particularly within the Lepontine and at the margins of the Prealps sublids. The Lepontine sublid, which is located at the narrowest part of the Central Alps, was simultaneously uplifted by tectonic underplating in the middle crust. It acquired the characteristics of a metamorphic core complex with a field of mostly W-moving fault blocks separated by listric normal faults that converge with the base of the brittle lid. It became a pull-apart segment of the dextral system of the Insubric line. Adjacent to the west (external side), the Prealps sublid continued dextral translation along its southern margin, the Giffre transverse zone. At the same time, complementary sinistral shearing along its northeastern margin, the Kander line, augmented axial stretching.

Below the lid, the middle and lower crust shortened disharmonically. According to seismic evidence and material balance considerations there was considerable imbrication of peels of lower crust and upper mantle in the Western Alps, whereas in the Central Alps, in addition to stacking in the middle crust, eclogitization and subduction of crustal masses seem to have prevailed.

In spite of the efforts to tie the sundry sets of data into a coherent and crudely quantified kinematic model, many problems remain. For instance: what did transpression look like in the Meso- and Eo-Alps if the Rhine line was straight at the beginning of Neoalpine movements? How may the middle to late Miocene Jura be added to this scheme? There are still lots of loose ends to play with.

References

Bayer, R., Cazes, M., Dal Piaz, G.V., Damotte, B., Elter, G., Gosso, G., Hirn, A., Lanza, R., Lombardo, B., Mugnier, J.-L., Nicolas, A., Nicolich, R., Polino, R., Roure, F., Sacchi, R., Scarascia, S., Tabacco, I., Tapponnier, P., Tardy, M., Taylor, M., Thouvenot, F., Torreilles, G. & Villien, A. 1987. Premiers résultats de la traversé des Alpes occidentales par sismique reflexion verticale (Programme ECORS-CROP). *Comptes Rendus de l'Académie des Sciences, Paris,* **305**, Série II, 1461-1470.

Bearth, P. 1956. Geologische Beobachtungen im Grenzgebiet der lepontinischen und penninischen Alpen. *Eclogae Geologicae Helvetiae*, **49**, 279-290.

Bernoulli, D., Bertotti, G. & Zingg, A. 1989. Northward thrusting of the Gonfolite Lombarda (South-Alpine Molasse) onto the Mesozoic sequence of the Lombardian Alps: Implications for the deformation history of the Southern Alps. *Eclogae Geologicae Helvetiae*, **82**, 841-856.

Coward, M. P. & Dietrich, D. 1989. Alpine tectonics: an overview. *In:* Coward, M. P., Dietrich, D. & Park, R. G. (eds) *Geological Society of London Special Publication*, **45**, 1-29.

Davis, G. A. 1988. Rapid upward transport of mid-crustal mylonitic gneisses in the footwall of a Miocene detachment fault, Whipple Mountains, southeastern California. *Geologische Rundschau*, **77**, 191-209.

Dewey, J. F., Helman, M. L., Turco, E., Hutton, D.H.W. & Knott, S. D. 1989. Kinematics of the western Mediterranean. *In:* Coward, M. P., Dietrich, D. & Park, R. G. (eds) *Geological Society of London Special Publication*, **45**, 265-283.

——, Pitman, W. C., Ryan, W. B. F. & Bonin, J. 1973. Plate tectonics and the evolution of the Alpine system. *Geological Society of America Bulletin*, **84**, 3137-3184.

Dietrich, D., & Casey, M. 1989. A new tectonic model for the Helvetic nappes. *In:* Coward, M. P., Dietrich, D. & Park, R. G. (eds) *Geological Society of London Special Publication*, **45**, 47-63.

Dokka, R. K., Mahaffie, M. J. & Snoke, A. W. 1986. Thermochronologic evidence of major tectonic denudation. *Tectonics*, **5**, 995-1006.

Elter, P. & Pertusati, P. 1973. Considerazioni sul limite Alpi-Appennino e sulle sue relazioni con l'arco delle Alpi occidentali. *Memorie della Società Geologica Italiana*, **12**, 359-375.

Ferrazzini, B. & Schuler, P. 1979. Eine Abwicklungskarte des Helvetikums zwischen Rhone und Reuss. *Eclogae Geologicae Helvetiae*, **72**, 439-453.

Frei, W., Heitzmann, P., Lehner, P., Muller, St., Olivier, R., Pfiffner, A., Steck, A. & Valasek, P. 1989. Geotraverses across the Swiss Alps. *Nature*, **340**, 544-548.

Gelati, R., Napolitano, A. & Valdisturlo, A. 1988. La gonfolite Lombarda, stratigrafia e significato nella evoluzione oligo-miocenica del margine sudalpino. *Rivista italiana di Paleontologia (Stratigrafia)*, **94**, 285-332.

Gidon, M. 1976. Carte géologique simplifiée des Alpes occidentales du Léman à Digne, 1 : 250 000. *Bureau de Recherches Géologiques et Minières*, éditions Didier-Richard.

Giger, M. & Hurford, A. J. 1989. The Tertiary intrusives north of the Insubric Line (Central Alps): Their Tertiary uplift, erosion, redeposition and burial in the South-Alpine foreland (Como, Northern Italy). *Eclogae Geologicae Helvetiae*, **82**, 857-866.

Hurford, A. J., Flisch, M. & Jäger, E. 1989. Unravelling the thermo-tectonic evolution of the Alps: a contribution from fission track analysis and mica dating. *In:* Coward, M. P., Dietrich, D. & Park, R. G. (eds) *Geological Society of London Special Publication*, **45**, 369-398.

Kligfield, R., Geiser, P. & Geiser, J. 1986. Construction of geologic cross sections using microcomputer systems. *Geobyte*, **2**, 60-67.

Laubscher, H. P. 1970. Bewegung und Wärme in der alpinen Orogenese. *Schweizerische mineralogische und petrographische Mitteilungen*, **53**, 565-596.

—— 1982. A northern hinge zone of the arc of the western Alps. *Eclogae Geologicae Helvetiae*, **75**, 233-246.

—— 1983. Detachment, shear and compression in the Central Alps. *Geological Society of America Memoir*, **158**, 191-211.

—— 1988. Material balance in Alpine orogeny. *Geological Society of America Bulletin*, **100**, 1313-1328.

—— 1989. The tectonics of the southern Alps and the Austroalpine nappes: a comparison. *In*: Coward, M. P., Dietrich, D. & Park, R. G. (eds) *Geological Society of London Special Publication*, **45**, 229-242.

——1990a. The problem of the deep structure of the Southern Alps: 3-D material balance considerations and regional consequences. *Tectonophysics*, **176**, 103-121

——1990b. Deep seismic data from the central Alps: mass distributions and their kinematics. *Mémoires de la Société Géologique de France* (in press).

Lugeon, M. 1902. Les grandes nappes de recouvrement des Alpes du Chablais et de la Suisse. *Bulletin de la Société Géologique de France*, **4**, **1**, 723-825.

Mancktelow, N. 1985. The Simplon Line: A major displacement zone in the western Lepontine Alps. *Eclogae Geologicae Helvetiae*, **78**, 73-96.

Merle, O., Cobbold, P. R.& Schmid, S. 1989. Tertiary kinematics in the Lepontine dome. *In:* Coward, M. P., Dietrich, D. & Park, R. G. (eds) *Geological Society of London Special Publication*, **45**, 113-134.

—— & Le Gal, Ph. 1988. Post-amphibolitic westward thrusting and fold vergence in the Ticino domain. *Eclogae geologicae Helvetiae*, **81**, 215-226.

Pieri, M. & Groppi, G. 1981. Subsurface geological structure of the Po plain, Italy. *Centro Nazionale Ricerche, Progetto Finalizzato Geodinamica*, 414.

Ramsay, J. G. 1989. Fold and fault geometry in the western Helvetic nappes of Switzerland and France. *In:* Coward, M. P., Dietrich, D. & Park, R. G. (eds) *Geological Society of London Special Publication*, **45**, 33-45.

Roure, F., Polino, R. & Nicolich, R. 1989. Poinçonnement, rétrocharriages et chevauchements post-basculement dans les Alpes occidentales: évolution intracontinentale d'une chaîne de collision. *Comptes Rendus de l'Académie des Sciences de Paris*, **309**, Série II, 283-290.

Schmid, S. M., Aebli, H. R., Heller, F.& Zingg, A. 1989. The role of the Periadriatic line in the tectonic evolution of the Alps. *In:* Coward, M. P., Dietrich, D. & Park, R. G. (eds) *Geological Society of London Special Publication*, **45**, 153-171.

——& Haas, R.. 1989. Transition from near-surface thrusting to intrabasement décollement, Schlinig thrust, Eastern Alps. *Tectonics*, **8**, 697-718.

Schreyer, W. 1988. Subduction of continental crust to mantle depths: petrological evidence. *Episodes*, **11**, 97 - 104.

Schweizerische Geologische Kommission (ed.). 1934. St-Maurice. *Geological Atlas of Switzerland 1:25 000*, Sheet **8**. Kümmerly and Frey, Bern.

—— 1937. Saxon-Morcles, *Geological Atlas of Switzerland 1:25 000*, Sheet **10**. Kümmerly and Frey, Bern.

—— 1951. Finhaut. *Geological Atlas of Switzerland 1:25 000*, Sheet **24**. Kümmerly and Frey, Bern.

—— 1980. Geologische Karte der Schweiz 1 : 500 000. Bundesamt für Landestopographie, Wabern.

Steck, A. 1984. Structures de déformations tertiaires dans les Alpes centrales (transversale Aar-Simplon-Ossola). *Eclogae Geologicae Helvetiae*, **77**, 55-100.

—— 1987. Le massif du Simplon-Réflexions sur la cinématique des nappes de gneiss. *Schweizerische mineralogische und petrographische Mitteilungen*, **67**, 27-45.

Suppe, J. 1983. Geometry and kinematics of fault-bend folding. *American Journal of Science*, **283**, 684-721.

Thomas, W. A. 1990. Transverse zones in thrust belts. *Eclogae Geologicae Helvetiae*, **83** (in press).

Trümpy, R. 1969. Die helvetischen Decken der Ostschweiz: Versuch einer palinspastischen Korrelation und Ansätze zu einer kinematischen Analyse. *Eclogae Geologicae Helvetiae*, **62**,105-142.

—— 1980. An outline of the geology of Switzerland. *Geology of Switzerland- a guide-book*, 5 - 104. Schweiz. Geol. Kommission, Wepf & Co. Publ., Basel/New York. 334p.

Wheeler, R. L. 1980. Cross-strike structural discontinuities: possible exploration tool for natural gas in Appalachian overthrust belt. *The American Association of Petroleum Geologists Bulletin*, **64**, 2166-2178.

Withjack, M.O.& Scheiner, C. 1982. Fault patterns associated with domes-an experimental and analytical study. *The American Association of Petroleum Geologists Bulletin*, **66**, 302-316.

Zwahlen, P. 1986. Die Kandertal-Störung, eine transversale Diskontinuität im Bau der Helvetischen Decken. Ph. D. Diss. Bern, Universitätsdruckerei Bern,103p.

Structural evolution of the western Chartreuse fold and thrust system, NW French Subalpine chains

Robert W.H. Butler*

Department of Earth Sciences, The Open University, Walton Hall, Milton Keynes, MK7 6AA, UK

Abstract: Structural styles in the outer parts of the Alps of SE France, in common with much of the Tethyan province, involve displacements on thrust faults together with important fold structures. These deform the pre-existing sedimentary pile which contained normal faults and displays significant lateral stratigraphic variations. The underlying crystalline basement is not involved suggesting important detachment at the base of the Mesozoic cover. Abundant field outcrops are used to construct a regional cross-section through the Chartreuse district of the Subalpine fold and thrust belt, across the outermost structures of the western Alps. Thrusting directions are estimated as being ESE-WNW, determined from the spherical mean azimuth (106°) of the wide range of orientations of slip axis indicators (striae, shear fibres etc.). However, the range in orientations of slip axis indicators probably does not reflect regional non-plane strain but combinations of rotations of blocks about vertical axes within broader fault zones (but not thrust sheets) and of earlier thrusts about horizontal axes during the development of thrust anticlines. The fold-thrust belt (Mesozoic shelf sediments and Tertiary molasse) shows a gross imbricate geometry but folding plays an important role, particularly preceding thrust ramp development, which cannot be explained by simple fault-bend fold models. Locally ramps coincide with abrupt changes in Mesozoic stratigraphy suggesting local normal fault controls. Thrust anticlines show complex geometry developed in part by fault bend folding but substantially modified by large numbers of minor thrusts concentrated in the forelimbs.

The Subalpine chains of France are a classic foreland thrust belt, formed on the outer, European margin of the western Alpine mountain belt. These structures are particularly well exposed in the segment from Haut Giffre on the Franco-Swiss border to Diois (Fig. 1). Formed in late Miocene times, the thrust belt represents the last 20-30 km of Alpine shortening (Butler 1989a). Since then the region has experienced several kilometres uplift caused by isostatic rebound during erosion and unloading in the tectonic hinterland. Deep gorge sections and over 2 km vertical relief combine to provide excellent surface control on sub-surface structure. A compilation of drill-hole, gravity and seismic refraction data by Ménard (1980) provides a depth to basement map for the region. A recently acquired deep seismic reflection line (Bayer *et al.* 1987) and associated experiments (e.g. Mugnier *et al.* 1987) show that basement is not involved directly beneath the thrust belt although it has been carried up on thrusts further east.

The Subalpine chains themselves, and the more outlying Jura hills, are formed by folded and thrust Mesozoic shelf sediments together with vestiges of the overlying foredeep basin sequences. The shelf units are up to 4 km thick, although more commonly lie between 2-3 km, and are made up of alternations of limestones and shales with several regionally continuous carbonate platforms. The Tertiary foredeep sediments in contrast are classic molasse sandstones, grits and conglomerates mostly deposited in a tide-dominated shallow marine environment (Demarcq 1970; Nicolet 1979). Following the cessation of regional sedimentation in the middle Cretaceous (Lemoine *et al.* 1986) the Subalpine shelf probably experienced two main episodes of burial; during the deposition of foredeep sediments (Oligo-Miocene) and during the subsequent thrusting in late Miocene times. Folding and thrusting effects both Mesozoic shelf sediments and Tertiary molasse deposits throughout most of the Subalpine chains and all of the Jura. Those thrusts which climb through the stratigraphy emerge into the molasse basin although many structures are blind, being imaged on seismic reflection profiles (e.g. Charollais *et al.* 1977) or predicted by section construction (e.g. Butler 1989a). Only in the north of the region, towards their continuation into the Helvetic nappes of Switzerland, were the Subalpine Mesozoic rocks enveloped by the far-travelled Prealpine thrust sheets.

The combination of significant vertical relief and continuous gorge sections makes the Subalpine chains a particularly good natural laboratory to study the evolution of fold-thrust complexes. This contribution discusses the structural evolution of a small part of this thrust belt, in the western Chartreuse massif which lies between the cities of Grenoble and Chambéry (Fig. 1). This region contains the southern termination of the vestigial Swiss foredeep basin, contained between converging Jura hills and the main Subalpine chains. It is possible therefore to examine a range of thrust structures

*Present address: Department of Earth Sciences, University of Leeds, Leeds, LS2 9JT, UK.

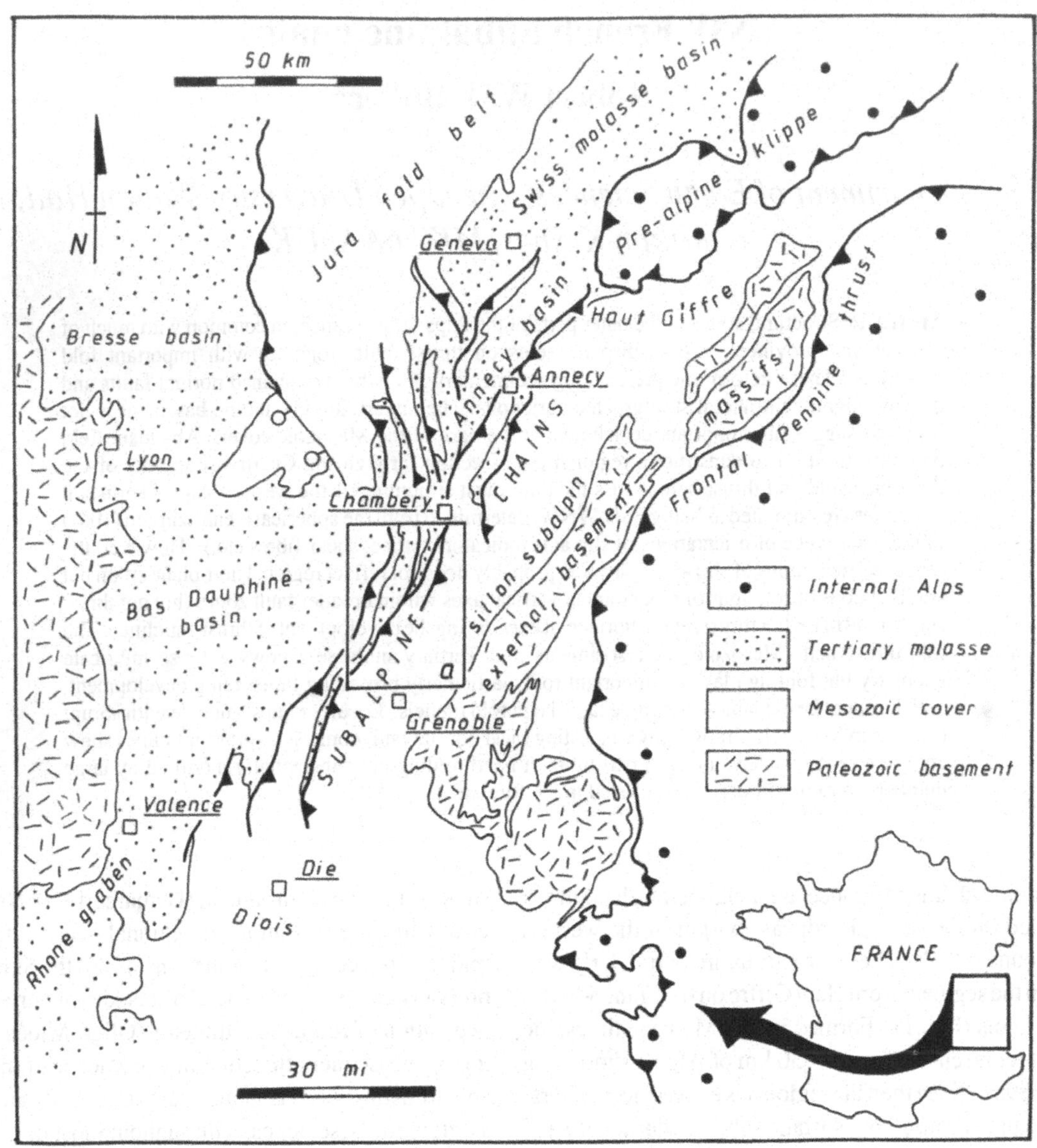

Figure 1. Location map of the Subalpine chains in the external part of the western Alps of France. The boxed area (a) is the Chartreuse massif (see Fig. 2).

from the most outlying (western) into the main thrust belt along a single transect. The problems to be addressed include thrust localization, the variations in 'structural style' and regional deformation patterns on a large scale.

Stratigraphy and regional structure

The basic geology of the Chartreuse (Fig. 2), southern Jura and surrounding regions has been well-known for many years (e.g. Gignoux & Moret 1944) and has formed part of several excellent review articles (e.g. Ramsay 1963; Debelmas & Lemoine 1970; Debelmas & Kerckhove 1980). Structural information is provided by Gidon (1964, 1981, 1988), Doudoux *et al.* (1982) and Siddans (1983). Balanced cross-sections have been constructed through the Chartreuse using angular fault-bend folding models (Suppe 1983) by Mugnier *et al.* (1987). Cross-sections through surrounding regions are provided by Doudoux *et al.* (1982) and Butler (1989a,b). The area is covered by 1:50,000 geological maps published by BRGM (sheets: Domène, 1969; Montmélian, 1969; Voiron, 1970; Grenoble, 1978; see also 1:250,000 Lyon, 1980). The explanatory booklets provide additional stratigraphic details.

The Chartreuse area has recently been studied using explosion seismology, as discussed by Thouvenot & Ménard (1990). This experiment used a single shot point to the north of the Chartreuse, recorded along two longitudinal lines running sub-parallel to strike and two fans which arc across strike. They obtained wide angle reflections from the assumed top of basement and from Mesozoic limestones. From

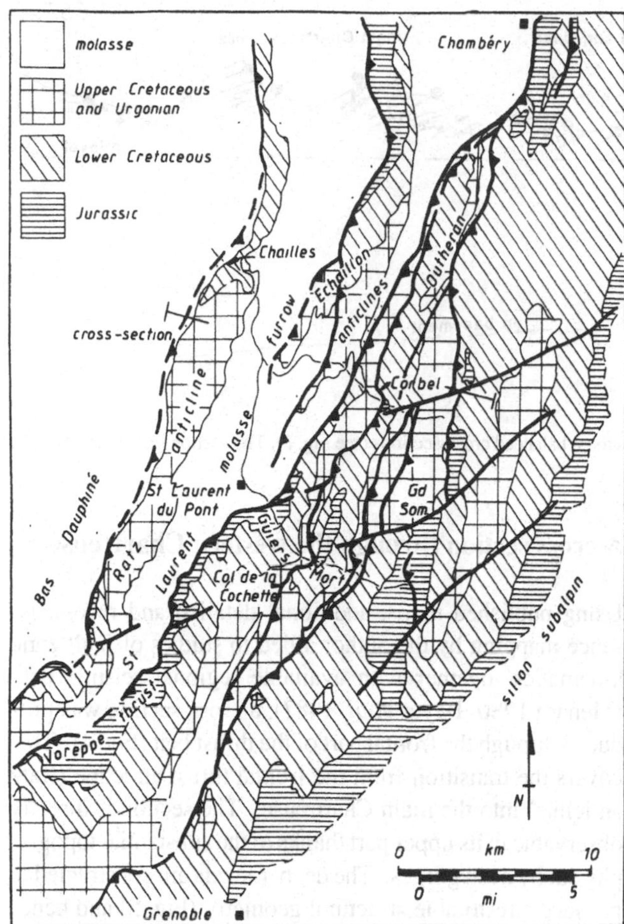

Figure 2. Geological and location map of the Chartreuse massif of the Subalpine chains (location on Fig. 1). The location of the geological profile (Fig. 4) and balanced cross-section (Fig. 5) is indicated. The Guiers Vif river runs just south of Corbel.

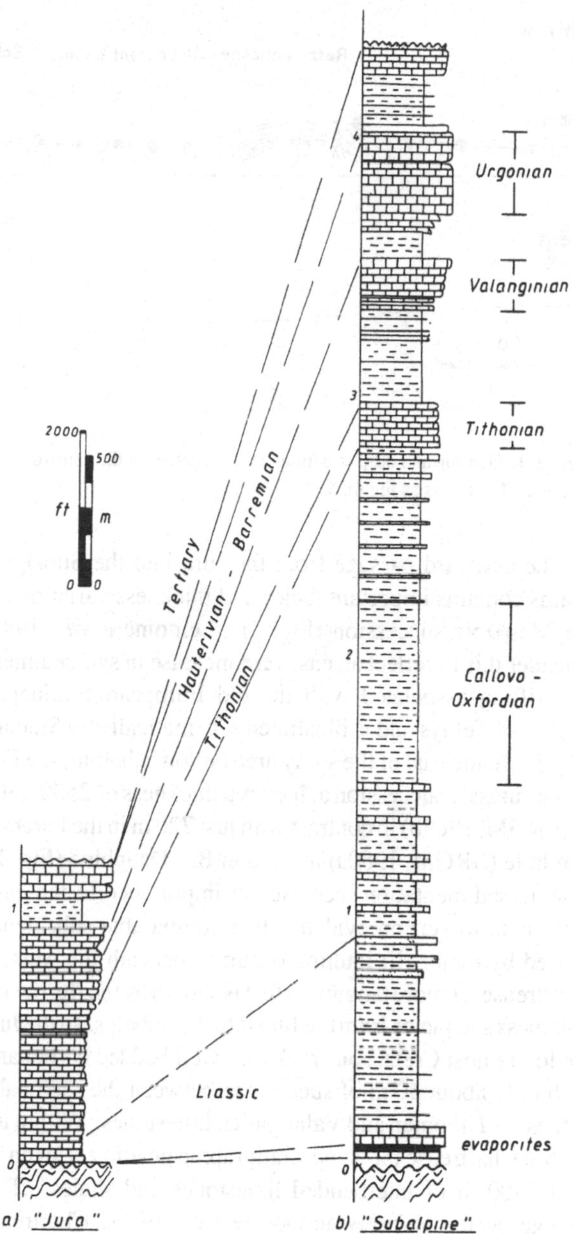

Figure 3. Simplified stratigraphic columns for the Jura (**a**) and Subalpine chains (**b**) in the Chartreuse district of the external Alps. The brick ornament represents carbonates, the pecked symbol is shale and stippled beds are siliciclastics. The numbers adjacent to each column represent the cumulative thickness in kilometres.

these data Thouvenot & Menard suggest major basement involvement in the thrusting, both passively as a step developed during the Mesozoic rifting history (discussed below) and by active WNW-directed thrusting. However, as the authors point out, the experimental design makes the correlation, and therefore determination of position and orientation, of reflectors rather difficult. The greatest limitations arise from the use of a single shot-point which prohibits back-shooting along the longitudinal sections. Wide angle experiments are best suited to regions with laterally persistant structure, not an attribute of the Subalpine chains (e.g. Doudoux *et al.* 1982). The authors have not published ray-tracing computations to support their seismic models. Therefore their interpretations which correlate particular limestone formations and basement steps with reflections are only weakly supported by the data. Nevertheless, the general position for top basement is used in construction of the regional section (Fig. 4).

Nature of the pre-thrusting template

In the west the Jura platform is dominated by laterally persistant carbonates with a cumulative thickness of about 2 km (Fig. 3). Commonly the upper part of this sequence is missing, having been eroded through the Palaeogene. This unconformity is spatially related to a suite of extensional faults which define the eastern margin of the Rhone-Bresse basin and probably represents footwall uplift in response to this normal faulting. Oligocene basins are also preserved within the thrust belt and, as we will see, the controlling normal faults influenced the subsequent development of thrust structures. Different levels of top basement detected by Thouvenot & Ménard's (1990) seismic experiment may reflect normal fault blocks in the subsurface. These faults are overlain stratigraphically by the true foredeep fill; there is an additional unconformity at the base of the Miocene molasse succession.

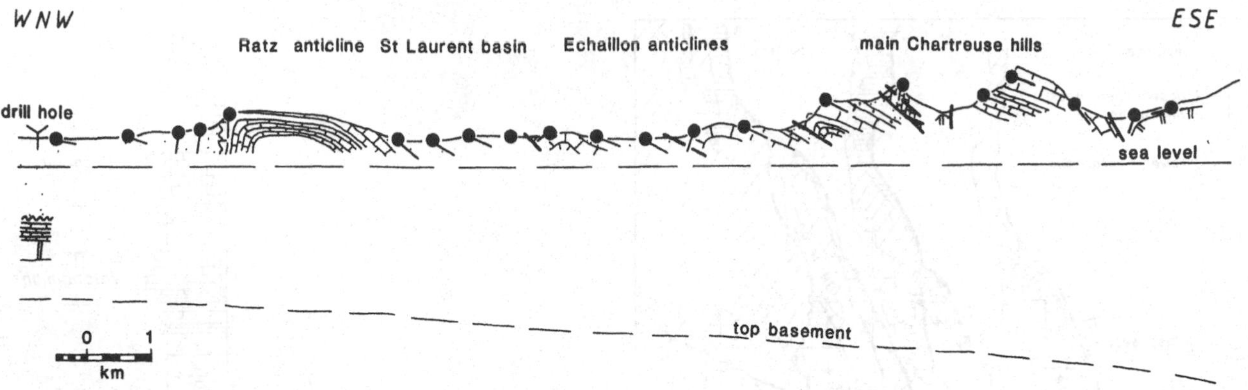

Figure 4. Outcrop and deep structural data available for the construction of regional balanced cross-sections (see Fig. 5). The section line is indicated on Figure 2. The key is as Figure 5.

The eastward passage from the Jura into the Subalpine chains contains important facies and thickness variations in the Mesozoic succession (Fig. 3). Lemoine *et al.* (1986) consider this to reflect an eastward increase in syn-sedimentary rifting associated with the old European continental margin of Tethys and is illustrated schematically by Siddans (1983). In the east of the study area (sillon subalpin, see Fig. 1) the Jurassic succession achieves a thickness of 2500-2800 m (BRGM, 1969b) in contrast with just 720 m in the Faramas borehole (BRGM, 1980a) located in Bas Dauphine (Fig. 1). Pelagic sedimentation increases in importance to the east; there is however, no evidence that comparable subsidence caused by major rift faulting occurred beneath the western Chartreuse. A more tangible effect is shown by the Berriasian thickness and facies from the Jura into the Subalps. In the Jura the lowermost Cretaceous rocks are well bedded carbonates with only about 50 m of succession between the regionally extensive Tithonian and Valanginian limestones. Within the central Chartreuse the same stratigraphic position is taken by up to 500 m of interbedded limestones and shales. This change occurs abruptly across the front of the Chartreuse hills. Although Lemoine *et al.* (1986) consider the lower Cretaceous to be a 'post-rift' sequence presumably deposited during thermal subsidence, the radical thickness changes across a short horizontal distance, together with local evidence for slumping (in the Guiers Mort valley, Fig. 2), suggests active extensional faulting continuing into Berriasian times. There is little evidence for radically differential subsidence across the Chartreuse region following the Berriasian although ongoing sequence stratigraphic studies on Cretacerous units may elucidate this. Much of the evidence has been removed by the Palaeogene uplift which effected the Jura but which was less marked to the east.

The net result of the various Mesozoic subsidence histories and differential uplift was to generate a laterally variable multi-layer sequence formed by 'competent' limestones and 'incompetent' shales with the local offsets caused by normal faults. This type of template, rather than the idealized layer-cake stratigraphies adopted for many foreland thrust belts, is characteristic of much of the Tethyan province (e.g. Lemoine *et al.* 1986; Butler 1989b).

A cross section through the western Chartreuse

Using published maps, additional detailed and reconnaissance mapping by the author allied to studies of fault zone kinematics together with available regional seismic data (Ménard 1980; Bayer *et al.* 1987), a cross-section was produced through the frontal part of the thrust belt (Fig. 4). This covers the transition from the frontal thrust anticline (Ratz anticline) into the main Chartreuse. The section is directly observable in its upper part thanks to the substantial topography cut by deep gorges. The deeper levels are constructed to produce a restorable structural geometry (Fig. 5) and hence the section runs WNW-ESE, parallel to the fault movement axes defined by shear fibres (Ramsay 1980), slickensides and grooves on fault surfaces (Hancock & Barka 1987). These kinematic data are discussed later. The background strain state of thrust sheets is very low with no appreciable cleavage generation or penetrative fracturing within the massive carbonates. As discussed later, more intense deformation occurs on the (western) fore-limbs of folds and around thrust zones but these regions are of restricted extent. For the most part simple line-length restoration methods (e.g. Goguel 1963; Dahlstrom 1969) can be employed. Parts of the Chartreuse massif contain NE-SW dextral oblique-slip fault zones (Goguel 1948; Gidon 1964) which locally generate patches of non-plane strain within the thrust belt. The cross-section (Fig. 5) was constructed away from these areas, concentrating on the western Chartreuse although the continuation of the topographic profile east of the Entremont thrust crosses an oblique-slip fault, in structures not considered here. Thrusting in the more outlying structures illustrated on Figure 4 (west of the Entremont thrust) is considered to have been plane strain.

In contrast to sections through some other foreland thrust belts, the western Chartreuse shows broad, open fold forms (Fig. 4) rather than the geometric kink-bands predicted for some fault bend structures (see Gidon 1988). This observation is crucial in chosing the appropriate method of section construction. The kink-band model popularized by Suppe (1983) has been applied to the Chartreuse (Mugnier *et al.* 1987). This approach has not been adopted here because it

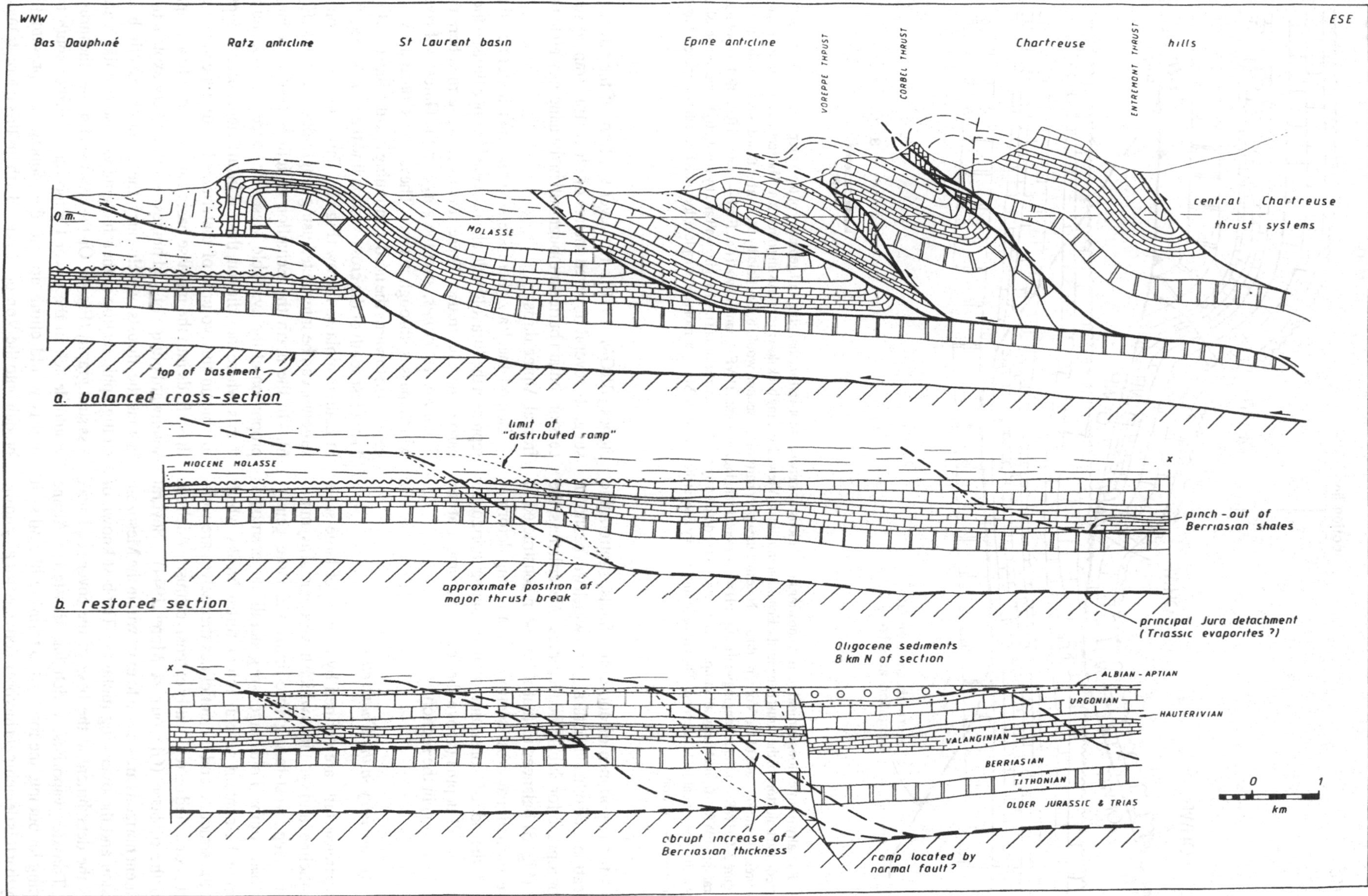

Figure 5. Balanced and restored section through the western Chartreuse, section line marked on Figure 2. Note that the two parts of the restored section join up (at x).

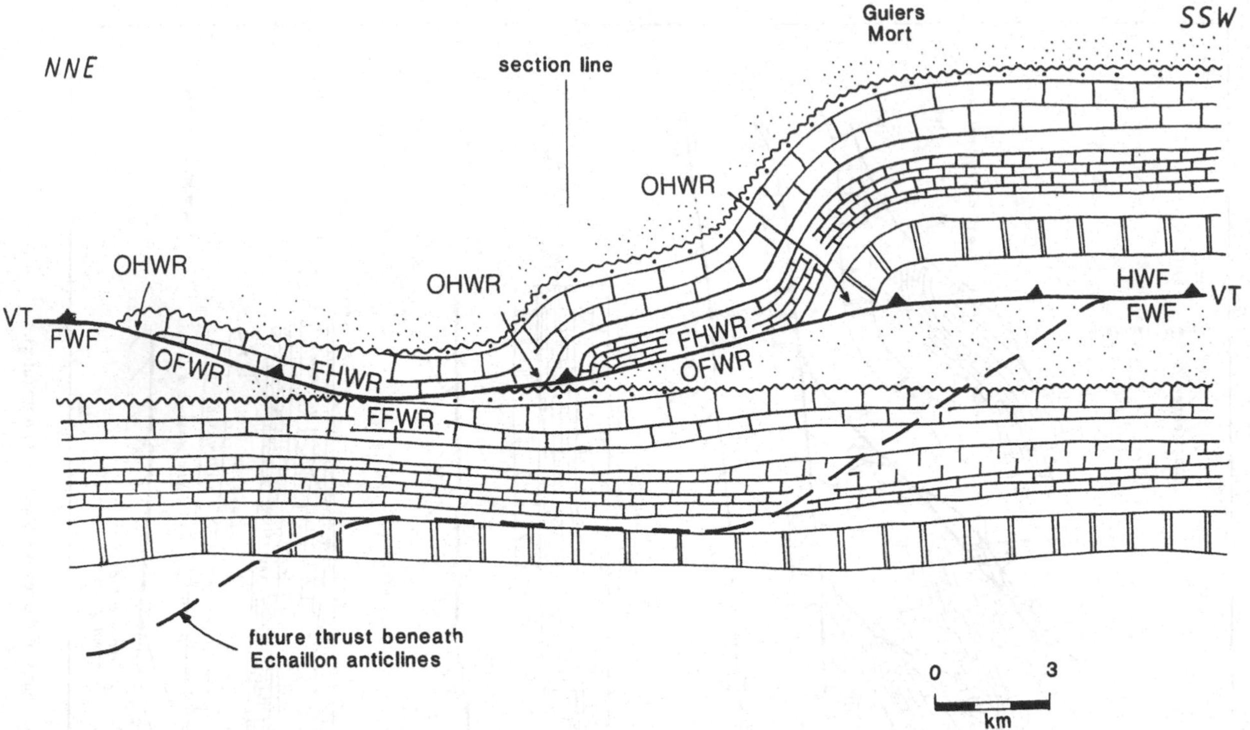

Figure 6. Partially restored longitudinal section showing the relationship between footwall and hangingwall geometry of the Voreppe thrust (VT) together with the predicted trajectory of the thrust beneath the Echaillon anticlines. Viewed looking back down the thrusting direction. The section runs approximately along the present outcrop trace of the Voreppe thrust. Note that the upper parts of the section would be overlain by the Corbel thrust sheet (not illustrated). Key as Figure 5. OHWR—oblique hangingwall ramp; FHWR—frontal hangingwall ramp; HWF—hangingwall flat; FWF—footwall flat; OFWR—oblique footwall ramp; FFWR—frontal footwall ramp. These elements are defined by bedding orientations in the hangingwall and by the stratigraphic relationships in the footwall. No vertical scale intended. The unconformity (wavy line) at the base of the Miocene molasse (stippled) does not reflect thrust-related erosion prior to molasse deposition.

obscures the importance of additional faults, stratigraphic complications and other structural styles which will be shown to be appropriate for Subalpine tectonics. Nevertheless, the section (Fig. 5) balances in that it has been constructed to preserve formational cross-sectional area during deformation with the added constraint of maintaining horizontal shortening at all depths (above the local detachment at the base of the cover) in the section.

Structural geometry in the Chartreuse

The Chartreuse hills are defined by a series of large-scale broad anticlines and synclines which have a general asymmetry with steep, west-facing fore-limbs and more gently-dipping back-limbs. Fold hinges are generally rather smooth. The forelimbs contain several minor thrust breaks which disrupt the structure and the anticlines themselves are well-known to be carried on regionally extensive thrusts which cut up into the molasse (Gignoux & Moret 1944). Several detachment horizons are known to exist within the Mesozoic succession and the overlying molasse. The best known of these is the detachment at the base of the cover sequence, within Triassic evaporites, which has transferred Alpine shortening far out into the foreland to form the Jura hills. It is likely that this detachment has been activated to carry up the Ratz anticline (Figs 2 & 4). The oldest rocks exposed in the core of the structure are Upper Jurassic (Tithonian) limestones (BRGM, 1970). However the form of the anticline suggests several hundred metres of older stratigraphic succession remain buried although still in the hangingwall to the frontal Alpine thrust.

Behind the Ratz there are several other anticlines that represent the Jura fold belt, ahead of the Voreppe thrust which carries the Chartreuse. These structures are termed here the *Echaillon anticlines*, after the gorge on the Guiers Vif river which passes through them. These folds become well-developed towards the north continuing into the Jura (see Fig. 1) but just south of the cross-section line they vanish into the St Laurent molasse furrow (Fig. 2). This lateral variation in structure can be interpreted using longitudinal sections (Fig. 6). In the north the Echaillon anticlines contain a large proportion of the available Mesozoic succession but southwards, where they die out, there is insufficient space at depth to contain this volume of material. The disappearance of these folds into the molasse basin cannot be due to a simple southward plunge of fold hinges. A more likely explanation is that the thrusts which carry these anticlines climb up stratigraphic section in their hangingwalls so that they carry less stratigraphy (Fig. 6). On the cross-section line the thrusts would detach in the lower Cretaceous. Further south the thrusts would climb up into the molasse and presumably branch onto the Voreppe thrust. Perhaps this lateral transfer of thrust displacement provides a model for the southward termination of the Jura fold belt as a separate belt of structures

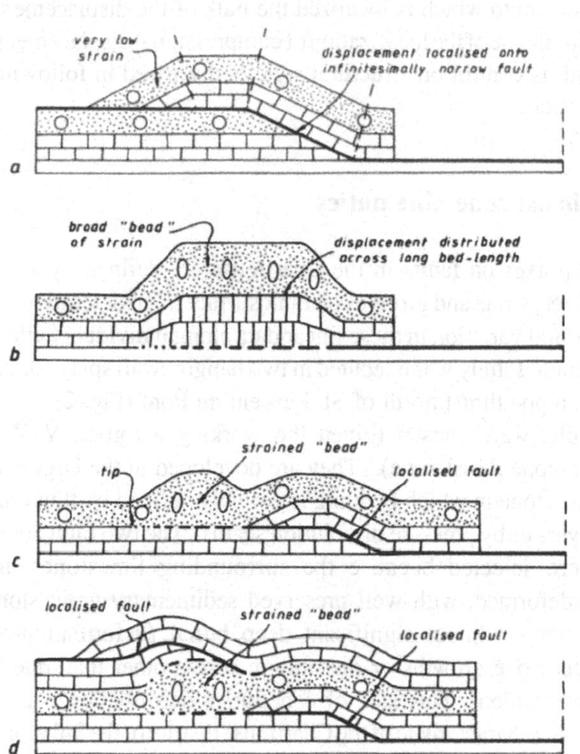

Figure 7. Variations in structural style which may exist in the Alpine thrust belt. **(a)** The classic fault bend fold (Suppe 1983) where displacements localize onto a discrete thrust. **(b)** A broad bead of strain developed above a detachment. 'Displacement' is distributed through a large volume of rock. **(c)** Combinations of fault-bend folding and distributed strain at different levels in the thrust belt. This geometry is the simple tip-line strain pattern (e.g. Williams & Chapman 1983). **(d)** A derivation of (c) where structural style alternates between strain and discrete thrusting. This is the most common deformation pattern in the thrust belt, in conjunction with folding.

distinct from the Subalpine chains. The activation of a detachment at the base of the Cretaceous is suggested to form the Echaillon anticlines. Certainly these folds do not contain any older (i.e. Jurassic) rocks on the section line (Fig. 5). This horizon was not activated further west since this region contains the facies change in the Berriasian from limestones in the west to interbedded limestones and shales where activated as a detachment in the east.

Figure 6 can be used qualitatively to detect lateral variations in the displacement on the Voreppe thrust. In the north hangingwall and footwall ramps for adjacent stratigraphic levels are juxtaposed implying relatively low displacements. This is consistent with the interpretation of sub-surface structure along the section line (Fig. 5). However, in the south the Voreppe thrust displays a hangingwall flat in mid-upper Jurassic rocks on a footwall flat about 1500 m up into the Miocene molasse (BRGM 1978) suggesting more substantial displacements. This geometric variation along strike is consistent with the southern segment of the Voreppe thrust accommodating extra slip which in the north is taken up on the Echaillon structures. This geometry supports the idea of thrust branching proposed above. Regionally it appears that displacements across the southern Jura are transferred onto the Subalpine chains from north to south.

Thouvenot & Ménard (1990) predict the presence of a long slab of Miocene molasse lying in the footwall to their 'Median overthrust' (equivalent to the Corbel thrust in this paper) based on correlation of a seismic reflector with the Urgonian limestone now exposed ahead of this thrust. The large displacements on the Corbel thrust required by Thouvenot & Menard's model are not compatible with the regional geology. This thrust terminates in a lateral tip c. 30 km south of the Chartreuse (see section in Butler 1989b). The interpretation favoured here interprets seismic reflector in the footwall to the Corbel thrust as being the Tithonian limestone, thereby reducing the required displacement.

Fold-thrust relationships

The hangingwall anticlines within the Chartreuse fold-thrust belt have a characteristic form whereby the forelimbs are steeply dipping, rather than showing the conventional moderate (30°) dips predicted by simple fault-bend models (e.g. Suppe's (1983) mode 1 folds). The back-limb dips of, e.g. the Ratz anticline (Fig. 4) are consistent with these rocks overlying a footwall ramp (i.e. <45°). The forelimb dips vertically, with an interlimb angle of 90°. This suggests an additional strain component, rather than fault slip alone.

The simple fault-bend fold model which dates from Rich (1934), formalized as the mode 1 structures of Suppe (1983) and since followed by numerous geologists (e.g. Crane 1987; De Paor 1988) is but one extreme of tectonic behaviour in a foreland thrust belt. In it shortening is highly localized onto a discrete fault surface throughout the evolution of the structure. The range of geometries which can be produced is extremely small, being a simple function of fault shape and

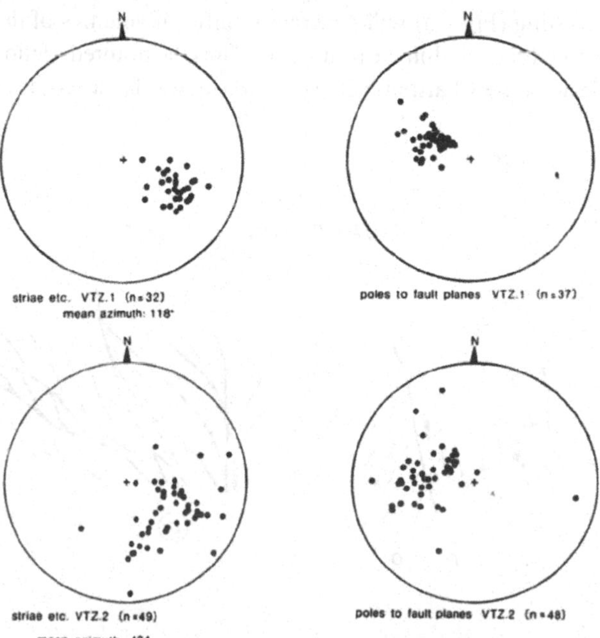

Figure 8. Stereonets of striae and shear fibres together with the poles to fault planes measured on two splays (VTZ.1 & VTZ.2) from the Voreppe thrust. The data come from a road section c. 1 km east of St. Christophe en Guiers (on the D520c). These and the following stereoplots are upper hemishere, equal angle were calculated together with the statistics using the programme STEREO, marketed by Rockware Inc.

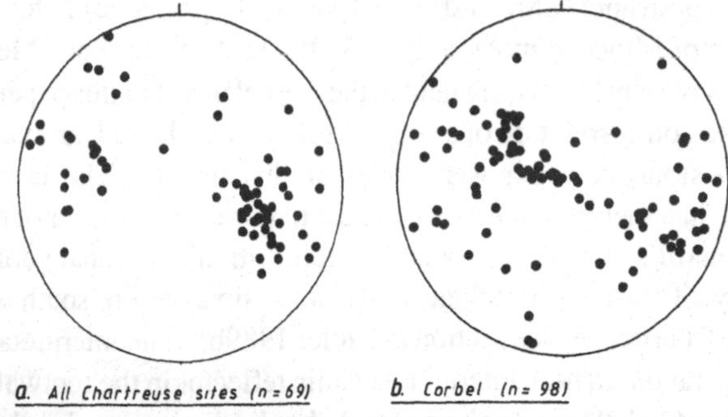

Figure 9. Stereonets of shear fibres and striations found on fault planes in the study area. **(a)** a selection of all measured fault planes from the Chartreuse; the spherical mean azimuth of these data is 106°. **(b)** a detailed study (omitted from (a)) of the Corbel area. The spread on a vertical WNW—ESE great circle is interpreted as due to folding of early thrusts around the anticline.

finite displacement (Fig. 7a). An alternative extreme is that the shortening is distributed across a wide region of a particular bed with time, generating a bead of strain without rupture (Fig. 7b). In reality, certainly for the Chartreuse thrust systems and other parts of the Subalpine chains, displacement localization lies between these end-members. The migrating tip model of Williams & Chapman (1983) is another example where displacements are distributed through a portion of a bed ahead of a localized single thrust surface (Fig. 7c). This does not imply a vertical limit of shortening since this would lead to the geologically incompatible situation of more shortening at depth in a section than at the surface. Theoretically the pattern of localized and distributed displacement could alternate through a multilayer sequence (Fig. 7d; see Butler in press).

Any deviation away from the highly localized models of thrusting (Fig. 7a) will generate significant volumes of deformed rock within the thrust belt. Thus the restored section through the Chartreuse (Fig. 5b) illustrates thrust ramps as being broad zones which ultimately contain the major thrust break onto which is localized the bulk of the displacement. This theme of the localization of compressional displacements and its control on structural style is discussed in following sections.

Thrust zone kinematics

Slip axes on faults in the Chartreuse are defined by shear fibres, striae and grooves on fault surfaces. To determine the natural variation in these linear data along individual faults a detailed study was executed in two hangingwall splays of the Voreppe thrust north of St. Laurent du Pont (Fig. 2). Two faults were chosen (given the working acronym VTZ = Voreppe thrust zone). They are developed in the Urgonian limestone in which bedding is variably defined by thin mud layers or by pressure dissolution seams. The two fault zones were selected because the surrounding limestones are undeformed, with well preserved sedimentary grain stone textures without significant deep burial (deformation-related) overgrowths or dissolution seams other than due to compaction. This lack of penetrative deformation is a regional characteristic of the Chartreuse district of the Subalpine thrust belt requiring rather simple translation patterns on thrusts. It is difficult to retain intact thrust sheets without keeping the deformation regionally plane strain. Therefore, any variations in the orientations of recorded slip axis indicators are likely to reflect processes operating within the fault zones (e.g. minor rotations of entirely fault-bounded blocks) rather than regionally non-parallel thrust motions.

VTZ.1 lies in the lower part of the road section section and is composed of a series of subparallel splays from a single bedding plane defined by dissolution seams. VTZ.2 lies 5m above VTZ.1 and is more complex. It is defined by a 1 m zone of lozenge blocks bounded by two master faults which appear

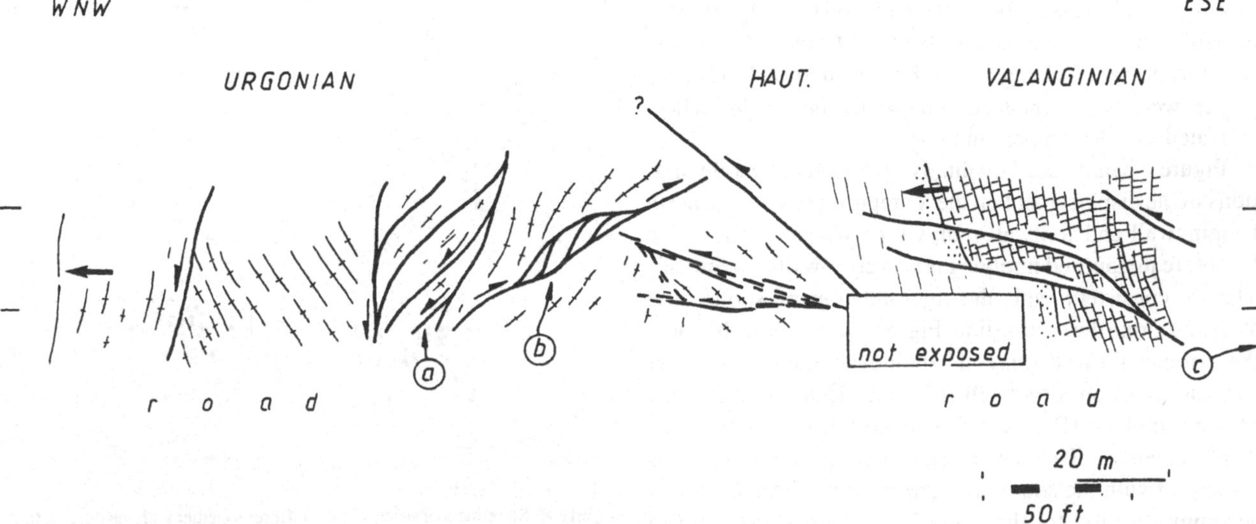

Figure 10. Sketch section of a road cut (c. 1 km west of Corbel village on the D45) through the forelimb of the anticline at Corbel (see Figs 2 & 4 for location of the structure). The large arrows represent stratigraphic way-up. The slip sense on thrusts is determined from small scale structures. Discussion of the deformation mechanisms on these and other thrusts in the Chartreuse is reserved for a companion paper.

to have localized along mud layers within the limestone. VTZ.2 could be termed a micro-duplex, defined by sub-parallel roof and floor thrusts. The orientations of slip axes defined by the linear elements listed above were measured together with the subsidiary fault plane on which they lay and grouped for each fault zone (VTZ.1 and VTZ.2). The data are presented here on upper hemisphere equal-angle stereoplots computed using the programme STEREO ™, marketed by Rockware (Fig. 8) with axes and fault planes grouped in pairs. Spherical mean lineation azimuths were also calculated using the STEREO ™ package.

VTZ.1 lineations are tightly grouped around the mean 118° azimuth but there is substantial scatter in VTZ.2 lineations. The same degree of spread is reflected in the orientations of subsidiary fault planes. Refolding cannot be invoked as both fault zones are sub-parallel. Therefore these variations reflect the natural variations in slip axis along the two fault zones. Notice that the mean lineation azimuths of the two faults are similar (118 and 126°). Subsidiary fault orientations suggest that VTZ.1 is defined primarily by frontal ramp splays while VTZ.2 contains far more oblique-lateral ramps. However, ramp orientations alone cannot explain the spread in transport axes in VTZ.2 because slip should still be parallel, regardless of the angle of incidence between thrusting and ramps. Passive rotation due to folding above oblique ramps can be rejected because the upper (roof) thrust is not folded. The prefered explanation is that the fault-bounded lozenges of limestone in VTZ.2 have experienced rotations about vertical axes during thrusting, achieved by simultaneous or alternating movement on the two bounding faults. The lozenges show substantial veining and fracturing indicating that significant strains, required to accommodate rotation, have been accumulated, but only within the fault zone. Pfiffner & Burkhard (1987) discuss the general problems of fault zone kinematics and the slip on individual surfaces.

Clearly natural fault zones can show substantial variety in apparent movement axis, particularly when earlier formed kinematic indicators are reoriented by later fault events. However, the mean transport axis for the two faults studies (VTZ.1 and VTZ.2) are surprisingly consistent. Figure 9a is a stereoplot of slip axes measured on thrusts throughout the Chartreuse (excluding the detailed studies of Figs 8 & 9b). Regionally the Chartreuse can be described as being a series of fault bounded blocks rather like VTZ.2. However, the rocks between the faults, in contrast to VTZ.2 are barely deformed suggesting generally parallel thrusting without rotations about vertical axes. Exceptions may exist adjacent to the regional oblique faults which transect the Chartreuse (Fig. 2) but these have not been considered here and are avoided by the regional section (Fig. 5). Palaeomagnetic studies are awaited which would demonstrate such rotations. The interpretation favoured here is that the Chartreuse experienced broadly plane strain thrusting on a 106° axis, the mean azimuth of slip axes for the region (Fig. 9a).

The regional dataset for the Chartreuse (Fig. 9a) shows substantial spread along a vertical great circle parallel to the presumed transport axis. The same trend is also illustrated by

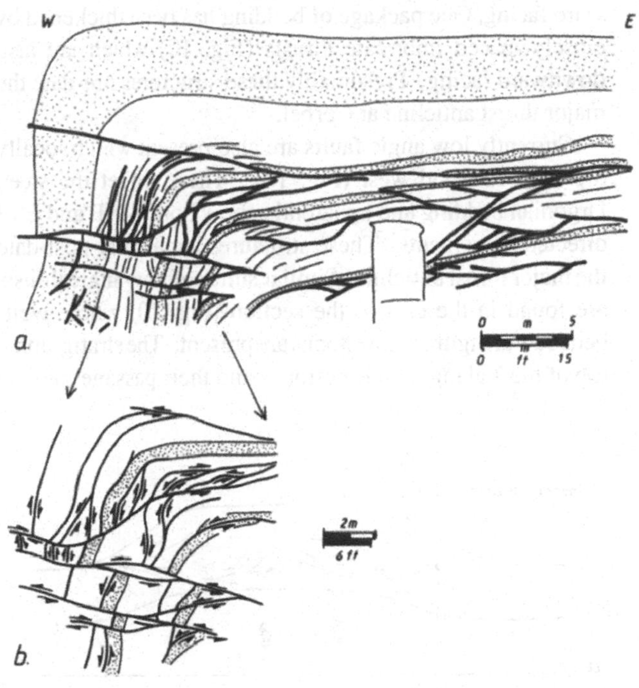

Figure 11. (**a**) Sketch section through the Ratz anticline at Chailles (Fig. 2), based on road cuttings along route N6. The rocks are well-bedded limestones (the 'marbre bâtard' of local texts) of upper Jurassic—lower Cretaceous age. Particularly prominent limestone beds are stippled. (**b**) is a detail of the forelimb area. Slip senses on faults were determined from small-scale structures (shear fibres, etc). Notice the early 'downward-facing' thrusts are cross-cut by gently-dipping faults.

the detailed study of the Corbel road section (Fig. 9b). These data will be discussed after considering the structure of this section.

Structural evolution of thrust anticlines

The interpretation of the Chartreuse anticlines as being tip or thrust-propagation folds (Williams & Chapman 1983; Suppe 1983) requires that oversteepening of the forelimbs should occur before movement up the thrust ramp. A series of detailed traverses to examine meso-scale structural histories of minor thrusts was made through well exposed regions. Two examples are discussed here.

The Corbel thrust zone

The Corbel thrust zone lies within the Chartreuse hills. In essence it consists of several thrust splays which break up the forelimb of a major anticline (Gidon 1988; see Fig. 5a). A road section provides a near continuous section through part of the structure (Fig. 10) which is described from left (west) to right (east).

Bedding within the massive Urgonian limestones is steeply dipping. However, there are discordances between packages of Urgonian, commonly between 10-15 m thick. These have the geometry of thrust cut-offs, originally gently cutting up section towards the west. These geometries are now downward facing. One package of bedding has been thickened by closely-spaced imbricate thrusts (Fig. 10) which are also downward facing. Presumably these structures pre-date the major thrust anticline at Corbel.

Currently low angle faults are also present which locally dip both east and west (Fig. 10). These offset the steep Urgonian bedding and apparently show both WNE and ESE directed movements. These structures apparently post-date the major thrust anticline. Similar suites of low-angle thrusts are found in the east of the section (Fig. 10) where well-bedded Valanginian limestones are present. The stratigraphic top of the Valanginian limestones and their passage into the Hauterivian shales is preserved, offset by small fault zones with about 5 m of WNW-directed offset. The distance between the Valanginian and Urgonian limestones is just 15 m along the road section, in contrast to the usual 100-150 m found in the Chartreuse area. The Valanginian limestones at outcrop probably lie in the hangingwall to another E-dipping fault which has carried them almost onto the Urgonian. The implied offset on this fault would be about 100 m. As with the smaller thrusts, this fault would post-date the main Corbel anticline.

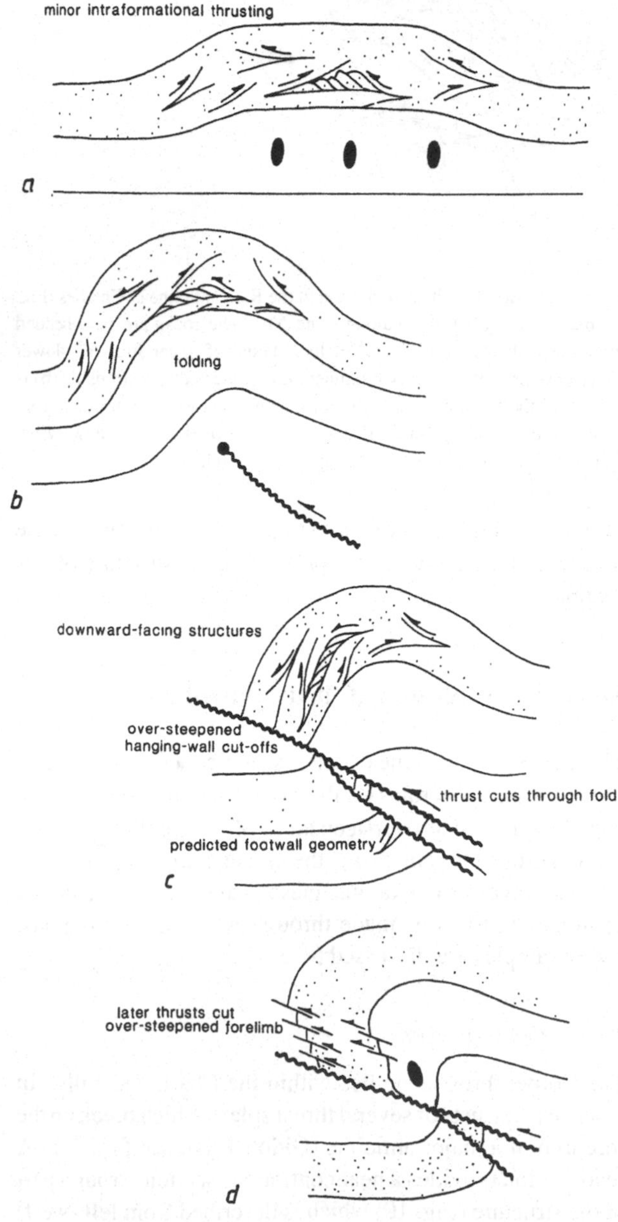

Figure 12. The proposed model of structural sequence (a-d in time) for a typical thrust-fold structure in the Chartreuse. The stippled layer represents a competent limestone unit (e.g. the Urgonian) overlying a less competent shale (e.g. the Hauterivian).

The complexity of structural evolution at Corbel is reflected in the distribution of the orientation of linear markers (striations, shear fibres) found on fault planes (Fig. 8b). These lie crudely along a vertical plane parallel to the regional thrusting direction but plunge by varying degrees towards the WNW and ESE. This population includes early thrust lineations which now lie on downward facing structures on the forelimb of the major Corbel anticline together with gently-dipping, post-fold thrusts which dip ESE and break up the forelimb.

The Ratz anticline

Complete exposure through the forelimb of the Ratz anticline is provided by the Guiers gorge at Chailles (Fig. 2). The forelimb of the anticline (Fig. 11) is cut by thrusts which generally offset towards WNW. At the present level of exposure many of these faults originate from the back-limb and cut across the hinge zone. Again these faults apparently post-date the Ratz anticline.

Within the forelimb, particularly the Valanginian limestones which dominate the present exposure levels, there are small faults which cut gently across the now-steeply dipping bedding (Fig. 11). These faults follow bedding surfaces and climb across to form ramps. Individual displacements are probably minor. The faults probably formed prior to the steepening of the Ratz forelimb, in an analogous position to those in the forelimb of the Corbel fold (Fig. 10).

A general model of Chartreuse thrust evolution

Both the Corbel and Ratz anticlines show the same sequence of structural development, summarised schematically in Figure 12. Bedding sub-parallel thrusting, directed both towards the foreland and hinterland, has thickened the limestone layers (Urgonian in the case of the Corbel and the 'marbre bâtard' of the Ratz anticline) while these formations were still sub-horizontal (i.e. before folding). Collectively these thrusts accomodate layer-parallel shortening. Subsequently bedding was steepened to be near vertical in the forelimbs of the major fold structures and then the forelimb was dissected by low angle faults which, although they extend bedding, are probably thrusts.

There are similarities in the finite geometry of some Chartreuse folds and the mode 2 hangingwall anticlines of Suppe (1983). Both the natural and theoretical geometries show tight interlimb angles. Indeed Suppe's model can be described as a distributed shear in the hangingwall which operates between the thrust and the axial plane of the anticline. A theoretical fold grows by the distributed shear operating during the early initiation of a thrust ramp, the fold being passively carried along the thrust surface once it has grown through the structure. However, the interplay between distributed shear and localized displacements on major thrusts is rather more complex in the natural examples of the Chartreuse. The minor thrusts which form before and after the steepening of the forelimbs, essentially accommodating distributed shear away from the major thrust planes, are commonly isolated with minor individual displacements. These faults may link kinematically, through zones of more distributed strain marked by weak cleavage development in shale formations, as illustrated hypothetically on Figure 7d. Collectively these faults are important in shaping the fold structures and in controlling the fore-limb dips. They provide a significant degree of freedom to the simple fault-bend fold models of Suppe (1983) and provide an explanation for a range of fore-limb dips which are not controlled by simple kink-band hinges above major ramp inflections.

The control of pre-existing structures

The regional balanced section (Fig. 5) suggests that some of the thrust anticlines coincide with major variations in the stratigraphic thicknesses of some Mesozoic formations. This suggests that the sites of thrust ramps may be influenced by pre-existing structures within sedimentary basins. The Corbel thrust zone coincides with a major eastward increase in the thickness of Berriasian shales which can be seen from surface geology (BRGM, 1969b). The hangingwall anticline contains a normal fault, downthrowing the Urgonian limestones to the east on the side of the Montagne d'Outheran (BRGM, 1969b). The map shows the fault off-setting Oligocene sandstones, probably part of the old Bresse-Rhone fault systems (Demarcq 1970) which pre-dated the thrust structures (i.e. pre-late Miocene) of the Chartreuse. Although this fault lies within the thickened pile of Berriasian shales, and so cannot have controlled their thickness variations directly, the two features may be related. On the restored section (Fig. 5b) thickness variations within the Berriasian are shown to be controlled by a fault splaying from, and lying to the west of, the Outheran fault.

The normal faults are not readily recognized within the Corbel thrust zone. Segments of the faults may have been reactivated by thrusting, or have been dismembered. However, it is possible that the distributed ramp zone represented by the Corbel thrust and the fore-limb of the Corbel anticline may have localized at the basement step generated by the pre-existing normal fault. These steps may become more marked in the northern Chartreuse where offsets in the top basement are suggested by the available seismic data (Thouvenot & Ménard 1990).

Discussion

The Chartreuse thrust systems apparently operated with a complex pattern of straining and highly localized displacements. Clearly to restore accurately any individual hangingwall anticlines, as is in vogue in some parts of the hydrocarbon exploration industry (e.g. Crane 1987; De Paor 1988; Kligfield *et al.* 1987), these sequences must be unravelled in the reverse order to which they were imposed. For example, the cut-off angles of hangingwall ramps will become severely modified by the generation of late 'short-cut' thrusts. In discussion of Crane's (1987) paper, Rowan & Ratliff (1988) point out that the forelimbs of hangingwall anticlines are commonly the mostly highly deformed parts of thrust belts. This observation is born out in the Chartreuse (see also Gidon 1988). The regional absence of activated slip surfaces on the back-limbs of Chartreuse folds suggests that the hindward propagation of ramp-related strains as flexural slip is not a significant feature (cf. Elliott 1976; Fischer & Coward 1982). Therefore Figure 5 balances while maintaining vertical trail lines (Butler in press). Clearly the deformation associated with ramp climb has been localized into the hangingwall ramp area. Thus many simple geometric models (e.g. De Paor 1988; Kligfield *et al.* 1987; Crane 1987; Mugnier *et al.* 1987) of thrust geometry, which use simple fault-bend folding and the hindward transmission of simple shear, are not appropriate for the analysis of Chartreuse structures nor, in the author's experience, to any of the Alpine-Tethyan thrust belts.

Figure 5 illustrates that thrust zones which dismember the forelimbs of major anticlines can be broad. Cooper & Trayner (1986) point out that the thrust trajectory on restored sections through such structures will have little mechanical significance because the final fault break developed after some deformation. Figure 5b illustrates the frontal ramps as being zones of deformation, largely, but not exclusively accomodated by large numbers of individually discrete faults. These ramps are linked by detachment horizons developed near the base of the Mesozoic cover and also along various weak horizons, such as the Berriasian where sufficiently shaley and within some levels of the molasse. Some of the thrust ramps are localized at pre-existing normal faults which can be detected through stratigraphic variations. There is no evidence for reactivation of these pre-existing fault structures from depth or any active involvement of the basement until further east in the thrust belt. Structural style is complex and variable, apparently controlled by the pre-existing basin architecture, facies-thickness variations, lithology and pre-existing faults.

This contribution is part of a study of Subalpine thrust system geometry and its 3-D evolution, funded by a NERC research grant (GR3/6172). I thank colleagues involved in this and the companion studies of fault-rock evolution, fluid migration and the region's thermal/diagenetic history: Sue Bowler, Maurice Tucker, Alastair Welbon and Gerald Roberts. Andy Tindle is acknowledged for computation support.

References

Bayer, R., Cazes, M., Dal Piaz, G.V., Damotte, B., Elter, G., Hirn, A., Lanza, R., Lomardo, B., Mugnier, J-L., Nicolas, A., Nicolich, R., Polino, R., Roure, F., Sacchi, R., Scarascia, S., Tobacco, I., Tapponnier, P., Tardy, M., Taylor, M., Thouvenet, F., Torrielles, G. & Villien, A. 1987. Premiers resultats de la traversée des Alpes occidentales par sismique réflexion verticale (Programme ECORS-CROP). *Comptes Rendus de l'Academie des Sciences, Paris*, **305**, 1461-70.

Butler R.W.H. 1989a. The geometry of crustal shortening in the western Alps. *In*: Sengor, A.M.C. (ed.) *Tectonic evolution of the Tethyan Region*. Proceedings of NATO Advanced Study Institute, **C239**, 43-76.

—— 1989b. The influence of pre-existing basin structure on thrust system evolution in the western Alps. *In:* Williams, G.D. & Cooper, M.A. (eds) *Inversion Tectonics*. Geological Society of London Special Publication, **44**,105-22.

—— in press. Evolution of Alpine fold-thrust complexes: a linked kinematic approach. *In:* Mitra, S. (ed.) *Structural geology of fold and thrust belts (Elliott volume)*. Johns Hopkins University Press.

BRGM 1969a. Carte géologique de la France à 1:50000, feuille Domène. *Bureau des Récherches Géologiques et Minières, Orléans*

—— 1969b. Carte géologique de la France à 1:50000, feuille Montmélian, *Bureau des Récherches Géologiques et Minières, Orléans*

—— 1970. Carte géologique de la France à 1:50000, feuille Voiron, *Bureau des Récherches Géologiques et Minières, Orléans*

—— 1978. Carte géologique de la France à 1:50000, feuille Grenoble, *Bureau des Récherches Géologiques et Minières, Orléans*

—— 1980a. Carte géologiques de la France à 1:50000, feuille La-Tour-du-Pin, *Bureau des Récherches Géologiques et Minières, Orléans*

—— 1980b. Carte géologique de la France à 1:250000, feuille Lyon, *Bureau des Récherches Géologiques et Minières, Orléans*

Charollais, J., Pairis, J-L. & Rosset, J. 1977. Compte rendu de l'excursion de la Societé Géologique Suisse en Haute Savoie (France) du10 au 12 Octobre 1976. *Eclogae Geologicae Helvetiae*, **70**, 253-85.

Cooper, M. A. & Trayner, P. M. 1986. Thrust-surface geometry: implications for thrust belt evolution and section balancing techniques. *Journal of Structural Geology*, **8**, 305-12.

Crane, R. C. 1987. Use of fault cut-offs and bed travel distance in balanced cross-sections. *Journal of Structural Geology*, **9**, 243-6.

Dahlstrom, C.D.A. 1969. Balanced cross-sections. *Canadian Journal of Earth Science*, **6**, 743-57.

Debelmas, J. & Kerchove, C. 1980. Les Alpes franco-italiènnes. *Géologie Alpine*, **56**, 21-58.

Debelmas, J. & Lemoine, M. 1970. The western Alps: Palaeogeography and structure. *Earth Science Reviews*, **6**, 221-56.

Demarcq, G. 1970. Etude stratigraphique du Miocene rhodanien. *Memoires de la Bureau des Récherches Géologiques et Minières, Orléans*, **61**.

De Paor, D. G. 1988. Balanced section in thrust belts. Part 1: construction. *American Association of Petroleum Geologists Bulletin*, **72**, 73-90.

Doudoux, B., Mercier, B. & Tardy, M. 1982. Une interprétation nouvelle de la structure des massifs subalpins savoyards (Alpes occidentales): Nappes de charriage oligocènes et déformations superposées. *Comptes Rendus de l'Academie des Sciences, Paris*, **295**, 63-8.

Elliott, D. 1976. The energy balance and deformation mechanisms of thrust sheets. *Philosophical transactions of the Royal Society of London*, **A283**, 289-312.

Fischer, M. W. & Coward, M. P. 1982. Strains and folds within thrust sheets: an analysis of the Heilam sheet, northwest Scotland. *Tectonophysics*, **88**, 291-312.

Gidon, M. 1964. Nouvelle contribution à l'étude du massif de la Grande Chartreuse et de ses relations avec les régions avoisinantes. *Géologie Alpine*, **40**, 187-205.

—— 1981. Les déformations de la couverture des Alpes occidentales externes dans la région de Grenoble: leurs rapports avec celles du socle. *Comptes Rendus de l'Academie des Sciences, Paris*, **292**, 1057-60.

—— 1988. L'anatomie des zones de chevauchement du massif de la Chartreuse (Chaines subalpines septentrionales, Isere, France). *Géologie Alpine*, **64**, 27-47.

Gignoux, M. & Moret, L. 1944. *Géologie Dauphinoise (ou l'initiation à la géologie par l'étude des environs de Grenoble)*. B. Arthaud, Grenoble, 425p.

Goguel, J. 1948. Le rôle des failles de décrochement dans le massif de la Grande Chartreuse. *Bulletin de la Société Géologique de la France*, **18**, 277-235.

—— 1963. *Tectonics* (English ed.). Freeman, San Fransisco, 384p.

Hancock, P. L. & Barka, A. A. 1987. Kinematic indicators on active normal faults in western Turkey. *Journal of Structural Geology*, **9**, 573-84.

Kligfield, R., Geiser, P. & Geiser, J. 1987. Construction of geologic cross-sections using microcomputer systems. *Geobyte*, **1**, 60-6.

Knipe, R.J. 1985. Footwall geometry and the rheology of thrust sheets. *Journal of Structural Geology*, **7**, 1-10.

Lemoine, M., Bas, T., Arnaud-VanneauU, A., Arnaud, H., Dumont, T., Gidon, M., Bourbob, M., De Graciansky, P. C., Rudkiewsky, J. L., Megard-Galli J. & Tricart, P. 1986. The continental margin of the Mesozoic Tethys in the Western Alps. *Marine and Petroleum Geology*, **3**, 179-99.

Menard, G. 1980. Profondeur du socle antétriassique dans le sud-est de la France. *Comptes Rendus de l'Academie de Sciences, Paris*, **290**, 299-302.

Mugnier, J. L., Arpin, R. & Thouvenot, F. 1987. Coupes equilibrées à travers le massif subalpin de la Chartreuse. *Geodinamica Acta*, **1**, 125-37.

Nicolett, C. 1979. Le Bas Dauphine septentrionale, étude stratigraphique et sédimentologique. Thèse 3ème cycle, University of Grenoble.

Pfiffner, O. A. & Burkhard, M. 1987. Determination of paleo-stress axes orientations from fault, twin and earthquake data. *Annales Tectonicae*, **1**, 48-57.

Ramsay, J. G. 1963. Stratigraphy, structure and metamorphism in the western Alps. *Proceeding of the Geological Association*, **74**, 357-91.

—— 1980. The crack seal mechanism of rock deformation. *Nature*, **284**, 135-9.

Rich, J. 1934. Mechanics of low angle overthrust faulting as illustrated by the Cumberland block, Virginia, Kentucky, Tennessee. *American Association of Petroleum Geologists Bulletin*, **18**, 1584-96.

Rowan, M. G. & Ratliff, R.A. 1988. Use of fault cut-offs and bed travel distance in balanced cross-sections: Discussion 1. *Journal of Structural Geology*, **10**, 311-6.

Siddans, A.W.B. 1983. Finite strain patterns in some Alpine nappes. *Journal of Structural Geology*, **5**, 441-8.

Suppe, J., 1983. Geometry and kinematics of fault bend folding. *American Journal of Science*, **283**, 684-721.

Thouvenot, F. & Menard, G. 1990. Allochthony of the Chartreuse Subalpine massif: explosion-seismology constraints. *Journal of Structural Geology*, **12**, 113-121.

Williams, G. D. & Chapman, T. 1983. Strain developed in the hangingwall of thrusts due to their slip-propagation rate: a dislocation model. *Journal of Structural Geology*, **5**, 563-72.

Kinematics of a transverse zone in the Southern Alps, Italy

Gregor Schönborn

Geological Institute of Basel University, Bernoullistrasse 32, CH - 4056 Basel, Switzerland

Abstract: Transverse zones cut the south-verging thrust belt of the Southern Alps into different segments. Detailed mapping together with computer-aided construction of balanced cross-sections has enabled the kinematic models of the structures east and west of the Ballabio-Barzio transverse zone to be expanded into the complex transverse zone itself. Kinematic analysis reveals three distinct stages of basement and sediment thrusting of 17 km, 25 km and >20 km shortening. The shortening for each stage is the same in all cross-sections. In the Ballabio-Barzio zone, an inherited normal fault caused the thrust of stage 1 to ramp onto the Upper Triassic and the thrust of stage 2 to split into different branches. East of a reactivated sinistral north-south trending tear fault, wedging of the Middle Triassic units led to backthrusting of the Upper Triassic sheet above. The triangular tip of a basement fault-bend fold produced lateral escape on both sides contemporaneous with southward thrusting. Stage 3 deformation was an out-of-sequence thrust, the frontal fault-bend fold of which was hindered by an oblique lineament. The subsequent stages of a thrust breaking through the hinge of this fold, overturning and flattening the cut off frontal limb, are shown by cross-sections through the frontal range at different distances to the oblique lineament.

The Insubric line is a steeply north dipping east-west trending fault zone, separating the Southern Alps from the rest of the Alps (Fig. 1). The Lombardian part of the Southern Alps is a south-verging thrust belt consisting of sediments and basement slivers, segmented by north-south striking transverse zones. Alpine metamorphism is absent to very low grade in the deeper parts and brittle deformation dominates everywhere.

Fundamental work in the Lombardian Alps was done by Dutch geologists (e.g. De Sitter & De Sitter-Koomans 1949). They advocated the concept of vertical basement uplift and gravitational sediment sliding and crustal shortening was not assumed at all until the late 1970s (Gaetani & Jadoul 1979; Castellarin 1978). Publication of seismic lines (Pieri & Groppi 1981) revealed a foothills belt deeply buried under post-kinematic basin fill. Laubscher (1985) constructed balanced cross-sections to estimate the amount of shortening of the exposed part of the Southern Alps, whereas Roeder (1989; in press) linked them to the foothills belt.

This work is part of a Swiss National Science Foundation project aimed at clarifying the kinematic role of the transverse zones in the Southern Alps. As this required knowledge of detailed tectonics not available in current publications, mapping at 1:5000 and 1:10000 of the whole area was carried out. In order to maximize ease and accuracy of cross-section construction, computer-aided balancing with the preservation of bedlength and thickness (GEOSEC - 20™©, GeoLogic Systems, Inc., Kligfield *et al.* 1986) was applied.

Balanced cross-sections through the structures east and west of the Ballabio-Barzio transverse zone (Fig. 2) were constructed involving both detached basement and sediments. These have been published previously (Schönborn 1990) but serve as a starting point for unravelling the kinematics of the transverse zone. Therefore they are presented briefly here.

The kinematic model of the transverse zone introduced here is quantitatively consistent with the development of the Grigna (in the west) and the Valtorta segment (in the east) and accounts for many features that have so far resisted explanation.

Methodology

Unravelling the Lombardian Alps means dealing with at least six stages of brittle deformation: Permian normal faulting expressed as ENE oriented graben and pull-aparts (isopach maps of the Permian graben fill indicate this orientation, De Sitter & De Sitter-Koomans 1949); Liassic normal faulting due to E-W oriented extension; compressional stages 1 to 3 (Cretaceous to Miocene?); a late Miocene compressional stage documented in the foothills below the Po plain. Additionally, the thickness of some Triassic strata changes up to hundreds of metres within short distances. Therefore quantitative kinematic models will hardly ever be accurate, but they are still very useful for understanding complex tectonics, since local observations are often incoherent, ambiguous, apparently contradictory or simply absent.

A lot of local data have been collected, classified into categories of significance and most of them incorporated into a 2-D kinematic model, which then was extended into the third dimension.

Stratigraphy

As summarized in Figure 3, the area investigated involves a Hercynian and older metamorphic basement, Permian clastics

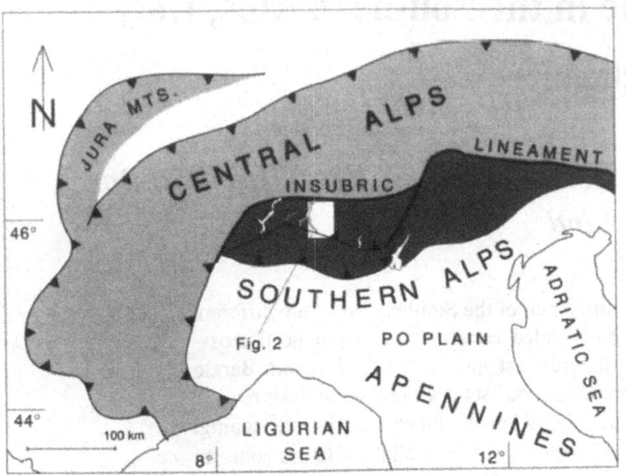

Figure 1. Location of the Southern Alps (dark) and Fig. 2a (rectangle) with respect to the rest of the Alps.

and volcanics, partly as graben fill, beneath the Lower Triassic shales and evaporites. Well bedded Anisian limestones and Ladinian platform carbonates (Esino Formation) form a competent layer (referred to as 'Middle Triassic' in this paper), that is overlain by Carnian carbonates, sandstones, shales and evaporites. The nearly 1.5 km thick layer of Norian Dolomia Principale comprises limestones formed, like the Esino Formation, in internal basins, where shallow-water carbonate sedimentation could not balance the very rapid subsidence. This layer, called here 'Upper Triassic', is capped by lower Rhaetian shales. Distinct thickness changes of the Lower Liassic siliceous limestones (Moltrasio Formation: 0 m on the Mt. Cornizzolo; 700 m on the Albenza ridge more than 1 km further west) record large-scale normal faulting preceding the opening of the Tethys. The condensed silica-rich pelagic sequence of Middle and Late Jurassic age is covered by the Maiolica Formation, a white layer of calcareous nannoplankton. The deposition of hemipelagic shales (Scaglia) during the middle Cretaceous was stopped by flysch sedimentation. These flysche document Late Cretaceous tectonic activity in the Southern or the Austroalpine Alps north of the outcrops along the margin of the Po plain (Bernoulli & Winkler 1990). The more than 5 km thick Mesozoic strata in the area discussed decrease southward (well data, compare Pieri & Groppi 1981). For an overview of the Jurassic stratigraphy see Winterer & Bosellini (1981).

Structural style

In the Triassic strata, classical ramp-flat tectonics developed in relation to the stratigraphy: ramps (22 - 28°) in the two rigid platform carbonates of Middle and Late Triassic age and flats in the decollement horizons of the Lower Triassic, the Carnian and the lower Rhaetian. The detachment climbs through the Jurassic and the lowermost Cretaceous strata (Maiolica Formation), but turns subparallel to the lower to middle Cretaceous (flat in the Marne di Bruntino?). For the construction of cross-sections it is important to emphasize that the vertical projection of structures across one of the important decollement horizons is rather risky and requires lateral information or kinematic assumptions. Because the exposed basement shows brittle deformation (Schönborn & Laubscher 1986) and the geometry of the antiforms corresponds to the formulae of Suppe (1983), the ramp-flat concept was also applied for the construction of the deeper parts of the sections, although here deformation may be more ductile. The calculated basement shortening, however, is in excellent agreement with that calculated for the sediments.

Various quantitative and qualitative criteria were applied to determine the conditions under which thrusting took place along the Orobic thrust. These criteria include quartz deformation, mineral assemblages (prograde and retrograde), style of large-scale deformation with respect to the lithologies, Illite-crystallinity measured on newly formed minerals. These indicate lowest grade metamorphic conditions (high anchizone) for the deformation of the basement front during stage 1.

Inherited faults

The South Alpine thrust belt is divided by transverse zones into compartments of different structures. Probably developing in the Late Triassic to Early Liassic as normal faults preceding the opening of the Tethys, they acted as zones of mechanical weakness and led to the segmentation during subsequent thrusting (e.g. Schumacher 1990; Bertotti 1990; and Bernoulli 1964 for the Lugano line; or Lichtensteiger 1986 for the Lecco line). In the Bergamasc Alps, sedimentological data account only for a narrow, east-west trending strip in the southernmost mountain range. The strike of those faults further north is doubtful.

Concerning the Ballabio-Barzio transverse zone, there is evidence for Late Permian (± ENE-WSW trending faults, near Introbio), Ladinian (±N-S, east of Mt. Grigna-N, Gaetani *et al.* 1986), late Norian (N-S, west of Mt. Sodadura, Jadoul 1985), Early Liassic (N-S, east of the Resegone range, Jadoul & Doniselli 1986) and Toarcian (± N-S, in the Barro - Erve area, Gaetani & Poliani, 1978) tectonic activity.

Marked ramp-flat tectonics and large inherited faults lead to the observed distinct transverse zones. Insights into the kinematic processes in this area may be helpful for understanding transverse zones in general.

Transport directions

All compressional stages are roughly south-verging, indicated by the large E-W trending structures: the Orobic and the Gallinera thrusts active during stage 1, the en-echelon aligned Orobic, Trabuchello and Cedegolo anticlines thrust during stage 2, the frontal range flexure between Como and Lake Iseo folded during stage 3 and the subsequent foothills belt. Other large-scale structures (e.g. the Giudicarie/Trompia system, Laubscher 1990) support a roughly southward transport, whereas small-scale structures adopting local geometries may differ from this general trend.

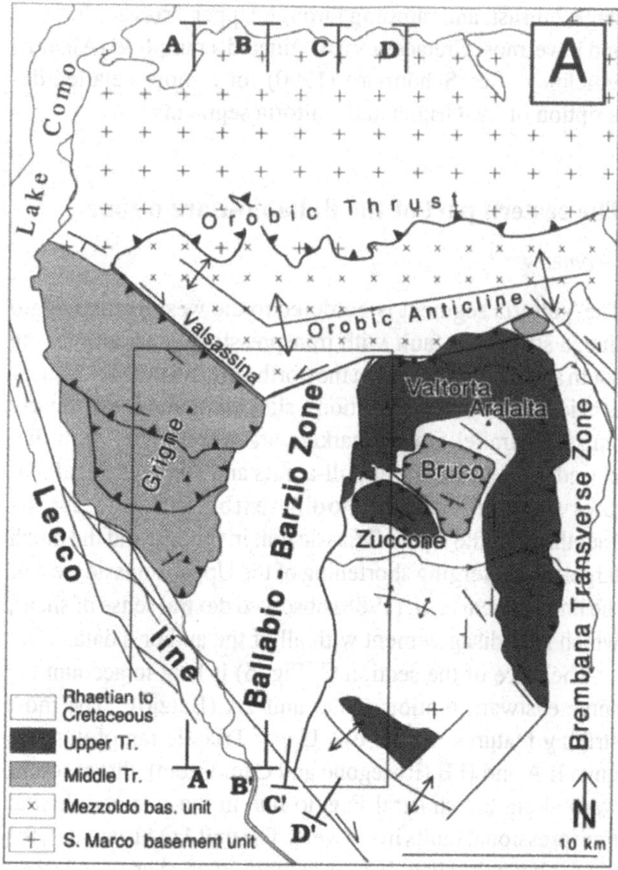

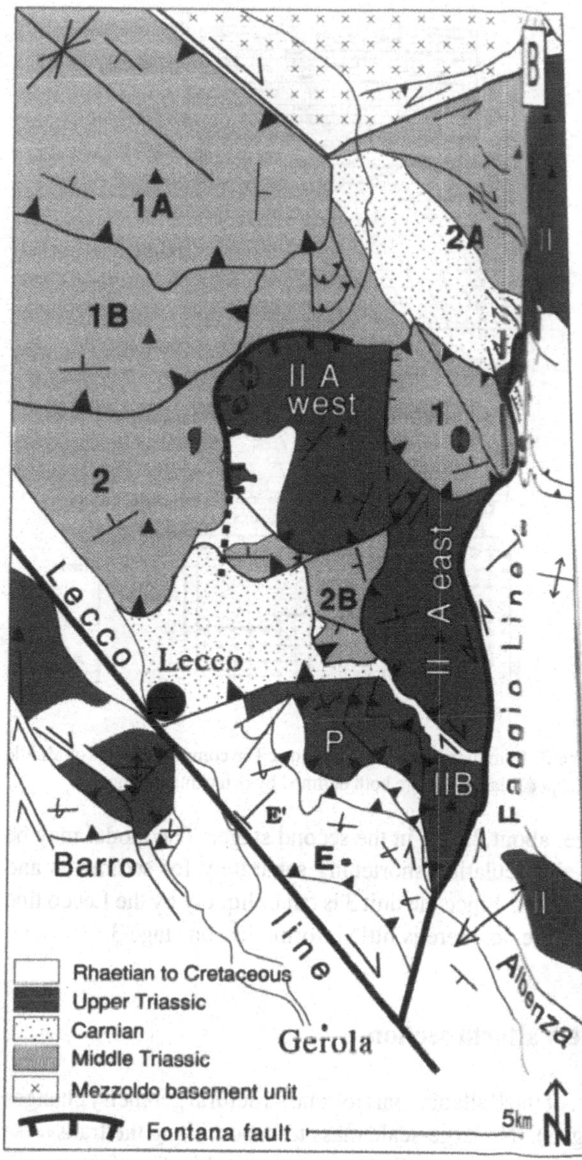

Figure 2. Simplified geological map of the Grigna - Valtorta region. The Grigna and Valtorta segments (Fig. 2a) are separated by the Ballabio-Barzio transverse zone (Fig. 2b). Permian and lowermost Triassic are included in the basement. Location of Figure 2b (rectangle) on Figure 2a. Area of Mt. Barro after Lichtensteiger (1986). Middle Triassic unit 1A = Grigna-N, 1B = Grigna-S, 1 = Muschiada, 2 = Coltignone, 2A = Barzio, 2B = Erna; Upper Triassic unit II = Aralalta-Albenza, II A west = Due Mani, II A east = Resegone, II B = Camozzera, P = Il Pizzo, E = Jurassic of Erve.

The en-echelon aligned basement anticlines are interpreted here as ramp-folds, directed by inherited Permian extensional structures. The rotation and plunge of the Orobic anticline's fold axis near its western termination (Fig. 2a) is in agreement with the model by Snedden & Spang (1990) for decreasing displacement of a fault-bend fold which is laterally replaced by another. Accentuation of the anticlines by dextral transpression, however, cannot be excluded.

The palaeo-stress analysis by Zanchi *et al.* (1988) also manifests southward transport with local deflections in the transverse zone.

The Grigna section

In the Grigna mountains east of Lake Como (Fig. 2), a pile of three thrust sheets of Middle Triassic is well exposed, a fourth one is hidden below the Carnian beds of Lecco. The Grigna segment is the clearest in Lombardy and therefore may serve as a guide for understanding more complex structures. Laubscher's (1985) interpretation is only slightly modified by this work.

The current cross-section is depicted in Figure 4c. At their northern end the Middle Triassic unit 1A (Grigna-N) lies in stratigraphic contact with the basement, as Gaetani (1982) pointed out in a study of a highway tunnel. This is fundamental for the model, because unit 1A has to be the sedimentary cover of the upper basement unit (San Marco unit, Schönborn 1990). This facilitates balancing of the detached basement slivers with respect to the sediments. The geometries suggest a similar relationship between the lower basement unit (Mezzoldo unit) and the Middle Triassic unit 2 (Coltignone).

In a first stage the upper basement unit and first the Middle Triassic unit 1A (Grigna-N), then the unit 1B (Grigna-S) were moved south along the stage 1 thrust (Orobic thrust, Figs 4a & 4b). This thrust was subsequently folded by the stage 2 thrust (Coltignone thrust), which transported the lower basement unit and the Middle Triassic unit 2 (Coltignone) on top of the unit 3 (Lecco). Shortening is about 17 km in the first

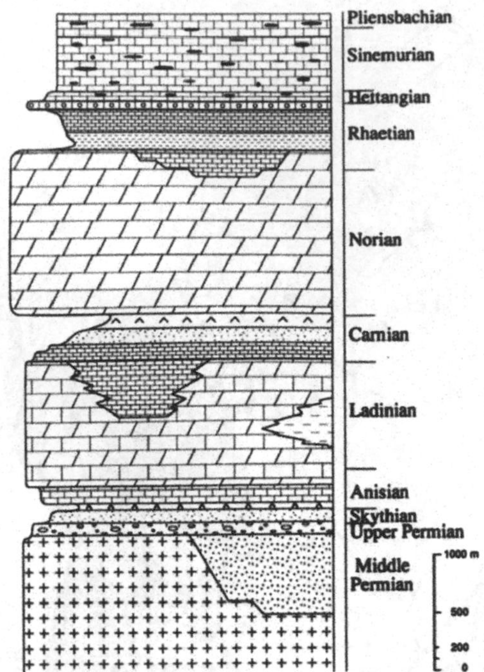

Figure 3. Summary stratigraphic column. The competent layers of Middle and Upper Triassic age are both outlined by detachment horizons.

stage, about 25 km in the second stage. The model may be tested calculating shortening separately for sediments and basement. Since the unit 3 is cut obliquely by the Lecco line in Figure 4c, there is little information on stage 3.

The Valtorta section

East of the Ballabio-Barzio zone structural geometry changes (Fig. 5). No large-scale mass transport along the transverse zone beyond the studied area is observable, therefore overall shortening in the east and the west has to be identical.

During stage 1 the Orobic thrust ramped through both the Middle and Upper Triassic layers, without a flat in the Carnian horizon between. The Middle Triassic unit 1 (Bruco) is an equivalent of 1A and 1B, and also remnants of the Upper Triassic unit above (I, Zuccone unit, not exposed in the Grigna segment) may be observed.

The kinematics of the second stage were complicated by backthrusting of the Upper Triassic sheet II (Aralalta, Figs 5b & 5c). After ramping through basement and Middle Triassic, the latter by two in-sequence thrusts, shortening was split into a south-verging thrust (12 km), some internal deformation of the Upper Triassic and a backthrust (about 5.5 km). Therefore the units 2A and 2B beneath the Upper Triassic II formed a wedge. In order to accommodate backthrusting of the unit II, the Middle Triassic unit 1 (Bruco) above is dissected in the model by an out-of-sequence thrust, creating the unit 1' ('Superbruco').

The backthrust was concealed by the Valtorta fault. Along this sinistral strike-slip fault eastward motion of the Valtorta segment took place.

The out-of-sequence thrust active during stage 3 cut the stage 2 thrust, and ramping through Upper Triassic, Jurassic and lowermost Cretaceous it originated a rampfold (Albenza anticline). See Schönborn (1990) for a more detailed description of the Grigna and Valtorta segments.

The eastern part of the Ballabio-Barzio zone

Geometry

The Valtorta segment is bordered to the west by the Faggio line, a strike-slip fault with transpressional character in the south and transtensional in the north. Sinistral sense of shear is evidenced by some striations, slickolites and steps, clearly sinistral parallel faults (markers are more than 100 m displaced), the geometry of pull-aparts and compressional features where the line splits into different branches (Fig. 2b), the backthrust of the Upper Triassic unit in the east and the much larger south-verging shortening of the Upper Triassic west of the line. Zanchi *et al.* (1988) observed dextral sense of shear, which is in disagreement with all of the author's data.

The trace of the section C (Fig. 6) is bent to account for some eastward motion of the unit 2A (Barzio). The most striking features are the two Upper Triassic rampfolds, the units II A and II B (Resegone and Camozzera), thrust southward along the sinistral Faggio line in the east and dextral transpressional faults in the west. The unit 1 (Muschiada), the block above the sheet II A, is composed of a very condensed Middle Triassic sequence and topped at Mt. Muschiada itself by a small remnant of the Upper Triassic sheet above (unit I). The cross-section (Fig. 6) is simplified for better clarity: the Middle Triassic is too thick.

The SW dipping Middle Triassic unit 2A is separated from the lower basement unit by the Valtorta fault, which has a marked vertical component of displacement there. SE trending dextral strike-slip faults in this unit, partly in the form of flower structures (e.g. east of the village of Barzio) may be interpreted as former continuations of the Valsassina line, but cut off and transported eastward by the contemporaneously active Valtorta fault (Fig. 2b).

Kinematics

Figure 6a shows the restored version of the section C. The Upper Triassic units II A and II B are interpreted as being equivalent to the unit II (Aralalta-Albenza) from further east. The link is enabled by correlating the Albenza - Upper Triassic (southern part of II) with the Camozzera - Upper Triassic (II B): the overlying Rhaetian to Cretaceous strata of both units are continous (compare Figs 10 a & b). Unit II A is the northern continuation of II B, and restoring the different Upper Triassic units levels up exactly the northern terminations of the units II A and II B in the west and II in the east (compare Fig. 9c and the Upper Triassic ramp on Fig. 9b). The Middle Triassic Muschiada unit (1 on section C) is explained as lateral continuation of the Bruco unit (1 on section D) because both are thrust onto the Upper Triassic level II (the correlation is covered up by subsequent rotation and backthrusting of the eastern segment). Lying on top of

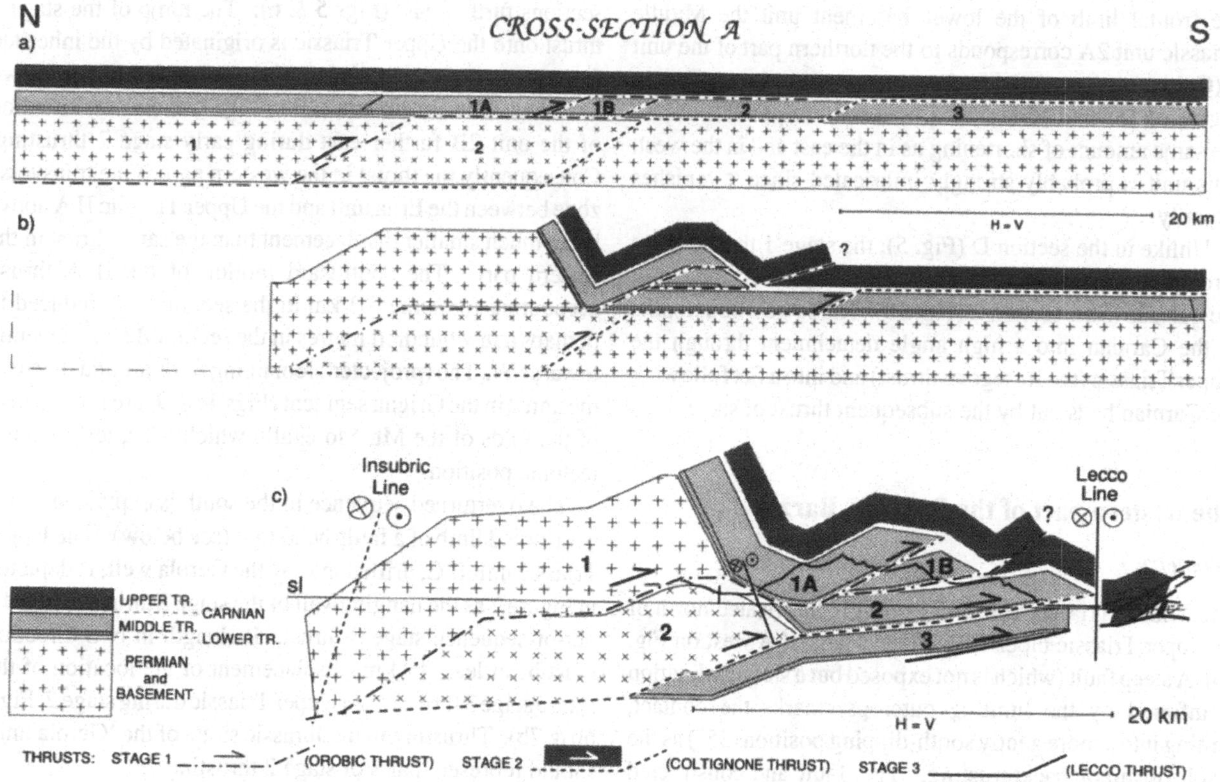

Figure 4. Cross-section A. A kinematic model of the Grigna mountains, modified after Schönborn (1990). In a first stage the upper basement unit, the Middle Triassic unit 1A and then 1B were transported along the Orobic thrust. In stage 2 the lower basement unit (Orobic anticline) and the Middle Triassic unit 2 were thrust on top of the unit 3. Shortening of the first stage is 17 km, of the second stage 25.5 km. Basement unit 1 = San Marco, 2 = Mezzoldo; Middle Triassic unit 1A = Grigna-N, 1B = Grigna-S, 2 = Coltignone, 3 = Lecco.

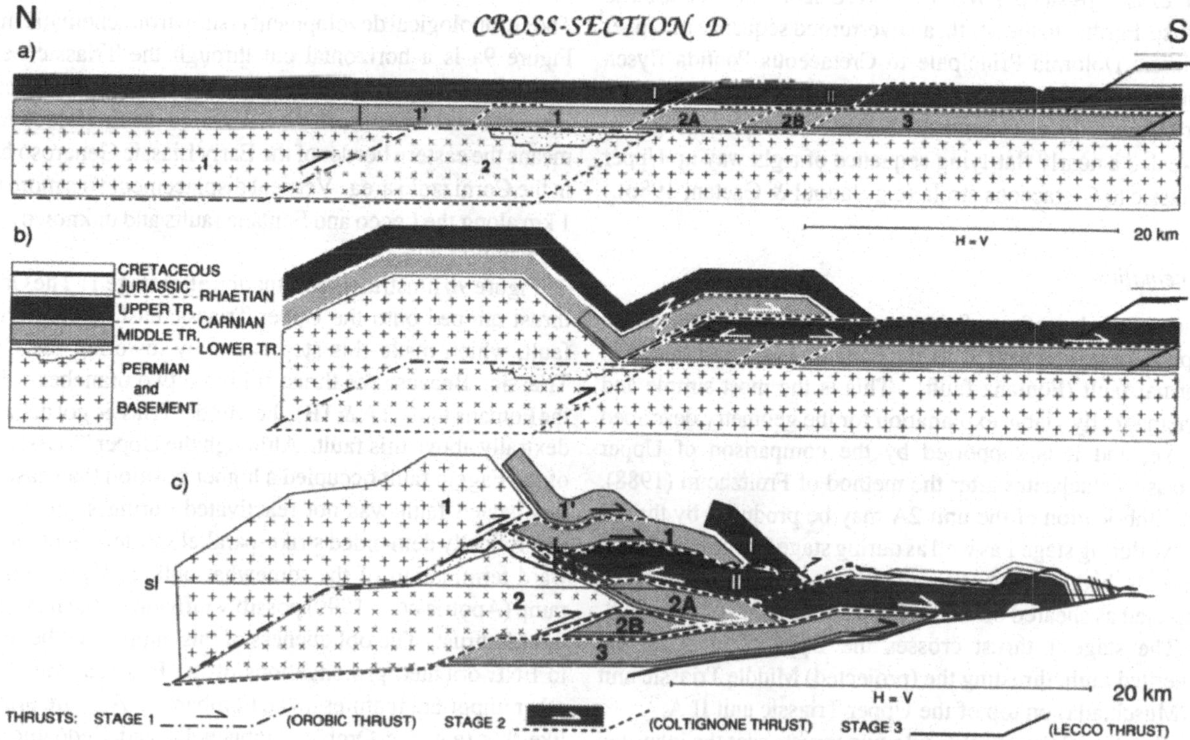

Figure 5. Cross-section D. A kinematic model of the Valtorta segment, modified after Schönborn (1990). Dashed lines are thrusts active in the subsequent stages. (a) Palinspastic configuration before thrusting. (b) Situation after stage 1. Arrows indicate the relative movements of the next stage. The Upper Triassic sheet II is thrust southward as well as northward. (c) Current cross-section after stage 3. The units 2A and 2B form a wedge below the backthrust Upper Triassic unit II. Shortening as in Figure 4. Basement unit 1 = San Marco, 2 = Mezzoldo; Middle Triassic unit 1 = Bruco, 1' = 'Superbruco', 2A = Barzio, 2B = 'Infrabarzio', 3 = Lecco; Upper Triassic unit I = Zuccone, II = Aralalta-Albenza, II C = 'Gerola'.

the frontal limb of the lower basement unit the Middle Triassic unit 2A corresponds to the northern part of the unit 2 (Coltignone), occupying the same position in the Grigna segment. The unit 2B is not exposed but is required to ensure the same amount of shortening as in the east and in the west. This unit is probably strongly imbricated since it vanishes laterally.

Unlike in the section D (Fig. 5), the stage 1 thrust ramps through the Upper Triassic after a low angle detachment in the Carnian in the section C (Fig. 6b). The stage 2 thrust splits in the Carnian into a high angle detachment through the Upper Triassic (the Resegone thrust) and into a flat following the Carnian beds cut by the subsequent thrust of stage 3.

The western part of the Ballabio-Barzio zone

Geometry

The Middle Triassic unit 2A is strongly imbricated north of an Upper Triassic block called Due Mani (II A west on Fig. 2b). A steep fault (which is not exposed but a steep inclination is inferred by the limiting outcrops) marks the contact, turning into a more gently south-dipping position (35°) as the result of minor backthrusting. This fault and constructed cross-sections (the base of the Upper Triassic is exposed further south) point out that this block does not lie on top of Middle Triassic 2A, but more or less on the same level. It was thrust southward onto two Middle Triassic blocks that are connected to the sheet 2 further west. The eastern one, called Mt. Erna, dips steeply WSW - SSW, discordant with its basal thrust. Further to the south, an overturned sequence of Upper Triassic Dolomia Principale to Cretaceous Pontida flysch with many south to west-verging thrusts is one of the major problems of local tectonics. A well near Gerola (Fig. 2b) revealed a nearly flat-lying sequence of right way up Upper Triassic to Cretaceous strata (e.g. Jadoul & Gaetani 1986).

Kinematics

The restored version of the model (Fig. 7a) interprets the Upper Triassic unit II A as the hangingwall of an inherited normal fault (Fontana fault). This is the most simple and kinematically viable explanation for the geometry described above, and it is supported by the comparison of Upper Triassic cataclasites after the method of Froitzheim (1988). The imbrication of the unit 2A may be produced by thrusts active during stage 1 as well as during stage 2. The Erna block with its bedding cut discordantly by the basal thrust is depicted as sheared off and rotated frontal limb of the unit 2.

The stage 1 thrust crosses the Upper Triassic at the inherited fault, thrusting the (projected) Middle Triassic unit 1 (Muschiada) on top of the Upper Triassic unit II A.

The stage 2 thrust splits into two thrusts near the inherited fault: one going on horizontally and then ramping through the Middle Triassic below the unit 2B, the other leading up immediately and thrusting Upper Triassic II A onto II B (Fig. 7b).

This section reveals the reason for the evolution of the sections further east (Figs 5 & 6): The ramp of the stage 1 thrust onto the Upper Triassic is originated by the inherited fault, as well as the splitting of the stage 2 thrust into two branches. The Erna block (2B of Fig. 7) shows the situation of the unit 2B further east during early stage 2 thrusting. Consequently the thrust in the western part of the transverse zone between the Erna unit and the Upper Triassic II A above had a much smaller displacement than the same thrust in the eastern part. The southward motion of the II A thrust, accounting for some 3.3 km in the section C, is reduced to perhaps a few hundred metres in the section B and vanishing towards W. The (projected) remaining folds around the tip of the unit 2 in the Grigna segment (Figs 4c & 9) are reminiscent of the folds of the Mt. Moregallo which occupies a similar tectonic position.

The overturned sequence in the south is explained as the overturned limb of a fault-bend fold (see below). The Upper Triassic unit II C, drilled into by the Gerola well, is depicted in Figure 7 as the hangingwall of the stage 2 thrust, cut by the out-of-sequence stage 3 thrust. Its length in N-S direction must be at least 12 km (displacement of the location of the future stage 3 thrust in the Upper Triassic during stage 2, Figs. 6b & 7b). Thrusting in the Jurassic strata of the 'Gerola unit' should represent parts of stage 2 thrusting.

The cross-sections A to D are summarized in a block diagram (Fig. 8) to clarify the 3-D geometries.

Chronological development

The chronological development is shown on schematic maps. Figure 9a is a horizontal cut through the Triassic before thrusting: a ridge of Middle Triassic strata was bordered by inherited normal faults to the SW and to the E. 'Lecco fault' means the eastern border of the Early Liassic Generoso basin in the Cornizzolo area. Vertical displacement was more than 1 km along the Lecco and Fontana faults and unknown along the Faggio fault.

Figure 9b illustrates the situation after stage 1: The Orobic thrust ramped onto the Upper Triassic east of the Fontana fault, which made this step easier by lowering the Upper Triassic. Because the thrust split into two branches west of the Fontana fault (1A & 1B), the Middle Triassic got displaced dextrally above this fault. Although the Upper Triassic west of the Faggio fault occupied a higher position than east of it, this healed fault was not reactivated during stage 1. The theoretically demanded strike-parallel shortening on the forward termination of the somewhat oblique Upper Triassic ramp (Apotria *et al.* 1990) possibly influenced the forming of the 1A thrust. The obliqueness of this ramp could be related to ENE oriented prolongations of the Fontana fault. Also other important features in the Lombardian Alps are oriented like this (e.g. the Orobic, Trabuchello and Cedegolo anticlines, the Clusone fault and the unnamed one northeast of San Pellegrino). Liassic normal faulting following an inherited Permian fault system and influencing Alpine thrusting could be an explanation for this phenomenon.

Figure 9c elucidates stage 2: The triangular shape of the

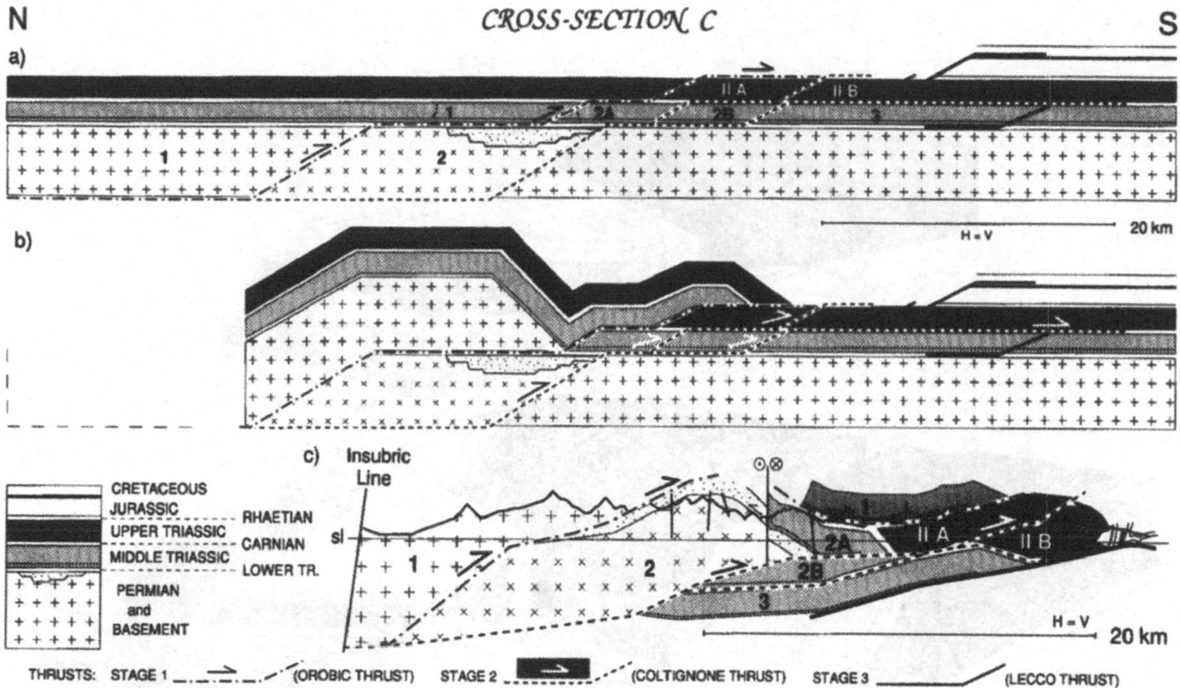

Figure 6. Cross-section C. A kinematic model of the eastern part of the Ballabio-Barzio transverse zone. (**a**) Restored section. (**b**) Situation after stage 1. Arrows indicate movements of stage 2. (**c**) Current cross-section. Minor backthrusting of unit 2A is located between the Permian and the basement. Some eastward motion and extension along strike of unit 2A has to be added. Unit 1 is too thick and the Upper Triassic unit I above not depicted, neither is unit II C in Figure 6c. Shortening as in Figure 4. Basement unit 1 = San Marco, 2 = Mezzoldo; Middle Triassic unit 1 = Muschiada, 2A = Barzio, 2B = 'Infrabarzio', 3 = Lecco; Upper Triassic unit II A = Resegone, II B = Camozzera, II C = 'Gerola'.

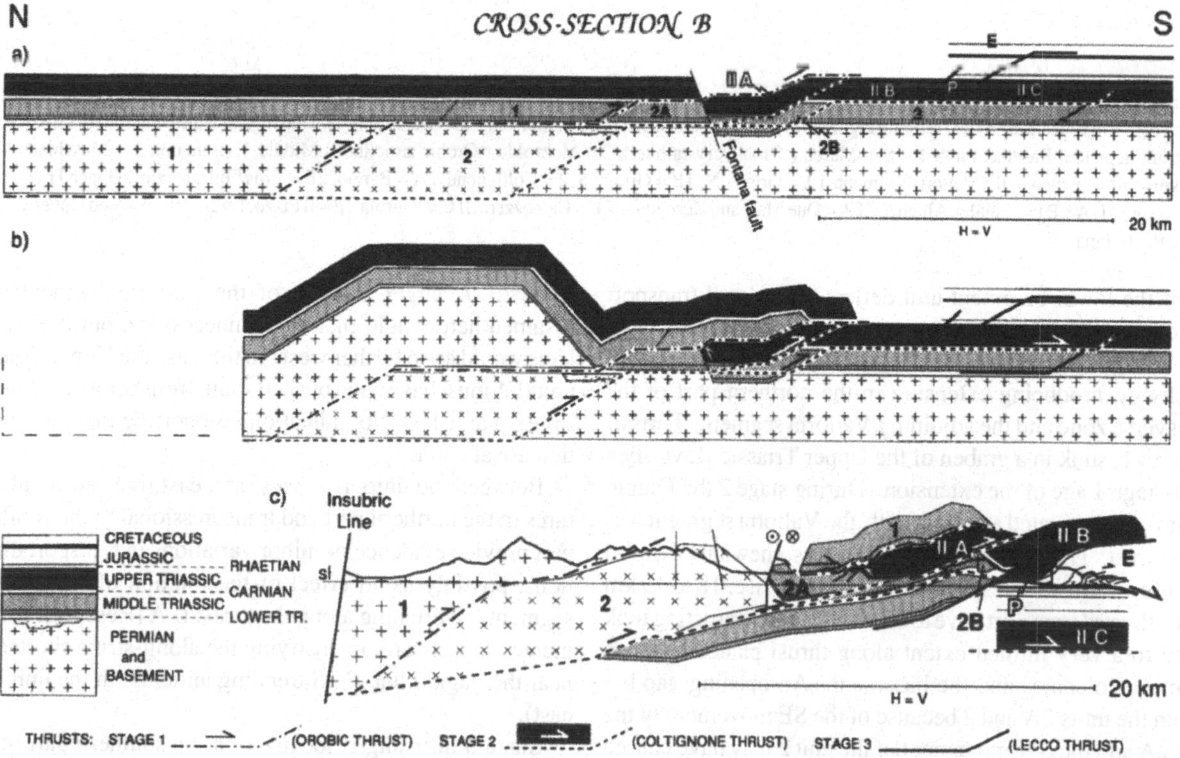

Figure 7. Cross-section B. A kinematic model of the western part of the Ballabio-Barzio transverse zone. The structures further east (Figs 5 & 6) were originated at the inherited Fontana normal fault: the ramp of the stage 1-thrust onto the Upper Triassic, the splitting of the stage 2-thrust and internal thrusting of the Upper Triassic during stage 2. The unit II C represents the hangingwall of the stage 2 thrust, cut by the out-of-sequence stage 3-thrust. Shortening as in Figure 4. Basement unit 1 = San Marco unit, 2 = Mezzoldo; Middle Triassic unit 1 = Muschiada, 2A = Barzio, 2B = 'Infrabarzio', 3 = Lecco; Upper Triassic unit II A = Due Mani, II B = Camozzera, II C = 'Gerola'. P = Upper Triassic of Il Pizzo, E = Jurassic of Erve.

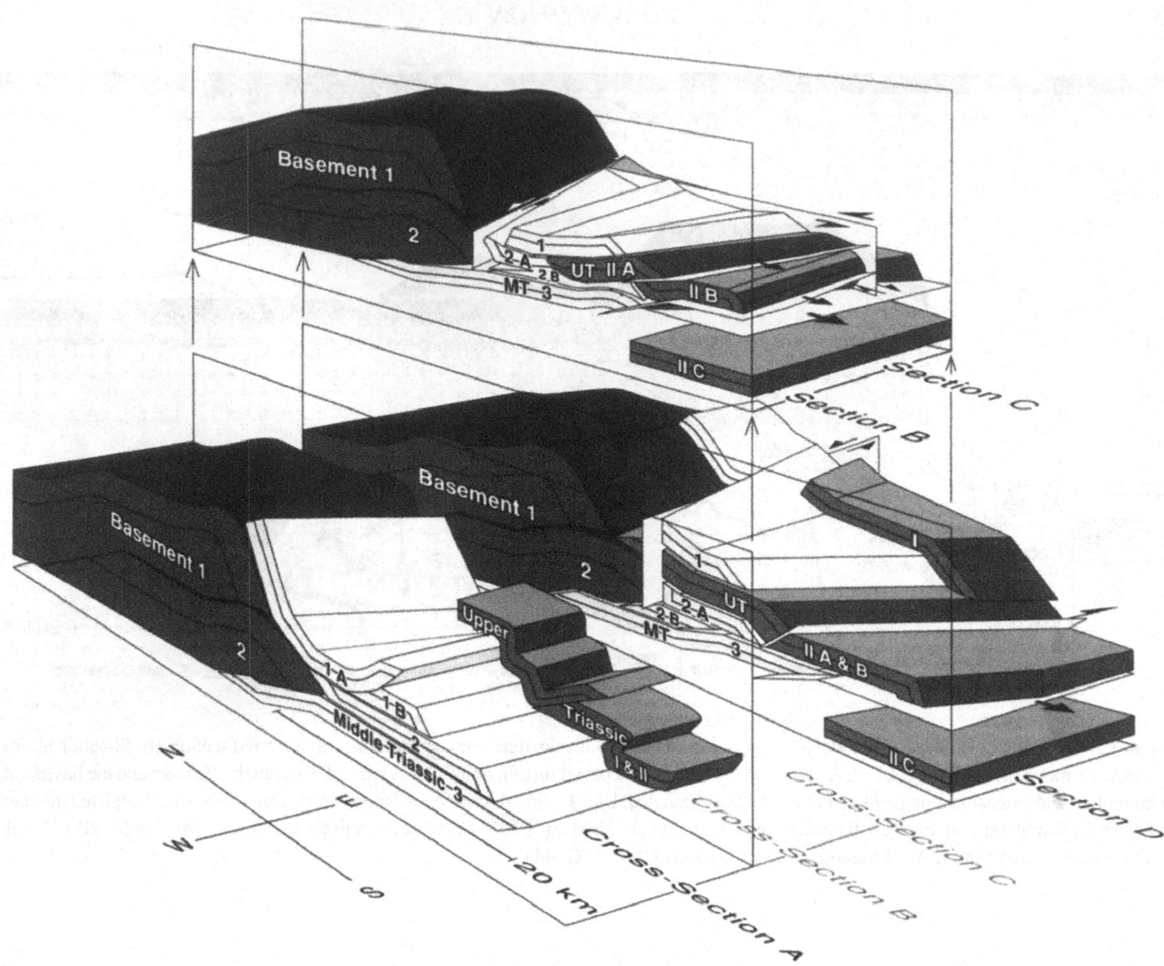

Figure 8. Simplified block diagram of the Grigna segment to the left, the Valtorta segment to the right and the uplifted Ballabio-Barzio transverse zone in between. Cross-sections A to D refer to Figures 4, 7, 6 & 5. The Upper Triassic unit I in the transverse zone and the frontal fault-bend fold of cross-section A are not depicted. Basement unit 1 = San Marco ('Orobic crystalline'), 2 = Mezzoldo ('Orobic anticline'); Middle Triassic unit 1 = Muschiada in the transverse zone, Bruco in the Valtorta segment, 1A = Grigna-N, 1B = Grigna-S, 2 = Coltignone, 2A = Barzio, 2B = Erna/'Infrabarzio'; Upper Triassic unit I = Zuccone, II (A&B) = Aralalta-Albenza, II A = Due Mani and Resegone, II B = Camozzera, II C = 'Gerola', P = Il Pizzo, I & II = not exposed Upper Triassic above the Grigna.

tip of the lower basement unit deflected the local transport directions. The Grigna segment was compressed towards SW (Grigna syncline), the Valtorta segment was rotated clockwise producing extension in the northern part of the transverse zone and the adjoining Valtorta segment. Parts of the unit 1, sunk in a graben of the Upper Triassic II, verify a post-stage 1 age of the extension. During stage 2 the Faggio fault was reactivated as a tear fault: the Valtorta segment was backthrust. The unit 2A was transported somewhat eastward and northward with respect to the basement (i. e. 18 to 19 km of southward transport have to be added). Backthrusting took place to a very limited extent along thrust planes between Permian volcanics and the basement. An opening gap between the units 2A and 2 because of the SE movement of the unit 2A and the SW movement of the unit 2 may have caused the eastward dip of the unit 1A (above) near the transverse zone. The swinging of the front of the unit 2A (shaded barbs) below the Upper Triassic indicates the area of existence of the unit 2B below. This probably strongly imbricated unit is necessary to ensure the shortening of the Middle Triassic level 2. A larger rotation of the Valtorta segment than assumed here would make 2B unnecessary, but then problems would arise further west: in this case the Upper Triassic unit II A must have been pushed south from below the Middle Triassic unit 1A. No indications supporting this interpretation are at hand.

Between the units II A west and east transtensional features in the northern part and transpressional in the southern part provide evidence of minor variations in transport directions, possibly as an effect of the rotation of the Valtorta segment. Shortening along the thrust II A (Resegone thrust) increased eastward, intensifying the along-strike shortening near the Faggio line (N-S trending anticline in the unit II A east).

Stage 3 thrusting is localized in the southern part of the map. Here the influence of a transverse zone on thrusting can be observed directly. Figure 10 exhibits the frontal parts of the discussed sections. Further away from the oblique Lecco lineament a fault-bend fold (or broken through fault-propagation fold, which can hardly be distinguished after some

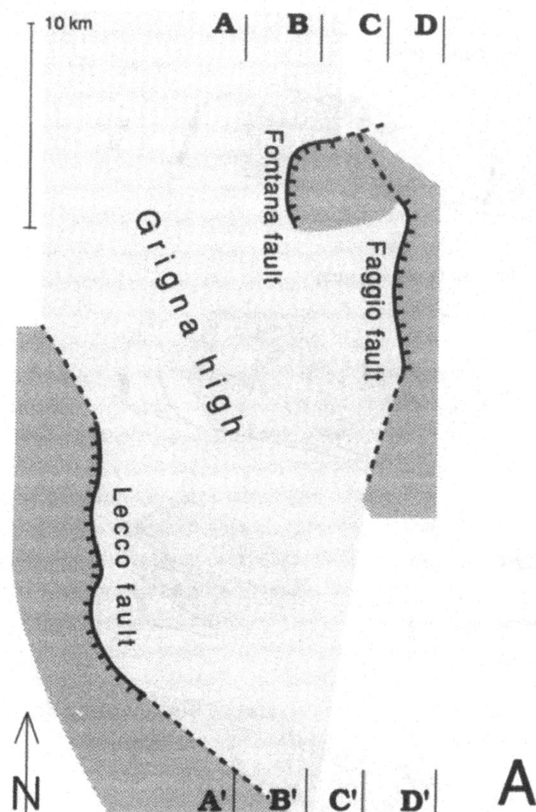

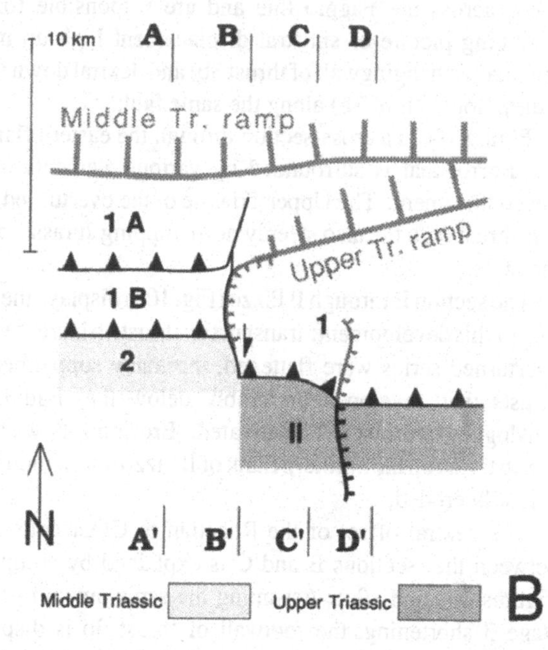

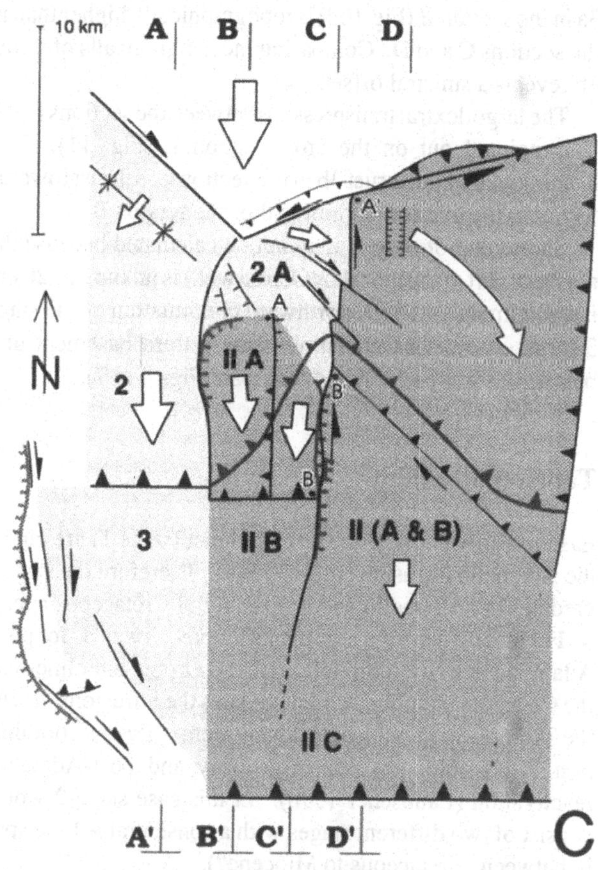

Figure 9. Schematic chronological development of the stages 1 and 2 on a map.
(a) Horizontal cut through the Triassic before thrusting. A ridge of Middle Triassic strata was bordered by Liassic normal faults bringing down the Upper Triassic. 'Lecco fault' means the eastern margin of the Generoso basin in the Cornizzolo area (simplified after Lichtensteiger 1986). Cross-sections A to D refer to the Figures 4a, 7a, 6a & 5a.
(b) Situation after stage 1; shaded tectonic features are overlain by the Middle Triassic of the hangingwall (white). The oblique Upper Triassic ramp originated near the inherited Fontana fault.
(c) Tectonics of the second stage. The triangular tip of the basement caused lateral escape to the SE and to the SW, contemporaneous to the much larger southward transport. The clockwise rotation of the Valtorta segment (right side) produced extension in the northern part of the transverse zone and along strike compression in the Brembo valley to the east. The Upper Triassic units II A were thrust south along the Faggio line, which was reactivated as a sinistral tear fault. Wedging of the unit 2A and the underlying unit 2B is manifested by backthrusting of the Upper Triassic unit II (A & B) and to a smaller extent of the unit 2A. Shaded barbs indicate the front of the unit 2A below the Upper Triassic. The points A and A', B and B' were neighbours before stage 2 thrusting. The passively transported and rotated units of the first stage are not drafted. During the third stage the units 3, II B and II (A & B) were thrust onto unit II C (compare Fig. 8). Middle Triassic unit 1 = Muschiada - Bruco, 1A = Grigna-N, 1B = Grigna-S, 2 = Coltignone, 2A = Barzio, 3 = Lecco; Upper Triassic unit II (A & B) = Aralalta - Albenza, II A = Resegone and Due Mani, II B = Camozzera, II C = 'Gerola'.

transport) with a steep southern limb is the sign of a ramp through the Upper Triassic to lowermost Cretaceous strata (Fig. 10a). The unexposed stage 3 thrust (Lecco thrust) turned into a flat in the lower to middle Cretaceous, as it is indicated by the fold style and the thrusts further south. West of the Lecco lineament (e.g. near the Mt. Cornizzolo) structures look similar to Figure10a.

In the section C through the Mt. Camozzera (Fig. 10b) transport along the thrust below the rampfold, 3a, was stopped by the Lecco lineament and another thrust, 3b, broke through the hinge of the fold, slightly overturning the frontal limb as shown by the steeply north dipping Jurassic beds. Minor backthrusts are demanded by the geometry and can actually be observed in the field. The changes between 10a and 10b occur across the Faggio line and are responsible for the intriguing picture of sinistral displacement high up in the mountains (hangingwall of thrust 3b) and dextral down in the valley (footwall of 3b) along the same fault.

Figure 10c is a cross-section through the eastern flank of Mt. Barro, that is surrounded by various elements of the Lecco lineament. The Upper Triassic of the overturned limb was thrust onto the also steeply north dipping Jurassic along thrust 3c.

The section B through Il Pizzo (Fig. 10d) displays the final step of this development: transport on thrust 3b increased, the overturned series were flattened, the many supplementary thrusts that developed preferably below the rigid layers (analogues to thrust 3c) inactivated. Erosion follows about the 3b thrust on the northern flank of Il Pizzo, the hangingwall is totally eroded.

The dextral offset of the Rhaetian to Cretaceous strata between the sections B and C is explained by comparing Figures 10b and 10d: Assuming the same amount of total stage 3 shortening, the footwall of thrust 3b is displaced dextrally as shortening along thrust 3b in Figure 10b is larger than along thrust 3a which is further east. Additionally thrust 3a in the section B (Fig. 10d) is topographically higher than in the sections C and D. Comparing the hangingwalls of thrust 3b reveals a sinistral offset.

The large dextral transpression between the sections B and C is pointed out on the cross-section E (Fig. 11). The hangingwall of the thrust 3b in the section C is thrust towards SW onto the overturned footwall of the section B.

Shortening of stage 3 may not be calculated because the northern part of the unit II C (Gerola well) is unknown. 20 km is a minimum value. A northward continuation of the stage 3 thrust into the basement, creating a third basement unit below may not be excluded (compare Figs 4 to 7).

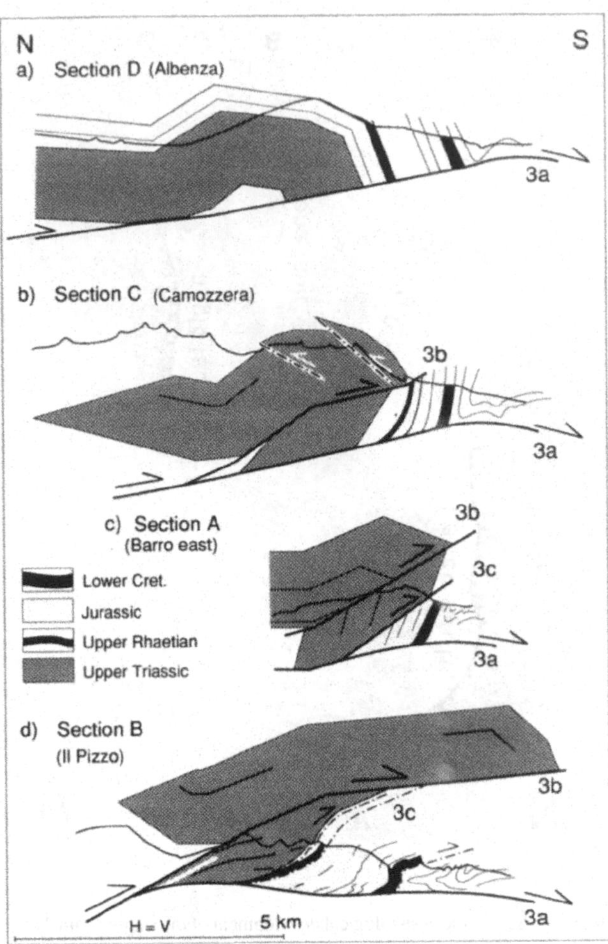

Figure 10. Development of the fault-bend fold of stage 3. (**a**) The rampfold above the stage 3 thrust. West of the Lecco lineament structures are similar. (**b**) Because displacement along the first thrust surface (3a) was hindered by the nearer Lecco lineament, thrust 3b broke through the hinge of the fold and overturned its footwall. (**c**) Another thrust (3c) transported the Upper Triassic of the overturned limb onto the Jurassic of the footwall. Modified after Lichtensteiger (1986). (**d**) Analogous thrusts to the 3c thrust developed beneath the competent layers and were cut by the 3b thrust, along which further transport flattened out the overturned limb. If total shortening of stage 3 remained the same, the footwall of thrust 3b was displaced dextrally from Figures 10 a - d, whereas the hangingwall was displaced sinistrally. These features can be observed in the field. Minor parallel faults of the Lecco line are not shown.

Time relationships

Eastern equivalents of the Orobic thrust (stage 1) are cut by the Adamello intrusions (43 - 30 Ma). Therefore the stage 1 predates the Adamello and is possibly of Cretaceous age.

For stage 2 there are arguments for pre- as well as for post-Adamello activity (correlation of the Orobic anticline with the Cedegolo anticline, which predates the intrusions or with the Trompia high, which postdates them). Even a combination is possible: pre-Adamello array and post-Adamello reactivation (Laubscher 1990). In this case stage 2 would consist of two different stages with a considerable time span in between (Cretaceous to Miocene?).

The stage 3 thrust can be traced westward beneath the rampfold of the frontal range of Mt. Cornizzolo and further west. A possible link to the Miocene tectonics in the southern Ticino (Bernoulli *et al.* 1989) may only be assumed but not proved at the moment.

The Late Miocene thrusts below the Po plain (Pieri & Groppi 1981) have to be projected below the discussed structures. Roeder (in press) interprets them as footwall imbricates of a main thrust corresponding to a combination of the stage 2 and 3 thrusts of this paper.

Conclusions

The transverse zone was brought into existence at an inherited normal fault that gave rise for the stage 1 thrust to ramp onto the Upper Triassic.

The expression of stage 2 tectonics in the basement are en-echelon aligned rampfolds. The triangular tip of one of them produced lateral escape to the SW and to the SE, contemporaneous to a much larger southward transport. The stage 2 thrust split into two branches near the mentioned inherited fault, imbricating the Middle Triassic to the east. Another inherited normal fault, oriented N-S, was reactivated as a tear

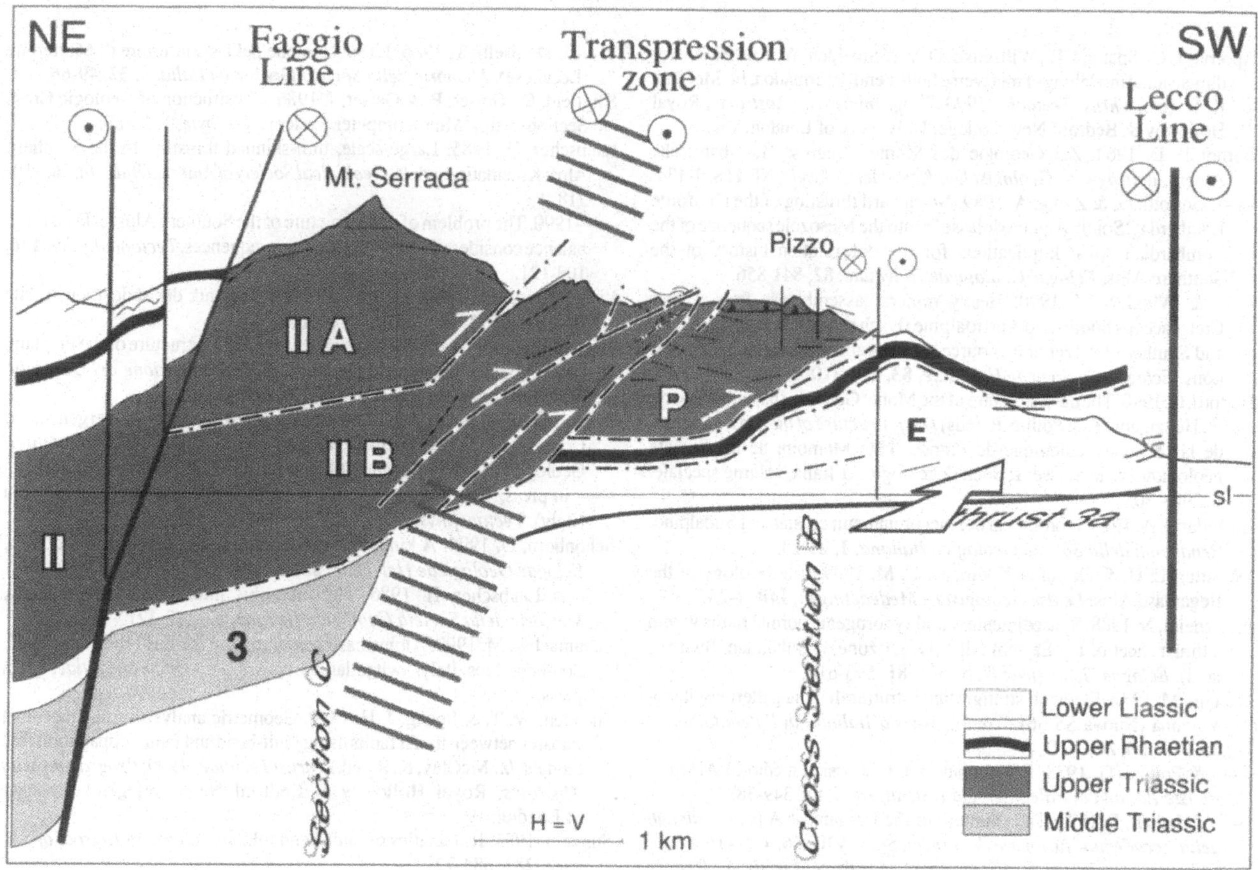

Figure 11. Cross-section E. SW trending cross-section in the southern part of the transverse zone. The hangingwall of the thrust 3b in Figure 10b (II A & II B) was thrust onto the overturned footwall of this thrust in Figure 10d (P & E). The corresponding hangingwall is eroded on the southwestern side. Parallel faults of the dextral Lecco line intersect the overturned strata. Upper Triassic unit II = Aralalta-Albenza, II A = Resegone, II B = Camozzera, P = Pizzo; E = Jurassic of Erve.

fault and the area further east backthrust.

The fault-bend fold of the stage 3 thrust was hindered by an oblique lineament (Lecco line). A thrust broke through the hinge of the fold and overturned the frontal limb. The out-of-sequence stage 3 thrust cut off the hangingwall of the stage 2 thrust now lying below.

Shortening is some 17 km during the first, 25 km during the second and at least 20 km during the third stage.

This transverse zones illustrates the severe influence of oblique inherited faults on subsequent thrusting: they may provoke additional ramps (Fontana fault) or more complex thrust systems, trying to overcome the obstacle (Lecco fault). Additionally all sorts of oblique and lateral ramps may originate near such faults, depending on fault geometry and stratigraphy. Healed inherited faults oriented parallel to the transport direction may be not reactivated until deflection of the local S_1-pattern produces favourable conditions (Mohr criterion) for reactivation as strike-slip or transpressional faults. Every lateral movement, although small compared to the main transport direction causes along strike compression and extension, which in turn has the same effect as inherited unevenness. In this way an inherited fault or other unevenness is able to propagate its consequences.

In the case of the Ballabio-Barzio transverse zone the oblique orientation of Permian graben with respect to subsequent thrusting led to en-echelon aligned rampfolds, which in turn produced lateral escape, which had the effect of (1) S_1-deflection and in this way enabled reactivation of the Faggio fault as a tear fault and therefore backthrusting of the Valtorta segment and (2) lateral extension in the northern part of the transverse zone, along strike compression in the Brembo valley and possibly in the southern part of the transverse zone.

This work benefited from discussions and field trips with H. Laubscher, M. Schumacher, Th. Noack and many others. Th. Widmer is thanked especially for offering his field data. The paper has been improved by the advice of R. Zoetemeijer, W. Sassi and an anonymous reviewer as well as by the editorial help of K. McClay. Support by the Swiss National Foundation, grant numbers 20 - 4798.85 and 20 - 25560.88 is gratefully acknowledged.

References

Apotria, T. G., Spang, J. H., Wiltschko, D. V. & Snedden, W. T. 1990. Three-dimensional modelling of transverse fault-bend deformation. *In:* McClay, K. R. (ed.) *Thrust Tectonics 1990, Programme with Abstracts*, Royal Holloway & Bedford New College, University of London, 18.

Bernoulli, D. 1964. Zur Geologie des Monte Generoso (Lombardische Alpen). *Beiträge zur Geologischen Karte der Schweiz*, NF **118**, 1-134.

——, Bertotti, G. & Zingg, A. 1989. Northward thrusting of the Gonfolite Lombarda ('South-Alpine Molasse') onto the Mesozoic sequence of the Lombardian Alps: Implications for the deformation history of the Southern Alps. *Eclogae Geologicae Helvetiae*, **82**, 841-856.

—— & Winkler, W. 1990. Heavy mineral assemblages from Upper Cretacaceous South- and Austroalpine flysch sequences (Northern Italy and Southern Switzerland): source terranes and paleotectonic implications. *Eclogae Geologicae Helvetiae*, **83**, 287-310.

Bertotti, G. 1990. The deep structure of the Monte Generoso basin. *In:* Roure, F., Heitzmann, P. & Polino, R. (eds) *Deep structure of the Alps*. Memoire de la Société géologique de France, **156**; Memoire de la Société géologique de la Suisse, **1**; Società geologica d'Italia, volume speciale, **1**, 290 - 303.

Castellarin, A. 1978. Il problema dei raccorciamenti crostali nel Sudalpino. *Rendiconti della Società Geologica Italiana*, **1**, 21-23.

De Sitter, L. U. & De Sitter-Koomans, C. M. 1949. The Geology of the Begamasc Alps. *Leidse Geologische Mededelingen*, **14B**, 1-257.

Froitzheim, N. 1988. Synsedimentary and synorogenic normal faults within a thrust sheet of the Eastern Alps (Ortler zone, Garubünden, Switzerland). *Eclogae Geologicae Helvetiae*, **81**, 593-610.

Gaetani, M. 1982. Elementi stratigrafici e strutturali della galleria Bellano-Varenna (Nuova SS 36) (Como). *Rivista Italiana di Paleontologia e Stratigrafia*, **88**, 1-10.

—— & Poliani, G. 1978. Il Toarciano e il Giurassico medio in Albenza. *Rivista Italiana di Paleontologia e Stratigrafia*, **84**, 349-382.

—— & Jadoul, F. 1979. The structure of the Bergamasc Alps. *Rendiconti della Accademia Nazionale dei Lincei*, Serie VIII, **66**, 411-416.

—— Gianotti, R., Jadoul, F., Ciarapica, G., Cirilli, S., Lualdi, A., Passeri, L., Pellegrini, M. & Tannoia, G. 1986. Carbonifero superiore, Permiano e Triassico nell'area lariana. *Memorie della Società Geologica Italiana*, **32**, 5-48.

Jadoul, F. 1985. Stratigrafia e paleografia del Norico nelle Prealpi Bergamasce occidentali. *Rivista Italiana di Paleontologia e Stratigrafia*, **91**, 479-512.

—— & Gaetani, M. 1986. L'assetto strutturale del settore Lariano centro-meridionale. *Memorie della Società Geologica Italiana*, **32**, 123-131.

—— & Doniselli, T. 1986. La successione del Lias inferiore di Morterone (Lecchese). *Memorie della Società Geologica Italiana*, **32**, 49-66.

Kligfield, R., Geiser, P. & Geiser, J. 1986. Construction of Geologic Cross Sections using Microcomputer Systems. *Geobyte*, **1**, 60-66.

Laubscher, H. 1985. Large-scale, thin-skinned thrusting in the Southern Alps: Kinematic models. *Geological Society of America Bulletin*, **96**, 710-718.

—— 1990. The problem of deep structure of the Southern Alps: 3-D material balance considerations and regional consequences. *Tectonophysics*, **176**, 103-121.

Lichtensteiger, T. 1986. Stratigraphie und Tektonik der südöstlichen Alta Brianza. *Ph.D. thesis*, Basel, (unpubl.).

Pieri, M. & Groppi, G. 1981. Subsurface geological structure of the Po plain, Italy. Progetto Finalizzato Geodinamica, *Pubblicazione del Consiglio Nazionale delle Ricerche*, **414**, 1-13.

Roeder, D. 1989. South-Alpine thrusting and trans-Alpine convergence. *In:* Coward, M. P., Dietrich, D. & Park, R. G. (eds) *Alpine Tectonics*. Geological Society Special Publication, **45**, 211-227.

—— in press. Thrusting and wedge growth, Southern Alps of Lombardia (Italy). *Tectonophysics*.

Schönborn, G. 1990. A kinematic model of the western Bergamasc Alps. *Eclogae Geologicae Helvetiae*, **83**, in press.

—— & Laubscher, H. 1986. The suborobic imbrications near Taceno. *Memorie della Società Geologica Italiana*, **32**, 113-121.

Schumacher, M. 1990. Alpine basement thrusts in the Eastern Seengebirge, Southern Alps, Italy/Switzerland. *Eclogae Geologicae Helvetiae*, **83**, in press.

Snedden, W. T. & Spang, J. H. 1990. Geometric analysis of displacement transfer between thrust faults using fault-bend and fault-propagation fold models. *In:* McClay, K. R. (ed.) *Thrust Tectonics 1990, Programme with Abstracts*, Royal Holloway & Bedford New College, University of London, 19.

Suppe, J. 1983. Kinematics of fault-bend folding. *American Journal of Science*, **283**, 684-721.

Winterer, E. L. & Bosellini, A. 1981. Subsidence and sedimentation on Jurassic platform margin. *American Association of Petroleum Geologists Bulletin*, **65**, 394-421.

Zanchi, A., Forcella, F., Jadoul, F., Bernini, M., Bersezio, R., Fornaciari, M., Rossetti, R. & Torazzi, S. 1988. The Faggio-Morterone transverse line: Mesoscopic analysis and kinematic implications. *Rendiconti della Società Geologica Italiana*, **11**, 279-286.

Hangingwall geometry of overthrusts emanating from ductile decollements

Peter Jordan & Thomas Noack

Geological Institute of Basel University, Bernoullistrasse 32, CH-4056 Basel, Switzerland

Abstract: The hangingwall geometry of thrusts emanating from decollement-type sole thrusts strongly depends on the position, the thickness and the deformation style of the sole thrust and the way the ductile basal layers are included in the hangingwall geometry. Analytical models for some characteristic overthrust geometries are formulated and compared to field examples from the Jura mountains. All these models are area balanced and within the portions which deform in a brittle fashion are also line length balanced. Due to a wedge-like inclusion of ductile material in the thrust, overthrusts nucleating from thick sole thrusts have back limbs that are longer and flatter than the ramp. Furthermore, these backlimbs and the respective hinges rotate during deformation and the flat-ramp hinge migrates in the opposite sense than bulk shear movement. Due to hinge rotation and migration, internal destruction of those back limb areas that migrate through the hinges is kept to a minimum. In these models, important portions of the backlimb do not even migrate through the flat-ramp hinge.

A decollement is characterized by a stratal controlled sole thrust localized in particularly soft or incompetent horizons (e.g. Ramsay & Huber 1987, p. 518). In many cases, these horizons, e.g. evaporites, behave in a ductile manner while the overlaying strata, e.g. carbonates or clastic sediments, behave in a brittle fashion (Laubscher 1961, 1973, 1986; Davis & Engelder 1985; Malavieille & Ritz 1989). Decollement in ductile layers has been recognized by Buxtorf as early as 1907. Consequently, with the inclusion of ductile basal layers in the hangingwall of thrust systems, the amount of thickening or thinning has often been overestimated. Ductile materials have commonly been used to fill gaps between the basement and deformed cover without regard to mass balance or thrust kinematics (Fig. 1a). In order to keep mass balance, other authors try to fill these gaps with duplexes and basement structures (e.g. Boyer & Elliot 1982; Mitra 1986,1988; Gürler *et al.* 1987; Bitterli 1988). Their constructions are based on a sharp sole fault at the base or within the ductile layers, and a mass and line length conserving *dip domain structure* (Suppe 1983, Groshong & Usdansky 1988) in the hangingwall, viz. the ductile layers are assumed to behave the same way as all other overlying brittle strata (Fig. 1b).

Natural examples show that sole thrusts are often not sharp faults but shear zones of a certain thickness (e.g. Jordan & Nüesch 1989). Based on recent field work in the Jura and on geometrical models, this paper will demonstrate that there are accumulations and rolling-out of weak basal materials in decollement-related overthrusts. However, they are in concordance with overall mass balance, and they are controlled by the thickness of the sole thrust, its localization within the ductile layers, its deformation style and the way the ductile material within and above the sole thrust is included in overthrusting. These local changes in thickness of ductile layers in the vicinity of thrusts have an important influence on the geometry and the internal deformation of the hangingwall. For instance, they cause a hangingwall geometry that is not (directly) related to the geometry of the footwall ramp as it is assumed by Suppe (1983), Jamison (1987) and Mitra & Namson (1989), who use a sharp, infinitesimally thin sole thrust for their constructions.

First attempts to quantify the geometry and deformation of overthrusts emanating from thick ductile sole thrusts were made by Suter (1981) and Taboada *et al.* (1990). In the following, it will be shown that the nearly infinite range of natural decollement-thrust geometries (e.g. Fig. 2e) can be regarded as combinations of four major types (Figs 2a – d) that differ in the localization and the deformation style of the sole thrust and the hangingwall geometry.

The Type 1 Thrust is associated with a 'Top Decollement' (Fig. 2a), characterized by a sharp (infinitesimally thin) sole thrust at the top of the weak layer. No weak material is included in the thrust. The hangingwall is assumed to deform in a geometry according to Suppe (1983). This type of deformation is quite common in the frontal duplex zone of eastern Jura (e.g. Bitterli 1988).

The Type 2 Thrust is related to a 'Simple Shear Decollement' (Fig. 2b), introduced by Malavieille & Ritz (1989). It is characterized by homogeneous layer-parallel simple shear in a thick ductile sole thrust. The ductile material is squeezed between the competent backlimb and the undeformed footwall, producing a backlimb that is considerably flatter than expected from the ramp angle. The horizontal crest and the forelimb still exhibit the angular relationship to the ramp angle indicated by the model of Suppe (1983). Malavieille & Ritz (1989) suggest that the Arc de Digne nappe of southern France is a good example for this geometry.

The Type 3 Thrust is associated with a 'Basal Decollement'

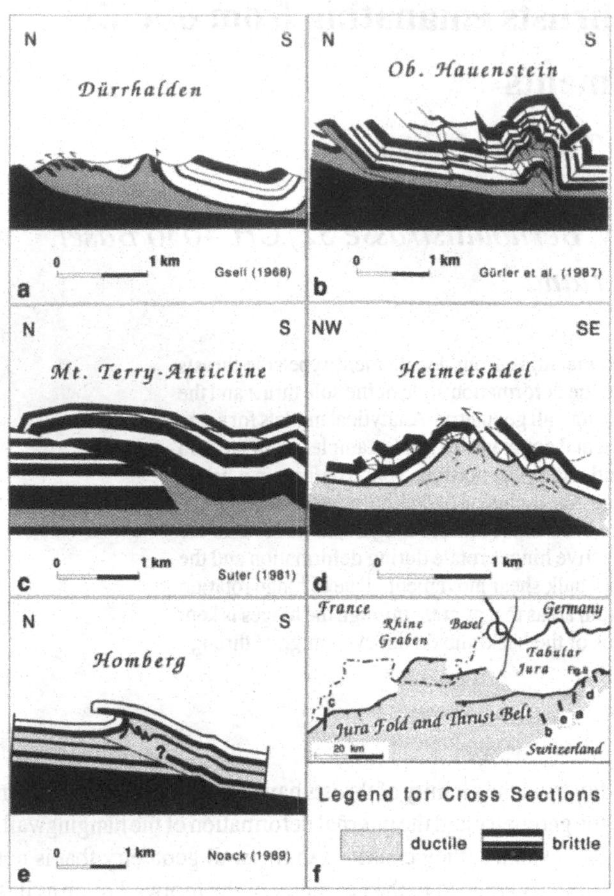

Figure 1. Interpretations of thrusts related to ductile decollements (examples from the Swiss Jura): (**a**) huge accumulations of evaporites as assumed by Gsell (1968); (**b**) extrapolation of the dip domain structure to the depth (indicated by arrow) of the Passwang fault related fold (after Gürler *et al.* 1987); (**c**) wedge-like involvement of ductile strata in thrust related to basal decollement: the Mt. Terry fault related fold (after Suter 1981); Underthrust cushions: (**d**) Heimetsädel anticline; (**e**) Homberg anticline with sharp frontal shear zone-type Homberg thrust (after Noack 1989). (**f**) Location of cross sections.

(Fig. 2c), first proposed by Suter (1981). It is characterized by a sharp sole thrust at the bottom of the weak basal layer combined with a wedge-shaped inclusion of the ductile layer in the thrust. As a consequence of the additional material squeezed in the ramp, the backlimb is even flatter than in Type 2-thrust while the amount of layer-normal shortening is smaller. The cross-section of Mt. Terry Thrust by Suter (1981) is a good example for this type of thrust (Fig. 1c).

The Type 4 Geometry is produced by an 'Underthrust Cushion Decollement' (Fig. 2d) that is characterized by two 'sole thrusts' within the weak basal layer (cf. *delamination*, Geiser 1988). The material between these two thrusts is squeezed under the overlying structure while the stiff layers above the upper thrust are shortened by overthrusting. In the example shown, the upper shear zone is localized at the top of the weak basal layer, and the material between the two shear zones is shortened parallel to the sole thrust to the same extent as the overlying segment. The resulting accumulation is assumed to be symmetrical and bulge-like. There are, of course, an infinite number of variations concerning the localization of the two 'sole thrusts', the deformation within the resulting segments and the final shape of the resulting cushion. As a consequence of the underthrusting, line lengths and layer thicknesses in the competent hangingwall are no longer conserved. There are various possibilities of internal shear. In Figure 2d, vertical simple shear in the competent units is assumed. Good examples for Type 4-decollements are the Heimetsädel (near Kienberg) and the Homberg (north of Olten) anticlines in the eastern Jura (Figs 1d – e).

Generally, the geometry of natural sole thrusts is a mixture between the types discussed above ('combined decollements') and thrusts emanating from such sole thrusts are consequently a combination of the four types. A possible combination of Types 1 to 3 is shown in Figure 2e. In the following, analytical models for Type 2, Type 3 and for thrusts nucleating from combined decollements are presented. Type 1 follows the rules of Suppe (1983) and Type 4 is not well enough constrained to be formulated in a simple and useful model, as is the situation when (except in Types 1 & 4) the

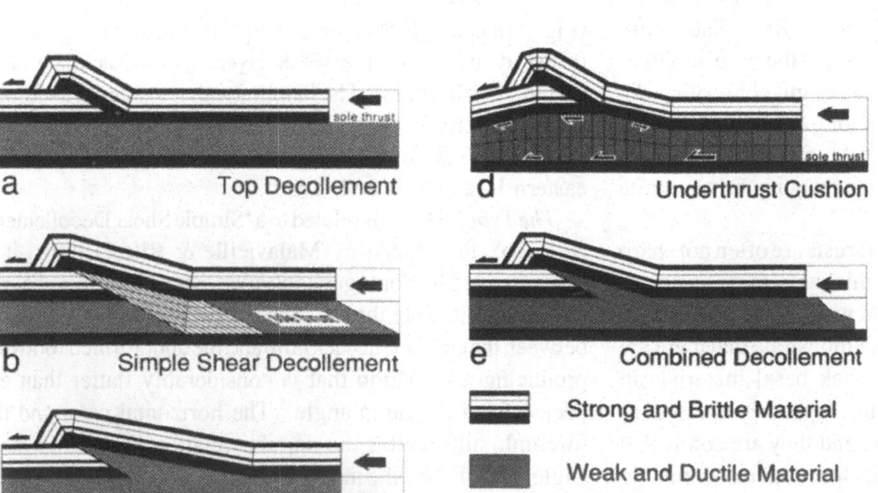

Figure 2. Thrusts related to various types of decollements (for discussion see text). In (b) and (d) the finite deformation of an initially orthogonal and equilateral net is indicated.

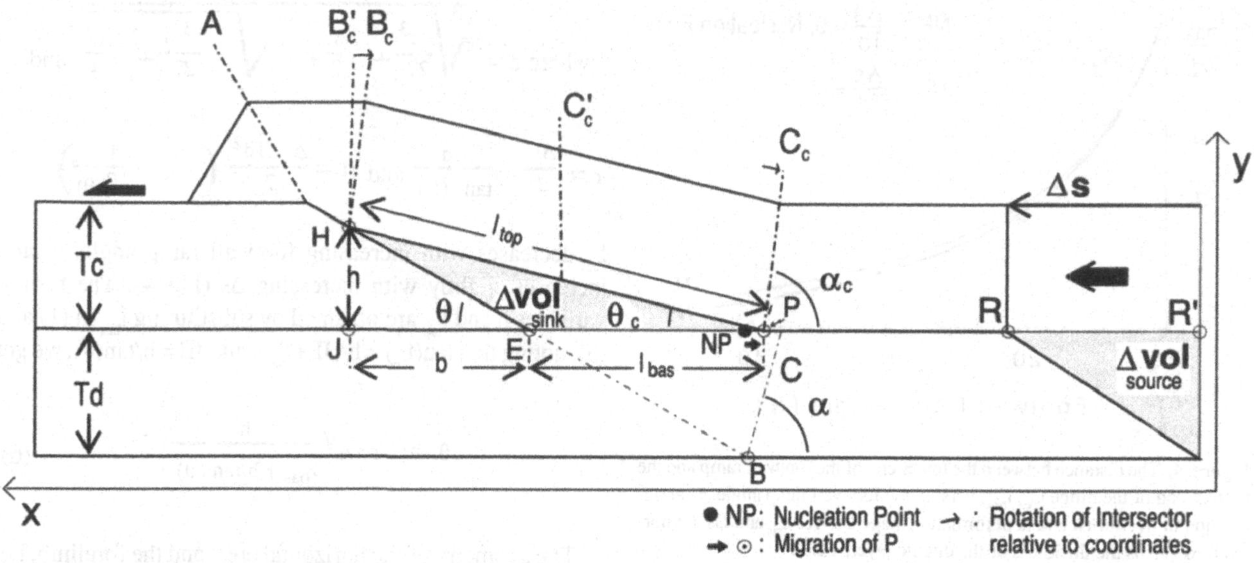

Figure 3. Model and geometry of a thrust related to simple shear decollement (Type 2) (for discussion see text, explanation of symbols in Table 1). The orientation of the passively rotated hinge C_c' gives the amount of internal shear in the stiff backlimb as the hinge was originally oriented perpendicularly to layering.

base of the stiff hangingwall segment has reached the upper footwall flat. Nevertheless, possible models for this latter situation are discussed.

Table 1 Symbols used in text and equations.

Symbol	Meaning
Ac, Bc, Cc	hinges of hangingwall above ductile basal layer
Ac´, Bc´, Cc´	inactive (passively transported) hinges of hangingwall
Bc*	hinge between backlimb and foreland dipping crest in 'shear model'
b	distance JE = h/ tan (θ)
C	ramp-flat hinge of footwall = imaginary flat-ramp bisector in basal ductile layer
h	distance HJ = ltop . sin (θc)
l_{bas}	$=f(\Delta s)$; distance EP; distance along initial position of top of basal ductile layer between footwall ramp and basal point of hangingwall Cc-hinge
l_{front}	length of 'sharp frontal shear zone'
l_{top}	distance HP; length of the top of basal ductile layer in backlimb
NP	nucleation point of initial hangingwall Cc-hinge (= P for Δs=0)
Tc	thickness of (brittle) hangingwall layers above ductile basal layer
Td	thickness of ductile basal layer
Ti	thickness of partial layer i
α	dip of C-hinge of footwall
αc	= 0.5.(π-θ); dip of Cc-hinge of hangingwall
γ	= Δs/Td; shear parallel to decollement
γi	shear of partial layer i
γ*	additional shear in ramp triangle parallel to Cc
Δa	= α - αc = 0.5. (θ - θc)
Δs	shortening of all competent hangingwall layers
Δvol	volume of source = volume of sink
θ	ramp angle of footwall
θc	dip of backlimb = apparent ramp angle of hangingwall

Ramp models

All models are formulated in two dimensions, i.e. in the transport direction assuming plane strain (cf. Suppe 1983). The ramp models for Type 2 and 3 thrusts and thrusts nucleating from combined decollements describe the situation from thrust nucleation until the arrival of the base of the brittle layer at the upper footwall flat.

Thrust related to a simple shear decollement (Type 2)

This model assumes that (Fig. 3): (1) all volumes are conserved; (2) within the competent layers, bed lengths and thicknesses are conserved; and (3) there is no displacement along the interfaces between ductile and competent layers. The thrust fault that cuts the competent layers at an angle θ is suggested to have propagated rapidly at the very first step of deformation. Subsequently the hangingwall starts to thrust over the footwall in such a way that the volume of the ductile strata remains constant, i. e. the 'volume loss', Δvol_{source}, is compensated by a 'volume gain', Δvol_{sink} (Fig. 3), therefore:

$$\underbrace{0.5 \cdot \Delta s \cdot Td}_{source} = \underbrace{0.5 \cdot l_{bas} \cdot h}_{sink} \tag{1}$$

where Δs is the shortening in the competent strata, Td the thickness of the ductile layer, l_{bas} the distance between P and E, (the lower end of the footwall ramp of the competent layers), and h the vertical component of the displacement vector EH. During thrusting, the stiff hangingwall is rotated by the angle θ_c around point P, resulting in a hinge C_c. In order to keep layer thickness constant, C_c has to be the bisector of the angle $(\pi-\theta_c)$. In contrast to the model of Taboada *et*

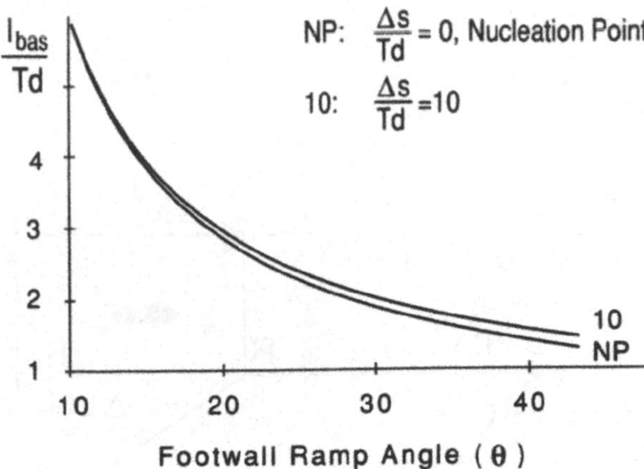

Figure 4. The distance between the lower end of the footwall ramp and the bottom end of the hinge C_c, l_{bas}, versus the footwall ramp angle, θ, at nucleation of overthrust and at deformation stage Ds=10. l_{bas} and Ds are normalized versus the thickness of the ductile layer, Td.

al. (1990) for the same geometrical setting, in the present model, the hinge C_c is not parallel to the hinge C (that bisects the angle $(\pi-\theta)$) between the ramp and the flat at the base of the decollement nor do the two hinges intersect with each other at the interface of the two layers. Noting that $b = h/\tan(\theta)$, from equation (1),

$$b=(\Delta s\cdot Td) / (\tan(\theta)\cdot l_{bas}) \tag{2}$$

From the requirement of constant bed length of the competent units, HPR = ER´, and,

$$l_{top} = l_{bas} + \Delta s \qquad \text{(note that } l_{bas} = f(\Delta s)) \tag{3}$$

From the Pythagoras' theorem:

$$(l_{top})^2 = (l_{bas} + b)^2 + (h)^2$$

or, substituting from equation (1), (2) & (3)

$$(l_{bas} + \Delta s)^2 = \left(l_{bas} + \frac{\Delta s\cdot Td}{l_{bas}\cdot \tan(\theta)}\right)^2 + \left(\frac{\Delta s\cdot Td}{l_{bas}}\right)^2$$

and

$$l_{bas}^3 + \underbrace{\left(\frac{\Delta s}{2} - \frac{Td^2}{\tan(\theta)}\right)}_{r} l_{bas}^2 \underbrace{- \frac{\Delta s\cdot Td^2}{2}\left(1 + \frac{1}{\tan^2(\theta)}\right)}_{t} = 0 \tag{4}$$

Using CARDAN's formula for cubic equations we get

$$l_{bas} = -\frac{r}{3} + \frac{r^2}{9a} + a \tag{5}$$

where $a = \sqrt[3]{\frac{-r^3}{27} + \frac{t}{2} + \sqrt{-\frac{r^3\cdot t}{27} + \frac{t^2}{4}}}$ and

$r = \frac{\Delta s}{2} - \frac{Td}{\tan(\theta)}$ and $t = \frac{\Delta s\cdot Td^2}{2}\left(1 + \frac{1}{\tan^2(\theta)}\right)$

l_{bas} decreases with increasing footwall ramp angle, θ, and increases slightly with increasing Δs (Fig. 4). The further variables h and l_{top} are obtained by substituting l_{bas} in (1) and (3), noting that $\tan(\theta_c) = h/(JE+l_{bas})$, and $JE = h/\tan(\theta)$, we get

$$\theta_c = \arctan\left(\frac{h}{l_{bas} + h/\tan(\theta)}\right) \tag{6}$$

The geometry of the horizontal crest and the forelimb, i.e. the part of the hangingwall to the left of hinge B_c (Fig. 3) is controlled by the ramp angle only and has, therefore, a Suppe (1983)-type shape. The backlimb, however, is much flatter than expected from the Suppe model. During progressive hangingwall displacement, the backlimb of a Type 2 thrust is rotated to become increasingly steeper until the base of the

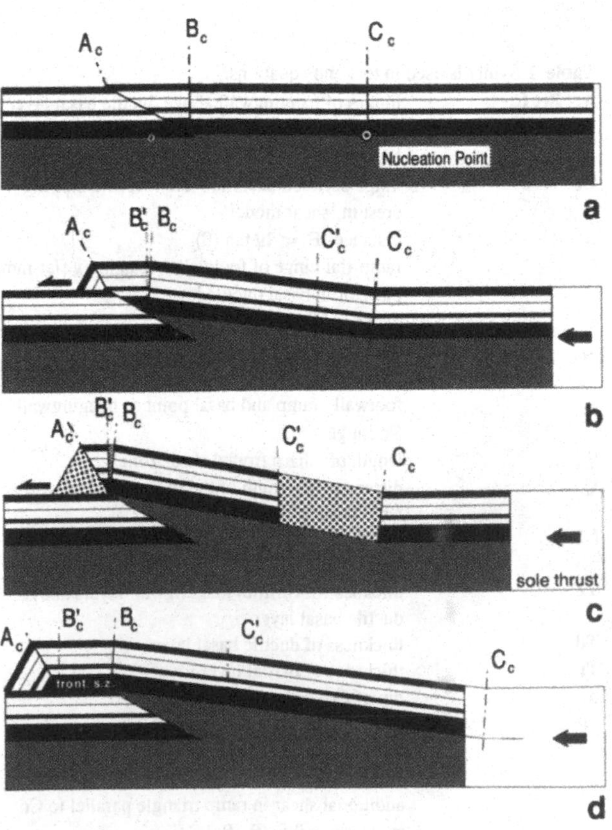

Figure 5. Kinematic sequence of a thrust related to a basal decollement (Type 3). (a) after nucleation; (b) ramp situation; (c) base of stiff backlimb has reached top of footwall (dotted domains have passed through hinges); (d) sharp frontal shear zone-type development after backlimb has reached flat.

competent layers reaches the upper footwall flat. The dip of hinges B_c and C_c are not fixed but rotate with increasing Δs. In addition, the lower end of hinge C_c (point P) migrates with progressive deformation, in the opposite sense to the thrust movement, as a function of l_{bas} (Fig. 3). The location of the nucleation point NP of the hinge C_c, which is the initial value of l_{bas}, is determined from equation (5) with $\Delta s = 0$ (Fig. 4):

$$l_{bas\,(NP)} = Td/\tan(\theta)$$

An important feature of Type 2 decollements is that the backlimb domain between hinge C_c' and B_c is deformed only by layer-parallel slip (cf. Fig. 5c), whereas the entire backlimb of a classical fault-bend-fold undergoes severe deformation when being moved through the basal hinge (B bisector of Suppe 1983).

The present model shows some important differences to the model of Taboada *et al.* (1990) for the same geometrical setting. The deformation matrix for the ductile basal layers of the present model is derived and outlined in the appendix.

Thrust related to a basal decollement (Type 3)

The model for this type (Fig. 2c) has the same constraints as the Type 2 model. It diverges, however, from the former model by a sharp sole thrust at the bottom of the ductile layer. Consequently, the Δvol_{source} is doubled in respect to the former model. This modifies equation (1) to

$$\underset{\text{source}}{\Delta s \cdot Td} = \underset{\text{sink}}{0.5 \cdot l_{bas} \cdot h} \qquad (7)$$

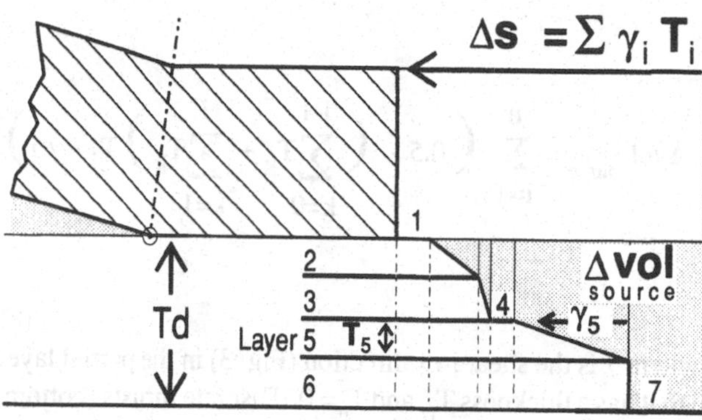

Figure 6. $Dvol_{source}$ resulting from variable shear in a ductile sole thrust. For layer 5, T_5 & γ_5 are indicated. Layer 6 is not sheared, and 'layers' 1, 4 & 7 are sharp sole faults with T_i=0.

Using equations (7) and (2) to (5) we get a kinematic sequence as shown in Figure 5(a to c). For identical values of Δs, Td and θ, l_{bas} is doubled and tan (θ_c) is halved in respect to the former model. Internal deformation in the gently dipping backlimb is less severe than in the Type 2 model. Figure 5c shows the domains of the backlimb that have passed through a hinge zone when the bottom of the brittle domain reaches the top of the footwall.

Thrust related to a combined decollement

This model has the same constraints as the former two models and is based again on equations (1) to (5). It assumes, however, a variable shear profile in the basal strata that range from sharp sole faults ('layers' 1, 4, & 7 in Fig. 6) to layers sheared to various degrees (layers 2, 3 & 5) and undeformed interlayers (6). The resulting 'volume loss', Δvol_{source}, has to be calculated layer by layer (or thrust plane) from top to bottom:

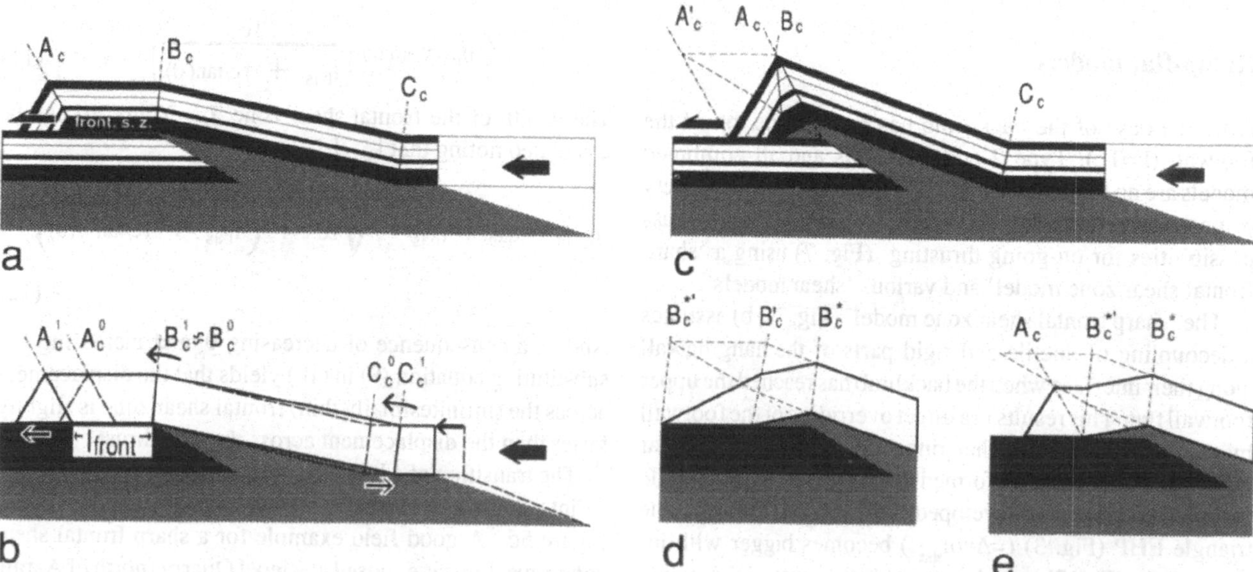

Figure 7. Plausible models for thrust development after arrival of the base of the stiff backlimb at the top of the footwall: **(a)** 'sharp frontal shear zone model'; **(b)** geometry of 'sharp frontal shear zone model' and kinematic sequence from step 0 (C_c^0 etc.) to step 1 (C_c^1 etc.). The rotation of B_c is indicated by an arrow as the rotation is to small to be drawn. C_c and B_c rotate parallel to each other. Notice backward migration of C_c hinge. 'Shear models': **(c)** hangingwall passes through the hinge A_c that is fixed. **(d)** hangingwall passes through new hinge B_c* that originates from B_c (after Taboada *et al.* 1990). **(e)** Face of hangingwall is equal to ramp situation (Suppe (1983)-type, cf. Fig. 3), rest of hangingwall passes through new hinge B_c* that nucleates from B_c (after Malavieille & Ritz 1989).

$$\Delta vol_{source} = \sum_{i=1}^{n} \left(0.5 \cdot \left(\sum_{k=0}^{i-1} T_k + \sum_{k=1}^{i} T_k \right) T_i \cdot \gamma_i \right) \quad (8)$$

where γ_i is the shear in x direction (Fig. 3) in the partial layer i that has a thickness T_i, and $T_0 = 0$. Discrete thrusts (bottom, intra and top decollements) have to be treated as partial layers (with $T_i \cdot \gamma_i$ = amount of displacement along thrust; and, for the summing-up of thicknesses, $T_i = 0$). The specific 'volume loss' is normalized by

$$fk = \frac{\Delta vol_{source}}{Td \Delta s}, \text{ where } Td = \sum_{i=1}^{n} T_i \text{ and } \Delta s = \sum_{i=1}^{n} T_i \gamma^i \quad (9a)$$

in a modified equation (1):

$$\underset{source}{fk \cdot \Delta s \cdot Td} = \underset{sink}{0.5 \cdot l_{bas} \cdot h} \quad (9b)$$

For an overthrust related to a top decollement (Type 1) fk=0, for one related to a simple shear decollement (Type 2) fk=0.5 and for one related to a basal decollement (Type 3) fk=1. Figure 2e gives an example for a combined decollement calculated with equations (8), (9) & (2) – (5).

Ramp-flat models

After the base of the back limb has reached the top of the footwall (h=Tc), Type 2 and 3 models and all combined models are no longer constrained. There is an infinite variety of possible geometries. In the following, we will discuss possibilities for on-going thrusting (Fig. 7) using a 'sharp frontal shear zone model' and various 'shear models'.

The 'sharp frontal shear zone model' (Fig. 7a, b) assumes a decoupling of ductile and rigid parts of the hangingwall along their interface when the backlimb has reached the upper footwall flat. This results in a direct overriding of the footwall by the brittle part of the hangingwall along a frontal shear zone which is commonly formed as a very thin slice of ductile material, that might be developed as a mylonite (Fig. 7d). The triangle EHP (Fig. 3) (=Δvol_{sink}) becomes bigger with increasing Δs (Fig. 7b) as almost no ductile material is transferred to the upper flat. Noting that h= EH $\cdot \sin(\theta)$=Tc is constant, equation (9b) yields

$$l_{bas} = EP = 2 \cdot fk \cdot \Delta s \cdot Td/Tc \quad . \quad (10)$$

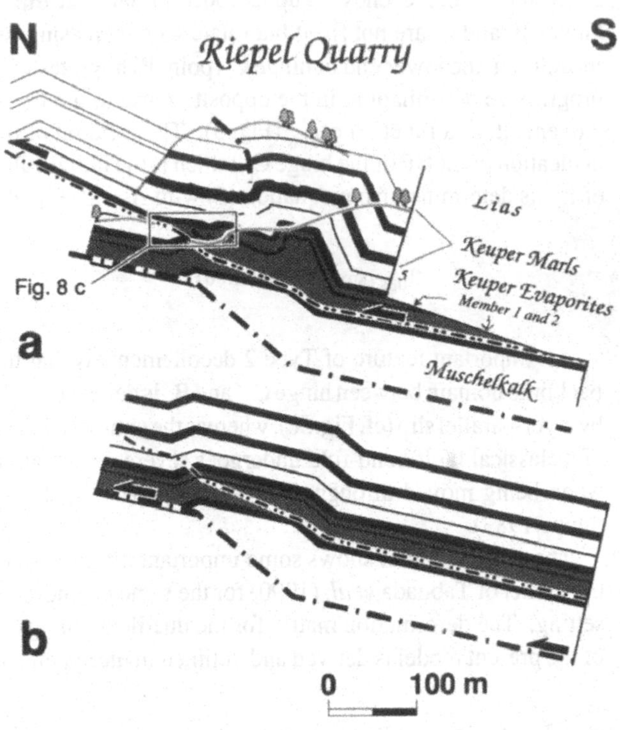

Figure 8. Riepel quarry (for location see Fig. 1): **(a)** cross section showing a sharp frontal shear zone-type thrust and an accumulation of ductile evaporites (members 1 and 2) in front of the footwall thrust; **(b)** restoration of latest deformation step (cross section constructed and restored using Geosec™, © Geo-Logic Systems Inc. 1989).

i.e. with increasing Δs, the hinge C_c migrates towards the hinterland. The backlimb becomes flatter, while the hinges B_c and C_c become slightly steeper. Noting that h = Tc, equation (6) yields

$$\theta_c = \arctan\left(\frac{Tc}{l_{bas} + Tc/\tan(\theta)} \right) \quad (11)$$

The length of the frontal shear zone, l_{front}, (Fig. 7b) can be calculated noting that $l_{front} + l_{top} = \Delta s + l_{bas}$

$$l_{front} = \Delta s + l_{bas} - \sqrt{Tc^2 + \left(l_{bas} + Tc/\tan(\theta) \right)^2} \quad (12)$$

And as a consequence of decreasing θ_c and increasing l_{bas}, substituting equation (9) in (10) yields that the displacement across the (infinitesimally thin) frontal shear zone is slightly faster than the displacement across the sole thrust.

The transition of a basal decollement related thrust (Type 3) into a sharp frontal shear zone-type thrust is shown in Figure 5d. A good field example for a sharp frontal shear zone-type thrust is exposed at Riepel Quarry (north of Aarau, Fig. 8). The deformation within the triangle (prism) below the backlimb is obviously compressive and accretionary while the frontal shear zone is developed as a shear zone.

Introduced by Malavieille & Ritz (1989), the 'shear models' discuss possibilities of transferring ductile material onto

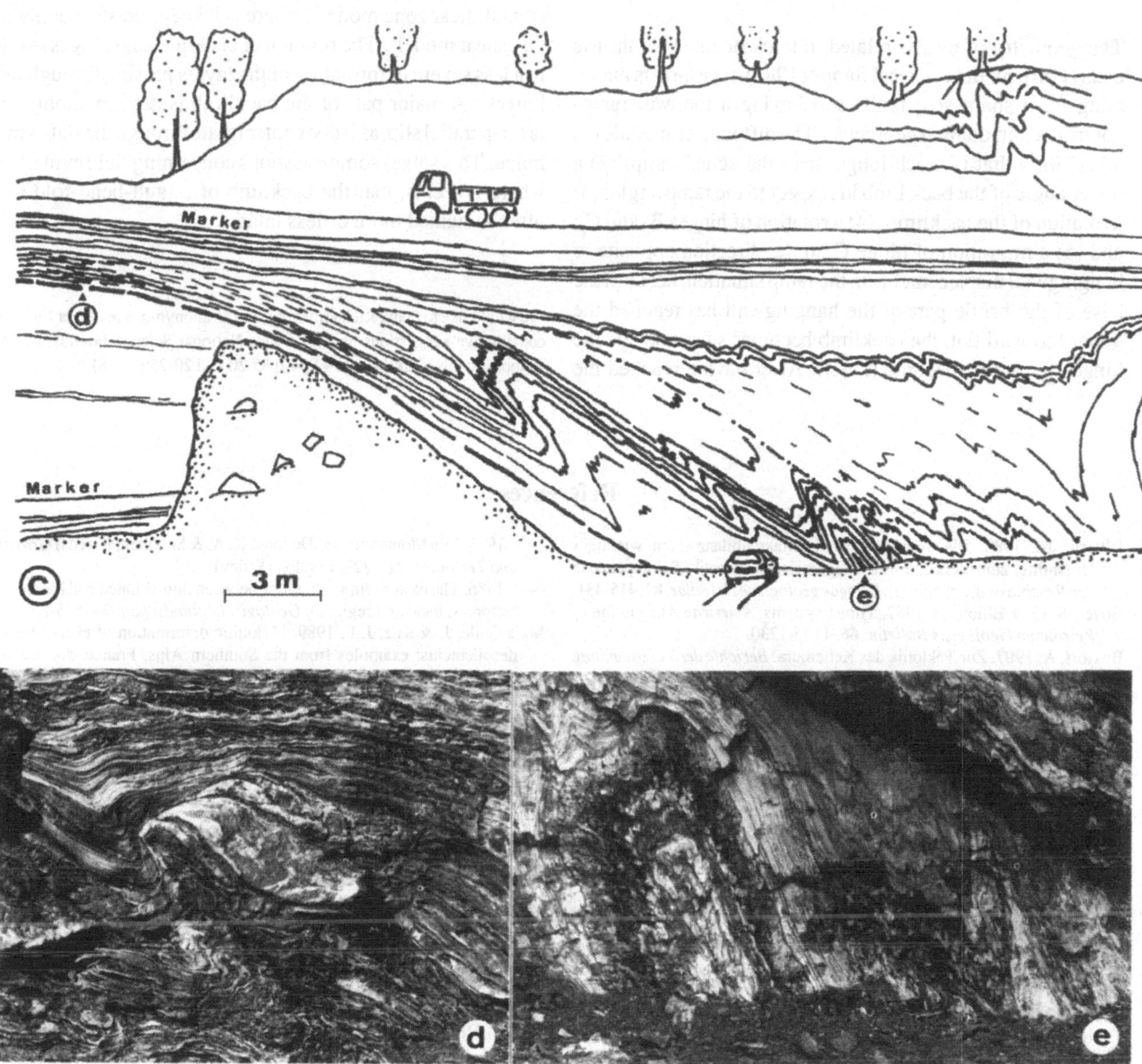

Figure 8 (cont'd.). (c) Line drawing of panorama picture (ca. 30°, cf. Fig. 8a) of lowest level of Riepel quarry showing footwall ramp (left), and accumulation of hangingwall evaporites (right). Top right shows north limb of the small fold in the southern part of Riepel Quarry (cf. Fig. 8a). (d) Detail of mylonite shear zone ('sharp frontal shear zone', size indicated by coin); (e) detail of fold in the accumulation domain in front of footwall thrust (size indicated by hammer).

the upper footwall flat. Geometrically, they are all based on a folding down of the crest and the upper backlimb parallel to different hinges (A_c or B_c*, Fig. 7c-d) that are all bisectors of the 'initial', unfolded geometry. All 'shear models' conserve the volume, however, line lengths and bed thicknesses of one or the other segment of the brittle layers have urgently to undergo changes in order to keep geometrical constraints. Kinematically, the three models differ in the function of the hinges.

In the model introduced here (Fig. 7c), the thrust front passes through the fixed hinge A_c. The hinge B_c, initially a bisector inherited from the ramp geometry, is deflected while passing through A_c. This model results in a steep forelimb and a high crest. Line lengths and bed thicknesses are conserved in the forelimb but not in the domain between A_c and B_c. In contrast, the model proposed by Taboada *et al.* (1990) (Fig. 7d) is based on a new hinge, B_c*, between the backlimb and the crest segment. The new hinge develops from the hinge B_c (initially a bisector) when the latter reaches the flat. The hinge B_c migrates further while B_c* is fixed (however, it rotates in time). This model results in a flat and far ranging forelimb and a foreland sloping crest segment. Line lengths and bed thicknesses are conserved within the crest segment but not in the forelimb. The last model (Fig. 7e), proposed by Malavieille & Ritz (1989), is a modification of the model of Taboada *et al.* (1990). Shear is restricted to the part between B_c* and the tip of the ductile thrust triangle. The thrust face, i.e. the frontal forelimb and the frontal crest (flat) are Suppe (1983)-type and are conserved from the preceding ramp thrusting.

Conclusions

The geometry of thrusts related to thick ductile sole thrusts clearly differ from classical Suppe (1983)-type thrusts nucleating from sharp sole faults and having a footwall ramp-controlled dip domain geometry. The differences include (1) a backlimb that is much longer than the actual ramp; (2) a lower angle of the back limb in respect to the ramp angle; (3) a rotation of the backlimb; (4) a rotation of hinges B_c and C_c; and (5) a migration of hinge C_c in the direction opposite to hangingwall displacement. In the ramp situation, i.e. until the base of the brittle part of the hangingwall has reached the upper footwall flat, the backlimb becomes steeper while the hinges B_c and C_c become flatter. After having reached the footwall flat, the backlimb becomes again flatter in the 'sharp frontal shear zone model', whereas it keeps on steepening in the 'shear model'. The rotation of backlimb and hinges result in a less severe deformation of the layers passing through the hinges. A major part of the backlimb is deformed only by layer-parallel slip, as it does not migrate through the flat-ramp hinge. This solves some questions concerning field evidence, where it seems, that the backlimb of a fault-bend-fold has often remained more or less intact.

We would like to thank Rick Groshong and an anonymous reviewer for their constructive suggestions and the Swiss National Science Foundation for supporting this work (grants 4.904-0.85.20 and 20-25560.88).

References

Bitterli, Th. 1988. Die dreidimensionale Massenbilanz - ein wichtiges Hilfsmittel zum Verständnis der regionalen Kinematik (Schuppenzone von Reigoldswil, Faltenjura). *Eclogae geologicae Helvetiae*, **81**, 415-431.

Boyer, S. E. & Elliott, D. 1982. Thrust systems. *American Association of Petroleum Geologists Bulletin*, **66**, 1196-1230.

Buxtorf, A. 1907. Zur Tektonik des Kettenjura. *Berichte der Versammlung der Oberrheinischen geologischen Vereinigung*, **30/40**, 29-38.

Davis, M.D. & Engelder, T. 1985. The role of salt in fold-and-thrust belts. *Tectonophysics*, **119**, 67-88

Geiser, P. A. 1988. The role of kinematics in the construction and analysis of geological cross-sections in deformed terranes. *In:* Mitra, G. & Wojtal, S. (eds) *Geometries and mechanisms of thrusting with special reference to the Appalachians*. Geological Society of America Special Paper, **222**, 47-76.

Groshong, R. H. & Usdansky, S. I. 1988. Kinematic models of plane-roofed duplex styles. *In:* Mitra, G. & Wojtal, S. (eds) *Geometries and mechanisms of thrusting with special reference to the Appalachians*. Geological Society of America Special Paper, **222**, 197-206.

Gsell, F. J. 1968. Geologie des Falten- und Tafeljuras zwischen Aare und Witnau und Betrachtungen zur Tektonik des Ostjuras zwischen unterem Hauenstein im Westen und der Aare im Osten. *Ph.D. thesis, Universität Zürich (Switzerland)* (unpubl.).

Gürler, B., Hauber, L. & Schwander, M. 1987. Die Geologie der Umgebung von Basel mit Hinweisen über die Nutzungsmöglichkeiten der Erdwärme. *Beiträge zur geologischen Karte der Schweiz*, N.F. **160**

Jamison, W. R. 1987. Geometric analysis of fold development in overthrust terranes. *Journal of Structural Geology*, **9** , 207-219.

Jordan, P. & Nüesch, R. 1989. Deformation structures in the Muschelkalk Anhydrites of the Schafisheim Well (Jura Overthrust, Northern Switzerland). *Eclogae geologicae Helvetiae*, **80**, 429-454.

Laubscher, H. P. 1961. Die Fernschubhypothese der Jurafaltung. *Eclogae geologicae Helvetiae*, **54**, 221-282 .

—— 1973. Jura Mountains. *In:* De Jong, K. A. & Scholten, R. (eds) *Gravity and Tectonics*, 217-227, London (Wiley).

—— 1986. The eastern Jura: Relations between thin-skinned and basement tectonics, local and regional. *Geologische Rundschau*, **73**, 535-553.

Malavieille, J. & Ritz, J. F. 1989. Mylonitic deformation of evaporites in decollements: examples from the Southern Alps, France. *Journal of Structural Geology*, **11**, 583-590.

Mitra, S. 1986. Duplex structures and imbricate thrust systems: Geometry, structural position, and hydrocarbon potential. *American Association of Petroleum Geologists Bulletin*, **70**, 1087-1112

—— 1988. Three-dimensional geometry and kinematic evolution of the Pine Mountain thrust system, southern Appalachians. *Geological Society of America Bulletin*, **100** , 72-95.

—— & Namson, J. 1989. Equal-area balancing. *American Journal of Science*, **289**, 563-599.

Noack, Th. 1989. Computergestützte Modellierung geologischer Strukturen im östlichen Jura: Konstruktion balancierter Profile, Gravimetrie, Refraktionsseismik. *Ph.D. thesis Universität Basel (Switzerland)* (unpubl.).

Ramsay, J. G. & Huber, M. I. 1983. *The techniques of modern structural geology, vol 1: Strain analysis*. Academic Press. London

—— & Huber, M. I. 1987. *The techniques of modern structural geology, vol. 2: folds and fractures*. Academic Press. London

Suppe, J. 1983. Geometry and kinematics of fault-bend-folding. *American Journal of Science*, **283**, 684-721.

Suter, M. 1981. Strukturelles Querprofil durch den nordwestlichen Faltenjura, Mt.-Terry-Randüberschiebung-Freiberge. *Eclogae geologicae Helvetiae*, **74**, 255-275.

Taboada, A., Ritz, J. F. & Malavieille, J. 1990. Effect of ramp geometry on deformation in a ductile decollement level. *Journal of Structural Geology*, **12**, 297-302.

Appendix

State of deformation in the ductile layers of a Type 2 decollement:

In a reference system normal and parallel to the original layering (Fig. 3), the deformation of the ductile basal layer in the ramp triangle HPB (Fig. 3) can be formulated by the superposition of two simple shear movements. The first is parallel to the decollement and is characterized by

$\gamma = \Delta s/Td$.

The second is parallel to the hinges C_c and B_c of the respective deformation step and is characterized by

$\gamma^* = 2\cdot\sin(\theta_c)/(1+\cos(\theta_c)) = 2\cdot\tan(\theta_c/2)$.

where $\theta_c = f(g)$ (equation (2, 5 & 6)). The finite strain as shown in Fig. 2b resulting from a given $g = \Delta s/Td$ is (cf. Ramsay & Huber 1983, p. 290):

$$\begin{pmatrix} x' \\ y' \end{pmatrix} = \begin{pmatrix} 1 + \gamma^* \sin(\alpha_c)\cos(\alpha_c) & -\gamma^* \cos^2(\alpha_c) \\ \gamma^* \sin^2(\alpha_c) & 1 - \gamma^* \sin(\alpha_c)\cos(\alpha_c) \end{pmatrix} \begin{pmatrix} 1 & \gamma \\ 0 & 1 \end{pmatrix} \begin{pmatrix} x \\ y \end{pmatrix}$$

Noting that $\alpha_c = \pi/2 - \theta_c/2$ (angle of bisector C_c)

$$\begin{pmatrix} x' \\ y' \end{pmatrix} = \begin{pmatrix} 2 - \cos(\theta_c) & \gamma(2 - \cos(\theta_c)) - \dfrac{(1 - \cos(\theta_c))^2}{\sin(\theta_c)} \\ \sin(\theta_c) & \gamma \sin(\theta_c) + \cos(\theta_c) \end{pmatrix} \begin{pmatrix} x \\ y \end{pmatrix}$$

The Venetian Alps thrust belt

C. Doglioni

Dipartimento di Scienze Geologiche e Paleontologiche, Università di Ferrara, 44100 Italy

Abstract: The Venetian part of the Southern Alps (N Italy) is a Neogene south-vergent thrust belt. The minimum shortening of the chain is 30 km. The thrusts trend N60°-80°E and show an inherited N10°W-N10°E normal fault pattern of the Mesozoic continental margin. These earlier features strongly controlled the evolution of the following oblique thrust belt. Structural undulations along strike of folds and thrusts occur in correspondence to Mesozoic faults, thickness and facies variations. The thrusts are arranged in an imbricate fan geometry. A frontal triangle zone laterally ends at transfer faults. Earlier stages of the thrust belt were characterized by frontal triangle zones which have later been involved and cut by the progression of the internal thrusts.

This paper describes the main structural grain of the Venetian Alps (Figs 1 & 2), a part of the Southern Alps, south of the Insubric Lineament. It is a SSE-vergent fold and thrust belt of mainly Neogene age (Dal Piaz 1912; Leonardi 1965; Castellarin 1979; Laubscher 1985, Massari *et al.* 1986; Roeder 1989; Doglioni 1987) probably produced by the dextral transpression in the central-eastern Alps (Laubscher 1983). The chain (Fig. 3) deformed a pre-existing Mesozoic passive continental margin (Aubouin 1963; Bosellini 1965; 1973; Bernoulli *et al.* 1979; Winterer & Bosellini 1981).

The study area has been shortened mainly during Neogene times (Venzo 1939; Massari *et al.* 1986; Doglioni 1987) and not deformed by the Dinaric chain which constituted the unfolded foreland during Palaeogene times (Doglioni & Bosellini 1987).

Inherited structures

The Southern Alps were part of a Mesozoic continental margin, according to stratigraphic analysis, i.e. facies and thicknesses changes (Bernoulli *et al.* 1979; Winterer & Bosellini 1981). The area can be divided into three main structural sectors during the Mesozoic. These are, from west to east: the Trento Platform, the Belluno Basin and the Friuli Platform. True Mesozoic normal faults (i.e. Liassic) have been documented at the western border of the Trento Platform (Castellarin 1972; Doglioni & Bosellini 1987) and at its eastern margin (Winterer & Bosellini 1981; Bosellini *et al.* 1981; Bosellini & Doglioni 1986; Masetti & Bianchin 1987, Doglioni & Neri 1988).

The Mesozoic normal faults trend mainly N10°W-N10°E. It can be argued that the Trento Platform, the Belluno Basin and the Friuli Platform were bounded by crustal normal faults, mainly N-S trending, acting at different times and with different displacements during Jurassic time and during at least the Early Cretaceous (Fig. 4). The main Mesozoic tectonic features bordering the Belluno Basin are from west to east: the eastern margin of the Asiago Plateau, the Seren (Graben) Valley, the Cismon Valley alignment (clearly seen on satellite images), the Passo Rolle Line, the S Gregorio alignment, and to the east the Col delle Tosatte - Fadalto alignment (Figs 4 & 5). The Mesozoic alignments probably used inherited Variscan discontinuities as well. Platform and basinal Mesozoic facies do not coincide everywhere with the old horst and graben structure, i.e. the drowned Trento Platform, which acts as a horst with reduced basinal sequences after the Middle Jurassic until the Late Cretaceous.

Structure of the thrust belt

The geometry of the thrust belt is that of an imbricate fan (Fig. 3) with a main envelope angle produced by the thrust slices close to 7° (critical taper of wedge) according to the model of Platt (1988). The main thrusts are in order from the internal parts to the foreland, the Valsugana Line, the Belluno Line, the Moline Line, the Tezze Line and the Bassano Line. The thrust belt is not cylindrical in shape and the strain continuously changes along strike. In a map view of the area (Figs 2 & 5), the thrusts show an anastomosing pattern along strike, maintaining constant shortening which can conservatively be calculated as 30 km (Fig. 3).

The structural evolution of the thrust belt shows a general rejuvenation from the internal thrust to the external ones. However the internal thrust sheets seem to have been reactivated also in recent times (Sleyko *et al.* 1987). The crystalline basement outcrops in the hangingwall to the Valsugana thrust and is composed of Variscan metamorphosed greenschist facies rocks intruded by Late Carboniferous granitic bodies. Basement depth in the Venetian Plain is inferred by magnetic data (Cassano *et al.* 1986) and by the assumption of the general hinterland dipping monocline typical of thrust belts. This is consistent with the southward rising of the basement discovered in the Assunta Well at 4747 m where Late Triassic dolomites onlap a Late Ordovician

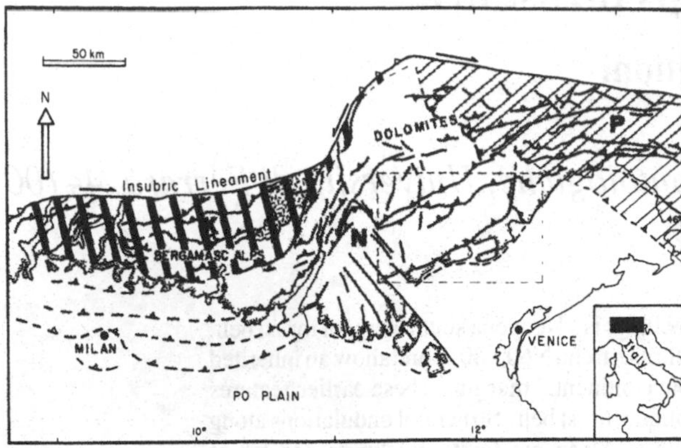

Figure 1. The Southern Alps underwent compressions at different times and in different areas. K: area of Late Cretaceous—Palaeogene compression; P: Palaeogene—Early Neogene compression (Dinarides); N: Palaeogene—Neogene compression (Southern Alps). The rectangle indicates the study area which has been deformed only by the SSE-vergent Neogene South Alpine thrust belt and was located in the foreland of the Palaeogene WSW-vergent Dinaric thrust belt. Note in the eastern side the overlapping between Southern Alps and Dinarides.

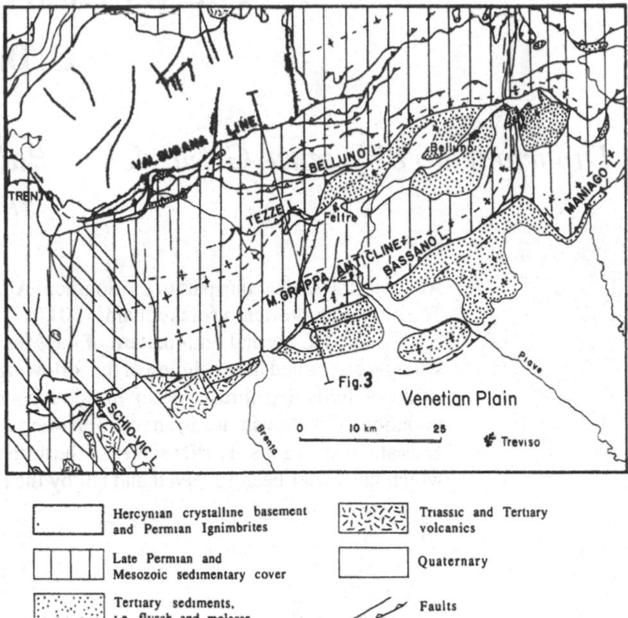

Figure 2. Simplified tectonic map of the Venetian Alps.

granite (Pieri & Groppi 1981). The basement is clearly involved in the Valsugana Line, but balanced cross-sections would indicate a wider involvement by southern thrusts as well. Significant thickness variations of the sedimentary cover and minor dips of the thrusts cannot be excluded and these could considerably increase the amount of shortening along the thrust belt (Roeder 1989). The main decollement of the thrust belt appears to be located in the basement (15-20 km in depth) beneath the Dolomites as suggested by the construction of balanced cross-sections (Doglioni 1987) and by focal mechanisms of earthquakes indicating low angle thrust planes (i.e. the Siusi event, Slejko *et al.* 1987). Triangle zones are present along the Valsugana thrust where the basement is sometimes wedged within the sedimentary cover, or it produces a triangle zone in the Valsugana Valley where the Valsugana thrust faces a north-vergent basement involved backthrust, to the north of the Asiago Plateau. Major undulations along the Valsugana thrust occur as a result of inherited features, i.e. the sinistral N-10°E striking transpressive undulation of Borgo Valsugana which occurs at an inherited structural high as indicated by the reduced thickness of the sedimentary cover. The thrust has a ramp trajectory and has not assumed a staircase geometry in the sedimentary cover, probably due to the difficulty in generating flexural slip folds.

Within the sedimentary cover the thrusts are characterized by cut-off angles ranging between 5° and 45°. Preferential decollement layers are the Tertiary Possagno Marls, the Late Cretaceous Scaglia Rossa, and other, buried levels within the Late Permian and Triassic sequences. The thrust planes assume steeper angles when a footwall syncline is present. Footwall synclines are well developed in Cretaceous pelagic thin-bedded rocks (Biancone and Scaglia Rossa) whose folding is accomodated by intense flexural slip. Chevron folds are particularly common in these two formations and their amplitude and wavelength decrease away from the thrust planes.

The frontal part of the thrust belt is characterized by a triangle zone (Figs 2 & 3) which generates a general southward dipping monocline characteristic of the Venetian foothills between Bassano and Vittorio Veneto. The frontal triangle zone is the most peculiar structure of the foreland and its presence is indicated by: (1) the general absence of an important thrust at the base of the mountains (Monte Grappa - Visentin Anticline); (2) the necessity of a thrust at the base of the anticline to resolve the volume problem of the structural high; (3) the south-dipping monocline in the frontal part of the chain which is typical for the triangle zones (i.e. Bally *et al.* 1966; Jones 1982; Boyer & Elliott 1982); (4) the presence of north-vergent backthrusts (i.e. in the Possagno and Follina areas, Braga 1970; Zanferrari *et al.* 1982). The interpretation presented in Figure 3 of the triangle zone is only one possibility: i.e. there are not clear indications of the southward continuity of the decollement necessary to adsorb the amount of displacement. This could be in turn expressed through pressure solution cleavage or more probably in antiformal stack duplexes within the anticline core of the triangle zone.

It is interesting to note that a similar triangle zone has been reported for the northern part of the Alps in the Bavarian foreland (Müller *et al.* 1988). The triangle zone between Bassano (Schio?) and Vittorio Veneto seems to be connected with a ramp-flat geometry of the deep-seated blind thrust which generated a thrust-propagation fold (the Monte Grappa-Visentin Anticline). This was active at least during Late Miocene times because Tortonian and Messinian sediments onlap with a gradually smaller inclination the southern limb of the anticline (Massari *et al.* 1986). Sequence boundaries in the southern fold limb are marked by angular unconformities with decreasing angles toward the foredeep suggesting the coeval activity of the frontal fold (Monte Grappa - Visentin Anticline). It is clear that the unconformities are angular only along dip where the frontal fold is perpendicular to the assumed regional maximum Neogene stress (σ_1: N20°-30°W) and the fold axis has a 'cylindrical' trend. Where there are structural undulations in the fold axis (i.e. the sinistral transpressive zones of Valdobbiadene-Cornuda and the greater Fadalto alignment) the unconformities are marked by angular relationships along both dip and strike. In summary, structures control the nature of the unconformities. A growth fold, with constant horizontal axis, generated by pure compression produces angular unconformities only along dip, whereas a growth fold generated by transpression produces angular unconformities both along dip and strike.

To the north, the Belluno Line may have been a blind thrust generating a triangle zone during earlier stages of the deformation, later rising at the surface in the northern limb of the Belluno Syncline. This is supported by the steep attitude of the northern limb of the Belluno Syncline which is difficult to explain geometrically as a simple footwall syncline. It is also notable that triangle zones mainly occur in the Mesozoic Belluno Basin (Fig. 5), rather than in the neighbouring platforms. In fact the Belluno Syncline is developed in the deepest structural zone with thick basinal lithofacies. This earlier structural situation had an important influence in the morphology and source areas of the hydrographic pattern during the Late Miocene.

Timing of the thrust belt

In the Venetian segment of the Alps the thrusts became active during Late Oligocene to Quaternary times. Tortonian sandstones are thrust by the Valsugana Line (Venzo 1939) and Pliocene shales are folded along the frontal triangle zone. Moreover Messinian-Pliocene onlap geometries in the southern border of the chain support a mainly Neogene age of deformation. The extension of the unconformities within the molasse is a function of the thrust belt structure and reflects areas of stronger uplift. A problem is represented by the style and timing of the orogenic evolution: Has the thrust belt grown southward in a regular fashion by continuous creep since Late Oligocene time and do the unconformities record moments of sea level fall (low stand)? Or was the chain generated step by step so that the unconformities simply mark moments of tectonic activity? In general, plates move with a regular velocity suggesting that in areas of deformation the tectonic evolution should follow an almost constant activity. If the tectonic evolution generated with a constant regularity and tectonic activity had a wavelength too short or too long with respect to the eustatic sea level changes, then an interesting problem appears in dating the thrust belt: the timing of the tectonic activity has been considered as the time missing at regional unconformities and the age of coarse-grained sediment supply (i.e. the Messinian conglomerate) onlapping the

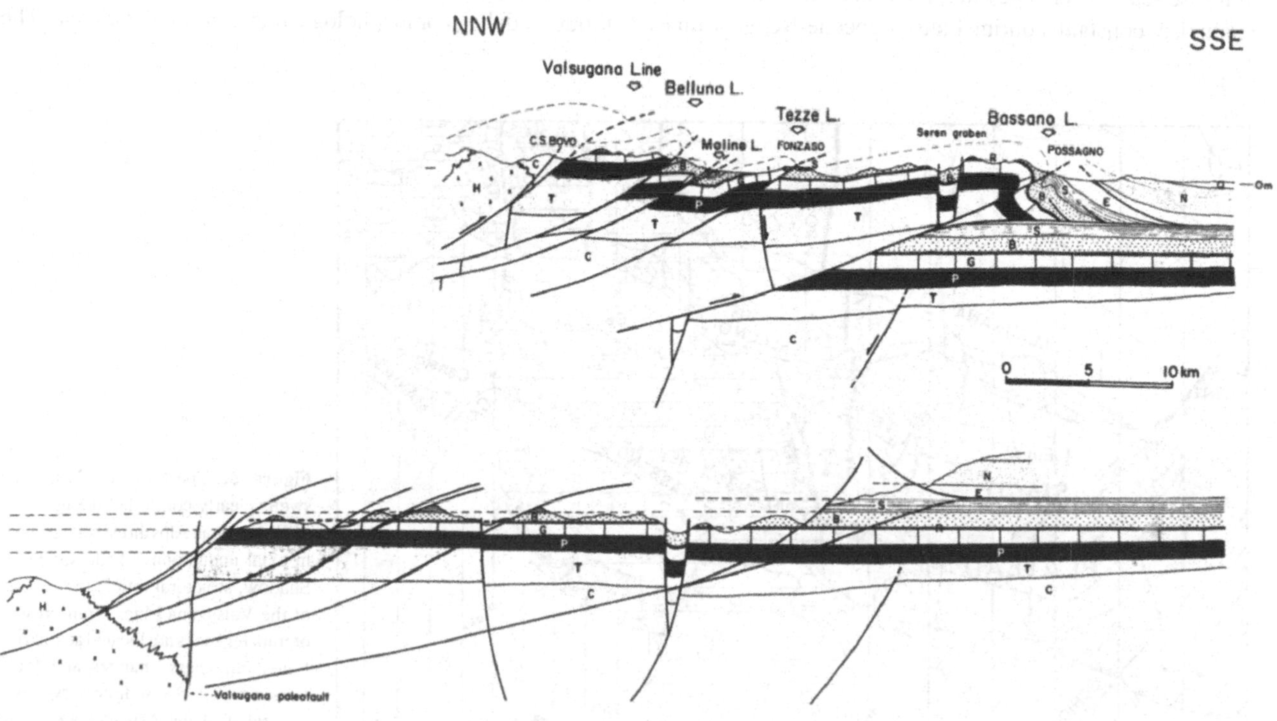

Figure 3. Balanced cross section across the Venetian Alps. See Figure 2 for location. Horizontal scale = vertical scale. Legend: C, crystalline basement; H, Late Hercynian granite; T, Late Permian-Lower and Middle Triassic formations; P, Late Triassic (Dolomia Principale); G, Liassic platform facies (Calcari Grigi) gradually southward passing to Liassic-Dogger basinal facies in the Venetian Plain (Soverzene Formation, Igne Formation, Vajont Limestone); R, Dogger-Malm basinal facies (Lower and Upper Rosso Ammonitico, Fonzaso Formation); B, Early Cretaceous (Biancone); S, Late Cretaceous (Scaglia Rossa); E, Palaeogene (Possagno Marls, etc.); N, Late Oligocene—Neogene Molasse; Q, Quaternary.

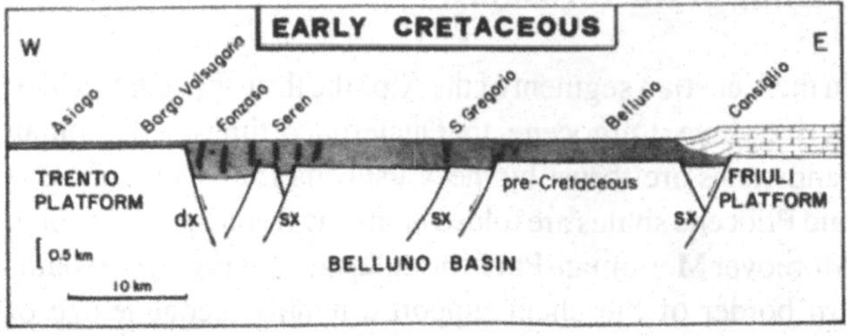

Figure 4. Interpreted W—E Early Cretaceous cross section from the Asiago plateau to the Cansiglio plateau, showing the coeval extensional tensional tectonics, which produced differential subsidence in the area. Note the basinward prograding carbonate platform in the eastern side (M. Cavallo Limestone) and the different thicknesses in the basinal sedimentation (Biancone). Neptunian dykes (vertically shaded) characterize the zones of tensional tectonics (i.e. the Arsiè Lake). The Neogene deformation has been strongly influenced by the pre-existing structural and stratigraphic geometries: dx and sx are dextral or sinistral transpressive or transfer zones which emplaced at the pre-existing anisotropies. Compare the general tectonics of the area with the inherited Mesozoic structural background (Fig. 5).

discontinuity. However, if the unconformities recorded only moments of general lowstand (in this case global or confined to the salinity crisis in the Mediterranean area) then one could argue that the thrust belt developed more gradually and that the sea level oscillations orchestrated the arrangement of the syntectonic sedimentary sequences. The different interpretations of the unconformities and conglomeratic supply in the molassic sequences allow different tectonic reconstructions. With the sea level change interpretation (Vail *et al.* 1977) the chain rises constantly during Late Oligocene-Neogene times, but if we assume the unconformities and conglomeratic supply to be tectonic-related, then episodic tectonic activity existed, which is in contrast to the regular activity of the frontal growth fold and the general plate motion.

In the Belluno syncline an angular unconformity marks the Early Eocene Flysch - Late Oligicene Molasse contact. Moreover the Flysch seems to onlap the northern limb of the Belluno syncline and to the west the Seren Valley alignment. Consequently it is not possible to exclude the interpretation that the area underwent compressive tectonics during Palaeogene times.

Interference pattern

The basement and the sedimentary cover of the region were broken by N-S trending normal faults and N60°-90°E transfer faults (the palaeo-Valsugana Line ? Fig. 3) during the Late Permian-Mesozoic rifting phases. These features have been cut, re-used or deformed during the Alpine inversion. Local structural undulations in the general N60°-80°E trend of the chain (fold axis, direction of thrust planes, etc.) everywhere occur in correspondence to inherited features in the basement and in the Mesozoic sedimentary cover which occurs in approximately N-S trending basins and swells. The present tectonic configuration is due to the inherited Mesozoic structure. The structural evolution of the area followed boundary features as transfer zones at horst margins (i.e. at the Trento Platform and Friuli Platform margins) which have influenced the geomorphological evolution of the area. The

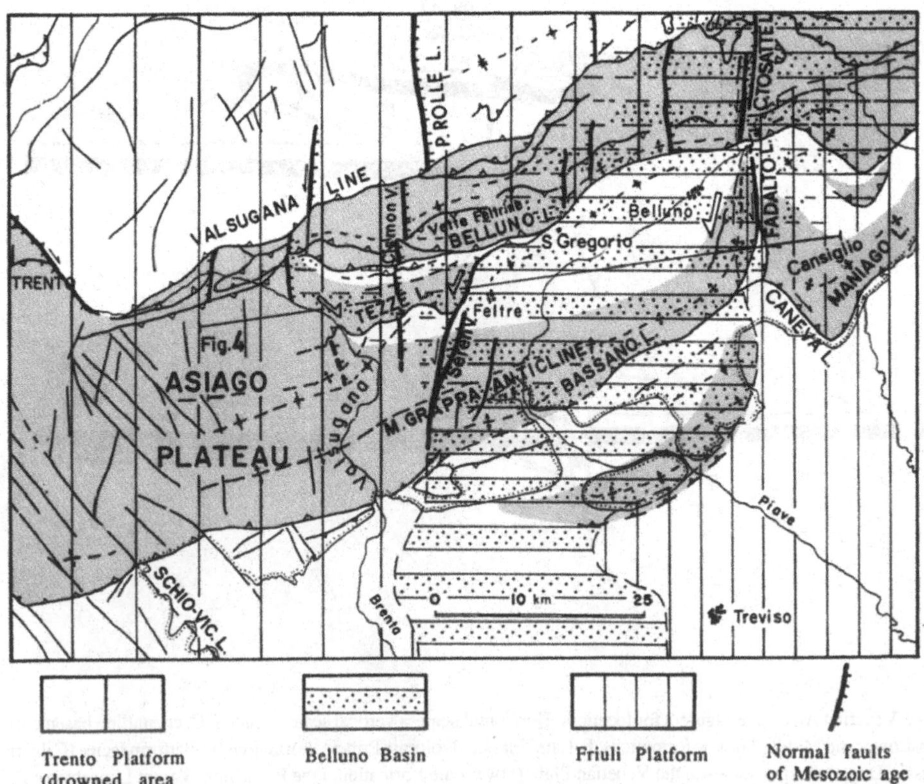

Figure 5. The Mesozoic basins and swells clearly controlled the geometry of the Venetian Alps thrust belt, resulting in axial undulations of the structures. Shadow, structural reliefs to the south of the Valsugana Line. Transpressive or transfer zones are located at inherited faults, thickness changes and facies transitions of the sedimentary cover. The relief of the Asiago Plateau—M. Grappa Anticline suddenly decreases in size when the deformation enters the Belluno Basin and eastward becomes a sinistral transpressive zone at the intersection with the Friuli Platform. Note how the structural reliefs are more diffuse within the inherited Belluno Basin. Compare Figure 4.

Asiago pop-up (Barbieri 1987) constitutes the western part of the study area, and is a wide plateau formed on the inherited Trento Platform (horst). The deformation at the western end of the Valsugana Line is transferred in the Trento area through the dextral transpressive Calisio Line (Fig. 2) to the sinistral transpressive Giudicarie Belt to the west. This undulation similarly runs around a minor inherited Late Palaeozoic and Mesozoic horst, within the wider Trento Platform. On the basis of thickness and facies changes the amplitude of the Trento Platform was probably wider in the east during Jurassic times (Seren Valley) and probably retreated (as horst, with basinal facies) by about 10 km during Cretaceous times (eastern margin of the Asiago Plateau, Valsugana Valley). The Tezze Line develops at this final eastern margin of the Trento horst and undulates in an oblique and lateral ramp (sinistral transpression) at the intersection with the inherited Seren Valley alignment (Fig. 5). The Alpine deformation within the Belluno Basin is more diffuse, characterized by a major number of thrust planes and reduced wavelength folds with respect to the lateral platform areas (Fig. 5). The Belluno Line mainly develops to the east of the Trento Platform. It branches the Valsugana thrust and shows an eastward increase in displacement and amplitude of the fault-propagation folding and fault-bend folding in the hangingwall. Commonly, the inherited tensional Mesozoic areas have been reactivated in transpressive zones (Fig. 4) and are transfer zones between two different styles of deformation. For instance the N-vergent backthrust in the hangingwall of the Belluno Line ends at the western margin of the Vette Feltrine at the intersection with the inherited tensional zone of the Cismon Valley (Fig. 5). The Caneva Line and the Fadalto Line are two respectively dextral and sinistral transpressive zones at the eastern margin of the Belluno Basin (Fig. 5). The Caneva Line represents the eastern transfer fault of the frontal triangle zone (Fig. 5). The Fadalto transpression was emplaced at the western termination of the Friuli Platform. The study area was located in the foredeep of the Dinaric thrust belt during Palaeogene times (Fig. 1) and suffered subsidence due to the load of the WSW-vergent Dinaric thrust sheets. A regional ENE-dipping monocline developed at that time and was inherited and involved in the younger SSE-vergent Neogene Southalpine deformation. The variations along strike of the deformation are reflected also in the Neogene and Quaternary foreland basin.

Conclusions

The good outcrops and the clear interference between inherited features and Alpine tectonics make the Venetian Alps a classic example of a thrust belt. Earlier Mesozoic features strongly influenced the evolution of the chain. Any kind of structural undulation along strike of the thrust belt is associated with pre-existing synsedimentary faults, thickness and/or facies variations in the sedimentary cover. The thrusts are arranged in an imbricate fan geometry and show a frontal triangle zone which was probably present at earlier stages of the thrust belt in more internal zones. The variations along strike of the deformation are reflected also in the Neogene and Quaternary molasse. The frontal triangle zone appears to be a growth fold from Late Oligocene to Quaternary time because clastic sedimentation on the southern limb of the anticline shows onlap geometries and reduced thicknesses. According to this progressive evolution of the thrust belt, the unconformities within the molasse could record lowstands of eustatic cycles.

Thanks to R. Crane and three anonymous referees who revised the paper. I also acknowledge K. R. McClay, A. Bally, G. Barbieri, D. Bernoulli, A. Bosellini, J. Channell, G.V. Dal Piaz, H. Laubscher, D. Masetti, F. Massari, I. Moretti, J. Platt, E. Semenza and E. Zappaterra for useful comments and discussions. The research was supported by an MPI grant.

References

Auboin, J. 1963. Essai sur la paléogéographie post-triasique et l'évolution secondaire et tertiaire du versant sud des Alpes orientales (Alpes méridionales; Lombardie et Vénétie, Italie; Slovénie occidentale, Yougoslavie). *Bulletin Société Géologique de France*, 7, **5**, 5, 730-766.

Bally, A. W., Gordy, P. L. & Stewart, G. A. 1966. Structure, seismic data and orogenic evolution of Southern Canadian Rocky Mountains. *Bulletin of Canadian Petroleum Geology*, **14**, 337-381.

Barbieri, G. 1987. Lineamenti tettonici degli Altipiani Trentini e Vicentini tra Folgaria e Asiago (Prealpi Venete). *Memorie Scienze Geologiche*, **39**, 257-264.

Bernoulli, D., Caron, C., Homewood, P., Kalin, O. & Von Stuijvenberg, J. 1979. Evolution of continental margins in the Alps. *Schweizerische Mineralogische und Petrographische Mitteilungen*, **59**, 165-170.

Bosellini, A. 1965. Lineamenti strutturali delle Alpi Meridionali durante il Permo-Trias. *Memorie Museo Storia Naturale Venezia Tridentina*, **15**, 3, 1-72.

—— 1973. Modello geodinamico e paleotettonico delle Alpi Meridionali durante il Giurassico-Cretacico. Sue possibili applicazioni agli Appennini. *In:* Accordi, B. (ed.) *Moderne vedute sulla geologia dell'Appennino*, Accademia Nazionale Lincei, **183**, 163-205.

—— & Doglioni, C. 1986. Inherited structures in the hangingwall of the Valsugana Overthrust (Southern Alps, Northern Italy). *Journal of Structural Geology*, **8**, 5, 581-583.

Boyer, S. E. & Elliott, D. 1982. Thrust systems. *American Association of Petroleum Geologists Bulletin*, **66**, 1196-1230.

Braga, G. P. 1970. L'assetto tettonico dei dintorni di Possagno (Trevigiano occidentale). *Accademia Nazionale Lincei*, **48**, 451-455.

Buxtorf, A. 1916. Prognosen und Befunde beim Hauensteinbasis und Grenchenbergtunnel und die Bedeutung der letzteren für die Geologie des Juragebirges. *Verhandlungen der Naturforschenden Gesellschaft in Basel*, **27**, 184-254.

Cassano, E., Anelli, L., Fichera, R. & Cappelli, V. 1986. Pianura Padana, Interpretazione integrata di dati geofisici e geologici. *Agip*, int. report, 1-27.

Castellarin, A. 1972. Evoluzione paleotettonica sinsedimentaria del limite tra 'piattaforma veneta' e 'bacino lombardo', a nord di Riva del Garda. *Giornale Geologia*, **37** (1970), s. 2, 1, 11-212.

—— 1979. Il problema dei raccorciamenti crostali nel sudalpino. *Rendiconti Società Geologica Italiana*, **1** (1978), 21-23.

Dal Piaz, G. 1912. Studi geotettonici sulle Alpi Orientali (Regione tra il Brenta ed il Lago di Santa Croce). *Memorie Istituto Geologia R. Università di Padova*, **1**, 1-196.

Doglioni, C. 1987. Tectonics of the Dolomites (Southern Alps, Northern Italy). *Journal of Structural Geology*, **9**, 181-193.

—— & Bosellini, A. 1987. Eoalpine and mesoalpine tectonics in the Southern Alps. *Geologische Rundschau*, **76**, 3, 735-754.

—— & Neri, C. 1988. Anisian tectonics in the Passo Rolle area. *Rendiconti Società Geologica Italiana*, **11**, 197-204.

Jones, P. B. 1982. Oil and gas beneath east-dipping underthrust faults in the Alberta foothills. *In:* Powers, R. B. (ed.) Geologic Studies of the Cordilleran Thrust Belt. *Rocky Mountain Association of Geologists*, 61-74.

Laubscher, H. P. 1983. The Late Alpine (Periadriatic intrusions and the Insubric Line. *Memorie Società Geologica Italiana*, **26**, 21-30.

—— 1985. Large scale, thin skinned thrusting in the Southern Alps: kinematic models. *Geological Society of America Bulletin*, **96**, 710-718.

Leonardi, P. 1965. Tectonics and Tectonogenesis of the Dolomites. *Eclogae Geologicae Helvetiae*, **58**, 1, 49-62.

Masetti, D. & Bianchin, G. 1987. Geologia del Gruppo della Schiara. *Memorie Scienze Geologiche*, **39**, 187-212.

Massari, F., Grandesso, P., Stefani, C. & Jobstraibizer, P. G. 1986. A small polyhistory foreland basin evolving in a context of oblique convergence: the Venetian basin (Chattian to Recent, Southern Alps, Italy). *International Association of Sedimentologists Special Publication*, **8**, 141-168.

Müller, M., Nieberding, F. & Wanninger, A. 1988. Tectonic style and pressure distribution at the northern margin of the Alps between Lake Constance and River Inn. *Geologische Rundschau*, **77**, 787-796.

Pieri, M. & Groppi, G. 1981. Subsurface geological structure of the Po plain, Italy. *Progetto Finalizzato Geodinamica*, **414**, 1-13.

Platt, J. P. 1988. The mechanics of frontal imbrication: a first order analysis. *Geologische Rundschau*, **77**, 577-589.

Roeder, D. 1989. South-Alpine thrusting and trans-Alpine convergence. *In:* Coward, M. P., Dietrich, D. & Park, R. G. (eds) *Alpine Tectonics*, Geological Society Special Publication, **45**, 211-227.

Slejko, D., Carulli, G. B., Carraro, F., Castaldini, D., Cavallin, A., Doglioni, C., Iliceto, V., Nicolich, R., Rebez, R., Semenza, E. Zanferrari, A. & Zanolla, C. 1987. Modello sismotettonico dell'Italia nord-orientale. *CNR, GNDT, Rendiconto*, **1**, 1-82.

Vail, P. R., Mitchum, R. M. Jr. & Thompson, S. III 1977. Seismic stratigraphy and Global Changes of Sea Level, Part 4: Global Cycles of Relative Changes of Sea Level. *In:* Payton, C. E. (ed.) *Seismic Stratigraphy - applications to hydrocarbon exploration*. American Association of Petroleum Geologists Memoir, **26**, 83-98.

Venzo, S. 1939. Nuovo lembo tortoniano strizzato tra le filladi ed il permiano a Strigno di Valsugana (Trentino meridionale orientale). *Bollettino Società Geologica Italiana*, **58**, 1, 175-185.

Winterer, E. L. & Bosellini, A. 1981. Subsidence and Sedimentation on a Jurassic Passive Continental Margin (Southern Alps, Italy). *American Association of Petroleum Geologists Bulletin*, **65**, 394-421.

Zanferrari, A., Bollettinari, G., Carobene, L., Carton, A., Carulli, G. B., Castaldini, D., Cavallin, A., Panizza, M., Pellegrini, G. B., Pianetti, F. & Sauro, U. 1982. Evoluzione neotettonica dell'Italia nord-orientale. *Memorie Scienze Geologiche*, **35**, 355-376.

PART SIX

Himalayas

Thrust geometries, interferences and rotations in the Northwest Himalaya

P. J. Treloar[1,2], M. P. Coward[1], A. F. Chambers[2], C. N. Izatt[3] & K. C. Jackson[1]

[1]*Department of Geology, Imperial College, London, SW7 2BP, UK*
[2]*School of Geological Sciences, Kingston Polytechnic, Kingston-upon-Thames, Surrey, KT1 2EE, UK*
[3]*International Exploration, British Gas, 59 Bryanston Street, London, W1A 2AZ, UK*

Abstract: In North Pakistan the dominant transport direction throughout Himalayan collision has been to the S or SSE. Southward propagation of thrusts within the thickened Indian Plate has, however, been impeded by interference with SW-verging thrusts in Kashmir, at the western end of the main Himalayan oroclinal chain. As a result of this interference, thrusts within both the Pakistani and Kashmiri systems have become pinned at their lateral terminations. Lineation and palaeomagnetic data document substantial rotations of whole thrust sheets, of up to 40° around the pinned terminations, anticlockwise in Pakistan and clockwise in Kashmir. Although such rotations are best seen within the Pliocene to Recent structures of the external zones, similar rotations can be determined within Oligocene structures in the internal zones. The NW Himalayan syntaxes are crustal scale folds which have grown within the zone of convergence between the two thrust systems. The main Himalayan thrust system is interpreted as having been pinned within the Himalayan chain, rather than at its western termination, due to early thickening of the northwestern Indian Plate having acted as a mechanical impediment to the lateral propagation of the main Himalayan thrusts.

The tectonic evolution of the Himalayan-Tibetan system of thickened continental crust is the result of the Tertiary collision between India and Asia. Following breakup of Gondwanaland, India commenced its northward movement at about 80 Ma ago. Sea floor palaeomagnetic stripe data show this movement to have been at 15-20 cm/yr. Following continental collision during the early Tertiary, at about 52 Ma (Patriat & Achache 1984) or 45 Ma (Dewey *et al.* 1989) Ma, this rate decreased to about 5 cm/yr. As a result of the cumulative post-collisional northward drift, the Himalayan-Tibet system must have accommodated over 2000 km of shortening through a combination of subduction of continental India under Asia (Argand 1924; Powell & Conaghan 1973; Butler & Coward 1989), thickening of the leading edge of India by thrusting (Coward & Butler 1985; Mattauer 1986), homogeneous thickening of Tibet and regions to the north (England & Houseman 1986; Dewey *et al.* 1988) and some form of lateral eastward extrusion of Tibet (Molnar & Tapponier 1985; Tapponier *et al.* 1986; England & Molnar 1990). Here we deal with aspects of the geometry and timing of the thrusting which accommodated thickening of the northern part of the Indian Plate.

The India-Asia collision is not a simple orthogonal one. Since collision India has rotated anti-clockwise with respect to Asia (Klootwijk *et al.* 1985), with the result that some 1000 km more shortening has occurred in the eastern Himalaya than in the west (Dewey *et al.* 1989). The structural effects of this westward decrease in continental subduction are clearly seen. The main Himalayan chain has an oroclinal form (Fig. 1) in which the strike of the major thrusts swings from E-W in Nepal to NW-SE in Kashmir. As the transport direction is everywhere approximately perpendicular to the strike of the thrusts (Brunel 1986), movement directions within the main chain, as indicated by stretching lineations and fault plane solution data (Seeber *et al.* 1981), are strongly divergent along the length of the chain. As a consequence of the divergence in transport direction, together with the reduction in the amount of transport towards the northwest, the Indian Plate thrust sheets of the main Himalayan chain show increasingly strong clockwise rotations towards their northwestern tips. Palaeomagnetic evidence for such clockwise rotations have been described by Klootwijk *et al.* (1985) and Bossart *et al.* (1990).

In North Pakistan the main Himalayan-age thrusts are ENE-striking with a dominantly southerly or SSE transport direction (Coward *et al.* 1988). This is in direct contrast to the NW-striking, SW-verging thrusts immediately to the the east in NW India. Consequently there is a strong convergence of movement directions in the NW Himalaya which is marked, within the zone of convergence, by the development of the large antiformal folds of the Northwest Himalayan syntaxes.

This paper aims to summarize the tectonic evolution of the Pakistan Himalaya and to show how lineation and palaeomagnetic data may be used to determine regional scale rotations. These data can then be used to constrain how the two converging thrust directions interfere and generate a range of regional structures.

Geology of North Pakistan

In Pakistan, the Indian and Asian plates are separated by the

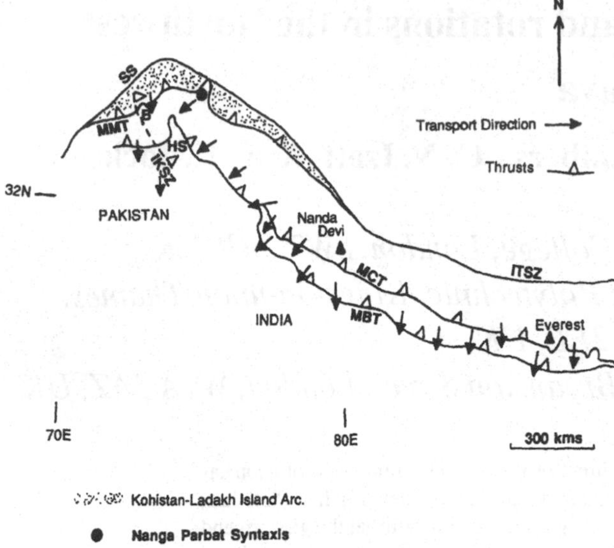

Figure 1. Simplified structural map of the Himalayan arc showing the location of the northwestern syntaxes, the radial nature of the movement direction (after Brunel 1986) that results in the thrust interferences in the northwestern part of the arc, and the location of the currently active Indus Kohistan seismic zone (IKSZ). ITSZ - Indus Tsangbo suture zone. MBT - Main Boundary thrust; MCT - Main Central thrust; MMT - Main Mantle thrust; SS - Shyok suture; HS - Hazara syntaxis.

rocks of the Kohistan Island Arc (Fig. 2). Kohistan was sutured to the Asian Plate along the Northern, or Shyok, Suture at about 100 Ma (Pudsey *et al.* 1986; Treloar *et al.* 1989a) well before the onset of Himalayan collision. Himalayan collision in Pakistan was thus between Kohistan and India, with Kohistan thrust southwards onto the Indian Plate along the Main Mantle thrust (MMT), which is the true westward continuation of the Indus-Tsangbo Suture Zone.

During collision the leading edge of the Indian Plate was subducted beneath Kohistan where it was thickened by a combination of processes including ductile shearing, recumbent folding and backthrusting (Butler & Coward 1989; Treloar *et al.* 1989b). In restoring the imbricated Indian Plate sequences, Coward & Butler (1985) estimated a total shortening of about 470 km. As the imbricate stack they restored is composed dominantly of cover rocks, with little or no basement involvement, Coward & Butler suggested that cover and basement sequences had been decoupled early in Himalayan collision with the equivalent of 470 km of basement being subducted and thickened underneath Kohistan. These decoupled cover and basement sequences were re-imbricated later in the Himalayan deformation history (Treloar *et al.* 1989b).

The Indian Plate rocks can be divided into three: internal crystalline thrust sheets, external thrust sheets and molasse

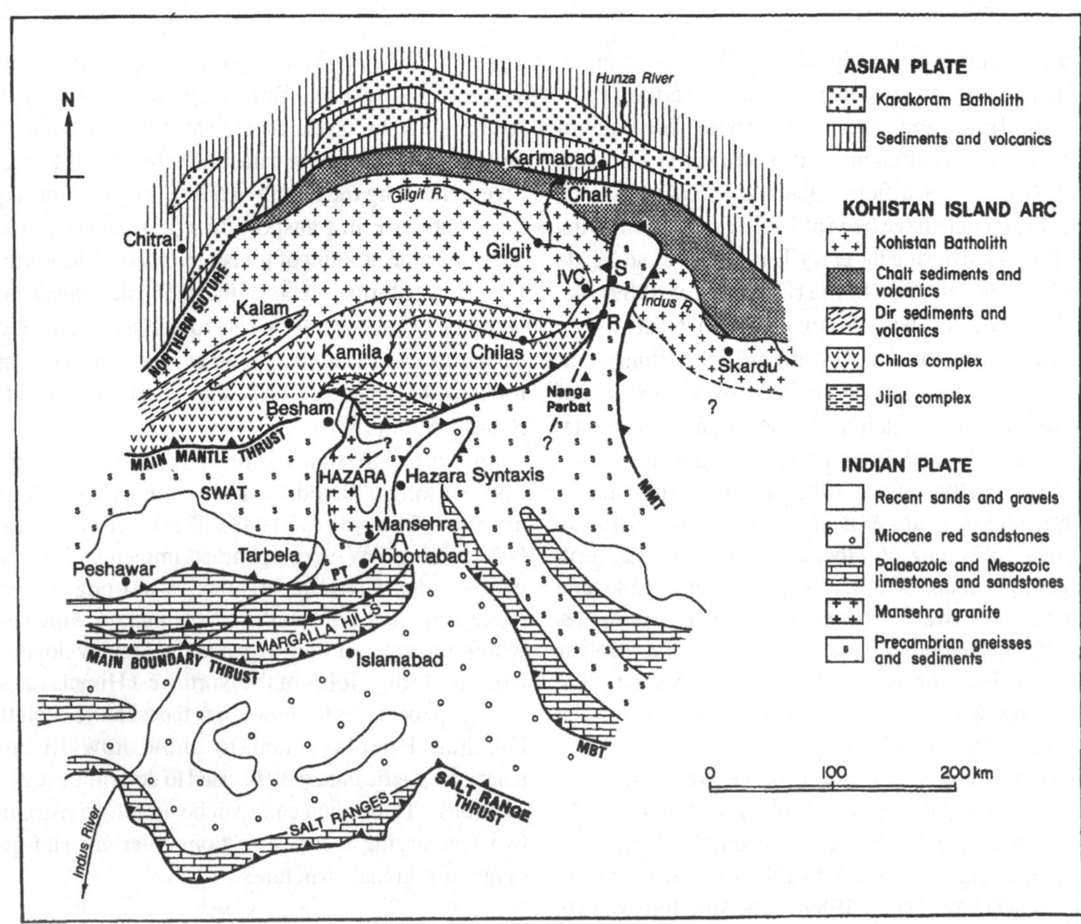

Figure 2. Geological map of northern Pakistan showing the location of the major thrusts in the Indian Plate and of the Nanga Parbat and Hazara syntaxes, and the separation of the Indian and Asian plates by the Kohistan island arc. R-Raikot; S-Sassi; PT - Panjal thrust; MBT - Main Boundary thrust.

bearing thrusts (Fig. 2). The crystalline rocks of the internal zones, including those rocks metamorphosed during the Himalayan orogeny, outcrop between the MMT and the Panjal thrust. To the south, between the Panjal thrust and the Main Boundary thrust (MBT) outcrop weakly to unmetamorphosed sediments that include Precambrian slates as well as carbonate-rich rocks of Lower Palaeozoic to Eocene age. By analogy with the main Himalayan chain these form the 'Lesser Himalayan' sequences. The Lesser Himalayan rocks are thrust southwards along the MBT onto Tertiary-aged fluviatile molasse sediments of the foreland basins.

Although some Palaeocene to Eocene aged molasse is preserved within the core of the Hazara syntaxis (Bossart & Ottiger 1989), much of the molasse, which has been dated by magneto-stratigraphic methods (Johnson *et al.* 1979, 1982; Opdyke *et al.* 1979), is of Miocene and younger age (Table 1). There is an apparent sedimentation gap between the Eocene molasse and the base of the Miocene-aged Murree Formation with little evidence for Oligocene deposition. Within the Miocene and younger sediments there is a clear southward migration of the sedimentary depocentre. Both the southward migration of the depocentre and the hiatus in sedimentation during the Oligocene are features recorded throughout the length of the Himalayan chain.

Group	Sub-Group	Formation	Age (Ma)
Siwalik	Upper Siwalik	Soan	5.0 - 0.0
Siwalik	Middle Siwalik	Dhok Pathan	8.5 - 5.0
Siwalik	Middle Siwalik	Nagri	10.6 - 8.5
Siwalik	Lower Siwalik	Chinji	14.2 - 10.6
Siwalik	Lower Siwalik	Kamlial	18.0 - 14.2
Rawalpindi	Murree		>18.0

Table 1. Stratigraphy of the Miocene to Recent foreland basin molasse sediments of northern Pakistan (after Burbank & Beck 1989, and references therein).

History of Indian plate deformation

a) The internal zone deformation

The earliest recorded deformations are within the internal

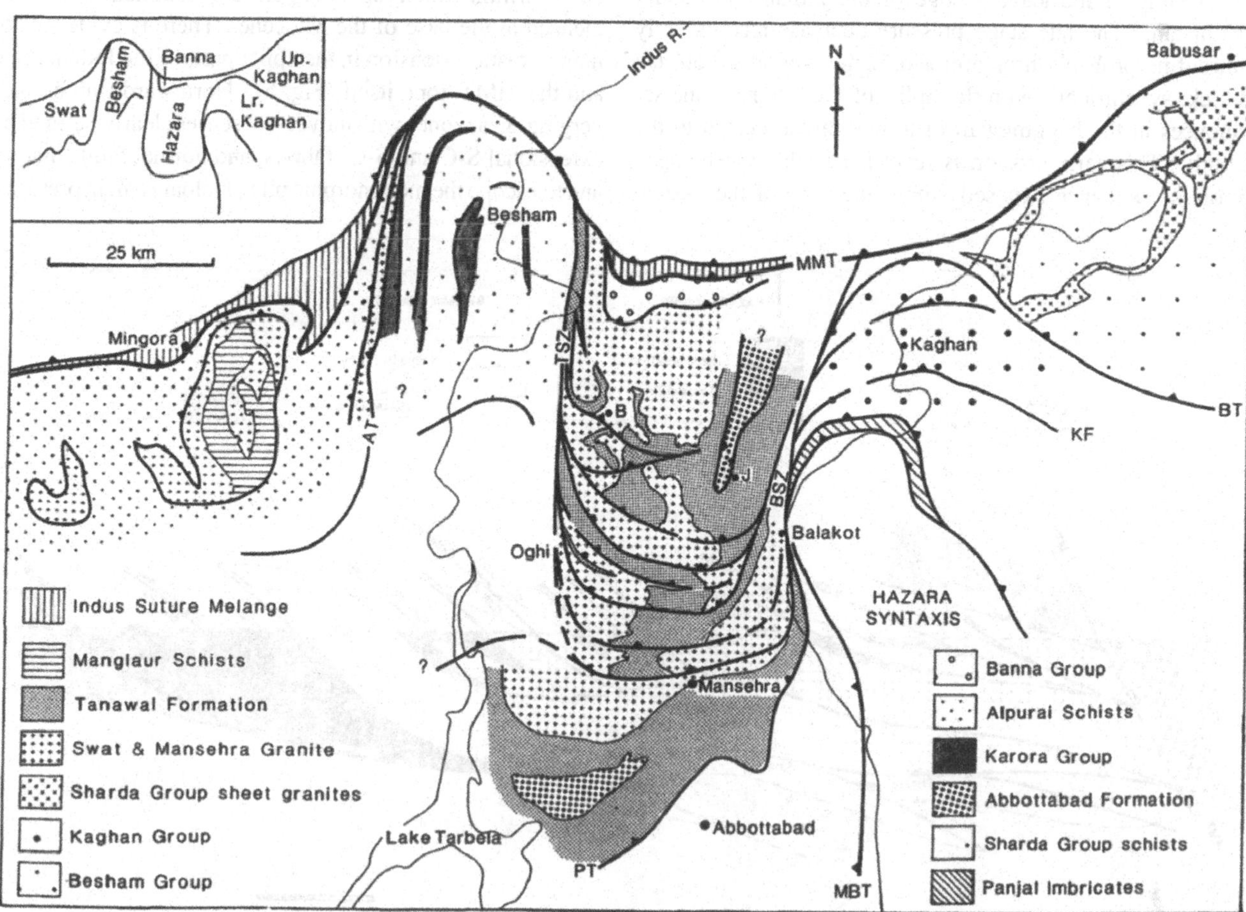

Figure 3. Geological map of the northern part of the Indian Plate, south of the Main Mantle thrust, between the Swat and Kaghan valleys. The inset shows how the region is divided into a series of large scale thrust nappes. Place names: B = Batagram, J = Jabori. Thrusts and shear zones: AT = Alpurai thrust; BT = Batal thrust; KF = Khannian fault; MBT = Main Boundary thrust; MMT = Main Mantle thrust; MT = Mansehra thrust; PT = Panjal thrust; BSZ and TSZ = Balakot and Thakot shear zones.

zones within which we can identify two distinct periods of shortening. The first of these was synchronous with the subduction of India under Kohistan (Treloar *et al.* 1989b), and involved the early decoupling of cover and basement sequences associated with intense ductile shearing and mylonitization. Porphyroblast-fabric relationships, including spectacularly spiralled garnets, show that metamorphism was synchronous with the ductile deformation (Treloar *et al.* 1989b & c). Within garnets in calcareous schists in the Swat Valley, ilmenite inclusions in garnet cores and rutile inclusions in their rims and in the groundmass, show that prograde metamorphism, to a peak at about 625±50°C at 10±2 kb was along a path of increasing pressure. Gibbs' method analysis of samples from both the Hazara and Kaghan valley districts confirm the overall core to rim pressure increase within the Swat Valley. However, a reverse Ca zonation in near rim regions of garnet profiles is consistent with a late stage pressure release. Evidence for this is also indicated from samples elsewhere in the crystalline zones. In one sample from the Kaghan Valley the mantling of rutile by ilmenite in a garnet-plagioclase-kyanite bearing rock clearly demonstrates this.

The P-T-t paths for the internal zones are thus in accord with prograde metamorphism during active subduction and thickening of the leading edge of the Indian Plate under Kohistan. The late stage pressure decrease records early unroofing, probably through erosion of the overthrust Kohistan block, synchronous with the uplift of the internal zone sequences in the hangingwall of a new thrust located to the south. This early erosion is recorded in the Eocene-aged molasse sequence exposed within the core of the Hazara syntaxis. Ar-Ar hornblende ages of 39±2 Ma (Treloar & Rex 1990a & b; Lawrence *et al.* 1990) show that metamorphism and early stages of post-metamorphic cooling through ca. 500°C predated 40 Ma.

The metamorphic pile was subsequently disrupted during a second deformation event. During this, the decoupled cover and basement sequences were re-imbricated within a series of lithologically distinct thrust nappes (Fig. 3) stacked within a S-verging crustal scale thrust stack (Coward *et al.* 1988; Treloar *et al.* 1989c & d). Sharp metamorphic breaks mark the boundaries between these nappes, each of which are internally imbricated by thrusts that placed already cooled but originally higher grade rocks on top of lower grade ones, generating an overall tectonic inversion of the metamorphic sequence (Fig. 4). Mica K-Ar ages (Treloar & Rex 1990a & b) from shear zones that stack these nappes show that thrust stacking pre-dated 30 Ma, although when it commenced is uncertain.

Post-metamorphic cooling histories show a period of rapid cooling of about 300°C in <10 Ma during the Oligocene to early Miocene which equates to a phase of rapid unroofing and exhumation through a combination of erosion and extensional normal faulting (Treloar & Rex 1990a). Erosion is recorded within the molasse sediments of the foreland basins within which the main period of sedimentation commenced at the base of the Miocene. There is evidence for major crustal extension in the upper parts of the Indian Plate and the MMT zone itself (Fig. 5). Northward (hinterland) verging shear zones with movement senses clearly defined by extensional S-C and S-C' fabrics, and normal faults disrupt and telescope the metamorphic pile (Treloar *et al.* in press, a).

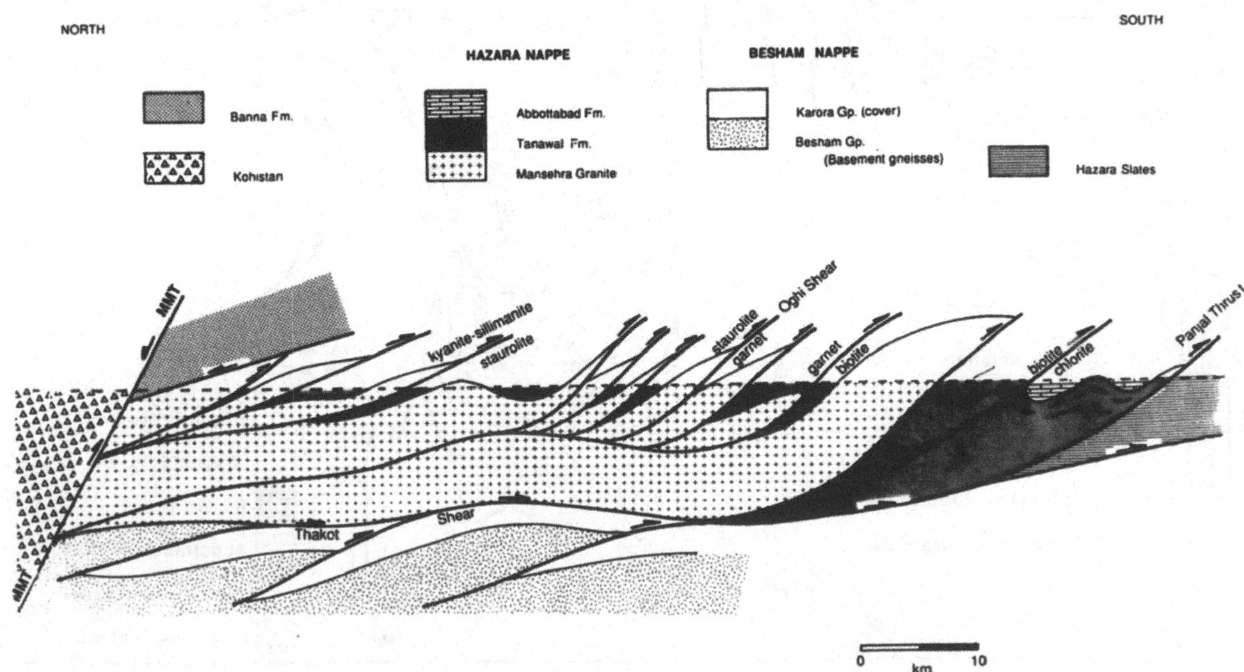

Figure 4. Section across the Hazara nappe showing the internal imbrication within the nappe and its structural relationships with the underlying Besham nappe and overlying low grade rocks of the Banna nappe. MMT - Main Mantle thrust. The Panjal thrust forms the effective sole thrust for the two internal zone deformations. Note the upward and northward increase in metamorphic grade, the result of post-metamorphic stacking and disruption of the metamorphic pile.

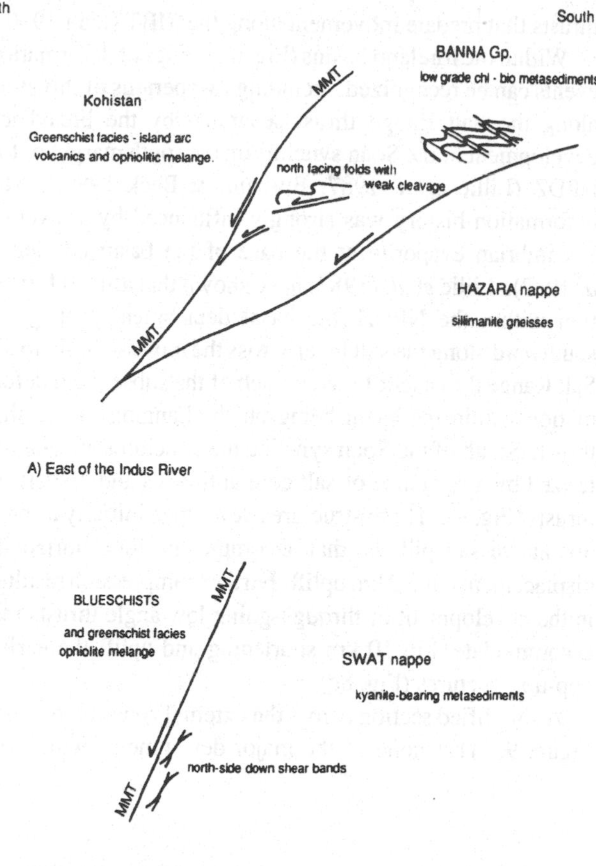

Figure 5. Sections across the Main Mantle thrust zone (**a**) east of the Indus river and (**b**) west of the river showing the disposition of low grade rocks in the hanging walls of late north-verging extensional normal faults each of which have high grade rocks in their hanging walls.

The chlorite-bearing greenschist facies rocks of the Banna nappe (Treloar *et al.* 1989b) are separated from the structurally underlying sillimanite gneisses of the Hazara nappe by one such shear zone (Figs 3 & 4). The MMT itself, formerly a S-vergent thrust, became a N-side down extensional fault with blueschist and greenschist rocks on its hangingwall and amphibolite grade metasediments on its footwall. Fission track data (Zeitler 1985) date this extension. In the Swat valley zircon ages at >45 Ma are some 20 Ma older on the MMT hangingwall than on the footwall where they are at <25 Ma. By contrast there is no significant difference in apatite ages across the MMT. On both hangingwall and footwall these are at about 16-20 Ma, implying that extension continued until after 23 Ma but ended before ca. 20 Ma.

Considerable tectonic uplift must have accompanied the exhumation, during the course of which up to 20 km of material was removed. Although some of this uplift may have been isostatic in nature, much must have been through uplift in the hangingwall of a major S-verging thrust. The obvious candidate is the Panjal Thrust, which already formed the effective sole thrust to the post-metamorphic thrust stack. Movement along the Panjal thrust must therefore have been (dis?)-continuous for over 20 Ma until about 20 Ma ago. This continued southward thrusting maintained the critical wedge dimensions of the mountain front during erosion. This model is analogous to those proposed for the Main Central thrust zone in Nepal and India (Royden & Burchfiel 1987; Searle & Rex 1989) where high level extension is synchronous with S-verging thrusting.

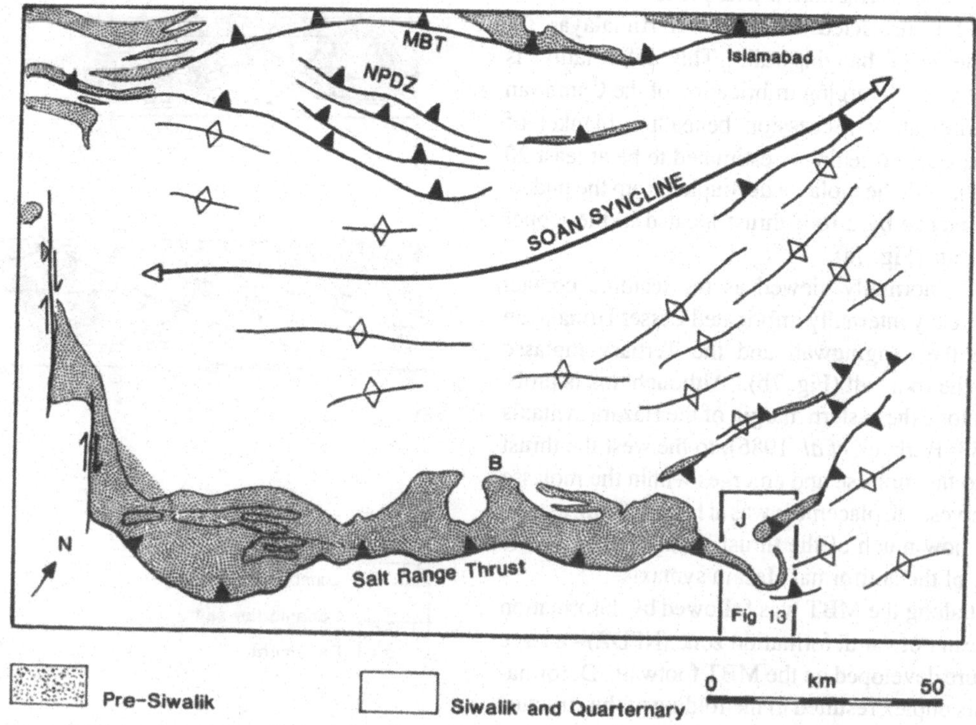

Figure 6. Simplified map of the geology of the foreland basins of northern Pakistan, after Butler *et al.* (1987); and Baker *et al.* (1988). (B) Bhaun. (J) Eastern Salt Ranges. MBT - Main Boundary thrust. NPDZ - North Potwar deformation zone.

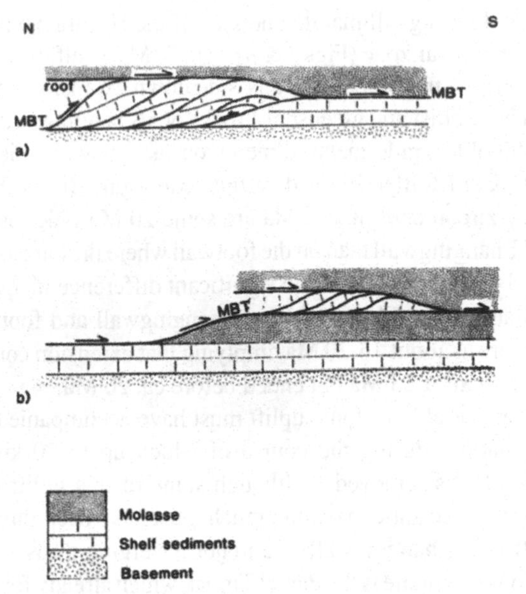

Figure 7. Structural evolution of the Lesser Himalaya and the Main Boundary thrust. (a) imbrication of the Cambrian to Eocene sediments on the Main Boundary thrust (MBT) hangingwall underneath a roof thrust along which the overlying Miocene molasse sediments were decoupled. (**b**) subsequent translation of the imbricated sediments southward along the Main Boundary thrust across the molasse basin.

b) The external zone deformation

Deformation within the external zones (Figs 2 & 6) can be divided into three phases: pre-, syn- and post-movement along the MBT. Deformation which predated movement along the MBT is restricted to the Lesser Himalayan sequences on the MBT hangingwall. This deformation is characterized by the S-verging imbrication of the Cambrian to Eocene sedimentary succession beneath a blanket of Miocene molasse. Shortening is estimated to be at least 20 km (Izatt 1990), with the molasse decoupled from the underlying shelf sequence by a roof thrust located in the Upper Eocene sediments (Fig. 7a).

The MBT is normally viewed as the tectonic contact between the already internally imbricated Lesser Himalayan sequences on the hangingwall and the Tertiary molasse sequences on the footwall (Fig. 7b). Although this is probably the case along the eastern margin of the Hazara syntaxis (Burbank 1983; Burbank *et al.* 1986), to the west the thrust cuts down into the molasse and emerges within the molasse basins themselves. Displacement was at least 120 km, as may be judged by how much of the thrust is exposed along the western flank of the antiformal Hazara syntaxis.

Movement along the MBT was followed by deformation within the North Potwar deformation zone (NPDZ) - a large duplex structure developed on the MBT footwall. Deformation within this duplex resulted in the folding of the structurally overlying thrust sequence. Much of the backsteepening of thrusts and apparent backthrusting within the Margalla Hills (Coward *et al.* 1988) is the result of this folding of thrusts that predate movement along the MBT (Izatt 1990).

Within the foreland basins (Fig. 6) a series of deformation events can be recognized, including two periods of thrusting along the Salt Range thrust separated by the breakback development of the Soan syncline on the south margin of the NPDZ (Lillie *et al.* 1987; Burbank & Beck 1989). The deformation history was strongly influenced by a layer of Eocambrian evaporite at the base of the basin (Butler *et al.*1987). Lillie *et al.* (1987) have shown that after deformation within the NPDZ, the basal detachment propagated southward along the salt layer across the molasse basin to the Salt Range thrust (SRT) with much of the subsequent deformation within the basin being on the hangingwall of that thrust. South of the Soan syncline the structures are characterized by a sequence of salt core antiforms and associated thrusts (Fig. 8). These structures developed initially as pop-ups above salt pillows that accommodate little horizontal displacement but 1-2 km uplift. Further compression resulted in the development of through-going low angle thrusts that accommodate 8 to 10 km shortening and uplift the earlier pop-up structures (Fig. 8c).

A simplified section across the external zones is shown in Figure 9. That none of the major detachments within the

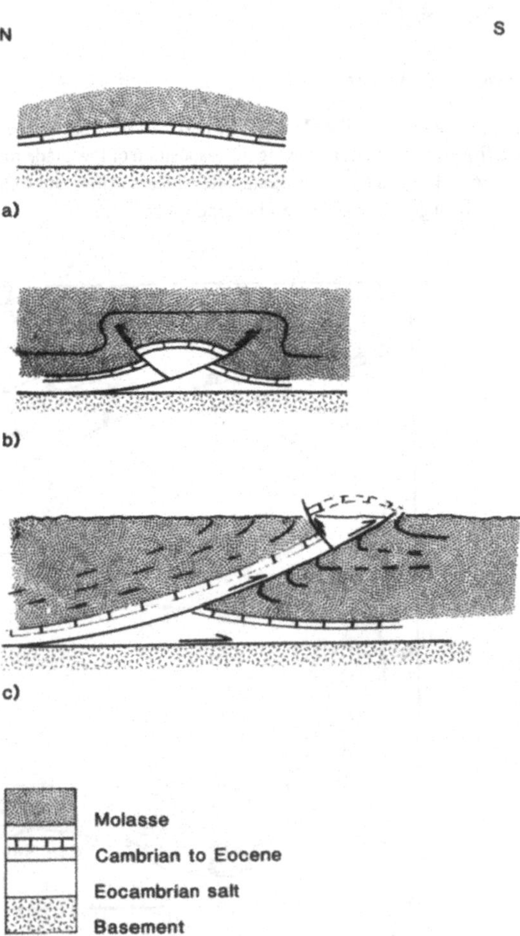

Figure 8. Generation of structures south of the Soan syncline. (a) early salt cored pillows. (b) pop-ups developed above the salt pillows (e.g. the Qazian fold). (c) late through going thrusting that uplifts the pop-up folds (e.g. the Domeli thrust) and generated salt cored antiformal structures at the surface.

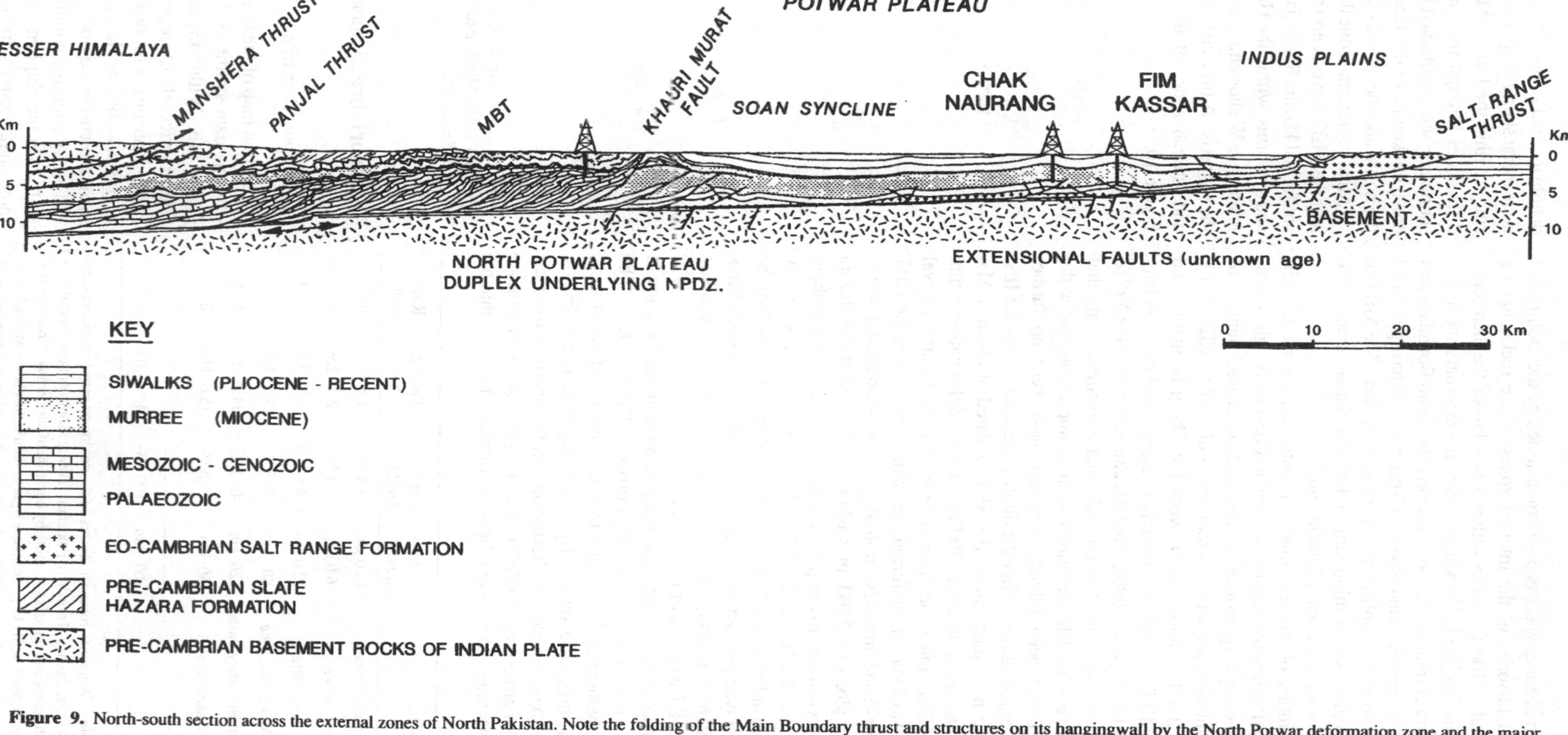

Figure 9. North-south section across the external zones of North Pakistan. Note the folding of the Main Boundary thrust and structures on its hangingwall by the North Potwar deformation zone and the major detachment within the foreland basin molasse.

external zones was steep is demonstrated by the cooling and uplift histories of the internal zones where rocks, having cooled to 100°C, were within 3 to 4 km of the then topographic surface by 20 Ma, or before the external zone deformation commenced. Consequently the Main Boundary and Salt Range thrusts and associated structures can have generated little hinterland uplift and subsequent erosion. The MBT in particular, with a minimum of 120 km displacement, have been a sub-horizontal detachment.

Timing of the deformation events, and the age of individual structures within the foreland basins, is tightly constrained by magnetostratigraphic and fission track analysis of the molasse sequences (Johnson *et al.* 1979, 1982, 1986). Structural relationships between the Chinji Formation and the MBT suggest that movement along the thrust was later than 10 Ma (Izatt 1990). That structures related to the NPDZ fold the MBT, implies not only that movement along that thrust predated deformation within the duplex, but that, as the MBT had been folded, there can have been no further movement along it after the duplex had formed. As the first phase movement along the SRT is dated at about 5 Ma, displacement along the MBT followed by deformation within the NPDZ must both have occurred within the time interval of 10 to 5 Ma. This inferred age for movement along the MBT as pre-5 and probably pre-8 Ma is earlier than similar movement along the MBT in Kashmir to the east of the Hazara syntaxis where major pulses of movement are dated at about 5 and 2 Ma (Burbank 1983). The Soan syncline is precisely dated at between 2.1 and 1.9 Ma. Deformation south of the Soan syncline post-dates deposition of the late Pliocene Dhok Pathan Formation, and thus dates the late stage movements on the SRT as younger than 2 Ma.

Amounts of shortening have been estimated for many of the external zone structures (Table 2). The magnetostratigraphic data permit estimates of when these structures were active. By combining these two estimates, approximate rates of shortening can be determined for the major structures (Table 2). It is not yet possible to estimate shortening amounts and hence deformation rates within the ductile internal zones. Within the external zones, however, it is clear that shortening rates have varied from >4 cm/yr to <1 cm/yr within the last 10 Ma. These variations may reflect differences through time in the way that total Himalayan shortening was partitioned between thickening of the Indian crust, thickening of the Tibetan crust and the Asiatic crust to the north, and the lateral extrusion of Tibet.

All of the internal zone thrusts and shears, the Panjal thrust and the MBT have been folded by the development of the antiformal Hazara Syntaxis. The large N-trending antiforms and synforms within the Hazara, Besham and Swat districts relate to E-W shortening associated with the development of the syntaxis. A time chart of events within both the external and internal zones of the Pakistan Himalaya is shown in Figure 10.

Structure	Shortening amount	Shortening cumulative	Timing Ma	Rate cm/yr
Salt Range thrust	12.0 km	4.6 %	1.8 - 0.0	0.67
Soan syncline	6.0 km	2.3 %	2.1 - 1.9	3.00
Salt Range thrust	12.0 km	4.6 %	4.0 - 3.0	1.20
N. Potwar def. zone	80 km	30.8 %	5.5 - 4.0	5.33
Main Boundary thrust	120 km	46.2 %	11.0 - 5.5	1.85
MBT hangingwall	30 km	11.5 %	15.0 -11.0	0.75
	260 km	100 %	15.0 - 0.0	1.73

Table 2. Amounts, timing and rates of shortening on major structures within the external zones of northern Pakistan. Average shortening rates are derived by dividing amount of shortening by duration of shortening. Errors on these rates are necessarily large and these figures should only be considered as very approximate indicators of the variations in shortening rate throughout the deformation history.

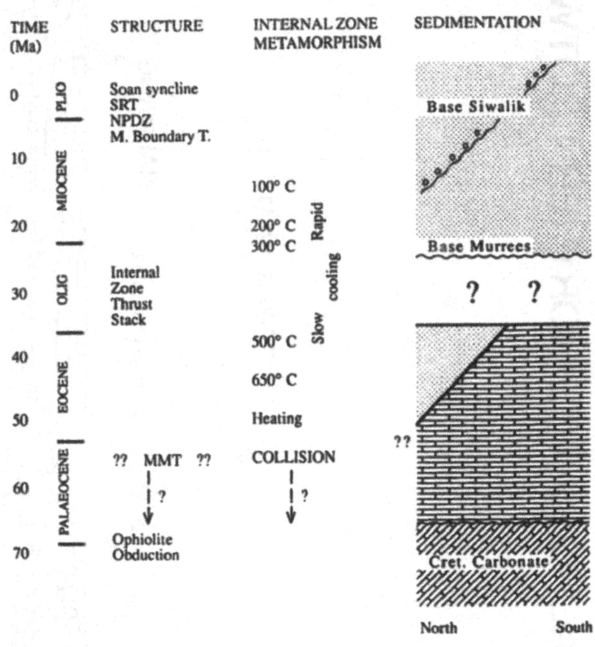

Figure 10. Time chart to show the sequence of events within the Indian Plate. MMT - Main Mantle thrust; NPDZ - North Potwar deformation zone; SRT - Salt Range thrust.

Transport directions and rotations

a) Transport directions

A lineation map of the internal zones is shown in Figure 11a with the data synthesised on a lineation trajectory map in Figure 11b. Within the crystalline internal zones, the main linear fabrics reflect the post-metamorphic imbrication of the Indian Plate on the footwall of the MMT. West of the Hazara syntaxis, within the Hazara, Besham and Swat nappes, movement directions are dominantly towards the south (SSW to SSE). By contrast, within and to the east of the syntaxis, as well as in the Kaghan valley nappes, linear fabrics have a more southwesterly trend. Within individual thrust sheets, for example the Besham sheet, movement directions may

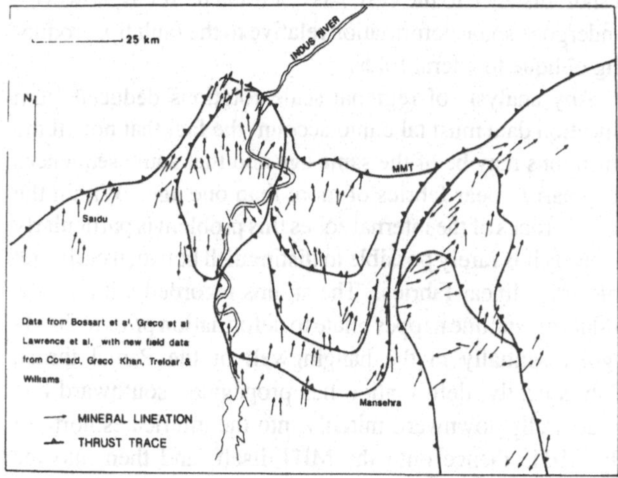

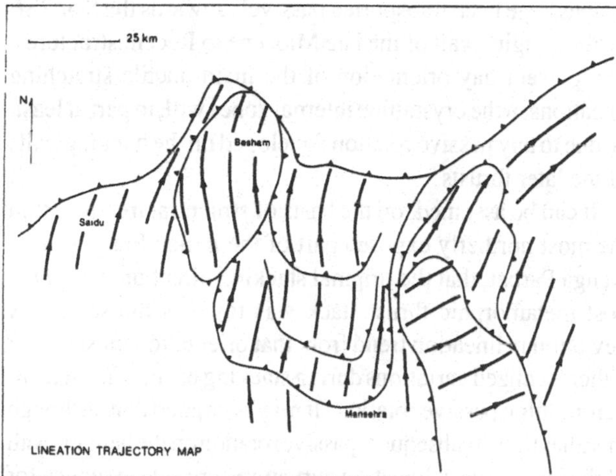

Figure 11. (a). Map to show the orientation of stretching lineations within the internal zones of the Indian Plate MMT - Main Mantle thrust; (b). Map to show movement trajectories within the internal zones based on the lineation data in Figure 10a.

vary, the lineation trajectories often defining curved traces. Further to the northeast, in the core of the Nanga Parbat syntaxis the lineation trend is more nearly southerly (Coward 1986; Treloar *et al.* in press, b).

Variations in thrust direction are seen further south along the MBT where, from W to E, transport directions swing from towards 210° at Attock, to towards 140° at Murree (Fig. 12). This demonstrates a divergence in thrust direction along the MBT similar to that demonstrated by Brunel (1986) for similar age structures on the other side of the Hazara syntaxis in the main Himalayan chain, although with an opposite sense.

b) Palaeomagnetic rotation data

To the west of the Hazara syntaxis, in the Potwar Plateau and Salt Ranges of North Pakistan, anticlockwise rotations (Opdyke *et al.* 1982) accompanied the thrusting and folding that occurred in the last 6 Ma (Burbank & Beck 1989). Thus a 35° rotation in the Bhaun area (Fig. 6) can be correlated with motion on the SRT at 5 Ma; and an approximate 40° rotation with renewed thrusting in the eastern Salt Ranges at around

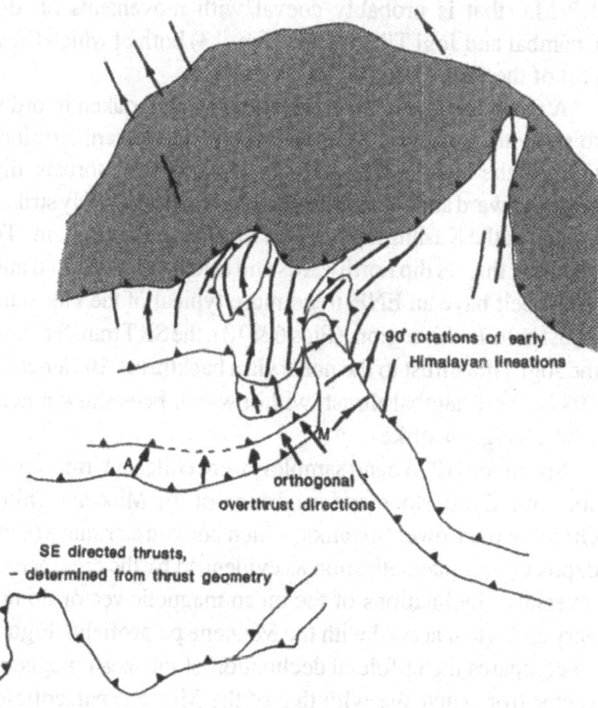

Figure 12. Map of the northern part of the Indian Plate to show the full range of movement directions from Nanga Parbat south to the external zone faults.

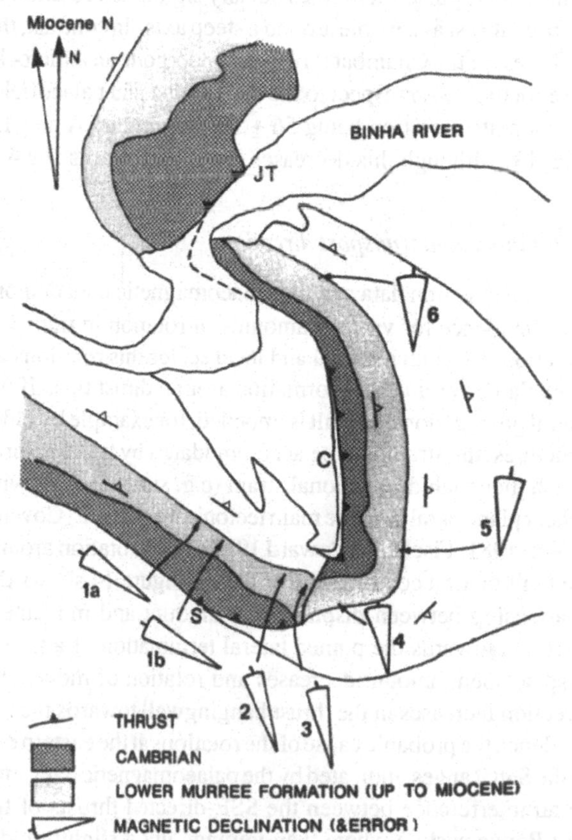

Figure 13. Geological map of the eastern termination of the Salt Range thrust (S) showing its relationship to the Chambal (C) and Jogi Tilla (JT) thrusts. For each of sites la to 6, open arrows (with 95% confidence cones) mark the magnetic declinations. Arrows in the top right corner show the difference between the Miocene palaeofield declination and the present.

1.5 Ma, that is probably coeval with movements on the Chambal and Jogi Tilla thrusts (Fig. 13) both of which form part of the SRT system.

A palaeomagnetic investigation was undertaken in order to study the complex deformation near the eastern termination of the Salt Range thrust. To the east, thrusts dip northeastward and folds and thrusts have northwesterly strikes typical of the Kashmiri part of the main Himalayan chain. To the west, thrusts dip northwards and folds within the fold and thrust belt have an ENE-trend more typical of the Pakistani thrust belt. In this region, (Figs 6 & 13), the SRT transfers into the Jogi Tilla thrust to the north via a backthrust (Butler *et al.* 1987), the Chambal thrust, within which beds show a near 180° change in strike.

Specimens (236 hand samples) were collected from seven sites in red mudstone and sandstone of the Miocene Chinji Group of the Lower Siwaliks, which contain a primary (syn-depositional) magnetisation as evidenced by the presence of reversals. Inclinations of the mean magnetic vector do not vary and are in accord with the Miocene palaeofield. Figure 13 compares the unfolded declination of the mean magnetic vector from each site with that of the Miocene palaeofield. Within the back-thrust (sites 4 - 6), the large changes in strike are not reflected in the magnetic declinations, showing that no rotation other than simple tilting around the present strike has occurred. Furthermore, the similarity between the declinations and the palaeofield rules out any significant rotation of the back thrust as a whole around a steep axis. In contrast, the SRT west of the Chambal thrust has undergone an anticlockwise rotation with respect to the back thrust and palaeofield, the estimated rotation being 56 ± 10° at section 1A and 1B (Fig. 13), although this decreases eastward towards site 4.

c) Rotations and transport directions

Both the lineation data and the palaeomagnetic data demonstrate evidence for varying amounts of rotation in transport direction. On both regional and local scales this rotation can be explained as due to deformation around thrust tips. If the lateral propagation of a fault is impeded, for example by older structures, the strain will be accommodated by tip structures which may include rotational strain (e.g. simple shear) with a shear plane parallel to the main tectonic movement (Coward & Kim 1981; Fischer & Coward 1982), body rotation around the fault tip or a combination of these. Figure 14 shows the relationship between displacement amount and movement direction towards the pinned lateral termination of a thrust. Displacement amount decreases and rotation of movement direction increases in the thrust hangingwall towards the tip.

Hence the probable cause of the rotations at the eastern end of the Salt Ranges, indicated by the palaeomagnetic data, may be an interference between the SSE-directed thrusts of the Salt Range system where they impinge upon slightly older NW-trending Himalayan structures (Butler *et al.* 1987). Given the short and recent periods of thrust motion, the more than 50° rotation of the pinned SRT is indeed spectacular. As can be theroretically predicted, the rotation decreases rapidly to the west (Opdyke *et al.* 1982). In contrast the Chambal thrust, a short distance to the east, has not rotated. It has, however, undergone some deformation relative to the fault tip, producing oblique to lateral folds.

Any analysis of regional scale rotations deduced from lineation data must take into account the fact that not all the lineations may be of the same age and that some sequences may carry linear fabrics of more than one age. Within the ductile rocks of the internal zones this problem is particularly acute as it is rarely possible to distinguish between early and late stage linear fabrics. The strains recorded within individual crystalline nappes relate to deformation prior to 20 Ma ago, essentially in the hangingwall of the Panjal thrust. Subsequently, deformation has propagated southward and structurally downward initially into the imbricates north of the MBT, thence onto the MBT itself, and then into the foreland basins. From, at the latest, 15 Ma ago the crystalline nappes have been transported passively towards the S or SSE on the hangingwall of the late Miocene to Recent structures. The present day orientation of the main ductile stretching lineations in the crystalline internal zones will, in part at least, be due to any passive rotation developed on the hangingwalls of the later thrusts.

It can be assumed, on the basis of structural trends within the most northerly exposed part of the Indian Plate, around Nanga Parbat, that the original stacking direction within the post metamorphic thrust stack was towards the south. A deviation in lineation trend from that orientation must reflect either localized variations during stacking or significant post-deformation passive rotation. It may be argued that, although there has been a subsequent passive rotation on the hangingwall of the later thrusts, lineation curvatures provide evidence for

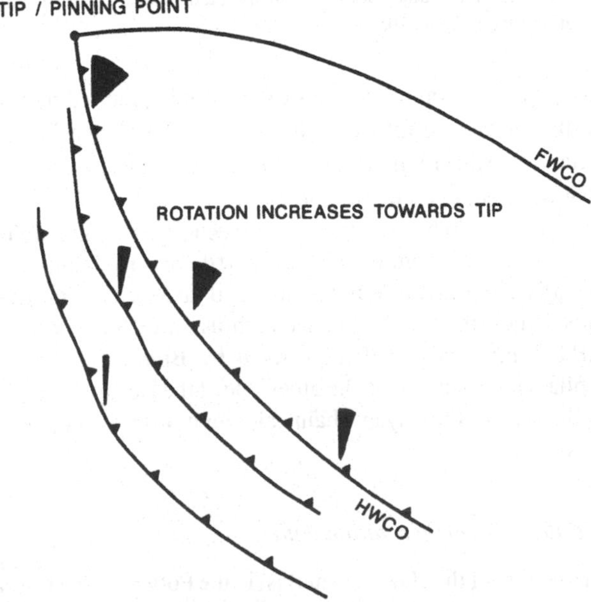

Figure 14. Diagram showing the expected relationships between displacement amounts, rotations of movement direction, dextral shearing and crustal scale folding at the lateral termination of a thrust. FWCO and HWCO - footwall and hangingwall cut offs.

early rotations within these nappes during the post-metamorphic thrust stacking period. Hence curvature of lineation trend within the Besham nappe is consistent with an anticlockwise rotation, whereas curvatures within the Kaghan valley nappes are consistent with clockwise rotations. That the situation is doubtless more complicated than this is indicated by the way that individual thrusts within the Hazara nappe show evidence for locally differing movement directions.

To the west of the Hazara syntaxis, the post 15 Ma bulk transport directions on the SRT, MBT and associated imbricates are to the SSE. As shown above, palaeomagnetic rotation data and kinematic indicators demonstrate that towards the eastern tip of these structures movement directions rotate, with an anticlockwise sense, towards the SE or ESE. To the east of the Hazara syntaxis the situation is the exact opposite. Lineations of all ages have a SW or SSW orientation (Brunel, 1986). Klootwijk *et al.* (1985) have presented palaeomagnetic evidence for late Tertiary clockwise rotations of about 45° in the Indian Plate rocks on the hangingwall of the MBT in Kashmir, to the west of the Hazara syntaxis. Within the core of that syntaxis, Bossart *et al.* (1990) have demonstrated similar magnitude rotations for Eocene-aged molasse on the footwall of the MBT. These clockwise rotations are consistent with a model in which displacement amounts decrease towards the western termination of the main Himalayan oroclinal thrust system. There is thus clear evidence that the crystalline thrust sheets have undergone passive rotations during the last 15 Ma in the hangingwall of the Miocene to Recent thrusts as a result of the pinning of these thrusts near their lateral terminations. These late rotations imply that Himalayan transport directions have varied through time, as indicated in Figure 15, and that, consequently, simple two dimensional attempts at balanced structural restorations may not be valid.

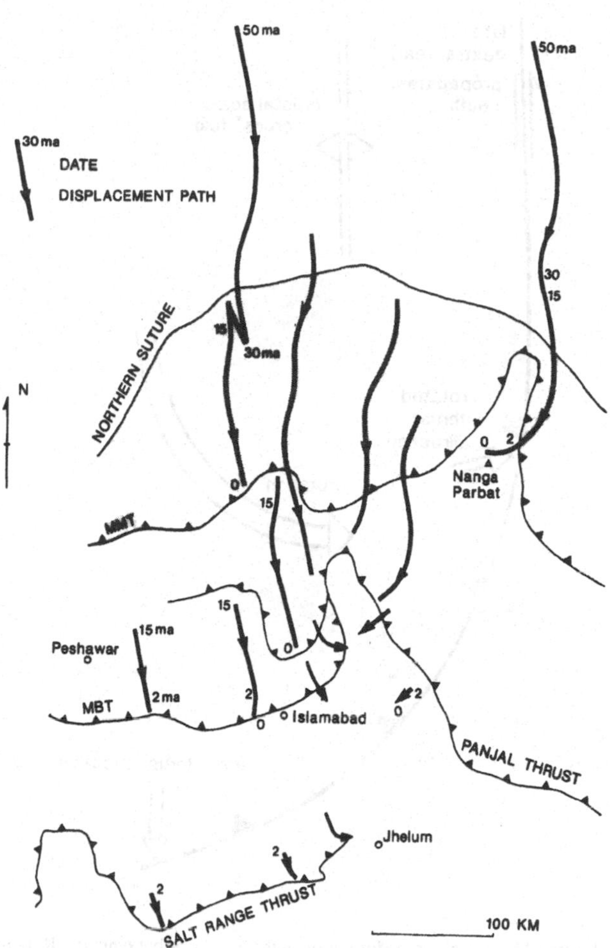

Figure 15. Schematic diagram to show the paths followed through time for points within the northern part of the Indian Plate MBT - Main Boundary thrust. MMT - Main Mantle thrust.

Interference structures in the Northwest Himalaya

The NW Himalaya are thus clearly a zone of convergence between the SW-verging thrusts in Kashmir and the S or SSE-verging thrusts in Pakistan. The changes in movement direction in both systems (clockwise towards the western tips of the Kashmir-Himalayan thrust system and anticlockwise towards the eastern tips of the Pakistan thrust system) are predicted by models in which each system is pinned by the other with rotation developing around that pinning point and increasing towards it. Structures developed in the convergence zone are naturally complex and have long been recognized. The Hazara and Nanga Parbat syntaxes (Fig. 2), two large scale antiformal folds, were recognized by Wadia (1931, 1932). Here we summarize their essential structural geometries and show how differences in their evolutionary history can be explained by different sequences of events within the interfering thrust system.

a) The Nanga Parbat syntaxis

The Nanga Parbat syntaxis marks the most northerly exposed part of the Indian Plate internal zones. Here Indian Plate gneisses outcrop within a structural half window flanked on either side by the basic rocks of the Kohistan Island Arc. The MMT is folded up and around the syntaxis which has the overall form of a crustal scale antiform. Within the syntaxis the main gneissic fabrics form a penetrative S-verging, L-S fabric developed during southward thrusting of Kohistan over India. These early fabrics have subsequently undergone an E-W shortening that resulted in the formation of a series of N-plunging folds with amplitude and half wavelengths in excess of 10 km. Along the western margin of the syntaxis uplift is accommodated within a major fault zone by a combination of NW-verging thrusting and dextral strike-slip faulting with the latter dominant within the northern part of the fault zone (Butler & Prior 1988a, b; Butler *et al.* 1989; Madin *et al.* 1989). The respective roles of folding within the syntaxis and faulting on the western margin in accommodating total uplift is discussed by Treloar *et al.* (in press, b). Ar-Ar, K-Ar and fission track data (Treloar *et al.* 1989a; in press, b; Zeitler 1985; Zeitler *et al.* 1989) demonstrate that syntaxial

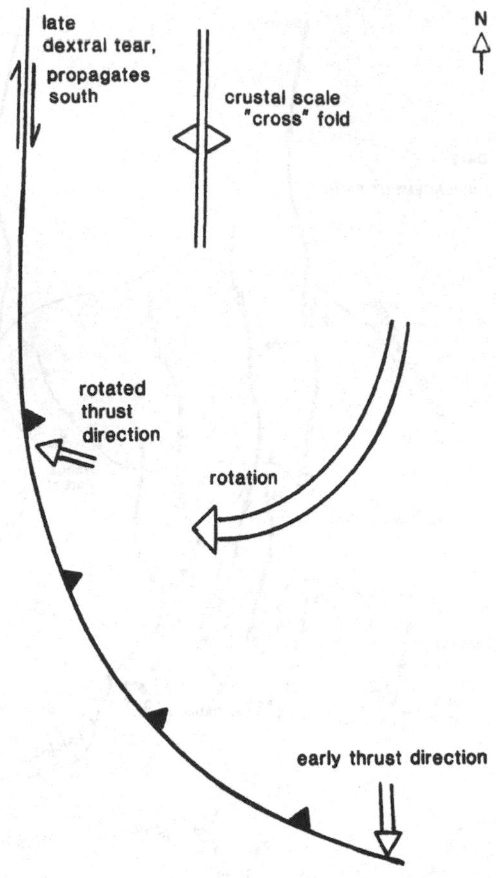

Figure 16. Model for the formation of the Nanga Parbat syntaxis. Rotation of the movement vector towards the thrust tip resulted in NW- verging thrusting along the western margin. These thrusts have been overprinted by dextral strike-slip structures that propagated southward along the western margin and which are related to tear faulting near the thrust tip. As strain became distributed throughout the hangingwall of the thrust, the rotation of the shortening direction generated large scale folds within the syntaxis.

uplift has occurred over a period of about 10 Ma. Fold amplification rate has been exponential, becoming explosive in the last 1 Ma.

The combination of deformation features (rotations, NW-directed thrusting, dextral strike-slip faulting and crustal scale folding) seen within the syntaxis is consistent with its development at the lateral termination of one of the main SW-verging Himalayan thrusts (Fig. 16). For the Nanga Parbat syntaxis the strong S-verging early L-S fabrics support an early phase of deformation within the S or SSE-verging Pakistan thrust system with later deformations related to tip strains within the SW-verging Kashmiri system.

b) The Hazara syntaxis

The structure and geometry of the northern part of the Hazara syntaxis has been described in detail by Bossart *et al.* (1988, 1990). Essentially the structure is a large, N-plunging antiformal fold with curvilinear axes giving it a sheath-type geometry. Within the core of the syntaxis, foreland basin molasse sequences of Palaeocene to Miocene age have been shortened and domed, and the structurally overlying Main Boundary and Panjal thrusts have been upfolded around the syntaxial margins. Growth of the Hazara syntaxis occurred during the early Pliocene (Burbank 1983). Bossart *et al.* (1988) suggested that the syntaxis grew as a result of an anticlockwise rotation of the movement direction from SW- to SSE-directed thrusting. Two supporting lines of evidence were presented for this. Firstly, quartz fibres showed anticlockwise curvatures and, secondly, late stage sinistral strike-slip faulting along the western margin was attributed to the late SSE-directed movements. Note that the sinistral ductile mylonite zone described by Bossart *et al.* (1988) as being related to the later stages of their anticlockwise rotation, is an earlier structure interpreted as one of the ductile shears involved in the post-metamorphic thrust stacking of the internal zones (Treloar *et al.* 1989c & d). Along with other similar aged shears this mylonite was subsequently passively folded by growth of the Hazara syntaxis. The palaeomagnetic documentation of a clockwise rotation of 45° within the core of the syntaxis (Bossart *et al.* 1990) is apparently contradictory to the anticlockwise rotation model.

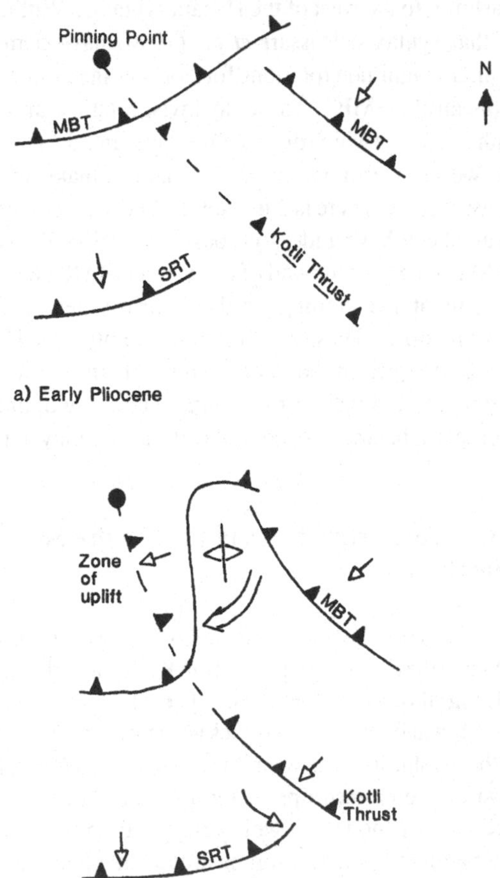

Figure 17. Model for the formation of the Hazara syntaxis. **(a)** Early Pliocene deformation along the Main Boundary thrust (MBT) in the east and the Salt Range thrust (SRT) in the west. (note how movement had ceased already on the MBT in the west) together with the initiation of the unexposed SW-verging Kotli thrust, pinned at its northwest termination; **(b)** Late Pliocene to Recent deformation in the hangingwall of the pinned Kotli thrust generates clockwise rotation in the footwall of the MBT and folding of the MBT. Note anticlockwise rotation at the eastern tip of the Salt Range thrust and neotectonic uplift along the Kotli thrust - Indus Kohistan seismic zone.

In the light of field observations by the authors, it is possible to interpret the totality of the data presented by Bossart and his co-workers as reflecting the interference between two active and converging thrust wedges. The SSE-vergent Pakistan thrust wedge propagated rapidly southwards during the Pliocene due to the presence of Eocambrian evaporites within the Potwar region. The southwestward advance of the Kashmiri thrust wedge was thus inhibited by the rapidly thickening Pakistani wedge, particularly in the north. This interference caused pinning and clockwise rotation in the core of the syntaxis as blind thrusts attempted to propagate beneath the Pakistani thrust system (Fig. 17). The emergent Kotli thrust represents such a structure, the northwest prolongation of which terminates as a large tip fold that uplifts the western limb of the Hazara syntaxis and folds the MBT. Present day SW-verging thrusting along this margin is indicated by seismicity studies (Seeber & Armbruster 1979).

Thus the syntaxial bend of the MBT is not the result of an anticlockwise rotation of movement direction, rather it results from the folding, uplift and erosion of the MBT by structures propagating simultaneously, but convergently, in its footwall. There is no need to invoke significant Pliocene to Recent sinistral strike-slip movements along the western margin of the syntaxis (cf. Bossart *et al.* 1988), as both the steepness and the N-S strike of the thrust reflect the magnitude of the folding of the MBT during syntaxial growth. It must be noted that the Pliocene thrust systems to the east of the syntaxis have different geometries from those to the west. On the eastern margin of the syntaxis, the Pliocene-aged structure was the MBT. On the western margin of the syntaxis, however, the simultaneously propagating structures were those in the footwall of the MBT, such as the SRT and related breakback structures.

c) Structures west of the Hazara syntaxis

West of the Hazara syntaxis, within the region of the Indus and Swat valleys, the S-verging thrusts of the post-metamorphic thrust stack have been deformed by a series of large scale N-trending antiformal and synformal folds, with half wavelengths of up to 40 km. Among these is the domal Besham antiform, which folds the MMT and within the core of which basement gneisses of the Besham nappe are exposed. These structures represent a change in the regional shortening and transport direction late in the tectonic history, from early S-directed thrusting to late stage E-W directed folding and shortening. The age of this deformation is uncertain, but it does deform the N-directed 20 Ma old extensional structures developed in the upper part of the Indian Plate. Apatite ages from two samples (Zeitler 1985) from the core of the Besham Dome, at about 4 Ma, are younger than ages of 16-20 Ma in the surrounding areas. These young ages indicate that the growth of the Besham fold is Recent, possibly within the last 5 Ma. As such, it is probable that these structures are related to the growth of the Nanga Parbat and Hazara syntaxes. Indeed the Besham dome may be a proto-syntaxis itself.

An important structure in this regard is the Indus Kohistan seismic zone (IKSZ), a NE-dipping fault zone imaged

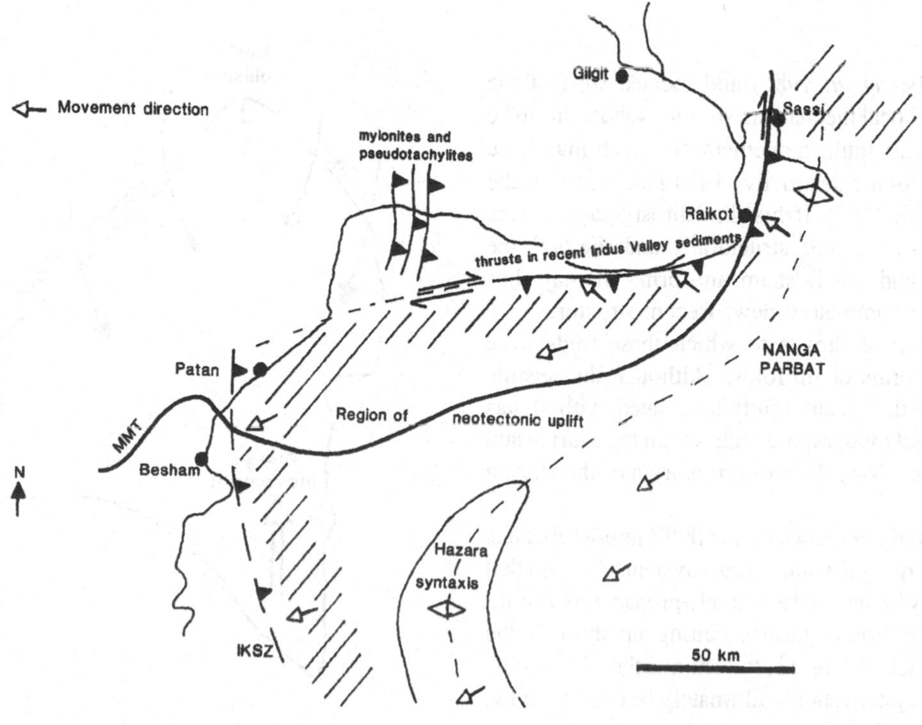

Figure 18. Sketch map showing how a zone of neotectonic uplift is located within the hangingwall of the Indus Kohistan seismic zone (IKSZ). This topographic high, and its northward extension into the Nanga Parbat region marks the western end of the main Himalayan mountain front characterized by elevations of greater than 4 km.

PLEISTOCENE (1.0 - 0 Ma).	Frontal Himalayan thrusts Uplift of Muree-Pattan zone in IKSZ hangingwall Explosive amplification of Nanga Parbat crustal scale fold
PLIO-PLEISTOCENE (2.0-0.5 Ma).	Salt Range thrusts (>25 km displacement) 30-40° anticlockwise rotations in eastern Salt Ranges
PLIOCENE (ca. 2.0 Ma)	Folding on Potwar Plateau
PLIOCENE (5.0 - 2.5 Ma)	Growth of Hazara syntaxis and of Besham antiform Accelerating growth of Nanga Parbat Fold Early rotations (ca. 35°) on Salt Range thrusts
UPPER MIOCENE (9.0 - 5.0 Ma)	Imbrication of Siwaliks in northen Potwar in MBT footwall Incipient development of Nanga Parbat fold
MIDDLE to UPPER MIOCENE (15.0 - 9.0 Ma)	SSE displacement of external thrust sheets (Nathia Gali and MBT) Pinning of thrust sheets (e.g. Panjal sheet in west) Clockwise rotation in Kashmir
LATE OLIGOCENE TO EARLY MIOCENE (30.0 - 20.0 Ma)	Extension in Upper Parts of Indian Plate. Beginning of erosion and of foreland basin molasse sedimentation. Anticlockwise rotation on Panjal thrust (=sole thrust)
LATE EOCENE to EARLY OLIGOCENE (40.0 - 30.0 Ma)	Post-metamorphic stacking in internal zones Rotation due to early stages of thrust pinning.
EARLY to MID EOCENE (50.0 - 40.0 Ma)	Collision and thickening/subduction of Indian Plate Main phase metamorphism Early erosion and Eocene age molasse sediments.

Table 3. Time chart of events on the Indian Plate in the Pakistan Himalaya.

seismically (Seeber *et al.* 1981) and located west of the Hazara syntaxis. Focal mechanism solutions show this to be a SW-verging thrust fault, movement on which may have been responsible for the destructive 1974 Pattan earthquake (Jackson & Yielding 1983). If this blind thrust detaches in the mid to lower crust, it should structurally underlie both the Hazara syntaxis and the Besham antiform. It may thus represent either a completely new, Recent structure, or a Recent reactivation of thrusts to which those folds were hangingwall antiforms or tip folds. Although the seismic zone is unexposed, Recent uplift associated with it has generated a marked topographic high within the Murree and Abbottabad region along the western margin of the Hazara syntaxis (Fig. 18)

The southwesterly vergence on the IKSZ implies that it is part of the SW-verging Kashmir thrust system. As such that system can clearly be seen to be actively propagating toward the SW, with the frontal thrusts cutting up through the Pakistan thrust stack. Were this to continue the whole SSE-verging Pakistan system should ultimately be overthrust by, and incorporated within, the dominant SW-verging system. In the interim a whole series of complex interference structures may be expected to develop, the locus of which should migrate towards the southwest. In the northeast, at Nanga Parbat, SW directed movements, clockwise rotations and dextral strike-slip faulting indicate that the earlier S-verging thrust system has been completely overprinted by the SW-verging one. Towards the southwest, in the Hazara syntaxis, there was Pliocene-aged interference with folding of the MBT and clockwise rotations developed on the MBT footwall. To the west of the Hazara syntaxis, in the Besham region, the early S-verging thrust structures were domed during the Pliocene by E-W shortening developed in response to newly initiated SW-verging thrusting. To the south, in the Salt Ranges, anticlockwise rotations are developing within the SSE-verging thrust wedge as the thrusts are pinned by the interfering SW-verging structures.

d) Timing

The time intervals over which thrusting, thrust interference and rotation, and syntaxial growth within the convergent zone have developed is indicated in Table 3. Within Pakistan the main S-verging thrust system has been operative, since collision, for over 50 Ma. Interference between the two thrust systems with pinning of thrusts and associated rotations around the pinned thrust tips has been ongoing for at least 30 Ma. The earliest evidence for such rotation is in the curved lineation trajectories within some of the internal zone crystalline nappes, implying that the imbricating thrusts were interfering across the zone of convergence within 20 Ma of collision. The spectacular rotations of over 40°, and the growth and explosive amplification of the crustal scale syntaxial folds are, however, features developed within the Pliocene to Recent.

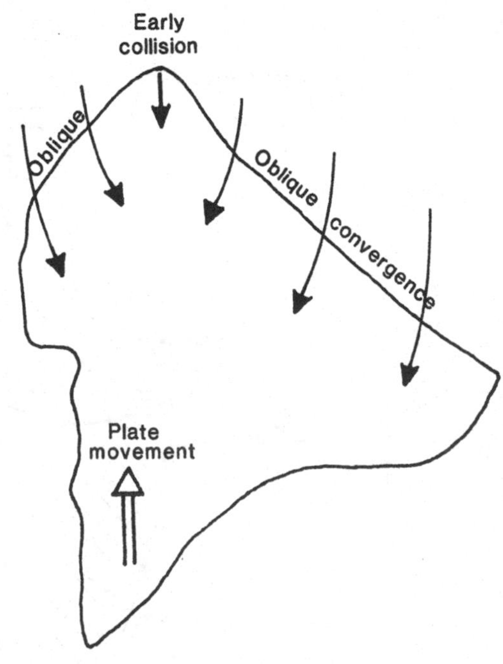

Figure 19. Diagram (modified after Klootwijk & Bingham 1980, and Klootwijk *et al.* 1985) to show the shape of pre-collisional India, in particular the two oblique ramps along the northern margin and the predicted rotation of thrust sheets emplaced onto those ramps.

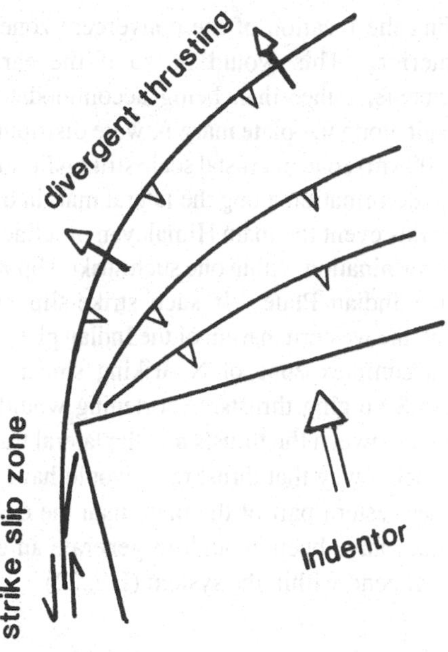

Figure 20. Diagram showing the relationship of an oroclinal thrust system to a rigid indentor, after Bale (1986). Note the curved nature of the thrusts which verge radially outwards, and the development of strike slip faults and lateral ramps which mark the thrust terminations at the arc margins.

Discussion

Throughout the period of Himalayan collision, thrust movement within the Pakistan Himalaya has been towards the S or SSE. Some variation in movement direction through time may be due to regional scale rotations of the Indian Plate resulting in a swing in the orthogonal thrust directions. Hence the anticlockwise rotation of the Indian Plate (Dewey *et al.* 1989) may have resulted in some anticlockwise rotation of the S-verging thrust sheets. Such an effect was invoked by Bossart *et al.* (1990) to explain the apparent anticlockwise rotation along the western margin of the Hazara syntaxis. However, these rotations are insufficient to explain the magnitude of the Pliocene to Recent anticlockwise rotations in the western Salt Ranges. Regionally the present day movement direction within the external zones of both the Pakistan Himalaya and the central part of the Himalayan orocline is to the SSE, which is consistent with the current NNW directed motion of the Indian Plate.

A number of factors may explain the oroclinal shape of the main Himalayan chain. Klootwijk & Bingham (1980) showed pre-collisional India to have been a diamond shape indenter with its northern margin composed of two segments each oblique to the convergence direction. When collision occurred, both these segments would have acted as oblique ramps, with thrust sheets rotating as they climbed onto them (Fig. 19). For a purely orthogonal system, Bale (1986) modelled thrust systems using a rigid indenter driving into a sand box, and showed that thrusts developed in front of the indenter develop an arcuate form with thrust directions verging radially outwards (Fig. 20). Within the Himalaya the natural radiation of thrust propagation direction would reinforce the rotations generated through oblique collision and would be further reinforced by the general rotation of India with greater subduction and shortening to the east than the west.

What the above does not explain is why the entire thrust belt from Pakistan east to India does not form a single curvilinear sequence, but has a strongly convergent zone developed within it. Nor does it explain why the main Himalayan thrusts appear to pin within the chain rather than its western end. There are a number of possible reasons for this. Firstly, the NW Himalaya is atypical with the Kohistan Island Arc trapped between the Indian and Asian Plates. Does the pinning of the main Himalayan chain at its western end reflect the inclusion of Kohistan within the orogenic system and its localized effects on crustal thickness and/or lithospheric structure, and the regional collisional history?

Secondly, it is possible that, due to the shape of the pre-collisional north Indian margin, collision in the NW Himalaya was earlier than elsewhere in the chain. Evidence for this includes plate reconstruction based on palaeomagnetic data (Besse & Courtillot 1988), the early age of metamorphism (pre-45 Ma) that followed collision and crustal thickening in the northwest Himalaya and the syn-collisional anticlockwise rotation of the Indian Plate. If collision was early then much of the crustal thickening in the northwest, including the obduction of the at least 20 km thick Kohistan sequence, should have been earlier there than in the main part of the Himalayan chain. Once thrusting commenced in the main part of the Himalayan chain at a slightly later date (circa 45 Ma?; Dewey *et al.* 1988, 1989) the lateral propagation of the main Himalayan thrusts, already likely to have an oroclinal form, could have been mechanically impeded by the presence of already thickened crust and lithosphere at their western terminations (Fig. 21). Once initiated, the convergent zone would have become intensified both because the anticlockwise rotation of India increased shortening rates in the east and because of the natural rotation of thrust sheets emplaced onto an oblique margin.

Any alternative model to the above must relate the location of the oroclinal thrust terminations as well as the location of the syntaxes within a zone of convergence within, rather than at the western margin of, the Himalayan chain. Were there major pre-existing structures within the Indian Plate which could have controlled the location of the zone of convergence and oroclinal thrust terminations? The western margin of continental India is currently marked by a series of N-S trending structures which extend from North Afghanistan south through the Chaman fault zone in western Pakistan into the Owen Fracture Zone in the Indian Ocean. These structures have been described as defining a sinistral continental transform along the western edge of the Indian Plate (Dewey *et al.* 1989) although there is no convincing field evidence for substantial displacements along any of them. The Chaman fault zone, for example, is a Recent structure with only limited sinistral displacement (Izatt 1990) which cannot be a continental transform. It may instead be a continental escape structure analogous to those described in Tibet (Molnar & Tapponier 1985) or East Anatolia (Dewey *et al.* 1986).

If the western margin of the Indian Plate, did carry a major component of sinistral strike-slip displacement, associated structures within the Indian Plate may have been important in

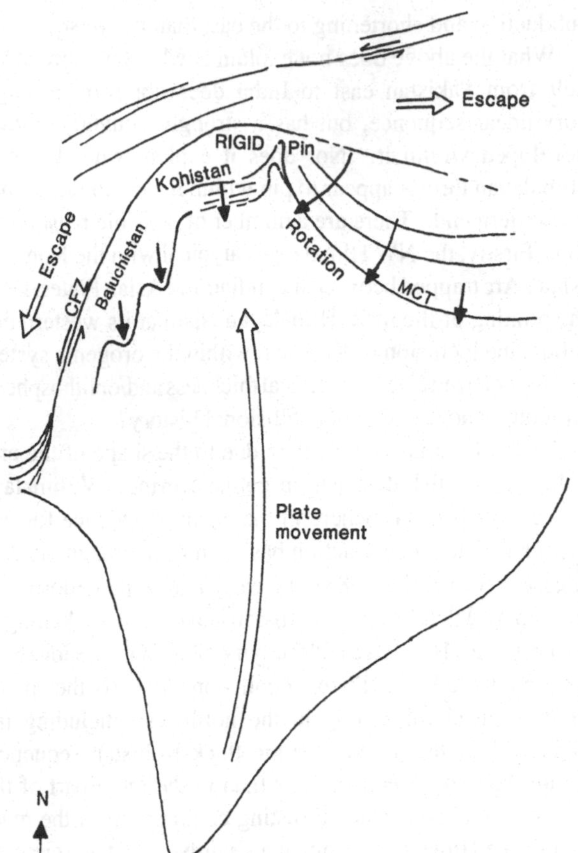

Figure 21. Schematic model to show the evolution of the Himalayan chain. Oblique convergence resulted in earlier collision to the northwest than to the east. Thickened crust in the northwest, enhanced by the obduction of the Kohistan island arc, mechanically impeded lateral propagation of the main Himalayan thrusts, pinning them at their western terminations. Anticlockwise rotation of the Indian Plate increases displacement on the main Himalayan thrusts towards the east, enhancing the orocline and hence the zone of convergence in the northwest Himalaya. The Chaman fault zone (CFZ) is here interpreted as a Recent continental escape structure.

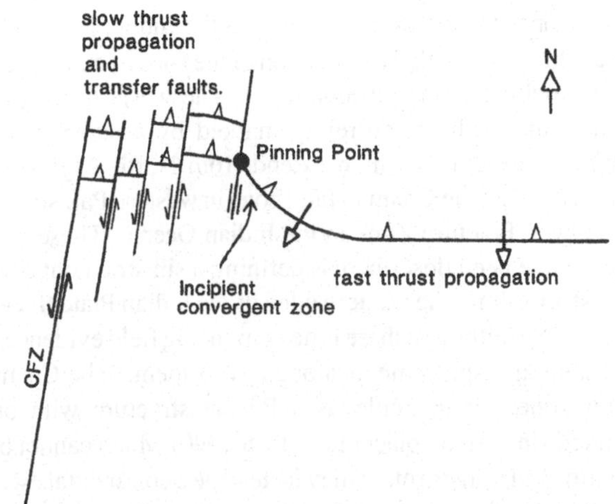

Figure 22. Diagram showing the relationship between strike-slip faulting and thrusting near the western margin of the Indian Plate, together with the incipient development of the convergent zone where the main Himalayan thrusts pin within the zone of strike-slip faulting.

controlling the location of the convergent zone within the plate interior. This would be so if the early sinistral displacements, rather than being accommodated within a single fault along the plate margin, were distributed across a wide (>200 km) zone of crustal scale strike-slip structures, all related to deformation along the lateral margin of the Indian Plate. In this event the main Himalayan orocline could have found its termination within one such strike-slip zone located within the Indian Plate. If such strike-slip faulting was important, the western margin of the Indian plate would have formed a complex zone of N-striking sinistral strike-slip faults and S-verging thrusts. Shortening would have been partitioned between the thrusts and the lateral transfer structures in such a way that thrust rates would have been lower within the western part of the plate than the east, which in itself would have been enough to generate an early proto-convergent bend within the system (Fig. 22).

Conclusions

The thrust system of the NW Himalaya is different from that of the main Himalayan chain in a number of respects. Collision appears to have been earlier in the northwest than to the east with early crustal thickening and metamorphism predating 40 Ma, with no evidence for the late metamorphism and leucogranite magmatism widely developed within the Indian and Nepalese Himalaya. Collision in the northwest is complicated by the obduction of the Kohistan island arc, a sheet at least 20 km thick, onto the leading edge of the Indian Plate. Movement is dominantly to the S or SSE, although with evidence of anticlockwise rotations towards the eastern ends of the main thrusts within the Pakistan thrust wedge.

Within the crystalline internal zones, metamorphism was synchronous with early phases of deformation related to crustal thickening, although the metamorphic pile was subsequently imbricated during a post-metamorphic phase of thrust stacking which generated a tectonic inversion of the metamorphic pile. There is no evidence here for syn-tectonic metamorphic inversions in thrust footwalls as are associated with the Main Central thrust in India and Nepal. The metamorphic pile was unroofed during the late-Oligocene to early Miocene due to a combination of erosion with resultant molasse deposition in the foreland basins, and extension in the upper parts of the Indian Plate and within the Main Mantle thrust zone synchronous with continued uplift along the Panjal thrust.

That less than 4 km of uplift and erosion has affected the internal zones during the last 20 Ma, implies that the Miocene and younger thrusts and detachments must have been shallow dipping to sub-horizontal, by contrast with similar age thrusts in the main Himalayan chain which do appear to be more steeply dipping. Deformation within the internal zones has been strongly controlled by the presence of a layer of Eocambrian salt at the base of the foreland basin sediments, generating a whole sequence of salt cored folds within the foreland fold and thrust belt.

Palaeomagnetic and kinematic data demonstrate anticlockwise rotations at the eastern terminations of the Pakistani thrust system. These rotations are the converse of the clockwise rotations recorded at the western terminations of the thrusts of the main oroclinal Himalayan thrust system. Both sets of rotations are the result of the pinning of each thrust system due to interference by the other. The two interfering thrust systems meet in a major zone of convergence, characterized by the growth of the crustal scale northwest Himalayan syntaxes. Continuing syntaxial growth, rotation, and the southwestward propagation of blind thrusts of the Kashmiri system into the Pakistani system demonstrate that interference and convergence are active to the present day.

Although the oroclinal shape of the main Himalayan thrust system can be predicted by experimental modelling, the pinning of the major Himalayan thrusts within the Himalayan chain, rather than at its western termination, is probably the result of oblique convergence. As a result of this, collision and associated crustal thickening, enhanced by the obduction of Kohistan onto the Indian Plate, was earlier in the northwest than the east. This thickened crust would have acted as a mechanical impediment to the propagation of the slightly younger thrusts of the main Himalayan system. Although early collision and crustal thickening in the northwest Himalaya probably controlled the pinning of the main oroclinal thrusts within the length of the Himalayan chain, strike-slip faulting near the western edge of the Indian Plate margin could have had a significant effect in controlling the exact location of the convergent zone. Once the zone of convergence was initiated it would rapidly have become an increasingly important feature due to the effects of oblique convergence, together with the continued anticlockwise rotation of the Indian Plate, with greater shortening in the east than the west.

The Imperial College Himalayan Research Group has been largely supported by NERC grant GR3/6113 to MPC and PJT. In addition we acknowledge a DENI Research studentship to AFC, a BP Research studentship to CNI, and a BP funded postdoctoral fellowship to KCJ. We acknowledge discussions with many colleagues, especially R.W.H. Butler, M. P. Williams, and R. J. Lillie, and reviews from R. A. Gayer and J. F. Dewey.

References

Argand, E. 1924. La tectonique de l'Asie. *13th International Geological Congress*, **1**, 170-372.

Baker, D. M., Lillie, R. J., Yeats, R. S., Johnson, G. D., Yousuf, M. & Zamin, MA.S.H. 1988. Development of the Himalayan frontal thrust zone: Salt Range, Pakistan. *Geology*, **16**, 3-7.

Bale, P. 1986. Tectonique Cadomienne en Bretagne Nord: interaction decrochement chevauchment: champs de deformation et modelisations experimentales. *Unpublished Ph.D. thesis, University of Rennes*, 361p.

Besse, J. & Courtillot, V. 1988. Palaeogeographic maps of the continents bordering the Indian Ocean since the early Jurassic. *Journal of Geophysical Research*, **93**, 11791-11808.

Bossart, P., Dietrich, D., Greco, A., Ottiger, R. & Ramsay, J. G. 1988. The tectonic structure of the Hazara-Kashmir Syntaxis, southern Himalayas, Pakistan. *Tectonics*, **7**, 273-297.

—— & Ottiger, R. 1989. Rocks of the Murree Formation in northern Pakistan: indicators of a descending foreland basin of late Paleocene to middle Eocene age. *Eclogae Geologicae Helvetiae*, **82**, 133-165.

Bossart, P., Ottiger, R. & Heller, F. 1990. Rock magnetic properties and structural developments in the core of the Hazara-Kashmir Syntaxis, NE Pakistan. *Tectonics*, **9**, 103-121.

Brunel, M. 1986. Ductile thrusting in the Himalayas: ductile thrusting and shear sense criteria. *Tectonics*, **5**, 247-265.

Burbank, D. W. 1983. The chronology of intermontane basin development in the northwestern Himalaya and the evolution of the Northwest syntaxis. *Earth and Planetary Science Letters*, **64**, 77-92.

—— & Beck, R. A. 1989. Early Pliocene uplift of the Salt Range; temporal constraints on thrust wedge development, northwest Himalaya, Pakistan. *Geological Society of America Special Paper*, **232**, 113-128.

——, Raynolds, R.G.H. & Johnson, G. D. 1986. Late Cenozoic tectonics and sedimentation in the northwestern Himalayan foredeep; II, eastern limb of the northwest syntaxis and regional synthesis. *In*: Allen, P. & Homewood, P. (eds) *Foreland Basins*, International Association of Sedimentologists Special Publication, **8**, Blackwells, 293-306.

Butler, R.W.H. & Coward, M. P. 1989. Crustal scale thrusting and continental subduction during Himalayan collision tectonics of the NW Indian Plate. *In*: Sengor, A.M.C. (ed) *Tectonic evolution of the Tethyan region* NATO ASI Series. C, **259**, 387-413.

——, Coward, M. P., Harwood, G. M. & Knipe, R. J. 1987. Salt: its control on thrust geometry, structural style and gravitational collapse along the Himalayan mountain front in the Salt Range of northern Pakistan. *In*: J. J. O'Brien & I. Lerche (eds) *Dynamical geology of salt and related structures*. Academic Press, 399-418.

—— & Prior, D. J. 1988a. Tectonic controls on the uplift of Nanga Parbat, Pakistan Himalayas. *Nature*, **333**, 247-250.

—— & Prior, D. J. 1988b. Anatomy of a continental subduction zone. *Geologische Rundschau*, **77**, 239-255.

——, Prior, D. J. & Knipe, R. J. 1989. Neotectonics of the Nanga Parbat syntaxis, Pakistan, and crustal stacking in the northwest Himalaya. *Earth and Planetary Science Letters*, **94**, 329-343.

Coward, M. P. 1986. A section through the Nanga Parbat syntaxis, Indus valley, Kohistan. *Geological Bulletin of the University of Peshawar*, **18**, 147-152.

—— & Butler, R.W.H. 1985. Thrust tectonics and the deep structure of the Pakistan Himalaya. *Geology*, **13**, 417-420.

——, Butler, R.W.H., Chambers, A. F., Graham, R. H., Izatt, C. N., Khan, M. A., Knipe, R. J., Prior, D. J., Treloar, P. J. & Williams, M. P. 1988. Folding and imbrication of the Indian crust during Himalayan collision. *Philosophical Transactions of the Royal Society of London, Series A*, **326**, 377-391.

—— & Kim, J. H. 1981. Strain within thrust sheets. *In*: McClay, K. R. & Price, N. J. (eds) *Thrust and Nappe Tectonics*. Geological Society of London Special Publication **9**, 275-292.

Dewey, J. F., Cande, S. & Pitman, W. C. 1989. Tectonic evolution of the India/Eurasia collision zone. *Eclogae Geologicae Helvetiae*, **82**, 717-734.

——, Shackleton, R. M., Chang Chengfa & Sun Yujin. 1988. The tectonic evolution of the Tibetan plateau. *Philosophical Transactions of the Royal Society of London, Series A*, **327**, 379-413.

——, Hempton, M. R., Kidd, W.S.F., Saroglu, F. & Sengor, A.M.C. 1986. Shortening of continental lithosphere: the neotectonics of Eastern Anatolia - a young collision zone. *In*: Coward, M. P. & Ries, A. C. (eds) *Collision Tectonics*. Geological Society of London Special Publication, **19**, 3-36.

England, P. C. & Houseman, G. A. 1986. Finite strain calculations of continental deformation. II. Application to the India-Asia plate collision. *Journal of Geophysical Research*, **91**, 3664-3676.

—— & Molnar, P. 1990. Right lateral shear and rotation as the explanation for strike slip faulting in eastern Tibet. *Nature*, **344**, 140-142.

Fischer, M. W. & Coward, M. P. 1982. Strain within thrust sheets: the Heilam sheet of NW Scotland. *Tectonophysics*, **88**, 291-312.

Izatt, C. N. 1990. Variations in thrust front geometry across the Potwar Plateau and Hazara/Kalachitta hill ranges, northern Pakistan. Unpublished Ph.D. thesis, University of London, 353p.

Jackson, J. & Yielding, G. 1983. The seismicity of Kohistan, Pakistan: source studies of the Hamran (1972.9.3), Darel (1981.9.12) and Patan (1974.12.28) earthquakes. *Tectonophysics*, **91**, 15-28.

Johnson, G. D., Johnson, N. M., Opdyke, N. D. & Tahirkelli, R.A.K. 1979. Magnetic reversal stratigraphy and sedimentary tectonics of the Upper Siwalik Group, eastern Salt Ranges and southwestern Kashmir. *In*: Farah, A. & De Jong, K. A. (eds) *Geodynamics of Pakistan*, Geological Survey of Pakistan, Quetta, 149-165.

——, Raynolds, R.G.H. and Burbank, D. W. 1986. Late Cenozoic tectonics and sedimentation in the northwestern Himalayan foredeep; I, Thrust ramping and associated deformation in the Potwar region. *In*: Allen, P. & Homewood, P. (eds) *Foreland Basins*, International Association of Sedimentologists Special Publication, **8**, Blackwells 273-291.

——, Zeitler, P. K., Naeser, C. W., Johnson, N.M ., Summers, D. M., Frost, C. D., Opdyke, N. D. & Tahirkelli, R.A.K. 1982. The occurence and fission track ages of late Neogene and Quarternary volcanic sediments, Siwalik Group, northern Pakistan. *Palaeogeography, Palaeoclimatology and Palaeoecology*, **37**, 67-93.

Klootwijk, C. T. & Bingham, D. K. 1980 The extent of greater Asia. III. Palaeomagnetic data from the Tibetan sedimentary series, Hakkhola region, Nepal Himalaya. *Earth and Planetary Science Letters*, **51**, 381-405.

——, Conaghan P. J. & Powell, C.McA. 1985. The Himalayan arc; large scale continental subduction, oroclinal bending and backarc spreading. *Earth and Planetary Science Letters*, **75**, 167-183

Lawrence, R. D., Kazmi, A. H. & Snee, L. W. 1990. Geological setting of the emerald deposits of Swat, North Pakistan. *In*: Kazmi A. H. & Snee, L. W. (eds) *Emeralds of Pakistan: Geology, gemmology, and genesis*. Van Nostrand Reinhold, New York, 1-25.

Lillie, R. J., Johnson, G. D., Yousuf, M., Seamin, A.S.H. & Yeats, R. S. 1987. Structural development within the Himalayan fold-and-thrust belt of Pakistan. *In*: Beaumont, C. & Tankard, A. J. (eds) *Sedimentary basins and basin forming mechanisms'*. Canadian Society of Petroleum Geologists Memoir, **12**, 379-392.

Madin, I. P., Lawrence, R. D. & Ur-Rehman, S. 1989. The northwestern Nanga Parbat - Haramosh massif: evidence for crustal uplift at the northwestern corner of the Indian craton. *Geological Society of America, Special Paper*, **232**, 169-182.

Mattauer, M. 1986. Intracontinental subduction, crust-mantle decollement and crustal-stacking wedge in the Himalayas and other collision belts. *In*: Coward, M. P. & Ries, A. C. (eds) *Collision Tectonics*, Geological Society of London Special Publication, **19**, 37-50.

Molnar P. & Tapponier, P. 1985. Cenozoic tectonics of Asia: effects of a continental collision. *Science*, **189**, 419-426.

Opdyke, N. D., Lindsay, E. H., Johnson, G. D., Johnson, N. M., Tahirkelli, R.A.K. & Mirza, M.A. 1979. Magnetic polarity stratigraphy and vertebrate palaeontology of the Upper Siwalik Subgroup of northern Pakistan. *Palaeogeography, Palaeoclimatolgy, Palaeocology*, **27**, 1-34.

——, Johnson, N. M., Johnson, G. D., Lindsay, E. H. & Tahirkelli, R.A.K. 1982. Palaeomagnetism of the Middle Siwalik Formations of northern Pakistan and rotation of the Salt Range decollement. *Palaeogeography, Palaeoclimatolgy, Palaeocology*, **37**, 1-15.

Patriat, P. & Achache, J. 1984. India-Eurasia collision chronology has implications for crustal shortening and driving mechanisms of plates. *Nature*, **311**, 615-621.

Powell, C. M. & Conaghan, P. J. 1973. Plate tectonics and the Himalayas. *Earth and Planetary Science Letters*, **20**, 1-12.

Pudsey, C. J., Coward, M. P., Luff, I. W., Shackleton, R. M., Windley, B. F. & Jan, M. Q. 1986. The collision zone between the Kohistan arc and the Asian Plate in NW Pakistan. *Transactions of the Royal Society of Edinburgh, Earth Sciences*, **76**, 463-479.

Royden, L. H. & Burchfiel, B. C. 1987. Thin skinned N-S extension within the convergent Himalayan region: gravitational collapse of a Miocene topographic front. *In*: Coward, M. P., Dewey, J. F. & Hancock, P. L. (eds) *Extension Tectonics*. Geological Society of London Special Publication, **28**, 611-619.

Searle, M. P. & Rex, A. J. 1989. Thermal model for the Zanskar Himalaya. *Journal of Metamorphic Geology*, **7**, 127-134.

Seeber, L. & Armbruster, J. G. 1979. Seismicity in the Hazara arc in northern Pakistan: decollement versus basement faulting. *In*: Farah, A. & De Jong, K. A. (eds) *Geodynamics of Pakistan*, Geological Survey of Pakistan, Quetta, 131-142.

——, Armbruster, J. G. & Quittmeyer, R. C. 1981. Seismicity and continental subduction in the Himalayan arc. *American Geophysical Union, Geodynamics Series*, **5**, 215-242.

Tapponier, P., Peltzer, G. & Armijo, R. 1986. On the mechanics of the collision between India and Asia. *In:* Coward, M. P. & Ries, A. C. (eds) *Collision Tectonics*. Geological Society of London Special Publication, **19**, 115-157.

Treloar, P. J., Rex, D. C., Guise, P. G., Coward, M. P., Searle, M. P., Petterson, M. G., Windley, B. F., Jan, M. Q. & Luff, I. W. 1989a. K-Ar and Ar-Ar geochronology of the Himalayan collision in NW Pakistan: constraints on the timing of collision, deformation, metamorphism and uplift. *Tectonics*, **8**, 881-909.

——, Coward, M. P., Williams, M. P. & Khan, M. A. 1989b. Basement cover imbrication south of the Main Mantle Thrust, North Pakistan. *Geological Society of America Special Paper*, **232**, 137-152.

——, Broughton, R. D., Williams, M. P., Coward, M. P. & Windley, B. F. 1989c. Deformation, metamorphism and imbrication of the Indian Plate south of the Main Mantle Thrust, North Pakistan. *Journal of Metamorphic Geology*, **7**, 111-125.

——, Williams, M. P. & Coward, M. P. 1989d. Metamorphism and crustal stacking in the north Indian Plate, North Pakistan. *Tectonophysics*, 165, 167-184.

—— & Rex, D. C. 1990a Cooling and uplift histories of the crystalline thrust stack of the Indian Plate internal zones west of Nanga Parbat, Pakistan Himalaya. *Tectonophysics*, **180**, 323-349.

—— & Rex, D. C., 1990b. Post-metamorphic cooling history of the Indian Plate crystalline thrust stack, Pakistan Himalaya. *Journal of the Geological Society of London*, **147**, 735-738.

——, Potts, G. J. & Wheeler, J. Structural evolution and asymmetric uplift of the Nanga Parbat syntaxis, Pakistan Himalaya. *Geologische Rundschau*, (in press, b).

——, Rex, D. C. & Williams, M. P. The role of erosion and extension in unroofing the Indian Plate thrust stack, Pakistan Himalaya. *Geological Magazine*, (in press a).

Wadia, D. N. 1931. The syntaxis of the northwest Himalaya: its rocks, tectonics and orogeny. *Records of the Geological Survey of India*, **65**, 189-220

—— 1932. Note on the geology of Nanga Parbat (Mt.Diamir) and adjoining portions of Chilas, Gilgit district, Kashmir. *Records of the Geological Survey of India*, **66**, 212-234.

Zeitler, P. K. 1985. Cooling history of the NW Himalaya. *Tectonics*, **4**, 127-151.

——, Chamberlain, C. P. & Jan, M. Q. 1989. Geochronology and temperature history of the Nanga Parbat-Haramosh massif, Pakistan. *Geological Society of America Special Paper*, **232**, 1-22.

Balanced and retrodeformed geological cross-section from the frontal Sulaiman Lobe, Pakistan: Duplex development in thick strata along the western margin of the Indian Plate

I.A.K. Jadoon[1]*, R. D. Lawrence[2] & R. J. Lillie[2]

[1] Department of Earth Sciences, Quaid-i-Azam University, Islamabad, Pakistan
[2] Department of Geosciences, Oregon State University, Corvallis, Oregon, 97331-5506, USA

Abstract: A balanced structural cross-section has been constructed integrating seismic reflection profiles, drillhole, surface geology, and Landsat data across the tectonically active frontal Sulaiman fold belt in the western Himalayas. Restoration of the section provides information regarding the chronology of structures, structural style, sequence of thrusting, and the amount of shortening. General structural form evidenced by gentle topography and a broad fold belt is similar to that of other mountain belts underlain by a weak detachment. A sequence of about 10 km of dominantly platform (>7 km) and molasse strata thickens tectonically to about 15 km, 129 km north of the southwards verging deformation front. Nearly all of the 10 km thick stratigraphic sequence has been detached at the deformation front. Structural style is that of a foreland-verging duplex separated from the roof sequence by a passive-roofthrust in thick Cretaceous shale. This structure is expressed at the surface by fault-related folds. Toward the northerly hinterland, progressively older rocks are present at the surface in the hinge zones of the anticlines. They have been uplifted by duplexing several kilometres higher than their regional stratigraphic level. The passive-roof thrust has not been cross-cut by backthrusts, and it is present over a distance of 60 km along the line of section. Progressive deformation reveals a series of structural and geometric features including: (1) broad concentric folding at the fault tip; (2) development of a passive roof and duplex sequence by forward propagation of floor and roof thrusts; (3) forward propagation of the duplex as critical taper is achieved; and (4) tear faults and extensional normal faults within the overthrust wedge. A retrodeformed cross-section shows that about 76 km of orogenic contraction in the cover sequence has occurred across the frontal 129 km of the Sulaiman fold belt.

The Himalayan mountain system represents an active continent-continent collision zone extending westward from Burma through northern India and Nepal into Pakistan (Gansser 1981). The broad Sulaiman fold belt is developed by transpression as a result of the left-lateral strike-slip motion along the Chaman fault zone and southward thrusting along the western boundary of the Indian subcontinent (Figs 1 & 2; Sarwar & De Jong 1979; Lawrence *et al.* 1981; Farah *et al.* 1984; Quittmeyer *et al.* 1984). The frontal part of the Sulaiman fold belt is seismically active (Quittmeyer *et al.* 1979, 1984); however, the stratigraphy is not disrupted by any thrust faults that break the surface (Fig. 3). The style of deformation from the western Sulaiman fold-and-thrust belt is reported to be of forward propagating duplexes developed in a piggy-back fashion (Banks & Warburton 1986). Southward migration of the deformation front shed erosion products into the active Sulaiman foredeep, where 7 km of molasse strata are currently present in the Sibi molasse basin (Banks & Warburton 1986).

Recent studies constrained by seismic reflection and borehole data in the North American Cordillera, Appalachians, Alps, Himalayas, and Taiwan have provided insight into the mechanism of deformation and geometry of structures in the frontal part of collision zones (Rich 1934; Dahlstrom 1969a, 1970; Suppe 1980, 1983; Laubscher 1981; Acharyya & Ray 1982; Bachman *et al.* 1982; Jones 1982; Davis *et al.* 1983; Davis & Engelder 1985; Banks & Warburton 1986; Boyer 1986; Mitra 1986; Lillie *et al.* 1987; Jaume & Lillie 1988; McDougal 1989; Izatt 1990). Studies of active mountain belts (i.e. Himalaya and Taiwan) are important because they provide constraints on collisional processes that are unavailable in ancient mountain belts. In this study seismic reflection and well data, available from the frontal part of the active Sulaiman fold belt are integrated with surface geology and Landsat data (Figs 3 & 4) in order to: (1) determine the thickness and nature of the overthrust wedge; (2) analyse the structural style and nature of deformation; (3) study progressive deformation and its effects on the basin geometry; and (4) estimate the amount of compressional shortening in the cover of Phanerozoic sediments.

Tectonic setting and stratigraphy

The Himalayan mountain belt changes trend from northwest-southeast in India to northeast-southwest in Pakistan (Fig. 1).

* Present address: Department of Geosciences, Oregon State University, Corvallis, Oregon, 97331-5506, USA

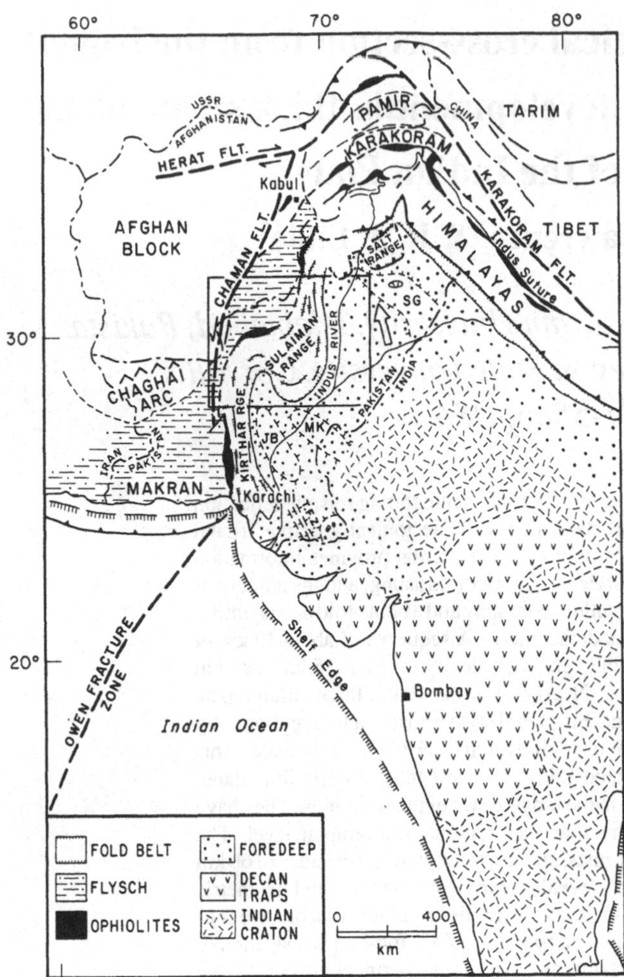

Figure 1. Tectonic features along the western edge of the Indian subcontinent. Area of Figure 2 shown by rectangle. JB=Jacobabad high; MK= Mari/Khandkot/Khandkot high; TFZ=Talhar fault zone; SG=Sargodha high.

Typical of the foreland part of the northwestern Himalaya in Pakistan are two broad lobate features - the Salt Range/ Potwar Plateau and the Sulaiman fold belt. Their lobate geometry is interpreted to be the result of rapid southward translation along a weak decollement of the tear fault bounded thrust sheets (Sarwar & De Jong 1979; Seeber *et al.* 1981). This is similar to the foreland translation of the Pine Mountain thrust block of the Central Appalachians (Rich 1934; Harris & Milici 1977) and the Jura Mountains of Europe (Laubscher 1981). Deformation is progressively younger toward the foreland, as constrained by magnetic stratigraphy (Johnson *et al.* 1982; Raynolds & Johnson 1985) and neotectonic activity in the Salt Range/Potwar Plateau (Yeats *et al.* 1984; Yeats 1989). In the Sulaiman fold belt progressive deformation is evidenced by structural style (Hunting Survey Corporation 1961; Kazmi & Rana 1982), a prominent topographic front, and seismicity over the frontal folds (Quittmeyer *et al.* 1979, 1984).

Unlike the Salt Range/Potwar Plateau that is associated with the main zone of Himalayan convergence, the Sulaiman fold belt is located along a zone of transpression in the northwestern part of the Indian subcontinent (Fig. 1). The broad Sulaiman fold belt is bounded to the west and north by the left-lateral strike-slip Chaman fault zone (Fig. 2). The foredeep basin to the east and south of the active Sulaiman Lobe is formed mainly as a result of tectonic compression between the Indian plate and the Afghan block (Fig. 2). The initial event of collision is manifested by the emplacement of the Muslimbagh ophiolite between Late Cretaceous and Early Eocene times (Allemann 1979). An unconformity between Cretaceous and Palaeocene rocks in the Attock Cherat Ranges north of the Potwar Plateau (Yeats & Hussain 1987) extends all the way to the Loralai valley of the Sulaiman

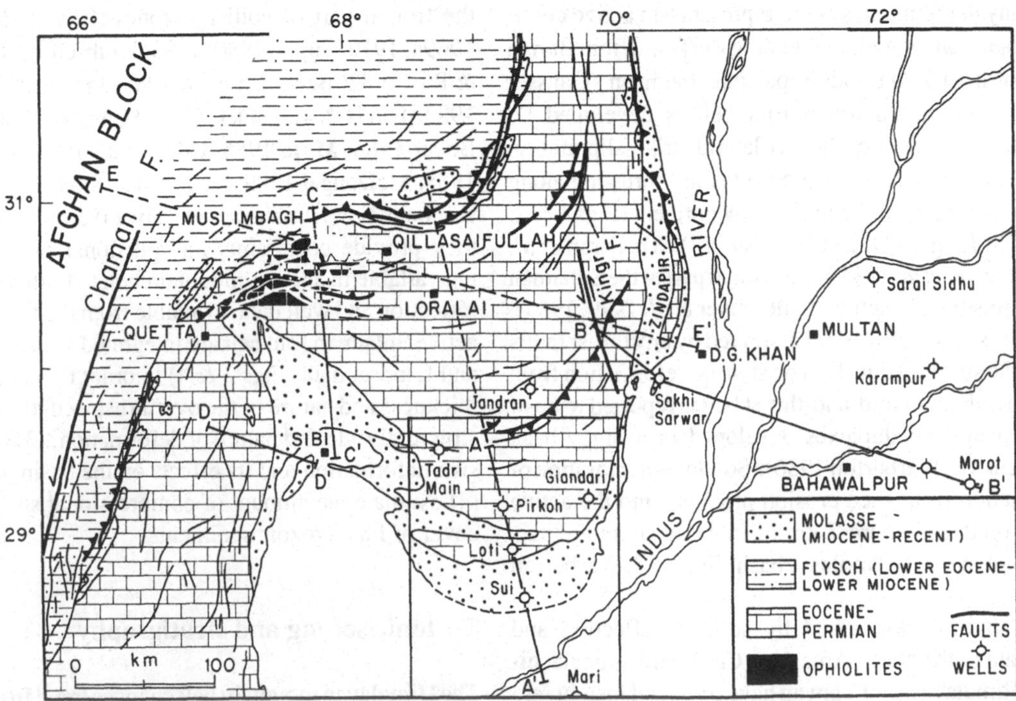

Figure 2. Generalized tectonic map of the Sulaiman fold belt in Pakistan (after Kazmi & Rana 1982). Area of investigation (Fig. 3) is shown by rectangle. Line AA′ is part of this study, while line BB′ is constructed by Humayun *et al.* (1990). Cross-sections CC′ and DD′ are shown in Figure 5 (after Banks & Warburton 1986). Line EE′ shows Bouguer gravity profile (Khurshid pers. comm.).

Range (Hunting Survey Corporation 1961). Renewed southward thrusting since Late Oligocene-Early Miocene constantly reworked the molasse strata migrating the Indus basin farther east and south (Banks & Warburton 1986; Waheed *et al.* 1988). This is similar to the southward migration of the active foredeep basins of the Ganges plain in India and the Jhelum plain in Pakistan (Acharyya & Ray 1982; Raiverman *et al.* 1983; Johnson *et al.* 1985; Raynolds & Johnson 1985).

The main structural elements in the Sulaiman fold belt are east-west trending arcuate folds and faults which rotate rapidly to a north-south direction along the margin of the active fold belt (Fig. 2). Imbricate faults are visible at the surface only in the north (Hunting Survey Corporation 1961; Kazmi & Rana 1982). They gradually disappear toward the

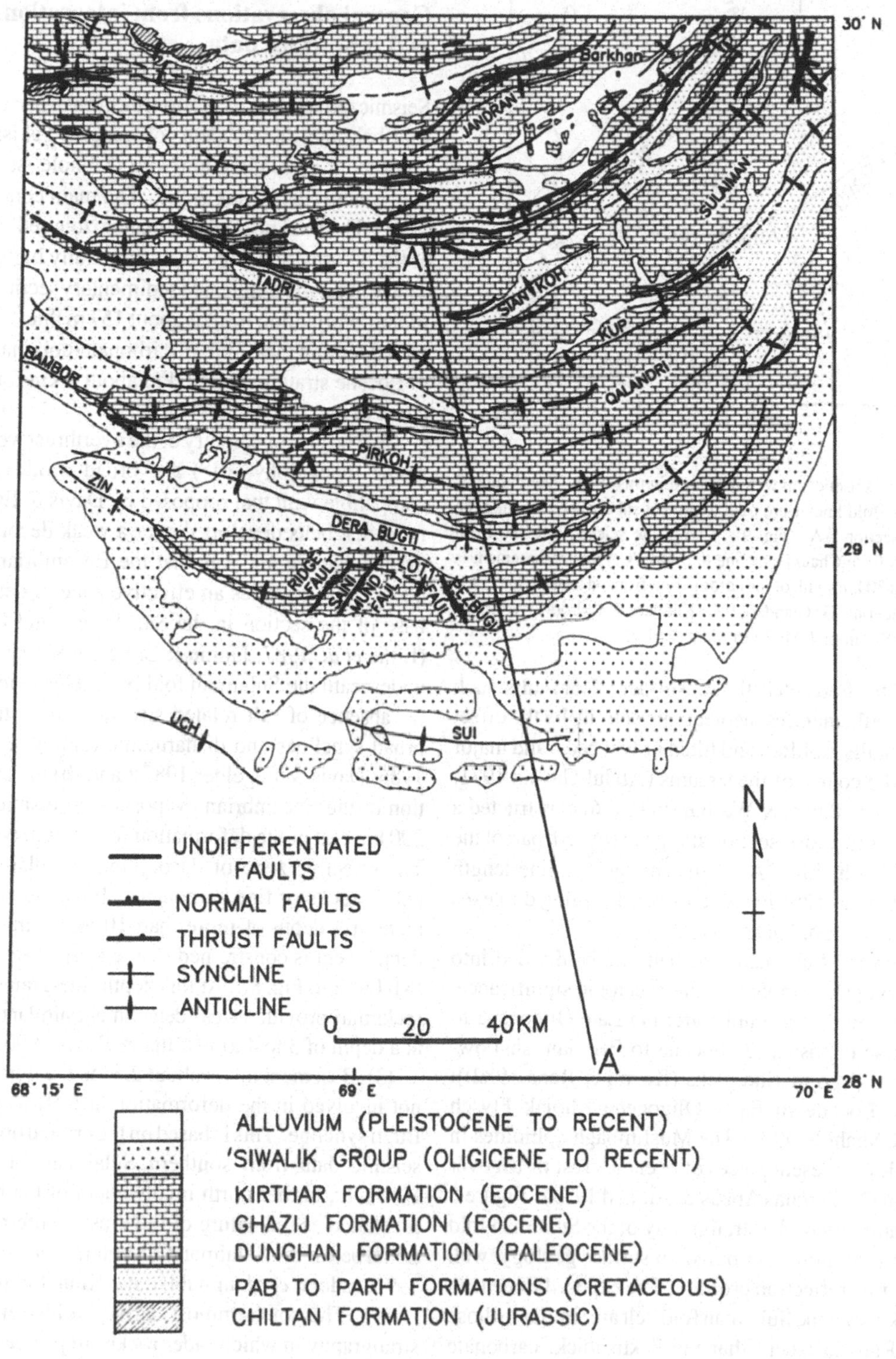

Figure 3. Generalized geological map of southern Sulaiman lobe. Mapping is compiled from the unpublished maps of the Oil and Gas Development Corporation (OGDC), the Geological Survey of Pakistan (GSP), the Hunting Survey Corporation (1961), and from Landsat images (1:125,000) supplied by Earth Satellite Corporation. Available seismic reflection data is shown in Figure 4. Deformed and retrodeformed section (AA′) is shown in Figure 9.

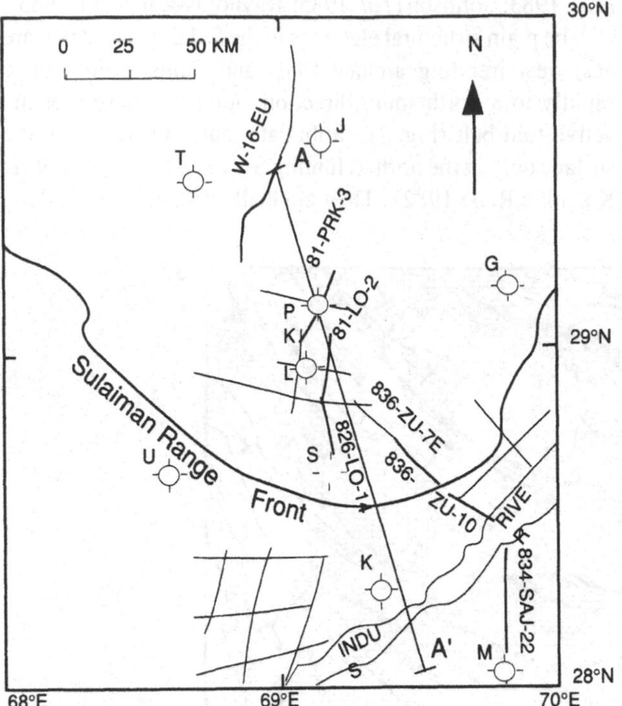

Figure 4. Map of seismic and well data used in this study. See Figures 2 & 3 for location. Bold lines were used to project subsurface data onto the balanced cross-section AA′ (Fig. 9). Composite seismic line shown in Figure 8 extends from Chauki syncline to Tadri. Crystalline basement is recognized about 30 km south of the deformation front on seismic line SAJ-22. Well abbreviations: G=Giandari-1; K=Kandkot-2; L=Loti-2; M=Mari-2; Pk=Pirkoh-2; S=Sui-1; T.M=Tadri Main; U=Uch-1.

frontal part of the fold belt in the subsurface. Tear faults, such as the Kingri fault, manifest neotectonic activity by the offset of fold axes, faults, uplifted and tilted gravel beds, and major bends along the course of the streams (Abdul-Gawad 1971; Rowlands 1978). Banks & Warburton (1986) constructed a balanced structural cross-section along the western part of the Sulaiman fold belt (Fig. 5A). This sugggests a line-length shortening of about 126 km along forward verging duplexes beneath a passive-roof backthrust.

Rocks from the Sulaiman fold belt can be divided into three main groups to emphasize their tectonic significance. From south to north these units are: (1) Late Oligocene to Recent molasse deposits; (2) Eocene to Permian, shallow-marine shelf to deep marine rocks (Kazmi & Rana 1982)); and (3) Late Eocene to Early Oligocene Khojak Flysch (Lawrence & Shahid 1990). The Muslimbagh ophiolites in the Zhob valley represent pieces of oceanic crust thrust over Maestrichtian shelf strata (Abbas & Ahmad 1979). Figure 6 provides a summary of the stratigraphy of the Sulaiman fold belt at the deformation front based on surface geology, well data, and seismic reflection profiles. The exposed Eocene to Permian rocks from the Sulaiman fold belt are similar to those of the Salt Range, except that the 7 km thick, carbonate dominated sequence is much thicker than that of the Salt Range. The Sulaiman fold belt with such a thick, sedimentary section yet with relatively high Bouguer gravity anomalies is interpreted to overly an extended crust (Lillie *et al.* 1989; Jadoon *et al.* 1989; Jadoon *et al.* unpubl. data; Khurshid pers. comm.).

Seismicity from the frontal part of the Sulaiman fold belt (Quittmeyer *et al.* 1984), along with multiple unconformities in the Siwalik molasse (Iqbal & Shah 1980; Banks & Warburton 1986) illustrate the ongoing deformation that is taken up by broad folds along the southern Sulaiman front.

General observations from integration of surface and subsurface data

Seismic reflection profiles from the frontal part of the Sulaiman fold belt and the adjacent foredeep in Pakistan have been interpreted in conjunction with drillhole, surface geology, and Landsat data (Figs 7-10). The main conclusions are:

(1) The thickness of the Phanerozoic sedimentary wedge at the deformation front is exceptionally high, about 10 km. The structurally duplicated sedimentary section 114 km north of the deformation front is about 15 km thick. This thickness includes more than 7 km of carbonate-dominated Palaeozoic to Eocene strata (compared to about 1 km for the same age strata in the Salt Range).

(2) The gross geometry of the overthrust wedge, including gentle topography (<1°) and broad width (> 250 km), is compatible with that proposed by Davis & Engelder (1985) for thrust belts developed over a weak decollement. However, evidence suggests that the Eocambrian evaporite sequence that provides an effective zone of decoupling at the base of the section in the Salt Range and Potwar Plateau (Lillie *et al.* 1987; Jaume & Lillie 1988) may not be present underneath the Sulaiman fold belt. This evidence includes: (a) absence of salt related structures (e.g. tight anticlines, broad synclines and disharmonic folding; Lillie & Yousaf 1986; Davis & Engelder 1985); and (b) the closest observation of the Eocambrian evaporites in seismic lines is about 200 km east of the deformation front (Humayon *et al.* 1990). The effective zone of decoupling in Sulaiman may be in pelitic rocks or fine carbonates above the crystalline basement at a depth of more than 10 to 15 km. Depth of the detachment is constrained by the seismic reflection profiles (81-LO-2 in Fig. 8). At this depth fine-grained sedimentary rocks may provide a weak detachment similar to the evaporites at a depth of 3 to 4 km (Lillie & Davis 1990).

(3) Basement dip is about 2.5° to the north. Basement is not involved in the deformation at least as far back as the Bugti syncline. This is based on the critical observation of the seismic data from southern Sulaiman foredeep (Fig. 4). However, farther north involvement of the basement is not precluded as the nature of the crust is inferred to be transitional below the Sulaiman fold belt (Jadoon *et al.* 1989; Izattt 1990, Jadoon *et al.* unpubl. data; Khurshid pers. comm.).

(4) The southernmost surface folds reflect a coherent stratigraphy in which older rocks are progressively exposed in the cores of more northerly, tighter anticlines (molasse in Sui anticline, Eocene in Loti and Pirkoh, Palaeocene in Kurdan, and Cretaceous in Tadri, Figs 3 & 9). The crests of these folds are cut only by small-scale normal faults. Seismic reflection profiles confirm that these surface folds reflect

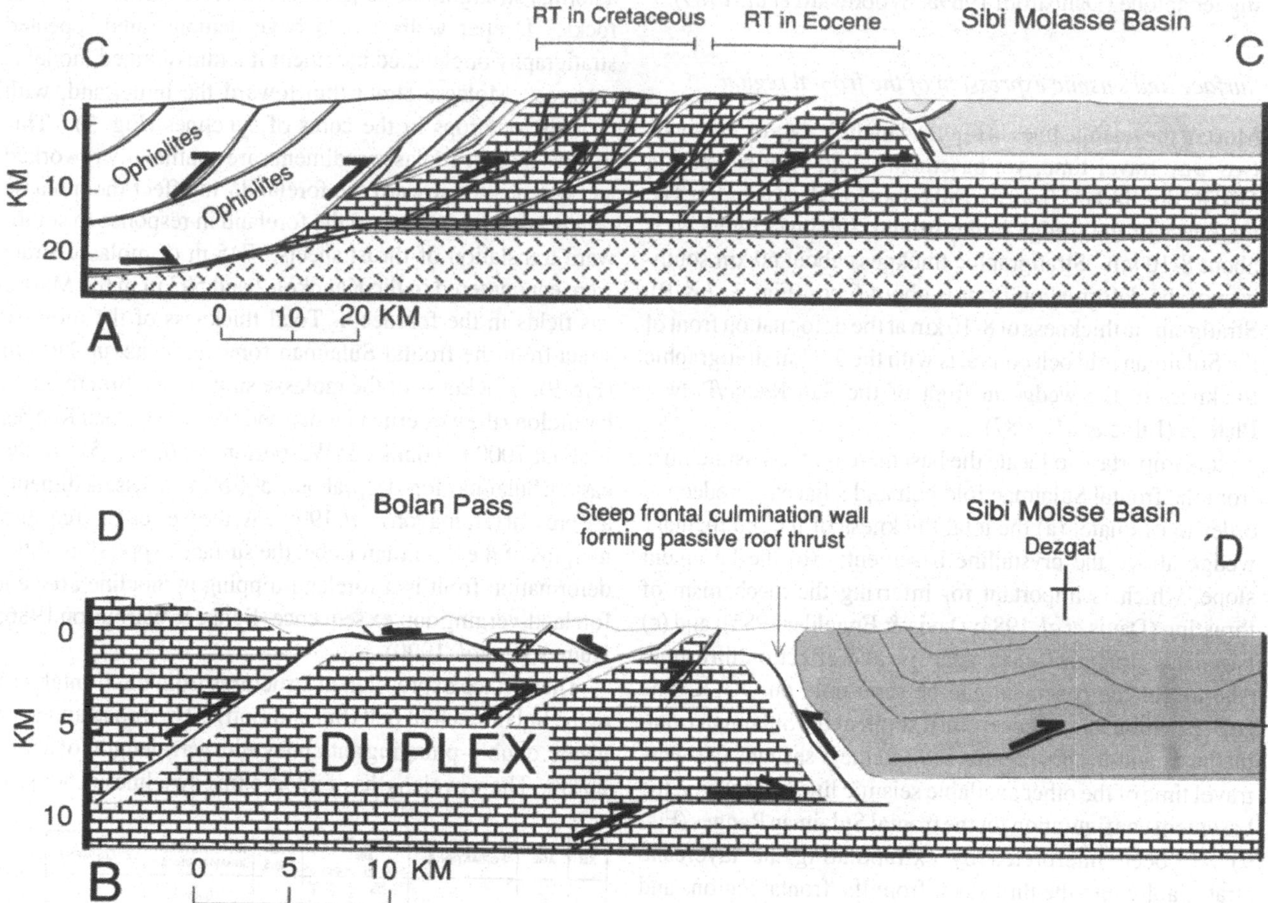

Figure 5. Structural cross-sections from the western Sulaiman fold belt CC´ and northern Kirthar Ranges DD´ (modified from Banks & Warburton 1986). (a) Passive-roof duplex geometry with a floor thrust at the base of the sedimentary section and a passive-roof thrust in the Ghazij (Eocene) and Goru (Cretaceous) Formations. (b) An antiformal stack duplex and the associated foredeep with 7 km of molasse sediments. A forward facing monocline is the surface expression of the duplex. See Figure 2 for location of cross-sections. Sand and clear pattern represents Molasse (Neogene) and Platform strata (Eocene to Cretaceous) that are a part of the roof-sequence. Brick pattern representing Jurassic to Precambrian rocks is a part of the duplex-sequence. Random lines show crystalline basement.

thrust faults and duplexes at depth. Seismic reflection profiles also show that northwards from the Bugti syncline rocks exposed at the surface are structurally elevated by the overthickened, active wedge. The resultant structural relief is 4 to 8 km from south to north.

(5) Overall structural style is of forward-verging duplexes bounded between a floor thrust near the base of the sedimentary section and a passive-roof thrust in thick Cretaceous shales. Fault related folds are exposed at the surface. Frontal broad and gentle folds (Sui and Loti), wavelength about 20 km, may be primarily formed as a result of ductile flow of material in the core of the anticlines at a depth of about 10 km (seismic line 81-LO-2 in Fig. 8).

(6) Total shortening parallel to the direction of tectonic transport along the duplex structures and the broad frontal folds is estimated as 76 km. Only a fraction of shortening (< 1 km) is accommodated by the surficial frontal folds (Sui and Loti), over a distance of about 55 km.

The details of these structures are discussed below in the context of seismic reflection profiles and the structural cross-section A-A' (Figs 7, 8 & 9). This is followed by discussion of the style of deformation and deposition in the foredeep basin, progressive deformation and crustal shortening.

Balanced structural cross-section

Drillhole data from the frontal part of the Sulaiman fold belt and adjacent foredeep (Fig. 4), have been provided to Oregon State University by the Hydrocarbon Development Institute of Pakistan (HDIP), Texaco and Amoco. A composite seismic line (bold lines in Fig. 4) has been constructed to project subsurface data onto a 174 km balanced structural cross-section (A-A' in Figs 3 & 9). A balanced structural cross-section is one that can be retrodeformed; thus it provides an opportunity to evaluate if the solution is geologically reasonable (Bally *et al.* 1966; Dahlstrom 1969a, 1970; Gwinn 1970; Woodward *et al.* 1989). A line length balancing technique is applied to the section except at the base of the broad frontal folds where the technique is invalid due to the ductile flow of the rocks in the core of the gentle anticline. The section was balanced in this basal zone using an area balancing technique (Dahlstrom 1969a; Woodward *et al.* 1989).

Surface and seismic expression of the frontal region

Most of the seismic lines in Figure 4 include data of 5 seconds two way travel time, yet basement could only be seen in OGDC line SAJ-22 from the Sulaiman foredeep (Fig. 7). The thickness of the wedge in the frontal Sulaiman foredeep is about 8-10 km. Stratigraphic thickness about 60 km south-east of the deformation front is about 6 km (Figs 3, 7 & 9). Stratigraphic thickness of 8-10 km at the deformation front of the Sulaiman fold belt contrasts with the 2-3 km stratigraphic thickness of the wedge in front of the Salt Range/Potwar Plateau (Lillie *et al.* 1987).

It is important to locate the basement on the seismic lines from the frontal Sulaiman fold belt and adjacent foredeep in order to evaluate: (a) the total thickness of the sedimentary wedge above the crystalline basement; (b) the basement slope, which is important for inferring the mechanism of thrusting (Davis *et al.* 1983; Davis & Engelder 1985); and (c) basement structures, their genesis, and effect on thrusting. The top of the basement can be seen only on OGDC line SAJ-22 within the foredeep. Just south of the Sulaiman front, the basement reaches depths beyond the 5 seconds two-way travel time of the other available seismic lines. However, the basement configuration for the frontal Sulaiman Ranges (Fig. 9) has been interpreted by extrapolating the layercake stratigraphy into the thrust belt from the frontal regions and adjacent foredeep using the seismic lines (ZU-10, ZU-7/7E, LO-14 in Figs 4, 7 & 8). Given an average basement slope (ß) of 2.5°, PreCambrian to Quaternary rocks that are about 6 km thick in the foredeep at Mari (Fig. 9) thickens stratigraphically to about 10 km at the deformation front. It attains a thickness of about 15 km below the Tadri structure 114 km north of the deformation front. Planar stratigraphy and gentle and broad structures (Sui and Loti folds), as far north as the Bugti syncline are inferred to reflect planar basement. However, the presence of rift related features of the Tethyan margin is not precluded, as transitional crust of about 25 km is inferred by Bouguer gravity modelling at the deformation front (Jadoon *et al.* unpubl. data).

Prominent reflections from the sedimentary wedge come from Eocene and Palaeocene limestones, Cretaceous sandstone and limestone, and from the top and base of the massive Jurassic limestone. At the surface progressively older strata are exposed towards the hinterland in the cores of doubly plunging anticlines (Fig. 3). Boreholes through the frontal folds, including Pirkoh anticline (Figs 2, 4 & 9), penetrate a normal stratigraphic sequence for about 3000 m and reach upper Cretaceous rocks. In the Mari gas field in the foredeep, a normal stratigraphic sequence of about 3300 m from Siwalik to Cretaceous (lower Goru) has been drilled. This sequence includes about 2100 m of Cretaceous, 800 m of Eocene and Palaeocene, and about 500 m of molasse strata (Kamran & Ranki 1987). All units of the carbonate-dominated sequence thicken to the north on the seismic lines except the Cretaceous. North of Pirkoh anticline, two wells drilled by Amoco on the Tadri and Jandran structures respectively, penetrated a normal stratigraphic sequences of Cretaceous and Jurassic rocks. Deeper wells would have demonstrated repeated stratigraphy or elevated basement if a thin-skinned model is incorrect. Molasse strata thin toward the hinterland, with sporadic outcrops in the cores of synclines (Fig. 9). This implies that the molasse sediments are continually reworked and redeposited toward the foreland. In effect the molasse sediments migrate toward the foreland in response to southward translation of thrust sheets. 745 m of molasse strata were encountered in the Kandkot-2 and 593 m in the Mari-2 gas fields in the foredeep. Total thickness of the molasse strata from the frontal Sulaiman foredeep is about 2400 m (Fig. 9). Thickness of the molasse strata in the Sibi molasse basin along the western Sulaiman and northern Kirthar Ranges is about 7000 m (Banks & Warburton 1986; Fig. 5). In the eastern Sulaiman foredeep, about 3500 m of molasse sediments are present (Humayon *et al.* 1990). At the western and eastern margins of the Sulaiman Lobe, the surface expression of the deformation front is a foreland-dipping monocline above a foreland-verging duplex sequence (Banks & Warburton 1986; Humayon *et al.* 1990).

The surface expression of structures from the frontal part of the Sulaiman fold belt (Sui and Loti) is two broad and very gentle, doubly-plunging anticlines with wavelengths of about 20 km. The wavelengths of intervening synclines is 6-8 km.

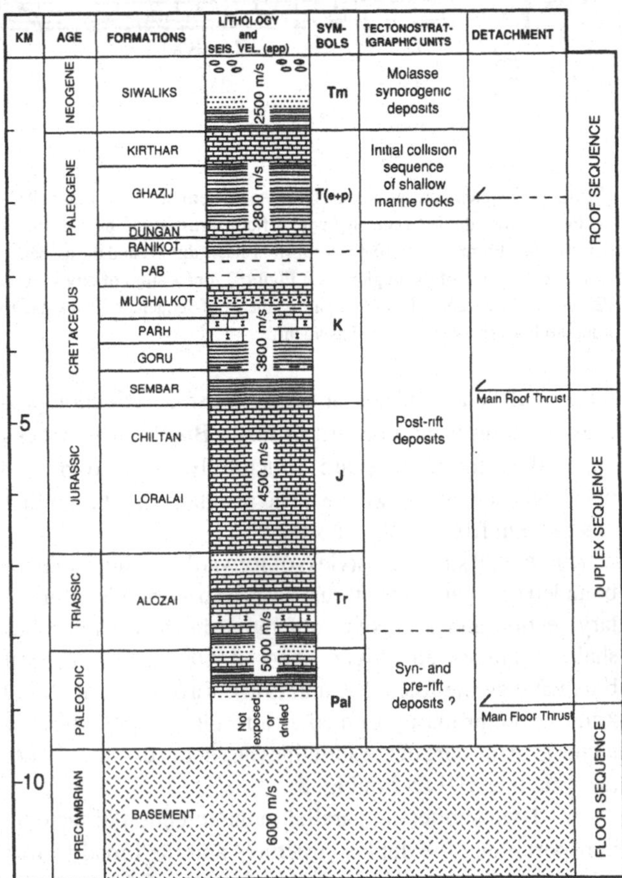

Figure 6. Simplified stratigraphic column of frontal Sulaiman fold belt. Approximate seismic velocities are estimates based on thicknesses from the well data, sonic logs, and converting stacking velocities from seismic lines to interval velocities. Detachment horizons are shown with a duplex sequence below and a roof sequence above Cretaceous shales.

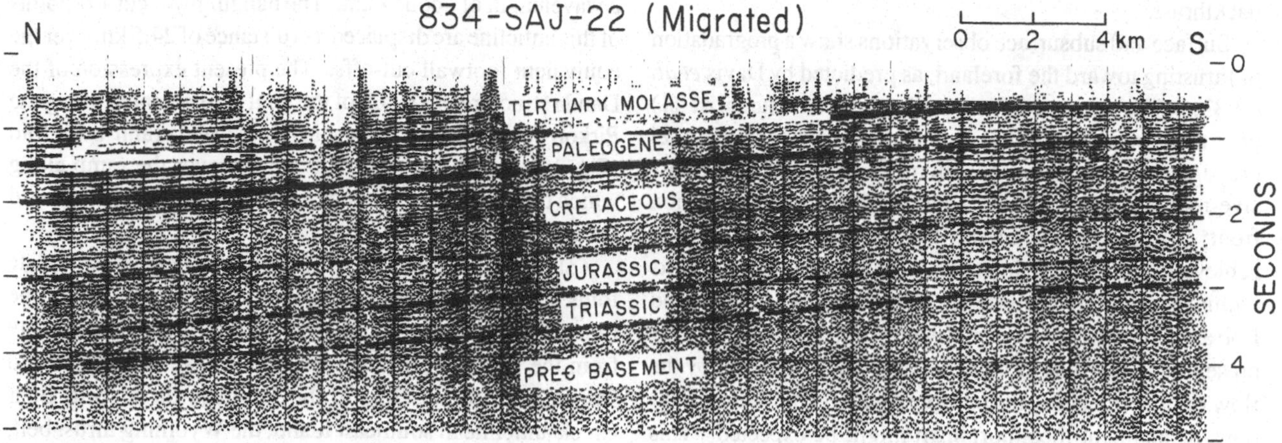

Figure 7. Seismic line about 30 km south of the deformation front showing crystalline basement at a depth of 3-4 seconds on two way travel time (about 6-8 km depth). See Figure 4 for location. Line SAJ-22 is 8-40 Hz, migrated vibroseis source, recorded in 1983 by OGDC and processed by Petty-Ray Geophysical Company.

Surface expression of the southerly limb of the third folded structure (Pirkoh) is a foreland-dipping monocline with dips between 35° and 70°. The top of this structure has almost horizontal strata over a distance of about 16 km (Figs 8 & 9). Further north, less open folds of smaller wavelength exist. Only Siwalik, Eocene and Palaeocene rocks are exposed along the line of the section. From south to north, progressively older strata occupy the hinge zones of the anticlines as structurally deeper levels are exposed. In addition to folds, normal and tear(?) faults are exposed at the surface. The Ridge, Saini Mund, and Pezbogi Nala faults on the Loti structure have a dip-slip offset of about 20 m and possibly displace the axis of the Loti syncline several hundred metres by a strike-slip component (Fig. 3).

The seismic expression of structures along the line of cross-section is of broad concentric anticlines (Sui and Loti) and more complex structures to the north (Fig. 8). The Sui and Loti folds maintain their layer parallel thickness and wavelength on the seismic lines, unlike typical concentric folds (Dahlstrom 1969b), where anticlines become tighter and synclines become broader at depth. It is inferred that the space in the cores of these anticlines is occupied by the ductile flow of material as a result of tectonic compression of the southward-propagating thrust sheet at depths of 10 to 15 km. Concentric folding is seen as deep as 5 seconds of two way travel time data on the seismic lines across the Sui and Loti structures (Figs 8 & 9). Basement is expected between 5 and 6 seconds. This implies that virtually all the Phanerozoic section is detached from the crystalline basement, with the decollement near the base of the wedge. North of the Bugti syncline, exposed rocks have a structural relief of about 4 km in the Pirkoh and about 8 km in the Tadri structures (Figs 8 & 9). The style of deformation above a detachment in Cretaceous shales is of passive folding in the roof thrust sheet; below, it is a duplex sequence of Jurassic and older rocks. A passive-roof backthrust in the Cretaceous Sembar shale accommodates forward movement of this duplex sequence (Figs 6, 8 & 9). Consequently, there is not a thrust fault at the surface in the tectonically thickened wedge from the frontal part of the Sulaiman fold belt. A similar style of deformation is reported from the western (Banks & Warburton 1986; Fig. 5) and eastern (Humayon *et al.* 1990) Sulaiman Range, the Kohat Plateau in the Trans-Indus Salt Range (Ahmed 1989; McDougal 1989), the northern Potwar Plateau (Lillie *et al.* 1987; Jaswal *et al.* 1990) and from other foreland fold and thrust belts (i.e. Canadian Cordillera, Price 1981, 1986; Jones 1982; Appalachians, Boyer & Elliot 1982; Mitra 1986; the Scottish Highlands, Elliot & Johnson 1980; the Pyrenees, Hossack *et al.* 1984; Williams 1985; the Alps, Boyer & Elliot 1982; Papua New Guinea, Hobson 1986; and the Taiwan thrust belt, Suppe 1980, 1983).

The deformed section is about 129 km long and restores to an undeformed length of about 205 km which gives a shortening of about 76 km (Fig. 9). This is very unevenly divided between the duplexes (75 km shortening), and broad Sui and Loti frontal anticlines (<1 km shortening). The central Sulaiman shortening determined here is similar to the 95 km of shortening found by Banks & Warburton (1986) for the equivalent portion of the western Sulaiman Lobe and also to the 70 km of shortening in the Kohat Plateau south of the Main Boundary thrust (McDougal 1989).

Structural style and development

Style of deformation

The Sulaiman lobe is an actively deforming fold belt that thickens northwestwards over a basement slope of 2.5°. About 10 km of undeformed platform and molasse strata, as measured at the deformation front, are thickened in a thrust wedge to about 15 km, 129 km north of the deformation front. This elevation is interpreted in this paper as due to thin-skinned structural duplication. However, the major thrust faults that are responsible for this thickening of the wedge do not crop out at the surface. Balanced and retrodeformed cross-sections based on seismic control show that the style of deformation is a duplex sequence of massive Jurassic limestone and older rocks probably detached from the crystalline basement along a sole thrust, and a roof-sequence of thick Cretaceous shales overlying these duplexes on a passive backthrust.

Surface and subsurface observations show a progradation of thrusting toward the foreland, as predicted by Davis *et al.* (1983). The interpreted chronology of structures is: (1) growth of broad, concentric fault tip folds in the foreland; (2) propagation of the basal decollement and uplift of the passive-roof sequence above a backthrust near deformation front; (3) propagation of the duplexes as critical taper is achieved; and (4) tear faults and extensional normal faults within the overthickened wedge. The currently active Sui and Loti anticlines in front of the forward verging duplexes are the present fault tip folds. Their cores are filled primarily by the flow of material at depths of more than 10 km where conditions of incipient metamorphism might be expected. This suggests that the folding precedes thrusting in the frontal Sulaiman fold belt. The extensional normal faults (for example, Pezbugi fault) are considered typical flexural structures formed above the neutral plane of folding and may initiate simultaneously with the initial folding.

The structural style of the roof-sequence is of fault-related folds (hybrid folds of Mitra, 1986) of variable tightness, symmetry, and extent as a result of ramp spacing, relative displacement along adjacent thrusts, degree of overlap, and final position of the cut-off point with respect to the next ramp. Hybrid folds from the frontal Sulaiman fold belt, using the terminology of Boyer & Elliot (1982), Butler (1982), Suppe (1983), Boyer (1986), and Mitra (1986), are classified as: (1) fault-bend folds; (2) leading edge, ramp-overlap anticlines; (3) intraplate anticlines; and (4) overlapping ramp anticlines. Specific examples of each type of structure are discussed below.

(1) Fault-bend fold: The Pirkoh anticline is a foreland verging fault-bend (box) fold. It has a broad hinge zone which exposes flat-lying, Eocene Pirkoh limestone at its core over a distance of about 16 km. This reflects the considerable displacement of the hangingwall beyond the footwall cutoff point. Displacement between cut-off points is about 20.5 km along a horse with a total length of about 45 km. The surface expression of the southern limb is of a monocline with dips of 35°-70°S (Fig. 9). The northern limb of Pirkoh, over the ramp, is overlapped by another fault-bend fold (Danda anticline), and so is not exposed at the surface. The topographic slope along the line of section north of Pirkoh is towards the hinterland.

Pirkoh anticline is the youngest fault-bend fold in this part of the actively deforming Sulaiman fold belt. The locus of shallow seismicity in the Sulaiman lobe is the Bugti syncline just south of Pirkoh anticline (Quittmeyer *et al.* 1979; Quittmeyer *et al.* 1984). This corresponds closely with the tip of the fault beneath the fold (Fig. 9). Epicentres of two events of magnitude between 6-7 are located on Loti and Sui anticlines (Quittmeyer *et al.* 1979). Determination of the depths of these seismic events could best determine if the active deformation is now concentrated along the decollement below and ahead of the Pirkoh duplex or along the roofthrust or otherwise along basement related faults.

(2) Leading edge, ramp overlap anticlines: Danda anticline, north of Pirkoh, is a comparatively tight structure with a wavelength of about 4 km. The hangingwall cut-off points of this anticline are displaced to a distance of 24.5 km over the equivalent footwall cut-offs. The present expression of the Danda structure is a result of the total displacement along the Pirkoh duplex below and ahead of the Danda fault-bend fold (Fig. 10-e), and the degree of overlap over the ramp along which the Pirkoh duplex steps upsection to propagate toward the foreland.

(3) Intraplate folds: Boyer (1986) describes how intraplate folds accommodate shortening strain within the body of the thrust sheet and are commonly cored by faults that propagate from the basal thrust fault. Examples have been presented from Elk Horn anticline in the Montana thrust belt, Bear Creek anticline in southeast Idaho, the Wyoming thrust belt, and from the Pennsylvania Valley and Ridge Province (Boyer 1986). The Kurdan anticline, with a wavelength of about 8-10 km and a steeper northern limb, is interpreted as an intraplate fold cored by a hinterland verging, passive backthrust that propagates from a within stack decollement surface (Fig. 10d). A shortening of about 3.8 km has occurred as a result of forward wedging of Triassic and older strata underneath the passive-backthrust (Fig. 9c).

(4) Overlapping ramp anticlines (anticlinal stack): In the core of the Tadri anticline Cretaceous rocks are exposed (Fig. 3), with an uplift of about 8 km. This is interpreted to be as a result of differential displacement along 2 horses (Fig. 9). Out of the total shortening of about 26.3 km, 10 km has occurred within the lower, and younger, horse (Fig. 10b-c). The result is completely overlapping ramp anticlines, as envisioned by Mitra (1986).

Passive-roof sequence and back thrust

A passive-roof sequence is a normal stratigraphic sequence separated from the duplex sequence by a roof thrust that remains stationary above a foreward propagating thrust sheet (Banks & Warburton 1986). To a minor extent shortening of the duplex sequence is accommodated in the roof sequence by uplift and folding. Banks & Warburton (1986) suggest that a roof-sequence, instead of extending over a large number of duplex horses, may be imbricated, thus equal amount of roof sequence is removed primarily by erosion (see their Fig. 7). How far a passive-roof sequence may extend has not yet been resolved. In Taiwan the length of passive-roof sequence is about 14 km (Suppe 1980). Jones (1982) predicts that a 50 km of relative backthrust sequence may have extended over the Alberta foothills. A retrodeformed geological cross-section from the thick Papua New Guinea thrust belt shows a major passive-roof sequence of more than 100 km (Hobson 1986). The Brooks Range of Alaska provides another example of a continuous passive-roof sequence that may extend several hundred of kilometres across the regional strike (I. R. Vann pers. comm., cited in Banks & Warburton 1986).

In the frontal Sulaiman fold belt, the passive-roof sequence of Cretaceous and younger rocks covers the entire area studied on a continuous roof backthrust. Along the line of section studied, the preserved extent of this roof sequence is about 60 km. Information from the Sulaiman fold belt

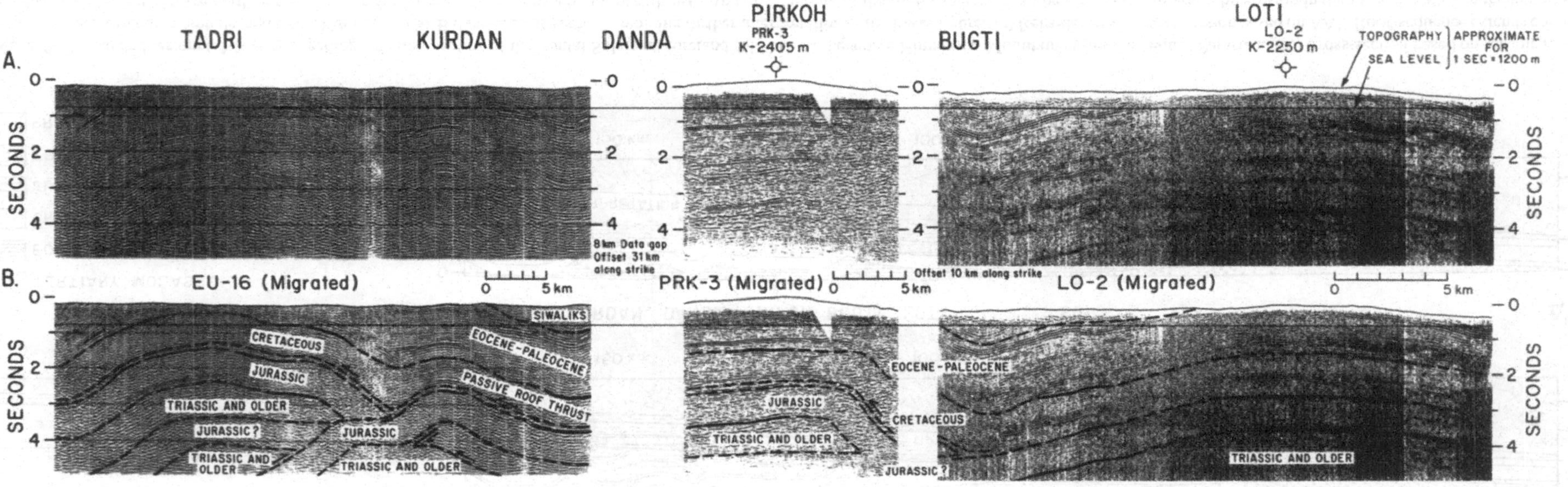

gure 8. Composite uninterpreted and interpreted seismic line from Chauki syncline to Tadri. Tertiary shallow marine rocks and Siwaliks are exposed at the surface. Basement is below 5 seconds on 2-way travel time. e section shows duplex bounded by a passive-roof thrust in Cretaceous shales and a floor thrust probably just above crystalline basement. The tip of the blind thrust extends below the Loti anticline (Fig. 9). Note that ncentric folding is the structural style of the broad Loti anticline; the space in the core of the broad folds (Loti and Sui) maybe filled by ductile flow or fine structures (small scale imbricates and duplex) within a decollement ne at a depth of more than 10 km. Line 81-LO-2 is 24-fold, migrated dynamite source, recorded and processed in 1981 by OGDC. Line 816-PRK-3 is 24 fold, migrated dynamite source, recorded by OGDC in December 80 to January 1981 and processed by Geophysical Service Inc. Azaiba, Oman. Line W16-EU is 10-40 Hz, migrated vibroseis source, recorded and processed by Western Geophysical Company of America in 1975. Lines tied along strike. See Figure 8 for geological details of seismic data gap between lines W16-EU and 816-PRK-3. Note that the horizontal scale differs between all the lines.

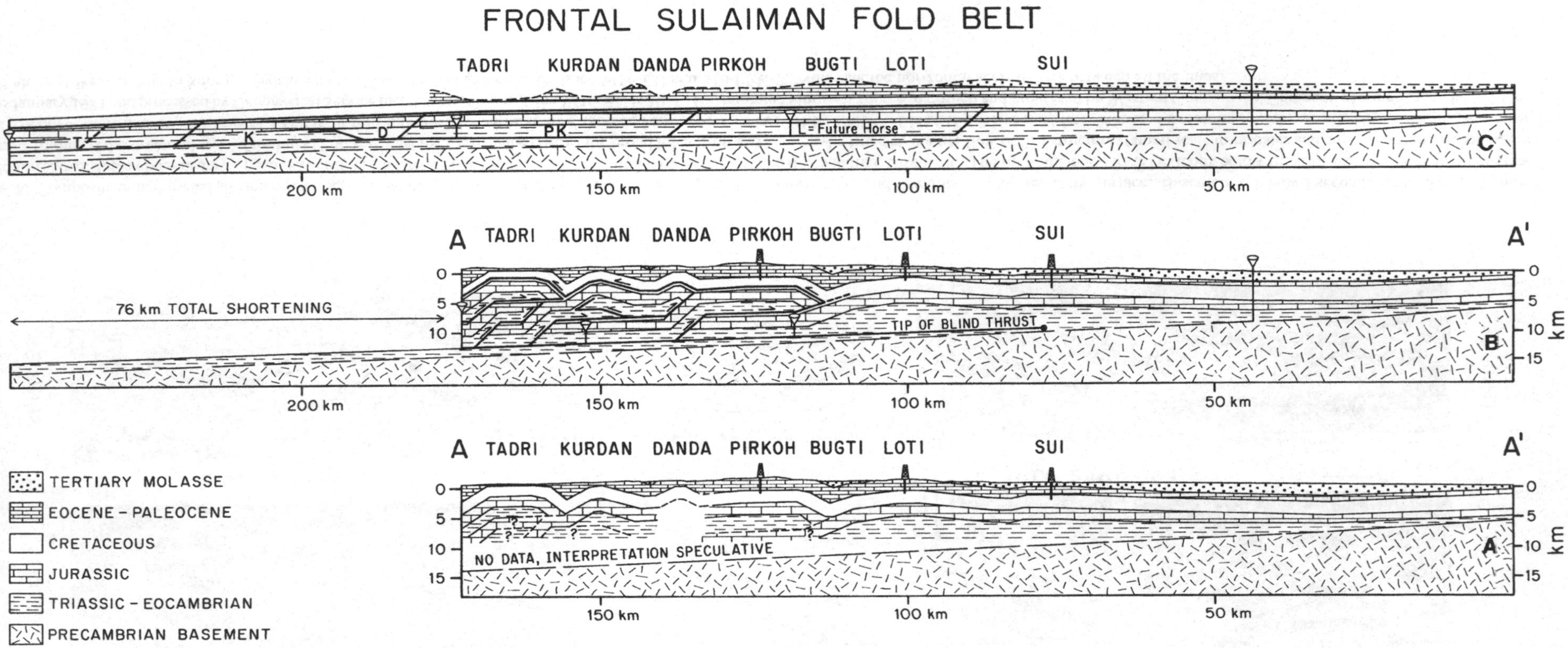

Figure 9. Actual and restored, NNW-SSE geological cross-section of the frontal Sulaiman foreland fold belt of the active Himalayan Mountain system in central Pakistan. **(a)** Cross-section based on seismic reflection profiles, surface geology, borehole, and LANDSAT data. **(b)** Balanced cross-section involving further interpretation of the lower figure. **(c)** Retrodeformed cross-section based on AA'. Roof-sequence extends continually for a distance of about 150 km north of the tip of the duplex and is not cut by a major backthrust. An equal amount of the roof-sequence must be removed primarally by erosion in the Loralai valley in the hinterland. In the current balanced section (top figure) only the shortening associated with the folds in the roof-sequence is shown. Seismic data has been projected from the bold lines in Figure 4 on to the cross-section AA′ in Figure 3. Letters identifying the individual horses in the duplex sequence are from the individual mountains (shown on Fig. 3), formed by the duplex propagation. From south to north these mountains are L, Loti; PK, Pirkoh; D, Danda Range; K, Kurdan Range; and T, Tadri.

(Jadoon unpubl. data) shows that roof sequence extends over a distance of about 150 km northwards from the tip of the duplex in the Bugti syncline and no major cross-cutting backthrust has yet been discovered. Finally, base of the roof sequence (thick Cretaceous shale) is exposed at the surface in the broad Loralai valley where majority of the roof sequence is removed by erosion. This example suggests that in the initial stages of its evolution, a passive roof sequence may extend over several duplex horses without overstep backthrusts cross-cutting the roof sequence. This idea is based on the absence of a major backthrust and presence of hinterland and foreland verging faults with minor displacement which propagate out of synclines in the roof sequence north of this line (Jadoon work in progress).

Duplex development, orogenic contraction, and deposition in the foredeep molasse basin

Sequential restoration of the balanced structural cross-section from the active Sulaiman fold-and-thrust belt provides an opportunity to unravel the progressive deformation and provide information on the deposition, uplift, and forward migration of the foredeep basin. The following is a chronological description of the evolution along the cross-section A-A' in Figure 10.

(1) Erosion of molasse and platform sediments from southward migrating thrust sheets in the Loralai and Kohlu areas north of the studied section (Fig. 2) developed a molasse basin that thinned toward the foreland. Deposition continued until depth to the basement became sufficient for the decollement to propagate southward. This initiated a thrust sheet of massive Jurassic limestone and older strata, bounded between a floor and a roof thrust, that stepped up-section and slid along thick Cretaceous shale at the base of the roof sequence for 16.25 km (Fig. 10a-b).

(2) Surface expression of the duplex became a foreland dipping monocline. Location of the ramp in this case is arbitrary, positioned only for balancing purposes. The displacement of 16.25 km (10-b) within this duplex is the amount of shortening along line A-A'. With uplift, the foredeep basin migrated further south, and reworking of the molasse sediments thickened the foredeep wedge. A topographic slope of 2.8° was produced on the section and may have provided critical taper. The second thrust sheet (T) stepped upsection below the tip of the first duplex and flattened along the shale horizon at the base of the Cretaceous (Fig. 10b-c).

(3) A displacement of 10 km of thrust slice T produced an antiformal stack, Tadri, and a 6.5 km deep molasse basin filled with reworked molasse eroded from structures north of the section. A modern example of this geometry exists in the northern Kirthar and western Sulaiman ranges, where Jurassic limestone is exposed 9 km above its stratigraphic level with a foredeep Sibi basin that contains 7 km of molasse strata (Banks & Warburton 1986; Fig. 5). Development of the Tadri antiformal stack and extreme steepening of the passive-roofthrust impeded backthrust motion. Continuous uplift allowed erosion through the deformed molasse strata and into the Eocene Kirthar limestone. When sufficient topographic slope was developed (4° in this section), the frictional resistance at the base and top of the thrust wedge was overcome and the roofthrust propagated southward (Fig. 10c).

(4) Within the duplex sequence, wedging of an intraplate thrust (between K and D and below the passive-roofthrust) produced the hinterland verging Kurdan fault-bend fold (Fig. 10d). The fault in the core of the Kurdan anticline remained passive during 3.75 km of foreland directed displacement of the Kurdan intraplate thrust sheet (K) along the basal detachment. Uplift and translation of the Kurdan structure shifted the deformation front farther south. A modern analogue to this situation is the Sibi molasse basin in front of the western Sulaiman/Kirthar ranges (Fig. 5), and the foredeep in front of the eastern Sulaiman ranges.

(5) Propagation of the sole detachment continued in front of the antiformal stack. The length of the horse (K + D) being about 42 km. This detached sequence stepped upsection and was translated along a flat for a distance of about 25 km (Fig. 10d-e). Successively older rocks were exposed in the cores of the fault-related anticlines. The top of Cretaceous sandstone in the Tadri structure was raised 8 km from its regional stratigraphic position. The decollement in all these stages remains at a depth of 12-15 km.

(6) The Pirkoh thrust slice (PK) stepped-up southwards in front of the Danda (D) monocline (Fig. 10e-f). This horse was translated for a distance of about 20 km to form the broad, fault-bend Pirkoh anticline. The surface expression of the duplex was that of a broad monocline. The topographic slope was about 1° to the south and the total displacement within the duplex sequence was about 75 km. The depositional axis of the molasse basin migrated farther toward the foreland throughout compression.

(7) The present day geometry developed by very gentle concentric folding (L and S) in front of Pirkoh anticline (Fig. 10g). We suggest that the space in the cores of these anticlines is occupied by ductile flow of fine-grained sedimentary rocks at a depth of more than 10 km and probably involving substantial pressure solution. Extensional normal faults with a component of strike-slip displacement are the dominant surface structures along the hinge zones of the Pirkoh and Loti anticlines. The deformed section of 129 km length within the duplex sequence restores to an undeformed length of 205 km (Fig. 9). Total displacement is 76 km at the present day. As discussed earlier, an equal amount of roof sequence must have been removed, primarily by erosion, in the broad Loralai valley. In this paper only the part of the shortening that was taken-up by folds in the roof sequence is balanced (Fig. 9c).

Conclusions

Surface and subsurface data from the southern Sulaiman Lobe have been integrated to look into the structural evolution of the active Himalayan fold belt along the western margin of the Indian subcontinent. The important conclusions are summarized as follows:

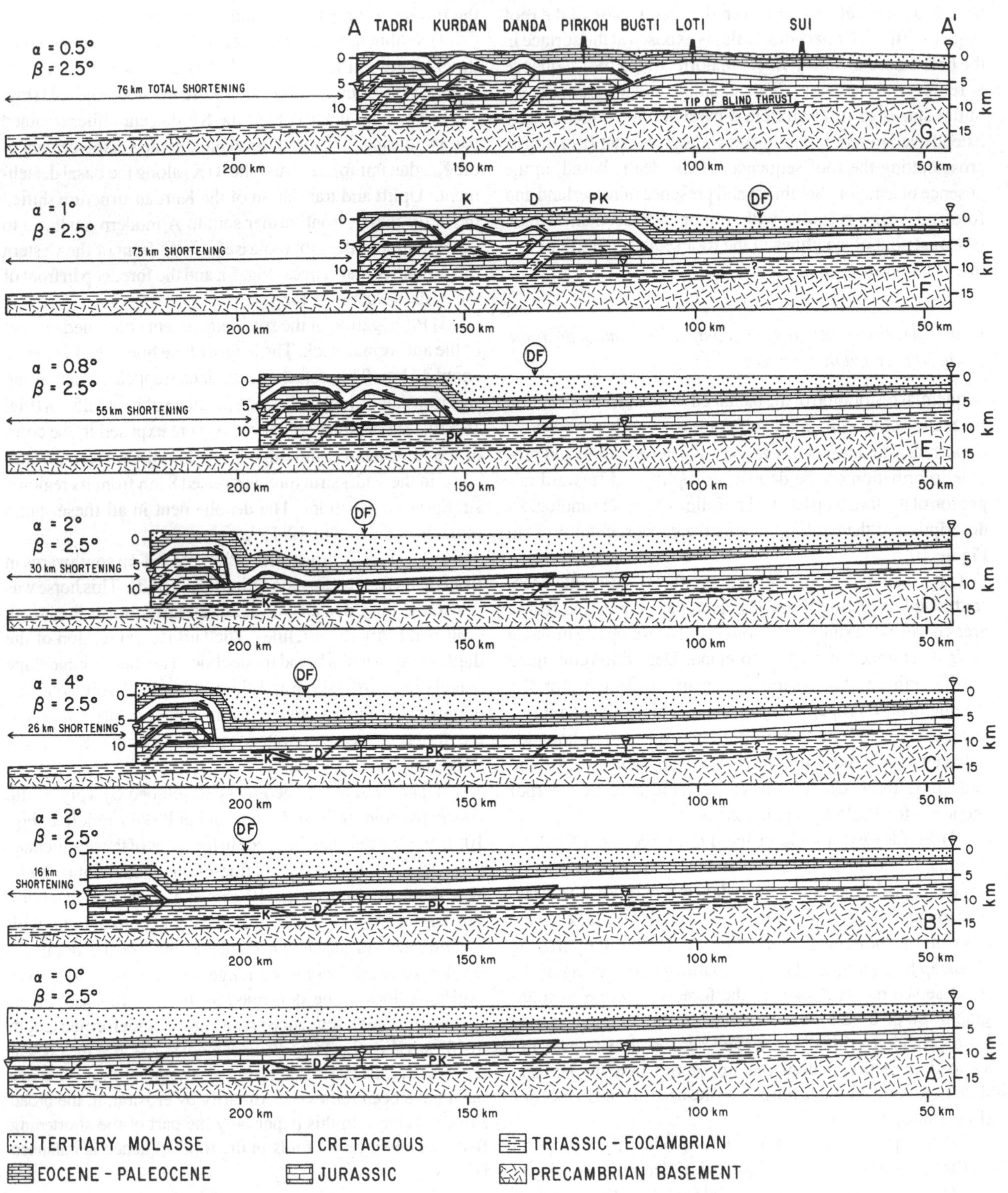

Figure 10. Palinspastic restoration of duplex development in the frontal part of the active Sulaiman fold belt along the line of cross-section AA´ (Fig. 3). Area balancing is done below the Triassic over the two frontal broad folds (Fig. 9). The current basement slope (β) 2.5° is considered to remain constant in the reconstruction. Topographic slope (α) changes at each step to create the suitable taper to overcome the frictional resistance at the base of the wedge allowing the duplex to propagate towards the foreland. At each stage the deformation front progressively moves towards the foreland and continental molasse strata are constantly reworked to thicken the foredeep wedge. See text (points 1 - 7) for discussion of A to G.

(1) The gentler surface topography (<1°) and broad width (>250 km) of the Sulaiman fold belt is similar to other mountain belts underlain by a weak decollement.

(2) The compressive deformation along the frontal part of the Sulaiman fold belt is accomodated by a duplex whose floor thrust is above the crystalline basement and roof thrust is in thick Cretaceous shales. The surface expression of deformation in the duplex is fault-related folds (Pirkoh, Danda, Kurdan, and Tadri), where exposed rocks at the surface are structurally uplifted 4-8 km above their regional stratigraphic level.

(3) The roof sequence is not breached along the cross-section, suggesting that a major passive-roof thrust extends over a 60 km length along the cross-section.

(4) Progressive structural development is as follows: (a) concentric folding behind the fault tip (Sui and Loti anticlines); (b) the development of a passive-roof duplex at the deformation front; (c) forward propagation of the duplex to produce a variety of structural geometries. From south to north, these features include a fault-bend fold (Pirkoh), a leading-edge ramp-overlap anticline (Danda), an intraplate fold (Kurdan), and an anticlinal stack (Tadri). Molasse sediments have been continually reworked and the depositional axis of the foredeep basin migrated southward due to southward migration of the deformation front.

(5) Planispastic restoration indicates a shortening of 75 km in the duplex sequence and 76 km in the entire deformed 129 km section. The restored section has a length of 205 km.

(6) Only a fraction of the shortening, <1 km, is taken up by the broad, frontal Sui and Loti anticlines. These folds, extending over a distance of 55 km in front of the main mountain belt, are concentric in the seismic lines as deep as 5 seconds two way travel time on seismic data. The basement is expected between five and six seconds on two way travel time data. This suggests that the space in the core of these anticlines is primarily filled by ductile flow of material at depths of about 10 km, as a result of tectonic compression; implying that tip-line folding preceeds faulting in the southern Sulaiman fold belt.

This work in the Sulaiman fold belt is part of a cooperative project between Oregon State University (OSU) and Hydrocarbon Development Institute of Pakistan (HDIP) and is supported by NSF grants INT-86-09914 and EAR-8816962. Additional support was provided by Amoco, Texaco Overseas Petroleum Company, and Mobil Oil Company. Ishtiaq Jadoon at OSU is supported by USAID. We would like to thank the Oil and Gas Development Corporation (OGDC) and Amoco for releasing seismic and well data for this project. Cooperation from Geological Survey of Pakistan (GSP) is also acknowledged. We thank to various individuals for logistical help and stimulating discusssion during the course of this study, particularly Hilal Raza, Riaz Ahmed, Manshoor Ali, Dan Davis, Ghazanfer Abbas, Mansoor Humayun, Jalil Ahmed, Amjid Cheema, Gary Huftile, Mirza Baig, and Mazhar Qayum. Critical review by J. Warburton and C. Izattt greatly improved the original manuscript.

References

Abbas, G. & Ahmad. 1979. The Muslimbagh Ophiolites. *In:* Farah, A. & De Jong, K.A. (eds) *Geodynamics of Pakistan*. Geological Survey of Pakistan, Quetta, 243-250.

Abdul-Gawad, M. 1971. Wrench movement in the Baluchistan and relation to Himalayan-Indian Ocean Tectonics. *Geological Society of America Bulletin*, **82**, 1235-1250.

Acharyya, S. K. & Ray, K. K. 1982. Hydrocarbon possibilities of concealed Mesozoic-Paleogene sediments below Himalayan nappes, reappraisal. *American Association of Petroleum Geologists Bulletin*, **66**, 57-70.

Ahmed, I. 1989. Sedimentology and structure of the south-west Kohat Plateau. Ph.D. thesis, Department of Earth Sciences, University of Cambridge (unpubl.).

Allemann, F. 1979. Time of emplacement of the Zhob valley ophiolites and Bela ophiolites. *In:* Farah, A. & De Jong, K. A. (eds) *Geodynamics of Pakistan*. Geological Survey of Pakistan, Quetta, 215-242.

Bachmann, G. H. Dohr, G. & Muller, M. 1982. Exploration in a classic thrust belt and its foreland: Bavarian Alps Germany. *American Association of Petroleum Geologists Bulletin*, **66**, 2529-2542.

Bally, A. W., Gordy, P. L. & Stewart, G. A. 1966. Structure, seismic data, and orogenic evolution of southern Canadian Rocky Mountains. *Bulletin of Canadian Petroleum Geology*, **14**, 337-381.

Banks, C. J. & Warburton, J. 1986. 'Passive-roof' duplex geometry in the frontal structures of the Kirthar and Sulaiman mountain belt, Pakistan. *Journal of Structural Geology*, **8**, 229-237.

Boyer, S. E. & Elliot, D. 1982. Thrust Systems. *American Association of Petroleum Geologists Bulletin*, **66**, 1196-1230.

—— 1986. Styles of folding within thrust sheets: Examples from the Appalachian and Rocky Mountains of the USA and Canada. *Journal of Structural Geology*, **8**, 325-340.

Butler, R.W.H. 1982. The terminology of structures in thrust belts. *Journal of Structural Geology*, **4**, 239-245.

Dahlstrom, C.D.A. 1969a. Balanced cross-sections. *Canadian Journal of Earth Science*, **6**, 743-757.

—— 1969b. The upper detachment in concentric folding. *Bulletin of Canadian Petroleum Geology*. **17**, 326-346.

—— 1970. Structural geology in the eastern margin of the Canadian Rocky Mountains. *Bulletin of Canadian Petroleum Geology*. **18**, 332-406.

Davis, D., Suppe, J. & Dahlen, F. A. 1983. Mechanics of fold-and-thrust belts and accretionary wedges. *Journal of Geophysical Research*, **88**, 1153-1172.

—— & Engelder, T. 1985. The role of salt in fold-and-thrust belts. *Tectonophysics*, **119**, 67-88.

Elliot, D. & Johnson, M. R. W. 1980. Structural evolution in the northern part of the Moine thrust belt, NW Scotland. *Royal Society of Edinburgh Transactions*, **71**, 69-96.

Farah, A., Abbas, G., De Jong, K. A. & Lawrence, R. D. 1984. Evolution of the lithosphere in Pakistan. *Tectonophysics*, **105**, 207- 227.

Gansser, A. 1981. The geodynamic history of the Himalaya. *In:* Gupta, H. K. & Delany, F. (eds) *Zagros, Hindu Kush, Himalaya: Geodynamic evolution*. American Geophysical Union, Geodynamic Series, **3**, 111-121.

Gwinn, V. E. 1970. Kinematic patterns and estimates of lateral shortening, Valley and Ridge and Great Valley Provinces, central Appalachians, south-central Pennsylvania. *In:* Fisher, G. W. *et al.* (eds) *Studies in Appalachian Geology, Central and Southern*, Interscience, New York, 127-146.

Harris, L. D. & Milici, R. C. 1977. Characteristics of thin-skinned style of deformation in the southern Appalachian and potential hydrocarbon traps. United States Geological Survey, professional paper, **1018**, 40p.

Hobson, D. M. 1986. A thin skinned model for the Papuan thrust belt and some implications for hydrocarbon exploration. *Australian Petroleum Exploration Association Journal*, **26**, 214-224.

Hossack, J. R., Deramond, J. & Graham, R. H. 1984. The geological structure and development of the Pyrenees (abst). *Toulouse Conference on Thrusting and Deformation, Abstracts of Programs*, 46-47.

Humayon, M., Lillie, R. J. & Lawrence, R. D. 1990. Structural interpretation of eastern Sulaiman foldbelt and foredeep, Pakistan. *Tectonics* (in press).

Hunting Survey Corporation. 1961. *Reconnaissance Geology of part of West Pakistan, A Columbo Plan Cooperative Project*. Government of Canada, Geodynamic, Toronto, 550p.

Iqbal, M.W.A. & Shah, S.M.I. 1980. *Records of the Geological Survey of Pakistan: A guide to the stratigraphy of Pakistan*. 37p.

Izattt, C. N. 1990. Variation in thrust front geometry across the Potwar Plateau and Hazara/Kalachitta hill ranges, northern Pakistan. Ph.D thesis, Imperial College of Science, Technology and Medicine, University of London, 353p (unpubl.).

Jadoon, I.A.K., Lillie, R. J., Humayon, M., Lawrence, R. D., Ali, S. M. & Cheema, A. 1989. Mechanism of deformation and the nature of the crust underneath the Himalayan foreland fold-and-thrust belts in Pakistan. EOS. *Transaction, American Geophysical Union*, **70**, 1372-1373.

Jaswal, T., Lillie, R. J. & Lawrence, R. D. 1990. Structure and evolution of the Durnal oil field, Nothern Potwar Deformed Zone, Pakistan. *American Association of Petroleum Geologists Bulletin* (in press).

Jaume, S. C. & Lillie, R. J. 1988. Mechanics of the Salt Range-Potwar Plateau, Pakistan. A fold and thrust belt underlain by evaporites. *Tectonics*, **7**, 57-71.

Johnson, N. M., Opdyke, N. D., Johnson, G. D., Lindsay, E. H. & Tahirkheli, R.A.K. 1982. Magnetic polarity stratigraphy and ages of the Siwalik Group rocks of the Potwar Plateau, Pakistan. *Paleogeography, Paleoclimatology, Paleoecology*, **37**, 17-42.

——, Stix, J., Tauxe, L., Cerveny, P. F. & Tahirkheli, R.A.J. 1985. Paleoclimatic chronology, fluvial processses, and tectonic implication of the Siwalik deposits near Chinji Village, Pakistan. *Journal of Geology*, **93**, 27-40.

Jones, P. B. 1982. *Oil and gas beneath east-dipping thrust faults in the Alberta Foothills. In:* Rocky Mountain Association of Geologists Guidebook, Denver, Colorado, 61-74.

Kamran, M. & Ranki, U. 1987. Pakistan well data. Hydrocarbon Development Institute of Pakistan, unpublished report.

Kazmi, A. H. & Rana, R. A. 1982. Tectonic map of Pakistan, 1:2,000,000. Geological Survey of Pakistan, Quetta.

Laubscher, H. P. 1981. The 3D propagation of decollement in the Jura. *In:* McClay, K. R. & Price, N. J. (eds) *Thrust and Nappe Tectonics*. Geological Society of London Special Publication, **9**, 311-318.

Lawrence, R. D., Khan, S. H., De Jong, K. A., Farah, A. & Yeats, R. S. 1981. Thrust and strike-slip fault interaction along the Chaman fault zone, Pakistan. *In:* McClay, K. R. & Price, N. J. (eds) *Thrust and Nappe Tectonics*, Geological Society of London Special Publication, **9**, 363-370.

—— & —— 1990. Structural reconnaissance of Khojak flysch, Pakistan and Afghanistan (in press).

Lillie, R. J. & Yousaf, M. 1986. Modern analogs for some midcrustal reflections observed beneath collisional mountain belts. *In:* Barazangi, M. & Brown, L. (eds) *Reflection Seismology: The Continental Crust*, American Geophysical Union, Geodynamic Series, **14**, 55-65.

——, Johnson, G. D., Yousaf, M., Zamin, A. S. H. & Yeats, R. S. 1987. Structural development within the Himalayan foreland fold-and-thrust belt of Pakistan, *In:* Beaumont, C. & Tankand, A. J. (eds) *Sedimentary basins and basin-forming mechanisms*. Canadian Society of Petroleum Geologists Memoir, **12**, 379-392.

——, Lawrence, R. D., Humayon, M. & Jadoon, I.A.K. 1989. The Sulaiman thrust lobe of Pakistan: Early stage underthrusting of the Mesozoic rifted margin of the Indian subcontinent. *Geological Society of America*, Abstract with program, 318.

—— & Davis, D. M. 1990. Structure and mechanics of the foldbelts of Pakistan. *Thrust Tectonics 1990, Programme with Abstracts*, Royal Holloway and Bedford New College, University of London (4-7 April 1990), 69.

McDougal, J. 1989. Geology and Geophysics of the foreland fold-thrust belt of northwestern Pakistan. Ph.D. thesis, Oregon State University, 140p (unpubl.).

Mitra, S. 1986. Duplex structures and imbricate thrust systems: Geometry, structural position, and hydrocarbon potential. *American Association of Petroleum Geologists Bulletin*, **70**,1087-1112.

Price, R. A. 1981. The Cordilleran foreland thrust and fold belt in the southern Canadian Rocky Mountains. *In:* McClay, K. R. & Price, N. J. (eds) *Thrust and Nappe Tectonics*, Geological Society of London Special Publication, **9**, 427-448.

—— 1986. The southeastern Canadian Cordillera: thrust faulting, tectonic wedging, and delamination of the lithosphere. *Journal of Structural Geology*, **8**, 239-254.

Quittmeyer, R. C., Farah, A. & Jacob, K. H. 1979. The seismicity of Pakistan and its relation to surface faults. *In:* Farah, A. & De Jong, K. A. (eds) *Geodynamics of Pakistan*, Geological Survey of Pakistan, Quetta, 351-358.

——, Kaffa, A. A. & Armbruster, J. G. 1984. Focal mechanism and depths of earthquakes in Central Pakistan: A tectonic interpretation. *Journal of Geophysical Research*, **89**, 2459-2470.

Raiverman, V., Kunte, S. V. & Mudherjea, A. 1983. Basin geometry, Cenozoic sedimentation, and hydrocarbon prospects in northwestern Himalaya and Indo-Gangetic plains. *Petroleum Asia Journal*, **6**, 67-92.

Raynolds, R.G.H. & Johnson, G. D. 1985. Rates of Neogene depositional and deformational processes, northwest Himalayan foredeep margin, Pakistan. *In:* Snelling, N. J. (ed.) *The chronology of the Geological Record*. Geological Society of London Memoir, **10**, 297-311.

Rich, J. L. 1934. Mechanics of low-angle overthrust faulting as illustrated by the Cumberland thrust block, Virginia, Kentucky, Tennessee. *American Association of Petroleum Geologists Bulletin*, **18**, 1584-1596.

Rowlands, D. 1978. The structure and seismicity of a portion of southern Sulaiman Ranges, Pakistan. *Tectonophysics*, **51**, 41-56.

Sarwar, G., & De Jong, K. A. 1979. Arcs, oroclines, syntaxes: The curvature of mountain belts in Pakistan. *In:* Farah, A. & De Jong, K. A. (eds) *Geodynamics of Pakistan*, Geological Survey of Pakistan, Quetta, 351-358.

Seeber, L., Armbruster, J. G. & Quittmeyer, R. C. 1981. Seismicity and continental subduction in the Himalayan Arc. *In:* Gupta, H. K. & Delany, F. M. (eds) *Zagros, Hindukush, Himalaya, Geodynamic Evolution*, American Geophysical Union, Geodynamic Series, **3**, 215-242.

Suppe, J. 1980. Imbricated structure of western foothills belt, south central Taiwan. *Petroleum Geology of Taiwan*, **17**, 1-16.

—— 1983. Geometry and kinematics of fault bend folding. *American Journal of Science*, 283, 684-721.

Waheed, A., Wells, N. A., Ahmad, N. & Tabutt, K. D. 1988. Paleocurrents along an obliquely convergent plate boundary, Sulaiman fold belt, south west Himalayas, West central Pakistan. Abs. *American Association of Petroleum Geologists Bulletin*, **72**, 255-25.

Williams, G. D.1985. Thrust Tectonics in the south central Pyrenees. *Journal of Structural Geology*, **7**, 11-17.

Woodward, N. B., Boyer, S. E. & Suppe, J. 1989. Balanced geological cross-sections: An essential technique in geological research and exploration. American Geophysical Union, Short course in geology, **6**, 132p.

Yeats, R. S., Khan, S. H. & Akhtar. M. 1984. Late Quaternary deformation of the Salt Range of Pakistan. *Geological Society of America Bulletin*, **95**, 958-966.

—— & Hussain, A.1987. Timing of structural events in the Himalayan foothills of northwestern Pakistan. *Geological Society of America Bulletin*, **99**, 161-176.-176.

—— 1989. Folds and blind thrusts in the sub-Himalaya and 1905 Kangra earthquake (M=8). *EOS* Transactions, American Geophysical Union, **70**, 1368.

PART SEVEN

NW American Cordillera

The Monashee decollement of the southern Canadian Cordillera: a crustal-scale shear zone linking the Rocky Mountain Foreland belt to lower crust beneath accreted terranes

Richard L. Brown,[1] Sharon D. Carr,[1] Bradford J. Johnson,[1] Vicki J. Coleman,[1] Frederick A. Cook[2] & John L. Varsek[2]

[1]*Department of Earth Sciences, Carleton University, & Ottawa-Carleton Geoscience Centre, Ottawa, Ontario, Canada K1S 5B6*

[2]*Department of Geology and Geophysics, The University of Calgary, Calgary, Alberta, Canada T2N 1N4*

Abstract: The Monashee decollement, a crustal-scale shear zone in the hinterland (Omineca belt) of the southern Canadian Cordillera is interpreted as correlating with the sole thrust of the Rocky Mountain Foreland belt. LITHOPROBE seismic reflection data indicate that the shear zone is rooted in the lower crust beneath the Intermontane Superterrane. The decollement is exposed on the margins of the Monashee complex, an elongate domal culmination within the Omineca belt. This complex includes Early Proterozoic gneisses of North American basement. The hangingwall of the decollement consists of deformed and metamorphosed North American continental margin and oceanic accreted terranes of the Selkirk allochthon. Layered reflections observed in LITHOPROBE seismic reflection data correlate with sheared and transposed rocks in both the hangingwall and the footwall of the Monashee decollement; the decollement is imaged as reflections, cutoffs and correlatable reflection segments. Field relations, kinematics and geochronology indicate that the shear zone has experienced a complex history that included protracted easterly directed shear of hangingwall rocks and imbrication of basement rocks in its footwall. U-Pb zircon dating of syn- and post-kinematic leucogranites in the decollement zone demonstrates that the final stages of thrusting occurred in the Late Palaeocene. The cessation of thrusting and the onset of crustal extension in the southern Omineca belt correspond to the end of thrusting in the Foreland belt.

Easterly directed thrust faults in the Rocky Mountain Foreland belt merge with a sole thrust above Precambrian basement rocks of the North American craton (Bally *et al.* 1966; Price & Mountjoy 1970; Cook *et al.* 1988). These basement rocks remained rigid and uninvolved in the thin-skinned deformation of the cover rocks as far west as the southern Rocky Mountain Trench (Fig. 1). Supracrustal shortening in the Foreland belt at latitude 52°N is approximately 50%, amounting to an easterly displacement of 200 km relative to a fixed reference in the underlying basement (Price & Mountjoy 1970).

Basement involvement in thrust deformation west of the Rocky Mountain Trench is evident from surface geology in the Omineca belt, the metamorphic hinterland of the orogen (Fig. 1; Simony *et al.* 1980; McDonough & Simony 1988). Brown *et al.* (1986) and Monger *et al.* (1986) proposed models that accommodate the thin-skinned shortening in the Foreland belt by development of a duplex in basement rocks beneath the Omineca belt. Balanced sections have been drawn in which shortening of the basement beneath the Omineca belt is assumed to have accommodated at least half of the shortening in the Foreland belt (Brown *et al.* 1986; Journeay 1986; Murphy 1987; Parrish *et al.* 1988). The remainder is thought to have been transferred eastward by displacement of cover rocks on a structurally higher detachment surface called the Monashee decollement.

This paper briefly reviews the structural history of the Monashee decollement and offers an interpretation of its subsurface extent that is consistent with seismic reflection data from the LITHOPROBE southern Canadian Cordillera transect. The purpose is to demonstrate that the Monashee decollement is an important part of a thrust system that can be traced in the subsurface as far east as the Rocky Mountain Trench and at least as far west as the central Intermontane belt. Correlation of the sole thrust of the Foreland belt with the Monashee decollement is explicit in this interpretation. The implications of this interpretation are examined and it is suggested that North American basement extends much farther west than previously recognized (Armstrong *et al.* 1977; Armstrong 1988).

Geological setting

The Mesozoic and Tertiary tectonic history of the southern Canadian Cordillera involved four interrelated processes:

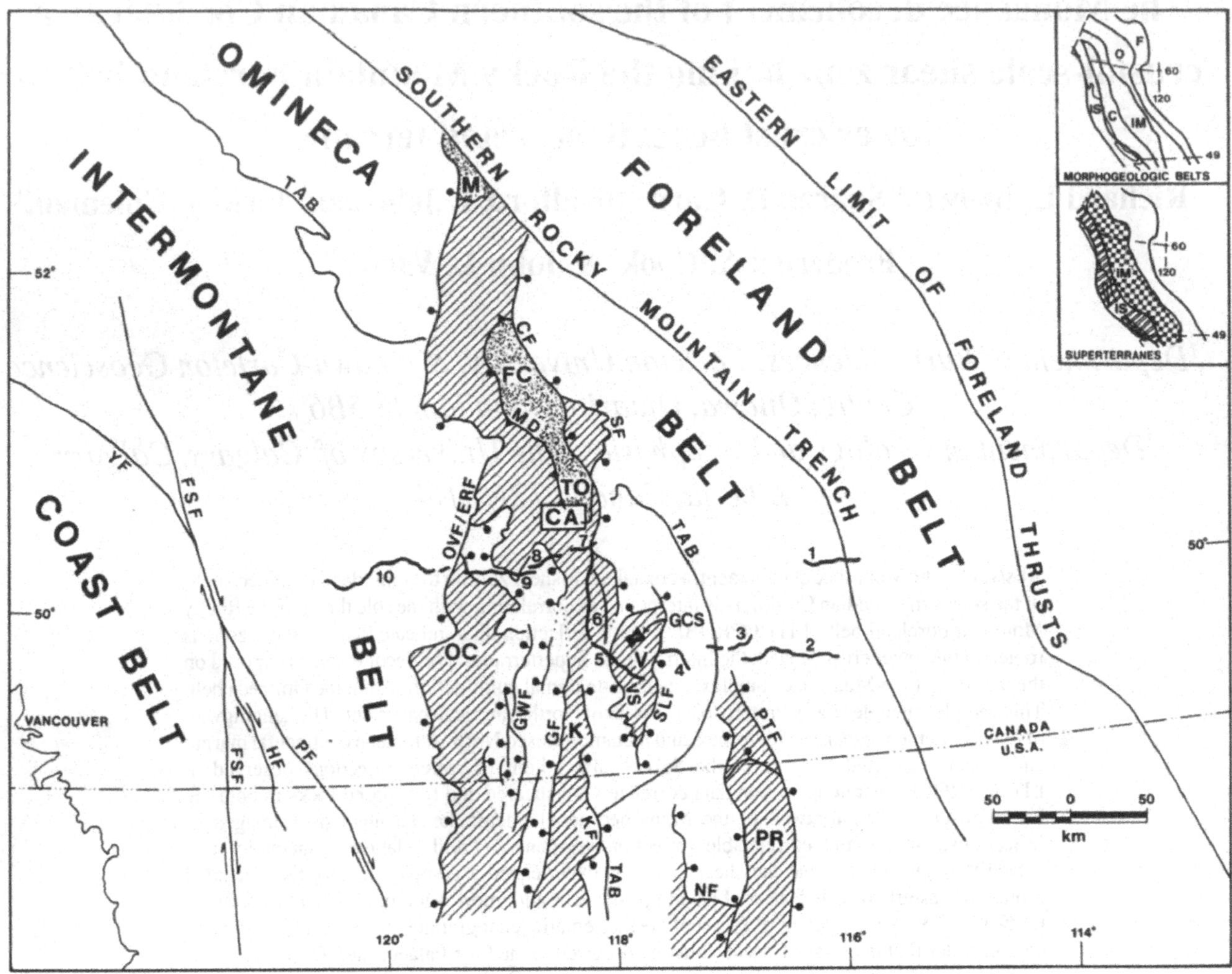

Figure 1. Tectonic map of southern Omineca Belt, modified from Parrish *et al.* (1988). Insets locate the morphogeological belts of the Cordillera and the Intermontane (IM) and Insular (IS) superterranes. The Omineca Belt is bounded on the east by the southern Rocky Mountain Trench and on the west by the Okanagan Valley-Eagle River normal fault system (OVF/ERF) and terrane accretion boundary (TAB). Intermontane Superterrane lies west of terrane accretion boundary. Diagonal stripe pattern indicates the Shuswap complex. Basement rocks (dark grey pattern) are exposed in the Frenchman Cap (FC) and Thor-Odin (TO) culminations of the Monashee complex and in the Malton complex (M). Other high-grade culminations within the Shuswap complex are individually named the Kettle-Grand Forks (K), Okanagan (OC), Priest River (PR) and Valhalla (V) complexes. Mesozoic-Palaeocene thrust faults are the Monashee decollement (MD) and Gwillim Creek shear zones (GCS). Late Mesozoic-Tertiary strike-slip faults are the Fraser-Straight Creek (FSF), Yalakom (YF), Hozameen (HF) and Pasayten (PF) faults. Eocene normal faults are the Columbia River fault (CF), Granby fault (GF), Greenwood fault system (GWF), Kettle fault (KF), Newport fault (NF), Okanagan Valley-Eagle River fault system (OVF/ERF), Purcell Trench fault (PTF), Slocan Lake fault (SLF) and Valkyr shear zone (VS). Numerals correspond to LITHOPROBE seismic reflection lines shown in Figure 4. Box labelled CA locates Cariboo Alp (see Figs 3 & 5).

accretion of outboard terranes to the western margin of North America, protracted compression and crustal thickening, strike-slip faulting and crustal extension (Monger *et al.* 1982; Gabrielse 1985; Brown *et al.* 1986; Parrish *et al.* 1988). This history is reflected in the distribution of the five morphogeological belts that parallel the structural grain of the orogen (Monger *et al.* 1982; Gabrielse & Yorath 1989; Fig. 1, inset). The Foreland belt consists of rocks of the Proterozoic to Palaeozoic North American pericratonic prism and of the Mesozoic to early Tertiary foreland basin; these rocks were imbricated by Mesozoic to early Tertiary thin-skinned thrusts and folds. The Omineca belt is the metamorphic and plutonic hinterland to the Foreland belt and has in large part been unroofed by tectonic removal of low-grade cover on crustal-scale Eocene extension faults. The Intermontane belt comprises weakly metamorphosed late Palaeozoic to Early Jurassic oceanic and island arc terranes. The Coast belt primarily consists of Cretaceous to early Tertiary plutons with metamorphic pendants, while the Insular belt consists of Palaeozoic to Jurassic volcanic, plutonic, metamorphic and sedimentary rocks.

Arc and oceanic terranes were amalgamated to form the Intermontane Superterrane and the Insular Superterrane (Fig. 1, inset), which were accreted to the western North American margin in the Jurassic and Cretaceous (Monger *et al.* 1982; Price *et al.* 1985). The accreted terranes lie predominantly outboard of the Omineca belt, although relatively thin slivers are preserved at high structural levels within the Omineca belt.

The present crustal structure of the southern Omineca belt is a result of compressional crustal thickening and subsequent extension. During the compressional phase, a thick composite sheet of internally deforming cover rocks, the Selkirk allochthon (Read & Brown 1981), was transported eastward over basement on the Monashee decollement (Brown 1980; Read & Brown 1981). The Selkirk allochthon includes (i) Proterozoic to late Palaeozoic metasedimentary rocks of the palaeocontinental margin, (ii) deformed and metamorphosed late Palaeozoic to Early Jurassic volcanic and sedimentary rocks distal to North America, and (iii) obducted and complexly infolded late Palaeozoic to Early Jurassic remnants of accreted terranes that collectively form the Intermontane Superterrane.

During Eocene crustal extension the Selkirk allochthon was dissected by generally north-trending low- to moderate-angle plastic-brittle normal fault systems (Columbia River - Standfast Creek, Granby, Greenwood, Kettle, Newport, Okanagan Valley-Eagle River, Purcell Trench and Valkyr-Slocan Lake, Fig. 1). The Shuswap complex, a Cordilleran metamorphic core complex (Coney 1980; Armstrong 1982), consists of rocks that were buried during compression and were exhumed by these normal fault systems. These high-grade rocks are exposed as elongate domal culminations, the Kettle-Grand Forks, Monashee, Okanagan, Priest River and Valhalla complexes, which are collectively called the Shuswap complex (Fig. 1). Beneath the normal fault systems that bound these complexes, upper-amphibolite-facies rocks preserve evidence of Mesozoic to Palaeocene compression, are overprinted by fabrics and mineral assemblages that were generated during extensional ductile deformation (Carr *et al.* 1987; Parrish *et al.* 1988; Johnson 1988a, 1989b, 1989c) and are characterized by an early Tertiary cooling history (Armstrong 1982; R. L. Armstrong, unpubl. data, 1984; Parrish *et al.* 1988 and references therein). In contrast, the hangingwall rocks consist of unmetamorphosed sedimentary and volcanic strata and rocks that were metamorphosed in the Middle Jurassic during accretion of the Intermontane Superterrane; these rocks were uplifted to high structural levels and cooled before the Late Cretaceous (Read & Wheeler 1976; Parrish & Wheeler 1983; R. L. Armstrong, unpubl. data, 1984; Parrish *et al.* 1988 and references therein).

Basement rocks are exposed in the Monashee complex (Frenchman Cap and Thor-Odin culminations) and in the Malton complex (Fig. 1). These include Early Proterozoic gneisses that are similar in age to autochthonous North American basement (Duncan 1984; Evenchick *et al.* 1984; Parrish & Ross 1989, 1990; Parkinson 1990). The Monashee complex was overthrust by Late Proterozoic strata of the Windermere Supergroup, which are of North American provenance and were deposited outboard (west) of the present exposure of the Monashee complex. These relationships lend support to the interpretation that the basement within the Monashee complex was part of the North American craton.

The Monashee decollement

The Monashee decollement is a crustal-scale easterly vergent and westerly rooted ductile compressional shear zone along which cover rocks of the Selkirk allochthon are detached from parautochthonous basement. The decollement is exposed as a 1-2 km thick zone of intense ductile strain over a strike length of >200 km on the flanks of Frenchman Cap and Thor-Odin culminations (Fig. 1). Precambrian basement and unconformably overlying Proterozoic to Cambrian cover rocks in the footwall (Hoy & Godwin 1988; Parrish & Scammell 1988 and references therein) are exposed in a tectonic window, the Monashee complex (Read & Brown 1981; Brown & Journeay 1987). The Monashee decollement is truncated on the eastern margin of the Monashee complex by the east-dipping early Tertiary Columbia River normal fault (Fig. 1, Lane 1984).

The Monashee decollement is part of a thrust system that accommodated thick-skinned crustal shortening; it is the roof thrust to an antiformal basement duplex (Brown *et al.* 1986, Fig. 2; Monger *et al.* 1986). Rocks of the Selkirk allochthon are inferred to be duplicated by blind thrusts in the hangingwall (Fig. 4). Displacements on these structures, which may be out of sequence, fed into the Monashee decollement.

Kinematics

Independent and consistent kinematic indicators are well preserved along the length of the decollement and have been studied in outcrop, hand sample and thin section (Brown 1980; Journeay 1986; Journeay & Brown 1986; Scammell 1986; Coleman 1987, 1989; Bosdachin 1989; Harrap 1990). Criteria including C-S fabrics (Berthe *et al.* 1979), strain-insensitive foliation (Means 1981; Simpson & Schmid 1983), asymmetrical extensional shears (Platt & Vissers 1980; White *et al.* 1980; Platt 1984), rotated porphyroclasts (Passchier & Simpson 1986), mica fish (Lister & Snoke 1984), asymmetrical foliation boudinage and asymmetrical pull-aparts (Hanmer 1984, 1986) consistently record a northeasterly directed sense of shear (Fig. 2).

Timing

Journeay (1986) documented a zone of inverted low-pressure metamorphic isograds in the footwall of the decollement that overprint previously quenched high-pressure assemblages. He interpreted these data as evidence that the Monashee decollement experienced a two-stage history of movement to which he tentatively assigned Middle Jurassic and Cretaceous-Palaeocene ages (see Brown & Journeay 1987). The inverted metamorphism implies that the footwall must have been cooler than the overthrust Selkirk allochthon during the final stages of movement and that thrusting was followed by rapid uplift and cooling to allow preservation of the inverted isograds.

At present, there is no direct geochronological evidence for mid-Mesozoic motion on the decollement. Carr (Carr

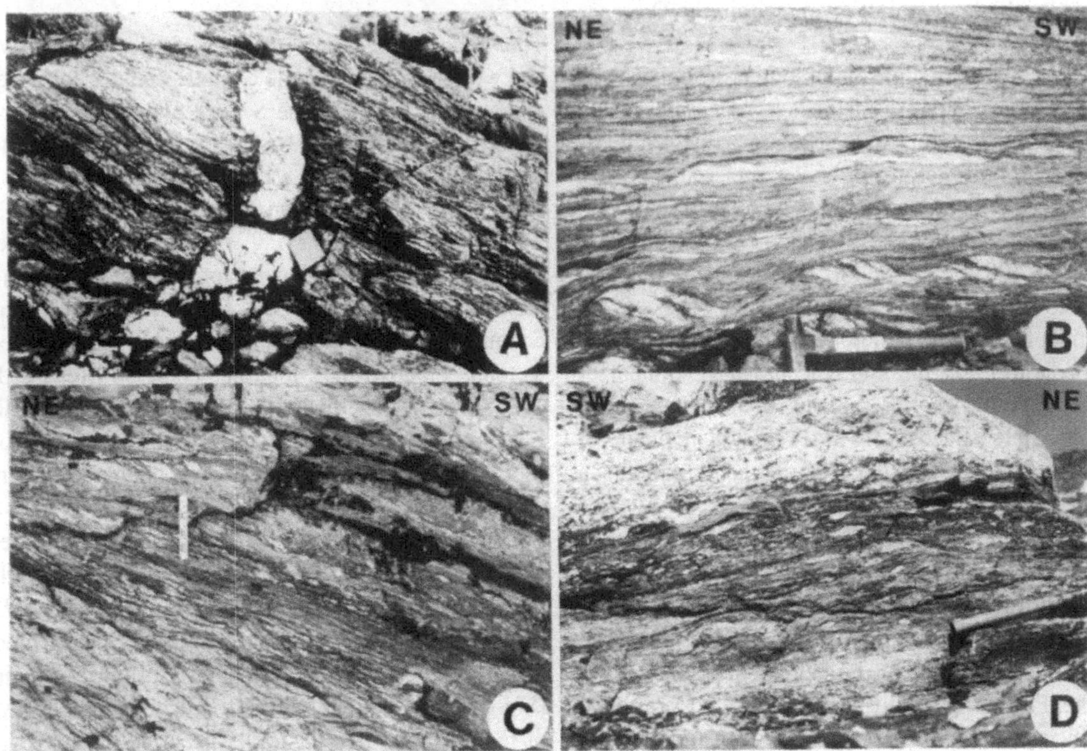

Figure 2. (**a**) A 58 Ma post-kinematic pegmatite dyke in Cariboo duplex cross–cuts the sheared footwall rocks of the Monashee decollement. See text for further explanation and Figure 3 for location. (**b**) Asymmetrical extensional shears and back-rotated asymmetrical boudins within sheared basement gneisses of the Cariboo duplex indicate an upper-member-to-the-northeast sense of shear. See Figure 1 for location. (**c**) Sheared gneisses in footwall of Monashee decollement on west flank of Frenchman Cap culmination. Asymmetric leucosome pods and shear foliation indicate an upper-member-to-the-northeast sense of shear. See Figure 1 for location. (**d**) Sheared gneisses in the hangingwall of the Monashee decollement on the west flank of Thor-Odin culmination. C-S fabrics and asymmetrical extensional shears indicate an upper-member-to-the-northeast sense of shear. Light-coloured layer is concordant sheared pegmatite. See Figure 1 for location.

Figure 3. View looking north through Cariboo Alp to Thor-Odin culmination (TO) of Monashee complex, showing imbricated slices of basement (B) and darker grey cover of the Cariboo duplex. Note undeformed pegmatite dykes (P). ST = sole thrust.

Additional material from *THRUST TECTONICS*,
ISBN 978-0-412-43900-1, is available at http://extras.springer.com

EXTRAS ONLINE

1989, 1990; Carr & Brown 1990) demonstrated that the Monashee decollement on the southern flank of Thor-Odin was active in the Late Palaeocene and that motion ceased by 58 Ma. Timing constraints are based on U-Pb zircon dating of deformed sillimanite-bearing syn-kinematic leucogranites in the Monashee decollement (62 + 0.3 & 59 + 0.3 Ma) and a post-kinematic cross-cutting pegmatite dyke (58 $\pm$ 0.5 Ma) (Figs 2 & 3).

Shortening in the foreland continued into the Palaeocene (Price & Mountjoy 1970; Donelick & Beaumont 1990). Therefore, the kinematic and geochronological data from the Omineca belt lend strong support to the proposal that the Monashee decollement was kinematically and temporally linked to the Foreland belt (Brown & Read 1983; Brown *et al.* 1986). The following section reviews deep seismic reflection data that further corroborate this interpretation.

LITHOPROBE seismic reflection data

LITHOPROBE deep seismic reflection data have been recorded in a series of profiles that extend from the Foreland belt to the Pacific Ocean (Clowes *et al.* 1987; 1988 transect data). The data presented in Figure 4 constitute only that part of the transect bearing directly on the nature of the reflection characteristics of the Monashee decollement and its extent to the east and west. These data have been processed into stacked sections using standard procedures, which include post-stack migrations and coherency filtering, and are plotted with no vertical exaggeration for a velocity of 6.0 km/s.

The Monashee decollement is exposed ~10 km north of LITHOPROBE lines 6 & 7 on the southern flank of Thor-Odin culmination, where it is a highly strained zone up to 2 km thick (Duncan 1984; Coleman 1989; Fig. 1). At Cariboo Alp (Figs 2 & 3), the decollement zone is a duplex with imbricated horses of basement and cover gneisses (Coleman 1989; Fig. 5). There is a strain gradient in the hangingwall where gneisses are transposed into parallelism with the decollement in a zone approximately 5 km thick. The trace of the decollement has been projected, using well constrained surface geology, into the intersection of lines 6 & 7 (Figs 1 & 4) at approximately 2 s (two way travel time), where it is correlated with a zone of layered reflections (Fig. 4). Above and below this zone are domains of reflections with different apparent dip and characteristic spacing and intensity; truncations of reflections are imaged in places along the boundary. Apparent southward dips on line 6 and westward dips on line 7 indicate a southwesterly dip for the decollement, consistent with surface geology. Reflections correlated with the decollement dip moderately to the southwest across lines 7 & 8 to about 5 s and continue westward on lines 9 & 10, where they become subhorizontal at about 25 km depth (~8 s) on the central part of line 10 (Fig. 4). This suggests that the Monashee decollement and related structures are listric into the lower crust. Antiformal geometries in the hangingwall of the Monashee decollement that occur on lines 6 & 9 are juxtaposed against underlying throughgoing layers (Fig. 4). These may be interpreted as hangingwall anticlines located at the leading edge of blind thrust sheets; displacement on such thrusts would have fed into the Monashee decollement.

Exposed basement rocks beneath the Monashee decollement are imbricated and isoclinally folded, as described above for the Cariboo Alp locality (Fig. 5; Reesor & Moore 1971; Duncan 1984; Journeay 1986). Reflections from beneath the decollement in lines 6, 7 & 8 appear to have structural configurations that are similar to such imbrications.

The crust between the inferred position of the Monashee decollement and the reflection MOHO tapers westward from the east end of line 7 to the central part of line 10 (Fig. 4). The Monashee decollement cannot be traced beyond the centre of line 10 with confidence, because interpretation of subhorizontal reflections in the western part of line 10 is ambiguous. Displacement on Eocene normal faults, which penetrate to mid-crustal levels (Okanagan Valley-Eagle River fault system, Figs 1 & 4), must be balanced somewhere by stretching in the lower crust (Johnson & Brown 1990), and therefore these subhorizontal reflections may have been produced by Eocene fabrics. In any case, if the crust beneath the Monashee decollement is North American basement, then the reflection geometry suggests that the attenuated margin of North America continues westward beneath the accreted terranes at least as far as the central part of line 10. This implies that the Intermontane Superterrane must have been emplaced as thin crustal slices.

The trace of the Monashee decollement in line 6 extends southward down to approximately 7 s (~21 km depth), where it is projected into line 5 east of Lower Arrow Lake (Fig. 4). The decollement is not as readily discernible in lines 5 & 4 as it is in line 6. However, it is clear that reflections are nearly horizontal where the decollement continues into line 5 from line 6, and these reflections arch over to become gently east-dipping beneath line 4. These reflection patterns are consistent with the domal geometry of the Valhalla complex known from geological constraints, and it can be assumed that the decollement continues eastward beneath the Valhalla complex and that it is truncated in the footwall of the Early Eocene Slocan Lake normal fault at approximately 7 s (~21 km depth) (Fig. 4).

The Slocan Lake fault is a moderately east-dipping normal fault with a brittle hangingwall and a mylonitic footwall (Parrish 1984; Carr *et al.* 1987). The fault has been interpreted on the basis of both geological and seismic data as a crustal-scale extensional shear zone (Cook *et al.* 1987, 1988; Parrish *et al.* 1988). Carr (1986) estimates displacement of 10-20 km along the central part of the fault.

It is proposed that the sole thrust of the Foreland belt was the eastern continuation of the Monashee decollement; the sole thrust coincides with the near-basement reflector (NBR, Fig. 4), which is typically 1-2 km above cratonic basement (Bally *et al.* 1966; Cook *et al.* 1988; Eaton & Cook 1988). At the west end of line 2 and at the east end of line 3, the sole thrust is cut by the Slocan Lake fault at ~30 km depth (Cook *et al.* 1988). Figure 4 shows that the amount of separation between the inferred cutoffs of the sole thrust and the Monashee decollement along the projected trace of the Slocan Lake fault is approximately 17 km. The estimates of displacement on

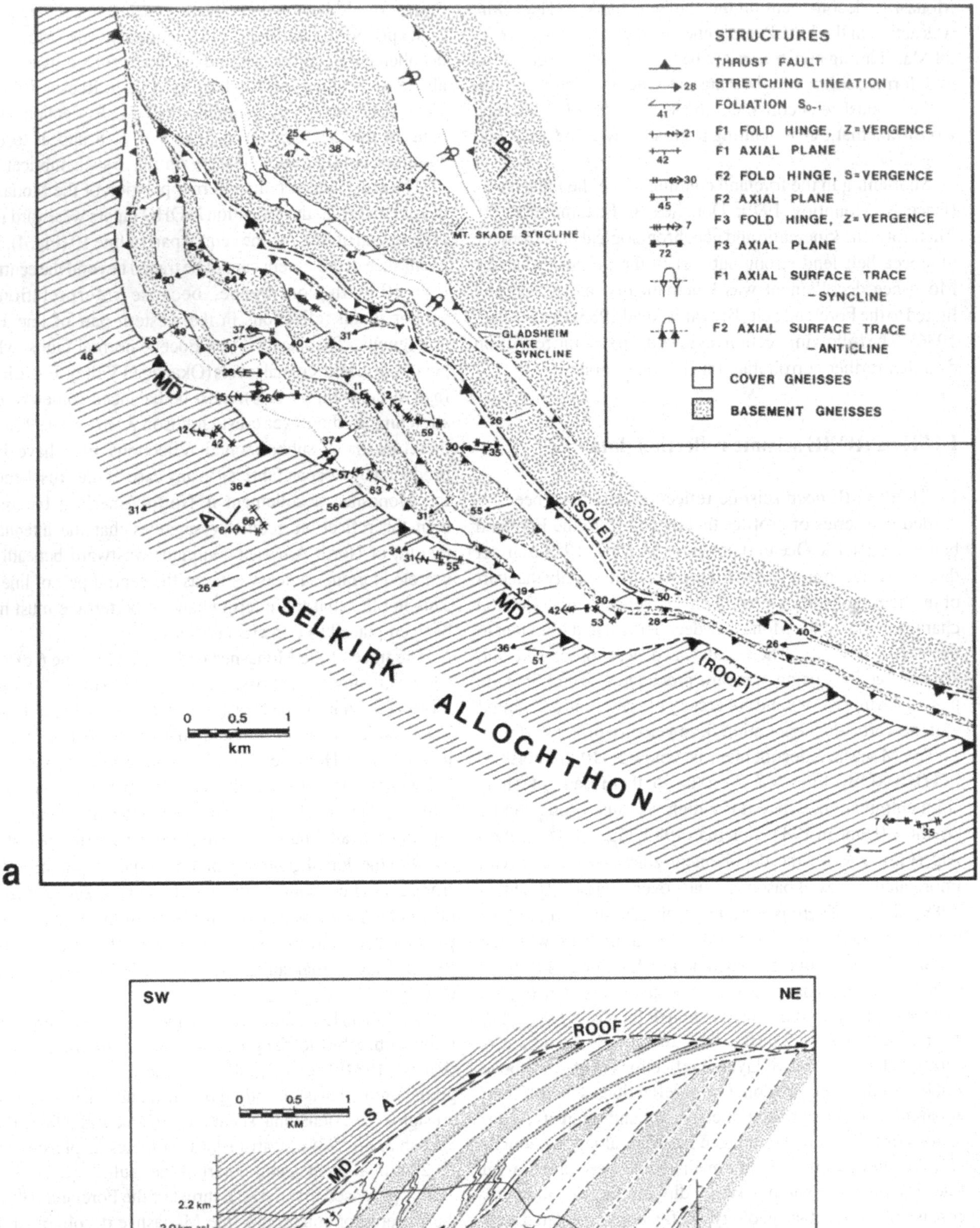

STRUCTURES

THRUST FAULT

STRETCHING LINEATION

FOLIATION S_{0-1}

F1 FOLD HINGE, Z=VERGENCE

F1 AXIAL PLANE

F2 FOLD HINGE, S=VERGENCE

F2 AXIAL PLANE

F3 FOLD HINGE, Z=VERGENCE

F3 AXIAL PLANE

F1 AXIAL SURFACE TRACE – SYNCLINE

F2 AXIAL SURFACE TRACE – ANTICLINE

COVER GNEISSES

BASEMENT GNEISSES

MT. SKADE SYNCLINE

GLADSHEIM LAKE SYNCLINE

MD

(SOLE)

(ROOF)

SELKIRK ALLOCHTHON

A

B

0 0.5 1 km

a

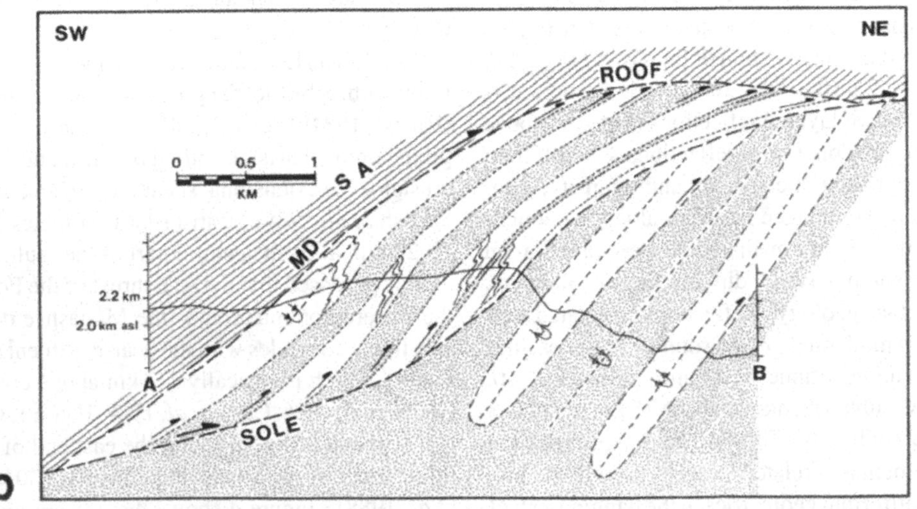

Figure 5. Cariboo duplex: (**a**) map and (**b**) cross section illustrate basement slices in footwall of Monashee decollement. See Figure 3 for view of part of Cariboo Alp and Figure 1 for location.

the Slocan Lake fault, together with similar geochronological and kinematic signatures are compelling evidence for correlating the sole thrust of the Foreland belt in the hangingwall of the Slocan Lake fault with the Monashee decollement.

Conclusions

The Monashee decollement is a crustal-scale compressional shear zone of great areal extent. It carried middle and upper crustal rocks in its hangingwall, including strata of North American cratonic provenance and amalgamated slices of westerly derived accreted terranes.

The Monashee decollement extends eastward in the subsurface to the Rocky Mountain Trench and is inferred to be the westward continuation of the sole thrust of the Foreland belt. It extends westward in the middle crust beneath the Intermontane Superterrane and can be traced into the lower crust, where its westward continuation becomes ambiguous.

Basement gneisses of probable North American cratonic origin occur in the southern Omineca belt and are projected westward in the subsurface beneath the Intermontane belt. This implies that accreted rocks of the Intermontane Superterrane are thin crustal slices that have been transported northeasterly onto the attenuated margin of North America.

Middle to lower crust beneath the Monashee decollement in the Omineca belt does not appear to have been involved in significant penetrative Tertiary extensional strain. However, strong subhorizontal reflections near the base of the crust beneath the western Intermontane belt (Fig. 4, line 10) may be a result of superimposed extensional shearing and could be the locus for balancing extension observed at higher crustal levels.

This paper has made use of seismic data produced by the Canadian LITHOPROBE project under the direction of Dr. R. Clowes. The interpretation of the data presented in this paper has been carried out independently by the authors, but we readily acknowledge the contribution of other members of the interpretation team. In particular, we have benefited from discussions with Peter Carroll, Murray Journeay, Ernie Kanasewich, Randy Parrish, Ray Price and Carl Spencer. The seismic field work was carried out by Sonix Exploration of Calgary, with initial processing by Western Geophysical of Calgary. Final processing and data preparation were done at the University of Calgary LITHOPROBE seismic processing facility.

Geological field work by RLB, SDC, BJJ and VJC has been funded by Natural Sciences and Engineering Research Council of Canada (NSERC) grant A2693, a LITHOPROBE supporting geoscience grant and Energy, Mines and Resources Canada research agreements to RLB. Also gratefully acknowledged are Geological Society of America research grants in 1987 and 1988 and Geological Survey of Canada (GSC) field support to SDC, British Columbia geoscience research grant RG88-02 to BJJ and a Texaco Canada Resources geological research grant to VJC. SDC, BJJ and VJC would like to thank Randy Parrish and the GSC Geochronology Section for the privilege of working in the geochronology laboratory. FAC and JLV are supported through NSERC operating grant A2623 to FAC. Thanks to L. Hardy for help with preparation of the manuscript. The paper is designated LITHOPROBE contribution number 186.

References

Armstrong, R. L. 1982. Cordilleran metamorphic core complexes-from Arizona to southern Canada. *Annual Reviews of Earth and Planetary Science*, **10**, 129-154.

—— 1988. Mesozoic and early Cenozoic magmatic evolution of the Canadian Cordillera. *Geological Society of America Special Paper*, **218**, 55-91.

——, Taubeneck, W. H. & Hales, P. O. 1977. Rb-Sr and K-Ar geochronometry of Mesozoic granitic rocks and their Sr isotopic composition, Oregon, Washington, and Idaho. *Geological Society of America Bulletin*, **88**, 397-411.

Bally, A. W., Gordy, P. L. & Stewart, G. A. 1966. Structure, seismic data, and orogenic evolution of southern Canadian Rockies. *Bulletin of Canadian Petroleum Geology*, **14**, 337-381.

Berthe, D., Choukroune, P. & Jegouzo, P. 1979. Orthogneiss, mylonite and non-coaxial deformation of granites: the example of the South Amorican Shear Zone. *Journal of Structural Geology*, **1**, 31-42.

Bosdachin, R. 1989. Structure, stratigraphy, and tectonothermal evolution of the central western flank of the Monashee complex, southeastern British Columbia. M.Sc. thesis, Carleton University (unpubl.).

Brown, R. L. 1980. Frenchman Cap dome, Shuswap complex, British Columbia: a progress report. *In: Current Research, Part A*. Geological Survey of Canada, Paper **80-1A**, 47-51.

—— & Journeay, J. M. 1987. Tectonic denudation of the Shuswap metamorphic terrane of southeastern British Columbia. *Geology*, **15**, 142-146.

—— & Read, P. B. 1983. Shuswap terrane of British Columbia: a Mesozoic 'core complex'. *Geology*, **11**, 164-168.

——, Journeay, J. M., Lane, L. S., Murphy, D. C. & Rees, C. J. 1986. Obduction, backfolding and piggyback thrusting in the metamorphic hinterland of the southeastern Canadian Cordillera. *Journal of Structural Geology*, **8**, 255-268.

Carr, S. D. 1986. The Valkyr shear zone and the Slocan Lake fault zone: Eocene structures that bound the Valhalla gneiss complex, southeastern British Columbia. M.Sc. thesis, Carleton University (unpubl.).

—— 1989. Implications of Early Eocene Ladybird granite in the Thor-Odin-Pinnacles area, southern British Columbia. *In: Current Research, Part E*. Geological Survey of Canada, Paper **89-1E**, 69-77.

—— 1990. Late Cretaceous-early Tertiary tectonic evolution of the southern Omineca belt, Canadian Cordillera. Ph.D. thesis, Carleton University (unpubl.).

—— & Brown, R. L. 1990. Southern Cordilleran LITHOPROBE transect: evidence for crustal thickening and collapse of the Omineca belt, Valhalla complex to Okanagan Valley, British Columbia. *Geological Association of Canada-Mineralogical Association of Canada, Annual Meeting, Program with Abstracts*, **15**, A22.

——, Parrish, R. & Brown, R. L. 1987. Eocene structural development of the Valhalla complex, southeastern British Columbia. *Tectonics*, **6**, 175-196.

Clowes, R. M., Brandon, M. T., Green, A. G., Yorath, C. J., Sutherland Brown, A., Kanasewich, E. R. & Spencer, C. 1987. LITHOPROBE - southern Vancouver Island: Cenozoic subduction complex imaged by deep seismic reflections. *Canadian Journal of Earth Sciences*, **24**, 31-51.

Coleman, V. J. 1987. A metamorphic and kinematic study of the Monashee decollement, southeastern British Columbia. B.Sc. thesis, Carleton University (unpubl.).

—— 1989. The Cariboo duplex at the southern boundary of the Monashee complex, Shuswap terrane, southern B.C. *In: Current Research, Part A*. Geological Survey of Canada, Paper **89-1A**, 89-93.

Coney, P. J. 1980. Cordilleran metamorphic core complexes: an overview. *In*: Crittenden, M. D., Jr., Coney, P. J. & Davis, G. H. (eds) *Cordilleran Metamorphic Core Complexes*. Geological Society of America Memoir, **153**, 7-31.

Cook, F. A., Green, A. G., Simony, P. S., Price, R. A., Parrish, R., Milkereit, B., Gordy, P. L., Brown, R. L. Coflin, K. C. & Patenaude, C. 1987. LITHOPROBE southern Canadian Cordilleran transect: Rocky Mountain thrust belt to Valhalla gneiss complex. *Geophysical Journal of the Royal Astronomical Society*, **89**, 91-98.

—— 1988. LITHOPROBE seismic reflection structure of the southeastern Canadian Cordillera: initial results. *Tectonics*, **7**, 157-180.

Donelick, R. A. & Beaumont, C. 1990. Late Cretaceous-early Tertiary cooling of the Brazeau thrust sheet, Central Foothills, Alberta: implications for the timing of the end of Laramide thrusting in the region. *Geological Association of Canada-Mineralogical Association of Canada, Annual Meeting, Program with Abstracts*, **15**, A34.

Duncan, I. J. 1984. Structural evolution of the Thor-Odin gneiss dome. *Tectonophysics*, **101**, 87-130.

Eaton, D. W. S. & Cook, F. A. 1988. LITHOPROBE seismic reflection imaging of Rocky Mountain structures east of Canal Flats, British Columbia. *Canadian Journal of Earth Sciences*, **25**, 1339-1348.

Evenchick, C. A., Parrish, R. R. & Gabrielse, H. 1984. Precambrian gneiss and Late Proterozoic sedimentation in north-central British Columbia. *Geology*, **12**, 233-237.

Gabrielse, H. 1985. Major dextral transcurrent displacements along the Northern Rocky Mountain Trench and related lineaments in north-central British Columbia. *Geological Society of America Bulletin*, **96**, 1-14.

—— & Yorath, C. J. 1989. DNAG #4. The Cordilleran orogen in Canada. *Geoscience Canada*, **16**, 67-83.

Hanmer, S. K. 1984. The potential use of planar and elliptical structures as indicators of strain regime and kinematics of tectonic flow. *In: Current Research, Part B*. Geological Survey of Canada, Paper **84-1B**, 133-142.

—— 1986. Asymmetric pull-aparts and foliation fish as kinematic indicators. *Journal of Structural Geology*, **8**, 111-122.

Harrap, R. 1990. Stratigraphy and structure of the Monashee terrane in the Mount English area, west of Revelstoke, B.C. M.Sc. thesis, Carleton University (unpubl.).

Hoy, T. & Godwin, C. I. 1988. Significance of a Cambrian date from galena lead-isotope data for the stratiform Cottonbelt deposit in the Monashee complex, southeastern British Columbia. *Canadian Journal of Earth Sciences*, **25**, 1534-1541.

Johnson, B. J. 1988a. Progress report: stratigraphy and structure of the Shuswap metamorphic complex in the Hunters Range, eastern Shuswap Highland (82L). *In: Geological Fieldwork, 1987*. British Columbia Ministry of Energy, Mines and Petroleum Resources, Paper **1988-1**, 55-58.

—— 1989b. Geology of the west margin of the Shuswap terrane near Sicamous: implications for Tertiary extensional tectonics (82L, M). *In: Geological Fieldwork, 1988*. British Columbia Ministry of Energy, Mines and Petroleum Resources, Paper **1989-1**, 49-54.

—— 1989b. Structural style of crustal extension in a cross-section through the Shuswap complex, southeastern British Columbia. *Geological Society of America, Abstracts with Programs*, **21**, 98.

—— & Brown, R. L. 1990. Crustal structure of the Omineca belt adjacent to line 19 of LITHOPROBE southern Canadian Cordillera transect. *Geological Association of Canada-Mineralogical Association of Canada, Annual Meeting, Program with Abstracts*, **15**, A66.

Journeay, J. M. 1986. Stratigraphy, internal strain and thermo-tectonic evolution of northern Frenchman Cap dome: an exhumed duplex structure, Omineca hinterland, S.E. Canadian Cordillera. Ph.D. thesis, Queen's University (unpubl.).

—— & Brown, R. L. 1986. Major tectonic boundaries of the Omineca belt in southern British Columbia: a progress report. *In: Current Research, Part A*. Geological Survey of Canada, Paper **86-1A**.

Lane, L. S. 1984. Brittle deformation in the Columbia River fault zone near Revelstoke, southeastern British Columbia. *Canadian Journal of Earth Sciences*, **21**, 584-598.

Lister, G. S. & Snoke, A. W. 1984. S-C mylonites. *Journal of Structural Geology*, **6**, 617-638.

McDonough, M. R. & Simony, P. S. 1988. Structural evolution of basement gneisses and Hadrynian cover, Bulldog Creek area, Rocky Mountains, British Columbia. *Canadian Journal of Earth Sciences*, **25**, 1687-1702.

Means, W. D. 1981. The concept of steady-state foliation. *Tectonophysics*, **78**, 179-199.

Monger, J.W.H., Price, R. A. & Tempelman-Kluit, D. J. 1982. Tectonic accretion and the origin of the two major metamorphic and plutonic welts in the Canadian Cordillera. *Geology*, **10**, 1261-1266.

—— (principal compiler), Woodsworth, G. J., Price, R. A., Clowes, R. N., Riddihough, R. P., Currie, R., Hoy, T., Preto, V. A. G., Simony, P. S., Snavely, P. D. & Yorath, C. J. 1986. Transect B2, southern Canadian Cordillera, North American continent-ocean transects. Decade of North American Geology Program, *Geological Society of America*.

Murphy, D. C. 1987. Suprastructure/infrastructure transition, east-central Cariboo Mountains, British Columbia: geometry, kinematics and tectonic implications. *Journal of Structural Geology*, **9**, 13-30.

Parkinson, D. 1990. Age and isotopic character of basement gneisses in the southern Monashee complex, southeastern B.C. *Geological Association of Canada - Mineralogical Association of Canada, Annual Meeting, Program with Abstracts*, **15**, A101.

Parrish, R. 1984. Slocan Lake fault: a low angle fault zone bounding the Valhalla gneiss complex, Nelson map area, southern British Columbia. *In: Current Research, Part A*. Geological Survey of Canada, Paper **84-1A**, 323-330.

Parrish, R. R. & Ross, G. 1989. Precambrian basement in the southern Canadian Cordillera: the cratonic connection. LITHOPROBE Cordilleran Workshop Summary, 72-77.

—— 1990. Review of the Precambrian basement of the Cordillera. *Geological Association of Canada-Mineralogical Association of Canada, Annual Meeting, Program with Abstracts*, **15**, A101.

Parrish, R. & Scammell, R. 1988. The age of the Mount Copeland syenite gneiss and its metamorphic zircons, Monashee complex, southeastern British Columbia. *Geological Survey of Canada, Paper* **88-2**, 21-28.

Parrish, R. R. & Wheeler, J. O. 1983. U-Pb zircon age of the Kuskanax batholith, southeastern British Columbia. *Canadian Journal of Earth Sciences*, **20**, 1751-1756.

——, Carr, S. D. & Parkinson, D. L. 1988. Extensional tectonics of the southern Omineca belt, British Columbia and Washington. *Tectonics*, **7**, 181-212.

Passchier, C. W. & Simpson, C. 1986. Porphyroclast systems as kinematic indicators. *Journal of Structural Geology*, **8**, 831-843.

Platt, J. P. 1984. Secondary cleavages in ductile shear zones. *Journal of Structural Geology*, **6**, 439-442.

—— & Vissers, R. L. M. 1980. Extensional structures in anisotropic rocks. *Journal of Structural Geology*, **2**, 397-410.

Price, R. A. & Mountjoy, E. W. 1970. Geologic structure of the Canadian Rocky Mountains between Bow and Athabaska rivers - a progress report. *Geological Association of Canada Special Paper*, **6**, 7-26.

——, Monger, J. W. H. & Roddick, J. A. 1985. Cordilleran cross-section, Calgary to Vancouver. *In:* Tempelman-Kluit, D. J. (ed.) *Field Guides to Geology and Mineral Deposits in the Southern Canadian Cordillera*. Geological Society of America, Cordilleran Section Guidebook, Vancouver, British Columbia, 3-1 to 3-85.

Read, P. B. & Brown, R. L. 1981. Columbia River fault zone: southeastern margin of the Shuswap and Monashee complexes, southern British Columbia. *Canadian Journal of Earth Sciences*, **18**, 1127-1145.

—— & Wheeler, J. O. 1976. Geology of Lardeau west-half map area. Geological Survey of Canada, Open File Map 288.

Reesor, J. E. & Moore, J. M., Jr. 1971. Petrology and structure of Thor-Odin gneiss dome, Shuswap metamorphic complex, British Columbia. *Geological Survey of Canada Bulletin* **195**.

Scammell, R. J. 1986. Stratigraphy, structure and metamorphism of the north flank of the Monashee complex, southeastern British Columbia: a record of the Proterozoic crustal thickening. M.Sc. thesis, Carleton University (unpubl.).

Simony, P. S., Ghent, E. D., Craw, D., Mitchell, W. & Robbins, D. B. 1980. Structural and metamorphic evolution of northeast flank of Shuswap complex, southern Canoe River area, British Columbia. *In:* Crittenden, M. D., Jr., Coney, P. J. & Davis, G. H. (eds) *Cordilleran Metamorphic Core Complexes*. Geological Society of America Memoir, **153**, 445-461.

Simpson, C. & Schmid, S. M. 1983. An evaluation of criteria to deduce the sense of movement in sheared rocks. *Geological Society of America Bulletin*, **94**, 1281-1288.

White, W. H., Burrows, S. E., Carreras, J., Shaw, N. D. & Humphreys, F. J. 1980. On mylonites in ductile shear zones. *Journal of Structural Geology*, **2**, 175-187.

The Skeena fold belt: a link between the Coast Plutonic Complex, the Omineca belt and the Rocky Mountain fold and thrust belt

Carol A. Evenchick

Geological Survey of Canada, 100 West Pender St., Vancouver, British Columbia, Canada, V6B 1R8

Abstract: The northern Canadian Cordillera contains a second fold and thrust belt west of the well-known Rocky Mountain fold and thrust belt and the metamorphic/plutonic Omineca belt. The Skeena fold belt occupies one quarter of the width of the Cordillera, and merges to the west with a second plutonic/metamorphic belt, the Coast Plutonic Complex. The Skeena fold belt has many features common to thin-skinned fold and thrust belts, including: low-angle thrust faults which sole into a detachment; a wide variety of fold styles depending on the rock type; a minimum of 44% shortening; a foreland basin that was cannibalized by continued deformation; termination in a frontal triangle zone; and a hinterland of metamorphic and plutonic rocks. The Skeena fold belt is similar in many respects to the Rocky Mountain fold and thrust belt, but rather than involving a continental terrace wedge, it deformed a terrane of Devonian to Lower Jurassic strata (Stikinia) which accreted to North America in the early Mesozoic. Also deformed is the Jurassic to Cretaceous clastic succession which overlies Stikinia, and the Cretaceous clastic succession of the associated foreland basin.

The Skeena fold belt represents as much as 160 km of northeasterly shortening which occurred between latest Jurassic (?) and early Tertiary time. The shortening was broadly contemporaneous with crustal thickening and plutonism in the Coast Plutonic Complex and Omineca belt, dextral strike-slip on faults east of the Skeena fold belt, and development of the Rocky Mountain fold and thrust belt. Generally concurrent Cordillera-wide deformation suggests that there was a structural link between all of these zones.

The Canadian Cordillera includes a collage of terranes which accreted to the western margin of North America in Mesozoic time (Fig. 1; Concy *et al.* 1980; Monger 1984). Some terranes were pericratonic, and merely slipped along the margin of North America. Others, however, were exotic and assembled into larger terranes prior to their accretion to North America. Their accretion as 'superterranes' had a profound tectonic response. Another framework for Cordilleran geology is that of morphogeological belts (Gabrielse *et al.* in press) which, defined on stratigraphy, structure, metamorphic grade, and physiography, expresses the cumulative geological history. While a terrane is defined by a characteristic stratigraphy, the process of superterrane accretion or tightening is inferred to have produced the major metamorphic and plutonic belts, the Coast and Omineca belts (Fig. 1; Monger *et al.* 1982). Accretion of the Intermontane Superterrane (Fig. 1) to North America in the mid-Mesozoic was followed in the Late Cretaceous by either accretion of the Insular Superterrane (Monger *et al.* 1982), or terrane tightening above a magmatic arc (van der Heyden 1989).

East (cratonward) of each metamorphic/plutonic belt is a belt of lower grade, stratified rocks with fewer intrusive rocks (Foreland and Intermontane belts; Fig. 1). The well-known Rocky Mountain fold and thrust belt (Foreland belt) consists of an eastward tapering wedge of miogeoclinal, platformal and foreland basin successions which were deposited on cratonic North America between Precambrian and early Tertiary time (e.g. Bally *et al.* 1966; Price 1981). Northeast verging thrust faults and associated folds in the southern Rockies accommodated a minimum of 170 km shortening (Price 1981) between Late Jurassic and early Tertiary time.

The Intermontane belt (Fig. 1) is defined by low-lying topography, sub-greenschist facies volcanic and sedimentary strata, and I-type plutonic rocks associated with Mesozoic and Cenozoic volcanism (Gabrielse *et al.* in press). In the terrane framework, the volcanic and oceanic strata comprise several accreted terranes. They are overlain by Mesozoic and Cenozoic clastic successions which postdate amalgamation of the terranes (overlap assemblages on Fig. 1). A large proportion of the belt is underlain by lower Palaeozoic platformal carbonates, mid-Palaeozoic to lower Middle Jurassic island-arc volcanic rocks, and related sedimentary and plutonic rocks of Stikinia (Monger 1984). Except for structures alluded to in an overview by Wheeler *et al.* (1972), the significance of structures in much of the northern Intermontane belt have been underestimated in regional syntheses (e.g. Souther & Armstrong 1966; Monger & Price 1979; Monger *et al.* 1982). Most of Stikinia was assumed to have acted as a rigid block during accretion of the superterranes. Recent mapping, however, has elucidated the nature and significance of a regional fold and thrust belt which occupies most of the width of the northern Intermontane belt (Evenchick 1991a). This is one of three companion papers on the Skeena fold belt. The first (Evenchick 1991a) describes its regional extent, geometry, and evolution; the second (Evenchick 1991b) presents arguments for the involvement of pre Middle

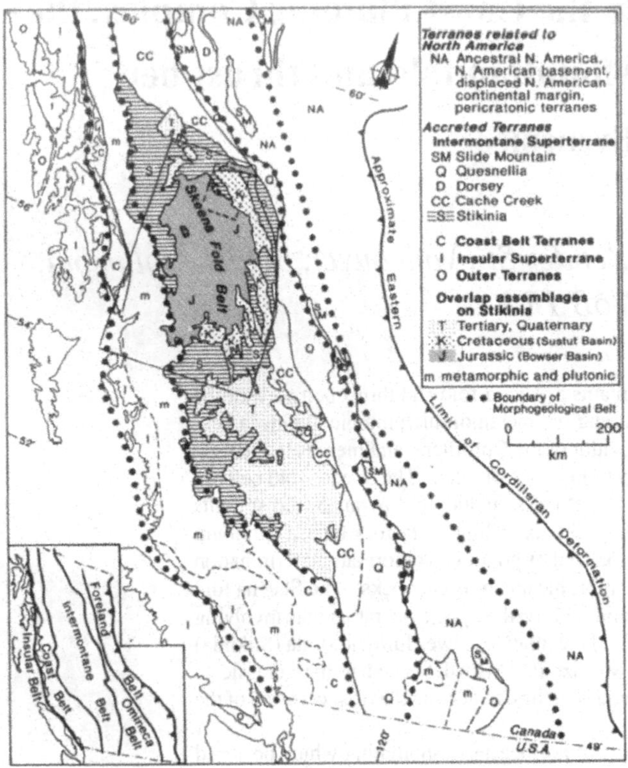

Figure 1. Generalized terrane map of the Canadian Cordillera south of 60°N. NA includes cratonic North America, and its displaced miogeocline. Slide Mountain and Cache Creek terranes are characterized by oceanic rocks of Palaeozoic to early Mesozoic age. Quesnellia, Stikinia, and Insular terranes are dominantly arc volcanic rocks and volcaniclastics of Palaeozoic to early Mesozoic age. Only overlap assemblages on Stikinia are shown. Modified after Wheeler *et al.* (1988) and Gabrielse and Yorath (1989). The location of Figure 2 is outlined by a dashed line, and Figure 4 by a solid line. Inset are the morphogeological belts, after Gabrielse & Yorath (1989).

Jurassic strata on the west side of the fold belt. The present paper is a synthesis of the fold belt, discusses its relationship with its metamorphic/plutonic hinterland, and speculates on the linkage between zones of contemporaneous deformation across the Cordillera. These papers are based on twelve months of regional mapping at the northeast margin of the fold belt and a compilation of the contractional structures.

Regional geology of the Skeena fold belt

The Skeena fold belt is bounded on the west by the Coast belt, on the north and east by a triangle zone at the front of deformation, and on the south by the limit of strata which display folds prominently (Evenchick 1991a). Although the fold belt may continue to the south, structures are obscured by poor exposure, the presence of massive volcanic rocks which display folds poorly, and overlapping younger strata.

Strata involved in the Skeena fold belt are assemblages of Stikinia, as well as overlying Jurassic and Cretaceous clastic successions called Bowser Lake and Sustut Groups. Most Stikinian rocks at the northeast margin of the fold belt are Lower to Middle Jurassic volcanic rocks of the Hazelton Group (Tipper & Richards 1976) and sedimentary rocks of the Spatsizi Group (Thomson *et al.* 1986). Thick, massive flows in the Hazelton Group are the most competent units in the fold belt. Thin, laterally discontinuous and incompetent layers, as well as large facies changes (e.g. Anderson & Thorkelson 1990) result in a mechanical stratigraphy which is laterally variable.

In Middle Jurassic time Stikinia was overlain by a widespread clastic succession called the Bowser Lake Group. The group is Middle Jurassic to Early Cretaceous in age, and is divided into a lower marine unit (Ashman Formation), and an upper unit of shallow marine to nonmarine facies (Figs 3 & 4; Tipper & Richards 1976; Gabrielse & Tipper 1984; Evenchick 1989; Cookenboo & Bustin 1989; MacLeod & Hills 1990; Evenchick & Green 1990). There is no single complete section of Bowser Lake Group, but mapping of several partial sections suggests that it is at least 3500 m thick. The Ashman Formation is characterized by incompetent siltstone and fine grained sandstone, with competent medium grained sandstone and conglomerate beds The lack of marker beds prohibits large-scale structural analysis. In the undivided shallow marine to nonmarine facies there is a wide variety of proportions and thicknesses of sandstone and conglomerate beds They control both the style and scale of folds, and provide a means of estimating the amount of shortening.

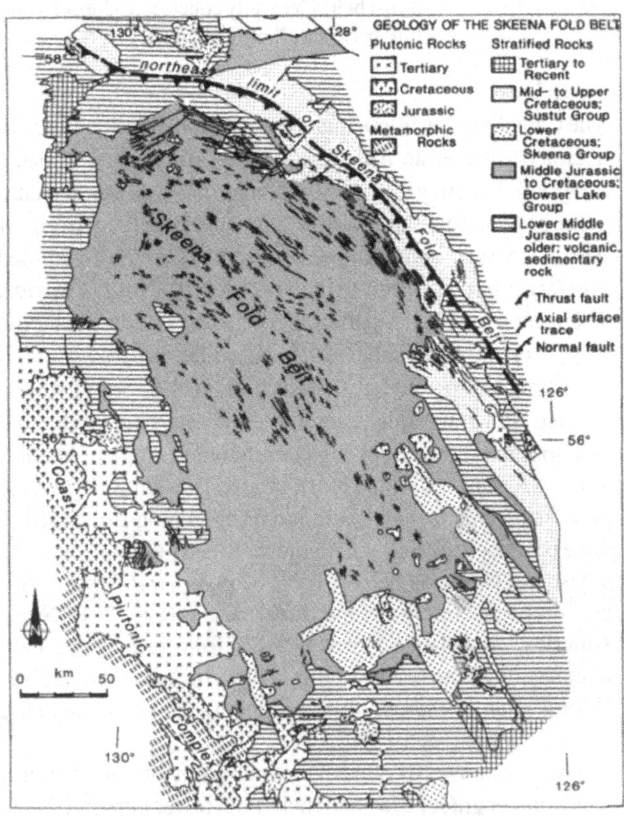

Figure 2. Geology of the Skeena fold belt, modified after Wheeler & McFeely (1987). Structures are compiled from the following maps: Geological Survey of Canada 1957; Richards 1975, 1980; Eisbacher 1974; Souther 1972; Tipper 1976; Read 1983; Woodsworth *et al.* 1985; Grove 1986; Evenchick 1988b, 1989; Britton 1989; Evenchick & Green 1990.

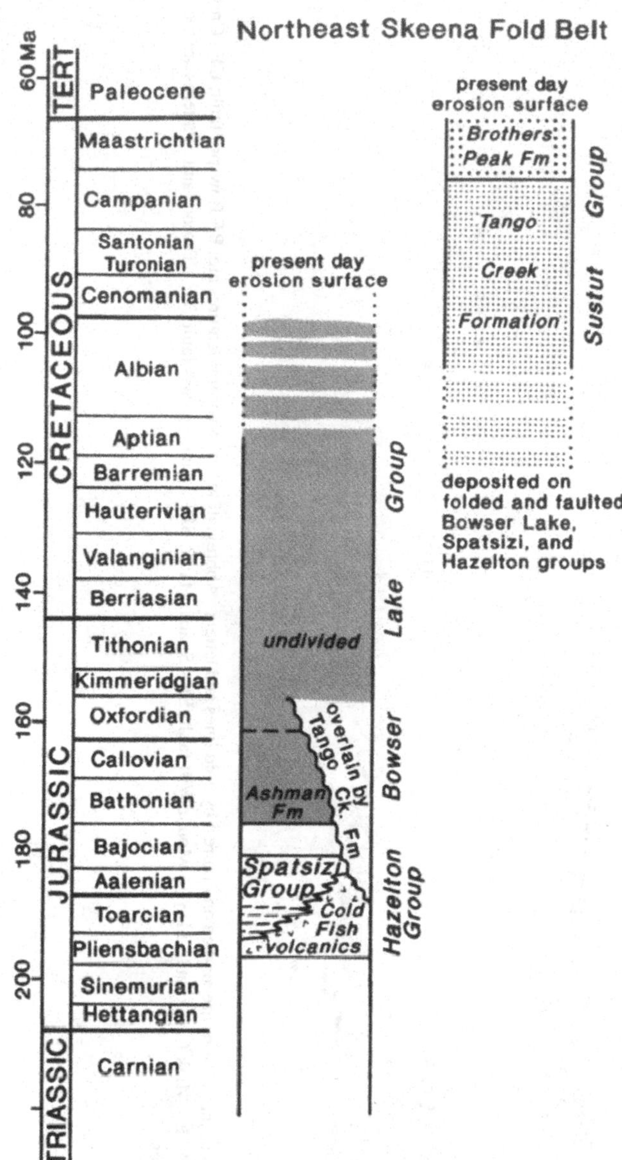

Figure 3. Stratigraphic units of the Skeena fold belt, combined from information in Tipper & Richards 1976; Thomson *et al.* 1986; Evenchick 1987; Cookenboo & Bustin 1989; MacLeod & Hills 1990; Sweet & Evenchick 1990. Gaps in the patterns represent uncertainties in the upper age limit of the Bowser Lake Group and the lower age limit of the Sustut Group.

Deposition in the Sustut Basin (Fig. 1) overlapped the northeast margin of the Bowser Basin in mid- to Late Cretaceous time (Fig. 3). At least 2000 m of dominantly nonmarine sandstone, siltstone, conglomerate and tuff is divided into the Tango Creek and Brothers Peak Formations (Figs 3 & 4; Eisbacher 1974). These units, which form the Sustut Group, were derived from the southwest and deposited broadly during formation of the Skeena fold belt. The Sustut Basin is therefore interpreted as the foreland basin (in part) to the evolving Skeena fold belt (Evenchick 1991a).

The northeast margin of the fold belt

Structural style

The scale and geometry of the fold belt are illustrated in maps and cross-sections of its northeast margin (Figs 4 & 5). The dominant structures are northwest trending folds that are either upright, or have west dipping axial surfaces; many are northeast verging (Figs 4, 6, 7 & 8). Folds plunge gently, except where they interfere with local folds and as a result plunge up to 35°. They are commonly hundreds of metres in amplitude and wavelength, and their geometry varies from open to tight, with both angular and rounded hinges. Chevron, box and disharmonic folds are also common. Folds in the volcanic rocks and immediately overlying strata are an order of magnitude larger in wavelength and amplitude than those in the Bowser Lake Group (Fig. 5).

Klippen of volcanic rocks on Tango Creek Formation and thrust faults with volcanic rocks in the hanging wall, and Spatsizi or Bowser Lake Group in the footwall (e.g. Fig. 6), show that the volcanic rocks (i.e. Stikinia) were intimately involved in contractional deformation. The large amount of

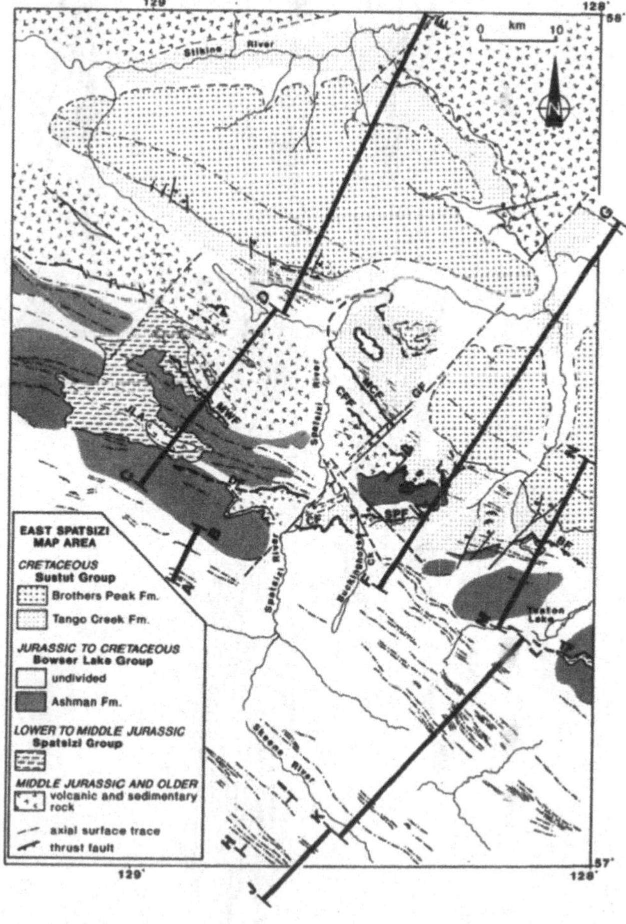

Figure 4. Geology of east Spatsizi area, at the northeast margin of the Skeena Fold Belt. Lines of cross-sections are shown by letters. Abbreviations of structural features are: **BF** Bowsprit fault; **CF** - Crescent fault; **CFF** - Cold Fish fault; **DF** - Denkladia fault; **GF** - Griffith fault; **JLA** - Joan Lake Anticline; **MCF** - Mink Creek fault; **MWF** - Mount Will fault; **SF** - Spatsizi fault; **SPF** - Sunday Pass fault; **TF** - Tuaton fault.

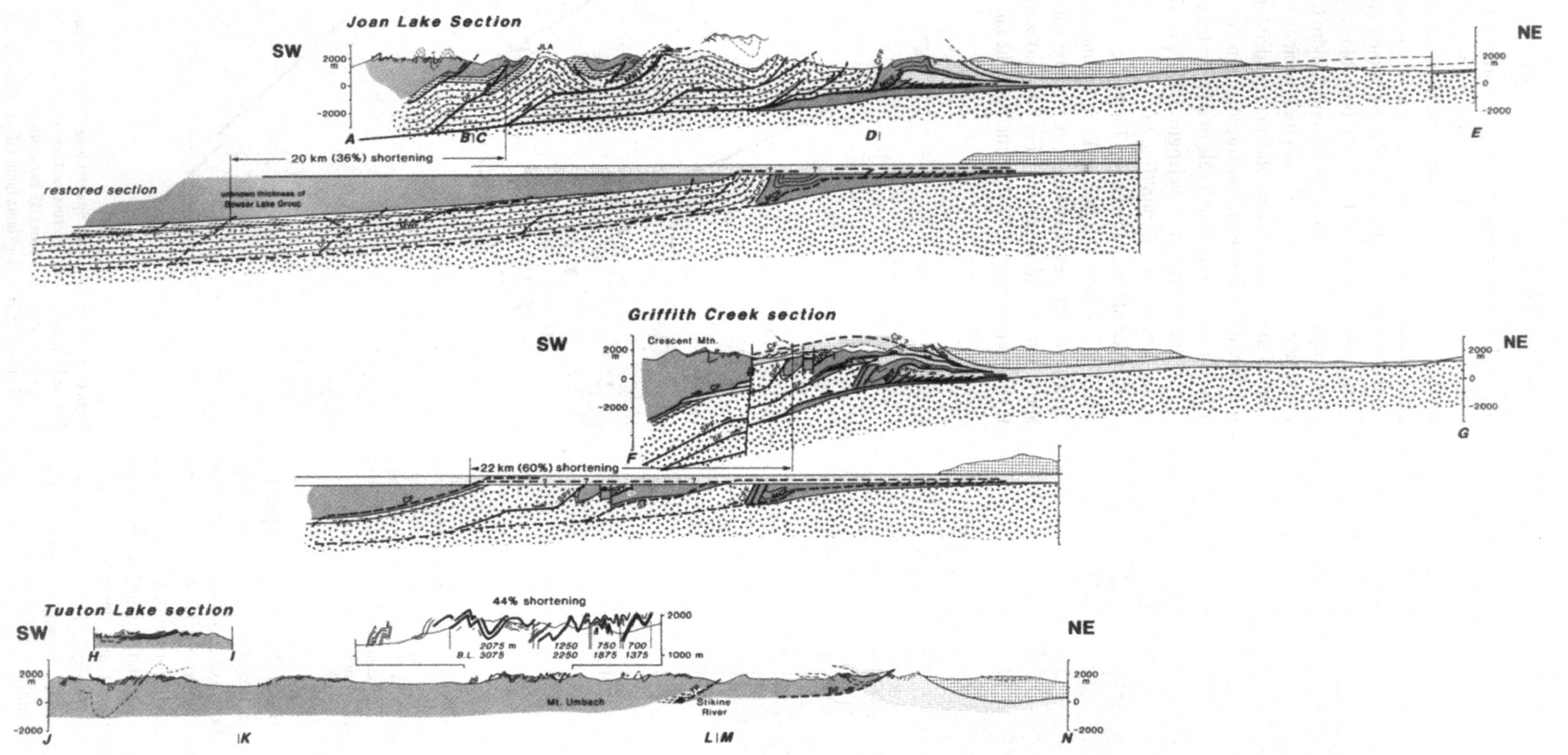

Figure 5. Balanced and restored cross sections of the Skeena fold belt. Locations of the sections are marked on Figure 4 by bold lines and letters. Abbreviations of structural features are: **BF** Bowsprit fault; **CF** - Crescent fault; **CFF** - Cold Fish fault; **DF** - Denkladia fault; **JLA** - Joan Lake Anticline; **MCF** - Mink Creek fault; **MWF** - Mount Will fault; **SF** - Spatsizi fault; **SPF** - Sunday Pass fault; **TF** - Tuaton fault. Patterns are the same as for Figure 4. Lines in the volcanic rocks in the Joan Lake section are form lines.

Figure 6. Tight fold in Hazelton Group volcanics in the hanging wall of the Mount Will fault. Recessive strata in the saddle to the right are Spatsizi Group (SG) overlying the volcanics. The thrust fault is marked by the bold dashed line, and the contact between the Hazelton Group and the Spatsizi Group by the thin solid line. Viewed to the northwest.

Figure 7. Folds and hanging-wall cutoff on the Mount Will fault, in the Spatsizi Group. View to northwest.

Figure 8. Northeast verging chevron folds in siltstone and sandstone of the Bowser Lake Group. View to northwest.

shortening, dominance of northeast verging and upright structures, and similarity between structures in the volcanic rocks and typical fold and thrust belt structures suggest that thrust faults in the volcanics sole into a basal detachment in Stikinia, and that the belt is 'thin skinned' (Evenchick 1991a).

Three structural levels of the Skeena fold belt are shown in Figure 5. The Joan Lake section (ABCDE) crosses the Sustut Basin, and the most extensive exposures of volcanic rocks. The Griffith Creek section (FG) crosses the Sustut Basin and a large region of structurally thickened Tango Creek Formation. The Tuaton Lake section (JKLMN) is dominated by folded Bowser Lake Group, which has only local markers.

The cross-sections also display three zones of common structural style. On its northeast side the Sustut Group unconformably overlies Stikinia and strata are flat lying or gently warped. The second zone is characterized by a gradual southwestward steepening of bedding to form a northeast dipping monocline of Brothers Peak Formation. Up to 200 m of Tango Creek Formation conformably underlies the Brothers Peak Formation, and continues downwards into at least 2000 m of structurally thickened Tango Creek Formation. This zone is characterized by northeast verging thrust faults and folds which, in the northeast, dip northeast below the monocline, a geometry similar to that of the triangle zone at the front of deformation in the Rocky Mountain fold and thrust belt (Jones 1982; Price 1986). Southwest of the triangle zone is a third zone, a 20 km wide belt of folded and thrust faulted volcanic rocks and Bowser Lake Group. To the southwest, folded Bowser Lake Group extends another 170 km across the Bowser Basin (Fig. 2).

The cross-sections only cover the northeast margin of the fold belt, but a regional compilation (Fig. 2) illustrates that contractional structures continue far to the west. Structures in the volcanic rocks west of the basin are cryptic, but folds in overlying strata indicate that the Skeena fold belt continues to the east margin of the Coast Plutonic Complex where it is engulfed by post-tectonic, Tertiary plutons (Evenchick 1991b).

Constraints on the timing of deformation

Stratigraphic and palaeontological data for the Sustut Group, summarized in Figure 9, constrain the timing of deformation. In one region, thrust faults and folds in pre-Sustut strata are unconformably overlain by Sustut Group (Fig. 9; Evenchick 1987). Because the youngest strata below the sub-Sustut unconformity are Oxfordian in age (H. W. Tipper, pers. comm. 1989), and because the basal Sustut Group is likely to be Aptian or Albian in age (Sweet & Evenchick 1990; A. Sweet, pers. comm. 1990), the oldest structures of the northeastern fold belt formed between Oxfordian and mid-Cretaceous time.

Palaeocurrent and provenance studies by Eisbacher (1974) illustrate an association between the Sustut Basin and the Skeena fold belt (Fig. 9). They show that rivers flowed from the east (Omineca belt), carrying the first miogeoclinal and metamorphic detritus to the northern Intermontane belt. During deposition of the Tango Creek Formation a western source is indicated by a change in palaeocurrents and a prominence of chert clasts, assumed to have been eroded from the Bowser Lake Group. This is inferred to be the first clastic record of the evolving Skeena fold belt (Eisbacher 1974; Evenchick 1991a). Clasts in the Brothers Peak Formation were derived from the east and west, and denote continued input from the fold belt. Deformation of the Brothers Peak Formation indicates that shortening continued to latest Cretaceous or early Tertiary time.

Regional relationships at the south and west margins of the fold belt place local constraints on the timing of deformation. For example, on the southwest margin of the fold belt, widespread Tertiary plutons intrude contractional structures in the Bowser Lake Group and Middle Jurassic granite (54°N to 55°N, Fig. 2; Woodsworth *et al.* 1985; age of granite from P. van der Heyden, pers. comm. 1990), indicating that folds and thrust faults formed between Late Jurassic and Tertiary time. In the southeast (55°N to 56°N, 126°W to 128°W; Fig. 2), Late Cretaceous to Eocene postkinematic plutons (Richards 1980; T. Richards, pers. comm. 1989), show that some of the folding apparently ended before the Late Cretaceous. However, cross-cutting plutons only limit the age of those particular structures, and do not preclude younger detachments carrying the plutons. The regional relationships provide examples which fall within the Late Jurassic (?) to latest Cretaceous or early Tertiary age for the fold belt that was determined from relationships in the Sustut Group. There is, however, no conclusive evidence to demonstrate or eliminate Late Jurassic and earliest Cretaceous deformation.

Shortening and variations in structural style

Estimates of shortening have been made at the northeast margin of the fold belt (Evenchick 1991a). They were derived by constructing balanced cross-sections (following Dahlstrom 1969) and restored sections in regions where several stratigraphic units provide control on the position of cut-offs of contacts by faults. Where all of the shortening is within the Bowser Lake Group, it was estimated by measuring the bed lengths of prominent local markers around folds where constant bed thickness is preserved. Details of the sections are given in Evenchick (1991a). In the absence of seismic and well data, the sections rely on a combination of down plunge projection and extrapolation of relationships among the structural levels exposed by gently plunging structures, and large dip-slip faults (Fig. 4). Additional factors guiding construction of the sections include: (1) the Sustut Group is pinned to Stikinia on the east side of the Sustut Basin; (2) the southwest side of the Sustut Basin is the location of a triangle zone which is inferred to lie above the northeast limit of deformation; (3) the basal detachment is assumed to dip gently southwest from the triangle zone (see discussion in Evenchick 1991a); (4) fold geometry in the volcanics and the position of the folds with respect to known detachments are consistent with the hypothesis that they are a result of displacement along underlying thrust faults which have a ramp-flat geometry.

In the Joan Lake section, the leading edge of the Denkladia thrust sheet has been displaced 20 km northeastward relative

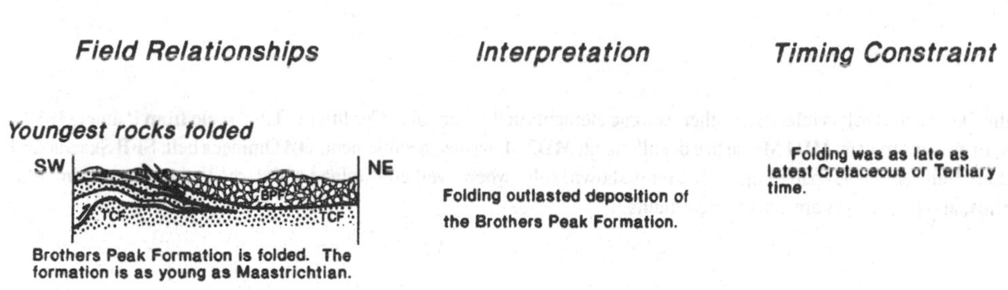

Field Relationships
Interpretation
Timing Constraint
Youngest rocks folded
SW
NE
BPF
TCF
TCF
Brothers Peak Formation is folded. The formation is as young as Maastrichtian.
Folding outlasted deposition of the Brothers Peak Formation.
Folding was as late as latest Cretaceous or Tertiary time.

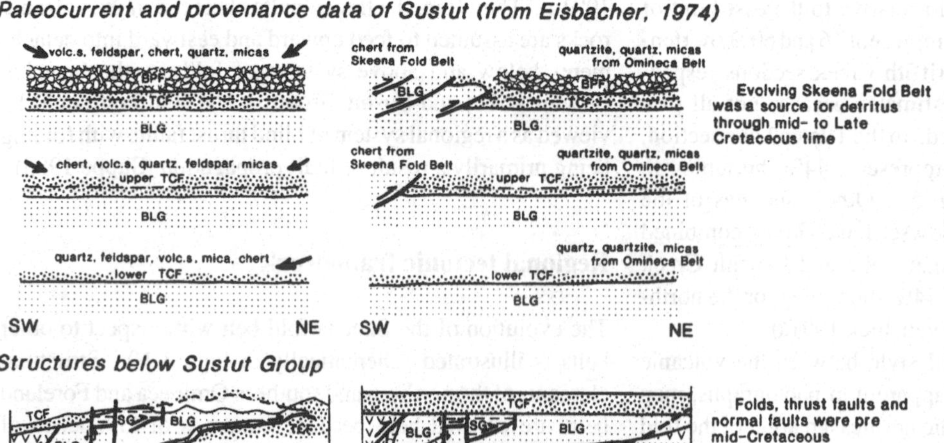

Figure 9. Summary of constraints on the timing of deformation in the Skeena fold belt. Data for the sedimentological interpretation are from Eisbacher (1974). Abbreviations: **SG** Spatsizi Group; **BLG** Bowser Lake Group; **TCF** Tango Creek Formation; **BPF** Brothers Peak Formation; v volcanic rocks. Full arrows denote palaeocurrent directions, and half arrows thrust faults.

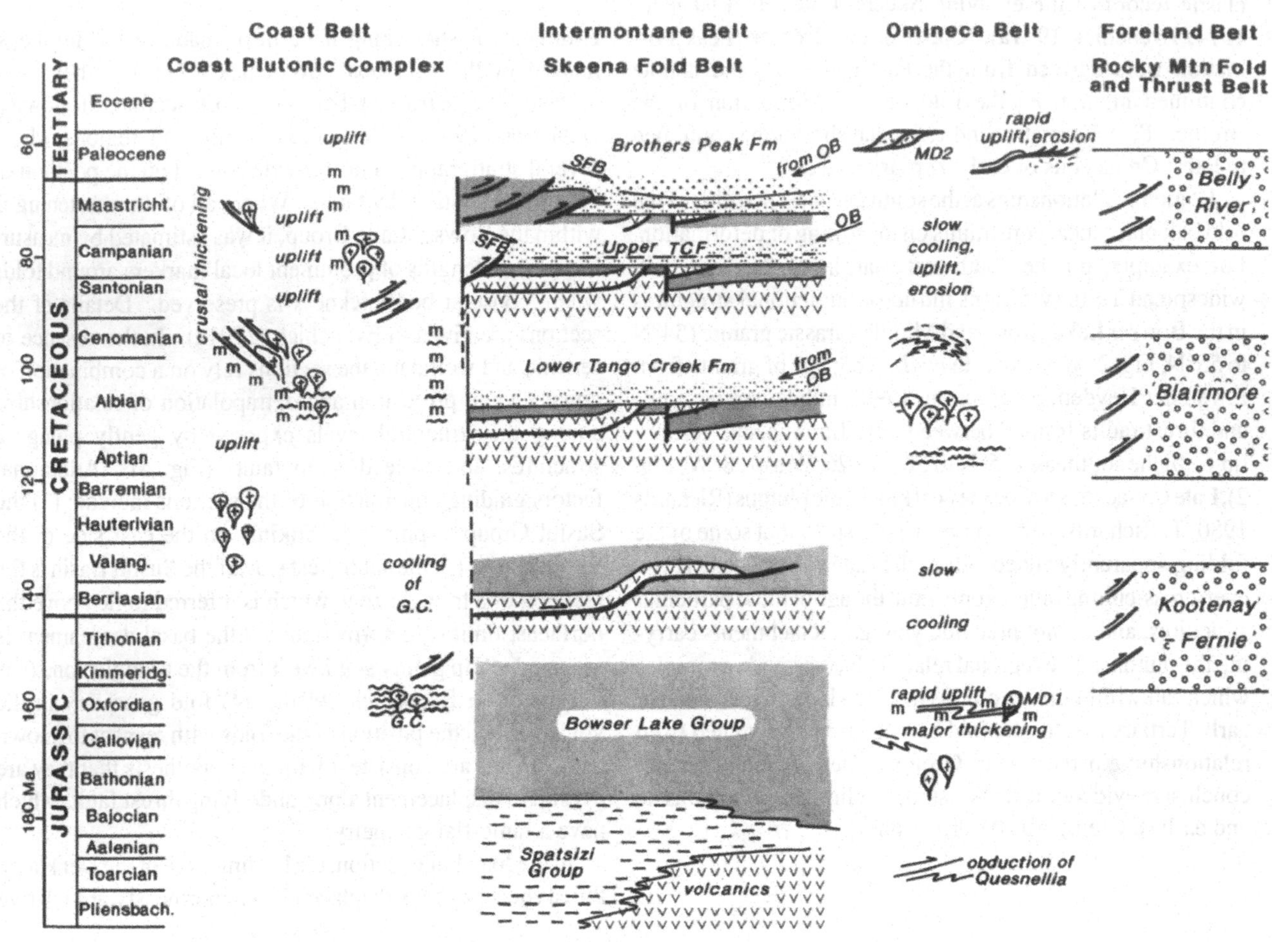

Figure 10. Summary of evolution of the Skeena fold belt in relation to other tectonic elements in the Canadian Cordillera. Time scale from Palmer (1983). Abbreviations: **G.C.** Gamsby Complex; m metamorphism; **MD1** Monashee decollement; **MD2** Monashee decollement; **OB** Omineca belt; **SFB** Skeena fold belt. Bars on the left side of Intermontane Belt reflect the span of ages for events shown, solid where well constrained and dashed where uncertain. Full arrows are from sources of clastic detritus, and half arrows are active thrust faults.

to the northeast limit of deformation. In the Griffith Creek section, the leading edge of the Crescent thrust sheet has been displaced 22 km northeastward relative to the east limit of deformation. These estimates represent 36 and 60% shortening for the Joan Lake and Griffith Creek sections respectively. They are minimum estimates because not all pre-Sustut structures can be restored. In the Tuaton Lake section, folding near Mount Umbach represents 44% shortening in the Bowser Lake Group (Fig. 5). Other estimates of the shortening by folding in the Bowser Lake Group, combined with the shortening in the Joan Lake and Griffith Creek section result in an average of 44% shortening for the northeast margin of the fold belt (Evenchick 1991a).

The difference in structural style between the volcanic rocks and overlying strata is apparent in photographs, map pattern and cross-sections of the northeast margin of the fold belt. The volcanic rocks appear to be deformed primarily by thrust faults, and form folds which are an order of magnitude larger than the numerous detached, disharmonic folds in the Bowser Lake Group. Although the style of deformation of the volcanic rocks and overlying strata varies, both units display generally the same amount of shortening (Evenchick 1991a). Therefore, the thrust faults observed in the volcanic rocks are assumed to feed upward and eastward into detachments below and above systems of folds in the Spatsizi, Bowser Lake, and Sustut Groups. The fold belt could be viewed as a regional system of blind thrust faults with folding being primarily a result of fault propagation (Suppe 1985).

Regional tectonic framework

The evolution of the Skeena fold belt with respect to other belts is illustrated schematically in Figure 10. Structural elements of the northern and southern Omineca and Foreland belts are enlisted here because hundreds of kilometres of dextral strike slip displaced the Skeena fold belt northward relative to the Foreland belt and parts of the Omineca belt before, during, and after development of the fold belt (Gabrielse 1985). In detail, the nature of the belts varies widely along

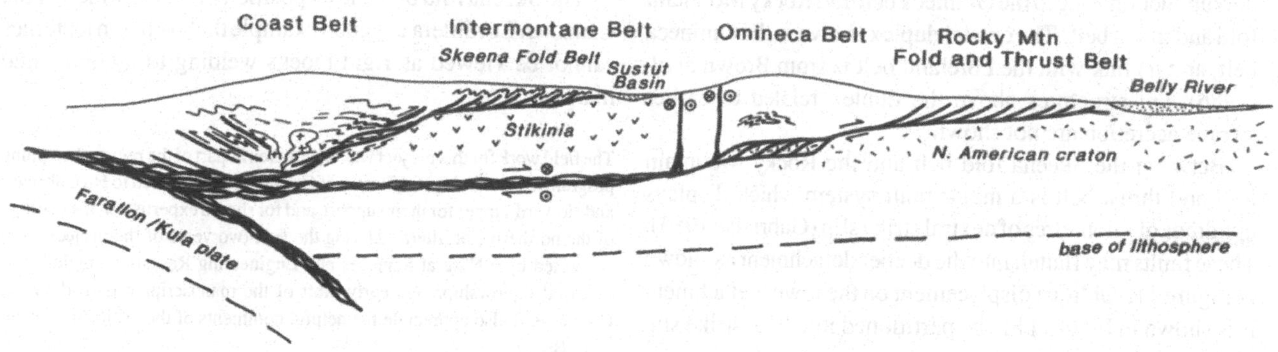

Figure 11. Schematic diagram of the Late Cretaceous tectonic framework, illustrating a possible tectonic link between the Skeena fold belt, and the Coast, Omineca and Foreland belts.

strike, but their general characteristics apply to this discussion.

Latest Jurassic folding in the Skeena fold belt, if present, was coeval with early crustal thickening along the east side of the Coast Plutonic Complex (van der Heyden 1989), crustal thickening, metamorphism, and uplift in the Omineca belt (Archibald *et al.* 1983; Brown *et al.* 1986), and the beginning of deposition of the Kootenay - Fernie clastic wedge in the Foreland belt. The clastic wedge deposits in the Foreland belt are attributed to the load of thrust sheets in the Rocky Mountain fold and thrust belt (Price 1973, 1981). Although Jurassic structures in the Skeena fold belt cannot be demonstrated, at least some shortening occurred before mid-Cretaceous time.

A mid-Cretaceous to earliest Tertiary age for the Skeena fold belt is indicated by many regional relationships. Mid-Cretaceous time was also a period of major crustal thickening in the Coast Plutonic Complex (Crawford *et al.* 1987; van der Heyden 1989). Crustal thickening by west verging thrust faults on the west side of the belt was accompanied by amphibolite facies metamorphism and emplacement of synkinematic plutons. Later, east verging thrust faults were active on the east side of the belt. Many plutons were emplaced in the Omineca belt at this time. Regional anticlinoria in the northern Omineca belt are a result of Early to mid-Cretaceous shortening (Evenchick 1988a). The southern Omineca belt underwent slow cooling, uplift and erosion, with deformation at the deepest levels (Archibald *et al.* 1983). In the Foreland belt, deposition of the Blairmore clastic wedge resulted from continued shortening in the Rockies.

Crustal thickening, metamorphism, and uplift in the Coast Plutonic Complex extended into earliest Tertiary time (Crawford *et al.* 1987; van der Heyden 1989). To the east, folded strata of the Brothers Peak Formation demonstrates that deformation in the Skeena fold belt lasted into latest Cretaceous or early Tertiary time. The last displacement on contractional shear zones in the southern Omineca belt was during the Palaeocene (Journeay 1989; Brown & Carr 1990). The shortening was transmitted eastward into the Rockies (Archibald *et al.* 1984; Brown *et al.* 1986) at a time broadly corresponding to deposition of the last foreland basin deposits (Belly River Formation) in the Foreland belt.

Discussion and conclusions

The Skeena fold belt is a regional fold and thrust belt that spans most of the Intermontane belt. It is characterized by structures typical of fold and thrust belts, and formed in Late Jurassic (?) to earliest Tertiary time. From the surface geometry of the Skeena fold belt, shortening is inferred to sole into a basal detachment that roots farther west in the Coast Plutonic Complex, where major crustal thickening was contemporaneous with development of the fold belt. Moreover, if the estimate of shortening at the northeast margin of the fold belt can be applied across the fold belt, then the Skeena fold belt represents about 160 km of shortening of the Bowser Basin and Stikinia. Furthermore, 160 km of Stikinia representing the footwall of the basal detachment must be found in or under the Coast Plutonic Complex.

The structural and temporal associations between the Skeena fold belt and the Coast Plutonic Complex suggest that their evolution was kinematically linked. The spatial and kinematic relationships between foreland and metamorphic/plutonic hinterland are similar along many ancient convergent margins, including the Omineca belt and Rocky Mountain fold and thrust belt (Price 1981; Archibald *et al.* 1984; Brown *et al.* 1986); the Asiak thrust-fold belt and Hepburn metamorphic internal zone of Wopmay Orogen (e.g. Hoffman *et al.* 1988); Appalachian Orogen (e.g. Hatcher 1981); the Himalayas (e.g. Gansser 1964; Mattauer 1986) and the Alps (Trumpy 1980). In the Canadian Cordillera, however, shortening in the Skeena fold belt and its metamorphic/plutonic hinterland was contemporaneous with shortening in the Foreland belt and its metamorphic/plutonic hinterland. This relationship invites speculation on the gross structure of the Cordillera.

The double pair of fold belt and hinterland permits postulation of a structural link between all belts of the Cordillera, illustrated schematically in Figure 11. Two systems of detachments shown in Figure 11 provide a mechanism for transferring displacement from the subduction zone to the east side of the Coast Plutonic Complex , and to the relatively high structural level of the Skeena fold belt. East verging structures observed on the east side of the complex (Fig. 10), and associated metamorphism and plutonism may be part of that system. As well, a deeper detachment may have led from the subduction zone to the Omineca belt and Rocky Mountain fold and thrust belt. The crustal duplex shown in the Omineca belt, and its link with the Foreland belt is from Brown *et al.* (1986), but structures above the duplex related to earlier terrane accretion are not shown.

Between the Skeena fold belt and the Rocky Mountain fold and thrust belt is a major fault system which displays hundreds of kilometres of dextral strike slip (Gabrielse 1985). These faults may flatten into the deeper detachment as shown in Figure 11. Oblique displacement on the lower detachment is is shown in Figure 11 to be partitioned into both strike slip and thrust components.

The Skeena fold belt and its position in the evolution of the Canadian Cordillera is a good example that collision tectonics cannot be viewed as rigid blocks welding to a continental margin.

The field work for this project was undertaken as part of the regional mapping program of the Geological Survey of Canada. I am indebted to Hu Gabrielse and Howard Tipper for their support, and for shared expertise on the geology of the northern Cordillera. During the first two years of the project I was supported by a Natural Sciences and Engineering Research Council Post Doctoral Fellowship. An early draft of the manuscript was read by Hu Gabrielse. I also appreciate the helpful comments of the critical reviewer, R. L. Brown.

References

Anderson, R. G. & Thorkelson, D. J. 1990. Mesozoic stratigraphy and setting for some mineral deposits in Iskut River map area, northwestern British Columbia. *In:* Current Research, Part E, *Geological Survey of Canada*, Paper 90-1E, 131-139.

Archibald, D. A., Glover, J. K., Price, R. A., Farrar, E. & Carmichael, D. M. 1983. Geochronology and tectonic implications of magmatism and metamorphism, southern Kootenay arc and neighbouring regions, southeastern British Columbia - I. Jurassic to mid-Cretaceous. *Canadian Journal of Earth Sciences*, **20**, 1891-1913.

——, Krogh, T. E., Armstrong, R. L. & Farrar, E. 1984. Geochronology and tectonic implications of magmatism and metamorphism, southern Kootenay arc and neighbouring regions, southeastern British Columbia - II. Mid-Cretaceous to Eocene. *Canadian Journal of Earth Sciences*, **21**, 567-584.

Bally, A. W., Gordey, P. L. & Stewart, G. A. 1966. Structure, seismic data, and orogenic evolution of southern Canadian Rocky Mountains. *Bulletin of Canadian Petroleum Geology*, **14**, 337-381.

Britton, J. M. (compiler) 1989. Geology and Mineral Deposits of the Unuk Area. *British Columbia Ministry of Energy, Mines, and Petroleum Resources*, Open File Map 1989-10.

Brown, R. L., Journeay, J. M., Lane, L. S., Murphy, D. C. & Rees, C. J. 1986. Obduction, backfolding and piggyback thrusting in the metamorphic hinterland of the southeastern Canadian Cordillera. *Journal of Structural Geology*, **8**, 255-268.

—— & Carr, S. D. in press. Lithospheric thickening and orogenic collapse within the Canadian Cordillera. *Pacific Rim 90 Congress.*

Coney, P., Jones, D. L. & Monger, J.W.H. 1980. Cordilleran Suspect Terranes. *Nature*, **288**, 329-333.

Cookenboo, H. & Bustin, R. M. 1989. Jura-Cretaceous (Oxfordian to Cenomanian) stratigraphy of the north-central Bowser Basin, northern British Columbia. *Canadian Journal of Earth Sciences*, **26**, 1001-1012.

Crawford, M. L., Hollister, L. S. & Woodsworth, G. J. 1987. Crustal deformation and regional metamorphism across a terrane boundary, Coast Plutonic Complex, British Columbia. *Tectonics*, **6**, 343-361.

Dahlstrom, C.D.A. 1969. Balanced cross-sections. *Canadian Journal of Earth Sciences*, **6**, 743-757.

Eisbacher, G. H. 1974. Sedimentary and tectonic evolution of the Sustut and Sifton basins, north-central British Columbia. *Geological Survey of Canada*, Paper 73-31.

Evenchick, C. A. 1987. Stratigraphy and structure of the northeast margin of the Bowser Basin, Spatsizi map area, north-central British Columbia. *In:* Current Research, Part A, *Geological Survey of Canada*, Paper 87-1A, 719-726.

——1988a. Stratigraphy, metamorphism, structure, and their tectonic implications in the Sifton and Deserters ranges, Cassiar and northern Rocky mountains, northern British Columbia. *Geological Survey of Canada*, Bulletin 376.

—— 1988b. Structural style and stratigraphy in northeast Bowser and Sustut basins, north-central British Columbia. *In:* Current Research, Part E, *Geological Survey of Canada*, Paper 88-1E, 91-95.

—— 1989. Stratigraphy and structure in east Spatsizi map area, north-central British Columbia. *In:* Current Research, Part E, *Geological Survey of Canada*, Paper 89-1E, 133-138.

—— 1991 a. Geometry, evolution, and tectonic framework of the Skeena Fold Belt, north-central British Columbia. *Tectonics* (in press).

—— 1991 b. Structural relationships of the Skeena Fold Belt on the west side of the Bowser Basin, northwest British Columbia. *Canadian Journal of Earth Sciences* (in press).

—— & Green, G. M. 1990. Structural style and stratigraphy of southwest Spatsizi map area. *In:* Current Research, Part F, *Geological Survey of Canada*, Paper 90-1F, 135-144.

Gabrielse, H. 1985. Major dextral transcurrent displacements along the Northern Rocky Mountain Trench and related lineaments in north-central British Columbia. *Geological Society of America Bulletin*, **96**, 1-14.

—— & Tipper, H. W. 1984. Bedrock geology of Spatsizi map area (104H). *Geological Survey of Canada*, Open File 1005.

—— & Yorath, C. J. 1989. DNAG#4. The Cordilleran Orogen in Canada. *Geoscience Canada*, 16, 67-83.

——, Monger, J.W.H., Wheeler, J. O. & Yorath, C. J. in press. Morphogeological Belts, Tectonic Assemblages and Terranes. In: Chapter 2, Tectonic Framework. *In:* Gabrielse, H. & Yorath, C. J. (eds), The Cordilleran Orogen: Canada, *Geological Survey of Canada*, Geology of Canada no. 4.

Gansser, A. 1964. *Geology of the Himalayas. Wiley Interscience*, London, 289p.

Geological Survey of Canada, 1957. Stikine River Area, Cassiar District, British Columbia. *Geological Survey of Canada*, Map 9-1957.

Grove, E. W. 1986. Geology and Mineral Deposits of the Unuk River - Salmon River - Anyox Area. *British Columbia Ministry of Energy, Mines, and Petroleum Resources*, Bulletin 63.

Hatcher, R. Jr., 1981. Thrusts and nappes in the North American Appalachian Orogen. *In:* McClay, K. R. & Price, N. J. (eds) *Thrust and Nappe Tectonics*. Geological Society of London Special Publication, **9**, 491-500.

Hoffman, P. F., Tirrul, R., King, J. E., St-Onge, M. R. & Lucas, S. B. 1988. Axial projections and modes of crustal thickening, eastern Wopmay orogen, northwest Canadian shield. *In:* Clark Jr., S.P., Burchfiel, B.C. & Suppe, J. (eds) *Processes in Continental Lithospheric Deformation*. Geological Society of America, Special Publication, **218**, 1-30.

Jones, P. B. 1982. Oil and gas beneath east-dipping underthrust faults in the Alberta Foothills. *In:* Powers, R.B. (ed.) *Studies of the Cordilleran Thrust Belt*. Rocky Mountain Association of Geologists, 61-74.

Journeay, J. M. 1989. The Shuswap Thrust: Sole fault to a metamorphic-plutonic complex of Late Cretaceous-Early Tertiary age, southern Omineca belt, B.C. *In: Geological Society of America Abstracts with Programs*, **21**, 99.

MacLeod, S. & Hills, L. V. 1990. Sedimentology and palaeontology of Late Jurassic (Oxfordian) to Early Cretaceous (Aptian) strata, northern Bowser Basin. *In: Geological Association of Canada, Mineralogical Association of Canada, Program with Abstracts*, **15**, A80.

Mattauer, J. 1986. Intracontinental subduction, crust-mantle decollement and crustal-stacking wedge in the Himalayas and other collision belts. *In:* Coward, M. P. & Ries, A. C. (eds) *Collision Tectonics*, Geological Society of London, Special Publication **19**, 37-50.

Monger, J.W.H. 1984. Cordilleran Tectonics: a Canadian perspective. *Bulletin of the Geological Society of France*, **7**, 255-278.

—— & Price, R. A. 1979. Geodynamic evolution of the Canadian Cordillera-progress and problems. *Canadian Journal of Earth Sciences*, **16**, 770-791.

——, Price, R. A. & Tempelman-Kluit, D. J. 1982. Tectonic accretion and the origin of the two major metamorphic and plutonic welts in the Canadian Cordillera. *Geology*, **10**, 70-75.

Palmer, A. R. (compiler). 1983. The Decade of North American Geology 1983 time scale. *Geology*, **11**, 503-504.

Price, R. A. 1973. Large-scale gravitational flow of supracrustal rocks, southern Canadian Rockies. *In:* De Jong, K. A. & Scholten, R. (eds) *Gravity and Tectonics*. Wiley, New York, 491-502.

—— 1981. The Cordilleran foreland thrust and fold belt in the southern Canadian Rocky Mountains. *In:* McClay, K. R. & Price, N. J. (eds) *Thrust and Nappe Tectonics*. Geological Society of London Special Publication, **9**, 427-448.

—— 1986. The southeastern Canadian Cordillera: thrust faulting, tectonic wedging, and delamination of the lithosphere. *Journal of Structural Geology*, **8**, 239-254.

Read, P. B. 1983. Geology, Classy Creek (104J/2E) and Stikine Canyon (104J/1W). *Geological Survey of Canada*, Open File 940.

Richards, T. A. 1975. Geology of McConnell Creek (94D\E) map area. *Geological Survey of Canada*, Open File 342.

—— 1980. Geology of Hazelton (93M) map area. *Geological Survey of Canada*, Open File 720.

Sweet, A. R. & Evenchick, C. A. 1990. Ages and depositional environments of the Sustut Group, north-central British Columbia. *In: Geological Association of Canada, Mineralogical Association of Canada, Program with Abstracts*, **15**, A127.

Souther, J. G. 1972. Telegraph Creek map-area, British Columbia. *Geological Survey of Canada*, Paper 71-44.

—— & Armstrong, J. E. 1966. North Central Belt of the Cordillera of British Columbia. *In: Canadian Institute of Mining and Metallurgy, Special Volume* **8**, 171-184.

Suppe, J. 1985. *Principles of Structural Geology*. Prentice-Hall Inc., 537p.

Thomson, R. C., Smith, P. L. & Tipper, H. W. 1986. Lower to Middle Jurassic (Pliensbachian to Bajocian) stratigraphy of the northern Spatsizi area, north-central British Columbia. *Canadian Journal of Earth Sciences*, **23**, 1963-1973.

Tipper, H. W. (compiler). 1976. Smithers, B. C. 93L. *Geological Survey of Canada*, Open File 351.

—— & Richards, T. A. 1976. Jurassic stratigraphy and history of north-central British Columbia. *Geological Survey of Canada Bulletin*, 270.

Trumpy, R. 1980. An outline of the geology of Switzerland. *In:* Trumpy, R. (ed.) *Geology of Switzerland A Guide-Book* 26th International Geological Congress, Guidebook G10, 7-102.

van der Heyden, P. 1989. U-Pb and K-Ar Geochronometry of the Coast Plutonic Complex, 53°N to 54°N, British Columbia, and Implications for the Insular-Intermontane Superterrane boundary. Unpublished Ph.D. thesis, *University of British Columbia*.

Wheeler, J. O., Aitken, J. D., Berry, M. J., Gabrielse, H., Hutchison, W. W., Jacoby, W. R., Monger, J.W.H., Niblett, E. R., Norris, D. K., Price, R. A. & Stacey. 1972. The Cordilleran Structural Province. *In:* Price, R.A. & Douglas, R.J.W. (eds) *Variations in Tectonic Styles in Canada*. Geological Association of Canada Special Paper, **11**, 1-82.

—— & McFeely, P. (compilers). 1987. Tectonic Assemblage Map of the Canadian Cordillera and adjacent parts of the United States of America. *Geological Survey of Canada*, Open File 1565.

——, Brookfield, A. J., Gabrielse, H., Monger, J.W.H., Tipper, H. W. & Woodsworth, G. J. (compilers). 1988. Terrane Map of the Canadian Cordillera. *Geological Survey of Canada*, Open File 1894.

Woodsworth, G. J., Hill, M. L. & van der Heyden, P. 1985. Preliminary Geologic Map of Terrace (NTS 103 I East Half) Map Area, British Columbia. *Geological Survey of Canada*, Open File 1136.

——, Anderson, R. G. & Armstrong, R. L. 1989. Plutonic regimes in the Canadian Cordillera. *Geological Survey of Canada Open File*, 1983.

Geometric evidence for synchronous thrusting in the southern Alberta and northwest Montana thrust belts

Steven E. Boyer

Department of Geological Sciences, University of Washington, Seattle, Washington 98195, USA

Abstract: Any model of thrust kinematics or mechanics must account for the hitherto unexplained presence of imbricate stacks on the forelimbs and crests of folds and structural culminations. Although map-scale patterns of folded thrusts and the sedimentary record of foredeeps seem to conclusively argue for a relatively simple hinterland-to-foreland ('piggy-back') progression of thrusting, such a simple sequence cannot produce such crestal imbricate stacks. The only model compatible with existing structural and stratigraphic evidence is one in which thrusts at depth develop in a cratonward progression but remain active during the initiation of younger thrusts. Thus at any one time two or more thrusts may be simultaneously active. This model differs from previous ones in which older thrusts were assumed to become deactivated as new thrusts propagated toward the foreland. Synchronous thrusting in the subsurface produces 'out-of-sequence' thrusting at the Earth's surface. The adoption of a synchronous thrust model over the classic 'piggy-back' model has important implications for mechanical models of thrusting, cross-section balancing, and hydrocarbon exploration in thrust belts and leads to more realistic models of duplex fault zone development. First, while not proving the critically-tapered wedge model of thrust mechanics, the observations demonstrate that the model is viable. Second, since most cross-section balancing techniques include the assumption of piggy-back thrusting, 'out-of-sequence' and synchronous thrusts must be identified if a section is to be correctly balanced. Third, models of hydrocarbon generation based on a cratonward progression of thrusting may lead to inaccurate conclusions if applied to regions of synchronous thrusting; appraisal of basins and prospects based on synchronous thrust models produce results considerably different from those generated assuming simple sequential thrust development. Lastly, existing duplex models cannot adequately explain the large variation in the geometry of duplex fault zones. However, synchronous thrusting, if halted at various stages, can produce a large variety of typically observed duplex geometries.

There have been two principal schools of thought regarding the sequence of thrust imbrication within thrust and fold belts; one maintaining that thrusts generally are younger toward the foreland (Bally *et al.* 1966; Dahlstrom 1970; Royse *et al.* 1975) and a second holding that thrusts develop in a foreland-to-hinterland sequence (Milici 1975; Mudge 1970). Until recently the theory of cratonward thrust imbrication, with allowances for minor 'out-of-sequence' thrusting (Dahlstrom 1970) has become the most widely accepted, having taken on the role of an *axiom* in the study of thrust kinematics. In hinterland-to-foreland progression of thrusting, older thrusts become inactive and are carried passively, in a so-called 'piggy-back' fashion, upon the backs of underlying younger thrusts. Many thrust models, including those for duplex development (Boyer 1978; Boyer & Elliott 1982) have been built upon this foundation, as the arguments for a hinterland-to-foreland thrust progression are based on convincing stratigraphic and geometric evidence (see discussion in the following section). However, much structural and stratigraphic evidence has come to light which casts doubt on the simple piggy-back thrust model (Ori & Friend 1984; Searle 1985; Butler 1987; Roeder 1988; Morley 1988).

The conclusion that thrusts develop in a relatively straight forward hinterland-to-foreland, progression has at least two major shortcomings: (1) it is incompatible with compelling new mechanical models for the emplacement of thrust belts, which require synchronous, 'out-of-sequence' thrusting, thrust reactivation and/or duplex underplating to maintain critical taper (Davis *et al.* 1983; Boyer & Geiser 1987; Platt 1988; Dahlen & Suppe 1988; Morley 1988) and (2) a piggy-back sequence cannot produce many observed structural geometries. This paper, dealing with the latter, presents observations which indicate that synchronous thrusting in the Canadian and US Rockies is more common than has previously been recognized. Therefore simple models of piggy-back thrusting may not have universal applicability. If these conclusions are correct, they have considerable implications for the mechanics of thrust emplacement, the application of cross-section balancing techniques, and patterns of hydrocarbon generation and entrapment in thrust belts.

Existing criteria for the determination of thrust sequence

Most patterns of synorogenic sedimentation suggest a cratonward progression of thrusting. Armstrong & Oriel (1965) postulated a west-to-east progression of thrusting in the Idaho-Wyoming-Utah thrust belt, commencing in latest Jurassic and terminating in early Eocene. They based their conclusions on observations that from west to east successively younger rocks were involved in thrusting and that during the Cretaceous period the axis of maximum deposition migrated eastward (Armstrong & Oriel 1965, p. 1854-6).

Presumably this eastward migration of the depositional axis accompanied the eastward advance or growth of the thrust belt. Oriel & Armstrong (1966, p. 2618) further noted that for Mesozoic to early Tertiary formations 'The westernmost known occurrence and coarsest facies of each (syntectonic) unit lie in a belt east of the coarsest facies of the next older unit.' Although there is much debate concerning the absolute ages of the various thrusts (Heller *et al.* 1986), most workers appear to accept a west-to-east progression for the US Rocky Mountain thrust belts (Royse *et al.* 1975; Wiltschko & Dorr 1983). In the Canadian foothills the eastward migration of Mesozoic to Palaeocene synorogenic foredeeps has led to similar conclusions (Bally *et al.* 1966, p. 365-369; Price 1973, 1981).

Arguments for a hinterland-to-foreland progression of thrusts are fortified by geometric evidence, most importantly patterns of folded thrusts (Verrall 1968; Jones 1971). The existence of folded thrusts was noted at least as early as the 1930s by geologists working in the Canadian Rocky Mountains (Hake *et al.* 1935, 1942; Hume 1941; Hage 1942; and Deiss, 1943). Scott (1951), attributed the folded thrusts to two distinct phases of deformation: (1) low-angle thrust faulting, followed by (2) regional-scale folding and accompanying 'adjustment faulting' and reactivation of the steeper portions of folded thrusts. However, extensive drilling and seismic acquisition, associated with petroleum exploration in the Canadian Rockies, and field observations have led to the conclusion that folded thrusts are best explained by a cratonward progression of thrusting in which older thrusts are carried piggy-back and folded over ramps in lower, more frontal thrusts (Bally *et al.*, 1966; Verrall 1968; Dahlstrom 1970; Jones 1971).

Despite the geometric and stratigraphic evidence for a foreland progession of thrusting there has long existed a school of thought maintaining that mechanical arguments and geometric evidence suggest the opposite sense of propagation (Mudge 1970; Milici 1975). Mechanical theory (Hubbert & Rubey, 1959) has indicated that for tabular thrust sheets to be emplaced by horizontal compression they must be limited to lengths deemed by many to be unrealistically short. This has led to the suggestion that gravity is the only viable driving force for the emplacement of thrust sheets (Milici 1975; Mudge 1970; LeMoine 1973; Scholten 1973; Choukroune & Seguret 1973). In a surficial gravity slide the imbricates often break backward from the toe, and if such a mechanism operated in the emplacement of thrust sheets it was thought that thrust belts must also develop in a foreland-to-hinterland sequence. These conclusions have been supported by patterns of thrust intersection which suggest that more frontal thrusts have been truncated by younger thrusts toward the hinterland (Milici 1975).

The *mechanical arguments* for foreland-to-hinterland thrust sequences must now be considered less reliable in light of recent work suggesting that thrust sheets can indeed be emplaced by horizontal compression if a critical taper angle is maintained (Davis *et al.* 1983). However, the *physical field evidence* for hindward thrust sequences cannot be as easily dismissed. Where the youngest synorogenic sediments are preserved in orogenic belts, many thrusts can be shown to be either be late and out-of-sequence or to have been periodically reactivated (Lamerson 1982; Hurst & Steidtmann 1986). In agreement with the stratigraphic evidence, the use of balanced cross sections also indicates that imbricate stacks are often generated by 'out-of-sequence' thrusting (Delphia & Bombolakis 1988). Many out-of-sequence imbricate stacks occur at the tops of ramps in thrust sheets and have been produced in experiments (Morse 1977, Fig. 3). Also out-of-sequence thrusts may develop at the base of a ramp as a means of decreasing the ramp angle (Serra 1977; Knipe 1985).

It should be noted that the term 'out-of-sequence' thrust is used by proponents of piggy-back thrusting to refer to those thrusts which disobey the 'rules' of thrust development and it has been suggested that the use of this term be discontinued (Geiser, pers. comm. 1990). Such thrust imbricates are usually thought to be of minor significance (Dahlstrom 1970, p. 354; Woodward 1987). This is probably true for portions of the Canadian and US Rockies, including the southern portion of the Utah-Wyoming thrust belt. However, patterns of imbrication, especially the locus of imbricate faults on the crests of folds, in the southern Alberta and northwest Montana thrust belts, will be used in subsequent sections to argue that such out-of-sequence motions are not incidental and that they are symptomatic of synchronous motion on two or more major thrusts at depth.

Imbricate stacks situated on the forelimbs of folds

Stacks of imbricate thrusts are often found on the crests or limbs of structural culminations (Figs 1, 4b, 5 & Table 1). Several examples gleaned from published sections are listed in Table 1. All these structures share common features: (1) the upper fault in the sequence emanates from the backlimb of the underlying fold and is usually quite planar; (2) structurally lower faults on the forelimb display a decreasing radius of curvature with depth (i.e. the lower faults in the sequence are more tightly folded than the higher faults); and (3) strata on the backlimb of the fold display little or no evidence of disruption or thrust repetition.

The geometry of crestal imbricate stacks have been described in the older thrust literature. Most commonly mentioned are 'back-limb' imbricates, or imbricates which overlie the hinterland-dipping limbs of folds. It has been supposed for some time that 'backlimb' imbricates often form sequentially from foreland to hinterland, or in a so-called 'out-of-sequence' order, on the back limbs of existing structural culminations (Douglas 1950; Bally *et al.* 1966). Douglas (1950, p. 88-95) maintained that field observations and map patterns supported a model in which the backlimb faults formed simultaneously with the underlying anticlines. Hume (1957, p. 408-409) followed Douglas (1950) in arguing for simultaneous folding and imbrication. Douglas (1950, p. 89) attributed back-limb out-of-sequence thrusts to greater interbed slip within the west limbs of folds, forming 'out-of-the-syncline' thrusts (Hake *et al.* 1942; Dahlstrom 1970). However, in most of the structures discussed in this paper, the

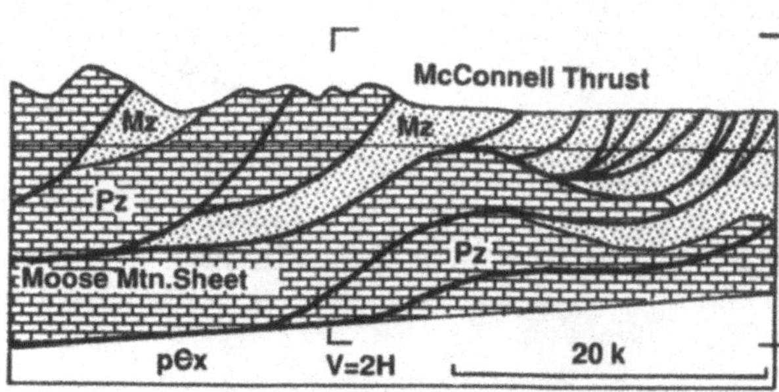

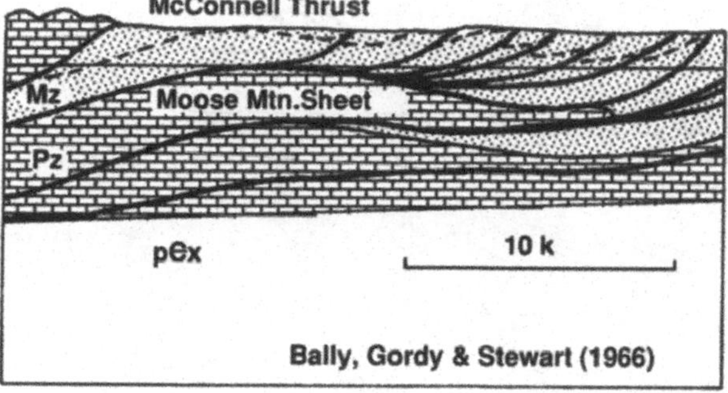

Figure 1. Imbricate stack overlying the frontal limb a structural culmination in the southern Alberta thrust belt. The upper section extends westward from the Bow Valley structure and is drawn with 2:1 vertical exaggeration. The lower section is an enlargement of the culmination (location indicated by box in upper figure) with no vertical exaggeration. Note that the structure on the west limb of the structure is relatively simple.

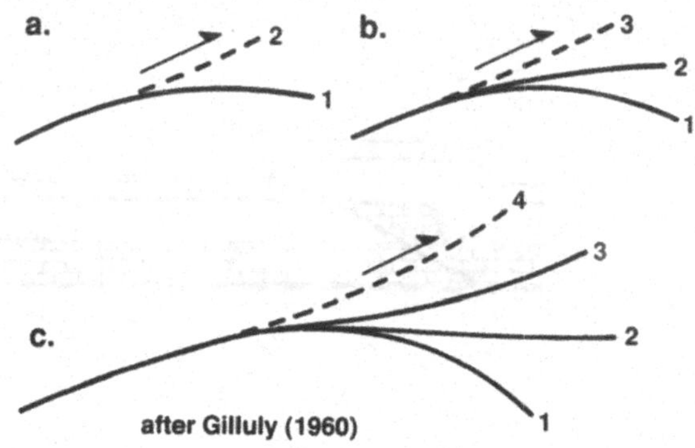

Figure 2. Gilluly's model of simultaneous thrusting and folding.
(a). As thrust '1' is folded by movement on the leading thrust, the frontal limb is locked while displacement is free to continue on the backlimb of the anticline. Continued movement on the backlimb portion of the fault initiates a new imbricate '2' which breaks across the crest of the structure.
(b). Continued folding inactivates the forelimb portion of fault '2', while synchronous motion on the backlimb thrust initiates a new imbricate '3'.
(c). The process is repeated as fault three is folded and imbricate '4' is formed.
In this simple model all imbricates merge at a single branch line at the crest of the anticline.

cumulative displacement on the crestal imbricates is on the order of several kilometres, an amount that greatly exceeds any bedding-plane slip that would accompany crowding in the troughs of adjacent synclines.

Gilluly (1960) provided a more satisfactory explanation for crestal imbricate stacks: simultaneous thrusting and folding (Fig. 2). This interpretation differs from the treatment of the above workers in that the upper detachment and associated imbricates were considered to be major thrust systems in their own right, rather than simply accommodating synclinal crowding and out-of-the-syncline, bedding-parallel shear. Gilluly's synchronous thrust/fold model was contested by Dahlstrom (1970, p. 349-350 & Fig. 20) and Boyer & Elliott (1982, p. 1216 & Fig. 27), the latter maintaining that the same geometry could be produced by the folding of a pre-existing imbricate fan. The best criterion for determining the relative timing of structures within a 'folded' imbricate stack is not the variation in fold intensity between the upper and lower imbricates, but rather the position of the branch lines of the various imbricates; in most structures of this type the imbricates share a single branch line near the crest of the fold (see Fig. 1 and examples cited in Table 1), implying a 'causal relationship between the two phenomenon' (Dahlstrom 1970, p. 350).

By incorporating modern concepts of thrust geometry, Gilluly's (1960) model can be adapted to explain the frontal-limb imbrication of the Bow Valley culmination (Fig.1) and

Table 1. Structures possessing forelimb imbricate stacks suggestive of synchronous thrusting*

Structure	*Location*	*Reference*
Hot Springs window	Southern Appalachians, USA	Oriel (1951)
Mountain City Window	Southern Appalachians, USA	Rodgers (1970, p. 172-173)
Valley & Ridge Province	Appalachians, Penna., USA	Nickelsen (1988)
Goat Ridge window	Basin & Range, Nevada, USA	Gilluly (1960)*
Engadine window	Central Alps	Oriel (1951, p. 24-27)
Savanna Creek field	Rocky Mtns., Alberta, Canada	Hennessey (1975)
Bow Valley structure	Rocky Mtns., Alberta, Canada	Bally et al. (1966)
Waterton field	Rocky Mtns., Alberta, Canada	Dahlstrom (1969, Fig. 10a)
Painter Reservoir field	Rocky Mtns., Wyoming, USA	Lamerson (1982, Fig. 12)
Tunp thrust	Rocky Mtns., Wyoming, USA	Lamerson (1982, Plate 8)
Pineview field	Rocky Mtns., Wyoming, USA	Lamerson (1982)
Stewart Peak culmination	Rocky Mtns., Wyoming, USA	Lageson (1984)
Northern Sawtooths	Rocky Mtns., Montana, USA	Mudge & Earhart (1983)

* Also see numerous additional examples suggested by Gilluly (1960, p. 76-77). This table refers to structures that resemble results produced by the synchronous thrust model, but these structures, with the exception of the Goat Ridge window, were not usually interpreted as such.

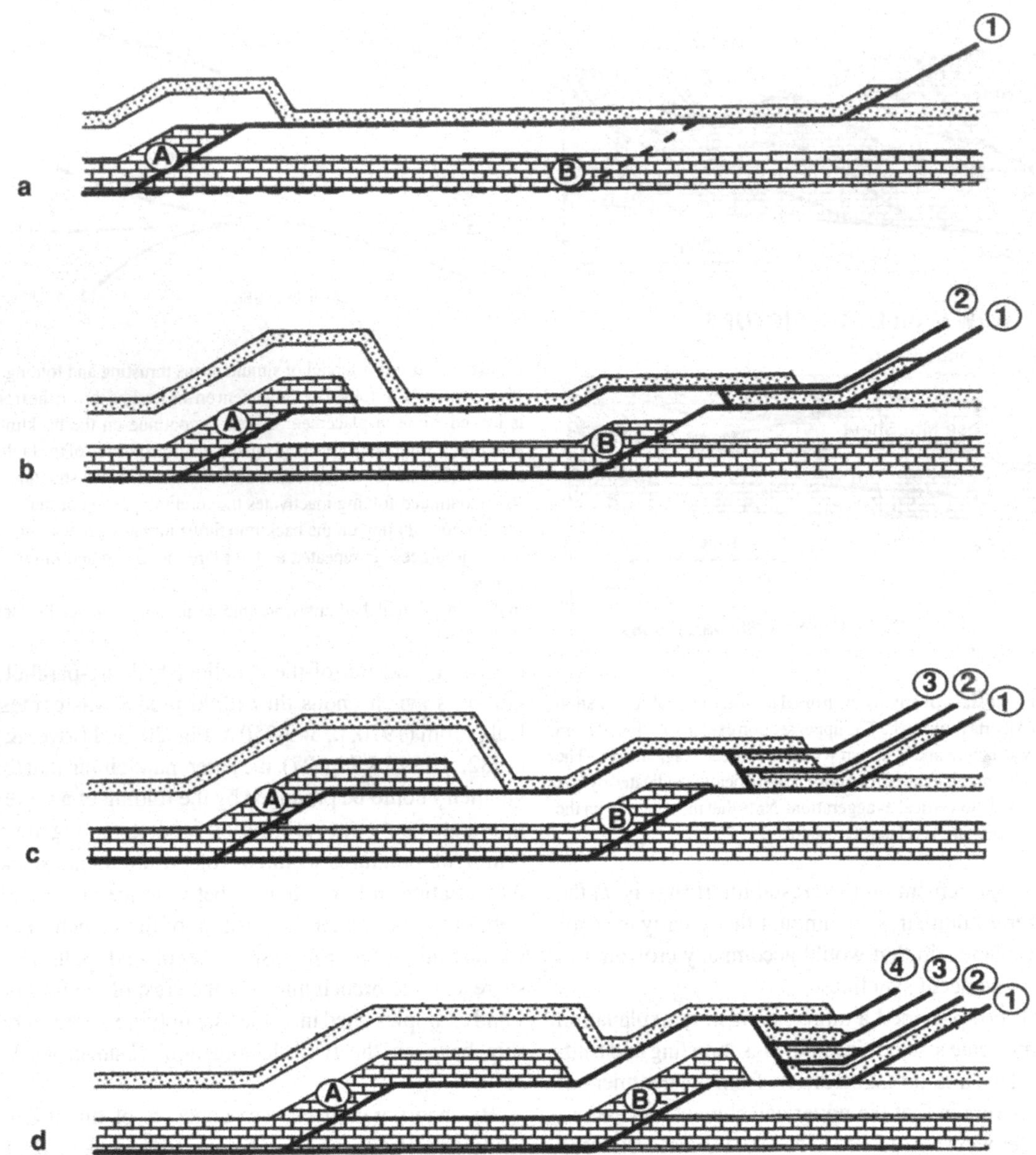

Figure 3. Gilluly's model of simultaneous folding and thrusting (Fig. 2) provided the inspiration for this synchronous thrust model. **(a)** After formation of the initial thrust ('A') a new thrust ('B', dashed) propagates toward the foreland. **(b)** Movement on thrust 'B' folds the upper flat of thrust 'A' deactivating imbricate '1'. Renewed movement on thrust 'A' breaks across the crest of the frontal fold, forming an out-of-sequence imbricate, '2' **(c & d)** The process continues with additional out-of-sequence imbricates, '3' and '4', being formed as motion alternates between thrusts 'A' and 'B' With each stage of displacement on thrust 'B' the amplitude of the frontal ramp anticline increases and new out-of-sequence develop in conjunction with continued motion on thrust 'A'. See text for additional discussion. Although the model invokes alternating motion between thrusts 'A' and 'B', the two thrusts will appear to have been synchronous when viewed in the context of geological time.

similar structures throughout the greater Rocky Mountain region (Boyer 1986a). None of the examples that will be discussed can easily be explained by folding of existing imbricates as suggested by piggy-back thrust models (Dahlstrom 1970; Boyer & Elliott 1982).

Simultaneous thrust model

In order to explain the locus of imbricate stacks on the crests and forelimbs of underlying culminations and to rectify the disparate observations of the hindward and cratonward schools of thrust progression, a model was prepared based on Gilluly's (1960) concept of synchronous folding and thrusting (Boyer, 1986a). It was assumed that both schools of thought were correct in their observations but partially in error in their conclusions.

The model was drawn using the kink domain geometry and flat - ramp - flat thrust/fold geometries (Rich 1934; Douglas 1950; Boyer & Elliott 1982) and the equations of Suppe (1983). In Figure 3a an imbricate (dashed) is about to form after some initial movement on an older thrust. Movement on the younger frontal thrust forms a low-amplitude ramp anticline which in turn folds the existing thrust. As it is folded, that portion of the older thrust which overlies the frontal limb of the leading ramp anticline is inactivated (Figs 3b & 3c). However, continued movement on the higher thrust produces an out-of-sequence imbricate which breaks across the crest of the frontal anticline. As this process proceeds, the amplitude of the frontal anticline grows and additional out-of-sequence imbricates propagate from the crest of the structure. Out-of-sequence imbrication continues until the frontal ramp anticline attains a maximum amplitude, a condition which is met as the rear kink plane of the ramp anticline has climbed unto the upper flat (Fig. 3d). From this stage forward, simultaneous motion on the two thrusts will proceed without the introduction of additional crestal imbrication and the youngest and highest imbricate will remain active. All imbricate faults occur on the forelimb of the structure; compare the model (Fig. 3d) with actual examples (Fig. 1 & Table 1).

Note that in this model, movement alternated between the two major thrusts, so the thrusts are not strictly synchronous. However, when averaged over geological time the motions of the two thrusts will appear to be simultaneous. Such alternating stick-slip motion probably also applies to many other thrusts that have been described as being synchronous.

Dahlstrom (1970, p. 354) has noted that regional and local thrust sequences are not entirely compatible. However, out-of-sequence imbrications may not be incompatible with the propagation patterns of regional thrusts when viewed in the light of the simultaneous thrust model just described. Note that 'in-sequence' initiation of thrusts at depth, combined with continued motion on all thrusts, leads to 'out-of-sequence' imbrication at shallow or surface levels (Fig. 3). In this model hinterland-to-foreland thrust propagation, synchronous thrusting, and out-of-sequence imbrication are all compatible.

Application of the synchronous thrust model to the southern Alberta & northwest Montana thrust belts

As summarized in Table 1, crestal and forelimb imbricate fans commonly overlie folds and thrust culminations in many thrust belts. These features are especially common in the Rocky Mountains of southern Alberta and the northwest Montana Sawtooth Ranges. Applying the synchronous thrust model (Fig. 3) to these thrust terranes one can draw conclusions concerning their probable kinematic development

Northern Sawtooth Range, Montana, USA

At the northern termination of the Sawtooth Range in northwest Montana a crestal imbricate stack overlies the frontal limb of a fault-cored culmination (Fig. 4). On the west limb of the structure is a normal stratigraphic sequence of Mississippian, and unconformably overlying Jurassic and lower Cretaceous strata. Indeed, owing to the lack of duplicated strata on the west flank of the culmination, Mudge & Earhart (1983) showed no detachment within the Jurassic of the west limb. However, in order to allow restoration of the tectonic shortening on the east limb of the structure, such a detachment is certainly required.

If the model (Fig. 3) applies to the origin of the northern Sawtooth structure, the numerous major thrusts within the Sawtooths must have been synchronously active. The presence of the forelimb imbricate stack suggests that the upper detachment in the Jurassic section must have been active during the growth of the culmination. Since this detachment roots westward in Palaeozoic carbonates, thrusting on western imbricates must have occurred synchronously with faults of Palaeozoic strata coring the frontal culmination to the east.

Southern Alberta, Canada

A section through the Lewis thrust sheet and the Waterton gas field in Alberta, Canada (Fig. 5), displays geometries strikingly similar to those mapped by Mudge & Earhart (1983) in the northern Sawtooth Range. The Lewis thrust sheet of Middle Proterozoic metasedimentary strata is folded into a broad syncline, the west and east limbs of the syncline having been uplifted by imbrication of Palaeozoic carbonates within two underlying duplex fault zones (Bally *et al.* 1966; Gordy *et al.* 1977, pp. 25 & 31). The two duplex structures are on trend with the two antiformal features that emerge southward from beneath the Lewis sheet to form the northern Sawtooths.

In order to understand the kinematic development of the two culminations sequentially restored cross sections were constructed by reversing the process utilized in the model (Fig. 3). Assuming that the least curved faults emanating from the backlimb of the frontal culmination were the last to form, these faults were restored first. Using balancing arguments, Dahlstrom (1969, p. 750) demonstrated that imbricates of Mesozoic rocks overlying Waterton field must be separated from the underlying duplex of Palaeozoic carbonates by a detachment; i.e. shortening within the Mesozoic strata is far in excess of that within the Palaeozoic sequence. Since the Mesozoic imbricates branch from a regional detachment within the Jurassic section, and this detachment in turn roots westward into the trailing duplex, restoration of the frontal imbricates required partial restoration of imbricates within the trailing duplex (Figs 5a & 5b). At the same time as the highest imbricates of the frontal structure were restored several of the underlying imbricates of Palaeozoic carbonates were also restored. This process decreased the amplitude of the frontal structure so that in the next stage (Figs 5b & 5c) the highest imbricate of the frontal duplex and an additional imbricate of the trailing duplex could be restored. The next sequence (from Figs 5c to 5d) was performed by continuing the process of removing the highest forelimb imbricates while restoring footwall imbricates to decrease the amplitude of the frontal duplex. This restoration is a simplification; to show the complete evolution of the section one would need to construct as many sequential sections as there are imbricates overlying the frontal structure, approximately twelve.

Note that this procedure reversed the process utilized in the model (Fig. 3). Although the 'real' example is considerably

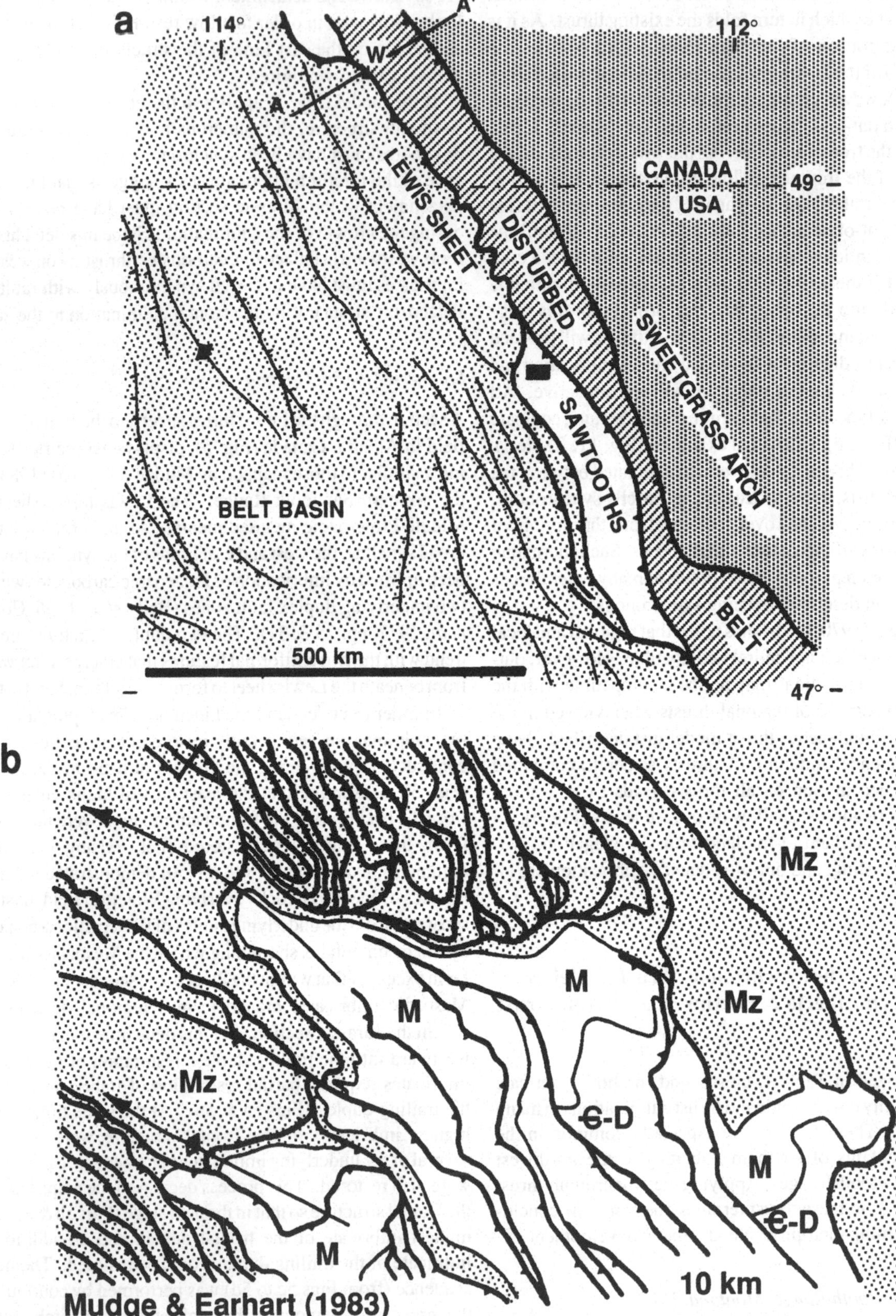

Figure 4. Imbricate stacks overlying the frontal culmination of the northern Sawtooth Range. The location of the map 'b' is indicated by the black rectangle on map 'a'. Application of the synchronous thrust model (Fig. 3) suggests that the folded detachment separating the Mesozoic (Jurassic and Cretaceous) sequence (Mz) from underlying Mississippian carbonates (M) was active in conjunction thrusts of Mississippian and Cambro-Devonian (C-D) which fill the core of the north-plunging structure. Movement on imbricates within the frontal anticlinorium (Palaeozoic strata) increased the amplitude of the anticlinorium. As the anticliorium grew portions of the overlying Jurassic-level detachment were deactivated and abandoned as the imbricate faults which overlie the northeast limb of the structure. Since the upper detachment roots westward in Palaeozoic carbonates this implies synchronous motion on faults of the eastern and western Sawtooth Ranges. The structure of the northern Sawtooths is similar to that at Waterton gas field ('W' in Fig. 4a). The sequential

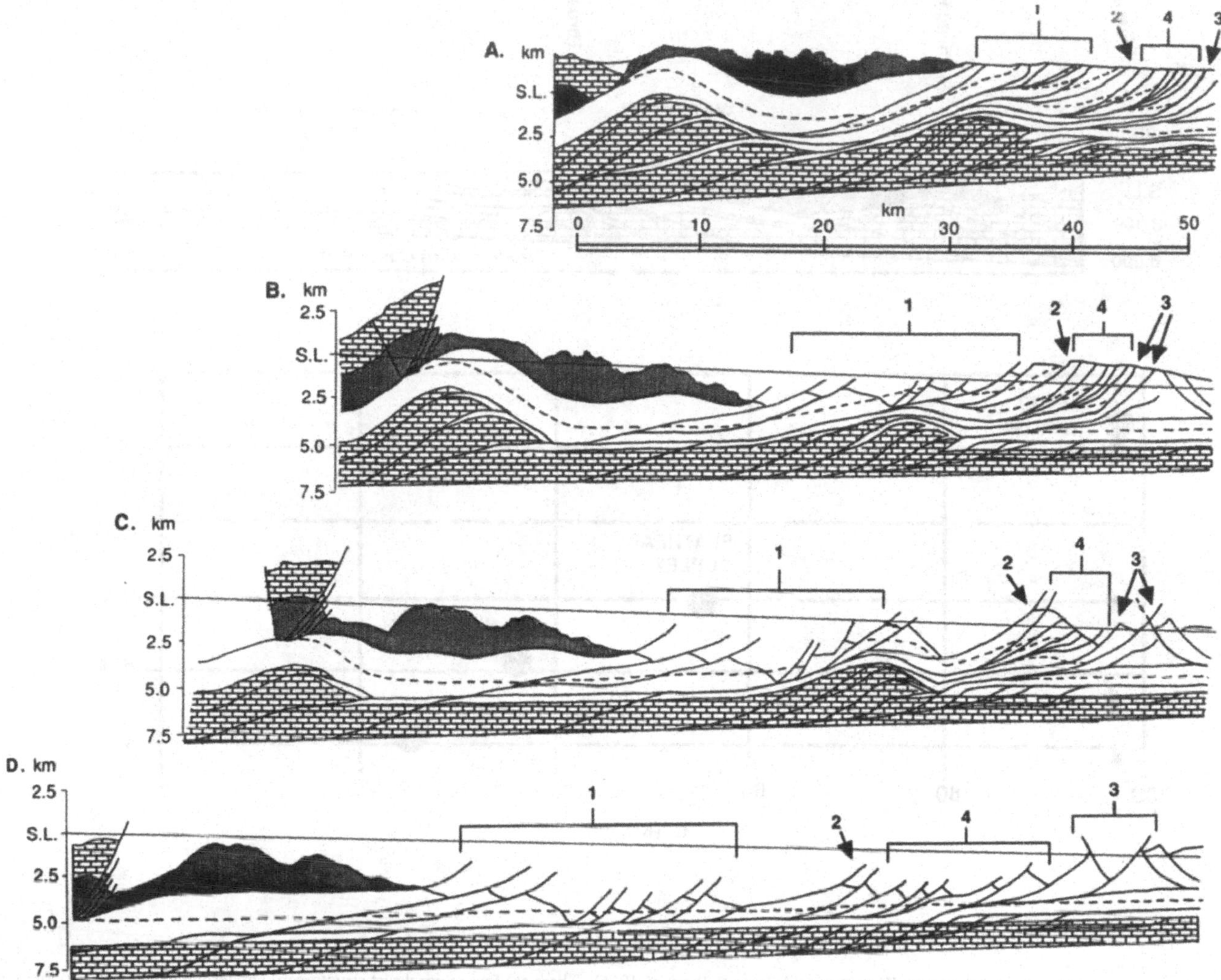

Figure 5. (a) The Lewis thrust sheet is folded by two duplexes of Palaeozoic carbonates, Waterton gas field to the east and the Flathead duplex to the west beneath the trailing edge of the Lewis thrust. These two structures lie on strike with the two culminations at the northern termination of the Sawtooth Range (Fig. 4). The position of the section is shown in Figure 4a. Waterton field and the frontal culmination of the Sawtooths (Fig. 4b) are similar in form, both with imbricate stacks overlying their forelimbs; the Sawtooth culmination can be considered a surface analogue to the Waterton field duplex. (b) The section was restored by reversing the process used in producing the model of Figure 3. The structurally highest and presumably the youngest imbricates, '1' were removed first. Then the amplitude of the leading duplex was decreased by restoring 4 frontal imbricates of the duplex. This process unfolded one or more overlying imbricates. Since the imbricates in Mesozoic section merge at depth with a Jurassic-level detachment, which roots westward into the Flathead duplex, imbricates of the Flathead duplex must be restored simultaneously with the imbricates overlying the frontal duplex. (c) The process is continued by removing the displacement on imbricate '2', while simlultaneously restoring an additional imbricate of the Flathead duplex. Also the amplitude of the Waterton duplex was further decreased by partially restoring imbricates '3'. (d) Between stages 'C' and 'D', the remaining duplex imbricates of Waterton structure were restored and the displacements on imbricates '4' were removed. Sequential restoration of this section using the synchronous thrust model indicates that the two duplex fault zones grew simultaneously.

more complex that the model, one can see that structure is broadly similar to the model which was produced by synchronous thrusting. The presence of the imbricate stacks on the forelimbs of the Waterton and frontal Sawtooth structures is thus interpreted to mean that major thrusts within the Canadian Front Ranges (on trend with the trailing duplex) and beneath the Foothills (on strike with the leading duplex) were simultaneously active. Although the southern Canadian Rocky Mountain and northwest Montana thrust belts may have initiated by a hinterland-to-foreland propagation of thrusting, the model requires that the earliest formed thrusts remain active, rather than being deactivated as previously inferred.

The same conclusions are reached by analysis of a section through the Bow Valley structure immediately northwest of Calgary (Fig. 1). Here all imbricates lie upon the crest and forelimb of the structure. The upper detachment within the Jurassic must have been active during emplacement of the Moose Mountain sheets. Since the upper detachment joins with the deeper portions of the McConnell thrust to the west, application of the synchronous thrust model (Fig. 3) requires the McConnell remained active or was reactivated during growth of the Bow Valley structure.

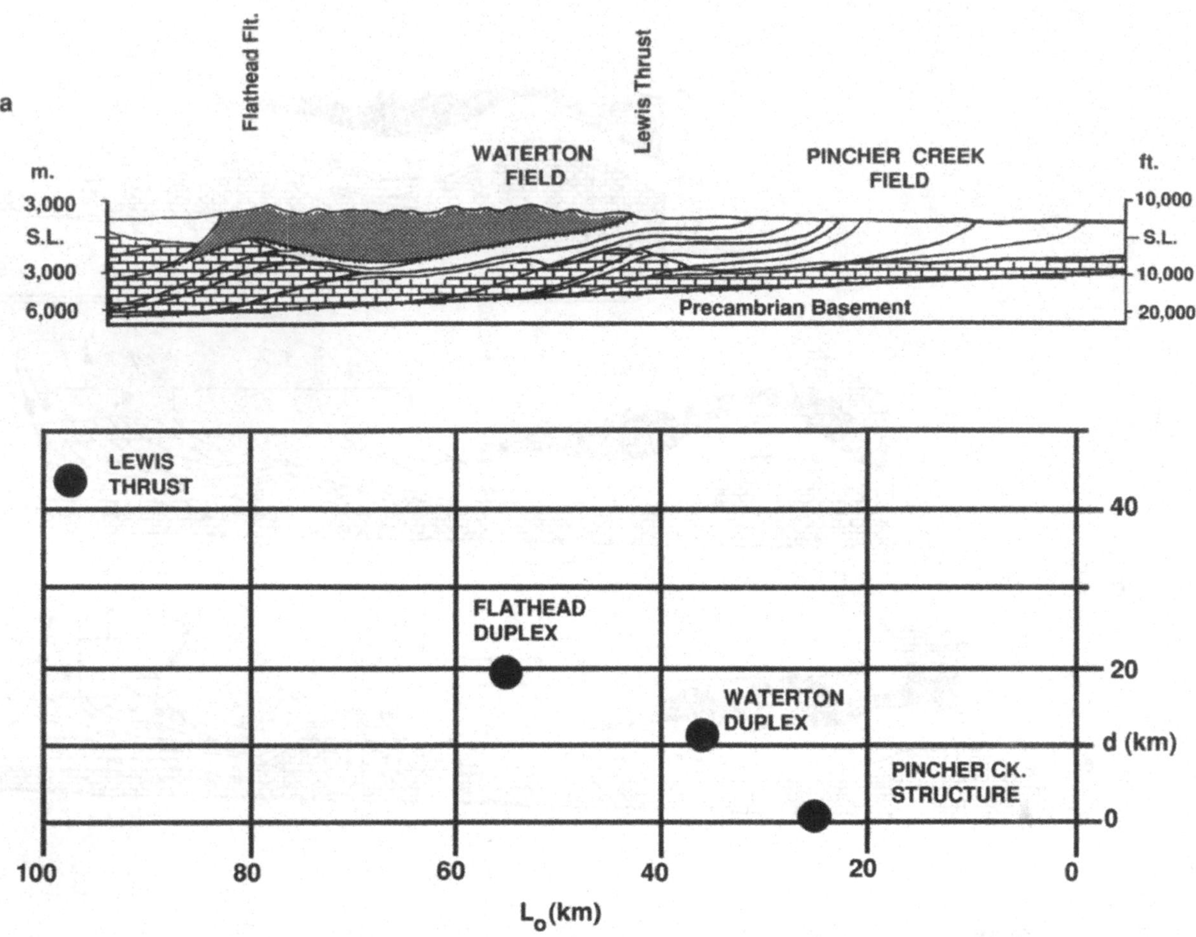

Figure 6. (**a**) Cross section through Waterton field (from Bally *et al.* 1966). There are four major thrust structures: 1) underlying Pincher Creek field is a frontal thrust with displacement of of less than 1 km, 2) Waterton field duplex whose imbricates have a combined shortening of approximately 11 km, 3) the Flathead duplex accounting for 20 km shortening, and 4) the Lewis thrust with a minimum displacement of 42 km. The amount of shortening was estimated from the sections of Bally *et al.* (1966). (**b**). The graph shows the amount of displacement or shortening on each of these four structures, plotted versus distance L_0. L_0 is the distance to each structure, measured in the restored state from the right end of the section. The systematic hindward increase in displacement suggests that existing western thrusts, such as the Lewis, remained active as new thrusts propagated toward the foreland.

Corroborating evidence for synchronous thrusting

Evidence for simultaneous motions on two or more major thrusts is corroborated by cross-strike displacement profiles (Fig. 6) and strike-parallel displacement transfer (Fig. 7).

Cross-strike displacement profiles

As shown in the model (Fig. 3), if the earliest formed thrust remains active as new imbricates propagate toward the foreland, the earliest thrust should have greater displacement. Inherent in this argument is the assumption that thrust displacement rates have remained constant during the life of a thrust belt. (Of course one could also argue that a hinterland increase in displacement may reflect higher rates of translation earlier in the thrust history.) A hinterland-to-foreland decrease in displacement is indicated for a cross-sections through the Canadian Rockies (Fig. 6), strongly suggesting that all the thrusts remained active, rather than being sequentially deactivated, as younger thrusts propagated from hinterland to foreland.

Displacement transfer

If one assumes that displacement in the northwest Montana and southern Alberta thrust belts is relatively constant along-strike and proceeded at constant rates, one is led to the possible conclusion that emplacement of the Lewis thrust sheet, in the vicinity of the US/Canada border, was balanced by shortening of the Palaeozoic strata in the the Sawtooth Range to the south. Displacement on the Lewis is on the order of 51 km (Price 1965, p. 115) to 65 km (Mudge & Earhart 1980, p. 12-13), with the maximum displacement occurring immediately south of the Canadian border. Approximately 250 km to the north the Lewis terminates in a fold complex at Mt. Kidd (Price *et al.* 1972; Stockmal 1979; see Fig. 11 in Boyer, 1986b) and the southern termination of the thrust lies in a fold approximately 170 km to the south (Mudge & Earhart 1980, 1983). In the central segment of the Sawtooth Range shortening in Palaeozoic sequence is on the order of 35

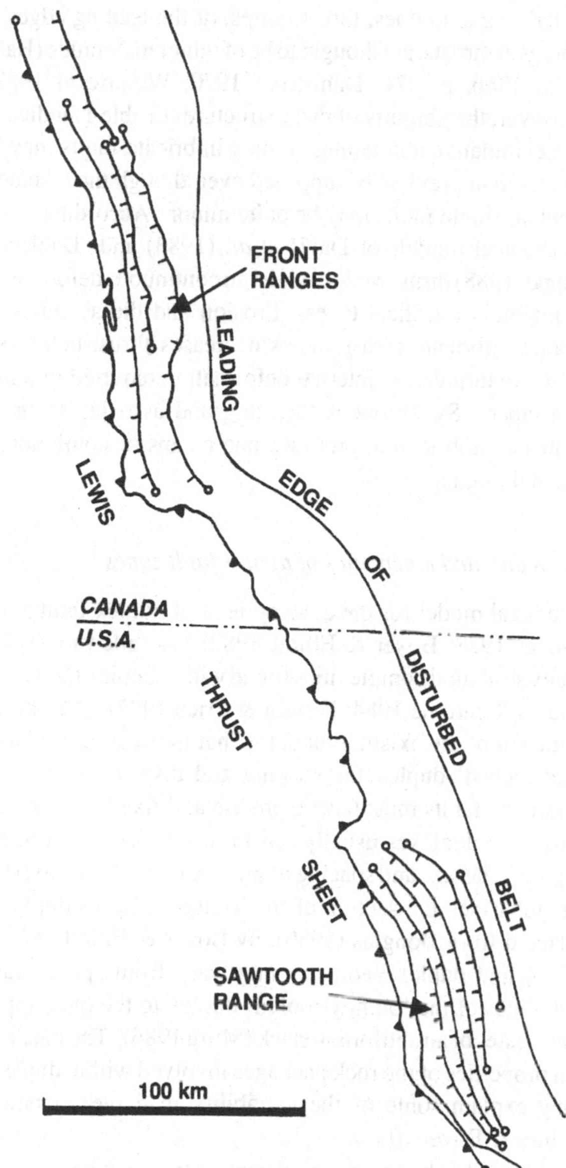

Figure 7. Transfer of displacement between thrust faults shortening the Palaeozoic sequence in the Sawtooth Range of Montana and the Lewis thrust transporting Middle Proterozoic Belt/Purcell Supergroup metasedimentary rocks. Thrusts of the Sawtooths terminate in folds south of Glacier National Park, while the Lewis begins in a fold and increases displacement northward toward the Canadian border. The US/Canada border is the approximate trace of a mirror plane of symmetry, to the north of which the Lewis loses displacement as the Front Range thrusts increase in shortening (see Dahlstrom 1970, Figs 27 & 46). Displacement transfer between the Lewis thrust and the Sawtooth and Front Ranges implies that the Lewis was active synchronously with imbricate shortening of Palaeozoic rocks in these ranges.

to 45 km (based on the sections of Mudge & Earhart 1983, and Mitra 1986). To the north, thrusts of the Sawtooths lose displacement into a number of fault-propagation folds (Fig. 4b; also see map of Mudge & Earhart 1983). The northward loss of 35 to 45 km of shortening in the Sawtooth Range is transferred into an additional 40 km of displacement on the Lewis, implying synchronous movement on the Lewis thrust and imbricate faults of the Palaeozoic carbonates to the south (Fig. 7).

Discussion

The synchronous thrust model (Fig. 3) explains many of the conflicts that have arisen among various schools of thought concerning thrust sequence. In the model, a hinterland-to-foreland sequence of propagation at depth, combined with continuing motion of existing thrusts, produces 'out-of-sequence' movements at more shallow levels. Working primarily with surface data Milici (1975) was most likely to observe 'out-of-sequence' geometries, the shallow manifestation of synchronous thrusting, whereas Dahlstrom (1970) and Bally *et al.* (1966), utilizing abundant subsurface control (seismic and well data), were more likely to be influenced by the geometry of deeper structures which suggest a foreland progression. Thus, both the cratonward and 'hindward' schools of thrust kinematics may have been in part correct in their observations and conclusions, as they were dealing with different parts of the same puzzle. Synchronous thrusting is compatible with observations that have been incorporated into the existing model of simple hinterland-to-foreland thrusting: folded thrusts, 'backlimb' thrusts, interaction of laterally propagating thrusts, multiple-stage thrusting at two structural levels, and the 'incidental' reactivation of existing thrusts.

Lateral thrust propagation

Jones (1971, p. 304) noted that apparent truncation patterns in map view led Choquette (1959) and Hunt (1956) to argue for a east-to-west sequence of thrusting. Building upon the arguments of Douglas (1958, p. 130) he argued that due to differences in propagation rate and displacement laterally propagating thrusts, developed in a hinterland-to-foreland fashion may intersect laterally to give the misleading appearance of the reverse sequence. Applying the concepts of Douglas (1958) and Jones (1971) one could argue that the southward propagating Lewis thrust and the northward propagating Sawtooth thrusts bypassed each other, while in mirror image to the north the Lewis bent to the west of the southward propagating thrusts of the Alberta Front Ranges. Does this mean that out-of-sequence and synchronous thrusting is limited to *narrow zones of displacement transfer* between overlapping faults? No, as the zones of overlap are far from limited; transfer between the Lewis and Front Range thrusts occurs over 175 km in Alberta (Dahlstrom 1970, Fig. 27) and 130 km between the Lewis and Sawtooth thrusts in northwest Montana (Mudge & Earhart 1980, Fig. 1). Thus the synchronous thrust model is compatible with the lateral thrust-propagation model of Douglas (1958) and Jones (1971): in any given cross section two or more thrusts must experience simultaneous motion.

Multiple-stage thrusting

This interpretation of imbricate stacks overlying structural culminations differs from that of Bally *et al.* (1966), who hinted at the relationship between structural culminations and intense imbrication in the overlying section: 'It appears that areas marked by a multitude of Mesozoic imbrications indicate the presence of carbonate sheets stacked on top of each other without intervening Mesozoic clastic strata.' In their interpretation of the Bow Valley structure they appealed to a two-stage thrusting wherein the detachment at within the Jurassic was active first, imbricating the Mesozoic sequence from footwall Palaeozoic carbonates (Bally *et al.* 1966, p.350-351, Fig.13, phases I & II). Subsequently faults broke from a sole thrust at the base of the Palaeozoic sequence and cut up section through the pre-existing upper detachment (Bally *et al.* 1966, p.350-351, Fig.13, phase III).

As an alternative to the sequential model of Bally *et al.*(1966) this paper argues that the imbricated Mesozoic clastics are found in conjunction with underlying multiply imbricated Palaeozoic carbonates because the Jurassic-level detachment was active during footwall imbrication as illustrated in the synchronous thrust model (Fig. 3). Furthermore, if the so-called 'back-limb thrusts' and the imbricates on the forelimbs of the structures are part of the same system, as proposed here, the combined displacement on these systems may not be as insignificant as previously supposed.

Thrust reactivation

One's intuition suggests that it is unrealistic to expect older thrusts to be deactivated as movement is transferred to younger thrusts: 'That the pre-existent movement planes in the deformed mass would remain totally inert during this transportation is inherently unlikely: some unsystematic incidental movement is to be expected.' (Dahlstrom 1970; p. 355). Periodic reactivation would likely result in out-of-sequence and backlimb thrusting. The thrust model of this paper invokes a continuous thrust reactivation, but the periods of reactivation are so closely spaced in time as to give the appearance of synchronous thrusting. Although the model involves alternating motion between two thrusts, when viewed in the context of geological time the motions can be thought of as simultaneous.

Implications

'Rules' of thrusting, the understanding of thrust-belt geometries and the hydrocarbon exploration strategies over the past 30 years have generally been based on the hinterland-to-foreland or 'piggy-back' model of thrusting. However, if the forelimb imbricates in the Rocky Mountains were formed in a systematic fashion indicative of synchronous thrusting, and if such structures are common in other orogens, various implications arise for the mechanical models of thrust sheet emplacement, the kinematic development of duplex fault zones, the interpretation of thrust sequence in other thrust belts, the balancing of cross sections, and the modelling of hydrocarbon generation and entrapment in thrust belts.

Mechanical models of thrust emplacement

Out-of-sequence imbricates, whether they be associated with underlying anticlines, thrust ramps, or the leading edges of emergent thrusts, are thought to be of minor magnitude (Bally *et al.* 1966, p. 371; Dahlstrom 1970; Woodward 1987). However, the ubiquity of these structures (Table 1) indicates that cumulative shortening of such imbricate faults may be greater than previously supposed even though the displacement on single faults may be quite minor. According to the mechanical models of Davis *et al.* (1983) and Dahlen & Suppe (1988) thrust wedges undergo continuous deformation to maintain a critical taper. Erosion and the accretion of frontal imbricate thrust sheets decreases thrust-belt taper, which in turn drives internal deformation required to maintain taper. Synchronous thrusting and associated out-of-sequence imbrication provides one means of continuously maintaining taper.

Geometry and kinematics of duplex fault zones

The ideal model for the development of duplex fault zones (Boyer 1978; Boyer & Elliott 1982) has failed to explain many structural complexities found within duplex fault zones (Davis & Jardine 1984; Fermor & Price 1987). A common criticism of the existing model is that in order to produce a 'flat-topped' duplex the spacing and displacement of the imbricate faults must have a precise and fixed relationship. Observed duplexes usually fail to match the model in this regard. The size and spacing of imbricates is often extremely irregular (see the section of the Waterton Lakes duplex as adapted from Douglas (1950), by Boyer & Elliott 1982, p. 1206) and duplex geometry may vary from spaced ramp anticlines of the 'bumpy-roofed' duplex to the overlapping imbricates of an antiformal stack (Mitra 1986). The mechanical properties of the rock packages involved within duplexes may explain some of the variability in duplex geometry (Mitra & Boyer 1986).

Alternatively, these various geometries can be produced as a continuous process during synchronous thrusting (Fig. 8). If thrust displacement ceases early in the development of synchronous thrust duplexes, a bumpy-roof results. Should shortening continue or be resumed a flat-roof duplex may be produced. Additional shortening would pile the imbricates to form an antiformal stack. Note that synchronous-thrust duplexes can easily be distinguished from 'standard' duplexes produced by a simple hinterland-to-foreland sequence of imbrication; the former will have imbricate faults in the hanging wall of the roof thrust (Fig. 8) whereas the latter do not (compare Fig. 8 of this paper with Fig. 19 of Boyer & Elliott 1982).

Sequence of thrusting in other thrust belts

The simultaneous thrust/fold model of Gilluly (1960; see Figs 2 & 3 of this paper) is substantiated by the existence of eye-lid windows. Oriel's (1951) description of the Hot Springs eye-lid window (North Carolina) indicates that it could be interpreted using Gilluly's model, and Rodgers (1970, p. 170) argued for a break-back sequence above possible eye-lid windows of the southern Appalachians, such

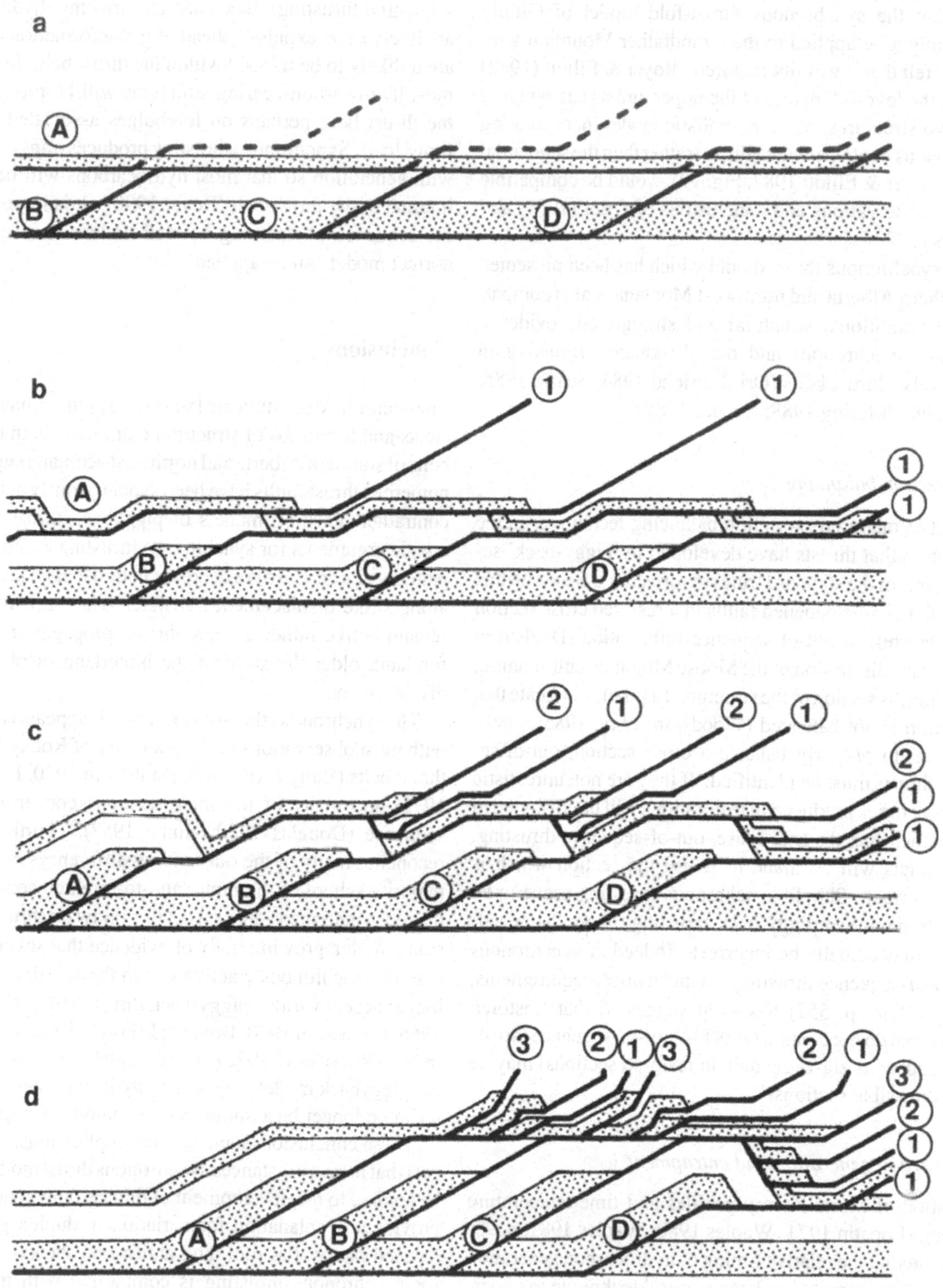

Figure 8. Formation of a duplex fault zone by synchronous thrusting. The model of Figure 3 has been applied to four imbricate faults. With continuous movement on the four thrusts the amplitude of each ramp anticline increases and younger out-of-sequence imbricate faults break across the crest of each structure. Note that in the initial stages of displacement (b) ramp anticlines are spaced, but with increased shortening synclines tighten (c) and eventually the synclines disappear as a flat-topped duplex is formed (d). With additional displacement the flat-topped duplex of section d would be transformed into an antiformal stack (Boyer & Elliott 1982, Fig. 12; Mitra 1986, Fig. 4). The numbers appended to the out-of-sequence imbricates indicate the relative ages of these faults, '1' being the oldest and '3' the youngest. Note that the imbricates comprise three distinct packages, each of which contains imbricates that become younger toward the hinterland. A 'piggy-back' duplex can be distinguished from a 'synchronous-thrust' duplex by the lack or presence of minor imbricate faults in the hangingwall of the roof thrust. Compare the evolution of the synchronous thrust duplex with that of a 'normal' duplex (Fig. 19 of Boyer & Elliott 1982).

as the Mountain City window. Rodgers (1970, p. 173) also noted that the synchronous thrust-fold model of Gilluly (1960) might be applied to the Grandfather Mountain window but felt that it was not required. Boyer & Elliott (1982) ignored the 'eye-lid' nature of the upper thrusts surrounding these two structures. A more realistic model incorporating synchronous duplex models (Fig. 8) rather than the sequential model (Boyer & Elliott 1982, Fig. 19), would be compatible with the observations and conclusions of Oriel (1951) and Rodgers (1970).

The synchronous thrust model which has been presented for southern Alberta and northwest Montana is also compatible with additional structural and stratigraphic evidence indicating synchronous and out-of-sequence thrusting in several other thrust belts (Ori & Friend 1984; Searle 1985; Butler 1987; Morley 1988; Roeder 1988).

Cross-section balancing

Inherent in many cross-section balancing techniques is the assumption that thrusts have developed in 'piggy-back' sequence, one of the standard 'rules of thrusting'. Based on this assumption, oversteepened faults in a restored cross section may either reflect out-of-sequence imbrication (Dahlstrom 1970, see his discussion of the Moose Mountain culmination, p. 353, and his section of the structure, Fig. 16) or indicate that the section is not balanced (Woodward *et al.* 1985; 1989). Therefore, to properly balance a cross section, out-of-sequence thrusts must be identified. If they are not, unrealistic fault trajectories produced in restorations will be attributed to error, rather than the true cause, out-of-sequence thrusting, and attempts will be made to revise the section which is already correct. Should synchronous thrusting prove to be more common than piggy-back thrusting, many 'balanced' sections may actually be incorrect. Indeed, if synchronous and out-of-sequence thrusting are mechanical requirements, Morley (1988, p. 557) has even suggested that 'restored sections that do not contain OOST's (out-of-sequence thrusts which appear as zig-zag patten in restored sections) may be the inadmissible sections!'

Hydrocarbon generation and entrapment

Utilization of burial history profiles and time-temperature indexing (Lopatin 1971; Waples 1980; Waples 1981, p. 95-106) to predict the relative timing of hydrocarbon generation and trap development has been a useful technique in understanding existing hydrocarbon-bearing structures and in the exploration for new fields (Warner 1983). However, the accuracy of the technique is dependent upon a correct knowledge of the sequence of thrusting (Boyer 1990). Most models of hydrocarbon generation and entrapment in thrust belts assume a simple hinterland-to-foreland sequence of thrusting, a model that has been reinforced by more than 30 years of thrust belt exploration. However, all other factors being equal, synchronous thrusting will produce a different timing between hydrocarbon generation and entrapment than will sequential thrusting. In sequential thrusting, hydrocarbons are likely to be expulsed ahead of the deformation front and are unlikely to be trapped within the thrust belt. In this case most hydrocarbon-bearing structures will be found outside the thrust belt, perhaps on forebulges associated with the thrust load. Synchronous thrusting produces traps concurrent with generation so that most hydrocarbons will be trapped internal to the thrust-belt (Boyer 1990). It is important that the sequence of thrusting be well understood, so that the correct model can be applied.

Conclusions

The occurrence of imbricate faults sited predominantly on the crests and forelimbs of structural culminations in the thrust belts of southern Alberta and northwest Montana suggest that numerous thrust faults have been synchronously active. This contradicts existing models of piggyback thrusting. Kinematic arguments for synchronous thrusting are fortified by patterns of a cross-strike decrease in thrust displacement and along-strike displacement transfer. When existing thrusts remain active rather as new thrusts propagate toward the foreland, older thrusts near the hinterland display greater displacement.

The synchronous thrust model (Fig. 3) appears compatible with most observations of the geometry of Rocky Mountain thrust belts (Bally *et al.* 1966; Dahlstrom 1970; Royse *et al.* 1975) and many of the interpretations concerning thrust sequence (Douglas 1950; Hume 1957). Furthermore, it reconciles many of the outstanding differences between the majority school of the hinterland-to-foreland sequence and the much maligned adherents of the hindward thrust progression. With a growing body of evidence that several thrusts may be synchronously active within thrust belts, we can no longer accept various piggy-back thrust models (Bally *et al.* 1966; Dahlstrom 1970; Boyer & Elliott 1982) as constituting 'rules' (Royse *et al.* 1975) of thrust belt behaviour. Although the piggyback model may still apply in a number of settings it can no longer be assumed to have universal applicability.

These conclusions have several implications. They suggest that in many instances synchronous thrust models should be applied to the development of duplex fault zones, as they provide an explanation for variation in duplex geometries (spaced ramps, flat-roof, and antiformal stacks). Evidence for synchronous thrusting is compatible with mechanical models of thrust emplacement (Davis *et al.* 1983), which require that thrust belts shorten internally to balance decreased taper resulting from erosion and frontal accretion. There are economic implications of this model as well. Most basin modelling and thermal maturation profiling routines in thrust belts are based on the piggyback model and therefore need to be modified when applied to regions of synchronous thrusting.

The observations which led to this paper resulted from discussions with Ricardo Presnell during the early 1980s, when he and I were employed by SOHIO (Standard Oil Production Co., now BP Exploration) in Denver. While we were attempting to interpret a number of seismic lines and tie geological maps to subsurface data, I was struck by the odd appearance of a 'folded' imbricate stack (Fig. 5, as mapped by Mudge & Earhart 1983) which brought to mind the work of Gilluly (1960) in Nevada. Those observations led to this paper. I am grateful for nine years of employment in the petroleum industry which permitted me to work throughout the thrust belts of the western US. Most beneficial has been the fieldwork conducted while employed by ARCO during 1978 and 1979 and the seismic interpretation and cross-section construction performed with SOHIO from 1981 to 1987.

I wish to acknowledge my gratitude to all my industry friends and colleagues with whom I worked during this period, as much of my past and present thrust-belt research has benefited from this collaboration. Observations made in this paper served as the nucleus for research directed toward evaluating the applicability of the 'critically-tapered-wedge' model to thrust tectonics in the Rocky Mountains. I gratefully acknowledge the National Science Foundation which is supporting this research under grants EAR-88-03623 and EAR-90-04303. I wish to thank Bob Holdsworth and Rob Butler, who reviewed the manuscript and provided much constructive criticism. I've attempted to incorporate as many of their suggestions as possible and hope that the result is a more comprehensible paper.

References

Armstrong, F. C. & Oriel, S. S. 1965. Tectonic development of Idaho-Wyoming thrust belt. *American Association of Petroleum Geologists Bulletin*, **49**, 1847-1866.

Bally, A. W., Gordy, P. L. & Stewart, G. A. 1966. Structure, seismic data, and orogenic evolution of southern Canadian Rocky Mountains. *Bulletin of Canadian Petroleum Geology*, **14**, 337-381.

Boyer, S. E. 1978. Structure and origin of the Grandfather Mountain window, North Carolina. *Ph.D. thesis, Johns Hopkins University.*

—— 1986a. Geometric evidence for simultaneous, rather than sequential, movement on major thrusts; implications for duplex development. *Geological Society of America Abstracts with Programs*, **18**, 549.

—— 1986b. Styles of folding within thrust sheets: examples from the Appalachian and Rocky Mountains of the USA and Canada. *Journal of Structural Geology*, **8**, 325-339.

—— 1990. New models for thrust kinematics and implications of the generation, migration, and entrapment of hydrocarbons in fold and thrust belts (abst.). *American Association of Petroleum Geologists Bulletin*, **74**, 617.

—— & Elliott, D. 1982. Thrust systems. *American Association of Petroleum Geologists*, **66**, 1196-1230.

—— & Geiser, P. 1987. Sequential development of thrust belts: implications for mechanics and cross section balancing. *Geological Society of America Abstracts with Programs*, **19**, 597.

Butler, R.W.H. 1987. Thrust sequences. *Journal of the Geological Society, London*, **144**, 619-634.

Choquette, A. L. 1959. Theoretical approach to foothills and mountain deformation of western Canada. *Alberta Society of Petroleum Geologists Journal*, **7**, 234-237.

Choukroune, P. & Seguret, M. 1973. Tectonics in the Pyrenees: role of compression and gravity. *In*: De Jong, K. A. & Scholten, R. (eds) *Gravity and Tectonics*. Wiley, New York, 141-156.

Dahlen, F. A. & Suppe, J. 1988. Mechanics, growth, and erosion ;of mountain belts. *Geological Society of America Special Paper*, **218**, 161-178.

Dahlstrom, C.D.A. 1969. Balanced cross sections. *Canadian Journal of Earth Sciences*, **6**, 743-757.

—— 1970. Structural geology in the eastern margin of the Canadian Rocky Mountains. *Bulletin of Canadian Petroleum Geology*, **18**, 332-406.

Davis, D., Suppe, J. & Dahlen, F. A. 1983. Mechanics of fold-and-thrust belts and accretionary wedges. *Journal of Geophysical Research*, **88**, 1153-1172.

Davis, G. A. & Jardine, E. A. 1984. Preliminary studies of the geometry and kinematics of the Lewis allochthon, Saint Mary Lake to Yellow Mountain, Glacier National Park, Montana. *Montana Geological Society Field Conference & Symposium, Northwest Montana & Adjacent Canada*, 201-209.

Deiss, C. F. 1943. Stratigraphy and structure of southwest Saypo quadrangle, Montana. *Geological Society of America Bulletin*, **54**, 205-262.

Delphia, J. G. & Bombolakis, E. G. 1988. Sequential development of a frontal ramp, imbricates, and a major fold in the Kemmerer region of the Wyoming thrust belt: *Geological Society of America Special Paper*, **222**, 207-222.

Douglas, R.J.W. 1950. Callum Creek, Langford Creek, and Gap map areas, Alberta. *Geological Survey of Canada Memoir 255*, 124p.

—— 1958. Mount Head map-area, Alberta. *Geological Survey of Canada Memoir 291*.

Erdman, O. A., Belot, R. E. & Slemko, W. 1953. Pincher Creek area, Alberta. *Alberta Society of Petroleum Geologists 3rd Annual Field Conference and Symposium*, 139-157.

Fermor, P. R. & Price, R. A. 1987. Multiduplex structure along the base of the Lewis thrust sheet in the southern Canadian Rockies. *Bulletin of Canadian Petroleum Geology*, **35**, 159-185.

Fox, F. G. 1959. Structure and accumulation of hydrocarbons in southern foothills, Alberta, Canada. *Bulletin of the American Association of Petroleum Geologists*, **43**, 992-1025.

Gallup, W. B. 1955. Geology of the Pincher Creek gas and naptha field and its regional implications. *Billings Geological Society 6th Annual Field Conference*, 150-159.

Gordy, P. L., Frey, F. R. & Norris, D. K. 1977. Geological field guide for the CSPG 1977 Waterton-Glacier Park Field Conference. *Canadian Society of Petroleum Geologists*, Calgary.

Gilluly, J. 1960. A folded thrust in Nevada - inferences as to time relations between folding and faulting. *American Journal of Science*, **258-A**, 68-79.

Hage, C. O. 1942. Folded thrust faults in Alberta Foothills west of Turner Valley. *Royal Society of Canada Transactions*, 3rd series, section 4, **36**, 67-78.

Hake, B. F., Willis, R. & Addison, C. C. 1942. Folded thrust faults in the Foothills of Alberta. *Geological Society of America Bulletin*, **53**, 291-334.

Heller, P. L., Bowdler, S. S., Chambers, H. P., Coogan, J. C., Hagen, E. S., Shuster, M. W., Winslow, N. S. & Lawton, T. F. 1986. Time of initial thrusting in the Sevier orogenic belt, Idaho-Wyoming and Utah. *Geology*, **14**, 388-391.

Hennessey, W. J. 1975. A brief history of Savanna Creek gas field. *Canadian Society of Petroleum Geology Guidebook, Exploration Up-date '75*, 18-21.

Hubbert, M. K. & Rubey, W. W. 1959. Role of fluid pressure in mechanics of overthrust faulting: I. Mechanics of fluid-filled porous solids and its applications to overthrust faulting. *Geological Society of America Bulletin*, **70**, 115-166.

Hume, G. S. 1957. Fault structure in the Foothills and eastern Rocky Mountains of southern Alberta. *Geological Society of America Bulletin*, **68**, 395-412.

Hunt, C. W. 1956. Panther dome: a minor orogen of the Canadian cordillera. *Alberta Society of Petroleum Geologists 6th Annual Field Conference Guidebook*, 44-55.

Hurst, D. J. & Steidtmann, J. R. 1986. Stratigraphy and tectonic significance of the Tunp conglomerate in the Fossil Basin, southwest Wyoming. *The Mountain Geologist*, **23**, 6-13.

Jones, P. B. 1971. Folded faults and sequence of thrusting in Alberta foothills. *American Association of Petroleum Geologists Bulletin*, **55**, 292-306.

Knipe, R. J. 1985. Footwall geometry and the rheology of thrust sheets. *Journal of Structural Geology*, **7**, 1-10.

Lageson, D. R. 1984. Structural geology of Stewart Peak culmination, Idaho-Wyoming thrust belt. *American Association of Petroleum Geologists Bulletin*, **68**, 401-416.

Lamerson, P. R. 1982. The Fossil basin and its relationship to the Absaroka thrust system, Wyoming and Utah. *Rocky Mountain Association of Geologists, Geologic Studies of the Cordilleran thrust belt*, 279-340.

LeMoine M. 1973. About gravity gliding tectonics in the western Alps: *In*: De Jong, K. A. & Scholten, R. (eds) *Gravity and Tectonics*. Wiley, New York, 201-216.

Lopatin, N. V. 1971. Temperature and geologic time as factors in coalification (in Russian). *Izv. Akad. Nauk SSSR, Seriya geologicheskaya*, **3**, 95-106.

Milici, R. C. 1975. Structural patterns in the southern Appalachians: evidence for a gravity slide mechanism for the Alleghanian deformation. *Geological Society of America Bulletin*, **86**, 1316-1320.

Mitra, G. & Boyer, S. E. Energy balance and deformation mechanisms of duplexes. *Journal of Structural Geology*, **8**, 291-304.

Mitra, S. 1986. Duplex structures and imbricate thrust systems: geometry, structural position, and hydrocarbon potential. *American Association of Petroleum Geologists Bulletin*, **70**, 1087-1112.

Morley, C. K. 1988. Out-of-sequence thrusts. *Tectonics*, **7**, 539-561.

Morse, J. 1977. Deformation in ramp regions of overthrust faults: experiments with small-scale rock models. *Wyoming Geological Association 29th Annual Field Conference Guidebook*, 457-470.

Mudge, M. R. 1970. Origin of the Disturbed Belt in northwestern Montana. *Geological Society of America Bulletin*, **81**, 377-392.

—— & Earhart, R. L. 1980. The Lewis thrust fault and related structures in the Disturbed Belt, northwestern Montana. *US Geological Survey Professional Paper 1174* , 18p.

Mudge, M. & Earhart, R. L. 1983. Bedrock geologic map of part of the northern Disturbed Belt, Lewis and Clark, Teton, Pondera, Glacier, Flathead, Cascade, and Powell Counties, Montana. *US Geological Survey Map I-1375*, scale 1:125,000.

Nickelsen, R. P. 1988. Structural evolution of folded thrusts and duplexes on a first-order anticlinorium in the Valley and Ridge Province of Pennsylvania. *Geological Society of America Special Paper*, **222**, 89-106.

Ori, G. G. & P. F Friend. 1984. Sedimentary basins formed and carried piggyback on active thrust sheets. *Geology*, **12**, 475-478.

Oriel, S. S. 1951. Structure of the Hot Springs window, Madison County, North Carolina. *American Journal of Science*, **249**, 1-30.

—— & Armstrong, F. C. 1966. Times of thrusting in Idaho-Wyoming thrust belt: Reply. *American Association of Petroleum Geologists Bulletin*, **50**, 2614-2621.

Platt, J. P. 1988. The mechanics of frontal imbrication: a first-order analysis. *Geologische Rundschau*, **77**, 577-589.

Price, R. A. 1965. Flathead map-area, British Columbia and Alberta. *Geological Survey of Canada Memoir 336*, 221p.

—— 1973. Large-scale gravitational flow of supracrustal rocks, southern Canadian Rockies: *In*: De Jong, K. A. & Scholten, R. (eds) *Gravity and Tectonics*. Wiley, New York, 491-502.

—— 1981. The southern foreland thrust and fold belt in the southern Canadian Rocky Mountains: *In*: McClay, K. R. & Price, N. J. (eds) *Thrust and Nappe Tectonics*, Geological Society of London Special Publication, **9**, 427- 448.

—— *et al.* 1972. The Canadian Rockies and tectonic evolution of the southeastern Canadian Cordillera. *XXIV International Geological Congress*, Excursion AC 15.

Rich, J. L. 1934. Mechanics of low-angle overthrust faulting illustrated by Cumberland thrust block, Virginia, Kentucky and Tennessee. *American Association of Petroleum Geologists Bulletin*, **18**, 1584-1596.

Rodgers, J. 1970. *The tectonics of the Appalachians*. John Wiley & Sons, New York, 271p.

Roeder, D. 1988. Andean-age structure of Eastern Cordillera (Province of La Paz, Bolivia). *Tectonics*, **7**, 23-39.

Royse, F., Jr., Warner, M. & Reese, D. L. 1975. Thrust belt structural geometry and related stratigraphic problems, Wyoming-Idaho-northern Utah: *In*: Bolyard, D.W. (ed.) Deep Drilling Frontiers of the Central Rocky Mountains. *Rocky Mountain Association of Geologists*, 41-54.

Scholten, R. 1973. Gravitational mechanisms in the northern Rocky Mountains of the United States: *In*: De Jong, K. A. & Scholten, R. (eds) *Gravity and Tectonics*. Wiley, New York, 473-489.

Scott, J. C. 1951. Folded faults in Rocky Foothills of Alberta, Canada. *American Association of Petroleum Geologists Bulletin*, **35**, 2316-2347.

Searle, M. P. 1985. Sequence of thrusting and origin of culminations in the northern and central Oman Mountains. *Journal of Structural Geology*, **7**, 129-143.

Serra, S. 1977. Styles of deformation in ramp regions of overthrust faults. *Wyoming Geological Association 29th Annual Field Conference Guidebook*, 487-498.

Stockmal, G. S. 1979. Structural geology of the northern termination of the Lewis thrust, Front Ranges, southern Canadian Rocky Mountains. M.Sc. thesis, The University of Calgary, Alberta (unpubl.).

Suppe, J. 1983. Geometry and kinematics of fault-bend folding. *American Journal of Science*, **283**, 684-721.

Verrall, P. 1968. Observations on geological structure between the Bow and North Saskatchewan rivers. *Alberta Society of Petroleum Geologists 16th Annual Field Conference Guidebook*, 106-118.

Waples, D. 1980. Time and temperature in petroleum exploration: application of Lopatin's method to petroleum exploration. *American Association of Petroleum Geologists Bulletin*, **64**, 916-926.

—— 1981. *Organic geochemistry for exploration geologists*. Burgess Publishing Co., 95-106.

Warner, M. A. 1983. Source and time of generation of hydrocarbons in the Fossil Basin, western Wyoming thrust belt. *Rocky Mountain Association of Geologists Geologic Studies of the Cordilleran thrust belt*, 805-815.

Wiltschko, D. V. & Dorr, J. A., Jr. 1983. Timing of deformation in overthrust belt and foreland of Idaho, Wyoming, and Utah. *American Association of Petroleum Geologists Bulletin*, **67**, 1304-1322.

Woodward, N. B. 1987. Geological applicability of critical-wedge thrust-belt models. *Geological Society of America Bulletin*, **99**, 827-832.

——, Boyer, S. E. & Suppe, J. 1985. An outline of balanced cross-sections. *University of Tennessee Studies in Geology*, **11**, 2nd edition, 170p.

——, —— & ——. 1989. Balanced geological cross-sections. *American Geophysical Union Short Course in Geology*, **6**, 132p.

The analysis of fracture systems in subsurface thrust structures from the Foothills of the Canadian Rockies

Mark Cooper

BP Resources Canada Limited, #2100, 855-2nd Street SW, Calgary, Alberta, Canada T2P 4J9

Abstract: The Foothills of the Canadian Rocky Mountains are one of the classic examples of a thin-skinned fold-thrust belt. A regionally significant detachment in the shales of the Jurassic Fernie Formation separates the intensely folded and imbricated clastics of the Jurassic and Cretaceous from the thrust structures in the prospective dolomite reservoirs of the Triassic and Upper Palaeozoic. As the foothills are traced northwards from Alberta to British Columbia the amount of displacement on the thrusts in the foothills decreases. Occasionally, far-travelled thrust sheets achieve sufficient elevation above regional to cause outcrop of the reservoir units. In these structures, which are analogues of the prospective structures at depth, the mesoscopic structures and fabrics can be studied; the opportunities to examine the structures in the subsurface by remote techniques has been much more limited. The advent of new well-logging technology such as the Formation Microscanner allows the geometry of subsurface foothills structures to be analysed in considerable detail. The exact geometric relationships of both bedding and fracture systems can be elucidated in-situ and correlated with core data. Two case studies using these techniques are presented; a tip fold from the British Columbia Foothills and a ramp anticline from the Alberta Foothills. The analysis reveals a systematic relationship between fracture set orientation and density to both structural geometry and lithological variations.

The Foothills of the Canadian Rocky Mountains have been explored for oil and gas since the early part of this century. This has resulted in an extensive literature on the fold and thrust belt which includes some of the classic papers on thrust geometry and techniques of analysis (e.g. Douglas 1950, 1958; Bally *et al.* 1966; Dahlstrom 1969, 1970; Price & Mountjoy 1970 and Thompson 1979). The Canadian Rockies rarely exceed 200 km in width and form part of the 5000 km long North American Cordilleran fold and thrust belt that extends from SE California to Alaska. Development of the fold and thrust belt commenced in the Late Jurassic and terminated in the Eocene, and is conventionally regarded as the result of the accretion of a series of exotic terranes to the Pacific Margin of North America (Monger *et al.* 1982). A good, recent review of the fold and thrust belt is given by McMechan & Thompson (1989).

The Foothills belt lies between the Front Ranges, a series of large thrust sheets of Palaeozoic rocks to the west, and the largely undeformed Western Canadian Sedimentary Basin to the east (Figs 1 & 2). The eastern boundary is characterized by a frontal monocline and/or a triangle zone (Jones 1982) which becomes a less reliable indicator of the limit of deformation as the belt is traced northwards. The boundary with the Front Ranges is characterized by a major displacement thrust with Palaeozoic and/or Proterozoic strata in the hangingwall (e.g. the McConnell and Lewis thrusts, Fig. 1). Within the Foothills belt thrust sheets with outcrops of Upper Palaeozoic rocks are locally observed (e.g. the Livingstone and Brazeau Thrusts, Fig. 1) but surface structure is dominated by numerous folds and small displacement imbricate thrusts within the Mesozoic clastics. The intensity of the folding and the magnitude of the thrust displacements generally decreases to the north.

In Alberta the closely spaced folds and imbricates in the Mesozoic clastic rocks obscure the simpler geometry of the Palaeozoic carbonate thrust sheets in the subsurface. The abundance of seismic and well data from gas exploration activities provides the constraints to develop quite detailed geometric models of these structures. The Palaeozoic carbonate thrust sheets tend to be widely spaced, relatively simple structures developed above shallow angle ramps and often have significant displacements. This is in contrast to the closely spaced, steeply dipping ramp anticlines and tip folds of the Mesozoic and implies a detachment in the Jurassic Fernie Formation between these two structural regimes.

As the foothills are traced northwards into British Columbia (Fig. 2), the amount of displacement on the thrusts gradually decreases, thrusts at surface are less common and folds predominate (McMechan & Thompson 1989). The change in style is gradual and is concomitant with a more diffuse eastern limit of deformation. Small displacement thrusts at depth persist much further to the east but have little expression at the surface. In the southern foothills of NE British Columbia, the primary exploration targets are the carbonates of the Upper Triassic which are deformed into a series of northeast vergent folds with steeply dipping northeast limbs that are locally overturned (Barss & Montandon 1981).

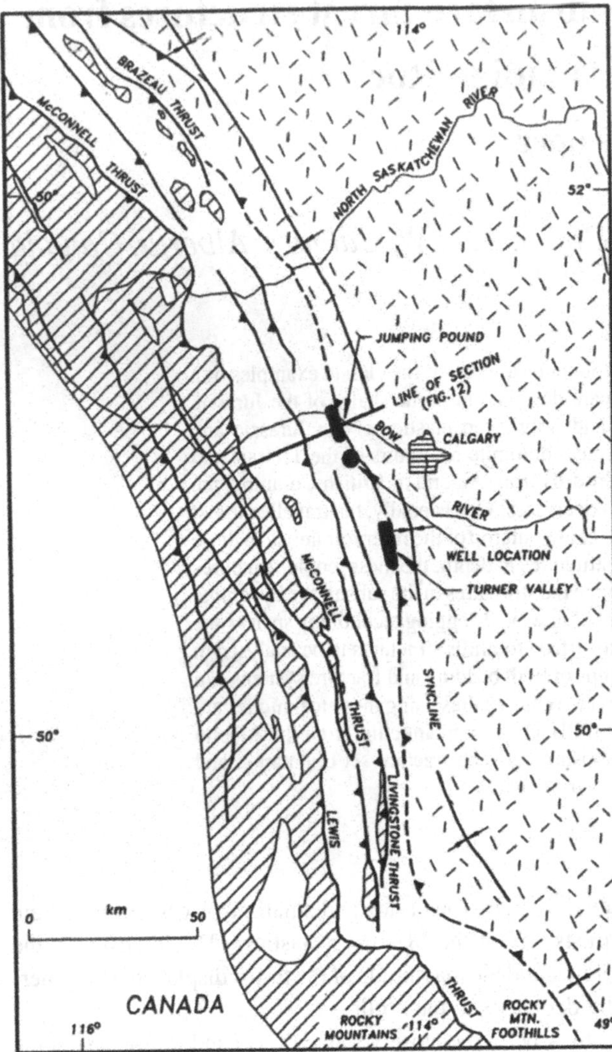

Figure 1. Simplified geological map of the Rocky Mountain Foothills of southern and central Alberta. The major structural features are shown on the map together with the locations of Figure 12 and the Example 2 well. The stratigraphic subdivisions shown on the map are defined on Figure 13.

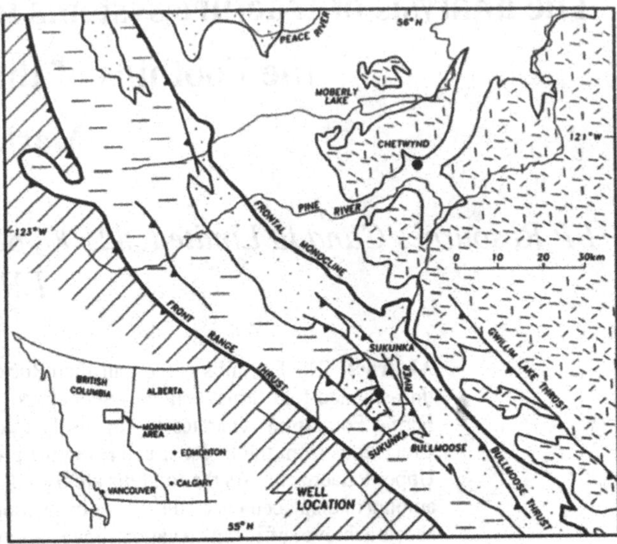

Figure 2. Simplified geological map of the Monkman area showing the location of the Example 1 well. The stratigraphic subdivisions used on the map are defined on Figure 3. The narrow foothills belt in this area lies between the Front Range Thrust and the Frontal Monocline.

Fracture development in thrust structures

Several workers have studied the geometry of fractures associated with thrusts and thrust related folds. Muecke & Charlesworth (1966) described folded Cardium sandstones from the Canadian Rocky Mountain foothills that contain three fracture sets perpendicular to bedding; one set parallels the orogenic strike and the other sets form a conjugate set with an acute bisectrix perpendicular to the fold axis. Norris (1971) has described the Type 1 and Type 2 fractures of Stearns (1968) in the northeast facing Castle River fold in the lower Cretaceous Blairmore Formation from the foothills of southern Alberta. Similar fracture patterns have also been noted by Price (1967) from a variety of folds and lithologies in the same general area. The nomenclature of Stearns (1968) was developed primarily on the Teton Anticline in the Sawtooth Range thrust belt in Montana; the Type 1 fractures comprise an extension fracture set perpendicular to the fold axis with an associated conjugate shear set whose acute bisectrix coincides with the orientation of the extension fracture set. The Type 2 fractures of Stearns (1968) comprise similar sets of fractures except that the extension fracture set is parallel to the fold axis and perpendicular to bedding. Stearns (1968) notes that the Type 2 fractures tend to occur in localized swarms whereas the Type 1 fractures are more uniformly distributed. Friedman (1969) described fractures associated with the development of the Owl Ridge thrust fault in Southern California, where two of the four fracture sets parallel the thrust fault. The other two sets have the same strike but a shallower dip and are interpreted as a conjugate set of shear fractures. McQuillan (1973, 1974) analysed the development of fractures associated with thrust related folds in the Zagros Mountains of SW Iran. When the fracture patterns were studied on air photos for one of these folds a simple orthogonal pattern of fractures oriented parallel and perpendicular to the fold axis emerges. These fractures are assumed to be perpendicular to bedding. The density of these fractures is higher in the plunging axis of the fold and in the crestal regions. In the field small scale fractures show no clear orientation distribution; McQuillan considered that the small scale fractures were earlier than the folding but that the larger scale fractures were intimately related to the development of the folds. Similar orthogonal patterns to those described by McQuillan (1974) have also been recorded in the Valley and Ridge of the Appalachians (Wiltschko *et al.* 1985; Mitra 1988). The state of strain in thrust sheets has been studied extensively, both theoretically (Berger & Johnson 1980; Sanderson 1982) and in the field (Allmendinger 1982; Spang & Groshong 1981; Spang *et al.* 1981; Wiltschko *et al.* 1985). Allmendinger (1982) noted that rotation of beds into overturned fold limbs can cause compression at a high angle to bedding to be superimposed on earlier layer parallel compression. Wiltschko *et al.* (1985) measured strain by calcite twin lamellae in the Mississippian Newman Limestone of the Pine Mountain block which shows early layer parallel shortening and subsequent inter-bed shear. Spang & Groshong (1981), also working in the Appalachians, concluded that the maximum compressive strain was always sub-parallel to the layering and perpendicular to the fold axis with the maximum extension sub-parallel to the fold axis. A very similar observation was made by Spang *et al.* (1981) above thrust ramps in Devonian and Mississippian carbonates of the Front Ranges in the Canadian Rockies where the maximum principal strain (averaging 6%) is layer parallel or at a small angle to bedding. This provides particularly useful and relevant data on the state of strain in structures analogous to those in the foothills described in this paper.

LITHOLOGY	FORMATION THICKNESS	AGE	MAP
	DUNVEGAN 220m	UPPER CRETACEOUS	
	CRUISER 170m		
	GOODRICH 300m		
	HASLER 270m		
	BOULDER CREEK 160m		
	HULCROSS 100m		
	GATES 200m	LOWER CRETACEOUS	
	MOOSEBAR 220m		
	GETHING 400m		
	CADOMIN 40m		
	NIKANASSIN 1700m		
	FERNIE 300m	JURASSIC	
	NORDEGG 15m		
	PARDONET 55m	TRIASSIC	
	BALDONNEL 65m		
	CHARLIE LAKE 350m		
	HALFWAY 70m		
	TOAD-GRAYLING 450m		
	BELLOY 30m	PERMIAN	
	DEBOLT	MISSISSIPPIAN	
	SHUNDA		
	PEKISKO		
	BANFF 200m		
	EXSHAW 10m		
	WABAMUN 100–350m	DEVONIAN	

Figure 3. Mesozoic and Upper Palaeozoic stratigraphy of the Monkman area. The shading within the map column shows the stratigraphic subdivisions used in Figure 2.

Example 1: Fracture development in a tip fold

Regional setting

This example is located in the southern portion of the NE British Columbia sector of the foothills in the Monkman area (Fig. 2) where the prolific gas fields of Sukunka and Bullmoose occur (Barss & Montandon 1981). The reservoir units in this area are the dolomites of the Upper Triassic Pardonet and Baldonnel Formations (Fig. 3). The Triassic is deformed by a series of thrust imbricates, which detach in the anhydrite beds of the Charlie Lake Formation, and tip (fault propagation) folds; typical sections through these structures are given by Barss & Montandon (1981). The tip folds have up to 800 m of amplitude and usually involve the Halfway Formation (Fig. 3) which is penetrated in structure in some of the wells. The geometric constraints provided by the well and seismic data suggest that the detachment is within the shales of the Toad-Grayling Formation (Fig. 3). The seismic mapping of the Triassic structures is complicated by the presence of detachments within the numerous shale units of the Upper Cretaceous section (McMechan 1985, and Fig. 3) and the strongly deformed sands and shales of the Nikanassin Group. The detailed 1:50000 scale maps of the area produced by the British Columbia Department of Mines (Kilby & Wrightson 1987) illustrate the low amplitude, short wavelength folds developed in the Cretaceous above some of these detachments. The shallower deformation makes it difficult to extrapolate surface structure to depth with any confidence; it also distorts the seismic ray paths thus making seismic imaging difficult.

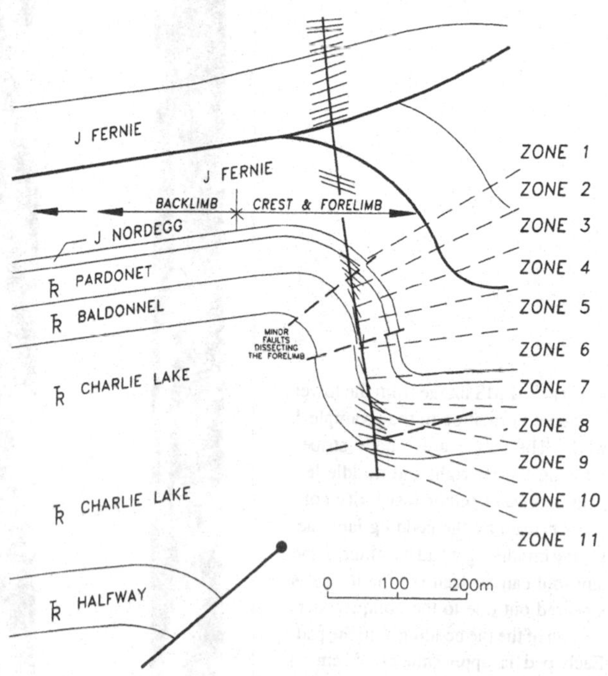

Figure 4. Detailed structural cross-section through the lower portion of the Example 1 well showing dip data, faults and stratigraphic boundaries. The subdivision of the structure into the 11 dip domains (structural zones)

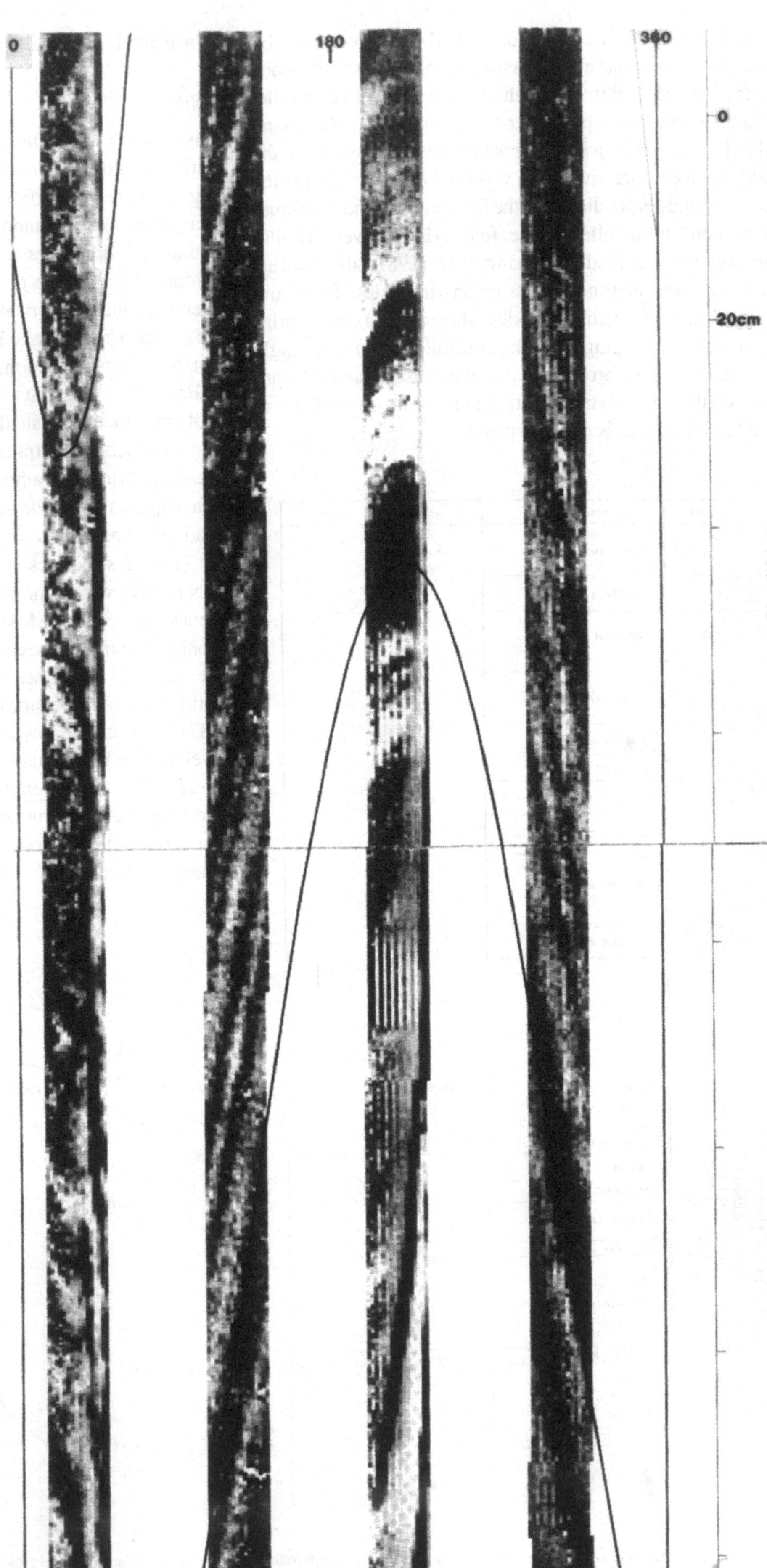

Figure 5. FMS image from the lower Pardonet Formation in the Example 1 well. The subtle grey shade stripes crossing the far right and middle left pads represent minor resistivity contrasts caused by the bedding laminae. On the middle right pad the hinge of the sinusoid can be seen but the image is smeared out due to the oblique intersection of the the bedding with the pad. Each pad is approximately 5 cm in width. The solid sinusoidal curve is an example of a correlated bedding event; the true dip in this particular segment of the well is 80° northeast. The location of this image is shown on Figure 10.

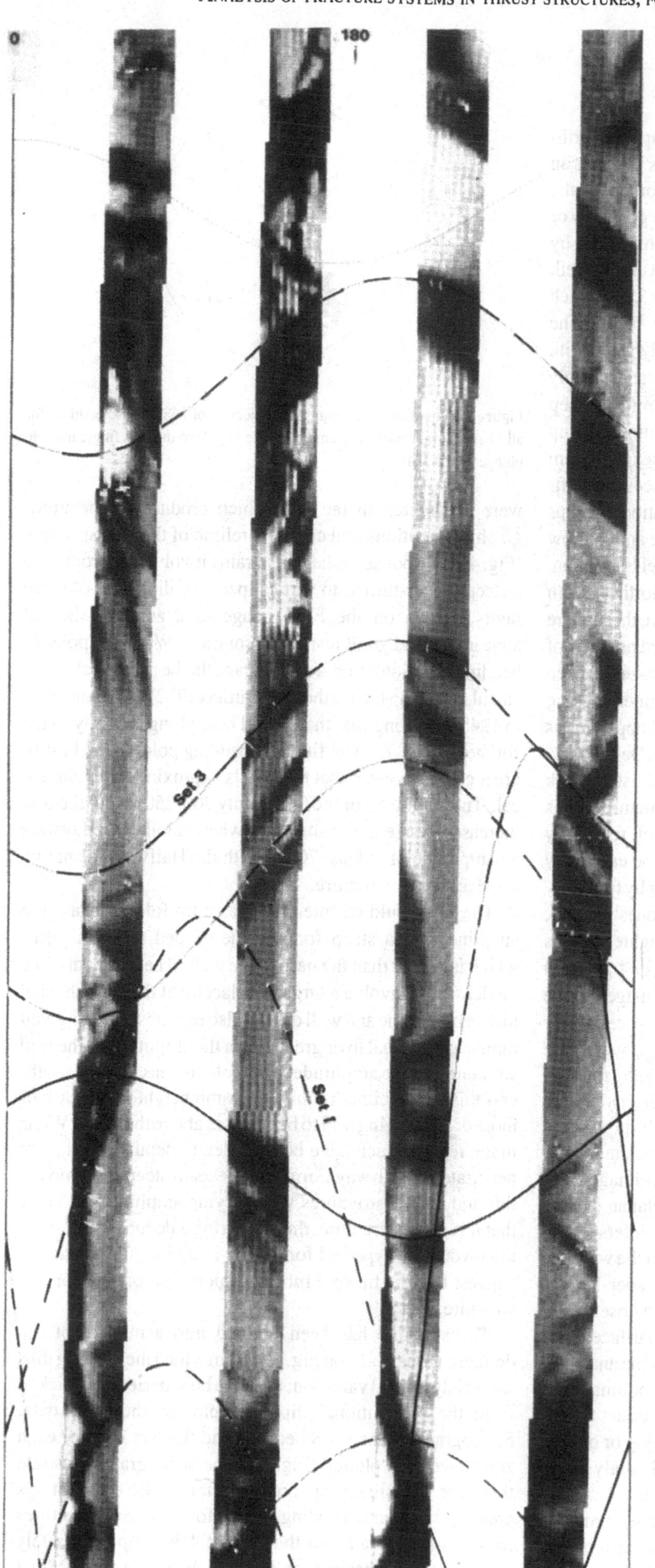

Figure 6. FMS image from the upper Pardonet Formation in Example 1. The solid sinusoidal curves are bedding and the dashed sinusoidal curves are fractures which are labelled to differentiate Set 1 and Set 2. Note that the fractures are defined by resistivity events that are highly conductive (black). The location of the image is shown on Figure 10.

Fold and fracture geometry

The well described here penetrated the steeply dipping northeast limb of one of the Triassic tip folds and thus provides an excellent opportunity to examine the fracture geometry in the most deformed portion of a tip fold (Fig. 4). The geometry of the crest and backlimb of the structure is constrained by seismic data and a vertical seismic profile shot in the well. The forelimb geometry was confirmed by the core which clearly contained steep northeasterly dipping beds. The northeast dip of the strata becomes increasingly steep with depth, then becomes vertical to over-turned and steeply dipping to the southwest before gradually returning to steep north east dips as the well penetrated the forelimb of the tip fold. These high angle dips are beyond the maximum correlation angle of conventional dipmeter processing software. Towards the base of the Pardonet Formation, the dips begin to shallow once more becoming moderate and shallow dipping to the northeast and, within the Baldonnel Formation, show progressively shallower dips towards the northeast with depth. In the underlying Charlie Lake Formation, the dips are only 5°-10° to the northeast (Fig. 4). A detailed analysis of the geometry of the tip fold structure in this well was undertaken using the Schlumberger Formation Microscanner logging tool (FMS). The tool has been used for several applications in structurally less complex settings (Serra 1989; Dennis *et al.* 1987; Fissinger & Gyllenstein 1986) but no published work exists on FMS analysis of fractures in thrust structures. This tool makes a wellbore wall image or map of resistivity variations (Figs 5 & 6). These variations can be caused by beds of varying porosity and hence resistivity, or by fractures, which if open, appear as highly conductive zones (Fig. 6). Highly resistive rock (tight) will appear in lighter shades whereas highly conductive rock (porous and/or fractured) will appear black. Using a workstation the image can be analysed by searching for valid correlations of resistivity character between the four strips of the image which are produced by the four pads of the tool and provide approximately 40% coverage of the wellbore. Valid correlations are considered to be between bands of similar resistivity shading, or their contacts, on all four pads. The correlation should also honour the angle of the band as it crosses the four image strips produced by the four pads. An inclined planar surface intersecting a pseudo-circular bore-hole will intersect the wellbore wall image in an elliptical trace. When the wellbore wall image is displayed on a conventional paper log the ellipse is transformed into a sinusoidal curve whose amplitude is proportional to the apparent dip of the surface in the wellbore (Figs 5 & 6). The workstation enables the analyst to make correlations after which the dips are automatically calculated, displayed and stored. As a final product the dip data can be output as a conventional dipmeter log or on disk as ASCII files for further processing and analysis by stereographic projections. The detailed geometry of the fold (Fig. 4) was constructed by applying the dip domain method of Suppe (1983) to the dip data, well formation tops and known stratigraphic thicknesses. A number of minor faults were intersected in the well which produce minor stratigraphic repetitions and cut the forelimb of the fold structure (Fig. 4); a response to the high strains involved in producing a steeply over-turned to vertical panel of dip. One of these faults is seen on the FMS image as a zone of sheared appearance and good resistivity contrast. When the poles to bedding are plotted on a stereogram the best fit great circle calculated using the method of Mardia (1972) has a pole of 6° to 124° indicating that the tip fold here plunges gently to the southeast (Fig. 7). The fit of the bedding poles to the best fit great circle suggests that the fold is approximately cylindrical. The amplitude of the fold is only 200-250 m, but the fold increases in size to the southeast where another well proves an amplitude of at least 700 m with the Halfway Formation involved in the structure.

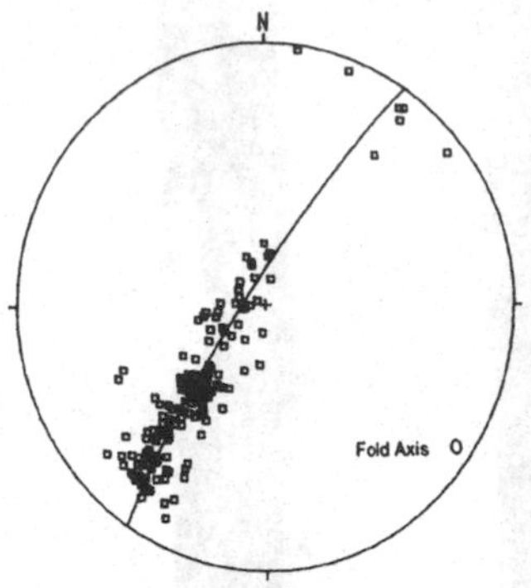

Figure 7. Equal-area stereographic projection of 168 poles to bedding for all 11 structural zones in Example 1. The pi pole to the best fit great circle plunges 6° to 124°.

The fold could be interpreted as a tip fold or as a ramp anticline with a steep forelimb developed above a thrust which is deeper than the base of the well. The ramp anticline model would involve a larger displacement than is indicated to date by seismic and well data. It also requires a hangingwall ramp cut-off equal to or greater than the amplitude of the fold structure. As the amplitude of the fold increases to the southeast this would either require the ramp height to increase or more of the hangingwall to be elevated above the ramp. Wells in the region which have been drilled to depths adequate to penetrate the Halfway Formation beneath steep forelimbs of this and similar structures with varying amplitudes indicate that it is still in structure, thus requiring a deeper detachment than would be expected for a ramp anticline. These factors suggest that the tip fold model is more appropriate for this structure.

The structure has been divided into a number of dip domains (zones 1-11 on Fig. 4) within which the bedding dips show only limited variation; the zone boundaries were picked from the conventional dipmeter plot of the FMS data. Stereograms of the FMS bedding and fracture data for each zone have been plotted (Fig. 8). The stereograms illustrate that three fracture sets are present in the well; Set 1 fractures are perpendicular to bedding and the fold axis, Set 2 fractures have a strike parallel to that of Set 1 but dip moderately northwest and are thus at an angle to the fold axis and Set 3

Table 1. Example 1 - mean orientations and statistics for structural zones

Zone	Bedding				Set 3 Fractures				Set 1 Fractures				Set 2 Fractures			
	Az	Dip	Rbar	#	Az	Dip	Rbar	#	Az	Dip	Rbar	#	Az	Dip	Rbar	#
1	042	34	0.9961	7	246	58	0.9562	10	309	87	0.7353	5				
2	039	47	0.9948	14	234	45	0.9402	18	296	85	0.6181	6				
2c	041	45	0.9958	6	227	46	0.9850	32	105	87		2				
3	044	51	0.9907	20	232	41	0.9876	3	306	85	0.3858	10				
4	038	61	0.9950	15					135	83		2				
5	036	77	0.7628	14					291	81	0.9896	3				
6	211	81	0.9517	5					285	80	0.9502	7	315	48	0.9661	11
7	039	63	0.9780	24					314	82	0.9631	11	338	35	0.9572	8
8	040	34	0.9960	25												
9	043	31	0.9909	1	235	49	0.8138	7	128	84		1				
10	070	14	0.9768	2					302	83	0.7025	11	315	37		1
11	134	7	0.9836	13					308	81	0.9532	6				

fractures are orthogonal to bedding but have a strike parallel to the fold axis. The fracture data from the core (in zone 2) are also included and show a similar distribution of fractures to the FMS data. The raw data are difficult to interpret and to clarify geometric relationships the mean orientations of bedding and of each fracture set have been calculated using the statistical methods of Mardia (1972) and Woodcock (1977); the results are summarized in Table 1. In each of the dip domain zones the mean azimuth and dip are given with the number of readings and the 'Rbar' value. The Rbar parameter is derived from the spherical variance of the data set and provides a measure of the dispersion of the data about the mean vector, a high value indicates low dispersion. With the exception of the Set 1 fractures in Zones 1 and 3 all the Rbar values exceed the critical values at a 99% level of significance; this indicates that, with these exceptions, all of the data sets are not uniformly distributed but show clustering about the mean (Davis 1983). The relationships of the fracture sets to the fold are conveniently illustrated by plotting the mean orientations of the fracture sets as great circles together with the best fit great circle to the bedding poles (the theoretical *ac* plane of the fold) and the calculated fold axis (Fig. 9). The stereonet derived by this method from Zone 2 illustrates how the Set 3 fractures tend to be orthogonal to bedding (parallel to the fold axis) and how the Set 1 fractures are closely parallel to the theoretical *ac* plane. In Zones 5-7 the Set 1 fractures appear to diverge further from the theoretical *ac* plane of the structure (see Fig. 8). As these zones cover most of the steeply dipping part of the structure this may be due to increased strain on the steep limb. However, the apparent differences in the mean orientations are probably caused by the small number of readings for which the means are being calculated (Table 1).

The Set 3 fractures in Zone 2 are dispersed over approximately 60° of arc (Fig. 8) which suggests that they are the

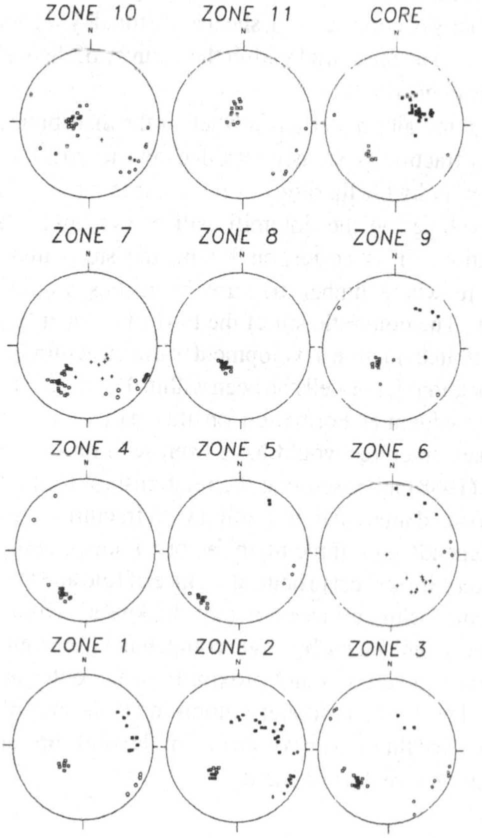

ZONE 10 ZONE 11 CORE

ZONE 7 ZONE 8 ZONE 9

ZONE 4 ZONE 5 ZONE 6

ZONE 1 ZONE 2 ZONE 3

Figure 8. Equal-area stereographic projections for the 11 structural zones in the well. Bedding poles; open squares. Set 1 fracture s; open circles. Set 2 fractures;crosses. Set 3 fractures; filled squares. The data from the core cut

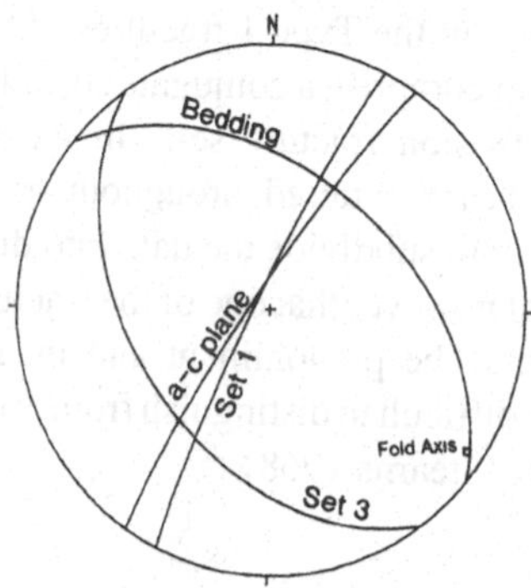

Figure 9. Simplified equal-area stereographic projection of the data from Zone 2. The mean orientations of the bedding, Set 1 fractures and Set 2 fractures are plotted as great circles together with the fold axis and the

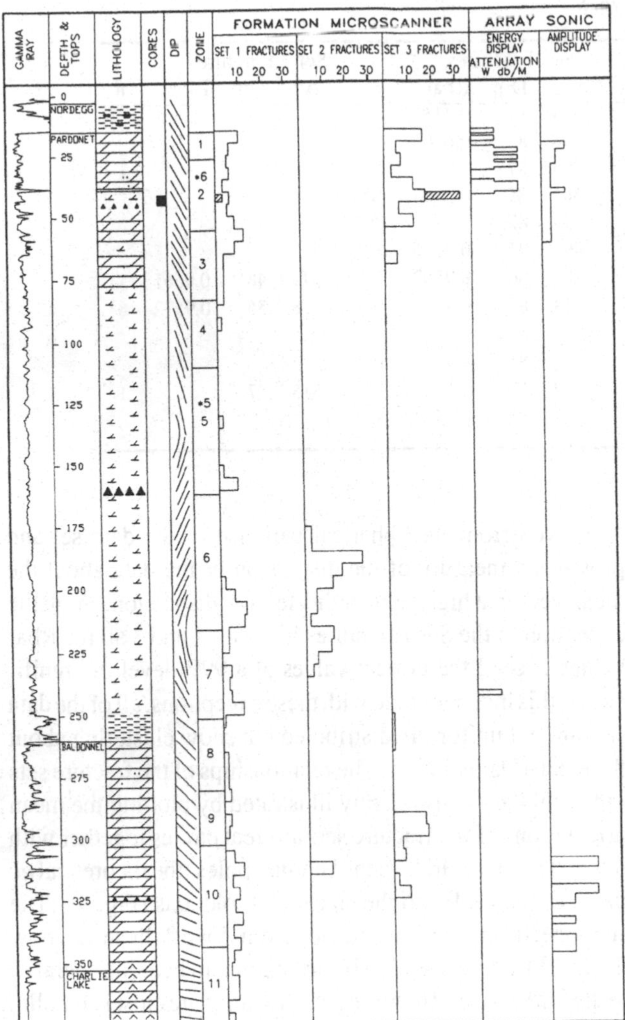

Figure 10. Summary log of the Example 1 well showing the distribution of the three fracture sets in relation to lithology and bedding dip. The location of the structural zone boundaries and the FMS images presented in Figures 5 & 6 are also indicated by *5 and *6 in the zone column.

Type 2 fractures of Stearns (1968) comprising a conjugate set of shear fractures and an extension fracture set that bisects the acute angle of the conjugate set. The data from Zone 1 could be interpreted in a similar fashion though the conjugate fracture sets are missing. The Set 1 fractures in several zones e.g. Zones 10 and 11 also show a spread in an arcuate pattern and could represent the Type 1 fractures of Stearns (1968); these also would comprise a conjugate shear fracture set and a bisecting extension fracture set. In most of the zones however the data are scattered throughout the arcuate pattern and it is difficult to subdivide the data into discrete clusters. It is not unusual however that one of the fracture sets of Type 1 or Type 2 may be predominant and in many cases the patterns can be difficult to distinguish from orthogonal *ac* and *bc* fracture sets (Stearns 1968).

Fracture distribution

The three sets of fractures illustrated by the stereographic analysis were then identified throughout the well to deter-

within the fold structure (Fig. 10). The fractures measured using the workstation from the FMS log are clearly parallel to many other fractures whose orientation was not measured. By carefully examining the FMS log it is possible to determine the number of fractures of each type within every 5 m interval of the well. These data are displayed on the fracture log together with lithology, gamma-ray log, cored intervals, and dips (Fig. 10). The intervals identified by the array sonic tool as being heavily fractured are also indicated on the log and correlate well with the fractured intervals determined from the FMS log. Unfortunately in this well only a limited amount of core was cut and as a result it is difficult to convincingly calibrate the log identification of fractures by core studies. Over the cored interval it appears that the FMS tool results in a slight overestimate of the number of Set 1 fractures but grossly underestimates the number of Set 3 fractures (Fig. 10). The Set 3 fractures, when studied in the core, are partially sealed by quartz/calcite mineralization and as a result will have little or no resistivity contrast to the FMS tool. The open fractures are readily seen because of the focusing effect on the electric current of the highly conductive fractures filled with drilling fluid; this also tends to exaggerate the width of the fracture.

The fracture distribution detected by the array sonic tool is also shown on the fracture log (Fig. 10); the details of the interpretation of this tool are discussed later with reference to Example 2. The array sonic tool is a more sophisticated version of the conventional sonic logging tool in that it allows the velocity of compression, shear and Stoneley waves (Sheriff 1973) to be measured within the vicinity of the well bore, (Morris *et al.* 1984).

What remains uncertain is whether the distribution of the various fracture types is controlled by the position on the structure or by the lithology as it is clear that fewer fractures are developed in the dolomitic silt of the lower Pardonet Formation. This corresponds with the steep limb of the structure where higher fracture intensities would be expected. The dolomitic silt of the lower Pardonet Formation has inhibited fracture development in this structurally favourable location. If the well had been within the cleaner dolomites of the Baldonnel Formation on this part of the structure abundant fractures would have been seen on the FMS log. Mitra (1988) proposed that the relationship of lithotectonic position and internal deformation with fracture development was dependent on three main factors; relative competency, position with respect to neutral surface of fold and the number of flexural slip surfaces per unit thickness. High fracture volume is enhanced by high competency, low number of flexural slip surfaces and proximity to the outer arc of the fold. The lower Pardonet dolomitic silt is very flaggy in outcrop and thus may have sufficient flexural slip surfaces to inhibit fracture development.

Origin of the fractures

In their study of the Uinta Basin in Utah Narr & Currie (1982) suggested that burial with hydrostatic pore fluid pressure will ing. The development of fractures is also enhanced during subsequent uplift by the presence of high pore fluid pressures. Engelder (1985) defines hydraulic fractures as those caused by abnormally high pore fluid pressures during burial, and considers that tectonic fractures develop in response to high pore pressures induced by tectonic compaction. All of the post-Triassic sequence has been deposited in foreland basins related to exotic terrane accretion (Monger *et al.* 1982; Cant & Stockmal 1989). The stress regime throughout the post-Triassic history has thus been similar; there will, however, have been relatively short lived peaks of higher differential stress during development of the fold and thrust structures in the region.

High pore fluid pressures are justified on the premise that the high sedimentation rates in the foreland basin (approximately 7 km of Jurassic and Cretaceous sediments, Kalkreuth & McMechan 1988) will facilitate the development of high pressures. Supporting evidence is provided by the fluid pressure in the Triassic Halfway Formation which is almost twice normal hydrostatic. The preservation of the anomalously high fluid pressures in the Halfway is due to the isolated nature of the sand bodies in this formation which are enclosed in tight shales. The development of the structures elevated the Triassic 2-2.5 km above regional. Later uplift is regarded as being driven by isostatic re-equilibration and erosion of the overburden. During deformation high pore fluid pressures developed inducing the tectonic fractures observed in the structure. The Set 1 fractures are almost certainly due to a component of along-strike extension during the deformation, a common phenomenon in many thrust belts (Hossack 1978). The Set 1 open fractures are Stearns (1968) Type 1 fractures, which were probably filled with gas thus inhibiting mineralization, and developed late in the deformation history, also as a result of high tectonically induced pore fluid pressures. The origin of the Set 2 fractures is at present unknown and is difficult to put into the overall kinematics of development of this fold.

The Set 3 fractures could superficially be interpreted as joints which formed as a regional set prior to Laramide thrusting. However, the parallelism of strike with the fold axis suggests a relationship with the deformation and where examined in core the rocks part readily along these partially mineralized fractures and reveal a down-dip mineral fibre lineation. This suggests that they are shear fractures which developed during folding and are sub-parallel to the Laramide thrusts. The Set 3 fractures are considered to be Stearns (1968) Type 2 shear fractures related to folding during the thrust movements; they probably developed early in the deformation history but may have been reactivated as the northeast fold limb rotated and steepened. The other possible origin for the fractures is as a set of radial fractures in the fold; however, such fractures should be tensional rather than showing evidence of shear.

The distribution and geometric relationship of fracture Sets 1 & 3 to the tip fold are illustrated in Figure 11, which shows that these fracture sets produce a pseudo-orthogonal fracture system similar to those described by Price (1967), and Mitra (1988). Good analogues to this structure have been analysed by Almendinger (1982) and Mitra (1987). Allmendinger (1982) described the microstructures associated with two, asymmetric, overturned folds which complicate the geometry of the Meade thrust sheet in the Idaho Wyoming thrust belt. A fracture set is locally developed perpendicular to the axis of the folds similar to Set 1 in Example 1. The geometry of the folds is very similar to the Monkman structure and appear to be tip folds cut by faults on the steep overturned limbs. In the upright west dipping fold limb, compression is parallel or sub-parallel to bedding with extension either perpendicular to bedding or parallel to the fold axis. In the hinge zones these simple relationships are complicated by the rotation of the beds. On the overturned limbs compression both parallel and at high angles to bedding is seen. This is interpreted as layer parallel shortening early in the development of the fold followed by high angle compression following rotation into the overturned limb, which has been rotated 90°-150° through the regional stress field. Allmendinger claims that the asymmetry of these folds was initiated in the earliest stages of folding; this is also clear from the geometric models of Suppe (1983). However the contention that the hinge need not therefore migrate (Allmendinger 1982) is incorrect as parts of beds which initially are located on the backlimb of the fold will pass through the hinge and become part of the forelimb as folding proceeds (see Suppe 1983). Mitra (1987) described a model for the development of the Wills Mountain anticline in the Appalachians which he interprets as a tip fold subsequently displaced onto the upper detachment beyond the top of the ramp. As the tip fold developed the forelimbs steepened and are thinned by bedding normal shortening, pressure solution on bed parallel stylolites and conjugate extension fractures;

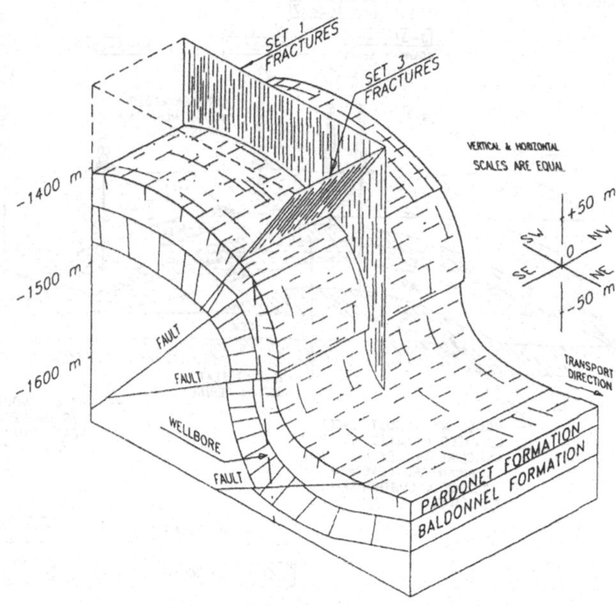

Figure 11. Three-dimensional block diagram to illustrate the relationship of the Set 1 and Set 3 fractures to the fold structure. The Set 2 fractures have been omitted for clarity and the Set 1 and Set 3 fracture sets have been shown as orthogonal sets

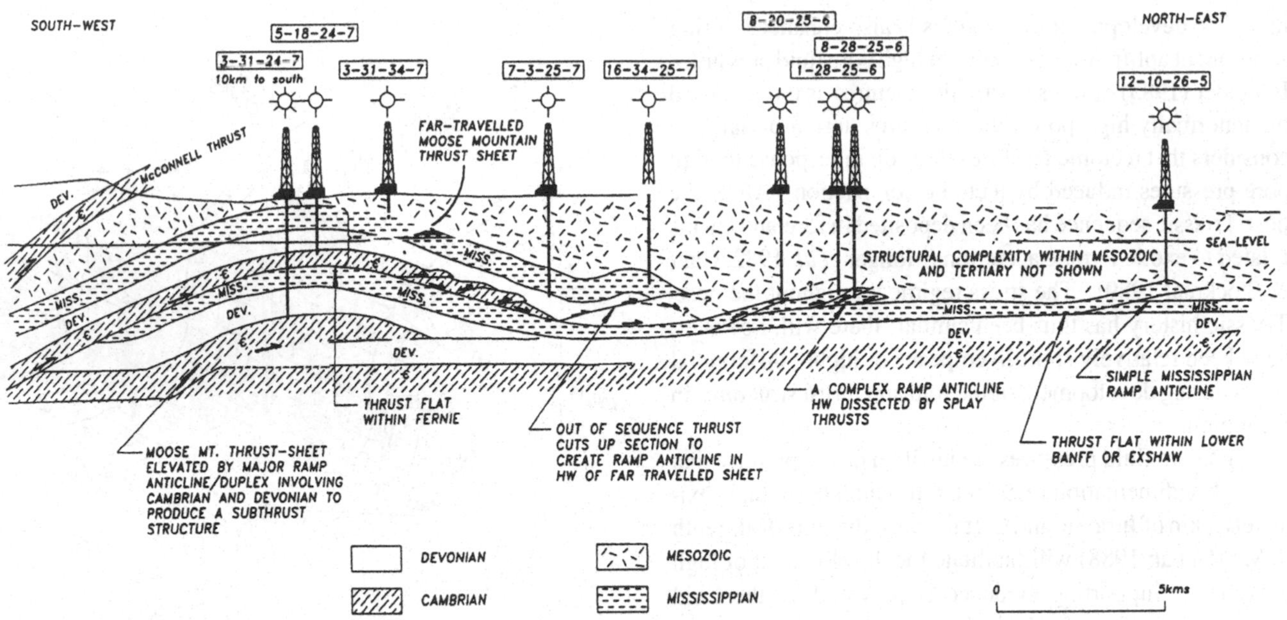

Figure 12. Generic cross-section through the Alberta Foothills to illustrate the various structural types in the Upper Palaeozoic carbonates. All the wells shown are west of the 5th meridian. The location of the section is shown on Figure 1 (based on a section by Ray Widdowson).

AGE		FORMATION OR GROUP	LITHOLOGY	MAP
TERTIARY		PASKAPOO		
CRETACEOUS	UPPER	EDMONTON		— 1000m
		BELLY RIVER		— 2000m
		WAPIABI		
		CARDIUM		
		BLACKSTONE		
	LOWER	BLAIRMORE		— 3000m
MISSISSIPPIAN		FERNIE & NORDEGG		
		TURNER VALLEY		
		SHUNDA		
		PEKISKO		
		BANFF		

Figure 13. Typical stratigraphic column for the Alberta Foothills. The shading in the right hand column of the figure shows the stratigraphic subdivisions used in Figure 1.

has shown much higher fracture density on the steeply dipping west limb than on the gently dipping east limb.

Example 2: Fracture geometry in a ramp anticline

Regional setting

The second example is located in the Alberta Foothills to the west of Calgary (Fig. 1). The productive Palaeozoic carbonate structures fall into four main types; simple ramp anticlines, ramp anticlines segmented by hangingwall shortcut faults, which may be out-of-sequence, hangingwall ramp anticlines at the leading edges of large displacement thrust sheets and sub-thrust structures, ramp anticlines and duplexes beneath large displacement thrust sheets. These structural types are illustrated in Figure 12, a generic section based on structures in the foothills west of Calgary.

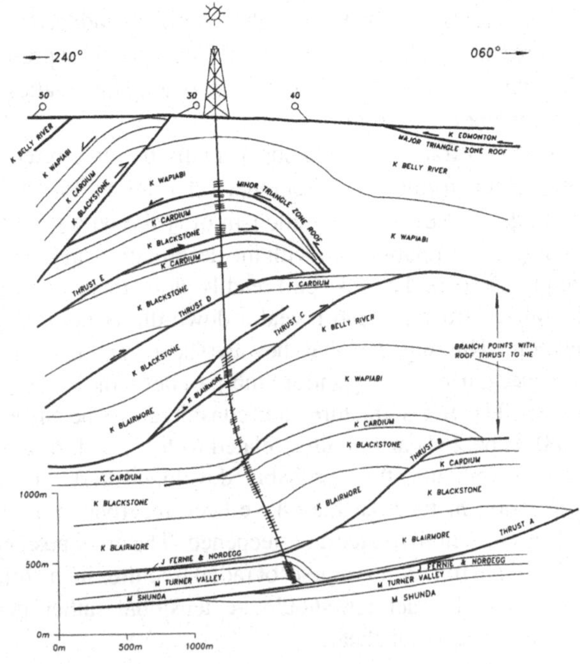

Figure 14. Detailed structural cross-section through the Example 2 well showing dip data, stratigraphic tops and major faults. The structure of the Mesozoic rocks in the well is discussed in the text.

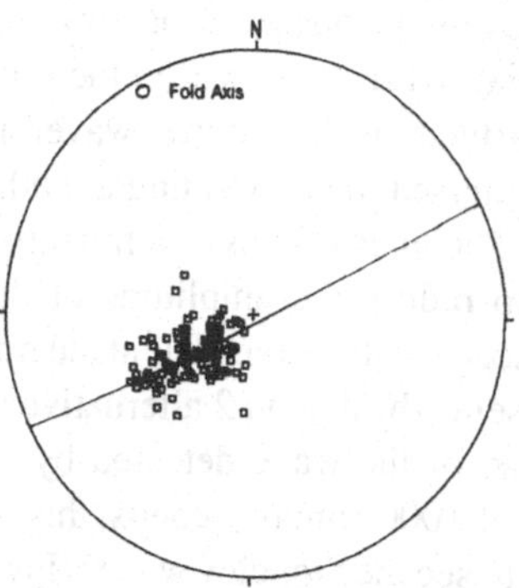

Figure 15. Equal area stereographic projection of 120 poles to bedding from the Mississippian section of the Example 2 well. The spherical variance of the data is 0.0233 reflecting the strongly unimodal distribution of the data. A best-fit great circle can be fitted to the data and indicates a fold axis plunging 3° to 334°.

Fold and fracture geometry

The structure is constrained by an extensive seismic database and a number of wells, one of which is described here. The deformation in the Mesozoic section is detached from that at the Mississippian level by a major regional detachment within the Jurassic Fernie Formation (Fig. 13). The structure is located beneath the frontal monocline of the Alberta Foothills, in an analogous structural position to many of the major producing fields in Alberta, for example, Jumping Pound and Turner Valley. The structure of the Mesozoic rocks in this area is dominated by two major stacked ramp anticlines both of which detach in the Fernie Formation (Fig. 14, Thrusts B & C). Beneath the frontal monocline the two ramp anticlines are stacked one on top of each other producing the sudden extreme structural elevation of up to 1500 m

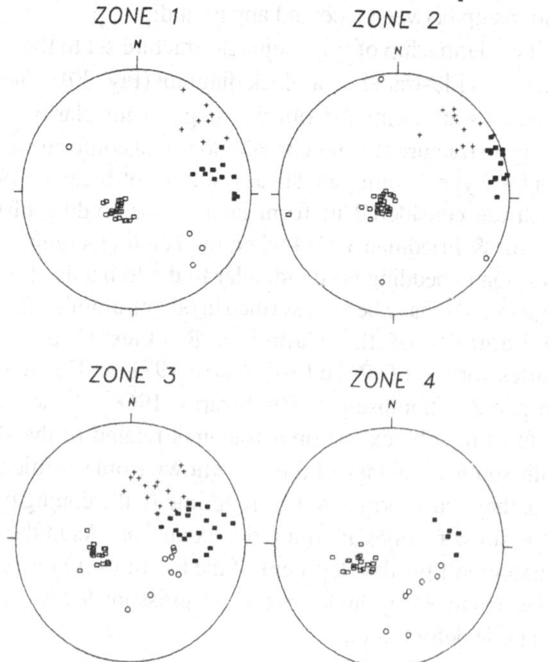

Figure 16. Equal area stereographic projections of bedding and fracture set data from Zones 1 to 4 in the Example 2 well. The data for Zone 5 is not included as there are few fractures in that interval. Bedding; open squares. Set 1 fractures; open circles. Set 2 fractures; crosses. Set 3 fractures; filled

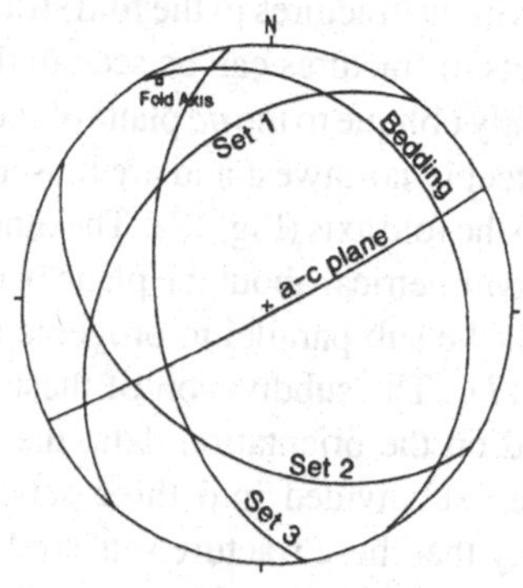

Figure 17. Simplified equal area stereographic projection of the data from Zone 3. The mean orientations of the bedding and the three fracture sets are plotted as great circles together with the fold axis and the theoretical *ac* plane of the fold.

observed at thefrontal monocline. The lower ramp anticline involves a stratigraphic section from the Fernie Formation up to the Cretaceous Belly River Formation (Fig. 13). The Mississippian level structure has only a small thrust displacement of about 500 m producing a structural elevation of some 350 m above regional. Both Mesozoic imbricates merge in a common roof thrust at the top of the Belly River Formation creating a triangle zone whose roof thrust emerges to outcrop within the frontal monocline (Fig. 13). The upper Mesozoic imbricate has some internal complications with repetition of the Cardium and Blackstone Formations (Fig. 14, thrusts D & E). They appear to constitute a minor triangle zone within the upper sheet whose upper detachment is near the base of the Wapiabi Formation. The structure of the Mississippian appears to be a simple ramp anticline approximately 5 km in strike length. The lower detachment within the Mississippian Shunda Formation and the upper detachment is within the Jurassic Fernie Formation, a regionally significant detachment throughout the Alberta Foothills. The structure dies out rapidly to the south of the well but displacement is transferred onto another Mississippian ramp anticline which is en-echelon and slightly to the east of this structure. The transfer of displacement from one fault structure to another is common in many thrust belts and is usually termed a transfer zone. An important point is that the offset of the two structures is not due to the presence of a lateral ramp.

The core and FMS data clearly indicates that the well penetrates the forelimb of the Mississippian structure (Fig. 14). Analysis of the bedding dip data from the Mississippian reflects this with a strongly unimodal distribution (Fig. 15 & Woodcock 1977). A best-fit great circle for the poles to bedding suggests a fold axis that plunges 3° to 332°; although, given the lack of data from the crest and backlimb, this is unreliable. Mapping based on the available seismic data suggest that the fold plunges sharply to the south-southeast from the well. The orientation of the fracture sets has been determined by measurements of the cores and from the FMS. These data have been analysed by the same methods as described for Example 1. In Example 2 the structure has been subdivided into five dip domains or zones the relationships of the fractures to the fold structure (Figs 16 & 17). Three sets of fractures can be seen in the well; Set 1 has a strike slightly oblique to the *ac* plane of the fold but dips moderately to steeply northwest and are thus approximately perpendicular to the fold axis (Fig. 17). The other two sets are approximately symmetrical about the plane perpendicular to bedding and a strike sub-parallel to orogenic strike and the fold axis (Fig. 17). The subdivision of these two sets was arbitrarily based on the orientation data; the data could as easily have been subdivided into three sets. The scatter suggests strongly that these fracture sets are Stearns (1968) Type 2 fractures; a conjugate set of fractures with an extension fracture set along the acute bisectrix which is sub-parallel to the fold axis.

Fracture distribution and origin

The distribution of the fractures in the Mississippian rocks penetrated by the well is summarized in Figure 18, which shows that the fractures occur throughout the well but become less frequent in the lower part of the well. This diagram

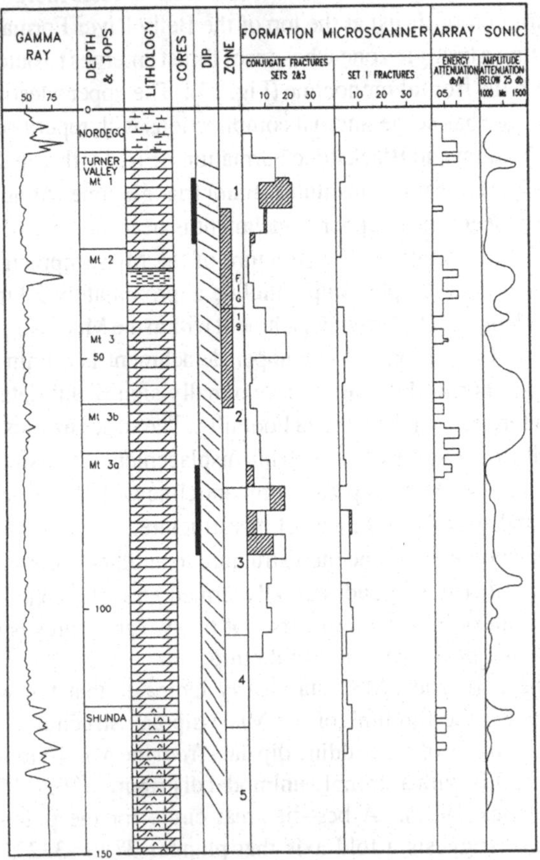

Figure 18. Summary log of the Example 2 well showing the distribution of the three fracture sets in relation to lithology and bedding dip. The subdivision of the Turner Valley Formation follows the scheme of Rupp (1969). The location of the structural zone boundaries and the portion of the well illustrated by Figure 19 are indicated in the zone column. The number of fractures of each set are recorded as histograms with a 5 m sample interval; the number of fractures measured in the cored intervals are indicated by cross-hatching.

also summarizes the response of the array sonic tool to the fractured intervals. Fractures within the rock will produce characteristic effects on the source waveform that can be detected and analysed thus providing a method for fracture identification. The main effects of a fracture are to slow the source wave, to reduce the amplitude of the wave and to attenuate the energy of the wave. The standard Schlumberger processing presents the data in 2 alternative ways. The first shows the phase of the wave detected by Receiver 1 to a recording time of 10000 microseconds; this is a long enough recording time to see the Stoneley waves (Fig. 19). Fractures tend to produce a characteristic chevron pattern due to reflection at the fracture interface. The energy attenuation can also be calculated and is displayed in Figure 19; this is expressed in decibels per metre and the attenuation effect decreases with time as the slower waves are detected. The other display shows phase and amplitude plotted against recording time, (Fig. 19) fractured zones cause a phase shift to the right due to the slowing of the waveform and hence a later recording time at the receiver. The amplitude display shows attenuation of the amplitude in fractured zones. The more fractured the rock the more attenuated the wave energy becomes. This tool correlates quite well with the FMS fracture logs (Figs 10 & 18).

From the perspective of hydrocarbon production the actual number of fractures that intersect the wellbore is the most important factor. However, this does not yield the true fracture spacing as this is dependent on the angular relationships between the wellbore and the fractures; a fracture with a dip azimuth parallel to the wellbore has much less chance of being intersected by the well than one which is at a high angle to the wellbore. The true fracture spacing will thus be a simple function of apparent fracture spacing identical to the relationship between true and apparent dip.

The relationship of the conjugate fracture set to the ramp anticline is illustrated by a block diagram (Fig. 20). The Set 1 fractures are omitted from the diagram for clarity. The conjugate fractures form in orientations that conform closely with the Type 2 conjugate shear fractures of Stearns (1968) which are considered to form later in the folding history (Stearns & Friedman 1972) when the bending strains cause extension in bedding perpendicular to the fold axis. Similar conjugate sets have been described in other Laramide folds in the Foothills of the Canadian Rockies (Muecke & Charlesworth 1966; Price 1967; Norris 1971) and in Montana (Burger & Thompson 1970; Stearns 1968). The Set 1 fractures may be extensional fractures related to the sharp south-southeast plunge of the fold known from seismic data. If so, they must form at a different time to the conjugate set as the stress regimes are mutually exclusive. As in the case of Example 1 the development of the fractures is considered to be enhanced by high pore fluid pressure levels during Laramide deformation.

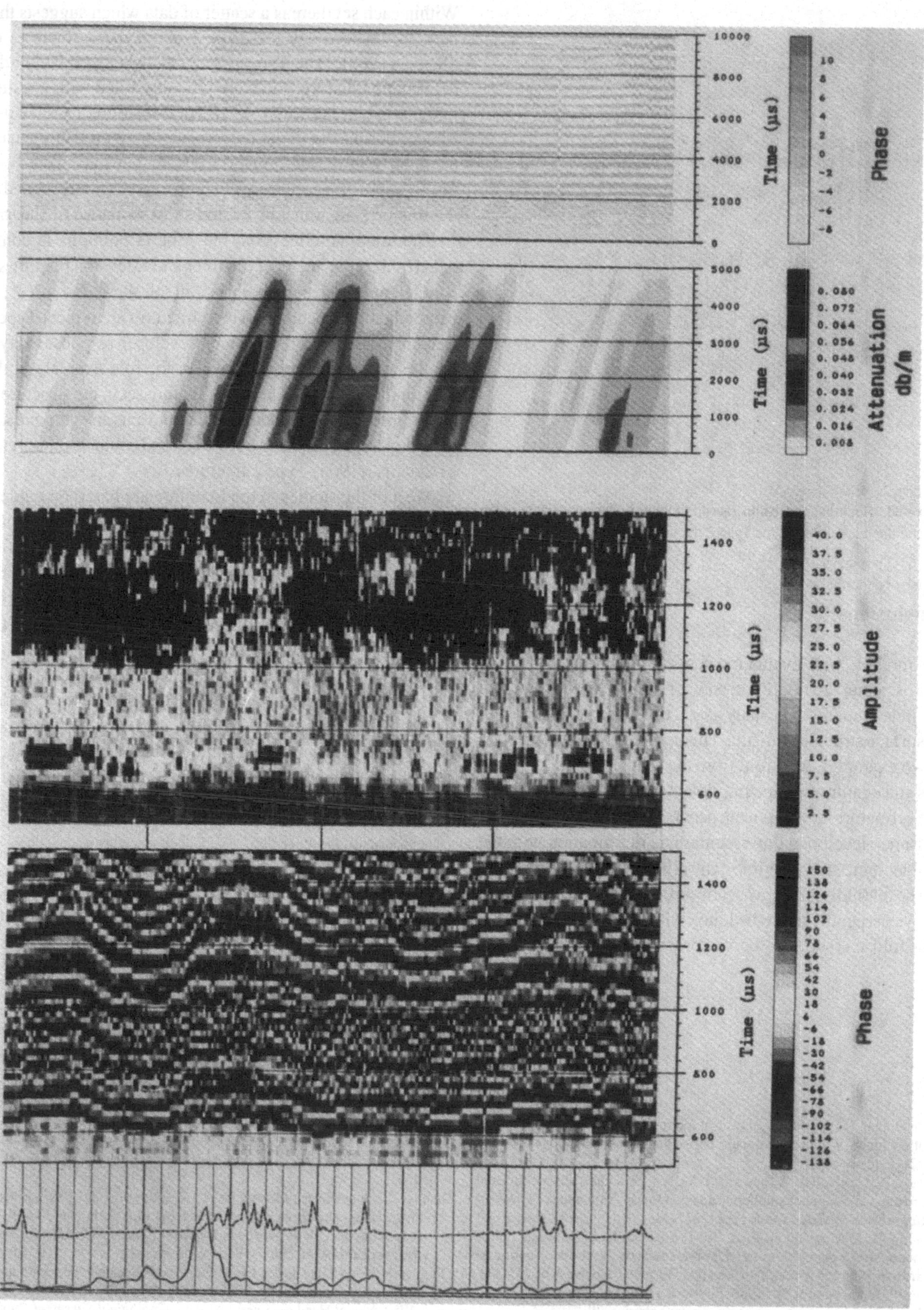

Figure 19. An example of the data output from the array sonic logging tool showing the effect of the fractures on the phase, amplitude and energy attenuation of the source wavelet. The right hand phase column is recorded for a longer time and shows the chevron pattern effect on the phase of the Stonely waves.

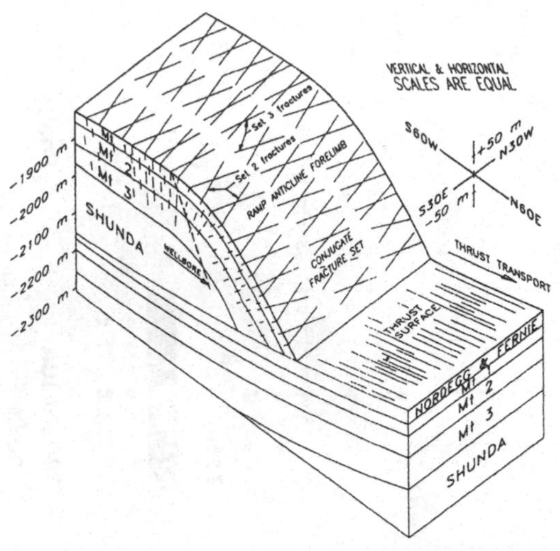

Figure 20. Block diagram to illustrate the three-dimensional structure of the fold and the relationship of the Type 2 fractures to the fold axis.

Conclusions

The modern well evaluation tools such as the Formation Microscanner, Array Sonic and sophisticated core analysis facilitate quantitative analysis of the fracture systems of subsurface structures in-situ. This approach has been applied to two examples, a tip fold from the British Columbia Foothills and a ramp anticline from the Alberta Foothills. In both cases, fracture development occurs due to high pore fluid pressures developed during Laramide deformation with the local tectonic stress regime controlling fracture orientation.

The tip fold contains a pseudo-orthogonal fracture system, always perpendicular to bedding, with one set perpendicular to the fold axis and the other major set parallel to the fold axis. Within each set there is a scatter of data which suggests that they are respectively the Type 1 and Type 2 fractures of Stearns (1968). The Type 1 fractures developed late in the deformation history, and were probably filled with gas thus inhibiting mineralization. The Type 2 fractures are related to folding during the thrust movements; they probably developed early in the deformation history but may have been reactivated as the northeast fold limb rotated and steepened as the fractures are mineralized and show evidence of shear.

The ramp anticline from the Alberta Foothills is dominated by a conjugate set of fractures and extension fracture set parallel to the acute bisectrix; all of these fractures are perpendicular to bedding. These fractures are typical Type 2 fractures of Stearns (1968) which develop in response to layer parallel extension perpendicular to the fold axis late in the deformation history. The other fracture set in the structure is approximately perpendicular to the fold axis and perpendicular to bedding. The data are scattered and could be interpreted as Stearns' (1968) Type 1 fractures.

Most of the studies in the literature are based on outcrop data and suggest that the model of Stearns (1968) is appropriate, although, in many cases the fracture sets are simplified to pseudo-orthogonal sets parallel and perpendicular to the fold axis. Few of the studies quantitatively analyse the relationship of fracture orientation and intensity to the mesoscopic folds to whose development they are related. This paper has presented two examples for which such an analysis has been undertaken. The development of predictive, empirical models of fracture geometry in subsurface thrust structures must await more studies of the type discussed here which quantitatively describe the density and geometry of the fracture sets.

I wish to thank BP Resources Canada Ltd. for permission to publish the work presented in this paper, my colleagues at BP Canada for discussions on the ideas presented and Peter Fermor and an anonymous reviewer for their valuble comments on the original draft. The figures were prepared by Les Neumann and Kevin Franck.

References

Allmendinger, R. W. 1982. Analysis of microstructures in the Meade Plate of the Idaho-Wyoming Foreland Thrust belt, USA. *Tectonophysics*, **85**, 221-251.

Bally, A. W. Gordy, P. L. & Stewart, G. A. 1966. Structure, seismic data and orogenic evolution of southern Canadian Rocky Mountains. *Bulletin of Canadian Petroleum Geology*, **14**, 337-381.

Barss, D. L. & Montandon, F. A. 1981. Sukunka-Bullmoose gas fields: models for a developing trend in the southern Foothills of northeast British Columbia. *Bulletin of Canadian Petroleum Geology*, **29**, 293-333.

Berger, P. & Johnson, A. M. 1980. First-order analysis of deformation of a thrust sheet moving over a ramp. *Tectonophysics*, **70**, T9-T24.

Burger, H. R. & Thompson, M. D. 1970. Fracture analysis of the Carmichael Peak Anticline, Madison County, Montana. *Geological Society of America Bulletin*, **81**, 1831-1836.

Cant, D. J. & Stockmal, G. S. 1989. The Alberta foreland basin: relationship between stratiraphy and Cordilleran terrane-accretion events. *Canadian Journal of Earth Sciences*, **26**, 1964-1975

Dahlstrom, C.D.A. 1969. Balanced cross-sections. *Canadian Journal of Earth*

—— 1970. Structural Geology in the eastern margin of the Canadian Rocky Mountains. *Bulletin of Canadian Petroleum Geology*, **18**, 332-406.

Davis, D. C. 1983. *Statistics and Data Analysis in Geology*. Wiley, New York, 646p.

Dennis, B., Standen, E., Georgi, D. T. & Calow, G. O. 1987. Fracture Identification and Productivity Predictions in a Carbonate Reef Complex. *62nd Society of Petroleum Engineers Annual Technical Conference and Exhibition, Dallas*. Paper **SPE 16808**.

Douglas, R.J.W. 1950. Callum Creek, Langford Creek and Gap map-areas, Alberta. *Geological Survey of Canada Memoir*, **255**, 124p.

—— 1958. Mount Head map-area, Alberta. *Geological Survey of Canada Memoir*, **291**, 241p.

Engelder, T. 1985. Loading paths to joint propagation during a tectonic cycle: an example from the Appalachian Plateau, *USA Journal of Structural Geology*, **7**, 459-476.

Fissinger, M. R. & Gyllenstein, A. 1986. Fracture Detection in North Sea Reservoirs. *Transactions of the Society of Petroleum Well Log Analysts, 10th European Formation Evaluation Symposium, Aberdeen, Scotland,*

Friedman, M. 1969. Structural analysis of fractures in cores from Saticoy Field, Ventura County, California. *American Association of Petroleum Geologists Bulletin*, **53**, 367-389.

Hossack, J. R. 1978. The correction of stratigraphic sections for tectonic finite strain in the Bygdin area, Norway. *Journal of the Geological Society of London*, **135**, 229-242.

Jones, P. B. 1982. Oil and gas beneath east-dipping underthrust faults in the Alberta Foothills. *In*: Powers, R.B. (ed.), *Studies of the Cordilleran Thrust Belt*. Rocky Mountain Association of Geologists, 61-74.

Kalkeuth, W. & McMechan, M. E.,1988. Burial history and thermal maturity, Rocky Mountain Front Ranges, Foothills and Foreland, east-central British Columbia and adjacent Alberta, Canada. *American Association of Petroleum Geologists Bulletin*, **72**, 1395-1410.

Kilby, W. E. & Wrightson, C. B. 1987. Bullmoose mapping and compilation project (93P/3,4). *British Columbia Ministry of Energy, Mines and Petroleum Resources, Geological Fieldwork 1986*, Paper **1987-1**, 373-378.

Mardia, K. V. 1972. *Statistics of Directional Data*. Academic Press, London, 357p.

McMechan, M. E. 1985. Low-taper triangle zone geometry: an interpretation for the Rocky Mountain Foothills, Pine Pass-Peace River area, British Columbia. *Bulletin of Canadian Petroleum Geology*, **33**, 31-38.

—— & Thompson, R.I. 1989. Structural style and history of the Rocky Mountain fold and thrust belt. *In:* Ricketts, B. D. (ed.) *Western Canada Sedimentary Basin; A Case History*. Canadian Society of Petroleum Geologists, 47-72.

McQuillan, H. 1973. Small-Scale Fracture Density in the Asmari Formation of Southwest Iran and its Relation to Bed Thickness and Structural Setting. *American Association of Petroleum Geologists Bulletin*, **57**, 2367-2385.

—— 1974. Fracture Patterns on Kuh-e Asmari Anticline, Southwest Iran. *American Association of Petroleum Geologists Bulletin*, **58**, 236-246.

Mitra, S. 1987. Regional variations in deformation mechanisms and structural styles in the central Appalachian orogenic belt. *Geological Society of America Bulletin*, **98**, 596-590.

—— 1988. Effects of Deformation Mechanisms on Reservoir Potential in Central Appalachian Overthrust Belt. *American Association of Petroleum Geologists Bulletin*, **72**, 536-554.

Monger, J.W.H., Price, R. A. & Tempelman-Kluit, D. J. 1982. Tectonic accretion and the origin of two major metamorphic and plutonic welts in the Canadian Cordillera. *Geology*, **10**, 70-75.

Morris, C. F., Little, T. M., & Letton, W. III. 1984. A New Sonic Array Tool for Full Waveform Logging. *Society of Petroleum Engineers Annual Technical Conference and Exhibition, Houston*, Paper **SPE 13285**.

Muecke, G. K. & Charlesworth, H.A.K. 1966. Jointing in folded Cardium Sandstones along the Bow River, Alberta. *Canadian Journal of Earth Sciences*, **3**, 579-596.

Narr, W. & Currie, J. B. 1982. Origin of fracture porosity - Example from Altamont field, Utah. *American Association of Petroleum Geologists Bulletin*, **66**, 1231-1247.

Norris, D. K. 1971. Comparative study of the Castle River and other folds in the Eastern Cordillera of Canada. *Geological Survey of Canada Bulletin*, **205**, 58p.

Price, R. A. 1967. The tectonic significance of mesoscopic subfabrics in the southern Rocky Mountains of Alberta and British Columbia. *Canadian Journal of Earth Sciences*, **4**, 39-70.

—— & Mountjoy, E. W. 1970. Geologic structure of the Canadian Rocky Mountains between Bow and Athabasca river, a progress report. *In:* Wheeler, J. O. (ed.), *Structure of the Southern Canadian Cordillera*. Geological Association of Canada, Special Publication, **6**, 7-25.

Rupp, A. W. 1969. Turner Valley Formation of the Jumping Pound area, Foothills, southern Alberta. *Bulletin of Canadian Petroleum Geology*, **17**, 460-485.

Sanderson, D. J. 1982. Models of strain variation in nappes and thrust sheets a review. *Tectonophysics*, **88**, 201-233.

Serra, O. 1989. *Formation Microscanner Image Interpretation*. Schlumberger Education Services. 117p.

Sheriff, R. E. 1973. *Encyclopedic Dictionary of Exploration Geophysics*. Society of Exploration Geophysicists, Tulsa Oklahoma, 266p.

Spang, J. H. & Groshong Jr., R. H., 1981. Deformation mechanisms and strain history of a minor fold from the Appalachian Valley and Ridge Province. *Tectonophysics*, **72**, 323-342.

——, Wolcott, T. L. & Serra, S. 1981. Strain in the ramp regions of two minor thrusts, southern Canadian Rocky Mountains. *In:* Carter, N. L. *et al.* (eds) *Mechanical Behaviour of crustal rocks. The Handin Volume*, American Geophysical Union, Geophysical Monograph, **24**, 243-250.

Stearns, D. W. 1968. Certain aspects of fracture in naturally deformed rocks. *In:* Rieker, R. E. (ed.) National Science Foundation Advanced Science Seminar in Rock Mechanics. *Special Report, Air Force Cambridge Research Laboratories*, Bedford, Mass., **AD66993751**, 97-118.

—— & Friedman, M. 1972. Reservoirs in Fractured Rock. *American Association of Petroleum Geologists Memoir*,**16**, 82-100.

Suppe, J. 1983. Geometry and kinematics of fault bend folding. *American Journal of Science*, **283**, 684-721.

Thompson, R. I. 1979. A structural interpretation across part of the northern Rocky Mountains, British Columbia, Canada. *Canadian Journal of Earth Sciences*, **16**, 1228-1241.

Wiltschko, D. V., Medwedeff, D. A. & Millson, H. E. 1985. Distribution and mechanisms of strain within rocks on the Northwest ramp of Pine Mountain block, southern Appalachian foreland: a field test of Theory. *Geological Society of America Bulletin*, **96**, 426-435.

Woodcock, N. H. 1977. Specification of fabric shapes using the eigenvalue method. *Geological Society of America Bulletin*, **88**, 1231-1236.

Thrust tectonics and Cretaceous intracontinental shortening in southeast Alaska

C. M. Rubin & J. B. Saleeby

Division of Geological and Planetary Sciences, California Institute of Technology, Pasadena, California 91125, USA

Abstract: An imbricate thrust belt extending along strike for more than 2000 km overprints the tectonic boundary between two of the largest allochthonous crustal fragments (Intermontane and Insular superterranes) in the North American Cordillera, and affects rocks west of the Coast Plutonic Complex in southeast Alaska and western British Columbia. Deformation was broadly coeval with mid-Cretaceous magmatism and involved the emplacement of west-directed thrust nappes over a structurally intact and relatively unmetamorphosed basement. The Palaeozoic and lower Mesozoic Alexander terrane forms structural basement for much of the thrust belt along a moderately northeast-dipping decollement.

There were two main episodes of mid-Cretaceous deformation, which were contemporaneous with the emplacement of tabular plutonic bodies. Older structures record ductile southwest-vergent folding and faulting, and regional metamorphism, associated with a well-developed axial-planar foliation. Second-generation structures include southwest-directed thrust faults that juxtapose contrasting metamorphic grades and refold earlier structures.

Structural, stratigraphical and geochronologic data suggest that regional-scale deformation in southeast Alaska occurred between 113 Ma and 89 Ma. Deformation involved the imbrication of marginal basin(s) and a magmatic arc, overprinting the older tectonic boundary between the Insular superterrane and the late Mesozoic western margin of North America (i.e. the Intermontane superterrane). Contractional deformation along the length of the thrust belt was broadly coeval with arc magmatism, and thus records intra-arc tectonism. Late Palaeocene to Early Eocene deformation and uplift may mark the transition from contractional to extensional tectonism, and perhaps records the collapse of tectonically thickened crust.

The western margin of the Coast Plutonic Complex of the North American Cordillera comprises a northwest-trending zone from western British Columbia to northern southeast Alaska (latitude 49-59°; Fig. 1). The belt is characterized by widespread Cretaceous crustal shortening and metamorphism, and deep erosion (Crawford *et al.* 1976; Rubin *et al.* 1990a). Thrust sheets in the belt show extensive basement involvement, and include juvenile ensimatic elements of the Alexander terrane and continent-derived North American slope-and-rise deposits. The involvement of crystalline basement in thrust sheets and large thrust displacements imply that deformation affected the upper and middle crust, and not just a thin, detached sedimentary cover. This paper develops evidence of significant crustal thickening in the late Mesozoic, and draw analogies to similar orogenic belts such as the Andes (Suárez *et al.* 1983), where an anomalously thick crust existed by late Mesozoic time.

The tectonic evolution of the western margin of the Coast Plutonic Complex in western Canada and southeast Alaska has been discussed by numerous workers (Hollister 1982; Monger *et al.* 1982; Crawford *et al.* 1987; Rubin *et al.* 1990a). Two principal tectonic interpretations have been proposed to explain the extensive late Mesozoic deformation that extends for over 2000 km, mostly along the western side of the Coast Plutonic Complex: (1) collision of allochthonous terranes, in this case Alexander terrane and Wrangellia, against the western margin of Mesozoic North America; and (2) intra-arc deformation (Rubin *et al.* 1990a). Most workers agree, however, that the mid-Cretaceous and younger calc-alkaline plutons originated in a convergent-margin setting (Armstrong 1988). This paper presents an overview of mid-Cretaceous structural history and propose a model for the deformational framework of the western metamorphic belt of the Coast Plutonic Complex in southern southeast Alaska, between latitudes 55° and 56° N (Fig. 1).

Tectonic framework

The western North American Cordillera has been subdivided into two N-NW-trending lithotectonic belts, the Intermontane and Insular superterranes (Figs 1 & 2; terranes I and II of Monger *et al.* 1982). The Intermontane superterrane consists of lower Palaeozoic continent-derived slope- and- rise deposits, and upper Palaeozoic to lower Mesozoic ensimatic arc assemblages that probably formed adjacent to North America (see summaries in Miller 1987; Rubin *et al.* 1990b). The superterrane was accreted to North America in Early to Middle Jurassic time (Monger *et al.* 1982), and thus formed the western margin of the North American continent during

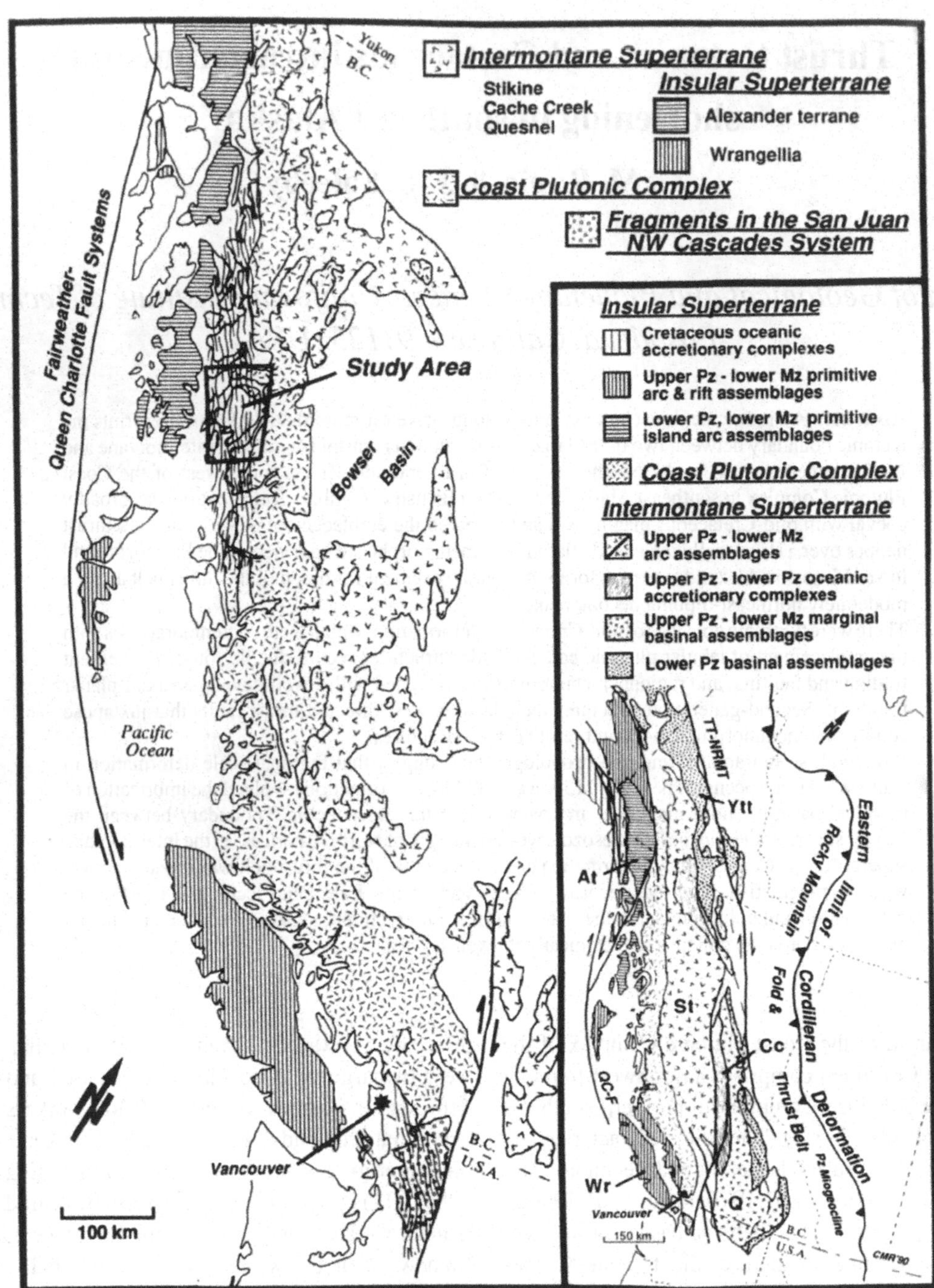

Figure 1. Generalized distribution of the Intermontane and Insular superterranes and the miogeocline of the northern Cordillera (modified after Monger *et al.* 1982; Monger & Berg 1987; Wheeler *et al.*,1988.; Miller 1987). Zone of mid-Cretaceous and younger deformation shown in wavy pattern. Also shown are some large-strike-slip faults; SC-FF = Straight Creek-Frazer River fault system. AT = Alexander terrane; Wr = Wrangellia. Inset shows superterranes and intervening Coast Plutonic Complex (CPC); TT-NRMT: Tintina-northern Rocky Mountain trench; QC-F = Queen Charlotte -Fairweather fault system.

Cretaceous time. The Insular superterrane, consisting of juvenile, mantle-derived volcanic arc and rift assemblages, is separated from the Intermontane superterrane to the west by the Mesozoic and early Cenozoic Coast Plutonic Complex (Fig. 1). The Insular superterrane has a strike length of 2000 km; apparently it had no palaeogeographic relation to North America until the Cretaceous (Monger *et al.* 1982; Saleeby 1983; Gehrels & Saleeby 1987a).

The tectonic boundary between the two superterranes generally coincides with the the Coast Plutonic Complex (Fig. 1). A regionally extensive mid- to Late Cretaceous thrust belt places rocks of the Intermontane superterrane above the eastern margin of the Insular superterrane. Different structural levels are exposed throughout the belt, from high-grade gneiss on the east, to low-grade phyllite on the west, making this an excellent area to study thrust tectonics.

Geological and stratigraphic setting

The Alexander terrane is structural basement to the thrust belt; it consists of a structurally intact lower Palaeozoic ensimatic arc sequence overlain by middle Palaeozoic clastic and carbonate strata (Fig. 2, Gehrels & Saleeby 1987a). The Palaeozoic rocks are capped by an Upper Triassic rift assemblage (Fig. 2, Gehrels & Saleeby 1987b). In most areas, rocks of the Alexander terrane are only slightly deformed and are not highly metamorphosed (Gehrels & Saleeby 1987), except near its eastern boundary where the Alexander terrane is affected by late Mesozoic deformational. Upper Jurassic and Lower Cretaceous marine pyroclastic and basinal strata of the Gravina sequence depositionally overlie the Alexander terrane are (Fig. 3, Berg *et al.* 1988), and constitute the remnants of an oceanic island arc that was constructed on the Alexander terrane (Rubin & McClelland 1989; Rubin & Saleeby unpubl. data). Middle Palaeozoic and lower Mesozoic rocks of the Alava sequence structurally overlie Alexander terrane and Gravina sequence rocks (Fig. 2). Locally, turbidite and channel-fill deposits of the Gravina sequence unconformably overlie the Alava sequence and thus form an overlap between the Alexander terrane and the Alava sequence. The Alava sequence consists of three lithotectonic units: (1) Pennsylvanian and Permian mafic metavolcanic tuff and flows, and massive crinoidal marble interbedded with black phyllite; (2) localized sequences of quartzite, mafic metavolcanic rocks, and marble; and (3) upper Middle Triassic carbonaceous marble, mafic metavolcanic tuff, breccia, and flows. The original stratigraphical relations and distribution of the Alava sequence are obscured by mid-Cretaceous and younger deformation, making stratigraphical correlation with adjacent terranes uncertain. Lithostratigraphy, age, and tectonic setting suggest that the Alava sequence is similar to the upper Palaeozoic and lower Mesozoic Stikine terrane (Intermontane superterrane; Rubin & Saleeby 1988). The presence of quartzite precludes a stratigraphical correlation with juvenile, age-equivalent elements of Wrangellia.

The Kah Shakes sequence occupies even higher structural levels in the northern portions of the thrust belt and consists of lower Palaeozoic quartzite, marble, amphibolite, calc-silicate rocks, and mid-Palaeozoic orthogneiss (Fig. 2). Detrital zircon from quartz-rich meta-psammitic rocks yield Proterozoic upper intercept U-Pb ages (Saleeby & Rubin 1989; Gehrels *et al.* 1990). Precambrian inheritance in zircon collected from orthogneiss implies an old continental source for some of the terrigenous rocks in the Kah Shakes sequence. The older portions of the Kah Shakes sequence represent outer shelf- and continental-rise deposits derived from the North American continent. This sequence may have been the depositional basement to the Alava sequence. Negative ε_{Nd} values and high initial Sr values for correlative rocks to the north support the interpretation that the Kah Shakes sequence is derived, in part, from old continental crust (Samson *et al.* 1989) and may be stratigraphicalally equivalent to the Yukon-Tanana terrane (Fig. 2).

Rocks in the thrust belt were intruded by mafic-ultramafic complexes, epidote-bearing tonalite, quartz diorite and granodiorite plutons, sills and dikes that range in age from ≈ 110 to 89 Ma (U-Pb ages on zircon, Rubin & Saleeby 1987a,

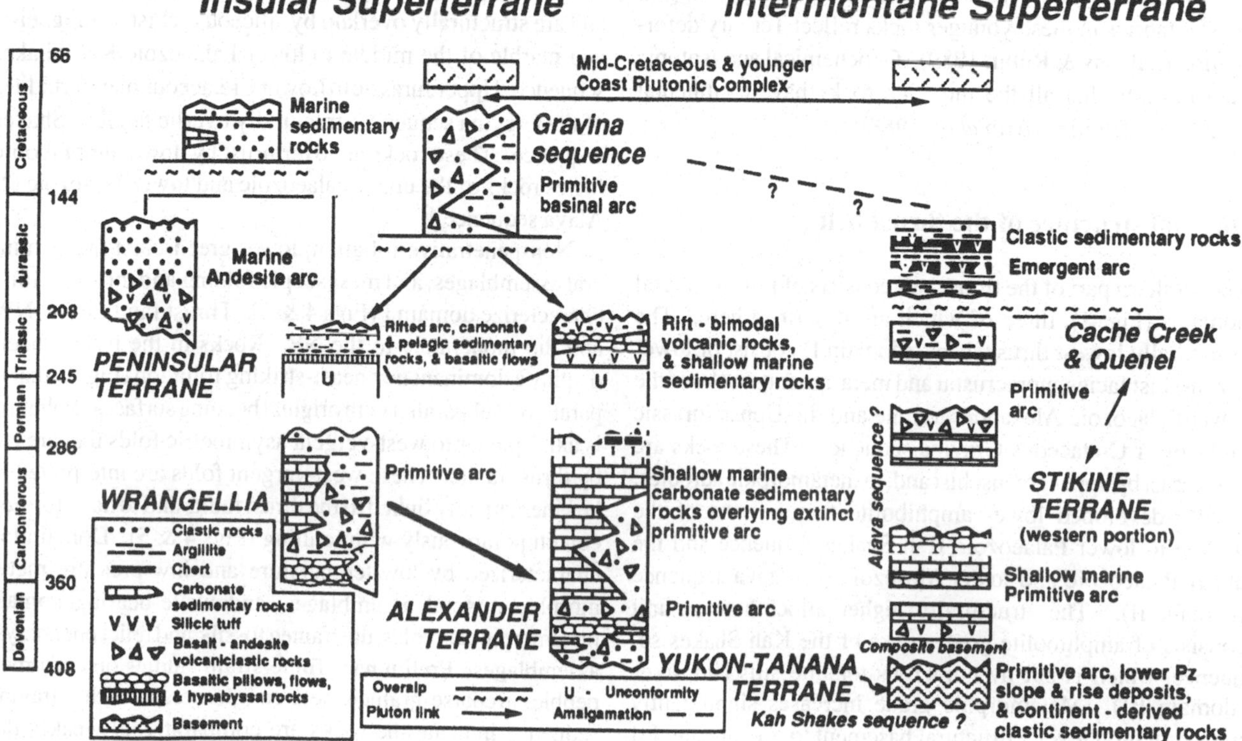

Figure 2. Petrotectonic and structural histories of the Intermontane and Insular superterranes of British Columbia and southern Alaska (modified from

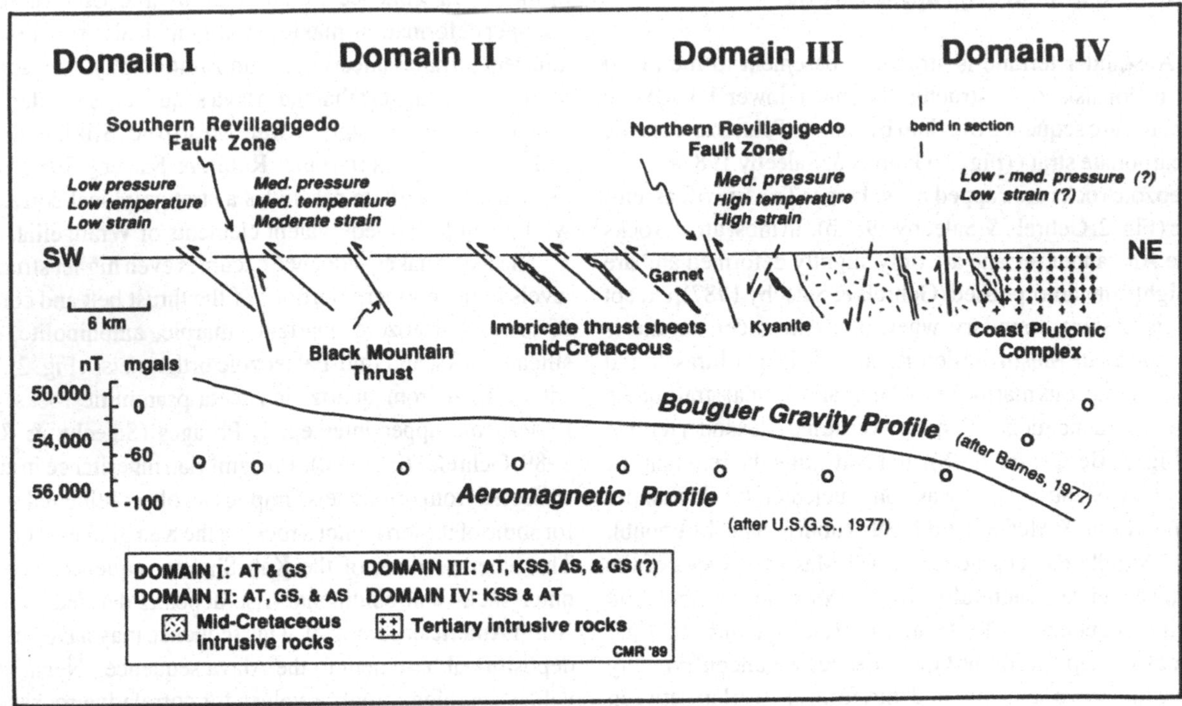

Figure 3. Geological domains, aeromagnetic, and gravity profiles across the western metamorphic belt of the Coast Plutonic Complex showing the mid-Cretaceous structural domains. Late Paleocene-Eocene fabrics do not strongly overprint this area. AT = Alexander terrane; GS = Gravina sequence; AS = Alava sequence; KSS = Kah Shakes sequence.

b; Arth *et al.* 1988). These ≈100 Ma or younger plutons crosscut the earlier regional metamorphic fabric; 95-100 Ma plutons are cut by thrust faults, whereas 90 Ma plutons crosscut thrust faults. A belt of Early Palaeocene to Early Eocene plutons occurs east of the thrust belt (Smith & Diggles 1981); fabrics in these younger rocks reflect Tertiary deformation (Saleeby & Rubin 1989). Geochemical and isotopic data indicate that all the intrusive rocks have continental magmatic affinities (Arth *et al.* 1988).

Internal structure of the thrust belt

The northern part of the thrust belt consists of four structural domains (Fig. 3), three of which are described here. The structurally lowest thrust sheets (domain I) consist of lower greenschist facies supracrustal and meta-igneous rocks of the lower Palaeozoic Alexander terrane and the Upper Jurassic and Lower Cretaceous Gravina sequence. These rocks are overthrust by upper greenschist and, in metamorphic aureoles, locally developed lower amphibolite facies rocks of the middle to lower Palaeozoic Kah Shakes sequence and the upper Palaeozoic and lower Mesozoic (?) Alava sequence (domain II). The structurally higher allochthonous unit consists of amphibolite grade rocks of the Kah Shakes sequence, which is cut by numerous tonalite sills and dikes (domain III). Metamorphic grade increases significantly with structural level. Structural basement to domain I and II is the Alexander terrane; higher nappes contain slices of basement consisting of the Alexander terrane rocks and the Kah Shakes sequence.

Structural styles and the stacking order of thrust sheets are different in the southern portion of the thrust belt. There, structurally lowest thrust sheets consist of lower Paleozoic supracrustal and meta-igneous rocks of the Alexander terrane and are structurally overlain by siliceous schist, orthogneiss, and marble of the middle to lower Palaeozoic Kah Shakes sequence. Upper Jurassic to Lower Cretaceous metaturbidites of the Gravina sequence structurally overlie the Kah Shakes sequence. These rocks are overthrust by lower amphibolite facies rocks of the upper Palaeozoic and lower Mesozoic (?) Alava sequence.

Non-penetrative foliation, lower greenschist facies mineral assemblages, and mesoscopic folding and thrust faulting characterize domain I (Figs 4 & 5). Thrust faults strike NW and dip moderately to the NE. Rocks in the thrust sheets display a dominant northeast-striking foliation (Fig. 6) that is parallel or subparallel with original bedding surfaces. Foliation is axial-planar to west-vergent asymmetric folds that are cut by thrust faults. These west-vergent folds are interpreted to be kinematically linked to the thrust faults and to have formed contemporaneously with faulting (Figs 4 & 5). Domain I is characterized by low-temperature and low-pressure metamorphic mineral assemblages, andalusite-bearing contact metamorphic aureoles, unstrained fossils, and relict phenocryst assemblages. Preliminary finite-strain studies on deformed pebbles in coarse-grained meta-conglomerates of the Gravina sequence indicate the rocks are comparatively weakly deformed; R_{XZ} ratios are 5:1 or smaller (C. Rubin unpubl. data). Based on these data, rocks in domain I show little finite strain and did not experience temperatures greater than 400° C or pressures above 3 kb.

Domain II lies structurally above domain I and is composed largely of imbricate thrust sheets consisting of lower Palaeozoic and lower Mesozoic supracrustal rocks of the Alava sequence, Upper Jurassic (?) and lower Cretaceous metasedimentary rocks of the Gravina sequence, and local lower Palaeozoic orthogneiss of the Alexander terrane. Some of these rocks (i.e. Alava sequence) may be thrust slices of Intermontane superterrane rocks (Fig. 3). The Alava and Gravina sequences on central Revillagigedo Island form an imbricate series of thrust sheets with intervening thrust slabs of Kah Shakes (Figs 4 & 5b). Stratigraphic control is based on isotopically dated (Pb-U zircon) granitic clasts in laterally continuous Gravina sequence metaturbidites, distinctive Permian fossiliferous marble and Middle Triassic carbonaceous marble horizons in the Alava sequence (Silberling *et al.* 1982). The southern Revillagigedo thrust separates domains I and II (Figs 3 & 5), ramps to deeper structural and stratigraphical levels, and dips moderately towards the east. Meta-igneous pebble- to cobble-conglomerate in the Gravina sequence shows flattening fabrics that

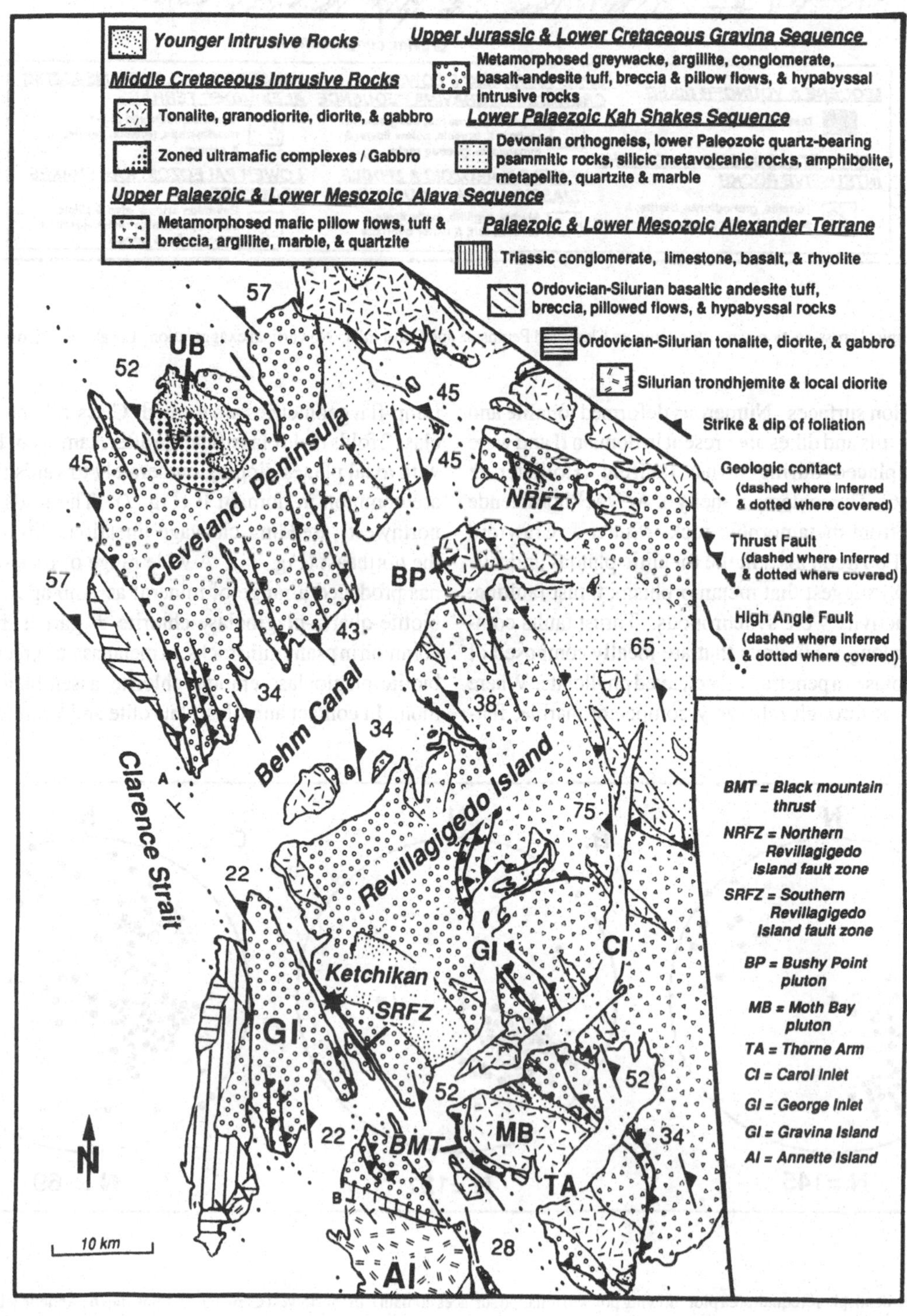

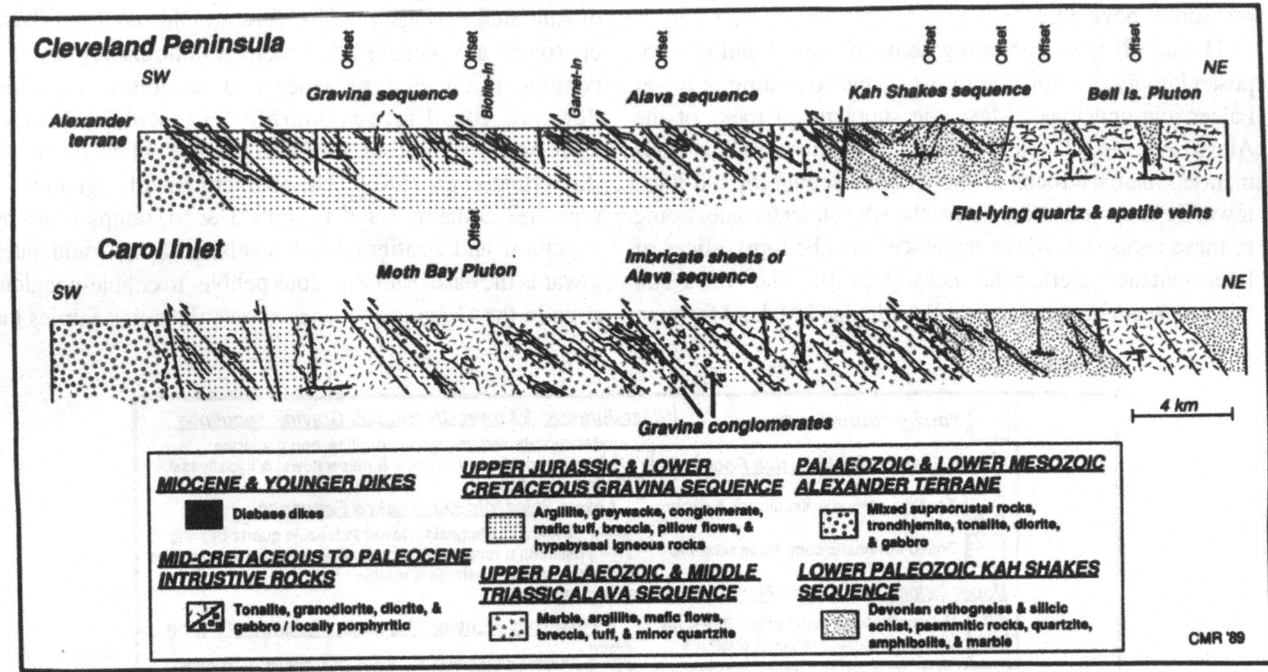

Figure 5. Geological cross-section across the southern Cleveland Peninsula and Carol Inlet. No vertical exaggeration. Locations of cross-sections shown on Figure 4.

parallel foliation surfaces. Numerous deformed tonalite and quartz diorite sills and dikes are present in domain II and were probably emplaced during thrust faulting (Hollister & Crawford 1986). The contemporaneity of Ar^{40}/Ar^{39} hornblende cooling ages from metamorphic rocks (Sutter & Crawford 1985) and U-Pb zircon ages in the tonalitic plutons and sills (Saleeby 1988) suggest that metamorphism, thrust faulting, and igneous activity were synchronous. Thrust faults commonly break along weak units that are locally composed of marble phaccoids in a penetratively cleaved phyllonite. Where thrust faults cut through relatively competent marble, flat-ramp-flat geometries are formed. Class 1C and similar-style class 2 folds (fold classification after Ramsay & Huber 1987), with axial planes dipping moderately towards the northeast, are common in domain II (Fig. 7). The folds verge to the northwest. Bedding-cleavage lineations dip moderately to the northeast (Fig. 7a). A wide range of rock compositions has produced a range of mineral assemblages: white mica-biotite-quartz-plagioclase-chlorite ± garnet ± calcite are common in psammitic rocks; in metabasalts, green amphibole-biotite-plagioclase-epidote-chlorite assemblages are common. In contact aureoles, staurolite and kyanite are present.

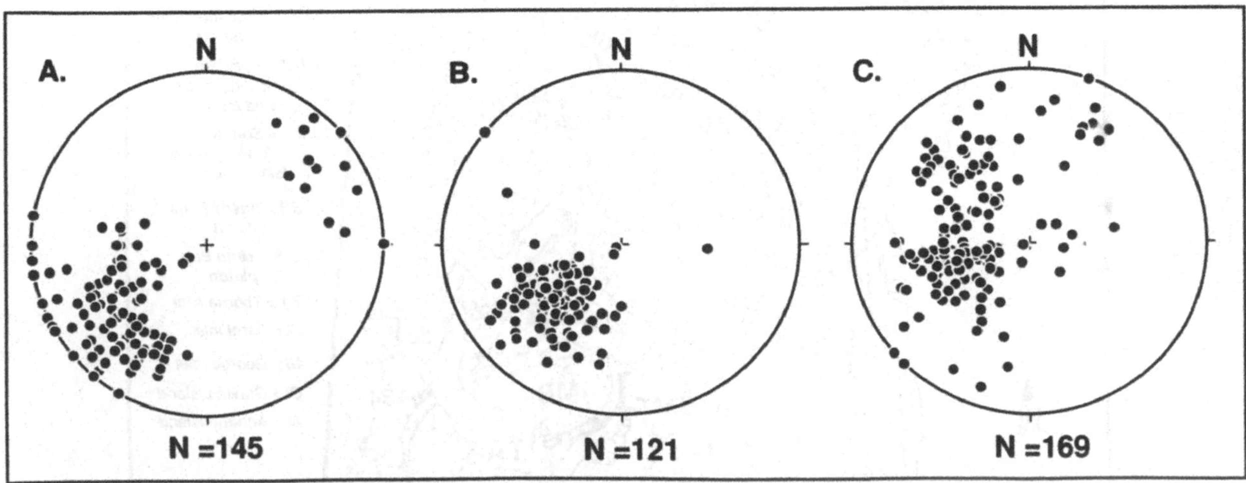

Figure 6. Lower-hemisphere, equal-area plot showing poles to foliation surfaces domain I, (a) northwest Cleveland Peninsula, (b) Annette Island, (c) Gravina Island.

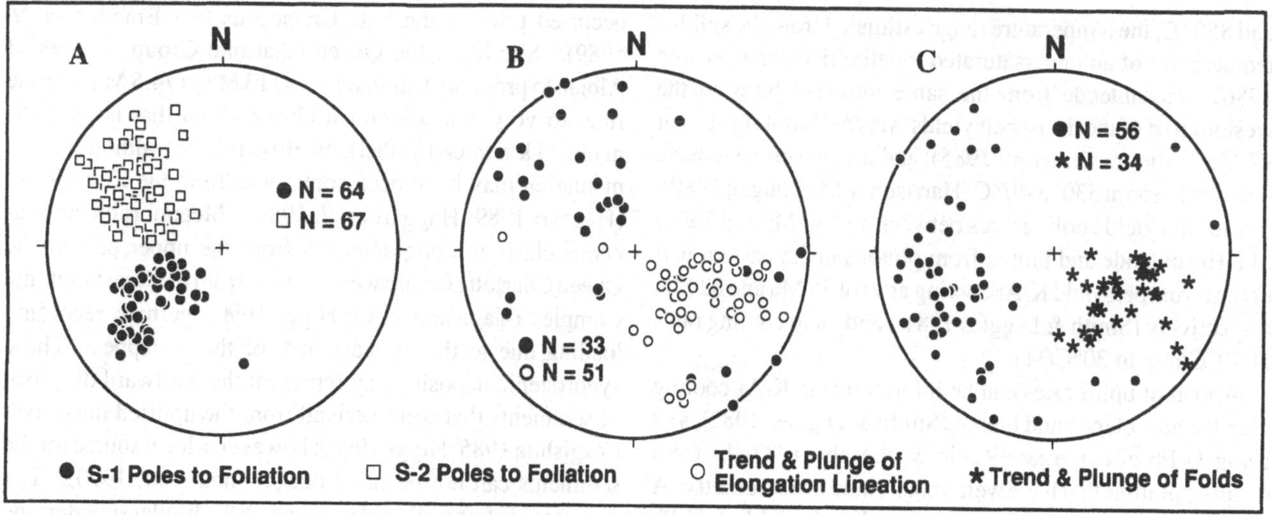

Figure 7. Lower-hemisphere, equal-area plot showing, (a) poles to foliation and cleavage in domain II, (b) poles to foliation and trend and plunge of small-scale folds in domain II, and (c) poles to foliation and trend and plunge of elongation lineation in domain III.

In summary, domain II is characterized by upper greenschist facies; locally amphibolite assemblages are present. Metamorphism occurred at pressures ranging as high as 5 to 6 kb.

The structurally higher thrust sheets, exposed in domain III, are internally complex. Domain III is characterized by polydeformed, isoclinally folded, amphibolite-facies supracrustal and meta-igneous rocks (Fig. 3). This domain contains large complex crystalline nappes. Characteristic features include well-developed moderately-dipping elongation lineation that parallels fold axes (Fig. 7b, c), highly attenuated fold limbs, and amphibolite facies metamorphism that is locally replaced by retrograde assemblages. Similar-style class 2 and class 3 folds are common in this domain (fold classification after Ramsay & Huber 1987). Metamorphic mineral assemblages include hornblende-plagioclase ± biotite ± garnet in mafic gneiss, quartz-plagioclase-biotite-kyanite in pelitic strata, and quartz-plagioclase-biotite-orthoclase-garnet in medium-grained psammatic gneiss. Steep north- to northeast-dipping ductile thrust faults crosscut earlier regional fabrics. These fault zones have a strongly developed foliation that transposes earlier fabrics (Figs 3 & 4; e.g. the Black Mountain thrust, northern and southern Revillagigedo fault zones). The orientation of the faults at depth is not accurately known. A consistent sense of motion can be inferred using kinematic indicators (e.g. s-c fabrics, fold vergence) and the juxtaposition of rocks with differing metamorphic grades (Fig. 3). Thrust faults, such as the northern Revillagigedo fault zone (Figs 3 & 4), have a minimum separation of 10 km, based on contrasting metamorphic grades. Late Cretaceous uplift and erosion of the western part of the orogenic belt were most likely accommodated along these thrust faults. In the southern and eastern portions of the thrust belt, subsequent Palaeocene and younger deformation has re-oriented mid-Cretaceous fabrics. Here, deep structural levels of the west-vergent thrust belt were transported upwards along steeply dipping east-vergent faults.

Timing of mid-Cretaceous orogenic events

Age constraints for the timing of tectonism come from plutons which pre-date deformation and others which post-date deformation; some are syntectonic. The earliest phase of thrust faulting affects domains I and II. Here, the a 101 Ma hornblende-bearing tonalite and granodiorite (U-Pb zircon age on the Moth Bay pluton, Rubin & Saleeby 1987) was emplaced across previously imbricated thrust slices of the Alexander terrane, and the Kah Shakes and Gravina sequences. The pluton crosscuts the dominant northeast striking foliation of the metasedimentary country rocks; foliated xenoliths occur in the pluton. Along the southern margin of the Moth Bay pluton, a southwest-vergent thrust fault crosscuts the pluton. These relations suggest that deformation was active before and after its emplacement. Deformed intrusive rocks of similar age that intrude the Alexander terrane are exposed to the west (Saleeby unpubl. data) and indicate that thrust faulting was active prior to 100 Ma. Thrust faulting continued to ≈ 90 Ma (U-Pb zircon), the age of the youngest plutons affected by thrust-related deformation; however, most 90 Ma plutons are not deformed and crosscut Cretaceous structural fabrics. Hornblende K-Ar cooling ages as young as 70 Ma for these plutons bracket the end of metamorphism and associated deformation (Smith & Diggles 1981). The age of the youngest strata (Gravina sequence) involved in thrust faulting is Albian (Berg *et al.* 1972). Thus, much of the deformation occurred between the Albian and the Turonian, approximately 113 Ma to 89 Ma (time scale after Harland *et al.* 1982).

Uplift, erosion and foredeep development

Cooling rates for the Moth Bay pluton can be calculated using U-Pb data obtained from the analyses of zircon, and from K/Ar and Ar^{40}/Ar^{39} data from hornblende and biotite. A 101 Ma zircon age dates crystallization that occurred between 1000°

and 850° C, the temperature range estimated from the solidus temperature of an undersaturated tonalite (Huang & Wyllie 1986). Hornblende from the same intrusive body in the western part of the thrust belt yields Ar^{40}/Ar^{39} cooling ages of 97 Ma (Sutter & Crawford 1985), and are presumed to have retained argon at 530° ± 40° C (Harrison & McDougall 1980). These data yield cooling rates between 112°C/Ma and 75°C/Ma Hornblende and biotite from plutons in the eastern part of the thrust belt yield K/Ar cooling ages of 89 Ma and 82 Ma, respectively (Smith & Diggles 1981), and yield cooling rates of 70°C/m.y to 30°C/Ma

Apparent uplift rates can be inferred using K/Ar cooling ages for hornblende and biotite (Smith & Diggles 1982), and using U-Pb zircon ages (Rubin & Saleeby 1987a,b) from tonalitic plutons that have well-constrained geobarometry. A tabular tonalite pluton (Bushy Point pluton) yielding U-Pb ages of 95 Ma is inferred to have crystallized at a depth of ≈ 25 km. The pressure estimate is based on Al^T (total aluminium) content in hornblende (Hammarstrom & Zen 1986) and the experimental calibration of the Al^T in the hornblende geobarometer (Johnson & Rutherford 1989). Hornblende and biotite from the surrounding country rock yield K/Ar cooling ages of 83.0 Ma ± 2.5 and 80.2 ± 4 Ma, and are presumed to have retained argon at 280° ± 40° C and 530° ± 40° C, respectively (Harrison 1981; Harrison & McDougall 1980). Assuming a mid-Cretaceous geothermal gradient of ≈25°C/km, these data imply an apparent uplift rate of ≈0.9 km/Ma To the south, the inferred mid-Cretaceous average uplift rate is ≈0.58mm/yr (Zen 1988), which is comparable with the uplift rates inferred for the Ketchikan area. These rates are comparable with modern orogenic uplift and erosion rates (see Dahlen & Suppe 1988).

Synorogenic to post-orogenic marine and nonmarine foredeep sequences may be represented by Upper Cretaceous parts of the Queen Charlotte and Nanaimo Groups in western British Columbia. The Nanaimo Group consists of Turonian to Maastrichtian (≈90 Ma to 65 Ma) marine and non-marine sandstone, conglomerate, and mudstone, approximately 2.5 km thick, that unconformably overlies rocks of Wrangellia (Muller & Jeletsky 1970; England 1989). The Upper Cretaceous Nanaimo conglomerates contain metamorphic clasts that were probably derived from thrust sheets exposed to the east, indicating that both metamorphism and thrust faulting occurred prior to the Late Cretaceous (see Brandon *et al.* 1989). Similarly, the Queen Charlotte Group consists of Albian to probably Campanian (≈113 Ma to 74.5 Ma) marine fine- to very coarse-grained clastic strata that is approximately 3 km thick (Sutherland-Brown 1968), although sedimentation may have been continuous from the Hauterivian (Haggart 1989; Haggart *et al.* 1989). Metamorphic and igneous clasts in conglomerates from the upper part of the Queen Charlotte Group were derived from the Coast Plutonic Complex (Yagishita 1985; Higgs 1990); perhaps recording loading due to the emplacement of thrust nappes. These synorogenic deposits may represent the westward dispersal of sediments that were derived from the uplifted thrust belt (Yagishita 1985; Higgs 1990); however, a local source for the sediments can not be ruled out (Gamba *et al.* 1990). The presence of strong lithosphere beneath the Insular superterrane may have prevented the development of deep basins; thus shallow foredeep basins were formed as the crust was flexed by the weight of the thrust sheets.

Implications for basement tectonics and convergent margin tectonics

Major differences between the structural styles present in southeast Alaska and mountain belts characterized by thin-skin deformation (e.g. fold and thrust belt of the Canadian Rockies) reflect fundamental differences in basement deformation, level of crustal exposure during deformation, and subsequent tectonic evolution. Mid-Cretaceous thrust faults in southeast Alaska dip between 40° and 70°, and affect predominantly crystalline basement, and are perhaps analogous to present-day deformation in the high Andes (Stauder 1975; Suárez *et al.* 1983). Steeply dipping thrust faults and focal depths between 15 and 25 km characterize the central Andes (Chinn & Isacks 1983; Stauder 1975; Suárez *et al.* 1983). Most earthquakes within the central Andes occur within basement and show little evidence of thin-skin deformation. Locally, however, some steeply dipping recent thrust faults recognized in the high Andes (e.g. Pariahuana fault) have shallower seismisity, with seismic activity no deeper than 20 km (Philip & Mégard 1977). In places, contemporaneous minor folds formed parallel to the fault trace (Philip &

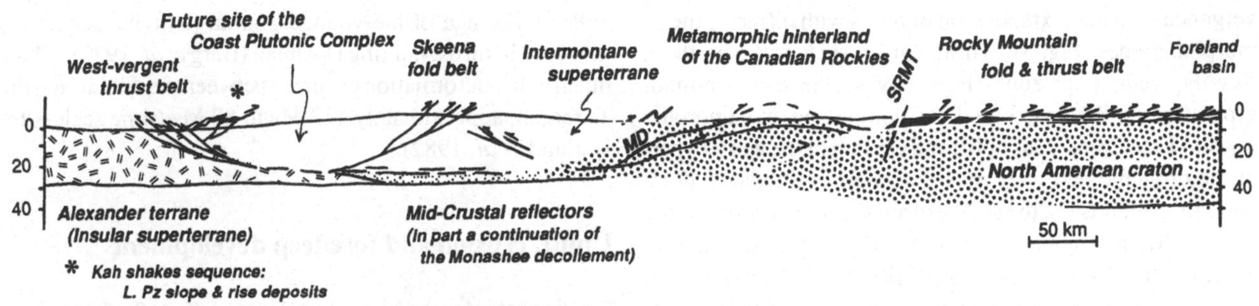

Figure 8. Simplified cross-section across the western Cordillera at the end of the Cretaceous compressive phase. Topography is current present day erosion-level. SRMT = southern Rocky Mountain trench; MD = Monashee decollement. Strike-slip faults normal to cross-section not included. Rocky Mountain thrust belt simplified from Price & Mountjoy (1970); metamorphic hinterland of the Rocky Mountain thrust belt simplified from Brown *et al.* (1976); Skeena

Mégard 1977). Contractional deformation may be due to a combination of factors which include, high convergence rates (Pardo-Casas & Molnar 1987), the age of the subducted ocean floor (Molnar & Atwater 1983), the angle of the subducting slab (Wdowinski *et al.* 1989), and the mechanical coupling between the South American and Nazca plates (Uyeda & Kanamori 1979). Such faulting is analogous to mid-Cretaceous thrust faulting in southeast Alaska, where thrust faults dip steeply and extend into crystalline basement. Mid-Cretaceous thrust faults, such as the Black Mountain thrust (Fig. 4), probably formed in a comparable tectonic setting, possibly recording coupled subduction between the North American and Pacific/Farallon plates. The high Andes have also undergone very different styles of deformation during their tectonic evolution, in which back-folding or 'retrocharriage' occurred during the later stages of thrust faulting. In this context, the westward dipping faults and foliation surfaces in the eastern portion of the thrust belt in southeast Alaska may be analogous to the modern day structural reversals seen in the Andes. Although the analogy between the mid-Cretaceous thrust system and the Andes is imperfect, the mid-Cretaceous thrust system involves basement deformation, whereas much of the deformation to the east (i.e. east-vergent thrust faulting in the Bowser basin) may have occurred aseismically along low-angle decollements. By analogy with the Andes, mid-Cretaceous thrust faulting was driven by strongly coupled subduction in which deformation is superposed on either an older collisional (Saleeby & Rubin 1990), a strike-slip boundary (McClelland & Gehrels in press) or a combination of the two. The presence of coeval igneous activity, convergence between continental fragments, and the transport of crystalline slabs along thrust faults in the mid-Cretaceous support such an analogy with the Andes.

Beneath most of the western wall rocks of the Coast Plutonic Complex, the present-day crustal thickness is less than 30 km (Barnes 1972a, b, 1977), and locally may be as thin as 25 km (Yorath *et al.* 1985). Emplacement pressure of ≈6-7 kb for 100 Ma plutons (Hammarstrom & Zen 1986; Rubin & Saleeby 1988) suggest the crust was at least 25 km thick at this time. Thrust faulting during the mid-Cretaceous (90 Ma) may have thickened the crust to 50 or 60 km. This estimate for mid-Cretaceous crustal thickness is consistent with the present-day thickness of 25 to 30 km, plus the amount of crust that must have been removed to expose Cretaceous mid-crustal rocks. This is in accord with crustal thicknesses in regions which have undergone similar styles of crustal thickening, such as the hinterland of the Canadian Rockies (Monger & Price 1979; Monger *et al.* 1985).

Uplift and erosion of the thrust belt in an extensional setting during Palaeocene and Eocene time subsequently thinned the crust; the available constraints on present crustal structure and magnitude of Tertiary extension, however, are poorly known. Rapid early Tertiary uplift rates of about 2 km/ma have been inferred for the western flank of the Coast Plutonic Complex (Hollister 1982; Harrison *et al.* 1979), although the uplift has also been ascribed to contraction (Hollister 1982; Crawford *et al.* 1987). Extensional tectonism is documented on the west flank of the Coast Plutonic Complex, where thrust-related fabrics are truncated by east-dipping Tertiary normal faults (McClelland *et al.* 1990). Abundant mid-Cretaceous and younger sills and plutons, and a moderate geothermal gradient resulted in hot, thermally weakened crust which effectively reduced the overall strength of the crust. By analogy with regions of active extension in the high Andes, tectonically thickened mid-Cretaceous crust may have collapsed and subsided due the gravitational potential of a high mountain belt.

Locally, the emplacement of arc-derived calc-alkaline magmas was coeval with mid-Cretaceous deformation. Thus, mid-Cretaceous contraction occurred in an intra-arc setting. Addition of voluminous early- to mid-Tertiary plutons coincided with and outlasted erosion, uplift, and inferred extension of the tectonically thickened and thermally weakened crust. The mid-Cretaceous to Early Tertiary tectonic history reflects contractional and extensional events that occurred in a convergent-margin setting. Changes in overall plate convergence rates and the rate of subduction may explain differing structural styles (see discussion in Royden & Burchfiel 1989). High convergence rates combined with strongly coupled subduction are characterized by compressional arc and back-arc regimes, and are typical of continental convergent margins (Uyeda & Kanamori 1979), whereas extension may have resulted from decoupling of the down-going slab from the overriding plate. Slower subduction rates reduce horizontal compressive stresses required to support high topography, and may result in topographic collapse by normal faulting (Dalmayrac & Molnar 1983). Mid-Cretaceous tectonism coincides with an abrupt increase of convergence rates (>100 km/Ma) between the Farallon and North American plates (Engebretson *et al.* 1985); the intense deformation in the northwestern Cordillera may have been related to rapid subduction. The correlation of rapid subduction with intracontinental tectonism has also been proposed in the Andes (see discussion in Pardo-Casas & Molnar 1987).

Controversy exists concerning the role of collision of allochthonous crustal elements (e.g. the Alexander terrane and Wrangellia) in late Mesozoic deformation of the northwest Cordillera (Monger *et al.* 1982; Crawford *et al.* 1987; Pavlis 1989; Rubin *et al.* 1990a). The absence of an accretionary wedge and high-pressure metamorphic terranes, combined with the lack of evidence for late Mesozoic oceanic crust between the Insular and Intermontane superterranes at these latitudes, are important constraints that argue against collisional tectonic models. There is no general agreement concerning plate tectonic models that explain mid-Cretaceous and younger deformation, yet deformation and igneous activity must have occurred in a magmatic arc setting. Thus, the collision of exotic tectonic elements and the concurrent closure of an late Mesozoic oceanic basin is not required to explain the stratigraphical, structural, and metamorphic relations seen in southeast Alaska and western British Columbia. Terrane accretion did not play a major role in mid-Cretaceous deformation and mountain building. Mid-Cretaceous deformation was broadly coeval with crustal shortening to the east in Bowser Basin (Skeena fold belt of Evenchick 1991, this volume) and to the south in the southern portions of the

Coast Plutonic Complex (Journeay 1990). The coincidence in timing of deformation in the western Canadian Cordillera suggest that there is a detachment which underlies all of these zones and which may continue as far east as the Rocky Mountain fold and thrust belt (Fig. 8). The fault that underlies continental slope-and-rise deposits west of the Coast Plutonic Complex may be the western continuation of the Monashee decollement, which separates North American-type crust with Palaezoic to Mesozoic terranes (Fig. 8). This interpretation is supported by layered seismic reflectors near the southern boundary of the Coast Plutonic Complex which have been correlated to the Monashee decollement (Brown *et al.* 1991, this volume). The remarkable similarity of timing of deformation across the western Cordillera suggests tectonism was a widespread intracontinental event. Deformation formed in response to east-west compression over a distance of at least 600 km across the Northern Cordillera, and developed in a structural setting much like that of the modern Andes.

C.M.R. is grateful for long discussions with Bill McClelland, Meghan Miller, Jim Monger, and Margi Rusmore on Cordilleran tectonics. Field and laboratory work for southeast Alaska regional studies has been supported by US National Science Foundation Grants EAR 86-05386 and EAR 88-03834 (to JBS); additional support (to CMR) was provided by Geological Society of America Penrose Grants, a Sigma-Xi grant-in-aid, and by the US Geological Survey, Alaska Branch. Some of the work reported here was part of a doctoral thesis by C.M.R. at Caltech. The stereonet program was graciously provided by Rick Allmendinger. Informal reviews by Meghan Miller were quite helpful. Hugh Gabrielse, Ken McClay, and Bob Thompson provided very helpful and critical reviews.

References

Anderson, R. G. 1989. A stratigraphical, plutonic, and structural framework for the Iskut River map area, northwestern British Columbia. *In: Current research, Part E.* Geological Survey of Canada Paper, **89-1E**, 145-154.

Armstrong, R. L. 1988. Mesozoic and early Cenozoic magmatic evolution of the Canadian Cordillera. *Geological Society of America Special Paper*, **218**, 55-91.

Arth, J. G., Barker, F. & Stern, T. W. 1988. Coast Batholith and Taku plutons near Ketchikan, Alaska: Petrography, geochronology, geochemistry, and isotopic character. *American Journal of Science*, **288-A**, 461-489.

Barnes, D. F. 1972a Simple bouguer gravity anomaly map of Prince Rupert 1:250,000 quadrangle, southeastern Alaska, showing station locations, anomaly values, and generalized 10-milligal contours. *US Geological Survey Geophysical Open-File Map*, 1 sheet.

—— 1972b. Simple bouguer gravity anomaly map of Ketchikan 1:250,000 quadrangle, southeastern Alaska, showing station locations, anomaly values, and generalized 10-milligal contours. *US Geological Survey Geophysical Open-File Map*, 1 sheet.

—— 1977. Bouguer gravity map of Alaska: *US Geological Survey Geophysical Investigations Map* **GP-913**, 1 sheet, scale 1:2,500,000.

Berg, H. C., Jones, D. L. & Richter, D. H. 1972. Gravina-Nutzotin belt: Tectonic significance of an upper Mesozoic sedimentary and volcanic sequence in southern and southeastern Alaska. *US Geological Survey Professional Paper* **800-D**, 1-24.

——, Elliott, R. L. & Koch, R. D.1988a. Geological map of the Ketchikan and Prince Rupert quadrangles, Alaska. *US Geological Survey Miscellaneous Investigations Series*, Map 1-1807, 1 sheet, scale 1:250,000.

Brandon, M. T., Cowan, D. S. & Vance, J. A. 1988. The Late Cretaceous San Juan thrust system, San Juan Islands, Washington. *Geological Society of America Special Paper*, **221**, 81p.

Brown, R. L., Journeay, J. M., Lane, J. M., Murphy, D. C. & Rees, C. J. 1986. Obduction, backfolding and piggyback thrusting in the metamorphic hinterland of the southeastern Canadian Cordillera. *Journal of Structural Geology*, **8**, 255-258.

Chinn, D. S. & Isacks, B. L. 1983. Accurate source depths and focal mechanisms of shallow earthquakes in western South America and the New Hebrides Island arc. *Tectonics*, **2**, 529-653.

Crawford, M. L., Hollister, L. S. & Woodsworth, G. J. 1987. Crustal deformation and regional metamorphism across a terrane boundary, Coast Plutonic Complex, British Columbia. *Tectonics*, **6**, 343-361.

Engebretson, D. C., Cox, A. & Gordon, R. G. 1985. Relative motions between oceanic and continental plates in the Pacific basin. *Geological Society of America Special Paper*, **206**, 59p.

England, T.D.J. 1989. Lithostratigraphy of the Nanaimo Group, Georgia Basin, southwestern British Columbia. *In: Current Research, Part E,* Geological Survey of Canada, **89-1E**, 197-206.

Dahlen, F. A. & Suppe, J. 1988. Mechanics, growth, and erosion of mountain belts. *Geological Society of America Special Paper*, **218**, 161-178.

Dalmayrac, B. & Molnar, P. 1981. Parallel thrust and normal faulting in Peru and constraints on the state of stress. *Earth and Planetary Science Letters*, **55**, 473-481.

Gamba, C. A., Indrelid, J. & Taite, S. 1990 Sedimentology of the Upper Cretaceous Queen Charlotte Group, with special reference to the Honna Formation, Queen Charlotte Islands, British Columbia. *Geological Survey of Canada*, **90-1E**, 67-73.

Garner, M. C., Bergman. S. C., Cushing, G. W., MacKevett, E. M., Plafker, G., Campbell, R. B., Dodds, C. J., McClelland, W. C. & Mueller, P. A. 1988. Pennsylvanian pluton stitching of Wrangellia and the Alexander terrane, Wrangell Mountains, Alaska. *Geology*, **16**, 967-971.

Gehrels, G. E. & Saleeby, J. B. 1987a. Geological framework, tectonic evolution, and displacement history of the Alexander terrane. *Tectonics*, **6**, 151-173.

—— & Saleeby J. B. 1987b. Geology of Prince of Wales Island, southeastern Alaska. *Geological Society of America Bulletin*, **98**, 123-137.

——, McClelland, W. C., Samson, S. D., Patchett, P. J. & Jackson, J. L. 1990. Ancient continental margin assemblage in the northern Coast Mountains, southeast Alaska and northwest Canada. *Geology*, **18**, 208-211.

Harland, W. B., Cox, A. V., Lewellyn, P. G., Pickton, C.A.G., Smith, A. G. & Walters, R. 1982. *A Geological time scale*. Cambridge University Press, Cambridge, England, 131p.

Haggart, J. W. 1989. Reconnaissance stratigraphy and biochronology of the Lower Cretaceous Longarm Formation, Queen Charlotte Islands, British Columbia. *Geological Survey of Canada Paper*, **89-1H**, 39-46.

——, Lewis, P. D. & Hickson, C. J. 1989. Stratigraphy and structure of Cretaceous strata, Long Inlet, Queen Charlotte Islands, British Columbia. *Geological Survey of Canada Paper*, **89-1H**, 65-72.

Hammarstrom, J. M. & Zen, E. 1986. Aluminum in hornblende: An empirical igneous geobarometer. *American Mineralogist*, **71**, 1297-1313.

Harrison, T. M. 1981. The diffusion of ^{40}Ar in hornblende. *Contributions to Mineralogy and Petrology*, **78**, 324-331.

—— & McDougall I. 1980. Investigations of an intrusive contact, northwest Nelson, New Zealand-I. Thermal, chronological, and isotopic constraints. *Geochemica et Cosmochimica Acta*, **44**, 1985-2003.

——, Armstrong, R. L., Naeser, C. W., & Harakal, J. E. 1979. Geochronology and thermal history of the Coast Plutonic Complex, near Prince Rupert, British Columbia. *Canadian Journal of Earth Sciences*, **16**, 400-410.

Higgs, R. 1990. Sedimentology and tectonic implications of Cretaceous fan-delta conglomerates, Queen Charlotte Islands, Canada. *Sedimentology*, **37**, 83-103.

Hollister, L. S. 1982. Metamorphic evidence for rapid uplift of a portion of the central gneiss complex, Coast Mountains, B.C. *Canadian Mineralogist*, **20**, 319-332.

—— & Crawford, M. L. 1986. Melt-enhanced deformation: A major tectonic process. *Geology*, **14**, 358-561.

Huang, W. L. & Wyllie, P. J. 1986. Phase relationships of gabbro-tonalite-granite-water at 15 kb with applications to differentiation and anatexis. *American Mineralogist*, **71**, 301-316.

Johnson, M. C. & Rutherford, M. J. 1989. Experimental calibration of aluminium-in-hornblende geobarometer with application to Long Valley caldera (California) volcanic rocks. *Geology*, **17**, 837-841.

Jordan, T., Isacks, B., Allmendinger, R., Brewer, J., Ramos, V. & Ando, C. 1983. Andean tectonics related to the geometry of the subducted plate. *Geological Society of America Bulletin*, **94**, 341-346.

Journeay, J. M. 1990. A progress report on the structural and tectonic framework of the southern Coast Belt, British Columbia. *Geological Survey of Canada*, **90-1E**, 183-195.

McClelland, W. C., & Gehrels, G. E. 1990. Geology of the Duncan Canal shear zone and its bearing on the accretionary history of the Alexander terrane, southeastern Alaska. *Geological Society of America Bulletin* (in press).

——, Gehrels, G. E., Samson, S. D. & Patchett, P.J . 1990. Geological and structural relations along the western flank of the Coast Mountains batholith: Stikine River to Cape Fanshaw, central SE Alaska. *Geological Society of Canada Abstract with Programs*, **15**, 116.

Miller, M. M. 1987. Dispersed remnants of a northeast Pacific fringing arc - Upper Paleozoic island arc terrane of Permian McCloud faunal affinity, western US. *Tectonics*, **6**, 807-830.

Monger, J.W.H. & Berg, H. 1987. Lithotectonic terrane map of western Canada and southeastern Alaska. *US Geological Survey Miscellaneous Field Studies Map* MF 1874B, scale 1: 2,500,000.

——& Price, R. A. 1979. Geodynamic evolution of the Canadian Cordillera—progress and problems. *Canadian Journal of Earth Science*, **16**, 770-791.

——, Price, R. A. & Tempelman-Kluit, J. D. 1982. Tectonic accretion and the origin of the two major metamorphic and plutonic welts in the Canadian Cordillera. *Geology*, **10**, 70-75.

——, Clowes, R. M., Price, R. A., Simony, P. S., Riddihough, R. P. & Woodsworth, G. J. 1985 B-2 Juan de Fuca plate to the Alberta plains: Geological Society of America Decade of North American Geology Centennial Continent/Ocean Transect 7, scale 1:1,000,000, 21p.

Muller, J. E. & Jeletsky, J. A. 1970. Geology of the Upper Cretaceous Nanaimo Group, Vancouver Island and Gulf Islands, British Columbia. *Geological Survey of Canada Paper*, **69-27**, 77p.

Pardo-Casas. F. & Molnar, P. 1987. Relative motion of the Nazca (Farallon) and South America plates since the Late Cretaceous time. *Tectonics*, **6**, 233-248.

Pavlis, T. 1989. Middle Cretaceous orogenesis in the northern Cordillera: A Mediterranean analog. *Geology*, **17**, 947-950.

Philip, H. & Mégard, F. 1977. Structural analysis of the superficial deformation of the 1969 Pariahuanca earthquake (central Peru). *Tectonophysics*, **38**, 259-278.

Price, R. A. & Mountjoy, E. W. 1970. Geological structure of the Canadian Rocky Mountains between Bow and Athabaska rivers—A progress report. *Geological Association of Canada Special Paper*, **6**, 7-26.

Ramsay, J. G. & Huber, M. I. 1987. Techniques of modern structural geology, volume 2, Folds and Fractures. Academic Press, 700p.

Royden, L. & Burchfiel, B. C. 1989. Arc systematic variations in ;thrust belt style related to plate boundary processes? (The western Alps versus the Carpathians). *Tectonics*, **8**, 51-61.

Rubin, C. M & Saleeby, J. B. 1987a. The inner boundary zone of the Alexander terrane in southern SE Alaska: A newly discovered thrust belt. *Geological Society of America Abstracts with Programs*, **19**, 455.

—— & —— 1987b. The inner boundary zone of the Alexander terrane in southern SE Alaska, Part 1: Cleveland Peninsula to southern Revillagigedo Island. *Geological Society of America Abstracts with Programs*, **19**, 826.

——&—— 1988. A new perspective on what is the Taku terrane in southern SE Alaska. *Geological Society of America Abstracts with Programs*, **20**, 226.

—— & McClelland, W. C. 1989. The Gravina Belt: Remnants of a mid-Mesozoic oceanic arc in southern southeast Alaska. *Eos Transactions in the American Geophysical Union*, 1308.

——, Saleeby, J. B., Cowan, D. S., McGroder M. F. & Brandon, M. T. 1990a. Late Mesozoic compressional tectonism: Development of a west-vergent thrust system in the northwestern Cordillera. *Geology*, **18**, 276-280.

——, Miller, M. M. & Smith, G. M. 1990b. Tectonic development of Cordilleran mid-Paleozoic volcano-plutonic complexes. *In*: Harwood, D. S. & Miller, M. M. (eds) *The paleogeography of the Klamath Mountains, Sierra Nevada, and adjacent areas. Geological Society of America Special Paper* (in press).

Saleeby, J. B. 1983 Accretionary tectonics of the North American Cordillera: *Annual. Reviews in Earth and Planetary. Sciences*, **15**, 45-73.

—— 1988. The inner boundary of the Alexander terrane in southern SE Alaska: part II southern Revillagigedo Island (RI) to Cape Fox (CI). *Geological Society of America Abstracts with Programs*, 19. 828.

—— & Rubin, C. M. 1989. The western margin of the Coast Plutonic Complex (CPC) in southernmost SE Alaska. *Geological Society of America Abstracts with Programs*, 21, 139.

—— & Rubin, C. M. 1990. Tectonic intercalation of crystalline nappes and basinal overlap relations in southern SE Alaska: Implications for the initial stages of Insular superterrane accretion. *Eos Transactions in the American Geophysical Union* (in press).

Samson, S. D., McClelland, W. C., Patchett, P. J. & Gehrels, G. E. 1989. Nd isotopes and Phanerozoic crustal genesis in the Canadian Cordillera. *Nature*, **337**, 705-709.

Silberling, N. J., Wardlaw, B. R. & Berg, H. C. 1982. New paleontologic age determinations form the Taku terrane, Ketchikan area, southeastern Alaska. *US Geological Survey Circular* **844**, 117-119.

Smith, J. G. & Diggles, M. F. 1981. Potassium-argon determinations in the Ketchikan and Prince Rupert quadrangles, southeastern Alaska. *US Geological Survey Open-File Report* **78-73N**, 1-6.

Stauder, W. 1975. Subduction of the Nazca plate under Peru as evidenced by focal mechanisms and by seismicity. *Journal of Geophyical Research*, **80**, 1053-1064.

Suárez, G., Molnar, P. & Burchfiel, B. C. 1983. Seismicity, fault plane solutions, depth of faulting, and active tectonics of the Andes of Peru, Ecuador, and southern Columbia. *Journal of Geophysical Research*, **88**, 10403-10428.

Sutherland-Brown, A 1968. Geology of the Queen Charlotte Islands British Columbia. *British Columbia Department of Mines and Petroleum Resources Bulletin*, **54**, 226pp.

Sutter, J. F. & Crawford, M. L. 1985. Timing of metamorphism and uplift in the vicinity of Prince Rupert, British Columbia and Ketchikan, Alaska. *Geological Society of America Abstracts with Programs*, **17**, 411.

United States Geological Survey. 1977. Aeromagnetic map of the Ketchikan, Prince Rupert, and northeastern Craig quadrangles, Alaska. *US Geological Survey Open-File Report Map*, 77-359.

Uyeda, S.& Kanamori, H. 1979. Back, arc opening and the mode of subduction. *Journal of Geophysical Research*, **84**, 1049-1061.

Yagishita, K. 1985. Evolution of a provenance as revealed by petrographic analyses of Cretaceous formations in the Queen Charlotte Islands, British Columbia, Canada. *Sedimentology*, **32**, 671-684.

Yorath, C. J., Woodsworth, G. J., Riddihough, R. P., Currie, R. G., Hyndman, R. D., Rogers, G. C., Seemann, D. A. & Collins, A. D. 1985. Continent—Ocean transect B1: Intermontane Belt (Skeena Mountains) to Insular Belt (Queen Charlotte Islands): Geological Society of America Decade of North American Geology Centennial Continent/Ocean Transect 8, scale 1:1,000,000, 7p.

Wdowinski, S., O'Connell, J. & England, P. 1989. A continuum model of continental deformation above subduction zones: Application to the Andes and the Aegean. *Journal of Geophysical Research*, **94**, 10331-10346.

Wheeler, J. O., Brookfield, A. J., Gabrielse, H., Monger, J.W.H., Tipper, H. W. & Woodsworth, G. J. 1988. Terrane map of the Canadian Cordillera. Canadian Geological Survey Open-File Report 1894, scale 1:2,000,000

Zen, E. 1988. Tectonic significance of high-pressure plutonic rocks in the westerns Cordillera of North America. *In*: Ernst, W. G. (ed.) *Metamorphism and crustal evolution of the western United States, Rubey volume VII*, Prentice Hall, Englewood Cliffs, New Jersey, 41-68.

Glossary of thrust tectonics terms

K.R. McClay

Department of Geology, Royal Holloway and Bedford New College, University of London, Egham, Surrey, England, TW20 0EX

This glossary aims to illustrate, and to define where possible, some of the more widely used terms in thrust tectonics. It is presented on a thematic basis - individual thrust faults and related structures, thrust systems, thrust fault related folds, 3-D thrust geometries, thrust sequences, models of thrust systems, and thrusts in inversion tectonics. Fundamental terms are defined first, followed by an alphabetical listing of related structures. Where appropriate key references are given.

Since some of the best studied thrust terranes such as the Canadian Rocky Mountains, the Appalachians, the Pyrenees, and the Moine thrust zone are relatively high level foreland fold and thrust belts it is inevitable that much of thrust tectonics terminology is concerned with structures found in the external zones of these belts. These thrustbelts characteristically consist of platform sediments deformed by thrust faults which have a ramp-flat trajectories (Bally *et al.* 1966; Dahlstrom 1969, 1970; Price 1981; Rich 1934; and others). Steps in the thrust surface generate geometrically necessary folds in the hangingwall above (Fig. 1). Therefore much attention has been focussed upon hangingwall deformation in thrust belts and upon conceptual and geometric models to explain them.

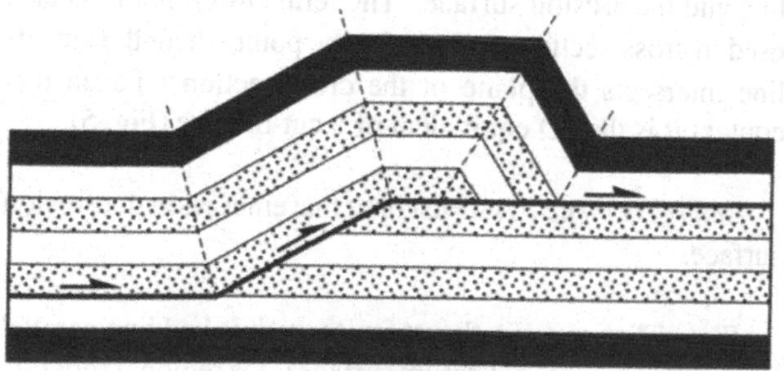

Figure 1. Idealised thrust fault showing kink band style folding in the hangingwall and no deformation in the footwall.

Many of the illustrations used in this glossary have been constructed using the programme *FaultII*™ (Wilkerson & Associates) that gives kink band geometries (e.g. Suppe 1983, 1985; Mitra 1986, 1990), commonly found in shallow thrustbelts. This, however, is not meant to imply that kink-band folding is the only geometric style found in high level thrustbelts (see Ramsay 1991, this volume). This glossary is not meant to be exhaustive but attempts to cover many of the terms used in the papers presented in this volume and those of thrust tectonics literature in general. The reader will recognise the difficulty in precisely defining many of the terms used in thrust tectonics as individual usages and preferences vary widely.

Thrust faults

Thrust fault: A contraction fault that shortens a datum surface, usually bedding in upper crustal rocks or a regional foliation surface in more highly metamorphosed rocks.

This section of the glossary defines terms applied to individual thrust faults (after McClay 1981; Butler 1982; Boyer & Elliott 1982; Diegel 1986).

Backthrust: A thrust fault which has an opposite vergence to that of the main thrust system or thrust belt (Fig. 2). Backthrusts are commonly *hinterland-vergent thrusts.*

Figure 2. Backthrust showing an opposite sense of vergence to that of the foreland vergent thrust system.

Blind thrust: A thrust fault that is not emergent - i.e. it remains buried such that the displacement on the thrust below is compensated by folding or cleavage development at a structurally higher level (Fig. 3) (cf. Thompson 1981).

Figure 3. Blind thrust system formed by a buried imbricate fan in which the overlying strata are shortened by folding.

Branch line: The line of intersection between two thrust sheets (Fig. 4a). For duplexes there are *trailing edge branch lines* and *leading edge branch lines* (see duplexes below). See Diegel (1986) for a full discussion of branch lines.

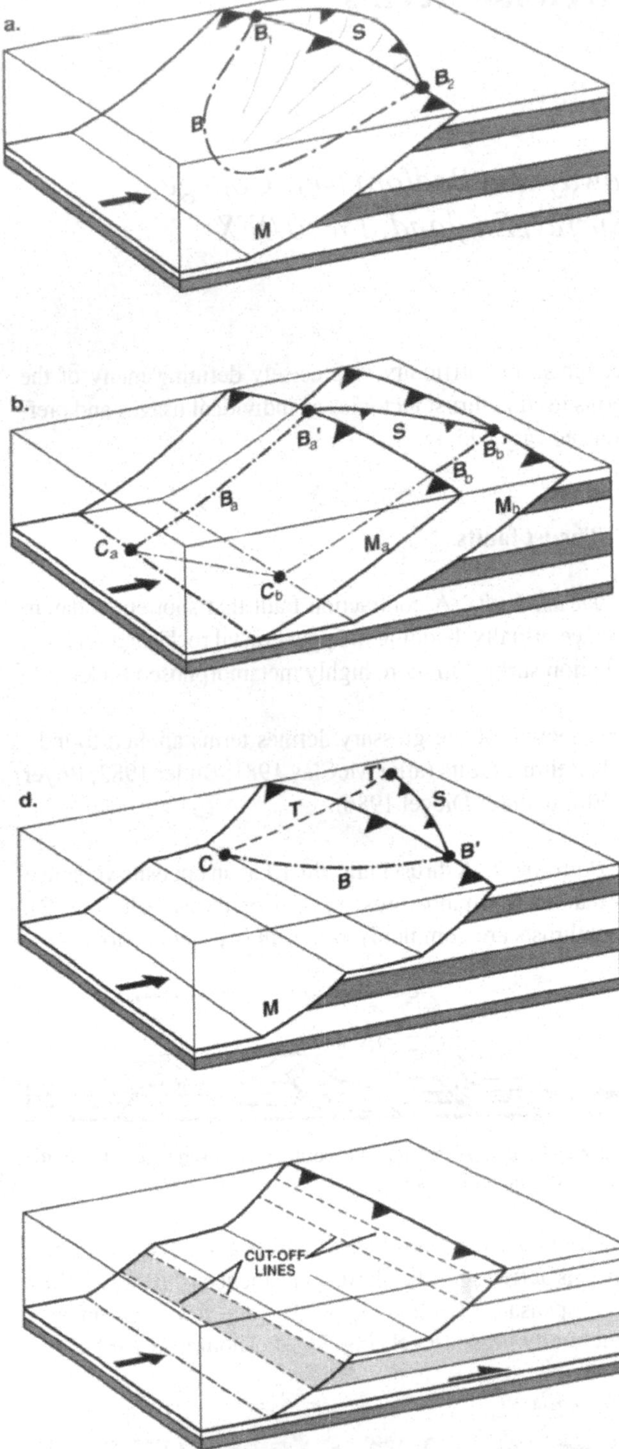

Figure 4. Thrust imbricates and splays showing branch lines, branch points and tip lines (adapted after Boyer & Elliott 1982). **(a)** *Rejoining imbricate* or *rejoining splay* **S** joins the *main fault* **M** along the *branch line* **B** which intersects the surface at two branch points $\mathbf{B_1}$ and $\mathbf{B_2}$. **(b)** *Connecting imbricate* or *connecting splay* **S** which joins two *main faults* $\mathbf{M_a}$ and $\mathbf{M_b}$ along *branch lines* $\mathbf{B_a}$ and $\mathbf{B_b}$. There are two *branch points* $\mathbf{B_a}'$ and $\mathbf{B_b}'$ and two *corners* $\mathbf{C_a}$ and $\mathbf{C_b}$. **(c)** *Isolated imbricate* or *isolated splay* **S** which intersects the *main fault* **M** along a *branch line* **B**. *Tip lines* $\mathbf{T_1}$ and $\mathbf{T_2}$ define the ends of the splay **S**. *Tip lines* $\mathbf{T_1}$ and $\mathbf{T_2}$ meet *branch line* **B** at *corners* $\mathbf{C_1}$ and $\mathbf{C_2}$. The *tip lines* intersect the surface at *tip points* $\mathbf{T_1}'$ and $\mathbf{T_2}'$. **(d)** *Diverging imbricate* or *diverging splay* **S** intersects the *main fault* **M** along *branch line* **B**. The *branch line* **B** has one *corner* **C** and a *branch point* **B'** where it intersects the surface. The *tip line* **T** has one *corner* **C** and a *tip point* **T'** where it intersects the surface.

Branch point: The point of intersection between a branch line and the erosion surface (Fig. 4b). The term *branch point* is also used in cross-section analysis for the point where the branch line intersects the plane of the cross-section - i.e. in this context it is the 2D equivalent of a branch line (Fig. 5).

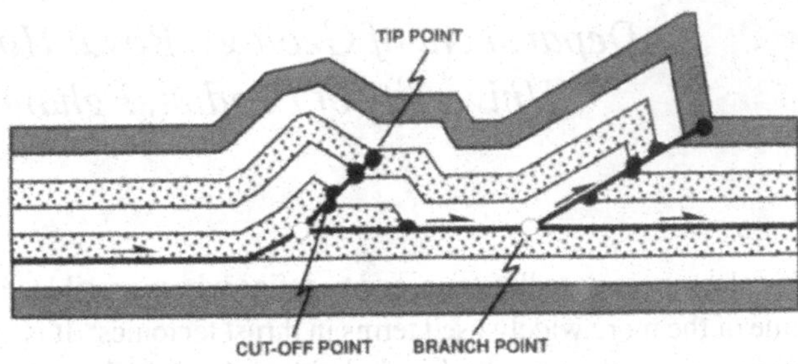

Figure 5. Cross-section illustrating branch points (white circles), cut-off points (black half-moons) and tip-points (black circles).

Cut-off line: The line of intersection between a thrust surface and a stratigraphic horizon (Fig. 6).

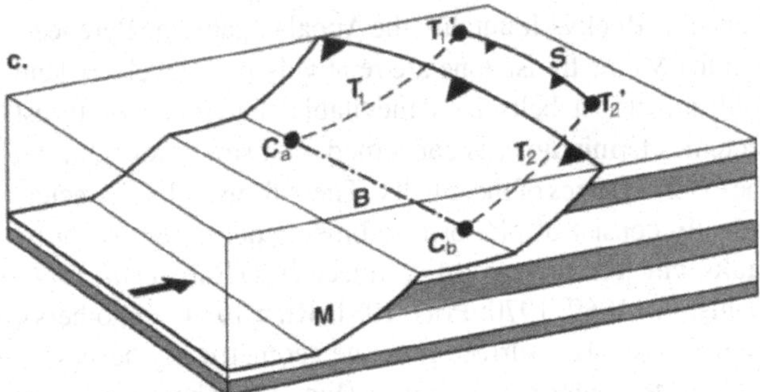

Figure 6. Footwall cut-off lines.

Cut-off point: The point of intersection between a cut-off line and the erosion surface. The term *cut-off point* is also used in cross-section analysis for the point where the cut-off line intersects the plane of the cross-section - i.e. in this context it is the 2D equivalent of a cut-off line (Fig. 5).

Emergent thrust: A thrust fault that emerges at the erosion surface.

Flat: That part of a thrust fault which is bedding parallel or parallel to a regional datum surface (cf. a regional foliation in metamorphic rocks) (Fig. 7).

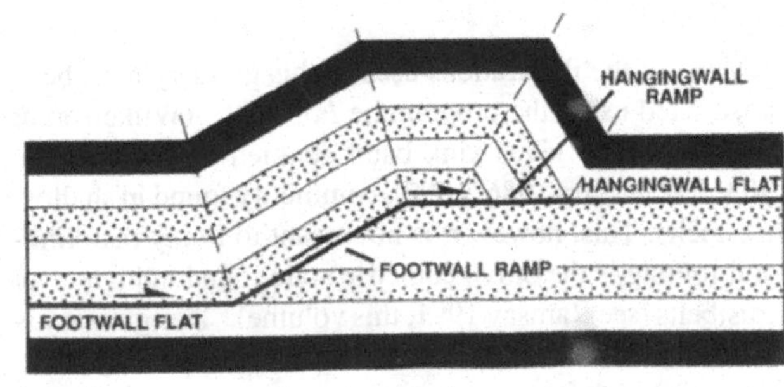

Figure 7. Thrust ramps in cross-section.

Footwall cut-off lines are the lines of intersection between a thrust surface and a stratigraphic horizon in the footwall of the thrust (Fig. 6) (Diegel 1986).

Footwall cut-off point: The point of intersection between a cut-off line in the footwall of the thrust and the erosion surface.

Footwall flat: That part of the footwall of a thrust fault where the thrust fault is bedding parallel or parallel to a regional datum surface (cf. a regional foliation in metamorphic rocks) (Fig. 7).

Footwall shortcut thrusts: A low angle thrust fault developed in the footwall of a steep thrust fault (Fig. 8). The resultant lower angle fault trajectory is kinematically and mechanically more feasible for large displacement than the high angle fault trajectory (see Knipe 1985; McClay & Buchanan 1991, this volume)

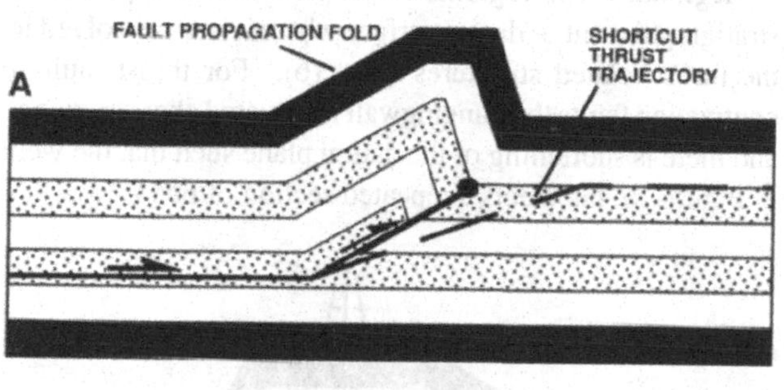

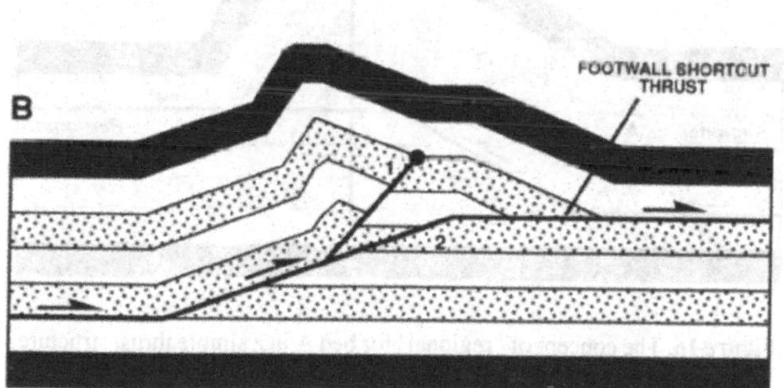

Figure 8. Footwall shortcut thrust. **(a)** Fault propagation fold with trajectory of incipient footwall shortcut thrust. **(b)** Final configuration after displacement on the footwall shortcut thrust.

Foreland-vergent thrust: A thrust fault that verges towards the undeformed foreland of the thrust belt (Fig. 9).

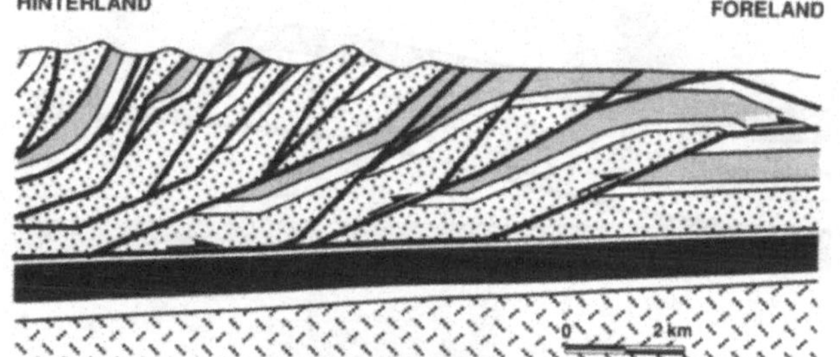

Figure 9. Foreland-vergent thrust system.

Frontal ramp: A ramp in the thrust surface that is perpendicular to the direction of transport of the thrust sheet (Fig. 10). Ramp angles are commonly between 10° and 30°.

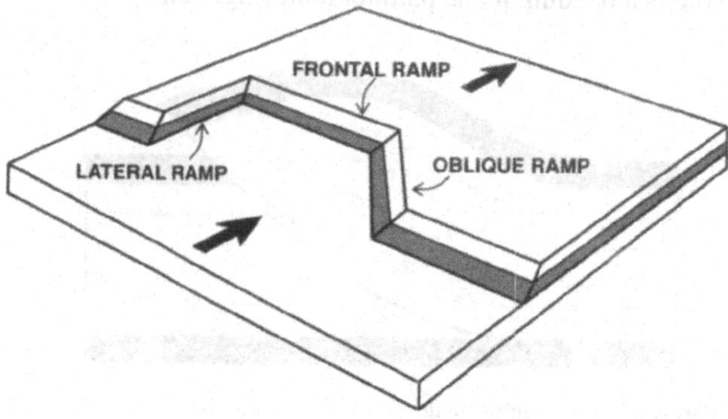

Figure 10. 3D thrust footwall ramp structures.

Hangingwall cut-off lines: The lines of intersection between a thrust surface and a stratigraphic horizon in the hangingwall of the thrust.

Hangingwall cut-off point: The point of intersection between a cut-off line in the hangingwall of the thrust and the erosion surface.

Hangingwall flat: That part of the hangingwall of a thrust fault where the thrust fault is bedding parallel or parallel to a regional datum surface (cf. a regional foliation in metamorphic rocks) (Fig. 7).

Hinterland-vergent thrust: A thrust fault that verges towards the hinterland of the orogen (i.e away from the undeformed foreland).

Klippe: An isolated portion of a thrust nappe or thrust sheet that is separated from the main part of the thrust nappe or thrust sheet as a result of erosion (Fig. 11).

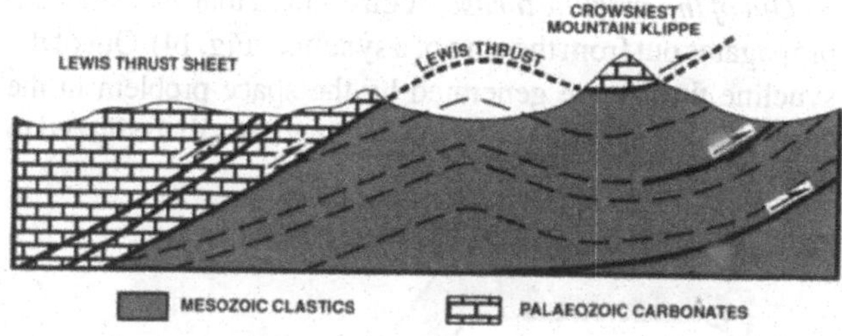

Figure 11. Klippe (adapted after McClay & Insley 1986).

Lateral ramp: A ramp in the thrust surface that is parallel to the direction of transport of the thrust sheet (Fig. 10). Ramp angles are generally between 10° and 30°. (Note that if the lateral structure is vertical then it becomes a thrust transport parallel tear or strike-slip fault and should not be termed a lateral ramp).

Listric thrust fault: A concave upwards thrust fault such that the upper section is a steep high angle contraction fault, the middle section is a medium angle contraction fault and the sole is a bedding plane parallel fault (Fig. 12).

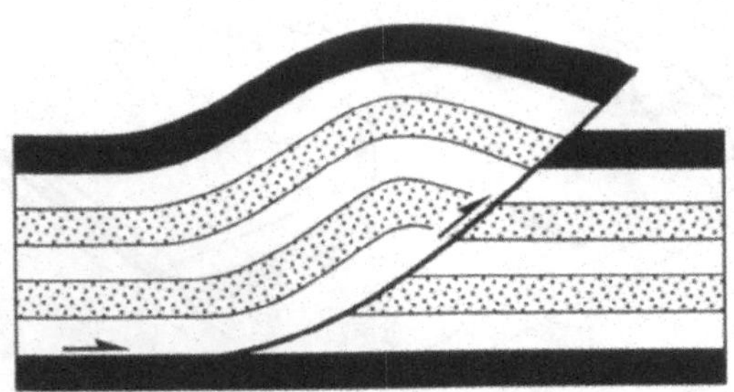

Figure 12. Listric thrust fault.

Oblique ramp: A ramp in the thrust surface that is oblique to the direction of transport of the thrust sheet (Fig. 10). Ramp angles are generally between 10° and 30°.

Out of the graben thrusts: Thrust faults that propagate outwards from a graben structure as a result of the contraction and inversion of a pre-existing extensional structure (Fig. 13) (see McClay & Buchanan 1991, this volume).

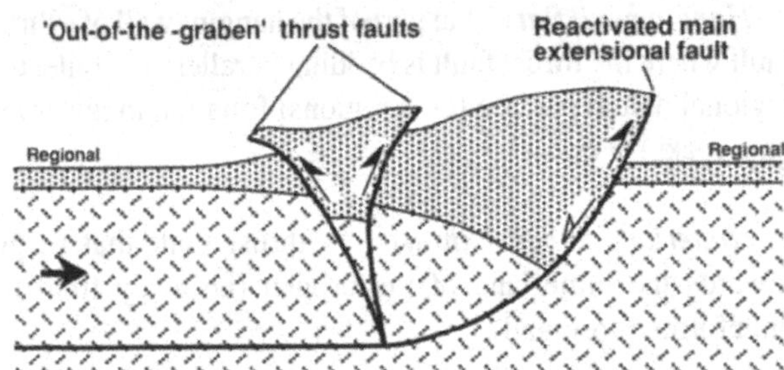

Figure 13. 'Out-of-the-graben' thrusts formed by inversion of a crestal collapse graben structure (McClay & Buchanan, 1991, this volume).

Out of the syncline thrust: A thrust fault that nucleates and propagates out from the core of a syncline (Fig. 14). Out of the syncline thrusts are generated by the space problem in the cores of tight synclines and may not necessarily be linked to other thrusts.

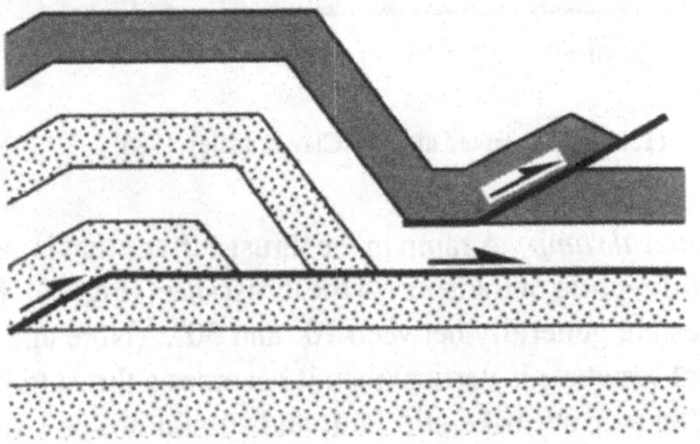

Figure 14. 'Out-of-the-syncline' thrust.

Pop -up: A section of hangingwall strata that has been uplifted by the combination of a foreland vergent thrust and a hinterland vergent thrust (Fig. 15).

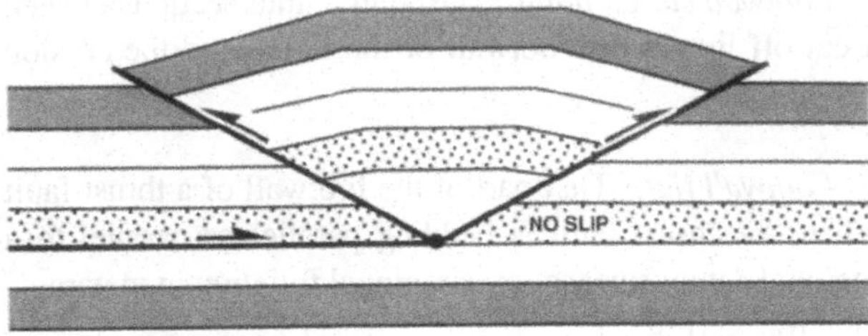

Figure 15. 'Pop-up' structure.

Ramp: That part of a thrust fault that cuts across bedding (or the appropriate datum plane-see above) (Fig. 7). Ramps may be divided into *hangingwall ramps* and *footwall ramps* (Fig. 7). Ramps commonly have angles between 10° and 30°.

Regional: The regional is the elevation of a particular stratigraphic unit or datum surface where it is not involved in the thrust related structures (Fig. 16). For thrust faults / contraction faults the hangingwall is elevated above regional and there is shortening of the datum plane such that the well in Figure 16 intersects a repeated section of bed A.

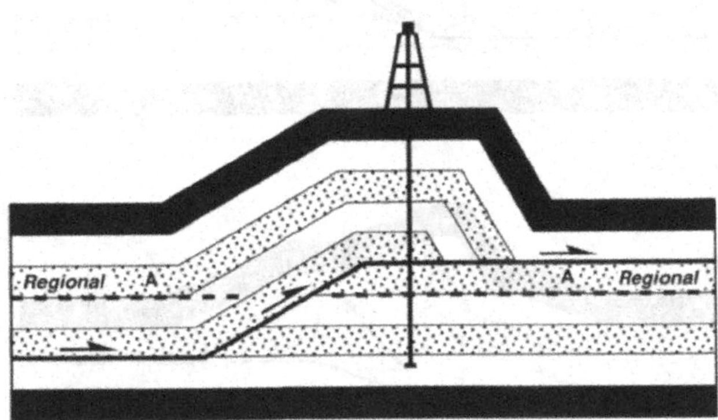

Figure 16. The concept of "regional" for bed A in a simple thrust structure.

Smooth trajectory thrust: A thrust fault with a trajectory that is smoothly varying and does not have a staircase form (Fig. 17). Smooth trajectory thrusts are found in higher grade metamorphic rocks (Cooper & Trayner 1986; McClay 1987) where ductile penetrative strains are developed within the thrust sheet.

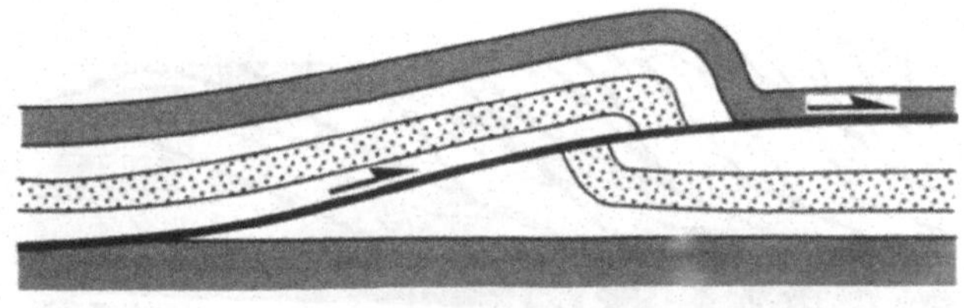

Figure 17. Smooth trajectory thrust.

Sole thrust: The lowermost thrust common to a thrust system (may also be termed a *floor thrust* - see thrust systems below).

Splay: A secondary thrust fault (i.e. smaller in size and displacement) that emerges from a main thrust fault. Boyer & Elliott (1982) define four types of splays (Fig. 4).

Rejoining splay (Fig. 4a), *Connecting splay* (Fig. 4b), *Isolated splay* (Fig. 4c), *and Divergent splay* (Fig. 4d).

Thrust nappe: A large thrust sheet which may have been generated from a recumbent fold in which the lower limb has been faulted out to form the sole thrust of the nappe (Fig. 18). Thrust nappes may also be generated from detachment thrusting and from inversion structures (cf. from inversion of ramp-flat extensional fault systems - see McClay & Buchanan 1991, this volume).

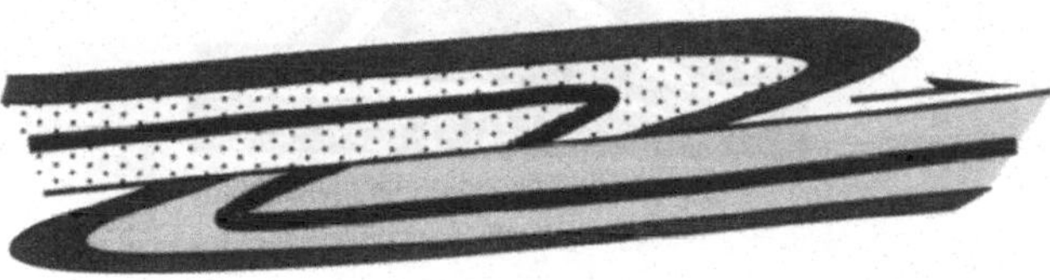

Figure 18. Thrust nappe developed by thrusting out the lower limb of a recumbent overfold.

Thrust sheet: A volume of rock bounded below by a thrust fault.

Thrust trajectory: The path that the thrust surface takes across the stratigraphy (often displayed on cross sections and restored sections). In high-level foreland fold and thrust belts thrust faults are commonly described as having *staircase* or *stair-step* trajectories (Fig. 19) typically with long bedding-parallel fault surfaces (*flats*) separated by shorter, high-angle fault segments that cut across the bedding (*ramps*).

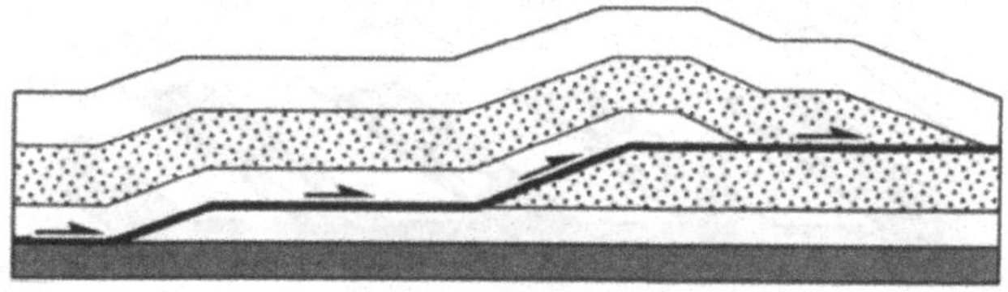

Figure 19. Staircase thrust trajectory consisting of ramps and flats.

Thrust vergence: The direction towards which the hangingwall of the thrust fault has moved relative to the footwall.

Tip line: The edge of a thrust fault where displacement dies to zero (Fig. 4c).

Tip point: The point of intersection between a tip line and the erosion surface (Fig. 4c). The term *tip point* is also used in cross-section analysis for the point where the tip line intersects the plane of the cross-section - i.e. in this context the tip point is the 2D equivalent of a tip line (Fig. 5).

Thrust systems

Thrust system: A zone of closely related thrusts that are geometrically, kinematically and mechanically linked.

This section of the glossary deals with *linked* thrust faults that form thrust systems. The terminology for thrust systems stems from Dahlstrom (1970), Boyer & Elliott (1982), Mitra (1986) and modified by Woodward *et al.* (1989). Thrust systems include duplexes, imbricate thrust systems and triangle zones (Fig. 20 - next page).

Duplexes

Duplex: An array of thrust horses bounded by a *floor thrust* (i.e. sole thrust) at the base and by a *roof thrust* at the top (Figs 20 & 21).

The stacking of the horses and hence the duplex shape depends upon the ramp angle, thrust spacing, and displacement on individual link thrusts. Models for duplex formation (Boyer & Elliott 1982; Mitra 1986) generally assume a 'forward-breaking' thrust sequence (see *thrust sequences* below). Mitra (1986) revised Boyer & Elliott's (1982) classification of duplexes and proposed a threefold classification (Fig. 21) consisting of -

1) Independent ramp anticlines and hinterland sloping duplexes (Fig. 21a);
2) True duplexes (Fig. 21b);
3) Overlapping ramp anticlines (Fig. 21c).

For independent ramp anticlines the final spacing between the thrusts is much greater than the displacement on the individual thrusts and the structure formed consists of independent ramp anticlines separated by broad synclines (Fig. 21a). Hinterland sloping duplexes (Fig. 21a) are formed where the initial spacing of thrust faults is small and displacement on individual thrusts is small such that, at the contact between horses, the roof thrust slopes towards the hinterland (Mitra 1986). True duplexes (such as those modelled by Boyer & Elliott (1982)) are formed by a particular combination of final thrust spacing, ramp angle and ramp height such that parts of all of the link thrusts and roof thrust are parallel to the frontal ramp of the duplex (Fig. 21b). Overlapping ramp anticlines are formed where the crests of successive ramp anticlines partially or totally overlap (Fig. 21c). A system of completely overlapping ramp anticlines in which the trailing branch lines are coincident is termed an antiformal stack (Fig. 21c).

Mitra (1986) further subdivided true duplexes depending upon their position with repect to larger thrusts (Fig. 21b). Duplexes may occur in the footwall to a ramp anticline, in the hangingwall to a ramp anticline and in front of a ramp anticline (Fig. 21b).

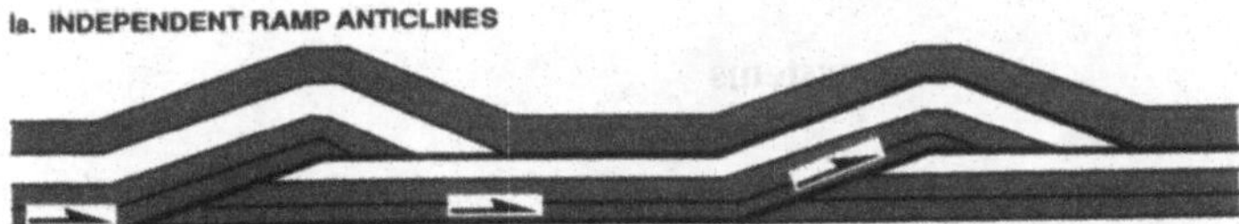

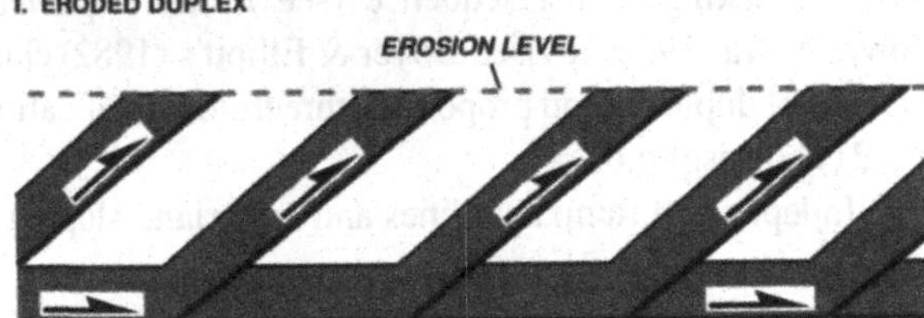

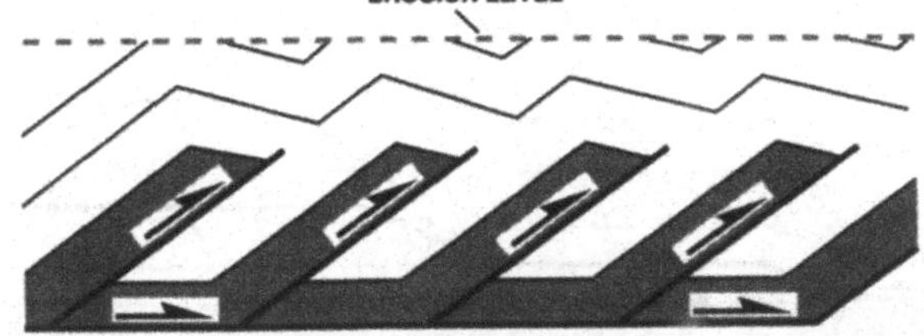

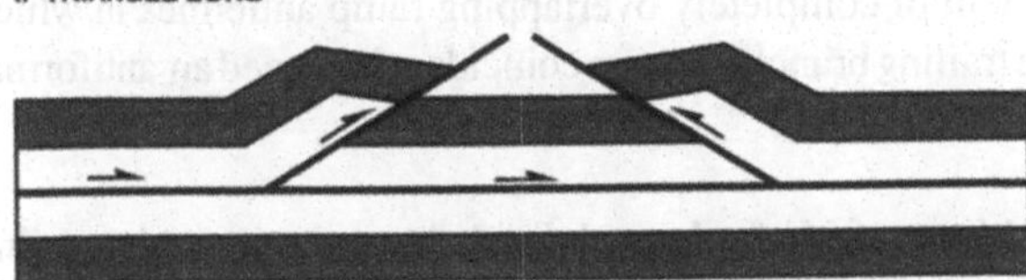

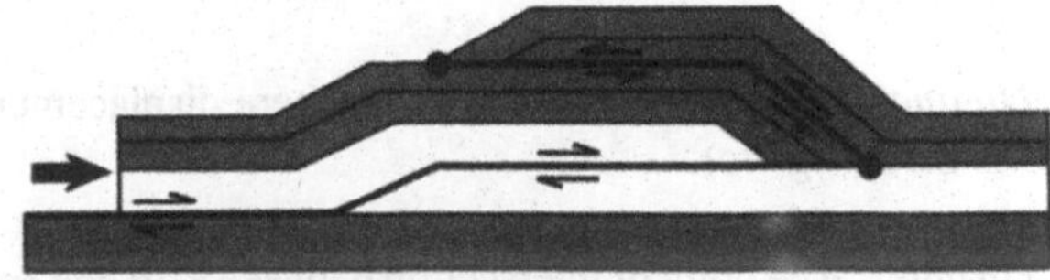

Figure 20. Thrust systems. **(a)** Duplexes. **(b)** Imbricate systems (schematic). **(c)** Triangle zones. (adapted after Boyer & Elliott 1982; Mitra 1986; and Woodward *et al.* 1989).

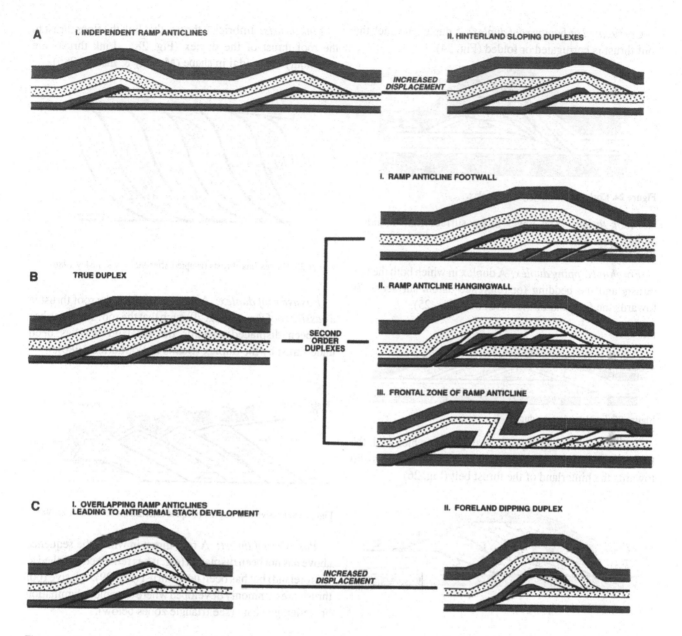

Figure 21. Duplex classification (modified after Mitra 1986). **(a)** Independent ramp anticlines and hinterland dipping duplexes. **(b)** True duplexes with second order duplexes. **(c)** Overlapping ramp anticlines which produce antiformal stacks and, with increased displacement, foreland dipping duplexes.

Antiformal stack: A duplex formed by overlapping ramp anticlines which have coincident trailing branch lines (Fig. 22). The individual horses are stacked up on top of each other such that they form an antiform.

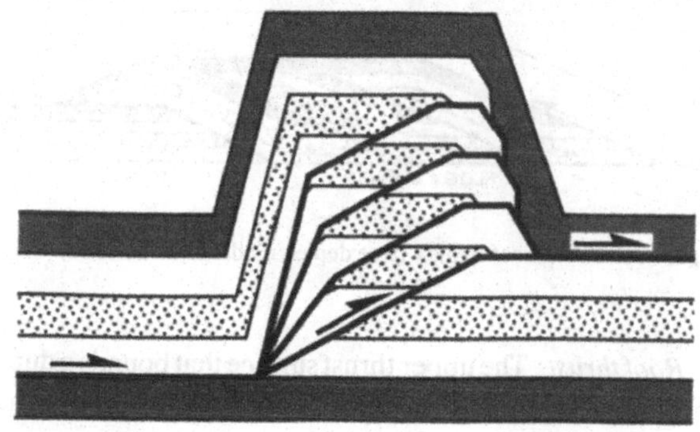

Figure 22. Antiformal stack.

Breached duplex: A duplex in which 'out of sequence movement' on the link thrusts have breached or cut through the roof thrust (Fig. 23). Butler (1987) discusses breaching of duplex structures.

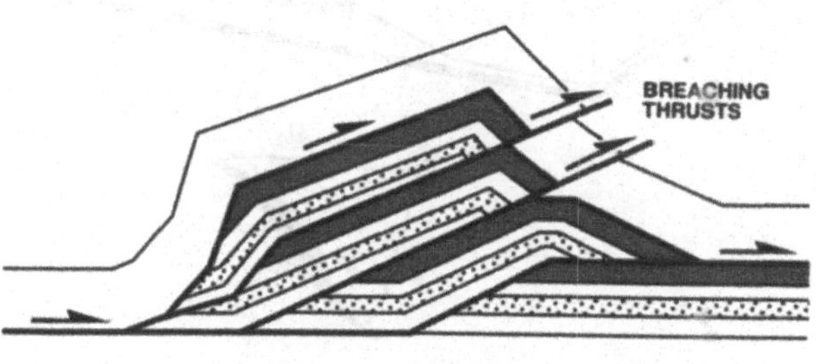

Figure 23. Duplex breached by reactivation of the link thrusts which displace the original duplex roof thrust.

Corrugated or bumpy roof duplex: A duplex in which the roof thrust is corrugated or folded (Fig. 24).

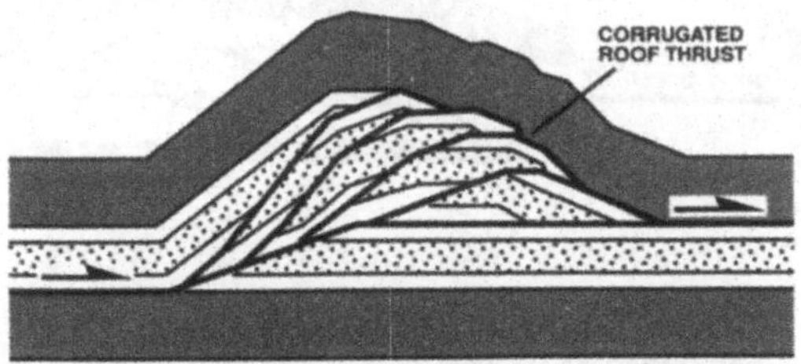

Figure 24. Corrugated or'bumpy roof' duplex.

Floor thrust: The lower thrust surface that bounds a duplex (Fig. 21).

Foreland dipping duplex; A duplex in which both the link thrusts and the bedding (or reference datum surfaces) dip towards the foreland of the thrust belt (Fig. 25).

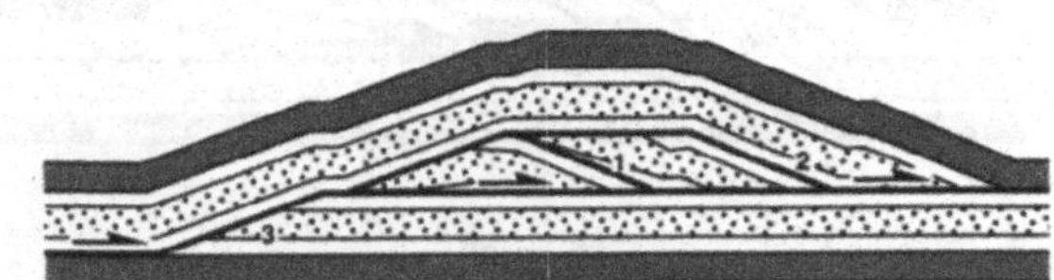

Figure 25. Foreland dipping duplex.

Hinterland dipping duplex: A duplex in which both the link thrusts and the bedding (or reference datum surfaces) dip towards the hinterland of the thrust belt (Fig. 26)

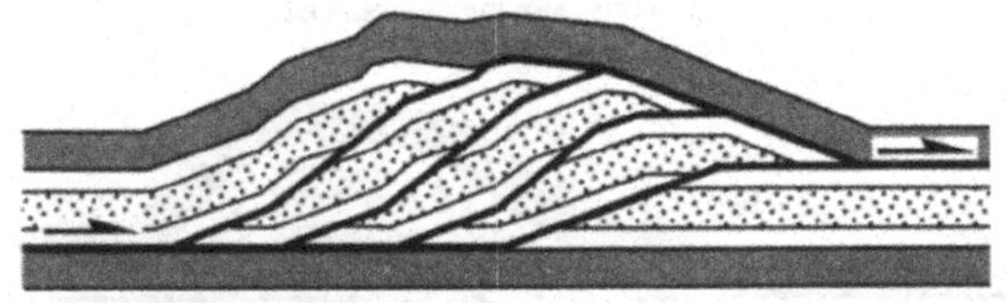

Figure 26. Hinterland dipping duplex.

Horse: A volume of rock completely surrounded (bounded) by thrust faults (Fig. 27).

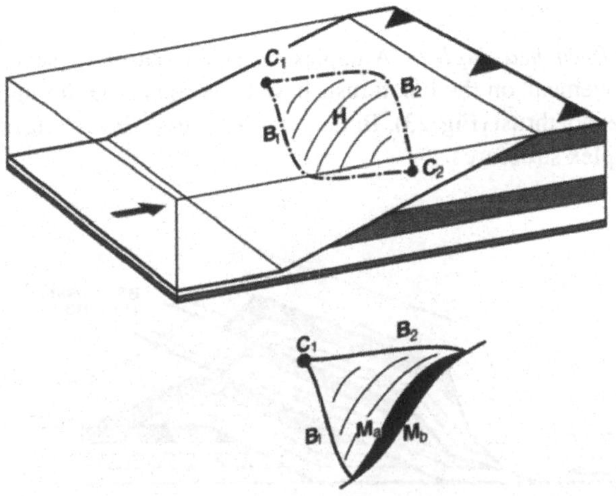

Figure 27. *Horse* **H** - a volume of rock enclosed by thrust faults. $\mathbf{B_1}$ and $\mathbf{B_2}$ are *branch lines.* $\mathbf{C_1}$ and $\mathbf{C_2}$ are *corner points* and $\mathbf{M_1}$ and $\mathbf{M_2}$ are thrust surfaces bounding the horse (adapted after Boyer & Elliott 1982).

Link thrusts: Imbricate thrusts that link the floor thrust to the roof thrust of the duplex (Fig. 28). Link thrusts are commonly sigmoidal in shape (McClay & Insley 1986).

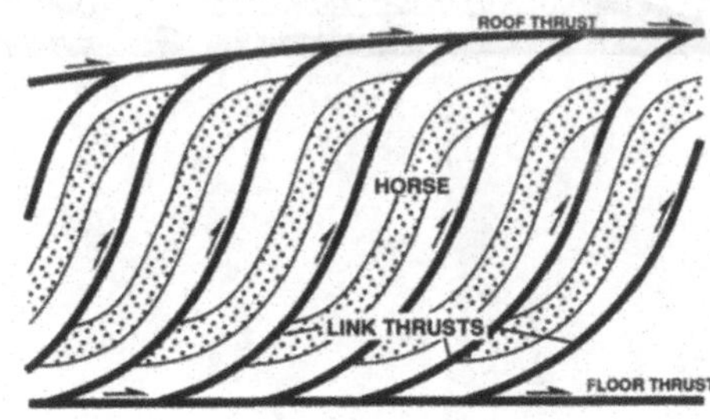

Figure 28. Duplex link thrusts (adapted after McClay & Insley 1986).

Passive roof duplex: A duplex in which the roof thrust is a *passive roof thrust* (Fig. 29) such that the roof sequence has not been displaced towards the foreland but has been underthrust by the duplex (Banks & Warburton 1986).

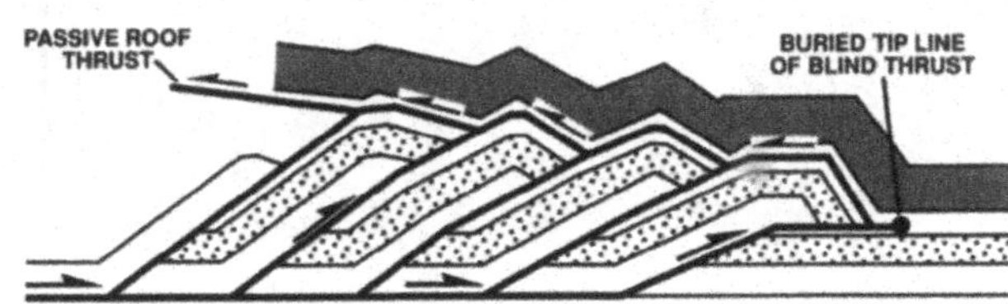

Figure 29. Passive roof duplex (adapted after Banks & Warburton 1986).

Passive roof thrust: A roof thrust in which the sequence above has not been displaced (e.g. it has remained attached to the foreland) but has been underthrust (Fig. 29). Passive roof thrusts are commonly developed where tectonic delamination or wedging occurs (see triangle zones below).

Planar roof duplex: A duplex in which the roof thrust is planar except where it is folded over the trailing ramp and over the leading ramp (Fig. 30). Groshong & Usdansky (1988) demonstrate that such a geometry is a result of a special combination of duplex thrust spacing and displacement.

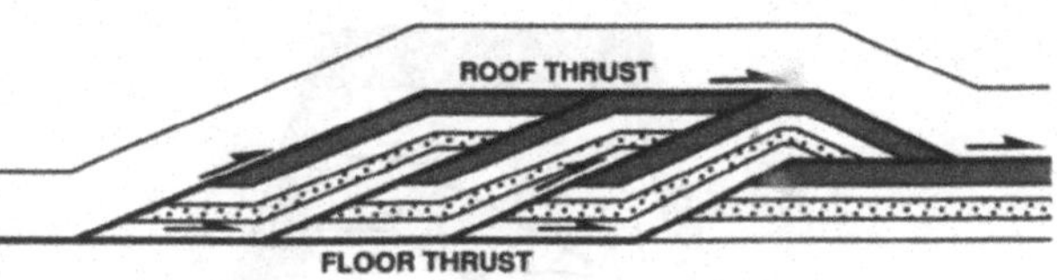

Figure 30. Planar roof duplex (true duplex model of Mitra 1986).

Roof thrust: The upper thrust surface that bounds a duplex (Fig. 30). Roof thrusts may be smooth or folded by movement on underlying thrusts of the duplex.

Smooth roof duplex: A duplex in which the roof thrust varies smoothly (Fig. 31) (see McClay & Insley 1986; Tanner 1991, this volume). Smoothly varying roof thrust geometry may be interpreted as indicating synchronous thrust movement (McClay & Insley 1986).

Figure 31. Smooth roof duplex where the roof thrust varies smoothly without folding by the underlying link thrusts (adapted after McClay & Insley 1986).

Truncated duplex: A duplex that is beheaded or truncated by an out of sequence thrust (Fig. 32).

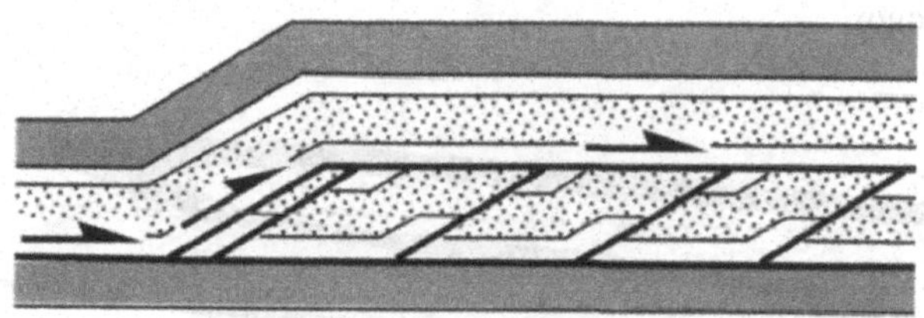

Figure 32. Truncated duplex in which the upper section (leading branch lines) has been removed by an 'out-of-sequence' thrust overriding the duplex.

Imbricate thrust systems

Imbricate thrust system: A closely related branching array of thrusts such that the thrust sheets overlap like roof tiles (Fig. 20).

Imbricate thrust systems may be formed a system of overlapping fault propagation folds (tip line folds - see *fault related folds and folding* below) as shown in Figure 33. Imbricate fans may also form from duplexes which have the leading branch lines eroded (Fig. 20). Boyer & Elliott (1982) point out the difficulty in distinguishing between imbricate systems formed from duplexes which have had the leading branch lines eroded and those imbricate systems formed from a branching array of thrusts that die out into tip lines and which have been subsequently eroded (Fig. 20).

Blind imbricate complex: An imbricate fan that remains buried such that the displacement on the imbricate faults below is compensated at a higher structural level by folding, cleavage development or another set of structures having a different style (Fig. 20) (see also Thompson 1981).

Imbricate fan: A system of linked, emergent thrusts that diverge upwards from a *sole thrust* (or floor thrust) (Fig. 33).

Leading imbricate fan: An imbricate fan that has most of its displacement on the leading (lowermost) thrust (Fig. 20).

Trailing imbricate fan: An imbricate fan that has most of its displacement on the trailing (highest) thrust (Fig. 20).

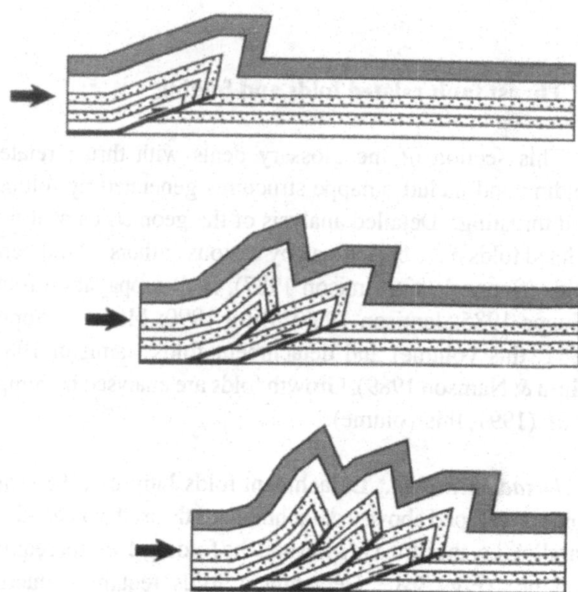

Figure 33. Imbricate fan formed from an array of overlapping fault-propagation folds (adapted after Mitra 1990).

Triangle zones

The term *'Triangle zone'* was first used to describe the thrustbelt termination in the southern Canadian Rocky Mountains (e.g. Price 1981). There it is a zone of opposed thrust dips, at the external margin of the thrust belt and often with a duplex or antiformal stack in the axial part. This is more correctly described as a *passive roof duplex* (Fig. 29). Such ' triangle zones' are basically intercutaneous wedges (Price 1986). A second usage of the term *triangle zone* refers to a combination of two thrusts with the same basal detachment and with opposing vergence such that they form a triangular zone (see below).

Intercutaneous thrust wedge: A thrust bounded wedge bounded by a sole or floor thrust at the base and by a passive roof thrust at the top (see Price 1986)

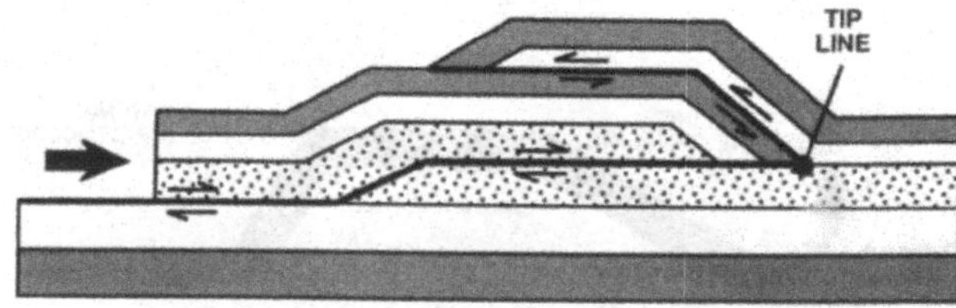

Figure 34. Intercutaneous thrust wedge (adapted after Price 1986).

Triangle zone: A combination of two thrusts with the same basal detachment and with opposing vergence such that they form a triangular zone (Fig. 35).

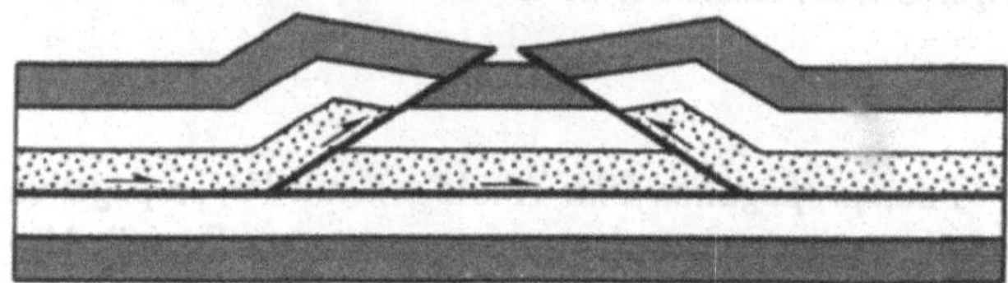

Figure 35. Triangle zone.

Thrust fault related folds and folding

This section of the glossary deals with thrust related folding and includes nappe structures generated by folding and thrusting. Detailed analysis of the geometries of thrust related folds have been given by various authors - fault bend folds (Suppe 1983; Jamison 1987), fault-propagation folds (Suppe 1985; Jamison 1987; Mitra 1990; Mosar & Suppe 1991, this volume) and detachment folds (Jamison 1987; Mitra & Namson 1989). Growth folds are analysed by Suppe *et al.* (1991, this volume).

Detachment fold: Detachment folds Jamison (1987) are folds developed above a detachment or thrust that is bedding parallel (i.e. the thrust is a flat and the folding does not require a ramp) (Fig. 36). Detachment folds require a ductile decollement layer (e.g. salt or shale) which can infill the space generated at the base of the fold (Fig. 36). Detachment folds are rootless and commonly disharmonic.

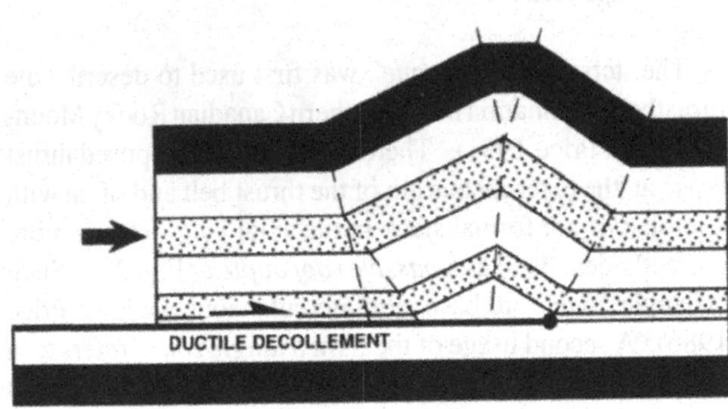

Figure 36. Detachment fold.

Fault-bend fold: A fold generated by movement of a thrust sheet over a ramp (Fig. 37) (analysed in detail by Suppe (1983).

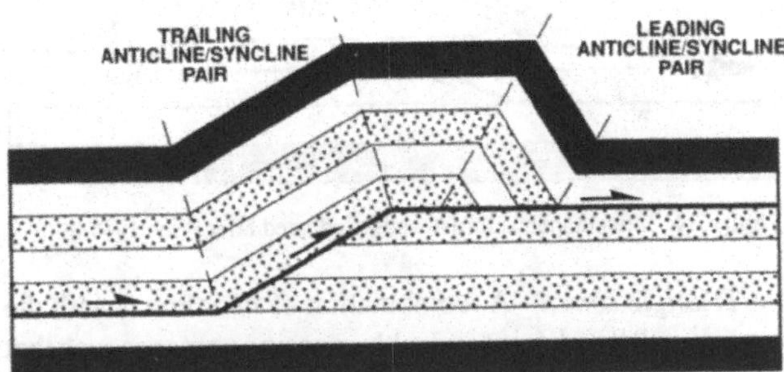

Figure 37. Fault-bend fold with leading and trailing anticline-syncline pairs.

Fault propagation fold: A fold generated by propagation of a thrust tip up a ramp into undeformed strata (Fig. 38). Also known as a *tip line fold.*

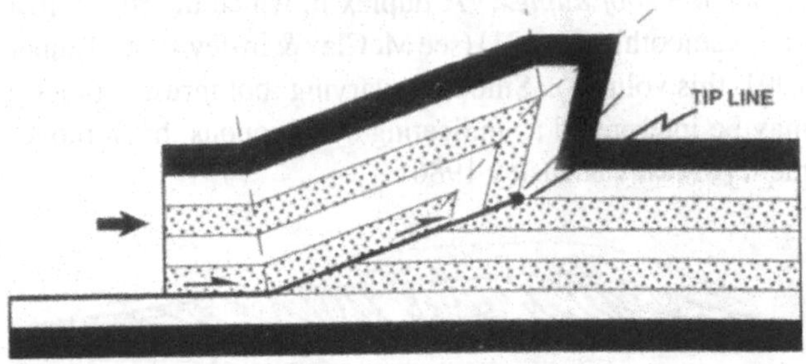

Figure 38. Fault-propagation fold.

Fold nappe: A nappe formed by a large recumbent overfold (Fig. 39) in which the underlimb is highly attenuated (cf. the Helvetic nappes, Ramsay 1981; Dietrich & Casey 1989)

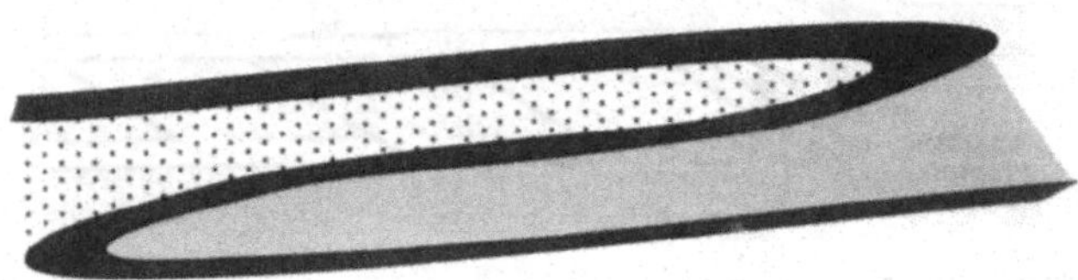

Figure 39. Fold nappe.

Footwall syncline: A syncline that is developed in the footwall to a thrust sheet below a ramp. Footwall synclines may develop by frictional drag below the thrust or develop from fault propagation folds.

Growth fold: A fold that develops in sedimentary strata at the same time as they are being deposited (Fig.40) (see Suppe *et al.* 1991, this volume). Growth folds (*growth anticlines*) may develop above the tip line of a thrust fault that is propagating upwards into the sedimentary section as it is being deposited. Strata typically thin onto the crest of the *growth anticline*.

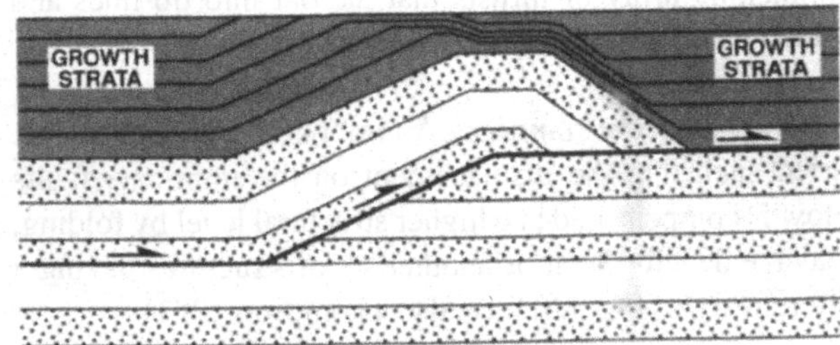

Figure 40. Growth fault-bend fold (after Zoetemeijer & Sassi 1991, this volume).

Growth strata: Strata that are deposited on a growth fold system as it develops (Fig. 40) and hence they record the evolution of the fold (see Suppe *et al.* 1991, this volume).

Lift-off fold: Lift-off folds (Mitra & Namson 1989) are detachment folds whereby the beds and the detachment are isoclinally folded in the core of the anticline (Fig. 41). Lift-off folds require a ductile decollement layer such as salt or shale which can flow from the core of the fold.

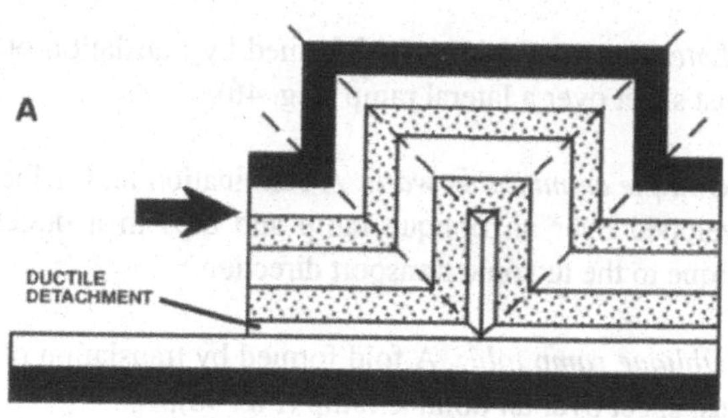

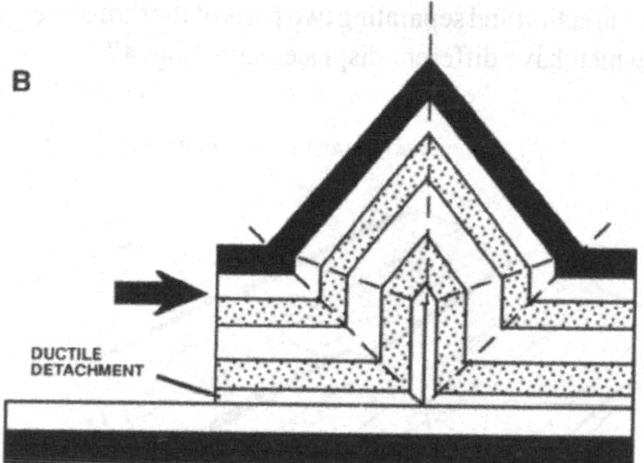

Figure 41. Lift-off folds in which the detachment is isoclinally folded in the core of the fold. A ductile detachment layer is needed in order to permit flow of ductile material from the collapsed fold core. **(a)** Box lift-off fold. **(b)** Chevron lift-off fold. (adapted after Mitra & Namson 1989)

Overthrust shear: A term used for thrust or fold nappes that have been subjected to bulk shear strain generally in the direction of nappe transport. See Ramsay *et al.* (1983), Dietrich & Casey (1989), and Rowan & Kligfield (1991, this volume) for discussions of overthrust shear and nappe emplacement.

Ramp anticline: An anticline in the hangingwall of a thrust generated by movement of the thrust sheet up and over a ramp in the footwall (Fig. 30).

Ramp syncline: A syncline in the hangingwall of a thrust generated by movement of the thrust sheet up and over a ramp in the footwall (Fig. 37). Ramp anticlines and ramp synclines occur in geometrically and kinematically linked pairs - a *leading anticline-syncline pair* and a *trailing anticline-syncline pair* (Fig. 37).

Thrust nappe: A large thrust sheet commonly with significant displacement (e.g. the Moine nappe). A thrust nappe may be generated from a recumbent fold in which the lower limb has been faulted out to form the sole thrust of the nappe (Fig. 18).

Transported fault-propagation fold: A fault propagation fold that has been transported by thrust that has broken through onto an upper flat (Fig. 42).

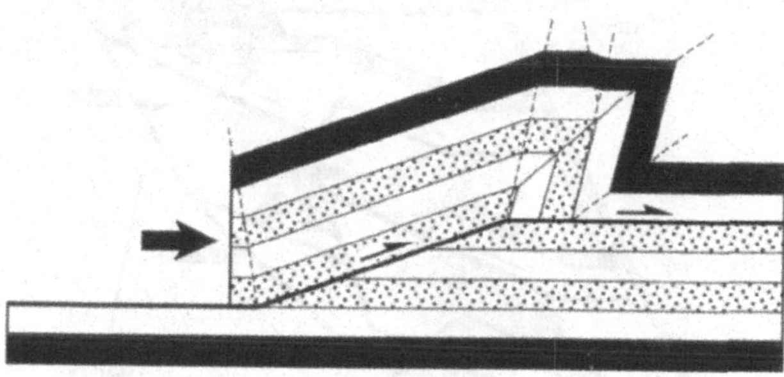

Figure 42. Transported fault propagation fold.

Translated (or transported) detachment fold: A detachment fold that is transported by a thoroughgoing thrust such that it is displaced from its point of formation (e.g. up a footwall ramp) (Fig. 43). Mitra (1990) discusses the differences between translated (transported) fault-propagation folds and translated (transported) detachment folds.

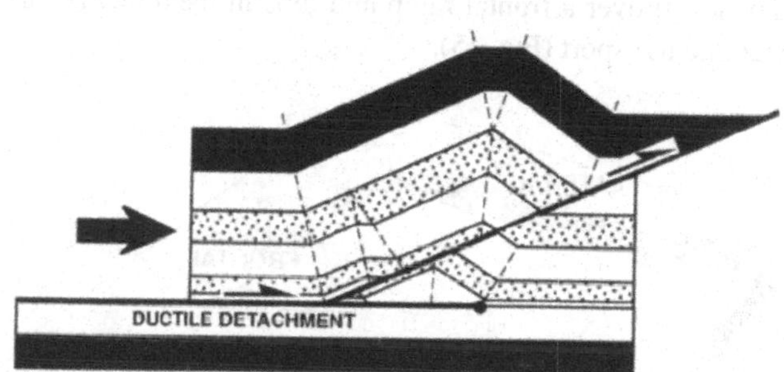

Figure 43. Transported detachment fold (adapted after Mitra 1990).

Thrust structures in 3D

The movement of a thrust sheet over a corrugated surface will generate flat topped anticlines and domes. These are termed culminations (Dahlstrom 1970; Butler 1982) and the limbs of these structures are termed culmination walls.

Culmination: An anticline or dome with four way closure generated by movement of the thrust sheet over underlying ramps.

Displacement transfer zone: The zone where displacement is transferred from one thrust to another (Dahlstrom 1970). Displacement transfer zones may occur by simple *en-echelon overlap* of thrust faults (Fig. 44) or by *tear faults* parallel to the direction of tectonic transport (see *tear faults* - below).

Dorsal culmination wall: The limb which is developed over the rear ramp of the culmination and dips in a direction opposite (i.e. 180°) to the tectonic transport direction (Fig. 45).

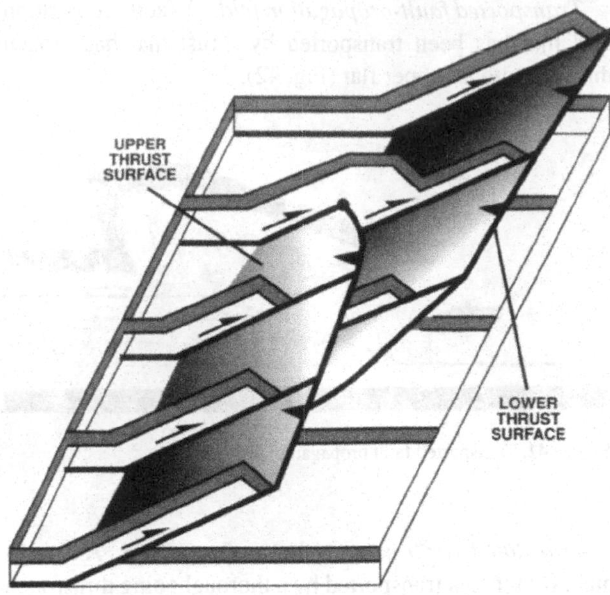

Figure 44. Displacement transfer by *en echelon* overlap of thrust faults.

Frontal culmination wall: A culmination limb which is developed over a frontal ramp and dips in the direction of tectonic transport (Fig. 45).

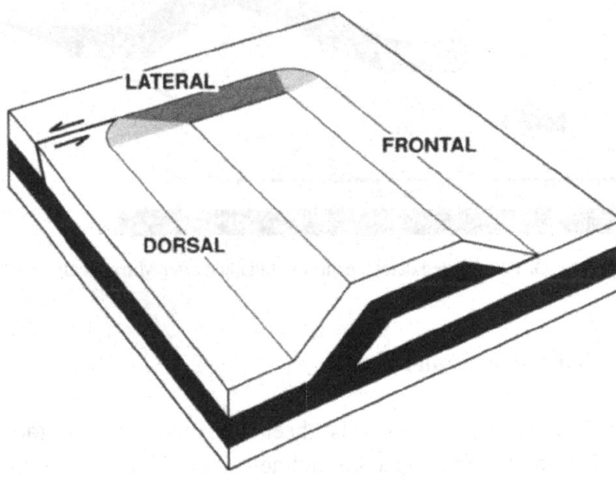

Figure 45. Culmination walls.

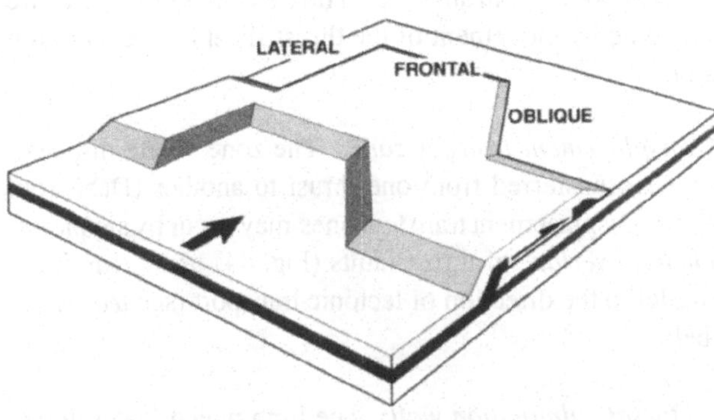

Figure 46. Ramp related folds in the hangingwall of a thrust system.

Frontal ramp fold: A fold formed by translation of the thrust sheet over a frontal ramp (Fig. 46).

Lateral culmination wall: A culmination limb which is developed over a lateral ramp and dips in a direction 90° to the tectonic transport direction (Fig. 45).

Lateral ramp fold: A fold formed by translation of the thrust sheet over a lateral ramp (Fig. 46).

Oblique culmination wall: A culmination limb which is developed over an oblique ramp and dips in a direction oblique to the tectonic transport direction.

Oblique ramp fold: A fold formed by translation of the thrust sheet over an oblique ramp (Fig. 46).

Tear fault. A strike-slip fault parallel to the thrust transport direction and separating two parts of the thrust sheet each of which have different displacements (Fig. 47).

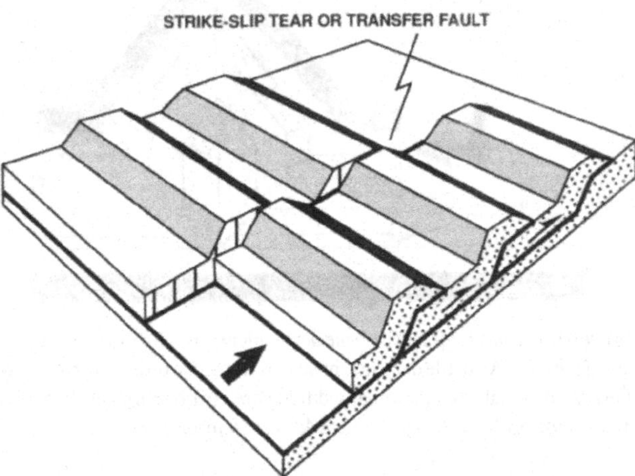

Figure 47. Tear fault parallel to the thrust transport direction and separating two parts of the thrust sheet each of which have different displacements.

Window: A hole within a thrust sheet whereby the hangingwall section of a thrust sheet has been removed by erosion thus exposing the underlying structures. Windows commonly form through erosion of culminations such that the underlying thrust system (commonly imbricates or duplexes) is revealed. Boyer & Elliott (1982) illustrate different types of windows formed by erosion of various duplex forms.

Thrust sequences

Thrust sequence: The sequence in which new thrust faults develop.

The sequence of development of thrust faults within a thrust belt or thrust system is an important parameter needed for the interpretation of both the geometry and the kinematic evolution of a thrust belt. It is essential for the construction of balanced and restored sections (Boyer and Elliott 1982; Boyer 1991, this volume; Butler 1987; Morley 1988; Suppe 1985; Woodward *et al.* 1989). A long accepted paradigm for thrust tectonics is that, in foreland fold and thrust belts, thrust faults develop sequentially in a sequence that both nucleates in a forward-breaking sequence and verges towards the foreland (Dahlstrom 1970; Bally *et al.* 1966; Boyer & Elliott 1982; Butler 1982). Recently these basic 'rules' of thrust tectonics have been challenged (see Boyer 1991, this volume; and Tanner 1991, this volume).

Breaching: Breaching occurs where an early formed thrust is cut by later thrusts (Butler 1987) (Fig. 48). It is a term describing the local geometric relationships between thrusts. The term may be applied to a *breached duplex* where the link thrusts do not join or anastomose with the roof thrust but cut and displace it (Fig. 23).

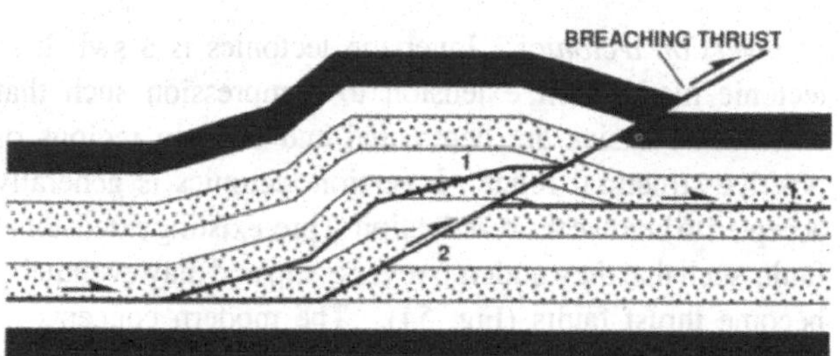

Figure 48. Breached thrust - an early formed thrust cut by later thrust. Numbers indicate sequence of faulting.

Break-back sequence: The sequence of thrusting where new (younger) thrusts nucleate in the hangingwalls of older thrusts and verge in the same direction as the older thrusts (Fig. 49).

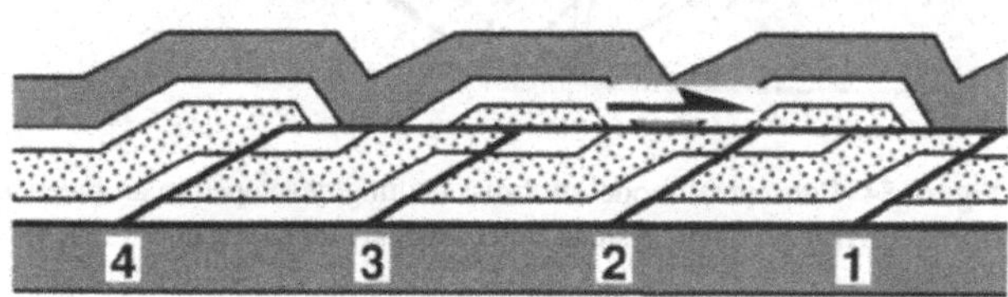

Figure 49. Break-back thrust sequence. Numbers indicate sequence of faulting.

Forward-breaking sequence: The sequence of thrusting in which new (younger) thrust faults nucleate in the footwalls of older thrusts and verge in the same direction as the older thrusts (Fig. 50).

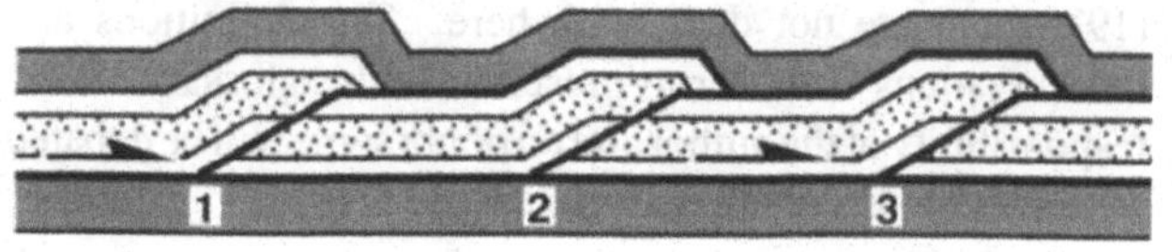

Figure 50. Forward-breaking or "piggy-back thrust" sequence. Numbers indicate sequence of faulting.

In sequence thrusting: A thrust sequence that has formed progressively and in order in one direction (i.e either a forward-breaking sequence or a break-back sequence). Figures 49 & 50 show in-sequence thrusting.

Out-of-sequence thrusting The opposite to 'in-sequence thrusting'. Thrust faulting which develops in a sequence other than in sequence (Fig. 51). Break-back sequences of thrusts have commonly been called out-of sequence thrusts but the term should be more appropriately used to describe thrust sequences which do not conform to a either a progressive forward-breaking or break-back sequence (Fig. 51). Out of sequence thrusts commonly cut through and displace pre-existing thrusts. Morley (1988) and Butler (1987) discuss out of sequence thrusting.

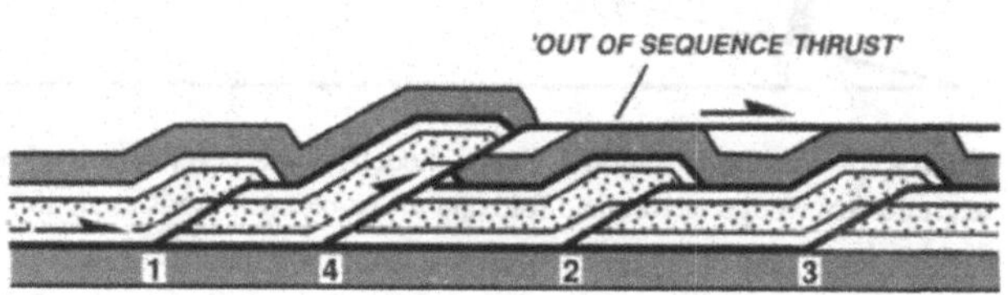

Figure 51. "Out-of-sequence" thrust cutting into a foreland-vergent thrust system. Numbers indicate sequence of faulting.

Piggy-back thrust sequence: Piggy-back thrust sequence occurs when topographically higher but older thrusts are carried by lower younger thrusts (Fig. 50). This is essentially the same as a *forward-breaking* thrust sequence.

Synchronous thrusting: Synchronous thrusting occurs when two or more thrusts move together (i.e. not in a piggy-back sequence). Boyer (1991, this volume) postulates that synchronous thrust movement may be significant in the Rocky Mountains of western North America.

Models for thrust faulting

The critical Coulomb wedge model

Wedge type models for the development of fold and thrust belts were first analysed by Elliott (1976) and Chapple (1978). The critical taper model or the critical Coulomb wedge model incorporating brittle frictional rheology for the emplacement of accretionary wedges and foreland fold and thrust belts was postulated by Davis et al. (1983) and subsequently developed by Dahlen and others (see review by Dahlen 1990). The critical wedge model assumes that the accretionary wedge or foreland fold and thrust belt has a triangular cross-section, a characteristic surface slope α and a basal decollement dip of β (Fig. 52). The wedge is composed of material that has a Navier-Coulomb rheology. The wedge forms by pushing an initial layered sequence from behind such that thrusts propagate in sequence (towards the foreland) until a critical taper ($\alpha + \beta$) is attained at such time that subsequent deformation involves transport of the whole wedge along the basal decollement or the wedge continues to grow in a self-similar fashion by addition of new material at the wedge toe but maintaining the critical taper. Critical wedge models have used to explain the development of accretionary prisms and foreland fold and thrust belts (e.g. Dahlen & Suppe 1988; Dahlen 1990; Willett 1991, this volume; Liu *et al.* 1991, this volume) but have been criticised by Woodward (1987).

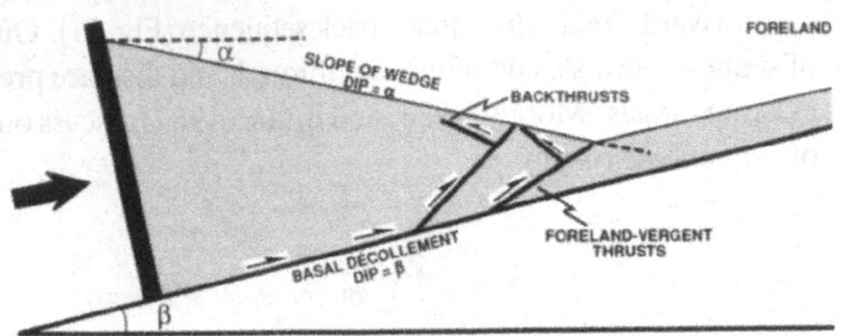

Figure 52. Idealised Coulomb wedge model for the development of an accretionary wedge or foreland fold and thrust belt (cf. Davis et al. 1983; Dahlen 1990).

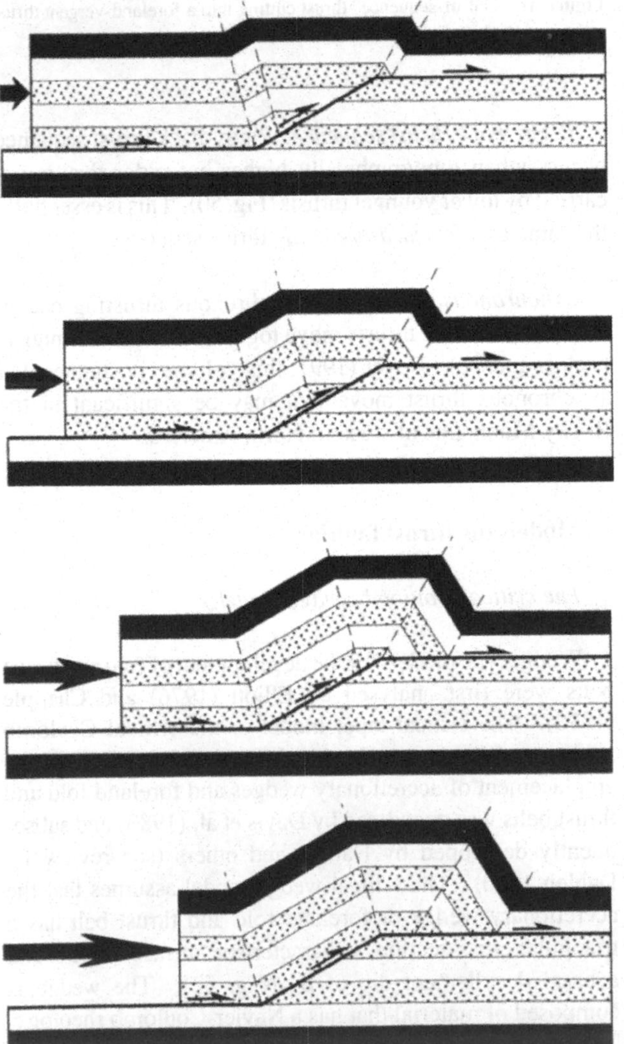

Figure 53. Self-similar growth model for the development of a fault-bend fold.

Self-similar thrust models

Self-similar model: It is commonly assumed that spatial variations in thrust system geometries reflect temporal variations in the evolution of the structure (Fischer & Woodward 1991, this volume). This is the basis for the self-similar growth model for thrust structures such as fault-bend folds or fault-propagation folds where, through time, the structure grows in amplitude as a simple geometric progression of the first formed structural feature (Fig. 53). This model assumes that the structure does not change form with time (e.g. evolve from a detachment fold to a fault propagation fold) - an assumption challenged by Fischer & Woodward (1991, this volume).

Inversion

Inversion tectonics: Inversion tectonics is a switch in tectonic mode from extension to compression such that extensional basins are contracted and become regions of positive structural relief. Inversion tectonics is generally accepted to involve the reactivation of pre-existing extensional faults such that they undergo reverse slip and may eventually become thrust faults (Fig. 54). The modern concepts of inversion tectonics are discussed in Cooper & Williams (1989). Thrust structures in inverted basins include footwall shortcut thrusts and 'out-of-the-graben' thrusts (Fig. 13).

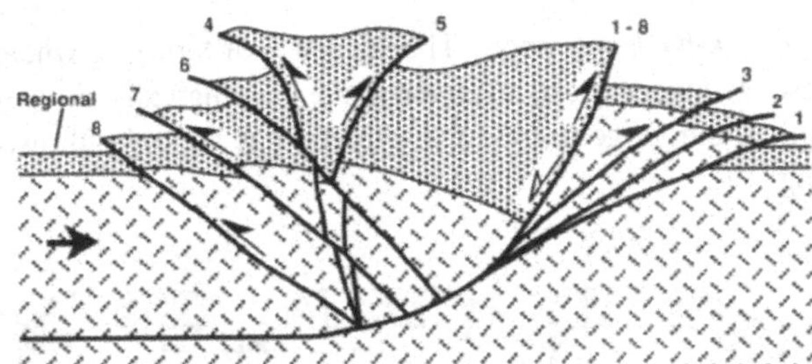

Figure 54 Conceptual model of thrust faulting developed in an inverted listric extensional fault system (after McClay & Buchanan 1991, this volume).

Concluding statements

In this glossary it has not been possible to review or illustrate all of the terms that have been applied to fold and thrust belts - that would be the subject of an entire book. Techniques of analysis such as the construction and restoration of balanced sections have been recently reviewed by Geiser (1988), Mitra & Namson (1989) and Woodward *et al.* (1989) and are not dealt with here. The definitions and interpretations of the terms presented in this glossary are the responsibility of the author and no doubt some will be debated as to exact interpretation and usage.

This glossary was compiled with the help of Tim Dooley and Joaquina Alvarez-Marron who are thanked for assistance with the illustrations. I. Davison, J. Alvarez-Marron, R. Knipe and C. Banks are thanked for critical reviews and comments.

References

Bally, A.W., Gordy, P.L., & Stewart, G.A., 1966, Structure, seismic data, and orogenic evolution of southern Canadian Rocky Mountains. *Bulletin of Canadian Petroleum Geology*, **14**, 337-381.

Banks, C.J. & Warburton, J. 1986. 'Passive-roof' duplex geometry in the frontal structures of the Kirthar and Sulaiman mountain belts, Pakistan. *Journal of Structural Geology*, **8**, 229-238.

Boyer, S.E. 1991. Geometric evidence for synchronous thrusting in the southern Alberta and northwest Montana thrust belts, this volume.

Boyer, S. E., & Elliott, D. 1982. Thrust systems. *American Association of Petroleum Geologists Bulletin*, **66**, 1196-1230.

Butler, R.W.H. 1982. The terminology of structures in thrust belts. *Journal of Structural Geology*, **4**, 239-246.

Butler, R.W.H. 1987. Thrust sequences. *Journal of the Geological Society, London* , **144**, 619-634.

Chapple, W.M. 1978. Mechanics of thin-skinned fold-and-thrust belts. *Geological Society of America Bulletin*, **89**, 1189-1198.

Cooper, M.A. & Trayner, P.M. 1986. Thrust-surface geometry: implications for thrust-belt evolution and section-balancing techniques. *Journal of Structural Geology*, **8**, 305-312.

Cooper, M. A. & Williams, G. D. (eds), 1989. *Inversion Tectonics*. Geological Society of London Special Publication, **44**, 376 p.

Dahlen, F.A. 1990. Critical taper model of fold-and-thrust belts and accretionary wedges. *Annual Reviews of Earth and Planetary Sciences*, **18**, 55-99.

Dahlen, F.A., & Suppe, J. 1988. Mechanics, growth, and erosion ;of mountain belts. *Geological Society of America Special Paper*, **218**, 161-178.

Dahlstrom, C.D.A. 1969. Balanced cross sections. *Canadian Journal of Earth Sciences*, **6**, 743-757.

Dahlstrom, C.D.A. 1970. Structural geology in the eastern margin of the Canadian Rocky Mountains. *Bulletin of Canadian Petroleum Geology*, **18**, 332-406.

Davis, D., Suppe, J. & Dahlen, F.A. 1983. Mechanics of fold-and-thrust belts and accretionary wedges. *Journal of Geophysical Research*, **94**, 10,347-54.

Diegel, F.A. 1986. Topological constraints on imbricate thrust networks, examples from the Mountain City window, Tennessee, U.S.A. *Journal of Structural Geology*, **8**, 269-280.

Dietrich, D. & Casey, M. 1989. A new tectonic model for the Helvetic nappes. *In:* Coward, M. P., Dietrich, D. & Park, R. G. (eds). *Alpine Tectonics*, Geological Society of London Special Publication, **45**, 47-63.

Elliott, D.E. 1976. The energy balance and deformation mechanisms of thrust sheets. *Philosophical Transactions of the Royal Society of London*, **283**, 289-312.

Fischer, M. P. & Woodward, N. B. 1991. The geometric evolution of foreland thrust systems, this volume.

Geiser, P. A. 1988. The role of kinematics in the construction and analysis of geological cross sections in deformed terranes. In: Mitra, G. & Wojtal, S. (eds) Geometries and mechanisms of thrusting with special reference to the Appalachians, *Geological Society of America, Special paper*, **222**, 47-76.

Groshong, R. H. & Usdansky, S. I. 1988. Kinematic models of plane-roofed duplex styles. In: Mitra, G. & Wojtal, S. (eds) Geometries and mechanisms of thrusting with special reference to the Appalachians. *Geological Society of America, Special paper*, **222**, 197-206.

Jamison, W.R. 1987. Geometric analysis of fold development in overthrust terranes. *Journal of Structural Geology*, **9**, 207-219.

Knipe, R.J. 1985. Footwall geometry and the rheology of thrust sheets. *Journal of Structural Geology*, **7**, 1-10.

Liu Huiqi, McClay, K. R. & Powell, D. 1991. Physical models of thrust wedges, this volume.

McClay, K.R. 1981. What is a thrust? What is a nappe? In: McClay, K.R. & Price, N.J. (eds). *Thrust and Nappe Tectonics*, Geological Society of London Special Publication, **9**, 7-12.

McClay, K.R. 1987. The mapping of geological structures. Geological Society of London Handbook Series, Open University Press, 164p.

McClay, K.R. & Insley, M.W. 1986. Duplex structures in the Lewis thrust sheet, Crowsnest Pass, Rocky Mountains, Alberta, Canada. *Journal of Structural Geology*, **8**, 911-922.

McClay, K.R. & Buchanan, P.G. 1991. Thrust faults in inverted extensional basins, this volume.

Mitra, S. 1986. Duplex structures and imbricate thrust systems: geometry, structural position, and hydrocarbon potential. *American Association of Petroleum Geologists Bulletin*, **70**, 1087-1112.

Mitra, S. 1990. Fault-propagation folds: geometry, kinematic evolution, and hydrocarbon traps. *American Association of Petroleum Geologists Bulletin*, **74**, 921-945.

Mitra, S. & Namson, J. 1989. Equal-area balancing. *American Journal of Science*, **289**, 563-599.

Morley, C.K. 1988. Out-of-sequence thrusts. *Tectonics*, **7**, 539-561.

Mosar, J. & Suppe, J. 1991. Role of shear in fault-propagation folding, this volume.

Price, R.A. 1981. The Cordilleran foreland thrust and fold belt in the southern Canadian Rocky Mountains. In: McClay, K.R. & Price, N.J. (eds). *Thrust and Nappe Tectonics*, Geological Society of London Special Publication, **9**, 427-448.

Price, R.A. 1986. The southeastern Canadian Cordillera: thrust faulting, tectonic wedging and delamination of the lithosphere. *Journal of Structural Geology*, **8**, 239-254.

Ramsay, J.G. 1981. Tectonics of the Helvetic Nappes. In: McClay, K.R. & Price, N.J. (eds). *Thrust and Nappe Tectonics*, Geological Society of London Special Publication, **9**, 293-310.

Ramsay, J.G. 1991. Some geometric problems of ramp-flat thrust models, this volume.

Ramsay, J.G. Casey, M. & Kligfield, R. 1983. Role of shear in development of the Helvetic fold-thrust belt of Switzerland. *Geology*, **11**, 439-442.

Rich, J.L. 1934. Mechanics of low-angle overthrust faulting illustrated by Cumberland thrust block, Virginia, Kentucky and Tennessee. *American Association of Petroleum Geologists Bulletin*, **18**, 1584-1596.

Rowan, M.G. & Kligfield, R. 1991. Kinematics of large-scale asymmetric buckle folds in overthrust shear: an example from the Helvetic nappes, this volume.

Suppe, J. 1983. Geometry and kinematics of fault-bend folding. *American Journal of Science* , **283**, 684-721.

Suppe, J. 1985. Principles of structural geology. Prentice Hall, New Jersey, 537p.

Suppe, J., Chou, G. T. & Hook, S. C. 1991. Rates of folding and faulting determined from growth strata, this volume.

Tanner, P.W.G. 1991. The duplex model: Implications from a study of flexural-slip duplexes, this volume.

Thompson, R.I. 1981. The nature and significance of large 'blind' thrusts within the northern Rocky Mountains of Canada. In: McClay, K.R. & Price, N.J. (eds). *Thrust and Nappe Tectonics*, Geological Society of London Special Publication, **9**, 449-462.

Willett, S. D. 1991. Dynamic and kinematic growth and change of a Coulomb wedge, this volume.

Woodward, N.B. 1987. Geological applicability of critical-wedge thrust-belt models. *Geological Society of America Bulletin*, **99**, 827-832.

Woodward, N.B., Boyer, S.E., & Suppe, J. 1989. Balanced geological cross-sections. *American Geophysical Union Short Course in Geology*, **6**, 132p.

Index

References to figures appear in *italics*; those to tables in **bold**

D

Q

R

S

T

U

V

W

X

Y

Z